POLYMERIC BIOMATERIALS

Structure and Function

VOLUME 1

T0341112

Polymeric Biomaterials

Polymeric Biomaterials: Structure and Function, Volume 1

Polymeric Biomaterials: Medicinal and Pharmaceutical Applications, Volume 2

POLYMERIC BIOMATERIALS

Structure and Function

VOLUME 1

Founding Editor
Severian Dumitriu

Editor
Valentin Popa

CRC Press
Taylor & Francis Group
Boca Raton   London   New York

CRC Press is an imprint of the
Taylor & Francis Group, an **informa** business

CRC Press
Taylor & Francis Group
6000 Broken Sound Parkway NW, Suite 300
Boca Raton, FL 33487-2742

First issued in paperback 2020

© 2013 by Taylor & Francis Group, LLC
CRC Press is an imprint of Taylor & Francis Group, an Informa business

No claim to original U.S. Government works

Version Date: 20120726

ISBN 13: 978-0-367-61797-4 (pbk)
ISBN 13: 978-1-4200-9470-1 (hbk)

Library of Congress Cataloging-in-Publication Data

Polymeric biomaterials / editors, Severian Dumitriu and Valentin Popa.
    p. ; cm.
  Includes bibliographical references and index.
  ISBN 978-1-4200-9470-1 (v. 1 : alk. paper) -- ISBN 978-1-4200-9468-8 (v. 2 : alk. paper)
  I. Dumitriu, Severian, 1939- II. Popa, Valentin I.
  [DNLM: 1. Polymers. 2. Biocompatible Materials--therapeutic use. 3. Regenerative Medicine--methods. QT 37.5.P7]

  610.28--dc23   2012029709

Visit the Taylor & Francis Web site at
http://www.taylorandfrancis.com

and the CRC Press Web site at
http://www.crcpress.com

Contents

Supplementary Resources Disclaimer

Additional resources were previously made available for this title on CD. However, as CD has become a less accessible format, all resources have been moved to a more convenient online download option.

You can find these resources available here: www.routledge.com/9781420094701

Please note: Where this title mentions the associated disc, please use the downloadable resources instead.

Preface

The field of biomaterials has developed rapidly because of the continuous and ever-expanding practical needs of medicine and health-care practice. There are currently thousands of medical devices, diagnostic products, and disposables on the market, and the range of applications continues to grow. In addition to traditional medical devices, diagnostic products, pharmaceutical preparations, and health-care disposables, the list of biomaterials applications includes smart delivery systems for drugs, tissue cultures, engineered tissues, and hybrid organs.

Undoubtedly, biomaterials have had a major impact on the practice of contemporary medicine and patient care, resulting in both saving and improving the quality of lives of humans and animals. Modern biomaterials practice is continuing to develop into a major interdisciplinary effort involving chemists, biologists, engineers, and physicians. It also takes advantage of developments in the traditional, nonmedical materials field, and much progress has been made since the beginning of the research in biomaterials that made possible the creation of a high-quality and much improved variety of devices, implants (permanent or temporary), and drug carrier devices. All of these now display a greater than ever biocompatibility and biofunctionality. The variety of chemical substances used in these materials is currently very broad, and most biomedical applications are associated with various polymers and materials based on them.

The pace of research in the field of polymeric biomaterials is so fast that two editions of *Polymeric Biomaterials* have already been edited by Severian Dumitriu. Due to the interest generated and the success of these books, Severian was working on a third edition. Unfortunately, he passed away before this could be finalized. Many of the scientists who accepted his invitation to cooperate for this new edition agreed to contribute to the book in memory of the contribution that Severian made to the field of polymeric biomaterials. Together with Daniela, his beloved daughter, and Barbara Glunn and Jessika Vakili from Taylor & Francis Group, we decided to continue the work and finalize this book.

This book is organized in two volumes consisting of 53 chapters that systematically provide the latest developments in different aspects of polymeric biomaterials. Thus, we can mention contributions to the field of synthesis and applications of polymers such as polyesters, poly(vinyl alcohol), polyphosphazenes, elastomers, bioceramics, blends or composites, enzymatic synthesis, along with natural ones such as mucoadhesives, chitin, chitosan, lignin, carbohydrates derivatives, heparin, etc.

Drugs carriers and delivery systems, gene and nucleic acids delivery represent other subjects of some chapters, dealing with both supports (biodegradable and biocompatible) and techniques (nanoparticles, electrospinning, photo- and pH responsive polymers, hydrogels, lipid-core micelles, biomimetic systems, medical devices) aspects. In some cases, biomaterials can be synthesized, modified, and processed by different methods to ensure biocompatibility and biodegradability to be used as membranes, composites, scaffolds, and implants. Some examples of specific utilizations of polymeric biomaterials are presented, such as orthopedic surgery, bone regeneration, wound healing, dental and maxillofacial surgery applications, artificial joints, diabetes, anticancer agents and cancer therapy, modification of living cells, myocardial tissue engineering—repair and reconstruction, and bioartificial organs.

Publishing this book was accomplished with the contributions of renowned scientists from all over the world. They are all experts in their particular field of biomaterials research and have made high-level contributions to various fields of research. We are very grateful to these scientists for their willingness to contribute to this reference work as well as for their engagement. Without their commitment and enthusiasm, it would not have been possible to compile such a book.

I am also grateful to the publisher for recognizing the demand for such a book, for taking the risk to bring out such a book, and for realizing the excellent quality of the publication.

I would like to thank Daniela for her inestimable help and assistance. I dedicate this book to memory of Severian, one of my best friends.

Last but not least, I would like to thank my family for their patience. I sincerely apologize for the many hours I spent in the preparation of this book, which kept me away from them.

This book is a very useful tool for many scientists, physicians, pharmacists, engineers, and other experts in a variety of disciplines, both in academe and industry. It may not only be useful for research and development but may also be suitable for teaching.

This book has a companion CD that contains color figures as noted at the applicable text figures.

Valentin I. Popa

Acknowledgments

My father was passionate about polymeric biomaterials. He was very happy when this project was planned with Taylor & Francis Group. He had worked tirelessly toward this. He would have loved to have seen this book published, but destiny willed otherwise.

I am extremely grateful to Professor Popa for accepting to serve as the editor, to all the authors for their precious contributions, and to the staff at Taylor & Francis Group.

The positive response from the authors to pursue their contribution to this book was amazing and is testimony of their appreciation for the scientific contribution that my father made to the field of polymeric biomaterials.

I trust the book is of great quality and reflects the efforts and dedication that have been put into it by my father and all the contributors.

My small contribution to this book is dedicated to the memory of my parents, Severian and Maria, for their unconditional love and for being the best teachers ever. And to finish on a positive note, I want to cite one quote of Dr. Seuss that I particularly like:

"Don't cry because it's over. Smile because it happened".

Daniela Dumitriu

Editors

Severian Dumitriu (deceased) was a research professor, Department of Chemical Engineering, University of Sherbrooke, Quebec, Canada. He edited several books, including *Polymeric Biomaterials*, second edition, *Polysaccharides in Medicinal Applications*, and *Polysaccharides: Structural Diversity and Functional Versatility* (all three titles were published by Taylor & Francis Group [previously Marcel Dekker]), and authored or coauthored over 190 professional papers and book chapters in the fields of polymer and cellulose chemistry, polyfunctional initiators, and bioactive polymers. He also held 15 international patents. Professor Dumitriu received his BSc (1959) and MS (1961) in chemical engineering and his PhD (1971) in macromolecular chemistry from the Polytechnic Institute of Jassy, Romania. Upon completing his doctorate, he worked with Professor G. Smets at the Catholic University of Louvain, Belgium, and was a research associate at the University of Pisa, Italy; the Hebrew University Medical School, Jerusalem, Israel; and the University of Paris, South France.

Valentin I. Popa earned his BSc and MSc in chemical engineering (1969) and PhD in the field of polysaccharide chemistry (1976) from Polytechnic Institute of Iasi, Romania. He was awarded the Romanian Academy Prize for his contributions in the field of seaweed chemistry (1976). He has published more than 500 papers in the following fields: wood chemistry and biotechnology, biomass complex processing, biosynthesis and biodegradation of natural compounds, allelochemicals, bioadhesives, and bioremediation. He is also the author or coauthor of 37 books or book chapters. Dr. Popa holds six patents and has been involved in many Romanian and European research projects as scientific manager. He was visiting scientist or visiting professor at Academy of Sciences (Seoul, Korea, 1972), Technical University of Helsinki (Finland, 1978), Institute of Biotechnology (Vienna, Austria, 1995), Research Institute for Pulp and Paper (Braila, Romania, 1976), "Petru Poni" Institute of Macromolecular Chemistry (Iasi, Romania, 1985, 1986), Université de Sherbrooke and University McGill (Canada, 2003), STFI–Packforsk (now known as Innventia, Stockholm, Sweden, 2008), and Institute of Wood Chemistry (Riga, Latvia, 2009). Dr. Popa is a member of the International Lignin Institute, International Association of Scientific Papermakers, International Academy of Wood Science, Romanian Academy for Technical Sciences, and American Chemical Society. He is also a professor of wood chemistry and biotechnology in "Gheorghe Asachi" Technical University of Iasi, PhD supervisor (30 students defended their theses), and editor-in-chief of *Cellulose Chemistry and Technology*.

Contributors

Harry R. Allcock
Department of Chemistry
The Pennsylvania State University
University Park, Pennsylvania

Pascal Auroy
Université d'Auvergne
Clermont-Ferrand, France

Andreas Bernkop-Schnürch
Institute of Pharmacy
University of Innsbruck
Innsbruck, Austria

Martine Bonnaure-Mallet
UFR Odontologie Equipe de Microbiologie
   EA 1254
Université Européenne de Bretagne
Rennes, France

Rhiannon Braund
School of Pharmacy
University of Otago
Dunedin, New Zealand

Toby D. Brown
Institute for Health and Biomedical
   Innovation
Queensland University of Technology
Brisbane, Queensland, Australia

Cesare Cametti
Department of Physics
University of Rome "La Sapienza"
and
CNR-INFM-SOFT
Rome, Italy

Dominique Chauvel-Lebret
Université Européenne de Bretagne
UFR d'Odontologie-UMR CNRS 6226
   Sciences Chimiques de Rennes-UR1,
   CHU-pole d'Odontologie et de chirurgie
   Buccale
Rennes, France

Weiliam Chen
Department of Surgery
Division of Wound Healing and Regenerative
   Medicine
School of Medicine
New York University
New York, New York

Federica Chiellini
Laboratory of Bioactive Polymeric Materials
   for Biomedical and Environmental
   Application (BIOLab)
Department of Chemistry and Industrial
   Chemistry
University of Pisa
Pisa, Italy

Valeria Chiono
Department of Mechanical and Aerospace
   Engineering
Politecnico di Torino
Torino, Italy

Gianluca Ciardelli
Department of Mechanical and Aerospace
   Engineering
Politecnico di Torino
Torino, Italy

E.C. Combe
School of Dentistry
University of Minnesota
Minneapolis, Minnesota

Paul D. Dalton
Institute for Health and Biomedical Innovation
Queensland University of Technology
Brisbane, Queensland, Australia

Meng Deng
Institute for Regenerative Engineering
Raymond and Beverly Sackler Center for
   Biological, Physical and Engineering
   Sciences
The University of Connecticut
Storrs, Connecticut

Jacques Desbrieres
Department of Physics and Chemistry
    of Polymers
Pau and Adour Countries University
Pau, France

Alberto Dessy
Laboratory of Bioactive Polymeric Materials
    for Biomedical and Environmental
    Application (BIOLab)
Department of Chemistry and Industrial
    Chemistry
University of Pisa
Pisa, Italy

Abraham J. Domb
Faculty of Medicine
School of Pharmacy
Institute of Drug Research
The Hebrew University of Jerusalem
Jerusalem, Israel

Cesare Errico
Laboratory of Bioactive Polymeric Materials
    for Biomedical and Environmental
    Application (BIOLab)
Department of Chemistry and Industrial
    Chemistry
University of Pisa
Pisa, Italy

Christine Falabella
Department of Biomedical
    Engineering
Stony Brook University
State University of New York
New York, New York

Mar Fernández
Institute for Health and Biomedical
    Innovation
Queensland University of Technology
Kelvin Grove, Queensland, Australia

Piergiorgio Gentile
Department of Mechanical and Aerospace
    Engineering
Politecnico di Torino
Torino, Italy

Catherine Gkioni
Department of Periodontology and Biomaterials
Radboud University Nijmegen Medical Center
Nijmegen, the Netherlands

Takao Hanawa
Institute of Biomaterials and Bioengineering
Tokyo Medical and Dental University
Tokyo, Japan

Dietmar W. Hutmacher
Institute for Health and Biomedical Innovation
Queensland University of Technology
Brisbane, Queensland, Australia

Florin Dan Irimie
Department of Biochemistry and Biochemical
    Engineering
"Babeş-Bolyai" University
Cluj-Napoca, Romania

Kazuhiko Ishihara
Department of Materials Engineering
and
Department of Bioengineering
School of Engineering
The University of Tokyo
Tokyo, Japan

John Jansen
Department of Periodontology and Biomaterials
Radboud University Nijmegen Medical Center
Nijmegen, the Netherlands

Christine Jérôme
Center for Education and
    Research on Macromolecules
University of Liège
Liège, Belgium

Toshiyuki Kanamori
National Institute of Advanced
    Industrial Science and
    Technology
Tsukuba, Japan

Toshihiro Kasuga
Graduate School of Engineering
Nagoya Institute of Technology
Nagoya, Japan

Wahid Khan
Institute of Drug Research (IDR)
School of Pharmacy-Faculty of Medicine
The Hebrew University
    of Jerusalem
Jerusalem, Israel

Menno L.W. Knetsch
Department of Biomedical
    Engineering/Biomaterials Science
Maastricht University
Maastricht, the Netherlands

Masanori Kobayashi
Department of Biomedical Engineering
Daido University
Nagoya, Japan

Sangamesh G. Kumbar
Department of Orthopaedic Surgery, Chemical,
    Materials and Biomolecular Engineering
Institute for Regenerative Engineering
Raymond and Beverly Sackler Center for
    Biological, Physical and Engineering
    Sciences
The University of Connecticut
Storrs, Connecticut

Masayuki Kyomoto
Department of Materials Engineering
School of Engineering
and
Science for Joint Reconstruction
Graduate School of Medicine
The University of Tokyo
and
Research Department
Kyocera Medical Corporation
Tokyo, Japan

Cato T. Laurencin
Connecticut Institute for Clinical and
    Translational Science
and
Institute for Regenerative Engineering
Raymond and Beverly Sackler Center for
    Biological, Physical and Engineering
    Sciences
The University of Connecticut
Storrs, Connecticut

Philippe Lecomte
Center for Education and Research on
    Macromolecules
University of Liège
Liège, Belgium

Sander Leeuwenburgh
Department of Periodontology
    and Biomaterials
Radboud University Nijmegen Medical Center
Nijmegen, the Netherlands

Natalie J. Medlicott
School of Pharmacy
University of Otago
Dunedin, New Zealand

Franco Maria Montevecchi
Department of Mechanical and Aerospace
    Engineering
Politecnico di Torino
Torino, Italy

Toru Moro
Science for Joint Reconstruction
Graduate School of Medicine
The University of Tokyo
Tokyo, Japan

Vijay Kumar Nandagiri
Department of Mechanical and Aerospace
    Engineering
Politecnico di Torino
Torino, Italy

Akiko Obata
Graduate School of Engineering
Nagoya Institute of Technology
Nagoya, Japan

Csaba Paizs
Department of Biochemistry and Biochemical
    Engineering
"Babeş-Bolyai" University
Cluj-Napoca, Romania

Gaio Paradossi
Department of Chemical Sciences
    and Technologies
University of Rome Tor Vergata
Rome, Italy

Juan Parra
Institute for Health and Biomedical
    Innovation
Queensland University of Technology
Kelvin Grove, Queensland, Australia

Anna Maria Piras
Laboratory of Bioactive Polymeric Materials
    for Biomedical and Environmental
    Application (BIOLab)
Department of Chemistry and Industrial
    Chemistry
University of Pisa
Pisa, Italy

Valentin I. Popa
Department of Natural and Synthetic
    Polymers
Faculty of Chemical Engineering and
    Environmental Protection
"Gheorghe Asachi" Technical University
    of Iasi
Iasi, Romania

Gema Rodríguez-Crespo
Institute for Health and Biomedical
    Innovation
Queensland University of Technology
Kelvin Grove, Queensland, Australia

Luis M. Rodríguez-Lorenzo
Institute for Health and Biomedical
    Innovation
Queensland University of Technology
Kelvin Grove, Queensland, Australia

Julio San Román
Institute for Health and Biomedical
    Innovation
Queensland University of Technology
Kelvin Grove, Queensland, Australia

Alina Sionkowska
Faculty of Chemistry
Nicolaus Copernicus University
Torun, Poland

Shinji Sugiura
National Institute of Advanced Industrial
    Science and Technology
Tsukuba, Japan

Kimio Sumaru
National Institute of Advanced Industrial
    Science and Technology
Tsukuba, Japan

Toshiyuki Takagi
National Institute of Advanced
    Industrial Science and
    Technology
Tsukuba, Japan

Monica Ioana Tosa
Department of Biochemistry
    and Biochemical Engineering
"Babeş-Bolyai" University
Cluj-Napoca, Romania

Cedryck Vaquette
Institute for Health and Biomedical
    Innovation
Queensland University of Technology
Brisbane, Queensland, Australia

Blanca Vázquez
Institute for Health and Biomedical
    Innovation
Queensland University of Technology
Kelvin Grove, Queensland, Australia

Lihui Weng
Department of Radiology
University of Minnesota
Minneapolis, Minnesota

1 Synthesis and Fabrication of Polyesters as Biomaterials

Philippe Lecomte and Christine Jérôme

CONTENTS

1.1 INTRODUCTION

Nowadays, biomaterials are produced from a wide range of polymers. Among them, biodegradable and biocompatible aliphatic polyesters occupy a key position. Their chemical structure can be easily modified, which allows the tailoring of important properties such as bioadherence, mechanical properties, and kinetics of biodegradation. Aliphatic polyesters are thus widely used in biomedical applications as implants, as scaffolds in tissue engineering, and as carriers for drug delivery.

This chapter aims at reviewing the most important techniques used for the synthesis of aliphatic polyesters in biomedical applications. The simplest approach relies on step-growth polymerization (Figure 1.1, routes a and b). Nevertheless, the control imparted to the polymerization is limited by this technique, and the synthesis of high molar mass polyesters is difficult. These drawbacks can be tackled by implementing another technique, the ring-opening polymerization of cyclic esters (Figure 1.1, route c). Under appropriate conditions, this polymerization is living and enables the synthesis of controlled high molar mass aliphatic polyesters. The chemical structure of the chain-ends can be controlled and it is thus possible to functionalize them on demand. Besides, aliphatic polyesters are produced by the ring-opening polymerization of cyclic esters at the industrial scale. Moreover, the livingness of the ring-opening polymerization of cyclic esters opens up the possibility to prepare various architectures such as star-shaped, comb-shaped, hyperbranched polymers and networks. The versatility of the ring-opening polymerization of cyclic esters thus allows fine tailoring of the properties of aliphatic polyesters in view of biomedical applications. A very important section of this chapter will thus be dedicated to this technique. It must be noted that other polymerization

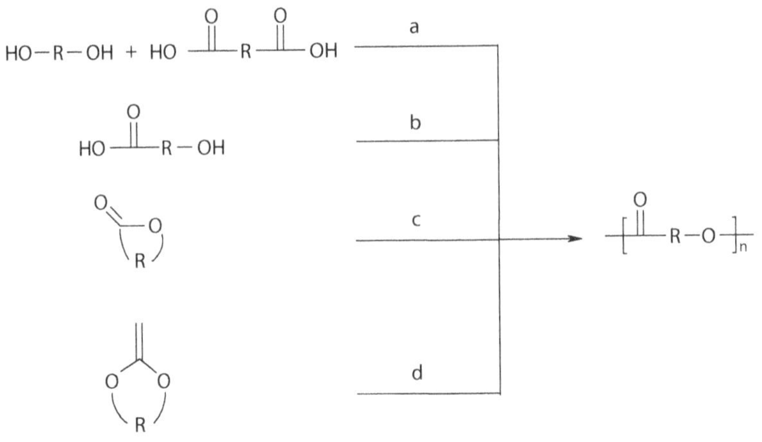

FIGURE 1.1 Main techniques for the synthesis of aliphatic polyesters.

techniques were investigated even though they were not so popular. In this regard, special attention will be paid to the ring-opening polymerization of cyclic ketene acetals (Figure 1.1, route d).

For biomedical applications, it is mandatory that aliphatic polyesters exhibit a very high purity and are not contaminated by potentially toxic impurities. Unfortunately, usual polymerization techniques are based on catalysts and initiators made up of toxic metals such as tin and aluminum, especially as far as the ring-opening polymerization of cyclic esters is concerned. Special attention will thus be paid to the recent advances in the implementation of polymerization processes based on less toxic metals or, even better, on metal-free processes.

1.2 STEP-GROWTH POLYMERIZATION

A straightforward approach for the synthesis of aliphatic polyesters is based on the step-growth polymerization of a mixture of diacids and diols or, more directly, of hydroxy acids, by implementing an esterification reaction. This technique allows the synthesis of a very wide range of aliphatic polyesters because of the easy synthesis of various diols, diacids, and hydroxy acids. Some of these can even be obtained from renewable resources, and bio-based aliphatic polyesters can then be produced. Step-growth polymerization has been in use for a long time, and hence we will not develop further the theory of polycondensation. It is just worth noting that, recently, Kricheldorf revised the concept of kinetically controlled (Kricheldorf and Schwarz 2003) and thermodynamically controlled (Kricheldorf 2003) step-growth polymerization.

The main limitation of all step-growth polymerizations remains the difficult synthesis of high molar masses. Indeed, the synthesis of high molar mass polyesters by step-growth polymerization requires to reach high conversions very close to 100% and to stay close to the ideal 1/1 stoichiometry between alcohol and acid functions.

The synthesis of polyhydroxyalkanoates (PHAs) by step-growth polymerization is a naturally occurring process (Lu et al. 2009). These aliphatic polyesters are produced by microorganisms as an intracellular reserve of carbon storage compounds and energy (Lu et al. 2009). Poly(3-(R)-hydroxybutyrate) (PHB) is a typical example, but PHAs with a huge variety of chemical structures can be synthesized from a very wide range of hydroxyalkanoates. The potential of PHAs has been assessed for several biomedical applications such as controlled release, tissue engineering, surgical sutures, and wound dressings (Williams et al. 1999).

The limitations of step-growth polymerization urged chemists to search for more efficient techniques for the synthesis of high molar mass aliphatic polyesters with a fine control of the chemical structure and the molar mass. This goal was achieved by implementing the ring-opening polymerization of cyclic esters, as will be shown in the next section.

1.3 RING-OPENING POLYMERIZATION OF CYCLIC ESTERS

Ring-opening polymerization is a very popular technique to synthesize aliphatic polyesters in view of biomedical applications. The main reason for its success lies in the very easy synthesis of aliphatic polyesters with high and controlled molar masses (Stridsberg et al. 2002, Penczek et al. 2007). The main limitation remains, for the time being, the scarcity of cyclic esters, despite the recent progresses being made in the last few years, especially compared to diacids, diols, and hydroxyacids used in step-growth polymerization.

The ring-opening polymerization of cyclic esters has been in use for a long time. van Natta et al. (1934) already reported in 1934 the ring-opening polymerization of ε-caprolactone (εCL). Nowadays, polylactide (PLA) and poly(ε-caprolactone) (PCL) are produced at the industrial scale (Figure 1.2).

PCL is a semicrystalline biodegradable and biocompatible polyester ($T_m = 60°C$; $T_g = -60°C$) (Woodruff and Hutmacher 2010). Nevertheless, the degradation of PCL is slow, making it suitable in the field of drug delivery applications for long-term applications (Sinha et al. 2004). PLA and copolymers of lactide and glycolide (PLGA) have the advantage to be more hydrophilic than PCL and thus to degrade faster. Polylactide contains a chiral center (R or S), and its properties depend on tacticity. It is indeed well known that isotactic PLA is semicrystalline, whereas atactic PLA is amorphous. Semicrystalline PLA exhibits interesting mechanical properties but at the expense of biodegradability. Isotactic PLLA is synthesized by the polymerization of enantiomerically pure L-lactide, whereas atactic PLLA is obtained from a racemic mixture of L- and D-lactides.

The ring-opening polymerization of εCL and LA is carried out at the industrial level. Interestingly enough, PLA is a bio-based polyester produced from agricultural renewable resources such as starch, whereas biodegradable PCL is produced from oil.

Although εCL, GA, and LA are very important monomers, other cyclic esters can be polymerized by ring-opening depending on their size. First, there is an important question to determine whether a cyclic ester can be polymerized or not. The answer to this question can be found by considering thermodynamic data. According to the microreversibility rule, there is a competition between polymerization and depolymerization. Table 1.1 shows the values of the enthalpy (ΔH_p) and entropy (ΔS_p) of polymerization, the monomer concentration at equilibrium ($[M]_{eq}$), and the ceiling temperature (T_c) (Duda et al. 2005).

Table 1.1 shows that the polymerization of low-membered cyclic esters is an enthalpy-driven process (Penczek et al. 2000, Duda et al. 2005). The release of the ring strain is obviously a key parameter in favor of polymerization. Nevertheless, the behavior of five-membered lactones is completely different, as shown by the high-ceiling temperature, because depolymerization is faster than polymerization. Although the ring-opening polymerization of γBL is difficult, it is possible to

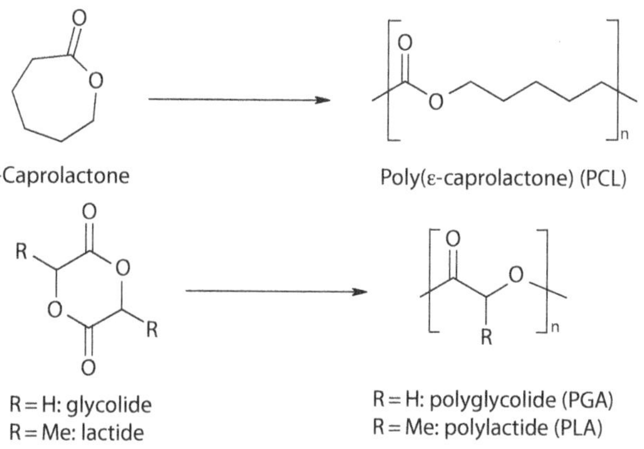

R = H: glycolide
R = Me: lactide

R = H: polyglycolide (PGA)
R = Me: polylactide (PLA)

FIGURE 1.2 Polymerization of ε-caprolactone (εCL) and lactide (LA) and glycolide (GA).

TABLE 1.1
Thermodynamics for the Polymerization of Cyclic Esters

Monomer	Ring Size	Monomer Polymer States	ΔH_p (kJ/mol)	ΔS_p (J/mol K)$_p$	$[M]_{eq}$ (mol/L)	T_c (K)
βPL	4	1c	−82.3	−74	3.9×10^{-10}	1112
γBL	5	1c	5.1	−29.9	3.3×10^{-3}	−171
LA	6	88	−22.9	−41	5.5×10^{-2}	914
GA	6	11	−13.8	−45	2.5	520
VL	6	Lc	−27.4	−65	3.9×10^{-1}	—
εCL	7	1c	−28.8	−53.9	5.3×10^{-2}	534

TABLE 1.2
Nomenclature of Cyclic Esters

Ring Size		IUPAC Name	Usual Name
4	βPL	oxetan-2-one	β-propionolactone
5	γBL	dihydrofuran-2(3*H*)-one	γ-butyrolactone
6	δVL	tetrahydro-2*H*-pyran-2-one	δ-valerolactone
6	LA	lactide	3,6-dimethyl-1,4-dioxane-2,5-dione
6	GA	glycolide	1,4-dioxane-2,5-dione
7	εCL	oxepan-2-one	ε-caprolactone

obtain low molar mass oligomers under suitable conditions and also to copolymerize γBL with other cyclic esters (Penczek et al. 2000).

It is worth noting that lactide, caprolactone, and glycolide are not the names recommended by IUPAC. Table 1.2 shows the usual and official names for several cyclic esters. For the sake of simplicity, the usual names will be mentioned in this review to avoid confusion.

So many initiators and catalysts have been used for the ring-opening polymerization of cyclic esters for the last few years that it is impossible to list all. This review will thus focus on the most popular processes with special attention being paid to the livingness and kinetics of polymerization, and the contamination of aliphatic polyesters by toxic residues not tolerated in the frame of biomedical applications.

1.3.1 ANIONIC POLYMERIZATION

Metal alkoxides are nucleophilic species prone to initiate the ring-opening polymerization of cyclic esters. Polymerization takes place usually by the cleavage of the acyl-oxygen bond. Figure 1.3 shows the initiation and propagation steps in the case of polymerization of εCL.

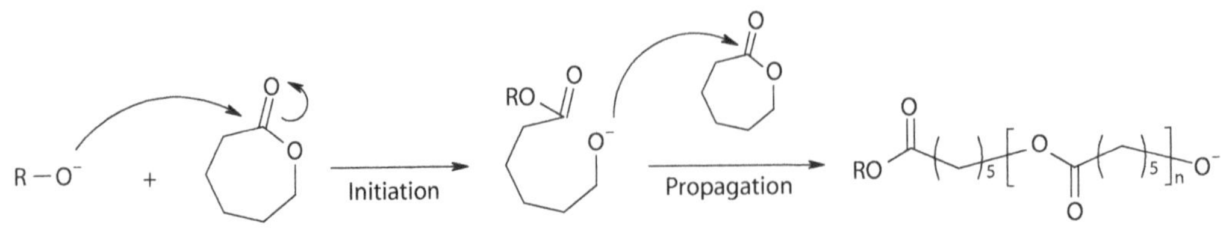

FIGURE 1.3   Anionic ring-opening polymerization of εCL.

The mechanism shown in Figure 1.3 assumes that polymerization is living and takes place only by initiation and propagation reactions according to Szwarc (1956), which means that no termination and transfer reactions are present. This view is unfortunately oversimplified because termination and transfer reactions are often observed. Indeed, anionic species are deactivated by traces of water or by other protic substances, which is nothing but a termination reaction. It is thus necessary to perform these anionic polymerizations with carefully purified monomers under strictly anhydrous conditions. Besides, although metal alkoxides react as nucleophiles, they react also as bases. For instance, potassium tert-butoxide reacts with εCL to form the corresponding potassium enolate.

Transfer reactions by transesterification reactions are commonly observed and are at the origin of loss of control of polymerization (Figures 1.4 and 1.5). Indeed, alkoxides are often too reactive to be selective, and they can react not only with the ester functions of monomers but also with the ester functions present all along the polymer chains. Depending upon whether alkoxide and ester are located on the same chain or not, transesterification reaction can be intramolecular or intermolecular. Intramolecular transesterification reactions result in a decrease of the molar mass and in the formation of cyclic oligomers (Figure 1.4). Conversely, intermolecular transesterification does not result in a decrease of the number-average molar mass but rather results in the reshuffling of the length of chains and thus in the modification of the polydispersity index (Figure 1.5). For the sake of simplicity, the mechanism of εCL is shown in Figures 1.4 and 1.5, but it can be extended to other cyclic esters as well.

Special attention has to be paid to the four-membered lactones because of their unusual behavior.

Indeed, polymerization of β-lactones by the mechanism based on the cleavage of the acyl-oxygen bond is disfavored, which was accounted for by stereo-electronic effects (Coulembier et al. 2006a). Another mechanism based on the scission of the alkyl-oxygen bond is then observed (Figure 1.6).

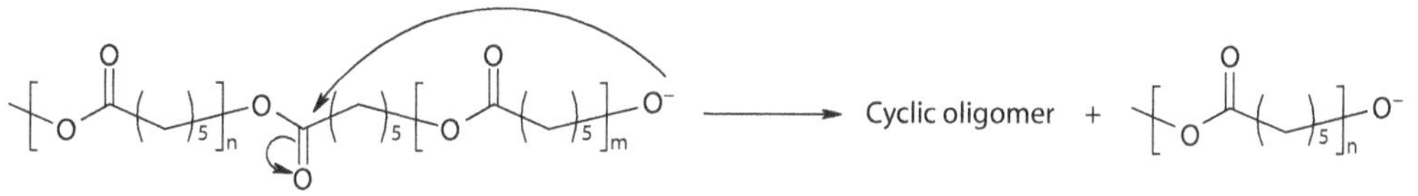

FIGURE 1.4 Intramolecular transesterification reactions during the polymerization of εCL.

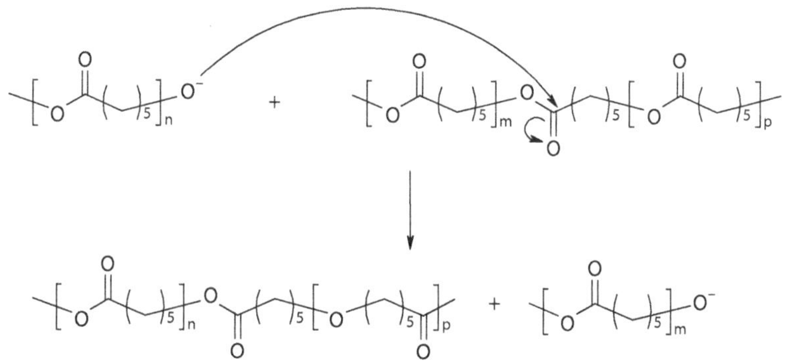

FIGURE 1.5 Intermolecular transesterification reactions during the polymerization of εCL.

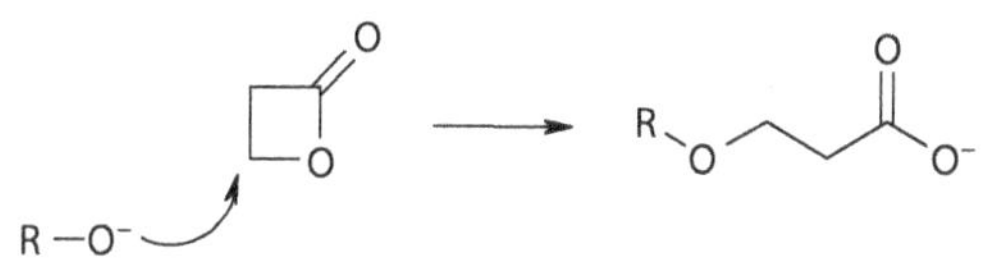

FIGURE 1.6 Ring-opening polymerization of β-lactones by cleavage of the alkyl-oxygen bond.

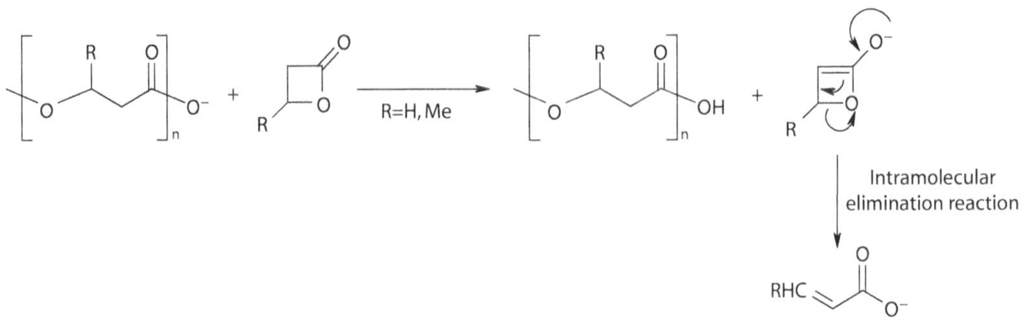

FIGURE 1.7   Transfer reaction for the polymerization of β-lactones.

Interestingly enough, ring-opening affords a carboxylate anion rather than an alkoxide. This polymerization can thus also be initiated by carboxylic salts even though they are less nucleophilic than alkoxides (Penczek 2007).

It is worth noting that another transfer reaction mechanism shown in Figure 1.7 is reported as far as β-lactones are concerned (Penczek 2007).

1.3.2   COORDINATION POLYMERIZATION

Anionic metal alkoxides are too reactive to be selective and transfer transesterification reactions are very difficult to get rid of, at the expense of control of polymerization. In order to improve the control of polymerization, it is mandatory to use less reactive initiators. This can be achieved by playing with steric and electronic effects (Lecomte and Jérôme 2004). The use of cumbersome ligands is the first possible route to obtain more selective species. The most common approach relies on the use of less electrophilic metals. The works of Teyssié and coworkers who used bimetallic μ-oxo-alkoxides as initiators have to be mentioned (Hamitou et al. 1973, Ouhadi et al. 1976). Since then, a wide range of transition metals have been investigated.

Aluminum alkoxides occupy a key position because of the very good control imparted to the polymerization of cyclic esters. Commercially available aluminum triisopropoxide is a widely used aluminum alkoxide. Firstly, the three alkoxide bonds initiate polymerization. Nevertheless, the behavior of aluminum triisopropoxide is complicated due to its aggregation in solution. Indeed, aluminum isopropoxide exists as a mixture of trimers (A_3) and tetramers (A_4; Ropson et al. 1995). As long as polymerization of εCL is carried out by aluminum isopropoxide in toluene at 0°C, only A_3 is prone to initiate polymerization. The interconversion of A_4 into A_3 is slow compared to the kinetics of polymerization, and A_4 does not initiate polymerization under these conditions. Conversely, polymerization of lactide is carried out at a higher temperature, typically at 70°C. The interconversion of A_4 into A_3 is faster, and all species initiate polymerization. It is thus very important to take into account the equilibrium between A_3 and A_4 species to determine the theoretical molar mass on the basis of the monomer-to-initiator molar ratio. Last but not least, it is worth noting that aluminum alkoxides can be prepared in the laboratory by the reaction of any alcohol with triethylaluminum or aluminum isopropoxide (Dubois et al. 1989).

The mechanism of polymerization is quite similar to that of anionic polymerization. Nevertheless, one has to take into account the coordination of the metal with oxygen atoms. The mechanism shown in Figure 1.8 is referred in the state of the art as the coordination–insertion mechanism. Firstly, the alkoxide (RO-M) coordinates with the carbonyl of the cyclic ester, followed by the nucleophilic addition of the alkoxide onto the electrophilic C=O bond. Thereafter, ring-opening takes place by an elimination reaction, resulting in the cleavage of the acyl-oxygen bond and in the formation of a new alkoxide. The initiator can contain one functional group, provided it is tolerated by aluminum alkoxide species (Dubois et al. 1989).

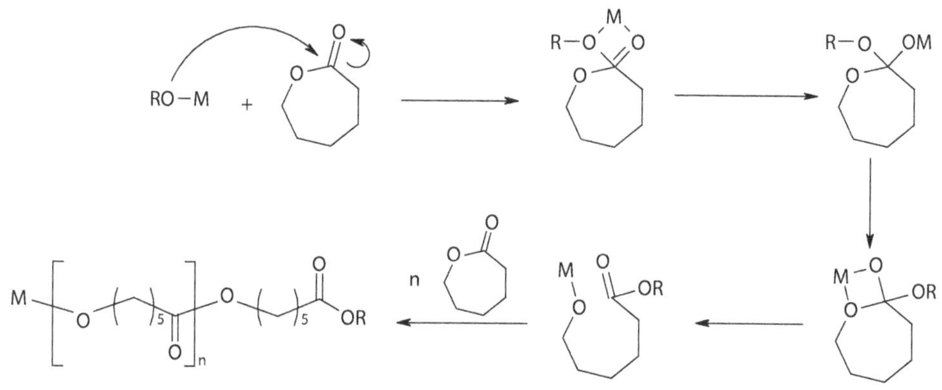

FIGURE 1.8 Coordination–insertion mechanism for the ring-opening polymerization of cyclic esters.

During the propagation of the ring-opening polymerization of cyclic esters initiated by aluminum alkoxides, aggregation can again take place depending upon stereoelectronic factors, as shown by NMR spectroscopy and kinetic studies (Duda and Penczek 1991). For instance, during the polymerization of εCL, the three-arm growing species is a unimer when initiated by A_3 and is a trimer when initiated by Et_2AlOR (Ropson et al. 1995).

Although aluminum alkoxides exerted an excellent control to the polymerization of cyclic esters, they are suspected to be involved in Alzheimer's disease. Moreover, catalytical remnants are very difficult to withdraw and these toxicity issues are thus a huge limitation for biomedical applications.

Many researchers used tin(II) bis-(2-ethylhexanoate), also commonly referred to as tin octoate, instead of aluminum alkoxides because of its recognition by the American Food and Drug Administration (FDA). Another reason for the popularity of tin(II) bis-(2-ethylhexanoate) is its lower sensitivity to water and other impurities. It is thus easier to achieve polymerization in laboratories or in industry. Nevertheless, it was found that tin(II) bis-(2-ethylhexanoate) is also cytotoxic and should ideally be avoided for the synthesis of biomaterials.

Tin(II) bis-(2-ethylhexanoate) does not contain any alkoxide bond, and the mechanism remained unclear for a long time and several proposals were reported. In 1998, Penczek and coworkers reported that tin alkoxides are formed in the polymerization medium by the reaction of tin(II) bis-(2-ethylhexanoate) with traces of water or with any other protic impurities. Polymerization thus takes place by the usual coordination–insertion mechanism at least as long as it is carried out in THF at 80°C (Kowalski et al. 1998). This proposal was substantiated by kinetic measurements (Kowalski et al. 1998, 2000a), analysis of the chain-ends by MALDI-TOF (Kowalski et al. 2000b), and proton-trapping agent experiments (Majerska et al. 2000).

It is worth noting that other tin-based initiators are also used. Hedrick and coworkers proposed to use tin trifluoromethanesulfonate to improve the kinetics of polymerization (Möller et al. 2000). Another possibility relies on the use of Sn(II) alkoxides (Duda et al. 2000, Kowalski et al. 2000c) and Sn(IV) alkoxides (Kricheldorf and Eggerstedt 1998, Kricheldorf et al. 2001, Kricheldorf 2004). Penczek and coworkers reported that the polymerization of lactide, initiated by $Sn(OBu)_2$, is under control in a range of molar masses up to 10^6 g/mol (Duda et al. 2000, Kowalski et al. 2000c). Nevertheless, whatever the tin derivative used to initiate polymerization, toxicity remains an issue. Albertsson and coworkers proposed a smart approach to extract tin derivatives (Stjerndahl et al. 2007). They initiated the polymerization of cyclic esters by tin(IV) alkoxides before adding 1,2-ethanedithiol to the polymerization medium to afford a sulfur-containing dibutyltin derivative, which can be extracted because of its high solubility in organic solvents. For example, this process allowed synthesizing a sample of PCL contaminated by only 23 ppm of tin residues (Stjerndahl et al. 2007).

If ring-opening polymerization can be carried out in bulk and in organic solvents, supercritical carbon dioxide is also a valuable medium. The technology based on supercritical carbon dioxide is particularly interesting in the field of biomaterials because of the remarkable extraction properties of this medium with the prospect to purify aliphatic polyesters. Ring-opening polymerization of εCL in supercritical carbon dioxide was first reported by Mingotaud and coworkers (Mingotaud et al. 1999, 2000). Then, Jérôme and coworkers observed that tin(IV) alkoxides initiate the controlled ring-opening polymerization of εCL in supercritical CO_2 (Stassin et al. 2001). Slow kinetics was accounted for by the reversible reaction of tin alkoxides and carbon dioxide (Stassin and Jérôme 2002). PCL is not soluble in supercritical carbon dioxide, and the polymerization of εCL is thus heterogeneous. Nevertheless, in the presence of a suitable surfactant, nanoparticles can be obtained (Stassin and Jérôme 2004). Last but not least, supercritical CO_2 was used to withdraw quantitatively unconverted monomer and metallic remnants.

The issues related to the toxicity of tin and aluminum derivatives urged researchers to initiate the polymerization of cyclic esters by alkoxides based on less toxic metals. In this regard, bismuth (Kim et al. 2004), magnesium (Shueh et al. 2004, Yu et al. 2005b), and calcium (Zhong et al. 2001, Westerhausen et al. 2003) alkoxides are reported. The mechanism remains the usual coordination–insertion mechanism. The kinetics and control of polymerization depend on the nature of the metal and on its ligands.

So many metallic derivatives have been used to initiate or catalyze the ring-opening polymerization of cyclic esters that it is almost impossible to list all. Among these derivatives, zinc octoate (Libiszowski et al. 2002, Kowalski et al. 2007), aluminum acetyl acetonate (Kowalski et al. 2007), scandium trifluoromethanesulfonate (Möller et al. 2000, Nomura et al. 2000), and scandium trifluoromethanesulfonimide [$Sc(NTf_2)_3$] (Oshimura and Takasu 2010) can be mentioned. Special attention has to be paid to lanthanides alkoxides (Metal = Er, Sm, Dy, La) as initiators because of the very fast kinetics of polymerization (McLain and Drysdale 1992). Shen et al. (1996) showed that the increase of steric hindrance of the ligand disfavors transesterification reactions. Yasuda and coworkers polymerized εCL by $SmOEt(C_5Me_5)_2(OEt_2)$, [$YOMe(C_5H5)_2]_2$, and $YOMe(C_5Me_5)_2(THF)$ (Yamashita et al. 1996). In 1996, yttrium isopropoxide was obtained "in situ" by the reaction of isopropanol and yttrium tris(2,6-di-tert-butylphenolate) (Stevels et al. 1996a,b). Then, a similar approach was implemented by Jérôme and Spitz who synthesized $Y(OiPr)_3$ (Martin et al. 2000, 2003a) and $Nd(OiPr)_3$ (Tortosa et al. 2001) by the reaction of isopropanol with $Y[N(SiMe_3)_2]_3$ and $Nd[N(SiMe_3)_2]_3$, respectively. In 2003, the polymerization of ε-caprolactone was carried out by $La(OiPr)_3$ (Save et al. 2002) and $M(BH_4)_3(THF)_3$ (M = Nd, La, Sm) (Guillaume et al. 2003, Palard et al. 2005). Cyclic esters other than εCL can be polymerized by the same family of lanthanide alkoxides (Yamashita et al. 1996). Last but not least, Jérôme and coworkers grafted yttrium isopropoxide onto a porous silica surface (Martin et al. 2003b,c). Two methods of immobilization were reported. Firstly, the hydroxyl groups located on the surface of silica were allowed to react with an excess of $Y[N(SiMe_3)_2]_3$ into silylamido groups, which were finally allowed to react with 2-propanol to obtain yttrium alkoxides. Secondly, $Y[N(SiMe_3)_2]_3$ was made to react with less than three equivalents of 2-propanol into an yttrium alkoxide, which was then grafted onto the surface. Hamaide and coworkers supported alkoxides based on other metals (Al, Zr, Sm, and Nd) onto silica and alumina (Miola-Delaite et al. 2000).

1.3.3 METAL-FREE RING-OPENING POLYMERIZATION

Aluminum and tin alkoxides being too toxic for biomedical applications, chemists investigated original metal-free processes for the ring-opening polymerization of cyclic esters. Their strategy relies on the initiation of the polymerization by nucleophilic species such as alcohols and amines. Nevertheless, these species are in general not nucleophilic enough to react with cyclic esters. Nevertheless, there are exceptions to this general rule. For instance, highly reactive β-lactones are polymerized by nucleophilic amines in the absence of any catalyst. Polymerization initiated by tertiary amines is known as zwitterionic polymerization in the literature (Löfgren et al. 1995). The mechanism shown in Figure 1.9 is based on the reaction of tertiary amines and cyclic esters

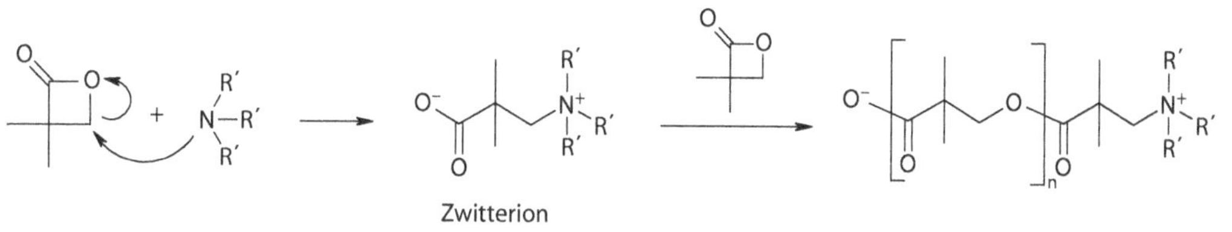

FIGURE 1.9 Zwitterionic polymerization of pivalolactone.

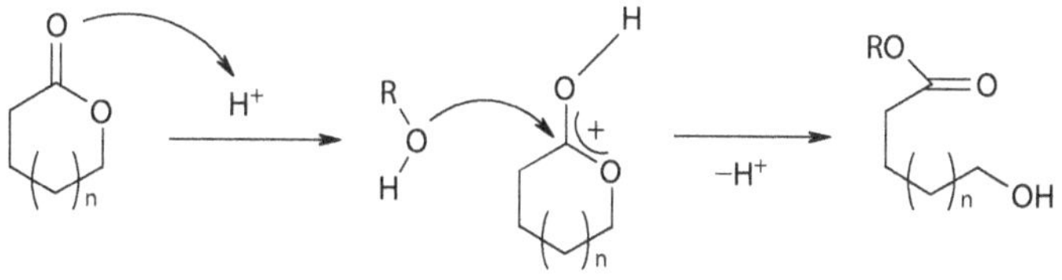

FIGURE 1.10 Ring-opening polymerization of cyclic esters catalyzed by acids and initiated by alcohols.

with a zwitterionic species made up of an ammonium cation and a carboxylate anion. Interestingly enough, Kricheldorf et al. (2005) mentioned the possibility that chain extension takes place by step-growth polycondensation at least at some stage of polymerization.

As a rule, alcohols and amines are in general not nucleophilic enough to react with cyclic esters and the reaction has thus to be catalyzed. In order to do so, two main strategies might be implemented. They rely on the activation of either the monomer or the initiator. Last but not least, some processes combine both mechanisms of activation.

Cyclic esters can be activated by either acids or nucleophiles to allow their reaction with alcohols and amines in the frame of a polymerization process. As far as acids are concerned as catalysts, the mechanism of ring-opening polymerization relies on the activation of cyclic esters by protonation of the exocyclic oxygen of cyclic diesters, which facilitates the reaction with the nucleophilic species, that can be the initiator during initiation or the hydroxyl-end capped chain during propagation (Figure 1.10). Polymerization takes place by the scission of the oxygen-acyl bond. In 2000, an example of ring-opening polymerization of εCL and δVL was reported by Endo and coworkers by using alcohol as an initiator and HCl.Et$_2$O as a catalyst (Shibasaki et al. 2000). Polymerization was under control but the molar mass did not exceed 15,000 g/mol. It is worth noting that, later on, Jérôme and coworkers reported that molar masses up to 50,000 g/mol were obtained as far as PVL was polymerized (Lou et al. 2002a). Recently, the process was extended to a wider range of acids. Trifluoromethanesulfonic is an efficient catalyst for the controlled ring-opening polymerization of lactide (Bourissou et al. 2005) and ε-caprolactone (Basko and Kubisa 2006). Later, Bourrissou showed that trifluoromethanesulfonic acid can be substituted for less acidic methanesulfonic acid for the polymerization of εCL (Gazeau-Bureau et al. 2008). Polymerization of δVL was catalyzed by trifluoromethanesulfonimide according to Kakuchi et al. (2010). Interestingly enough, polymerization was under control and various functionalized alcohols were used as initiators. Very recently, Takasu and coworkers extended this strategy to nonafluorobutanesulfonimide for the polymerization of εCL (Oshimura et al. 2011). Interestingly enough, organic catalysts such as lactic acid (Casas et al. 2004, Persson et al. 2006), citric acid (Casas et al. 2004), fumaric acid (Sanda et al. 2002, Zeng et al. 2005), and amino acids (Casas et al. 2004) were also used. In addition, the acid catalyst can be supported on silica (Wilson and Jones 2004). Amino acids exhibit a particular behavior because they both catalyze and initiate polymerization (Liu and Liu 2004).

In the absence of any nucleophilic species such as an alcohol, the only nucleophilic species remaining in the system is the cyclic ester. This is the typical case of cationic ring-opening polymerization of cyclic esters, which has been in use for a long time. The cationic polymerization can be catalyzed not only by Bronsted acids but also by alkylating agents, acylating agents, and Lewis acids.

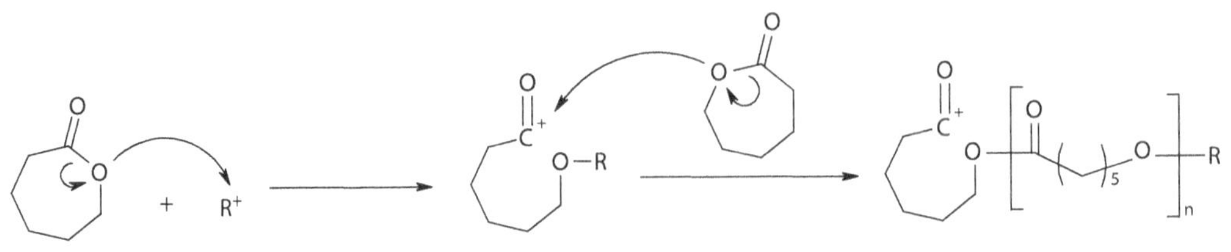

FIGURE 1.11 Mechanism for cationic ring-opening polymerization based on the reaction of the endocyclic oxygen.

For instance, in 1984, Penczek reported the cationic polymerization of εCL and βPL by acylating agents (Hofman et al. 1984). For a long time it was accepted that cationic ring-opening polymerization mediated by alkylating agents takes place by the mechanism shown in Figure 1.11, which is based on the reaction of the cation with the endocyclic oxygen followed by the cleavage of the acyl-oxygen bond. In 1984, Penczek (Hofman et al. 1984) and Kricheldorf et al. (1986) proposed that the cation reacts with the exocyclic oxygen rather than the endocyclic oxygen to afford a dialkoxycarbo-cationic species, which finally reacts by cleavage of the alkyl-oxygen bond (Figure 1.12). It is worth noting that both mechanisms shown in Figures 1.12 and 1.13 are observed when acylating agents are used (Slomkowski et al. 1985).

In recent years, the ring-opening polymerization of cyclic esters by nucleophilic catalysts has emerged as a very promising process under the impulse given by the group of Hedrick (Kamber et al. 2007). Table 1.3 shows several nucleophiles known to activate cyclic esters and thus prone to catalyze the ring-opening polymerization of cyclic esters initiated by alcohols by the mechanism shown in Figure 1.13. Among nucleophilic species, *N*-heterocyclic carbenes (Table 1.3, entries 1–7), amines (Table 1.3, entries 8–11), and phosphines (Table 1.3, entries 12–17) can be mentioned. Depending upon the cyclic ester that has to be polymerized, the nucleophilic catalyst must be carefully selected because its nucleophilicity and thus the kinetics of this ring-opening is influenced by steric and electronic effects. Interestingly enough, polymerization is under control. Although this approach is very promising, more work is needed to assess the impact of these catalysts on the biocompatibility of aliphatic polyesters.

Another mechanism relies on the activation of the initiator. Bases that are able to activate nucleophilic alcohols and catalyze the ring-opening polymerization of cyclic esters are shown in Table 1.4. TBD and DBU catalyze the ring-opening polymerization of cyclic esters (Table 1.4, entries 1, 2).

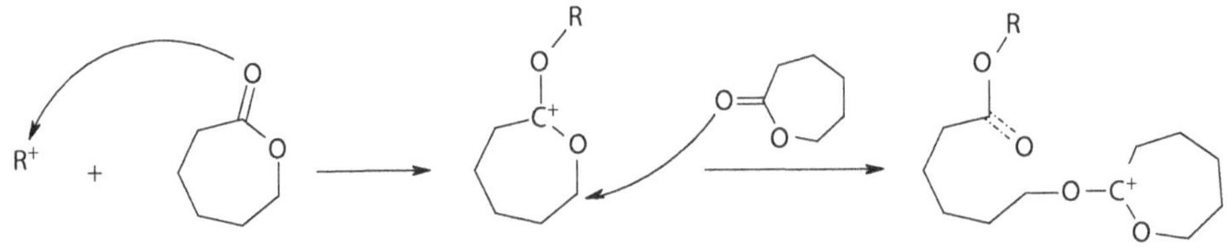

FIGURE 1.12 Mechanism for cationic ring-opening polymerization based on the reaction of the exocyclic oxygen.

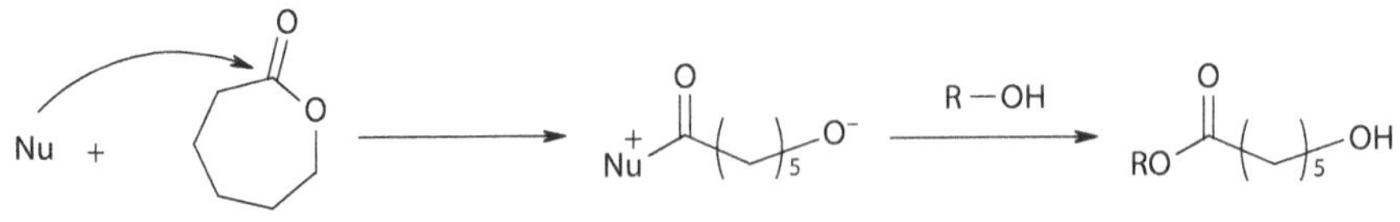

FIGURE 1.13 Activation of cyclic esters by nucleophiles.

TABLE 1.3

Nucleophilic Catalysts for the Ring-Opening Polymerization of Lactones

Entry	Organocatalyst	Monomer	Initiator	Reference
1	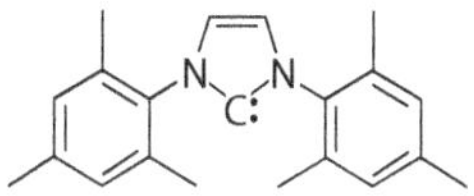	L-lactide	EtOH, pyrenebutanol	Connor et al. (2002)
		εCL	EtOH, pyrenebutanol	
		βBL	EtOH, pyrenebutanol	
		rac-lactide	—	Dove et al. (2006)
		meso-lactide	—	Dove et al. (2006)
2	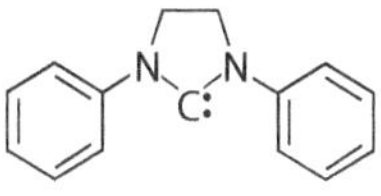	εCL	PhCH$_2$OH	Nyce et al. (2003)
		δVL	PhCH$_2$OH	
3	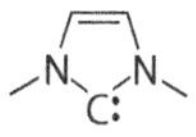	βBL	PhCH$_2$OH	Nyce et al. (2003)
		δVL	PhCH$_2$OH	
		εCL	PhCH$_2$OH	
4	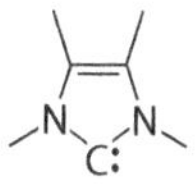	εCL	PhCH$_2$OH	Nyce et al. (2003)
5	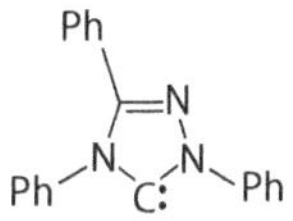	L-lactide	MeOH, pyrenebutanol, PEO-OH	Coulembier et al. (2005)
		βBL	pyrenebutanol	Coulembier et al. (2006b)
6	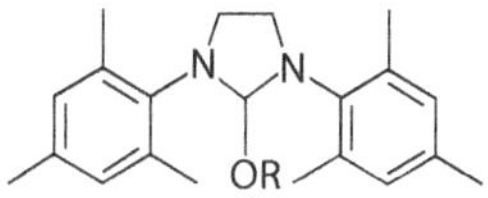	L-lactide	—	Connor et al. (2002)
7	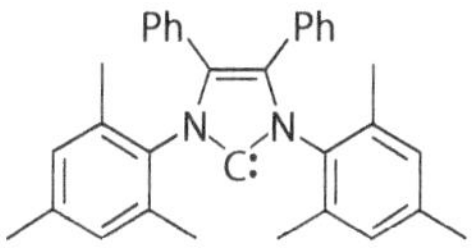	rac-lactide	—	Dove et al. (2006)
		meso-lactide	—	
7	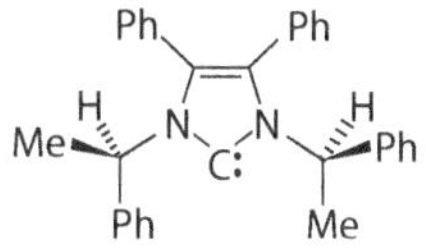	rac-lactide	—	Dove et al. (2006)
		meso-lactide	—	
8	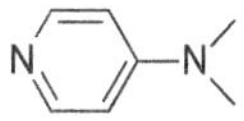 DMAP	Lactide	EtOH, PhCH$_2$OH	Nederberg et al. (2001)
9	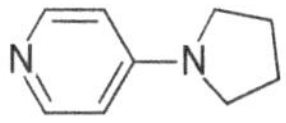	Lactide	EtOH, PhCH$_2$OH	Nederberg et al. (2001)
10	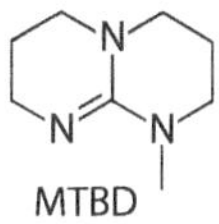 MTBD	Lactide	Pyrenebutanol	Pratt et al. (2006) and Lohmeijer et al. (2006)
11	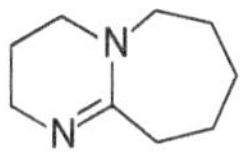 DBU	Lactide	Pyrenebutanol	Lohmeijer et al. (2006)

(continued)

TABLE 1.3 (continued)
Nucleophilic Catalysts for the Ring-Opening Polymerization of Lactones

Entry	Organocatalyst	Monomer	Initiator	Reference
12	PBu_3	Lactide	PhEtOH	Myers et al. (2002)
13	$PPhMe_2$	Lactide	PhEtOH	Myers et al. (2002)
14	PPh_3	Lactide	PhEtOH	Myers et al. (2002)
15		Lactide	PhEtOH	Myers et al. (2002)
16		Lactide	PhEtOH	Myers et al. (2002)
17		Lactide	PhEtOH	Myers et al. (2002)
18		Lactide	PhEtOH	Myers et al. (2002)

The mechanism is pseudoanionic and is based on the activation of the alcohol by attracting the proton. Phosphazene bases are also prone to catalyzing the ring-opening polymerization of cyclic esters (Zhang et al. 2007a). In particular, 2-*tert*-butylimino-2-diethylamino-1,3-dimethylper-hydrdro-1,3,2-diazaphosphorine (BEMP) is efficient (Table 1.4, entry 3). It is worth noting that BEMP exhibits a higher basicity (pK_{BH+} = 27.6) compared to DBU (pK_{BH+} = 24.3) and MTBD (pK_{BH+} = 25.4). As long as a dimeric phosphazene is used to catalyze the polymerization of racemic lactide (Table 1.4, entry 4), isotactic polylactide is obtained (Zhang et al. 2007b).

Several catalysts have the ability to activate both the monomer and the initiator at the same time (Table 1.5). 1,5,7-Triazabicyclo[4.4.0]dec-5-ene (TBD) is an efficient organocatalyst for the ring-opening polymerization of cyclic esters initiated by alcohols (Table 1.5, entry 1), and the polymerization takes place by dual activation of both the monomer and the initiator. The activation of the monomer can take place either by an acyl transfer mechanism or by a mechanism based on hydrogen bonding. This last mechanism was preferred as far as the polymerization of L-lactide is concerned (Chuma et al. 2008). As long as 1,8-diaza[5.4.0]bicycloundec-7-ene (DBU) and *N*-methylated TBD (MTBD) were used as catalysts instead of TBD, the polymerization of lactide was slower. Polymerization of CL and VL did not take place. Indeed, these experimental data can be easily understood if one takes into account that these catalysts can only activate the alcohol but not cyclic esters (Lohmeijer et al. 2006). However, a possible trick to overcome this issue relies on the addition of a thiourea activating cyclic esters by hydrogen

TABLE 1.4
Basic Catalysts for the Ring-Opening Polymerization of Lactones

Entry	Organocatalyst	Monomer	Initiator	Reference
1	MTBD	Lactide	1-Pyrenebutanol	Pratt et al. (2006) and Lohmeijer et al. (2006)
2	DBU	Lactide	1-Pyrenebutanol	Lohmeijer et al. (2006)
3	Et₂N–, tBu	L-lactide δVL εCL	1-Pyrenebutanol 1-Pyrenebutanol 1-Pyrenebutanol	Zhang et al. (2007a) Zhang et al. (2007a) Zhang et al. (2007a)
4	N—P=N=P—N—tBu	L-lactide	1-Pyrenebutanol	Zhang et al. (2007a)

bonding (Table 1.5, entry 2). Interestingly enough, both catalytical species can be combined in a single molecule. Indeed, Hedrick and coworkers reported that the polymerization of cyclic esters is efficiently catalyzed by 1-[3,5-bis(trifluoromethyl)phenyl]-3-[2-(dimethylamino)cyclohexyl]thiourea with a very good control of the molecular parameters (Table 1.5, entry 3; Dove et al. 2005). Polymerization takes place also in the presence of both 1-[3,5-bis(trifluoromethyl) phenyl]-3-cyclohexylthiourea and N,N-dimethylcyclohexanamine (Table 1.5, entry 4; Dove et al. 2005). Whenever only one of these is present, no polymer was obtained, in agreement with a dual activation mechanism.

1.3.4 ENZYMATIC RING-OPENING POLYMERIZATION

Enzymes are very promising nontoxic catalysts for the preparation of biomaterials. They are green catalysts obtained from renewable resources and are easily separated from polyesters. Among enzymes, lipases are known to catalyze the hydrolysis reaction of esters. Chemists used these natural enzymes to catalyze the reverse reaction, for example, the esterification reaction. The ring-opening polymerization of cyclic esters can be catalyzed by lipases as independently discovered in 1993 by the groups of Kobayashi (Uyama and Kobayashi 1993) and Knani et al. (1993). Since then, a wide variety of cyclic esters of different size were polymerized. Several reviews have been published on this topic (Gross et al. 2001, Varma et al. 2005, Albertsson and Srivastava 2008, Kobayashi and Makino 2009). Among the different lipases, *Candida antarctica* (lipase CA, CALB or Novozym 435) is widely used. An alcohol can purposely be added in the reaction medium to initiate polymerization. The course of polymerization is influenced by water.

TABLE 1.5
Dual Catalysts for the Ring-Opening Polymerization of Lactones

Entry	Organocatalyst	Monomer	Initiator	Reference
1	TBD	Lactide	Pyrenebutanol	Pratt et al. (2006)
		δVL	Pyrenebutanol	Lohmeijer et al. (2006)
		εCL	Pyrenebutanol	Lohmeijer et al. (2006)
2	DBU + thiourea (CF₃, F₃C)	δVL	Pyrenebutanol	Lohmeijer et al. (2006)
		εCL	Pyrenebutanol	Lohmeijer et al. (2006)
3	thiourea (CF₃, F₃C) rac NMe₂	Lactide	Pyrenebutanol	Dove et al. (2005)
4	thiourea (CF₃, F₃C) + cyclohexyl NMe₂	Lactide	Pyrenebutanol	Dove et al. (2005)

Indeed, a minimum amount of water needs to be bound to the surface of the enzyme to maintain its conformational flexibility and catalytic activity (Gross et al. 2001). Nevertheless, water also initiates this polymerization. The synthesis of high molar mass aliphatic polyesters by enzymatic ring-opening polymerization is thus difficult because it is not possible to carry out polymerization in strictly anhydrous conditions. In terms of the control of polymerization, enzymatic polymerization cannot compete with coordination or organocatalytic polymerization. Nevertheless, enzymatic polymerization has some practical interests. Polymerization proceeds under mild conditions in terms of pH, temperature, and pressure and can be carried out in bulk, in organic media, and even in supercritical carbon dioxide (Takamoto et al. 2001, Loeker et al. 2004). Moreover, enzymes are able to catalyze the ring-opening polymerization of large-membered cyclic esters, which are very difficult to polymerize by chemical catalysts and initiators such as metal alkoxides and organocatalysts (Duda et al. 2002).

The mechanism of enzymatic polymerization is based on the activation of the monomer and is very similar to the one already shown for nucleophilic organocatalysts. Briefly, a complex is formed between the enzyme and cyclic esters. The hydroxyl group of a serine residue of the active site of the enzyme reacts with cyclic esters affording an activated open form of the monomer. This activated intermediate then reacts with an alcohol, which can be the initiator or a hydroxyl-end capped chain. This mechanism is summarized in Figure 1.14.

1.3.5 POLYMERIZATION OF SUBSTITUTED AND FUNCTIONALIZED CYCLIC ESTERS

Although PCL, PLA, and PGA are widely used for biomedical applications, it is highly desirable to synthesize a wider range of aliphatic polyesters in view of the development of novel biomedical applications. Two main routes were investigated for this application: the first one relies on the direct

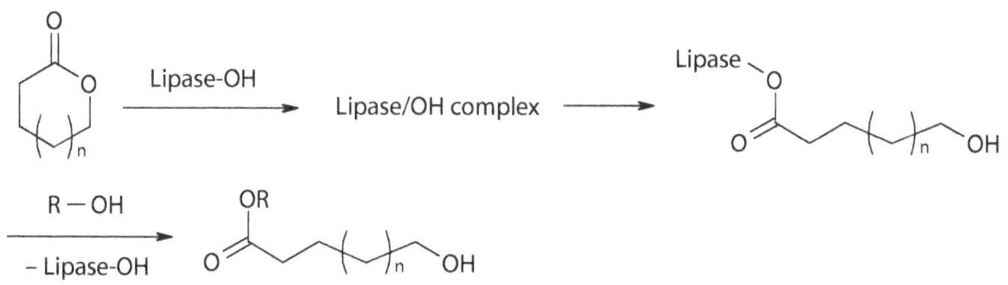

FIGURE 1.14 Enzymatic polymerization of cyclic esters.

chemical modification of aliphatic polyesters, and the second one is based on the synthesis and polymerization of substituted and/or functionalized cyclic esters.

The direct functionalization of aliphatic polyesters is the most straightforward route. Vert and coworkers functionalized PCL by a two-step process. Firstly, the metallation of PCL was carried out by lithium diisopropylamide in order to obtain a poly(enolate), which was then reacted with any electrophile such as naphthoyl chloride (Ponsart et al. 2000), benzylchloroformate (Ponsart et al. 2000), acetophenone (Ponsart et al. 2000), benzaldehyde (Ponsart et al. 2000), carbon dioxide (Gimenez et al. 2001), tritiated water (Ponsart et al. 2001), α-bromoacetoxy-ω-methoxy-poly(ethylene oxide) (Ponsart et al. 2002), and iodine (Nottelet et al. 2006; Figure 1.15).

This approach is very simple, and many functionalized PCL can be synthesized just by changing the nature of the electrophile. Moreover, the absence of any potentially toxic catalyst is a huge advantage. Unfortunately, this strategy is very touchy because degradation of aliphatic polyesters by nucleophilic enolates always takes place at a significant level. Under optimized conditions, degradation can only be limited. Although the process was extended to the functionalization of PLA, degradation is even more difficult to limit, this polymer being more sensitive than PCL. Moreover, the content of functionalization is quite low (<30%) even under optimized conditions.

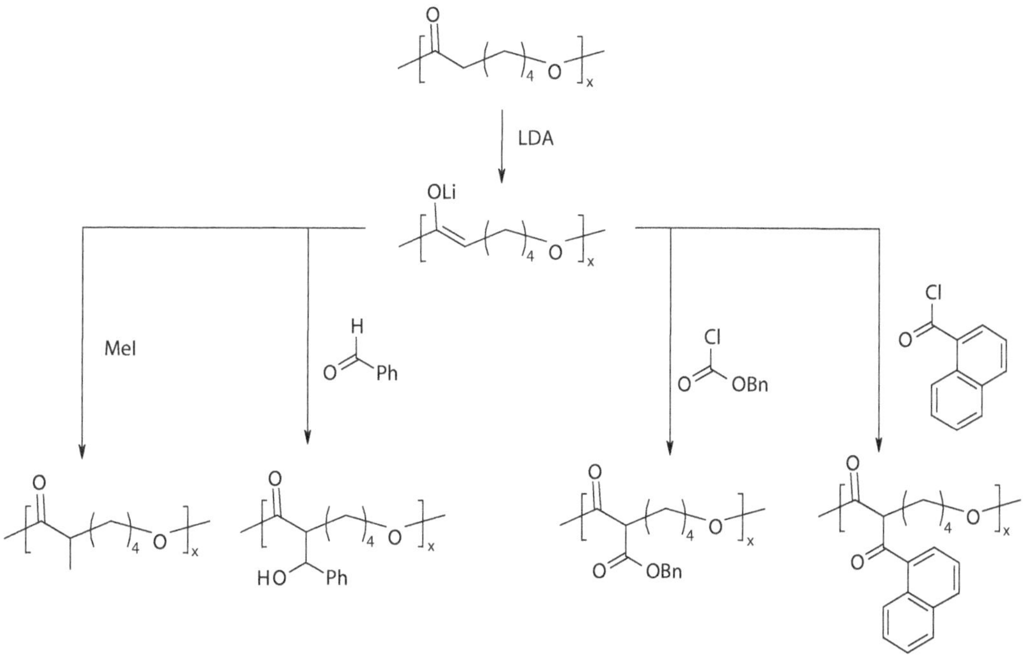

FIGURE 1.15 Functionalization of PCL according to Vert and coworkers.

The important limitations of the process shown in Figure 1.15 prompted chemists to investigate a less direct approach based on the synthesis and polymerization of functionalized cyclic esters (Lou et al. 2003). For that sake, two strategies can be applied. The first one relies on the insertion of the functional group inside the ring, and the second one is based on the substitution of the cyclic ester by a pendent functionalized group.

Table 1.6 shows the conditions used for the polymerization of cyclic esters bearing, inside the ring, functional groups such as an ether (Table 1.6, entries 1–3), a protected amine (Table 1.6, entries 4,5), an unsaturation (Table 1.6, entries 6,7), a ketone (Table 1.6, entry 8), and an amide (Table 1.6, entries 9–12) are present.

Table 1.7 shows cyclic esters bearing a substituent, functionalized or not. As functional groups, aromatics (Table 1.7, entry 1), chloride (Table 1.7, entries 2,3), bromide (Table 1.7, entries 4–6), iodide (Table 1.7, entry 7), alkyne (Table 1.7, entries 8, 9), alkene (Table 1.7, entries 10–15), and epoxide (Table 1.7, entry 16) can be mentioned.

Cyclic esters reported in Tables 1.6 and 1.7 are usually polymerized by tin and aluminum alkoxides. It is worth noting that 6,7-dihydro-2(3H)-oxepinone is an unusual case because this monomer can be polymerized by another mechanism, the ring-opening metathesis polymerization (ROMP) process by using the Schrock's catalyst (Figure 1.16; Lou et al. 2002c).

In terms of chemoselectivity, enzymes are appealing catalysts because they tolerate a wide range of functional groups and even epoxides (Table 1.7, entry 16). Unfortunately, tin and aluminum alkoxides are less tolerant and do not tolerate the presence of alcohols, carboxylic acids, and epoxides. Noteworthily, ketones are tolerated by tin (IV) alkoxides but not by aluminum alkoxides for reasons which still remain unclear (Latere et al. 2002). This issue was overcome by protecting these functional groups prior to polymerization. Table 1.8 shows lactones substituted by protected ketones (Table 1.8, entry 1), alcohols (Table 1.8, entries 2–6), diols (Table 1.8, entry 7), carboxylic acids (Table 1.8, entries 8–15), and amines (Table 1.8, entries 16–19). The choice of the protection group is essential for the success of this strategy. One the one hand, the protected functional groups have to be stable enough to avoid any degradation prior to polymerization and to allow obtaining the monomer in a very high purity, which is a prerequisite to achieve polymerization, at least as far as sensitive initiators are concerned, such as aluminum alkoxides. On the other hand, the protection group has to be removed after polymerization under conditions that prevent degradation from occurring. For this reason, benzylic alcohols and carboxylic esters are widely used because they can be deprotected under neutral conditions rather than in acidic conditions. As a rule, these two conditions are contradictory and it is not always that easy to find a satisfactory compromise between the stability of the monomer and the easy deprotection.

Many of the cyclic esters shown in Tables 1.6 through 1.8 are chiral and possess at least one chiral carbon (R or S). Although the stereoselective polymerization of lactide (Zhong et al. 2004) and β-butyrolactone (Amgoune et al. 2006, Carpentier 2010) is reported in the literature, the stereoselectivity of their polymerization is in general not discussed and racemic mixtures are polymerized. Obviously, more attention should be paid to these aspects in the future because stereoselective polymerization allows the synthesis of aliphatic polyesters with different tacticities, and thus of different properties.

Last but not least, it must be noted that substituted cyclic esters can sometimes be expensive because their synthesis can require several steps from commercially available compounds at the expense of the global yield, which can be low. It is beyond the scope of this chapter to describe all the syntheses of these functionalized cyclic esters, and the reader is invited to read the references given in Tables 1.6 through 1.8. Another limitation relies on the necessity to synthesize a new cyclic ester for any new functional aliphatic polyester.

The functionalization of aliphatic polyesters opens up new avenues in the field of biomaterials. The presence of functional groups allows modifying the biodegradation rate by changing the crystallinity and hydrophilicity or by adding functional groups such as carboxylic acids prone to catalyze the degradation. Interestingly enough, pH-sensitive aliphatic polyesters can be obtained by grafting carboxylic acids or amines along aliphatic polyesters. These groups can be under a neutral or an

TABLE 1.6
Ring-Opening of Cyclic Monoesters Functionalized inside the Ring

Entry	Monomer	Initiator	Reference
1		Bu$_2$SnO, Sn(Oct)$_2$	Mathisen and Albertsson (1989) and Mathisen et al. (1989)
		Al(OiPr)$_3$	Löfgren et al. (1994)
2		Et$_3$Al, H$_2$O	Shirahama et al. (1993, 1996)
3		Zn(II) L-lactate Al (OiPr)$_3$	Kricheldorf and Damrau (1998) and Raquez et al. (2000)
4		ROH/Sn(Oct)$_2$	Trollsas et al. (2000)
5		DBU	Kudoh et al. (2009)
6		Al(OiPr)$_3$	Lou et al. (2001)
7		Al(OiPr)$_3$ Schrock's catalyst	Lou et al. (2002b) Lou et al. (2002c)

(continued)

TABLE 1.6 (continued)
Ring-Opening of Cyclic Monoesters Functionalized inside the Ring

Entry	Monomer	Initiator	Reference
8		1-Phenyl-2-propanol, Sn(Oct)$_2$	Latere et al. (2002)
9		*Porcine pancreatic* lipase Lipase type XIII from *Pseudomonas species* Lipase from *Pseudomonas cepacia* Lipase type VII from *Candida rugosa*	Feng et al. (1999a, 2000) Feng et al. (1999a,b) Feng et al. (1999a,b) Feng et al. (1999b)
10		*Porcine pancreatic* lipase	Feng et al. (2000)
11		*Porcine pancreatic* lipase	Feng et al. (2000)
12		*Porcine pancreatic* lipase	Feng et al. (2000)

ionic form depending upon the pH and their pK$_a$, which affects their solubility in water. Another application relies on the use of these functional groups for the covalent grafting of biologically active molecules, drugs, targeting units, and other functional groups. For that sake, very efficient reactions have to be used under nondegrading conditions. It is not very easy to find a reaction that meets these criteria (Lecomte et al. 2006). Recently, it turned out that the click copper-catalyzed alkyne-azyde Huisgen's cycloaddition reaction (CuAAC) is particularly efficient (Parrish et al. 2005, Riva et al. 2007, Lecomte et al. 2008) in the presence of limited degradation (Lecomte et al. 2008). Moreover, the reaction can take place in water and in organic solvents. Nevertheless, the use of copper salts as catalysts is a severe limitation for biomedical applications. Although they are not as efficient as the CuAAC reaction, several other metal-free reactions have been used such as the esterification reaction between alcohols and carboxylic acids (Renard et al. 2003, Parrish and Emrick 2004), the ring-opening of epoxides by thiols (Lou et al. 2002b), the thiol-ene reaction (Rieger et al. 2005), and the coupling of ketones and oxyamines (Taniguchi et al. 2005, van Horn et al. 2008).

TABLE 1.7

Ring-Opening Polymerization of Functionalized Cyclic Esters

Entry	Monomer	Initiator	Reference
1	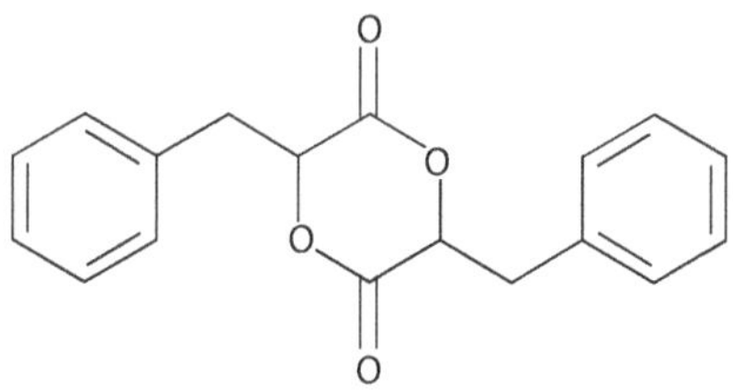	4-tBu-C_6H_4-CH_2OH, Sn(Oct)$_2$	Simmons and Baker (2001)
2	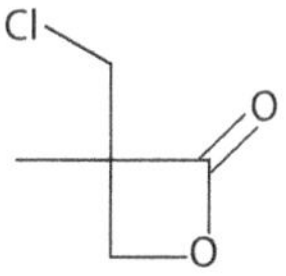	Sn(Oct)$_2$ Al(OiOr)$_3$ Pyridine, Et$_3$N CF$_3$CO$_2$H	Liu et al. (1999) Liu et al. (1999) Liu et al. (1999) Liu et al. (1999)
3	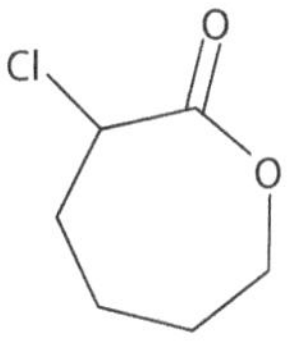	2,2-Dibutyl-2-stanna-1,3-dioxepane	Lenoir et al. (2004)
4	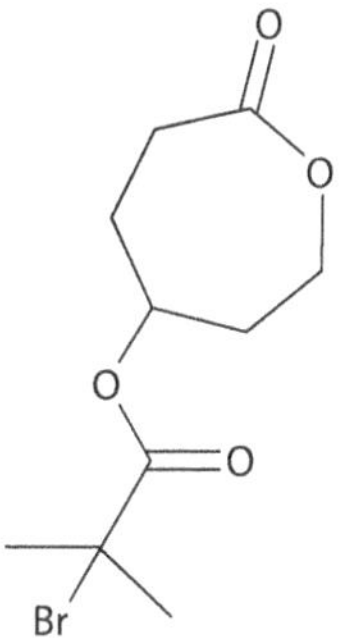	Al(OiPr)$_3$	Mecerreyes et al. (1999)
5	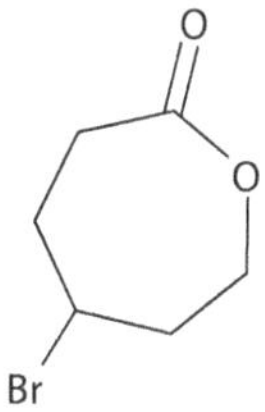	Al(OiPr)$_3$	Detrembleur et al. (2000)
6	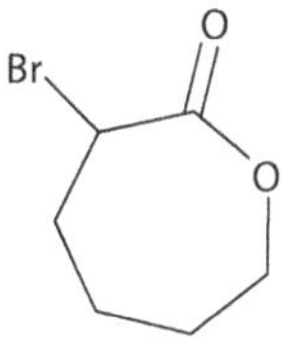	Al(OiPr)$_3$ (copolymerization with εCL)	Wang et al. (2005)

(continued)

TABLE 1.7 (continued)
Ring-Opening Polymerization of Functionalized Cyclic Esters

Entry	Monomer	Initiator	Reference
7		MeOH, Sn(Oct)$_2$	El Habnouni et al. (2009)
8		EtOH, Sn(OTf)$_2$	Parrish et al. (2005)
9		4-tBu-C$_6$H$_4$-CH$_2$OH, Sn(Oct)$_2$	Jiang et al. (2008)
10		Al(OiPr)$_3$	Mecerreyes et al. (2000a)
11		PhCH$_2$OH, Sn(Oct)$_2$	Mecerreyes et al. (2000b)
12		MeO-[Y] (copolymerization with βBL)	Ajellal et al. (2009)

TABLE 1.7 (continued)

Ring-Opening Polymerization of Functionalized Cyclic Esters

Entry	Monomer	Initiator	Reference
13	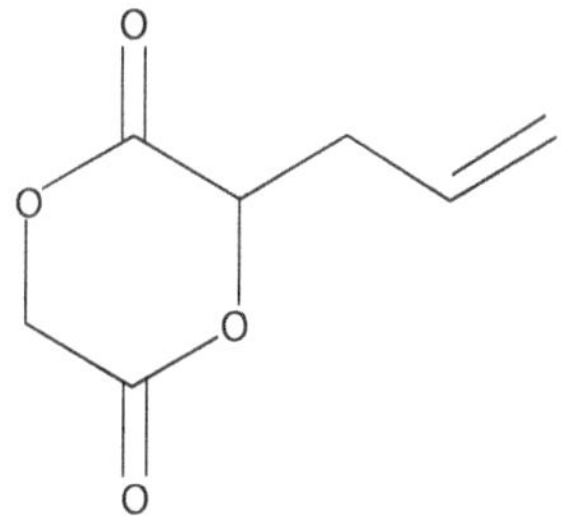	PhCH$_2$OH, Sn(Oct)$_2$	Leemhuis et al. (2008)
14	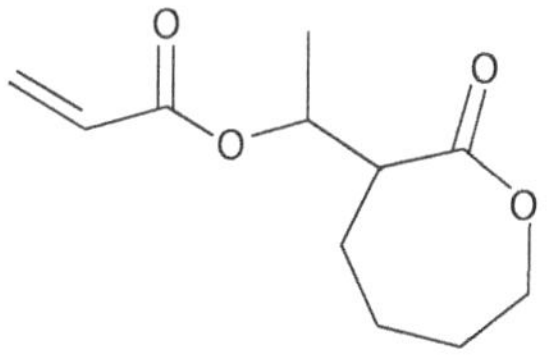	2,2-Dibutyl-2-stanna-1,3-dioxepane (block copolymerization with εCL)	Li et al. (2006)
15	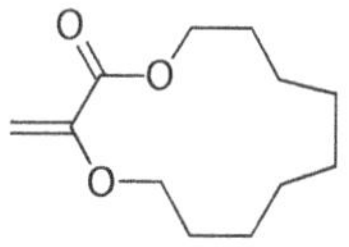	Novozym 435	Veld et al. (2007)
16	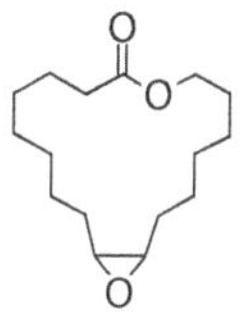	Novozym 435	Veld et al. (2007)

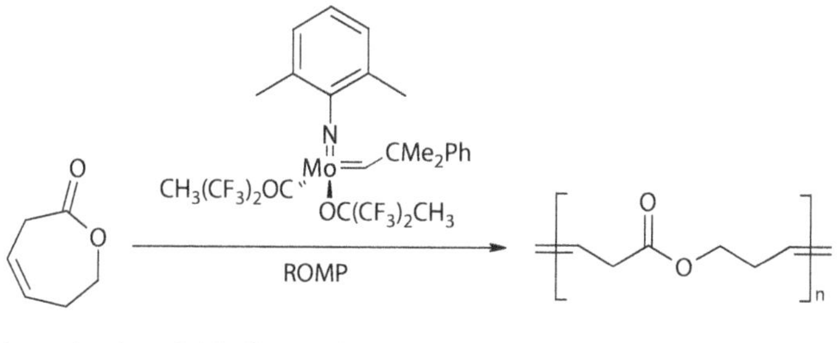

FIGURE 1.16 Polymerization of 6,7-dihydro-2(3*H*)-oxepinone by ROMP.

TABLE 1.8

Ring-Opening Polymerization of Cyclic Esters Bearing Protected Functional Groups

Entry	Monomer	Initiator	Reference
1		Al(OiPr)$_3$	Tian et al. (1997)
2		Sn(Oct)$_2$ (copolymerization with δVL and a dilactone)	Pitt et al. (1987)
3		Al(OiPr)$_3$ (copolymerization with εCL and TOSUO)	Stassin et al. (2000)
4		ROH, Sn(Oct)$_2$	Trollsas et al. (2000)
5		iPrOH,	Leemhuis et al. (2006)
6		iPrOH,	Leemhuis et al. (2006)

TABLE 1.8 (continued)
Ring-Opening Polymerization of Cyclic Esters Bearing Protected Functional Groups

Entry	Monomer	Initiator	Reference
7		ROH, Sn(Oct)$_2$	Trollsas et al. (2000)
8		ROH, Sn(Oct)$_2$	Trollsas et al. (2000)
9		ROH, Sn(Oct)$_2$	Trollsas et al. (2000)
10		Al(OiPr)$_3$ (copolymerization with εCL)	Lecomte et al. (2000)
11		Tetraethylammonium benzoate	Bizzari et al. (2002)

(continued)

TABLE 1.8 (continued)
Ring-Opening Polymerization of Cyclic Esters Bearing Protected Functional Groups

Entry	Monomer	Initiator	Reference
12		Tetraethylammonium benzoate	Barbaud et al. (2004)
13		PEO-OH, Sn(Oct)$_2$	Mahmud et al. (2006)
14		PhCH$_2$OH, Sn(Oct)$_2$	Gerhardt et al. (2007)
15		Al(OiPr)$_3$, Sn(Oct)$_2$, Et$_2$zn	Kimura et al. (1988)
16		Tetrabutylammonium acetate	Flétier et al. (1990)
17		iPrOH, Sn(Oct)$_2$	Blanquer et al. (2010)

TABLE 1.8 (continued)
Ring-Opening Polymerization of Cyclic Esters Bearing Protected Functional Groups

Entry	Monomer	Initiator	Reference
18		Sn(Oct)$_2$ (copolymerization with εCL)	Yan et al. (2010)
19		PhCH$_2$OH, Sn(Oct)$_2$	Gerhardt et al. (2007)

1.4 RADICAL RING-OPENING POLYMERIZATION OF CYCLIC KETENE ACETALS

Although the ring-opening polymerization of cyclic esters is a very widely used technique to prepare aliphatic polyesters, special attention has to be paid to a less common process based on the ring-opening polymerization of cyclic ketene acetals (Bailey 1985, Sanda and Endo 2001, Agarwal 2010). For instance, the polymerization of εCL and 2-methylidene-1,3-dioxepane affords an aliphatic polyester with the same chemical structure. Nevertheless, the polyester obtained from εCL is semicrystalline and the polyester obtained from 2-methylidene-1,3-dioxepane is completely amorphous due to a high degree of branching (Jin and Gonsalves 1997, Undin et al. 2010). The degradation rate of amorphous polymers being faster compared to semicrystalline polymers, the synthesis of amorphous aliphatic polyesters is promising in terms of biomedical applications.

The mechanism of the polymerization of 2-methylidene-1,3-dioxepane is shown in Figure 1.17. Radicals add onto the double bond to afford a new radical. This radical rearrange by ring-opening of the ketal into a new radical, which then propagate. The new C=O double bond is approximately more stable than 50 kcal than the starting C=C double bond. This energy gain, combined with the release of the ring-strain, is the driving force of polymerization (Bailey 1985).

Another possible mechanism is the direct radical polymerization without any ring-opening (Figure 1.18). The polymerization selectivity (ring-opening versus direct polymerization) depends on several parameters such as ring size, substituents, and temperature. As far as the seven-membered cyclic acetal 2-methylene-1,3-dioxepane is concerned, polymerization takes place with 100% of ring-opening at room temperature (Bailey et al. 1982). Conversely, ring-opening is accompanied by direct vinyl polymerization whenever five- and six-membered cyclic acetals are polymerized (Bailey 1985). It is also worth noting that the presence of substituents can affect the selectivity of polymerization. Indeed, substituents able to stabilize the radical formed after the ring-opening step disfavor the direct vinyl polymerization. For instance, polymerization of 2-methylidene-4-phenyl-1,3-dioxolane is fully selective with 100% ring-opening, whereas polymerization of unsubstituted five-membered cycles is usually not selective (Bailey 1985).

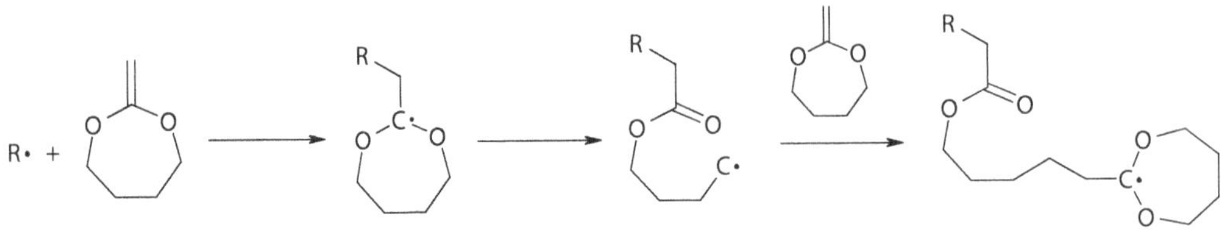

FIGURE 1.17   Polymerization of 2-methylidene-1,3-dioxepane.

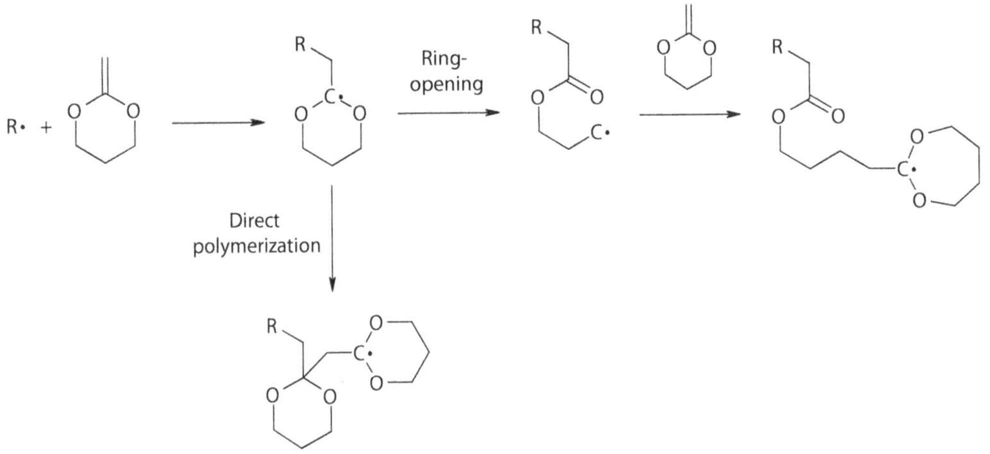

FIGURE 1.18   Polymerization of six-membered cyclic ketene acetals.

This technique of polymerization is not very popular because of its poor selectivity and the very difficult synthesis and purification of cyclic ketene acetals with moderate or even low yields.

1.5   MACROMOLECULAR ENGINEERING OF ALIPHATIC POLYESTERS

The properties of aliphatic polyesters have to be tailored on demand to develop novel biomedical applications. The previous sections of this review dealt with the main processes of polymerization and the examples given were limited to homopolymerizations. It was already shown that the functionalization of aliphatic polyesters is a first tool allowing the extension of the range of their properties. A second possible approach is based on copolymerization or in the modification of the architecture in the frame of macromolecular engineering.

1.5.1   COPOLYMERIZATION

The simplest technique of copolymerization is the polymerization of a mixture of, at least, two comonomers. The distribution of the two comonomers in the final polymer depends on their reactivity ratios. As far as copolymerization is random, the so-obtained copolyesters exhibit averaged properties of the corresponding homopolymers. The sequential polymerization of comonomers provides block copolymers, provided that the polymerization is living. In this review, it was shown that a plethora of initiators and catalysts allows the ring-opening polymerization of cyclic esters to be living. Accordingly, the literature reports a large number of examples dealing with the synthesis of block copolymers by this technique. Interestingly enough, block copolymers exhibit brand new properties compared to the corresponding homopolymers, and thus not just averaged properties as is the case for random copolymers. In many examples, the ring-opening polymerization of cyclic esters is just used to synthesize one block and the other block is obtained by another technique.

For example, PCL-b-PEO diblock copolymers are synthesized by successive polymerization of ethylene oxide and εCL (Kim et al. 2004, Bednarek and Kubisa 2005). These polymers are amphiphilic, PEO being soluble in water, which is not in the case of PCL. The self-association of these amphiphilic copolymers in water affords micelles or hollow spheres, which are used as drug carriers in drug delivery applications. It is worth recalling that PEO is not biodegradable but is biocompatible and bioresorbable provided its molar mass is low enough. A plethora of other works describes examples of fully degradable block copolyesters synthesized by ring-opening polymerization.

1.5.2 Modification of the Architecture

The synthesis of aliphatic polyesters with various architectures is achieved by using the ring-opening polymerization of cyclic esters, this technique being living under appropriate conditions.

A first example is given by star-branched polyesters, which show particular properties such as lower melt viscosities, lower crystallinity, and smaller hydrodynamic volume. Interestingly enough, star-shaped copolyesters contain a higher number of chain-ends compared to linear polymers. Whenever chemists desire to graft a drug, a targeting unit, or a probe, the use of star-shaped polyesters rather than linear polyesters is thus an easy trick to increase their number. Two main techniques allow synthesizing star-shaped copolyesters. In the frame of the arm-first technique, living chains are coupled onto multifunctional (>3) electrophiles (Tian et al. 1994). Conversely, the initiation of the ring-opening polymerization by multifunctional initiators is known as the core-first technique. The most usual conditions are based on the initiation of the polymerization of cyclic esters by polyols in the presence of tin octoate (Trollsas et al. 1998, Lang et al. 2002, Kricheldorf 2004, Choi et al. 2005). Another approach less commonly used relies on the initiation of the polymerization by a spirocyclic initiator (Kricheldorf and Lee 1996, Li et al. 2008). Finally, the arm-first and core-first techniques can be combined to synthesize star-shaped copolymers in order to increase further the range of properties of star-shaped copolymers (Van Butsele et al. 2006, Riva et al. 2011).

Graft polymers are other examples of polymers with a branched architecture (Dai et al. 2009). These polymers can be synthesized by three main methods. The polymerization of chains end-capped by a polymerizable unit, i.e., a macromonomer, is known as the "grafting through approach." The "grafting onto" process is based on the coupling of chains functionalized at one chain-end onto a backbone bearing several complementary functions. The last process, the "grafting from" technique, relies on the initiation of the polymerization by a macroinitiator bearing several initiating units. Hyperbranched aliphatic polyesters are obtained by the polymerization of AB_x inimers made up of an initiator and a polymerizable group. Typical examples are lactones substituted by unprotected alcohols (Liu et al. 1999a, Trollsas et al. 1999, Tasaka et al. 2001, Yu et al. 2005a, Parzuchowski et al. 2006).

The crystallinity, biodegradation rate, and the mechanical properties of biomaterials made up of aliphatic polyesters can be modified by implementing cross-linking reactions. For instance, Hedrick and coworkers reported the cross-linking of PCL bearing pendant acrylates (Mecerreyes et al. 2001). An intramolecular cross-linking takes place under diluted conditions and nanoparticles are then prepared. As far as the cross-linking is carried out at higher concentration, intermolecular cross-linking affords three-dimensional networks. The cross-linking can be carried out in the presence of radicals (Mecerreyes et al. 2001) or photochemically (Riva et al. 2007, Vaida et al. 2008). The cross-linking of linear or star-shaped aliphatic polyesters bearing an unsaturation at least at two chain-ends is also possible (Turunen et al. 2001, Kweon et al. 2003). The cross-linking can be carried out by processes based exclusively on ring-opening polymerization. In this respect, Albertsson and coworkers reported on the ring-opening polymerization of tetrafunctional bis-(ε-caprolactones) (Palmgren et al. 1997, Albertsson et al. 2000). Other cross-linking agents are made up of other polymerizable heterocycles, such as bis-carbonates (Grijpma et al. 1993) or lactones substituted by epoxides (Lowe et al. 2009). All these processes are based on the ring-opening polymerization technique. Nevertheless, networks can also be obtained by the polycondensation of comonomers, one of them being at least trifunctional (Kricheldorf and Fechner 2002, Theiler et al. 2010).

Finally, the cross-linking can be carried out by the coupling of telechelic polymers with polyesters bearing along the chains the complementary functional groups. Several coupling reactions are reported for that sake such as the Michael addition of amines onto acrylates (Theiler et al. 2010), the coupling of ketones and oxyamines (van Horn and Wooley 2007), the click copper(II) catalyzed azide-alkyne cycloaddition (Zednik et al. 2008), and the esterification reaction (Kricheldorf and Fechner 2001, Kricheldorf 2004, Theiler et al. 2010).

Finally, it is worth noting that other examples of architectures such as macrocycles (Li et al. 2006, Jeong et al. 2007, Lang et al. 2002, Hiskins and Grayson 2009, Misaka et al. 2009, Xie et al. 2009) can be found in the literature. Nevertheless, their impact on the field of biomaterial is quite limited and they will not be reported in this review. As far as macrocycles are concerned, the lack of applications as biomaterials can be accounted for by their difficult synthesis even though some recent progresses have been made in this field (Laurent and Grayson 2009).

1.6   CONCLUSIONS

The importance of biodegradable and biocompatible aliphatic polyesters as biomaterials and as environmentally friendly thermoplastics prompted researchers to develop efficient processes for their synthesis, mainly by step-growth polymerization and ring-opening polymerization of cyclic esters. Nowadays, owing to the impressive progresses achieved in this field, it is possible to synthesize aliphatic polyesters with high and controlled molar masses, to control the functionalities at the chain-ends and to graft functional groups all along the chains. The architecture of aliphatic polyesters can also be modified on demand in the frame of macromolecular engineering. Remarkably, ring-opening polymerization can be carried out by a very wide range of polymerization techniques such as anionic, cationic, coordination, organocatalytic, enzymatic, and radical polymerizations. The synthesis of aliphatic polyesters can thus be considered as a mature field. Currently, the challenge for a chemist remains to develop efficient processes for the synthesis of ultrapure aliphatic polyesters, thus noncontaminated by toxic catalytical residues. In this regard, tin(II) bis-(2-ethylhexanoate) is still widely used despite its known toxicity and very difficult extraction. Obviously, more work needs to be done despite the very important progresses being reported in the last few years. In the future, special attention will have to be paid to implement green processes, for instance, by avoiding organic solvents and to the synthesis of a wider range of bio-based aliphatic polyesters from new monomers produced from renewable resources.

ACKNOWLEDGMENTS

CERM is indebted to the "Belgian Science Policy" for general support in the frame of the "Interuniversity Attraction Poles Programme (IAP 6/27)—Functional Supramolecular Systems." P.L. is research associate by the "Fonds National pour la Recherche Scientifique" (FRS-FNRS).

REFERENCES

Agarwal, S. 2010. Chemistry, chances and limitations of the radical ring-opening polymerization of cyclic ketene acetals for the synthesis of degradable polyesters. *Polym. Chem.* 1: 953–954.

Ajellal, N., C.M. Thomas, and J.F. Carpentier. 2009. Functional syndiotactic poly(β-hydroxyalkanoate)s via stereoselective ring-opening copolymerization of rac-β-butyrolactone and rac-allyl-β-butyrolactone. *J. Polym. Sci. A Polym. Chem.* 47: 3177–3189.

Albertsson, A.C., U. Edlund, and K. Stridsberg. 2000. Controlled ring-opening polymerization of lactones and lactides. *Macromol. Symp.* 157: 39–46.

Albertsson, A.C. and R.K. Srivastava. 2008. Recent developments in enzyme-catalyzed ring-opening polymerization. *Adv. Drug Deliv. Rev.* 60: 1077–1093.

Amgoune, A., C.M. Thomas, S. Ilinca, T. Roisnel, and J.-F. Carpentier. 2006. Highly active, productive, and syndiospecific yttrium initiators for the polymerization of racemic β-butyrolactone. *Ang. Chem. Int. Ed.* 45: 2782–2784.

Bailey, W.J. 1985. Free-radical ring-opening polymerization. *Polymer J.* 17: 85–95.

Bailey, W.J., Z. Ni, and S.R. Wu. 1982. Synthesis of poly-ε-caprolactone via a free radical mechanism. Free radical ring opening polymerization of 2-methylene-1,3-dioxepane. *J. Polym. Sci. A Polym. Chem.* 20: 3021–3030.

Barbaud, C., F. Fay, F. Abdillah, S. Randriamahefa, and P. Guérin. 2004. Synthesis of new homopolyester and copolyesters by anionic ring-opening polymerization of α,α′,β-trisubstituted β-lactones. *Macromol. Chem. Phys.* 205: 199–207.

Basko, M. and P. Kubisa. 2006. Cationic copolymerization of ε-caprolactone and L,L-lactide by an activated monomer mechanism. *J. Polym. Sci. A Polym. Chem.* 44: 7071–7081.

Bednarek, M. and P. Kubisa. 2005. Copolymerization with the feeding of one of the comonomers: Cationic activated monomer copolymerization of ε-caprolactone with ethylene oxide. *J. Polym. Sci. A Polym. Chem.* 43: 3788–3796.

Bizzari, R., F. Chiellini, R. Solaro, E. Chiellini, S. Cammas-Marion, and P. Guerin. 2002. Synthesis and characterization of new malolactonate polymers and copolymers for biomedical applications. *Macromolecules* 35: 1215–1223.

Blanquer, S., J. Tailhades, V. Darcos, M. Amblard, J. Martinez, B. Nottelet, and J. Coudane. 2010. Easy synthesis and ring-opening polymerization of 5-Z-Amino-δ-valerolactone: New degradable amino-functionalized (co)polyesters. *J. Polym. Sci. A Polym. Chem.* 48: 5891–5898.

Bourissou, D., B. Martin-Vaca, A. Dumitrescu, M. Graullier, and F. Lacombe. 2005. Controlled cationic polymerization of lactide. *Macromolecules* 38: 9993–9998.

Carpentier, J.-F. 2010. Discrete metal catalysts for stereoselective ring-opening polymerization of chiral racemic β-lactones. *Macromol. Rapid Commun.* 31: 1696–1705.

Casas, J., P.V. Persson, T. Iversen, and A. Cordova. 2004. Direct organocatalytic ring-opening polymerizations of lactones. *Adv. Synth. Catal.* 346: 1087–1089.

Choi, J., I.K. Kim, and C.Y. Kwak. 2005. Synthesis and characterization of a series of star-branched poly(ε-caprolactone)s with the variation in arm numbers and lengths. *Polymer* 46: 9725–9735.

Chuma, A., H.W. Horn, W.C. Swope, R.C. Pratt, L. Zhang, B.G.G. Lohmeijer, C.G. Wade et al. 2008. The reaction mechanism for the organocatalytic ring-opening polymerization of L-Lactide using a guanidine-based catalyst: Hydrogen-bonded or covalently bound? *J. Am. Chem. Soc.* 130: 6749–6754.

Connor, E.F., G.W. Nyce, M. Myers, A. Möck, and J.L. Hedrick. 2002. First example of N-heterocyclic carbenes as catalysts for living polymerization: Organocatalytic ring-opening polymerization of cyclic esters. *J. Am. Chem. Soc.* 124: 914–915.

Coulembier, O., P. Degée, J.L. Hedrick, and P. Dubois. 2006a. From controlled ring-opening polymerization to biodegradable aliphatic polyester: Especially poly(β-malic acid) derivatives. *Prog. Polym. Sci.* 31: 723–747.

Coulembier, O., A.P. Dove, R.C. Pratt, A.C. Sentman, D.A. Culkin, L. Mespouille, P. Dubois et al. 2005. Latent, thermally activated organic catalysts for the on-demand living polymerization of lactide. *Ang. Chem. Int. Ed.* 44: 4964–4968.

Coulembier, O., B.G.G. Lohmeijer, A.P. Dove, R.C. Pratt, L. Mespouille, D.A. Culkin, S.J. Benight et al. 2006b. Alcohol adducts of *N*-heterocyclic carbenes: Latent catalysts for the thermally-controlled living polymerization of cyclic esters. *Macromolecules* 39: 5617–5628.

Dai, W., J. Zhu, A. Shangguan, and M. Lang. 2009. Synthesis, characterization and degradability of the comb-type poly(4-hydroxyl-ε-caprolactone-*co*-ε-caprolactone)-g-poly(L-lactide). *Eur. Polym. J.* 45: 1659–1667.

Detrembleur, C., M. Mazza, O. Halleux, P. Lecomte, D. Mecerreyes, J.L. Hedrick, and R. Jérôme. 2000. Ring-opening polymerization of γ-bromo-ε-caprolactone: A novel route to functionalized aliphatic polyesters. *Macromolecules* 33: 14–18.

Dove, A.P., H. Li, R.C. Pratt, B.G.G. Lohmeijer, D.A. Culkin, R.M. Waymouth, and J.L. Hedrick. 2006. Stereoselective polymerization of rac- and meso-lactide catalyzed by sterically encumbered *N*-heterocyclic carbenes. *Chem. Commun.* 2006: 2881–2883.

Dove, A.P., R.C. Pratt, B.G.G. Lohmeijer, R.M. Waymouth, and J.L. Hedrick. 2005. Thiourea-based bifunctional organocatalysis: Supramolecular recognition for living polymerization. *J. Am. Chem. Soc.* 127: 13798–13799.

Dubois, P., R. Jérôme, and P. Teyssié. 1989. Macromolecular engineering of polylactones and polylactides. I. End-functionalization of poly-ε-caprolactone. *Polym. Bull.* 22: 475–482.

Duda, A., A. Kowalski, S. Penczek, H. Uyama, and S. Kobayashi. 2002. Kinetics of the ring-opening polymerization of 6-, 7-, 9-, 12-, 13-, 16-, and 17-membered lactones. Comparison of chemical and enzymatic polymerizations. *Macromolecules* 35: 4266–4270.

Duda, A., J. Libiszowski, J. Mosnacek, and S. Penczek. 2005. Copolymerization of cyclic esters at the living polymer-monomer equilibrium. *Macromol. Symp.* 226: 109–119.

Duda, A. and S. Penczek. 1991. Anionic and pseudoanionic polymerization of ε-caprolactone. *Macromol. Symp.* 42/43: 135–143.

Duda, A., S. Penczek, A. Kowalski, and J. Libiszowski. 2000. Polymerization of ε-caprolactone and L,L-dilactide initiated with stannous octoate and stannous butoxide—A comparison. *Macromol. Symp.* 153: 41–53.

El Habnouni, S., V. Darcos, and J. Coudane. 2009. Synthesis and ring-opening polymerization of a new functional lactone, α-iodo-ε-caprolactone: A novel route to functionalized aliphatic polyesters. *Macromol. Rapid Commun.* 30: 165–169.

Feng, Y., D. Klee, H. Keul, and H. Höcker. 2000. Lipase-catalyzed ring-opening polymerization of morpholine-2,5-dione derivatives: A novel route to the synthesis of poly(ester amide)s. *Macromol. Chem. Phys.* 201: 2670–2675.

Feng, Y., J. Knüfermann, D. Klee, and H. Höcker. 1999a. Enzyme-catalyzed ring-opening polymerization of 3(*S*)-isopropylmorpholine-2,5-dione. *Macromol. Rapid Commun.* 20: 88–90.

Feng, Y., J. Knüfermann, D. Klee, and H. Höcker. 1999b. Lipase-catalyzed ring-opening polymerization of 3(*S*)-isopropylmorpholine-2,5-dione. *Macromol. Chem. Phys.* 200: 1506–1514.

Flétier, I., A. Le Borgne, and N. Spassky. 1990. Synthesis of functional polyesters derived from serine. *Polym. Bull.* 24: 349–353.

Gazeau-Bureau, S., D. Delcroix, B. Martin-Vaca, D. Bourissou, C. Navarro, and S. Magnet. 2008. Organo-catalyzed ROP of ε-caprolactone: Methanesulfonic acid competes with trifluoromethanesulfonic acid. *Macromolecules* 41: 3782–3784.

Gerhardt, W.W., D.E. Noga, K.I. Hardcastle, A.J. García, D.M. Collard, and M. Weck. 2007. Functional lactide monomers: Methodology and polymerization. *Biomacromolecules* 8: 1735–1742.

Gimenez, S., S. Ponsart, J. Coudane, and M. Vert. 2001. Synthesis, properties and in vitro degradation of carboxyl-bearing PCL. *J. Bioact. Compat. Polym.* 16: 32–46.

Grijpma, D.W., E. Kroeze, A.J. Nijenhuis, and A.J. Pennings. 1993. Poly(L-lactide) crosslinked with spiro-bis-dimethylene-carbonate. *Polymer* 34: 1496–1503.

Gross, R.A., A. Kumar, and B. Kalra. 2001. Polymer synthesis by in vitro enzyme catalysis. *Chem. Rev.* 101: 2097–2124.

Guillaume, S.M., M. Schappacher, and A. Soum. 2003. Polymerization of ε-caprolactone by Nd(BH$_4$)$_3$(THF)$_3$: Synthesis of hydroxytelechelic poly(ε-caprolactone). *Macromolecules* 36: 54–60.

Hamitou, A., R. Jérôme, A.J. Hubert, and P. Teyssié. 1973. A new catalyst for the ring-opening polymerization of lactones to polyesters. *Macromolecules* 6: 651–652.

Hiskins, J.N. and J.M. Grayson. 2009. Synthesis and degradation behavior of cyclic poly(ε-caprolactone). *Macromolecules* 42: 6406–6413.

Hofman, A., R. Szymanski, S. Slomkowski, and S. Penczek. 1984. Structure of active species in the cationic polymerization of β-propiolactone and ε-caprolactone. *Makromol. Chem.* 185: 655–667.

van Horn, B.A., R.K. Iha, and K.L. Wooley. 2008. Sequential and single-step, one-pot strategies for the transformation of hydrolytically degradable polyesters into multifunctional systems. *Macromolecules* 41: 1618–1626.

van Horn, B.A. and K.L. Wooley. 2007. Cross-linked and functionalized polyester materials constructed using ketoxime ether linkages. *Soft Matter* 3: 1032–1040.

Jeong, W., J.L. Hedrick, and R.M. Waymouth. 2007. Organic spirocyclic initiators for the ring-expansion polymerization of β-lactones. *J. Am. Chem. Soc.* 129: 8414–8415.

Jiang, X., E.B. Vogel, M.R. Smith III, and G.L. Baker. 2008. "Clickable" polyglycolides: Tunable synthons for thermoresponsive, degradable polymers. *Macromolecules* 41: 1937–1944.

Jin, S. and K.E. Gonsalves. 1997. A study of the mechanism of the free-radical ring-opening polymerization of 2-methylene-1,3-dioxepane. *Macromolecules* 30: 3104–3106.

Kakuchi, R., Y. Tsuji, K. Chiba, K. Fuchise, R. Sakai, T. Satoh, and T. Kakuchi. 2010. Controlled/living ring-opening polymerization of δ-valerolactone using triflylimide as an efficient cationic organocatalyst. *Macromolecules* 43: 7090–7094.

Kamber, N.E., W. Jeong, R.M. Waymouth, R.C. Pratt, B.G.G. Lohmeijer, and J.L. Hedrick. 2007. Organocatalytic ring-opening polymerization. *Chem. Rev.* 107: 5813–5840.

Kim, M.S., K.S. Seo, G. Khang, S.H. Cho, and H.B. Lee. 2004. Preparation of methoxy poly(ethylene glycol)/polyester diblock copolymers and examination of the gel-to-sol transition. *J. Polym. Sci. A Polym. Chem.* 42: 5784–5793.

Kimura, Y., K. Shirotani, H. Yamane, and T. Kitao. 1988. Ring-opening polymerization of 3(S)-[(benzyloxycarbonyl)methyl]-1,4-dioxane-2,5-dione: A new route to a poly(α-hydroxy acid) with pendant carboxyl groups. *Macromolecules* 21: 3338–3340.

Knani, D., A.L. Gutman, and D.H. Kohn. 1993. Enzymatic polyesterification in organic media. Enzyme-catalyzed synthesis of linear polyesters. I. Condensation polymerization of linear hydroxyester. II. Ring-opening polymerization of ε-caprolactone. *J. Polym. Sci. A Polym. Chem.* 31: 1221–1232.

Kobayashi, S. and A. Makino. 2009. Enzymatic polymer synthesis: An opportunity for green polymer chemistry. *Chem. Rev.* 109: 5288–5353.

Kowalski, A., A. Duda, and S. Penczek. 1998. Kinetics and mechanism of cyclic esters polymerization initiated with tin(II) octoate, 1 Polymerization of ε-caprolactone. *Macromol. Rapid Commun.* 19: 567–572.

Kowalski, A., A. Duda, and S. Penczek. 2000a. Kinetics and mechanism of cyclic esters polymerization initiated with tin(II) octoate, 3. Polymerization of L,L-dilactide. *Macromolecules* 33: 7359–7370.

Kowalski, A., A. Duda, and S. Penczek. 2000b. Mechanism of cyclic ester polymerization initiated with tin(II) octoate. 2. Macromolecules fitted with tin(II) alkoxide species observed directly in MALDI-TOF spectra. *Macromolecules* 33: 689–695.

Kowalski, A., J. Libiszowski, A. Duda, and S. Penczek. 2000c. Polymerization of L,L-dilactide initiated by tin(II) butoxide. *Macromolecules* 33: 1964–1971.

Kowalski, A., J. Libiszowski, K. Majerska, A. Duda, and S. Penczek. 2007. Kinetics and mechanism of ε-caprolactone and L,L-lactide polymerization coinitiated with zinc octoate or aluminum acetylacetonate: The next proofs for the general alkoxide mechanism and synthetic applications. *Polymer* 48: 3952–3960.

Kricheldorf, H.R. 2003. Macrocycles. 21. Role of ring-ring equilibria in thermodynamically controlled polycondensations. *Macromolecules* 36: 2302–2308.

Kricheldorf, H.R. 2004. Biodegradable polymers with variable architectures via ring-expansion polymerization. *J. Polym. Sci. A Polym. Chem.* 42: 4723–4742.

Kricheldorf, H.R., K. Ahrensdorf, and S. Rost. 2004. Star-shaped homo- and copolyesters derived from ε-caprolactone, L,L-lactide and trimethylene carbonate. *Macromol. Chem. Phys.* 205: 1602–1610.

Kricheldorf, H.R. and D.O. Damrau. 1998. Zn L-lactate-catalyzed polymerizations of 1,4-dioxan-2-one. *Macromol. Chem. Phys.* 199: 1089–1097.

Kricheldorf, H.R. and S. Eggerstedt. 1998. Macrocycles 2. Living macrocyclic polymerization of ε-caprolactone with 2,2-dibutyl-2-stanna-1,3-dioxepane as initiator. *Macromol. Chem. Phys.* 199: 283–290.

Kricheldorf, H.R. and B. Fechner. 2001. Polylactones. 51. Resorbable networks by combined ring-expansion polymerization and ring-opening polycondensation of ε-caprolactone or DL-lactide. *Macromolecules* 34: 3517–3521.

Kricheldorf, H.R. and B. Fechner. 2002. Polylactones. 59. Biodegradable networks via ring-expansion polymerization of lactones and lactides. *Biomacromolecules* 3: 691–695.

Kricheldorf, H.R., M. Garaleh, and G. Schwarz. 2005. Tertiary amine-initiated zwitterionic polymerization of pivalolactone—A reinvestigation by means of MALDI-TOF mass spectrometry. *J. Macromol. Sci. A Pure Appl. Chem.* 42: 139–148.

Kricheldorf, H.R., J.M. Jonte, and R. Dunsing. 1986. Polylactones. 7. The mechanism of cationic polymerization of β-propiolactone and ε-caprolactone. *Makromol. Chem.* 187: 771–785.

Kricheldorf, H.R. and S.R. Lee. 1996. Polylactones. 40. Nanopretzels by macrocyclic polymerization of lactones via a spirocyclic tin initiator derived from pentaerythritol. *Macromolecules* 29: 8669–8695.

Kricheldorf, H.R. and G. Schwarz. 2003. Cyclic polymers by kinetically controlled step-growth polymerization. *Macromol. Rapid Commun.* 24: 359–381.

Kricheldorf, H.R., A. Stricker, and D. Langanke. 2001. Polylactones, 50. The reactivity of cyclic and noncyclic dibutyltin bisalkoxides as initiators in the polymerization of lactones. *Macromol. Chem. Phys.* 202: 2525–2534.

Kudoh, R., A. Sudo, and T. Endo. 2009. Synthesis of eight-membered lactone having tertiary amine moiety by ring-expansion reaction of 1,3-benzoxazine and its anionic ring-opening polymerization behavior. *Macromolecules* 42: 2327–2329.

Kweon, H., M.K. Yoo, I.K. Park, T.H. Kim, H.C. Lee, H.S. Lee, J.S. Oh et al. 2003. A novel degradable polycaprolactone networks for tissue engineering. *Biomaterials* 24: 801–808.

Lang, M., R.P. Wong, and C.C. Chu. 2002. Synthesis and structural analysis of functionalized poly(ε-caprolactone)-based three arm star polymers. *J. Polym. Sci. A Polym. Chem.* 40: 1127–1141.

Latere, J.P., P. Lecomte, P. Dubois, and R. Jérôme. 2002. 2-Oxepane-1,5-dione: A precursor of a novel class of versatile semicrystalline biodegradable (co)polyester. *Macromolecules* 21: 7857–7859.

Laurent, B.A. and S.M. Grayson. 2009. Synthetic approaches for the preparation of cyclic polymers. *Chem. Soc. Rev.* 38: 2202–2213.

Lecomte, P., V. D'Aloia, M. Mazza, O. Halleux, S. Gautier, C. Detrembleur, and R. Jérôme. 2000. Synthesis of new hydrophilic γ-substituted poly-ε-caprolactones. Polymer Preprints. *Am. Chem. Soc.* 41(2): 1534–1535.

Lecomte, P. and R. Jérôme. 2004. Recent developments in controlled/living ring opening polymerization. In *Encyclopedia of Polymer Science and Technology*, ed. J. Kroschwitz, pp. 547–565. Hoboken, NJ: Wiley.

Lecomte, P., R. Riva, C. Jérôme, and R. Jérôme. 2008. Macromolecular engineering of biodegradable polyesters by ring-opening polymerization and click chemistry. *Macromol. Rapid Commun.* 29: 982–997.

Lecomte, P., R. Riva, S. Schmeits, J. Rieger, K. Van Butsele, C. Jérôme, and R. Jérôme. 2006. New prospects for the grafting of functional groups onto aliphatic polyesters. Ring-opening polymerization of α- or γ-substituted ε-caprolactone followed by chemical derivatization of the substituents. *Macromol. Symp.* 240: 157–165.

Leemhuis, M., N. Akeroyd, J.A.W. Kruijtzer, C.F. van Nostrum, and W.E. Hennink. 2008. Synthesis and characterization of allyl functionalized poly(α-hydroxy)acids and their further dihydroxylation and epoxidation. *Eur. Polym. J.* 44: 308–317.

Leemhuis, M., C.F. van Nostrum, J.A.W. Kruijtzer, Z.Y. Zhong, M.R. ten Breteler, P.J. Dijkstra, J. Feijen et al. 2006. Functionalized poly(α-hydroxy acid)s via ring-opening polymerization: Toward hydrophilic polyesters with pendant hydroxyl groups. *Macromolecules* 39: 3500–3508.

Lenoir, S., R. Riva, X. Lou, C. Detrembleur, R. Jérôme, and P. Lecomte. 2004. Ring-opening polymerization of α-chloro-ε-caprolactone and chemical modification of poly(α-chloro-ε-caprolactone) by atom transfer radical processes. *Macromolecules* 37: 4055–4061.

Li, H., A. Debuigne, R. Jérôme, and P. Lecomte. 2006. Synthesis of macrocyclic poly(ε-caprolactone) by intramolecular cross-linking of unsaturated end groups of chains precyclic by the initiation. *Ang. Chem. Int. Ed.* 45: 2264–2267.

Li, H., R. Riva, H.R. Kricheldorf, R. Jérôme, and P. Lecomte. 2008. Synthesis of eight and star-shaped poly(ε-caprolactone)s and their amphiphilic derivatives. *Chem. Eur. J.* 14: 358–368.

Libiszowki, J., A. Kowalski, A. Duda, and S. Penczek. 2002. Kinetics and mechanism of cyclic esters polymerization initiated with covalent metal carboxylates, 5. End-group studies in the model ε-caprolactone and L,L-dilactide/tin(II) and zinc octoate/butyl alcohol systems. *Macromol. Chem. Phys.* 203: 1694–1701.

Liu, J. and L. Liu. 2004. Ring-opening polymerization of ε-caprolactone initiated by natural amino acids. *Macromolecules* 37: 2674–2676.

Liu, M., N. Vladimirov, and J.M.J. Fréchet. 1999a. A new approach to hyperbranched polymers by ring-opening polymerization of an AB Monomer: 4-(2-hydroxyethyl)-ε-caprolactone. *Macromolecules* 32: 6881–6884.

Liu, X.Q., M.X. Wang, Z.C. Li, and F.M. Li. 1999b. Synthesis and ring-opening polymerization of α-chloromethyl-α-methyl-β-propiolactone. *Macromol. Chem. Phys.* 200: 468–473.

Loeker, F.C., C.J. Duxbury, R. Kumar, W. Gao, R.A. Gross, and S.M. Howdle. 2004. Enzyme-catalyzed ring-opening polymerization of ε-caprolactone in supercritical carbon dioxide. *Macromolecules* 37: 2450–2453.

Löfgren, A., A.C. Albertsson, P. Dubois, and R. Jérôme. 1995. Recent advances in ring-opening polymerization of lactones and related compounds. *J. Macromol. Sci. C Rev. Macromol. Chem. Phys.* 35: 379–418.

Löfgren, A., A.C. Albertsson, P. Dubois, R. Jérôme, and P. Teyssié. 1994. Synthesis and characterization of biodegradable homopolymers and block copolymers based on 1,5-dioxepan-2-one. *Macromolecules* 27: 5556–5562.

Lohmeijer, B.G.G., R.C. Pratt, F. Leibfarth, J.W. Logan, D.A. Long, A.P. Dove, F. Nederberg, J. Choi, C. Wade, R.M. Waymouth, and J.L. Hedrick. 2006. Guanidine and amidine organocatalysts for ring-opening polymerization of cyclic esters. *Macromolecules* 39: 8574–8583.

Lou, X., C. Detrembleur, and R. Jérôme. 2002a. Living cationic polymerization of δ-valerolactone and synthesis of high molecular weight homopolymer and asymmetric telechelic and block copolymer. *Macromolecules* 35: 1190–1195.

Lou, X., C. Detrembleur, and R. Jérôme. 2003. Novel aliphatic polyesters based on functional cyclic (di)esters. *Macromol. Rapid Commun.* 24: 161–172.

Lou, X., C. Detrembleur, P. Lecomte, and R. Jérôme. 2001. Living ring-opening (co)polymerization of 6,7-dihydro-2(5H)-oxepinone into unsaturated aliphatic polyesters. *Macromolecules* 34: 5806–5811.

Lou, X., C. Detrembleur, P. Lecomte, and R. Jérôme. 2002b. Controlled synthesis and chemical modification of unsaturated aliphatic (co)polyesters based on 6,7-dihydro-2(3H)-oxepinone. *J. Polym. Sci. A Polym. Chem.* 40: 2286–2297.

Lou, X., C. Detrembleur, P. Lecomte, and R. Jérôme. 2002c. Novel unsaturated ε-caprolactone polymerizable by ring-opening metathesis mechanisms. *e-polymers* 34: 1–12.

Lowe, J.R., W.B. Tolman, and M.A. Hillmyer. 2009. Oxidized dihydrocarvone as a renewable multifunctional monomer for the synthesis of shape memory polyesters. *Biomacromolecules* 10: 2003–2008.

Lu, J., R.C. Tappel, and C.T. Nomura. 2009. Mini-review: Biosynthesis of poly(hydroxyalkanoates). *J. Macromol. Sci. C Polym. Rev.* 49: 226–248.

Mahmud, A., X.B. Xiong, and A. Lavasanifar. 2006. Novel self-associating poly(ethylene oxide)-block-poly(ε-caprolactone) block copolymers with functional side groups on the polyester block for drug delivery. *Macromolecules* 39: 9419–9428.

Majerska, K., A. Duda, and S. Penczek. 2000. Kinetics and mechanism of cyclic esters polymerization initiated with tin(II) octoate, 4. Influence of proton trapping agents on the kinetics of ε-caprolactone and L,L-dilactide polymerization. *Macromol. Rapid Commun.* 21: 1327–1332.

Martin, E., P. Dubois, and R. Jérôme. 2000. Controlled ring-opening polymerization of ε-caprolactone promoted by "in situ" formed yttrium alkoxides. *Macromolecules* 33: 1530–1535.

Martin, E., P. Dubois, and R. Jérôme. 2003a. "In situ" formation of yttrium alkoxides: A versatile and efficient catalyst for the ROP of ε-caprolactone. *Macromolecules* 36: 5934–5941.

Martin, E., P. Dubois, and R. Jérôme. 2003b. Preparation of supported yttrium alkoxides as catalysts for the polymerization of lactones and oxirane. *J. Polym. Sci. A Polym. Chem.* 41: 569–578.

Martin, E., P. Dubois, and R. Jérôme. 2003c. Polymerization of ε-caprolactone initiated by Y alkoxide grafted onto porous silica. *Macromolecules* 36: 7094–7099.

Mathisen, T. and A.C. Albertsson. 1989. Polymerization of 1,5-dioxepan-2-one. 1. Synthesis and characterization of the monomer 1,5-dioxepan-2-one and its cyclic dimer 1,5,8,12-tetraoxacyclotetradecane-2,9-dione. *Macromolecules* 22: 3838–3842.

Mathisen, T., K. Masus, and A.C. Albertsson. 1989. Polymerization of 1,5-dioxepan-2-one. 2. Polymerization of 1,5-dioxepan-2-one and its cyclic dimer, including a new procedure for the synthesis of 1,5-dioxepan-2-one. *Macromolecules* 22: 3842–3846.

McLain, S.J. and N.E. Drysdale. 1992. Living ring-opening polymerization of ε-caprolactone by yttrium and lanthanide alkoxides. Polymer Preprints. *Am. Chem. Soc.* 33(1): 174–175.

Mecerreyes, D., B. Atthoff, K.A. Boduch, M. Trollsas, and J.L. Hedrick. 1999. Unimolecular combination of an atom transfer radical polymerization initiator and a lactone monomer as a route to new graft copolymers. *Macromolecules* 16: 5175–5182.

Mecerreyes, D., J. Humes, R.D. Miller, J.L. Hedrick, P. Lecomte, C. Detrembleur, and R. Jérôme. 2000a. First example of an unsymmetrical difunctional monomer polymerizable by two living/controlled methods. *Macromol. Rapid Commun.* 21: 779–784.

Mecerreyes, D., V. Lee., C.J. Hawker, J.L. Hedrick, A. Wursch, W. Volksen, T. Magbitang et al. 2001. A novel approach to functionalized nanoparticles: Self-crosslinking of macromolecules in ultradilute solution. *Adv. Mater.* 13: 204–208.

Mecerreyes, D., R.D. Miller, J.L. Hedrick, C. Detrembleur, and R. Jérôme. 2000b. Ring-opening polymerization of 6-hydroxynon-8-enoic acid lactone: Novel biodegradable copolymers containing allyl pendent groups. *J. Polym. Sci. A Polym. Chem.* 38: 870–875.

Mingotaud, A.-F., F. Cansell, N. Gilbert, and A. Soum. 1999. Cationic and anionic ring-opening polymerization in supercritical CO_2. Preliminary results. *Polym. J.* 31: 406–410.

Mingotaud, A.-F., F. Dargelas, and F. Cansell. 2000. Cationic and anionic ring-opening polymerization in supercritical CO_2. *Macromol. Symp.* 153: 77–86.

Miola-Delaite, C., E. Colomb, E. Pollet, and T. Hamaide. 2000. Anionic ring-opening polymerization of oxygenated heterocycles with supported zirconium and rare earth alkoxides as initiators in protic conditions towards a catalytic heterogeneous process. *Macromol. Symp.* 153: 275–286.

Misaka, H., R. Kakuchi, C. Zhang, R. Sakai, T. Satoh, and T. Kakuchi. 2009. Synthesis of well-defined macrocyclic poly(δ-valerolactone) by "click cyclization." *Macromolecules* 42: 5091–5096.

Möller, M., F. Nederberg, L.S. Lim, R. Kange, C.J. Hawker, J.L. Hedrick, Y. Gu et al. 2000. Sn(OTf)$_2$ and Sc(OTf)$_3$: Efficient and versatile catalysts for the controlled polymerization of lactones. *J. Polym. Sci. A Polym. Chem.* 38: 2067–2074.

Myers, M., E.F. Connor, T. Glauser, A. Möck, G. Nyce, and J.L. Hedrick. 2002. Phosphines: Nucleophilic organic catalysts for the controlled ring-opening polymerization of lactides. *J. Polym. Sci. A Polym. Chem.* 40: 844–851.

van Natta, F.J., J.W. Hill, and W.H. Carothers. 1934. Studies of polymerization and ring formation. XXIII. ε-Caprolactone and its polymers. *J. Am. Chem. Soc.* 56: 455–457.

Nederberg, F., E.F. Connor, M. Möller, T. Glauser, and J.L. Hedrick. 2001. New paradigms for organic catalysts: The first organocatalytic living polymerization. *Ang. Chem. Int. Ed.* 40: 2712–2715.

Nomura, N., A. Taira, T. Tomioka, and M. Okada. 2000. A catalytic approach for cationic living polymerization: Sc(OTf)$_3$-catalyzed ring-opening polymerization of lactones. *Macromolecules* 33: 1497–1499.

Nottelet, B., J. Coudane, and M. Vert. 2006. Synthesis of an x-ray opaque biodegradable copolyester by chemical modification of poly (ε-caprolactone). *Biomaterials* 27: 4948–4954.

Nyce, G.W., T. Glauser, E.F. Connor, A. Möck, R.M. Waymouth, and J.L. Hedrick. 2003. In situ generation of carbenes: A general and versatile platform for organocatalytic living polymerization. *J. Am. Chem. Soc.* 125: 3046–3056.

Oshimura, M. and A. Takasu. 2010. Controlled ring-opening polymerization of ε-caprolactone catalyzed by rare-earth perfluoroalkanesulfonates and perfluoroalkanesulfonimides. *Macromolecules* 43: 2283–2290.

Oshimura, M., T. Tang, and A. Takasu. 2011. Ring-opening polymerization of ε-caprolactone using perfluoroalkanesulfonates and perfluoroalkanesulfonimides as organic catalysts. *J. Polym. Sci. A Polym. Chem.* 49: 1210–1218.

Ouhadi, T., A. Hamitou, R. Jérôme, and P. Teyssié. 1976. Soluble bimetallic μ-oxoalkoxides. 8. Structure and kinetic behavior of the catalytic species in unsubstituted lactone ring-opening polymerization. *Macromolecules* 9: 927–931.

Palard, I., A. Soum, and S.M. Guillaume. 2005. Rare earth metal tris(borohydride) complexes as initiators for ε-caprolactone polymerization: General features and IR investigations of the process. *Macromolecules* 36: 54–60.

Palmgren, R., S. Karlsson, and A.C. Albertsson. 1997. Synthesis of degradable crosslinked polymers based on 1,5-dioxepan-2-one and crosslinker of bis-1-caprolactone type. *J. Polym. Sci. A Polym. Chem.* 35: 1635–1649.

Parrish, B., R.B. Breitenkamp, and T. Emrick. 2005. PEG- and peptide-grafted aliphatic polyesters by click chemistry. *J. Am. Chem. Soc.* 127: 7404–7410.

Parrish, B. and T. Emrick. 2004. Aliphatic polyesters withy pendant cyclopentene groups: Controlled synthesis and conversion to polyester-graft-PEG copolymers. *Macromolecules* 37: 5863–5865.

Parzuchowski, P.G., M. Grabowska, M. Tryznowski, and G. Rokicki. 2006. Synthesis of glycerol based hyperbranched polyesters with primary hydroxyl groups. *Macromolecules* 39: 7181–7186.

Penczek, S., T. Biela, and A. Duda. 2000. Living polymerization with reversible chain transfer and reversible deactivation: The case of cyclic esters. *Macromol. Rapid Commun.* 21: 941–950.

Penczek, S., M. Cypryk, A. Duda, P. Kubisa, and S. Slomkowski. 2007. Living ring-opening polymerizations of heterocyclic monomers. *Prog. Polym. Sci.* 32: 247–282.

Persson, P.V., J. Casas, T. Iversen, and A. Cordova. 2006. Direct organocatalytic chemoselective synthesis of a dendrimer-like star polyester. *Macromolecules* 39: 2819–2822.

Pitt, C.G., Z.H. Gu, P. Ingram, and R.W. Hendren. 1987. The synthesis of biodegradable polymers with functional side chains. *J. Polym. Sci. A Polym. Chem.* 25: 955–966.

Ponsart, S., J. Coudane, J. McGrath, and M. Vert. 2002. Study of the grafting of bromoacetylated α-hydroxy-ω-methoxypoly(ethyleneglycol) onto anionically activated poly(ε-caprolactone). *J. Bioact. Compat. Polym.* 17: 417–432.

Ponsart, S., J. Coudane, J.L. Morgat, and M. Vert. 2001. Synthesis of ^{3}H and fluorescence-labelled poly (DL-Lactic acid). *J. Labelled Comp. Radiopharm.* 44: 677–687.

Ponsart, S., J. Coudane, and M. Vert. 2000. A novel route to poly(ε-caprolactone)-based copolymers via anionic derivatization. *Biomacromolecules* 1: 275–281.

Pratt, R.C., B.G.G. Lohmeijer, D.A. Long, R.M. Waymouth, and J.L. Hedrick. 2006. Triazabicyclodecene: A simple bifunctional organocatalyst for acyl transfer and ring-opening polymerization of cyclic esters. *J. Am. Chem. Soc.* 128: 4556–4557.

Raquez, J.M., P. Degée, R. Narayan, and P. Dubois. 2000. "Coordination-insertion" ring-opening polymerization of 1,4-dioxan-2-one and controlled synthesis of diblock copolymers with ε-caprolactone. *Macromol. Rapid Commun.* 21: 1063–1071.

Raquez, J.M., P. Degée, R. Narayan, and P. Dubois. 2001. Some thermodynamic, kinetic, and mechanistic aspects of the ring-opening polymerization of 1,4-dioxan-2-one initiated by Al(OiPr)₃ in bulk. *Macromolecules* 34: 8419–8425.

Renard, E., C. Ternat, V. Langlois, and P. Guérin. 2003. Synthesis of graft bacterial polyesters for nanoparticles preparation. *Macromol. Biosci.* 3: 248–252.

Rieger, J., K. Van Butsele, P. Lecomte, C. Detrembleur, R. Jérôme, and C. Jérôme. 2005. Versatile functionalization and grafting of poly(ε-caprolactone) by Michael-type addition. *Chem. Commun.* 274–276.

Riva, R., W. Lazzari, L. Billiet, F. Du Prez, C. Jérôme, and P. Lecomte. 2011. Preparation of pH-sensitive star-shaped aliphatic polyesters as precursors of polymersomes. *J. Polym. Sci. A Polym. Chem.* 49: 1552–1563.

Riva, R., S. Schmeits, C. Jérôme, R. Jérôme, and P. Lecomte. 2007. Combination of ring-opening polymerization toward functionalization and grafting of poly(ε-caprolactone). *Macromolecules* 40: 796–803.

Ropson, N., P. Dubois, R. Jérôme, and P. Teyssié. 1995. Macromolecular engineering of polylactones and polylactides. 20. Effect of monomer, solvent, and initiator on the ring-opening polymerization as initiated with aluminum alkoxides. *Macromolecules* 28: 7589–7598.

Sanda, F. and T. Endo. 2001. Radical ring-opening polymerization. *J. Polym. Sci. A Polym. Chem.* 39: 265–276.

Sanda, F., H. Sanada, Y. Shibasaki, and T. Endo. 2002. Star polymer synthesis from ε-caprolactone utilizing polyol/protonic acid initiator. *Macromolecules* 35: 680–683.

Save, M., M. Schappacher, and A. Soum. 2002. Controlled ring-opening polymerization of lactones and lactides initiated by lanthanum isopropoxide. I. General aspects and kinetics. *Macromol. Chem. Phys.* 203: 889–899.

Shen, Y., Z. Shen, Y. Zhang, and K. Yao. 1996. Novel rare earth catalysts for the living polymerization and block copolymerization of ε-caprolactone. *Macromolecules* 29: 8289–8295.

Shibasaki, Y., H. Sanada, M. Yokoi, F. Sanda, and T. Endo. 2000. Activated monomer cationic polymerization of lactones and the application to well-defined block copolymer synthesis with seven-membered cyclic carbonate. *Macromolecules* 33: 4316–4320.

Shirahama, H., K. Mizuma, Y. Kawaguchi, M. Shomi, and H. Yasuda. 1993. Development of new biodegradable polymers. *Kobunshi Ronbunshu* 50: 821–835.

Shirahama, H., M. Shomi, M. Sakane, and H. Yasuda. 1996. Biodegradation of novel optically active polyesters synthesized by copolymerization of (*R*)-MOHEL with lactones. *Macromolecules* 29: 4821–4828.

Shueh, M.L., Y.S. Wang, B.H. Huang, C.Y. Kuo, and C.C. Lin. 2004. Reactions of 2,2′-methylenebis(4-chloro-6-isopropyl-3-methylphenol) and 2,2′-ethylenebis(4,6-di-*tert*-butylphenol) with Mg^nBr_2: Efficient catalysts for the ring-opening polymerization of ε-caprolactone and L-lactide. *Macromolecules* 37: 5155–5162.

Simmons, T.L. and G.L. Baker. 2001. Poly(phenyllactide): Synthesis, characterization and hydrolytic degradation. *Biomacromolecules* 2: 658–663.

Sinha, V. R., K. Bansal, R. Kaushik, R. Kumria, and A. Trehan. 2004. Poly-ε-caprolactone microspheres and nanospheres: An overview. *Int. J. Pharm.* 278: 1–23.

Slomkowski, S., R. Szymanski, and A. Hofman. 1985. Formation of the intermediate cyclic six-membered oxonium ion in the cationic polymerization of β-propiolactone initiated with $CH_3CO^+SbF_6^-$. *Makromol. Chem.* 186: 2283–2290.

Stassin, H., O. Halleux, P. Dubois, C. Detrembleur, P. Lecomte, and R. Jérôme. 2000. Ring-opening copolymerization of ε-caprolactone, γ-triethylsilyloxy-ε-caprolactone and γ-ethylene ketal- ε-caprolactone: A route to hetero-graft copolyesters. *Macromol. Symp.* 153: 27–39.

Stassin, F., O. Halleux, and R. Jérôme. 2001. Ring-opening polymerization of ε-caprolactone in supercritical carbon dioxide. *Macromolecules* 34: 775–781.

Stassin, F. and R. Jérôme. 2002. Effect of pressure and temperature upon tin alkoxide-promoted ring-opening polymerisation of ε-caprolactone in supercritical carbon dioxide. *Chem. Commun.* 232–233.

Stassin, F. and R. Jérôme. 2004. Contribution of supercritical CO_2 to the preparation of aliphatic polyesters and materials thereof. *Macromol. Symp.* 217: 135–146.

Stevels, W.M., M.J.K. Ankoné, P.J. Dijkstra, and J. Feijen. 1996a. A versatile and highly efficient catalyst system for the preparation of polyesters based on lanthanide tris(2,6-di-tert-butylphenolate)s and various alcohols. *Macromolecules* 29: 3332–3333.

Stevels, W.M., M.J.K. Ankoné, P.J. Dijkstra, and J. Feijen. 1996b. Kinetics and mechanism of ε-caprolactone polymerization using yttrium alkoxides as initiators. *Macromolecules* 29: 8296–8303.

Stjerndahl, A., A.F. Wistrand, and A.C. Albertsson. 2007. Industrial utilization of tin-initiated resorbable polymers: Synthesis on a large scale with a low amount of initiator residue. *Biomacromolecules* 8: 937–940.

Stridsberg, K.M., M. Ryner, and A.-C. Albertsson. 2002. Controlled ring-opening polymerization: Polymers with designed macromolecular architecture. *Adv. Polym. Sci.* 157: 42–139.

Szwarc, M. 1956. Living polymers. *Nature* 178: 1168–1169.

Takamoto, T., H. Uyama, and S. Kobayashi. 2001. Lipase-catalyzed synthesis of aliphatic polyesters in supercritical carbon dioxide. *e-polymers* 4:1–6.

Taniguchi, I., A.M. Mayes, E.W.L. Chan, and L.G. Griffith. 2005. A chemoselective approach to grafting biodegradable polyesters. *Macromolecules* 38: 216–219.

Tasaka, F., Y. Ohya, and T. Ouchi. 2001. One-pot synthesis of novel branched polylactide through the copolymerization of lactide with mevalonolactone. *Macromol. Rapid Commun.* 22: 820–824.

Theiler, S., M. Teske, H. Keul, K. Sternberg, and M. Möller. 2010. Synthesis, characterization and in vitro degradation of 3D-microstructured poly(ε-caprolactone) resins. *Polym. Chem.* 1: 1215–1225.

Tian, D., P. Dubois, C. Grandfils, and R. Jérôme. 1997. Ring-opening polymerization of 1,4,8-trioxaspiro[4.6]-9-undecanone: A new route to aliphatic polyesters bearing functional pendent groups. *Macromolecules* 30: 406–409.

Tian, D., P. Dubois, R. Jérôme, and P. Teyssié. 1994. Macromolecular engineering of polylactones and polylactides. 18. Synthesis of star-branched aliphatic polyesters bearing various functional end-groups. *Macromolecules* 27: 4134–4144.

Tortosa, K., T. Hamaide, C. Boisson, and R. Spitz. 2001. Homogeneous and heterogeneous polymerization of ε-caprolactone by neodymium alkoxides prepared in situ. *Macromol. Chem. Phys.* 202: 1156–1160.

Trollsas, M., C.J. Hawker, J.F. Remenar, J.L. Hedrick, M. Johansson, H. Ihre, and A. Hult. 1998. Highly branched radial block copolymers via dendritic initiation of aliphatic polyesters. *J. Polym. Sci. A Polym. Chem.* 36: 2793–2798.

Trollsas, M., V.Y. Lee, D. Mecerreyes, P. Löwenhielm, M. Möller, R.D. Miller, and J.L. Hedrick. 2000. Hydrophilic aliphatic polyesters: Design, synthesis, and ring-opening polymerization of functional cyclic esters. *Macromolecules* 33: 4619–4627.

Trollsas, M., P. Löwenhielm, V.Y. Lee, M. Möller, R.D. Miller, and J.L. Hedrick. 1999. New approach to hyper-branched polyesters: Self-condensing cyclic ester polymerization of bis(hydroxymethyl)-substituted ε-caprolactone. *Macromolecules* 32: 9062–9066.

Turunen, M.P.K., H. Korhonen, J. Tuominen, and J.V. Seppälä. 2001. Synthesis, characterization and crosslinking of functional star-shaped poly(ε-caprolactone). *Polym. Int.* 51: 92–100.

Undin, J., P. Plikk, A. Finne-Wistrand, and A.C. Albertsson. 2010. Synthesis of amorphous aliphatic polyester-ether homo- and copolymers by radical polymerization of ketene acetals. *J. Polym. Sci. A Polym. Chem.* 48: 4965–4973.

Uyama, H. and S. Kobayashi. 1993. Enzymatic ring-opening polymerization of lactones catalyzed by lipase. *Chem. Lett.* 7: 1149–1150.

Vaida, C., P. Mela, H. Keul, and M. Möller. 2008. 2D- and 3D-Microstructured biodegradable polyester resins. *J. Polym. Sci. A Polym. Chem.* 46: 6789–6800.

Van Butsele, K., F. Stoffelbach, R. Jérôme, and C. Jérôme. 2006. Synthesis of novel amphiphilic and pH-sensitive ABC miktoarm star terpolymers. *Macromolecules* 39: 5652–5656.

Varma, I.K., A.C. Albertsson, R. Rajkhowa, and R.K. Srivistava. 2005. Enzyme catalyzed synthesis of polyesters. *Prog. Polym. Sci.* 30: 949–981.

Veld, M.J., A.R.A. Palmans, and E.W. Meijer. 2007. Selective polymerization of functional monomers with Novozym 435. *J. Polym. Sci. A Polym. Chem.* 45: 5968–5978.

Wang, G., Y. Shi, Z. Fu, W. Yang, Q. Huang, and Y. Zhang. 2005. Controlled synthesis of poly(ε-caprolactone)-*graft*-polystyrene by atom transfer radical polymerization with poly(ε-caprolactone-co-α-bromo-ε-caprolactone) copolymer as macroinitiator. *Polymer* 46: 10601–10606.

Westerhausen, M., S. Schneiderbauer, A.N. Kneifel, Y. Söltl, P. Mayer, H. Nöth, Z. Zhong, P.J. Dijkstra, and J. Feijen. 2003. Organocalcium compounds with catalytic activity for the ring-opening polymerization of lactones. *Eur. J. Inorg. Chem.* 2003: 3432–3439.

Williams, S.F., D.P. Martin, D.M. Horowitz, and O.P. Peoples. 1999. PHA applications: Addressing the price performance issue I. Tissue engineering. *Int. J. Biol. Macromol.* 25: 111–121.

Wilson, B.C. and C.W. Jones. 2004. A recoverable, metal-free catalyst for the green polymerization of ε-caprolactone. *Macromolecules* 37: 9709–9714.

Woodruff, M.A. and W. Hutmacher. 2010. The return of a forgotten polymer-polycaprolactone in the 21st century. *Prog. Polym. Sci.* 35: 1217–1256.

Xie, M., J. Shi, L. Ding, J. Li, H. Han, and Y. Zhang. 2009. Cyclic poly(ε-caprolactone) synthesized by combination of ring-opening polymerization with ring-closing metathesis, ring closing enyne metathesis, or "click" reaction. *J. Polym. Sci. A Polym. Chem.* 47: 3022–3033.

Yamashita, M., Y. Takemoto, E. Ihara, and H. Yasuda. 1996. Organolanthanide-initiated living polymerization of ε-caprolactone, δ-valerolactone, and β-propiolactone. *Macromolecules* 29: 1798–1806.

Yan, J., Y. Zhang, Y. Xiao, Y. Zhang, and M.D. Lang. 2010. Novel poly(ε-caprolactone)s bearing amino groups: Synthesis, characterization and biotinylation. *React. Funct. Polym.* 70: 400–407.

Yu, X.H., J. Feng, and R.X. Zhuo. 2005a. Preparation of hyperbranched aliphatic polyester derived from functionalized 1,4-dioxan-2-one. *Macromolecules* 38: 6244–6247.

Yu, T.L., C.C. Wu, C.C. Chen, B.H. Huang, J. Wu, and C.C. Lin. 2005b. Catalysts for the ring-opening polymerization of ε-caprolactone and L-lactide and the mechanistic study. *Polymer* 46: 5909–5917.

Zednik, J., R. Riva, P. Lussis, C. Jérôme, R. Jérôme, and P. Lecomte. 2008. pH-responsive biodegradable amphiphilic networks. *Polymer* 49: 697–702.

Zeng, F., H. Lee, M. Chidiac, and C. Allen. 2005. Synthesis and characterization of six-arm star poly(δ-valerolactone)-*block*-methoxy poly(ethylene glycol) copolymers. *Biomacromolecules* 6: 2140–2149.

Zhang, L., F. Nederberg, J.M. Messman, R.C. Pratt, R.M. Waymouth, J.L. Hedrick, and C.G. Wade. 2007b. Organocatalytic stereoselective ring-opening polymerization of lactide with dimeric phosphazene bases. *J. Am. Chem. Soc.* 129: 12610–12611.

Zhang, L., F. Nederberg, R.C. Pratt, R.M. Waymouth, J.L. Hedrick, and C.G. Wade. 2007a. Phosphazene bases: A new category of organocatalysts for the living ring-opening polymerization of cyclic esters. *Macromolecules* 40: 4154–4158.

Zhong, Z., P.J. Dijkstra, C. Birg, M. Westerhausen, and J. Feijen. 2001. A novel and versatile calcium-based initiator system for the ring-opening polymerization of cyclic esters. *Macromolecules* 34: 3863–3868.

Zhong, Z., P.J. Dijkstra, and J. Feijen. 2004. Controlled synthesis of biodegradable lactide polymers and copolymers using novel in situ generated or single site stereoselective polymerization initiators. *J. Biomater. Sci. Polymer Ed.* 15: 929–946.

2 Hydrogels Formed by Cross-Linked Poly(Vinyl Alcohol)

Gaio Paradossi

CONTENTS

2.1 INTRODUCTION

Poly(vinyl alcohol) (PVA) has a long-lasting history in polymer science [1]. It was first synthesized in 1924 by Herrmann and Haechnel, and, during the time, the interest for this polymer has moved from applications mainly as material for the fabrication of adhesives to the use as component for the design of biomedical devices. The versatility of this synthetic polymer stems in its chemical and physical features as well as in the low cytotoxic impact of derived materials/devices when they interact with tissues and organs.

The hydrophilic nature of the polymer, due to the presence of a hydroxyl group in the repeating unit, can be considered a key feature. The alcoholic functionality is also responsible for the unusual chemical versatility of PVA as it allows the grafting of side chains with various chemical and physical characteristics, providing the tuning of the hydrophilicity of the polymer backbone and the possibility to formulate various cross-linking strategies. Backbone grafting is employed to introduce side chains on PVA able to form cross-links directly with other PVA chains or able to react selectively with complementary functional groups belonging to another PVA chain or to a different polymer. The combination of these features opens a vast number of networking possibilities leading to hydrogels based on new functionalities, degree of cross-linking, and 3D arrangements and pore interconnections. Biomedical applications, addressing soft tissues or cartilage replacements, often require an ad hoc design of hydrogels appropriate for not easily accessible physiological districts and of procedures that should be as less invasive as possible. The approaches described here have been developed to match such concepts with the final aim to provide new answers for the fast growing fields of tissue engineering and drug delivery.

Vinyl alcohol cannot be directly polymerized for the obtainment of PVA as the monomer exhibits cheto-enolic forms [1]. Bulk, solution, emulsion, and suspension radical polymerizations of vinyl acetate monomers are standard methods used to obtain the parent poly(vinyl acetate) polymer. After polymerization of vinyl acetate monomers, PVA is obtained by alkaline hydrolysis of the acetate groups of the precursor polymer. The deacetylation degree is a key parameter influencing the PVA behavior in solution. The already complex behavior of the starting material is further amplified when substituted/grafted PVA chains are considered.

Without any presumption of completeness, this chapter highlights recent results on chemical hydrogels based on PVA sorting out from the literature the most promising and attractive approaches for the obtainment of PVA-based networks without reporting on PVA copolymers as most of the considerations on PVA-based systems can be easily extended to chains having in the average repeating unit a vinyl alcohol residue.

The chapter is organized as follows: Section 2.2 reviews PVA-based chemical hydrogels formed by di-functional cross-linkers (with small and high molecular weights), Section 2.3 considers recent advancements in hydrogels synthesized by using functionalized PVA chains as precursors (macromers) for photo- and chemo-selective cross-linking processes. In this section, "click chemistry" as a method for the synthesis of chemical hydrogels is considered separately as one of the newest and most promising approaches in this field. Section 2.4 addresses hybrid hydrogels combining the features of PVA with other polymer components into a multifunctional matrix.

The synthetic approaches to the chemical gelation of PVA-based networks consist mainly of two synthetic routes: condensation, photo- or chemo-selective cross-linking of functionalized PVA macromers. Within the chemo-selective approach, synthetic strategies involving "click chemistry" [2–4] have been successfully used to design PVA chemical hydrogels. These different networking methods offer the possibility to fabricate PVA-based chemical hydrogels with a wide range of structural, mechanical, and dynamic features addressing different types of applications. In this chapter, we will confine the discussion to PVA hydrogels as materials for the design of polymer device for biomedicine. Properties and potential applications of these hydrogels will be considered along with their synthesis.

Characterization of PVA-based chemical hydrogels can be attempted using various physical methods. Some of them will be reported in this chapter when considering specific examples of PVA chemical networks. However, the analysis based on the concepts introduced by the Flory–Rehner swelling theory and the rubber elasticity theory [5] developed for a cross-linking process in the absence of solvent and further implemented by Peppas [6,7] for the case where the process is carried out in solution still remains a rule of thumb in network characterization. According to the earlier-mentioned approaches, swelling degree and elastic modulus are among the parameters characterizing the hydrogel behavior and, within the assumptions of the method, their combined use allows the determination of the effective degree of cross-linking, the value of the molecular weight between cross-links and the determination of the polymer–solvent interaction parameter, χ, of the Flory's mean-field polymer solution theory.

In the Flory–Rehner theory, the swelling pressure, Π_{swell}, is the sum of the uncoupled contributions π_{osm} and G, the osmotic contribution and the elastic modulus, respectively:

$$\Pi_{swell} = \pi_{osm} - G \tag{2.1}$$

The equilibrium condition is achieved by the hydrogel when these counter-acting contributions to the swelling balance each other, zeroing the equilibrium swelling pressure. This approach assumes that the network can be treated as a set of polymer chains in solution with the volume fractions in the relaxed and swollen states, φ_0 and φ_2, respectively. In terms of change of solvent chemical potential, this is described by the following equation:

$$-\left[\ln(1-\varphi_2)+\varphi_2+\chi\varphi_2^2\right] = V_1\left(\frac{v_e}{V_0}\right)\varphi_0\left[\left(\frac{\varphi_2}{\varphi_0}\right)^{1/3}-\frac{2}{f}\frac{\varphi_2}{\varphi_0}\right] \qquad (2.2)$$

where

(v_e/V_0) is the density of cross-links contributing to the elastic response of the hydrogel

χ is the polymer–solvent interaction parameter

According to this theory, it can be shown that the volumetric swelling ratio, Q, between the swollen gel volume and the dry volume in the case of high hydration ($Q > 10$) is inversely proportional to the cross-link density raised to the 3/5 power.

$$Q \propto \left(\frac{v_e}{V_0}\right)^{-3/5} \qquad (2.3)$$

Also, the storage modulus, G', is linked to the density of cross-links by the relation

$$\frac{v_e}{V_0} = \frac{G'}{RT}\left[\frac{\varphi_0}{\varphi_2}\right]^{1/3} \qquad (2.4)$$

Equations 2.2 through 2.4, based on the swelling and rubber elasticity theories, bear several assumptions: (1) uncoupling of the osmotic pressure from the elastic terms, (2) equivalence between the polymer–solvent interaction parameter in the gel and in the corresponding polymeric solution, and (3) a strict connection of the elastic force to individual cross-links (affine network model). Despite the criticisms accumulated over decades [8,9], the outlined approach is still a valid investigation tool in the chemistry of polymer networks as it gives direct evaluation of the network average degree of cross-links starting from easily accessible experimental parameters.

2.2 CROSS-LINKING PVA WITH DI-FUNCTIONAL MOLECULES

PVA can be cross-linked by condensation of dialdehydes, diisocyanates, or any molecule containing any two functionalities with hydroxyl groups of PVA chains. The cross-linker can be either low or high molecular weight molecules. The cross-links, formed by condensation reactions, will form hydrogels with different mechanical properties depending on the type of used di-functional molecules.

2.2.1 LOW MOLECULAR WEIGHT MOLECULES AS DI-FUNCTIONAL CROSS-LINKERS

Hydroxyl groups of PVA can be easily coupled with low molecular weight molecules bearing two functionalities. The most used cross-linkers are glutaraldehyde, esa-methylene-diisocyanate, and epichlorohydrin. The main drawback in the use of such cross-linkers is their toxicity, which can jeopardize the compatibility of the polymer moiety. This is caused by the slow release of cross-linker molecules after the gelation. For this reason, PVA hydrogels cross-linked with such molecules are usually not considered for biomedical applications. However, they can be regarded as models for studying dynamic processes under controlled structural conditions as the diffusion in confined environments and for probing the dependence of dynamic properties of polymer segments and water from the hydrogel architecture.

Following this line, PVA has been cross-linked with glutaraldehyde by an acetalization reaction. The diffusional properties of PVA chains have been studied during the hydrogel formation by pulse field gradient ^{1}H-NMR [10]. This approach allows the almost complete suppression of the diffusion of water molecules, making possible the monitoring of NMR parameters regarding the polymer as the spin–lattice and spin–spin relaxation times, T_1 and T_2, respectively, and the self-diffusion coefficient, D_{self}, during the gel formation. The study of the time decay of the amplitude of the echo signal revealed three self-diffusion coefficients as a consequence of the heterogeneity of the system. Among these, the smaller D_{self}, accounting for the slowest diffusion process, depends on the gelation time, whereas the other two D_{self} values were not influenced by the progressing of the cross-linking reaction. The different dynamic behaviors found for the PVA/glutaraldehyde hydrogels can be explained by the local inhomogeneities of the matrix and with the presence of polymer concentration gradients. Interestingly, the smaller D_{self} was characterized by a time dependence similar to the trend displayed by the viscosity versus time, with an induction lag independent from the gelation time, followed by a "secondary" gel regime characterized by a decrease of about two orders of magnitude of the D_{self} value. Further time evaluation of the viscosity data was not possible due to the abrupt increase of the viscosity close to the gel point.

Water dynamics in PVA-glutaraldehyde hydrogels was probed by quasi-elastic neutron scattering (QENS) showing a diffusionally quenched water due to the confinement present in the PVA matrix [11].

Short chain cross-linkers were used recently in an investigation aimed to find a suitable material among different polyalcohols to substitute natural lens [12]. The natural matrix constituting the lens is a hydrogel containing a high molecular weight protein as polymer component imparting a low elastic modulus and a high refractive index to the system. These features, together with transparency, are crucial for adapting the lens when it is necessary to focus from short to far distances and vice versa. In general, a low elastic modulus can be obtained with hydrogels based on synthetic polymers with a small cross-linking degree; on the contrary, a high refractive index is accomplished with a high polymer chain density in the hydrogel. PVA was one of the polyalcohols considered as material for lens design. A butylated PVA was dissolved in N-methylpyrrolidone for the cross-linking reaction with 1,4-butane diisocyanate and 1,12-dodecyldiisocyanate. The reaction yielded opaque hydrogels due to macroscopic inhomogeneities caused by the difficulty to have an efficient mixing of the polymer solution with the diisocyanates during the gelation process.

PVA networks were synthesized using epichlorohydrin as cross-linker. Mechanical and swelling experiments confirmed the high cross-linking degree. Crystallization was promoted by drying the PVA hydrogel at 90°C. The presence of chemical cross-links enhanced the crystalline zone of PVA hydrogels [13].

2.2.2 HIGH MOLECULAR WEIGHT MOLECULES (PVA MACROMERS) AS DI-FUNCTIONAL CROSS-LINKERS

In this section, we will discuss the exploitation of high molecular weight di-functional molecules used as cross-linkers according to the same synthetic approach outlined in Section 2.1. In particular, we will focus on the case of a PVA macromer with cross-linking functionalities at the ends of the molecules, which allows the networking of PVA chains in order to obtain a hydrogel without the use and the incorporation of external molecules in the polymer network as outlined in the previous section.

As any other vinyl polymers, also PVA backbone bears head-to-head sequences, although in much lower extent with respect to head-to-tail ones. Each head-to-head sequence contains a vicinal diol that can be selectively and quantitatively splitted by metaperiodate, according to a well-known reaction used in carbohydrate chemistry for the determination of the amount of equatorial hydroxyl groups in a saccharide residue [14].

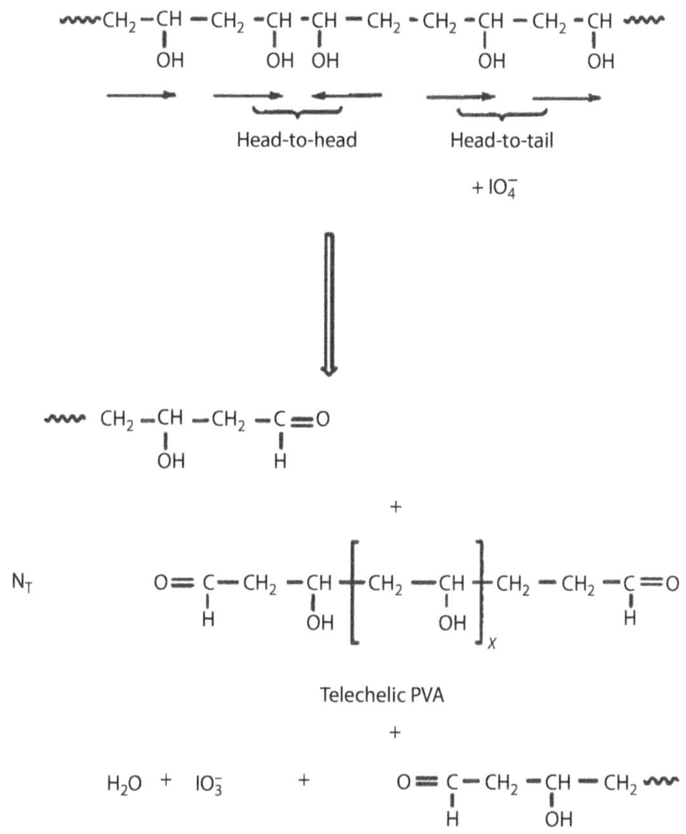

SCHEME 2.1 Oxidation of head-to-head sequences in PVA.

The by-product of the splitting reaction of PVA chains are shorter PVA chains bearing aldehydes as end groups, called telechelic PVA (see Scheme 2.1). The number of telechelic PVA macromers originating from the parent PVA chains is determined by the amount of head-to-head sequences present in the backbone, about 1%–2% (mol/mol) in commercially available samples. These PVA macromers with a molecular weight of about 2000 g/mol can be used as cross-linkers of chains of the same polymer via an acetalization in aqueous medium involving the aldehyde end groups of the telechelic PVA and the hydroxyl moiety of the PVA backbone, both on telechelic and on intact PVA chains. The main advantage of using reactive PVA macromers as cross-linkers is the absence of toxic external molecules participating to the network formation which may impact the biocompatibility of the resulting hydrogel.

The chain dynamics and the diffusive behavior of the confined water of this PVA-based hydrogel were investigated by dynamic light scattering (DLS) around the sol $\leftrightarrow$ gel transition and by QENS to probe the dynamics of the water confined in the hydrogel [15,16]. The DLS autocorrelation function of the scattered light intensity, $g_2(t)$, displays a different phenomenology whether the state of the investigated system is a polymer solution or a gel. In particular in a gel system, the $g_2(t)$ function does not decay to zero, i.e., complete loss of correlation, even extending the delay times over several decades. This is the evidence of the occurrence of an arrest of some of the diffusional processes taking place in the system. In a polymer system undergoing a sol $\leftrightarrow$ gel transition, a progressive quenching of long-range dynamics will occur during gelation.

The autocorrelation functions of the intensity of light scattered by PVA systems collected around the transition point clearly showed the passage from the solution to the gel state. The collected autocorrelation functions were analyzed by means of a fractal model in order to assess a correlation time, τ, accounting for the local collective mesh motions and a characteristic average length, δ, of

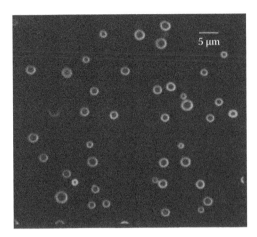

FIGURE 2.1 **(See companion CD for color figure.)** Laser scanning confocal microscopy image after tagging PVA shelled microballoons with fluorescein isothiocyanate (FITC).

the network meshes. These parameters changed during the network formation from about 300 to 100 μs and from 350 to 270 Å for τ and δ, respectively, when passing from the initial stage of the gel formation to the final structured hydrogels. The results were corroborated by the comparison with the equilibrium swelling and G' measurements (Equations 2.1 and 2.2).

A particular application of this reaction involving PVA reactive macromers is represented by the possibility to carry out the networking reaction at the water/air interface [17].

This is achieved by foaming the PVA aqueous solution by high shear stirring while the splitting reaction described in Scheme 2.1 produces the telechelic PVA macromers. In this case, the network formation is limited to the thin aqueous layer between the air contained inside the foam bubbles and the outside. The buildup of hydrated cross-linked PVA shell generates well-dispersed and stable microballoons with a gas (air)-filled core and with dimensions of about 2 μm with a standard deviation of ±1 μm. Confocal laser scanning microscopy is a powerful technique for characterizing such systems. As shown in Figure 2.1, fluorescein-labeled microballoons appear as fluorescent rings when focused on their equatorial plane.

Colloidal dispersions of microbubbles are currently used in the ultrasound imaging, i.e., in echography, as injectable ultrasound contrast agent (UCA). However, the marketed UCAs, based on lipidic shells, present some drawbacks as their short time life in the circulation stream and a poor control on the size distribution with shell diameters ranging from few to more than 10 μm. The lipidic shells, stabilized by a core filled by hydrophobic gasses as perfluorocarbons or sulfur hexafluoride, SF_6, offer a time life of few minutes after parenteral administration. The new PVA shelled microballoons have a shelf life of many months and a longer circulation life than the commercial microbubbles can be expected as well. The biointerface of the PVA-based microballoons has been investigated in terms of biocompatibility, localization of cancer cell lines, delivery of antitumor drugs as doxorubicin, and impact on their viability. This is a step further toward the design of a multifunctional device supporting diagnostic, i.e., ultrasound imaging, as well as therapeutic, i.e., localized drug release, activity.

2.3 PVA MACROMERS AS HYDROGEL BUILDING BLOCKS

In this section, we will consider first the photo-activated cross-linking process occurring on PVA macromers grafted with side chains with different length and chemical composition. In the second part, we will highlight PVA macromers cross-linked via a chemoselective process. Hydrogels obtained by click chemistry will be discussed in the third part.

2.3.1 PHOTO-CROSS-LINKED PVA HYDROGELS

In situ formation of polymer networks offers a number of advantages as (1) reduced invasiveness of surgical treatment as the polymeric precursors can be injected followed by a photo-curing process and (2) complete compliance of the liquid (solutions) polymeric precursors to adapt to the walls of the organ cavity and to its exact shape and dimensions for an efficient replacement. The recent advent of photo-initiators sensible to near UV or visible light and the possibility to convoy and confine light via thin fiber optic guides have contributed to finalize the application of photo-cross-linkable polymer solution for tissue replacement in in vivo treatments.

Modified PVA has been considered as a key polymer component in this synthetic route for the obtainment of chemically cross-linked hydrogels, based on the merging of features as hydrophilicity, good biocompatibility, and tissue-like elasticity. The strategies for introducing vinyl groups ready to photo-polymerize in the presence of a suitable initiator exploit the chemical versatility of PVA secondary –OH groups present as pendants in every repeating unit. In particular, we will examine two coupling reactions to graft PVA with photo-cross-linkable molecules:

1. Grafting acrylate and methacrylate functionalities on PVA by esterification of hydroxyl groups with the vinyl acid.
2. Conjugation of macromers containing a carboxylic function and a vinyl moiety as end groups to the hydroxyl moiety of PVA by activating the carboxylic group with the so-called zero-length cross-linkers, consisting of different types of carbodiimide, as dicyclohexyl carbodiimide (DCC) or 1-ethyl-3-(3-dimethylaminopropyl) carbodiimide (EDC). The comb-like structure resulting from the coupling of the macromers to the PVA chains is then cross-linked by photo-polymerization of the vinyl end groups contained in the side chains.

The networks based on both approaches have a common structural feature consisting in the formation of a photo-polymerized vinyl backbone zipping together two or more PVA chains. In the papers describing the synthetic routes (1) and (2), the vinyl backbone obtained after photo-polymerization is either a poly(acrylate) or a poly(methacrylate), according to the chemical composition of photo-polymerized group grafted to PVA chain. A schematic description of this structure is given in Scheme 2.2. The introduction of vinyl groups impacts with the hydrophilicity of PVA and limits its solubility in water. For this reason, the degree of substitution of grafted PVA is low, typically below 5%. However, this is not an obstacle to efficiently cross-link PVA chains due to the zipping mechanism at the base of the networking.

Hydrogels obtained from photo-polymerization of PVA macromers grafted with acrylate according to route (1) were comparatively studied with PVA chemically cross-linked with glutaraldehyde (see Section 2.1) and physically cross-linked hydrogel domains obtained by freeze–thaw cycles [18]. This work is a systematic study on the factors influencing the network structure and evidences that the macromer concentration and the type of networking process are key role factors influencing the

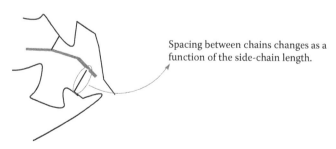

Spacing between chains changes as a function of the side-chain length.

SCHEME 2.2 (See companion CD for color figure.) Representation of the zipping of the PVA chains (in black) occurring during the photopolymerization of vinyl side chains.

network architecture. Equilibrium swelling ratio and compressive modulus of hydrogels based on acrylated PVA macromers with a degree of substitution of 7% were photo-polymerized at different concentrations. From Equation 2.2, the cross-linking density is readily obtained from the equilibrium degree of swelling, Q, with $Q = 1/\varphi_2$. The structural parameter M_c, i.e., the average molecular weight of the polymer chain between cross-linking points, is linked to the density of cross-links and can be determined according to

$$M_c = \frac{\rho_p}{(\nu_e/V_0) + (2\rho_p/M_n)} \tag{2.5}$$

where
  ρ_p is the density of the dry polymer
  M_n is the number average molecular weight of the polymer chains before the cross-linking

The higher value of M_c measured in hydrogels derived from acrylated PVA macromers with respect to the corresponding theoretical value, $M_{c,th}$, showed that the polymer network synthesized at lower concentration contained a larger number of chain loops formed by the reaction of intramolecular vinyl functionalities, not contributing to the elastic response to swelling of the hydrogels. Moreover, a comparative study of the observed equilibrium swelling degrees of PVA hydrogels obtained (1) by photo-polymerization of macromers with an acrylation degree of 7 mol%, (2) by freeze–thaw cycling with a final crystallinity of 2% determined by differential scanning calorimetry, and (3) by acetalization in the presence of 2 and 7 mol% glutaraldehyde, evidenced that for PVA hydrogels cross-linked in the presence of 7% glutaraldehyde, a Q value of 4.5 was obtained, lower than the Q measured for the network based on the acrylated PVA macromers. This finding was explained with the substantially different cross-linking mechanisms occurring in the two types of processes. At the base of the photo-polymerization, there is the chain polymerization propagating through the pendant vinyl groups attached to PVA backbones, whereas the acetalization reaction is based on the random condensation of fast diffusing glutaraldehyde molecules and of alcohol functionalities of the polymer. In the former case, networking will contain a considerable amount of loops and inhomogeneities, whereas the latter process will take place with a uniform cross-linker distribution in the reaction medium.

According to procedure (1), Mühlebach et al. [19] reported the modification of PVA with acrylic and methacrylic acid and with 2-vinyl-4,4-dimethylazlactone. Photoinitiated cross-linking was carried out by UV irradiation (280–320 nm) in the presence of Irgacure 2959, a photo-initiator soluble in water at concentration lower than 0.1%. Biomedical applications generally require highly hydrated polymer networks. In this respect, PVA is a suitable material for the design of polymer scaffolds as hydrophilicity is one of the main features of this polymer.

Acrylamide-grafted PVA chains with different molecular weights and degree of acetylation were prepared by reacting N-(2,2-dimethoxyethyl) acrylamide in water with PVA in the presence of HCl [20]. The study was aimed to the assessment of the dependence of the swelling and of the mechanical properties of the PVA-based network on structural parameters as degree of substitution (DS), molecular weight, degree of acetylation (DA), and modality of hydrogel formulation as the percent of macromer used in the photo-cross-linking reaction.

An interesting trans-esterification reaction proposed by Hennink et al. [21,22] between glycidyl methacrylate and dextran, a natural water-soluble polysaccharide, was efficiently applied to PVA. The trans-esterification of PVA with glycidyl methacrylate in dimethyl sulfoxide (DMSO) in the presence of 4-dimethylaminopyridine (DMAP) allowed the obtainment of a wide range of DS [23].

As these chemical hydrogels are designed for biomedicine, the focus was on methacryloyl substituted PVA macromers with low DS values, suitable for the formation of hydrogel with high hydration degree. Once the methacrylate substitution was introduced on PVA, the photo-cross-linking was accomplished by irradiation of UV light at 365 nm in the presence of Irgacure. The outlined strategy is suitable for the formation of biocompatible hydrogel in the vitreal cavity, providing an alternative vitreous substitute for vitrectomy usually consisting in the replacement of the vit-reous with silicone oil. With the method described earlier, many chemical PVA hydrogels were designed as scaffolds for tissue regeneration. This application requires hydrogels in a monolithic or "wall-to-wall" morphology. However, the possibility to fabricate nano- or micron-sized hydrogels, i.e., microgels, opens other fields of application as drug delivery. In the context, it was reported that a PVA-based chemical microgel can be obtained as a colloidal stable dispersion by carrying out the photo-cross-linking reaction of methacryloyl derivative of PVA with different DS in the presence of an aqueous solution of dextran with a molecular weight of about 40 kDa, Dextran T40. Due to their incompatibility, PVA and dextran form in water a microemulsion under high shear stirring. This is an application of the so-called water-in-water emulsion method. The absence of organic phases in the synthesis of microgels makes this method particularly suitable for designing polymer-based micro-devices for close contact with tissues or organs. The aqueous phase containing the methacryloyl PVA derivative is sensible to photo-cross-linking in the presence of the photoinitiator Irgacure. In this way, it was possible to obtain hydrogel beads based on cross-linked PVA with a diameter of about 2 μm and with a quite narrow distribution ($\sigma = \pm 1$ μm), potentially usable as injectable drug delivery system. The chemical versatility of PVA is also an asset for the surface coating of the microgel par-ticles. Succinoylation of the microbeads surface imparts a negative ζ potential, the electric potential at the microgel–liquid interface, allowing the loading of doxorubicin, a positively charged antitumor drug, by favorable charge interaction with microgel particles. The cargo amount can be as much as the 50% by weight of the dry microgel, providing a large dose range for treating tumor cells [24].

To enhance the biointerface features of these PVA-based microgels, other surface coatings were also applied as hyaluronate. Hyaluronic acid (HA) is a glycosaminoglycan contained in synovial fluids and cartilages of mammalians. It is a bioactive molecule as it participates to several cellular processes, besides having an important dumping and lubrication function in the articular joints. In 1990, HA was assessed as a ligand for CD 44 transmembrane receptor [25], a protein regulating bioadhesion and proliferation process and overexpressed by tumor cells. This has opened the pos-sibility to target tumor cells and to enhance the bioavailability of the antitumor drug delivered by an HA-functionalized device. The HA functionalization of PVA microgels surface is easily obtained by partial oxidation of the polysaccharide by metaperiodate, followed by condensation with the alcohol functionalities of the microgel particles [26].

The water-in-water emulsion polymerization is a suitable method to incorporate in the PVA network other vinyl comonomers. The obtainment of thermoresponsive spherical-shaped micro-gels using this method has been recently reported [27,28]. This was accomplished by carrying out the "water-in-water" emulsion polymerization method in the presence of N-isopropylacrylamide (NiPAAm) monomer. In this variant of the method, NiPAAm monomers distribute in both the dispersing phase water/dextran T40 and the disperse phase containing water/methacrylated PVA. The photo-polymerization incorporates NiPAAm in the PVA-rich phase, whereas oligomers of NiPAAm are formed in the dextran-rich phase. After polymerization, NiPAAm oligomers are washed out in the purification step. It is known that polymerized NiPAAm shows an LCST behav-ior and, correspondingly, cross-linked NiPAAm displays a volume-phase transition. The behavior of these PVA/poly(methacrylate-co-NiPAAm) microgel was studied with respect to their average size and size distribution, swelling, and release properties. It was observed that the stirring speed is a key parameter for controlling the amount of incorporated NiPAAm, the particle size and the sharpness of the volume-phase transition. The volume-phase transition temperature (VPTT) of the microgel was evaluated around 38°C and 34°C for microgels with a NiPAAm/methacrylate molar ratio of 0.8 and 2.4, respectively. A volume reduction of about 50% of the microgel particles was accompanied by a size decrease of the network meshes from 4 to 2 nm when temperature was brought from 25°C to 40°C, as determined by size exclusion experiments with fluorescein-tagged dextran, FITC-dextran, molecules. O-Succinoylated microgels were loaded with doxorubicin, a well-known antitumor drug, by exploiting the favorable electrostatic interaction between negatively charged microgel surface and positively charged doxorubicin. The drug release was enhanced at temperatures higher than the VPTT, i.e., 40°C, due to the microgel shrinking and the consequent increase of the surface/volume ratio.

Recently it has been shown [29] that structural and diffusion features of the PVA-based hydrogels can be tuned by choosing the cross-linking strategy. The network structure and the macromolecular drug release of hydrogels were discussed in terms of the cross-linking modality. Curing acryloyl-derivative of PVA (DS = 2%, deacetylation degree 83%) aqueous solutions by photo- or by redox reaction polymerizations yielded hydrogels with different behaviors in terms of swelling, release of FITC-dextrans with 20 and 4 kDa, used as model high molecular weight drugs, and polymer mass loss. The different curing methods gave indication of a higher extent of cross-linking in the case of the photo-polymerized PVA network. The two curing strategies can be used for tuning the release properties of PVA-based hydrogels based on radical cross-linking.

Synthetic routes based on approach (2) were reported by Anseth et al. [30] for obtaining bio-degradable PVA-based chemical hydrogels. In this work, a degradable oligoester derived from hydroxyethyl methacrylate, lactide and succinic anhydride, HEMA-$(Lac)_n$-Suc, was prepared (step 1) by ring-opening polymerization of molten lactide in the presence of stannous octoate at 110°C, followed by an esterification with succinic anhydride in DMSO using DMAP/pyridine as catalyst (step 2). This macromer is then grafted by N,N'-dicyclohexylcarbodiimide (DCC) chemistry to the PVA with DS ranging from about 1 to 15 mol% (see Scheme 2.3). Mass loss determinations

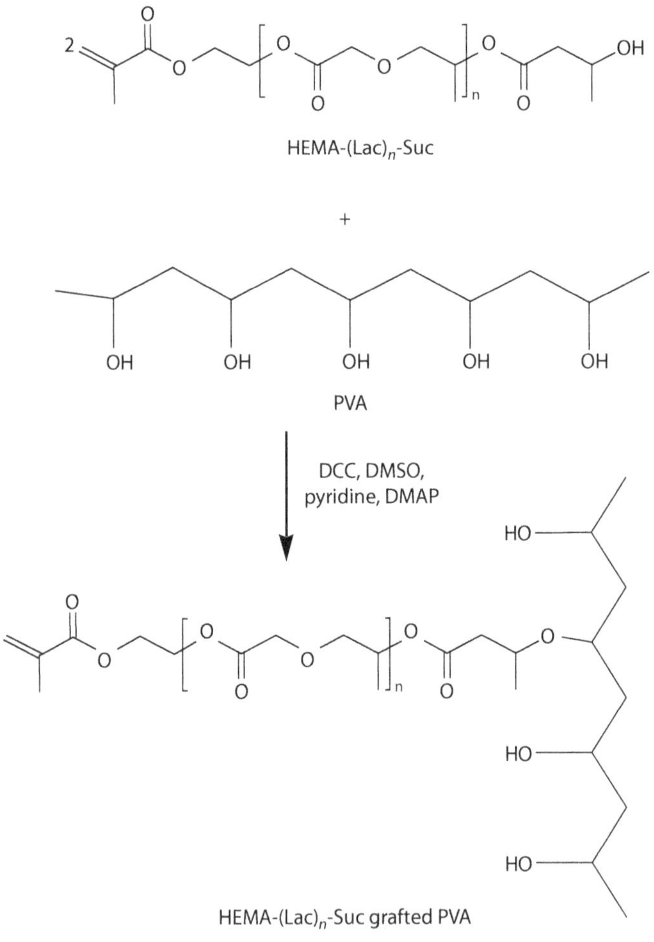

SCHEME 2.3 Synthesis of HEMA-$(Lac)_n$-Suc grafted PVA.

as a function of time showed the dependence of network degradation in phosphate-buffered solution at pH 7.4 (PBS) and at 37°C from the number of lactide residues and from the degree of substitution. Hydrolysis of lactide residues in the side chains was effective in controlling the network degradation. In addition, tuning of the hydrophobicity of the matrix was possible by controlling the degree of substitution of PVA chains as demonstrated by the improvement of valve interstitial cells bioadhesion to PVA grafted HEMA-(Lac)$_4$-Suc surface, characterized by the higher hydrophobicity among the investigated degradable scaffolds.

With a similar approach, degradable ester-acrylate chains terminating with a carboxylic acid group were coupled by DCC chemistry to yield ester-acrylate anhydride molecules. These anhydrides were conjugated with PVA hydroxyl moiety in DMSO in the presence of triethylamine [31], see Scheme 2.4.

The kinetic analysis of the hydrogel degradation caused by esters hydrolysis was treated as a pseudo-first-order process:

$$\frac{n_{ester}}{n_0} = \exp(-k't) \tag{2.6}$$

where

n_{ester} and n_0 are the number of ester blocks at time t and at $t = 0$, respectively
k' is the apparent kinetic constant

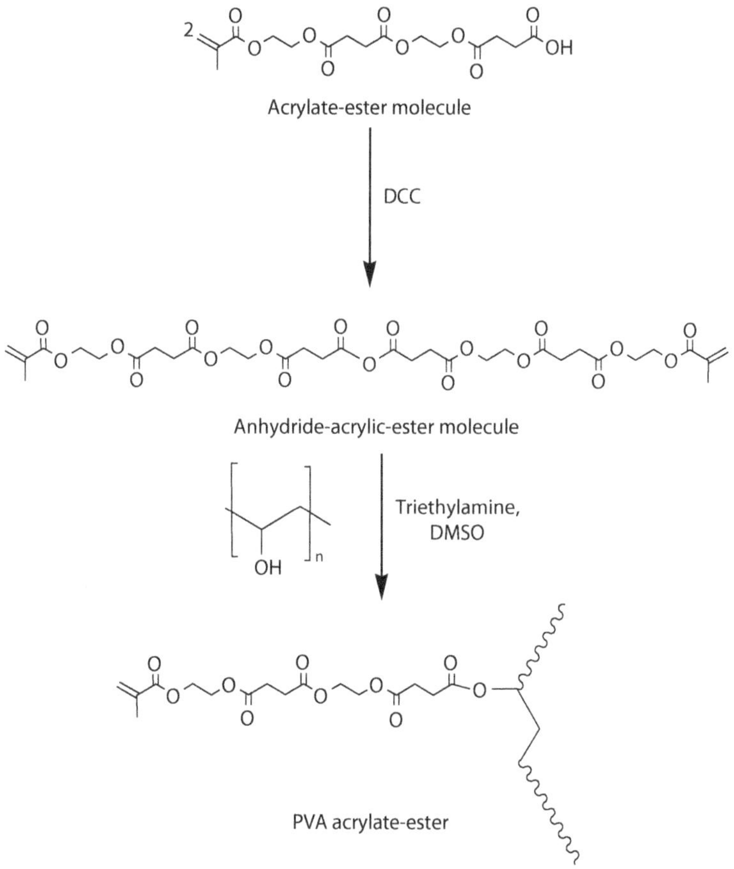

SCHEME 2.4   Synthesis of PVA acrylate-ester for photopolymerization.

For the structural features of these photo-polymerized hydrogels, there is a direct connection of the number of ester blocks with the cross-link density (ν_e/V_0). It is then possible to write Equation 2.3 in the form

$$Q \propto \exp\left(\frac{3}{5}k't\right) \tag{2.7}$$

2.3.2   HYDROGELS FABRICATED BY CHEMOSELECTIVE DERIVATIZATION OF PVA

A possible bias to photo-cross-linking is the production during the network formation of free radicals triggered by the photoinitiation of the cross-linking reaction that may damage the vitreal cavity tissues, although their presence should be expected in transient and low concentration. To circumvent the impact of cross-linking reaction based on radicals, Michael addition and more in general nucleophilic additions offer the possibility to cross-link in situ and in aqueous medium several polymers functionalized with suitable pairs of chemoselective coupling groups. In this way, the outlined drawbacks of the photo-polymerization can be circumvented. Similarly to the approach described in Section 2.3.1, (2), multifunctional PVA macromers were exploited for the obtainment of hydrogels with high elasticity. However, in this approach, replacing the photo-polymerization of vinyl moiety of the derivatized PVA, the cross-linking efficacy is based on the chemoselectivity of the substituents and directly linked to the kinetic and thermodynamic driving forces of the reactions involving the chemical functionalities born on the PVA chains.

Following this line, Ossipov et al. [32] prepared a set of PVA macromers bearing nucleophilic and electrophilic complementary functional groups. In this study, cysteine-functionalized or thiolated PVA were coupled with maleimide functionalized PVA chains obtaining networks with 1:1 stoichiometry of the cross-linking reaction. Derivatization of the hydroxyl groups of the PVA main chain with nucleophilic moieties was achieved by exploiting the chemical versatility of the alcohol functionality, coupled with known methods of peptide synthesis as 1,1′-carbonyldiimidazole (CDI) chemistry. This synthetic route allowed the activation of PVA hydroxyl groups and the coupling with molecules bearing appropriate groups, i.e., amines, thiols, and cysteine (see Scheme 2.5). Protection is often required on the nucleophilic groups to avoid the conjugation with the O-imidazolylcarbonyl PVA intermediate. The protection is removed in the presence of trifluoroacetic acid to allow coupling of the nucleophilic group with the complementary electrophilic functions and chain cross-linking.

The approach used to introduce electrophilic moieties on PVA is based [32] on the coupling of amino-derivatized PVA with N-hydroxysuccinimide ester of acids containing electrophilic groups as maleimide, acrylate, α-iodoacetyl functionalities (see Scheme 2.6). The elastic properties and the degree of swelling of the resulting hydrogels were, as expected, determined by the stability and density of the cross-links contributing to the network elasticity. The same group [33] designed hydrogel based on two PVA components, containing aldehydes and hydrazide side chains, respectively. In this study, encapsulation of murine neuroblastoma cells in the hydrogel was carried out during the cross-linking reaction occurring in aqueous medium at physiological condition of pH and temperature.

Electrophilic–nucleophilic chemoselective coupling, based on Michael addition of cysteine-modified PVA macromer on methacrylate-functionalized PVA, was also reported in the literature [34]. Telechelic PVA macromers prepared according to Scheme 2.2 were coupled with the amine group of cysteine, present in large excess, by reductive amination. The thiol end-capped PVA chains were then conjugated with a methacrylic derivative of PVA with DS of 4 mol%, yielding a hydrogel with thiol cross-links. The kinetics of the cross-links formation depended on the reactivity of the functional groups involved in the reaction. A possible drawback of this thiol containing cross-links could be the sensitivity of the hydrogel to oxidative conditions.

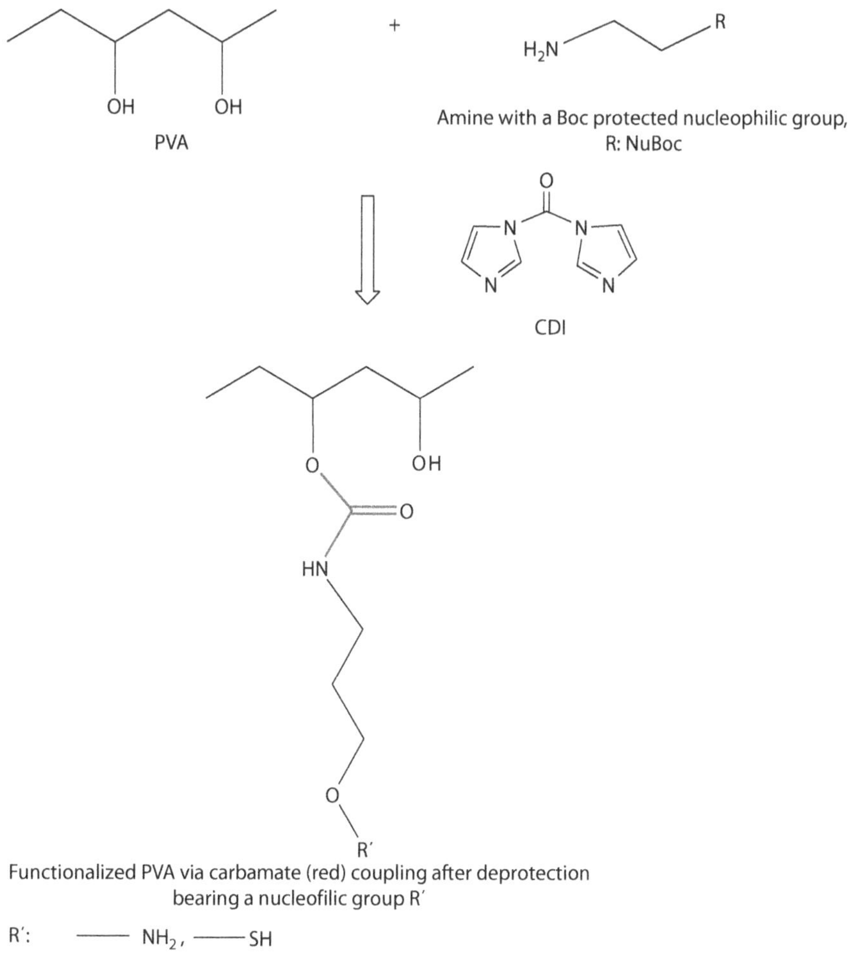
PVA
OH
OH
+
H_2N
R
Amine with a Boc protected nucleophilic group,
R: NuBoc
CDI
O
N
N
OH
O
HN
R′
Functionalized PVA via carbamate (red) coupling after deprotection bearing a nucleofilic group R′
R′: —— NH_2, —— SH

SCHEME 2.5 **(See companion CD for color figure.)** CDI chemistry on PVA.

2.3.3 PVA "Click" Hydrogels

Polymer chemistry used "click chemistry" since early stages of this highly selective synthetic pathway as an efficient cross-linking method for the formation of structurally controlled networks. The general features of a "click" reaction are (1) mild reaction conditions, (2) high yields, (3) stereospecificity, (4) absence of by-products, (5) no side reactions, and (6) a high thermodynamic driving force (usually more than 20 kcal/mol). Moreover, the possibility to carry out click reactions in water is fundamental for hydrogel fabrication. One of the most used click reactions, among the copper(I)-catalyzed Huisgen reactions, is the chemoselective 1,3-cycloaddition between alkynyl-functionalized and azido-functionalized macromers with the multiple formation of triazole cross-links. The general features of this reaction are reported in Scheme 2.7.

The first application of click chemistry to PVA appeared recently in the literature [35]. In this paper, click chemistry was exploited to construct hydrogels based on clicks between PVA chains bearing acetylenic and azido functionalities on the side chains grafted by means of CDI chemistry. The reported hydrogels are formed by activation of PVA chains with *O*-(imidazol-1-ylcarbonyl)-functionality. In a second step, the reaction with primary or secondary amines containing alkyne or azide functionalities allows the grafting of the PVA chains. In this way, the macromolecular precursors are ready to form PVA-based hydrogels by clicking the complementary functionalities grafted on the PVA chains. The design of this class of hydrogel should be bioorthogonal, i.e., the functionalities

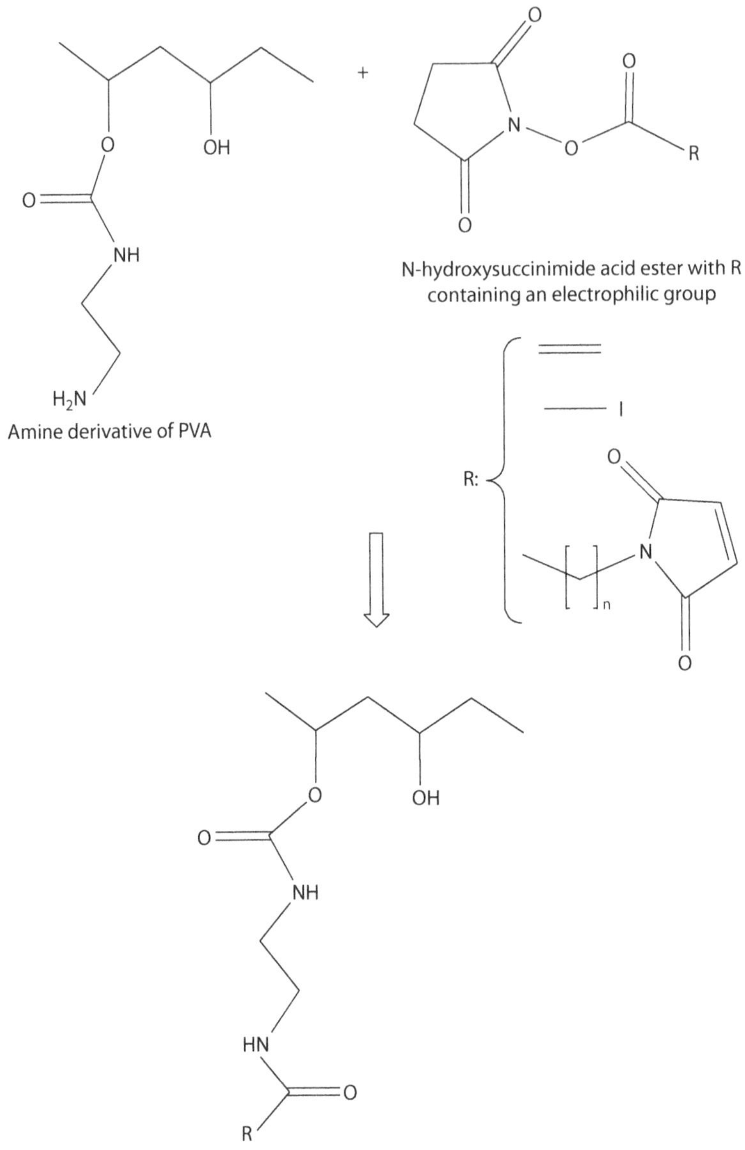

N-hydroxysuccinimide acid ester with R
containing an electrophilic group

Amine derivative of PVA

SCHEME 2.6   Electrophilic functionalization of PVA chains.

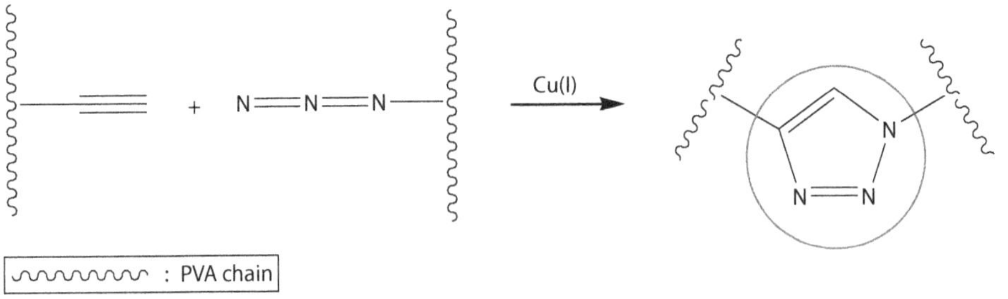

SCHEME 2.7   (See companion CD for color figure.) 1,3 Cycloaddition click reaction scheme for the formation of a triazole ring.

of PVA necessary to the "clicks" should be stable with respect to oxidation and hydrolysis pathways in order to avoid any release of molecular species toxic to cells, and specificity should be enhanced as much as possible with no side reactions involving metabolic or cellular materials. Moreover, the kinetics of the hydrogel formation should be compatible with the time necessary to mix the starting polymer components. The PVA "click" hydrogels described in the literature [35] do not exhaust the possibilities provided by this enhanced chemoselective approach and many novelties concerning hydrogel architectures and properties are to be expected in short time. Another "click chemistry" approach [35] leads to a PVA-poly(ethylene glycol) hybrid hydrogel. Poly(ethylene glycol) (PEG) was end-capped with azide functionalities, obtaining a telechelic diazide PEG spacer ready to click the alkyne-functionalized PVA by exploiting the high thermodynamic driving force of the click event. The hydrogels described in the paper provide a clear example of how "click chemistry" can be used to link structure and properties of the hydrogels. The outlined azide end-capped PEG is a two functional macromer, having two functional groups *per* polymer chain, $f = 2$, with a ratio $M/f = 1050$, where M is the polymer molecular weight. The same M/f value was obtained with an azide-functionalized PVA macromer with a DS of about 4.5%. However, "clicking" these types of macromers with the same alkyne-grafted PVA produced hydrogels with very different behaviors. Comparison of gelation time, gel fraction, storage, and loss moduli for mixtures of polymers indicates that sol $\leftrightarrow$ gel transition occurred in much less time when azide-functionalized PVA is clicked with the alkyne-derivative of PVA, hereafter called PVA-PVA. In a solution containing an equimolar amount of azide and alkyne with a fixed total polymer and catalyst concentration, the gel fraction, i.e., the dry polymer mass weighted after the hydrogel formation and swelling in water with respect to the initial polymer mass, for the PVA-PVA was twice the gel fraction found for PVA-PEG hydrogels, indicating that in the latter the occurrence of single chain links and intramolecular loops is more probable than in the former hydrogel in agreement with a drop of the elastic modulus, G', of the PVA-PEG hydrogel. According to Equations 2.5 and 2.6, the observed G' was converted into the cross-linking density, v_e/V_0, and linked to the molecular weight of the chain between cross-links, M_c. The molecular weight between cross-links found for the PVA-PVA hydrogel is lower than the M_c of PVA-PEG, indicating a more frequent interchain cross-linking with a larger number of chains participating to the elasticity of the network.

2.4 HYBRID HYDROGELS

The term hybrid hydrogels is used to indicate hydrated networks where two or more polymers participate to the network. Together with hydrogels entirely based on PVA, in the last decade many examples of hybrid hydrogels have been reported in the literature. In principle, a network based on more than one polymer component enables a larger number of functions with respect to one-polymer-based network. Hybrid hydrogels can be formed by chemically cross-linked polymer chains or by the interpenetration of networks, each one based on a single polymer component (IPN). Here, we will consider only those formed by the chemical cross-linking between the polymer components.

Chemo-selective and "click chemistry" approaches are commonly used to generate hybrid hydrogels, see, e.g., PEG-PVA hydrogel described in Section 2.3.3. Some examples of hybrid hydrogels have been already reviewed in the earlier sections. Coupling of nucleophilic with electrophilic polymer partners in which one is a derivatized PVA chain and the other is a modified or intact biopolymer, allows a number of hybrid hydrogels having different features as charge density, chain stiffness, biodegradability, and bioelimination properties.

In the design of hybrid hydrogels, polymer compatibility is an often overlooked issue, although being a key factor to avoid polymer phase segregation, breakdown of mechanical properties, low and/or uneven cross-link distribution. Once the compatibility of the polymers at a given mass or molar ratio is assessed, the fabrication of a multicomponent network is based on the functional groups to link. Hybrid hydrogels have been used for mimicking complex natural networks and in particular the extracellular matrix (ECM) with artificial scaffolds matching the manifold features required for the controlled growth of a cell line. ECM is a complex multicomponent system, containing low and high molecular weight constituents, produced by the cells themselves as suitable medium for metabolic processes involving bioadhesion and proliferation. Many biological key factors are involved in different steps of cell culturing, from the RGD tripeptide sequences contained in the fibronectin and enabling the bioadhesion process of the cells on the ECM matrix to cytokines and growth factors triggering and supporting the growth, respectively. The proliferation step of the cells requires a controlled hydrolysis of the polymer matrix promoted by the metallo-proteinases expressed by the cells in order to have available a larger, destructured portion of the ECM. The hydrolysis of the proteic part of the matrix is also paired by the progressive degradation of HA, a polysaccharide component of the ECM, driven by the hydrolytic activity of the hyaluronidases, a family of degradative enzymes also expressed by the cells hosted in the ECM. Only if this schematically described interplay between the host ECM and the guest cells is properly established, the growth of a cell colony leading to the final aim of the (re)generation of a tissue can be achieved.

An artificial scaffold for tissue regeneration should enable such complex interaction pattern, functions, and processes in order to be a sound substitute of ECM. A possible approach in view of the design of scaffolds for tissue regeneration is to favor cell growth incorporating some of the natural components in the artificially reproduced systems. HA, a key component in the ECM, is a candidate to be a polymeric partner in a PVA-based hybrid hydrogel. Ossipov et al. [32] focused on the coupling between a nucleophilic derivatized PVA with a partially oxidized HA. PVA backbone was modified with side chains containing nucleophilic functionalities as thiol, cysteine, 1,2-aminothiol, and aminoxy side chains prepared according to the strategy outlined in Scheme 2.5, or, alternatively, by introducing electrophilic functionalities in the PVA chains according to Scheme 2.6. Nucleophilic or electrophilic functionalized PVA chains were coupled with HA after partial oxidation with periodate. The action of periodate on HA consists in the specific oxidation of C2 and C3 carbons of the D-guluronic acid residue of the hyaluronic repeating unit (Scheme 2.8) into aldehyde groups.

The hybrid hydrogels obtained by the conjugation of oxidized HA with aminooxy-(PVAAO), semicarbazide-(PVASemi) and hydrazide-(PVAH) modified PVAs yielded strong hydrogels ranging

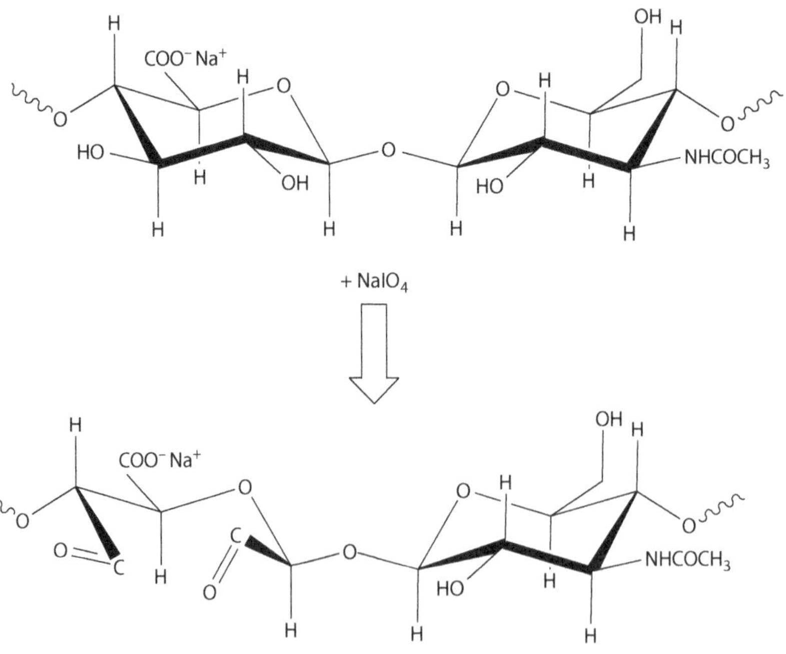

SCHEME 2.8 Oxidation of hyaluronate.

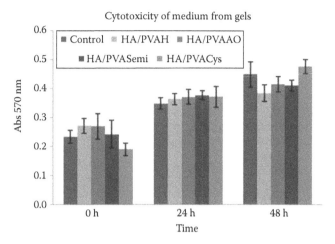

FIGURE 2.2 Cytotoxicity tests of different PVA-based/HA hybrid hydrogels on human dermal fibroblasts.

from compliable materials to more elastic networks. The chemoselective reaction responsible for the cross-link formation takes place with the incorporation of the cells allowing the intimate contact between the living material and the hydrogel constituents. Figure 2.2 shows the cytotoxic effect of these hybrid hydrogels tested on human dermal fibroblasts. The impact of the hydrogels on cell viability, monitored by MTT assay, was absent and the enzymatic degradation of HA by hyaluronidase was accomplished within 3 days for all of the studied hydrogels. The degradation of HA due to the hyaluronidase expressed by the cells incorporated in the PVA/HA co-hydrogels was not investigated in the paper.

Another hybrid hydrogel based on the conjugation of PVA with heparin, a natural polysaccharide belonging to the family of glycosaminoglycans, was described recently in the literature [36]. Methacrylated macromers of PVA and heparin, conjugated via photopolymerization of the vinyl moiety, yielded a co-hydrogel with mechanical properties similar to the network generated by the photo-cross-linking of methacrylated PVA and retained the biological activity characteristic of heparin as its growth factor activity. Tests, performed with BaF3 fibroblasts seeded on PVA/heparin co-hydrogels in the presence of 0.3 nM fibroblast growth factor FGF-2, showed increased cells proliferation when compared with methacrylated PVA hydrogels (Figure 2.3).

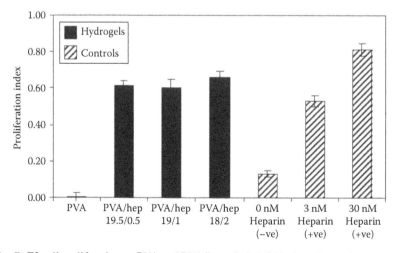

FIGURE 2.3 BaF3 cell proliferation on PVA and PVA/heparin hybrid hydrogels, and in response to negative (no heparin) and positive (3 and 30 nM heparin) controls, all in the presence of 0.3 nM FGF-2.

The effect was independent from the amount of heparin contained in the co-hydrogels, and this was interpreted as evidence that the optimal signaling activity was already obtained with the co-hydrogel incorporating the lowest amount of heparin.

The low molecular weight cross-linker, glutaraldehyde, as outlined in Section 2.2, was exploited for the networking of PVA with chitosan, a polysaccharide derived from the deacetylation of chitin, the second most abundant polysaccharide after cellulose. The bioactivity of this co-hydrogel was tested with respect to cell viability and cytotoxicity [37]. The use of glutaraldehyde, a highly reactive and toxic cross-linker, did not impact with the viability of the cells.

2.5   CONCLUDING REMARKS

The versatility of PVA is paramount in the fabrication of several chemical hydrogels enabling the obtainment of a variety of internal architectures and overall morphologies, including monolithic hydrogels, microgels, and microbubbles. This is combined with a favorable bio-interface provided by several PVA-based hydrogels, validated by in vitro and in vivo studies with various cell lines. Modulation of the properties and introduction of responsiveness to external stimuli are other assets of recently formulated chemical hydrogels based on PVA.

The picture emerging from a literature survey shows a constantly ongoing research activity on the synthesis of new hydrogels as well as on their characterization, including more advanced aspects as the study of the network internal dynamics and the diffusive behavior of confined water ranging from the molecular detail to the macroscopic level. Several, not less negligible, features are still to be explored and new PVA-based hydrogels soon will enrich the panorama of the potential tools available for biomedicine.

LIST OF SYMBOLS

D_{self}	self-diffusion coefficient
f	functionality of the cross-link
$g_2(t)$	autocorrelation function of the scattered intensity of light
G	elastic modulus
G'	storage modulus
k'	apparent kinetic constant
M_c	number average molecular weight between cross-links
M_n	number average molecular weight
n_{est}	number of ester blocks at time t
n_0	initial number of ester blocks
Q	swelling ratio
V_1	molar volume of the solvent (water)
VPTT	volume-phase transition temperature
χ	polymer–solvent interaction parameter
π_{osm}	osmotic pressure
Π_{swell}	swelling pressure
φ_0	polymer volume fraction in the relaxed state
φ_2	polymer volume fraction in the swollen state
v_e/V_0	density of cross-links

REFERENCES

1. Finch, C. A. 1973. *Polyvinyl Alcohol. Properties and Applications*. London, U.K.: Wiley.
2. Huisgen, R. 1989. Kinetics and reaction mechanisms: Selected examples from the experience of forty years. *Pure Appl. Chem.* 61:613–628.
3. Kolb, H. C., Finn, M. G., Sharpless, K. B. 2001. Click chemistry: Diverse chemical. Function from a few good reactions. *Angew. Chem. Int. Ed.* 40:2004–2021.
4. Evans, R. A. 2007. The rise of azide-alkyne 1,3-dipolar 'click' cycloaddition and its application to polymer science and surface modification. *Aust. J. Chem.* 60:384–395.
5. Flory, P. J. 1953. *Principles of Polymer Chemistry*. Ithaca, NY: Cornell University.
6. Peppas, N. A., Merril, E. W. 1976. Determination of interaction parameter χ_1, for poly(vinyl alcohol) and water in gels crosslinked from solutions. *J. Pol. Sci. Polym. Chem. Ed.* 14:459–464.
7. Peppas, N. A., Bar-Howell, B. D. 1986. Characterization of the cross-linked structure of hydrogels. In *Hydrogels in Medicine and Pharmacy*, ed. N. A. Peppas, pp. 27–55, Vol. 1. Boca Raton, FL: CRC Press.
8. De Gennes, P. G. 1979. *Scaling Concept in Polymer Physics*. Ithaca, NY: Cornell University Press.
9. Neuburger, N. A., Eichinger, B. E. 1988. Critical experimental test of the Flory-Rehner theory of swelling. *Macromolecules* 21:3060–3070.
10. Hansen, E. W., Bouzga, A. M., Sommer, B., Kvernberg, P. O. 2000. Crosslinking of PVA and glutaraldehyde in water monitored by viscosity and pulse field gradient NMR: A comparative study. *Polym. Adv. Technol.* 11:185–191.
11. Paradossi, G., Di Bari, M. T., Telling, M. T. F., Turtù, A., Cavalieri, F. 2001. Incoherent quasi-elastic neutron scattering study of chemical hydrogels based on poly (vinyl alcohol). *Physica B* 301:150–156.
12. de Groot, J. H., Spaans, C. J., van Calck, R. V., van Beijma, F. J., Norrby, S., Pennings, A. J. 2003. Hydrogels for an accommodating intraocular lens. An explorative study. *Biomacromolecules* 4:608–616.
13. Bo, J. 1992. Study on PVA hydrogel crosslinked by epichlorohydrin. *J. Appl. Polym. Sci.* 46:783–786.
14. Painter, T., Larsen, B. 1970. Formation of hemiacetals between neighbouring hexuronic acid residues during the periodate oxidation of alginate. *Acta Chem. Scand.* 24:813–33.
15. Barretta, P., Bordi, F., Rinaldi, C., Paradossi, G. 2000. A dynamic light scattering study of hydrogels based on telechelic PVA. *J. Phys. Chem. B* 104:11019–11026.
16. Paradossi, G., Cavalieri, F., Chiessi, E., Telling, M. T. F. 2003. Super-cooled water in PVA matrices: I. An incoherent quasi elastic neutron scattering (QENS) study. *J. Phys. Chem. B* 107:8363–8371.
17. Cavalieri, F., El Hamassi, A., Chiessi, E., Paradossi, G. 2005. Stable polymeric microballoons as multifunctional device for biomedical uses: Synthesis and characterization. *Langmuir* 21:8758–8764.
18. Martens, P., Anseth, K. S. 2000. Characterization of hydrogels formed from acrylate modified poly(vinyl alcohol) macromers. *Polymer* 41:7715–7722.
19. Mühlebach, A., Müller, B., Pharisa, C., Hofmann, M., Seiferling, B., Guerry, D. 1997. New water-soluble photo crosslinkable polymers based on modified poly(vinyl *alcohol*). *J. Polym. Sci. Part A Polym. Chem.* 35:3603–3611.
20. Martens, P., Blundo, J., Nilasaroya, A., Odell, R. A., Cooper-White, J., Poole-Warren, L. A. 2007. Effect of poly(vinyl alcohol) macromer chemistry and chain interactions on hydrogel mechanical properties. *Chem. Mater.* 19:2641–2648.
21. van Dijk-Wolthuis, W. N. E., Franssen, O., Talsma, H., van Stenbergen, M. J., Kettenes-van den Bosh, J. J., Hennink, W. E. 1995. Synthesis, characterization, and polymerization of glycidyl methacrylate derivatized dextran. *Macromolecules* 28: 6317–6322.
22. Meyvis, T. K. L., De Smedt, S. C., Demeester, J., Hennink, E. 1999. Rheological monitoring of long-term degrading polymer hydrogels. *J. Rheol.* 43: 933–950.
23. Cavalieri, F., Miano, F., D'Antona, P., Paradossi, G. 2004. Study of gelling behaviour of poly(vinyl alcohol)-methacrylate for in situ utilizations in tissue replacement and drug delivery. *Biomacromolecules* 5:2439–2446.
24. Cavalieri, F., Chiessi, E., Villa, R., Viganò, L., Zaffaroni, N., Telling, M. F., Paradossi, G. 2008. Novel PVA-based hydrogel microparticles for doxorubicin delivery. *Biomacromolecules* 9:1967–1973.
25. Aruffo, A., Stamenkovic, I., Melnick, M., Underhill, C. B., Seed, B. 1990. CD44 is the principal cell surface receptor for hyaluronate. *Cell* 61:1303–1313.
26. Cerroni, B., Chiessi, E., Margheritelli, S., Oddo, L., Paradossi, G. 2011. Polymer shelled microparticles for a targeted doxorubicin delivery in cancer therapy. *Biomacromolecules* 12:593–601.
27. Ghugare, S. V., Mozetic, P., Paradossi, G. 2009. Temperature-sensitive poly(vinyl alcohol)/poly(methacrylate-co-N-isopropyl acrylamide) microgels for doxorubicin delivery. *Biomacromolecules* 10:1589–1596.

28. Ghugare, S. V., Chiessi, E., Telling, M. T. F., Deriu, A., Gerelli, Y., Wuttke, J., Paradossi, G. 2010. Structure and dynamics of a thermoresponsive microgel around its volume phase transition temperature. *J. Phys. Chem. B* 114:10285–10293.

29. Mawad, D., Odell, R., Poole-Warren, L. A. 2009. Network structure and macromolecular drug release form poly(vinyl alcohol) hydrogels fabricated via two crosslinking strategies. *Int. J. Pharm.* 366:31–37.

30. Nuttelman, C. R., Henry, S. M., Anseth, K. S. 2002. Synthesis and characterization of photocrosslinkable, degradable poly(vinyl alcohol)-based tissue engineering scaffolds. *Biomaterials* 23:3617–3626.

31. Martens, P., Holland, T., Anseth, K. S. 2002. Synthesis and characterization of degradable hydrogels formed from acrylate modified poly(vinyl alcohol) macromers. *Polymer* 43:6093–6100.

32. Ossipov, D. A., Piskounova, S., Hilborn, J. 2008. Poly(vinyl alcohol) cross-linkers for in vivo injectable hydrogels. *Macromolecules* 41:3971–3982.

33. Ossipov, D. A., Brännvall, K., Forsberg-Nilsson, K., Hilborn, J. 2007. Formation of the first injectable poly(vinyl alcohol) hydrogel by mixing of functional PVA precursors. *J. Appl. Polym. Sci.* 106:60–70.

34. Tortora, M., Cavalieri, F., Chiessi, E., Paradossi, G. 2007. Michael-type addition reactions for the in situ formation of poly(vinyl alcohol)-based hydrogels. *Biomacromolecules* 8:209–214.

35. Ossipov, D. A., Hilborn, J. 2006. Poly(vinyl alcohol)-based hydrogels formed by "Click Chemistry." *Macromolecules* 39:1709–1718.

36. Nilasarova, A., Poole-warren, L. A., Whitelock, J. M., Jo Martens, P. 2008. Structural and functional characterization of poly (vinyl alcohol) and heparin hydrogels. *Biomaterials* 29:4658–4664.

37. Mansur, H. S., Costa, Jr., E., de, S., Mansur, A. A. P., Barbosa-Stancioli, E. F. 2009. Cytocompatibility evaluation in cell-culture systems of chemically crosslinked chitosan/PVA hydrogels. *Mater. Sci. Eng. C* 29:1574–1583.

3 Development and Evaluation of Poly(Vinyl Alcohol) Hydrogels as a Component of Hybrid Artificial Tissues for Orthopedics Surgery Application

Masanori Kobayashi

CONTENTS

3.1 INTRODUCTION

The area of research of polymeric biomaterials covers a wide range of fields from basic research to applied research. Due to the development of nanotechnology, biotechnology, and tissue engineering and the technical fusion of polymer sciences in recent years, the field of polymeric biomaterial science is expanding rapidly. In vivo compartmentalized organic macromolecules (extracellular matrix [ECM], proteins, polysaccharides, DNA, etc.) play an important role among cellular components, and it is not an exaggeration to say that the living body and functional expression as such depend on biological macromolecules. The essence of biomaterial research is to imitate the living body, to substitute vital functions, and to actively influence the living body. One can say that it is inevitable that research on polymer materials, the key elements that constitute the living body, is wide-ranging. Among the research and development of such biomaterials, particularly regarding the fields of application, eventually the most important point is a safe and reliable technical potential in order to meet the characteristics demanded in the desired medical field, as well as the specific performance of clinically applicable material and devices and their evaluation. Due to its excellent biocompatibility and mechanical properties, the biomedical application of poly(vinyl alcohol) hydrogels (PVA-H) as artificial artery, artificial cartilage, and artificial muscle has been made possible. Regarding the points mentioned earlier, PVA-H can be identified as a typical biomaterial for which no specific clinical application has not been found yet in spite of a lot of research and development being done in various fields for a long time. Since 1985, Oka et al. have been engaged in research on the clinical application of PVA-H in the orthopedic surgery field and have also performed many basic experiments on PVA-H, improved PVA-H, and developed PVA-H implants. The author also joined this research group around 1995 and has been engaged in various researches on the application of PVA-H. In this chapter, limiting this topic to the application of PVA-H in the field of orthopedic surgery, and focusing on the implant study examples that have been developed so far, the author will explain a series of findings with regard to the clinical application of PVA-H and its possibilities.

3.2   DEVELOPMENT OF PVA-H

PVA was developed in 1924 by Hermann et al. and is a synthetic fiber with various excellent mechanical properties. It is also the raw material of "vinylon," the first high-strength, high-modulus synthetic fiber that was developed in Japan. PVA-H is prepared from PVA. It is not only used in the manufacturing of high-strength, high-modulus fiber but also serves as raw material for films and acetal resin, as textile processing agent, adhesive agent, polyvinyl chloride polymerization stabilizer, and inorganic binder. Especially, since the excellent biocompatibility of PVA-H has been widely known, this has raised great expectations for its use as a biomaterial. Initially PVA-H was not found to be a strong gel, and it was also turbid. Then, Hyon, who was a collaboration researcher, and colleagues developed the freeze-thaw method [1,2]. Generally, hydrogels are defined as gels that contain water but are not soluble in water. Later on, Hyon et al. made a complete homogenous PVA solution by heating the mixture of PVA and water/dimethyl-sulfoxide (DMSO), after agitating under nitrogen current, leaving the solution in low temperature ($-20°C$) for 10 h, and promoted the crystallization and cross-linking of PVA molecules. This frozen gel was brought into contact with water to exchange DMSO in the PVA gel with water. Repeating this cryogels process by freeze-thaw cycling, a PVA-H with high mechanical strength, high water content, and excellent transparency was successfully produced. By using this technique, we have succeeded in the reinforcement of the mechanical properties of gels by only the arrangement of water content and polymerization of PVA-H without the chemical agent or other blend polymers. Adding the cross-linking by γ-radiation, we have characterized further improved mechanical properties and biocompatibility of PVA-H.

However, to develop PVA-H as an arthroplasty material, as shown in later Section 3.5, it is necessary for PVA-H to infiltrate into the titanium fiber mesh as a composite device so that it can be firmly fixed to the underlying bone. However, because of poor infiltration of PVA solution into the pores of titanium mesh using the low-temperature crystallization method, the filtration technique was improved by adopting high-pressure injection molding. The bonding strength of PVA-H and titanium fiber mesh interface was increased remarkably with this improved fabrication technique. Figure 3.1 shows the flow chart of the fabrication process of two types of PVA-H. Then, two types of PVA-H were selected and examined with the use of a medical device we tried to investigate.

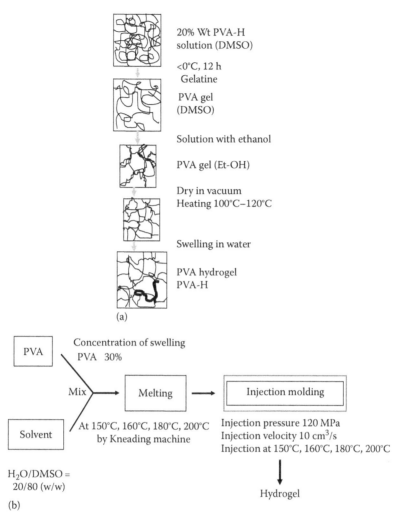

FIGURE 3.1   (See companion CD for color figure.) (a) Flow chart of the production process of poly(vinyl alcohol)-hydrogel (PVA-H) by freeze-thaw method. (b) Flow chart of the production process of poly(vinyl alcohol)-hydrogel (PVA-H) by injection molding method.

3.3   BIOCOMPATIBILITY OF PVA-H

It has already been reported that PVA-H has excellent biocompatibility; however, as the newly improved PVA-H was not tested, de novo basic biocompatibility experiments using animals were performed [3,4]. PVA-H was then implanted into various sites in rabbits and dogs, subcutaneously, intramuscular, etc., and follow-up after surgery confirmed superior biocompatibility. In order to make certain in vivo foreign-body immune reactions against PVA-H, fine particles of PVA-H and ultra-high molecular weight polyethylene (UHMWPE) were injected into the right and left knee joints, respectively, in the same rabbit. The PVA-H used were fine particles with a diameter of $100\,\mu m$, with UHMWPE particles as control. Figure 3.2 is a histological photograph of tissue observed 3 months after injection. In the periphery of the UHMWPE particles, accumulated macrophages and foreign-body giant cells, and remarkable foreign-body reactions were observed, whereas almost no reactions were seen around PVA gel particles on the opposite side. This bio-inert characteristic of this PVA-H is speculated to be due to the strong hydrophilic effects of this gel, which inhibits water circulation to the cells and makes it difficult for cells to adhere to its surface.

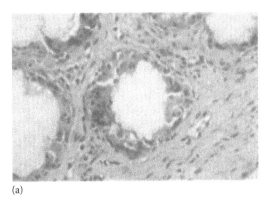
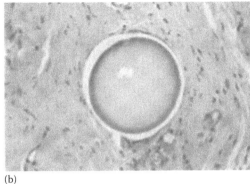

(a)   (b)

FIGURE 3.2   **(See companion CD for color figure.)** Histological appearance of the implant material particles and the surrounding tissue in rabbits knee joint (×100). (a) UHMWPE particle: many macrophage and giant cells are surrounding particle due to intense immunoreactions. (b) PVA-H particle: Immunoreactions were hardly observed.

The results of these basic experiments confirmed that PVA-H has favorable biocompatibility. Even in many animal experiments subsequently performed for the development of implants, no immune reactions were observed for PVA-H.

3.4   MECHANICAL PROPERTIES OF PVA-H

PVA-H has been chosen for investigation as it possesses several useful properties, including permeability, hydrophilicity, and low frictional function. It has been widely commercialized and studied in the medical industries for the production of membrane, gel, and films for an artificial pancreas, drug delivery system, and adhesion protection sheet [5–7].

However, the use of PVA-H as an application in the orthopedics surgery field has been thought to be limited because of its low mechanical strength and durability. Actually, PVA-H is not such a strong gel mechanically. In this chapter, a series of the development and evaluation concerning the mechanical properties of PVA-H is mentioned.

3.4.1   MECHANICAL PROPERTIES AND FABRICATION METHODS OF PVA-H

Traditional cross-linking methods have been used to synthesize PVA-H materials with improved mechanical properties; however, the chemical agents introduced in their preparation are often toxic, inevitably effecting the biocompatibility of PVA-H. On the other hand, polymer blending is a useful method of improving or modifying the physiochemical properties of polymer materials; blends between synthetic polymers and biopolymers are of particular interest to current hydrogel research. A polymer blend can be defined as a combination of two polymers without any chemical bonding between them. Some chemical interactions can occur between components, but it sometimes induces the deterioration of mechanical properties of gels.

In this academic background, Hyon et al. succeeded in gelling PVA by cryogenic crystallization using a freeze-thawing method at low temperatures with the organic solvent DMSO in the synthesis process, making it possible to prepare PVA of improved transparency and high strength. Subsequently, the research and development of all kinds of implants was performed using an improved version of this PVA-H, and it underwent improvements such as molecule cross-linking by γ-irradiation in order to achieve higher strength.

However, as shown earlier, in order to develop PVA-H as an artificial articular cartilage [4,8], PVA-H is required to infiltrate into titanium fiber mesh as a composite device to fix PVA-H firmly to the underlying bone. Hence, the high-pressure injection molding technique was used instead of the low-temperature crystallization method. Although the bonding strength of PVA-H and titanium fiber mesh interface increased remarkably with this improved technique, the mechanical properties of the PVA-H prepared by the new fabrication technique were found to be uncertain. Accordingly, we have investigated the mechanical properties of novel PVA-Hs consecutively [9]. Figure 3.3 shows the tensile strength test results of PVA-H (degree of polymerization: 8800; water content: 30%) developed in our laboratory. The mechanical strength and Young's modulus of gels by injection molding were all close to that of natural articular cartilage and PVA-H by freeze-thawing, the difference of heating temperature and water content also showed no significant influence. The rheological behavior is also very important in investigating the mechanical properties of hydrogels. Although its variability in measurement technique has been reported, we can evaluate

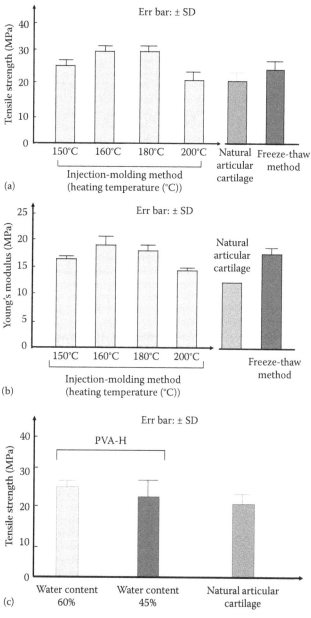

FIGURE 3.3 (See companion CD for color figure.) Mechanical properties of PVA-H prepared by injection molding method. (a) Influence of heating temperature on ultimate tensile strength (MPa). (b) Influence of heating temperature on Young's modulus (MPa). (c) Influence of water content on ultimate tensile strength (MPa).

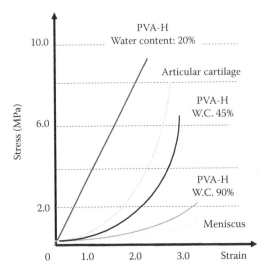

FIGURE 3.4 **(See companion CD for color figure.)** Representative stress–strain curves of various PVA-H samples and natural tissue specimens in the compression test.

the potentiality of its viscoelastic characteristics by comparing with that of natural human tissue [10]. Figure 3.4 shows the stress–strain curves in compression tests of various PVA-Hs in some mechanical tests. As for the stress–strain curves of PVA-H, the appearance of viscoelastic curve in high water content samples resembled that of natural tissue such as an articular cartilage, while the elastic response (or solid-like response) means high Young's modulus value dominated in low water content. For yield stress, since PVA-H has high viscosity and does not reveal the fracture, a definite yield point as observed in metals cannot be determined. Thus, with a gradually increasing load on the PVA-H sample with a loading speed of 5 mm/min, a compression test consisting of loading and unloading was repeatedly performed, and the load at which PVA-H deformation remained even without loading after releasing was defined as the yield point. Figure 3.5 shows the comparison of these yield stresses of PVA-H.

Figure 3.6 shows the results of stress–relaxation tests. PVA-H with a high water content tends to show more marked stress relaxation. The human menisci exhibited the most acute stress relaxation. The results

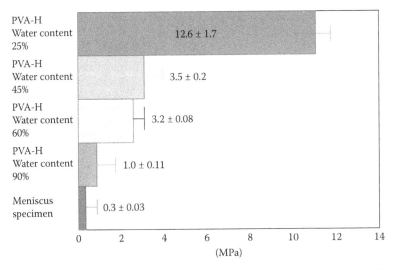

FIGURE 3.5 **(See companion CD for color figure.)** Comparison of yield stress value of various PVA-H and natural tissue specimens.

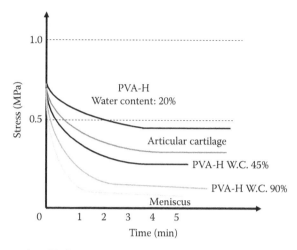

FIGURE 3.6   **(See companion CD for color figure.)** Comparison of the stress–relaxation curves of PVA-H and natural tissue specimens.

from mechanical testing indicated that changing the degree of cross-linking and water content could be utilized as a means of achieving the desired mechanical properties, including the viscoelastic characteristics of the PVA-H prepared by the freeze-thawing method or injection molding method.

3.4.2   TRIBOLOGY OF PVA-H

One of the important reasons that PVA-H has been widely studied as a biomedical application is its excellent lubrication function. Numerous authors have already reported the lubricative mechanism of PVA-H, which is said to be able to preserve the fluid film between articulating surface by elasto-hydro-dynamic lubrication mode under a certain loading condition. However, boundary lubrication or solid lubrication mode might also operate on the PVA-H as an articular cartilage substitute under the more severe high loading such as athletic dynamic motion. In this chapter, the tribological property of PVA-H obtained from our studies is described. The basic frictional function of PVA-H has been examined using an end face–type friction test machine for the purpose of developing artificial cartilage [10,11]. Figures 3.7 through 3.9 explain these friction tests. Figure 3.8 shows the frictional coefficient of various

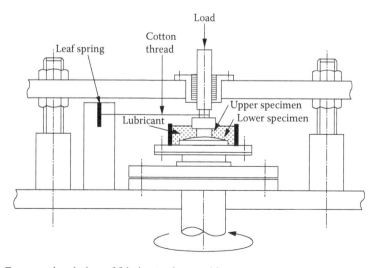

FIGURE 3.7   Cross-sectional view of friction testing machine.

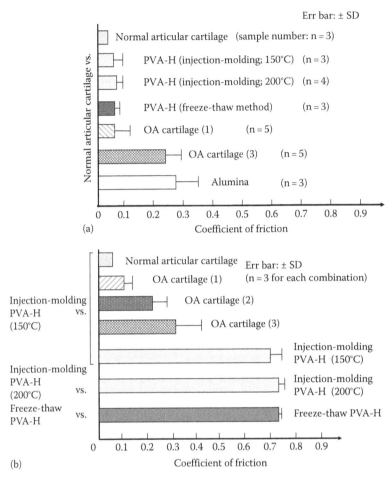

FIGURE 3.8 **(See companion CD for color figure.)** Histogram showing coefficients of friction for various combination. (a) Friction coefficients of various materials specimens against normal articular cartilage. (b) Friction coefficients of various materials specimens against PVA-H.

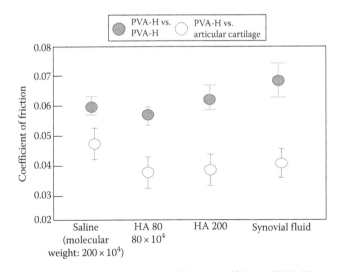

FIGURE 3.9 **(See companion CD for color figure.)** Frictional coefficient of PVA-H versus articular cartilage or PVA-H, and the influence of hyaluronic acid.

biomaterials against natural articular cartilage or PVA-H in synovial fluid simulated lubricants (saline with 0.375% hyaluronic acid, 3.0% albumin, and 0.5% γ-globulin). All friction coefficients of PVA-H against normal articular cartilage showed low values compared to that of articular cartilage versus articular cartilage, whereas the friction coefficients of PVA-H versus PVA-H was remarkably high. Figure 3.9 compares the friction coefficients of PVA-H versus articular cartilage, or PVA-H versus PVA-H with each different lubricant. This result demonstrates that the coefficient of PVA-H versus PVA-H shows high value against the dose of hyaluronic acid, while the lubrication of PVA-H versus articular cartilage improve according to a hyaluronic acid dose. These frictional experimental data show that the adhesion function of the lubricants such as hyaluronic acid, albumin, and γ-globulin on the PVA-H surface is inferior to the articular cartilage surface. Therefore, as another experimental approach for tribological function of PVA-H, the behavior of lubricating surfaces on two kinds of materials, natural articular cartilage and PVA-H, was observed by confocal laser scanning microscopy (CLSM) [12,13].

Each specimen was placed on the table, the synovial fluid added to its surface, pressed with a glass plate (0.15 mm thickness), and a load (12 N) applied for CLSM observation (Figure 3.10). The compression of natural articular cartilage by a glass plate, which was equivalent to physiological loading, caused the cartilage surface to exhibit two distinct areas: one in direct contact area with a glass plate and another with a fluid pool between the cartilage and the plate (Figure 3.11). Between these two areas, a third morphological area was observed, as shown in Figure 3.12. In the area in direct contact with the glass plate, the detailed morphology in the

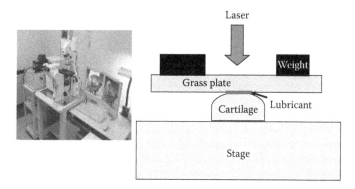

FIGURE 3.10 **(See companion CD for color figure.)** Photograph of real-time 1LM21 confocal laser scanning microscopy (CLSM) and schematic cross-sectional view of CLSM observation.

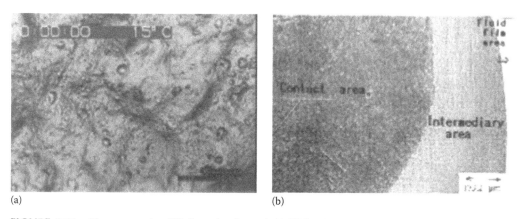

(a)   (b)

FIGURE 3.11 **(See companion CD for color figure.)** (a) CLSM image of normal articular cartilage surface without loading (×1400). (b) CLSM image of normal articular cartilage surface under the loading (×250). The solid contact area, fluid pooling area with synovial fluid and intermediary area with fringe-like pattern were observed.

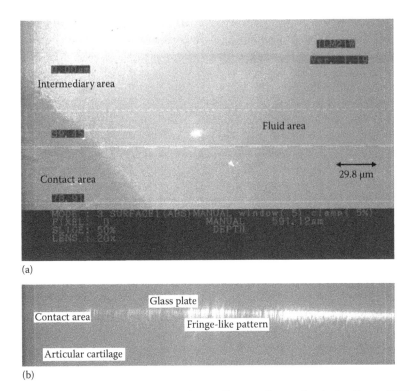

(a)

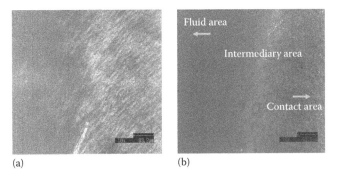

(b)

FIGURE 3.12   (a) CLSM image of normal articular cartilage surface under the loading (×1000). (b) Cross-sectional view of CLSM image (×1000).

articular cartilage surface disappeared due to compression. In the area of the fluid pool, the presence of a fluid membrane between the articular cartilage and the glass plate was confirmed, but the state of the cartilage surface was unclear. In the third area, a fringe-like pattern was observed around the contact area at higher magnification. In addition, the observation of this area by a particular optical-isolation method revealed a reflected image that corresponded to the third area, as shown in Figure 3.13. This finding indicates that the third area is composed of a liquid crystal structure. In contrast, the CLSM images of the PVA-H surface under loading were uniform, and distinguishing between the area of contact and the fluid pool was impossible, as shown in Figure 3.14. In the cross-sectional view, the grass plate and PVA-H surface were clear, but no images corresponding to the third area, which was seen in the articular cartilage

(a)   (b)

FIGURE 3.13   (See companion CD for color figure.) CLSM images with optical-isolator technique about the intermediary area. The reflection image (a) was obtained corresponding to the third area in the original CLSM image (b).

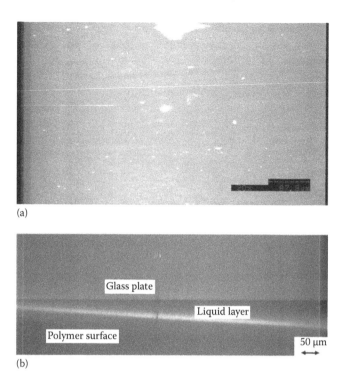

(a)

(b)

FIGURE 3.14 (a) CLSM image of PVA-H surface under the loading (×700). Third area and fluid pooling area were unclear. (b) Cross-sectional view of CLSM image (×700).

specimens, were observed. The optical-isolation technique also did not reveal any image. To discuss the excellent lubricative mechanism of natural synovial joints under the high loading conditions such as boundary lubrication, we must address the molecular nature of the protective film layer that covers the articular cartilage surface. Many investigators have already proposed the presence of this lubricin [14–16]. Our CLSM image on articular cartilage may indicate that liquid crystal region reflected the protective film layer on cartilage surface. While PVA-H was developed to have the same lubricative mechanism of natural joints, however, CLSM of PVA-H did not reveal the presence of a film layer. Even though further analysis and discussion may be demanded regarding these results, the finding means that PVA-H does not form a fluid film nor liquid crystal, as does natural articular cartilage, and PVA-H has the limitation as an artificial cartilage to imitate the lubricative mechanism of synovial joint cartilage.

3.5 APPLICATION FOR ARTHROPLASTY IMPLANT

3.5.1 DEVELOPMENT OF ARTIFICIAL ARTICULAR CARTILAGE

Arthroplasty using implants as used in total hip joint arthroplasty (THA) and total knee joint arthroplasty (TKA) is an established treatment for patients with severe arthrosis caused by diseases such as osteoarthritis (OA) and rheumatoid arthritis (RA). However, the tolerance of artificial joints to physiological loading is inferior to that of natural joints, and the number of revision operations performed to adjust for loosening or wear of these implants has been increasing. The long-term survival of THA and TKA is a significant problem in orthopedic surgery [17,18].

In contrast, it is known that natural joints have excellent lubrication. Though the mechanism involved is not still clear, tribological studies, biomechanical studies, and morphological studies have proposed various theories to clarify the mechanism behind the superior lubricative function of natural joints. If the mechanism of natural joint lubrication could be imitated by artificial materials, the properties of artificial joints can be much improved. In this respect, Oka et al. have attempted to develop an artificial articular cartilage based on PVA-H. Initially, we thought that the medical application as an artificial articular cartilage by PVA-H had a high potential as we were able to get favorable in vitro data [4,19] and many studies had already reported the superior lubricative properties of PVA-H [20,21]. However, as shown in previous section, the frictional test results of PVA-H exhibited high frictional coefficients and much wear, enough to predict the failure of arthroplasty by PVA-H. Although the mechanical property of PVA-H was improved for medical application, the use of lubrication remains a problem.

3.5.2 DEVELOPMENT OF ARTIFICIAL ARTICULAR CARTILAGE AS A HEMI-ARTHROPLASTY

The development of total arthroplasty by PVA-H has some key issues. However, as mentioned earlier, the friction coefficients of PVA-H against normal articular cartilage have shown low values in frictional tests (4.2: tribology of PVA-H). This finding encourages us to attempt to use PVA-H as a medical implant for partial hemi-arthroplasty replacement. Osteonecrosis of the femoral head is a disabling disease that can lead to destruction of the hip joint. The appropriate treatment, depending on the stage of the disease, remains controversial, yet progression to collapse of the femoral head often necessitates total hip replacement (THR). Considering the age of patients and the poor prognosis associated with THR, it is desirable to preserve as much of the joint as possible during treatment as same as the concept of an artificial cartilage for osteoarthritis [22–24].

We have developed an articular cartilage composite device with PVA-H, which have characterized the mechanical properties and lubricating functions, and a titanium fiber mesh, which is used to attach PVA-H firmly to underlying bone as shown in Figure 3.15 [25]. The PVA-H used was polymerization degree 8800, water content 30%, which corresponds to the mechanical properties of natural articular cartilage in vitro.

This composite has been implanted into the surface of canine femoral head for the purpose of simulating partial surface replacement hemi-arthroplasty, as shown in Figure 3.16. In addition to this PVA-H device, prostheses made from UHMWPE and alumina ceramics were also implanted as a control group to compare with the present implant material. The dogs with implant were allowed unrestrained weight-bearing and normal activity in their living area, and killed at time intervals of 1, 3, 6, and 12 months. Dogs with all three types of devices showed good locomotive function of the hip joint without dislocation, deformity, or limping during their survival periods. Concerning the endurance of the PVA-H prosthesis, the transected specimens showed good congruity of the PVA-H with the adjacent natural articular cartilage. Figure 3.17 shows the histological and radiographic findings at 12 months after operation. It shows that the reduction

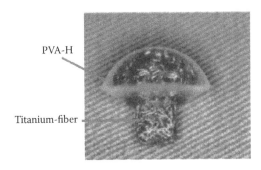

FIGURE 3.15 (See companion CD for color figure.) Photograph of a synthetic composite osteochondral device for partial replacement of the canine femoral head. The transparent portion is made of PVA-H.

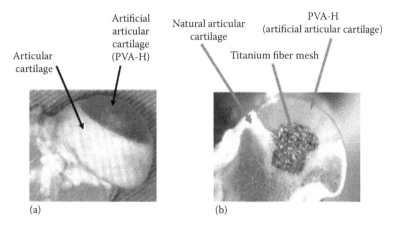

Articular cartilage

Artificial articular cartilage (PVA-H)

Natural articular cartilage

PVA-H (artificial articular cartilage)

Titanium fiber mesh

(a)   (b)

FIGURE 3.16 **(See companion CD for color figure.)** Photograph of a specimen of the femoral head with composite osteochondral device at 12 months after implant operation. (a) Macroscopic appearance; (b) cross-sectional appearance.

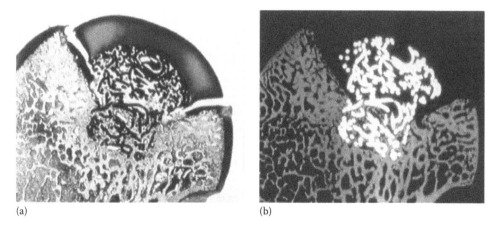

(a)   (b)

FIGURE 3.17 **(See companion CD for color figure.)** Cross-sectional view of a specimen of the femoral head with composite osteochondral device at 12 months after implant operation (×2.5). (a) Histological appearance (Giemusa stain). (b) Contact microradiograph appearance.

and position of femoral head were well maintained, and there was an increase in bone density around the stem portion of the titanium fiber mesh and good fixation to the surrounding bone. To evaluate the lubrication function of the PVA-H prosthesis, opposite canine acetabular articular cartilages were also reviewed macroscopically and histologically. No adverse responses, such as erosion or hemorrhage, were observed in the macroscopic appearance of the articular cartilage of acetabulum of the PVA-H group, though a slight synovitis and discoloration in the alumina group was observed. In the UHMWPE group, there was severe synovitis and erosion of the acetabular cartilage 6 months after the operation. Histological evaluation of acetabular articular cartilage was graded using the Mankin's score [26]. These results are summarized in Figure 3.18. The PVA-H group acetabular cartilage showed that structural integrity, surface regularity, and thickness were well maintained throughout the experimental periods, whereas the histological appearance of the other material groups deteriorated depending on the experimental periods. Although this study has limitations as an evaluation, we think these results demonstrate the superior potential of the clinical partial surface replacement hemi-arthroplasty using PVA-H articular cartilage.

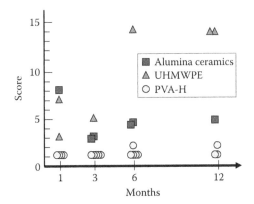

FIGURE 3.18   (See companion CD for color figure.) Mankin's score for the changes in acetabular cartilage in each group. Each point represents a single score for a joint.

3.6 DEVELOPMENT OF ARTIFICIAL MENISCUS

As well as the articular cartilage substitute for hemi-arthroplasty, the favorable lubricative function of PVA-H against articular cartilage seems promising for the development of artificial meniscus. Basic and biomechanical studies on the meniscus function of knee joint have suggested the important role of meniscus, such as stabilization of knee joint movements, shock absorption, and stress dispersion of loading, and there is an increasing understanding of the significance of meniscus [27–29]. Furthermore, many clinical reports have also indicated the articular degeneration cases in meniscetimized patients. Nowadays, it is well known that meniscectomy results in osteoarthritis changes in articular cartilage [30–32].

Thus, the treatment of meniscus injury has been changed from meniscectomy to repair. However, depending on the type of meniscus injury, meniscectomy cannot be avoidable. However, meniscus implant replacement is the most favorable treatment for the reconstruction of meniscus function. Thus, we have developed artificial meniscus using PVA-H [33,34]. The PVA-H used as an artificial meniscus should have two major characteristics: (1) mechanical properties and (2) tribological properties. The mechanical property of meniscus tissue is different from that of articular cartilage. The microstructural characteristics of the meniscus determine the mechanical properties. Meniscus are composed of 70% water and 30% organic matter. Type-I collagen fibers constitute 75% of the organic matter, which provides the primary meniscal structural scaffolding. This type-I collagen is one of the major differences between the menisci and hyaline or articular cartilage, which is composed of predominantly type-II collagen. The remaining organic matter is composed of proteoglycans. The proteoglycan aggregates make up only 1% of the wet weight of meniscus but contribute most to mechanical properties such as compressive stiffness, hydration, and viscoelasticity. The stress–strain curve and the stress–relaxation curve of human meniscus express the typical viscoelasticity. Although PVA-H cannot fully substitute all the mechanical functions of meniscus, the viscoelastic characteristics similar to human's meniscus could be provided by adjusting the water content of PVA-H in the gel production process. As for the lubricative function of PVA-H meniscus, previous studies have proved favorable tribology of the PVA-H versus articular cartilage. Based on these biomechanical data, we have selected high water content PVA-H (water content: 90%, degree of polymerization: 17,500) as an implant sample and performed the in vivo implantation experiment. The bilateral knees of rabbits were used, and PVA-H artificial meniscus replacement was performed in the lateral compartment of one knee, with the lateral meniscectomy in the other knee as a control (Figure 3.19). After operation, the rabbits were reared in cages, and time-dependently sacrificed after a certain period of time. The status of the knee joint cartilage and the degree of injury in the artificial meniscus were observed. The postoperative observation period was 6 months, 1 year, 18 months, and 2 years. Figure 3.20 shows the macroscopic state

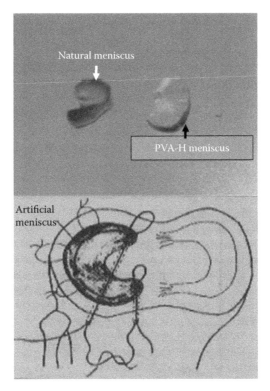

FIGURE 3.19 **(See companion CD for color figure.)** Photograph of PVA-H artificial meniscus and the schema of artificial meniscus replacement.

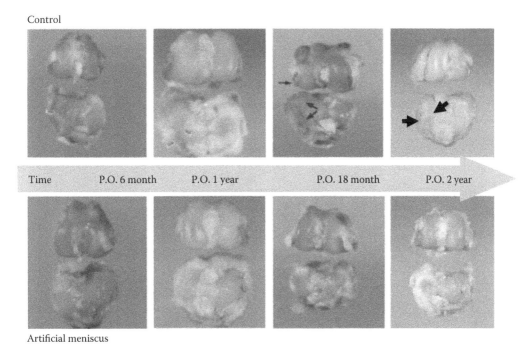

FIGURE 3.20 **(See companion CD for color figure.)** Macroscopic appearance of articular cartilage of rabbit's knee joint up to 2 years after operation. In both groups, lateral side of knee joint was operated. In the control group, OA change appears in post operative 6 months, remarkable change becomes obvious in P.O. 1 year.

of the articular cartilage surface of the femoral condyles and tibia in rabbits knees replaced by
PVA-H artificial meniscus and the control knees. Six months after the operation, the erosion and
degeneration in the articular cartilage were noted in both group knees, and the degree of degenerative change was more marked in the control group. After 1 year, no more progressive osteoarthritis (OA) changes were observed in the artificial meniscus group, whereas more severe OA-like
changes such as partial disappearances of the cartilage surface and bone changes such as elevated
osteophyte were observed in the control knee group. Two years after the operation, articular cartilage in the control group showed progressive OA changes. On the other hand, all PVA-H artificial
menisci were attached to tibia side, and no breakage, displacement, or deformation of PVA-H was
observed. In addition, the color and texture of PVA-H were not seriously different from intact
PVA-H. Most of the articular cartilage surfaces also exhibit normal condition or only slight erosion
in the artificial meniscus group. These macroscopic changes in articular cartilage were assessed
using the scoring system concerning the degree of OA progress as shown in Figure 3.21. In the
control group, OA progressed with time; however, slight cartilage damage remained without progression of degenerative change in the artificial meniscus group between the follow-up period. On
the other hand, all PVA-H artificial menisci were attached to peripheral tissue, and no breakage,
dislocation, or deformation of PVA-H was observed. The color and texture of gels were also not
markedly different from intact PVA-H (Figure 3.22).

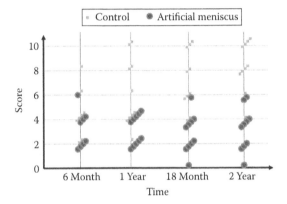

FIGURE 3.21   (See companion CD for color figure.) Macroscopic evaluation of articular cartilage according to classification by Chang.

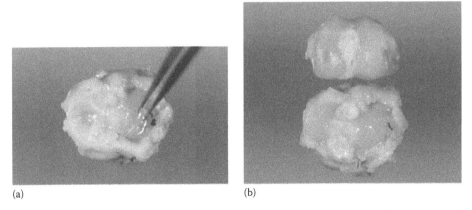

(a)   (b)

FIGURE 3.22   (See companion CD for color figure.) Macroscopic appearance of articular cartilage of
rabbit's knee joint in 2 years after operation in PVA-H meniscus replacement group. (a) Neither breakage nor
dislocation of PVA-H is observed. (b) Articular cartilages of knee joint are also intact.

Figure 3.23 shows the microscopic appearance of articular cartilage of both groups 2 years after the operation. In the control group, the cartilage layer was almost absent. In even survival cartilage layer, the stainability was poor and the presence of chondrocytes was not clear. In the PVA-H meniscus group, abrasion and regressive changes in the cartilage layer were observed in some parts, but the cartilage layer itself was almost similar to normal cartilage. These histological findings dependent on postoperative time were scored numerically as shown in Figure 3.24 by Mankin's scoring system. With the observation time, a significant difference appeared in the degree of progression of OA change in articular cartilage between the PVA-H artificial meniscus group and the control group.

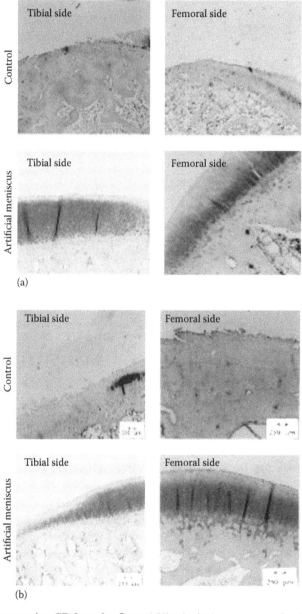

FIGURE 3.23 **(See companion CD for color figure.)** Histological appearance of articular cartilage of knee joint 2 years after operation (×20, Safranin-O stain). (a) P.O. 1 year; (b) P.O. 1.5 year;

(continued)

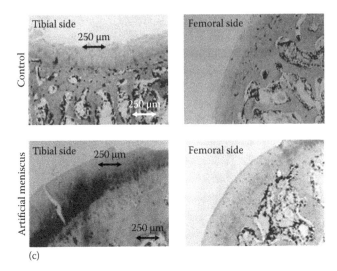

(c)

FIGURE 3.23 (continued) **(See companion CD for color figure.)** (c) P.O. 2 years.

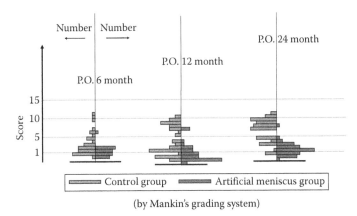

(by Mankin's grading system)

FIGURE 3.24 **(See companion CD for color figure.)** Histological evaluation of articular cartilage according to Mankin's grading system.

These results indicate that PVA-H with high water content is a promising material as an artificial meniscus implant that reconstruct meniscus function and prevent OA changes in articular cartilage. Further studies are needed to validate the PVA-H implant fixation method for clinical application.

3.7 DEVELOPMENT OF IMPLANT FOR SPINE SURGERY

3.7.1 DEVELOPMENT OF ARTIFICIAL INTERVERTEBRAL DISK

As regards the advantage of PVA-H that changing the degree of cross-linking and water content could be utilized as a means of achieving the desired mechanical properties of gel, the application as an intervertebral disk substitute that does not have lubrication issues would be favorable. Lesion of the intervertebral disk is one of the most common causes of spinal disorders. The function of the intervertebral disk is not only to stabilize the spine but also to give the spine its flexibility.

Spinal fusion is one of the most frequently performed procedures in spinal surgery. Spinal fusion, although it restores stability, does not preserve flexibility, and biomechanical and clinical studies have reported that it increases stress on the adjacent segments, leading to accelerated degeneration in the long term. Thus, development of an artificial intervertebral disk (AID) to restore both the stability and flexibility of the disordered spine is desirable [35,36].

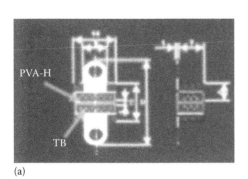
(a)

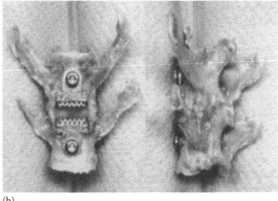
(b)

FIGURE 3.25 **(See companion CD for color figure.)** (a) Diagram of artificial intervertebral disk for beagle dog. (b) Macro appearance of the canine spine implanted with artificial intervertebral disk.

We have developed the AID made of PVA-H and titanium fiber mesh, and studied the mechanical and biological behavior in vitro and in vivo [37,38]. Figure 3.25 shows the diagram and macro appearance of AID for a beagle dog in animal experiment. A specially designed AID composed of titanium fiber mesh and PVA-H was manufactured for canine lumbar vertebrae. To attach PVA-H and titanium fiber mesh, PVA-H in fluid phase was permeated into the pores of titanium fiber mesh under low pressure conditions, and these two materials were mechanically bonded together by gelling the PVA. The degree of polymerization of PVA was 8800 and the water content of hydrogels was 30% and 50% according to the mechanical strength of canine lumbar intervertebral disk. Adult beagle dogs were operated, and the lumbar spine L5/6 intervertebral disk was replaced by AID. These dogs were sacrificed at 1, 3, 6, and 12 months after the operation, and the samples were evaluated by histological method and radiographic examination. All the dogs survived without any postoperative complication such as neuroparalysis. Figure 3.26 shows the radiographs of

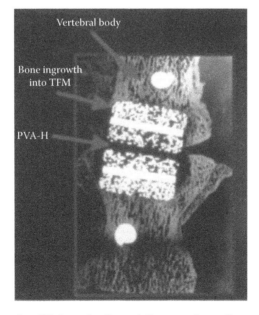

FIGURE 3.26 **(See companion CD for color figure.)** Contact microradiography findings of the artificial intervertebral disk implanted into dog 1 month after operation. Massive bone ingrowth into the pores of titanium fiber mesh was observed.

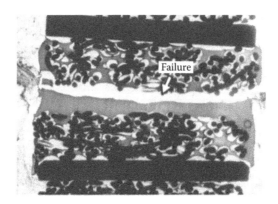

FIGURE 3.27 (See companion CD for color figure.) Micro-appearance of the interface between PVA-H and titanium fiber mesh of artificial intervertebral disk at 6 months after operation into canine spine.

the lumbar spine implanted with this AID prosthesis in 3 months after the operation. Massive bone ingrowth into the pores of the titanium fiber mesh was observed, with the implant attaching firmly to the vertebral body. However, prosthesis failure occurred in some samples at 6 and 12 months after the operation (Figure 3.27), which were observed at the interface between PVA-H and titanium fiber mesh. Furthermore, excessive osteophytes also were formed in the vertebral body implanted with AID.

These phenomena have been observed at postoperative more than 6 months. We presume that osteophyte was caused by the instability of spine with prosthesis due to the insufficient bonding strength of PVA-H with titanium fiber mesh and low mechanical strength of PVA-H. Interface failure between PVA-H and titanium fiber mesh occurred.

In order to improve the mechanical properties and strength of PVA-H and the bonding strength of the interface between titanium fiber mesh and PVA-H, we tried to use high-pressure injection molding in PVA-H synthesis process. Using an improved method, a higher concentration of a stronger PVA-H resulted. As a result, the AID with new PVA-H can solve these problems until postoperative 6 months in animal experiment. Further studies are needed to assess the durability of implant for this clinical application.

3.7.2 Application of Artificial Intervertebral Disk Nucleus Implant

Although the AID using PVA-H and titanium fiber mesh is useful, its treatment is invasive and tasks on the patient in the view of concrete orthopedic surgical technique. It is more desirable if an equivalent effective and the minimum invasive implant was achieved.

For the purpose of developing minimum-invasive treatment for spine disorders, Sou et al. studied the intervertebral disk nucleus replacement using PVA-H implant [39]. The mechanical properties and strength of PVA-H were controlled by changing the degree of polymerization and water content of hydrogel. An AID nucleus similar to the mechanical properties of natural disk nucleus has been developed. Figure 3.28 shows the histological appearance of a canine spine specimen 3 months after the intervertebral disk nucleus replacement. In vivo data displayed neither inflammation reaction nor displacement or breakage of PVA-H prosthesis. The minimum-invasive replacement using this PVA-H seems to be a promising technique in the treatment of intervertebral disk hernia. Further evaluations regarding the in vivo results in the long term are needed for clinical application.

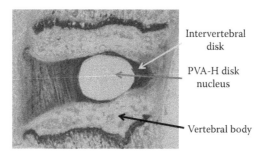

FIGURE 3.28 **(See companion CD for color figure.)** Histological appearance of the partial disk replacement by PVA-H artificial disk nucleus in an animal surgery.

3.8 FUTURE AND POTENTIAL OF PVA-H

PVA-H research and its development for clinical application as an orthopedics surgery implant have been outlined in this chapter so far. The biocompatibility and in vivo safety of PVA-H can be confirmed when taking into consideration previous studies. Furthermore, an advantage of PVA-H is that a tissue substitute with appropriate mechanical properties for the desired tissues can be prepared by adjusting the water content and the degree of polymerization.

Considering the other characteristic functions of PVA-H, such as permeability, hydrophilicity, and transparency, depending on ideas and ingenuity, PVA-H is an extremely promising material in not only orthopedic surgery but also other medical fields; however, in view of the new developments in current biomaterial research, the future development of PVA-H will highlight various problems in its application.

3.8.1 PROBLEMS WITH PVA-H FIXATION TECHNIQUES

The greatest challenge regarding the future clinical application of PVA-H, whereby a method that enables its fixation to living tissues is shown, has already been discussed earlier. PVA-H has strong bio-inert properties, that is, it hardly adheres or binds to the living body. As mentioned earlier, in order to couple PVA-H implants with a living bone, we tried to conjugate them with titanium fiber mesh (Figure 3.29) [40–43]. Other reports have already confirmed the growth of the newly formed bone into titanium mesh. If this method proves to be favorable to fix PVA-H and titanium mesh, then

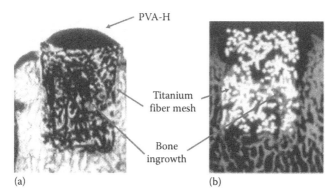

FIGURE 3.29 **(See companion CD for color figure.)** Histological appearance of composite device at 12 weeks after implantation of canine femoral condyle. Mature bone has grown into the pores of titanium fiber mesh. (a) Giemusa stain (×10); (b) contact microradiograph (×10).

we can expect this method to be in practical use in the future. In order to achieve strong binding between PVA-H and titanium mesh, a PVA solution injection molding method has been developed; however, the conjugation of PVA-H with titanium mesh is first and foremost similar to joining machine parts, and an uncertainty remains regarding its long-term fixation in vivo. However, as research on the bioactivation of biomaterial surfaces and the interaction between artificial material surfaces and the living body has made great progress in recent years, for this fixation method between the living body and PVA-H, the development of other adhesion techniques with the living body is greatly anticipated. As a result of a large number of researches, adhesion mechanisms between artificial material surfaces and the cells of living tissues are already being gradually elucidated and, using special surface treatment protein, adhesion has become possible even on biomaterial that was believed to be inert (stable) to living tissues; as a result, techniques enabling the adhesion of living cells to biomaterial are being reported [44,45]. The greatest advantage of this adhesion technique is that as long as the material surface has not lost the inducing functions of living tissue, restoration and recombination by biological reactions can be expected even in the case of rupture and damage to a portion of the adhesion site, which is a major stepping stone in solving the problem of long-term fixation. As regards PVA-H, the report that adhesion to soft tissues is possible using surface treatment has been addressed [46]. Other methods for bioactivation of the PVA-H gel surface are also considered.

3.8.2 FUTURE POSSIBILITIES OF PVA-H

Currently, not only in the field of orthopedic surgery, but in many fields for the in vivo restoration of soft tissues, tissue regeneration by regenerated medicine is attracting attention, and research is thriving [47–49]. If tissue regeneration should advance successfully, it would be an ideal treatment method as it assimilates with auto-tissue and has promising potential. However, specific control techniques by which prepared regenerated tissues up to the micro-cell level are identical to the desired tissues have not been established, and there is a long-term constant risk of carcinogenicity and teratoma. Furthermore, in addition to such overall problems of regenerated medicine, there are the unique problems regarding its application in the field of orthopedic surgery. The most important problems that are needed to be assessed are how to handle a mechanical environment, taking into consideration the biomechanics and the problem of a restoration period, considering the time required for recovery with reduced mobility. The mechanical properties of orthopedic surgery implants, which place a great deal of importance on biomechanical functions, present major problems. Even in cartilage regeneration, in which current research is more advanced, the fine viscoelastic properties of regenerating cartilage still often differ from the original cartilage. Furthermore, in the actual treatment of arthropathy, which replaces THA and TKA, not only the mechanical properties of the cartilage itself but also the tribology of joints and the shape of the sliding surface must be taken into consideration. It is widely known that pathological joint deformities can develop even in healthy normal cartilage as a result of a slight mechanical centralized load and distortion of the alignment of the small articular surface. Whether it is possible to prepare joints with an alignment that can maintain for a long period the same lubricating functions as healthy natural joints, similar to the problem with THA and TKA, greatly depends on the fixation of the regenerated cartilage over the whole articular surface. Another problem is the time needed for tissue regeneration and the recovery of motor functions. Orthopedic surgery aims at the recovery of motor functions, but this is difficult to achieve merely by surgical procedures using implants and scaffolding material alone. Exercise from an early stage that considers rehabilitation is necessary, and, from this aspect, the time required for tissue self-regeneration is an important factor in functional recovery. Elderly people often suffer from femoral neck fractures, and implant replacements, such as femoral head replacement, are performed early because of the reduced recovery functions of the elderly and the problem of timing constraints for the early recovery of motor functions. If regenerated medicine techniques advance in the future, such clinical issues will likely be overcome, but in the short term, the advantages of artificial materials, which can recover motor functions, can hardly be abandoned. On the other hand, it is expected that regenerated tissues derived from artificial materials, such as PVA-H, and tissue engineering will not only be perceived from the viewpoints of comparison and competition but will be expected to be developed as treatment methods using new implants that combine and compensate for the advantages and shortcomings of regenerated tissues derived from artificial implants and regenerative medicine. Research is being conducted to use PVA-H with high mechanical durability as scaffolding material with mechanical stability by using its properties as a gel in greater proportions, and hybridization with other polymer molecules [50,51]. In scaffolding material, the material is ultimately degraded and absorbed and substituted by regenerated tissue, and it is possible that by replacing a portion of the implant with regenerated tissue and leaving a portion of the implant as artificial material in the body, engineered tissues with additional new functions can be developed. For example, when using regenerated cartilage instead of TKA, it is difficult to construct and retain the shape of the knee joint surface directly, but it is possible to develop surface-substituted engineered artificial cartilage, which combines the mechanical properties and shape by constructing the shape of the original knee joint and covering the surface with regenerated cartilage using PVA-H, etc., which is bioactivated on the PVA-H surface mentioned earlier, as a matrix. In the future, together with developments in medical technology, not only in the field of orthopedic surgery, new methods for implant treatments with combined characteristics will be developed for each clinical department from the perspective of biomaterial and tissue engineering. As far as PVA-H is concerned, it will be necessary to include the required new functions using our accumulated knowledge and techniques, considering clinically specific treatment methods.

ACKNOWLEDGMENTS

This work was supported by the Department of Medical Engineering, Institute for Frontier Medical Sciences, Kyoto University. The author would like to express special thanks to all staff including Dr. S.-H. Hyon and Dr. Matsumura for their kind help and useful advice.

REFERENCES

1. S.H. Hyon, W.I. Cha, Y. Ikada; Preparation of transparent poly (vinyl alcohol) hydrogel. *Polym. Bull.* 1989: 22, 119–122.
2. W.I. Cha, S.H. Hyon, M. Oka, Y. Ikada; Mechanical and wear properties of poly (vinyl alcohol) hydrogels. *Macromol. Symp.* 1996: 109, 115–126.
3. J. Delecrin, M. Oka, P. Kumar et al.; Joint reactions against polymer particles: PVA-H versus UHMEPE. *J. Jpn. Ortho. Assoc.* 1990: 64, S1395.
4. T. Noguchi, T. Yamamuro, M. Oka. et al.; Poly (vinyl alcohol) hydrogel as an artificial articular cartilage: Evaluation of biocompatibility. *J. Appl. Biomater.* 1991: 2, 101–107.
5. N.A. Peppas, E.W. Merrill; Development of semicrystalline poly(vinyl alcohol) hydrogels for biomedical application. *J. Biomed. Mater. Res.* 1997: 11, 423–434.
6. T.H. Young, N.K. Yao, R.F. Chang, L.W. Chen; Evaluation of asymmetric poly (vinyl alcohol) membranes for use in artificial islets. *Biomaterials* 1996: 17, 2139–2145.
7. M. Kobayashi, J. Toguchida, M. Oka; Development of PVA-H shields with a high water content for tendon injury repair. *J. Hand Surg.* 2001: 26(B) 5, 436–440.
8. J. Nishiura, M. Oka, K. Sakaguchi et al.; Mechanical properties of PVA-H using injection molding method. *J. Jpn. Clin. Biomech.* 2001: 22, 129–133 (Japanese).
9. M. Kobayashi, M. Oka; Characterization of a poly(vinyl alcohol)-hydrogel artificial articular cartilage prepared by injection molding. *J. Biomater. Sci. Polym. Edn.* 2004: 15(6), 741–751.
10. M. Kobayashi, J. Toguchida, M. Oka; Development of an artificial meniscus using polyvinyl alcohol-hydrogel for early return to, and continuance of athletic life in sports persons with severe meniscus injury. I: Mechanical evaluation. *The Knee* 2003: 10, 47–51.
11. R. Hayashi, M. Oka, K. Ikeuchi et al.; Friction of artificial cartilage sliding against articular cartilage. *J. Jpn. Clin. Biomech.* 1999: 20, 307–313 (Japanese).

12. M. Kobayashi, J. Toguchida, M. Oka; Study on the lubrication mechanism of natural joint by confocal laser scanning microscopy. *J. Biomed. Mater. Res.* 2001: 55, 645–651.

13. M. Kobayashi, M. Oka; The lubricative function of artificial joint material surfaces by confocal laser scanning microscopy. Comparison with natural synovial joint surface. *Bio-Med. Mater. Eng.* 2003: 12, 429–437.

14. D.A. Swan, F.H. Silver, H.S. Slayter et al.; The molecular structure and lubricating activity of lubricin isolated from bovine and human synovial fluids. *Biochemistry* 1985: 225, 195–201.

15. J.E. Pickadr, E. Fisher, E. Ingham, J. Egan; Investigation into the effect of proteins and lipids on the frictional properties of articular cartilage. *Biomaterials* 1998: 19, 1807–1812.

16. T. Murakami; The adaptive multimode lubrication in biotribological systems. *Proc. Int. Tribol. Conf.* 1995: 3, 1981–1986.

17. Y. Matsuda, T. Yamamuro, R. Kasai et al.; Severe metallosis observed 17 years after replacement of the knee with a tumor prosthesis: A case report. *J. Long-Term Effects Med. Implants* 1992: 1, 295–303.

18. E.A. Salvati, C.N. Coornell; Long-term follow up of total hip replacement in patient with avascular necrosis. *Instr. Course Lect.* 1988: 37, 67–73.

19. M. Oka, Y.S. Chang, K. Ushio et al.; Synthetic osteochondral replacement of the femoral articular surface. *J. Bone Joint Surg. (B)* 1997: 79-B, 1003–1007.

20. P. Corkhill, A. Trevent, B. Tighe; The potential of hydrogels as synthetic articular cartilage. *Proc. Inst. Mech. Eng.* 1990: 204, 147–155.

21. T. Murakami, Y. Sawae, K. Nakajima; Evaluation of friction and wear characteristics of artificial cartilage materials. *J. Jpn. Clin. Biomech.* 1999: 20, 319–323 (Japanese).

22. J. Arict; Nontraumatic avascular necrosis of the femoral head: Past, present, and future. *Clin. Orthop.* 1992: 277, 12–21.

23. M.W. Hugerford, M.A. Mont, R. Scott et al.; Surface replacement hemiarthroplasty for the treatment of osteonecrosis of the femoral head. *J. Bone Joint Surg. (A)* 1998: 80-A, 1656–1664.

24. M. Singuler, T. Judel, T. Siguier et al.; Preliminary results of partial surface replacement of the femoral head in osteonecrosis. *J. Arthroplast.* 1999: 14, 45–51.

25. K. Ushio, M. Oka, S.H. Hyon et al.; Partial hemiarthroplasty for the treatment of osteonecrosis of the femoral head. *J. Bone-Joint. Surg. (B)* 2003: 85-B, 922–930.

26. H.J. Mankin, H. Dorfman, L. Lippiello, A. Zarins; Biochemical and metabolic abnormalities in articular cartilage from osteo-arthritic human hips. *J. Bone Joint Surg. (A)* 1971: 53-A, 523–537.

27. T. Brindle, J. Nyland, D.J. Johnson; The Meniscus: Review of basic principles with application to surgery and rehabilitation. *J. Athl. Train.* 2001: 36, 160–169.

28. E.L. Radin, F. Lamotte, P. Maquet; Role of the menisci in the distribution of stress in the knee. *Clin. Orthop.* 1984: 185, 290–294.

29. A.S. Voloshin, J. Wosk; Shock absorption of meniscectomized and painful knees: A comparative in vivo study. *J. Biomed. Eng.* 1983: 5, 157–161.

30. T.J. Fairbank; Knee joint changes after meniscectomy. *J. Bone Joint Surg. (B)* 1948: 30-B, 664–670.

31. W.L. Lanzer, G. Komenda; Changes in articular cartilage after meniscectomy. *Clin. Orthop.* 1990: 252, 41–48.

32. M.J. McNicholas, D.I. Roeley, D. Mcgurty et.al; Total meniscectomy in adolescence. A thirty-year follow up. *J. Bone Joint Surg. (B)* 2000: 82-B, 217–221.

33. M. Kobayashi, J. Toguchida, M. Oka; Preliminary study of poly(vinyl alcohol)-hydrogel (PVA-H) artificial meniscus. *Biomaterials* 2003: 24, 639–647.

34. M. Kobayashi, Y.S. Chang, M. Oka; A two year in vivo study of polyvinyl alcohol-hydrogel (PVA-H) artificial meniscus. *Biomaterials* 2005: 26, 3243–3248.

35. C.K. Lee; Accelerated degeneration of the segment adjacent to a lumber fusion. *Spine* 1988: 13, 375–377.

36. P. Enker, A. Steffee, C. Mcmillin et al.; Artificial disc replacement. *Spine* 1993: 18, 1061–1070.

37. K. Taniyama, M. Oka, Y.S. Hyon et al.; Evaluation of mechanical properties od canine artificial intervertebral disc. *J. Jpn. Clin. Biomech.* 2001: 22, 109–115 (Japanese).

38. S. Yura, M. Oka, K. Ushio et al.; Development of artificial intervertabral disc using poly(vinyl alcohol) hydrogels. *SIROT 99* (Sydney, Australia), Freund Publishing House Ltd., London, U.K., 1999, pp. 371–376.

39. K. So, M. Takemoto, S. Fujibayashi et al.; Anti degenerative effects of partial disc replacement in an animal surgery model. *Spine* 2007: 32, 1586–1591.

40. Y.S. Chang, M. Oka, M. Kobayashi et al.; Significance of interstitial bone ingrowth under load-bearing conditions: A comparison between solid and porous implant materials. *Biomaterials* 1996: 17, 1141–1148.

41. Y.S. Chang, M. Oka, M. Kobayashi et al.; Comparison of the bony ingrowth into an osetochondral defect and an artificial osteochondral composite device in load-bearing joints. *The Knee* 1998: 5, 205–213.
42. Y.S. Chang, M. Oka, M. Kobayashi et al.; Influence of various structure treatment on histological fixation of titanium implants. *J. Arthroplasty* 1998: 13, 816–825.
43. M. Kobayashi, M. Oka; Composite device for attachment of poly(vinyl alcohol)-hydrogel to underlying bone. *Artif. Organs* 2004: 28, 734–738.
44. C. Roberts, C.S. Chen, M. Mrksich et al.; Using mixed self-assembled monolayers presenting RGD and (EG) 3OH groups to characterize long-term attachment of bovine capillary endothelial cells to surfaces. *J. Am. Chem. Soc.* 1998: 120, 6548–6555.
45. M. Arnold, E.A. Cavalcanti-Adam, R. Glass et al.; Activation of integrin function by nanopatterned adhesive interfaces. *ChemPhysChem.* 2004: 5, 383–388.
46. T. Hayami, K. Matsumura, Y.S. Hyon et al.; The effect of ultra-thin coating of hydroxyapatite on artificial articular cartilage (PVA-H) for cell contracture. *Proceedings of the 35th Annual Meeting of Japanese Society for Clinical Biomechanics*, Osaka, Japan, 2008, p. 142 (Japanese).
47. C.J. Xian, B.K. Foster; Repair of injured articular and growth plate cartilage using mesenchymal stem cells and chondrogenic gene therapy. *Curr. Stem Cell Res. Ther.* 2006: 1, 213–229.
48. H.X. Song, F.B. Li, H.L. Shen et al.; Repairing articular cartilage defects with tissue- engineering cartilage in rabbits. *Chin. J. Traumatol.* 2006: 9, 266–271.
49. A. Heymer, G. Bradica, J. Eulert, U. Nöth; Multiphasic collagen fibre-PLA composites seeded with human mesenchymal stem cells for osteochondral defect repair: An in vitro study. *J. Tissue Eng. Regen. Med.* 2009: 3(5), 389–397.
50. Y. Uchino, S. Shimmura, H. Miyashita et al.; Amniotic membrane immobilized poly(vinyl alcohol) hybrid polymer as an artificial cornea scaffold that supports a stratified and differentiated corneal epithelium. *J. Biomed. Mater. Res. Appl. Biomater.* 2007: 81(1), 201–206.
51. C.R. Nuttelman, S.M. Henry, K.S. Anseth; Synthesis and characterization of photocrosslinkable, degradable poly(vinyl alcohol)-based tissue engineering scaffolds. *Biomaterials* 2002: 23, 3617–3626.

4 Polyphosphazenes as Biomaterials

Meng Deng, Cato T. Laurencin,
Harry R. Allcock, and Sangamesh G. Kumbar

CONTENTS

4.1 INTRODUCTION

Biomaterials are generally "substances other than food or drugs contained in therapeutic or diagnostic systems that are in contact with tissue or biological fluids" (Langer and Peppas 2003). Synthetic or natural materials are thus designed to interface with biology and have been widely used for a number of biomedical applications including drug delivery devices, temporary prostheses, and tissue engineering scaffolds (Langer and Peppas 2003; Ratner and Bryant 2004). In controlled drug delivery, a biodegradable matrix that can be loaded with different biologically active molecules is preferred to release various therapeutic agents in an erosion- or diffusion-controlled fashion or a combination of both. For tissue engineering applications, a three-dimensional (3D) biodegradable scaffold that mimics the properties of various tissues is required to support cellular activities and promote neo-tissue formation. Even though biomaterials have already contributed greatly to the improvement of human health, the need for better biomaterial systems is still increasing (Ratner and Bryant 2004).

Among the existing biomaterials, biodegradable polymeric materials have played a paramount role in the development of drug delivery systems and tissue engineering scaffolds due to their biodegradability and inherent biocompatibility (Peppas et al. 2003; Nair and Laurencin 2007). Biodegradable polymers including both natural and synthetic polymers degrade into low molecular weight fragments via hydrolysis or enzymolysis under physiological conditions (Nair and Laurencin 2007). Synthetic biodegradable polymers are attractive, since they offer unique advantages such as predictable properties and ease of tailoring for specific applications. Currently, some major classes of synthetic biodegradable polymers include polyesters, poly(α-amino acids), poly(ortho esters), and polyanhydrides (Nair and Laurencin 2007). Among them, aliphatic polyesters such as polymers of lactic (PLA) and glycolic (PGA) acid or their co-polymers poly(lactide-co-glycolide) (PLAGA) are the most commonly investigated biodegradable polymers. However, these polymers are associated with several distinct drawbacks. For example, they degrade via unstable backbone ester hydrolysis into the corresponding acids, which can compromise the structural integrity and potentially affect biocompatibility both in vitro and in vivo (Taylor et al. 1994; Agrawal and Athanasiou 1997; Böstman 1998; Fu et al. 2000). One example is that in anatomical sites such as articular cartilage, a significant accumulation of these acidic degradation products could occur and affect the cells and the tissues surrounding the implants (Taylor et al. 1994; Böstman 1998). Furthermore, it is possible that these acids can inactivate the sensitive molecules such as proteins used for drug delivery applications (Yang and Alexandridis 2000; Zhu et al. 2000). In order to overcome these limitations, PLA or PLAGA with high lactide content has been investigated for various medical applications. However, several clinical complications have been reported (Böstman 1998; Böstman and Pihlajamäki 2000; Landes et al. 2006). Additionally, the bulk erosion of polyesters often yields uncontrollable release kinetics and catastrophic implant failure, which is not desirable for drug delivery and tissue engineering applications. These studies have propelled researchers to search for and develop alternative biodegradable polymers with controlled degradation, nontoxic degradation products, appropriate physicochemical and biological properties that match the requirements of each specific medical application. Thus, polyphosphazene polymers that offer the combination of synthetic flexibility and modulated properties (i.e., controlled degradability and biocompatibility) are highly preferred for these biomedical applications.

This chapter provides the fundamental knowledge of polyphosphazene synthesis, and discusses the types of biodegradable polyphosphazenes, the mode of degradation mechanisms, and the inherent characteristics of biocompatibility. It further covers the medical applications of biodegradable polyphosphazenes with a specific focus on the use of polyphosphazenes as attractive candidate materials for developing drug delivery matrices and tissue engineering scaffolds.

4.2 SYNTHESIS OF POLYPHOSPHAZENES

Polyphosphazenes, a relatively new synthetic polymer class, are inorganic–organic hybrid polymers with a backbone of alternating phosphorus and nitrogen atoms and with each phosphorus atom bearing two organic or organometallic side groups (Allcock 1972b; Allcock et al. 1972; Allcock and Pucher 1991).

The general structure for polyphosphazenes is shown as Structure 4.1 (Allcock et al. 1977, 1994b; Nair et al. 2003). Nitrogen and phosphorus atoms on the backbone are linked by alternating single and double bonds. Each phosphorus atom is substituted with two side groups R (organic or organometallic or a combination of different functional groups). Generally polyphosphazenes are high molecular weight compounds, and their physicochemical and biological properties are greatly dependent on the nature and composition of the substituted side groups (Allcock 1967, 1972a; Allcock 2003). So far, more than 700 polyphosphazenes with diverse properties have been reported for a variety of applications including drug delivery and tissue engineering applications.

4.2.1 RING-OPENING POLYMERIZATION

The synthesis of polyphosphazenes was attempted by H.N. Stokes in 1895 via thermal ring-opening polymerization of hexachlorocyclotriphosphazene ($[NPCl_2]_3$) (2; Scheme 4.1). But the product obtained was an insoluble cross-linked rubbery material which was also hydrolytically labile and decomposed into phosphates, ammonia, and hydrochloric acid upon the contact with moisture. The polymer would

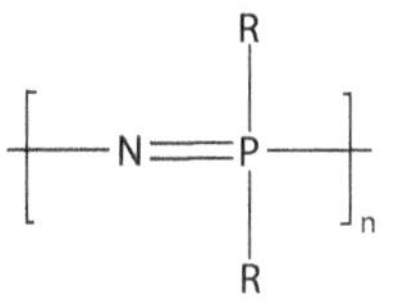

STRUCTURE 4.1 General structure of polyphosphazenes. R can be organic or organometallic or a combination of different functional groups.

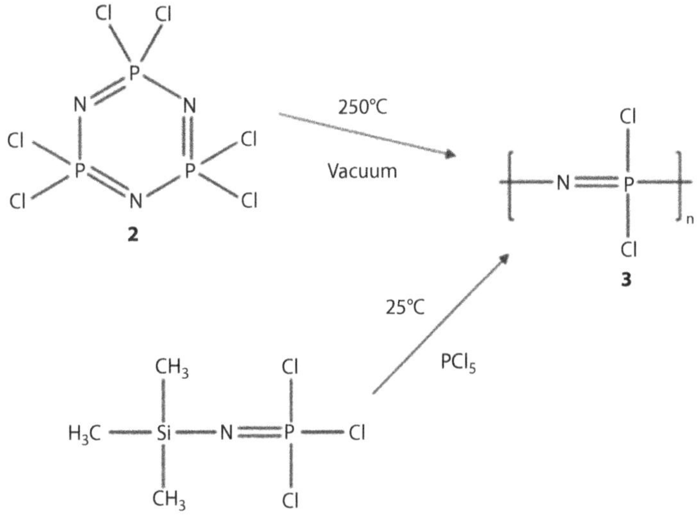

SCHEME 4.1 Synthesis of the prepolymer poly(dichlorophosphazene) (3) via thermal ring-opening polymerization of trimer hexachlorocyclotriphosphazene (2) or living cationic polymerization of a phosphoranamine monomer (4).

not find technological application for another 70 years because of its insolubility and hydrolytic instability (Allcock 1985). In 1965, Allcock and Kugel reported the first successful synthesis of linear
poly(dichlorophospazene) (**3**; Scheme 4.1) by carefully controlling the time and temperature for the
thermal ring-opening polymerization of trimer **2** (Allcock and Kugel 1965). Due to the highly reactive
and polar phosphorus-chlorine bonds (P-Cl), polymer **3** is hydrolytically unstable. Allcock's group
synthesized the first hydrolytically stable, soluble, and high molecular weight polyphosphazene by
replacing the chlorine atoms with organic or organometallic nucleophiles in 1966. Different classes
of polyphosphazenes were then prepared by replacing chlorine atoms of the polydichlorophosphazene
intermediate by alkoxide or aryloxide (Allcock et al. 1966), primary or secondary amines (Allcock
and Kugel 1966), or organometallic reagents (Allcock and Chu 1979). In addition to this versatile
synthetic flexibility, another great advantage of polyphosphazene chemistry lies in the existence of
small molecular analogs of the high polymer for use as reaction and structural models (Allcock 1979).

Through a melt polymerization process, the thermal ring-opening polymerization produces high
molecular weight polymer **3** with an average of ~15,000 repeating units (Allcock 2003). Briefly,
the trimer **2** is purified by recrystallization from heptane followed by sublimation. The sublimated
trimer **2** is then sealed in an evacuated glass tube and heated to 250°C for polymerization. The
polymerization is completed when the melt has become highly viscous. The sealed reaction tube
is then cooled to room temperature and broken carefully to harvest its contents. Pure polymer **3** is
isolated from the trimers and oligomers by sublimation. The polymerization step must be monitored
carefully because cross-linking can occur if the melt becomes too viscous. A high conversion from
cyclic trimer to polymer could also result in cross-linking. It is very important to limit the exposure
of the polymer synthesis to moisture because hydrolysis and cross-linking would occur. The melt
polymerization route generally produces polymer **3** with molecular weights over 1000 kDa.

4.2.1.1   Mechanism of Ring-Opening Polymerization

Ring-opening polymerizations are widely used to convert cyclic small molecules into linear polymers
(Allcock et al. 2003b; Odian 2004). The ring-opening polymerization for polyphosphazenes is a two-
step process, which involves the initial activation of trimer ring **2** and subsequent chain propagation
(Allcock 2003). At high temperature, ionization of a chlorine atom from the phosphorus atom occurs
and results in the formation a phosphazenium cation. In the second step, the formed phosphazenium cation attacks another trimer ring, which leads to chain cleavage of the attacked trimer and the consequent
transfer of positive charge to another phosphorus atom. The charge transfer process propagates to form
a linear chain. Since a ring structure is broken, the propagation rates are slow and often result in slower
molecular weight increases (Odian 2004). The chain continues to grow until all the trimers are used. If
the P-Cl unit in the middle portion of the polymer chain undergoes ionization, branching and growth
occur with the side chains. In addition, the presence of moisture results in cross-linking of the polymer.

4.2.1.2   Thermodynamics of Ring-Opening Polymerization

The ring-opening polymerization is governed by the fundamental thermodynamics. The change in
the Gibbs free energy (ΔG) for the polymerization reaction can be expressed as follows:

$$\Delta G = \Delta H - T\Delta S$$

where
   ΔH represents the enthalpy change
   ΔS represents the entropy change
   T is the reaction temperature

In general, the enthalpy change (ΔH) for most ring-opening polymerization is relatively small, since
the smaller rings, larger rings, and polymer chains are composed of the same repeating units as trimer ring. Specifically in the ring-opening polymerization for polyphosphazenes, the bond angle of the
cyclic trimers should be around 109.5°. However, in a six-membered ring, the –N=P–N=bond angle is

increased to 120°. As the trimer is converted to linear polymer, the strain is released and hence ΔH would be negative (Allcock 2003). Furthermore, the conversion of rigid trimer molecules to a flexible linear polymer contributes to a slight increase in the entropy. However, there is a decrease in the translational entropy when small molecules are converted into large macromolecules. Therefore, the temperature T is a critical factor in determining the reaction process. At lower or moderate temperatures, ΔH is dominant and ΔG is negative, which favors the polymerization. However, at higher temperatures, the contribution of TΔS becomes dominant and ΔG is positive, which results in depolymerization. The transition temperature at which a polymerization reaction shifts to depolymerization is called the "ceiling temperature."

4.2.2 Living Cationic Polymerization

Although ring-opening polymerization is the most fully developed and commonly used method to synthesize high molecular weight polymers (Allcock et al. 1977, 1994b; Nair et al. 2003), this process is often associated with little or no control over molecular weight and with high polydispersities. Thus, several attempts were made to synthesize polymer **3** with controlled molecular weight and small polydispersities (Neilson and Wisian-Neilson 1988; Potin and De Jaeger 1991). For example, De Jaeger et al. synthesized polymer **3** from dichlorophosphynoyliminotrichlorophosphorane $Cl_3P=N-P(O)Cl_2$ via condensation polymerization at 240°–290° with $POCl_3$ as by-products (Helioui et al. 1982; De Jaeger et al. 1983). A certain degree of molecular weight control was achieved, but the limitations included the necessity for high temperatures, the production of corrosive side products, and high polydispersities.

It was not until the early 1990s that Allcock, Manners, and co-workers reported a new method that allows the synthesis of the macromolecular intermediate with molecular weight control and small polydispersities by living cationic polymerization of phosphoranimines at ambient temperature (Scheme 4.1) (Honeyman et al. 1995; Allcock et al. 1996, 1997a,b; Nelson and Allcock 1997). It involves a catalyzed condensation of the monomer trichloro(trimethylsilyl)phosphoranimine, $(CH_3)_3 Si-N = PCl_3$ (**4**) with loss of $(CH_3)_3 SiCl$. Briefly, purified trichloro(trimethylsilyl)phosphoranimineeither in solid state or in solution (anhydrous dichloromethane or toluene) is treated with PCl_5 in a well-defined monomer/initiator ratio at 25°C for a variable period of time. In general, lower initiator concentration results in a smaller number of initiated chains and higher molecular weight of polymer **3**. Furthermore, increase of reaction temperature reduces the reaction time. Additionally, the molecular weight of the polymer **3** prepared in solution is higher with a smaller polydispersity than those prepared in solid state. Branching of polymer could also occur if the cationic charge is delocalized down the chain.

This living cationic polymerization method has also been used for the direct synthesis of polyphosphazenes by the PCl_5-induced polymerization of mono- and di-substituted organophosphoranimines, such as $PhCl_2P=NSiMe_3$, at ambient temperature (Allcock et al. 1997a). The chain length of the polymer **3** can be varied by changing the ratio of the monomer to PCl_5 catalyst. For example, block copolymers with other phosphazene monomers or with organic monomers have been formed through the living sites at the polymer chain ends in living cationic polymerization (Allcock et al. 1997b). In general, this route has several advantages over the thermal ring-opening route including lower polymerization temperature, better control over polymer molecular weight with a smaller polydispersity, and an active chain end that enables the synthesis of block copolymers and complex macromolecular architectures such as star and dendrimer geometries (Allcock 1967, 1972a; Honeyman et al. 1995; Allcock et al. 1996, 1997a,b; Nelson and Allcock 1997; Allcock 2003).

4.2.2.1 Reaction Mechanism of Living Cationic Polymerization

The reaction mechanism of living cationic polymerization has been reported (Honeyman et al. 1995; Allcock et al. 1996, 1997a,b; Nelson and Allcock 1997; Allcock 2003). This polymerization involves the addition of a cationic initiator, such as phosphorus pentachloride (PCl_5), to the monomer **4**. The initiator causes the displacement of the trimethylsilyl group. This creates an active site where polymer chain growth occurs with the addition of more monomers. The chain keeps growing until all the monomers are used and the reaction is terminated with the addition of a terminator.

4.2.3 MACROMOLECULAR SUBSTITUTION

As discussed earlier, the first step of polyphosphazene synthesis involves the synthesis of linear polymer **3**, which can be achieved via the thermal ring-opening polymerization of the commercially available cyclic trimer [NPCl$_2$]$_3$ or living cationic polymerization (Scheme 4.1). In a second step, the reactive chlorine atoms on the polymer **3** can be replaced via macromolecular substitution reactions by various organic nucleophiles including amino acid esters, alkoxy, aryloxy, or a combination of different functional groups as shown in Scheme 4.2.

The macromolecular substitution of polymer **3** makes polyphosphazene polymers a unique platform for biomaterial development. For example, in contrast to direct synthesis from monomer to polymer for many organic or inorganic macromolecules, polyphosphazene synthesis involves the replacement of chlorine atoms in a macromolecular intermediate polymer **3**. The two chlorine atoms in polymer **3** are susceptible to nucleophilic attack by many nucleophiles such as alkoxides, aryloxides, and amines (Scheme 4.2; Allcock 2003). Typically, about 30,000 chlorine atoms per molecule are replaced in this step. Macromolecular substitution reactions allow introducing multiple side groups that determine the polymer properties in a controlled manner. This unique synthetic versatility has led to the development of an interesting class of polymers with different physico-chemical and biological properties (Allcock 2003). In addition to single side group substitution, multiple substituents can be covalently linked to the polymer backbone. For example, substitution of two different types of side groups on the reactive intermediate polymer **3** can be achieved by sequential substitution, allowing the bulkier group to react first followed by smaller reactive group in solution phase during substitution (Allcock 2003). Secondary reactions can be further used to add different functionality to the side groups. Thus, compared to the conventional monomer-dependent

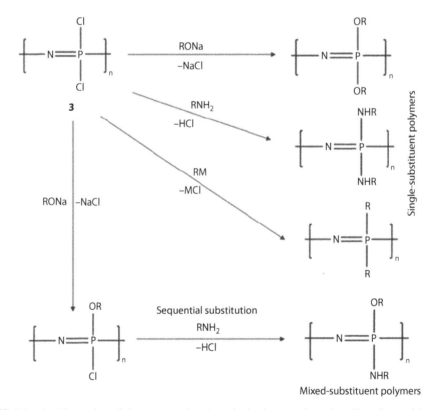

SCHEME 4.2 An illustration of the macromolecular substitution reactions that allow for a wide variety of organic side groups to be linked to the phosphorus atoms to form single-substituent polymers (simultaneous replacement of chlorine atoms) or mixed-substituent polymers (sequential substitution of chlorine atoms).

polymerizations, the macromolecular substitution is highly advantageous due to the systematic control of the material properties via side group chemistry rather than monomer polymerizability.

Another unique advantage of the macromolecular substitution reaction is the ability to introduce two or more different side groups in different compositions, which allows fine-tuning the polymer properties such as the degradation rate to suit a specific application. For example, in the case of a mixed-substituted polymer having two side groups, one of which is resistant to hydrolysis and the other being hydrolytically labile, the polymer erosion rate depends on the ratios of these two groups. Since many combinations of different types of side groups are accessible to be linked to the polymer backbone, it is possible to modulate the hydrolysis rate in a precisely controlled fashion. Allcock et al. have shown that side groups such as amino acid ester units (Allcock et al. 1994b), imidazolyl (Allcock et al. 1982b), glucosyl (Allcock and Scopelianos 1983), glyceryl (Allcock and Kwon 1988), and lactate or glycolate (Allcock et al. 1994c) groups sensitize the polymer to break down over periods of hours, days, months, or years at the physiological temperature depending on the specific molecular structure of the polymer.

4.2.4 OTHER SYNTHESIS METHODS

Several other polyphosphazene synthesis methods have also been reported in the literature for the generation of polymers with direct P–C bonds between the backbone and the sidegroups: (1) the ring-opening polymerization of phosphazene cyclic trimers that have alkyl or aryl units linked to the ring (Allcock 1992); (2) condensation reactions of organophosphoranimines (Neilson and Wisian-Neilson 1988; Montague and Matyjaszewski 1990); and (3) denitrogenation reactions of organophosphorus azides (Franz et al. 1994). For example, Neilson and Wisian-Neilson et al. reported the direct synthesis of poly(alkyl/aryl)phosphazenes with moderate molecular weights and polydispersities of 1.5–3.0 via the condensation polymerization of N-silylphosphoranimines at 200°C (Neilson and Wisian-Neilson 1988).

4.2.5 POLYPHOSPHAZENE ARCHITECTURE

So far, the macromolecular substitution of polymer **3** has allowed for over 250 different side groups to be covalently linked to the backbone via replacement of the labile chlorine atoms (Allcock 2003, 2010). This makes polyphosphazenes more versatile than many other polymer systems and has led to the formation of various different phosphazene polymer architectures such as linear polymers, branched or graft polymers, star polymers, and dendrimers (Scheme 4.3; Allcock 2003, 2010). Using the living cationic route shown in Scheme 4.1, block polyphosphazene copolymers have also been synthesized with various organic polymers including polystyrene (Allcock et al. 2004a; Chang et al. 2005), poly(methyl methacrylate) (Allcock et al. 2004b), poly(ethylene oxide) (Chang et al. 2002, 2003), and poly(dimethylsiloxane) (Prange and Allcock 1999; Allcock and Prange 2001). Furthermore, complex structures that involve hexachlorocyclotriphosphazene, such as cyclo-linear polymers, have also been synthesized with phosphazene rings linked between linear polymer backbones with polyethylene (Scheme 4.3) (Allcock and Kellam 2001; Allcock et al. 2001c). The cyclic phosphazenes can also be used as side chains for other polymers such as polystyrene (Allcock et al. 2001b), polynorbornene (Allcock et al. 2001d, 2003a), and polyoctenamer (Allcock et al. 2005). Finally, 3D cross-linked networks "cyclomatrix" can be fabricated to form tough materials with high melting point and hardness using the cyclic phosphazene ring through the multiple functional groups in the phosphazene system (Devadoss and Nair 1985; Zhang et al. 2005). In addition to phosphorus and nitrogen in the skeleton, a third element such as sulfur or carbon has also been incorporated in the main chain for further functionalizations (Scheme 4.3).

4.2.6 STRUCTURE–PROPERTY RELATIONSHIPS

The torsional and angular freedom within phosphorus-nitrogen backbone offers high molecular weight and high flexibility to the polyphosphazenes. The bond torsional mobility allows the polyphosphazene backbone to be stretched from a random coil. It is dependent on the electronic

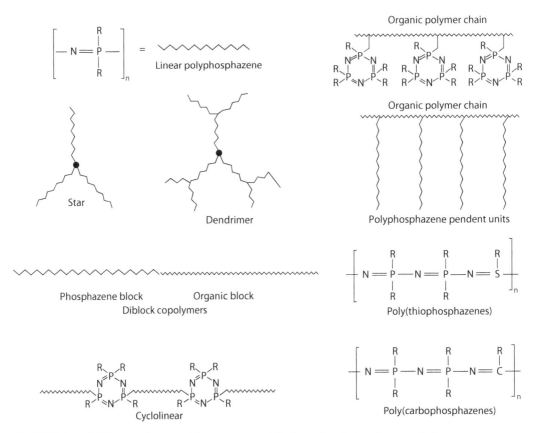

SCHEME 4.3 Different polymer architectures accessible through polyphosphazene platform.

character of the bond, bond angle and length, and the number, nature, and ratio of the attached side groups (Allcock 2003). In a repeating unit of $-P=N-$, each phosphorus atom provides five valence electrons while each nitrogen contributes another five. If the electron pairs are assigned in a sigma bond framework, there are four electrons that are unaccounted for. Three of these electrons are from the nitrogen atom and one is from the phosphorus atom. The electron from the phosphorus atom resides in a 3d orbital forming a "stabilized excited state." For the three electrons on the nitrogen atom, two of them are confined to a lone-pair orbital, which gives the polymer its basicity. The remaining electron from the nitrogen atom is in a 2p orbital. The pairing of these two electrons from the 2p and 3d orbitals results in an "out of plane" $d_\pi - p_\pi$ bond. Due to the orbital mismatch and nodes occurring at every phosphorus atom, delocalization is limited only over three skeletal atoms and long range delocalization does not occur. Furthermore, the 2p orbital can be brought into an overlapping position with a d-orbital at any torsion angle. Hence, the torsional barrier in the backbone bonds is much smaller when compared to $p_\pi - p_\pi$ double bond in organic molecules (Allcock 2003). The bond angle at the nitrogen atom in the backbone ranges from 130° to 160° while the bond angle at the phosphorus atom is 114° (Allcock 2003). The presence of the side groups in alternating atoms in the backbone contributes to the less density of the side groups in polyphosphazenes as compared to many other organic polymers and thus increases the backbone flexibility. Bond length of typical $-C=C-$ is ~1.54 Å whereas the bond length between nitrogen and phosphorus atoms ranges from 1.55 to 1.60 Å (Allcock 2003). The torsional barrier is further reduced due to the increased distance between the side groups and the alternating bond lengths in the backbone. Chatani et al. found a difference of 0.22 Å between the bonds in the polymer skeleton (Chatani and Yatsuyanagi 1987). However, other studies reported the values of bond differences to be 0.05 ~ 0.07 Å (Allcock 2003).

The overall physicochemical and biological properties of polyphosphazenes such as solubility, hydrophilicity/hydrophobicity, glass transition temperatures (T_g's), mechanical properties, hydrolytic sensitivity, and biocompatibility depend on the nature and composition of the side groups attached to the backbone. Bulky and inflexible side groups or polar or hydrogen bonding units increase the torsional barrier of the polymer backbone and result in a rigid polymer with high T_g's. Furthermore, the incorporation of side groups that impart intermolecular hydrogen bonding such as glycylglycine dipeptide units also leads to an increase in the torsional barrier of the motility. A large number of reagents such as alcohols, phenols, and amines have been used alone or in combination to synthesize polyphosphazenes with a wide range of properties (Allcock et al. 1977, 1994b; Allcock and Fuller 1980; Allcock and Pucher 1991; Crommen et al. 1992a,b, 1993; Laurencin et al. 1993, 1996; Qiu and Zhu 2000a; Gümüsdereliolu and Gür 2002; Nair et al. 2003, 2004; Conconi et al. 2004; Cui et al. 2004; Barrett et al. 2005; Yuan et al. 2005; Andrianov and Marin 2006; Sethuraman et al. 2006; Singh et al. 2006). Table 4.1 summarizes the

TABLE 4.1

Summary of Glass Transition Temperatures (T_g's) of Single-Substituent Polyphosphazenes Illustrating the Great Tunability of Physico-Chemical Properties by Modulating Side Group Chemistry

Polymer	Side Group	T_g (°C)
PNEG		−40
PNEA		−10
PNEL		15
PNEPhA		42
PNGEG		56
PNMEEP		−84
PNmPh		2

The selected polymers are (PNEG), poly[bis(ethyl glycinato)phosphazene]; (PNEA), poly[bis(ethyl alanato)phosphazene]; (PNEL), poly[bis(ethyl leucinato)phosphazene]; (PNEPhA), poly[bis(ethyl phenylalaninato) phosphazene]; (PNGEG), poly[bis(glycyl-ethylglycinato)phosphazene]; (PNMEEP), poly[bis(methoxyethoxyethoxy)phosphazene]; (PNmPh), poly[bis(p-methylphenoxy)phosphazene].

physicochemical properties such as T_g's of some of the representative polyphosphazenes for biomedical applications (Allcock et al. 1977, 1994b; Allcock and Fuller 1980; Allcock and Pucher 1991; Crommen et al. 1992a,b, 1993; Cohen et al. 1993; Laurencin et al. 1993, 1996; Schacht et al. 1996; Qiu and Zhu 2000a; Lee and Mooney 2001; Gümüsderaliolu and Gür 2002; Chaubal et al. 2003; Gunatillake and Adhikari 2003; Katti and Laurencin 2003; Conconi et al. 2004; Cui et al. 2004; Gleria and De Jaeger 2004; Kricheldorf et al. 2004; Nair et al. 2004; Barrett et al. 2005; Yuan et al. 2005; Andrianov and Marin 2006; Sethuraman et al. 2006; Singh et al. 2006). As seen from the table, a wide range of the glass transition temperature from −84°C to 56°C can be obtained by changing the side groups. For example, the polymer T_g is increased in the order of PNEG < PNEA < PNEL < PNEPhA, which is indicative of the bulkier side group resulting in a higher T_g.

4.3 POLYPHOSPHAZENES AS A VERSATILE CLASS OF BIOMEDICAL MATERIALS

As discussed previously, the side groups largely determine the material properties of polyphosphazenes. Polyphosphazenes have an inherently flexible backbone due to the bonding nature of the phosphorus-nitrogen atoms in the backbone. This flexibility offers great tunability of material properties. This great synthetic flexibility of polyphosphazenes enables the design and development of prospective biomedical materials with diverse properties. Most of the polyphosphazenes synthesized so far are hydrolytically stable. These polymers exhibit unique properties including unusual thermal properties (Sulkowski et al. 1997), optical properties (Allcock et al. 1998), electrolytic properties (Allcock et al. 2001a), and electroactivity (Li et al. 2001). Early reports also attempted to characterize tissue compatibility of several nondegradable fluorinated poly(organophosphazenes) (Wade et al. 1978; Welle et al. 2000).

Hydrolytically labile polymers that can degrade into nontoxic products are highly preferable as short-term medical implants for drug delivery and tissue engineering applications (Nair and Laurencin 2007). Due to their transient nature, biodegradable polyphosphazenes have attracted a great deal of attention as a unique class of biomedical materials. The phosphorus–nitrogen backbone can be rendered hydrolytical sensitivity by substituting with appropriate sidegroups (Allcock 2003). The polymers undergo bulk erosion or surface erosion or a combination of both generating nontoxic degradation products, such as phosphoric acid, ammonia, and the corresponding side groups. The degradation rate is greatly influenced by the type and composition of the substituted side groups (Andrianov and Marin 2006; Kumbar et al. 2006; Singh et al. 2006). It is possible to design polyphosphazenes that can degrade over a period of hours to days, months and years at body temperature depending on the structure and composition of the polymer (Allcock et al. 1977, 1994b; Allcock and Pucher 1991). In addition, the mode of side group attachment also plays an important role in determining the polymer degradability (Allcock 2003; Allcock et al. 2003c). Thus, a versatile array of biodegradable polyphosphazenes can be synthesized by modulating the side group chemistry. Specifically, this section is confined to various biodegradable polyphosphazenes (aminated polyphosphazenes and alkoxy-substituted polyphophazenes) and polyphosphazene blends/composites as well as their applications as drug delivery matrices and tissue engineering scaffolds.

4.3.1 BIODEGRADABLE POLYPHOSPHAZENES

Biodegradable polyphosphazenes can be synthesized by substituting chlorine atoms of the polymer **3** with hydrolytically labile groups such as amino acid esters (**5**) (Allcock et al. 1994b), imidazolyl (**6**) (Allcock et al. 1982b), glucosyl (**7**) (Allcock and Scopelianos 1983), glyceryl (**8**) (Allcock and Kwon 1988), and glycolate or lactate ester (**9**) (Allcock et al. 1994c), to name a few (Figure 4.1). In general, biodegradable polyphosphazenes alone can be classified into two categories based on the type of side group substituents: aminated polyphosphazenes (those substituted with amines of low acid dissociation constant pK_a) and alkoxy substituted polyphosphazenes (those substituted with activated alcohols). In addition, polyphosphazene blends/composites constitute another attractive class of biodegradable polymer systems for drug delivery and tissue engineering applications.

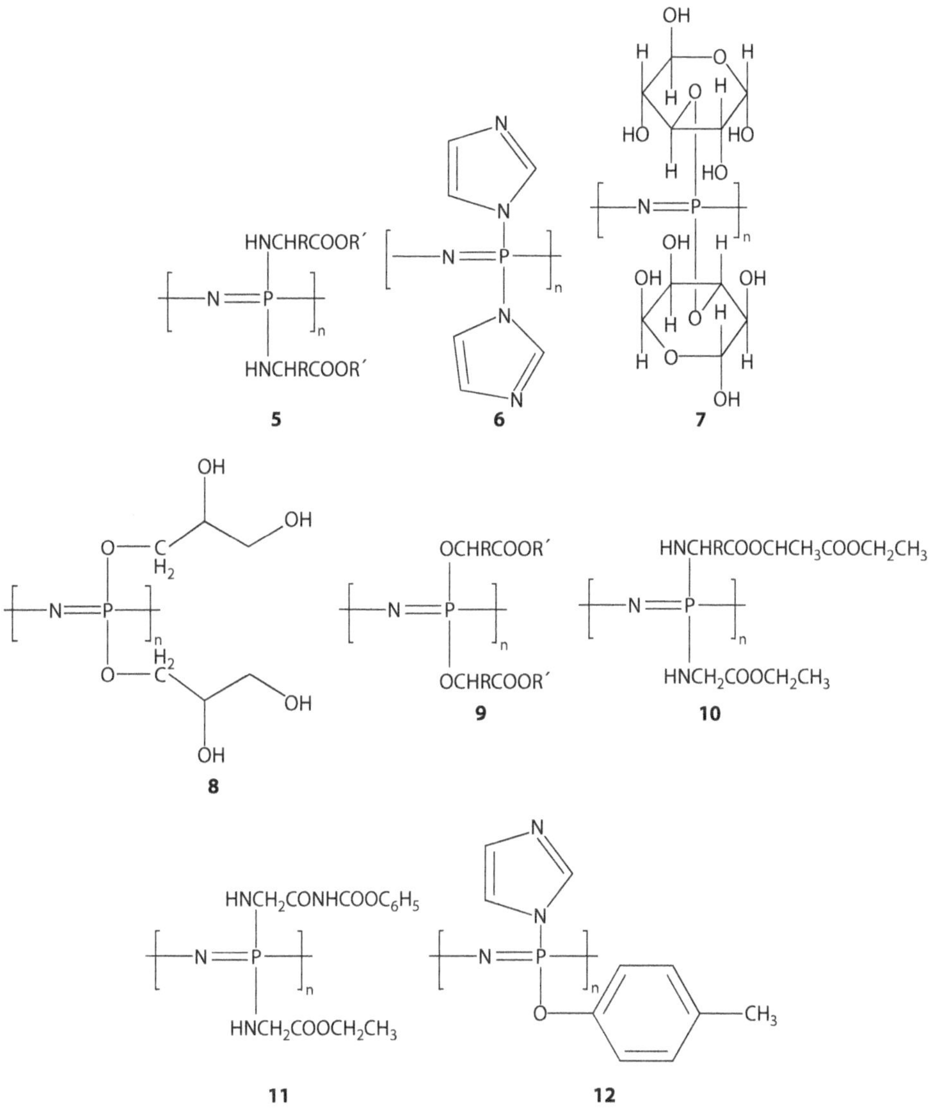

FIGURE 4.1 Structures of various biodegradable polyphosphazenes by substituting the chlorine atoms of poly(dichlorophospazene) (**3**) with hydrolytically labile groups such as amino acid esters (**5**), imidazolyl (**6**), glucosyl (**7**), glyceryl (**8**), and glycolate or lactate esters (**9**), to name a few. Polymers **10–12** are representative aminated polyphosphazenes.

4.3.1.1 Aminated Polyphosphazenes

Aminated polyphosphazenes constitute the largest and most extensively investigated class of biodegradable polyphosphazenes. In 1966 Allcock and Kugel reported the first successful synthesis of hydrolytically sensitive aminated polyphosphazenes (Allcock and Kugel 1966). Since then, combined efforts of Allcock and coworkers have led to the synthesis of a series of aminated polyphosphazenes. Among them, polyphosphazenes substituted with amino acid ester (**5**) and imidazole (**6**) side groups attracted much interest as candidate biomaterials due to their hydrolytic degradability and nontoxic degradation products.

The amino acid ester polyphosphazenes comprise the most extensively investigated biodegradable polyphosphazenes for biomedical applications due to their inherent biocompatibility and bioactivity. Allcock et al. synthesized a wide spectrum of amino acid ester–substituted polyphosphazenes as new biomaterials (Allcock et al. 1977, 1994b). Efforts were also made to increase the degradation rate of poly[(amino acid ester)phosphazenes] by incorporating side groups with hydrolytically sensitive ester functions (**10**, Figure 4.1; Crommen et al. 1992b). Furthermore, pH-sensitive polyphosphazenes were synthesized by incorporating aminoacethydroxamic acid as cosubstituents to poly[(amino acid ester)phosphazenes] (**11**, Figure 4.1; Qiu and Zhu 2000a). More recently, polyphosphazenes substituted with glycylglycine dipeptides have been designed and synthesized to render excellent degradability and hydrogen bond–forming ability (Krogman et al. 2007). Additionally, polyphosphazenes cosubstituted with the amino acid ester/purine or pyrimidine polyphosphazenes were reported to undergo hydrolytic degradation (Krogman et al. 2008a).

Specifically, the imidazole-substituted polymer (**6**) was found to be the most hydrolytically sensitive among the aminated phosphazenes synthesized. Furthermore, the hydrolysis rate of imidazole-substituted polyphosphazenes could be modulated by cosubstituting it with less hydrolytic sensitive groups such as *p*-methylphenoxy on the polymer backbone (**12**, Figure 4.1) (Allcock and Kugel 1966).

4.3.1.2 Alkoxy-Substituted Polyphosphazenes

In addition to the class of aminated phosphazenes, some alkoxy-substituted polyphosphazenes were also found to be hydrolytically labile (Allcock and Scopelianos 1983; Allcock and Kwon 1988; Allcock and Pucher 1991; Allcock et al. 1994c). For example, Allcock and Kwon synthesized the first glyceryl-substituted polyphosphazene (**8**) (Allcock and Kwon 1988). Hydrogels of this polymer were further synthesized by cross-linking with various cross-linking agents such as adipoyl chloride and hexamethylenediisocyanate. Moreover, polyphosphazenes substituted with glycosyl and methyl amino side groups were also found to be hydrolytically labile (Allcock and Pucher 1991). Additionally, Allcock et al. synthesized another alkoxy-substituted polyphosphazene bearing glycolic acid ester and lactic acid ester side groups (**9**) (Allcock et al. 1994c). These polymers are not crystalline and hydrolyze much faster than the commonly used biodegradable polyesters such as PLA and PGA.

4.3.1.3 Polyphosphazene Blends/Composites

The synthetic flexibility of polyphosphazenes offers great tunability in polymer properties. Although considerable advancement has been made in the development of polyphosphazenes with desirable properties, some challenges still remain especially in the development of ideal candidate materials with the right combination of desirable properties for a specific application. For example, the incorporation of bulkier side groups such as phenylalanine ethyl ester provides higher mechanical strength than other amino acid esters, but also results in a significant increase in polymer hydrophobicity, which substantially slows down the polymer degradation. The difficulties to attribute all the required physical, chemical, and biological properties to a single polymer necessitate the continued search for an ideal system (Deng et al. 2009).

Blending of two or more different polymers is a practical approach to develop novel biomaterials owing to the ease of blend preparation and the efficient control of blend properties via compositional changes (Utracki 1989). It synergistically combines the advantages of the polymer components. Moreover, the synthetic challenges of producing a copolymer are not encountered. Polymer blends/composites are of great commercial interest and occupy over 30% of the polymer market (Utracki 1989). For example, efforts have been made to improve the properties of polyesters by blending them with a variety of other polymers such as 2-methacryoloxyethyl phosphoroylcholine (PMPC) (Iwasaki et al. 2002), poly(vinyl alcohol) (Pitt et al. 1992), chitin (Mi et al. 2003), polycaprolactone (PCL) (Chen et al. 2003), and triblock copolymers of poly(ethylene oxide) (PEO) and poly(propylene oxide) (PPO), PEO – PPO – PEO (Park et al. 1992). In addition, the effect of enantiomeric polymer blending on degradation rate of polyesters has also been extensively investigated (Ikada and Tsuji 2000; Tsuji 2003).

In recent years, different blend systems of polyphosphazenes with a variety of classical synthetic biodegradable polymers including polyesters such as PLA, PLAGA, and polyanhydrides have been reported (Ibim et al. 1997; Qiu and Zhu 2000b; Ambrosio et al. 2002; Qiu 2002a,b; Nair et al. 2005; Krogman et al. 2007). Among them, polyphosphazene–polyester blends have several unique advantages compared to other polymer blend systems. In specific, the blend systems composed of biodegradable polyphosphazenes and polyesters are inherited with unique self-neutralizing ability where the polyphosphazene degradation products such as ammonium phosphate act as a buffer system to neutralize the acidic degradation products of polyesters. Furthermore, various functional groups can also be easily introduced to the blends through the versatile side group chemistry in polyphosphazenes, and thus the blend properties can be easily tailored for different biomedical applications. For example, carboxyl-substituted polyphosphazenes poly[bis(carboxylatephenoxy)phosphazene] (PCPP) have been shown to interact with simulated body fluid (SBF) and form a bone-like apatite layer on the surface (Brown et al. 2007). Such bioactivity would greatly improve the osteointegrativity of scaffolds/implants for bone tissue engineering (Kokubo and Takadama 2006). However, in order to achieve uniform and predictable properties of polyphosphazene–polyester blends, there are several design criteria on the molecular structure of polyphosphazenes: (1) the polymer should have appropriate functional groups that promote hydrogen bonding interactions with polyesters. This would allow the formation of miscible blends with improved physical, chemical, and biological properties (Deng et al. 2009); (2) the polymer should be biocompatible and should hydrolyze into nontoxic small molecules that can be metabolized or excreted. The long-term collaborative efforts between Laurencin and Allcock's groups have led to the development of a series of promising polyphosphazene–polyester blends for various biomedical applications based on the polymers substituted with amino acid esters, alkoxy, and aryloxy side groups (Figure 4.2). In addition, blends of polyphosphazenes alone have also been investigated for a number of biomedical applications.

4.3.1.3.1 Theory of Hydrogen-Bonded Polymer Blends

For biomedical applications, blend miscibility is a critical factor that affects the material performance. Uniform and predictable properties can only be achieved when blends are miscible. Partially miscible or immiscible blends are often characterized by polymer phase segregation, which adversely affects the material properties such as mechanical properties.

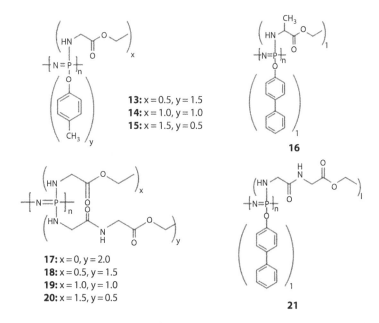

FIGURE 4.2 Structures of several representative polyphosphazenes that are substituted with various amino acid esters, alkoxy, and aryloxy side groups for the development of polyphosphazene-polyester blends.

Blend miscibility is obtained when the Gibbs free energy change of mixing is negative or $\Delta G_{mix} = \Delta H_{mix} - T\Delta S_{mix} < 0$. The miscibility of two polymers in a blend is generally a result of intermolecular interactions between the polymers such as through hydrogen bonding, dispersion forces, dipole–dipole interactions and/or van der Waals forces (Utracki 1989; Bates 1991; Coleman and Painter 1995; He et al. 2004).

For hydrogen-bonded polymer blends, ΔG_{mix} can be expressed in the following form:

$$\frac{\Delta G_{mix}}{RT} = \frac{\Phi_A}{M_A}\ln\Phi_A + \frac{\Phi_B}{M_B}\ln\Phi_B + \chi\Phi_A\Phi_B + \frac{\Delta G_H}{RT}$$

$$\chi = \frac{V_r}{RT}[\delta_A - \delta_B]^2$$

where Φ_i represents the volume fraction of polymer i in the blend, M_i is the degree of polymerization for polymer i (i = A or B). There are three components that contribute to ΔG_{mix}. The component $(\Phi_A/M_A)\ln\Phi_A + (\Phi_B/M_B)\ln\Phi_B$ represents a favorable contribution of combinatorial entropy; the term $\chi\Phi_A\Phi_B$ represents an unfavorable contribution of physical forces such as dispersion forces or van der Waals forces; the $\Delta G_H/RT$ represents a favorable contribution from hydrogen bonding. Since the value of combinatorial entropy is usually very small, blend miscibility is largely determined by the overall contributions from the $\chi\Phi_A\Phi_B$ and $\Delta G_H/RT$ components. When hydrogen bonding is absent, miscibility can only be achieved when the critical value of χ determined by the solubility parameters of the two polymers is extremely small. For example, χ must be less than 0.002 where only dispersive forces are present in Flory–Huggins model. When hydrogen bonding plays a dominant role, the contribution of $\Delta G_H/RT$ is greater than that of $\chi\Phi_A\Phi_B$ and completely miscible blends are obtained.

In the polyphosphazene–polyester blend systems, the synthetic flexibility of polyphosphazenes allows the incorporation of hydrogen bonding side groups on polymer backbone that interact with polyesters to favor blend miscibility.

4.3.1.3.2 Miscibility of Polyphosphazene Blends

The concept and feasibility of fabricating various blends of polyphosphazenes with polyesters was first demonstrated by Ibim et al. (1997). In the study, three different polyphosphazenes containing different molar ratios of ethyl glycinato to p-methylphenoxy (**13**, **14**, and **15**, Figure 4.2) were synthesized and blended with PLAGA (50:50) at a weight ratio of 50:50. All blends showed homogenous surface morphologies as evidenced by scanning electron microscopy (SEM) and their miscibility was confirmed by differential scanning calorimetry (DSC).

Qiu et al. have investigated the miscibility of a series of degradable blends of PNEG (**5**) (where R = H and R′ = CH_2CH_3) and polyesters or polyanhydrides (Qiu and Zhu 2000b; Qiu 2002a,b). In the study, different blends of PNEG with PLA, PLAGA (80:20), poly(sebacic anhydride) (PSA) and poly(sebacic anhydride-co-trimellitylimidoglycine)-block-poly(ethylene glycol) (30:50:20 by mole) (PSTP) were fabricated in various ratios using a solvent-mixing technique (Qiu and Zhu 2000b). The PNEG-PLA blends were completely immiscible, whereas the PNEG-PLAGA and PNEG-PSTP blends were found to be partially miscible as evaluated by DSC, Fourier transform infrared spectroscopy (FTIR), and phase contrast microscopy.

In order to develop high-strength candidate materials for load-bearing regenerative applications, several studies have indicated the feasibility of significantly increasing polyphosphazene hydrophobicity and mechanical properties by incorporating phenylphenoxy side groups (Singh et al. 2006; Sethuraman et al. 2010). Thus, Deng et al. (2008) fabricated blends of poly[(ethyl alanato)$_i$ (p-phenylphenoxy)$_i$phosphazene] (PNEAPhPh) (**16**, Figure 4.2) and PLAGA (85:15). Solvent effect on the miscibility of polyphosphazene-polyester blends was investigated for the first time and tetrahydrofuran (THF) was used to produce different blends at three weight ratios of PNEAPhPh:PLAGA namely

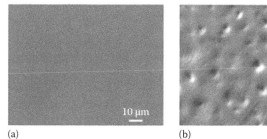
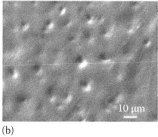
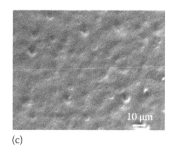

(a)   (b)   (c)

FIGURE 4.3 SEM micrographs showing the surface morphologies of different blends. (a) BLEND25; (b) BLEND50; and (c) BLEND75. BLEND25 showed uniform morphology indicating blend miscibility whereas the other two blends showed visible phase separation. (From *Biomaterials*, 29(3), Deng, M., Nair, L.S., Nukavarapu, S.P. et al., Miscibility and in vitro osteocompatibility of biodegradable blends of poly[(ethyl alanato) (p-phenyl phenoxy) phosphazene] and poly(lactic acid-glycolic acid), 337–349. Copyright 2008, with permission from Elsevier.)

25:75 (BLEND25), 50:50 (BLEND50) and 75:25 (BLEND75). Miscible blends were formed at lower polyphosphazene content as characterized by the single T_g value with smooth surface morphology. As shown in Figure 4.3, an increase in the polyphosphazene content of the blend caused phase separation due to the steric hindrance from the α-CH_3 groups in the alanine units and bulky phenylphenoxy side groups, as well as insufficient hydrogen bonding sites available to form miscible blends. This further indicated that the solvent has a limited effect on blend miscibility (Deng et al. 2008).

Blend miscibility is one of the prerequisites in applications that require controllable and predictable degradation pattern. Efforts have also been made to improve blend miscibility by introducing different side groups containing multiple hydrogen bonding sites in polyphosphazenes. In that direction, Krogman et al. (2009) developed polyphosphazenes having tris(hydroxymethyl)amino methane (Trisma or THAM) side groups and with cosubstitutents glycine ethyl ester and alanine ethyl ester. The THAM side group was linked to the polyphosphazene backbone via the amino function. The polymer has three pendent hydroxyl groups. The polymers form completely miscible blends with both PLAGA (50:50) and PLAGA (85:15) (Krogman et al. 2009). In another study, Krogman et al. have synthesized different glycylglycine dipeptide–based polyphosphazenes (**17–20**, Figure 4.2) and investigated the miscibility of their blends with both PLAGA (50:50) and PLAGA (85:15) (Krogman et al. 2007). Among them, polymer **18** was found to form completely miscible blends with both PLAGA, which was characterized by SEM, DSC, and attenuated total reflectance infrared spectroscopy (ATR-IR). It was found that the T_g for each blend was lower than that for each parent polymer, implying both polyphosphazenes and PLAGA in the blend acted as plasticizers for each other.

The aforementioned studies have highlighted the significant challenge in the development of an ideal polyphosphazene blend system with the desirable balanced properties such as miscibility and mechanical properties. More recently, Deng et al. (2010b,d) designed and fabricated blends of poly[(glycine ethyl glycinato)$_l$(p – phenylphenoxy)$_l$phosphazene] (PNGEGPhPh) (**21**, Figure 4.2) and PLAGA (50:50) in two different weight ratios (25:75 and 50:50, BLEND25 and BLEND50) (Deng et al. 2010b,d). The two side groups for the PNGEGPhPh were selected based on the following rationale. The glycylglycine dipeptide was shown to provide two proton donors per monomer for hydrogen bonding with PLAGA while the phenylphenoxy was beneficial for maintaining mechanical function and hydrophobicity of the polymer (Sethuraman et al. 2006, 2010; Singh et al. 2006; Krogman et al. 2007). Laurencin et al. demonstrated excellent osteocompatibility of phenylphenoxy cosubstituted polyphosphazenes both in vitro and in vivo (Sethuraman et al. 2004, 2006). In addition, a mixed-substituent polymer with both hydrolysis sensitive groups and hydrophobic groups provided efficient tunability to achieve a suitable degradation rate. Specifically the combination of the two side groups in a 50:50 ratio was expected to result in a complete degradation time in the range of 12–24 weeks. Furthermore, the glycylglycine dipeptide was recognized to be biocompatible, with degradation into glycine units, one of the common amino acids in the body (Adibi 1989). The miscibility of the blends was confirmed by a variety of analytical techniques including DSC, ATR-IR, SEM, and high-resolution solid-state nuclear magnetic resonance spectroscopy (NMR).

This mixed-substituent polymer resulted in a completely miscible blend system with the polyesters through the hydrogen bonding between the amide and amine protons in the dipeptide units of PNGEGPhPh and the carbonyl groups in PLAGA. For example, BLEND25 and BLEND50 showed a single T_g of ~36°C and ~40°C, whereas PLAGA and PNGEGPhPh showed a T_g of ~28°C and ~53°C, respectively. The single T_g value for both blends was intermediate between those of the two parent polymers indicating blend miscibility. Furthermore, for a miscible blend system, the T_g of the blend can be predicted by the Wood equation (Wood 1958): $T_g = w_1 T_{g1} + w_2 T_{g2}$ or by the Fox equation (Fox 1956): $T_g = T_{g1} T_{g2}/(w_1 T_{g2} + w_2 T_{g1})$ where w_i and T_{gi} are the weight fraction and the T_g of polymer i (i = 1, 2 designating polyphosphazene and PLAGA, respectively). The T_g's estimated using the Wood and Fox equations for Matrix1 and Matrix2 were 34°C and 40°C, and 32°C and 37°C, respectively, which are very close to the experimental values from DSC thermograms (Figure 4.4).

Recently, the synthesis of PLA grafted polyphosphazenes has been reported by Krogman et al. (2008c). In addition, novel polyphosphazene block copolymers such as polyphosphazene-block-polyester or -polycarbonate macromolecules have been recently synthesized (Krogman et al. 2008b; Weikel et al. 2010a). The lactic and glycolic acid blocks in the block copolymers could act as compatibilizers in the blend, thereby leading to complete miscibility. Efforts are currently underway to develop advanced polyphosphazene miscible blend systems using these polymeric architectures as prototypes as well as polymers with bioactive side groups for further functionalization modifications.

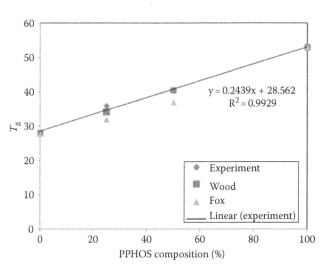

FIGURE 4.4 **(See companion CD for color figure.)** Miscibility of the polyphosphazene–polyester blends. Linear relationship between blend T_g and polyphosphazene composition was based on the experimental values from DSC thermograms, which was in line with Wood equation for miscible blends. Furthermore, the glass transition temperatures estimated using the Fox equation for the blends were very close to the experimental values from DSC thermograms. (From Deng, M., Nair, L.S., Nukavarapu, S.P. et al.: In situ porous structures: A unique polymer erosion mechanism in biodegradable dipeptide-based polyphosphazene and polyester blends producing matrices for regenerative engineering. *Adv. Funct. Mater.* 2010d. 20(17). 2794–2806. Copyright Wiley-VCH Verlag GmbH & Co. KGaA. With permission.)

In the case of blends composed of polyphosphazenes alone, Lemmouchi et al. prepared a blend (50/50, w/w) of PNEG with poly[(ethyl alanate)-*co*-(ethyl 2-(*O*-glycyl lactate))phosphazene] (3% gly-lact-OEt) (Lemmouchi et al. 1998). A single T_g was observed for the blend materials.

4.3.2 DEGRADATION OF BIODEGRADABLE POLYPHOSPHAZENES

4.3.2.1 Polymer Degradation Mechanisms

Allcock and coworkers first demonstrated that the incorporation of certain hydrolytically labile side groups on polymer backbone could confer biodegradability to polyphosphazenes (Allcock et al. 1982b). Since then, biodegradable polyphosphazenes have attracted a great deal of attention in biomedical applications. Biodegradation of hydrolytically sensitive polymers involves cleavage of hydrolytically labile bonds followed by polymer erosion. Although the polymer degradation mechanism varies with the chemical composition, the mechanism of hydrolytic degradation can be broadly classified into two types: bulk erosion and surface erosion. For example, synthetic biodegradable polyesters such as PLAGA are known to undergo bulk erosion whereas polyanhydrides such as poly(carboxyphenoxy propane) degrade through surface erosion. In bulk erosion, the polymer undergoes degradation with significant decrease in molecular weight and the corresponding material properties (such as mechanical properties) as a function of degradation time. As illustrated in Figure 4.5a, the matrix dimension remains constant until the structure fails catastrophically during hydrolytic degradation. In the case of surface erosion, degradation occurs at the implant surface with insignificant decrease in the molecular weight of the bulk material. The matrix becomes smaller but maintains its original geometric shape as a function of degradation time until the structure is completely eroded as indicated in Figure 4.5b. In comparison, biodegradable polyphosphazenes can degrade by a combination of both bulk and surface erosion (Laurencin et al. 1992). More recently, on the platform of biodegradable polyphosphazene blends, Deng et al. (2010d) demonstrated for the first time a unique polymer erosion process through which polymer matrices evolve

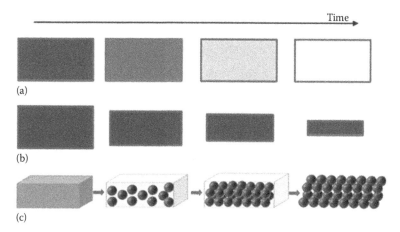

(a)

(b)

(c)

FIGURE 4.5 (**See companion CD for color figure.**) Schematic illustration of different types of polymer erosion. (a) Bulk erosion; (b) Surface erosion; and (c) A unique polymer erosion through which the polymer changes from a solid coherent film to an assemblage of microspheres with an interconnected porous structure. (From Deng, M., Nair, L.S., Nukavarapu, S.P. et al.: In situ porous structures: A unique polymer erosion mechanism in biodegradable dipeptide-based polyphosphazene and polyester blends producing matrices for regenerative engineering. *Adv. Funct. Mater.* 2010d. 20(17). 2794–2806. Copyright Wiley-VCH Verlag GmbH & Co. KGaA. With permission.)

from a solid coherent film to an assemblage of microspheres with an interconnected 3D porous structure (Figure 4.5c).

Polymer erosion results from polymer degradation. The process involved in the erosion of a degradable polymer generally include several phases: (1) hydration with or without swelling, (2) oligomers and monomers generation by water intrusion, (3) progressive degradation, and (4) oligomers and monomers release leading to mass loss. There are several important factors that affect the polymer degradation rate such as lability of the chemical bond, molecular weight, hydrophilicity/hydrophobicity, polymer matrix water permeability, solubility of the degradation products, environmental conditions (pH, temperature, in vitro, and in vivo, etc.), and copolymer composition. For example, PLAGA is known to degrade by simple hydrolysis of the ester bonds into lactic and glycolic acids. The degradation rate of polyphosphazenes mainly depends on several factors such as molecular weight of the polymer backbone, nature of the substituted side groups, composition and extent of side group substitution, or blend composition in the case of polyphosphazene blends (Andrianov and Marin 2006; Kumbar et al. 2006; Singh et al. 2006).

A general degradation pathway for biodegradable polyphosphazenes is shown in Scheme 4.4. During the polymer degradation, water molecules attack the organic side groups. The formation of P-OH units after the removal of side groups followed by the migration of the proton from oxygen to nitrogen sensitizes the polymer skeleton to hydrolysis. The polymer ultimately degrades into non-toxic degradation products comprised mainly of ammonia, phosphoric acid,

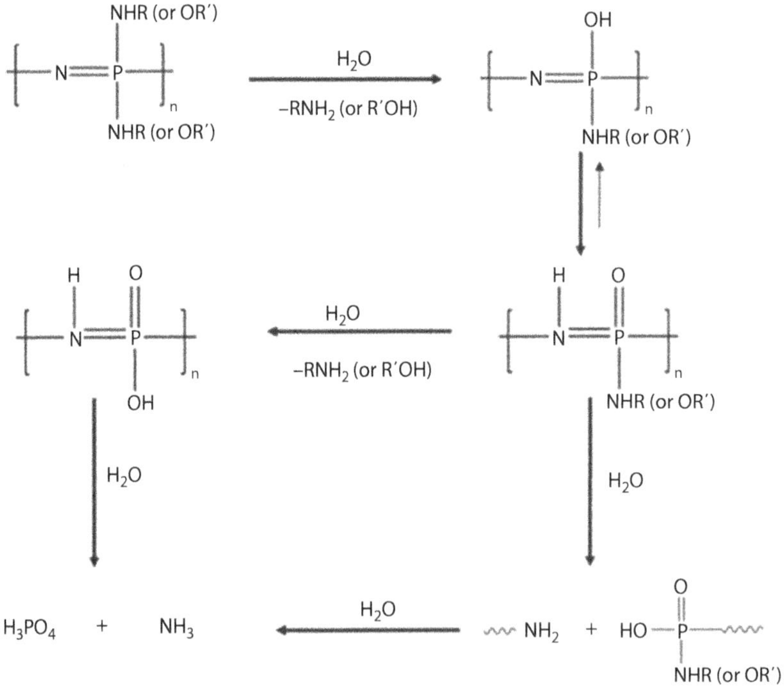

SCHEME 4.4 A general degradation pathway for biodegradable polyphosphazenes. The degradation is initiated by the water molecule attack on the organic side groups. The formation of P-OH units after the removal of side groups followed by the migration of the proton from oxygen to nitrogen sensitizes the polymer skeleton to hydrolysis. The polymer ultimately degrades into nontoxic degradation products comprised mainly of ammonium phosphate buffer and the corresponding side groups. The polymer degradation rate can also be altered by modulating the side group chemistry. For example, polyphosphazenes substituted with hydrolytically sensitive groups such as amino acid esters or imidazolyl degraded faster than the bulkier hydrophobic groups substituted polymers.

and the corresponding side groups (Andrianov and Marin 2006; Kumbar et al. 2006; Singh et al. 2006). In the case of a mixed-substituent polymer, the most hydrolytically sensitive group leaves the polyphosphazene backbone first followed by bulkier side groups such as *p*-methylphenoxy, *p*-phenylphenoxy, tyrosine, or pyrrolidone (Allcock et al. 1977, 1994b; Allcock and Fuller 1980; Allcock and Pucher 1991; Crommen et al. 1992a,b, 1993; Laurencin et al. 1993, 1996; Qiu and Zhu 2000a; Gümüsdereliolu and Gür 2002; Nair et al. 2003, 2004; Conconi et al. 2004; Cui et al. 2004; Barrett et al. 2005; Yuan et al. 2005; Andrianov and Marin 2006; Sethuraman et al. 2006; Singh et al. 2006). Degradation products such as ammonium phosphate constitute a natural buffer system for the microenvironment both in vitro and in vivo (Andrianov and Marin 2006; Kumbar et al. 2006; Singh et al. 2006).

4.3.2.2 Degradation of Aminated Polyphosphazenes

Allcock et al. extensively investigated the effect of different amino side groups on the hydrolytic behavior of polyphosphazenes as well as small molecular cyclic analogs (Allcock et al. 1982b). Among a series of synthesized aminated cyclophosphazenes, hexaaminocyclophosphazenes with amino, methyl amino, ethyl glycinato, methyl glycinamido, and imidazole groups were found to be soluble in water. However, a detectable rate of hydrolysis at 25°C was only observed for cyclophosphazene substituted with the imidazole group. Furthermore, all the aforementioned aminated phosphazenes experienced hydrolysis at 100°C as evidenced by ^{31}P NMR indicating that an increase in temperature increased the degradation rate of polyphosphazenes. Hydrolysis products of these compounds were found to be ammonia, phosphate, and the corresponding side group amine or amino acid. Under the same reaction condition, the sensitivity to hydrolysis for these compounds increased in the order of methyl amino-substituted cyclophosphazenes < amino-substituted < methyl glycinamido-substituted < ethyl glycinato-substituted < imidazole-substituted, as evidenced from changes in ^{31}P NMR spectra and recovered starting materials. Thus, highly hydrophilic imidazole groups can significantly increase the water permeability and the resulting polymer degradationrate. Two mechanistic hydrolysis pathways have been proposed for these compounds (Allcock et al. 1982b). The degradation of these compounds can be triggered through the protonation of atoms in the skeleton or in the side groups. Protonation of the skeletal nitrogen atom would directly result in ring cleavage. Ultimately this could lead to the conversion of the compound to ammonium ion, phosphoric acid, and the free amine salt. On the other hand, protonation of the side group nitrogen atom could take place, followed by nucleophilic attack by water at phosphorus, yielding a monohydroxycyclophosphazene. The monohydroxycyclophosphazene, by further hydrolysis, would undergo ring cleavage and eventual degradation. The detection of a monohydroxycyclophosphazene can be used as a key factor to identify this pathway. However, both mechanisms are blocked in strongly basic media. It has been further consolidated by experimental results where the rates of hydrolysis of all these phosphazenes were very slow in basic media containing 1 N or stronger sodium hydroxide. The overall hydrolysis rate of these phosphazenes in 1 N or stronger acid was higher than in water. Thus, acidic media accelerate the hydrolytic breakdown of these aminated phosphazenes compared to neutral or basic media.

Aminated polyphosphazenes followed the same general trend in hydrolytic sensitivity as the aforementioned small molecular analogs. The imidazole-substituted polymer (poly[bis(imidazolyl) phosphazene]) (**6**) was found to be the most hydrolytically labile among several synthesized aminated polyphosphazenes. The high hydrolytic sensitivity of imidazole-substituted polyphosphazenes could be further modulated by co-substituting it with less hydrolytically sensitive groups on the polyphosphazene backbone. This has been verified by the experiments where imidazolyl mixed-substituent polymer (poly[(imidazolyl)(methylamine)phosphazene]) underwent a slow hydrolysis over several months at 25°C in water. Similarly, Laurencin et al. evaluated the hydrolytic degradation of imidazole cosubstituted polyphosphazenes (poly[(imidazolyl)(*p*-methylphenoxy)phosphazene]) (**12**) (Laurencin et al. 1987). It was found that the relative amount of hydrolytically labile imidazolyl groups determined the degradation rate of the polymer. Degradation of 20% imidazole-substituted polyphosphazene was quite slow, with only 4% of the polymer degraded in 600 h. The contact angle measurement confirmed the hydrophobic characteristic of the polymer (Laurencin et al. 1993), which in turn decreased the water permeability of polymer matrix. In the case of 45% imidazole-substituted polymer, approximately 30% of the polymer degraded in approximately 300 h, whereas 80% imidazole-substituted polyphosphazene degraded in a few hours under ambient conditions.

Among the aminated polyphosphazenes, those substituted with amino acid ester groups have been extensively studied. In general, these macromolecules degrade through either bulk or surface erosion or a combination of both to body-friendly products such as amino acids and ammonium phosphate buffer, which ensures excellent biocompatibility for the degradation micro-environment. The selection of an amino acid ester as the substituent group rather than an amino acid is based on the following rationales: (1) a free carboxylic group in an amino acid would participate in chlorine substitution steps to form cross-links; (2) the free carboxylic groups in an amino acid could promote polymer degradation. Therefore, Allcock et al. synthesized a series of amino acid ester polyphosphazenes as potential biomaterials (Allcock et al. 1977). The polymers were synthesized by introducing different amino acid ester groups (5) (where R = H, CH_3, $CH_2CH(CH_3)_2$ or $CH_2C_6H_5$ and R' = CH_2CH_3) or an amino acid ester group first followed by treatment with methyl amino groups sequentially to tune the polymer degradation rate. All the polymers underwent a slow decrease in molecular weight both in solution and in the solid state at 25°C. In specific, PNEG underwent a decrease in chain length from 8,000 to 20,000 repeating units to less than 1000 repeating units in 15 days in the solid state at 25°C. Additionally the mixed-substituent polymer with methylamine side group underwent a chain shortening from about 8000 to 1000 repeating units in 25–30 days. This study clearly demonstrated that incorporation of suitable cosubstituents along with appropriate amino acid ester groups could efficiently modulate the polymer degradation pattern.

Allcock et al. (1994b) studied the influence of ester groups and α-substituents on the hydrolytic sensitivity of amino acid ester-substituted polyphosphazenes. Methyl, ethyl, *tert*-butyl and benzyl esters of glycine-, alanine-, valine- and phenylalanine-substituted polyphosphazenes were used for the study. The degradation studies were performed in deionized water and the molecular weight changes of these polymers were followed by gel permeation chromatography (GPC) analysis. The hydrolytic sensitivity of the polymers was found to increase as the size of the ester groups decreased. Thus, for the same amino acids, the sensitive to hydrolysis was found to increase in the order of benzyl esters < *tert*-butyl esters < ethyl esters < methyl esters. Similarly, the smaller the group linked to the α-carbon atom of the amino acid residue, the more sensitive was the polymer to hydrolysis. Thus, phenylalanato units were less sensitive than the valinato units, which were less sensitive than the alaninato groups, with the glycinato side groups being the most sensitive to hydrolysis. The hydrolysis products of these compounds were found to be phosphates, amino acid, the respective alcohol and ammonia. The physical properties of these polymers allows for facile processing and retainment of dimensional stability at physiological temperatures, making them prospective biomaterials for short-term medical applications as drug delivery matrices and tissue engineering scaffolds.

Allcock et al. (1994) investigated the hydrolytic degradation of three polyphosphazenes as potential matrices for controlled drug release. Those macromolecules were substituted with either the same amino acid residue or the same ester group, i.e., PNEG, PNEA (5, where R = CH_3 and R' = CH_2CH_3) and poly[bis(benzyl alanato)phosphazene] (PNBA) (5, where R = CH_3 and R' = $CH_2C_6H_5$). The polymer degradation rate at physiological pH was found to decrease in the order of ethyl glycinato-substituted > ethyl alanato-substituted > benzyl alanato-substituted polyphosphazenes. For the ethyl glycinato-substituted polymer, the small hydrogen atom at the α-position provides the least steric repulsion against hydrolytic attack on the phosphorus atom in the skeleton or on the side group P–N bonds. The decreased degradability of ethyl alanato and benzyl alanato compared to ethyl glycinato can be attributed to the relatively large methyl group in ethyl alanato-substituted polyphosphazenesand the bulky hydrophobic benzyl group in benzyl alanato-substituted polyphosphazenes. Thus, the substituents at the α-carbon of the amino acid

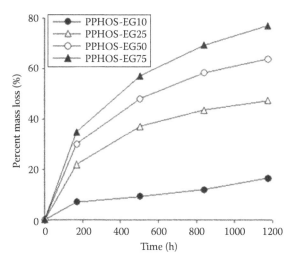

FIGURE 4.6 Percentage mass loss versus degradation time (h) of ethyl glycinato-substituted polyphosphazenes of different compositions in 0.1 M sodium phosphate buffer at 37°C and pH 7.4. The standard deviation (SD) of the mean is less than 3%. (From Laurencin, C.T., Norman, M.E., Elgendy, H.M. et al.: Use of polyphosphazenes for skeletal tissue regeneration. *J. Biomed. Mater. Res.* 1993. 27(7). 963–973. Copyright Wiley-VCH Verlag GmbH & Co. KGaA. With permission.)

ester side group appear to have more bearing on the hydrolysis rate of amino acid ester polymers than the types of ester groups. On the other hand, the ester groups at the chain end exhibited a significant effect on the polymer surface properties, such as hydrophilicity or hydrophobicity, as evidenced from contact angle measurements. Hydrophobic ester groups significantly decrease water penetration into the matrix and thereby retard the hydrolytic breakdown of these polymers. The hydrolysis products of these polymers were found to be phosphates, alanine or glycine, benzyl alcohol or ethanol, and ammonia.

Laurencin et al. demonstrated that the degradation rate of ethyl glycinato-substituted polyphosphazene could be efficiently modulated by incorporating a hydrophobic substituent, such as a methylphenoxy group, as co-substituent (Laurencin et al. 1993). Figure 4.6 shows the percentage mass loss versus degradation time of poly[(ethyl glycinato)(*p*-methylphenoxy)phosphazene] of different compositions. It can be seen that the degradation rate of the polymer increases with an increase in the percentage of the ethyl glycinato group. Furthermore, the degradation rate of aminated-polyphosphazene can also be tuned by varying the bulkiness and composition of the substituted side groups. For instance, among the four different L-alanine-substituted polyphosphazenes namely PNEA, poly[(ethyl alanato)$_1$(ethyl glycinato)$_1$phosphazene] (PNEAEG), poly[(ethyl alanato)$_1$(*p*-methylphenoxy)$_1$phosphazene] (PNEAmPh) and PNEAPhPh (**16**), the water contact angle increased in the order of PNEAEG < PNEA < PNEAmPh < PNEAPhPh. Water contact angles are the measure of surface hydrophilicity indicating the reduced polymer hydrophilicity with the increased side group bulkiness. Degradation studies carried out in phosphate buffer saline at 37°C showed a molecular weight loss pattern in the order of PNEAEG > PNEA > PNEAmPh > PNEAPhPh due to the differences in the bulkiness of the side groups and initial molecular weight of the polymer (Singh et al. 2006). SEM micrographs of polymer surface morphologies following hydrolytic degradation suggested that PNEA, PNEAmPh, and PNEAPhPh degraded via surface erosion mechanism, whereas the degradation of PNEAEG proceeded through a combination of bulk and surface erosion. In addition, the degradation kinetics of these polymers can also be modulated by adjusting the ratios of the two cosubstituents (Laurencin et al. 1993). These studies systematically demonstrated the versatility of polyphosphazene synthesis and the efficient control over polymer degradation by changing side group chemistry to suit the requirements of a specific biomedical application.

Ibim et al. (1996) studied the effect of pH on the degradation rate of poly[(ethyl glycinato)₁ (p-methylphenoxy)₁phosphazene] (PNEGmPh) (**14**). Degradation studies were performed in distilled water with the pH adjusted continuously to 2.0, 7.4, and 10.0. The polymer degradation was found to be very slow in neutral and basic solution (pH 7.4 and 10.0). In contrast, more than 80% of the polymer degraded in 35 days in acidic solution (pH 2.0).

So far, several hydrolysis routes have been proposed for poly[(amino acid ester)phosphazenes] (Allcock et al. 1982b, 1994b). The ester functionality could be involved in the breakdown of the −P=N− skeleton via three different mechanisms. In the first, hydrolysis of the ester unit by water molecules would lead to the formation of the corresponding polymer-bound amino acid. The carboxylic acid unit could then attack a nearby phosphorus atom in the polymer chain. This species would further react with water molecules to release the amino acid and to form a hydrolytically unstable phosphazane, which would ultimately break down to phosphates and ammonia. In a second mechanism, the presence of water could facilitate an attack on the polymer backbone by the ester functionality itself. Water molecules could then react with the unstable phosphorus–ester bond. As a result of this reaction, the amino acid would be released and a hydrolytically sensitive phosphazane could be formed. In a third mechanism, water molecules would displace the amino acid esters from the phosphorus atoms to form a hydroxyphosphazene. This species could rearrange in the presence of water to a phosphazane. The phosphazane would then react with water molecules to yield phosphates and ammonia. Since the evidence for all three mechanisms was found, it was suggested that all three could occur simultaneously, although it is still not clear which mechanism predominates (Allcock et al. 1982b, 1994b).

Since the biodegradation of poly[(amino acid ester)phosphazenes] could occur through hydrolysis of the amino acid ester function, leading to a carboxylic acid which may catalyze the polymer backbone breakdown, the polymer degradation rate could be increased by incorporating side groups with a hydrolytically sensitive ester functionality. Crommen et al. synthesized a series of polyphosphazenes bearing a number of hydrolytically labile substituents, such as ethyl 2-[O-glycyl] lactate (**10**) (where R = H) or ethyl 2-[O-alanyl] lactate (**10**) (where R = CH₃), and the remaining P–Cl groups of polymer **3** substituted by ethyl glycinate or ethyl phenylalanate moieties (Crommen et al. 1992b). The degradation of these polymers was carried out in phosphate buffer solution at pH 7.4. The in vitro degradation rate of these polymers was found to be directly dependent on the content of the depsipeptide side groups. The activated ester compounds were found to be more sensitive to hydrolysis than the normal alkyl esters. Thus, the incorporation of controlled amounts of a hydrolysis-sensitive amino acid ester is an elegant approach for controlling the degradation rate of amino acid ester–substituted polyphosphazenes.

Nair et al. (2006) investigated degradation behavior of two ethyl alanato based polymers namely poly[(ethyl alanato)₁(ethyl oxybenzoate)₁phosphazene] (PNEAEOB) and poly[(ethyl alanato)₁(propyl oxybenzoate)₁phosphazene] (PNEAPOB) in phosphate buffer (pH 7.4) at 37°C. PNEAPOB showed a faster degradation rate and higher water absorption compared to PNEAEOB. Both polymers were found to be insoluble in common organic solvents following hydrolysis presumably due to cross-linking reactions accompanying the degradation process.

Most aminated polyphosphazenes undergo faster degradation in acidic environment compared to physiological or basic conditions. Qiu and Zhu (2000a) prepared polymers (**11**) that degrade at a higher rate under physiological conditions, but are relatively stable at low pH. The polymer degradation was studied at different pH values at 37°C. It took only 12 h for the sample to dissolve completely in a pH 8.0 buffer, 1.5 days in pH 7.4, but more than 20 days in weakly acidic media (pH 5.0 and 6.0). A two-stage degradation mechanism was proposed for these polymers. In the first phase, the breakage of side groups played a prominent role, in that the benzyl ester of amino acethydroxamic acid was hydrolyzed to produce amino groups, probably accompanied by the hydrolysis of glycine ethyl ester to glycine. As a consequence, the polymer degraded to water-soluble polymeric products. In the second phase, the soluble polymeric products were further degraded by scission of the backbone into phosphates and ammonia.

4.3.2.3   Degradation of Alkoxy-Substituted Polyphosphazenes

Glyceryl-substituted polyphosphazenes undergo hydrolysis generating glycerol and phosphate in water at 100°C (Allcock and Kwon 1988). The hydrolysis was detectable after 12h and was complete after 150h. At 37°C the hydrolysis was essentially complete after 720h. The polymer with glycosyl and methyl amino substituents decomposed completely within 24–96h at 100°C in aqueous buffer solutions (pH 6.0, 7.0, and 8.0) (Allcock and Pucher 1991). Under physiological conditions the hydrolysis was much slower, with half-lives in the range of 165–175h. The degradation products of these polymers are presumed to be phosphate, glucose, ammonia, and methylamine. Another type of biodegradable alkoxy-substituted polyphosphazene consists of those with glycolic or lactic acid as side groups (Allcock et al. 1994c). These polymers (**9**) (where $R = H$ or CH_3 and $R' = CH_2CH_3$ or $CH_2C_6H_5$) undergo hydrolysis when exposed to water in the solid state as evidenced from ^{31}P NMR and GPC. Polymer **9** showed evidence of initial backbone hydrolysis at 320h (where $R = H$ and $R' = CH_2CH_3$), at 525h (where $R = CH_3$ and $R' = CH_2CH_3$), at 656h (where $R = H$ and $R' = CH_2C_6H_5$), and at 11:07h (where $R = CH_3$ and $R' = CH_2C_6H_5$), respectively. The fastest decomposition of polymer **9** (where $R = H$ and $R' = CH_2CH_3$) was attributed to the least sterically hindered groups present in it. Water molecules could displace ethyl glycolato side units or hydrolyze the ester function to the acid, and the pendent carboxylic acid could then induce skeletal degradation. Polymer **9** (where $R = CH_3$ and $R' = CH_2CH_3$) and polymer **9** (where $R = H$ and $R' = CH_2C_6H_5$) showed almost similar degradation profiles, much slower than polymer **9** (where $R = H$ and $R' = CH_2CH_3$). Both had either a methyl group or a bulky benzyl ester group at the α-position, which could hinder hydrolytic attack at the phosphorus atoms. Polymer **9** (where $R = CH_3$ and $R' = CH_2C_6H_5$) has both the hydrolysis-limiting factors of polymers **9** (where $R = CH_3$ and $R' = CH_2CH_3$) and polymer **9** (where $R = H$ and $R' = CH_2C_6H_5$), specifically the methyl group on the α-position as well as the benzyl ester group. The hydrolysis products of these compounds were identified as benzyl alcohol or ethanol, glycolic or lactic acid, phosphates, and ammonia.

4.3.2.4   Degradation of Polyphosphazene Blends/Composites

As discussed in the earlier sections, it is possible to design polyphosphazenes with modulated degradation rate through appropriate selection of side groups and their compositions. Polymeric blending is an alternative approach to develop novel biomaterials with controllable properties and degradation kinetics (Utracki 1989; Deng et al. 2009). A wide variety of biodegradable blend/composite materials can be developed by blending a biodegradable polyphosphazene with either another biodegradable polymer or with another biodegradable polyphosphazene having a different hydrolytic sensitivity. The advantage of the blending process is that a series of materials with variable degrees of degradability can be obtained from the two parent polymers (Utracki 1989; Deng et al. 2009).

The degradation studies performed by Ibim et al. demonstrated that blends degraded at different rates, implying that the blend properties can be easily tuned for a variety of different biomedical applications (Ibim et al. 1997).The follow-up study by Ambrosio et al. investigated the self-neutralizing ability of such blend systems and the possibility of maximizing the neutralization effect by optimizing the blend composition (Ambrosio et al. 2002). Three types of samples including PLAGA (50:50), PNEGmPh, and their blends were subjected to degradation in distilled water at pH 7.4 over a period of 40 weeks. The blend showed an intermediate degradation rate between those of the parent polymers, and the hydrolysis of polyphosphazenes could have been catalyzed by the acidic degradation products of PLAGA (Figure 4.7a). In addition, the amounts of neutralizing base (0.01N solution of sodium hydroxide) for titrating the medium back to pH 7.4 were measured for each type of sample throughout degradation. Results showed that the blend required significantly lower amounts of neutralizing base than PLAGA, which indicated that the blend system released less acidic degradation products due to the buffering effect of polyphosphazene degradation products (Figure 4.7b). Hence, these two earlier studies demonstrated the advantages of self-neutralizing blends and established the ground for developing polyphosphazene–polyester blends as promising candidate materials for controlled drug delivery and tissue engineering applications.

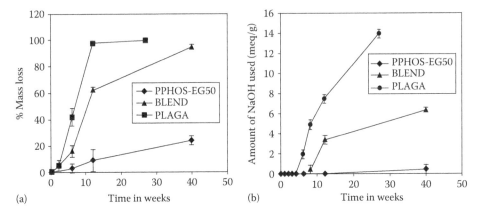

FIGURE 4.7 Degradation of PNEGmPh-PLAGA blend and parent polymers in pH 7.4 distilled water at 37°C over 12 weeks. (a) Percent mass loss; (b) Amount of NaOH (meq/g) used to neutralize the acid released during the degradation of the blend and its parent polymers. (From *Biomaterials*, 23(7), Ambrosio, A.M.A., Allcock, H.R., Katti, D.S., and Laurencin, C.T., Degradable polyphosphazene/poly([alpha]-hydroxyester) blends: Degradation studies, 1667–1672. Copyright 2002, with permission from Elsevier.)

Meanwhile, Qiu et al. performed degradation studies on slabs of PNEG-PLAGA and PNEG-PSTP blends in distilled water with or without *Rhizopus delemer* lipase at 37°C (Qiu 2002b). The blend of PNEG-PLAGA (70/30, w/w) took 120 days to disappear completely, while PNEG-PSTP (70/30, w/w) slabs needed only 20 days. The degradation rate of PNEG-PLAGA blends was strongly accelerated by the enzyme and the degree of enzymatic degradation depended on the weight percentage of PLAGA in the blend. In vivo degradation of the polymer blends was carried out by subcutaneously implanting them in the back of mice and some differences in degradation were observed compared to in vitro degradation. Such variation was attributed to more complex physiological environment, which indicated that the degradation of PNEG-PLAGA occurred through a combination of hydrolysis and enzymolysis, while the degradation of PNEG-PSTP proceeded through only hydrolysis. In addition, both in vitro and in vivo studies showed that the degradation rate of the blends could be easily controlled by adjusting the blend composition.

Krogman et al. (2009) blended PLAGA with polyphosphazenes containing THAM side groups and cosubstitutents such as glycine ethyl ester and alanine ethyl ester. However, the degradation rates of the aforementioned polymers were found to be very low. The dipeptide side groups impart hydrolytic sensitivity and significant hydrogen bonding capability to the polymers. Krogman et al. (2007) investigated the degradation behavior of blends of PLAGA and different glycylglycine ethyl ester–substituted polyphosphazenes (**17–20**). Hydrolysis studies showed the blends degraded at a slower rate than both parent polymers. The pH of the degradation medium containing the blends increased from 2.5 to 4.0, which was attributed to the buffering capacity of polyphosphazenes degradation products. This study has demonstrated the feasibility of fine-tuning the blend degradation profile by adjusting either the side group ratios of polyphosphazenes or blend composition. In another study, a class of biodegradable polyphosphazenes containing dipeptides alanylglycine ethyl ester, valinylglycine ethyl ester, and phenylalanylglycine ethyl ester were synthesized by Weikel et al. (2009). The polymers were found to be less sensitive to hydrolysis under neutral and basic conditions than acidic conditions (pH 4.0) (Weikel et al. 2009).

More recently, Deng et al. (2010d) demonstrated for the first time a unique degradation process for biodegradable biomaterials based on PNGEGPhPh-PLAGA blends both in vitro (phosphate-buffered saline and distilled water) and in vivo (physiological conditions using a rat subcutaneous implantation model). As shown in Figure 4.8, a unique polymer erosion process was evident in all the blend compositions in which a solid blend film degraded to form a self-assembled polyphosphazene sphere-based 3D porous structure with interconnected porosity.

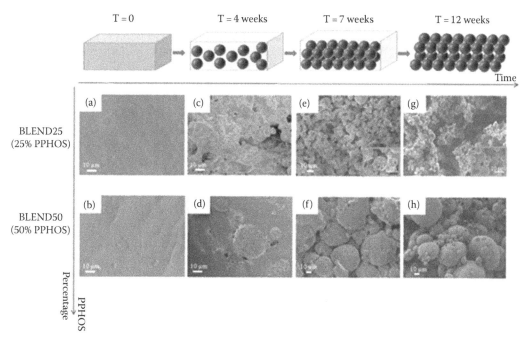

FIGURE 4.8 (See companion CD for color figure.) Surface morphologies of the blends during in vitro degradation at 37°C over 12 weeks as a function of time. (a, c, e, g): SEM images showing surface morphologies of BLEND25 following 0, 4, 7, and 12 weeks of in vitro degradation. The insets of (e) and (g) show the detailed 3D spherical structures. (b, d, f, h): SEM images showing surface morphologies of BLEND50 following 0, 4, 7, and 12 weeks of in vitro degradation. The unique polymer erosion of the blend system resulted in the change of matrix morphology from a solid coherent film to an assemblage of microspheres with interconnected porous structures characterized by macropores (10–100 μm) between polyphosphazene spheres as well as micro/nanopores on the sphere surface. (From Deng, M., Nair, L.S., Nukavarapu, S.P. et al.: Biomimetic, bioactive etheric polyphosphazene-poly(lactide-co-glycolide) blends for bone tissue engineering. *J. Biomed. Mater. Res. A* 2010c. 92A(1). 114–125. Copyright Wiley-VCH Verlag GmbH & Co. KGaA. With permission; Deng, M., Nair, L.S., Nukavarapu, S.P. et al.: In situ porous structures: A unique polymer erosion mechanism in biodegradable dipeptide-based polyphosphazene and polyester blends producing matrices for regenerative engineering. *Adv. Funct. Mater.* 2010d. 20(17). 2794–2806. Copyright Wiley-VCH Verlag GmbH & Co. KGaA. With permission.)

Specifically, the polyphosphazene cosubstituted with hydrophilic glycylglycine dipeptide ester and hydrophobic phenylphenoxy groups showed an exceptional ability to form inter- and intramolecular hydrogen bonds with a tunable degradation pattern. This copolymer resulted in a completely miscible blend system with the polyester through intermolecular hydrogen bonding. By altering the blend composition, it was possible to control blend physicochemical properties and degradation pattern. The blends showed a degradation rate in the order of BLEND50 < BLEND25 < PLAGA in vitro at 37°C over 12 weeks. Significantly higher pH values of degradation media were observed for blends compared to PLAGA confirming the neutralization of PLAGA acidic degradation by polyphosphazene hydrolysis products. In all the blend compositions both PLAGA and polyphosphazene components exhibited a similar molecular weight loss degradation pattern. For example, the number and weight average molecular weights (M_n and M_w) of both the PLAGA and polyphosphazene components decreased exponentially with degradation time throughout the subcutaneous implantation period. Furthermore, the apparent degradation rate, K, can be obtained according to the following exponential relationship between molecular weight and degradation time: $\lg M = \lg M_0 - Kt$. For example, the apparent degradation rates based on M_w were calculated to be 0.1014 week^{-1} and 0.0859 week^{-1} for the polyphosphazene component in BLEND25 and BLEND50, respectively. This indicates that the polyphosphazene degraded faster in BLEND25 due to higher PLAGA composition. A similar trend in the apparent degradation rates was found for the PLAGA component in pristine PLAGA, BLEND25, and BLEND50. This suggests that the PLAGA degraded the slowest in BLEND50 due to the higher polyphosphazene composition. These results agree well with in vitro findings that the PLAGA hydrolysis accelerated the polyphosphazene degradation, whereas the degradation of polyphosphazene retarded the hydrolysis of PLAGA. Characterization of in vivo degradation samples at 12 weeks revealed the formation of 3D interconnected porous structures with 82%–87% porosity. It was determined that the in situ porous structure resulted from inter- and intra-molecular hydrogen bonding interactions and variations in the degradation rates of both blend components promoting a three-stage degradation mechanism including intermolecular hydrogen bonding dominant state (Stage I, 0 week), intermediate state (Stage II, 0–4 weeks), and intramolecular hydrogen bonding dominant state (Stage III, 4–12 weeks).

Meanwhile, Weikel et al. synthesized polyphosphazenes cosubstituted with bioactive molecule such as choline chloride and with glycine ethyl ester, alanine ethyl ester, valine ethyl ester, or phenyl alanine ethyl ester (Weikel et al. 2010b). These choline-based polyphosphazenes were blended with PLAGA (50:50) and PLAGA (85:15) via solution casting techniques. It was found that the increased choline content allowed these polymers to hydrolyze into water-soluble products. Polymers were able to form miscible blends with both types of PLAGA. Hydrolysis experiments in aqueous media revealed that the polymer blends hydrolyzed to near-neutral pH media (~5.8 to 6.8).

For the blends of polyphosphazenes alone, Crommen et al. prepared blends of a hydrolytically sensitive depsipeptide-containing polymer with PNEG. It was found that the degradation rate of the blend increases as the composition of the depsipeptide-containing polymer increases (Crommen et al. 1993). Similarly, the degradation rate can also be adjusted by blending (50/50, w/w) poly[(ethyl glycinate)-co-(ethyl 2-(O-glycyl lactate))phosphazene] (3% gly-lact-OEt) with PNEA. However, it was found that the presence of the depsipeptide-substituted polymer only promoted the release of glycine and glycine ethyl ester but did not affect the release of the alanine side groups. Lemmouchi et al. observed a similar phenomenon for a blend (50/50, w/w) of PNEG with poly[(ethyl alanate)-co-(ethyl 2-(O-glycyl lactate))phosphazene] (3% gly-lact-OEt) (Lemmouchi et al. 1998). Only the release of alanine and alanine ethyl ester was significantly affected. Such observed degradation behavior of depsipeptide-containing polyphosphazene blends might be attributed to the intramolecular catalysis of the polymer by the pendant carboxylic acid.

4.3.3 Biocompatibility of Biodegradable Polyphosphazenes

Another important requirement for biodegradable materials in biomedical applications is biocompatibility. The efficacy of the drug delivery matrix to achieve controlled release kinetics as well as the tissue engineering scaffold to support cell growth and achieve functional tissue repair largely depends on the material biocompatibility.

4.3.3.1 Biocompatibility of Aminated Polyphosphazenes

Laurencin et al. first characterized the biocompatibility of the amino acid ester–substituted polyphosphazenes for tissue engineering applications (Laurencin et al. 1992). It was demonstrated that biodegradable PNEG supported the adhesion of primary rat osteoblasts (PRO) to the same extent as PLAGA and polyanhydrides. Since then, several amino acid ester polyphosphazenes have been investigated for their in vitro cytocompatibility. For example, both PNEAEOB and PNEAPOB supported adhesion and proliferation of PRO cells, and were found to be suitable as reinforcing materials for developing self-setting bone cements (Nair et al. 2006). Cytotoxicity study of the degradation media of poly[bis(ethyl 4-aminobutyro)phosphazene] was performed using Swiss 3T3 and HepG2 cell lines (Gümüsdereliolu and Gür 2002). It was found that the presence of degradation media did not affect the proliferation of both cell lines. These in vitro results confirmed the cytocompatibility of the polyphosphazene degradation products, and further supported them as potential biomaterials for in vivo applications (Gümüsdereliolu and Gür 2002). Early studies focused on characterizing these amino acid ester–substituted polyphosphazenes for their in vivo tissue compatibility. For instance, alanine ethyl ester–substituted polyphosphazenes namely PNEA, PNEAmPh, and PNEAPhPh were characterized for their in vivo tissue compatibility using a rat subcutaneous implantation model (Sethuraman et al. 2006). All these polymers initially provoked a mild to moderate immune response that subsided over time. After 12 weeks, a minimal immune response was found for PNEAmPh and PNEAPhPh which was characterized by the presence of few neutrophils, erythrocytes, and lymphocytes. Further in vitro osteocompatibility evaluation using PRO cells demonstrated that cells adhered, proliferated, and maintained their phenotype on polymers of PNEA, PNEAmPh, and PNEAPhPh (Sethuraman et al. 2010). Laurencin et al. (1998) also evaluated the biocompatibility of PNEGmPh-based bone grafts in New Zealand white rabbits. This polymer supported the bone growth with a mild tissue response when compared to PLAGA grafts. Histological evaluations revealed the presence of a thin layer of fibrous tissue at the implant interface and the adjacent bone at 1 week. After 12 weeks of implantation, the interface contained lamellar bone and showed the presence of fat and bone marrow cells adjacent to the implant, whereas PLAGA implants were surrounded by a thin fibrous membrane that consisted of a few giant cells, occasional lymphocytes, and plasma cells. The superior biocompatibility of these novel polymers was further supported by the absence of implant fragmentation, fat necrosis, or granulomas formation (Laurencin et al. 1998). These in vitro and in vivo characterizations demonstrated the excellent biocompatibility of polyphosphazenes. Therefore, the aminated polymers are actively pursued as biomaterials for biomedical applications.

4.3.3.2 Biocompatibility of Alkoxy-Substituted Polyphosphazenes

Alkoxy-substituted polyphosphazenes have not yet been extensively investigated as biocompatible materials, although they possess great potential due to the presence of reactive hydroxyl groups in many of them. For example, incorporation of alkoxy side groups to polymer backbone can modulate polymer surface hydrophobicity/hydrophilicity, which in turn affects the biocompatibility of the polyphosphazenes.

4.3.3.3 Biocompatibility of Polyphosphazene Blends/Composites

Since the first report on the biodegradable blends of poly[(amino acid ester)phosphazenes] and polyesters, a series of polyphosphazene blends have been developed and characterized for biocompatibility as prospective biomaterials. Most amino acid ester–based polyphosphazene blends were found to be biocompatible. Nair et al. blended PNEA with PLAGA (85:15) at two weight ratios of 1:3 and 1:1. Significantly higher PRO adhesion and proliferation were observed for both the blends compared to the parent polymers (Nair et al. 2005). Meanwhile, Qiu et al. have characterized a series of degradable blends comprising PNEG and polyesters or polyanhydrides for tissue biocompatibility using a rat subcutaneous implantation model (Qiu 2002a). The tissue biocompatibility of PNEG-PLAGA was found to be better than that of PNEG-PSTP. Moreover, the tissue biocompatibility of PNEG-PSTP blends was improved by increasing the weight percentage of PNEG in blends. These findings suggested the blends of PNEG-PLAGA or -PSTP may be used as drug delivery matrices or for other potential biomedical applications. Deng et al. (2008) developed high-strength materials for tissue engineering applications from the blends of PNEAPhPh and PLAGA at three weight ratios of 1:3, 1:1, and 3:1. In an in vitro osteocompatibility study using PRO culture, these blends not only supported the PRO proliferation to the same extent as PLAGA but also showed enhanced differentiation characterized by increased levels of alkaline phosphatase activity and mineralized matrix synthesis compared to parent polymers.

In addition, in an effort to develop biomimetic and bioactive materials for achieving better osteointegration, Deng et al. (2010c) also demonstrated that the novel etheric poly[(ethyl glycinato)$_1$(methoxyethoxyethoxy)$_1$phosphazene]-PLAGA blends were able to nucleate bone-like apatite via a biomimetic process. Enhanced phenotypic expression of the PRO was found on the blends compared to parent polymers. Krogman et al. evaluated the in vitro osteocompatibility of blends based on PLAGA and polyphosphazenes cosubstituted with both THAM and glycine or alanine ethyl esters (Krogman et al. 2009). The polymer blends showed good adhesion and proliferation of PRO during 14 days of culture. However, the dramatic decrease in cell number from 14 to 21 days was attributed to the different degradation kinetics of the parent polymers. This underlines the need to develop polymers with balanced properties in degradation rates and biocompatibility. In that direction, Deng et al. (2010a,b) demonstrated that blends of PLAGA and PNGEGPhPh resulted in improved osteoblast activity and biocompatibility compared to the polyester materials due to their appropriate degradation pattern and self-neutralizing ability. The blends demonstrated significantly higher osteoblast growth rates compared to PLAGA while maintaining osteoblast phenotype over a 21 day culture. In a 12 week rat subcutaneous implantation study, the dipeptide-based blends elicited minimal inflammatory response and exhibited superior biocompatibility as compared to PLAGA as shown in Figure 4.9. More recently, it was reported by Weikel et al. that the choline-based blends supported PRO cell growth with an elevated osteoblast phenotype expression compared to the polyester (Weikel et al. 2010b). Specifically, during the 21 day culture period, the elevated ALP expressions of osteoblast cells on choline-based blends suggested the progression of cell maturity, whereas osteoblasts on PLAGA were proliferating throughout the later time points. Choline-based blend system offers an advantage in the development of a scaffold material with balanced properties such as the elevated cellular activity from the release of choline and the increased buffering capacity from the hydrolysis of the polyphosphazene backbone.

Thus, polyphosphazene–polyester blend systems are promising candidate materials for biomedical applications. Current research efforts are focused on the design and development of polyphosphazene blends with other synthetic or natural biodegradable polymers. Attempts also focus on the incorporation of polyphosphazene block copolymers as compatibilizers to improve blend miscibility and other physicochemical and biological properties.

4.3.4 Biodegradable Polyphosphazenes for Biomedical Applications

Biodegradable polymer systems are widely investigated as controlled drug delivery matrices for enhancing the therapeutic performance of drugs or 3D engineered scaffolds for tissue repair and regeneration. The significant advantage of using biodegradable polymers is the ability to circumvent the long-term biocompatibility concerns associated with many of the nondegradable polymers. In addition to traditional polyesters, polyanhydrides, and poly(ortho esters), polyphosphazene polymers have recently shown exponentially growing interest as a biodegradable material platform for controlled drug release and tissue engineering applications as evidenced by some of the studies discussed next.

4.3.4.1 Drug Delivery

4.3.4.1.1 Matrix Fabrication

Most studies have used biodegradable polyphosphazenes as pellets or films for drug delivery applications. These matrices are usually fabricated via compression molding (Crommen et al. 1992b) or solvent casting due to great polymer solubility in a wide range of solvents (Laurencin et al. 1987, 1993; Allcock et al. 1994a; Ibim et al. 1996). In addition, Veronese et al. prepared naproxen-loaded microspheres of poly[(phenylalanine ethyl ester)(imidazolyl)phosphazene] with side group ratios of 71:29 and 80:20 by spray drying, emulsion/solvent evaporation, and emulsion/solvent evaporation-extraction (polymer dissolved in 50:50 ethanol:dichloromethane mixture and injected

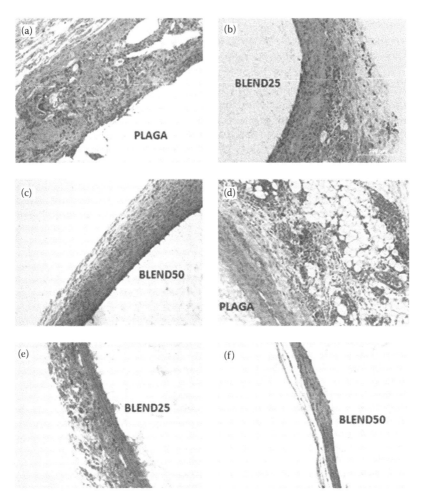

FIGURE 4.9 **(See companion CD for color figure.)** Representative photomicrographs of hematoxylin and eosin (H&E) stained section illustrating the chronic inflammatory response to PLAGA and blend matrices during the post-implantation period. After 2 weeks, (a) PLAGA, (b) BLEND25, (c) BLEND50; After 4 weeks, (d) PLAGA, (e) BLEND25, (f) BLEND50. Both blend matrices elicited less inflammatory responses than PLAGA. (From *Biomaterials*. 31(18), Deng, M., Nair, L.S., Nukavarapu, S.P. et al., Dipeptide-based polyphosphazene and polyester blends for bone tissue engineering, 4898–4908. Copyright 2010b, with permission from Elsevier.)

into phosphate buffer containing 5% Tween 80 as surfactant at pH 4.2) from a dichloromethane or acetone solution of the polymer (Veronese et al. 1998). Microspheres prepared by spray drying from a dichloromethane solution of the polymer yielded microcapsules with a hollow core, most of which were found to collapse when freeze-dried. Microspheres prepared by spray drying using acetone as the solvent maintained their spherical geometry with a narrow size distribution with a mean diameter of 2–5 μm. The solvent evaporation and emulsion/solvent evaporation-extraction technique yielded almost similar type of microspheres, with a larger size distribution of 25–80 μm with more than 70% of the particles in the size range of 40–70 μm. Similarly, Calceti et al. used suspension solvent evaporation, double emulsion-solvent evaporation, and suspension/double emulsion-solvent evaporation for the preparation of insulin-loaded polyphosphazene microspheres (Caliceti et al. 2000). SEM analysis demonstrated that all the preparation procedures produced spherical microparticles with a porous surface and a honeycomb internal structure. The emulsion techniques yielded microspheres with high porosity due to the elimination of the water microdrops of the double emulsion inner phase during hardening of the microspheres. However, the microspheres prepared by solvent evaporation also showed high porosity, indicating that microdrops of the continuous phase can be entrapped in the polymeric matrix during microparticle formation. All the three techniques yielded particles with a narrow size distribution, with the suspension solvent evaporation method yielding 90% of the particles in the range of 30–90 µm, double emulsion-solvent evaporation having 90% of the particles in the range of 10–60 µm and suspension/double emulsion-solvent evaporation having 90% of the particles in the range of 0–60 µm.

4.3.4.1.2 Drug Release Kinetics

Degradation studies of biodegradable polyphosphazenes clearly demonstrate that the polymer degradation rate can be efficiently tuned by incorporating different hydrophilic, hydrophobic, or bulky substituent groups. This, combined with the nontoxic degradation products of the biodegradable polyphosphazenes, distinguished them from other biodegradable polymers as attractive candidates for the controlled release of small and macromolecular drugs. The tunability in the degradation characteristics of these polymers is so versatile that it could offer the ability to formulate drug release systems based primarily on diffusion, erosion, or a combination of erosion and diffusion.

The first attempt in that direction was carried out by Allcock and co-workers by attaching various bioactive agents to a polyphosphazene backbone for subsequent hydrolytic release. Accordingly, steroidal residues such as desoxoestrone, estrone, 17β-estradiol, 17α-ethynylestradiol, estradiol 3-methyl ether, and 1,4-dihydroestradiol 3-methyl ether were attached to the polyphosphazene backbone by Allcock and Fuller via the sodium salt of the steroidal hydroxy function (Allcock and Fuller 1980). The degree of replacement of P–Cl bonds by P–OR bonds varied in the range of 0.5%–40% and the remaining chlorine atoms were replaced by methylamine, ethyl glycinate, or n-butylamine. Hydrolytically stable polymers were obtained when the steroidal units were linked to phosphorus through an aryloxy residue, whereas linkage through an alkoxy residue led to hydrolytically labile polymers. The use of ethyl glycinato residues as cosubstituents yielded polymers that underwent chain cleavage in aqueous media. However, detailed experiments on the degradation kinetics and release profiles of the attached steroids have not been carried out. Similarly, local anesthetics such as procaine, benzocaine, chloroprocaine, butyl-p-aminobenzoate, and 2-amino-4-picoline were attached to the polyphosphazene backbone by direct aminolysis in order to modify the duration of biological activity of these compounds (Allcock et al. 1982a). Preliminary studies indicated that the procaino-substituted high polymers underwent a slow hydrolysis in buffered aqueous media at pH 7.0. In another effort, Grolleman et al. developed a system consisting of 65%–90% of glycine ethyl ester and 10%–35% of a spacer group such as L-lysine ethyl ester linked to an anti-inflammatory agent, naproxen (Grolleman et al. 1986). The degradation kinetics and naproxen release studies exhibited a nearly zero-order release of the drug from the system. However, the amount of drug released was found to be very low.

One of the earlier studies reported by Laurencin et al. investigated the feasibility of imidazole co-substituted polyphosphazenes as a monolithic controlled drug delivery system (Laurencin et al. 1987). Imidazole and methylphenoxy co-substituted polyphosphazene matrices, polymer **12** with 20%, 45%, and 80% imidazole content, were used for the in vitro release of small molecules such as progesterone, p-nitroaniline, and macromolecule such as ^{14}C-labeled bovine serum albumin (BSA). The in vivo release of progesterone was studied via the subcutaneous implantation of matrices in Sprague–Dawley rats. The degradation rate of these matrices was discussed previously in Section 4.3.2.2. It was found that release of progesterone from 20% imidazole-substituted polyphosphazene matrices occurred over 10:00 h in vitro, whereas the in vivo study demonstrated an initial lag period of approximately 80 h followed by a period of almost constant release for 900 h. Furthermore, the release kinetics of 1% and 5% loaded p-nitroaniline from

20% imidazole-substituted polyphosphazene was determined as a function of square root of time. A diffusion-controlled release of the small molecules occurred over approximately 250 h (square root 15–16 h) and the release rate was found not to be very dependent on drug loading. In contrast to the low molecular weight release, the release profile for BSA-loaded 20% imidazole-substituted polyphosphazene was characterized by an initial burst release of the protein diffusing out of the matrix followed by a sustained release up to 500 h. The 25% burst release within the first 4 h was successfully eliminated by coating the matrix with a thin layer of the 20% imidazole-substituted polymer alone. It was further determined that the coated matrices released only 5% of protein during the first 4 h. Thus, prolonged release of low molecular weight molecules as well as macromolecules in combination with the inherent biodegradability and biocompatibility makes imidazole-substituted polymers (12) promising candidate matrices for developing drug delivery systems.

Allcock et al. investigated the potential of using biodegradable poly[(amino acid ester)phosphazenes] for the controlled release of small molecules in a monolithic system (Allcock et al. 1994a). Three different poly[(amino acid ester)phosphazenes], PNEG, PNEA, and PNBA, were examined as potential drug delivery matrices. The degradation rate of these polymers was discussed previously in Section 4.3.2.2. The diffusion rate of small molecules such as ethacrynic acid and Biebrich Scarlet was found to be dependent on the hydrolysis rate of the polymers. Bimodal release was observed for all three polymers: an initial slow release phase due to diffusion of the drug through the polymer matrix, and a second phase with an increased release rate owing to both diffusion and erosion, dominated by release through polymer erosion. Thus, fine-tuning the polyphosphazene structure with a specific combination of amino acid ester side groups could allow elegant control over the rate of polymer hydrolysis and the subsequent drug release.

It has been extensively demonstrated that the amino acid ester side groups confer biodegradability and biocompatibility to polyphosphazenes while the imidazole group can significantly increase the polymer degradation rate. Thus, mixed-substituent polymers with both amino acid ester and imidazole groups have been used for the controlled release of drugs to achieve desirable pharmaceutical dosages. Conforti et al. (1996) prepared a discoid matrix based on this mixed-substituent polymer with phenylalanine ethyl ester, imidazolyl group, and unsubstituted chlorine atoms in the side group ratio of 75:18:7 for the prolonged release of naproxen. The feasibility of modulating the drug release rate was demonstrated in vitro by either varying the drug loading or the matrix thickness. In general, thinner disk matrices showed a faster release rate than thicker matrices. The drug release from the thicker disks showed a release rate approaching zero-order kinetics. In vivo studies in a rat model further showed the maintenance of naproxen concentration in plasma within the therapeutic range for several weeks. The pharmacological efficacy of the systems was demonstrated in the inhibition of acute and chronic inflammatory diseases. In the acute inflammation model, the implantation of thin matrices, which showed fast release in vitro, resulted in significant inhibition of edema development, both 1 h and 1 week after matrix implantation. Interestingly, the implantation of thicker matrices resulted in a slow and steady drug release for inhibiting chronic adjuvant arthritis up to day 28. Although these discoidal polyphosphazene matrices showed promise as a controlled release system, the complexity associated with the need for implantation often limited their applications. Thus, polymer microspheres with the ability to be delivered by a syringe were developed for naproxen release. As mentioned earlier in Section 4.3.4.1.1, microspheres prepared by emulsion/solvent evaporation or emulsion/solvent evaporation-extraction showed a size distribution of 10–100 μm. About 80% loading of naproxen could be obtained in these microspheres (Veronese et al. 1998). The in vitro release of naproxen from microspheres produced by spray drying was very rapid and resulted in complete release of entrapped drug within 2 h irrespective of polymer composition and drug loading (Veronese et al. 1998). On the other hand, naproxen release from microspheres fabricated by solvent evaporation was slow and dependent on drug loading and polymer composition. The increase of the imidazole content of the polymer led to greater drug release due to the ease of polymer degradation and ease of water penetration into the matrix as a result of increased hydrophilicity. The rate of drug release also increased as the drug loading in the matrix increased. Thus, release from these microspheres was predominantly diffusion-controlled. In vivo release of the drug, through the implantation in rats, demonstrated a constant plasma concentration (suitable for anti-inflammatory activity) of naproxen maintained up to 400 h. Therefore, the drug release pattern from polyphosphazene matrices can be modulated by varying the polymer side group composition, method of microsphere fabrication, and drug loading.

Ibim et al. (1998) developed a musculoskeletal delivery system for the controlled release of the anti-inflammatory agent colchicines to joints based on two types of polyphosphazenes namely poly[(imidazolyl)$_{0.4}$(p-methylphenoxy)$_{1.6}$phosphazene] (**12**, 20% imidazole-substituted) and PNEGmPh (Ibim et al.). Polymer degradation and drug release studies were carried out using colchicine-loaded matrices. The degradation rate of PNEGmPh was much higher than that of polymer **12** at physiological pH. Colchicine release was found to be 20% and 60% from polymer **12** and PNEGmPh over the 21 day period of study, respectively. The release profile of colchicine from these polymers was found to be consistent with the degradation pattern of the respective polymers (i.e., the faster polymer degradation yielded the faster colchicine release), although the release appeared to proceed through a combination of erosion and diffusion mechanisms during the initial stages.

In another study, membranes and microspheres of polyphosphazenes PNEA and poly[(ethyl phenylalanato)$_{1.6}$(imidazolyl)$_{0.4}$phosphazene] were adopted for the treatment of periodontal diseases (Veronese et al. 1999). Drug release from polyphosphazene membranes was characterized both in vitro and in vivo. It was found that drug released at a rate that ensured therapeutic concentrations in the surrounding tissue. In vivo studies suggested that polyphosphazene membranes were much more effective in promoting the healing of rabbit tibia defects than polytetrafluoroethylene membranes. The drug release from polyphosphazene microspheres yielded local drug concentrations that are useful per se or when mixed with hydroxyapatite (HAp) for better bone formation.

Due to their desirable properties such as biodegradability and matrix permeability, biodegradable polyphosphazenes have also been extensively investigated as matrices for protein delivery. Andrianov et al. reviewed the advantages of using polyphosphazene matrices for the protein release (Andrianov and Payne 1998). For example, microspheres of poly[(ethyl phenylalanato)$_{1.6}$ (imidazolyl)$_{0.4}$phosphazene] were used for the controlled release of insulin (Caliceti et al. 2000). The microsphere preparation was discussed in Section 4.3.4.1.1. The in vitro release of insulin from these microspheres exhibited a bimodal profile, which was characterized by a burst release during the initial 2 h and a subsequent slow release in the range of 2–70 h. Subcutaneous administration of these microspheres prepared by suspension solvent evaporation and suspension/double emulsion-solvent evaporation to diabetic mice rapidly reduced glucose levels to 80%, but they lost most of their activity in 100 h. However, those obtained by double emulsion-solvent evaporation induced a remarkable decrease in glucose level and the activity was maintained for 10:00 h. The anti-insulin antibody production stimulated by all these preparations was found to increase constantly over a period of 8 weeks. This attests to the potential of these materials in the development of one-shot vaccines.

A polyphosphazene polymer (PNEGmPh) was also investigated by Ibim et al. as a substrate for the controlled release of macromolecules (Ibim et al. 1996). ^{14}C-labelled inulin, a polysaccharide with a molecular weight of 5000, was the macromolecule used for the study. Inulin was incorporated into the matrices by a solvent casting technique at 1%, 10%, and 40% loading. The in vitro degradation studies of these polymers were discussed in Section 4.3.2.2. The polymer degradation was accelerated with higher hydrophilic macromolecule inulin content due to increased water permeability. It was found that rate of inulin release from PNEGmPh matrices was highest at pH 2.0 followed by pH 10.0 and 7.4, which was in line with the trend of polymer degradation. Here, drug release also proceeded through a combination of erosion and diffusion mechanisms. Higher macromolecule loading increased the levels of initial drug burst release and resulted in increased amounts of inulin released due to the increased matrix hydrophilicity.

Qiu and Zhu (2000a) synthesized biodegradable polyphosphazenes containing the glycine ethyl ester and benzyl ester of aminoacethydroxamic acid as co-substituents by further modifying polymer PNEG. The polymers were found to undergo hydrolysis at a greater rate under physiological conditions, but were relatively stable at low pH. The potential of these polymers for macromolecular release was evaluated using polysaccharide fluorescein isothiocyanate dextran (FITC-dextran) and the protein myoglobin. The cumulative release of FITC-dextran from the polymer matrix was found to be greater at pH 7.4 and 8.0 than at pH 6.0 and 5.0. However, in the case of myoglobin, the matrix was found to retain the protein completely over 25 h at pH 5.0 and 6.0, whereas there was an appreciable protein release from the matrix at pH 7.4 and 8.0. This complete retention of the protein below pH 6.0 may be attributed to the formation of water-insoluble polyelectrolyte complexes between the protein and polyphosphazene below the isoelectric point (6.9) of the protein.

Oredein-McCoy et al. (2009) recently investigated the feasibility of utilizing polyphosphazene poly[(ethyl phenylalanato)$_{1.5}$(ethyl glycinato)$_{0.5}$phosphazene] microsphere sintered scaffold (Figure 4.10a) alone or incorporated with HAp (Figure 4.10b) to regulate the release of BSA. In the study, polyphosphazene and its composite microspheres with nanocrystalline HAp were fabricated via the emulsion-solvent evaporation technique. BSA-loaded microsphere matrices were created using a solvent/nonsolvent sintering method. Confocal microscopy verified the presence of FITC-BSA both in the adjoining regions between microspheres and within the surface of microspheres.

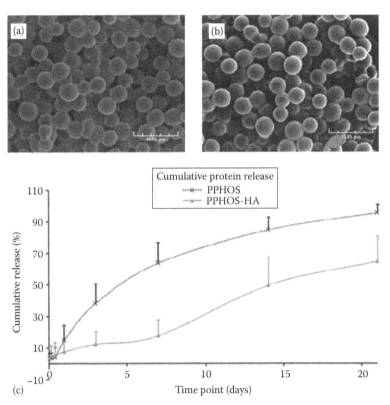

FIGURE 4.10 **(See companion CD for color figure.)** Polyphosphazene-based matrices for growth factor delivery. (a) SEM micrograph showing the morphology of polyphosphazene matrices via solvent/nonsolvent sintering; (b) SEM micrograph showing the morphology of polyphosphazene-HAp matrices via solvent/nonsolvent sintering; and (c) Cumulative BSA release. Zero order release of BSA was demonstrated over the course of 21 days from both the BSA-loaded polyphosphazene and BSA-loaded polyphosphazene-HAp composite scaffolds. Electrostatic interactions between HAp and BSA contributed to the controlled release curve shown here. (From Oredein-McCoy, O. et al., *J. Microencapsul.*, 26(6), 544, 2009, with permission from Informa Healthcare.)

It was found that for both polyphosphazene and its composite microsphere scaffolds, BSA was released over a 21 day period following a zero order release profile (Figure 4.10c). Furthermore, the presence of nanocrystalline HAp particles in the composite scaffolds improved the release profile pattern, which demonstrated the potential of HAp for controlled delivery of growth factors such as angiogenic factors. The results of this work have established the platform for future development of angiogenic factor-loaded polyphosphazene-based matrices to invoke neovascularization.

Thermosensitive polymer hydrogels have also attracted much attention as injected delivery systems of drugs, factors or cells for a variety of biomedical applications. Such polymer hydrogels show sol-gel transitions under certain stimuli such as temperature and pH. Most of them are formed via miscelle packing or hydrophobic interactions between polymer chains, and contain both hydrophilic and hydrophobic regions. Earlier literatures have reported several synthetic polymer hydrogels including N-isopropylacrylamide (NiPAAM), copolymers poly(lactide-co-glycolide)-poly(ethylene glycol)-poly(lactide-co-glycolide) (PLAGA-PEG-PLAGA), and poly(ethylene oxide)-poly(propylene oxide)-poly(ethylene oxide) (PEO-PPO-PEO) (Glatter et al. 1994; Chen and Hoffman 1995; Jeong et al. 1999; Zentner et al. 2001). Polyphosphazenes that respond to external stimuli such as temperature and pH can be designed and applied for controlled drug release (Allcock et al. 1988; Allcock and Ambrosio 1996; Lee et al. 2002; Zhang et al. 2006; Ghattas and Leroux 2009; Ilia 2009; Qiu and Zheng 2009; Bi et al. 2010). For example, amphiphilic graft polyphosphazene with oligopoly(N-isopropylacrylamide) and ethyl 4-aminobenzoate as side groups were synthesized by Qiu and coworkers with different hydrophobic/hydrophilic balances (Zhang et al. 2006). These polymers possess a lower critical solution temperature (LCST) below which they remain water-soluble and above which they precipitate out. Furthermore, pH-responsive hydrogels obtained by gamma irradiation of polyphosphazenes substituted with sodium oxybenzoate and methoxyethoxyethoxy side groups showed pH-dependent swelling behavior (Allcock and Ambrosio 1996). Lee et al. synthesized thermosensitive polyphosphophazene gels bearing α-amino-ω-methoxy PEG (AMPEG) and hydrophobic amino acid ethyl esters such as isoleucine ethyl ester, leucine ethyl ester and valine ethyl ester (Lee et al. 2002). By adjusting the nature and compositional ratios of the hydrophilic and hydrophobic moieties, the gelation properties of the polymers can be finely tuned (Lee and Song 2005). For example, a polyphosphazene gel that contains a greater hydrophobic component such as isoleucine ethyl ester (IleOEt) generally shows a higher gel viscosity and lower gelation temperature. On the other hand, a polyphosphazene polymer containing less hydrophobic or hydrolysis-sensitive groups such as ethyl-2-(O-glycyl) lactate (GlyLacOEt) tends to exhibit lower gel viscosities and higher gelation temperatures.

On the polyphosphazene platform, a variety of prodrugs or bioactive molecule conjugates have also been developed for controlled delivery. For example, drug conjugates of biodegradable polyphosphazene–platinum (II) with a wide range of molecular weights showed high tumor selectivity and excellent in vivo antitumor activity against both murine and human cancer cell lines than their drug alone counterparts (Jun et al. 2005). Such an observed high antitumor activity was achieved through the controlled release of the antitumor active platinum (II) moiety from the polyphosphazene backbone in the degradation process. Anticancer drugs such as paclitaxel and doxorubicin have also been successfully conjugated with polyphosphazenes for inhibiting local tumor growth (Chun et al. 2009a,b). The aqueous solutions of these conjugates showed a solgel transition behavior corresponding to temperature changes. The tumor-active drugs from the polymer-drug conjugate hydrogels were slowly released and effectively accumulated locally in the tumor sites. More recently, in an effort to induce osteogenic differentiation of mesenchymal stem cells (MSCs) for enhancing ectopic bone formation, an injectable and thermosensitive poly(organophosphazene)–RGD conjugate was successfully synthesized through a covalent amide linkage between a cell adhesion peptide, GRGDS, and carboxylic acid-terminated poly(organophosphazene) (Chun et al. 2009c). The polymer-GRGDS conjugates were in fluid state at room temperature and immediately gelled at body temperature. The rabbit mesenchymal stem cells (rMSCs) were cultivated in the polymer–GRGDS conjugate hydrogel constructs using an injection method into a nude mouse. rMSCs were found to express markers at mRNA level for all stages toward osteogenesis and showed a sharp increase of osteocalcin levels at 4 week post-induction.

Histological and immunohistochemical evaluations illustrated a significantly high level of mineralization and collagen type I expression by the rMSCs cultivated on the polymer–GRGDS conjugate hydrogels after 4 weeks. Furthermore, during the fabrication process of drug delivery matrices, bioactive agents are exposed to conditions like elevated temperature or harsh organic solvents which might compromise their bioactivity. It is possible to design hydrogels by the ionic cross-linking of a water-soluble polyphosphazenes that can eliminate the exposure of bioactive agents to the aforementioned harsh conditions. For instance, aqueous PCPP hydrogel microspheres can be obtained by cross-linking with Ca^{2+} ions (Andrianov et al. 1998). Such formulations allow a sustained protein release (Payne et al. 1995; Andrianov and Payne 1998; Andrianov et al. 2009). These biodegradable polyphosphazene matrices have shown tremendous potential as carriers for controlled release of bioactive agents.

Compared to the aforementioned polyphosphazenes, alkoxy-substituted polyphosphazenes have not been extensively investigated as drug delivery matrices. In fact, the presence of reactive hydroxyl groups in many of alkoxy-substituted polymers makes them great candidates for controlled drug release applications. Particularly, the glyceryl-substituted polyphosphazenes are very interesting, since the presence of hydroxyl groups in the side units can generate water solubility and can also provide potential sites either for cross-linking or for the attachment of bioactive agents (Allcock and Kwon 1988). The capacity of this polymer to cross-link and form hydrogels offers the possibility of using this system for the diffusion-controlled release of bioactive agents. Preliminary experiments with the diffusion release of tripelennamine hydrochloride from this polymer hydrogel suggested diffusion-controlled release followed by polymer hydrolysis. Furthermore, pharmaceutical agents or peptides can be linked to the glucosyl-substituted polyphosphazenes (Allcock and Pucher 1991). In vivo hydrolysis would release the glucose-bonded bioactive agents. In the case of polyphosphazenes with glycolic acid ester and lactic acid ester side groups, the lack of crystallinity of these polymers combined with the faster degradation rate compared to PGA or PLA makes them appropriate candidates for controlled drug release applications (Allcock et al. 1994c).

In addition, polyphosphazene blends/composites have also been explored as controlled drug delivery matrices due to the efficient control over the polymer properties and release profiles. Schacht et al. evaluated the potential of the blends of two polyphosphazenes with different degradation rates as a matrix for controlled drug release (Schacht et al. 1996, 1998). In vitro release of the antitumor agent mitomycin C (MMC) was studied using polyphosphazenes containing ethyl glycinate and phenylalanate side groups as well as blends of these two polymers. It was found that the MMC release was much faster from PNEG than from PNEPhA. For the blends, the drug release decreased with the increase of phenylalanato content, which may be attributed to the increase in hydrophobicity of the matrix and the possible decrease in the water permeability of the matrix. The release profile was found to be diffusion-controlled. In the case of thermosensitive polymer hydrogels, the precise control over the balance between hydrophilicity and hydrophobicity is required in order to have the desired gelation properties. However, complexity and many unexpected factors involved in polymer synthesis make such control unpredictable. Kang et al. (2005) demonstrated the efficacy of controlling the gelation behavior of hydrogels via the compositional regulation over the hydrophobic and hydrophilic components in polymer blends. In the study, both hard and soft polyphosphazenes were synthesized. A hard polymer such as $[NP(IleOEt)_{1.16}(AMPEG550)_{0.84}]_n$ exhibits a low gelation temperature and a high viscosity, whereas a soft one like $[NP(IleOEt)_{1.07}(GlyLacOEt)_{0.02}(AMPEG550)_{0.91}]_n$ shows a high gelation temperature and a low viscosity. By blending these two types of polyphosphazenes at an appropriate ratio, the blended aqueous solution was able to change into a transparent hydrogel that showed great strength at 37°C. According to DSC and IR, the two polymers blended homogenously and formed a thermosensitive injectable hydrogel with a T_{max} of 37°C–38°C. Recently, Potta et al. synthesized a class of injectable, self-cross-linkable, and thermosensitive polyphosphazene-based blends of functional thiolated and acrylated polymers to develop an ideal injectable carrier (Potta et al. 2010). The gel properties such as mechanical strength, inner 3D network, and degradation rate can be modulated by varying the degree of cross-linking between the thermosensitive and functional blended polymers.

In summary, the polyphosphazene polymers (aminated polyphosphazenes, alkoxy-substituted polyphosphazenes, and polyphosphazene blends/composites) constitute a very promising matrix platform capable of controlled delivery of both small molecules and macromolecules through the mechanisms of diffusion, erosion, alone or in combination.

4.3.4.2 Tissue Engineering

4.3.4.2.1 Scaffold Fabrication

Tissue loss or organ failure is one of the most devastating and costly problems in health care. A variety of musculoskeletal diseases including back pain, arthritis, osteoporosis, and bodily injuries are often reported among the U.S. population. Each year, more than 3 million musculoskeletal procedures are performed in the United States alone (Vacanti 1999). To overcome the limitations associated with the current graft options, tissue engineering has emerged as a promising strategy that provides de novo regenerated viable tissue substitutes. Tissue engineering has been defined as "the application of biological, chemical and engineering principles toward the repair, restoration or regeneration of tissues using biomaterials, cells, and factors alone or in combination" (Laurencin et al. 1999).

All approaches in tissue engineering involve one or more of the three key components: (1) 3D porous matrices with optimum properties to support and promote tissue regeneration, (2) a healthy population of harvested cells, and (3) biological growth factors that can induce cellular migration, differentiation, and tissue growth in vivo (Langer et al. 1993). Biodegradable scaffold plays crucial roles in scaffold-based tissue engineering approach. During the regeneration process, the biodegradable scaffold provides structural and mechanical restoration of the damaged tissues, gradually degrades into biocompatible products, presents an interconnected porous structure to accommodate cell infiltration and vascularization, and promote matrix synthesis (Laurencin et al. 1999; Langer 2000). The success of a 3D scaffold in tissue regeneration is largely dependent on its nature, composition, and associated properties. Biomaterials including synthetic biodegradable polymers and composites have shown great promise in a tissue engineering approach (Nair and Laurencin 2007).

Biodegradable polyphosphazenes have been identified as one of the important classes of biomaterials in tissue engineering applications (Ibim et al. 1997; Ambrosio et al. 2002; Nair et al. 2005; Krogman et al. 2007; Deng et al. 2010a). The synthetic flexibility of polyphosphazenes allows for the design and development of 3D scaffolds with the right combination of desirable physical, chemical, and biological properties for a specific tissue engineering application. A number of techniques including particulate leaching, microsphere sintering, and electrospinning have been developed to fabricate polyphosphazene-based porous scaffolds for a variety of tissue engineering applications based on aminated polyphosphazene, alkoxy-substituted polyphosphazenes, and polyphosphazene blends/composites (Laurencin et al. 1993, 1996; Ambrosio et al. 2003; Nair et al. 2004; Bhattacharyya et al. 2006, 2009; Brown et al. 2008; Nukavarapu et al. 2008, 2009). In addition, several in situ scaffold systems have also been developed on the polyphosphazene platform for tissue regeneration (Greish et al. 2005b; Deng et al. 2010d).

One of the early reports from Laurencin et al. detailed the fabrication of 3D porous scaffolds from PNEGmPh with interconnected porosity and an average pore diameter of 165 μm using a salt leaching technique for skeletal tissue regeneration (Laurencin et al. 1996). However, salt-leached scaffolds are often not applicable to load-bearing applications and associated with several limitations in generating the interconnected porous structures and eliminating the entrapped prorogens. Thus, mechanically competent 3D porous microsphere scaffolds have been developed by Laurencin et al. and showed great promise for regenerative applications (Borden et al. 2002a,b, 2003, 2004). In brief, these 3D microsphere scaffolds are fabricated from polymers such as PLAGA by heating the individual polymer microspheres above the polymer T_g in a steel mold of desired size and shape. Polymer becomes rubbery above the T_g and holds the microspheres together to form a 3D scaffold according to the mode shape with an interconnected porous structure at room temperature. It has been shown that the pore size and porosity of the scaffolds can be varied by changing the duration/temperature of sintering as well as the microsphere diameter. However, the applicability of heat sintering technique to polyphosphazene microspheres is limited due to its dependence on specific physicochemical properties such as T_g, crystallinity, viscosity, and surface tension of the polymer. Thus, a novel route based on a solvent/nonsolvent approach was adopted to create porous polymeric microsphere scaffolds suitable for tissue engineering purposes (Brown et al. 2008; Nukavarapu et al. 2008, 2009). As shown in Figure 4.11, sintering the microspheres was based on the partial dissolution of the microsphere surface in a solvent/nonsolvent composition at ambient conditions that establishes a balance between polymer dissolution and precipitation. At this intermediate state, the microsphere surfaces were susceptible to bonding with adjacent microspheres prior to precipitation of the intertwined surface chains and sintering of the microspheres. The resultant 3D scaffold properties such as the overall porosity, average pore diameter, and mechanical properties can be fine-tuned by changing the nature and composition of solvent/nonsolvent. Brown et al. (2008) used five different biodegradable polyphosphazenes exhibiting glass transition temperatures from −8°C to 41°C to study the versatility of the process (Figure 4.11). For example, to achieve effective

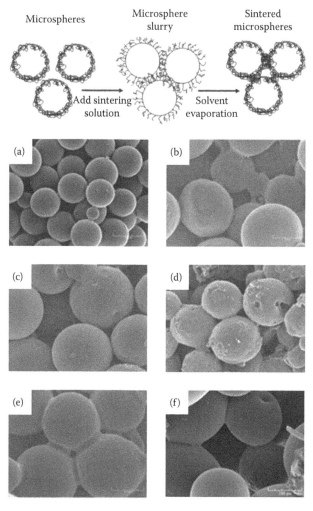

FIGURE 4.11 Schematic illustration of the solvent/nonsolvent sintering approach and SEM micrographs illustrating the flexibility of this dynamic solvent approach for creating 3D porous sintered microsphere scaffolds from a wide range of different polyphosphazene polymers where (a) PLAGA; (b) PNEA; (c) PNEPhA; (d) PNMV; (e) PNEAPhPh; and (f) PNEAmPh. (From Brown, J.L., Nair, L.S., and Laurencin, C.T.: Solvent/non-solvent sintering: A novel route to create porous microsphere scaffolds for tissue regeneration. *J. Biomed. Mater. Res. B. Appl. Biomater.*, 2008. 86B(2). 396–406. Copyright Wiley-VCH Verlag GmbH & Co. KGaA. With permission.)

sintering of PNEA, acetone was chosen as a solvent while hexane was a non-solvent. Similarly for poly[bis(methyl valinato)phosphazene] (PNMV) and PNEPhA, the combination of THF/hexane was used as the solvent/nonsolvent system. Thus fabricated polyphosphazene microsphere scaffolds exhibited a maximum interconnected porosity of 37.6% and a maximum compressive modulus of 94.3 MPa. Nukavarapu et al. developed composite microspheres from PNEPhA and 100 nm sized HAp with varying compositions in the range of 10%–30% (w/w) (Nukavarapu et al. 2008, 2009). These composite microspheres were sintered into 3D architecture using a similar solvent/non-solvent approach. These porous scaffolds exhibited mean pore diameters in the range of 86–145 µm and compressive moduli of 46–81 MPa.

Polymeric nanofibers due to their similarity to natural extracellular matrix (ECM) are attractive candidates as tissue engineering scaffolds. They can be fabricated using various processing techniques such as drawing (Ondarçuhu and Joachim 1998; Huang et al. 2003), self-assembly (Whitesides and Grzybowski 2002; Huang et al. 2003), template synthesis (Martin 1996; Huang et al. 2003), phase separation (Ma and Zhang 1999; Huang et al. 2003), and electrospinning (Huang et al. 2003). Among these, electrospinning is the most general and simplest technique that can be used to fabricate nanofibers from biodegradable polymers or polymer blends/composites (Huang et al. 2003; Kumbar et al. 2008; Nair and Laurencin 2008). Figure 4.12a shows a general representation of an electrospinning apparatus used to create polymeric nanofibers (Kumbar et al. 2008). In brief, a polymer jet is ejected from the surface of a charged polymer solution when the applied electric potential overcomes the surface tension. The ejected jet under the influence of applied electrical field travels rapidly to the collector and collects in the form of nonwoven web as the jet dries. Before reaching the collector the jet undergoes a series of electrically driven bending instabilities that results in a series of looping and spiraling motions. In an effort to minimize the instability due to the repulsive electrostatic forces, the jet elongates to undergo large amounts of plastic stretching that consequently leads to a significant reduction in its diameter and results in ultrathin fibers.

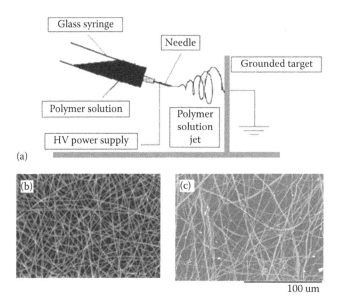

FIGURE 4.12 Electrospinning of polyphosphazene polymers for tissue engineering. (a) Schematic representation of an electrospinning apparatus to produce nonwoven nanofiber scaffolds, which consists of a high voltage supply (5–40 kV), a source electrode, a grounded collector electrode, and a capillary tube with a needle of small diameter; (b) SEM micrograph of electrospun PNmPh nanofibers (With permission from Nair, L. S., Bhattacharyya, S., Bender, J. D. et al., Fabrication and optimization of methylphenoxy substituted polyphosphazene nanofibers for biomedical applications, *Biomacromolecules*, 5(6), 2212–2220. Copyright 2004 American Chemical Society); and (c) SEM micrograph of electrospun PNGEGPhPh-PLAGA blend nanofibers.

With low-viscosity polymer solutions, the ejected jet may break down into droplets and result in electrospraying (Kumbar et al. 2007). Therefore, a suitable polymer solution viscosity is essential to fabricate nanofibers without any beads or beads-on-a-string appearance. For example, increase in viscosity or polymer concentration results in fiber diameter increase. By changing polymer concentration alone it is possible to fabricate the fiber diameters in the range of few nanometers to several micrometers while keeping other electrospinning parameters constant. In addition, the electrical potential applied also plays an important role in the fiber diameter obtained by electrospinning. So far, several poly[(amino acid ester)phosphazenes] have been successfully electrospun to produce ECM-mimicking nanofiber scaffolds for a variety of tissue engineering applications (Conconi et al. 2004, 2006; Nair et al. 2004; Bhattacharyya et al. 2006, 2009; Krogman et al. 2010; Lin et al. 2010). The morphologies of the PNmPh and PNGEGPhPh-PLAGA blend nanofibers under optimized electrospinning conditions are presented in Figure 4.12b and c, respectively.

In addition to the aforementioned preformed matrices, several composite matrices including self-setting polyphosphazene bone cements and in situ pore forming polyphosphazene blends have been developed to offer a great advantage in a clinical setup by forming the scaffolds in situ (Greish et al. 2005b; Deng et al. 2010d).

Bone is a natural composite principally composed of collagen fibrils and HAp (Rho et al. 1998). Collagen component provides bone with a structural framework, high tensile strength, and flexibility while ceramic HAp component offers stiffness and high compressive strength. Fabrication of a self-setting composite comprised of a polymer component and HAp closely mimics the bone composition while providing an easy way to form bone implants at the injury sites. Thus, a polyphosphazene composite cement system has been developed by exploiting the favorable interactions between polyphosphazene side groups and calcium phosphate ceramics. For example, the formation of composites has been achieved via an acid–base reaction of tetracalcium phosphate and anhydrous dicalcium phosphate in the presence of polyphosphazenes bearing alkyl ester containing side groups at physiological temperature (Greish et al. 2005b). Alkaline conditions resulted in the partial hydrolysis of alkyl ester side groups of PNEOB and PNPOB producing carboxylate groups on the polymer chain. Calcium ions in solution cross-linked these carboxylate groups resulting in the formation of a calcium cross-linked polyphosphazene surface layer, which in turn nucleated the depositing HAp on the surface (Greish et al. 2005b). Polymers such as PCPP and its water-soluble salts such as Na-PCPP or K-PCPP also favored the composite formation at low polymer concentrations under physiologic conditions (Greish et al. 2005a, 2006). Composites produced similarly using PNEAPhPh showed higher compressive strength than PNEA composites due to bulkier side groups (Greish et al. 2008). Such biomimetic deposition of HAp on the polymers closely resembles the mineralization of collagen, which enables the potential of developing composite implants for tissue repair and regeneration in vivo.

Recently, a dipeptide-based polyphosphazene blend system has demonstrated a unique erosion mechanism through which the polymer evolves from a solid coherent film to an assemblage of microspheres with interconnected porous structures (Deng et al. 2010d). This allows to develop and control 3D scaffolds for regenerative purposes with initial higher mechanical properties that reduces over time as polymer degrades while generating interconnected porous structures to allow tissue in-growth and matrix integration. Studies are currently underway to optimize dipeptide-based polyphosphazene side group composition to achieve higher mechanical properties suitable for load-bearing applications. We are also evaluating this blend system as a factor delivery vehicle and tissue engineering scaffold for regenerative applications.

4.3.4.2.2 Scaffold-Based Tissue Regeneration

So far, biodegradable polyphosphazene scaffolds have demonstrated great potential for the regeneration of different tissue types such as bone, tendon, blood vessel, and nerve, to name a few.

For example, more than 6.5 million fractures occur in the United States each year (Braddock et al. 2001). Reconstruction of large bone defects presents a significant medical challenge. The porous 3D scaffolds of PNEGmPh developed by Laurencin et al. using a salt leaching

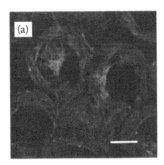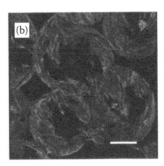

FIGURE 4.13 (**See companion CD for color figure.**) Cytoskeletal actin distribution of PRO grown on polyphosphazene-HAp composite microsphere scaffold for (a) 6 days; and (b) 12 days. Scale bar: 250 μm. After 12 days, cells covered the entire scaffold including the microsphere surface as well as the microsphere adjoining areas. (With permission from Nukavarapu, S.P., Kumbar, S.G., Brown, J.L. et al., Polyphosphazene/nano-hydroxyapatite composite microsphere scaffolds for bone tissue engineering, *Biomacromolecules*, 9(7), 1818–1825. Copyright 2008 American Chemical Society.)

technique as mentioned earlier in Section 4.3.4.2.1 supported cell growth within the pores as well as on surface of the polymer as early as day 1 (Laurencin et al. 1996). On these porous constructs, PRO cells proliferated in a steady state over the 21 day culture time whereas a decline in cell number was found for 2D films after 7 days. Thus, 3D porous constructs provide increased surface area and promote cell growth throughout the scaffold. In an effort to develop sintered polyphosphazene microsphere scaffolds for bone tissue engineering, Nukavarapu et al. prepared composite microspheres of PNEPhA with 100 nm sized HAp with varying HAp compositions between 10%–30% (w/w) (Nukavarapu et al. 2008). These composite microspheres were sintered into 3D architecture using a solvent/nonsolvent approach as discussed earlier in Section 4.3.4.2.1. The increased presence of cytoskeletal actin on the composite microsphere scaffolds suggested the progressive growth of PRO (Figure 4.13). Quantitative analysis confirmed that the composite scaffolds were able to support good osteoblast adhesion, proliferation, and alkaline phosphatase expression throughout the 21 day culture. In another study, El-Amin et al. evaluated the in vivo biocompatibility and degradation of 5 mm diameter plugs from PNEGmPh and PNEG in a rabbit metaphyseal distal femur defect model (El-Amin et al. 2007). Histological evaluations revealed that both polymers appeared to support bone growth comparable to the PLAGA control. Femurs with polyphosphazene implants showed evidence of bone in-growth and a mild fibrous response at 12 weeks.

Ambrosio et al. fabricated composites of PNEGmPh-HAp and evaluated their potential as scaffold materials for bone tissue engineering (Ambrosio et al. 2003). These novel composites retained their mechanical properties during degradation and supported osteoblast-like MC3T3-E1 cell proliferation. Such in vitro studies strongly support the use of polyphosphazene-based composites as promising new materials for developing biodegradable bone implants.

Recently, Deng et al. (2010d) fabricated the dipeptide-based polyphosphazene–polyester blends and investigated their potential to accommodate cell infiltration using a rat subcutaneous implantation model. It was found that the in situ formed 3D interconnected porous structures enabled cell infiltration and collagen tissue in-growth throughout the void space between spheres. The robust tissue in-growth of the dynamic pore forming scaffold attests to the utility of this system as a new strategy in regenerative medicine for developing solid matrices that balance degradation with tissue formation.

Nair et al. evaluated the potential of the PNmPh nanofiber matrices for tissue engineering applications. Preliminary results showed that these polyphosphazene fiber matrices supported the adhesion and proliferation of osteoblast-like MC3T3-E1 cells (Nair et al. 2004). PNmPh polymer can be used to modulate the physico-chemical properties (hydrophobicity/hydrophilicity, mechanical properties, and degradation pattern) and consequent regenerative efficacy of the mixed-substituent polyphosphazenes. Bhattacharyya et al. fabricated PNEA as well as PNEA-HAp composite nanofiber matrices as scaffolds for bone tissue engineering applications (Bhattacharyya et al. 2006, 2009). Such polyphosphazene nanofiber structures closely mimic the ECM architecture, and have shown improved cell performance over the conventional scaffold architectures (Nair et al. 2004; Bhattacharyya et al. 2006, 2009; Conconi et al. 2006). Conconi et al. fabricated electrospun scaffolds from poly[(ethyl phenylalanato)$_{1.4}$(ethyl glycinato)$_{0.6}$phosphazene] and its blends with PLA and PCL (Conconi et al. 2009). Osteoblast adhesion and growth were found to be less on electrospun polyphosphazene scaffolds than on electrospun PLA scaffolds. However, a synergic effect on cell proliferation was observed when cells were cultured on polyphosphazene-PLA blends. Although the poor mechanical properties limit their applications to repair load-bearing defects, the inherited buffering capacity of polyphosphazenes to neutralize the acidic degradation of PLA makes this system interesting materials to repair cranial and maxillofacial defects.

As discussed earlier, nanofiber matrices have shown great promise for regeneration purposes (Huang et al. 2003; Li and Xia 2004; Nair and Laurencin 2008). However, there are some significant challenges that need to be addressed in order to enable their load-bearing applications in tissue engineering. For example, nanofibers lack stability under compression, and hence may not be suitable to support load-bearing bone defects. (Li et al. 2002; Whitesides and Boncheva 2002; Ma 2008). With a more in-depth investigation into the constituents of bone and their unique properties, several scaffold platforms have been developed to closely mimic the inherent characteristics of the native bone.

Brown et al. (2010) recently developed a novel structure that combines the robust mechanical aspects of the sintered microsphere scaffold with a highly bioactive nanofiber structure to produce a composite scaffold that demonstrates an ability to mimic the mechanical environment of trabecular bone while also promoting the osteoinduction of osteoblast progenitor cells. Exploiting the chemistry of two biodegradable polymers, a 3D PLA nanofiber mesh was successfully incorporated within the void spaces between sintered PNEPhA microspheres (Figure 4.14a and b). The non-load-bearing fiber portion of these scaffolds is sufficiently porous to allow cell migration and ECM matrix production throughout the fibrous portion of the scaffold (Figure 4.14c and d). These composite nanofiber/microsphere scaffolds promote osteoinduction through focal adhesion kinase activity. Ultimately the focal adhesion kinase activity on the composite nanofiber/microsphere scaffolds demonstrated causality over the production of the mature osteoblast marker, osteocalcin, and the development of a calcified matrix (Figure 4.14e and f).The phenotype progression of osteoblast progenitor cells on the composite nanofiber/microsphere scaffolds illustrated a stronger and more rapid progression leading to fully matured osteoblasts by 21 days.

Inspired by the hierarchical structures that enable bone function, Deng (2010) recently developed a mechanically competent 3D scaffold mimicking the bone marrow cavity, as well as, the lamellar structure of bone by orienting electrospun polyphosphazene-polyester blend nanofibers in a concentric manner with an open central cavity. The 3D biomimetic scaffold exhibited a similar characteristic mechanical behavior to that of native bone. Compressive modulus of the scaffold was found to be within the range of human trabecular bone. To our knowledge this is the first mechanically competent 3D electrospun nanofiber scaffold with mechanical properties in the middle range of human trabecular bone. The potential of this scaffold for bone repair was further investigated by monitoring the cellular activity and mechanical performance over time using in vitro culture. These blend nanofiber matrices supported PRO adhesion, proliferation, and showed an elevated phenotype expression compared to PLAGA nanofibers. This biomimetic scaffold supported the robust PRO growth throughout the scaffold architecture and maintained osteoblast phenotype expression in vitro, which resulted in a similar cell-matrix organization to that of native bone and maintenance of structure integrity. When combined with the desirable polymer blend properties, the concentric open macrostructures of nanofibers that structurally and mechanically mimic the native bone can be a potential scaffold design for accelerated bone healing.

Besides bone repair, polyphosphazenes have also shown tremendous potential for regenerating other types of tissue such as nerve (Langone et al. 1995; Aldini et al. 1997; Zhang et al. 2009;

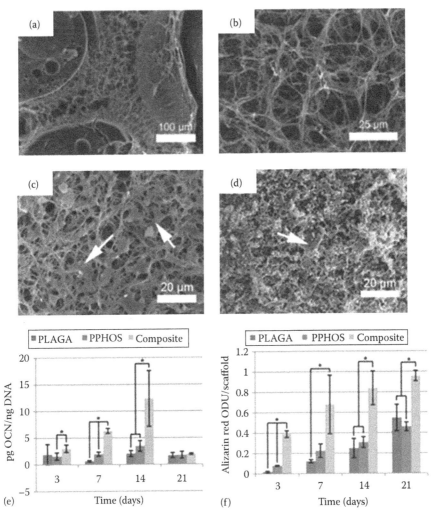

FIGURE 4.14   (See companion CD for color figure.) Polyphosphazene composite scaffolds bridging nano-fiber and microsphere architectures to improve bioactivity of mechanically competent constructs for bone regeneration. (a, b) SEM micrographs of the cross-sections of the composite nanofiber/microsphere scaffold demonstrating incorporation of nanofibers of less than 1 μm within the void space generated by sintering microspheres were (a) low magnification image and (b) high magnification image; (c, d) SEM micrographs of preosteoblasts in the interior of composite scaffolds after 3 days (c) and 14 days (d) of culture demonstrating the presence of preosteoblasts and the accumulation of matrix proteins in the interior of the composite scaffold during cell culture. The arrows point to representative cells. Notably, the fibrous portion of the scaffold has been extensively modified through the accumulation of ECM proteins after 14 days, to the extent that identifying the cells is difficult. (e, f) Phenotype evaluation of preosteoblasts seeded on composite scaffolds, (n = 4). Mature osteoblast marker such as osteocalcin (e) evaluated and normalized to total DNA demonstrates a peak at 14 days for the preosteoblasts seeded on the composite scaffold; and (f) Total scaffold calcification normalized to an unseeded construct demonstrates that the preosteoblasts seeded on the composite scaffold produced a substantially more mineralized matrix than did the preosteoblasts on control PLAGA and poly-phosphazene scaffold. (*) represents statistical significance, $p < 0.05$, from all other samples at a given time point. All error bars represent standard deviation. (From Brown, J.L., Peach, M.S., Nair, L.S., Kumbar, S.G., and Laurencin, C.T.: Composite scaffolds: Bridging nanofiber and microsphere architectures to improve bio-activity of mechanically competent constructs. *J. Biomed. Mater. Res. A.* 2010. 95(4). 1150–1158. Copyright Wiley-VCH Verlag GmbH & Co. KGaA. With permission.)

Deng et al. 2010a), vessel (Carampin et al. 2007; Deng et al. 2010a), and tendon (Huang et al. 2006; Deng et al. 2010a). For example, in vivo nerve repair studies were carried out by using tubes of poly[(ethylalanato)$_{1.4}$(imidazolyl)$_{0.6}$phosphazene] (Langone et al. 1995). Rat sciatic nerve stumps were transectioned and immediately sutured into the ends of 10 mm long polymer tubes. After implantation for 45 days, a tissue cable was found bridging the nerve stumps. Histological analysis illustrated that the tissue cable was mainly composed of a regenerated nerve fiber bundle. Another study using the conduits of alanine ester–substituted polyphosphazenes demonstrated a similar nerve regenerative capacity compared to autologous grafts in Wistar rats (Aldini et al. 1997). Recently, a biodegradable polyphosphazene cosubstituted with both aniline pentamer and glycine ethyl ester side groups showed an electrical conductivity of ∼2 × 10^{-5} S/cm in the semiconducting region (Zhang et al. 2009). The polymer supported the attachment and proliferation of RSC96 Schwann cells, suggesting that it is a potential scaffold material for peripheral nerve regeneration. As a first step to investigate the feasibility to construct human vessels or cardiac valve, Carampin et al. demonstrated that electrospun nanofibers of biodegradable poly[(ethyl phenylalanato)$_{1.4}$(ethyl glycinato)$_{0.6}$phosphazene] supported adhesion and growth of rat neuromicrovascular endothelial cells (Carampin et al. 2007). Nair et al. (2004) also showed that the PNmPh nanofiber matrices supported the adhesion of bovine coronary artery endothelial cells (BCAECs).

In summary, the synthetic versatility of polyphosphazene materials combined with advanced scaffold fabrication techniques hold great promise in developing optimal 3D structures with the right combination of physical, chemical, and biological properties to suit the requirements of a specific tissue engineering application.

4.4 CONCLUSIONS AND FUTURE TRENDS

The development of biodegradable materials has led to the synthesis of new polymers with diverse properties. The success of synthetic polymers as biomaterials for biomedical applications largely depends on their inherent physicochemical properties, biodegradability, and biocompatibility. Biodegradable polyphosphazenes are a unique class of polymers due to their synthetic flexibility and versatile adaptability for biomedical applications. They are high molecular weight polymers having an inorganic backbone of alternating phosphorus and nitrogen atoms bearing two side groups attached to each phosphorus atom. The synthetic flexibility of polymer backbone allows controlled tuning of physico-chemical properties, biodegradability, and biocompatibility via macromolecular substitutions of more than 250 organic nucleophiles. Specifically, the substitution of appropriate side groups such as amino acid esters or alkoxy substituents imparts biodegradability to polyphosphazenes. Biodegradable polyphosphazenes undergo degradation through surface erosion or bulk erosion or a combination of both into nontoxic and neutral pH degradation products. Furthermore, polyphosphazene blends/composites synergistically combine the advantages of the parent polymers. The blend properties can be fine-tuned to satisfy the requirements of a specific application via the blend compositional change. Additionally, the synthesis of novel polyphosphazene polyester or polycarbonate block copolymers enables the development of advanced polymeric blend systems using these polymeric architectures as prototypes. Thus, the polyphosphazene polymers based on polyphosphazenes alone and polyphosphazene blends/composites are an excellent material platform for developing controlled drug release matrices and tissue engineering scaffolds.

In controlled drug delivery, biodegradable polyphosphazenes are attractive materials due to their hydrolytic lability, nontoxic degradation products, excellent biocompatibility, ease of matrix fabrication, and matrix permeability. The controlled erosion rates and drug release profiles can also be tuned through macromolecular substitution. Studies have demonstrated that the release of small molecules and macromolecules from biodegradable polyphosphazene matrices occurs either by diffusion, erosion, or a combination of these mechanisms. Their success in both the in vitro and in vivo environments makes them potentially ideal candidates for the delivery of bioactive molecules. Future work will combine the versatile synthetic chemistry in polyphosphazenes with the advancement in matrix fabrication (i.e., nanoparticles preparation) to develop targeted drug delivery systems with controlled release kinetics.

In the field of tissue engineering, biodegradable polyphosphazenes have shown great promise in the regeneration of different tissue types via a scaffold-based tissue regeneration approach due to their excellent combination of physical, chemical, and biological properties. Recent studies have provided significant insights into the influence of topographic features in regulating cell behavior, including preferential adhesion, migration, proliferation, and expression of cell-specific phenotypes. With the advancement in micro- and nano-fabrication technology, biomaterials can be fabricated into various 3D structures that mimic the native tissue hierarchical structures. Reflecting the biological processes during native tissue development, it is critical to study and exploit the molecular mechanisms governing cell–cell and cell–material interaction at the micro- and nano-scale to generate tissue-inducing scaffolds and achieve facilitated tissue regeneration. Polyphosphazenes as a unique class of biomaterials, when appropriately integrated with the essential physical and chemical cues present in natural ECM, will ultimately enable successful regeneration of complex tissue systems using integrated graft systems in the near future.

ACKNOWLEDGMENTS

This work was supported by NIH RO1 EB004051, RO1 AR052536, and NSF EFRI-0736002. Dr. Laurencin was the recipient of Presidential Faculty Fellow Award from the National Science Foundation.

REFERENCES

Adibi, S. A. 1989. Glycyl-dipeptides: New substrates for protein nutrition. *J Lab Clin Med* 113 (6):665–673.
Agrawal, C. M. and Athanasiou, K. A. 1997. Technique to control pH in vicinity of biodegrading PLA-PGA implants. *J Biomed Mater Res* 38 (2):105–114.
Aldini, N. N., Fini, M., Rocca, M. et al. 1997. Peripheral nerve reconstruction with bioabsorbable polyphosphazene conduits. *J Bioact Compat Pol* 12 (1):3–13.
Allcock, H. R. 1967. *Heteroatom Ring Systems and Polymers*. New York: Academic Press.
Allcock, H. R. 1972a. *Phosphorus-Nitrogen Compounds. Cyclic, Linear and High Polymeric Systems*. New York: Academic Press.
Allcock, H. R. 1972b. Recent advances in phosphazene (phosphonitrilic) chemistry. *Chem Rev* 72 (4):315–356.
Allcock, H. R. 1979. Small-molecule phosphazene rings as models for high polymeric chains. *Acc Chem Res* 12 (10):351–358.
Allcock, H. R. 1985. Developments at the interface of inorganic, organic, and polymer chemistry. *Chem Eng News* 63 (11):22–36.
Allcock, H. R. 1992. Mechanisms and catalysis in cyclophosphazene polymerization. In *Catalysis in Polymer Synthesis*, eds., E. J. Vandenberg and J. C. Salamone, pp. 236–247. Washington, DC: American Chemical Society.
Allcock, H. R. 2003. *Chemistry and Applications of Polyphosphazenes*. Hoboken, NJ: Wiley Interscience.
Allcock, H. R. 2010. Hybrids of hybrids: Nano-scale combinations of polyphosphazenes with other materials. *Appl Organometal Chem* 24 (8):600–607.
Allcock, H. R. and Ambrosio, A. M. 1996. Synthesis and characterization of pH-sensitive poly (organophosphazene) hydrogels. *Biomaterials* 17 (23):2295–2302.
Allcock, H. R., Austin, P. E., and Neenan, T. X. 1982a. Phosphazene high polymers with bioactive substituent groups: Prospective anesthetic aminophosphazenes. *Macromolecules* 15 (3):689–693.
Allcock, H. R., Bender, J. D., Morford, R. V., and Berda, E. B. 2003a. Synthesis and characterization of novel solid polymer electrolytes based on poly(7-oxanorbornenes) with pendent oligoethyleneoxy-functionalized cyclotriphosphazenes. *Macromolecules* 36 (10):3563–3569.
Allcock, H. R. and Chu, C. T.-W. 1979. Reaction of phenyllithium with poly(dichlorophosphazene). *Macromolecules* 12 (4):551–555.

Allcock, H. R., Clay Kellam, E., and Morford, R. V. 2001a. Gel electrolytes from co-substituted oligoethylene-oxy/trifluoroethoxy linear polyphosphazenes. *Solid State Ionics* 143 (3–4):297–308.

Allcock, H. R., Cook, W. J., and Mack, D. P. 1972. Phosphonitrilic compounds. XV. High molecular weight poly[bis(amino)phosphazenes] and mixed-substituent poly(aminophosphazenes). *Inorg Chem* 11 (11):2584–2590.

Allcock, H. R., Crane, C. A., Morrissey, C. T. et al. 1996. Living cationic polymerization of phosphoranimines as an ambient temperature route to polyphosphazenes with controlled molecular weights. *Macromolecules* 29 (24):7740–7747.

Allcock, H. R. and Fuller, T. J. 1980. Phosphazene high polymers with steroidal side groups. *Macromolecules* 13 (6):1338–1345.

Allcock, H. R., Fuller, T. J., Mack, D. P., Matsumura, K., and Smeltz, K. M. 1977. Synthesis of poly[(amino acid alkyl ester)phosphazenes]. *Macromolecules* 10 (4):824–830.

Allcock, H. R., Fuller, T. J., and Matsumura, K. 1982b. Hydrolysis pathways for aminophosphazenes. *Inorg Chem* 21 (2):515–521.

Allcock, H. R., Hartle, T. J., Taylor, J. P., and Sunderland, N. J. 2001b. Organic polymers with cyclophosphazene side groups: Influence of the phosphazene on physical properties and thermolysis. *Macromolecules* 34 (12):3896–3904.

Allcock, H. R. and Kellam, E. C. 2001. Incorporation of cyclic phosphazene trimers into saturated and unsaturated ethylene-like polymer backbones. *Macromolecules* 35 (1):40–47.

Allcock, H. R., Kellam, E. C., and Hofmann, M. A. 2001c. Synthesis of cyclolinear phosphazene-containing polymers via ADMET polymerization. *Macromolecules* 34 (15):5140–5146.

Allcock, H. R. and Kugel, R. L. 1965. Synthesis of high polymeric alkoxy- and aryloxyphosphonitriles. *J Am Chem Soc* 87 (18):4216–4217.

Allcock, H. R. and Kugel, R. L. 1966. Phosphonitrilic compounds. VII. High molecular weight poly(diaminophosphazenes). *Inorg Chem* 5 (10):1716–1718.

Allcock, H. R., Kugel, R. L., and Valan, K. J. 1966. Phosphonitrilic compounds. VI. High molecular weight poly(alkoxy- and aryloxyphosphazenes). *Inorg Chem* 5 (10):1709–1715.

Allcock, H. R. and Kwon, S. 1988. Glyceryl polyphosphazenes: Synthesis, properties, and hydrolysis. *Macromolecules* 21 (7):1980–1985.

Allcock, H. E., Kwon, S., Riding, G. H., Fitzpatrick, R. J., and Bennett, J. L. 1988. Hydrophilic polyphosphazenes as hydrogels: Radiation cross-linking and hydrogel characteristics of poly [bis (methoxyethoxyethoxy)phosphazene]. *Biomaterials* 9 (6):509–513.

Allcock, H. R., Lampe, F. W., and Mark, J. E. 2003b. *Contemporary Polymer Chemistry*, 3rd edn. Upper Saddle River, NJ: Pearson Education, Inc (Pearson/Prentice Hall).

Allcock, H. R., Laredo, W. R., Kellam, E. C., and Morford, R. V. 2001d. Polynorbornenes bearing pendent cyclotriphosphazenes with oligoethyleneoxy side groups: Behavior as solid polymer electrolytes. *Macromolecules* 34 (4):787–794.

Allcock, H. R., Nelson, J. M., Reeves, S. D., Honeyman, C. H., and Manners, I. 1997a. Ambient-temperature direct synthesis of poly(organophosphazenes) via the living cationic polymerization of organo-substituted phosphoranimines. *Macromolecules* 30 (1):50–56.

Allcock, H. R., Powell, E. S., Chang, Y., and Kim, C. 2004a. Synthesis and micellar behavior of amphiphilic polystyrene-poly[bis(methoxyethoxyethoxy)phosphazene] block copolymers. *Macromolecules* 37 (19):7163–7167.

Allcock, H. R., Powell, E. S., Maher, A. E., and Berda, E. B. 2004b. Poly(methyl methacrylate)-graft-poly-[bis(trifluoroethoxy)phosphazene] copolymers: Synthesis, characterization, and effects of polyphosphazene incorporation. *Macromolecules* 37 (15):5824–5829.

Allcock, H. R. and Prange, R. 2001. Properties of poly(phosphazene-siloxane) block copolymers synthesized via telechelic polyphosphazenes and polysiloxane phosphoranimines. *Macromolecules* 34 (20):6858–6865.

Allcock, H. R. and Pucher, S. R. 1991. Polyphosphazenes with glucosyl and methylamino, trifluoroethoxy, phenoxy, or (methoxyethoxy)ethoxy side groups. *Macromolecules* 24 (1):23–34.

Allcock, H. R., Pucher, S. R., and Scopelianos, A. G. 1994a. Poly[(amino acid ester)phosphazenes] as substrates for the controlled release of small molecules. *Biomaterials* 15 (8):563–569.

Allcock, H. R., Pucher, S. R., and Scopelianos, A. G. 1994b. Poly[(amino acid ester)phosphazenes]: Synthesis, crystallinity, and hydrolytic sensitivity in solution and the solid state. *Macromolecules* 27 (5):1071–1075.

Allcock, H. R., Pucher, S. R., and Scopelianos, A. G. 1994c. Synthesis of poly(orgnaophosphazenes) with glycolic acid ester and lactic acid ester side groups: Prototypes for new bioerodible polymers. *Macromolecules* 27 (1):1–4.

Allcock, H. R., Ravikiran, R., and Olshavsky, M. A. 1998. Synthesis and characterization of hindered poly-phosphazenes via functionalized intermediates: Exploratory models for electro-optical materials. *Macromolecules* 31 (16):5206–5214.

Allcock, H. R., Reeves, S. D., Nelson, J. M., Crane, C. A., and Manners, I. 1997b. Polyphosphazene block copolymers via the controlled cationic, ambient temperature polymerization of phosphoranimines. *Macromolecules* 30 (7):2213–2215.

Allcock, H. R. and Scopelianos, A. G. 1983. Synthesis of sugar-substituted cyclic and polymeric phosphazenes and their oxidation, reduction, and acetylation reactions. *Macromolecules* 16 (5):715–719.

Allcock, H. R., Singh, A., Ambrosio, A. M., and Laredo, W. R. 2003c. Tyrosine-bearing polyphosphazenes. *Biomacromolecules* 4 (6):1646–1653.

Allcock, H. R., Welna, D. T., and Stone, D. A. 2005. Synthesis of pendent functionalized cyclotriphos-phazene polyoctenamers: Amphiphilic lithium ion conductive materials. *Macromolecules* 38 (25):10406–10412.

Ambrosio, A. M. A., Allcock, H. R., Katti, D. S., and Laurencin, C. T. 2002. Degradable polyphosphazene/poly([alpha]-hydroxyester) blends: Degradation studies. *Biomaterials* 23 (7):1667–1672.

Ambrosio, A. M., Sahota, J. S., Runge, C. et al. 2003. Novel polyphosphazene-hydroxyapatite composites as biomaterials. *IEEE Eng Med Biol Mag* 22 (5):18–26.

Andrianov, A. K., DeCollibus, D. P., Gillis, H. A. et al. 2009. Poly[di(carboxylatophenoxy)phosphazene] is a potent adjuvant for intradermal immunization. *Proc Natl Acad Sci USA* 106 (45):18936–18941.

Andrianov, A. K., Jianping, C., and Payne, L. G. 1998. Preparation of hydrogel microspheres by coacervation of aqueous polyphosphazene solutions. *Biomaterials* 19 (1–3):109–115.

Andrianov, A. K. and Marin, A. 2006. Degradation of polyaminophosphazenes: Effects of hydrolytic environment and polymer processing. *Biomacromolecules* 7 (5):1581–1586.

Andrianov, A. K. and Payne, L. G. 1998. Protein release from polyphosphazene matrices. *Adv Drug Deliv Rev* 31 (3):185–196.

Barrett, E. W., Phelps, M. V. B., Silva, R. J., Gaumond, R. P., and Allcock, H. R. 2005. Patterning poly(organophosphazenes) for selective cell adhesion applications. *Biomacromolecules* 6 (3):1689–1697.

Bates, F. S. 1991. Polymer-polymer phase behavior. *Science* 251 (4996):898–905.

Bhattacharyya, S., Kumbar, S. G., Khan, Y. M. et al. 2009. Biodegradable polyphosphazene-nanohydroxyapatite composite nanofibers: Scaffolds for bone tissue engineering. *J Biomed Nanotechnol* 5 (1):69–75.

Bhattacharyya, S., Nair, L. S., Singh, A. et al. 2006. Electrospinning of poly[bis(ethyl alanato) phosphazene] nanofibers *J Biomed Nanotechnol* 2 (1):36–45.

Bi, Y. M., Gong, X. Y., Wang, W. Z. et al. 2010. Synthesis and characterization of new biodegradable thermo-sensitive polyphosphazenes with lactic acid ester and methoxyethoxyethoxy side groups. *Chin Chem Lett* 21 (2):237–241.

Borden, M., Attawia, M., Khan, Y., El-Amin, S. F., and Laurencin, C. T. 2004. Tissue-engineered bone formation in vivo using a novel sintered polymeric microsphere matrix. *J Bone Joint Surg Br* 86 (8):1200–1208.

Borden, M., Attawia, M., Khan, Y., and Laurencin, C. T. 2002a. Tissue engineered microsphere-based matrices for bone repair: Design and evaluation. *Biomaterials* 23 (2):551–559.

Borden, M., Attawia, M., and Laurencin, C. T. 2002b. The sintered microsphere matrix for bone tissue engi-neering: In vitro osteoconductivity studies. *J Biomed Mater Res* 61 (3):421–429.

Borden, M., El-Amin, S. F., Attawia, M., and Laurencin, C. T. 2003. Structural and human cellular assessment of a novel microsphere-based tissue engineered scaffold for bone repair. *Biomaterials* 24 (4):597–609.

Böstman, O. M. 1998. Osteoarthritis of the ankle after foreign-body reaction to absorbable pins and screws: A three- to nine-year follow-up study. *J Bone Joint Surg Br* 80 (2):333–338.

Böstman, O. and Pihlajamäki, H. 2000. Clinical biocompatibility of biodegradable orthopaedic implants for internal fixation: A review. *Biomaterials* 21 (24):2615–2621.

Braddock, M., Houston, P., Campbell, C., and Ashcroft, P. 2001. Born again bone: Tissue engineering for bone repair. *News Physiol Sci* 16 (5):208–213.

Brown, J. L., Nair, L. S., Bender, J., Allcock, H. R., and Laurencin, C. T. 2007. The formation of an apatite coating on carboxylated polyphosphazenes via a biomimetic process. *Mater Lett* 61 (17):3692–3695.

Brown, J. L., Nair, L. S., and Laurencin, C. T. 2008. Solvent/non-solvent sintering: A novel route to cre-ate porous microsphere scaffolds for tissue regeneration. *J Biomed Mater Res B Appl Biomater* 86B (2):396–406.

Brown, J. L., Peach, M. S., Nair, L. S., Kumbar, S. G., and Laurencin, C. T. 2010. Composite scaffolds: Bridging nanofiber and microsphere architectures to improve bioactivity of mechanically competent con-structs. *J Biomed Mater Res A* 95 (4):1150–1158. DOI: 10.1002/jbm.a.32934.

Caliceti, P., Veronese, F. M., and Lora, S. 2000. Polyphosphazene microspheres for insulin delivery. *Int J Pharm* 211 (1–2):57–65.

Carampin, P., Conconi, M. T., Lora, S. et al. 2007. Electrospun polyphosphazene nanofibers for in vitro rat endothelial cells proliferation. *J Biomed Mater Res A* 80A (3):661–668.

Chang, Y., Powell, E. S., and Allcock, H. R. 2005. Environmentally responsive micelles from polystyrene–poly[bis(potassium carboxylatophenoxy)phosphazene] block copolymers. *J Polym Sci Polym Chem* 43 (13):2912–2920.

Chang, Y., Powell, E. S., Allcock, H. R., Park, S. M., and Kim, C. 2003. Thermosensitive behavior of poly(ethylene oxide)-poly[bis(methoxyethoxyethoxy)-phosphazene] block copolymers. *Macromolecules* 36 (7):2568–2570.

Chang, Y., Prange, R., Allcock, H. R., Lee, S. C., and Kim, C. 2002. Amphiphilic poly[bis(trifluoroethoxy) phosphazene]-poly(ethylene oxide) block copolymers: Synthesis and micellar characteristics. *Macromolecules* 35 (22):8556–8559.

Chatani, Y. and Yatsuyanagi, K. 1987. Structural studies of poly(phosphazenes). 1. Molecular and crystal structures of poly(dichlorophosphazene). *Macromolecules* 20 (5):1042–1045.

Chaubal, M. V., Gupta, A. S., Lopina, S. T., and Bruley, D. F. 2003. Polyphosphates and other phosphorus-containing polymers for drug delivery applications. *Crit Rev Ther Drug Carrier Syst* 20 (4):295–315.

Chen, C.-C., Chueh, J.-Y., Tseng, H., Huang, H.-M., and Lee, S.-Y. 2003. Preparation and characterization of biodegradable PLA polymeric blends. *Biomaterials* 24 (7):1167–1173.

Chen, G. and Hoffman, A. S. 1995. Graft copolymers that exhibit temperature-induced phase transitions over a wide range of pH. *Nature* 373 (6509):49–52.

Chun, C., Lee, S. M., Kim, C. W. et al. 2009a. Doxorubicin-polyphosphazene conjugate hydrogels for locally controlled delivery of cancer therapeutics. *Biomaterials* 30 (27):4752–4762.

Chun, C., Lee, S. M., Kim, S. Y., Yang, H. K., and Song, S.-C. 2009b. Thermosensitive poly(organophosphazene)-paclitaxel conjugate gels for antitumor applications. *Biomaterials* 30 (12):2349–2360.

Chun, C., Lim, H. J., Hong, K.-Y., Park, K.-H., and Song, S.-C. 2009c. The use of injectable, thermosensitive poly(organophosphazene)-RGD conjugates for the enhancement of mesenchymal stem cell osteogenic differentiation. *Biomaterials* 30 (31):6295–6308.

Cohen, S., Bano, M. C., Cima, L. G. et al. 1993. Design of synthetic polymeric structures for cell transplantation and tissue engineering. *Clin Mater* 13 (1–4):3–10.

Coleman, M. M. and Painter, P. C. 1995. Hydrogen bonded polymer blends. *Prog Polym Sci* 20 (1):1–59.

Conconi, M. T., Carampin, P., Lora, S., Grandi, C., and Parnigotto, P. P. 2009. Electrospun polyphosphazene nanofibers for in vitro osteoblast culture. In *Polyphosphazenes for Biomedical Applications*, ed., A. Andrianov. Hoboken, NJ: Wiley-Interscience. pp. 169–184.

Conconi, M. T., Lora, S., Baiguera, S. et al. 2004. In vitro culture of rat neuromicrovascular endothelial cells on polymeric scaffolds. *J Biomed Mater Res A* 71A (4):669–674.

Conconi, M. T., Lora, S., Menti, A. M., Carampin, P., and Parnigotto, P. P. 2006. In vitro evaluation of poly[bis(ethyl alanato)phosphazene] as a scaffold for bone tissue engineering. *Tissue Eng* 12 (4):811–819.

Conforti, A., Bertani, S., Lussignoli, S. et al. 1996. Anti-inflammatory activity of polyphosphazene-based naproxen slow-release systems. *J Pharm Pharmacol* 48 (5):468–473.

Crommen, J. H., Schacht, E. H., and Mense, E. H. 1992a. Biodegradable polymers. I. Synthesis of hydrolysis-sensitive poly[(organo)phosphazenes]. *Biomaterials* 13 (8):511–520.

Crommen, J. H., Schacht, E. H., and Mense, E. H. 1992b. Biodegradable polymers. II. Degradation characteristics of hydrolysis-sensitive poly[(organo)phosphazenes]. *Biomaterials* 13 (9):601–611.

Crommen, J., Vandorpe, J., and Schacht, E. 1993. Degradable polyphosphazenes for biomedical applications. *J Control Release* 24 (1–3):167–180.

Cui, Y., Zhao, X., Tang, X., and Luo, Y. 2004. Novel micro-crosslinked poly(organophosphazenes) with improved mechanical properties and controllable degradation rate as potential biodegradable matrix. *Biomaterials* 25 (3):451–457.

De Jaeger, R., Helioui, M., and Puscaric, E. 1983. Novel polychlorophosphazenes and process for their preparation. U.S. Patent 4377558.

Deng, M. 2010. Novel biocompatible polymeric blends for bone regeneration: Material and matrix design and development, PhD dissertation, Department of Chemical Engineering, University of Virginia, Charlottesville, VA.

Deng, M., Kumbar, S. G., Wan, Y. et al. 2010a. Polyphosphazene polymers for tissue engineering: An analysis of material synthesis, characterization and applications. *Soft Matter* 6:3119–3132.

Deng, M., Nair, L. S., Krogman, N. R., Allcock, H. R., and Laurencin, C. T. 2009. Biodegradable polyphosphazene blends for biomedical applications. In *Polyphosphazenes for Biomedical Applications*, ed., A. Andrianov. Hoboken, NJ: Wiley-Interscience. pp. 139–154.

Deng, M., Nair, L. S., Nukavarapu, S. P. et al. 2008. Miscibility and in vitro osteocompatibility of biodegradable blends of poly[(ethyl alanato) (p-phenyl phenoxy) phosphazene] and poly(lactic acid-glycolic acid). *Biomaterials* 29 (3):337–349.

Deng, M., Nair, L. S., Nukavarapu, S. P. et al. 2010b. Dipeptide-based polyphosphazene and polyester blends for bone tissue engineering. *Biomaterials* 31 (18):4898–4908.

Deng, M., Nair, L. S., Nukavarapu, S. P. et al. 2010c. Biomimetic, bioactive etheric polyphosphazene-poly(lactide-co-glycolide) blends for bone tissue engineering. *J Biomed Mater Res A* 92A (1):114–125.

Deng, M., Nair, L. S., Nukavarapu, S. P. et al. 2010d. In situ porous structures: A unique polymer erosion mechanism in biodegradable dipeptide-based polyphosphazene and polyester blends producing matrices for regenerative engineering. *Adv Funct Mater* 20 (17):2794–2806.

Devadoss, E. and Nair, C. P. R. 1985. A novel cyclomatrix polymer based on (hydroxy phenoxy) phosphazenes and 2-methyl aziridine: Some aspects of synthesis and adhesive heat resistance. *Polymer* 26 (12):1895–1900.

El-Amin, S. F., Kwon, M. S., Starnes, T., Allcock, H. R., and Laurencin, C. T. 2007. The biocompatibility of biodegradable glycine containing polyphosphazenes: A comparative study in bone. *J Inorg Organomet Polym Mater* 16 (4):387–396.

Fox, T. 1956. Influence of diluent and of copolymer composition on the glass temperature of a polymer system. *Bull Am Phys Soc* 1:123.

Franz, U., Nuyken, O., and Matyjaszewski, K. 1994. Synthesis and characterization of poly(phenyl-p-tolylphosphazene), prepared via in situ polymerization of phenyl-p-tolylphosphine azide. *Macromol Rapid Commun* 15 (2):169–174.

Fu, K., Pack, D. W., Klibanov, A. M., and Langer, R. 2000. Visual evidence of acidic environment within degrading poly(lactic-co-glycolic acid) (PLGA) microspheres. *Pharm Res* 17 (1):100–106.

Ghattas, D. and Leroux, J. 2009. Amphiphilic ionizable polyphosphazenes for the preparation of pH-responsive liposomes. In *Polyphosphazenes for Biomedical Applications*, ed., A. Andrianov. Hoboken, NJ: Wiley-Interscience. pp. 225–247.

Glatter, O., Scherf, G., Schillen, K., and Brown, W. 1994. Characterization of a poly(ethylene oxide)-poly(propylene oxide) triblock copolymer (EO27-PO39-EO27) in aqueous solution. *Macromolecules* 27 (21):6046–6054.

Gleria, M. and De Jaeger, R. 2004. *Phosphazenes: A Worldwide Insight*. New York: Nova Science Publishers.

Greish, Y. E., Bender, J. D., Lakshmi, S. et al. 2005a. Composite formation from hydroxyapatite with sodium and potassium salts of polyphosphazene. *J Mater Sci Mater Med* 16 (7):613–620.

Greish, Y. E., Bender, J. D., Lakshmi, S. et al. 2005b. Low temperature formation of hydroxyapatite-poly(alkyl oxybenzoate)phosphazene composites for biomedical applications. *Biomaterials* 26 (1):1–9.

Greish, Y. E., Bender, J. D., Lakshmi, S. et al. 2006. Formation of hydroxyapatite-polyphosphazene polymer composites at physiologic temperature. *J Biomed Mater Res A* 77 (2):416–425.

Greish, Y. E., Sturgeon, J. L., Singh, A. et al. 2008. Formation and properties of composites comprised of calcium-deficient hydroxyapatites and ethyl alanate polyphosphazenes. *J Mater Sci Mater Med* 19 (9):3153–3160.

Grolleman, C. W. J., de Visser, A. C., Wolke, J. G. C., van der Goot, H., and Timmerman, H. 1986. Studies on a bioerodible drug carrier system based on polyphosphazene Part I. Synthesis. *J Control Release* 3 (1–4):143–154.

Gümüsdereliolu, M. and Gür, A. 2002. Synthesis, characterization, in vitro degradation and cytotoxicity of poly[bis(ethyl 4-aminobutyro)phosphazene]. *React Funct Polym* 52 (2):71–80.

Gunatillake, P. A. and Adhikari, R. 2003. Biodegradable synthetic polymers for tissue engineering. *Eur Cell Mater* 5:1–16.

He, Y., Zhu, B., and Inoue, Y. 2004. Hydrogen bonds in polymer blends. *Prog Polym Sci* 29 (10):1021–1051.

Helioui, M., De Jaeger, R., Puskaric, E., and Heubel, J. 1982. Nouvelle préparation de polychlorophosphazènes linéaires. *Die Makromol. Chem.* 183 (5):1137–1143.

Honeyman, C. H., Manners, I., Morrissey, C. T., and Allcock, H. R. 1995. Ambient temperature synthesis of poly(dichlorophosphazene) with molecular weight control. *J Am Chem Soc* 117 (26):7035–7036.

Huang, D., Balian, G., and Chhabra, A. B. 2006. Tendon tissue engineering and gene transfer: The future of surgical treatment. *J Hand Surg* 31 (5):693–704.

Huang, Z. M., Zhang, Y. Z., Kotaki, M., and Ramakrishna, S. 2003. A review on polymer nanofibers by electrospinning and their applications in nanocomposites. *Compos Sci Technol* 63:2223–2253.

Ibim, S. E. M., Ambrosio, A. M. A., Kwon, M. S. et al. 1997. Novel polyphosphazene/poly(lactide-co-glycolide) blends: Miscibility and degradation studies. *Biomaterials* 18 (23):1565–1569.

Ibim, S. M., Ambrosio, A. A., Larrier, D., Allcock, H. R., and Laurencin, C. T. 1996. Controlled macromolecule release from poly(phosphazene) matrices. *J Control Release* 40 (1–2):31–39.

Ibim, S. M., El-Amin, S. F., Goad, M. E. P. et al. 1998. In vitro release of colchicine using poly(phosphazenes): The development of delivery systems for musculoskeletal use. *Pharmaceut Dev Tech* 3 (1):55–62.

Ikada, Y. and Tsuji, H. 2000. Biodegradable polyesters for medical and ecological applications. *Macromol Rapid Commun* 21 (3):117–132.

Ilia, G. 2009. Phosphorus containing hydrogels. *Polym Adv Technol* 20 (9):707–722.

Iwasaki, Y., Sawada, S.-I., Ishihara, K., Khang, G., and Lee, H. B. 2002. Reduction of surface-induced inflammatory reaction on PLGA/MPC polymer blend. *Biomaterials* 23 (18):3897–3903.

Jeong, B., Bae, Y. H., and Kim, S. W. 1999. Thermoreversible gelation of PEG-PLGA-PEG triblock copolymer aqueous solutions. *Macromolecules* 32 (21):7064–7069.

Jun, Y. J., Kim, J. I., Jun, M. J., and Sohn, Y. S. 2005. Selective tumor targeting by enhanced permeability and retention effect. Synthesis and antitumor activity of polyphosphazene-platinum (II) conjugates. *J Inorg Biochem* 99 (8):1593–1601.

Kang, G. D., Heo, J. Y., Jung, S. B., and Song, S. C. 2005. Controlling the thermosensitive gelation properties of poly(organophosphazenes) by blending. *Macromol Rapid Commun* 26 (20):1615–1618.

Katti, D. and Laurencin, C. T. 2003. Synthetic biomedical polymers for tissue engineering and drug delivery. In *Advanced Polymeric Materials: Structure Property Relationships*, eds., G. O. Shonaike and S. G. Advani. Boca Raton, FL: CRC Press. pp. 479–525.

Kokubo, T. and Takadama, H. 2006. How useful is SBF in predicting in vivo bone bioactivity? *Biomaterials* 27 (15):2907–2915.

Kricheldorf, H. R., Nuyken, O., and Swift, G. 2004. *Handbook of Polymer Synthesis: Second Edition (Plastics Engineering)*, 2nd edn. Boca Raton, FL: CRC Press.

Krogman, N. R., Hindenlang, M. D., Nair, L. S., Laurencin, C. T., and Allcock, H. R. 2008a. Synthesis of purine- and pyrimidine-containing polyphosphazenes: Physical properties and hydrolytic behavior. *Macromolecules* 41 (22):8467–8472.

Krogman, N. R., Singh, A., Nair, L. S., Laurencin, C. T., and Allcock, H. R. 2007. Miscibility of bioerodible polyphosphazene/poly(lactide-co-glycolide) blends. *Biomacromolecules* 8 (4):1306–1312.

Krogman, N. R., Steely, L., Hindenlang, M. D. et al. 2008b. Synthesis and characterization of polyphosphazene-block-polyester and polyphosphazene-block-polycarbonate macromolecules. *Macromolecules* 41 (4):1126–1130.

Krogman, N. R., Weikel, A. L., Kristhart, K. A. et al. 2009. The influence of side group modification in polyphosphazenes on hydrolysis and cell adhesion of blends with PLGA. *Biomaterials* 30 (17):3035–3041.

Krogman, N. R., Weikel, A. L., Nguyen, N. Q. et al. 2008c. Synthesis and characterization of new biomedical polymers: Serine- and threonine-containing polyphosphazenes and poly(l-lactic acid) grafted copolymers. *Macromolecules* 41 (21):7824–7828.

Krogman, N. R., Weikel, A. L., Nguyen, N. Q. et al. 2010. Hydrogen bonding in blends of polyesters with dipeptide-containing polyphosphazenes. *J Appl Polym Sci* 115 (1):431–437.

Kumbar, S. G., Bhattacharyya, S., Nukavarapu, S. P. et al. 2006. In vitro and in vivo characterization of biodegradable poly(organophosphazenes) for biomedical applications. *J Inorg Organomet Polym Mater* 16 (4):365–385.

Kumbar, S. G., Bhattacharyya, S., Sethuraman, S., and Laurencin, C. T. 2007. A preliminary report on a novel electrospray technique for nanoparticle based biomedical implants coating: Precision electrospraying. *J Biomed Mater Res B Appl Biomater.* 81B (1):91–103.

Kumbar, S. G., James, R., Nukavarapu, S. P., and Laurencin, C. T. 2008. Electrospun nanofiber scaffolds: Engineering soft tissues. *Biomed Mater* 3 (3):034002.

Landes, C. A., Ballon, A., and Roth, C. 2006. Maxillary and mandibular osteosyntheses with PLGA and P(L/DL)LA implants: A 5-year inpatient biocompatibility and degradation experience. *Plast Reconstr Surg* 117 (7):2347–2360.

Langer, R. 2000. Tissue engineering. *Mol Ther* 1 (1):12–15.

Langer, R. and Peppas, N. A. 2003. Advances in biomaterials, drug delivery, and bionanotechnology. *AIChE J* 49 (12):2990–3006.

Langer, R. and Vacanti, J. P. 1993. Tissue engineering. *Science* 260 (5110):920–926.

Langone, F., Lora, S., Veronese, F. M. et al. 1995. Peripheral nerve repair using a poly(organo)phosphazene tubular prosthesis. *Biomaterials* 16 (5):347–353.

Laurencin, C. T., Ambrosio, A. M. A., Bauer, T. W. et al. 1998. The biocompatibility of polyphophazenes. Evaluation in bone. Paper read at *Society for Biomaterials. 24th Annual Meeting in Conjunction with 30th International Symposium*, San Diego, CA. p. 436.

Laurencin, C. T., Ambrosio, A. M. A., Borden, M. D., and Cooper, J. A. 1999. Tissue engineering: Orthopedic applications. *Annu Rev Biomed Eng* 1 (1):19–46.

Laurencin, C. T., El-Amin, S. F., Ibim, S. E. et al. 1996. A highly porous 3-dimensional polyphosphazene polymer matrix for skeletal tissue regeneration. *J Biomed Mater Res* 30 (2):133–138.

Laurencin, C. T., Koh, H. J., Neenan, T. X., Allcock, H. R., and Langer, R. 1987. Controlled release using a new bioerodible polyphosphazene matrix system. *J Biomed Mater Res* 21 (10):1231–1246.

Laurencin, C. T., Morris, C. D., Pierre-Jacques, H. et al. 1992. Osteoblast culture on bioerodible polymers: Studies of initial cell adhesion and spread. *Polym Adv Tech* 3 (6):359–364.

Laurencin, C. T., Norman, M. E., Elgendy, H. M. et al. 1993. Use of polyphosphazenes for skeletal tissue regeneration. *J Biomed Mater Res* 27 (7):963–973.

Lee, B. H., Lee, Y. M., Sohn, Y. S., and Song, S.-C. 2002. A thermosensitive poly(organophosphazene) gel. *Macromolecules* 35 (10):3876–3879.

Lee, K. Y. and Mooney, D. J. 2001. Hydrogels for tissue engineering. *Chem Rev* 101 (7):1869–1880.

Lee, S. B. and Song, S.-C. 2005. Hydrolysis-improved thermosensitive polyorganophosphazenes with alpha-amino-omega-methoxy-poly(ethylene glycol) and amino acid esters as side groups. *Polym Int* 54 (9):1225–1232.

Lemmouchi, Y., Schacht, E., and Dejardin, S. 1998. Biodegradable poly[(amino acid ester)phosphazenes] for biomedical applications. *J Bioact Compat Polym* 13 (1):4–18.

Li, W. J., Laurencin, C. T., Caterson, E. J., Tuan, R. S., and Ko, F. K. 2002. Electrospun nanofibrous structure: A novel scaffold for tissue engineering. *J Biomed Mater Res* 60 (4):613–621.

Li, Z., Li, J., and Qin, J. 2001. Synthesis of polyphosphazenes as potential photorefractive materials. *React Funct Polym* 48 (1–3):113–118.

Li, D. and Xia, Y. 2004. Electrospinning of nanofibers: Reinventing the wheel? *Adv Mater* 16 (14):1151–1170.

Lin, Y.-J., Cai, Q., Li, L. et al. 2010. Co-electrospun composite nanofibers of blends of polyamino acid ester-phosphazene and gelatin. *Polym Int* 59:610–616.

Ma, P. X. 2008. Biomimetic materials for tissue engineering. *Adv Drug Deliv Rev* 60 (2):184–198.

Ma, P. X. and Zhang, R. 1999. Synthetic nano-scale fibrous extracellular matrix. *J Biomed Mater Res* 46 (1):60–72.

Martin, C. R. 1996. Membrane-based synthesis of nanomaterials. *Chem Mater* 8 (8):1739–1746.

Mi, F.-L., Shyu, S.-S., Lin, Y.-M. et al. 2003. Chitin/PLGA blend microspheres as a biodegradable drug delivery system: A new delivery system for protein. *Biomaterials* 24 (27):5023–5036.

Montague, R. A. and Matyjaszewski, K. 1990. Synthesis of poly[bis(trifluoroethoxy)phosphazene] under mild conditions using a fluoride initiator. *J Am Chem Soc* 112 (18):6721–6723.

Nair, L. S., Allcock, H. R., and Laurencin, C. T. 2005. Biodegradable poly[bis(ethyl alanato)phosphazene]-poly(lactide-co-glycolide) blends: Miscibility and osteocompatibility evaluations. Paper read at *Materials Research Society Symposium*, Boston, MA. 844:Y9.7.1–Y9.7.7.

Nair, L. S., Bhattacharyya, S., Bender, J. D. et al. 2004. Fabrication and optimization of methylphenoxy substituted polyphosphazene nanofibers for biomedical applications. *Biomacromolecules* 5 (6):2212–2220.

Nair, L. S., Katti, D. S., and Laurencin, C. T. 2003. Biodegradable polyphosphazenes for drug delivery applications. *Adv Drug Deliv Rev* 55 (4):467–482.

Nair, L. S. and Laurencin, C. T. 2007. Biodegradable polymers as biomaterials. *Progr Polym Sci* 32 (8–9):762–798.

Nair, L. S. and Laurencin, C. T. 2008. Nanofibers and nanoparticles for orthopaedic surgery applications. *J Bone Joint Surg Am* 90 (Suppl. 1):128–131.

Nair, L. S., Lee, D. A., Bender, J. D. et al. 2006. Synthesis, characterization, and osteocompatibility evaluation of novel alanine-based polyphosphazenes. *J Biomed Mater Res A* 76 (1):206–213.

Neilson, R. H. and Wisian-Neilson, P. 1988. Poly(alkyl/arylphosphazenes) and their precursors. *Chem Rev* 88 (3):541–562.

Nelson, J. M. and Allcock, H. R. 1997. Synthesis of triarmed-star polyphosphazenes via the "living" cationic polymerization of phosphoranimines at ambient temperatures. *Macromolecules* 30 (6):1854–1856.

Nukavarapu, S. P., Kumbar, S. G., Allcock, H. R., and Laurencin, C. T. 2009. Biodegradable polyphosphazene scaffolds for tissue engineering. In *Polyphosphazenes for Biomedical Applications*, ed., A. Andrianov. Hoboken, NJ: Wiley-Interscience. pp. 117–138.

Nukavarapu, S. P., Kumbar, S. G., Brown, J. L. et al. 2008. Polyphosphazene/nano-hydroxyapatite composite microsphere scaffolds for bone tissue engineering. *Biomacromolecules* 9 (7):1818–1825.

Odian, G. 2004. *Principles of Polymerization*, 4th edn., Hoboken, NJ: John Wiley & Sons, Inc.

Ondarçuhu, T. and Joachim, C. 1998. Drawing a single nanofibre over hundreds of microns. *Europhys Lett* 42 (2):215.

Oredein-McCoy, O., Krogman, N. R., Weikel, A. L. et al. 2009. Novel factor-loaded polyphosphazene matrices: Potential for driving angiogenesis. *J Microencapsul* 26 (6):544–555.

Park, T. G., Cohen, S., and Langer, R. 1992. Poly(L-lactic acid)/pluronic blends: Characterization of phase separation behavior, degradation, and morphology and use as protein-releasing matrixes. *Macromolecules* 25 (1):116–122.

Payne, L. G., Jenkins, S. A., Andrianov, A., and Roberts, B. E. 1995. Water-soluble phosphazene polymers for parenteral and mucosal vaccine delivery. *Pharm Biotechnol* 6:473–493.

Peppas, N. A., Huang, Y., Torres-Lugo, M., Ward, J. H., and Zhang, J. 2003. Physicochemical foundations and structural design of hydrogels in medicine and biology. *Annu Rev Biomed Eng* 2 (1):9–29.

Pitt, G. G., Cha, Y., Shah, S. S., and Zhu, K. J. 1992. Blends of PVA and PGLA: Control of the permeability and degradability of hydrogels by blending. *J Control Release* 19 (1–3):189–199.

Potin, P. and De Jaeger, R. 1991. Polyphosphazenes: Synthesis, structures, properties, applications. *Eur Polym J* 27 (4–5):341–348.

Potta, T., Chun, C., and Song, S.-C. 2010. Injectable, dual cross-linkable polyphosphazene blend hydrogels. *Biomaterials* 31 (32):8107–8120.

Prange, R. and Allcock, H. R. 1999. Telechelic syntheses of the first phosphazene siloxane block copolymers. *Macromolecules* 32 (19):6390–6392.

Qiu, L. Y. 2002a. Degradation and tissue compatibility of polyphosphazene blend films in vivo. *Sheng Wu Yi Xue Gong Cheng Xue Za Zhi* 19 (2):191–195.

Qiu, L. Y. 2002b. In vitro and in vivo degradation study on novel blends composed of polyphosphazene and polyester or polyanhydride. *Polym Int* 51:481–487.

Qiu, L. and Zheng, C. 2009. Amphiphilic polyphosphazenes as drug carriers. In *Polyphosphazenes for Biomedical Applications*, ed., A. Andrianov. Hoboken, NJ: Wiley-Interscience. pp. 277–295.

Qiu, L. Y. and Zhu, K. J. 2000a. Novel biodegradable polyphosphazenes containing glycine ethyl ester and benzyl ester of amino acethydroxamic acid as cosubstituents. *J Appl Polym Sci* 77 (13):2987–2995.

Qiu, L. Y. and Zhu, K. J. 2000b. Novel blends of poly[bis(glycine ethyl ester) phosphazene] and polyesters or polyanhydrides: Compatibility and degradation characteristics *in vitro*. *Polym Int* 49 (11):1283–1288.

Ratner, B. D. and Bryant, S. J. 2004. Biomaterials: Where we have been and where we are going. *Annu Rev Biomed Eng* 6 (1):41–75.

Rho, J. -Y., Kuhn-Spearing, L., and Zioupos, P. 1998. Mechanical properties and the hierarchical structure of bone. *Med Eng Phys* 20 (2):92–102.

Schacht, E., Vandorpe, J., Dejardin, S., Lemmouchi, Y., and Seymour, L. 1996. Biomedical applications of degradable polyphosphazenes. *Biotechnol Bioeng* 52 (1):102–108.

Schacht, E., Vandorpe, J., Lemmouchi, Y., Dejardin, S., and Seymour, L. 1998. Degradable polyphosphazenes for biomedical applications. In *Frontiers in Biomedical Polymer Applications*, ed., R. Ottembrite. Lancaster, U.K.: Technomic. pp. 27–42.

Sethuraman, S., Nair, L. S., El-Amin, S. et al. 2006. In vivo biodegradability and biocompatibility evaluation of novel alanine ester based polyphosphazenes in a rat model. *J Biomed Mater Res A* 77 (4):679–687.

Sethuraman, S., Nair, L. S., El-Amin, S. et al. 2010. Mechanical properties and osteocompatibility of novel biodegradable alanine based polyphosphazenes: Side group effects. *Acta Biomaterialia* 6 (6):1931–1937.

Sethuraman, S., Nair, L. S., Singh, A. et al. 2004. Synthesis and evaluation of novel amino acid ester phenyl phenoxy polyphosphazene for bone tissue engineering. Paper read at *Transactions of the Fifth Combined Meeting of the Orthopaedic Research Societies of Canada, USA, Japan, and Europe*. Calgary, Alberta, Canada.

Singh, A., Krogman, N. R., Sethuraman, S. et al. 2006. Effect of side group chemistry on the properties of biodegradable L-alanine cosubstituted polyphosphazenes. *Biomacromolecules* 7 (3):914–918.

Sulkowski, W., Sulkowska, A., and Kireev, V. 1997. Synthesis and spectroscopic studies of polyorganophosphazenes containing binaphthalene groups. *J Mol Struct* 410–411:241–244.

Taylor, M. S., Daniels, A. U., Andriano, K. P., and Heller, J. 1994. Six bioabsorbable polymers: In vitro acute toxicity of accumulated degradation products. *J Appl Biomater* 5 (2):151–157.

Tsuji, H. 2003. In vitro hydrolysis of blends from enantiomeric poly(lactide)s. Part 4: Well-homo-crystallized blend and nonblended films. *Biomaterials* 24 (4):537–547.

Utracki, L. A. 1989. *Polymer Alloys and Blends Thermodynamics and Rheology*. Munich, Germany: Hanser.

Vacanti, J. P. 1999. Tissue engineering: The design and fabrication of living replacement devices for surgical reconstruction and transplantation. *Lancet* 354:S32.

Veronese, F. M., Marsilio, F., Caliceti, P. et al. 1998. Polyorganophosphazene microspheres for drug release: Polymer synthesis, microsphere preparation, in vitro and in vivo naproxen release. *J Control Release* 52 (3):227–237.

Veronese, F. M., Marsilio, F., Lora, S. et al. 1999. Polyphosphazene membranes and microspheres in periodontal diseases and implant surgery. *Biomaterials* 20 (1):91–98.

Wade, C. W. R., Gourlay, S., Rice, R. et al. 1978. Biocompatibility of eight poly(organophosphazenes). In *Organometallic Polymers*, eds., C. E. Carraher, J. E. Sheats and C. U. Pittman. New York: Academic. pp. 283–288.

Weikel, A. L., Cho, S. Y., Morozowich, N. L. et al. 2010a. Hydrolysable polylactide-polyphosphazene block copolymers for biomedical applications: Synthesis, characterization, and composites with poly(lactic-co-glycolic acid). *Polym Chem* 1 (9):1459–1466.

Weikel, A. L., Krogman, N. R., Nguyen, N. Q. et al. 2009. Polyphosphazenes that contain dipeptide side groups: Synthesis, characterization, and sensitivity to hydrolysis. *Macromolecules* 42 (3):636–639.

Weikel, A. L., Owens, S. G., Morozowich, N. L. et al. 2010b. Miscibility of choline-substituted polyphosphazenes with PLGA and osteoblast activity on resulting blends. *Biomaterials* 31 (33):8507–8515.

Welle, A., Grunze, M., and Tur, D. 2000. Blood compatibility of poly[bis(trifluoroethoxy)phosphazene]. *J Appl Med Polym* 4 (1):6–10.

Whitesides, G. M. and Boncheva, M. 2002. Beyond molecules: Self-assembly of mesoscopic and macroscopic components. *Proc Natl Acad Sci USA* 99 (8):4769–4774.

Whitesides, G. M. and Grzybowski, B. 2002. Self-assembly at all scales. *Science* 295 (5564):2418–2421.

Wood, L. A. 1958. Glass transition temperatures of copolymers. *J Polym Sci* 28 (117):319–330.

Yang, L. and Alexandridis, P. 2000. Physicochemical aspects of drug delivery and release from polymer-based colloids. *Curr Opin Colloid Interface Sci* 5 (1–2):132–143.

Yuan, W., Song, Q., Zhu, L. et al. 2005. Asymmetric penta-armed poly(e-caprolactone)s with short-chain phosphazene core: Synthesis, characterization, and in vitro degradation. *Polym Int* 54 (9):1262–1267.

Zentner, G. M., Rathi, R., Shih, C. et al. 2001. Biodegradable block copolymers for delivery of proteins and water-insoluble drugs. *J Control Release* 72 (1–3):203–215.

Zhang, T., Cai, Q., Wu, D. Z., and Jin, R. G. 2005. Phosphazene cyclomatrix network polymers: Some aspects of the synthesis, characterization, and flame-retardant mechanisms of polymer. *J Appl Polym Sci* 95 (4):880–889.

Zhang, J. X., Qiu, L. Y., Jin, Y., and Zhu, K. J. 2006. Multimorphological self-assemblies of amphiphilic graft polyphosphazenes with oligopoly(N-isopropylacrylamide) and ethyl 4-aminobenzoate as side groups. *Macromolecules* 39 (1):451–455.

Zhang, Q.-S., Yan, Y.-H., Li, S.-P., and Feng, T. 2009. Synthesis of a novel biodegradable and electroactive polyphosphazene for biomedical application. *Biomed Mater* 4 (3):035008.

Zhu, G., Mallery, S. R., and Schwendeman, S. P. 2000. Stabilization of proteins encapsulated in injectable poly (lactide-co-glycolide). *Nat Biotechnol* 18 (1):52–57.

5 Biodegradable Polymers as Drug Carrier Systems

Abraham J. Domb and Wahid Khan

CONTENTS

5.1 INTRODUCTION

In recent years, there has been a rapid growth in the area of drug discovery, facilitated by novel technologies such as combinatorial chemistry and high-throughput screening. These novel approaches have led to drugs, which are generally more potent and have poorer solubility than the drugs developed from traditional approaches of medicinal chemistry. The development of these complex drugs has triggered a more urgent focus on developing novel techniques to deliver these drugs more effectively and efficiently (Molinari 2009; Shanmugasundaram and Rigby 2009; Yi Mok et al. 2009). Secondly, there is a continuous advance in biotechnology and drug development which produces more pharmaceutically active agents that will be difficult to administer by conventional means (Shire 2009). Thus, there is an increased demand for controlled or site-specific delivery system, which thereby needs a material, i.e., biomaterial, to deliver the drug and to perform its intended work.

Over the past two decades, the use of polymeric materials has increased dramatically for biomedical applications. The most important biomedical applications of biodegradable polymers are in the areas of controlled drug delivery systems (Zheng et al. 2009; Bobo and Shelton 2010) and in the form of implants and devices for bone and cartilage repair (Jung et al. 2008; Puppi et al. 2010), ligament reconstruction (Ge et al. 2005; Sahoo et al. 2006, 2007; Frosch et al. 2009), surgical dressings (Izumi et al. 2007; Xie et al. 2008), dental repairs (El-Backly et al. 2008), artificial heart valves, contact lenses, cardiac pacemakers, vascular and bone grafts (Iwai et al. 2005; Shen et al. 2006; Ngiam et al. 2009), tracheal replacements (Villegas-Cabello et al. 1994; Kaschke et al. 1996), and organ regeneration (Seckel et al. 1984; Heng et al. 2005).

The purpose of this chapter is to review the chemistry, properties, and formulation procedures of the different biodegradable polymers available, and to describe some of their release kinetics, safety, and biocompatibility considerations related to these polymers.

5.2 BIODEGRADABLE POLYMERS

5.2.1 POLYESTERS

Polyesters represent a class of polymers with wide applications owing to their immense diversity and synthetic versatility. This class includes poly(α-hydroxy acids) and other ester polymers with or without oxygen atom adjacent to the α-carbon of the acid moiety (Farng and Sherman 2004). Among polyesters, the most extensively investigated polymers are the poly(α-hydroxy acids), which include poly(glycolic acid) (PGA), poly(lactic acid) (PLA), and their copolymer poly(lactide-co-glycolide) (PLGA; Puppi et al. 2010). Several other aliphatic polyesters are now attracting significant attention as biomaterials due to their good biocompatibility and controllable degradation profiles.

5.2.1.1 Lactide/Glycolide and Their Copolymers

These are thermoplastic, aliphatic polyesters having hydrolytically labile aliphatic ester linkages in their backbone and hence are biodegradable. Linear polyesters of lactide and glycolide have been used for more than four decades for a variety of medical applications. Extensive research has been devoted to the use of these polymers as carriers for controlled drug delivery of a wide range of bioactive agents for human and animal use. They have been used for the delivery of steroids (Ju et al. 2009), anticancer agents (Danhier et al. 2009; Mukerjee and Vishwanatha 2009; Park et al. 2009), peptides and proteins (Santander-Ortega et al. 2009), antimicrobial agents (Kim et al. 2004; Xue et al. 2006; Nahar and Jain 2009), anesthetics, growth factors (Wenk et al. 2009), vaccines (Quintilio et al. 2009), and in tissue engineering (Zhang et al. 2009). Injectable formulations containing microspheres of lactide/glycolide polymers have received the most attention and many products (Table 5.1) are there in market.

TABLE 5.1
Commercially Available Biodegradable Polymeric Systems

Type of Polymer	Name of the Product	Dosage Form	Active Ingredients	Reference
1. Polyesters				
Lactide and glycolide copolymers	Lupron depot	Microspheres	Leoprolide	Hild et al. (2001), Wheeler et al. (1993), Dlugi et al. (1990), Kappy et al. (1989)
	Nutropin depot	Microspheres	Somatropin	Kemp et al. (2004), Silverman et al. (2002), Cook et al. (2002)
	Risperdal consta	Microspheres	Triptorelin	Gefvert et al. (2005), Bobo and Shelton (2010)
Polycaprolactones	Capronor	Subdermal capsule	Levonorgestrel	Anonymous (1991), Darney et al. (1989), Anonymous (1985), Ory et al. (1983)
	SynBiosys™	Drug delivery system	Peptides, antibiotics	Steendam et al. (2006), Gillissen et al. (2006)
Polycarbonates	Maxon™, BioSyn™	Absorbable sutures	—	Sonmez et al. (2009), Rodeheaver et al. (1996)
2. Poly(amides)				
3. Poly(phosphate esters)	Paclimer®	Microspheres	Paclitaxel	Pradilla et al. (2006), Armstrong et al. (2006), Lapidus et al. (2004), Harper et al. (1999)
4. Polyphosphazenes	—	Micelles	Indomethacin	Zhang et al. (2007)
5. Poly(orthoesters)	Alzamer®	Implant	—	Pinholt et al. (1991), Heller (1990)
6. Polyanhydrides	Gliadel®	Wafer implant	Carmustine	Kleinberg et al. (2004), Whittle et al. (2003)

5.2.1.1.1 Synthesis

Aliphatic polyesters can be synthesized by two mechanisms: (1) the step-growth polymerization (polycondensation) and (2) the ring-opening polymerization(ROP) (polyaddition). When polyesters are synthesized by the polycondensation, polymerization technique relies on the condensation of hydroxyl acids or of the mixtures of diacids and diols. But the major drawbacks of polycondensation mechanism are the high temperatures and long reaction time, and it is difficult to achieve high molecular weight polymers by this route (Edlund and Albertsson 2003; Jerome and Lecomte 2008).

ROP of cyclic lactones has been developed as the most effective polymerization route to develop high molecular weight homo- and co-polyesters. This process is free of polycondensation limitations and is thus preferred. Linear lactide and glycolide-based polymers are most commonly synthesized by ROP (Scheme 5.1) of lactide and glycolide by heating at 140°C–180°C for 2–10 h using catalyst/ initiators. A broad range of anionic, cationic, and coordinative initiators or catalysts have been reported for the ROP (Dittrich and Schultz 1971; Nijenhuis et al. 1992; Urakami and Guan 2008).

Tin octoate ($Sn(OCT)_2$) and alkoxides were the most widely used organometallic mediators for the ROP of lactones even if novel powerful metal-free catalytic systems are emerging.

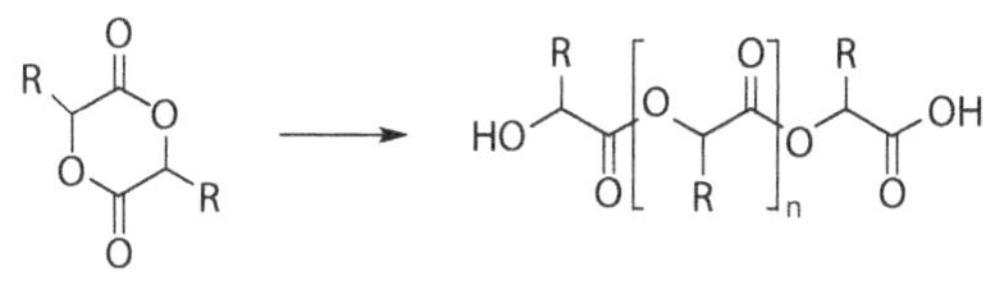

SCHEME 5.1 Scheme for ROP of lactide/glycolide.

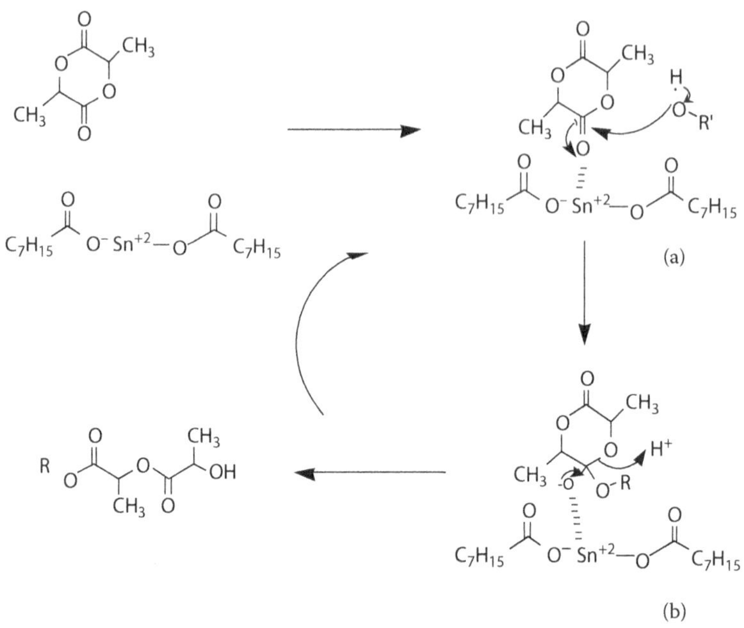

SCHEME 5.2 Hypothetical mechanism of ROP using tin octoate as catalyst. (a) Activated ester-carbonyl group in the dilactone, and (b) unstable intermediate which opens to the ester alcohol. (From Kissel, T. et al., *J. Control Release*, 16(1–2), 27, 1991.)

Yet, the common polymerization catalysts are tin derivatives such as tin octoate or tin hexanoate. A hypothetical mechanism of the ROP of lactide using a tin catalyst was suggested by Kissel et al. (1991) (Scheme 5.2).

In this mechanism, the Lewis acid character of the tin catalyst activates the ester-carbonyl group in the dilactone (a, in Scheme 5.2). The activated species reacts with the alcohol initiator to form an unstable intermediate which opens to the ester alcohol (b, in Scheme 5.2). The propagation reaction proceeds by tin catalyst activation of another dilactone carbonyl group and reaction with the hydroxyl end group. Tin catalyst activation of a carbonyl ester in the polymer chain results is an undesirable depolymerization. The hydroxyl initiator is incorporated in the polymer at its carbonyl end group. Lauryl alcohol is generally added to control the molecular weight (Dittrich and Schultz 1971; Nijenhuis et al. 1992). Polymers with a molecular weight as high as 500,000 can be obtained by the melt process when high-purity monomers (>99.9%) are used. Since Sn(OCT)$_2$ promoted polymerization are hardly controlled, a variety of other organometallic derivatives, particularly metal alkoxides, are continuously tested as initiators in the polymerization of lactides. The aluminum alkoxide initiators are the most versatile and readily available. The high selectivity of aluminum alkoxides is the major reason for their choice to produce well-defined polyesters and molecular weight is very well controlled. The very good control imparted to ROP by aluminum alkoxides is a unique platform for the macromolecular engineering of aliphatic polyesters, for instance making comb-like, star-shaped, graft, and hyper-branched polyesters. Both monoalkoxides (R$_2$-AlOR) and trialkoxides (Al(OR)$_3$) were used. By using only one form of the trialkoxide isopropyloxode (Al(O-iPr)$_3$), namely, its trimer, a perfect control of polymerization can be achieved (Ma et al. 2005).

This type of initiator can also polymerize other cyclic monomers, such as adipic anhydride and trimethylene carbonate (TMC) (Mecerreyes et al. 1999). For safety reasons, residual aluminum should be removed from the material before use in food or biomedical applications. Aluminum can be efficiently extracted by EDTA complexation, tin (IV) alkoxides is rather used as substitute for aluminum alkoxides, although the polydispersity of the chains is then higher (~1.5).

5.2.1.1.2 Polymer Properties

The polymer characteristics are affected by the comonomer composition, the polymer architecture, and molecular weight. The crystallinity of the polymer, an important factor in polymer biodegradation, varies with the stereo regularity of the polymer. Polyglycolide is a highly crystalline polymer (45%–55% crystallinity) and therefore exhibits a high tensile modulus and very low solubility in organic solvents. The melting point of the polymer is greater than 200°C. Unlike glycolide, lactide is a chiral molecule and exists in two optically active forms: L-lactide and D-lactide. The polymerization of these monomers leads to the formation of semi-crystalline polymers. The racemic DL polymer is less crystalline than the D- or L-lactide homopolymers (Nair and Laurencin 2007).

A comprehensive review on the mechanical properties of several biodegradable materials used in orthopedic devices has been published (Daniels et al. 1990). Polyglycolide shows excellent mechanical properties due to its high crystallinity. Due to its good initial mechanical properties, polyglycolides have been investigated as bone internal fixation devices, Biofix® (Steinmann et al. 1990). The thermal properties, tensile properties, and the flextural storage modulus as a function of temperature were determined. The following polymers were compared: poly(L-lactic acid), poly(D,L-lactic acid), poly(glycolic acid), poly(caprolactone), poly(hydroxybutyrate), and copolymers with hydroxyvaleric acid, and poly(trimethylene carbonate) (PTMC). The thermal and mechanical properties of several of the polymers tested are summarized in Table 5.2 (Engelberg and Kohn 1991). Different ratios of poly(lactide-co-glycolides) have been commercially developed and are being investigated for a wide range of biomedical applications according to mechanical properties and rate of degradation (Nair and Laurencin 2007).

The irradiated polymers continue to decrease in molecular weight during storage at room temperature. This decline in molecular weight affects the mechanical properties and the release rate from the polymers. Sterilization using γ-irradiation decreases the polymer molecular weight by 30%–40% (Montanari et al. 1998). PLA and its copolymers with less than 50% glycolic acid content

TABLE 5.2
Thermal and Mechanical Properties of Biodegradable Polyesters

Polymer	Mw	T_g (°C)	T_m (°C)	Tensile Strength (MPa)	Tensile Modulus (MPa)	Elongation Yield (%)	Elongation Break (%)
Poly(lactic acid)							
L-PLA	50,000	54	170	28	1200	3.7	6.0
L-PLA	100,000	58	159	50	2700	2.6	3.3
L-PLA	300,000	59	178	48	3000	1.8	2.0
DL-PLA	107,000	51	—	29	1900	4.0	5.0
Poly(glycolic acid)							
PGA	50,000	35	210	NA	NA	NA	NA
Poly(β-hydroxybutirate)							
PHB	370,000	1	171	36	2500	2.2	2.5
P(HB-11%HV)	529,000	2	145	20	1100	5.5	17
Poly(e-caprolactone)							
PCL	44,000	−62	57	16	400	7.0	80
Poly(trimethylene carbonate)							
PTC	48,000	−15	—	0.5	3	20	160
Poly(ortho esters)							
t-CDM:1,6-HD 35:65	99,000	55	—	20	820	4.1	220
t-CDM:1,6-HD 70:30	101,000	84	—	19	800	4.1	180

Source: Engelberg, I. and Kohn, J., *Biomaterials*, 12(3), 292, 1991.

are soluble in common solvents such as chlorinated hydrocarbons, tetrahydrofuran, and ethyl acetate. For poly(D,L-lactide-co-glycolide), study was reported aiming to determine solubility parameters for different proportions of lactide to glycolide (Schenderlein et al. 2004). The solubility parameters were in the range of 16.2–16.8, which is comparable to those of polystyrene and polyisoprene.

Frank et al. studied the effect of the chemical nature of the drug on matrix degradation and drug release behavior of degradable polymers, using lidocaine as a model drug in base and salt forms. The study showed that the drug in the base form has a substantial effect on the release characteristics, through an accelerating effect on matrix degradation. Drug release study shows that lidocaine salt follows a three-phase release pattern, in contrast to the biphasic release of the lidocaine base (Frank et al. 2005). Blends of low and high molecular weight poly(D,L-lactide) were studied as carriers for drugs. As expected, the addition of low molecular weight polymer accelerated the release of drugs from the blend formulations (Bodmeier et al. 1989).

Various other applications of polyester were reported in literature according to polymer properties. Intranasal administration of polylactide nanoparticles containing thyrotropin-releasing hormone as a viable means to suppress seizures and perhaps epileptogenesis and become the lead compound for intranasal anticonvulsant nanoparticle therapeutics has been reported. Results provide proof of principle that intranasal delivery of sustained-release nanoparticles which may be neuroprotective and suppress seizures and epileptogenesis (Kubek et al. 2009; Veronesi et al. 2009). Not only this, diastereoisomeric complexes of insulin with D-poly(lactic acid) or stereocomplexes of D- and L-PLA entrapping insulin were reported (Slager and Domb 2002). The complexes were spontaneously formed when insulin and D-PLA were mixed together in acetonitrile solution. These macromolecular stereocomplexes may form the onset of the development of a new generation of controlled release systems for peptides and proteins, by molecular complexation with enantiomeric polymers.

5.2.1.1.3 Biodegradation

The biodegradation of polyesters was reviewed by Lewis (1990), Gopferich (1997), and Holland et al. (1986). The molecular weight and polydispersity as well as the crystallinity and morphology of the polymers are important factors in polymer biodegradation. Polyesters mainly undergo bulk erosion, i.e., the polymeric matrices degrade all over their cross-section and have erosion kinetics that are nonlinear and usually characterized by a discontinuity (Vert 2005). The factors that may affect the polylactide degradation include chemical and configurational structure, molecular weight and distribution, fabrication conditions, site of implantation, physical factors, and degradation conditions (Nakamura et al. 1989). The degradation of semi-crystalline polymers proceeds in two phases; in the first phase, the amorphous regions are hydrolyzed and then the crystalline regions. The polymers degraded by bulk hydrolysis of the ester bonds, which resulted in a decrease in molecular with no weight loss.

A comprehensive investigation on the hydrolysis of lactide polymers was described by Vert et al. (1991). In these studies, a standardized set of experiments was designed. All specimens were prepared in a similar way and allowed to hydrolyze at 37°C in distilled water or isotonic phosphate buffer. The changes in the polymer during hydrolysis were monitored by weighing for water uptake and weight loss, gel permeation chromatography (GPC) for molecular weight change, and differential scanning calorimetry (DSC) and X-ray scattering for thermal properties and crystallinity change, potentiometry and enzymatic assays for pH change and lactic acid release, and dynamic mechanical tests for changes in mechanical properties. The polymer, semi-crystalline PLA, lost about 50% of its mechanical strength after 18 weeks in buffer, with no weight loss until about 30 weeks of hydrolysis. The degradation of branched PLA was characterized as bulk erosion, like the linear polymers (Kissel et al. 1991).

The biodegradation of branched PLA with glucose or macromolecular polyol in rats was determined by weight loss (Kissel et al. 1991). In vitro degradation was essentially the same, indicating minimal involvement of enzymatic degradation. The branched materials degraded much faster than the reference linear PLA. On the contrary, the linear PLA had a higher water uptake than the branched polymers. For example, after 36 weeks, the linear PLA contained about 21 wt% of water while the corresponding branched PLA contained only 2%. No adequate explanation was given for this phenomenon.

The degradation of several aliphatic polyesters in the form of microspheres in phosphate buffer solution at 37°C and 85°C was reported. Lower molecular weight polymers degraded faster than higher molecular weight polymers. Degradation at 85°C resulted in a similar degradation profile but faster. The biodegradation of low molecular weight PLA used in tablets for oral delivery of drugs was also studied (Moll and Koller 1990).

5.2.1.2 Polycaprolactones

The successful use of lactide and glycolide polymers in absorbable drug delivery systems and medical devices and absorbable sutures, encouraged the evaluation of other polyesters for this purpose. The most studied polymers in this category are the polycaprolactones, polyhydroxy-butyrates, and polymers of other α-hydroxy acids. Among them, polycaprolactones is semi-crystalline polyester; due to its slow degradation and biocompatibility, it offers an attractive material for long-term biomedical and drug delivery applications, e.g., fibers of copolymer of PGA and poly(ε-caprolactone) (PCL) are commercially available as monofilament suture (Monoacryl®) (Sinha et al. 2004).

5.2.1.2.1 Synthesis

PCL is a semi-crystalline polyester and is of great interest as it can be obtained by the ROP of a relatively cheap monomeric unit "ε-caprolactone." PCL has been synthesized from the anionic, cationic, or coordination polymerization of ε-caprolactone. The synthesis of polycaprolactones has been reviewed extensively (Pitt 1990; Perrin and English 1997; Zhang and Guo 2003). A schematic description of caprolactone polymerization using these three types of initiators is shown in Scheme 5.3 (Pitt 1990).

Various initiators and polymerization conditions were reported for each type of catalyst. Effective anionic reaction systems are tertiary amines, alkali metal alkoxides, and carboxylates in tetrahydrofuran, toluene, and benzene. The anionic method of polymerization is most useful for the synthesis of low molecular weight hydroxy-terminated oligomers and polymers. Kinetics of the anionic polymerization of ε-caprolactone initiated with $(CH_3)_3SiO^-K^+$ and carried out in THF solution has been reported in the temperature range from 0°C to 20°C (Sosnowski et al. 1991).

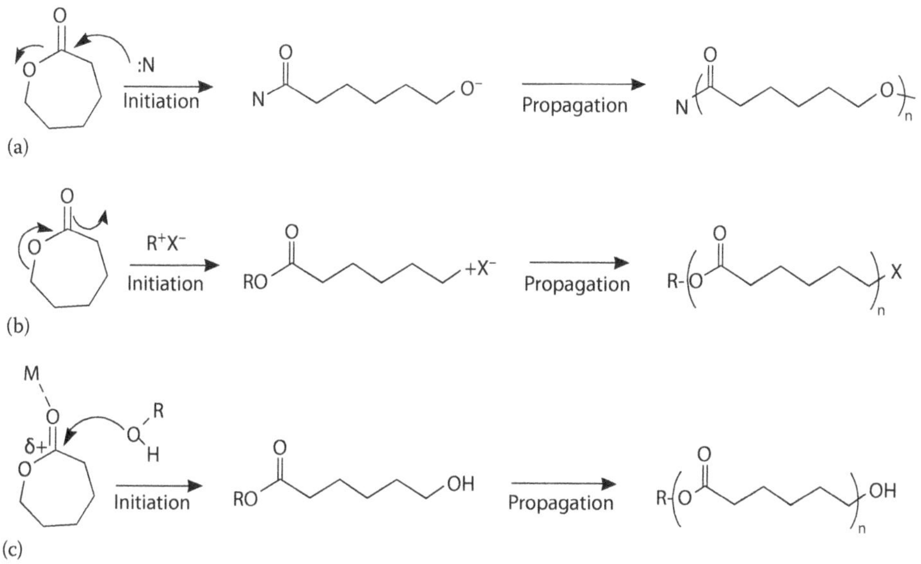

SCHEME 5.3 Polymerization of ε-caprolactone using (a) anionic, (b) cationic, and (c) coordination catalysts.

Cationic polymerization (Ludvig and Belenkaya 1974) is affected by cationic catalysts, which include protic acids, Lewis acids, acylating agents, and alkylating agents. The agents $FeCl_3$, BF_3, Et_2O, alkyl sulfonate, and trimethylsilyl triflate have been used in 1,2-dichloroethane at 50°C to yield polymers with a molecular weight range of 15,000–50,000. High molecular weight homopolymers and random copolymers with lactides and other lactones were obtained using coordination catalysts such as di-n-butyl zinc, stannous octoate, and alkoxides and halides of Al, Sn, Mg, and Ti. Polymerization occurs at 120°C under argon to yield polymers with a narrow molecular weight distribution (Mw/Mn = 1.1) and molecular weights above 50,000 (Hamitou et al. 1977).

In another study, kinetic and PMR spectroscopic methods have been used to investigate the cationic polymerization of ε-caprolactone initiated by $(C_2H_5)_3O^+SbF^-$ in the presence of ethylene glycol at 20°C–60°C. Alcohols markedly increase the rate of polymerization of ε-caprolactone. The difference in the basicity of the hydroxyl groups of the compounds formed in the initial linking of the monomer causes preferential growth of the chain at one of the ends. In such systems, the molecular mass of the polymer is determined by the monomer:addition ratio (Belen'Kaya et al. 1982).

In contrast to random copolymers, block and graft multicomponent systems are most often multiphase materials with properties that are different from the homopolymers or random copolymers. They are useful in improving the phase morphology, the interfacial adhesion and, accordingly, the ultimate mechanical properties of immiscible polymer blends. As an example, block copolymerization of ε-caprolactone (CL) and lactides (LA) allows the permeability of the PCL to be combined with the rapid biodegradation of PLA. Formation of large amount of homo PLA is observed and has been attributed to the increase in the mean degree in association with aluminum alkoxide in toluene from 1 to 3 in the presence of CL and LA, respectively. The homopolymer formation can be prevented by the addition of a small amount of an alcohol, like 2-propanol, or the use of Al derivative that bears only one alkoxide group (Endo et al. 1987).

The ROP of ε-caprolactone initiated by novel single lanthanide tris(4-*tert*-butylphenolate)s [Ln(OTBP)$_3$] is reported. Single-component La(OTBP)$_3$ can effectively prepare polycaprolactone with over 90% yield under quite mild conditions: molar ratio of ε-caprolactone to initiator is 1000, 60°C, 2 h in toluene (Susperregui et al. 2010). In another work, phosphotungstic acid (PTA) as a novel initiator was reported for the ROP of ε-caprolactone. It was found that PTA was an efficient initiator and ROP can be readily initiated by PTA at room temperature to form PCL with narrow molecular weight distribution. Polymerization mechanism study indicates that the polymerization proceeds via acyl-oxygen bond cleavage (Cheng et al. 2010).

5.2.1.2.2 Polymer Properties

PCL is a semi-crystalline polymer having glass transition temperature of −62°C and melting point ranging between 57°C and 64°C, depending upon the crystalline nature of PCL. PCL is of great research interest owing to its ease of processability and solubility in a wide range of organic solvents. PCL is soluble in chloroform, dichloromethane, carbon tetrachloride, benzene, toluene, cyclohexanone, and 2-nitropropane at room temperature. It has a low solubility in acetone, 2-butanone, ethyl acetate, dimethylformamide, and acetonitrile and is insoluble in alcohol, petroleum ether, and diethyl ether. PCL can be blended with other polymers to improve stress crack resistance, dyeability, and adhesion. Copolymerization with lactide increases the T_g with the increase in the lactide content in the polymer (Shindler et al. 1977; Sinha et al. 2004). The crystallinity of the polymer decreases with the increase in polymer molecular weight; polymer of 5,000 is 80% crystalline, whereas the 60,000 polymer is 45% crystalline (Pitt et al. 1981a).

5.2.1.2.3 Biodegradation

The biodegradation of PCL has been extensively studied in the past 40 years and several reviews are available (Pitt 1990; Sinha et al. 2004). Like the lactide polymers, PCL and its copolymers degrade both in vitro and in vivo by bulk hydrolysis of hydrolytically labile aliphatic ester linkages, with the degradation rate affected by the size and shape of the device and additives (Pitt et al. 1981a). The polymers degrade in two phases. In the first phase, a random hydrolytic chain scission occurs, which results in a reduction of the polymer molecular weight. In the second phase, the low molecular fragments and the small polymer particles are carried away from the site of implantation by solubilization in the body fluids or by phagocytosis, which results in a weight loss. Complete degradation and elimination of PCL homopolymers may last for 2–4 years. The degradation rate is significantly increased by copolymerization or blending with lactide and glycolide. The rate of degradation can also be increased by the addition of oleic acid or tertiary amines to the polymer which catalyzes the chain hydrolysis (Pitt et al. 1981b; Woodward et al. 1985). Due to the slow degradation, high permeability (to many drugs), and nontoxicity, PCL was initially investigated as a long-term drug/vaccine delivery vehicle, one, e.g., is long-term contraceptive device Capronors for the long-term zero-order release of levonorgestrel (Darney et al. 1989).

5.2.1.3 Poly(β-Hydroxybutyrate)

These are naturally occurring biodegradable polyesters produced by many bacteria as their energy source. Poly(β-hydroxybutyrate) (PHB) was discovered in 1920 as produced by the bacteria *Bacillus megaterium* (Nair and Laurencin 2007).

5.2.1.3.1 Synthesis

PHB is made by a controlled bacterial fermentation. The producing organism occurs naturally (Arun et al. 2009). An optically active copolymer of 3-hydroxybutirate (3HB) and 3-hydroxyvalerate (3HV) has been produced from propanonic acid or pentanoic acid by *Alcaligenes eutrophus* (Holmes 1985):

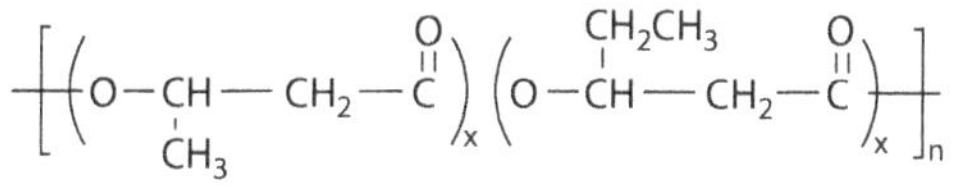

3HB   3HV

The copolymer compositions (0–95 mole% 3HV content) can be controlled by the composition of the carbon sources. Random copolymers of 3HB and 4HV were produced from 4-hydroxybutyric acid and butyric acids by *Alcaligenes eutrophus* (Doi et al. 1988; Kunioka et al. 1988):

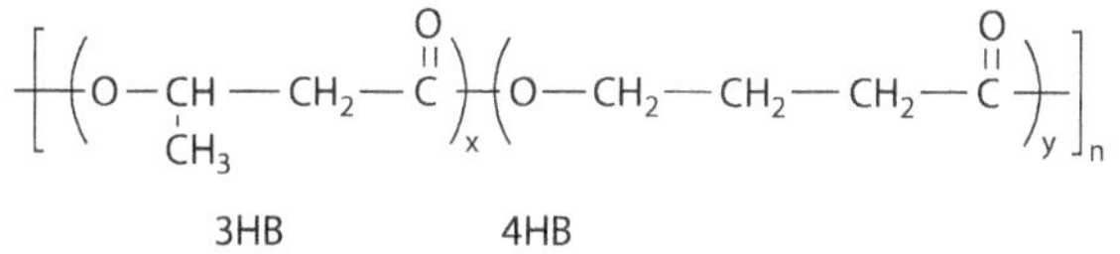

3HB   4HB

Studies aimed at screening and identifying a potential PHB accumulating *Bacillus* sp. and optimization of media parameters for increased PHB production by the strain have been reported (Pal et al. 2009). PHB production was found to be comparable to most of the *Bacillus* sp. reported to date. Whereas some studies aim to study PHB formation, and degradation in sequencing batch biofilm reactor has been reported. These reactors were operated in cycles comprising three individual phases: mixed fill, aeration, and draw. A synthetic substrate solution with acetate and ammonium was used. PHB was formed during the aeration phase immediately after acetate depletion and was subsequently consumed for biomass growth, owing to the high oxygen concentration in the reactor (Nogueira et al. 2009).

5.2.1.3.2 Properties and Biodegradation

The polymers are characterized as having a high molecular weight (>100,000, [n] > 3 dL/g) with a narrow polydispersity and a crystallinity of around 50%. The melting point depends on the polymer composition; P(3HB) homopolymer melts at 177°C with a T_g at 9°C, the 91:9 copolymer with 4HB melts at 159°C, and the 1:1 copolymer with 3HV melts at 91°C. The PHB properties in the living cells of *Alcaligenes eutrophus* were determined using X-ray and variable-temperature ^{13}C NMR relaxation studies (Amor et al. 1991). PHB is an amorphous elastomer with a T_g around $-40°C$ in its "native" state within the granules. The biodegradation of these polymers in soil and activated sludge show the rate of degradation to be in the following order:

$$P(3HB\text{-co-}9\%\ 4HB) > P(3HB) = P(3HB\text{-co-}50\%\ 3HV)$$

The hydrolytic degradation of HB polymers was studied (Brandl et al. 1988). Microspheres degraded slowly in phosphate buffer at 85°C and after 5 months 20%–40% of the polymer eroded under these conditions. Copolymers having a higher fraction of 3-HV and low molecular weight polymers were more susceptible to hydrolysis.

The mass loss of this polymer follows zero-order release kinetics, and this property along with its hydrophobic nature indicates that this polymer primarily undergoes surface erosion. This property makes it an ideal candidate for developing drug delivery vehicles that can achieve zero-order drug release (Nair and Laurencin 2007). Attempts are currently underway to increase the rate of degradation of these polymers by blending them with more hydrophilic polymers or other low molecular weight additives to increase water penetration and facilitate degradation (Chen et al. 2006). In another biomedical application, the effects of biodegradable tubular conduit made from PHB scaffold with predominantly unidirectional fiber orientation and supplemented with cultured adult Schwann cells on axonal regeneration after cervical spinal cord injury in adult rats were studied by Novikova et al. The results demonstrated that a PHB scaffold promotes attachment, proliferation, and survival of adult Schwann cells and supports marked axonal regeneration within the graft (Novikova et al. 2008).

5.2.1.4 Polycarbonates

Poly(ethylene carbonate), poly(propylene carbonate), and PTMC have been tested as biodegradable carriers for the drug delivery (Kawaguchi et al. 1982). They are linear thermoplastic polyesters of carbonic acid with aliphatic dihydroxy compounds.

5.2.1.4.1 Synthesis

The polymers are synthesized from the reaction of dihydroxy compounds with phosgene or with bischloroformates of aliphatic dihydroxy compounds by transesterification, and by polymerization of cyclic carbonates. These polymers have been synthesized from carbon dioxide and the corresponding epoxides in the presence of organometallic compounds as initiators (Kawaguchi et al. 1983).

$$HO-CH_2-\overset{\overset{\textstyle R}{\mid}}{CH}-OH + COCl_2 \longrightarrow \left(CH_2-\overset{\overset{\textstyle R}{\mid}}{CH}-O-\overset{\overset{\textstyle O}{\overset{\|}{}}}{C}-O\right)_n$$

$$R = H, CH_3$$

High molecular weight flexible PTMC can be obtained by the ROP of TMC. Being elastomeric aliphatic polyester with excellent flexibility and poor mechanical strength, PTMC has been investigated as a candidate implant material for soft tissue regeneration and as a suitable material for developing drug delivery vehicles (Nair and Laurencin 2007).

5.2.1.4.2 Biodegradation

Since the carbonate linkage may be labile to hydrolysis, the biodegradability of polycarbonates has been studied (Kawaguchi et al. 1982, 1983). Poly(ethylene carbonate) is biodegraded by surface erosion without significant change in the molecular weight of the residual polymer mass. The main degradation product of poly(ethylene carbonate) in aqueous systems is ethylene glycol, formed presumably by hydrolysis of ethylene carbonate (Stoll et al. 2001).

Pellets of poly(ethylene carbonate) and poly(propylene carbonate) were implanted into the peritoneal cavity of rats and the toxicity and weight loss of polymer pellets were determined. Poly(ethylene carbonate) was completely eliminated 15 days post-implantation, whereas poly(propylene carbonate) remained intact after 60 days. When pellets of the polymers were incubated in phosphate buffer pH 7.4 at 37°C, both polymers did not degrade even after 40 days. These data indicate that poly(ethylene carbonate) was degraded by enzymes. No visible inflammatory reaction was noted at the implantation sites.

PTMC undergoes surface degradation with the rate of in vivo degradation found to be much higher than in vitro degradation. This is presumably due to the contribution of in vivo enzymatic degradation process (Zhang et al. 2006). Copolymers of aliphatic carbonates and lactide showed excellent biocompatibility and mechanical properties. Block copolymers of TMC and lactide were synthesized from the reaction of the monomers with stannous octoate as catalyst at 160°C for 16 h:

$$\left[\begin{array}{c} CH-\overset{\overset{O}{\|}}{C}-O \\ | \\ CH_3 \end{array}\right]_n \left[CH_2-CH_2-CH_2-O-\overset{\overset{O}{\|}}{C}-O\right]_m$$

PLA   TMC

The polymers were soluble in common organic solvents and had a weight average molecular weight of 90,000. They completely degraded in vivo in 1 year (Katz et al. 1985).

5.2.1.5 Other Polyesters

Poly(p-dioxanone) is clinically used as an alternative to poly(lactide) in absorbable sutures with similar properties to poly(lactide) with the advantage of better irradiation stability during sterilization (Bezwada et al. 1990).

$$\left[O-CH_2-CH_2-O-CH_2-\overset{\overset{O}{\|}}{C}\right]_n$$

Poly(p-dioxanone)

5.2.1.5.1 Poly(p-Dioxanone)

The preparation of mono- and multifilament sutures incorporating ibuprofen as an anti-inflammatory agent was reported using p-dioxanone (Zurita et al. 2006). The kinetics of both the loading process and the release in a Sorensen's medium at 37°C has been investigated. Besides this, to produce an estrogenic and bactericidal biomaterial for the treatment of infected nonunions or bone defects, a synthetic degradable block copolymer of poly-D,L-lactic acid segments with randomly inserted p-dioxanone and polyethylene glycol segments was mixed with recombinant human BMP-2 (rhBMP-2) and antibiotics at high concentration. The in vitro elution profile of an antibiotic (teicoplanin) from the polymer, the effects of antibiotics on the bone-inducing capacity of rhBMP-2 or on ectopic new bone formation induced by the rhBMP, and the ability of the polymer to repair bone in a rat cranial defect model were determined. It was observed that the in vivo performance of pellets with antibiotics and rhBMP-2 revealed no significant change in bone yield within the ossicles after 3 weeks. Additionally, the biological activity of rhBMP-2 was retained irrespective of the presence of antibiotics. The data provide important insights into the fabrication of implants that provide efficacious delivery of rhBMP-2 using the lowest possible dose of this expensive osteoinductive protein. This information will be of value for the clinical use of BMPs (Kato et al. 2006a,b; Suzuki et al. 2006).

Biodegradable polymers derived from naturally occurring, multifunctional hydroxy acids and amino acid have been investigated by Lenz and Guerin (1983). The monomers, malic acid, and aspartic acid were polymerized into a polyester or polyamide using a ROP process as follows:

$$\begin{array}{ccc} \underset{\substack{|\\X-CH\\|\\CH_2\\|\\COOH}}{COOH} \rightarrow \rightarrow & \underset{\substack{|\\X-CH\\|\\CH_2\\|\\COOH}}{COOR} \rightarrow & \underset{\substack{|\\CH-X\\|\\CH_2-\overset{\overset{O}{\|}}{C}}}{COOR} \rightarrow & \left[\begin{array}{c} COOR \\ | \\ X-CH-CH_2-\overset{\overset{O}{\|}}{C} \end{array}\right]_n \end{array}$$

X = O (ester), X = NH (amide)

The molecular weights of the polymers were highly dependent on the purity of the cyclic monomers. Polymers of 50,000 were obtained in 93% polymerization yield when very pure monomers were used. Both polymers are water soluble; however, crystallinity and controlled number of acid groups in the polymer chains can be used to alter water swellability and solubility of the polymers. Biocompatibility evaluation in mice indicated that poly (β-malic acid) is nontoxic (Lenz and Guerin 1983).

A series of polyesters, poly(propylene fumarate) (PPF), based on the reaction product of fumaric acid (Domb et al. 1990, 1996), and tartaric acid with different aliphatic diols have also been evaluated as drug carriers. The polymers were synthesized by a direct melt condensation of the acids with the diols with acid catalysis.

5.2.2 Poly(Amides)

A poly(amide) is a polymer-containing monomer of amide joined by peptide bonds. Poly(amides) are commonly used in textiles, automotives, carpet, and sportswear due to their extreme durability and strength. The utilization of amide-based polymers, especially natural proteins, in the preparation of biodegradable matrices has been extensively investigated (Kemnitzer and Kohn 1997). Microcapsules and microspheres of cross-linked collagen (Sehgal and Srinivasan 2009), gelatin (Cao et al. 2009), and albumin (Mathew et al. 2009; Khan and Kumar 2011) have been used for drug delivery.

Poly(amides) occur naturally (proteins, such as wool and silk) or can be made artificially (nylons, and sodium poly(aspartate)) through step-growth polymerization or solid-phase synthesis. Synthetic routes for new classes of biodegradable polymers derived from amino acids having alternating anhydride or ester and amide bonds in the polymer backbone were described by Domb (1990). This approach is useful in the synthesis of degradable polymeric drugs, when bioactive amino acids are used, and possibly in the design of polymers with improved mechanical strength and biocompatibility.

Various poly(amides) were reported, e.g., poly(L-glutamic acid) is a poly(amide) composed of naturally occurring L-glutamic acid residues linked together through amide bonds. It is degraded into monomeric L-glutamic acid by lysosomal enzymes, thus making it an ideal candidate as biodegradable biomaterial. The polymer is highly charged at physiological pH and has been identified as a unique gene/plasmid delivery vehicle (Nicol et al. 2002). Poly(aspartic acid) is synthesized from aspartic acid by thermal polymerization. It is a highly water-soluble ionic polymer with a carboxylate content much higher than poly(glutamic acid) (Nair and Laurencin 2007). Poly(L-lysine) is a small polypeptide of the essential amino acid L-lysine. Recently, the poly-L-lysine-coated PLGA microspheres containing retinoic acid have been explored for nerve tissue engineering. Embryonic carcinoma cells were seeded on them and found to exhibit differentiation into neural cells (Nojehdehian et al. 2009).

The natural amino acid L-tyrosine is a major nutrient having a phenolic hydroxyl group. This feature makes it possible to use derivatives of tyrosine dipeptide as a motif to generate diphenolic monomers, which are important building blocks for the design of biodegradable polymers. Using this approach, a wide variety of polymers have been synthesized (Bourke and Kohn 2003).

Bailey et al. reported two-dimensional thin films consisting of homopolymer and discrete compositional blends of tyrosine-derived polycarbonates to elucidate the nature of different cell responses that were measured in vitro. The data suggest strongly that optimal composition and processing conditions can significantly affect the acute inflammatory and extracellular matrix production responses (Bailey et al. 2006).

Poly(amides) are ubiquitous, naturally occurring biodegradable polymers; however, their application as a biomaterial has been limited due to immunogenicity (Nair and Laurencin 2007). To overcome these limitations, attempts have been made to develop pseudopoly(amino acids) composed of amino acids linked by non-amide bonds such as esters, imino carbonates, and carbonates.

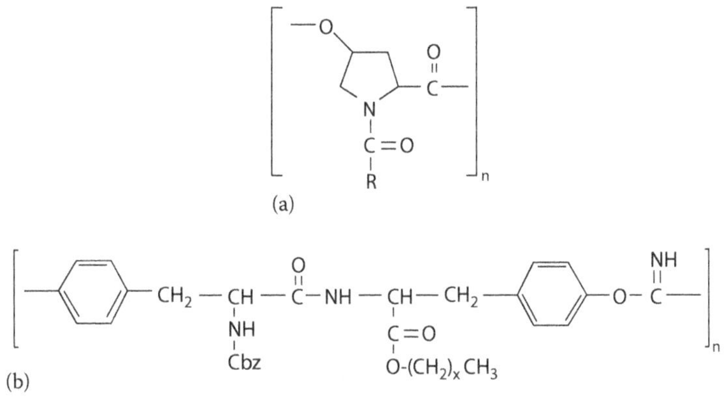

(a)

(b)

SCHEME 5.4 Molecular structure of pseudo(poly amino acids). (a) Poly(N-acylhydroxyproline esters); (b) Tyrosine-derived polyiminocarbonate, poly(CTTE), x = 1; poly(CTH), x = 4; poly(CTTP), x = 15.

Pseudopoly(amino acids) as an approach for biomaterials based on amino acids have been first suggested by Kohn and Langer (1987) who prepared a polyester from N-protected *trans*-4-hydroxy-L-proline and poly(iminocarbonate) from tyrosine dipeptide as monomeric starting material. The structures of poly(N-acylhydroxyproline esters) and a homologous series of tyrosine derived polymers are described in Scheme 5.4 (Kemnitzer and Kohn 1997).

The properties, biodegradability, drug release, and biocompatibility of this class of polymers have been reviewed (Kemnitzer and Kohn 1997). Random copolymers of the α-amino acids N-(3-hydroxypropyl)-L-glutamine and L-leucine were synthesized and used as carriers for naltrexone (Bennett et al. 1991). Naltrexone was covalently bound through the 3-phenolic or the 14-tertiary hydroxyls to the polymer hydroxyl side chains via a carbonate bond. Naltrexone was released from the polymer in a relatively constant way for 30 days, both in vitro and in vivo experiments in rats (Bennett et al. 1991).

5.2.3 POLY(PHOSPHATE ESTERS)

Poly(phosphate ester), polyphosphonates, and polyphosphazenes are three classes of phosphorus-containing polymers that have received wide attention over the past decade for their utility in biomedicine. Significant research in this area has led to niche polymers with morphologies ranging from viscous gels to amorphous microparticles for utility in drug delivery (Chaubal et al. 2003).

Poly(phosphate esters) were synthesized from the reaction of ethyl or phenyl phosphorodichloridates and various dialcohols including bisphenol A and poly(ethylene glycol) of various molecular weights.

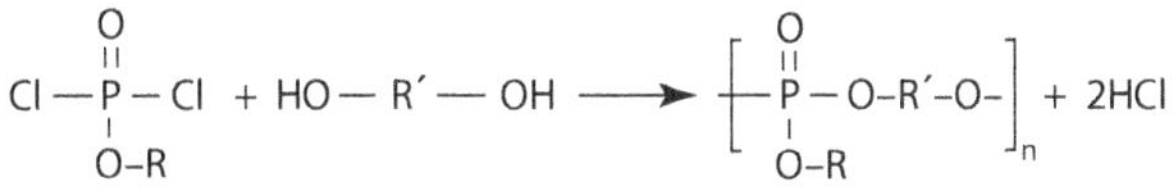

Interfacial condensation using a phase transfer catalyst and bisphenol A as comonomer yielded polymers with a weight average molecular weight around 36,000.

Poly(phosphate ester) polymers are among the attractive candidates, in the view of their high biocompatibility, adjustable biodegradability, flexibility in coupling fragile biomolecules under physiological conditions, and a wide variety of physicochemical properties (Wang et al. 2001). As an important class of eminent biomaterials, polyphosphates have good biocompatibility, biodegradability probably due to the structural similarity to naturally occurring nucleic and teichoic acids. They can be degraded naturally into harmless low molecular weight products through hydrolysis or enzymatic digestion of phosphate linkages under physiological conditions. In recent years, some linear random and block copolymers with phosphate units have been synthesized and applied to drug and gene delivery. Some biocompatible and biodegradable hydrogels based on polyphosphates have also been produced for cell encapsulations (Liu et al. 2010a). Poly(phosphate ester) has been studied as a potential biodegradable matrix for drug delivery (Wen et al. 2004). The pentavalency of phosphorus offers the potential for covalent linking of the drug (Chaubal et al. 2003).

The polymers based on bisphenol A release drugs for a long period of time; 8%–20% cortisone was released after 75 days in buffer solution. The degradation rate depends on the nature of the polymer side chain; polymers with phenyl side chains degrade much slower than those containing ethyl or ethoxyethyl side chains (Leong 1995). The in vivo degradation in rabbits was faster than in vitro.

A biodegradable polyphosphoester, poly[[(cholesteryl oxocarbonylamido ethyl) methyl bis(ethylene) ammonium iodide] ethyl phosphate] (PCEP), was synthesized and investigated for gene delivery (Wen et al. 2004). Results obtained suggest the potential of this polyphosphoester for naked DNA-based gene therapy (Huang et al. 2004). Cellular-specific micellar systems from functional amphiphilic block copolymers are attractive for targeted intracellular drug delivery. Micelles based on diblock copolymer of poly(ethyl ethylene phosphate) and poly(ε-caprolactone) were developed by Wang et al. The surface functionalized micellar system is promising for specific anticancer drug transportation and intracellular drug release (Wang et al. 2008).

5.2.4 POLYPHOSPHAZENES

The uniqueness of the polyphosphazenes stems from its inorganic backbone (N=P) with certain organic side groups which can be hydrolyzed to phosphate and ammonia. Several polymer structures have been used as matrix carriers for drugs or as a hydrolyzable polymeric drug, where the drug is covalently bound to the polymer backbone and released from the polymer by hydrolysis (Allcock and Kwon 1986). Various comprehensive reports on the synthesis, characterization, and medical applications of polyphosphazenes were published by Allcock (Allcock 1976; Allcock and Ambrosio 1996; Allcock et al. 2003, 2007).

The polymers are most commonly synthesized by a substitution reaction of the reactive poly(dichlorophosphazene) with a wide range of reactive nucleophils such as amines, alkoxides, and organometallic molecules (Scheme 5.5). The reaction is carried out in general at room temperature in tetrahydrofuran, or aromatic hydrocarbon solutions. Polymers containing mixed substituent can be obtained from the sequential or simultaneous reaction with several nucleophils (Scheme 5.5):

$$RX = RONa,\ RNH_2,\ C_6H_5ONa$$

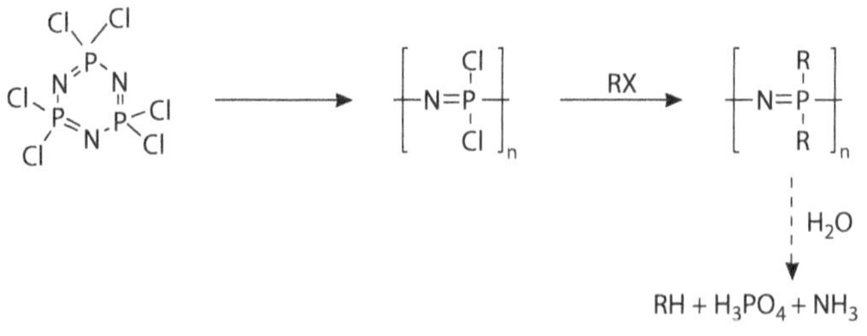

SCHEME 5.5 Synthesis and hydrolysis of polyphosphazenes. (From Allcock, H.R. and Kugel, R.L., *Inorg. Chem.*, 5, 1716, 1966.)

The properties of the polymers depend on the nature of the side groups. Hydrolytically degradable polyphosphazene were obtained when amino acid and imidazole derivatives were used as substituent. The first bioerodible polymer was the ethylglycinato derivative (R=NHCH$_2$COOEt) which hydrolytically degrade to ethanol, glycine, phosphate, and ammonia (Allcock et al. 1977).

Imidazolyl substituted polyphosphazene are hydrolytically unstable and hydrolyze in room moisture. The rate of hydrolysis can be slowed by the incorporation of hydrophobic side groups such as phenoxy or methylphenoxy groups. The in vivo and in vitro release of progesterone and bovine serum albumin (BSA) from imidazole, 4-methylphenoxy substituted polyphosphazene matrices were reported (Laurencin et al. 1993). Almost 90% of the loaded progesterone was released in 30 days with about 60% released in 8 days when placed in phosphate buffer pH 7.4 at 37°C.

Tyrosine-functionalized polyphosphazenes were synthesized by Allcock et al. (2003) and their hydrolytic stability, pH-sensitive behavior, and hydrogel-forming capabilities were investigated. The physical and chemical properties of the polymers varied with the type of linkage between the tyrosine unit and phosphazene backbone. The rate of hydrolysis was dependent on the ratio of the two side groups, the slowest rate being associated with the highest concentration of tyrosine. The hydrolysis products were identified as phosphates, tyrosine, glycine, ammonia, and ethanol derived from the ester group.

Amino acid ester substituted polyphosphazene have been used for controlled release of the covalently bounded anti-inflammatory agent, naproxen. Steroids having hydroxyl group were bound to the polymer chain through the hydroxyl group (Allcock and Fuller 1980). These amino acid ester substituted polyphosphazenes are attractive candidates because of their biocompatibility, controllable hydrolytic degradation rates, and nontoxic degradation products. These polymers demonstrated excellent tissue compatibility and in vivo biodegradability and can be potential candidates for various biomedical applications (Sethuraman et al. 2006), e.g., in vivo orthopedic applications (Nukavarapu et al. 2008).

Zheng et al. have reported a new class of amphiphilic methoxy-poly(ethylene glycol) grafted polyphosphazene. Doxorubicin was encapsulated into polymeric micelles derived from these copolymers. Findings suggest that these copolymers can encapsulate water-insoluble anti-cancer agents and contribute to improve drug sensitivity of adriamycin-resistant cell line (Zheng et al. 2009). In another study, Zheng et al. have reported the construction of self-assembled nanoparticles from novel pH-sensitive amphiphilic polyphosphazenes. These nanoparticles provide fast pH-responsive drug release and have the capability to disturb endosomal membranes. The pH-responsive nanoparticles designed in this study have achieved their potential as a drug delivery system for tumor therapy applications (Zheng et al. 2011).

5.2.5 POLY(ORTHOESTERS)

Poly(orthoesters) (POEs) are amorphous hydrophobic polymers containing hydrolytically labile, acid-sensitive, backbone linkages. These were developed by ALIZA Corporation (Alzamer®) as hydrophobic surface eroding polymers particularly for drug delivery applications. POEs were invented during the pursuit of developing a bioerodible polymer, subdermally implantable that would release contraceptive steroids by close to zero-order kinetics for at least 6 months (Benagiano and Gabelnick 1979). An additional objective was that the polymer erosion and drug release should be concomitant so that no polymer remnants are present in the tissue after the entire drug has been released.

The objectives mentioned earlier could only be met if the polymer was truly surface eroding. For a surface eroding polymer, the erosion process at the surface of the polymer should be much faster than in the interior of the device. To exhibit such a phenomenon, the polymer has to be extremely hydrophobic with very labile linkages. Hence, it was envisioned that polymeric devices with an orthoester linkage in the backbone, which is an acid-sensitive linkage, could provide a surface-eroding polymer if the interior of the matrix is buffered with basic salts.

By using diols with varying levels of chain flexibility, the rate of degradation for these polymers, pH sensitivity, and glass transition temperatures can be controlled. To date four different classes of POEs have been developed (POE I, II, III, and IV). Few orthopedic applications of this class of polymers have been explored, but major use of POEs has been limited to drug delivery systems (Sokolsky-Papkov et al. 2007).

5.2.5.1 Synthesis

The first POEs were reported by Choi and Heller (1979), assigned to Alza Corporation. These proprietary polymers were first designated as Chronomer and later as Alzamer. They were prepared by a transesterification reaction (Choi and Heller 1979). The general synthesis involved the heating of the reaction mixture to 110°C–115°C for about 1.5–2 h and then further heated at 180°C and 0.01 T for 24 h. The synthesis of such polymers, with minimal amount of cross-linkage, requires an orthoester starting material in which one alkoxy group has a greatly reduced reactivity. This can be achieved by using a cyclic structure as shown earlier.

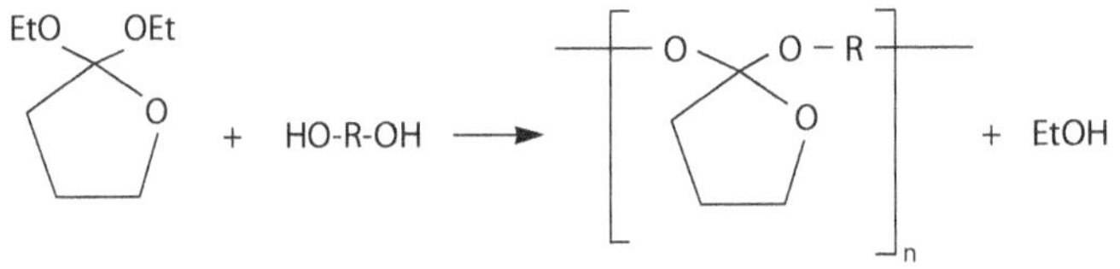

Hydrolysis of these polymers regenerates the diol and γ-butyrolactone. The γ-butyrolactone rapidly hydrolyses to γ-hydroxybutyric acid. The production of γ-hydroxybutyric acid would further catalyze the breakdown of orthoester linkages leading to bulk erosion of the matrix. Thus, it was decided to incorporate basic salts into the polymer matrix to neutralize the generated acid and keep the hydrolysis process under control (Heller 1990). This polymer system was originally designated as Chronomer@ but is now marketed as Alzamere. The polymer has been used in the treatment of burns, in the delivery of the narcotic antagonist naltrexone, and in the delivery of the contraceptive steroids norethindrone and levonorgestrel (Heller 1990).

Subsequently, another family of POEs were developed that were not related to Alzamer. These polymers are prepared by the addition of polyols to diketene acetals. The general reaction can be schematically represented as follows:

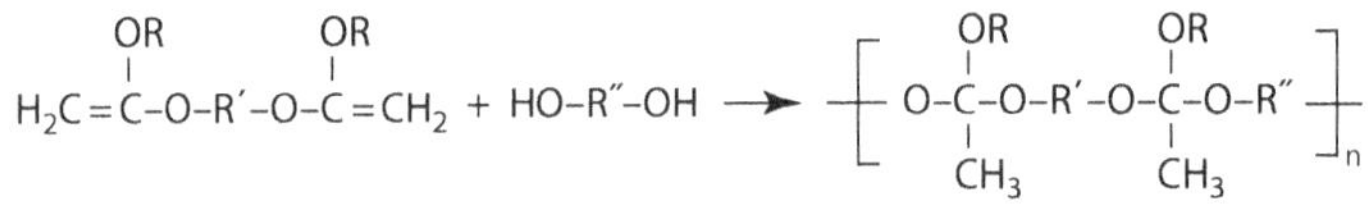

Initial work was conducted with the monomer, diketene acetal, derived from pentaerythritol, and 1,6 hexanediol. The reaction is exothermic and proceeds to completion virtually instantaneously. The reaction is as follows:

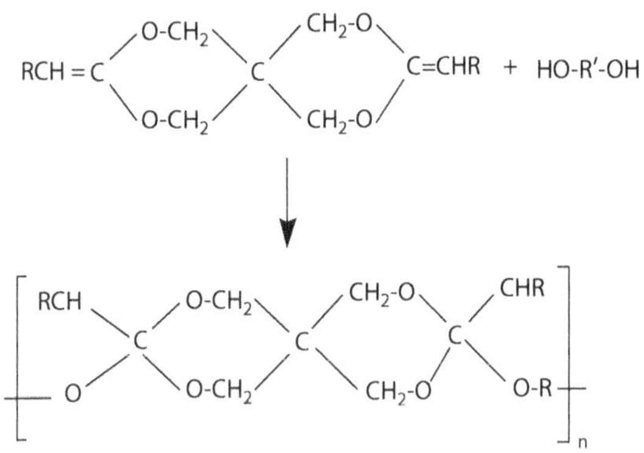

where R = H, for 3,9-bis(methylene 2,4,8, 10-tetra oxaspiro [5,5] undecane). The diols investigated were 1,6 hexanediol *trans*-1,4 cyclohexane dimethanol, 1,6-cyclohexanediol, ethylene glycol, and biophenol A.

5.2.5.2 Polymer Properties

The molecular weight of POEs was significantly dependent on the type of diol and catalyst used for synthesis. A linear, flexible diol like 1,6 hexane diol gave molecular weights greater than 200 K, whereas bisphenol A in the presence of catalyst gave molecular weight around only 10,000 (Heller et al. 1983).

The mechanical properties of the linear POEs can be varied over a large range by selecting various compositions of diols. It was shown that the glass transition temperature of the polymer prepared from DETOSU can be varied from 25°C to 110°C by simply changing the amount of 1,6 hexanediol in *trans*-1,4 cyclohexane dimethanol from 100% to 0% (Heller et al. 1985). There seems to be a linearly decreasing relationship between the T_g and percentage of 1,6 hexanediol. One could take advantage of the aforementioned relationship in selecting the polymer for in vivo applications because in vivo the T_g of the polymer would drop due to imbitions of water. This can result in the loss of stiffness and rigidity of the polymer.

5.2.5.3 Cross-Linked Poly(Orthoesters)

To prepare a cross-linked polymer, there should be at least one monomer that has functionality greater than 2. In the case of POEs, it is possible for a ketene acetal or an alcohol to have a functionality greater than 2. Due to the difficulty in preparing trifunctional ketene acetals, triols were used to prepare the cross-linked polymer.

In general, the synthesis of cross-linked polymers was synthesized by reacting prepolymer with the triols or a mixture of diols and triols (Scheme 5.6; Heller et al. 1987).

The prepolymer is an acetal with a diol and is a viscous liquid at room temperature. Thus, the compound of interest could be incorporated into the prepolymer along with the triol and the mixture can be cross-linked at temperatures as low as 40°C. This can be a good method for incorporating thermo labile drugs. However, one should be cautious with using compounds with hydroxyl functionality.

Another family of POEs can be prepared by reacting a triol with two vicinal hydroxyl groups and one removed by at least three methylene groups with an alkyl orthoacetate (Heller et al. 1990). The intermediate does not have to be isolated and continuous reaction produces a polymer. The use of flexible triols such as 1,2,6-hexanetriol produces highly flexible polymers that have ointment-like properties even at relatively high molecular weights. However, properties such as

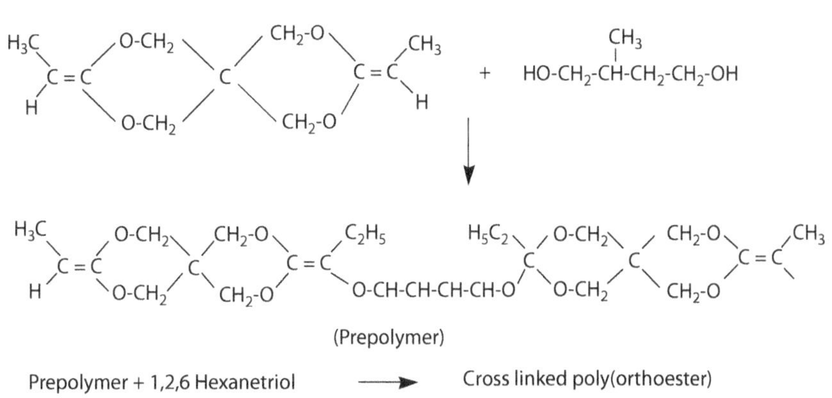

(Prepolymer)

Prepolymer + 1,2,6 Hexanetriol ⟶ Cross linked poly(orthoester)

SCHEME 5.6 Synthesis of cross-linked poly(orthoesters).

viscosity and hydrophobicity can be readily varied by controlling molecular weight and the size of the alkyl group R′.

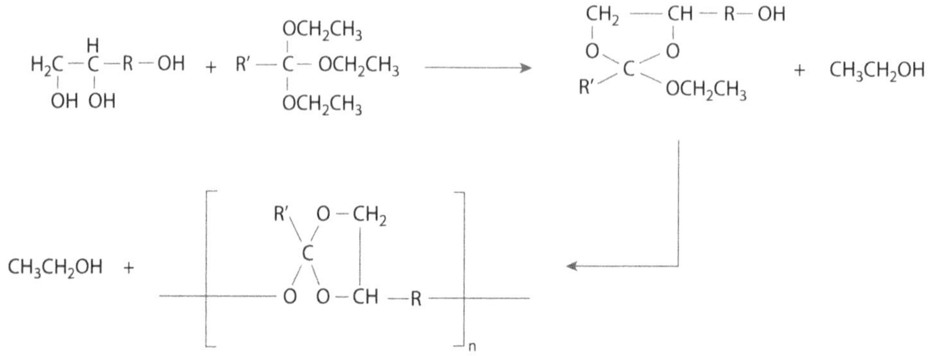

In another experiment, Heller et al. (1992) replaced the flexible triol to a rigid one such as 1,1,4-cyclohexanetrimethanol to obtain a solid polymer, as shown next:

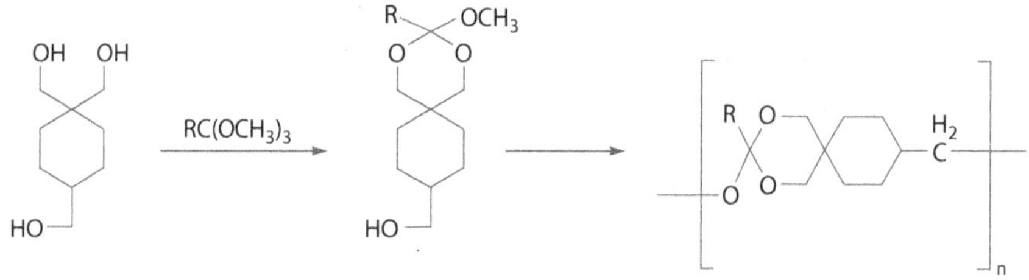

5.2.5.4   Polymer Hydrolysis

The primary mechanism for the degradation of POEs is via hydrolysis. Depending on the reactants used during the synthesis of the polymer, the hydrolysis products are diol or mixture of diols, and pentaerythritol dipropionate or diacetate if 3,9-bis (methylene-2,4,8,10-tetraoxaspiro [5,5] undecane) was used (Sparer et al. 1984). The pentaerythritol esters hydrolyze at a slower rate to pentaerythritol and the corresponding acetic or propionic acid. The sequence of reaction is as follows:

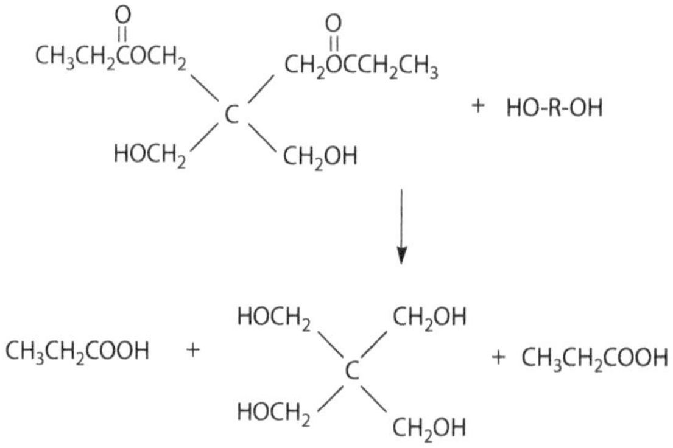

The difference in the sensitivity of the hydrolysis of orthoester linkages in acid versus alkaline medium has been used to the advantage in designing the orthoester-based delivery systems.

This preferential sensitivity is used by incorporating acid anhydrides into the matrix to accelerate the rate of hydrolysis. While, on the other hand, a base is used to stabilize the interior of the matrix.

5.2.5.5 Polymer Processing

The orthoester linkage is inherently unstable in the presence of water. However, because of the polymer's highly hydrophobic nature, they can be stored without careful exclusion of moisture. Even though the polymer is relatively stable in trace amounts of moisture, it is unstable to heat and undergoes disproportionation to an alcohol and ketene acetal. The combination of moisture and heat can be fatal for the processing of POEs, which are designed to erode within days. Thus, if injection molding is necessary to fabricate the device, then moisture must be rigorously excluded during fabrication. One should also consider the interaction between the incorporated anhydride as catalysts, the polymer, and drug during the thermal processing.

Deng et al. reported the use of laboratory-scale spray-congealing equipment to fabricate injectable microparticles consisting of poly(orthoester) and bupivacaine. Characterizations were performed to determine the chemico-physical properties of poly(orthoester) before and after microparticle fabrication. The microparticles have demonstrated the potential to be used for long-acting post-surgery pain management by local injection (Deng et al. 2003).

The effect of local delivery of indomethacin by a bioerodible poly(orthoester) on the reossification of segmental defects of the radius in rats was studied. The results of this study suggest that the poly(orthoester) may be used as a bioerodible system for local delivery of indomethacin to inhibit reossification of skeletal defects without tissue reaction, unabsorbed carrier, or systemic effects (Solheim et al. 1995).

5.2.6 POLYANHYDRIDES

The invention and development of polyanhydrides dates as far back as 1900s. In 1909, Bucher and Slade reported the development of aromatic polyanhydrides composed of isophthallic acid and terephthallic acid (Bucher and Slade 1909). Subsequently, Hill and Carothers (1932) reported a series of aliphatic polyanhydrides. Systematic development of polyanhydrides, as substitutes for polyesters in textile applications, was undertaken initially by Conix et al. (1958). They prepared and studied a number of aromatic and heterocyclic polyanhydrides. It has been recognized that the matrix should undergo heterogeneous degradation to maximize the control over release process. These polymers were not suitable for textile applications because of the extreme reactivity of anhydride linkage toward water. In early 1980s, Rosen et al. (1983) envisioned the use of hydrophobic polyanhydrides in designing the surface eroding matrix for applications in controlled drug delivery. The fact that they are extremely hydrophobic and hydrolytically unstable renders them useful in drug delivery applications (Leong et al. 1985, 1986). Therefore, although originally developed for the textile industry, polyanhydrides found use in biomedical applications due to their biodegradability and excellent biocompatibility (Katti et al. 2002).

Efficient control over various physicochemical properties, such as biodegradability and biocompatibility, can be achieved for polyanhydrides due to the availability of a wide variety of diacid monomers as well as by copolymerization of these monomers. Biodegradation of these polymers takes place by the hydrolysis of the anhydride bonds and the polymer undergoes predominantly surface erosion, a desired property to attain near zero-order drug release profile (Katti et al. 2002). The properties of major classes of anhydrides and the impact of geometry on degradation and erosion are discussed by Gopferich and Tessmar (2002).

5.2.6.1 Synthesis

5.2.6.1.1 Melt Polycondensation

The majority of the polyanhydrides are prepared by melt polycondensation. The sequence of reaction involves first the conversion of a dicarboxylic acid monomer into a prepolymer consisting of a mixed anhydride of the diacid with acetic anhydride. This is achieved by simply refluxing the diacid monomer with acetic anhydride for a specified length of time. The polymer is obtained subsequently by heating the prepolymer under vacuum to eliminate the acetic anhydride (Yoda 1963):

$$HOOC\text{-}(CH_2)_8\text{-}COOH \quad + \quad (CH_3CO)_2O$$

$$\downarrow Reflux$$

$$\underset{CH_3\text{-}\overset{O}{\overset{\|}{C}}\text{-}O\text{-}\overset{O}{\overset{\|}{C}}\text{-}(CH_2)_8\text{-}\overset{O}{\overset{\|}{C}}\text{-}O\text{-}\overset{O}{\overset{\|}{C}}\text{-}CH_3}{}$$

$$10^{-4}\,mm\,Hg \quad \Big| \quad 180°C, 90\,min$$

$$\left[-O\text{-}\overset{O}{\overset{\|}{C}}\text{-}(CH_2)_8\text{-}\overset{O}{\overset{\|}{C}}\text{-}O\text{-} \right]_n$$

The polyanhydride thus obtained was of low molecular weight. For most of the practical applications, high molecular weight polyanhydrides are desirable. Hence, a systematic study was undertaken to determine the factors that affected the polymer molecular weight (Domb and Langer 1987). It was found that the critical factors were monomer purity, reaction time and temperature, and an efficient system to remove the byproduct, acetic anhydride is required. The highest molecular weight polymers were obtained using pure isolated prepolymers and heating them at 180°C for 90 min with a vacuum of 10^{-4} mmHg, using a dry ice/acetone trap. Significantly higher molecular weights were obtained in shorter times by using coordination catalysts such as cadmium acetate, earth metal oxides, and $ZnEt_2.H_2O$ (Domb and Langer 1987). The weight average molecular weight varied from 90,000 to 240,000 when the concentration of cadmium acetate was changed from 0.5 to 3 mole%, with the reaction time of less than 1 h. Other catalysts, such as titanium and iron, inhibited the polymerization reaction and the polymers were dark brown. Acidic catalyst, such as p-toluene sulfonic acid, did not show any effect on polymer molecular weight while the basic catalyst, 4-dimethyl amino pyridine, caused a decrease in molecular weight.

5.2.6.1.2 Solution Polymerization

Synthesis of polyanhydrides, using melt polycondensation, is useful to obtain high molecular weight polymers but is not useful if the monomers are thermolabile. Hence, methods were developed to synthesize polyanhydrides under ambient conditions for heat-sensitive monomers such as dipeptides and therapeutically active diacids.

The solution polymerization is carried out by Schotten–Baumann technique. In this method, the solution of diacid chloride is added dropwise into an ice-cooled solution of a dicarboxylic acid. The reaction is facilitated by using an acid acceptor such as triethylamine. Polymerization takes place instantly on contact of the monomers and is essentially complete within 1 h. The solvents employed can be a single solvent or a mixture of solvents such as dichloromethane, chloroform, benzene, and ethyl ether. It was found that the order of addition is very important in obtaining relatively high molecular weight polyanhydrides. Addition of a diacid solution dropwise to the diacid chloride solution consistently produced high molecular weight polymers (Subramanyam and Pinkus 1985).

$$HOOC\text{-}R\text{-}COOH \ + \ ClOC\text{-}R'\text{-}COCl \ \xrightarrow{\text{Base}} \ \text{-}(R\text{-}C(O)\text{-}O\text{-}C(O)\text{-}R'\text{-}C(O)\text{-}O\text{-}C(O))\text{-} \ + \ Base\text{-}HCl$$

The drawback of this homogeneous Schotten–Baumann condensation reaction in solution is that the diacid chloride monomer should be of very high purity. An alternate approach was the conversion of dicarboxylic acid monomer into the polyanhydride using a dehydrative coupling agent under ambient conditions. The dehydrative coupling agent, N'N bis[2-oxo-3-oxazolidinyl]phosphonic chloride, was the most effective in forming polyanhydrides with the degree of polymerization around 20 (Leong et al. 1987). It is essential that the catalyst be ground into fine particles before use and should be freshly prepared. A disadvantage of this method is that the final product contains polymerization byproducts, which have to be removed by washing with protic solvents such as methanol or cold dilute hydrochloric acid. The washing by protic solvents may evoke some hydrolysis of the polymer.

Coupling agents such as phosgene and diphosgene could also be used for the polyanhydride formation. Polymerization of sebacic acid using either phosgene or diphosgene as coupling agents with the amine-based heterogeneous acid acceptor, poly(4-vinyl pyridine), produced higher molecular weights in comparison to non-amine heterogeneous base K_2CO_3 (Domb et al. 1988).

5.2.6.1.3   Ring-Opening Polymerization

ROP offers an alternate approach to the synthesis of polyanhydrides used for medical applications. Albertsson and co-workers prepared adipic acid polyanhydride from cyclic adipic anhydride (oxepane-2,7-dione) using cationic (e.g., $AlCl_3$ and $BF_3.(C_2H_5)_2O$), anionic (e.g., $CH_3COO^-K^+$ and NaH), and coordination-type inhibitors such as stannous-2-ethylhexanoate and dibutyltin oxide (Albertsson and Lundmark 1988; Lundmark et al. 1991). ROP takes place in two steps: (1) preparation of the cyclic monomer and (2) polymerization of the cyclic monomers (Kumar et al. 2005).

5.2.6.2   Polymer Properties

Almost all polyanhydrides show some degree of crystallinity as manifested by their crystalline melting points. An in-depth X-ray diffraction analysis was conducted with the homopolymers of sebacic acid (SA), bis(carboxyphenoxy)propane (CPP), bis(carboxyphenoxy)hexane (CPH) and fumaric acid, and the copolymers of SA with CPP, CPH, and fumaric acid. The results indicated that the homopolymers were highly crystalline and the crystallinity of the copolymers was determined, in most cases, by the monomer of highest concentration. Copolymers with a composition close to 1:1 were essentially amorphous (Mathiowitz et al. 1990).

The melting point, as determined by differential scanning calorimetry, of these aromatic polyanhydrides is much higher than the aliphatic polyanhydrides. The melting point of the aliphatic-aromatic copolyanhydrides is proportional to the aromatic content. For this type of copolymers there is characteristically a minimum T_m between 5 and 20 mole% of the lower melting component (Domb and Langer 1987).

The majority of polyanhydrides dissolve in solvents such as dichloromethane and chloroform. However, the aromatic polyanhydrides display much lower solubility than the aliphatic polyanhydrides. In an attempt to improve the solubility and decrease the T_m, copolymers of two different aromatic monomers were prepared. These copolymers displayed a substantial decrease in T_m and an increase in solubility than the corresponding homopolymers of aromatic diacids (Domb 1992).

The data on mechanical properties of polyanhydrides are very limited. The fibers of poly[1,2-bis(p-carboxyphenoxy) ethane anhydride] showed a tensile strength of $40\,kg/mm^2$ with an elongation of 17.2% and a Young's modulus of $505\,kg/mm^2$. A systematic study on the tensile strength of the copolymers of CPP and SA showed that increasing the CPP content in the copolymer or the molecular weight of the copolymer increased the tensile strength. Unsaturated polyanhydrides of the structure $[-(OOC-CH=CH-CO)_x-(OOC-R-CO)_y]_n$ were developed to improve the mechanical properties of the polymers. The advantage of the unsaturated polyanhydrides is that it can undergo secondary polymerization of the double bonds to create a cross-linked matrix (Domb et al. 1991b).

5.2.6.3   Polymer Hydrolysis

Anhydride linkage is extremely susceptible to hydrolysis in the presence of moisture to generate the dicarboxylic acids (Domb and Nudelman 1995). Hydrolysis of monomeric anhydrides is catalyzed by both acid and base, and the hydrolytic degradation rate of polyanhydrides increases with an increase in pH. It is believed that the poor solubility of the oligomeric products, under low pH conditions, formed at the surface of the matrix impede the degradation of the core. In general, the hydrophobic polymers such as P(CPP) and P(CPH) display constant erosion kinetics. The degradation rates of the polyanhydrides can be altered in a number of ways. The degradation rates can be enhanced by incorporating the aliphatic monomer, such as sebacic acid, into the polymer. The degradation can be slowed by increasing the methylene groups into the backbone of the polymer. For example, in the case of poly[bis(p-carboxyphenoxy)alkane] series, increasing the methylene groups from 1 to 6 increased the hydrophobicity of the polymer and the erosion rates underwent a decrease of 3 orders of magnitude.

To achieve a variety of degradation rates, aliphatic-aromatic homopolyanhydrides of the structure $-(OOC-C_6H_4-O(CH_2)_x-CO-)_n$ were prepared with "x" varying from 1 to 10. Increasing the value of "x" decreases the erosion rates (Domb et al. 1989). Increased erosion rates were also observed when poly(sebacic acid) was branched with either 1,3,5 tricarboxylic acid or low molecular weight poly(acrylic acid) (Maniar et al. 1990).

Apart from the reactivity of anhydride linkage toward water, aliphatic polyanhydrides and their copolymers are found to undergo self-depolymerization under anhydrous conditions in the solid state and in solution (Domb and Langer 1989). The depolymerization reaction mainly affects the high molecular weight fraction of the polymer. Aromatic homopolymers show no sign of depolymerization when stored under anhydrous conditions. The depolymerization rate is found to follow a first-order kinetics, accelerate with temperature, and increase with polarity of the solvent (Domb and Langer 1989). The depolymerized polymer can be repolymerized to yield the original polymer, suggesting that inter or intramolecular anhydride interchange takes place during depolymerization.

In order to tailor the erosion rate of polyanhydrides while retaining their surface erosion characteristics, new three-component polyanhydrides of sebacic acid, 1,3-bis(p-carboxyphenoxy)propane and poly(ethylene glycol), were synthesized by Hou et al. The incorporation of poly(ethylene glycol) into traditional two-component polyanhydrides retains their surface erosion properties while making the erosion rate tuneable. The new polyanhydrides hold potential for drug delivery applications (Hou et al. 2007).

5.2.6.4 Polymer Processing

Drug incorporated matrix can be formulated either by compression or injection molding. The polymer and drug can be ground in a Micro Mill grinder, sieved in to a particle size range of 90–120 μm, and can be pressed into circular discs using Carver press. Alternatively, the drug can be mixed into the molten polymer to form small chips of drug–polymer conjugate. These chips are fed into the injection molder to mold the drug/polymer matrix into the desired shaped device. One must consider the thermal stability of the polymer and the potential chemical interaction between drug and polymer at high temperatures of injection molding.

The preferred method of drug delivery, in many instances, is by injection. This requires the development of microcapsule or microspheres of the drug. Several different techniques have been developed for the preparation of microspheres from polyanhydrides, which includes "Hot-melt' microencapsulation (Mathiowitz and Langer 1987) and solvent removal technique (Mathiowitz et al. 1988).

5.2.6.5 Other Polyanhydrides

In addition to the previously discussed aliphatic, aromatic, and the copolyanhydrides of the respective diacids, several other modifications of the backbone of the polyanhydrides have been reported. These new polyanhydrides were developed to improve their physicochemical, mechanical, thermal, and hydrolytic properties.

One of these new polyanhydrides is polyanhydride-imides, also referred to as copolyimides (Domb 1990). They showed good thermal resistance but were essentially insoluble in most organic solvents. In an attempt to improve the solubility in more polar solvents, polyanhydrides were synthesized using an imide-diacids containing aliphatic aromatic characteristics. A systemic study was reported in which the starting monomers, imide-diacids, were prepared from aromatic acid anhydrides and α-amino acids. Varying the number of methylenic units in the α-amino acids provided the variability in the aliphatic character of the aromatic-aliphatic monomer. A typical example of such an aromatic-aliphatic monomer is the one obtained by the reaction of trimellitic anhydride with glycine:

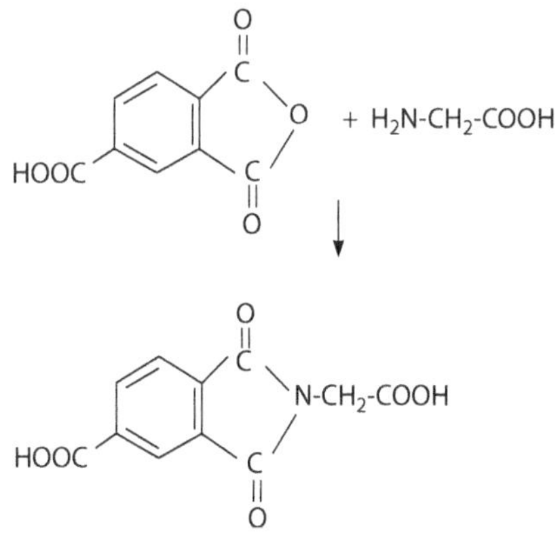

For the synthesis of polyanhydride, the aliphatic-aromatic diacid is first converted to the diacetyl derivative by refluxing the diacid in the presence of excess acetic anhydride. The diacetyl derivative is polymerized either by melt polycondensation or in solution. The polyanhydride-imides thus obtained are very soluble in polar organic solvents. However, they showed melt transitions at temperatures of 245°C and above. Along with that, insufficient data are available on the hydrolytic stability of these materials, rendering them questionable materials as a carrier in drug delivery systems.

Another class of polyanhydrides is based on natural fatty acids. The dimers of oleic acid and eurucic acid are liquid oils containing two carboxylic acids available for anhydride polymerization. The homopolymers are viscous liquids. Copolymerization with increasing amounts of sebacic acid forms solid polymers with increasing melting points as a function of SA content. The polymers are soluble in chlorinated hydrocarbons, tetrahydrofuran, 2-butanone, and acetone (Domb and Maniar 1993).

Polyanhydrides synthesized from nonlinear hydrophobic fatty acid esters, based on ricinoleic, maleic acid and sebacic acid, possessed the desired physicochemical properties such as low melting point, hydrophobicity and flexibility to the polymer formed in addition to biocompatibility and biodegradability. The polymers were synthesized by melt condensation to yield film-forming polymers with molecular weights exceeding 100,000 (Teomim et al. 1999).

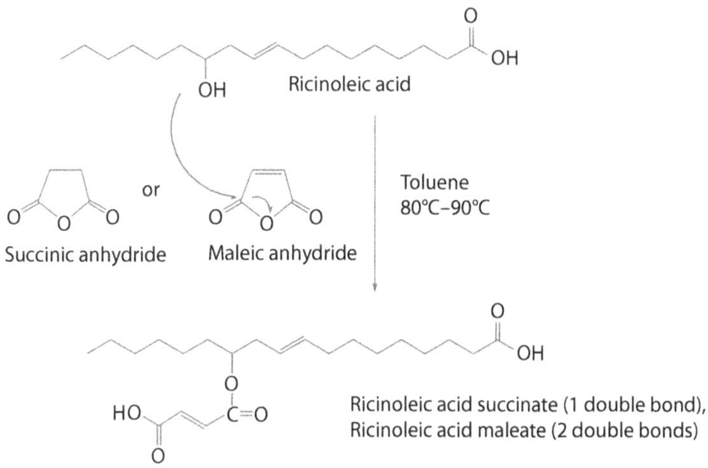

The properties of polyanhydrides were modified by the incorporation on long chain fatty acid terminals such as stearic acid in the polymer composition, which alters its hydrophobicity and decreases its degradation rate (Teomim and Domb 1999). Since natural fatty acids are monofunctional, they would act as polymerization chain terminators and control the molecular weight. A detailed analysis of the polymerization reaction shows that up to about 10 mole% content of stearic acid, the final product is essentially a stearic acid terminated polymer. Whereas higher amounts of acetyl stearate in the reaction mixture resulted in the formation of increasing amounts of stearic anhydride byproduct with minimal effect on the polymer molecular weight, which remains in the range of 5000. Physical mixtures of polyanhydrides with triglycerides and fatty acids or alcohols did not form uniform blends.

The potential of poly 1,3-bis-(p-carboxyphenoxy)propane-co-sebacic acid (p(CPP:SA)) microspheres was investigated for controlled delivery of basal insulin. The results indicate that CPP:SA microspheres controlled insulin release in vitro and in vivo over a month and the released insulin was conformationally and chemically stable and bioactive (Manoharan and Singh 2009).

Poly(1,3-bis-(p-carboxyphenoxy propane)-co-(sebacic anhydride) (p(CPP-SA)) was used to prepare microspheres, with human serum albumin (HSA) as the model protein. The structural integrity of HSA extracted from microspheres was detected by gel permeation chromatography, compared with native HSA. The results showed HSA remained its molecule weight after encapsulated (Sun et al. 2009).

Slow drug release in osteomyelitis treatment is an important biomedical problem. Therefore, a sustained-release bead system consisting of gentamicin sulfate in biodegradable poly(dimer acid-tetradecandioic acid) copolymer [P(DA-TA), WDA: WTA = 50: 50] is prepared by melt casting for osteomyelitis treatment. In vitro bacteriostatic activity studies demonstrated that the beads possessed desired bacteriostatic activity for *Staphylococcus aureus* and *Escherichia coli*, which are common bacteria for infections in bone (Guo et al. 2007).

In another application, novel amphiphilic biodegradable systems based on polyanhydrides for the stabilization and sustained release of peptides and proteins were reported (Torres et al. 2007). These results indicated polyanhydride carriers for the stabilization and sustained release of therapeutic peptides and proteins.

A series of biodegradable pasty poly(ester anhydride)s were prepared from alkanedicarboxylic acids and ricinoleic acid and its oligomers by transesterification-repolymerization method by Krasko and Domb. Pasty polymers can be mixed with drugs at room temperature and injected to tissue for delivery of drugs, particularly for heat-sensitive drugs. The polymers release model drugs for a few weeks while being degraded to their fatty acid counterparts. Copolymerization of alkanedicarboxylic acids with ricinoleic acid resulted in pasty biodegradable polymers useful as injectable carriers for drugs (Krasko and Domb 2007).

Recently as another application of polyanhydrides, Agueros et al. have reported the effect of the combination between 2-hydroxypropyl-beta-cyclodextrin and bioadhesive polyanhydride nanoparticles on the encapsulation and intestinal permeability of paclitaxel. A solid inclusion complex between drug and cyclodextrin was prepared by an evaporation method. The complex was incorporated in polyanhydride nanoparticles. The association between these three components would induce a positive effect over the intestinal permeability of paclitaxel (Agueros et al. 2009).

5.3 BIOCOMPATIBILITY AND TOXICITY

In all the potential uses of polymeric material, a direct contact between the polymer and biological tissues is evident. Therefore, for the eventual human application of these biomedical implants and devices, an adequate testing for safety and biocompatibility of the specific matrix polymer used in each case is essential.

Whenever a synthetic polymer material is to be utilized in vivo, the possible tissue–implant interactions must be taken into consideration. In the case of biodegradable matrices, not only the possible toxicity of the polymer has to be evaluated, but also the potential toxicity of its degradation products. The last section of this chapter reviews existing data about the biocompatibility and toxicity of the different polymers actually available for biomedical applications.

5.3.1 Polyesters

5.3.1.1 Lactide/Glycolide Copolymers

Many biodegradable polymers have been developed for controlled drug delivery. The plethora of drug therapies and types of drugs demand different formulations, fabrications conditions, and release kinetics. Wen et al. have reported that no one single polymer can satisfy all the requirements (Wen et al. 2003).

The feasibility of lactide/glycolide polymers as excipients for the controlled release of bioactive agents is well proven, and they are the most widely investigated biodegradable polymers for drug delivery. Most of the research work on the use of lactide/glycolide polymers as matrices for delivery systems has been focused on the development of injectable microspheres formulations, although implantable rod and pellet devices are also being investigated.

The lactide/glycolide copolymers have been subjected to extensive animal and human trials without any significant harmful side-effects (Lewis 1990). No evidence of inflammatory response, irritation, or other adverse effects has been reported upon implantation of lactide/glycolide polymer devices.

Many conventional pharmaceutical agents formulated in lactide/glycolide polymer matrices have been widely studied since almost three decades. One of the most successful lactide/glycolide drug delivery formulations, in terms of clinical results obtained, is the steroid-loaded injectable microspheres for the controlled release of contraceptives (Beck et al. 1981, 1983a,b). Many animal and clinical trials with these systems were performed showing very good biocompatibility. The success of the steroid microsphere system based on lactide/glycolide matrices is probably due to the combination of several factors: the reproducibility of the microencapsulation process, the in vivo drug-release performance, reliability in the treatment procedure, and the safety of the polymer.

Lactide/glycolide implants containing Naltrexone and other narcotic antagonist agents have also been extensively studied (Schwope et al. 1975; Chiang et al. 1984). In one of these studies describing the clinical evaluation of a bead preparation containing 70% naltrexone and 30% of a 90:10 lactide/glycolide copolymer, a local inflammatory reaction at the site of implantation was reported in two of three subjects after subcutaneous implantation of the beads containing the drug (Chiang et al. 1984). This finding prevented further clinical testing of that particular formulation. No similar problems were reported with other lactide/glycolide polymer preparations, and that incident was related to some unique aspect of that product (Chiang et al. 1984).

Good biocompatibility data were also reported with lactide/glycolide copolymer matrices containing antineoplastic drugs, antibiotics, and anti-inflammatory compounds (Lewis 1990). The delivery of therapeutic molecules to the brain has been limited in part due to the presence of the blood–brain barrier. The biocompatibility of lactide/glycolide copolymer in the brain was examined regarding the gliotic response following implants of lactide/glycolide copolymer into the brains of rats. It was found that lactide/glycolide copolymer is well tolerated following implantation into the CNS and that the astrocytic response to lactide/glycolide copolymer is largely a consequence of the mechanical trauma that occurs during surgery (Emerich et al. 1999).

Studies carried out for the encapsulation of bioactive macromolecules as proteins, peptides, and antigens showed mixed biocompatibility results. Serious problems in achieving long-term release have been reported in several cases where the macromolecules lost bioactivity in vivo after a few days, as in the case of lactide/glycolide copolymer containing growth hormone. A complex interaction apparently occurs in vivo between the acidic polymer and the hormone. In vitro release studies have shown that growth hormone can become insoluble when incorporated in poly(lactide) films.

Similar problems in maintaining biological activity for longer than 5–10 days were also observed with interferon-lactide/glycolide polymer formulation (Lewis 1990). On the other hand, promising results were obtained with luteinizing hormone releasing hormone (LHRH) incorporated in lactide/glycolide polymer showing long-term delivery of the macromolecule (Asch et al. 1985a,b). In general, hydrophilic polypeptides of low molecular weight (<5000) are considered quite stable in the presence of lactide/glycolide excipients and their acidic bioerosion byproducts.

Lactide/glycolide polymer implants, in the form of microbeads and pellets, containing insulin were reported to be effective in lowering blood glucose levels in diabetic rats for about 2 weeks, and no adverse effects, inflammation at the implant site, or deactivation of the macromolecule were observed (Kwong et al. 1986).

Chen and Singh have designed injectable controlled release polymer formulations for growth hormone using triblock copolymer PLGA-PEG-PLGA. The triblock copolymer used in this study was able to control the release of incorporated growth hormone in vitro and in vivo for longer duration. Both in vitro and in vivo results support the biocompatible nature of these polymer delivery systems (Chen and Singh 2008).

Bunger et al. have determined the feasibility and biocompatibility of a sirolimus eluting biodegradable poly(L-lactide) stent for peripheral vascular application. Histological analysis showed that these stents were without any signs of excessive recoiling or collapse i.e., they showed sufficient mechanical stability. The vascular inflammation against these stents was reduced by incorporation sirolimus in these stents. No systemic toxicity or thrombotic complications was observed with these stents (Bunger et al. 2007).

Aliphatic polyesters were tested for proliferation of human fibroblasts. Results obtained provide the non-in vitro toxicity of the materials and of the processing method for the scaffolds preparation in tissue engineering (Vaquette et al. 2006). Since biocompatibility of polylactic acid and polyglycolic acid copolymers is affected by chemical composition, molecular weight, cell environment, and by the methods of polymerization and processing, Di Toro et al. (2004) have determined all these factors and reported that these materials do not show any negative influence on cell proliferation and differentiation.

5.3.1.2 Poly(Caprolactone)

The biocompatibility and toxicity of poly(caprolactone) have mostly been tested in conjunction with evaluations of Capronor™, which is an implantable 1 year contraceptive delivery system composed of a levonorgestrel-ethyl oleate slurry within a poly(caprolactone) capsule. In a preliminary 90-day toxicology study of Capronor in female rats and guinea pigs, except for a bland response at the implant site and a minimal tissue encapsulating reaction, no toxic effects were observed (Pitt and Schindler 1984).

The Capronor contraceptive delivery system was also tested implanted in rats and monkeys in 2 year period (Pitt 1990). The results of this study based on animal clinical and physical data as blood and urine analysis, ophthalmoscopic tests, and histopathology after necropsy showed no significant differences between the test and control groups.

Biocompatibility and toxicology of poly(caprolactone), which was synthesized using $Ti(OBu)_4$ as an initiator, were determined (Chen et al. 2000). The results obtained showed that poly(caprolactone) films have a mild inflammatory reaction in the early days of planting and after 3 months the inflammation disappeared. They conclude that poly(caprolactone) material possesses good biocompatibility and has a good developmental prospect.

Byrro et al. reported microspheres of poly(caprolactone) containing prednisolone acetate. There microspheres were developed and evaluated as an injectable extended-release formulation. The local biocompatibility of the system was verified by histopathological analysis of the deployment region. Preliminary in vivo study showed the absence of local toxicity and confirmed the prolonged release profile of drug, suggesting the viability of the developed system for the treatment of orbital inflammatory diseases (Byrro et al. 2009).

Poly(caprolactone)-poly(ethylene glycol) copolymers are synthetic biomedical materials with amphiphilicity, controlled biodegradability, and great biocompatibility according to Wei et al. (2009). They have great potential application in the fields of nanotechnology, tissue engineering, pharmaceutics, and medicinal chemistry.

5.3.1.3 Polycarbonates

Biodegradation and biocompatibility of poly(ethylene carbonate) were examined using an in vivo cage implant system by Dadsetan et al. Exudate analysis showed the degradation products were biocompatible and induced minimal inflammatory and wound healing responses(Dadsetan et al. 2003).

To evaluate the technical feasibility of poly(ethylene carbonate), for injectable in situ forming drug delivery systems, the physical properties of poly(ethylene carbonate) solutions were characterized. This study demonstrated that in situ depot forming systems can be obtained from poly(ethylene carbonate) solutions (Liu et al. 2010b).

A study exploring the utility of poly(ethylene carbonate) as coating material for drug-eluting stents under in vitro conditions was reported. Poly(ethylene carbonate) (Mw 242 kDa) was found to be an amorphous polymer with thermoelastic properties. Tensile testing revealed a stress to strain failure of more than 600%. These properties are thought to be advantageous for expanding coated stents. In vitro cytotoxicity tests showed excellent cytocompatibility of poly(ethylene carbonate) (Unger et al. 2007).

5.3.2 POLY(AMIDES)

5.3.2.1 Natural Polymers

The use of natural biodegradable polymers to deliver drugs continues to be an area of active research despite the advent of synthetic biodegradable polymers. Natural polymers remain attractive primarily because they are natural products of living organisms, readily available, relatively inexpensive, and capable of a multitude of chemical modifications (Bogdansky 1990).

Most of the investigations of natural polymers as matrices in drug delivery systems have focused on the use of proteins (polypeptides or polyamides) as gelatin, collagen, and albumin. Collagens are the structural building blocks of the body. They are the most abundant group of organic macromolecules in an organism and serve important. Collagen is a major structural protein found in animal tissues where it is normally present in the form aligned fibers. The main applications of collagen are collagen shields in ophthalmology, sponges for burns/wounds, minipellets for protein delivery, and as basic matrices for cell culture systems in bone, tendon, and peripheral nerve repair (Sano et al. 1998). Collagen can be processed into a number of forms such as sheets, films, tubes, rods, sponges, powders, injectable solutions, and dispersions. In most drug delivery systems made of collagen, in vivo absorption of collagen is controlled by the use of cross-linking agents such as glutaraldehyde. In order to render collagen suitable for tissue engineering applications, the mechanical strength of collagen must be enhanced (Friess 1998; Sano et al. 1998; Lee et al. 2001).

Albumin in microspheres form was approved by the U.S. Food and Drug Administration in 1972 for clinical use (Schafer et al. 1994) and has subsequently found many applications in diagnostics and therapeutics. As a result of this, more than 100 diagnostic agents and drugs have been incorporated into albumin microspheres (Dandagi et al. 2006). Noncollagenous proteins, particularly albumin and to a lesser extent gelatin, continue to be developed as drug delivery vehicles (Adhirajan et al. 2009; Ohta et al. 2009; Samad et al. 2009). The exploitable features of albumin include its reported biodegradation into natural products and lack of toxicity. Although many examples on the use of albumin microspheres were reported in the literature, there are only a few studies describing gelatin systems. Compared with albumin, gelatin offers the advantages of a good history in parenteral formulations and lower antigenicity. A biocompatibility study with gelatin microspheres reported no untoward effects when injected intravenously into mice over a 12 week period.

5.3.2.2   Pseudopoly(Amino Acids)

Pseudopoly(amino acids) belonging to the poly(amides) group represents one of the new and, therefore, less advanced biodegradable polymers for medical use. Only a few of them have been synthesized and characterized. Preliminary results showed no gross toxicity or tissue incompatibility upon subcutaneous implantation for present available pseudopoly(amino acids), and they are actively investigated in several laboratories for medical applications ranging from biodegradable bone nails to implantable adjuvant (Kemnitzer and Kohn 1997).

In a biocompatibility study for the cornea bioassay, implantation of four small pieces of the polymer into four rabbit corneas elicited on pathological response in three of them and a very mild inflammatory response in one cornea (Yu 1988). Histological examination of the corneas 4 weeks after implantation showed no invading blood vessels or migrating inflammatory cells in the area around the implants (Langer et al. 1981). In another biocompatibility study, 10mg implants of poly(N-palmitoylhydroxyproline ester) were inserted subcutaneously in mice in the dorsal area of the animals between the dermis and the adipose tissue layer. The mice responses were examined up to 1 year post-implantation (Yu 1988). No inflammatory reaction at the implantation site was observed in any of the mice during the first 7 weeks. A thin but distinct layer of fibrous connective tissue appeared around the implant by week 14 with a few multinucleated giant cells associated with the fibrous connective tissue. Animal necropsy performed at 16 and 56 weeks post-implantation showed no histopathological abnormalities (Yu 1988).

These preliminary biocompatibility results from two animal models, rabbit and mouse, indicate that the poly(N-palmitoylhydroxyproline ester) elicits a very mild, local tissue reaction typical of a foreign body response, resulting in the encapsulation of the foreign biomaterial without evidence for any significant pathological abnormalities. However, additional safety and toxicity tests need to be performed to evaluate possible systemic toxic effects, allergic reactions, or any mutagenic, teratogenic, and carcinogenic activities.

Pseudopoly(amino acids) containing aromatic side chains of tyrosine in their backbone structure were also developed to investigate whether biodegradable polymers that incorporate aromatic components would combine a high degree of biocompatibility with a high degree of mechanical strength (Kemnitzer and Kohn 1997). Ertel et al. (1995) prepared a new pseudopoly(amino acid) and determined the performance of this orthopedic implant material.

5.3.3   POLYPHOSPHAZENES

Two different types of polyphosphazenes are of interest as bioinert materials: those with strongly hydrophobic surfaces characteristics and those with hydrophilic surfaces. Polyphosphazenes bearing fluoroalkoxy side groups are some of the most hydrophobic synthetic polymers known. Such polymers are as hydrophobic as poly(tetrafluoroethylene) (Teflon), but unlike Teflon, polyphosphazenes of this type are flexible or elastomeric, easy to prepare, and they can be used as coatings for other materials.

Biocompatibility and safety testing of these polymers by subcutaneous implantation in animals have shown minimal tissue response, similar in fact to the response reported with Teflon (Wade et al. 1978). The connection between hydrophobicity and tissue compatibility has been noted for classical organic polymers (Lyman and Knutson 1980). Thus, these hydrophobic polyphosphazenes have been mentioned as good candidates for use in heart valves, heart pumps, blood vessel prostheses, or as coating materials for pacemakers or other implantable devices; however, more in vivo testing and clinical trials are needed.

In their bioerosion reactions, polyphosphazenes display a uniqueness that stems from the presence of the inorganic backbone, which in the presence of appropriate side groups is capable of undergoing facile hydrolysis to phosphate and ammonia. The phosphate can be metabolized and the ammonia excreted. Theoretically, if side groups attached to the polymer are released by the same process being excretable or metabolizable, then the polymer can be eroded under hydrolytic conditions without the danger of a toxic response. Polyphosphazenes of this type are potential candidates as erodible biostructural materials for sutures, or as matrices for controlled delivery of drugs (Kohn and Langer 1987).

Polyphosphazenes containing amino acid ester side groups were the first bioerodible polyphosphazenes synthesized (Allcock et al. 1977). They are solid materials which erode hydrolytically to ethanol, glycine, phosphate, and ammonia. These polymers were tested in subcutaneous tissue response experiments showing no evidences of irritation, cell toxicity, giant cell formation, or tissue inflammation (Allcock and Kwon 1986). Imidazoyl groups linked to polyphosphazene chains are also hydrolyzed very easily showing good biocompatibility (Laurencin et al. 1993).

By changing the molar ratio of hydrophilic and hydrophobic segments, a series of novel amphiphilic graft polyphosphazenes was synthesized (Qiu and Yan 2009). Doxorubicin was physically loaded into micelles prepared by an O/W emulsion method. The results of cytotoxicity study showed that amphiphilic graft polyphosphazenes were biocompatible while doxorubicin-loaded micelles achieved comparable cytotoxicity with that of free doxorubicin.

5.3.4 POLY(ORTHOESTERS)

Polymers with ester linkages in their main chain comprise a family of polymers with immense diversity and versatility. Biodegradability is sought after in a wide range of applications, above all in the preparation of environmentally friendly polymers and biomedical materials for temporary surgical use and in drug delivery (Edlund and Albertsson 2003). As mentioned previously, the Chronomer® poly(orthoester) material from Alza Corp. or Alzamer has been investigated as bioerodible inserts for the delivery of the narcotic antagonist naltrexone and for the delivery of the contraceptive steroid norethisterone. The steroidal implant was tested in two separate human clinical trials causing local tissue irritation, and therefore further work with this formulation was discontinued (Solheim et al. 1995). The reasons for the local irritation were never properly elucidated. New types of POEswere developed but no data on the biocompatibility and safety of these materials were reported.

5.3.5 POLYANHYDRIDES

Polyanhydrides are a novel class of biodegradable polymers under development as vehicles for the release of bioactive molecules including drug peptides and proteins. A series of biocompatibility studies reported on several polyanhydrides have shown them to be nonmutagenic and nontoxic (Leong et al. 1986). In vitro tests measuring teratogenic potential were also negative. Growth of two types of mammalian cells in tissue culture was also not affected by the polyanhydride polymers (Leong et al. 1986); both the cellular doubling time and cellular morphology were unchanged when either bovine aorta endothelial cells or smooth muscle cells were grown directly on the polymeric substrate.

Subcutaneous implantation in rats of high doses of the 20:80 copolymer of bis-(p-carboxyphenoxy)propane (CPP) and sebacic acid (SA) for up to 8 weeks indicated relatively minimal tissue irritation with no evidence of local or systemic toxicity (Laurencin et al. 1990). Since this polymer was designed to be used clinically to deliver an anticancer agent directly into the brain for the treatment of brain neoplasms, its biocompatibility in rat brain was also studied (Tamargo et al. 1989). The tissue reaction of the polymer was compared to the reaction observed with two control materials used in surgery, oxidized cellulose absorbable hemostat (Surgicel®, Johnson and Johnson) and with absorbable gelatin sponge (Gelfoam®, Upjohn). The inflammatory reaction of the polymer was intermediate between the controls (Tamargo et al. 1989). A closely related polyanhydride copolymer poly(CPP)-SA 50:50 was also implanted in rabbit brains and was found to be essentially equivalent to Gelfoam in terms of biocompatibility evaluations (Brem et al. 1989). In a similar study conducted in monkey brains, no abnormalities were noted in the CT scans and magnetic resonance image, nor in the blood chemistry or hematology evaluations (Brem et al. 1988a). No systemic effects of the implants were observed on histological examinations of any of the tissues tested (Brem et al. 1988b).

Different classes of polyanhydrides have been synthesized and are undergoing extensive preclinical testing, including a wide range of biocompatibility studies. Examples of these new materials are polymers of sebacic acid poly(SA), and copolymers 1:1 of SA with fatty acid dimer (FAD), fumaric acid (FA) and isophthalic acid poly(FAD:SA), poly(FA:SA), poly(ISO:SA). These materials were implanted in rabbits intramuscularly, subcutaneously and in the cornea, and ocular and muscle irritation studies were performed compared to the material controls Gelfoam, Surgicel, and Vycryl® a synthetic absorbable suture (Ethicon) (Rock et al. 1991). Detailed observations of toxicity, bleeding, selling, or infection of the implantation site were conducted daily. At the end of the study, the animals were sacrificed and a gross necropsy examination of the tissues surrounding the implant site was performed. No significant clinical signs or abnormalities of the incision sites were observed during the study period (4 weeks). No meaningful differences could be seen in reaction between the various polymer implants tested and the control materials (Rock et al. 1991).

In the rabbit cornea bioassay, no evidence of inflammatory response was observed with any of the implants at any time. On an average, the bulk of the polymers disappeared completely between 7 and 14 days after the implantation (Rock et al. 1991). The cornea is a very sensitive indicator of inflammatory reactions (Aronson and Horton 1971; Henkind 1978). The rabbit cornea possesses clear advantages over other implant sites for studying implant–host interactions, due to the easy accessibility for frequent observations without having to gain surgical access to the implantation site; the transparency and avascularity of the cornea also enable the observer to distinguish among the different inflammatory characteristics such as edema, cellular infiltration, and ingrowths of blood vessels from the periphery of the cornea or neovascularization, which are strong indications that the biomaterial under testing is unsuitable for implantation (Langer et al. 1981).

In similar animal experiments in which polyanhydride matrices containing tumor angiogenic factor (TAF) were implanted in rabbit cornea, a significant vascularization response was observed without edema or white cells. Moreover, and most importantly from the biocompatibility standpoint, polymer matrices without incorporated TAF showed no adverse vascular response (Langer et al. 1981, 1985).

The biocompatibility of polyanhydrides based on ricinoleic acid as compared to Vicryl™ surgical suture and sham surgery was tested in rats (Teomim et al. 1999). No evidence for tissue necrosis was detected in any of the treated animals upon evaluation of the tissues 21 days post-implantation. No indication for post-implantation test site contamination was noted. Generally, in all tested groups there was a clear indication of time-related healing process (i.e., time-related reduction in the incidence and severity of necrosis and acute to subacute inflammatory reaction, associated with increased fibroplasia). The tissue reaction in the different treatment groups showed a clear trend of healing with time. In particular, comparison of the tissue reactivity, along the three different time periods (3, 7, and 21 days), indicated that the only noted remnants of tissue reaction at the 21 day time period were minimal subacute inflammation and mild fibrosis. In comparison, under the same conditions, Vicryl™ implant induced minimal fibrosis associated with the presence of minimal quantity of giant cells and encapsulated foreign material.

Based on the biocompatibility and safety preclinical studies carried out in rats (Tamargo et al. 1989; Laurencin et al. 1990), rabbits (Brem et al. 1989), and monkeys (Brem et al. 1988a,b) reviewed here showing quite acceptability of the polyanhydrides for human use, a Phase I/II clinical protocol was instituted (Domb et al. 1991a). In these clinical trials, a polyanhydride dosage form (Gliadel®) consisting of wafer polymer implants of poly(CPP-SA) 20:80 and containing the chemotherapeutic agent Carmustine (BCNU) were used for the treatment of glioblastoma multiforme, a universally fatal form of brain cancer. In these studies, up to eight of these wafer implants were placed to line the surgical cavity created during the surgical debulking of the brain tumor in patients undergoing a second operation for surgical debulking of either a Grade III or

IV anaplastic astrocytoma. In keeping with the results of the earlier preclinical studies suggesting a lack of toxicity, no central or systemic toxicity of the treatment was observed during the course of treating 21 patients under this protocol. Phase III human clinical trials have demonstrated that site-specific delivery of BCNU from a P(CPP:SA)20:80 wafer (Gliadel®) in patients with recurring brain cancer (glioblastoma multiforme) significantly prolongs patients survival (Brem et al. 1995). Gliadel® finally won approval from the FDA adjust therapy for the treatment of brain tumors.

Biocompatibility of an injectable gelling polymeric device for the controlled release of gentamicin sulfate in the treatment of invasive bacterial infections in bone of male Wister rats. The biodegradable delivery carrier, poly(sebacic-co-ricinoleic-ester-anhydride), designated as p(SA:RA), was injected, with and without gentamicin, into the tibial canal. Rats were killed 3 weeks later and the tibiae were processed histologically. The local tissue reaction to the polymer with or without antibiotic consisted mainly of mild reactive fibroplasia/fibrosis and mild to moderate increased reactive bone formation. No evidence for any active inflammatory response to the polymer were observed; thus, the injection of p(SA:RA) was well tolerated (Brin et al. 2009).

Low molecular weight hydroxy fatty acid based polyanhydrides were synthesized by Jain et al. Control over drug release was accessed with drugs featuring different aqueous solubility, i.e., methotrexate (hydrophobic) and 5-fluorouracil (hydrophilic). Due to their low melting temperatures, they can be injected locally (SC or intratumorally) to from regional in situ depot and have a great potential as a drug carrier for localized delivery of anticancer drugs. These polymers were found good to control the release of the entrapped drug and were found biocompatible in a preliminary in vivo study (Jain et al. 2008).

Cell–polymer interactions in subcutaneous and bony tissue were examined for a novel class of in situ forming and surface eroding polyanhydride networks. Specifically, photopolymerized disks of several polyanhydride compositions were implanted subcutaneously in rats, and the tissue was analyzed for an inflammatory response (Poshusta et al. 2003).

REFERENCES

Adhirajan, N., N. Shanmugasundaram, S. Shanmuganathan, and M. Babu. 2009. Functionally modified gelatin microspheres impregnated collagen scaffold as novel wound dressing to attenuate the proteases and bacterial growth. *Eur J Pharm Sci* 36 (2–3):235–245.

Agueros, M., L. Ruiz-Gaton, C. Vauthier, K. Bouchemal, S. Espuelas, G. Ponchel, and J. M. Irache. 2009. Combined hydroxypropyl-beta-cyclodextrin and poly(anhydride) nanoparticles improve the oral permeability of paclitaxel. *Eur J Pharm Sci* 38 (4):405–413.

Albertsson, A. C. and S. Lundmark. 1988. Synthesis of poly(adipic anhydride) by use of ketene. 25 (3):247–258.

Allcock, H. R. 1976. Polyphosphazenes: New polymers with inorganic backbone atoms. *Science* 193 (4259):1214–1219.

Allcock, H. R. and A. M. Ambrosio. 1996. Synthesis and characterization of pH-sensitive poly(organophosphazene) hydrogels. *Biomaterials* 17 (23):2295–2302.

Allcock, H. R. and T. J. Fuller. 1980. Phosphazene high polymers with steroidal side groups. *Macromolecules* 13 (6):1338–1345.

Allcock, H. R., T. J. Fuller, D. P. Mack, K. Matsumura, and K. M. Smeltz. 1977. Synthesis of Poly[(amino acid alkyl ester)phosphazenes]. *Macromolecules* 10 (4):824–830.

Allcock, H. R. and R. L. Kugel. 1966. High molecular weight poly(diaminophosphazenes). *Inorg Chem* 5:1716.

Allcock, H. R. and S. Kwon. 1986. Covalent linkage of proteins to surface-modified poly(organophosphazenes): Immobilization of glucose-6-phosphate dehydrogenase and trypsin. *Macromolecules* 19 (6):1502–1508.

Allcock, H. R., A. Singh, A. M. Ambrosio, and W. R. Laredo. 2003. Tyrosine-bearing polyphosphazenes. *Biomacromolecules* 4 (6):1646–1653.

Allcock, H. R., L. B. Steely, S. H. Kim, J. H. Kim, and B. K. Kang. 2007. Plasma surface functionalization of poly[bis(2,2,2-trifluoroethoxy)phosphazene] films and nanofibers. *Langmuir* 23 (15):8103–8107.

Amor, S. R., T. Rayment, and J. K. M. Sanders. 1991. Polyhydroxybutyrate in vivo: NMR and X-ray diffraction characterisation of the elastomeric state. *Macromolecules* 24:4583–4588.

Anonymous. 1985. Capronor. *Hypotenuse*, May–Jun:2–5.

Anonymous. 1991. Phase II—Clinical trial with biodegradable subdermal contraceptive implant Capronor (4.0-cm single implant). Indian Council of Medical Research Task Force on Hormonal Contraception. *Contraception* 44 (4):409–417.

Armstrong, D. K., G. F. Fleming, M. Markman, and H. H. Bailey. 2006. A phase I trial of intraperitoneal sustained-release paclitaxel microspheres (Paclimer) in recurrent ovarian cancer: A Gynecologic Oncology Group study. *Gynecol Oncol* 103 (2):391–396.

Aronson, S. B. and R. C. Horton. 1971. Mechanisms of the host response in the eye. VII. The normal rabbit eye in anterior ocular inflammation. *Arch Ophthalmol* 85 (3):306–308.

Arun, A., R. Arthi, V. Shanmugabalaji, and M. Eyini. 2009. Microbial production of poly-beta-hydroxy-butyrate by marine microbes isolated from various marine environments. *Bioresour Technol* 100 (7):2320–2323.

Asch, R. H., F. J. Rojas, A. Bartke, A. V. Schally, T. R. Tice, H. G. Klemcke, T. M. Siler-Khodr, R. E. Bray, and M. P. Hogan. 1985a. Prolonged suppression of plasma LH levels in male rats after a single injection of an LH-RH agonist in poly(DL-lactide-co-glycolide) microcapsules. *J Androl* 6 (2):83–88.

Asch, R. H., F. J. Rojas, T. R. Tice, and A. V. Schally. 1985b. Studies of a controlled-release microcapsule formulation of an LH-RH agonist (D-Trp-6-LH-RH) in the rhesus monkey menstrual cycle. *Int J Fertil* 30 (2):19–26.

Bailey, L. O., M. L. Becker, J. S. Stephens, N. D. Gallant, C. M. Mahoney, N. R. Washburn, A. Rege, J. Kohn, and E. J. Amis. 2006. Cellular response to phase-separated blends of tyrosine-derived polycarbonates. *J Biomed Mater Res A* 76 (3):491–502.

Beck, L. R., C. E. Flowers, V. Z. Pope, W. H. Wilborn, and T. R. Tice. 1983a. Clinical evaluation of an improved injectable microcapsule contraceptive system. *Am J Obstet Gynecol* 147 (7):815–821.

Beck, L. R., V. Z. Pope, C. E. Flowers, D. R. Cowsar, T. R. Tice, D. H. Lewis, R. L. Dunn, A. B. Moore, and R. M. Gilley. 1983b. Poly(DL-lactide-co-glycolide)/norethisterone microcapsules: An injectable biode-gradable contraceptive. *Biol Reprod* 28 (1):186–195.

Beck, L. R., R. A. Ramos, C. E. Flowers, G. Z. Lopez, D. H. Lewis, and D. R. Cowsar. 1981. Clinical evaluation of injectable biodegradable contraceptive system. *Am J Obstet Gynecol* 140 (7):799–806.

Belen'Kaya, B. G., Ye B. Lyudvig, A. L. Izyumnikov, and Yu I. Kul'Velis. 1982. Aspects of the cationic polym-erization of [epsilon]-caprolactone in the presence of alcohols. *Polym Sci U.S.S.R.* 24 (2):306–313.

Benagiano, G. and H. L. Gabelnick. 1979. Biodegradable systems for the sustained release of fertility-regulating agents. *J Steroid Biochem* 11 (1B):449–455.

Bennett, D. B., X. Li, N. W. Adams, S. W. Kim, C. J. T. Hoes, and J. Feijen. 1991. Biodegradable polymeric prodrugs of naltrexone. *J Control Release* 16 (1–2):43–52.

Bezwada, R. S., S. W. Shalaby, H. D. Newman, and A. Kafrauy. 1990. Bioabsorbable copolymers of p-dioxane and lactide for surgical devices. *Trans Soc Biomater* 13:194.

Bobo, W. V. and R. C. Shelton. 2010. Risperidone long-acting injectable (Risperdal Consta) for maintenance treatment in patients with bipolar disorder. *Expert Rev Neurother* 10 (11):1637–1658.

Bodmeier, R., K. H. Oh, and H. Chen. 1989. The effect of the addition of low molecular weight poly(DL-lactide) on drug release from biodegradable poly(DL-lactide) drug delivery systems. *Int J Pharm* 51 (1):1–8.

Bogdansky, S. 1990. Natural polymers as drug delivery systems, In *Biodegradable Polymers as Drug Delivery Systems*. M. Chasin and R. S. Langer, eds. New York: Marcel Dekker. pp. 231–259.

Bourke, S. L. and J. Kohn. 2003. Polymers derived from the amino acid -tyrosine: Polycarbonates, polyarylates and copolymers with poly(ethylene glycol). *Adv Drug Deliv Rev* 55 (4):447–466.

Brandl, H., R. A. Gross, R. W. Lenz, and R. C. Fuller. 1988. *Pseudomonas oleovorans* as a source of poly(β-hydroxyalkanoates) for potential applications as biodegradable polyesters. *Appl Environ Microbiol* 54 (8):1977–1982.

Brem, H., H. Ahn, R. J. Tamargo, M. Pinn, and M. Chasin. 1988a. A biodegradable polymer for intracranial drug delivery: A radiological study in primates. *Am Assoc Neurol Surg* 24:349.

Brem, H., A. Kader, J. I. Epstein, R. J. Tamargo, A. J. Domb, R. Langer, and K. W. Leong. 1989. Biocompatibility of a biodegradable, controlled-release polymer in the rabbit brain. *Sel Cancer Ther* 5 (2):55–65.

Brem, H., S. Piantadosi, P. C. Burger, M. Walker, R. Selker, N. A. Vick, K. Black et al. 1995. Placebo-controlled trial of safety and efficacy of intraoperative controlled delivery by biodegradable polymers of chemother-apy for recurrent gliomas. The Polymer-brain Tumor Treatment Group. *Lancet* 345 (8956):1008–1012.

Brem, H., R. J. Tamargo, M. Pinn, and M. Chasin. 1988b. Biocompatibility of BCNU-loaded biodegradable polymer: A toxicity study in primates. *Am Assoc Neurol Surg* 24:381.

Brin, Y. S., A. Nyska, A. J. Domb, J. Golenser, B. Mizrahi, and M. Nyska. 2009. Biocompatibility of a poly-meric implant for the treatment of osteomyelitis. *J Biomater Sci Polym Ed* 20 (7–8):1081–1090.

Bucher, J. E. and W.C. Slade. 1909. The anhydrides of isophthalic and terephthalic acids. *J Am Chem Soc* 31:1319–1321.

Bunger, C. M., N. Grabow, K. Sternberg, C. Kroger, L. Ketner, K. P. Schmitz, H. J. Kreutzer, H. Ince, C. A. Nienaber, E. Klar, and W. Schareck. 2007. Sirolimus-eluting biodegradable poly-L-lactide stent for peripheral vascular application: A preliminary study in porcine carotid arteries. *J Surg Res* 139 (1):77–82.

Byrro, R. M., D. Miyashita, V. B. Albuquerque, A. A. Velasco e Cruz, and S. Cunha Junior Ada. 2009. Biodegradable systems containing prednisolone acetate for orbital administration. *Arq Bras Oftalmol* 72 (4):444–450.

Cao, F. L., Y. W. Xi, L. Tang, A. H. Yu, and G. X. Zhai. 2009. Preparation and characterization of curcumin loaded gelatin microspheres for lung targeting. *Zhong Yao Cai* 32 (3):423–426.

Chaubal, M. V., A. S. Gupta, S. T. Lopina, and D. F. Bruley. 2003. Polyphosphates and other phosphorus-containing polymers for drug delivery applications. *Crit Rev Ther Drug Carrier Syst* 20 (4):295–315.

Chen, C., L. Dong, and P. H. F. Yu. 2006. Characterization and properties of biodegradable poly(hydroxyalkanoates) and 4,4-dihydroxydiphenylpropane blends: Intermolecular hydrogen bonds, miscibility and crystallization. *Eur Polym J* 42:2838–2848.

Chen, J., C. Huang, and Z. Chen. 2000. Study on the biocompatibility and toxicology of biomaterials—Poly(epsilon-caprolactone). *Sheng Wu Yi Xue Gong Cheng Xue Za Zhi* 17 (4):380–382.

Chen, S. and J. Singh. 2008. Controlled release of growth hormone from thermosensitive triblock copolymer systems: In vitro and in vivo evaluation. *Int J Pharm* 352 (1–2):58–65.

Cheng, G., X. Fan, W. Pan, and Y Liu. 2010. Ring-opening polymerization of ε-caprolactone initiated by heteropolyacid. *J Polym Res* 17 (6):847–851.

Chiang, C. N., L. E. Hollister, A. Kishimoto, and G. Barnett. 1984. Kinetics of a naltrexone sustained-release preparation. *Clin Pharmacol Ther* 36 (5):704–708.

Choi, N. S. and J. Heller. 1979. Erodible agent releasing device comprising poly(orthoesters) and poly(orthocarbonates). In *US4138344* USA.

Conix, A. 1958. Aromatic polyanhydrides, a new class of high melting fiber-forming polymers. *J Polym Sci* 29:343–353.

Cook, D. M., B. M. Biller, M. L. Vance, A. R. Hoffman, L. S. Phillips, K. M. Ford, D. P. Benziger et al. 2002. The pharmacokinetic and pharmacodynamic characteristics of a long-acting growth hormone (GH) preparation (nutropin depot) in GH-deficient adults. *J Clin Endocrinol Metab* 87 (10):4508–4514.

Dadsetan, M., E. M. Christenson, F. Unger, M. Ausborn, T. Kissel, A. Hiltner, and J. M. Anderson. 2003. In vivo biocompatibility and biodegradation of poly(ethylene carbonate). *J Control Release* 93 (3):259–270.

Dandagi, P. M., V. S. Mastiholimath, M. B. Patil, and M. K. Gupta. 2006. Biodegradable microparticulate system of captopril. *Int J Pharm* 307 (1):83–88.

Danhier, F., B. Vroman, N. Lecouturier, N. Crokart, V. Pourcelle, H. Freichels, C. Jerome, J. Marchand-Brynaert, O. Feron, and V. Preat. 2009. Targeting of tumor endothelium by RGD-grafted PLGA-nanoparticles loaded with paclitaxel. *J Control Release* 140 (2):166–173.

Daniels, A. U., M. K. Chang, and K. P. Andriano. 1990. Mechanical properties of biodegradable polymers and composites proposed for internal fixation of bone. *J Appl Biomater* 1 (1):57–78.

Darney, P. D., S. E. Monroe, C. M. Klaisle, and A. Alvarado. 1989. Clinical evaluation of the Capronor contraceptive implant: Preliminary report. *Am J Obstet Gynecol* 160 (5 Pt 2):1292–1295.

Deng, J. S., L. Li, Y. Tian, E. Ginsburg, M. Widman, and A. Myers. 2003. In vitro characterization of polyorthoester microparticles containing bupivacaine. *Pharm Dev Technol* 8 (1):31–38.

Di Toro, R., V. Betti, and S. Spampinato. 2004. Biocompatibility and integrin-mediated adhesion of human osteoblasts to poly(DL-lactide-co-glycolide) copolymers. *Eur J Pharm Sci* 21 (2–3):161–169.

Dittrich, V. W. and R. C. Schultz. 1971. Kinetics and mechanism of the ring opening polymerization of L-lactide. *Angew Makromol Chem* 15:109.

Dlugi, A. M., J. D. Miller, and J. Knittle. 1990. Lupron depot (leuprolide acetate for depot suspension) in the treatment of endometriosis: A randomized, placebo-controlled, double-blind study. Lupron Study Group. *Fertil Steril* 54 (3):419–427.

Doi, Y., A. Tamaki, M. Kunioka, and K. Soga. 1988. Production of copolyesters of 3-hydroxybutirate and 3-hydroxyvalerate by *Alcanigenes eutrophus* from butyric and pentanoic acids. *Appl Microbiol Biotech* 28:330.

Domb, A. J. 1990. Biodegradable polymers derived from amino acids. *Biomaterials* 11 (9):686–689.

Domb, A. 1992. Synthesis and characterization of biodegradable aromatic anhydride copolymers. *Macromolecules* 25 (1):12–17.

Domb, A., C. Gallardo, and R. Langer. 1989. Poly(anhydrides) 3. Poly(anhydrides) based on aliphatic-aromatic diacids. *Macromolecules* 22:3200–3204.

Domb, A. and R. Langer. 1987. Polyanhydrides: I. preparation of high molecular weight polyanhydrides. *J Polym Sci* 25:3373–3386.

Domb, A. J. and R. Langer. 1989. Solid state and solution stability of poly(anhydrides) and poly(esters). *Macromolecules* 22:2117–2122.

Domb, A. J., C. T. Laurencin, O. Israeli, T. N. Gerhart, and R. Langer. 1990. The formation of propylene fumarate oligomers for use in bioerodible bone cement composites. *J Polym Sci A: Polym Chem* 28:973–985.

Domb, A. J. and M. Maniar. 1993. Absorbable biopolymers derived from dimer fatty acids. *J Polym Sci Polym Chem* 31 (5):1275.

Domb, A., M. Maniar, S. Bogdansky, and M. Chasin. 1991a. Drug delivery to the brain using polymers. *Crit Rev Ther Drug Carrier Syst* 8 (1):1–17.

Domb, A. J., N. Manor, and O. Elmalak. 1996. Biodegradable bone cement compositions based on acrylate and epoxide terminated poly(propylene fumarate) oligomers and calcium salt compositions. *Biomaterials* 17 (4):411–417.

Domb, A., E. Mathiowitz, E. Ron, S. Giannos, and R. Langer. 1991b. Polyanhydrides IV: Unsaturated and cross-linked polyanhydrides. *J Polym Sci* 29:571–579.

Domb, A. J. and R. Nudelman. 1995. In vivo and in vitro elimination of aliphatic polyanhydrides. *Biomaterials* 16 (4):319–323.

Domb, A., E. Ron, and R. Langer. 1988. Polyanhydrides II. One step polymerization using phosgene or diphosgene as coupling agents. *Macromolecules* 21:1925–1929.

Edlund, U. and A. C. Albertsson. 2003. Polyesters based on diacid monomers. *Adv Drug Deliv Rev* 55 (4):585–609.

El-Backly, R. M., A. G. Massoud, A. M. El-Badry, R. A. Sherif, and M. K. Marei. 2008. Regeneration of dentine/pulp-like tissue using a dental pulp stem cell/poly(lactic-co-glycolic) acid scaffold construct in New Zealand white rabbits. *Aust Endod J* 34 (2):52–67.

Emerich, D. F., M. A. Tracy, K. L. Ward, M. Figueiredo, R. Qian, C. Henschel, and R. T. Bartus. 1999. Biocompatibility of poly (DL-lactide-co-glycolide) microspheres implanted into the brain. *Cell Transplant* 8 (1):47–58.

Endo, M., T. Aida, and S. Inoue. 1987. Immortal polymerization of .epsilon.-caprolactone initiated by aluminum porphyrin in the presence of alcohol. *Macromolecules* 20 (12):2982–2988.

Engelberg, I. and J. Kohn. 1991. Physico-mechanical properties of degradable polymers used in medical applications: A comparative study. *Biomaterials* 12 (3):292–304.

Ertel, S. I., J. Kohn, M. C. Zimmerman, and J. R. Parsons. 1995. Evaluation of poly(DTH carbonate), a tyrosine-derived degradable polymer, for orthopedic applications. *J Biomed Mater Res* 29 (11):1337–1348.

Farng, E. and O. Sherman. 2004. Meniscal repair devices: A clinical and biomechanical literature review. *Arthroscopy* 20 (3):273–286.

Frank, A., S. K. Rath, and S. S. Venkatraman. 2005. Controlled release from bioerodible polymers: Effect of drug type and polymer composition. *J Control Release* 102 (2):333–344.

Friess, W. 1998. Collagen—Biomaterial for drug delivery. *Eur J Pharm Biopharm* 45 (2):113–136.

Frosch, K. H., T. Sawallich, G. Schutze, A. Losch, T. Walde, P. Balcarek, F. Konietschke, and K. M. Sturmer. 2009. Magnetic resonance imaging analysis of the bioabsorbable Milagro interference screw for graft fixation in anterior cruciate ligament reconstruction. *Strategies Trauma Limb Reconstr* 4 (2):73–79.

Ge, Z., J. C. Goh, L. Wang, E. P. Tan, and E. H. Lee. 2005. Characterization of knitted polymeric scaffolds for potential use in ligament tissue engineering. *J Biomater Sci Polym Ed* 16 (9):1179–1192.

Gefvert, O., B. Eriksson, P. Persson, L. Helldin, A. Bjorner, E. Mannaert, B. Remmerie, M. Eerdekens, and S. Nyberg. 2005. Pharmacokinetics and D2 receptor occupancy of long-acting injectable risperidone (Risperdal Consta) in patients with schizophrenia. *Int J Neuropsychopharmacol* 8 (1):27–36.

Gillissen, M., R. Steendam, A. van der Laan, and E. Tijsma. 2006. Development of doxycycline-eluting delivery systems based on SynBiosys biodegradable multi-block copolymers. *J Control Release* 116 (2):e90–e92.

Gopferich, A. 1997. Mechanisms of polymer degradation and elimination. In *Handbook of Biodegradable Polymers*, J. Kost, D. M. Wiseman, and A. J. Domb, eds., pp. 451–471. Amsterdam, the Netherlands: Harwood Academic.

Gopferich, A. and J. Tessmar. 2002. Polyanhydride degradation and erosion. *Adv Drug Deliv Rev* 54 (7):911–931.

Guo, W., Z. Shi, R. Guo, and R. Sun. 2007. Preparation of gentamicin sulfate-polyanhydride sustained-release beads and in vitro bacteriostatic activity studies. *Sheng Wu Yi Xue Gong Cheng Xue Za Zhi* 24 (2):360–362, 384.

Hamitou, A., T. Ouhadi, R. Jerome, and P. Teyssie. 1977. Soluble bimetallic μ-oxoalkoxides. VII. Characteristics and mechanism of ring-opening polymerization of lactones. *J Polym Sci A: Polym Chem* 15:865–873.

Harper, E., W. Dang, R. G. Lapidus, and R. I. Garver, Jr. 1999. Enhanced efficacy of a novel controlled release paclitaxel formulation (PACLIMER delivery system) for local-regional therapy of lung cancer tumor nodules in mice. *Clin Cancer Res* 5 (12):4242–4248.

Heller, J. 1990. Development of poly(ortho esters): A historical overview. *Biomaterials* 11 (9):659–665.

Heller, J., B. K. Fritzinger, S. Y. Ng, and D. W. H. Pennale. 1985. In vitro and in vivo release of levonorgestrel from poly(ortho esters): II. Crosslinked polymers. *J Control Release* 1 (3):233–238.

Heller, J., S. Y. Ng, and B. K. Fritzinger. 1992. Synthesis and characterization of a new family of poly(ortho esters). *Macromolecules* 25 (13):3362–3364.

Heller, J., S. Y. Ng, B. K. Fritzinger, and K. V. Roskos. 1990. Controlled drug release from bioerodible hydrophobic ointments. *Biomaterials* 11 (4):235–237.

Heller, J., S. Y. Ng, D. W. Penhale, B. K. Fritzinger, L. M. Sanders, R. A. Burns, M. G. Gaynon, and S. S. Bhosale. 1987. Use of poly(ortho esters) for the controlled release of 5-fluorouracyl and a LHRH analogue. *J Control Release* 6 (1):217–224.

Heller, J., D. W. Penhale, B. K. Fritzinger, J. E. Rose, and R. F. Helwing. 1983. Controlled release of contraceptive steroids from biodegradable poly (ortho esters). *Contracept Deliv Syst* 4 (1):43–53.

Heng, B. C., H. Liu, and T. Cao. 2005. Scaffold implants for the controlled release of heparan sulfate (HS) and other glycosaminoglycan (GAG) species: This could facilitate the homing of adult stem cells for tissue/organ regeneration. *Med Hypotheses* 65 (2):414–415.

Henkind, P. 1978. Ocular neovascularization. *Am J Ophthalmol* 85:287–301.

Hild, S. A., M. L. Meistrich, R. P. Blye, and J. R. Reel. 2001. Lupron depot prevention of antispermatogenic/antifertility activity of the indenopyridine, CDB-4022, in the rat. *Biol Reprod* 65 (1):165–172.

Hill, J. W. and W. H. Carothers. 1932. Studies of polymerization and ring formation. XIV. A linear superpolyanhydride and a cyclic dimeric anhydride from sebacic acid. *J Am Chem Soc* 54:1569.

Holland, S. J., B. J. Tighe, and P. L. Gould. 1986. Polymers for biodegradable medical devices. 1. The potential of polyesters as controlled macromolecular release systems. *J Control Release* 4 (3):155–180.

Holmes, P. A. 1985. Application of PHB-a microbially produced biodegradable thermoplastic. *Phys Technol* 16:32–36.

Hou, S., L. K. McCauley, and P. X. Ma. 2007. Synthesis and erosion properties of PEG-containing polyanhydrides. *Macromol Biosci* 7 (5):620–628.

Huang, S. W., J. Wang, P. C. Zhang, H. Q. Mao, R. X. Zhuo, and K. W. Leong. 2004. Water-soluble and non-ionic polyphosphoester: Synthesis, degradation, biocompatibility and enhancement of gene expression in mouse muscle. *Biomacromolecules* 5 (2):306–311.

Iwai, S., Y. Sawa, S. Taketani, K. Torikai, K. Hirakawa, and H. Matsuda. 2005. Novel tissue-engineered biodegradable material for reconstruction of vascular wall. *Ann Thorac Surg* 80 (5):1821–1827.

Izumi, Y., M. Gika, N. Shinya, S. Miyabashira, T. Imamura, C. Nozaki, M. Kawamura, and K. Kobayashi. 2007. Hemostatic efficacy of a recombinant thrombin-coated polyglycolic acid sheet coupled with liquid fibrinogen, evaluated in a canine model of pulmonary arterial hemorrhage. *J Trauma* 63 (4):783–787; discussion 787.

Jain, J. P., S. Modi, and N. Kumar. 2008. Hydroxy fatty acid based polyanhydride as drug delivery system: Synthesis, characterization, in vitro degradation, drug release, and biocompatibility. *J Biomed Mater Res A* 84 (3):740–752.

Jerome, C. and P. Lecomte. 2008. Recent advances in the synthesis of aliphatic polyesters by ring-opening polymerization. *Adv Drug Deliv Rev* 60 (9):1056–1076.

Ju, Y. M., B. Yu, L. West, Y. Moussy, and F. Moussy. 2009. A dexamethasone-loaded PLGA microspheres/collagen scaffold composite for implantable glucose sensors. *J Biomed Mater Res A* 93 (1):200–210.

Jung, Y., M. S. Park, J. W. Lee, Y. H. Kim, and S. H. Kim. 2008. Cartilage regeneration with highly-elastic three-dimensional scaffolds prepared from biodegradable poly(L-lactide-co-epsilon-caprolactone). *Biomaterials* 29 (35):4630–4636.

Kappy, M., T. Stuart, A. Perelman, and R. Clemons. 1989. Suppression of gonadotropin secretion by a long-acting gonadotropin-releasing hormone analog (leuprolide acetate, Lupron Depot) in children with precocious puberty. *J Clin Endocrinol Metab* 69 (5):1087–1089.

Kaschke, O., H. J. Gerhardt, K. Bohm, M. Wenzel, and H. Planck. 1996. Experimental in vitro and in vivo studies of epithelium formation on biomaterials seeded with isolated respiratory cells. *J Invest Surg* 9 (2):59–79.

Kato, M., T. Namikawa, H. Terai, M. Hoshino, S. Miyamoto, and K. Takaoka. 2006a. Ectopic bone formation in mice associated with a lactic acid/dioxanone/ethylene glycol copolymer-tricalcium phosphate composite with added recombinant human bone morphogenetic protein-2. *Biomaterials* 27 (21):3927–3933.

Kato, M., H. Toyoda, T. Namikawa, M. Hoshino, H. Terai, S. Miyamoto, and K. Takaoka. 2006b. Optimized use of a biodegradable polymer as a carrier material for the local delivery of recombinant human bone morphogenetic protein-2 (rhBMP-2). *Biomaterials* 27 (9):2035–2041.

Katti, D. S., S. Lakshmi, R. Langer, and C. T. Laurencin. 2002. Toxicity, biodegradation and elimination of polyanhydrides. *Adv Drug Deliv Rev* 54 (7):933–961.

Katz, A., D. P. Mukherjee, A. L. Kaganov, and S. Gordon. 1985. A new synthetic monofilament absorbable suture made from polytrimethylene carbonate. *Surg Gynecol Obstet* 161:213–222.

Kawaguchi, T., M. Nakano, K. Juni, S. Inoue, and Y. Yoshida. 1982. Release profiles of 5-fluorouracil and its derivatives from polycarbonate matrices in vitro. *Chem Pharm Bull* 30 1517–1520.

Kawaguchi, T., M. Nakano, K. Juni, S. Inoue, and Y. Yoshida. 1983. Examination of biodegradability of poly(ethylene carbonate) and poly(propylene carbonate) in the peritoneal cavity in rats. *Chem Pharm Bull (Tokyo)* 31 (4):1400–1403.

Kemnitzer, J. and J. Kohn. 1997. Degradable polymers derived from the amino acid L-tyrosine. In *Handbook of Biodegradable Polymers*, J. Kost, A. J. Domb, and D. M. Weiseman, eds., pp. 251–272. Amsterdam, the Netherlands: Hardwood academic publishers.

Kemp, S. F., P. J. Fielder, K. M. Attie, S. L. Blethen, E. O. Reiter, K. M. Ford, M. Marian, L. N. Dao, H. J. Lee, and P. Saenger. 2004. Pharmacokinetic and pharmacodynamic characteristics of a long-acting growth hormone (GH) preparation (nutropin depot) in GH-deficient children. *J Clin Endocrinol Metab* 89 (7):3234–3240.

Khan, W. and N. Kumar. 2011. Drug targeting to macrophages using paromomycin-loaded albumin microspheres for treatment of visceral leishmaniasis: An in vitro evaluation. *J Drug Target* 19 (4):239–250.

Kim, K., Y. K. Luu, C. Chang, D. Fang, B. S. Hsiao, B. Chu, and M. Hadjiargyrou. 2004. Incorporation and controlled release of a hydrophilic antibiotic using poly(lactide-co-glycolide)-based electrospun nanofibrous scaffolds. *J Control Release* 98 (1):47–56.

Kissel, T., Z. Brich, S. Bantle, I. Lancranjan, F. Nimmerfall, and P. Vit. 1991. Parenteral depot-systems on the basis of biodegradable polyesters. *J Control Release* 16 (1–2):27–41.

Kleinberg, L. R., J. Weingart, P. Burger, K. Carson, S. A. Grossman, K. Li, A. Olivi, M. D. Wharam, and H. Brem. 2004. Clinical course and pathologic findings after Gliadel and radiotherapy for newly diagnosed malignant glioma: Implications for patient management. *Cancer Invest* 22 (1):1–9.

Kohn, J. and R. Langer. 1987. Polymerization reactions involving the side chains of.alpha.-L-amino acids. *J Am Chem Soc* 109 (3):817–820.

Krasko, M. Y. and A. J. Domb. 2007. Pasty injectable biodegradable polymers derived from natural acids. *J Biomed Mater Res A* 83 (4):1138–1145.

Kubek, M. J., A. J. Domb, and M. C. Veronesi. 2009. Attenuation of kindled seizures by intranasal delivery of neuropeptide-loaded nanoparticles. *Neurotherapeutics* 6 (2):359–371.

Kumar, N., A. C. Albertsson, U. Edlund, D. Teomim, R. Aliza, and A. J. Domb. 2005. *Polyanhydrides*, *Biopolymers Online*. New York: Wiley-VCH Verlag GmbH & Co. KGaA.

Kunioka, M., Y. Nakamura, and Y. Doi. 1988. New bacterial copolyester produced in alcanigenes eutrophus from organic acids. *Polym Commun* 29:174–176.

Kwong, A. K., S. Chou, A. M. Sun, M. V. Sefton, and M. F. A. Goosen. 1986. In vitro and in vivo release of insulin from poly(lactic acid) microbeads and pellets. *J Control Release* 4 (1):47–62.

Langer, R., H. Brem, and D. Tapper. 1981. Biocompatibility of polymeric delivery systems for macromolecules. *J Biomed Mater Res* 15 (2):267–277.

Langer, R., D. Lund, K. Leong, and J. Folkman. 1985. Controlled release of macromolecules: Biological studies. *J Control Release* 2:331–341.

Lapidus, R. G., W. Dang, D. M. Rosen, A. M. Gady, Y. Zabelinka, R. O'Meally, T. L. DeWeese, and S. R. Denmeade. 2004. Anti-tumor effect of combination therapy with intratumoral controlled-release paclitaxel (PACLIMER microspheres) and radiation. *Prostate* 58 (3):291–298.

Laurencin, C., A. Domb, C. Morris, V. Brown, M. Chasin, R. McConnell, N. Lange, and R. Langer. 1990. Poly(anhydride) administration in high doses in vivo: Studies of biocompatibility and toxicology. *J Biomed Mater Res* 24 (11):1463–1481.

Laurencin, C. T., M. E. Norman, H. M. Elgendy, S. F. el-Amin, H. R. Allcock, S. R. Pucher, and A. A. Ambrosio. 1993. Use of polyphosphazenes for skeletal tissue regeneration. *J Biomed Mater Res.* 27 (7):963–973.

Lee, C. H., A. Singla, and Y. Lee. 2001. Biomedical applications of collagen. *Int J Pharm* 221 (1–2):1–22.

Lenz, R. W. and P. Guerin. 1983. Functional polyesters and polyamides for medical applications of biodegradable polymers. In *Polymers in Medicine*, E. Chiellini and P. Giusti, eds. New York: Plenum Press. pp. 219–230.

Leong, K. W. 1995. Alternative materials for fracture fixation. *Connect Tissue Res* 31 (4):S69–S75.

Leong, K. W., B. C. Brott, and R. Langer. 1985. Bioerodible polyanhydrides as drug-carrier matrices. I: Characterization, degradation, and release characteristics. *J Biomed Mater Res* 19 (8):941–955.

Leong, K. W., P. D. D'Amore, M. Marletta, and R. Langer. 1986. Bioerodible polyanhydrides as drug-carrier matrices. II. Biocompatibility and chemical reactivity. *J Biomed Mater Res* 20 (1):51–64.

Leong, K., V. Simonte, and R. Langer. 1987. Synthesis of polyanhydrides: Melt-polycondensation, dehydrochlorination, and dehydrative coupling. *Macromolecules* 20:705–712.

Lewis, D. H. 1990. Controlled release of bioactive agents from lactide/glycolide polymers. In *Biodegradable Polymers as Drug Delivery Systems*, M. Chasin and R. Langer, eds. New York: Marcel Dekker. pp. 1–41.

Liu, J., W. Huang, Y. Pang, X. Zhu, Y. Zhou, and D. Yan. 2010a. Hyperbranched polyphosphates for drug delivery application: Design, synthesis, and in vitro evaluation. *Biomacromolecules* 11 (6):1564–1570.

Liu, Y., A. Kemmer, K. Keim, C. Curdy, H. Petersen, and T. Kissel. 2010b. Poly(ethylene carbonate) as a surface-eroding biomaterial for in situ forming parenteral drug delivery systems: A feasibility study. *Eur J Pharm Biopharm* 76 (2):222–229.

Ludvig, E. B. and B. G. Belenkaya. 1974. Investigation of the mechanism of cationic polymerization of Îμ-caprolactone. *J Macromol Sci A Pure Appl Chem* 8 (4):819–828.

Lundmark, S., M. SjÖling, and A. C. Albertsson. 1991. Polymerization of oxepan-2,7-dione in solution and synthesis of block copolymers of oxepan-2,7-dione and 2-oxepanone. *J Macromol Sci A* 28 (1):15–29.

Lyman, D. J. and K. Knutson. 1980. Chemical, physical, and mechanical aspects of blood compatibility. In *Biomedical Polymers*, E. Goldberg and A. Nakajima, eds., pp. 1–30. New York: Academic Press.

Ma, H., G. Melillo, L. Oliva, T. P. Spaniol, U. Englert, and J. Okuda. 2005. Aluminium alkyl complexes supported by [OSSO] type bisphenolato ligands: Synthesis, characterization and living polymerization of rac-lactide. *Dalton Trans* 4:721–727.

Maniar, M., X. D. Xie, and A. J. Domb. 1990. Polyanhydrides. V. Branched polyanhydrides. *Biomaterials* 11 (9):690–694.

Manoharan, C. and J. Singh. 2009. Evaluation of polyanhydride microspheres for basal insulin delivery: Effect of copolymer composition and zinc salt on encapsulation, in vitro release, stability, in vivo absorption and bioactivity in diabetic rats. *J Pharm Sci* 98 (11):4237–4250.

Mathew, S. T., S. G. Devi, V. V. Prasanth, and B. Vinod. 2009. Formulation and in vitro-in vivo evaluation of ketoprofen-loaded albumin microspheres for intramuscular administration. *J Microencapsul* 26 (5):456–469.

Mathiowitz, E. and R. Langer. 1987. Polyanhydride microspheres as drug carriers I. Hot-melt microencapsulation. *J Control Release* 5 (1):13–22.

Mathiowitz, E., E. Ron, G. Mathiowtiz, C. Amato, and R. Langer. 1990. Morphological characterization of bioerodible polymers. I. Crystallinity of poly(anhydride) copolymers. *Macromolecules* 23:3212–3218.

Mathiowitz, E., M. Saltzman, A. Domb, P. Dor, and R. Langer. 1988. Polyanhydride microspheres as drug carriers. II. Microencapsulation by solvent removal. *J Appl Polym Sci* 35:755–774.

Mecerreyes, D., R. Jérôme, and P. Dubois. 1999. Novel macromolecular architectures based on aliphatic polyesters: Relevance of the "Coordination-Insertion" ring-opening polymerization. *Adv Polym Sci* 147:1–59.

Molinari, G. 2009. Natural products in drug discovery: Present status and perspectives. *Adv Exp Med Biol* 655:13–27.

Moll, F. and G. Koller. 1990. Biodegradable tablets having a matrix of low molecular weight poly-L-lactic acid and poly-D,L-lactic acid. *Arch Pharm* 323:887–888.

Montanari, L., M. Costantini, E. C. Signoretti, L. Valvo, M. Santucci, M. Bartolomei, P. Fattibene, S. Onori, A. Faucitano, B. Conti, and I. Genta. 1998. Gamma irradiation effects on poly(DL-lactictide-co-glycolide) microspheres. *J Control Release* 56 (1–3):219–229.

Mukerjee, A. and J. K. Vishwanatha. 2009. Formulation, characterization and evaluation of curcumin-loaded PLGA nanospheres for cancer therapy. *Anticancer Res* 29 (10):3867–3875.

Nahar, M. and N. K. Jain. 2009. Preparation, characterization and evaluation of targeting potential of amphotericin B-loaded engineered PLGA nanoparticles. *Pharm Res* 26 (12):2588–2598.

Nair, L. S. and C. T. Laurencin. 2007. Biodegradable polymers as biomaterials. *Prog Polym Sci* 32 (8–9):762–798.

Nakamura, T., S. Hitomi, S. Watanabe, Y. Shimizu, K. Jamshidi, S. H. Hyon, and Y. Ikada. 1989. Bioabsorption of polylactides with different molecular properties. *J Biomed Mater Res* 23 (10):1115–1130.

Ngiam, M., S. Liao, A. J. Patil, Z. Cheng, F. Yang, M. J. Gubler, S. Ramakrishna, and C. K. Chan. 2009. Fabrication of mineralized polymeric nanofibrous composites for bone graft materials. *Tissue Eng Part A* 15 (3):535–546.

Nicol, F., M. Wong, F. C. MacLaughlin, J. Perrard, E. Wilson, and J. L. Nordstrom. 2002. L-glutamate, an anionic polymer, enhances transgene expression for plasmids delivered by intramuscular injection with in vivo electroporation. *Gene Ther* 9:1351–1358.

Nijenhuis, A. J., D. W. Grijpma, and A. J. Pennings. 1992. Lewis acid catalyzed polymerization of L-lactide. kinetics and mechanism of the bulk polymerization. *Macromolecules* 25:6419–6424.

Nogueira, R., C. Alves, M. Matos, and A. G. Brito. 2009. Synthesis and degradation of poly-beta-hydroxybutyrate in a sequencing batch biofilm reactor. *Bioresour Technol* 100 (7):2106–2110.

Nojehdehian, H., F. Moztarzadeh, H. Baharvand, H. Nazarian, and M. Tahriri. 2009. Preparation and surface characterization of poly-L-lysine-coated PLGA microsphere scaffolds containing retinoic acid for nerve tissue engineering: In vitro study. *Colloids Surf B Biointerfaces* 73:23–29.

Novikova, L. N., J. Pettersson, M. Brohlin, M. Wiberg, and L. N. Novikov. 2008. Biodegradable poly-beta-hydroxybutyrate scaffold seeded with Schwann cells to promote spinal cord repair. *Biomaterials* 29 (9):1198–1206.

Nukavarapu, S. P., S. G. Kumbar, J. L. Brown, N. R. Krogman, A. L. Weikel, M. D. Hindenlang, L. S. Nair, H. R. Allcock, and C. T. Laurencin. 2008. Polyphosphazene/nano-hydroxyapatite composite microsphere scaffolds for bone tissue engineering. *Biomacromolecules* 9 (7):1818–1825.

Ohta, S., N. Nitta, A. Sonoda, A. Seko, T. Tanaka, M. Takahashi, Y. Kimura, Y. Tabata, and K. Murata. 2009. Cisplatin-conjugated degradable gelatin microspheres: Fundamental study in vitro. *Br J Radiol* 82 (977):380–385.

Ory, S. J., C. B. Hammond, S. G. Yancy, R. W. Hendren, and C. G. Pitt. 1983. The effect of a biodegradable contraceptive capsule (Capronor) containing levonorgestrel on gonadotropin, estrogen, and progesterone levels. *Am J Obstet Gynecol* 145 (5):600–605.

Pal, A., A. Prabhu, A. A. Kumar, B. Rajagopal, K. Dadhe, V. Ponnamma, and S. Shivakumar. 2009. Optimization of process parameters for maximum poly(-beta-)hydroxybutyrate (PHB) production by *Bacillus thuringiensis* IAM 12077. *Pol J Microbiol* 58 (2):149–154.

Park, J., P. M. Fong, J. Lu, K. S. Russell, C. J. Booth, W. M. Saltzman, and T. M. Fahmy. 2009. PEGylated PLGA nanoparticles for the improved delivery of doxorubicin. *Nanomedicine* 5 (4):410–418.

Perrin, D. E. and J. P. English. 1997. Polycaprolactone. In *Handbook of Biodegradable Polymers*, J. Kost, A. J. Domb, and D. M. Weiseman eds. Amsterdam, the Netherlands: Hardwood Academic Publishers. pp. 63–77.

Pinholt, E. M., E. Solheim, G. Bang, and E. Sudmann. 1991. Bone induction by composite of bioerodible polyorthoester and demineralized bone matrix in rats. *Acta Orthop Scand* 62 (5):476–480.

Pitt, C. G. 1990. Poly(ε-caprolactone) and its copolymers. In *Biodegradable Polymers as Drug Delivery Systems*, R. S. Langer and M. Chasin, eds. New York: Marcel Dekker. pp. 71–120.

Pitt, C. G., F. I. Chasalow, Y. M. Hibionada, D. M. Klimas, and A. Schindler. 1981a. Aliphatic polyesters. I. The degradation of poly(ε-caprolactone) in vivo. *J Appl Polym Sci* 26 (11):3779–3787.

Pitt, C. G., M. M. Gratzl, G. L. Kimmel, J. Surles, and A. Schindler. 1981b. Aliphatic polyesters II. The degradation of poly (DL-lactide), poly (epsilon-caprolactone), and their copolymers in vivo. *Biomaterials* 2 (4):215–20.

Pitt, C. G. and A. Schindler. 1984. Capronor™: A biodegradable delivery system for levonorgestrel. In *Long-Acting Contraceptive Delivery Systems*, A. Glodsmith, G. L. Zatuchni, J. D. Shelton, and J. J. Sciarra, eds. Philadelphia, PA: Harper and Row, pp. 48–63.

Poshusta, A. K., J. A. Burdick, D. J. Mortisen, R. F. Padera, D. Ruehlman, M. J. Yaszemski, and K. S. Anseth. 2003. Histocompatibility of photocrosslinked polyanhydrides: A novel in situ forming orthopaedic biomaterial. *J Biomed Mater Res A* 64 (1):62–69.

Pradilla, G., P. P. Wang, P. Gabikian, K. Li, C. A. Magee, K. A. Walter, and H. Brem. 2006. Local intracerebral administration of Paclitaxel with the paclimer delivery system: Toxicity study in a canine model. *J Neurooncol* 76 (2):131–138.

Puppi, D., F. Chiellini, A. M. Piras, and E. Chiellini. 2010. Polymeric materials for bone and cartilage repair. *Prog Polym Sci* 35 (4):403–440.

Qiu, L. Y. and M. Q. Yan. 2009. Constructing doxorubicin-loaded polymeric micelles through amphiphilic graft polyphosphazenes containing ethyl tryptophan and PEG segments. *Acta Biomater* 5 (6):2132–2141.

Quintilio, W., C. S. Takata, O. A. Sant'Anna, M. H. da Costa, and I. Raw. 2009. Evaluation of a diphtheria and tetanus PLGA microencapsulated vaccine formulation without stabilizers. *Curr Drug Deliv* 6 (3):297–304.

Rock, M., M. Green, C. Fait, R. Geil, J. Myer, M. Maniar, and A. Domb. 1991. Evaluation and comparison of biocompatibility of various classes of polyanhydrides. *Polym Preprints* 32:221–222.

Rodeheaver, G. T., K. A. Beltran, C. W. Green, B. C. Faulkner, B. M. Stiles, G. W. Stanimir, H. Traeland, G. M. Fried, H. C. Brown, and R. F. Edlich. 1996. Biomechanical and clinical performance of a new synthetic monofilament absorbable suture. *J Long Term Eff Med Implants* 6 (3–4):181–198.

Rosen, H. B., J. Chang, G. E. Wnek, R. J. Linhardt, and R. Langer. 1983. Bioerodible polyanhydrides for controlled drug delivery. *Biomaterials* 4 (2):131–133.

Sahoo, S., J. G. Cho-Hong, and T. Siew-Lok. 2007. Development of hybrid polymer scaffolds for potential applications in ligament and tendon tissue engineering. *Biomed Mater* 2 (3):169–173.

Sahoo, S., H. Ouyang, J. C. Goh, T. E. Tay, and S. L. Toh. 2006. Characterization of a novel polymeric scaffold for potential application in tendon/ligament tissue engineering. *Tissue Eng* 12 (1):91–99.

Samad, A., Y. Sultana, R. K. Khar, K. Chuttani, and A. K. Mishra. 2009. Gelatin microspheres of rifampicin cross-linked with sucrose using thermal gelation method for the treatment of tuberculosis. *J Microencapsul* 26 (1):83–89.

Sano, A., T. Hojo, M. Maeda, and K. Fujioka. 1998. Protein release from collagen matrices. *Adv Drug Deliv Rev* 31 (3):247–266.

Santander-Ortega, M. J., D. Bastos-Gonzalez, J. L. Ortega-Vinuesa, and M. J. Alonso. 2009. Insulin-loaded PLGA nanoparticles for oral administration: An in vitro physico-chemical characterization. *J Biomed Nanotechnol* 5 (1):45–53.

Schafer, V., H. von Briesen, H. Rubsamen-Waigmann, A. M. Steffan, C. Royer, and J. Kreuter. 1994. Phagocytosis and degradation of human serum albumin microspheres and nanoparticles in human macrophages. *J Microencapsul* 11 (3):261–269.

Schenderlein, S., M. Luck, and B. W. Muller. 2004. Partial solubility parameters of poly(D,L-lactide-co-glycolide). *Int J Pharm* 286 (1–2):19–26.

Schwope, A. D., D. L. Wise, and J. F. Howes. 1975. Development of polylactic/glycolic acid delivery systems for use in treatment of narcotic addiction. *Natl Inst Drug Abuse Res Monogr Ser* (4):13–18.

Seckel, B. R., T. H. Chiu, E. Nyilas, and R. L. Sidman. 1984. Nerve regeneration through synthetic biodegradable nerve guides: Regulation by the target organ. *Plast Reconstr Surg* 74 (2):173–181.

Sehgal, P. K. and A. Srinivasan. 2009. Collagen-coated microparticles in drug delivery. *Expert Opin Drug Deliv* 6 (7):687–695.

Sethuraman, S., L. S. Nair, S. El-Amin, R. Farrar, M. T. Nguyen, A. Singh, H. R. Allcock, Y. E. Greish, P. W. Brown, and C. T. Laurencin. 2006. In vivo biodegradability and biocompatibility evaluation of novel alanine ester based polyphosphazenes in a rat model. *J Biomed Mater Res A* 77 (4):679–687.

Shanmugasundaram, K. and A. C. Rigby. 2009. Exploring novel target space: A need to partner high throughput docking and ligand-based similarity searches? *Comb Chem High Throughput Screen* 12 (10):984–999.

Shen, J. Y., M. B. Chan-Park, B. He, A. P. Zhu, X. Zhu, R. W. Beuerman, E. B. Yang, W. Chen, and V. Chan. 2006. Three-dimensional microchannels in biodegradable polymeric films for control orientation and phenotype of vascular smooth muscle cells. *Tissue Eng* 12 (8):2229–2240.

Shindler, A., R. Jeffcoat, G. L. Kimmel, C. G. Pitt, M. E. Wall, and R. Zweidinger. 1977. Biodegradable polymers for sustained drug delivery. *Contemp. Top. Polym. Sci.* 2:251–289.

Shire, S. J. 2009. Formulation and manufacturability of biologics. *Curr Opin Biotechnol* 20 (6):708–714.

Silverman, B. L., S. L. Blethen, E. O. Reiter, K. M. Attie, R. B. Neuwirth, and K. M. Ford. 2002. A long-acting human growth hormone (Nutropin Depot): Efficacy and safety following two years of treatment in children with growth hormone deficiency. *J Pediatr Endocrinol Metab* 15 (Suppl 2):715–722.

Sinha, V. R., K. Bansal, R. Kaushik, R. Kumria, and A. Trehan. 2004. Poly-epsilon-caprolactone microspheres and nanospheres: An overview. *Int J Pharm* 278 (1):1–23.

Slager, J. and A. J. Domb. 2002. Stereocomplexes based on poly(lactic acid) and insulin: Formulation and release studies. *Biomaterials* 23 (22):4389–4396.

Sokolsky-Papkov, M., K. Agashi, A. Olaye, K. Shakesheff, and A. J. Domb. 2007. Polymer carriers for drug delivery in tissue engineering. *Adv Drug Deliv Rev* 59 (4–5):187–206.

Solheim, E., E. M. Pinholt, R. Andersen, G. Bang, and E. Sudmann. 1995. Local delivery of indomethacin by a polyorthoester inhibits reossification of experimental bone defects. *J Biomed Mater Res* 29 (9):1141–1146.

Sonmez, K., B. Bahar, R. Karabulut, O. Gulbahar, A. Poyraz, Z. Turkyilmaz, B. Sancak, and A. C. Basaklar. 2009. Effects of different suture materials on wound healing and infection in subcutaneous closure techniques. *B-ENT* 5 (3):149–152.

Sosnowski, S., S. Slornkowski, and S. Penczek. 1991. Kinetics of the anionic polymerization of caprolactone with K+—(dibenzo-18-crownd ether) counterion. Propagation via macroions and macroion pairs. *Die Makromolekulare Chemie* 192:735.

Sparer, Randall V., Shih Chung, Cheryl D. Ringeisen, and Kenneth J. Himmelstein. 1984. Controlled release from erod1ble poly(ortho ester) drug delivery systems. *J Control Release* 1 (1):23–32.

Steendam, R., A. van der Laan, and D. Hissink. 2006. Bioresorbable drug-eluting stent coating formulations based on SynBiosys biodegradable multi-block copolymers. *J Control Release* 116 (2):e94–e95.

Steinmann, R., H. Gerngross, and W. Hartel. 1990. [The use of bioresorbable implants (Biofix) in surgery. The indications, technic and initial clinical results]. *Aktuelle Traumatol* 20 (2):102–107.

Stoll, G. H., F. Nimmerfall, M. Acemoglu, D. Bodmer, S. Bantle, I. Muller, A. Mahl, M. Kolopp, and K. Tullberg. 2001. Poly(ethylene carbonate)s, part II: Degradation mechanisms and parenteral delivery of bioactive agents. *J Control Release* 76 (3):209–225.

Subramanyam, R., and A. G. Pinkus. 1985. Synthesis of poly(terephthalic anhydride) by hydrolysis of terephthaloyl chloride triethylamine intermediate adduct: Characterization of intermediate adduct. *J Macromol Sci Chem* A22 (1):23.

Sun, L., S. Zhou, W. Wang, Q. Su, X. Li, and J. Weng. 2009. Preparation and characterization of protein-loaded polyanhydride microspheres. *J Mater Sci Mater Med* 20 (10):2035–2042.

Susperregui, N., D. Delcroix, B. Martin-Vaca, D. Bourissou, and L. Maron. 2010. Ring-opening polymerization of epsilon-caprolactone catalyzed by sulfonic acids: Computational evidence for bifunctional activation. *J Org Chem* 75 (19):6581–6587.

Suzuki, A., H. Terai, H. Toyoda, T. Namikawa, Y. Yokota, T. Tsunoda, and K. Takaoka. 2006. A biodegradable delivery system for antibiotics and recombinant human bone morphogenetic protein-2: A potential treatment for infected bone defects. *J Orthop Res* 24 (3):327–332.

Tamargo, R. J., J. I. Epstein, C. S. Reinhard, M. Chasin, and H. Brem. 1989. Brain biocompatibility of a biodegradable, controlled-release polymer in rats. *J Biomed Mater Res* 23 (2):253–266.

Teomim, D. and A. J. Domb. 1999. Fatty acid terminated polyanhydrides. *J Polym Sci A Polym Chem* 37 (16):3337–3344.

Teomim, D., A. Nyska, and A. J. Domb. 1999. Ricinoleic acid-based biopolymers. *J Biomed Mater Res* 45 (3):258–267.

Torres, M. P., A. S. Determan, G. L. Anderson, S. K. Mallapragada, and B. Narasimhan. 2007. Amphiphilic polyanhydrides for protein stabilization and release. *Biomaterials* 28 (1):108–116.

Unger, F., U. Westedt, P. Hanefeld, R. Wombacher, S. Zimmermann, A. Greiner, M. Ausborn, and T. Kissel. 2007. Poly(ethylene carbonate): A thermoelastic and biodegradable biomaterial for drug eluting stent coatings? *J Control Release* 117 (3):312–321.

Urakami, H. and Z. Guan. 2008. Living ring-opening polymerization of a carbohydrate-derived lactone for the synthesis of protein-resistant biomaterials. *Biomacromolecules* 9 (2):592–597.

Vaquette, C., S. Fawzi-Grancher, P. Lavalle, C. Frochot, M. L. Viriot, S. Muller, and X. Wang. 2006. In vitro biocompatibility of different polyester membranes. *Biomed Mater Eng* 16 (4 Suppl):S131–S136.

Veronesi, M. C., Y. Aldouby, A. J. Domb, and M. J. Kubek. 2009. Thyrotropin-releasing hormone D,L polylactide nanoparticles (TRH-NPs) protect against glutamate toxicity in vitro and kindling development in vivo. *Brain Res* 1303:151–160.

Vert, M. 2005. Aliphatic polyesters: Great degradable polymers that cannot do everything. *Biomacromolecules* 6 (2):538–546.

Vert, M., S. Li, and H. Garreau. 1991. More about the degradation of LA/GA-derived matrices in aqueous media. *J Control Release* 16 (1–2):15–26.

Villegas-Cabello, O., J. L. Vazquez-Juarez, F. M. Gutierrez-Perez, R. F. Davila-Cordova, and C. Diaz-Montemayor. 1994. Staged replacement of the canine trachea with ringed polyethylene terephthalate grafts. *Thorac Cardiovasc Surg* 42 (5):302–305.

Wade, C. W. R., S. Gourlay, R. Rice, A. Hegyeli, R. Singler, and J. White. 1978. Biocompatibility of eight poly(organophosphazenes). In *Organometallic Polymers*, eds., J. E. Sheats, C. U. Pittmann, and C. E. Carraher, pp. 283–288. New York: Academic Press.

Wang, Y. C., X. Q. Liu, T. M. Sun, M. H. Xiong, and J. Wang. 2008. Functionalized micelles from block copolymer of polyphosphoester and poly(epsilon-caprolactone) for receptor-mediated drug delivery. *J Control Release* 128 (1):32–40.

Wang, S., A. C. Wan, X. Xu, S. Gao, H. Q. Mao, K. W. Leong, and H. Yu. 2001. A new nerve guide conduit material composed of a biodegradable poly(phosphoester). *Biomaterials* 22 (10):1157–1169.

Wei, X., C. Gong, M. Gou, S. Fu, Q. Guo, S. Shi, F. Luo, G. Guo, L. Qiu, and Z. Qian. 2009. Biodegradable poly(epsilon-caprolactone)-poly(ethylene glycol) copolymers as drug delivery system. *Int J Pharm* 381 (1):1–18.

Wen, J., G. J. Kim, and K. W. Leong. 2003. Poly(D,L-lactide-co-ethyl ethylene phosphate)s as new drug carriers. *J Control Release* 92 (1–2):39–48.

Wen, J., H. Q. Mao, W. Li, K. Y. Lin, and K. W. Leong. 2004. Biodegradable polyphosphoester micelles for gene delivery. *J Pharm Sci* 93 (8):2142–2157.

Wenk, E., A. J. Meinel, S. Wildy, H. P. Merkle, and L. Meinel. 2009. Microporous silk fibroin scaffolds embedding PLGA microparticles for controlled growth factor delivery in tissue engineering. *Biomaterials* 30 (13):2571–2581.

Wheeler, J. M., J. D. Knittle, and J. D. Miller. 1993. Depot leuprolide acetate versus danazol in the treatment of women with symptomatic endometriosis: A multicenter, double-blind randomized clinical trial. II. Assessment of safety. The Lupron Endometriosis Study Group. *Am J Obstet Gynecol* 169 (1):26–33.

Whittle, I. R., S. Lyles, and M. Walker. 2003. Gliadel therapy given for first resection of malignant glioma: A single centre study of the potential use of Gliadel. *Br J Neurosurg* 17 (4):352–354.

Woodward, S. C., P. S. Brewer, F. Moatamed, A. Schindler, and C. G. Pitt. 1985. The intracellular degradation of poly(epsilon-caprolactone). *J Biomed Mater Res* 19 (4):437–444.

Xie, H., Y. S. Khajanchee, and B. S. Shaffer. 2008. Chitosan hemostatic dressing for renal parenchymal wound sealing in a porcine model: Implications for laparoscopic partial nephrectomy technique. *JSLS* 12 (1):18–24.

Xue, J. M., C. H. Tan, and D. Lukito. 2006. Biodegradable polymer-silica xerogel composite microspheres for controlled release of gentamicin. *J Biomed Mater Res B Appl Biomater* 78 (2):417–422.

Yi Mok, N., J. Chadwick, K. A. Kellett, N. M. Hooper, A. P. Johnson, and C. W. Fishwick. 2009. Discovery of novel non-peptide inhibitors of BACE-1 using virtual high-throughput screening. *Bioorg Med Chem Lett* 19 (23):6770–6774.

Yoda, N. 1963. Synthesis of polyanhydrides. XII. *J Polym Sci* 1:1323–1338.

Yu, H. 1988. *Pseudopoly(Amino Acids): A Study of the Synthesis and Characterization of Polyesters Made from-l-Amino Acids*. Cambridge, MA: Massachusetts Institute of Technology.

Zhang, N. and S. Guo. 2003. [Studies on a kind of new biodegradable material—Polycaprolactone and developments in medical area]. *Sheng Wu Yi Xue Gong Cheng Xue Za Zhi* 20 (4):746–749.

Zhang, P., Z. Hong, T. Yu, X. Chen, and X. Jing. 2009. In vivo mineralization and osteogenesis of nanocomposite scaffold of poly(lactide-co-glycolide) and hydroxyapatite surface-grafted with poly(L-lactide). *Biomaterials* 30 (1):58–70.

Zhang, Z., R. Kuijer, S. K. Bulstra, D. W. Grijpma, and J. Feijen. 2006. The in vivo and in vitro degradation behavior of poly(trimethylene carbonate). *Biomaterials* 27 (9):1741–1748.

Zhang, J. X., M. Q. Yan, X. H. Li, L. Y. Qiu, X. D. Li, X. J. Li, Y. Jin, and K. J. Zhu. 2007. Local delivery of indomethacin to arthritis-bearing rats through polymeric micelles based on amphiphilic polyphosphazenes. *Pharm Res* 24 (10):1944–1953.

Zheng, C., L. Qiu, X. Yao, and K. Zhu. 2009. Novel micelles from graft polyphosphazenes as potential anti-cancer drug delivery systems: Drug encapsulation and in vitro evaluation. *Int J Pharm* 373 (1–2):133–140.

Zheng, C., J. Xu, X. Yao, and L. Qiu. 2011. Polyphosphazene nanoparticles for cytoplasmic release of doxorubicin with improved cytotoxicity against Dox-resistant tumor cells. *J Colloid Interface Sci* 355 (2):374–382.

Zurita, R., J. Puiggali, and A. Rodriguez-Galan. 2006. Loading and release of ibuprofen in multi- and monofilament surgical sutures. *Macromol Biosci* 6 (9):767–775.

6 Bioresorbable Hybrid Membranes for Bone Regeneration

Akiko Obata and Toshihiro Kasuga

CONTENTS

6.1 INTRODUCTION

Bioresorbable materials have been used as implants for bone reconstruction in clinical field. The ideal implant materials should fulfill several criteria. The materials are basically required to have biocompatibility (nontoxicity) and bioactivity. Bioactivity is regarded as an ability of bonding to the host bone directly in body in the research field of implant materials. Various types of bioceramics have been developed as the implant materials, particularly calcium phosphates have been reported to show an excellent bioactivity. In fact, bioresorbable polymer and calcium phosphate composite materials also have been investigated to be applied to the implant materials enhancing bone formation.

Poly(L-lactic acid) (PLLA) is one of the biodegradable polymers and has been applied clinically. Bos et al. reported tissue reactions to PLLA implanted subcutaneously in the backs of rats (Bos et al. 1991). In the report, PLLA showed the degradation pattern of the biodegradable polymers during 26 weeks and no acute or chronic inflammatory reaction to PLLA was observed during 143 weeks of implantation, except for the early part implant period. PLLA has been investigated for application to scaffold materials for bone reconstruction (Mainil-Varlet et al. 1997). PLLA–ceramics composites were developed for bone reconstruction and showed good mechanical and bone-forming properties (Kasuga et al. 2003, Zhang et al. 2004, Kim et al. 2006). Calcium carbonate, for example, is well known as a bioresorbable material, shows excellent bone-forming ability, and has been used clinically as a bone filler (Liao et al. 2000, Vago et al. 2002). PLLA/calcium carbonate (vaterite) composites showed excellent hydroxyapatite (HA)-forming ability, cellular compatibility, and bio-compatibility by in vitro and in vivo tests (Maeda et al. 2004, 2006).

Silicon species is one of the essential, trace elements for healthy bone and has been recently reported to stimulate functions of osteogenic cells (Carlisle 1986, Xynos et al. 2000). This has been supported by the reports about bioactive glasses, such as Bioglass® (45% SiO_2, 24.5% CaO, 24.5% Na_2O, 6% P_2O_5). Ionic dissolution products released from Bioglass stimulated osteogenic cells at the genetic level. Xynos et al. reported that some gene expressions in human osteoblasts were stimulated by silicon species in a culture medium, derived from Bioglass. They also reported that the expression of insulin-like growth factor II in human osteoblasts was enhanced by silicon species in the culture medium, which stimulated the cells to proliferate. Reffitt et al. reported that orthosilicic acid in a culture medium was effective to stimulate collagen type I synthesis and osteo-blastic differentiation in human osteoblast-like cells (Reffitt et al. 2003).

Human mesenchymal stem cells (MSCs) and osteoblasts (HOBs) are one of the important and useful cells in the field of bone tissue engineering. MSCs are capable of differentiating in vitro, at the minimum, into the bone, cartilage, fat, and mature stromal cell lineages. When MSCs are cultured on the surfaces of a ceramic block and a diffusion chamber, they can exhibit osteogenic potential resulting in bone formation in vivo. MSCs have been reported to differentiate into osteo-blasts and undergo mineralization when they are cultured in the presence of ascorbic acid (AsA), β-glycerophosphate (βGP), and dexamethasone (Dex). These supplements stimulate HOBs to min-eralize as well. AsA promotes synthesis and accumulation of type I collagen. βGP is used as an exogenous phosphate source for the cells to synthesize mineralized material. Dex is used to induce proliferation, osteoblast maturation, and extracellular matrix mineralization. Jaiswal et al., how-ever, reported that MSCs showed osteogenic differentiation in Dulbecco's modified Eagle's medium (DMEM) containing AsA, Dex, and βGP; moreover, the addition of AsA to MSCs in the absence of Dex and βGP did not cause osteogenic differentiation of the cells (Jaiswal et al. 1997). We used DMEM containing serum, antibiotics (penicillin/streptomycin), and AsA as the culture medium for in vitro tests using MSCs or HOBs to establish whether the addition of commonly used osteogenic factors is necessary for osteogenesis on our composite materials. Gough et al. reported that HOBs in medium unsupplemented Dex and βGP was stimulated to mineralize in response to culture on a bioactive glass in the SiO_2-CaO-P_2O_5 system and also in response to dissolution products, silicon and calcium species, from the glass (Gough et al. 2004). Jones et al. reported the mineralization of HOBs was stimulated by culturing an SiO_2-CaO bioactive glass in a medium unsupplemented osteogenic factor (Jones et al. 2007). This chapter shows various types of PLLA/vaterite compos-ites releasing silicon and calcium species that we have developed so far. The composites have been found to have good HA-forming ability and cellular and tissue compatibilities.

6.2   SILOXANE-DOPED POLY(L-LACTIC ACID)/CALCIUM CARBONATE (VATERITE) COMPOSITES (SiPVC)

Biodegradable composite materials releasing silicon and calcium species were prepared by mix-ing silane coupling agent, aminopropyltrietoxysilane (APTES), into PLLA. We hypothesized it is possible to get the materials silicon species are homogeneously mixed in by using APTES as it has active sites for making hybrids with polymers. Silicon species were expected to be released from the composite materials by the PLLA degradation. A composite containing no silicon species, PLLA/vaterite (PVC), was prepared to clarify effects of silicon species on cell functions by comparing cell behaviors on between the two types of composite materials.

6.2.1   Preparation of the Composites and bHA-Coating

SiPVC was prepared using PLLA (molecular weights, M_w; = 160 ± 20 kDa) dissolved in methy-lene chloride and aminopropyltrietoxysilane (APTES; $Si(OC_2H_2)_3(CH_2)_3NH_2$) hydrolyzed through

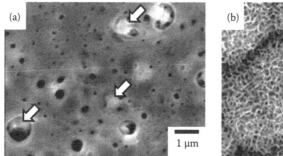
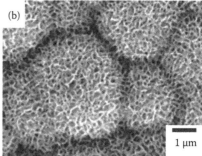

FIGURE 6.1 Surface morphologies of (a) SiPVC and (b) SiPVC+. SiPVC+ means SiPVC after soaking in SBF for 3 days to coat its surface with bone-like hydroxyapatite. Arrows indicate vaterite particles with about 500 µm in diameter.

addition of carbodiimide and ammonium solution. The PLLA solution and the APTES solution were mixed in weight ratio of PLLA:APTES = 1:0.67, and then added calcium carbonate (vaterite) in weight ratio of PLLA:vaterite = 1:0.5. SiPVC and PVC coated with bone-like HA (bHA) (denoted by SiPVC+ and PVC+, respectively) were prepared through immersion of the film in simulated body fluid (SBF) for 3 days. The bHA is deposited on some bioactive materials after immersing in SBF. SBF is a tris-buffer solution (pH 7.4) containing inorganic ions almost equal to those of human plasma (Kokubo et al. 1990). The bHA formed in SBF has a leaf-like crystalline shape, and therefore can ensure a large surface area for cellular spreading.

Figure 6.1 shows the surface morphologies of SiPVC and SiPVC+. Many micropores of 0.1–1 µm in diameter and vaterite particles of 0.5 µm in diameter (pointed by arrows in Figure 6.1a) were observed on SiPVC surface. On the other hand, leaf-like bHA crystals were deposited on SiPVC; the surface was completely covered with bHA. It has been observed that the thickness of SiPVC film and the bHA coating was about 10 and ~3 µm, respectively, and the bHA coating closely adhered to SiPVC surface (Maeda et al. 2006). The bHA on SiPVC has been confirmed to contain silicon species. The amount of silicon species in the bHA has been estimated to be 2.1 at% with an energy dispersive spectrometer (EDS).

6.2.2 CELLULAR COMPATIBILITY

Osteoblast-like cells (MC3T3-E1 cells) from rat calvaria were seeded onto SiPVC+ or PVC+ sterilized using ethylene oxide gas in a 24-well plate at a density of 60,000 cells per well. Alpha-modified minimum essential medium containing 10% fetal bovine serum (FBS) was used as the culture medium and changed every 2 days. Cells were cultured at 37°C in a humidified incubator with 5% CO_2, harvested after 1, 3, and 5 days using 0.1% actinase in PBS, and counted by a hemocytometer. Figure 6.2 shows the cellular numbers of MC3T3-E1 cells on SiPVC+ and PVC+ in a 24-well plate after incubation. The cells satisfactorily proliferated on both sample surfaces, and the cellular numbers on SiPVC+ were significantly higher than that on PVC+.

HOBs were seeded on SiPVC and PVC at a density of 10,000 cells per well, and then cultured for 21 days using medium, which is a Dulbecco's modified Eagles medium (DMEM) containing 10% FBS with 0.05 mM L-ascorbic acid-2-phosphate (AaAP) and 2% penicillin/streptomycin. The surfaces of dried samples were analyzed by laser Raman spectroscopy (LRS) and observed by field emission scanning electron microscopy (SEM) after coated with osmium. Figure 6.3 shows SEM micrographs of HOBs cultured on SiPVC+ and PVC+ for 21 days. Figure 6.3a shows an image of HOBs cultured on SiPVC+. The cells proliferated prosperously and appeared to elongate and formed rough shape and some agglomerates. Figure 6.3b shows that the cells cultured on PVC+. The cells appeared to elongate and aligned on the surface. No agglomerate was

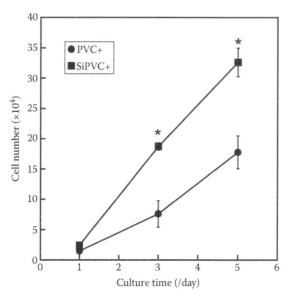

FIGURE 6.2 Numbers of MC3T3-E1 cells on SiPVC+ and PVC+ after 1, 3, and 5 day culturing. Cellular proliferation on SiPVC+ was better (*$p < 0.05$).

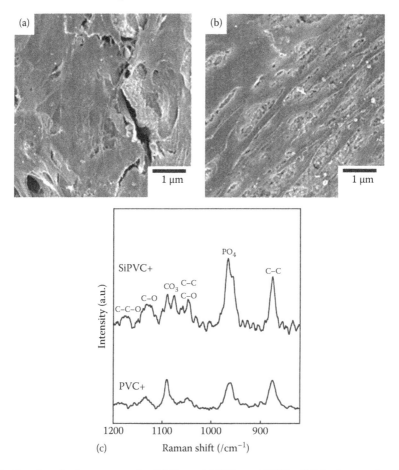

FIGURE 6.3 Results of culture tests using HOBs on SiPVC+ and PVC+. SEM micrographs of HOBs cultured on (a) SiPVC+ and (b) PVC+ for 21 days. (c) LRS spectra of (a) and (b).

observed in the cells. Figure 6.3c shows the LRS spectra of SiPVC+ and PVC+ surfaces after 21 days of HOBs culturing. A peak corresponding to orthophosphate appeared at approximately 965 cm^{-1} on each spectrum. The peak on the spectrum of SiPVC+ was higher in intensity and sharper in form than that of PVC+.

Bone nodule containing minerals forms after the maturation of extracellular matrix and the agglomeration of HOBs. Although no agglomerate was observed on the PVC+ surface after HOBs culturing, some agglomerates were observed on the SiPVC+ surface as shown in Figure 6.3a. In the result of LRS, the intensity of PO$_4$ peak on the spectrum of SiPVC+ after HOBs culturing was higher and sharper than that on the spectrum of PVC+. The strong and sharp PO$_4$ peak on the LRS spectrum of SiPVC after culturing is expected to originate from calcium phosphate precipitation in HOBs. According to the results of LRS, the agglomerates observed on the SiPVC+ surface are regarded to be bone nodules. The bone nodule formation in HOBs was enhanced on the SiPVC+ surface. It was reported that the bone nodule formation was enhanced on some bioactive glasses containing silica (Gough et al. 2004, Jones et al. 2007). It was also reported that calcium phosphate significantly precipitated in HOBs cultured on silicon-substituted HA (Thian et al. 2006). The effect of a trace amount of silicon species released from SiPVC or contained in bHA layer formed on SiPVC was expected to enhance the mineralization of HOBs.

MSCs are capable of differentiating in vitro into various tissues when placed in an appropriate environment. ALP activity in MSCs is known to increase during osteogenic differentiation. Jaiswal et al. reported that MSCs showed osteogenic differentiation in DMEM containing AsAP, Dex, and βGP, and that the addition of AsAP to MSCs in the absence of Dex and βGP did not cause osteogenic differentiation of the cells (Jaiswal et al. 1997). MSCs were seeded onto SiPVC+ and PVC+ in 24-well plates, at a density of 10,000 cells per well. The seeded cells were incubated in DMEM supplemented with only AsAP, not Dex nor βGP. Table 6.1 shows ALP activity and cell number of MSCs cultured on SiPVC+ and PVC+ for 21 days. Although the level of ALP activity in MSCs cultured on PVC+ remained unchanged, that on SiPVC+ significantly increased for 21 day culturing. The cells cultured on SiPVC+ showed larger number than that on PVC+ at all time points. The cell proliferation was enhanced on SiPVC+.

There are few reports showing the effects of silicon species on the activity of osteogenic differentiation of MSCs. Reffitt et al. reported that the collagen type 1 synthesis in MSCs was enhanced in the medium containing 10–50 μM of orthosilicic acid (Reffitt et al. 2003). The collagen type 1 synthesis by preosteoblasts or early undifferentiated osteoblast-like cells is regarded to represent bone formation (Boskey et al. 1999). The ALP activity in MSCs increased on SiPVC+; the osteogenic differentiation of MSCs was induced by the silicon species included in/released from SiPVC+.

The effects of silicon species on the cellular activities have been mainly reported to enhance the HOBs mineralization and bone formation. Silicon species also stimulate the MSC activity; especially they induce the osteogenic differentiation of MSCs without supplementing the

TABLE 6.1
ALP Activity and Cell Number of MSCs Cultured on SiPVC+ and PVC+ for 14 and 21 Days

	14 Days		21 Days	
	ALP	Number	ALP	Number
PVC+	2.89 ± 0.0202	2.07 ± 0.0822	2.98 ± 0.115	3.61 ± 0.975
SiPVC+	3.38 ± 0.250	6.59 ± 2.95	4.13 ± 0.125*	11.1 ± 1.35*

Unit ALP: p-nitrophenol nmol/min, number: ×10^4.
*$p < 0.05$, comparing the value between PVC+ and SiPVC+ on each stage.

osteogenic factors (Dex and βGP) in a culture medium. The mechanism for the effects of silicon species on the MSCs activity has not been clarified yet. Some gene-expressions in HOBs were reported to change by the addition of a trace amount of silicon species in the culture medium (Xynos et al. 2001). Therefore, a change in gene-expression is believed to occur in MSCs cultured on SiPVC+, which originates the stimulation to the proliferation and the osteogenic differentiation of MSCs.

6.3   SILOXANE-DOPED VATERITE PARTICLE AND ITS COMPOSITES WITH POLY(L-LACTIC ACID)

We have developed siloxane and PLLA hybrid materials and also successfully prepared the materials with film or hollow sphere shape, as aforementioned. The matrix of the material consists of a hybrid between amino groups in APTES and carboxyl groups in PLLA. The structure of the matrix may be seriously changed by changing the amount of APTES; therefore, it is difficult to drastically increase the silicon releasability of the material. It was assumed that vaterite particles containing siloxane should be very useful to prepare composite materials releasing silicon species continuously and controllably using various types of biodegradable materials.

A novel composite consisting of PLLA and calcium carbonate particles (vaterite) doped with silicon species was developed. The materials were successfully prepared using silane coupling agents by a carbonation process. It was expected that silicon species were gradually released from the composites through the degradation of PLLA matrix and vaterite and that the composites were able to show the chronic effects of silicon species on cellular activities.

6.3.1   Preparation of Siloxane-Doped Vaterite Particles

Silicon-doped vaterite (SiV) powders were prepared by a carbonation process with methanol. A sample of 150 g of $Ca(OH)_2$ was mixed with 30 or 60 mL of aminopropyltriethoxysilane (APTES) and 2000 mL of methanol with adding CO_2 gas for 75 min at a rate of 2000 mL/min. The resulting slurry was dried at 110°C, resulting in the SiV powders. The amount of silicon in SiV prepared using 30 and 60 mL of APTES was estimated to be approximately 1 and 3 wt%. SiV containing 1 and 3 wt% of silicon are hereafter denoted by SiV1 and SiV3, respectively.

Figure 6.4 shows SEM photographs of the SiV3 particles. The SiV3 particles have diameters of ~1 μm. These two types of particles consist of aggregates of nano-sized primary particles. The sizes of SiV particles are larger than that of vaterite. The results of the nitrogen gas adsorption method reveal that the specific surfaces of SiV3 and vaterite are 67.2 and 30.7 m²/g, respectively.

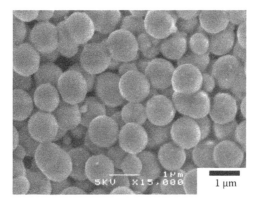

FIGURE 6.4   SEM photographs of SiV3 particles.

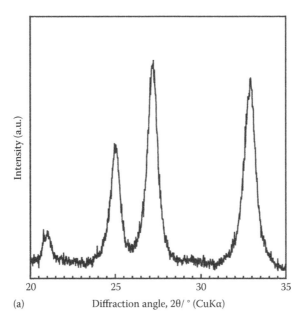

(a)   Diffraction angle, 2θ/ ° (CuKα)

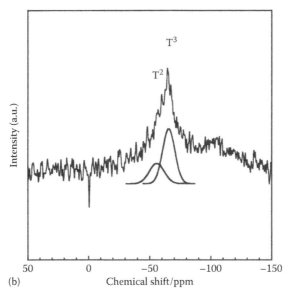

(b)   Chemical shift/ppm

FIGURE 6.5 Characterization of SiV particles: (a) XRD pattern and (b) ^{29}Si MAS-NMR spectrum. SiV consists predominantly of vaterite phase. No APTES exists as its monomer and siloxane was formed in SiV.

These results suggest that the conditions for the aggregation of the primary particles in the suspension are influenced by adding APTES. Figure 6.5a shows the x-ray diffraction (XRD) pattern of SiV. SiV consists predominantly of vaterite. The structure of APTES in SiV was further characterized by ^{29}Si magic angle spinning nuclear magnetic resonance (^{29}Si MAS-NMR) spectroscopy. Figure 6.5b shows the ^{29}Si MAS-NMR spectrum of SiV. A peak at −50 to −80 ppm is seen. The peak may be superimposed by those corresponding to T^2 [NH$_2$(CH$_2$)$_3$Si(OSi)$_2$ (OH, OC$_2$H$_5$)] and T^3 [NH$_2$(CH$_2$)$_3$Si(OSi)$_3$] (Yabuta et al. 2000). This indicates that no APTES in SiV exists as its monomer, and siloxane was formed by the condensation of APTES through the preparation of SiV.

6.3.2 PREPARATION OF SILOXANE-DOPED VATERITE/POLY(L-LACTIC ACID) COMPOSITE FILMS

A solution of PLLA (molecular weights, $M_w = 160 \pm 20\,kDa$) prepared by dissolving 1.5 g of PLLA in 20 mL of methylene chloride was used to disperse portions of 0.75 g of the vaterite or SiV powders under stirring for 30 min, resulting in the as-prepared composite. The composites prepared using vaterite, SiV1 and SiV3 were denoted by V-PLLA, SiV1-PLLA, and SiV3-PLLA, respectively. The composite films showed high flexibility. The surface observation of the film by SEM revealed that the vaterite or SiV particles dispersively existed in PLLA.

For estimation of the silicon releasabilities of SiV1-PLLA and SiV3-PLLA, silicon concentrations in α-minimum essential medium (α-MEM) containing 10% FBS after soaking SiV1-PLLA and SiV3-PLLA were evaluated by inductively coupled plasma atomic emission spectroscopy (ICP-AES). The medium was changed at days 1, 3, and 5 after soaking the sample and the silicon concentration in the harvested medium was evaluated. Figure 6.6 shows the silicon concentration in the culture medium after soaking SiV1-PLLA and SiV3-PLLA. Both the samples showed silicon releasability for 7 days of soaking in the culture medium. SiV1-PLLA and SiV3-PLLA release approximately 2.2 and 9.2 mg/L of silicon ions, respectively, within 1 day of soaking. The average value of the silicon released from SiV1-PLLA for 2–7 days of the incubation was estimated to be 0.21 mg/L/day, while that of SiV3-PLLA was 0.51 mg/L/day. The SiV particles existing on the composite sample surfaces are expected to be exposed to the culture medium and then immediately release ionic silicon species to the medium during the first day of the incubation. From day 2, the amounts of the released silicon species in both the samples significantly decreased. The amount of silicon species released from SiV1-PLLA and SiV3-PLLA decreased from 2.2 and 9.2 to 0.74 and 1.3 mg/L, respectively, at day 3. The releasing mechanism of the silicon species from day 2 may be different from that within day 1. From day 2, the SiV particles embedded in the PLLA matrix are gradually exposed to the culture medium through the PLLA degradation and then release a trace amount of silicon species to the culture medium. The silicon species release depends on the PLLA degradation from day 2 of incubation. The molecular weight of PLLA was reported to decrease after implanted in body and the degradation speed of PLLA depends on its molecular weight (Kenawy et al. 2002). There was no significant difference in M_w of the matrix PLLA among all of the composites. The M_w of V-PLLA,

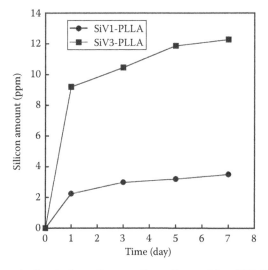

FIGURE 6.6 Silicon concentration in the culture medium after soaking SiV1-PLLA or SiV3-PLLA. The medium was changed at days 1, 3, and 5 after soaking of the sample and the total concentration in the medium harvested so far was evaluated.

SiV1-PLLA, and SiV3-PLLA were 189, 179, and 175 kDa, respectively. Therefore, the difference in the released amount of the silicon species between SiV1-PLLA and SiV3-PLLA is suggested to originate from the amount of APTES in SiV, not the speed of the degradation of the PLLA matrix. This indicates that the amount of silicon species released from SiV-PLLA composites is controllable by changing the amount of APTES in SiV. That is, to control the APTES content in SiV particles is regarded to be one of the most effective methods for controlling the silicon releasability of SiV-PLLA. In fact, the amount of the silicon species released from SiV3-PLLA per day was approximately 2.4 times as large as that released from SiV1-PLLA during day 2–7 of incubation. The value is evidently related to the difference in the amount of silicon species between SiV1 and SiV3.

6.3.3 Cellular Compatibility of SiV-PLLA Composites

The cellular proliferations on two types of the siloxane-containing samples were examined to investigate the effects on the cells of the silicon species amount in SiV-PLLA. Mouse osteoblast-like cells (MC3T3-E1 cells) were seeded onto the prepared samples in 24-well plates, at a density of 30,000 cells per well. The α-MEM containing 10% FBS was used as a culture medium. The medium was changed after 1 day of culturing and then changed every other day. Figure 6.7 shows the number of cells cultured on SiV1-PLLA and SiV3-PLLA for 7 days. The number of cells cultured on SiV3-PLLA is significantly higher than that on SiV1-PLLA at days 3, 5, and 7. The proliferation of MC3T3-E1 cells cultured on SiV3-PLLA, which released a larger amount of silicon species, was higher than that on SiV1-PLLA. On the other hand, the effects of the silicon species on the cellular functions may relate to their ionic states. The ionic state of the silicon species released from SiV-PLLA has been not found yet. Further works to clarify the ionic state and the correlation between the state and its effect on the cellular functions are needed.

For estimating alkaline phosphatase (ALP) activity in MC3T3-E1 cells, they were seeded onto the samples in 24-well plates, at a density of 10,000 cells per well. The culture medium contained no differentiation agent. A Thermanox® plastic plate was used as a control sample.

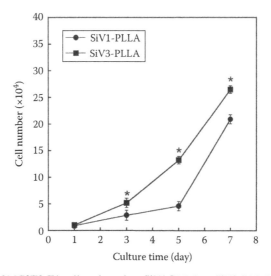

FIGURE 6.7 Numbers of MC3T3-E1 cells cultured on SiV1-PLLA or SiV3-PLLA for 7 days. The proliferation of MC3T3-E1 cells cultured on SiV3-PLLA, which released a larger amount of silicon species, was higher than that on SiV1-PLLA ($*p < 0.05$).

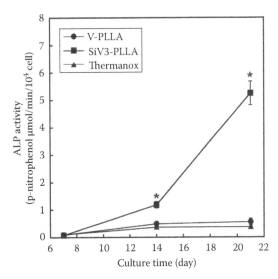

FIGURE 6.8   ALP activities in MC3T3-E1 cells cultured on V-PLLA, SiV3-PLLA, and Thermanox®. The levels of the activities of V-PLLA and Thermanox are almost same at all time points although those of SiV3-PLLA increased with prolongation of the culture time ($*p < 0.05$).

Figure 6.8 shows the ALP activities in MC3T3-E1 cells cultured on V-PLLA, SiV3-PLLA, and Thermanox. The activity levels of these samples increase during 21 days of culturing. In the case of V-PLLA and Thermanox, the levels of the activities are approximately 0.1–0.6 at all time points. On the other hand, the levels of SiV3-PLLA are approximately 1.2 at day 14 and approximately 5.3 at day 21. These values of SiV3-PLLA are significantly higher than those of V-PLLA at days 14 and 21. This indicates that the silicon species released from SiV3-PLLA induced the differentiation of MC3T3-E1 cells.

6.4   SILOXANE-DOPED VATERITE/POLY(L-LACTIC ACID) COMPOSITE MICROFIBER MEMBRANES

Three-dimensional scaffolds are common in biomaterials. Electrospinning is a process that can generate polymer fiber meshes with high porosity and flexibility, both essential components of tissue engineering (Still and von Recum 2008). In general, the process of electrospinning is mainly affected by both system parameters and process parameters. System parameters include polymer molecular weight (M_w) and polymer solution properties such as viscosity and conductivity. Process parameters, on the other hand, involve flow rate of the polymer solution, electric potential, and the distance between capillary and collector, among others (Still and von Recum 2008). It is possible to control the fiber diameter and the thickness of the fiber mesh scaffolds by optimizing these parameters. To use electrospun fiber meshes in bone tissue engineering applications, their various properties, including material, fiber orientation, porosity, and surface modification, must be considered. To optimize the mechanical and biomimetic properties of the fiber mesh, it is necessary to choose the appropriate materials, since the orientation and diameter of the fibers in electrospun scaffolds can influence cellular compatibility (Li et al. 2002, Zeng et al. 2003, Yang et al. 2005, Bandami et al. 2006, Xin et al. 2007). The fiber diameter in electrospun scaffolds, for example, has been reported to influence osteoblast proliferation and mineralization.

There has been increasing interest in the use of electrospun fibrous meshes as scaffolds in medicine. In particular, biocompatible polymers, such as PLGA and poly(caprolactone), have been applied as materials for the meshes, incorporated with various growth factors or DNA (Still and von Recum 2008). On the other hand, very few studies have been performed on fibrous meshes consisting of polymer/ceramics composites. Previously, PLGA and HA composite scaffolds, which were fabricated by electrospinning, contained only 5–10 wt% of HA nanocrystals (Nie and Wang 2007). A larger amount of calcium is needed to enhance the bone-forming ability of the electrospun fibrous mesh, and vaterite powders are effective materials for providing calcium.

The electrospun fiber mesh would contain a large volume of pores and would show flexibility, both essential components of biomaterials, such as bone fillers. The fibrous mesh containing a large amount of vaterite (up to 60 wt%) has been successfully prepared by kneading and subsequent electrospinning.

6.4.1 Preparation of SiV-PLLA Microfiber Membranes

The microfiber meshes were prepared by an electrospinning method. SiV powders were mixed with PLLA (PURASORB®; Mw; 260 kDa, PURAC, the Netherlands) under 200°C for 10 min with a kneader, resulting in the formation of the PLLA composites containing 60 wt% of the powders. The composites were dissolved in chloroform at 10 wt% to prepare the solution for electrospinning.

Figure 6.9a and b shows morphologies of the SiV-PLLA-mesh. Microfibers approximately 10 μm in diameter were observed at the SiV-PLLA-mesh surface and were found to twine one another. The size of gaps between the microfibers varied between 10 μm and several hundred micrometers, as shown in Figure 6.9a. The SiV particles observed at the microfiber surface were coated with a thin film of the PLLA matrix, as shown in Figure 6.9b. The particles were found to be packed closely in the microfiber from the fracture face of the microfiber. On the other hand, the V-PLLA mesh consisted of short microfibers approximately 7 μm in diameter. The prepared SiV-PLLA mesh showed flexibility without breaking, whereas the V-PLLA mesh was brittle and was easily broken by hand.

The prepared microfiber mesh was immersed in 10 mL of an aqueous solution containing ion concentrations 1.5 times those of SBF (1.5 SBF) at 37°C for 1 day. Figure 6.9c and d shows morphologies of the mesh after soaking. The particles were embedded in the PLLA matrix and after the immersion in 1.5 SBF for 1 day, the surfaces of the microfibers were coated with leaf-shaped deposits as shown in Figure 6.9d. No significant change was observed in the gap sizes between the

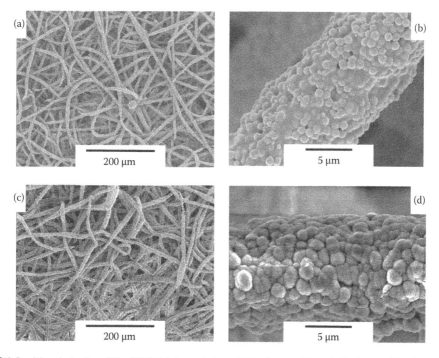

FIGURE 6.9 Morphologies of the SiV-PLLA-mesh (a and b) before and (c and d) after soaking in 1.5SBF for 1 day. The fibers have about 10 μm of diameter. The SiV particles are observed at the microfiber surfaces. After soaking, the surfaces of the microfibers were coated with leaf-shaped deposits, bone-like hydroxyapatite (bHA).

microfibers before and after the immersion. After the bHA coating, further testing demonstrated several cracks, but the coating did not peel off. The deposits were confirmed by XRD to be HA of approximately 1 μm in thickness. The release of silicon ion from SiV-PLLA mesh coated with bHA into the culture medium was evaluated over 5 days. The amount of the silicon species released from it was 0.68 mg/L after 1 day of soaking in the culture medium. The sample kept releasing the species; the amount in the culture medium was 0.39 mg/L at day 3 and 0.41 mg/L at day 5.

The flexibility of the mesh was felt to be a reflection of its structural properties such as the molecular weight of the PLLA matrix and structural changes in the interface between the matrix and the SiV powders. The structural analyses of SiV-PLLA were performed using ^{13}C cross polarization magic angle spinning nuclear magnetic resonance (^{13}C CP/MAS-NMR) and Fourier transform infrared reflection (FT-IR) spectroscopy. In the ^{13}C CP/MAS-NMR spectra of the prepared composites, the band of the carboxy peak shifted slightly and a new peak appeared at the low magnetic field area of the band, where carboxy groups form coordinate bonds with bivalent ions (Asada et al. 1991), as shown in Figure 6.10a. The peaks further separated into two peaks, Peaks A and B. Peak A was defined at one magnetic field in the band and Peak B as one magnetic field lower in the band. The intensity of

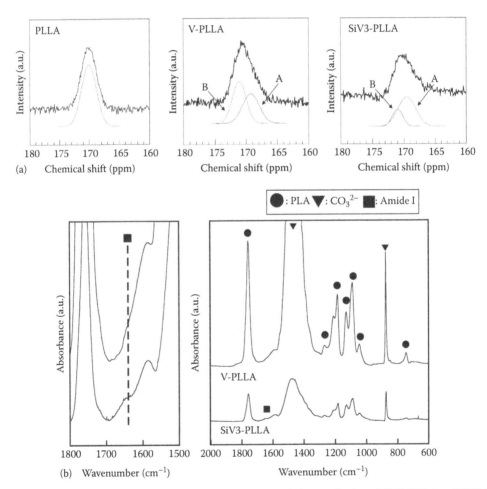

FIGURE 6.10 (See companion CD for color figure.) Results of characterization of SiV3-PLLA and V-PLLA. (a) ^{13}C CP/MAS-NMR spectra of the two samples and PLLA. The peaks further separated into two peaks, Peaks A and B. The intensity of Peak B in the spectrum of V-PLLA is larger than that of SiV3-PLLA. This indicates that more carboxy groups formed coordinate bonds with calcium ions in V-PLLA than SiV3-PLLA. (b) FT-IR spectra of the two samples. A weak peak appears at around 1650 cm^{-1} in the spectrum of SiV3-PLLA. This peak appears to originate from the amide bond between carboxy groups in PLLA and amino groups in APTES structure.

TABLE 6.2
Weight-Average Molecular Weight (Mw),
Number-Average Molecular Weight (Mn),
and Polydispersity (Mw/Mn) of PLLA and
PLLA Matrices in SiV3-PLLA or V-PLLA

	PLLA	V-PLLA	SiV3-PLLA
Mw (kDa)	259	17	185
Mn (kDa)	168	7	86
Mw/Mn	1.5	2.4	2.1

Peak B in the spectrum of V-PLLA is larger than that of SiV-PLLA. This indicates that more carboxy groups formed coordinate bonds with calcium ions in V-PLLA than SiV-PLLA and the formation of the coordinate bonds caused the decrease in the molecular weight. In the FT-IR spectrum of SiV-PLLA, a new weak peak was found at around $1650\,cm^{-1}$ except for peaks corresponding to PLLA and CO_3^{2-}, as shown in Figure 6.10b. This peak appeared to originate from the amide bond, which indicates that carboxy groups in PLLA structure form chemical bonds with amino groups in APTES structure.

The molecular weights of PLLA matrix in the prepared meshes were measured by gel permeation chromatography (GPC). Table 6.2 shows the weight-average molecular weight (Mw), number-average molecular weight (Mn), and polydispersity (Mw/Mn) of PLLA and PLLA matrix in SiV-PLLA or V-PLLA. Although the molecular weight of PLLA slightly decreased after kneading with SiV powders, it significantly decreased after kneading with vaterite powders. We suppose that the coordinate bonds between calcium ions and carboxy groups caused breaks in polymer chains. This may be the reason why the molecular weight of V-PLLA mesh was significantly smaller than that of SiV-PLLA one. The SiV particles were partially coated with siloxane derived from APTES; the amount of calcium ions explored to PLLA matrix was smaller comparing to that of vaterite. From these results of structural analyses, the SiV-PLLA mesh structure is suggested, as shown in Figure 6.11.

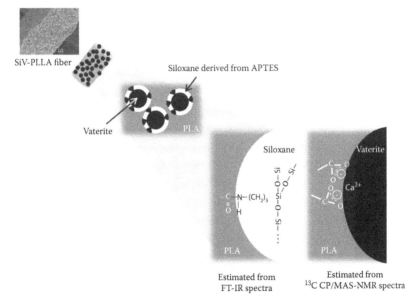

FIGURE 6.11 Schematic image of SiV-PLLA structure, based on the results of GPC, FT-IR, and ^{13}C CP/MAS-NMR. Carboxy groups in PLLA form coordinate bonds with calcium ions and also amide bonds with amino groups in siloxane derived from APTES on the SiV surface.

FIGURE 6.12   (See companion CD for color figure.) Histology of in vivo response to the bilayered meshes. Villanueva Goldner stains show new bone formed in the meshes at the center of holes 8 mm in diameter formed at frontal midline site of the rabbit calvaria. Asterisks indicate mineralized tissues. New bone formation was observed over almost the entire area of implanted Si-PVH/HA-mesh after 12 weeks.

6.4.2   TISSUE COMPATIBILITY AND BONE-FORMING ABILITY OF THE MEMBRANE

The tissue compatibility and bone-forming ability of the fiber meshes was evaluated by in vivo tests using rabbits. An 8 mm-diameter hole was drilled into the front midline of the animal's calvaria using a bone cutter and then covered with the bilayered meshes consisting of SiV-PLLA mesh and PLLA one. An 8 mm-diameter hole in bone is generally not regenerated if remained to be done. In our preliminary experiment, a PLGA mesh was implanted in the same position for 12 weeks and a new bone was not to be formed in/on the mesh at the center of an 8 mm hole. The bilayered mesh was fixed by putting between skin and bone. The meshes were implanted with SiV-PLLA mesh on the side in contact with the hole and dense PLLA mesh on the side in contact with the skin. Figure 6.12 shows the histology of in vivo response to the bilayered meshes and the new bone formation at the center of the holes. Villanueva Goldner stain showed that mineralized tissue was formed over almost the entire area of implanted SiV-PLLA mesh after 12 weeks. There was no tissue inflammation observed by histology. The bone formation started from the inside of SiV-PLLA mesh and not from the edge of the hole.

6.5   SUMMARY

This chapter described various types of PLLA–vaterite composite materials releasing silicon species that we have developed so far. These materials with various shapes have been prepared, such as film, bulk, hollow sphere, and fibrous membrane. We can choose the shape of the materials to adapt to the applications. Also, the shapability and mechanical elasticity of the materials enable us to allow good technical flexibility in the application. The effects of silicon species on the osteogenic cell functions have got much attention in the biomaterials for bone reconstruction and been gradually clarified in their mechanisms. We believe applying the silicon species is useful to develop materials static and chronic releasability of factors stimulating osteogenic cells and having good bone-forming ability.

ACKNOWLEDGMENTS

This chapter was based mainly on our projects: these works were supported in part by Grant-in-Aids for Scientific Research (B) (No. 14380398 and 20390497) from Japan Society for Promotion of Science, a Grant-in-Aid for Young Scientists (B) (No. 21700487) from The Ministry of Education, Culture, Sports, Science and Technology (MEXT), New Regional Consortium R&D Program from Ministry of Economy, Trade and Industry (METI), and a grant from Institute of Ceramics Research and Education (ICRE), NIT. We thank Dr. Yoshio Ota and Dr. Takashi Wakita for supporting our projects.

REFERENCES

Asada, M., Asada, N., Toyoda, A., Ando, I., and Kurosu, H. 1991. Side-chain structure of poly(methacrylic acid) and its zinc salts in the solid state as studied by high-resolution solid-state ^{13}C NMR spectroscopy. *J Mol Struct* 244:237–248.

Badami, A.S., Kreke, M.R., Thompson, M.S., Riffle, J.S., and Goldstein, A.S. 2006. Effect of fiber diameter on spreading, proliferation, and differentiation of osteoblastic cells on electrospun poly(lactic acid) substrates. *Biomaterials* 27:596–606.

Bos, R.R.M., Rozema, F.B., Boering, G. et al. 1991. Degradation of and tissue reaction to biodegradable poly(L-lactide) for use as internal fixation of fractures: A study in rats. *Biomaterials* 12:32–36.

Boskey, A.L., Wright, T.M., and Blank, R.D. 1999. Collagen and bone strength. *J Bone Miner Res* 14:330–335.

Carlisle, E.M. 1986. Silicon as an essential trace element in animal nutrition. In *Silicon Biochemistry*, eds. Evered, D. and O'Connor, M., pp. 123–139. Chichester, U.K.: Wiley.

Gough, J.E., Jones, J.R., and Hench, L.L. 2004. Nodule formation and mineralisation of human primary osteoblasts cultured on a porous bioactive glass scaffold. *Biomaterials* 25:2039–2046.

Jaiswal, N., Haynesworth, S.E., Caplan, A.I., and Bruder, S.P. 1997. Osteogenic differentiation of purified, culture-expanded human mesenchymal stem cells in vitro. *J Cell Biochem* 64:295–312.

Jones, J.R., Tsigkou, O., Coates, E.E., Stevens, M.M., Polak, J.M., and Hench, L.L. 2007. Extracellular matrix formation and mineralization on a phosphate-free porous bioactive glass scaffold using primary human osteoblast (HOB) cells. *Biomaterials* 28:1653–1663.

Kasuga, T., Maeda, H., Kato, K., Nogami, M., Hata, K., and Ueda, M. 2003. Preparation of poly(lactic acid) composites containing calcium carbonate (vaterite). *Biomaterials* 24:3247–3253.

Kenawy, E.R., Bowlin, G.L., Mansfield, K. et al. 2002. Release of tetracycline hydrochloride from electrospun poly(ethylene-co-vinylacetate), poly(lactic acid), and a blend. *J Control Release* 81:57–64.

Kim, S.S., Park, M.S., Jeon, O., Choi, C.Y., and Kim, B.S. 2006. Poly(lactide-co-glycolide)/hydroxyapatite composted scaffolds for bone tissue engineering. *Biomaterials* 27:1399–1409.

Kokubo, T., Kushitani, H., Sakka, S., Kitsugi, T., and Yamamuro, T. 1990. Solutions able to reproduce in vivo surface-structure changes in bioactive glass-ceramic A-W3. *J Biomed Mater Res* 24:721–734.

Li, W.J., Laurencin, C.T., Caterson, E.J., Tuan, R.S., and Ko, F.K. 2002. Electrospun nanofibrous structure: A novel scaffold for tissue engineering. *J Biomed Mater Res* 60:613–621.

Liao, H., Mutvei, H., Sjöström, M., Hammarström, L., and Li, J. 2000. Tissue responses to natural aragonite (*Margaritifera* shell) implants in vivo. *Biomaterials* 21:457–468.

Maeda, H., Kasuga, T., and Hench, L.L. 2006. Preparation of poly(L-lactic acid)-polysiloxane-calcium carbonate hybrid membranes for guided bone regeneration. *Biomaterials* 27:1216–1222.

Maeda, H., Kasuga, T., and Nogami, M. 2004. Bonelike apatite coating on skeleton of poly(lactic acid) composite sponge. *Mater Trans* 45:989–993.

Mainil-Varlet, P., Rahn, B., and Gogolewski, S. 1997. Long-term in vivo degradation and bone reaction to various polylactides: 1. One-year results. *Biomaterials* 18:257–266.

Nie, H. and Wang, C.H. 2007. Fabrication and characterization of PLGA/HAp composite scaffolds for delivery of BMP-2 plasmid DNA. *J Control Release* 120:111–121.

Reffitt, D.M., Ogston, N., Jugdaohsingh, R. et al. 2003. Orthosilicic acid stimulates collagen type 1 synthesis and osteoblastic differentiation in human osteoblast-like cells in vitro. *Bone* 32:127–135.

Still, T.J. and von Recum, H.A. 2008. Electrospinning: Applications in drug delivery and tissue engineering. *Biomaterials* 29:1989–2006.

Thian, E.S., Huang, J., Best, S.M., Barber, Z.H., Brooks, R.A., Rushton, N., and Bonfield, W. 2006. The response of osteoblasts to nanocrystalline silicon-substituted hydroxyapatite thin films. *Biomaterials* 27:2692–2698.

Vago, R., Plotquin, D., Bunin, A., Sinelnikov, I., Atar, D., and Itzhak, D. 2002. Hard tissue remodeling using biofabricated coralline biomaterials. *J Biochem Biophys Methods* 50:253–259.

Xin, X., Hussain, M., and Mao, J.J. 2007. Continuing differentiation of human mesenchymal stem cells and induced chondrogenic and osteogenic lineages in electrospun PLGA nanofiber scaffold. *Biomaterials* 28:316–325.

Xynos, I.D., Edgar, A.J., Buttery, L.D.K., Hench, L.L., and Polak, J.M. 2000. Ionic products of bioactive glass dissolution increase proliferation of human osteoblasts and induce insulin-like growth factor II mRNA expression and protein synthesis. *Biochem Biophys Res Commun* 276:461–465.

Xynos, I.D., Edgar, A.J., Buttery, L.D.K., Hench, L.L., and Polak, J.M. 2001. Gene-expression profiling of human osteoblasts following treatment with the ionic products of Bioglass® 45S5 dissolution. *J Biomed Mater Res* 55:151–157.

Yabuta, T., Tsuru, K., Hayakawa, S., Ohtsuki, C., and Osaka, A. 2000. Synthesis of bioactive organic-inorganic hybrids with γ-methacryloxypropyltrimethoxysilane. *J Sol-Gel Sci Technol* 19:745–748.

Yang, F., Murugan, R., Wang, S., and Ramakrishna, S. 2005. Electrospinning of nano/micro scale poly(L-lactic acid) aligned fibers and their potential in neural tissue engineering. *Biomaterials* 26:2603–2610.

Zeng, J., Xu, X., Chen, X. et al. 2003. Biodegradable electrospun fibers for drug delivery. *J Control Release* 92:227–231.

Zhang, K., Wang, Y., Hillmyer, M.A., and Francis, L.F. 2004. Processing and properties of porous poly(L-lactide)/bioactive glass composites. *Biomaterials* 25:2489–2500.

7 Mucoadhesive Polymers
Basics, Strategies, and Future Trends

Andreas Bernkop-Schnürch

CONTENTS

7.1 INTRODUCTION

In the early 1980s, academic research groups pioneered the concept of mucoadhesion as a new strategy in order to improve the therapeutic effect of various drugs. *Mucoadhesive polymers are able to adhere to the mucus gel layer* which covers various tissues of the body. These mucoadhesive properties are in many cases advantageous, rendering such polymers interesting tools for various pharmaceutical reasons:

1. Mediated by mucoadhesive polymers, the residence time of dosage forms on the mucosa can be prolonged, which allows a sustained drug release at a given target site in order to maximize the therapeutic effect. Robinson and Bologna, for instance, have reported that a polycarbophil gel is capable of remaining on the vaginal tissue for 3–4 days and thus provides an excellent vehicle for the delivery of drugs such as progesterone and nonoxynol-9 [1].
2. Furthermore, drug delivery systems can be localized on a certain surface area for purpose of local therapy or of drug liberation at the "absorption window." The absorption of riboflavin, for instance, which has its "absorption window" in the stomach as well as the small intestine, could be strongly improved in human volunteers by oral administration of mucoadhesive microspheres versus non-adhesive microspheres as illustrated in Figure 7.1 [2].
3. In addition, mucoadhesive polymers can guarantee an intimate contact with the absorption membrane providing the basis for a high-concentration gradient as driving force of drug absorption (IIIa), for the exclusion of a presystemic metabolism such as the degradation of orally given (poly)peptide drugs by luminally secreted intestinal enzymes (IIIb) [3,4], and for interactions of the polymer with the epithelium such as a permeation enhancing effect [e.g., 5,6] or the inhibition of brush border membrane-bound enzymes (IIIc) [7,8].

Because of all these benefits, research work in the field of mucoadhesive polymers has strongly increased within the last two decades, resulting in numerous promising ideas, strategies, systems,

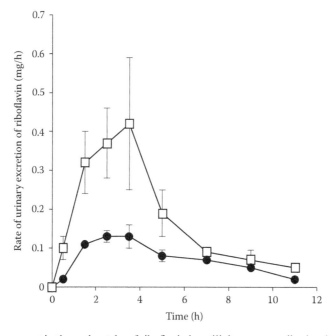

FIGURE 7.1   Improvement in the oral uptake of riboflavin by utilizing a mucoadhesive delivery system (□) in comparison to the same delivery system without mucoadhesive properties (●) in human volunteers. (Adapted from Akiyama, Y. et al., *J. Pharm. Pharmacol.*, 50, 159, 1998.)

and techniques based on a more and more profound basic knowledge. An overview reflecting the status quo as well as future trends concerning mucoadhesive polymers should provide a good platform for ongoing research and development in this field.

7.2 MUCUS GEL COMPOSITION

As mucoadhesive macromolecules adhere to the mucus gel layer, it is important to characterize first of all this polymeric network representing the target structure for mucoadhesive polymers. The most important component building up the mucus structure are glycoproteins with a relative molecular mass range of $1–40 \times 10^6$ Da. These so-called mucins possess a linear protein core, typically of high serine and threonine content that is glycosylated by oligosaccharide side chains that contain blood group structures. The protein core of many mucins exhibits furthermore N- and/or C-terminally located cysteine-rich subdomains, which are connected with each other via intra- and/or intermolecular disulfide bonds. This presumptive structure of the mucus is illustrated in Figure 7.2. The water content of the mucus gel has been determined to be 83% [9]. Generally, mucins may be classified into two classes, *secretory* (I) and *membrane-bound* (II) forms. Secretory mucins are secreted from mucosal absorptive epithelial cells as well as specialized goblet cells. They constitute the major component of mucus gels of the gastrointestinal, respiratory, ocular, and urogenital surface. The mucus layer based on secretory mucins represents not only a physical barrier but also a protective diffusion barrier [9,10]. Membrane-bound mucins possess a hydrophilic membrane-spanning domain and they are attached to cell surfaces. Up to now, eight different types of human mucins have been discovered and characterized, which are listed in Table 7.1.

Secreted mucins are continuously released from cells as well as glands undergoing immediately thereafter a polymerization process which is mainly based on an intermolecular disulfide bond formation. This so-formed mucus layer, on the other hand, is continuously eroded by enzymatic and mechanical challenges on the luminal surface. Although the turnover time of the mucus gel layer seems to be crucial in order to estimate for how long a mucoadhesive delivery system can remain on the mucosa in maximum, there is no accurate information on this time scale available. A clue was given by Lehr et al., who determined the turnover time of the mucus gel layer of chronically isolated intestinal loops in rats to be in the order of approximately 1–4 h [21]. Both mucus secretion and mucus erosion, however, are influenced by so many factors such as mucus secretagogues, mechanical stimuli, stress, calcium concentration, or the enzymatic activity of luminal proteases leading to a highly variable turnover time, which is therefore quite complex and difficult to evaluate.

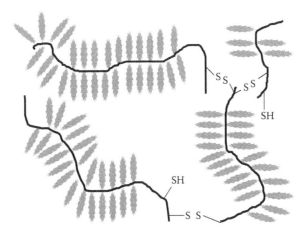

FIGURE 7.2 Schematic presentation of the three-dimensional network of the mucus gel layer; protein core: ▬▬ glycosidic side-chains: ⬬.

TABLE 7.1
Synopsis of Human Mucins

Mucin	Characteristics	High-Level Expression	Cysteine-Rich Subdomains	References
MUC1	Membrane-bound epithelial mucin	Breast, pancreas		[12]
MUC2	Secreted intestinal mucin	Intestine, tracheobronchus	Yes	[13]
MUC3	Secreted intestinal mucin	Intestine, gallbladder, pancreas	Yes	[14]
MUC4	Tracheobronchial secretory mucin	Tracheobronchus, colon, uterine endocervix		[15]
MUC5A/C	Secretory mucin	Tracheobronchus, stomach, ocular, uterine endocervix	Yes	[16]
MUC5B	Secretory mucin	Tracheobronchus, salivary	Yes	[17]
MUC6	Major secretory mucin of the stomach	Stomach, gallbladder	Yes	[18]
MUC7		Salivary		[19]
MUC8		Tracheobronchus, reproductive tract		[20]

Source: Campbell, B.J., Biochemical and functional aspects of mucus and mucin-type glycoproteins, bioadhesive drug delivery systems, E. Mathiowitz, Chickering, D.E. III, and Lehr, C.-M., eds., Marcel Dekker, New York, pp. 85–130, 1999.

7.3   PRINCIPLES OF MUCOADHESION

Since the concept of mucoadhesion has been introduced into the scientific literature, considerably many attempts have been undertaken in order to explain this phenomenon. So far, however, no generally accepted theory has been found. A reason for this situation can be seen in the fact that many parameters seem to have an impact on mucoadhesion (Figure 7.3), which makes a unique explanation impossible. Although there are many controversies in case of which theories should be favored, at least two basic steps are generally accepted. In step I, the *contact stage*, an intimate contact between the mucoadhesive and the mucus gel layer is formed. In step II, the *consolidation stage*, the adhesive joint is strengthened and consolidated, providing a prolonged adhesion.

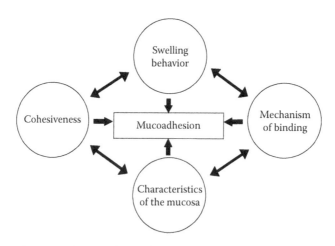

FIGURE 7.3   Schematic presentation of effects influencing mucoadhesion. (From Bernkop-Schnürch, A. and Steininger, S., *Int. J. Pharm.*, 194, 239, 2000.)

7.3.1 CHEMICAL PRINCIPLES

7.3.1.1 Formation of Non-Covalent Bonds

Non-covalent chemical bonds include hydrogen bonding (I), which is based on hydrophilic functional groups such as hydroxylic groups, carboxylic groups, amino groups, and sulfate groups, ionic interactions (II) such as the interaction of the cationic polymer chitosan with anionic sialic acid moieties of the mucus, and van der Waals forces (III) based on various dipole–dipole interactions.

7.3.1.2 Formation of Covalent Bonds

In contrast to secondary bonds, covalent bonds are much stronger and are not any more influenced by parameters such as ionic strength and pH value. Functional groups that are able to form covalent bonds to the mucus layer are overall thiol groups. The way how such functional groups can form covalent bonds with mucus glycoproteins is illustrated in Figure 7.4. Thiolated polymers are able to mimic the natural occurring mechanism how mucus glycoproteins are immobilized in the mucus.

7.3.2 PHYSICAL PRINCIPLES

7.3.2.1 Interpenetration

One theory in order to explain the phenomenon of mucoadhesion is based on a macromolecular interpenetration effect. The mucoadhesive macromolecules interpenetrate with mucus glycoproteins as illustrated in Figure 7.5. The resulting consolidation provides the formation of a strong stable mucoadhesive joint. The theory can be substantiated by the observation that chain flexibility favoring a polymeric interpenetration is a crucial parameter for mucoadhesion. The cross-linking of various polymers or the covalent attachment of large sized ligands [23] leading to a reduction in chain flexibility results in a strong decrease in mucoadhesive properties.

Rheological approaches—as discussed in detail later on—demonstrating a synergistic increase in the resistance to elastic deformation by mixing mucus with mucoadhesive polymer, i.e., the consolidation of the adhesive joint and attenuated total reflection Fourier transform infrared (ATR-FTIR) studies showing changes in the spectrum of a poly(acrylic acid) film because of interpenetrating mucin molecules within 6 min, provide further evidence for this theory [24].

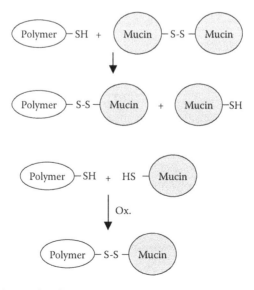

FIGURE 7.4   Thiolated polymers forming covalent bonds with the mucus layer.

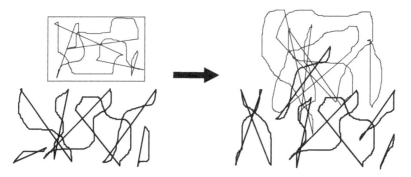

FIGURE 7.5   Interpenetration of a mucoadhesive matrix tablet (——) and the mucus gel layer (▬▬).

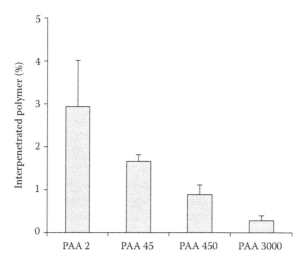

FIGURE 7.6   Size-dependent interpenetration of polyacrylates (PAA) of indicated molecular mass (2–3000 kDa) in a mucus layer. (Adapted from Imam, M.E. et al., *STP Pharma Sci.*, 13, 171, 2003.)

Recently, Imam et al. evaluated the degree of interpenetration of fluorescence labeled polyacrylates of increasing molecular mass with various mucus gel layers utilizing confocal laser microscopy and fluorescence quantification techniques [25]. Results of these studies are illustrated in Figure 7.6, demonstrating a clear correlation between the molecular mass of the mucoadhesive polymer and the degree of interpenetration.

7.3.2.2   Mucus Dehydration

Dehydration of a mucus gel layer increases its cohesive nature, which was shown by Mortazavi and Smart [26]. Dehydration essentially alters the physicochemical properties of a mucus gel layer, making it locally more cohesive and promoting the retention of a delivery system. The theory can be substantiated by studies with dialysis tubings. Bringing dry mucoadhesive polymers wrapped in dialysis tubings into contact with a mucus layer leads to its dehydration rapidly [26]. Dehydrating a mucus gel increases its cohesive nature and subsequently its adhesive behavior, which could be shown by tensile studies [26]. An objection to the theory that no interpenetration but exclusively mucus dehydration occurs is given by various rheological studies and tensile studies carried out by Caramella et al. [27], who observed a significant increase in the total work of adhesion (TWA) by mixing the polymer directly with mucin before tensile measurements. It is therefore likely that glycoproteins of the mucus are carried with the flow of water into the mucoadhesive polymer which leads to the already described interpenetration; an explanation which allows the combination of both the interpenetration and mucus dehydration theory.

7.3.2.3 Entanglements of Polymer Chains

The mucoadhesive as well as cohesive properties of polymers can also be explained by physical entanglements of polymer chains. The difference in cohesive as well as mucoadhesive properties of lyophilized and precipitated mucoadhesive polymers gives strong evidence for this theory. Precipitated polymers display high cohesive as well as mucoadhesive properties with a likely high number of polymer chain entanglements, whereas these properties are comparatively lower for the corresponding lyophilized polymers, which do not exhibit such a high extent of entanglements [28].

7.4 METHODS TO EVALUATE MUCOADHESIVE PROPERTIES

7.4.1 In Vitro Methods

The selection of the mucoadhesive material is the first step in developing a mucoadhesive drug delivery system. A screening of the adhesive properties of polymeric materials can be done by various in vitro methods such as visual tests, tensile studies, and rheological methods, which are often accompanied by additional spectroscopic techniques. Apart from these well-established methods which are described in detail later, some novel methods such as magnetic [29] and direct force measurement techniques [30] have been introduced into the literature as well.

7.4.1.1 Visual Tests

7.4.1.1.1 Rotating Cylinder

In order to evaluate the duration of binding to the mucosa as well as the cohesiveness of mucoadhesive polymers, the rotating cylinder seems to be an appropriate method. In particular, tablets consisting of the test polymer can be brought into contact with freshly excised intestinal mucosa (e.g., porcine), which has been spanned on a stainless steel cylinder (diameter: 4.4 cm; height: 5.1 cm; apparatus 4—cylinder, USP XXII). Thereafter, the cylinder is placed in the dissolution apparatus according to the USP containing an artificial gastric or intestinal fluid at 37°C. The experimental setup is illustrated in Figure 7.7. The fully immersed cylinder is agitated with 250 rpm. The time needed for detachment, disintegration, and/or erosion of test tablets can be determined visually [22].

7.4.1.1.2 Rinsed Channel

At this method, freshly excised mucosa is spread out on a lop-sided channel with the mucus gel layer facing upward and placed in a thermostatic chamber, which is kept at 37°C. After applying the test material on the mucosa, the rinse is flushed with an artificial gastric or intestinal fluid at a constant flow rate and the residence time of the mucoadhesive polymer is determined visually or quantified utilizing insoluble markers such as fluorescein diacetate [31]. The experimental setup is illustrated in Figure 7.8 [32,33].

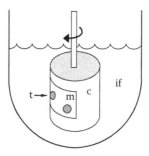

FIGURE 7.7 Schematic presentation of the test system used to evaluate the mucoadhesive properties of tablets based on various polymers. c, cylinder; if, intestinal fluid; m, porcine mucosa; t, tablet.

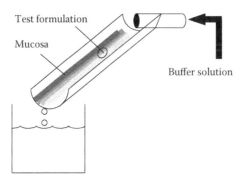

FIGURE 7.8 Experimental setup in order to evaluate the mucoadhesive properties of test formulations on a freshly excised mucosa spread out on a lop-sided channel.

7.4.1.2 Tensile Tests

7.4.1.2.1 Tensile Studies with Dry Polymer Compacts

Test polymers are thereby compressed to flat-faced discs. Tensiometer studies with these test discs are carried out on freshly excised mucosa. Test discs are therefore attached to the mucosa. After a certain contact time between test disc and mucosa, the mucosa is pulled at a certain rate (mm s^{-1}) from the disc. The TWA representing the area under the force/distance curve and the maximum detachment force (MDF) are determined. The experimental setup is illustrated in Figure 7.9. It represents one of the best established in vitro test systems which are used by numerous research groups [e.g., 23,34,35].

7.4.1.2.2 Tensile Studies with Hydrated Polymers

In order to minimize the influence of an "adhesion by hydration," tensile studies can also be carried out with hydrated polymers as described by Robinson et al. [36]. Hydrated test polymers are thereby spread in a uniform monolayer over excised mucosa which has been fixed on a flat surface. In an artificial gastric or intestinal fluid at 37°C, the hydrated polymer is brought in contact with the mucus layer of a second mucosa, which is fixed on a flat surface of a weight hanging on a balance. The TWA and MDF are then determined as described earlier.

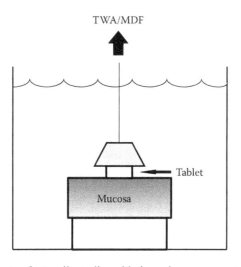

FIGURE 7.9 Experimental setup for tensile studies with dry polymer compacts.

7.4.1.2.3 Tensile Studies with Microspheres

Tensile studies as described earlier are not designed for measuring microscopic interactions such as those that may occur between microparticles and the mucus gel layer. Hence, a method was developed for measuring mucoadhesive properties of microspheres. In vivo interactions are thereby mimicked utilizing a miniature tissue chamber, which is heated by a water jacket. Thermoplastic microspheres are mounted to the tips of fine iron wires using a melting technique. The unloaded ends of the wires are then attached to a sample clip and suspended in the microbalance enclosure. The freshly excised section of mucosa is clamped in the buffer-filled chamber at 37°C and the microsphere is brought into contact with the tissue. To fracture the adhesive interactions, the tissue is pulled off the microsphere and certain mucoadhesive parameters are calculated and graphs of force versus position and time are plotted using appropriate software for the microbalance. The method provides valuable information concerning the adhesive properties of microspheres. So far, however, it is limited to microspheres not smaller than 300 μm [37].

7.4.1.3 Rheological Techniques

During the chain interpenetration of mucoadhesive polymers with mucin macromolecules, physical entanglements, conformational changes, and chemical interactions occur. Thereby, changes in the rheological behavior of the two macromolecular species are produced. An evaluation of the resulting synergistic increase in viscosity, which is supposed to be in many cases directly proportional to results obtained with tensile studies [e.g., 38], can be achieved by mixing the mucoadhesive polymer with mucus and measuring viscosity. The rheological behavior can be determined either by a classical rotational viscometry test at a certain shear rate or by dynamic oscillatory measurements, which give useful information about the structure of the polymer–mucin network [39].

7.4.1.4 Spectroscopic Techniques

Since mucoadhesive properties of polymers have been investigated, these tests were accompanied by additional spectroscopic analyses [40]. Kerr et al. [41], for instance, using ^{13}C nuclear magnetic resonance spectroscopy have provided evidence of hydrogen bonding between mucus glycoproteins and the carboxylic acid groups of polyacrylic acid. Moreover, Tobyn et al., using Fourier transform infrared spectroscopy, reported also interactions between the pig gastric mucus glycoproteins and the test mucoadhesive [42]. Jabbari et al. could confirm the chain interpenetration theory by investigating a mucin and polyacrylic acid interface via attenuated total reflection Fourier transform infrared spectroscopy (ATR-FTIR) [24]. Mortazavi et al., using infrared and ^{13}C-NMR spectroscopy, suggested the formation of hydrogen bonds between the mucoadhesive polyacrylic acid and the terminal sugar residues on the mucus glycoprotein [43]. In another study, the nature of interactions between the mucus gel and polyacrylic acid was investigated by tensile studies (I), dynamic oscillatory rheology (II), and ^{13}C-NMR spectroscopy (III) as well. The addition of hydrogen bond breaking agents resulted thereby in a decrease in mucoadhesive strength, a reduction in viscoelastic properties of polymer/mucin mixtures, and a positional change in the chemical shift of the polyacrylic acid signal [44].

7.4.2 IN VIVO METHODS

To date, the mucoadhesion of dosage forms on mucosal membranes has been evaluated in vivo by direct observation (I), by gamma scintigraphy (II), and by using insoluble markers (III). Thereby either the time period of mucoadhesion is determined or in case of the GI-tract to which extent the transit time of the mucoadhesive dosage form can be prolonged.

The direct observation offers the advantage that neither radionuclides nor insoluble markers are needed and that a pretreatment of the test formulation is in most cases not necessary. The technique can be used to evaluate mucoadhesion on various tissues in animal studies. Akiyama et al., for instance, administered mucoadhesive microspheres orally to rats. After 2.5 h, the extent of the adhesion of these microspheres to the gastric mucosa was evaluated visually, demonstrating a high amount of mucoadhesive microspheres being present in the stomach compared to non-adhesive microspheres [45]. For studies in humans, however, the direct observation is limited only to a few mucosal tissues such as the oral cavity. Bouckaert et al., for instance, determined the adhesion of tablets in the region of the upper canine. Test tablets were thereby fixed for 1 min with a slight manual pressure on the lip followed by moistening with the tongue to prevent sticking of the tablet to the lip. The adhesion time was defined as the time after which the mucoadhesive tablet was no longer visible under the lip upon control at 30 min interval [46]. In another study, which was carried out in the same way, volunteers participating in this study were asked to record the time and circumstances of the end of adhesion (erosion or detachment of the tablet) [47].

In contrast, no tissue limitations seem to exist for gamma scintigraphic methods. Using these techniques makes it even possible to evaluate the increase in the GI-transit time of mucoadhesive formulations. Radionuclides used for imaging studies include $^{99}Tc^m$, $^{111}In^m$, $^{113}In^m$, and $^{81}Kr^m$. Among them technetium-99m represents the most commonly used radionuclide, as it displays no beta or alpha radiation and a comparatively short half-life of 6.03 h. Indium-113m, which has an even shorter half-life of 1.7 h, has a different energy to technetium-99m and can therefore be used in double-labeling studies. For many applications, the longer lived isotope indium-111m (half-life 2.8 days) seems to be more appropriate. Whereas a strongly prolonged GI-transit time of mucoadhesive polymers was demonstrated in various animal studies [36], the same effect cannot in all studies be shown in human volunteers (effect: [2]; no effect: [48,49]).

7.5 MUCOADHESIVE POLYMERS

A systematic for mucoadhesive polymers can be based on their origin (e.g., natural–synthetic), the type of mucosa on which they are mainly applied (e.g., ocular–buccal), or their chemical structure (e.g., cellulose derivatives–polyacrylates). Apart from these approaches, mucoadhesive polymers can also be classified according to their mechanism of binding as shown within this chapter. An overview in the mucoadhesive properties of these different types of mucoadhesive polymers was provided by Grabovac et al. A rank order of the 10 most mucoadhesive polymers as listed in Table 7.2 was established by testing more than 50 polymers under standardized conditions [50].

TABLE 7.2
Overview on the Mucoadhesive Properties of the 10 Most Mucoadhesive Polymers

Polymer	pH	Total Work of Adhesion (µJ); Means ± SD (n = 5–8)
Chitosan-thiobutylamidine	pH 3 lyophilized	740.0 ± 146.7
Polyacrylic acid (450 kDa)-cysteine	pH 3 lyophilized	412.3 ± 27.3
Chitosan-thiobutylamidine	pH 6.5 precipitated	408.0 ± 67.9
Polycarbophil	pH 7 precipitated	342.0 ± 11.2
Carbopol 980	pH 7 precipitated	311.8 ± 42.0
Sodium carboxymethylcellulose-cysteine	pH 3 precipitated	261.7 ± 32.8
Carbopol 980	pH 3 precipitated	256.9 ± 49.0
Carbopol 974	pH 3 precipitated	219.3 ± 36.6
Polycarbophil-cysteine	pH 3 lyophilized	212.9 ± 14.4
Carbopol 974	pH 7 precipitated	211.2 ± 22.0

Source: Grabovac, V. et al., *Adv. Drug Deliv. Rev.*, 57, 1713, 2005.

7.5.1   Non-Covalent Binding Polymers

According to their surface charge which is important for the mechanism of adhesion, they can be divided into anionic, cationic, non-ionic, and ambiphilic polymers.

7.5.1.1   Anionic Polymers

For this group of polymers, mainly –COOH groups are responsible for their adhesion to the mucus gel layer. Carbonic acid moieties are supposed to form hydrogen bonds with hydroxyl groups of the oligosaccharide side chains on mucus proteins. Further anionic groups are sulfate as well as sulfonate moieties, which seem to be more of theoretical than of practical relevance. Important representatives of this group of mucoadhesive polymers are listed in Table 7.3. Among them one can find the most adhesive non-covalent binding polymers such as polyacrylates and NaCMC [e.g., 34,62].

TABLE 7.3
Anionic Mucoadhesive Polymers

Polymer	Chemical Structure	Additional Information	References
Alginate			[51–53]
Carbomer		Cross-linked with sucrose	[34,53,54]
Chitosan–EDTA		Optionally cross-linked with EDTA	[55,56]
Hyaluronic acid			[57–59]
NaCMC		0.3–1.0 carboxymethyl groups per glucose unit	[34,52,60]
Pectins		R = OH or methyl	[61,62]
Polycarbophil		Cross-linked with divinylglycol	[34,60,63,64]

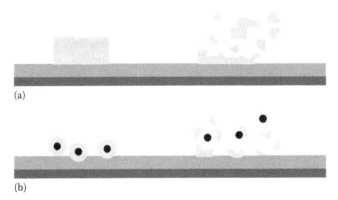

FIGURE 7.10   (See companion CD for color figure.) Adhesive bond failure in case of insufficient cohesive properties of the mucoadhesive polymer: (a) matrix tablets and (b) microspheres.

Because of their high charge density, these polymers display a high buffer capacity which might be beneficial for various reasons as discussed in Section 7.6.4. In contrast to non-ionic polymers, their swelling behavior, which is also crucial for mucoadhesion (Figure 7.3), strongly depends on the pH value. The lower the pH value, the lower is the swelling behavior leading to a quite insufficient adhesion in many cases. On the contrary, a too rapid swelling of such polymers at higher pH values can lead to an over-swelling, which causes a strong decrease in the cohesive properties of such polymers. Even if the polymer sticks to the mucus layer, in this case, the drug delivery system will not be any more mucoadhesive, as the adhesive bond fails within the mucoadhesive polymer itself. This effect is illustrated in Figure 7.10. A further drawback of anionic mucoadhesive polymers, however, is their incompatibility with multivalent cations such as Ca^{2+}, Mg^{2+}, and Fe^{3+}. In the presence of such cations, these polymers precipitate and/or coagulate [65], leading to a strong reduction in their adhesive properties.

7.5.1.2   Cationic Polymers

The strong mucoadhesion of cationic polymers can be explained by ionic interactions between these polymers and anionic substructures such as sialic acid moieties of the mucus gel layer. In particular, chitosan, which can be produced in high amounts for a reasonable price, seems to be a promising mucoadhesive excipient. Apart from its mucoadhesive properties, chitosan is also reported to display permeation enhancing properties [66,67]. The most important cationic mucoadhesive polymers are listed in Table 7.4. Their swelling behavior is strongly pH-dependent as well. In contrast to anionic polymers, however, their swelling behavior is improved at higher proton concentrations. Chitosan, for instance, is hydrated rapidly in the gastric fluid, leading to a strong reduction of its cohesive properties, whereas it does not swell at all at pH values above 6.5 causing a complete loss in its mucoadhesive properties.

7.5.1.3   Non-Ionic Polymers

The adhesion of anionic as well as cationic polymers strongly depends on the pH value of the surrounding fluid. On the contrary, non-ionic mucoadhesive polymers are mostly independent from this parameter. Whereas the formation of secondary chemical bonds due to ionic interactions can be completely excluded for this group of polymers, some of them such as poly(ethylene oxide) are capable of forming hydrogen bonds. Apart from these interactions, their adhesion to the mucosa seems to be rather based on an interpenetration followed by polymer chain entanglements. These theoretical considerations are in good accordance with mucoadhesion studies, demonstrating almost no adhesion of non-ionic polymers, if they are applied to the mucosa already in the completely hydrated form, whereas they are adhesive if applied in dry form [69]. Hence, non-ionic polymers are in most cases less adhesive than anionic as well as cationic

TABLE 7.4

Cationic Mucoadhesive Polymers

Polymer	Chemical Structure	Additional Information	References
Chitosan		Primary amino groups can be acetylated to some extent	[3]
Trimethylated chitosan			[68]
Polylysine			[61]

mucoadhesive polymers. Well-known representatives of this group of mucoadhesive polymers are listed in Table 7.5. In contrast to ionic polymers, non-ionic polymers are not influenced by electrolytes of the surrounding milieu. The addition of 0.9% NaCl, for instance, to carbomer leads to a tremendous decrease in its cohesiveness and subsequently to a strong reduction of its mucoadhesive properties, whereas these electrolytes have no influence on non-ionic mucoadhesive polymers.

7.5.1.4 Ambiphilic Polymers

Ambiphilic mucoadhesive polymers display cationic as well as anionic substructures on their polymer chains. On the one hand, mucoadhesion of the cationic polymers is referred to be caused by electrostatic interactions with negatively charged mucosal surfaces and for anionic polymers, on the other hand, mucoadhesion can be explained by the formation of hydrogen bonds of carboxylic acid groups with the mucus gel layer. The combination of positive as well as negative charges on the same polymer, however, seems to compensate both effects leading to strongly reduced adhesive properties of ambiphilic polymers. Mucoadhesion studies of chitosan–EDTA conjugates with increasing amounts of covalently attached EDTA can clearly show this effect. Whereas the exclusively anionic chitosan–EDTA conjugate (Table 7.3) exhibiting no remaining cationic moieties and the exclusively cationic polymer chitosan displayed the highest mucoadhesive properties, the mucoadhesion of chitosan–EDTA conjugates showing both cationic moieties of remaining primary amino groups and anionic moieties of already covalently attached EDTA was much lower. In addition, Lueßen et al. could show a strongly increased intestinal buserelin bioavailability in rats using chitosan HCl as mucoadhesive excipient. A mixture of this cationic polymer with the anionic polymer carbomer, however, led to a significantly reduced bioavailability of the therapeutic peptide [71]. Representatives of this type of mucoadhesive polymers are mainly proteins such as gelatin, which is reported as mucoadhesive in various studies [e.g., 62,72].

TABLE 7.5
Non-Ionic Mucoadhesive Polymers

Polymer	Chemical Structure	Additional Information	References
Hydroxypropyl-cellulose		R = H or hydroxypropyl	[35,70]
Hydroxypropyl-methylcellulose		R = H or methoxy or hydroxypropyl	[53,62]
Poly(ethylene oxide)			[35,53]
Poly(vinyl alcohol)			[54,62]
Poly(vinyl pyrrolidone)			[62]

Due to the combination of cationic as well as anionic mucoadhesive polymers leading to ionic interactions, however, the cohesiveness of delivery systems can be strongly improved [57,61]. If the adhesive bond of a delivery system fails rather within the mucoadhesive polymer itself, than between the polymer and the mucus layer as illustrated in Figure 7.10, this effect is more important than the mucoadhesive properties of the polymer in order to improve the adhesiveness of the whole dosage form.

7.5.2　Covalent Binding Polymers

Recently, a presumptive new generation of mucoadhesive polymers has been introduced into the pharmaceutical literature [73]. Whereas the attachment of mucoadhesive polymers to the mucus layer has to date been achieved by non-covalent bonds, these novel polymers are capable of forming covalent bonds [39,74]. The bridging structure most commonly encountered in biological systems—the disulfide bond—has thereby been discovered for the covalent adhesion of polymers to the mucus layer of the mucosa. *Thio*lated poly*mers* or so-called *thiomers* are mucoadhesive basis polymers, which display thiol-bearing side chains (Figure 7.11). Based on thiol/disulfide exchange reactions [75] as illustrated in Figure 7.12 and/or a simple oxidation process, disulfide bonds are formed between such polymers and cysteine-rich subdomains of mucus glycoproteins (see Table 7.1). Hence, thiomers mimic the natural mechanism of secreted mucus glycoproteins, which are also covalently anchored in the mucus layer by the formation of disulfide bonds. Due to the covalent attachment of

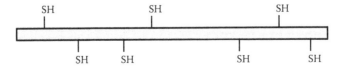

FIGURE 7.11　Thiomer (thiolated polymer); mucoadhesive polymers, which display thiol moieties bearing side chains.

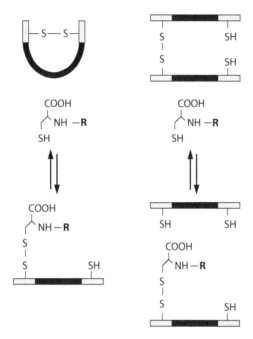

FIGURE 7.12 Schematic presentation of disulfide exchange reactions between a poly(peptide) and a cysteine derivative according to G. H. Snyder [76]. The poly(peptide) stands here for a mucin glycoprotein of the mucus and the cysteine derivative is a polymer–cysteine conjugate (R, mucoadhesive basis polymer). (Adapted from Bernkop-Schnürch, A. et al., *Pharm. Res.*, 16, 876, 1999.)

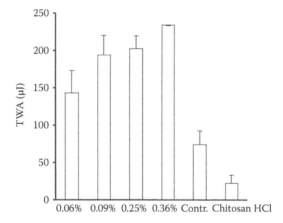

FIGURE 7.13 Comparison of the adhesive properties of chitosan-thiobutylamidine displaying increasing amounts of covalently attached thiol groups (%) and controls of unmodified chitosan. Represented values are means of the TWA determined in tensile studies with dry compacts of indicated test material. (From Roldo, M. et al., *Eur. J. Pharm. Biopharm.*, 57, 115, 2004.)

thioglycolic acid to chitosan, for instance, the adhesive properties of this polymer could be strongly increased [76]. Results of this study are shown in Figure 7.13. Apart from these improved muco-adhesive properties, which could meanwhile also be shown for various other thiolated polymers, thiomers exhibit strongly improved cohesive properties as well. Whereas, for example, tablets consisting of polycarbophil disintegrate within 2 h, tablets based on the corresponding thiolated polymer remain stable even for days in the disintegration apparatus according to the Pharmacopoeia Europea [77]. The result as shown in Figure 7.14 can be explained by the continuous oxidation of thiol moieties on thiomers which takes place in aqueous solutions at pH values above 5. Due to the

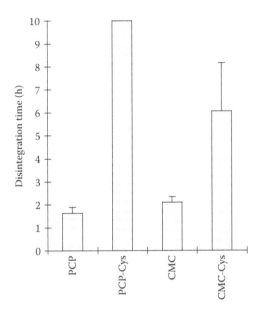

FIGURE 7.14 Comparison of the disintegration behavior of matrix-tablets (30 mg; 5 mm i.d.) containing indicated lyophilized polymers (PCP, polycarbophil; CMC, carboxymethylcellulose; PCP-Cys, polycarbophil–cysteine conjugate; CMC-Cys, carboxymethylcellulose–cysteine conjugate). Studies were carried out with a disintegration test apparatus (Pharm. Eur.) in 50 mM TBS pH 6.8 at 37°C. Indicated values are means ±SD of at least three experiments. Polycarbophil-cysteine tablets did not disintegrate even after 48 h of incubation. (From Bernkop-Schnürch, A. et al., *J. Control. Release*, 66, 39, 2000.)

high density of negative charges within anionic mucoadhesive polymers, they can also function as ion exchange resins displaying a high buffer capacity (see Section 7.6.4). According to this, the formation of disulfide bonds within polymer–cysteine conjugates can be controlled by adjusting the pH value of the system a priori. Adhesion of many quick swelling polymers, as already mentioned, is limited by an insufficient cohesion of the polymer resulting in a break within the polymer network rather than between the polymer and mucus layer (Figure 7.10). Although thiolated polymers are rapidly hydrated, they are able to form highly cohesive and viscoelastic gels due to the formation of additional disulfide bonds. The formation of an over-hydrated slippery mucilage can thereby be excluded completely. Using the rotating cylinder (Section 7.4.1.1) in order to evaluate the mucoadhesive properties of various formulations, for instance, revealed a comparatively much longer adhesion of tablets consisting of thiolated polymers [22]. Meanwhile, various anionic as well as cationic thiolated polymers have been synthesized as listed in Table 7.6. They all display strongly improved mucoadhesive properties compared to the corresponding unmodified polymers. The potential of thiolated polyacrylates could be demonstrated even in clinical trials. As illustrated in Figure 7.15, a sustained release of sodium fluorescein over several hours was achieved with thiolated polyacrylate minitablets adhering to the eye of volunteers [86].

7.6　MULTIFUNCTIONAL MUCOADHESIVE POLYMERS

In the 1980s, the adhesive properties of polymers played a central role in the field of mucoadhesion, whereas numerous scientists began in the 1990s to focus their interest also on additional features of mucoadhesive polymers. These properties include enzyme inhibition, permeation enhancement, high buffer capacity, and controlled drug release. For some of these properties, mucoadhesion is substantial, and for others various synergistic effects can be expected due to adhesion.

TABLE 7.6
Thiolated Mucoadhesive Polymers

Polymer	Chemical Structure	Additional Information	References
Chitosan-cysteine		21 up to 100 μmol thiol-groups per gram polymer	[78]
Chitosan-glutathione			[79]
Chitosan-thiobutylamidine			[80]
Chitosan-thioethylamidine			[81]
Chitosan-thioglycolic acid		11 up to 25 μmol thiol-groups per gram polymer	[82]

(*continued*)

TABLE 7.6 (continued)
Thiolated Mucoadhesive Polymers

Polymer	Chemical Structure	Additional Information	References
NaCMC-cysteine		22 up to 1280 μmol thiol-groups per gram polymer	[22,77]
Polyacrylate-cystamine		1 up to 20 μmol thiol-groups per gram polymer	[83]
Polyacrylate-cysteine		1 up to 142 μmol thiol-groups per gram polymer	[22,39,74,84,85]

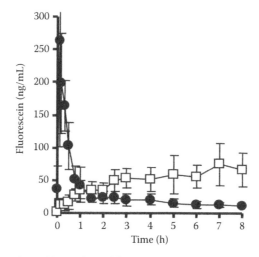

FIGURE 7.15 Prolonged ocular residence time of fluorescein being embedded in thiolated polyacrylate ocular inserts (□) in comparison to the same formulation but containing the unmodified polyacrylate (●). (Adapted from Hornof, M.D. et al., *J. Control. Release*, 89, 419, 2003.)

7.6.1 MUCOADHESION: ENZYME INHIBITION

Due to the great progress in biotechnology as well as gentechnology, the industry is capable of producing a large number of potential therapeutic peptides and proteins in commercial quantities. The majority of such drugs is most commonly administered by the parenteral routes that are often complex, difficult, and occasionally dangerous. Besides so-called alternative routes of application such as the nasal or transdermal route, there is no doubt that the peroral route is one of the most favored, as it offers the greatest ease of application. A presystemic metabolism of therapeutic peptides and proteins caused by proteolytic enzymes of the GI-tract, however, leads to a very poor bioavailability after oral dosing. Attempts to reduce this barrier include the use of analogs, prodrugs, formulations such as nanoparticles, microparticles, and liposomes that shield therapeutic peptides and proteins from luminal enzymatic attack, and the design of delivery systems targeting the colon where the proteolytic activity is relatively low. Moreover, considerable interest is shown in the use of muco-adhesive polymers, since due to such excipients various in vivo studies could demonstrate a significantly improved bioavailability of peptide and protein drugs after oral dosing. As formulations containing mucoadhesive polymers provide an intimate contact with the mucosa, a presystemic degradation of these drugs on the way between the delivery system and the absorbing membrane can be excluded. Takeuchi et al., for instance, demonstrated a significantly stronger reduction in the plasma calcium level after oral administration of calcitonin-loaded liposomes which were coated with a mucoadhesive polymer in comparison to the same formulation without the mucoadhesive coating [4].

In recent years, it could be demonstrated by various studies that certain mucoadhesive polymers display also an enzyme inhibitory effect. In particular, poly(acrylic acid) was shown to exhibit a pronounced inhibitory effect toward trypsin [e.g., 87,88]. Additionally, the immobilization of enzyme inhibitors to mucoadhesive polymers acting only in a very restricted area of the intestine seems to be a promising approach in order to improve their enzyme inhibitory properties. Due to the immobilization of the inhibitor, it remains concentrated on the polymer, which should certainly make a reduced share of this auxiliary agent in the dosage form sufficient. Side effects of the inhibitor such as systemic toxicity, a disturbed digestion of nutritive proteins and pancreatic hypersecretion caused by a luminal feedback regulation can be avoided. Hence, the covalent attachment of enzyme inhibitors to mucoadhesive polymers such as those shown in Figure 7.16 represents the combination of two favorable strategies for the oral administration of poly(peptide) drugs, offering additional advantages compared to a simple combination of both excipients without the covalent linkage. So far, various polymer–inhibitor conjugates have been generated as listed in Table 7.7. Their efficacy could be verified by in vivo studies in diabetic mice showing a significantly reduced glucose level after the oral administration of (PEGylated-)insulin tablets containing a polymer–inhibitor conjugate [96,97]. The results of this study are shown in Figure 7.17 [97]. In another study, the potential of the combination of mucoadhesive polymers and polymer–enzyme inhibitor conjugates was demonstrated by the improved oral uptake of salmon calcitonin in rats utilizing thiolated chitosan in combination with a chitosan–pepstatin conjugate as illustrated in Figure 7.18 [98].

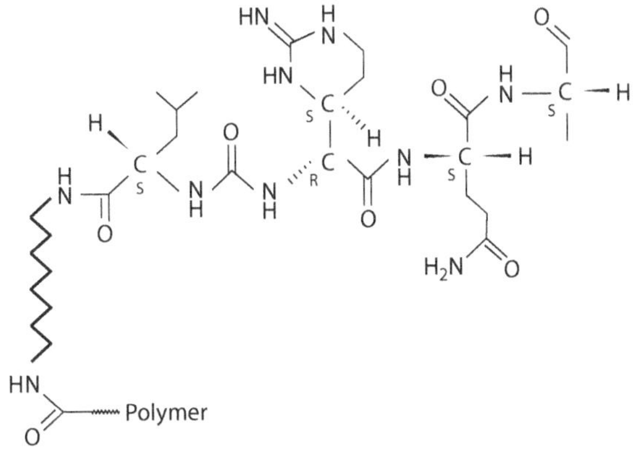

FIGURE 7.16 Example for a mucoadhesive polymer–inhibitor conjugate. The elastase inhibitor elastatinal is thereby covalently attached via a C8-spacer to polymers like polycarbophil or NaCMC. (From Bernkop-Schnürch, A. et al., *J. Control. Release*, 47, 113, 1997.)

TABLE 7.7

Comparison of Various Mucoadhesive Polymer–Inhibitor Conjugates

| | Inhibited Enzymes | | | |
Polymer–Inhibitor Conjugate	Based on Complexing Properties	Based on Competitive Inhibition	Mucoadhesive Properties	References
Carboxymethylcellulose–elastatinal conjugate		Elastase	+	[89]
Carboxymethylcellulose–pepstatin conjugate		Pepsin	n.d.	[90]
Chitosan–antipain conjugate		Trypsin	++	[91,92]
Chitosan–chymostatin conjugate		Chymotrypsin	n.d.	[92]
Chitosan–elastatinal conjugate		Elastase	n.d.	[92]
Chitosan–ACE conjugate		Trypsin, chymotrypsin, elastase	n.d.	[92]
Chitosan–EDTA	Aminopeptidase N, carboxypeptidase A/B		+++	[8,55,56]
Chitosan–EDTA–antipain conjugate	Aminopeptidase N, carboxypeptidase A/B	Trypsin	n.d.	[92]
Chitosan–EDTA–chymostatin conjugate	Aminopeptidase N, carboxypeptidase A/B	Chymotrypsin	n.d.	[92]
Chitosan–EDTA–elastatinal conjugate	Aminopeptidase N, carboxypeptidase A/B	Elastase	n.d.	[92]
Chitosan–EDTA–ACE conjugate	Aminopeptidase N, carboxypeptidase A/B	Trypsin, Chymotrypsin, Elastase	+	[92]
Chitosan–EDTA–BBI conjugate	Aminopeptidase N, carboxypeptidase A	Trypsin, Chymotrypsin, Elastase	+	[93]
Poly(acrylic acid)-Bowman–Birk inhibitor conjugate		Chymotrypsin	++	[94]
Poly(acrylic acid)–chymostatin conjugate		Chymotrypsin	+++	[23]
Poly(acrylic acid)–elastatinal conjugate		Elastase	+++	[89]
Polycarbophil–elastatinal conjugate		Elastase	+++	[89]
Poly(acrylic acid)–bacitracin conjugate	Aminopeptidase N		n.d.	[95]
Polycarbophil–cysteine	Carboxypeptidase A/B		++++	[85]

Mucoadhesive properties are classified in poor (+), good (++), very good (+++), and excellent (++++).
ACE, antipain/chymostatin/elastatinal; BBI, Bowman–Birk inhibitor.

7.6.2 MUCOADHESION: PERMEATION ENHANCEMENT

A number of mucoadhesive polymers have also promising effects on the modulation of the absorption barrier by opening of the intestinal intercellular junctions [64,99]. In contrast to permeation enhancers of low molecular size such as sodium salicylate and medium-chain glycerides [100,101], these polymers might not be absorbed from the intestine, which should exclude systemic side effects of these auxiliary agents. Permeation studies, for instance, carried out in Ussing chambers on Caco-2 monolayers demonstrated a strong permeation enhancing effect of chitosan and

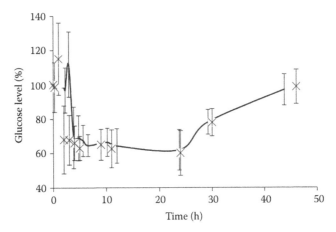

FIGURE 7.17 Decrease in blood glucose level in diabetic mice after oral administration of PEGylated insulin being incorporated in a thiolated polyacrylate. (Adapted from Caliceti, P. et al., *Eur. J. Pharm. Sci.*, 22, 315, 2004.)

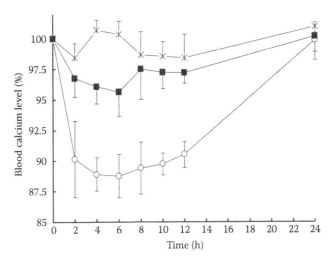

FIGURE 7.18 Comparison in the oral uptake of salmon calcitonin in rats determined via the decrease in blood calcium level after oral administration of the therapeutic peptide with chitosan (×), with chitosan/chitosan–inhibitor conjugate (■) and with thiolated chitosan/chitosan–inhibitor conjugate (o). (Adapted from Guggi, D. et al., *J. Control. Release*, 92, 125, 2003.)

carbomer [99,102]. This permeation enhancing effect on these mucoadhesive polymers can even be significantly improved due to the immobilization of cysteine on these polymers [5,74]. This improved permeation across the mucosa was accompanied by a decrease in the TEER, indicating a loosening of the tightness of intercellular junctions, i.e., the opening of the paracellular route across the epithelium for otherwise non-absorbable hydrophilic compounds such as peptides. Mechanisms that are responsible for the permeation enhancing effect of mucoadhesive polymers, however, are still unclear.

The permeation enhancing effect of cationic polymers such as chitosan might be based on the positive charges of these polymers which interact with the cell membrane, resulting in a structural reorganization of tight junction-associated proteins [6].

In case of thiolated polymers, protein tyrosine phosphatase might be involved in the underlying mechanism. This thiol-dependent enzyme mediates the closing process of tight junctions by dephosphorylation of tyrosine groups from the extracellular region [103]. The inhibition of PTP by specific inhibitors such as vanadate or pervanadate causes an enhanced opening of the tight junctions. As it is also inhibited by sulfhydryl compounds such as glutathione forming a mixed disulfide with Cys 215 [104], it is likely that thiolated polymers might also lead to an inhibition of this enzyme.

7.6.3 Mucoadhesion: Efflux Pump Inhibition

Transmembrane efflux pump proteins are one of the factors influencing and restricting the absorption of non-invasively administered drugs. These proteins can effectively alter the pharmacokinetics of various drugs such as anticancer drugs, immune suppressive drugs, and antibiotics. Efflux pump inhibitors have therefore gained a considerable attention as tool to improve clinical efficacy of drugs that act as substrates for efflux pumps. Among efflux pump inhibitors, polymeric excipients such as polyethyleneglycols, pluronics, and certain polysaccharides have gained considerable attention, as due to their comparatively high molecular mass these compounds are not absorbed from mucosal membranes. Consequently, systemic toxic side effects—as it is the case for low molecular mass efflux pump inhibitors—can be excluded for such mucoadhesive polymers. More recently, Werle et al. revealed efflux pump inhibitory capability of thiolated polymers [105]. The transmucosal transport of the P-gp substrate rhodamine 123, for instance, was strongly improved in the presence of thiolated chitosan. In the following, these in vitro results were confirmed by in vivo studies in rats. Föger et al. showed that the oral bioavailability of rhodamine 123 is even 3.0-fold improved when this model P-gp substrate is embedded in thiolated chitosan minitablets administered orally to rats (Figure 7.19) [106]. Furthermore, poly(acrylic acid)-cysteine showed to inhibit effectively Mrp2 efflux pump transporter, improving the permeation of sulforhodamine 101 4.67-fold [107]. In another study, tumor growth in rats was significantly reduced when paclitaxel was given orally in combination with a thiolated polymer [108]. In contrast to PEGs showing a P-gp inhibitory effect, which is completely independent from the molecular mass of the applied PEG [109], the inhibitory effect of thiolated polymers is dependent on their molecular mass. Thiolated polyacrylates of 250 kDa exhibit the highest P-pg inhibitory effect, for instance, whereas thiolated poly(acrylic acid) of higher and lower molecular mass displays significantly lower inhibitory activity. Studies performed in vivo in rats confirmed the efficacy of thiolated polyacrylates of 250 kDa as Mrp2 inhibitor [110].

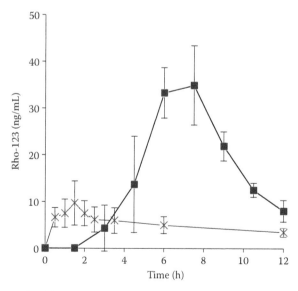

FIGURE 7.19 Efflux pump inhibitory effect of thiolated chitosan (×) in comparison to unmodified chitosan (■) in rats using rhodamine 123 as model Pgp-substrate. (Adapted from Föger, F. et al., *Biomaterials*, 27, 4250, 2006.)

The postulated mechanism of efflux pump inhibition is based on an interaction of thiolated polymers with the channel forming transmembrane domain of efflux pumps. Thiolated polymers seem to enter in the channel and likely form subsequently one or two disulfide bonds with cysteine subunits located within the channel. Due to this covalent interaction, the allosteric change of the transporter being essential to move drugs outside of the cell seems to be blocked [111].

7.6.4 MUCOADHESION: BUFFER SYSTEM

A further advantage of mucoadhesive polymers can be seen in the high buffer capacity of ionic polymers. As these polymers can act as ion exchange resins, they are able to maintain their pH value inside the polymeric network over a considerable period of time. Matrix tablets based on neutralized carbomer, for instance, can buffer the pH value inside the swollen carrier system even for hours in an artificial gastric fluid of pH 2 [112]. This high buffer capacity seems to be highly beneficial for various reasons. For example, the epidermal growth factor (EGF) is recognized as an important agent for acceleration of ulcer healing and has a peculiar biological property to repair tissue damage by an enhanced proliferation and differentiation of epithelial tissues [113]. Itoh and Matsuo demonstrated in a double-blind controlled clinical study the enhanced healing of rat gastric ulcers after oral administration of EGF. This effect could even be drastically increased by using the mucoadhesive polymer hydroxypropyl cellulose as drug carrier matrix [114]. As EGF is strongly degraded by pepsin [115], the use of mucoadhesive polymers providing an additional protective effect toward pepsinic degradation might therefore be helpful. It can be achieved by a comparatively higher pH value inside the drug carrier matrix based on its high buffer capacity at which penetrating pepsin is already inactive.

Apart from this likely advantage for the poly(peptide) administration, the high buffer capacity of neutralized anionic polymer might also be highly beneficial in treatment of *Helicobacter pylori* infection in peptic ulcer disease, as common antibiotics such as amoxicillin or metronidazole display poor stability in the acidic pH of the stomach. The incorporation of these therapeutic agents in mucoadhesive polymers displaying a high buffer capacity might improve their stability in the acidic milieu.

7.6.5 MUCOADHESION: CONTROLLED RELEASE

If the therapeutic agent is incorporated in the mucoadhesive polymer, the excipient can act both as mucoadhesive and as a matrix system providing a controlled drug release. The release behavior of drugs embedded in mucoadhesive polymers depends thereby mainly on their molecular size and charge. According to the equation determining the diffusion coefficient, in which the radius of a molecule indirectly correlates with the diffusion coefficient, small-sized drugs will be released faster than greater ones. Apart from their size, the release of therapeutic agents from mucoadhesive polymers can be simply controlled by raising or lowering the share of the polymer in the delivery system. Whereas a low amount of the mucoadhesive polymer in the carrier matrix can guarantee a rapid drug release, a sustained drug release can be provided by raising the share of the polymer in the delivery system [77]. In addition, the drug release from mucoadhesive polymers can also be controlled by the extent of cross-linking. The higher the polymer is cross-linked, the lower is the release rate of the drug. Such a cross-linking can be achieved by the formation of covalent bonds, for example, the cross-linking of gelatin with glutaraldehyde [72] or on the basis of ionic interactions. The release rate of insulin from matrix tablets consisting of the anionic mucoadhesive polymer carbomer (C934P), for instance, can be strongly reduced by the addition of the divalent cationic amino acid lysine [28]. In case of ionic drugs, a sustained release can also be guaranteed by the use of an ionic mucoadhesive polymer displaying the opposite charge of the therapeutic agent. On the basis of an ion exchange resin, for instance, a sustained release of the therapeutic agent α-lipoic acid over a period of more than 12h can be provided by the incorporation of this anionic drug in the cationic mucoadhesive polymer chitosan [116].

7.7   CONCLUDING REMARKS

Motivated by the great benefits that can be provided by mucoadhesive polymers such as a prolonged residence time and an intimate contact of the dosage form on the mucosa, considerably intensive research and development has been performed in this field since the concept of mucoadhesion has been pioneered in the early 1980s. Merits of these efforts are the establishment of various useful techniques in order to evaluate the mucoadhesive properties of different polymers as well as the design of novel polymers displaying improved mucoadhesive properties. In addition, the development of multifunctional mucoadhesive polymers exhibiting also features such as enzyme inhibitory properties, permeation enhancing effects, high buffer capacity, and the possibility to control the drug release made them even more promising auxiliary agents. Although there are already numerous formulations based on mucoadhesive polymers on the market, the number of delivery systems making use of these advantages will certainly increase in the near future.

REFERENCES

1. Robinson, J. R. and W. J. Bologna. 1994. Vaginal and reproductive system treatments using a bioadhesive polymer. *J. Control. Release* 28:87–94.
2. Akiyama, Y., N. Nagahara, E. Nara, M. Kitano, S. Iwasa, I. Yamamoto, J. Azuma, and Y. Ogawa. 1998. Evaluation of oral mucoadhesive microspheres in man on the basis of the pharmacokinetics of furosemide and riboflavin, compounds with limited gastrointestinal absorption sites. *J. Pharm. Pharmacol.* 50:159–166.
3. Takeuchi, H., H. Yamamoto, N. Toshiyuki, H. Tomoaki, and Y. Kawashima. 1996. Enteral absorption of insulin in rats from mucoadhesive chitosan-coated liposomes. *Pharm. Res.* 13:896–901.
4. Takeuchi, H., Y. Matsui, H. Yamamoto, and Y. Kawashima. 2003. Mucoadhesive properties of carbopol or chitosan-coated liposomes and their effectiveness in the oral administration of calcitonin to rats. *J. Control. Release* 86:235–242.
5. Clausen, A. and A. Bernkop-Schnürch. 2000. In vitro evaluation of the permeation enhancing effect of thiolated polycarbophil. *J. Pharm. Sci.* 89:1253–1261.
6. Schipper, N. G. M., S. Olsson, J. A. Hoogstraate, A. G. de Boer, K. M. Varum, and P. Artursson. 1997. Chitosans as absorption enhancers for poorly absorbable drugs 2: Mechanism of absorption enhancement. *Pharm. Res.* 14:923–929.
7. Lueßen, H. L., V. Bohner, D. Perard, P. Langguth, J. C. Verhoef, A. G. de Boer, H. P. Merkle, and H. E. Junginger. 1996. Mucoadhesive polymers in peroral peptide drug delivery. V. Effect of poly(acrylates) on the enzymatic degradation of peptide drugs by intestinal brush border membrane vesicles. *Int. J. Pharm.* 141:39–52.
8. Bernkop-Schnürch, A., C. Paikl, and C. Valenta. 1997. Novel bioadhesive chitosan-EDTA conjugate protects leucine enkephalin from degradation by aminopeptidase N. *Pharm. Res.* 14:917–922.
9. Matthes, I., F. Nimmerfall, and H. Sucker. 1992. Mucusmodelle zur Untersuchung von intestinalen Absorptionsmechanismen. *Pharmazie* 47:505–515.
10. Bernkop-Schnürch, A. and R. Fragner. 1996. Investigations into the diffusion behaviour of polypeptides in native intestinal mucus with regard to their peroral administration. *Pharm. Sci.* 2:361–363.
11. Campbell, B. J. 1999. Biochemical and functional aspects of mucus and mucin-type glycoproteins, bioadhesive drug delivery systems in *Bioadhesive Drug Delivery Systems: Fundamentals, Novel Approaches, and Developments*, E. Mathiowitz, D. E. Chickering III, and C.-M. Lehr, eds., pp. 85–130. New York: Marcel Dekker.
12. Gendler, S. J. and A. P. Spicer. 1995. Epithelial mucin genes. *Ann. Rev. Physiol.* 57:607–634.
13. Gum, J. R., J. W. Hicks, N. W. Toribara, E. M. Rothe, R. E. Lagace, and Y. S. Kim. 1992. The human MUC2 intestinal mucin has cysteine-rich subdomains located both upstream and downstream of its central repetitive region. *J. Biol. Chem.* 267:21375–21383.
14. Gum, J. R., J. W. Hicks, D. M. Swallow, R. L. Lagace, J. C. Byrd, D. T. A. Lamport, B. Siddiki, and Y. S. Kim. 1990. Molecular cloning of cDNAs derived from a novel human intestinal mucin gene. *Biochem. Biophys. Res. Commun.* 171:407–415.
15. Porchet, N., V. C. Nguyen, J. Dufosse, J. P. Audie, V. Guyonnet-Duperat, M. S. Gross, C. Denis, P. Degand, A. Bernheim, and J. P. Aubert. 1991. Molecular cloning and chromosomal localisation of a novel human tracheo-bronchial mucin cDNA containing tandemly repeated sequences of 48 base pairs. *Biochem. Biophys. Res. Commun.* 175:414–422.

16. Guyonnet-Duperat, V., J. P. Audie, V. Debailleul, A. Laine, M. P. Buisine, S. Galiegue-Zouitina, P. Pigny, P. Degand, J. P. Aubert, and N. Porchet. 1995. Characterisation of the human mucin gene MUC5AC: A consensus cysteine rich domain for 11p15 mucin genes? *Biochem. J.* 305:211–219.

17. Thornton, D. J., M. Howard, N. Khan, and J. K. Sheehan. 1997. Identification of two glycoforms of the MUC5B mucin in human respiratory mucus. Evidence for a cysteine-rich sequence repeated within the molecule. *J. Biol. Chem.* 272:9561–9566.

18. Toribara, N. W., A. M. Roberton, S. B. Ho, W. L. Kuo, E. T. Gum, J. R. Gum, J. C. Byrd, B. Siddiki, and Y. S. Kim. 1993. Human gastric mucin: Identification of unique species by expression cloning. *J. Biol. Chem.* 268:5879–5885.

19. Bobek, L. A., H. Tsai, A. R. Biesbrock, and M. J. Levine. 1993. Molecular cloning, sequence, and specificity of expression of the gene encoding the low molecular weight human salivary mucin (MUC7). *J. Biol. Chem.* 268:20563–20569.

20. Shankar, V., P. Pichan, R. L. Eddy, V. Tonk, N. Nowak, S. N. Sait, T. B. Shows, R. E. Shultz, G. Gotway, R. C. Elkins, M. S. Gilmore, and G. P. Sachdev. 1997. Chromosomal localisation of a human mucin gene (MUC8) and cloning of the cDNA corresponding to the carboxy terminus. *Am. J. Respir. Cell Mol. Biol.* 16:232–241.

21. Lehr, C.-M., F. G. J. Poelma, and H. E. Junginger. 1991. An estimate of turnover time of intestinal mucus gel layer in the rat in situ loop. *Int. J. Pharm.* 70:235–240.

22. Bernkop-Schnürch, A. and S. Steininger. 2000. Synthesis and characterisation of mucoadhesive thiolated polymers. *Int. J. Pharm.* 194:239–247.

23. Bernkop-Schnürch, A. and I. Apprich. 1997. Synthesis and evaluation of a modified mucoadhesive polymer protecting from α-chymotrypsinic degradation. *Int. J. Pharm.* 146:247–254.

24. Jabbari, E., N. Wisniewski, and N. A. Peppas. 1993. Evidence of mucoadhesion by chain interpenetration at a poly(acrylic acid)/mucin interface using ATR-FTIR spectroscopy. *J. Control. Release* 26:99–108.

25. Imam, M. E., M. Hornof, C. Valenta, G. Reznicek, and A. Bernkop-Schnürch. 2003. Evidence for the interpenetration of mucoadhesive polymers into the mucous gel layer. *STP Pharma. Sci.* 13:171–176.

26. Mortazavi, S. A. and J. D. Smart. 1993. An investigation into the role of water movement and mucus gel dehydration in mucoadhesion. *J. Control. Release* 25:197–203.

27. Caramella, C. M., S. Rossi, and M. C. Bonferoni. 1999. A rheological approach to explain the mucoadhesive behavior of polymer hydrogels in *Bioadhesive Drug Delivery Systems: Fundamentals*, E. Mathiowitz, D. E. Chickering III, and C.-M. Lehr, eds., pp. 25–65. New York: Marcel Dekker.

28. Bernkop-Schnürch, A., C. Humenberger, and C. Valenta. 1998. Basic studies on bioadhesive delivery systems for peptide and protein drugs. *Int. J. Pharm.* 165:217–225.

29. Hertzog, B. A. and E. Mathiowitz. 1999. Novel magnetic technique to measure bioadhesion *Novel Approaches, and Developments*, E. Mathiowitz, D. E. Chickering III, and C.-M. Lehr, eds., pp. 147–175. New York: Marcel Dekker.

30. Schneider, J. and M. Tirrell. 1999. Direct measurement of molecular-level forces and adhesion in biological systems *Novel Approaches, and Developments*, E. Mathiowitz, D. E. Chickering III, and C.-M. Lehr, eds., pp. 223–261. New York: Marcel Dekker.

31. Bernkop-Schnürch, A., A. Weithaler, K. Albrecht, and A. Greimel. 2006. Thiomers: Preparation and in vitro evaluation of a mucoadhesive nanoparticulate drug delivery system. *Int. J. Pharm.* 317:76–81.

32. Nielsen, L. S., L. Schubert, and J. Hansen. 1998. Bioadhesive drug delivery systems—I. Characterisation of mucoadhesive properties of systems based on glyceryl mono-oleate and glyceryl monolinoleate. *Eur. J. Pharm. Sci.* 6:231–239.

33. Rango Rao, K. V. and P. Buri. 1989. A novel in situ method to test polymers and coated microparticles for bioadhesion. *Int. J. Pharm.* 52:265–270.

34. Tobyn, M. J., J. R. Johnson, and P. W. Dettmar. 1996. Factors affecting in vitro gastric mucoadhesion II. Physical properties of polymers. *Eur. J. Pharm. Biopharm.* 42:56–61.

35. Mortazavi, S. A. and J. D. Smart. 1993. An *in-vitro* evaluation of mucosa-adhesion using tensile and shear stresses. *J. Pharm. Pharmacol.* 45 (suppl.):1108–1111.

36. Ch'ng, H. S., H. Park, P. Kelly, and J. R. Robinson. 1985. Bioadhesive polymers as platforms for oral controlled drug delivery II: Synthesis and evaluation of some swelling, water-insoluble bioadhesive polymers. *J. Pharm. Sci.* 74:399–405.

37. Chickering III, D. E., C. A. Santos, and E. Mathiowitz. 1999. Adaptation of a microbalance to measure bioadhesive properties of microspheres *Novel Approaches, and Developments*, E. Mathiowitz, D. E. Chickering III, and C.-M. Lehr, eds., pp. 131–147. New York: Marcel Dekker.

38. Hassan, E. E. and J. M. Gallo. 1990. A simple rheological method for the in vitro assessment of mucin-polymer bioadhesive bond strength. *Pharm. Res.* 7:491–495.

39. Leitner, V., G. F. Walker, and A. Bernkop-Schnürch. 2003. Thiolated polymers: Evidence for the formation of disulphide bonds with mucus glycoproteins. *Eur. J. Pharm. Biopharm.* 56:207–214.

40. Kellaway, I. W. 1990. In vitro test methods for the measurement of mucoadhesion, R. Gurny and H. E. Junginger. eds., pp. 86–97. Stuttgart, Germany: Wissenschaftliche Verlagsgesellschaft.

41. Kerr, L. J., I. W. Kellaway, C. Rowlands, and G. D. Parr. 1990. The influence of poly(acrylic) acids on the rheology of glycoprotein gels. *Proc. Int. Symp. Control. Release Bioact. Mater.* 122–123.

42. Tobyn, M. J., J. R. Johnson, and S. A. Gibson. 1992. Investigations into the role of hydrogen bonding in the interaction between mucoadhesives and mucin at gastric pH. *J. Pharm. Pharmacol.* 44 (suppl.):1048–1048.

43. Mortazavi, S. A., B. G. Carpenter, and J. D. Smart. 1993. An investigation into the nature of mucoadhesive interactions. *J. Pharm. Pharmacol.* 45 (suppl.): 1141.

44. Mortazavi, S. A. 1995. In vitro assessment of mucus/mucoadhesive interactions. *Int. J. Pharm.* 124:173–182.

45. Akiyama, Y. and N. Nagahara. 1999. Novel formulation approaches to oral mucoadhesive drug delivery systems in *Bioadhesive Drug Delivery Systems: Fundamentals, Novel Approaches, and Developments*, E. Mathiowitz, D. E. Chickering III, and C.-M. Lehr, eds., pp. 477–507. New York: Marcel Dekker.

46. Bouckaert, S., R. A. Lefebvre, and J.-P. Remon. 1993. *In vitro/in vivo* correlation of the bioadhesive properties of a buccal bioadhesive miconazole slow-release tablet. *Pharm. Res.* 10:853–856.

47. Bottenberg, P., R. Cleymaet, C. de Muynck, J.-P. Remon, D. Coomans, Y. Michotte, and D. Slop. 1991. Development and testing of bioadhesive, fluoride-containing slow-release tablets for oral use. *J. Pharm. Pharmacol.* 43:457–464.

48. Khosla, L. and S. S. Davis. 1987. The effect of polycarbophil on the gastric emptying of pellets. *J. Pharm. Pharmacol.* 39:47–49.

49. Harris, D., J. T. Fell, H. L. Sharma, and D. C. Taylor. 1990. GI transit of potential bioadhesive formulations in man: A scintigraphic study. *J. Control. Release* 12:45–53.

50. Grabovac, V., D. Guggi, and A. Bernkop-Schnürch. 2005. Comparison of the mucoadhesive properties of various polymers. *Adv. Drug Deliv. Rev.* 57:1713–1723.

51. Witschi, C. and R. J. Mrsny. 1999. In vitro evaluation of microparticles and polymer gels for use as nasal platforms for protein delivery. *Pharm. Res.* 16:382–390.

52. Evans, I. V. 1989. Mucilaginous substances from macroalgae: An overview. *Symp. Soc. Exp. Biol.* 43:455–461.

53. Mortazavi, S. A. and J. D. Smart. 1994. An in-vitro method for assessing the duration of mucoadhesion. *J. Control. Release* 31:207–217.

54. El Hameed, M. D. and I. W. Kellaway. 1997. Preparation and in vitro characterization of mucoadhesive polymeric microspheres as intra-nasal delivery systems. *Eur. J. Pharm. Biopharm.* 44:53–60.

55. Bernkop-Schnürch, A. and M. E. Krajicek. 1998. Mucoadhesive polymers as platforms for peroral peptide delivery and absorption: synthesis and evaluation of different chitosan-EDTA conjugates. *J. Control. Release* 50:215–223.

56. Bernkop-Schnürch, A. and J. Freudl. 1999. Comparative in vitro study of different chitosan-complexing agent conjugates. *Pharmazie* 54:369–371.

57. Takayama, K., M. Hirata, Y. Machida, T. Masada, T. Sannan, and T. Nagai. 1990. Effect of interpolymer complex formation on bioadhesive property and drug release phenomenon of compressed tablets consisting of chitosan and sodium hyaluronate. *Chem. Pharm. Bull.* 38:1993–1997.

58. Hadler, N. M., R. R. Dourmashikin, M. V. Nermut, and L. D. Williams. 1982. Ultrastructure of hyaluronic acid matrix. *Biochemistry* 79:307–309.

59. Sanzgiri, Y. D., E. M. Topp, L. Benedetti, and V. J. Stella. 1994. Evaluation of mucoadhesive properties of hyaluronic acid benzyl esters. *Int. J. Pharm.* 107:91–97.

60. Madsen, F., K. Eberth, and J. D. Smart. 1998. A rheological assessment of the nature of interactions between mucoadhesive polymers and a homogenised mucus gel. *Biomaterials* 19:1083–1092.

61. Liu, P. and T. R. Krishnan. 1999. Alginate-pectin-poly-L-lysine particulate as a potential controlled release formulation. *J. Pharm. Pharmacol.* 51:141–149.

62. Smart, J. D., I. W. Kellaway, and H. E. C. Worthington. 1984. An in vitro investigation of mucosa-adhesive materials for use in controlled drug delivery. *J. Pharm. Pharmacol.* 36:295–299.

63. Longer, M. A., H. S. Ch'ng, and J. R. Robinson. 1985. Bioadhesive polymers as platforms for oral controlled drug delivery III: Oral delivery of chlorothiazide using a bioadhesive polymer. *J. Pharm. Sci.* 74:406–411.

64. Lehr, C.-M., J. A. Bouwstra, E. H. Schacht, and H. E. Junginger. 1992. In vitro evaluation of mucoadhesive properties of chitosan and some other natural polymers. *Int. J. Pharm.* 78:43–48.

65. Valenta, C., B. Christen, and A. Bernkop-Schnürch. 1998. Chitosan-EDTA conjugate: A novel polymer for topical used gels. *J. Pharm. Pharmacol.* 50:445–452.

66. Artursson, P., T. Lindmark, S. S. Davis, and L. Illum. 1994. Effect of chitosan on the permeability of monolayers of intestinal epithelial cells (Caco-2). *Pharm. Res.* 11:1358–1361.

67. Lueßen, H. L., C.-O. Rentel, A. F. Kotzé, C.-M. Lehr, A. G. de Boer, J. C. Verhoef, and H. E. Junginger. 1997. Mucoadhesive polymers in peroral peptide drug delivery. IV. Polycarbophil and chitosan are potent enhancers of peptide transport across intestinal mucosae *in vitro*. *J. Control. Release* 45:15–23.

68. Thanou, M., B. I. Florea, M. W. Langemeÿer, J. C. Verhoef, and H. E. Junginger. 2000. N-trimethylated chitosan chloride (TMC) improves the intestinal permeation of the peptide drug buserelin in vitro (Caco-2 cells) and in vivo (rats). *Pharm. Res.* 17:27–31.

69. Lehr, C.-M. 1996. From sticky stuff to sweet receptors—Achievements, limits and novel approaches to bioadhesion. *Eur. J. Drug Metabol. Pharmacokinet.* 21:139–148.

70. Rillosi, M. and G. Buckton. 1995. Modelling mucoadhesion by use of surface energy terms obtained from the Lewis acid—Lewis base approach. II. Studies on anionic, cationic, and unionisable polymers, *Pharm. Res.* 12:669–675.

71. Lueßen, H. L., B. J. de Leeuw, M. W. Langemeyer, A. G. de Boer, J. C. Verhoef, and H. E. Junginger. 1996. Mucoadhesive polymers in peroral peptide drug delivery. VI. Carbomer and chitosan improve the intestinal absorption of the peptide drug buserelin in vivo. *Pharm. Res.* 13:1668–1672.

72. Matsuda, S., H. Iwata, N. Se, and Y. Ikada. 1999. Bioadhesion of gelatin films crosslinked with glutaraldehyde. *J. Biomed. Mater. Res.* 45:20–27.

73. Bernkop-Schnürch, A., V. Schwarz, and S. Steininger. 1999. Polymers with thiol groups: A new generation of mucoadhesive polymers? *Pharm. Res.* 16:876–881.

74. Bernkop-Schnürch, A. 2005. Thiomers: A new generation of mucoadhesive polymers. *Adv. Drug Deliv. Rev.* 57:1569–1582.

75. Snyder, G. H. 1987. Intramolecular disulfide loop formation in a peptide containing two cysteines. *Biochemistry* 26:688–694.

76. Roldo, M., M. Hornof, P. Caliceti, and A. Bernkop-Schnürch. 2004. Mucoadhesive thiolated chitosans as platforms for oral controlled drug delivery: Synthesis and in vitro evaluation. *Eur. J. Pharm. Biopharm.* 57:115–121.

77. Bernkop-Schnürch, A., S. Scholler, and R. G. Biebel. 2000. Development of controlled drug release systems based on polymer-cysteine conjugates. *J. Control. Release* 66:39–48.

78. Bernkop-Schnürch, A., U.-M. Brandt, and A. Clausen. 1999. Synthese und in vitro Evaluierung von Chitosan-Cystein Konjugaten. *Sci. Pharm.* 67:197–208.

79. Kafedjiiski, K., F. Föger, M. Werle, and A. Bernkop-Schnürch. 2005. Synthesis and in vitro evaluation of a novel chitosan-glutathione conjugate. *Pharm. Res.* 22:1480–1488.

80. Bernkop-Schnürch, A., M. Hornof, and T. Zoidl. 2003. Thiolated polymers—Thiomers: Modification of chitosan with 2-iminothiolane. *Int. J. Pharm.* 260:229–237.

81. Kafedjiiski, K., M. Hoffer, M. Werle, and A. Bernkop-Schnürch. 2006. Improved synthesis and in vitro characterization of chitosan-thioethylamidine conjugate. *Biomaterials* 27:127–135.

82. Kast, C. E. and A. Bernkop-Schnürch. 2001. Thiolated polymers—Thiomers: Development and in vitro evaluation of chitosan-thioglycolic acid conjugates. *Biomaterials* 22:2345–2352.

83. Kast, C. E. and A. Bernkop-Schnürch. 2002. Polymer-cystamine conjugates: New mucoadhesive excipients for drug delivery? *Int. J. Pharm.* 234:91–99.

84. Leitner, V., M. Marschütz, and A. Bernkop-Schnürch. 2003. Mucoadhesive and cohesive properties of poly(acrylic acid)-cysteine conjugates with regard to their molecular mass. *Eur. J. Pharm. Sci.* 18:89–96.

85. Bernkop-Schnürch, A. and S. Thaler. 2000. Polycarbophil-cysteine conjugates as platforms for peroral poly(peptide) delivery systems. *J. Pharm. Sci.* 89:901–909.

86. Hornof, M. D., W. Weyenberg, A. Ludwig, and A. Bernkop–Schnürch. 2003. A mucoadhesive ocular insert: Development and in vivo evaluation in humans. *J. Control. Release* 89:419–428.

87. Lueßen, H. L., J. C. Verhoef, G. Borchard, C.-M. Lehr, A. G. de Boer, and H. E. Junginger. 1995. Mucoadhesive polymers in peroral peptide drug delivery. II. Carbomer and polycarbophil are potent inhibitors of the intestinal proteolytic enzyme trypsin. *Pharm. Res.* 12:1293–1298.

88. Walker, G. F., R. Ledger, and I. G. Tucker. 1999. Carbomer inhibits tryptic proteolysis of luteinizing hormone-releasing hormone and N-α-benzoyl-L-arginine ethyl ester by binding the enzyme. *Pharm. Res.* 16:1074–1080.

89. Bernkop-Schnürch, A., G. Schwarz, and M. Kratzel. 1997. Modified mucoadhesive polymers for the peroral administration of mainly elastase degradable therapeutic poly(peptides). *J. Control. Release* 47:113–121.

90. Bernkop-Schnürch, A. and K. Dundalek. 1996. Novel bioadhesive drug delivery system protecting poly(peptides) from gastric enzymatic degradation. *Int. J. Pharm.* 138:75–83.

91. Bernkop-Schnürch, A., I. Bratengeyer, and C. Valenta. 1997. Development and in vitro evaluation of a drug delivery system protecting from trypsinic degradation. *Int. J. Pharm.* 157:17–25.

92. Bernkop-Schnürch, A. and A. Scerbe-Saiko. 1998. Synthesis and in vitro evaluation of chitosan-EDTA-protease-inhibitor conjugates which might be useful in oral delivery of peptides and proteins. *Pharm. Res.* 15:263–269.

93. Bernkop-Schnürch, A. and M. Pasta. 1998. Intestinal peptide and protein delivery: Novel bioadhesive drug carrier matrix shielding from enzymatic attack. *J. Pharm. Sci.* 87:430–434.

94. Bernkop-Schnürch, A. and N. C. Göckel. 1997. Development and analysis of a polymer protecting from luminal enzymatic degradation caused by α-chymotrypsin. *Drug Dev. Ind. Pharm.* 23:733–740.

95. Bernkop-Schnürch, A. and M. K. Marschütz. 1997. Development and in vitro evaluation of systems to protect peptide drugs from aminopeptidase N. *Pharm. Res.* 14:181–185.

96. Marschütz, M. K, P. Caliceti, and A. Bernkop-Schnürch. 2000. Design and in vivo evaluation of an oral delivery system for insulin. *Pharm. Res.* 17:1468–1474.

97. Caliceti, P., S. Salmaso, G. Walker, and A. Bernkop-Schnürch. 2004. Development and in vivo evaluation of an oral insulin-PEG delivery system. *Eur. J. Pharm. Sci.* 22:315–323.

98. Guggi, D., A. H. Krauland, and A. Bernkop-Schnürch. 2003. Systemic peptide delivery via the stomach: In vivo evaluation of an oral dosage form for salmon calcitonin. *J. Control. Release* 92:125–135.

99. Borchard, G., H. L. Lueßen, J. C. Verhoef, C.-M. Lehr, A. G. de Boer, and H. E. Junginger. 1996. The potential of mucoadhesive polymers in enhancing intestinal peptide drug absorption III: Effects of chitosan-glutamate and carbomer on epithelial tight junctions *in vitro. J. Control. Release* 39:131–138.

100. Lee, V. H. L. 1990. Protease inhibitors and permeation enhancers as approaches to modify peptide absorption. *J. Control. Release* 13:213–223.

101. Aungst, B. J., H. Saitoh, D. L. Burcham, S. M. Huang, S. A. Mousa, and M. A. Hussain. 1996. Enhancement of the intestinal absorption of peptides and non-peptides. *J. Control. Release* 41:19–31.

102. Illum, L., N. F. Farraj, and S. S. Davis. 1994. Chitosan as a novel nasal delivery system for peptide drugs. *Pharm. Res.* 11:1186–1189.

103. Rao, R. K., R. D. Baker, S. S. Baker, A. Gupta, and M. Holycross. 1997. Oxidant-induced disruption of intestinal epithelial barrier function: Role of tyrosine phosphorylation. *Am. J. Physiol.* 273:G812–G823.

104. Barret, W. C., J. P. DeGnore, S. Konig, H. M. Fales, Y. F. Keng, Y. Zhang, M. B. Yim, and P. B. Chock. 1999. Regulation of PTP1B via glutathionylation of the active site cysteine 215. *Biochemistry* 38:6699–6705.

105. Werle, M. and M. Hoffer. 2006. Glutathione and thiolated chitosan inhibit multidrug resistance P-glycoprotein activity in excised small intestine. *J. Control. Release* 111:41–46.

106. Föger, F., T. Schmitz, and A. Bernkop-Schnürch. 2006. In vivo evaluation of polymeric delivery systems for P-glycoprotein substrates. *Biomaterials* 27:4250–4255.

107. Bernkop-Schnürch, A. and V. Grabovac. 2006. Polymeric efflux pump inhibitors in oral drug delivery. *Am. J. Drug Deliv.* 4: 263–272.

108. Föger, F., S. Malaivijitnond, T. Wannaprasert, C. Huck, A. Bernkop-Schnürch, and M. Werle. 2008. Effect of a thiolated polymer on oral paclitaxel absorption and tumor growth in rats. *J. Drug Target* 16:149–155.

109. Shen, Q., Y. L. Lin, T. Handa, M. Doi, M. Sugie, K. Wakayama, N. Okada, T. Fujita, and A. Yamamoto 2006. Modulation of intestinal P-glycoprotein function by polyethylene glycols and their derivatives by in vitro transport and in situ absorption studies. *Int. J. Pharm.* 313:49–56.

110. Greindl, M., F. Föger, J. Hombach, and A. Bernkop-Schnürch. 2009. In vivo evaluation of thiolated poly(acrylic acid) as a drug absorption modulator for Mpr2 efflux pump substrates. *Eur. J. Pharm. Biopharm* 72:561–566.

111. Gottesman, M. M. and I. Pastan. 1988. The multidrug transporter, a double-edged sword. *J. Biol. Chem.* 263:12163–12166.

112. Bernkop-Schnürch, A. and B. Gilge. 2000. Anionic mucoadhesive polymers as auxiliary agents for the peroral administration of poly(peptide) drugs: Influence of the gastric fluid. *Drug Dev. Ind. Pharm.* 26:107–113.

113. Skov, O. P., S. S. Poulsen, P. Kirkegaard, and E. Nexo. 1984. Role of submandibular saliva and epidermal growth factor in gastric cytoprotection. *Gastroenterology* 87:103–108.

114. Itoh, M. and Y. Matsuo. 1994. Gastric ulcer treatment with intravenous human epidermal growth factor: A double-blind controlled clinical study. *J. Gastroenterol. Hepatol.* 9: S78–S83.

115. Slomiany, B. L., H. Nishikawa, J. Bilski, and A. Slomiany. 1990. Colloidal bismuth subcitrate inhibits peptic degradation of gastric mucus and epidermal growth factor in vitro. *Am. J. Gastroenterol.* 85:390–393.

116. Bernkop-Schnürch, A., H. Schuhbauer, and I. Pischel. 1999. α-Liponsäure(-Derivate) enthaltende Retardform, Deutsche Patentschrift 1999-09-30.

8 Biodegradable Polymeric/Ceramic Composite Scaffolds to Regenerate Bone Tissue

Catherine Gkioni, Sander Leeuwenburgh, and John Jansen

CONTENTS

8.1 INTRODUCTION

According to the U.S. Healthcare Cost and Utilization Project, nationwide inpatient statistics show that in 2004 more than 1,100,000 surgical procedures were conducted involving the partial excision of bone, bone grafting, and inpatient fracture repair [1]. The total estimated cost of those procedures was more than \$5 billion [1]. The majority of those procedures are due to congenital defects, bone tumors, or traumatic events in young patients and also from osteoarthritis in older patients [2].

The "gold standard" for bone grafting therapies is the use of autogenous bone, which is a part of healthy bone tissue as harvested from the same patient. It is the only solution that is effectively fulfilling all the requirements needed for the replacement of bone tissue [3]. Even though autogenous bone is osteogenic, osteoinductive, osteoconductive, and biocompatible, its use has a lot of drawbacks. As an extra surgery is required in order to harvest the bone, great risk exists for septic complications, viral transmission and morbidity at the site of the harvesting. The surgery can sometimes be painful, and the amount of the bone removed is always limited [2]. An alternative to the use of autogenous bone is the use of allografts (bone from another human donor) or xenografts (bone from animals). Allografts, however, are also of limited amount and there is always the risk of a viral or bacterial infection from the donor or disease transmission [2,3], while the disadvantages of using xenografts are similar to those of using allografts.

In this chapter, the major reasons that inspired scientists to develop polymer/ceramic composite biomaterials for bone substitution are discussed. To this end, a short introduction to the biology of bone including its mechanical and biological properties will provide the proper background to be able to understand the requirements for bone-substituting materials. Basic polymers and ceramics used for the manufacturing of bone-substituting composite biomaterials and their properties that are relevant to tissue engineering applications are being discussed, since the properties of the final composite greatly depend on the physicochemical characteristics of the single components. Biodegradable polymer/ceramic composites are considered to be a promising solution for bone regeneration, since the mechanical properties of these materials can be tailored to match the host tissue. Moreover, these temporary biomaterials can be replaced by newly formed bone upon gradual degradation of the original biomaterial. Therefore, this chapter will provide an overview of currently investigated biodegradable, polymeric/ceramic composite scaffolds that have been synthesized to replace bone tissue.

8.2 BONE STRUCTURE

Nature is the ultimate example of how material properties can be tailored to desire by combining single components. Wood, for example, is a composite material consisting of long cellulose fibers in a lignin matrix. Other examples of the composite structure of natural materials include corals as well as teeth and bone.

Bone is a very complex composite material with various functions. Among others it provides structural support for all the muscles, protects very sensitive organs (e.g., the brain and the eyes), and provides the body with minerals for metabolic reactions. The overall mechanical properties of different types of bone are not only a result of the synthesis of bones (a combination of two materials, i.e., an organic matrix and inorganic mineral crystals dispersed in this matrix) but also the way those materials are combined together from the order of some angstroms until the macro-level. Even though the structure remains always the same, the composition of bone is constantly changing through the life span of humans, affecting the mechanical properties. As mentioned before, bone mainly consists out of an organic collagen matrix and a dispersion of inorganic mineral crystals. The latter are plate-shaped carbonated apatite crystals of hexagonal crystallographic symmetry (Dahlite) [4]. The average dimensions of the crystals are 50 nm length and 25 nm wide with a thickness of 3 nm. Often a lot of impurities are present in the bone apatite such as HPO_4, Na, Mg, K, citrate, carbonate, and more [5].

The other major component of bone (collagen type I) has a fibrous structure that is created by the self-assembly of three collagen molecules into a triple helix [5]. The fibrils are ordered to sublayers that are parallel and have a gap of approximately 40 nm between the ends of the molecules. The crystals of the mineral actually grow in that gap and are in alignment from one fibril to the next. Five of these sublayers rotate one from each other approximately 30° and comprise a lamellar creating a plywood-like motif [6]. Lamellae (3–7 μm wide) are further arranged to osteons or Haversian systems and to woven bone. Finally woven bone and osteons combine together to give two types of bone, i.e., cortical and trabecular bone. These two types of bone are distinguished by their degree of porosity, density, and histological appearance of the tissue's microstructure [5].

Bone is mainly an anisotropic material [7,8] and it has a range of values for its mechanical properties rather than specific values. The Young's modulus is between 7 and 30 GPa; it has a tensile strength ranging between 50 and 150 GPa and elongation at fracture of about 1%–3%. From everything described earlier, it is quite clear that the mechanical properties of bone are the result of many factors, including the structure and the components of this fascinating composite material. The collagen in bone (an elastic hydrogel network) in combination with a hard inorganic material such as hydroxyapatite (HA) as well as the way they structurally combine together endow bone with unique mechanical properties such as low stiffness, resistance to tensile and compressive forces, and high fracture toughness [9].

8.3 BONE-SUBSTITUTING IMPLANTS

8.3.1 IMPLANT REQUIREMENTS FOR BONE SUBSTITUTION

When the biology of the natural bone is taken into respect, it is becoming quite clear that the implant meant to replace such a tissue should meet some standards. Mechanical properties close to that of healthy tissue and also the maintenance of these properties throughout the implantation period are one of the basic requirements that a bone replacing material or scaffold must possess in order to be inserted into the body. Another important requirement is the biocompatibility of the scaffold, which means that the scaffold should not elicit an inflammatory response when introduced in the body. This also implies that the material itself but also any degradation products should not be immunogenic and cytotoxic. A very important issue is the sterilization ability of the scaffold. The material should be easily sterilized not only on the surface but in the whole interior of the material in order to prevent infections. Another preferable characteristic of the scaffold, especially when intended for bone regeneration, is the porosity. The scaffold should have large interconnected pores to allow the migration of cells into the implant and promote vascularization of the newly formed tissue. Finally, all of the aforementioned characteristics should be combined in a way that leads to materials easily fabricated with validated production methods at low production costs [10].

8.3.2 CANDIDATE MATERIALS

A very wide range of materials has been used as implants through the ages. Archeologists have mentioned findings that people were using ox teeth, ivory, shells, coral, wood, human teeth from corpses, gold, and silver to replace missing human teeth or body parts [11]. It is quite obvious that as the living standards of society are growing higher and higher, the demand for better solutions is becoming more of a necessity. For bone replacement (such as hip-joint prostheses), metallic and ceramic implants are used most frequently, even though these materials are associated with a variety of problems. The most common problem, resulting from the interaction of the metal with bone and with the fluids of the body, is corrosion of the metal surface and the release of metal ions that can be detrimental for the organism at high concentrations [12]. Many metal alloys used for fracture repair contain Ni or Co ions, which cause allergic, toxic, and carcinogenic reactions [13]. Another drawback is that metallic materials are not resorbable, while ceramic scaffolds show very poor degradability [14,15].

Even though these materials are still widely used in clinics, there is a strong need for novel biomaterials with improved properties. Generally, metallic materials possess high moduli of elasticity, whereas ceramic materials have different values for tensile strength and brittleness compared to bone [16,17]. These differences in mechanical properties between implant materials and surrounding tissue are addressed by Wolff's law [16]. According to "Wolff's law," bone will remodel so that stress or strain is retained within specific levels, which means that that part of the mechanical load will be carried by the implant when a stiff metal implant is placed in bone, resulting into bone resorption and implant loosening [16]. In some cases, revision of the surgery is necessary for the scaffold to be fixated in position [17]. So, the mechanical properties of the material as implanted should not differ from the mechanical properties of the host tissue because this could lead to changes in the host tissue after implantation. The need for materials that combine mechanical properties similar to those of bone but also elicit biological reactions that can lead to a better bonding between the implant and the surrounding tissue is becoming more obvious. In order to fabricate materials that mimic bone's functions and mechanical properties, the understanding of the properties of the host tissue is very important.

8.3.3 Composite Approach toward Novel Bone-Substituting Biomaterials

During the past decades, much effort has been put in research on composite materials consisting of a polymeric matrix and a ceramic dispersed phase. By the combination of the two materials, scientists managed to eliminate many of the drawbacks that these materials might have when they stand on their own, thereby tailoring and enhancing the final properties of the composite material by controlling many of their properties such as the mechanical characteristics, bioactivity, and of course their degradation time.

Polymers, in general, lack the ability to bond with the surrounding tissue when implanted in the body. The basic concept for preparing polymer/ceramic composites is that the dispersed mineral phase will provide nucleation sites for HA formation in order to connect with the surrounding tissue by providing sites for the cells to attach [18]. While the polymer degrades gradually, the newly formed bone will replace the temporary implant and will provide the necessary mechanical stability. Among the many advantages of these composite materials is also the fact that ceramics in the polymer matrix can act as a buffer to the acidic microenvironment of the implant during polymer degradation [19]. This acidic environment is known to induce inflammation around the implantation site. Titration experiments [20] showed that carbonated calcium phosphates are able to buffer the acidic residues of polymer degradation and preserve the pH at a physiological value (pH 7.4) [20]. By a careful selection of the polymer and the ceramic phases, degradation times and mechanical properties of the scaffold can be controlled to a large extent [1]. Another characteristic is the improvement of the processing properties of pure ceramics since with the addition of a polymer the final material can be easily moldable and in some cases even injectable because of the elasticity caused by polymeric components [10,21].

In what follows, the biodegradable polymers and ceramics that are mostly used for the fabrication of biomaterials are discussed. Polymers that exist already in nature and are derived from natural sources are called natural polymers, whereas polymers synthesized in the lab are called synthetic polymers. Ceramics are categorized as bioactive (calcium phosphates and bioactive glasses) or bioinert ceramics. Subsequently, composite materials made of polymers (matrix) and ceramics (dispersed phase) mentioned earlier are discussed in the next section. The review ends with a short discussion about composite materials having ceramic matrices reinforced with polymers.

8.4 POLYMERS

The first polymeric biomaterials were developed under the great need for materials that could be used for repair, reconstruction, or regeneration of tissue during the Second World War in the decade 1940–1950 [22]. Biomaterials were first used for corneal substitution, hip-joint replacements, intraocular lenses, and blood contact devices [22,23]. There are a wide range of polymers used for replacement of diseased or damaged tissues. Owing to the tremendous progress in organic chemistry of the last century, it has become possible to synthesize copolymers, which can combine properties from the different polymers, such as degradability, biological response, and good mechanical properties. This widens the horizons even more and makes polymers a perfect candidate for composite scaffolds material. Polymers are mainly categorized as non-degradable or degradable polymers. An additional advantage of these polymers is that the fractions of the degradation can enter into the physiological cycles of the human metabolism and can be excreted without having any malicious effects for the body.

8.4.1 Degradation of Polymers

Polymers can degrade either by hydrolysis (passively) or by enzymatic reaction (actively). Enzymatic degradation only occurs to natural polymers [24], whereas passive degradation occurs by hydrolysis in the backbone of the polymer at a rate determined by the type of bonds that are affected by hydrolysis. Other factors can also affect the degradation rates (pH, catalytic phenomena, and the presence of functional groups in the polymer), but the most important ones are the composition and amount of water uptake. For example, the change in composition upon introduction of a second monomer in the polymer chain may lead to changes in the crystallinity or the glass transition of the material affecting the degradation rates. Hydrolysis is a reaction of water with the bonds from the polymer. Greater water concentrations will lead to faster degradation times; thus, hydrophilic polymers will tend to degrade much faster than hydrophobic ones [24].

8.4.2 Degradable Polymers of Natural Origin

Since collagen is the main polymer phase of bone, collagen has been the most popular natural polymer under investigation for the fabrication of bone substitutes. Even though at least 27 different types of collagen have been identified, bone and cartilage consist mainly of type I and II collagen. One molecule of collagen is structured from the arrangement of three polypeptide chains into a triple helix [25,26]. Collagen is highly biocompatible and degrades enzymatically and it can be processed also very easily into different forms such as sheets, tubes, sponges, and fibers [23]. The main disadvantage of collagen usage for scaffolds is that collagen can cause allergic reactions. There is also concern about the transmission of diseases, since the main collagen source is bovine or porcine. On the other hand, recombinant collagen often lacks the biological activity of native tissue since this type of collagen does not undergo posttranslational modifications [27].

Another common natural polymer used in tissue engineering is gelatin. Gelatin is produced from collagen and it can be derived by hydrolysis of collagen under acidic or basic conditions. It is considered to be a very good candidate for tissue engineering and it can replace collagen because it has a composition similar to collagen, while gelatin is not likely to transmit pathogens since it is a denatured protein [28].

Hyaluronic acid (HAc) is a polysaccharide that is also present in human organism, i.e., mainly found in the extracellular matrix. It is a glycosaminoglycan, which chemically consists of D-glucuronic acid and 2-acetamido-2-deoxy-D-glucose monosaccharides. The main disadvantages of this polysaccharide are the high solubility and the fast degradation times. Those two characteristics, though, can be tailored by cross-linking hyaluronic acid or by chemical modification leading to materials with controllable properties more suitable for scaffold fabrication [29,30].

Alginic acid is another natural polymer that is widely used for tissue engineering. Alginic acid is a linear polysaccharide and is composed of repeating units of D-mannuronic acid and L-guluronic acid. Its main source is brown algae and it is commercially available as sodium salt.

Alginate is widely used in the food industry as thickener or as emulsifying agents [31]. Due to the presence of carboxylate groups in its structure, it can form gels by cross-linking using divalent cations such as calcium [32]. Even though alginate is a natural polymer, it does not exist in the human body. Thus, its presence in the human body can always result in an immune response.

8.4.3 Synthetic Degradable Polymers

8.4.3.1 Polyesters

A very common polymer group used for fabricating scaffolds for tissue engineering or used as drug delivery system is that of aliphatic polyesters. Polyesters can be synthesized by the polycondensation of diacids and diols; self-polycondensation of hydroxyacids; or the ring-opening polymerization of cyclic diesters, lactones, glycolides, and lactides [10,23]. Since esterification is a reversible process, hydrolytic degradation of these polymers happens through a random scission of the ester bond yielding products that can be easily metabolized via natural procedures [33]. Some of the factors that can affect the degradation kinetics are the chemical composition and configurational structure, processing history, molar mass, polydispersity, crystallinity, morphology, presence of the original monomers, and the overall hydrophilicity [34].

8.4.3.1.1 Polyglycolic Acid

One of the polymers belonging to aliphatic polyesters is polyglycolide. Polyglycolide or poly (glycolic acid) (PGA) is a highly crystalline polymer with a high tensile modulus and a poor solubility in organic solvents. Right after implantation the polymer comes in contact with the body fluids and immediately degradation begins, resulting to a loss of strength in 1–2 months and a loss of mass after 6–12 months [35]. It is widely used for surgical sutures since the 1970s [36], but it also has been used as a material for devices for internal bone fixation [23].

8.4.3.1.2 Polylactic Acid

Another widely used aliphatic polyester is poly(lactic acid) (PLA). Due to the fact that lactide, the monomer from which PLA is synthesized, is a chiral molecule, the polymer exists in four different isotopes:

- Poly(L-lactic acid)
- Poly(D-lactic acid)
- Poly(DL-lactic acid)
- Meso-poly(lactic acid)

The L-isomer (PLLA) is also a crystalline polymer with almost 37% crystallinity. It is among the most used polymers for biomaterials because of the good mechanical properties, good tensile strength, low elongation, and high tensile modulus, and therefore, it has been used for load-bearing applications [37]. In poly(DL-lactic acid), the L- and D-units are randomly distributed in the polymeric chain resulting in a completely amorphous polymer with lower mechanical properties than to the L-isomer. Even though the polymer loses its strength in 1–2 months like PLLA, the loss in mass can take from 12 to 16 months [35].

8.4.3.1.3 Poly(Caprolactone)

Poly(caprolactone) (PCL) is also part of the polyester group. It is obtained by the polymerization of caprolactone, which is a very cheap starting material. PCL is semi-crystalline, with low tensile strength, high elongation at breakage, and it is soluble in most organic solvents [38]. It attracts the interest of most researchers because of its ability to blend perfectly with other polymers and it has a long degradation time (over 2 years), a property that makes PCL a perfect candidate for long-term drug delivery vehicles [35,39]. Copolymers of PCL with PLA have been investigated in order to achieve faster degradation times, whereas copolymers of PCL with PGA have been developed for the improvement of its mechanical properties [39].

As mentioned earlier, the overall goal is to find an optimal polymer with the best combination of properties. To this end, many of the aforementioned polymers are polymerized together to make copolymers with tailored properties. Poly(lactic-co-glycolic) acid is one of these copolymers. Degradation and mechanical properties change depending on the percentage of lactic and glycolic acid in the copolymer. Since lactic acid is more hydrophobic, copolymers with low lactic acid content tend to degrade more rapidly [40].

8.4.3.2 Polyethylene Glycol/Polyethylene Oxide

A very important polymer used widely in tissue engineering for drug delivery systems and scaffolds is polyethylene glycol (PEG) or polyethylene oxide (PEO) when it is of higher molecular weight. This polymer is extremely hydrophilic, it can degrade hydrolytically, and the products of the hydrolysis can be cleared by the kidneys [41]. A very important property of PEG is that it is not recognized by the molecules that provoke immune response, making it highly biocompatible [42]. It can be combined with a lot of polymers to form copolymers that adopt many of PEG's properties such as biocompatibility and biodegradability. Such a synthetic polymer is oligo(poly(ethylene glycol)fumarate) (OPF). This relatively new copolymer was synthesized by Jo et al. [43], by the copolymerization of PEG with the presence of fumaric acid, a naturally derived dicarboxylic acid. The main benefits of OPF hydrogels for tissue engineering applications include the ability to tailor the macromolecular structure in order to elicit different degradation rates as well as the good cell viability in vitro and tissue response in vivo [44,45].

8.5 CERAMICS

Ceramics are non-metallic inorganic materials that can have different degrees of crystallinity. In general, ceramic materials have a high compressive strength, but their resistance to deformation is limited resulting into a high brittleness [14].

8.5.1 Bioactive Ceramics

8.5.1.1 Calcium Phosphates

The class of materials that can create a firm bond with bone at the site of implantation, by the formation of a layer of HA, are called bioactive materials [46]. Calcium phosphates are a group of ceramics that are widely used as biomaterials because they resemble the ceramic part of bone, or they appear in human organisms at places of pathological calcifications such as urinary stones, dental calculi, and crystalluria [47].

Monocalcium phosphate monohydrate (MCPM, $Ca(H_2PO_4)_2 \cdot H_2O$) is the most water soluble compound of the calcium phosphate family. It is precipitated from highly acidic solutions, and when it is heated above 100°C it is transformed to anhydrous monocalcium phosphate. In general, MCPM is not found in biological calcifications and it is mostly used in self-hardening calcium phosphate cements (CPC) [48,49]. Dicalcium phosphate dihydrate (DCPD, $CaHPO_4 \cdot 2H_2O$) (also known as brushite) is a mineral often found in pathological calcifications. It can be easily precipitated from aqueous solutions, and when it is heated at temperatures above 80°C, it is transformed to its anhydrous component. Both DCPD and the anhydrous component are also used for the synthesis of CPC [50,51]. Octacalcium phosphate (OCP, $Ca_8(HPO_4)_2(PO_4)_4 \cdot 5H_2O$) is also found in humans as the stable phase of dental and urinary calculi. Its structure consists of apatitic layers separated by hydrated layers. It is supposed that OCP acts as a precursor in vivo for biological apatite formation. It is used in medicine as implant for bone substitution in various forms such as coatings or granules [52–54]. β-Tricalcium phosphate (β-TCP, β-$Ca_3(PO_4)_2$)

cannot be precipitated from aqueous solutions but it can be obtained by thermal decomposition of other calcium phosphates (such as calcium-deficient HA) and by calcining of bones. When it is heated above 1125°C, it transforms in α-TCP. Both β-TCP and α-TCP have the same chemical composition, but different crystal structure and thus solubility. β-TCP is more stable at room temperature and less soluble than α-TCP and it is usually mixed with HA to form biphasic calcium phosphate (BCP). Both phases of TCP and BCP are used widely for the preparation of bone-substituting products [47]. Amorphous calcium phosphate (ACP, $Ca_xH_y(PO_4)_z \cdot nH_2O$) is considered to be the first phase that precipitates in solutions containing calcium and phosphate ions. In humans it is often found in pathological calcifications of soft tissues such as heart valves [55]. It is widely used as dental and bone implant material [56].

Hydroxyapatite (HA, $Ca_5(PO_4)_3$) is a very stable component with low solubility. It can be precipitated in aqueous solutions by mixing exactly stoichiometric quantities of calcium and phosphate containing solutions at alkaline pH, followed by boiling in CO_2 free atmosphere [47]. Even though pure HA is never met in pathological calcifications, its similarity with the mineral found in bone and teeth makes it a perfect candidate for the use as bone-replacing material. It has been used in a variety of different forms such as coatings, granules, blocks, and as a component for bone cement [57].

8.5.1.2 Bioactive Glass

Another group of ceramics that is widely used in research for biomedical applications is bioactive glass. The first bioactive glass Bioglass®, a quaternary (SiO_2-CaO-P_2O_5-Na_2O) system, was introduced by L.L Hench [58] and was prepared by the quenching method. Extensive research has been performed since then and there are many glasses with various components that are used in the preparations of biomedical devices [59]. The problems generated by the high-temperature process and of the limited composition of these glasses can be overcome with the preparation of glasses using the sol–gel method [60,61]. Glasses made by the sol–gel method are more porous than the melt-derived ones. Due to their mesoporous structure, they tend to be more bioactive and more bioresorbable compared to the traditional glasses [62].

8.5.2 Bioinert Ceramics

Ceramic materials that are encapsulated by a fibrous tissue layer between the implant surface and surrounding bone tissue are called bioinert. Alumina (Al_2O_3) and zirconia (ZrO_2) are bioinert ceramic materials that are most frequently used for the manufacturing of biomedical devices. High-density and highly pure alumina ($α$-Al_2O_3) has been used for femoral heads of hip prostheses and acetabular cups. Zirconia is being used clinically as a material in the hip prosthesis due to its good mechanical properties such as high strength, low wear and high hardness [63].

8.6 COMPOSITE MATERIALS

8.6.1 Preparation Techniques

All of the aforementioned materials have been combined extensively by biomaterials scientists in search of the optimal combination of material properties in terms of mechanical and biological characteristics.

In case of a dispersed ceramic phase, this ceramic phase can be introduced into the organic matrix using either biomimetic, wet-chemical or physical mixing methods.

Biomimetic methods: Using this method, ceramic phases are formed in organic matrices according to synthesis strategies that resemble the way in which calcium phosphates are synthesized in nature. By soaking the matrix into saturated solutions (like simulated body fluids, SBF), the organic matrix is mineralized by precipitation of nano-crystallized minerals. Alternatively, these matrices can be calcified by alternate soaking in solutions rich in calcium or phosphate ions.

Wet-chemical methods: This method involves mixing of the polymer matrix in the solution in which the ceramic precursors are present either as dissolved salts or as precipitated particles. The advantage of this method is that it uses very small particles, which remain suspended in the solution, resulting in composite materials with a very good dispersion avoiding the formation of agglomerates.

Physical mixing: This method is the most common one and is used not only for particles but also for the preparation of fiber composites. The size and the shape of the dispersed phase of the composite material play a key role to the final mechanical properties of the composite. For example, fine particles tend to agglomerate by forming strong aggregates. It is very important to break down the agglomerates during processing and distribute the particles homogeneously in the matrix.

The packing behavior of a material is strongly dependent on particle size, shape, and surface characteristics. In general, bi-modal size distribution of the ceramic particles can be beneficial for composite preparation resulting into composites with high ceramic densities since the smaller particles occupy the space between the larger particles [64].

The first effort to create a material with a structure and mechanical properties similar to human bone was performed during the 1980s. This so-called Bone Analogue was introduced by Bonfield et al. [65,66], and it is a composite material consisting of a polymeric matrix of polyethylene (PE) mixed with HA crystals. This biomaterial is commercially available under the name HAPEX and it has been successfully used as middle-ear implant.

8.6.2 POLYMER/CERAMIC COMPOSITES DERIVED FROM NATURAL POLYMERS

8.6.2.1 Collagen-Based Composite Materials

Since collagen is the polymer found in human tissues, it is a logical consequence that a lot of work has been performed on composite materials consisting of a collagen matrix with a mineral dispersed phase. Collagen-HA composites show remarkable properties. They have been shown to be biocompatible both in humans and in animals; they accelerate osteogenesis and they revealed better osteoconductive properties in comparison to monolithic HA implants [67]. Yamauchi et al. [68], reported the fabrication of multilayer collagen/HA sheets, which present better stability, minimum swelling compared to the collagen reference, and better mechanical properties.

There are several collagen-HA composites commercially available. The commercialization of these materials means that they have been subjected to pre-clinical and clinical studies in animals and in humans and have been proved safe for use. Collapat II® (BioMet, Inc) is made from calf skin collagen of type I and synthetic HA granules. Another material derived from bovine collagen, HA, and bone marrow aspirate is Healos® (Depuy Orthopedics, Inc). The collagen consists of cross-linked fibers coated with the HA. Biostite® (Vebas) is a composite material made from equinine collagen of type I, synthetic HA and chondroitin-6-sulphate. Collagraft® (Neucoll, Inc and Zimmer, Inc) (a bovine collagen type I matrix mixed with HA and TCP granules and autologous bone marrow) is another commercially available material.

In a comparative study performed by Leupold et al. [69], three commercially available bone graft substitutes (including a ceramic one (ProOsteon®), demineralized bone matrix (DBX®) and Collagraft) were implanted in New Zealand white rabbits. After 3 months of implantation, Collagraft showed the most pronounced bony ingrowth compared to the other two and to empty controls.

Another interesting material comprised of recombinant human-like collagen with HA assembled into fibrils was dispersed into a PLA matrix and implanted into defects made in the radius of rabbits [70]. After 24 weeks of implantation, bone density increased, while cell ingrowth, new bone formation, scaffold degradation, and complete bone union were observed.

Serre et al. [71], studied the in vitro behavior of Biostite and compared it with the in vitro behavior of collagen sponges and HA crystals. The materials were analyzed with transmission electron microscopy and X-ray microanalysis to investigate the synthesis and calcification of organic matrix when cultured with human osteoblast-like cells. The composite biomaterial showed an earlier and more intense activity, while mineral deposits were observed on their surface.

Kikuchi et al. [72] developed a nanocomposite consisting of collagen and HA by the method of co-precipitation. Using $Ca(OH)_2$, H_3PO_4, and collagen solutions under alkaline conditions (pH 9) and at a temperature of 40°C, a bone-like nanostructured material was formed with HA crystals of up to 50 nm aligned along the collagen fibers similar to the nanostructure of bone. The collagen/HA blocks were compacted by removing water under cold isostatic pressure. After implantation of 12 weeks in the tibiae of Beagle dogs, the composite material was shown to be biodegradable and osteoconductive, but revealed a low resorption rate without significant tissue invasion in the implant due to its dense body. The same group [73] used glutaraldehyde cross-linking to the nanocomposites in order to improve the mechanical properties and the biodegradation rate. The results showed that biodegradability was affected by the addition of the glutaraldehyde as the resorption rate of the material was reduced with increasing concentration of the cross-linker, but the mechanical properties were improved.

In order to overcome the problems that were caused by the dense structure of the collagen/HA composite described earlier, Sotome et al. [74], tried to incorporate the collagen/HA nanocomposite into an alginate matrix. After implantation of the new composite in rats, the new material showed extensive bone formation and a satisfactory tissue invasion in the collagen-alginate/HA composite mainly due to swelling. Because of the different swelling ratios between the collagen/HA and the alginate, fragmentation was observed to occur in the implants. The tissue invaded the implant from the fragmentation sites and the collagen/HA composite was released from the implant inducing more bone formation around the material.

Song et al. [75] manufactured a porous composite HA/collagen membrane using a co-precipitation technique and a dynamic filtration technique. The membranes, designed for use as guided bone regeneration, were compared with a membrane made only out of collagen. The mechanical properties of the composite membrane (stiffness and tensile strength) and the enzymatic stability were significantly improved compared to the membranes made only from pure collagen.

Wu et al. [76] prepared collagen/HA microspheres with a size of 50–60 nm using a water-in-oil emulsion. After the preparation of the emulsion at 4°C, glutaraldehyde was used to cross-link collagen. The spheres were then cultured with rat osteoblasts. It was concluded that the composite microspheres can be an excellent carrier for osteoblast cells.

Joseph et al. [77] studied a bovine fibril type I collagen matrix reinforced with a BCP phase consisting of HA and TCP to a 60:40 ratio combined with a small amount of autogenous bone. The material was used instead of autogenous bone graft substitute into a spinal fusion model in canines for 12 months. The overall behavior of the composite was tested using biomechanical and biochemical methods. No differences were observed in the behavior of the composite and the control groups upon biomechanical testing, and they propose that the material is a good alternative to autologous bone.

Zou et al. [78] prepared a β-TCP/collagen scaffold by mechanically mixing the TCP powder with an acidic suspension of collagen. The TCP powder was prepared after a precipitation reaction and calcination at 900°C. After mixing the β-TCP powder with the suspension of the collagen, the material was cross-linked with glutaraldehyde and it was put into molds. β-TCP has a good biodegradability, making it a good candidate for use in composite materials. As a result, many research groups have also studied β-TCP/collagen materials [79–81], but the particle size was big (5–50 μm) and the mechanical mixing that was used for the preparation of the composite resulted in inhomogeneous distribution of the particles and resulted in a weak interaction between the particles and the matrix.

Another scaffold, as described by Shinji et al. [82], was made from porcine collagen sponge combined with OCP. The authors implanted collagen/β-TCP, collagen/HA, and their scaffold in critical sized defects in rat crania and compared the bone regeneration induced by the different scaffolds. Even though after week 4 of implantation there was no significant difference between the three types of implants, the authors report that at week 12 bone regeneration of collagen/OCP implant was significantly higher than those of the collagen/β-TCP, and collagen/HA implants. They also report that the collagen/OCP implant was more degradable.

Du et al. [83,84] developed a series of calcium phosphate/collagen composites by immersing commercially available type I collagen sheets into solutions containing calcium and phosphate ions. By controlling the reaction conditions, various materials were obtained including ACP/collagen and poorly crystalline carbonate apatite/collagen. The later ceramic is the one that resembles the mineral of bone. The nano-HA-sized composite was also implanted in a marrow cavity. The implant was shown to be bioactive and biodegradable and it had mechanical properties that could reach the lower limits of bone.

A different approach was followed by Pederson et al. [85], who developed a novel injectable biomaterial that consisted of a collagen matrix mixed with calcium and phosphate loaded liposomes. Phospholipids and phospholipids concentration was carefully chosen such as the liposomes would melt at body temperatures. In addition, the acidic collagen monomers self-assemble to a gel near the same temperatures. The combination of both the thermally gelling collagen and thermally triggered mineralization induced by the melting of the phospholipids resulted into a novel injectable scaffold that gels and mineralizes in situ triggered by an increase in temperature from room temperature to body temperature.

In a study performed by Eglin et al. [86], two composite materials were prepared consisting of the same polymer, i.e., type I collagen, and different reinforcing agents (CaO-P_2O_5-SiO_2 bioactive glass and silica particles derived from the sol–gel method). The materials were tested for their in vitro ability to induce HA mineralization and the results were compared with pure collagen, bioactive glass, silica and alumina/collagen scaffolds. It was also proven that both manufactured scaffolds induced the formation of HA after immersion in SBF. Surprisingly, collagen/silica particles induced mineralization even though the single components did not, which implies a synergistic role of collagen and silica that needs further investigation.

8.6.2.2 Miscellaneous Polymer/Ceramic Composites Derived from Natural Polymers

A three-dimensional porous alginate/HA scaffold with interconnected pores was manufactured by Lin et al. [87] using a three-step process. The final scaffolds were sponges with weight ratios of alginate:HA equal to 75/25 and 50/50 and were cross-linked with different divalent cations. The porous structure in the scaffolds was well defined and the average pore size was 150 μm with a porosity of 85%. The mechanical properties of the scaffolds were tested and a rat osteosarcoma cell line was used to evaluate the in vitro response. The results showed that the 50/50 scaffold had better mechanical properties than the 75/25 and the pure alginate that was used as a control. Both the composite scaffolds showed a better capacity for cell attachment than pure alginate.

Panzavolta et al. [88] prepared new porous composite scaffolds of gelatin containing α-TCP. The α-TCP particles partially hydrolyze during the foaming procedure to yield OCP. After immersion in phosphate buffer solution, the mineral undergoes further hydrolysis to give poorly crystalline carbonated apatite. To stabilize and prevent dissolution of gelatin, the authors cross-linked the scaffolds with genipin, a naturally occurring cross-linking agent. Another porous scaffold was prepared by Kim et al. [89]. The scaffold consisted of gelatin and up to 30 wt% of HA particles. The suspensions of gelatin and HA particles were first freeze-dried and then immersed in a solution containing a cross-linking agent, yielding cross-linked, highly porous composite scaffolds. The mechanical properties of the reinforced gelatin foam and the pure foam were tested and the results showed that the HA particles improved the compressive stress and the elastic modulus of the composites in both dry and wet states. The authors also made coatings on titanium disks with the composite gelatin/HA and measured the ALP activity of both the coatings and the composite foams. Both materials revealed enhanced ALP activity in comparison with scaffolds and coatings made out of pure gelatin.

8.6.3 SYNTHETIC POLYESTER-BASED POLYMER/CERAMIC COMPOSITES

Nazhat et al. [90] prepared a hot-pressed three-phase composite material combining a matrix of PDLA reinforced with HA particulates and semi-crystalline unidirectional fibers of PLLA. The authors studied the effect of the reinforcing phases on some of the mechanical properties and the glass transition temperature of the PDLA matrix. Following a similar pathway, Nehad et al. [91] prepared a HA/PLLA composite and evaluated how the hot-pressing method affects the structural characteristics of the material. They reported that there might be a change in bioresorption due to change in crystallinity of the PLLA during hot-pressing.

A composite porous material composed of HA/PDLLA was studied in terms of biocompatibility, osteoconductivity, and biodegradability by Hasegawa et al. [92]. Both the composite and the porous HA were implanted in rabbit femoral condyles. The composite scaffold was surrounded by bone in direct contact with the implant surface after 6 weeks of implantation. The amount of newly formed bone in the composite scaffold increased after 26 weeks and by that time newly formed bone was formed inside the pores and bone marrow was present. The bone in the HA scaffold did not show any significant increase after 12 weeks. The HA/PDLLA scaffold was resorbed much faster than pure HA, which was hardly resorbed throughout the implantation period.

Ignatius et al. [93] tested in vivo two composite materials consisting of α-TCP or GB14N glass and PLA polymer matrix and compared the results with an autogenous implant. The group tested the implants for up to 24 months implantation period in sheep. Both composite materials showed a good biocompatibility, osseous integration, and compression strengths in the range of autogenous grafts after 6 and 12 months of implantation. After 24 months of implantation, both implants presented an inflammation around the tissue and the mechanical properties were significantly decreased resulting from severe degradation of the polymer component.

PLA was also combined with heat-treated and non-heat-treated glass fibers of the system $50P_2O_5$-$40CaO$-$5Na_2O$-$5Fe_2O_3$ by Ahmed et al. [94]. The material had a flexural strength close to that of cortical bone, but the modulus values were much lower and after 6 weeks of degradation in distilled water it maintained 50% of the initial values. The authors concluded that the composite with the heat-treated fibers maintained its mechanical properties and showed higher cell viabilities mainly due to the slower degradation of the heat-treated fibers.

Kim et al. [95] prepared a porous HA scaffold coated with a composite solution of PCL and HA. The antibiotic tetracycline hydrochloride was entrapped in the coating layer resulting in a scaffold with drug release capability. They reported that the thickness of the coating played a more significant role than the HA:PCL ratio to the mechanical properties of the scaffold. Higher amounts of HA in the coating increased the biodegradation, while the antibiotic was released in an uncontrolled manner during the first 2 h, but after that period it was released with more sustained manner.

Causa et al. [96] prepared a composite material using PCL and HA. The mechanical performance of the composite material was improved compared to pure PCL. After seeding osteoblasts from human trabecular bone in the scaffolds, they showed that the composite materials had better osteoconductivity than the PCL alone. Guarino et al. [97], prepared a PCL matrix reinforced with two different phases. The biodegradable PLA fibers helped in maintaining the spaces required for cellular ingrowth, while the calcium phosphate particles reacted in the matrix and produced needle-like calcium-deficient HA that improved the mechanical properties.

A guided bone regeneration membrane comprising β-TCP and poly(L-lactide-co-glycolide-co-ε-caprolactone) (PLGC) or poly(L-lactide-co-ε-caprolactone) (PLCL) made by the heat-kneading method was described by Kikuchi et al. [98]. The scaffolds were first soaked in physiological and buffer saline and seemed to maintain their mechanical properties after 12 weeks. The composites were also implanted in the mandible of Beagle dogs for up to 6 weeks and showed good bone regeneration.

8.6.4 MISCELLANEOUS POLYMER/CERAMIC BIODEGRADABLE COMPOSITES

Song et al. [17] fabricated a pHEMA/carbonated apatite scaffold using a biomimetic process. The hydrogel was soaked in an acidic solution of HA in the presence of urea. As the solution was heated, urea started to decompose and during the change to a more basic pH partial hydrolysis of the hydrogel was achieved. The hydrolysis on the surface of the hydrogel promoted not only the nucleation and growth of a several microns thick surface layer of apatite but also into the whole interior.

Khan et al. [99] prepared a novel material by chemically binding nano-composite HA with polyurethane polymer. The HA nanoparticles were prepared by the sol–gel method and they were chemically bonded on the backbone of the polyurethane through solvent polymerization. This new material was then dissolved in DMF and the solution was electrospun to make nano-fibers. The material seemed to be very promising for the use in dental applications even though more in vivo and in vitro tests should be conducted. Bil et al. [100] created polyurethane/bioglass and polyurethane/PDLLA/bioglass composites by the slurry-dipping method. The bioglass coatings were very homogeneous on the surface of the polymer foams and after immersion in SBF there was a formation of a calcium phosphate layer within only 7 days, while the uncoated polyurethane foams did not show any calcium phosphate formation even after 21 days of immersion. The degradation rate of the coated polyurethane in SBF was slightly higher than the non-coated one.

An in vitro and in vivo study of a composite scaffold consisting of two PVA polymers and gelatin combined with HA ceramic in a weight ratio of 6:1:3 was conducted by Wang et al. [101]. The scaffold was synthesized by emulsification resulting in a highly interconnected porous network. The pores were about 100–500 μm allowing the ingrowth of fibrous tissue after implantation in rats.

A study by Leeuwenburgh et al. [102] involved the synthesis of a functionalized OPF hydrogel with finely dispersed calcium phosphates. The composite scaffolds were fabricated by synthesizing calcium phosphate nanocrystals directly in the OPF hydrogel by a wet-chemistry reaction. Even though the composite material had finely dispersed nano-apatite crystals throughout the polymer matrix and reduced swelling behavior, no mineralization was observed after 28 days in SBF.

The synthesis of a composite hydrogel/calcium phosphate material was reported by Demitreas et al. [103]. The authors fabricated a highly porous hydrogel scaffold based on poly-acrylamide and reinforced it with HA beads with size between 25 and 45 μm. The HA beads were added in the monomer and the polymerization occurred in the presence of the HA. The mechanical properties were improved and as reported the interconnectivity of the pores was maintained. Upon testing in vitro using cell culture, the material proved to be cytocompatible and did not release any cytotoxic compounds in the cell culture media. In Table 8.1, all the aforementioned materials are listed.

8.7 COMPOSITE CERAMIC/POLYMER DEGRADABLE BIOMATERIALS

All the aforementioned examples of composite scaffolds are mainly examples of polymeric matrices reinforced with a ceramic phase in the form of granules, particles fibers, or coatings. The opposite approach toward enhanced mechanical and biological properties involves functionalization of ceramic matrices with a polymer phase. The dispersed polymeric phase can be either in the form of fibers or in the form of particles to provide the final material with certain porosity.

TABLE 8.1

Composite Polymer-Ceramic Biomaterials for Bone Substitution

Polymer	Ceramic	Author	Reference	Description
PE	HA	Bonfield et al.	[65,66]	"The Bone Analogue" The first composite material manufactured for use as bone substitute
Collagen	HA	Yamauchi et al.	[68]	Multilayer collagen/HA sheets (2–10 layers)
Collagen (calf skin type I)	HA			Collapat II® (BioMet Inc) Commercially available
Collagen (bovine type I)	HA			Healos® (DePuy Orthopedics, Inc) Commercially available, contains also bone marrow aspirate
Collagen (equinine)	HA	Serre et al.	[71]	Biostite® (Vebas), commercially available material, contains also chondroitin-6-sulfate
Collagen (bovine type I)	HA+TCP	Leupold et al.	[69]	Collagraft® (Neucoll, Inc and Zimmer, Inc), commercially available material, contains also autologous bone marrow
Collagen (recombinant human like)	HA	Wang et al.	[70]	Fibrils of mineralized collagen in a PLA matrix
Collagen	HA	Kikuchi et al.	[72,73]	Bone-like nanocomposite material made by co-precipitation of HA on collagen fibrils
		Sotome et al.	[74]	The nanocomposite of Ref. [85] was introduced in an alginate matrix
Collagen	HA	Song et al.	[75]	Porous membranes
Collagen	HA	Wu et al.	[76]	Microspheres with the size of 50–60 nm cross-linked with glutaraldehyde
Collagen (bovine type I)	HA+TCP	Joseph et al.	[77]	Collagen fibrils combined with biphasic calcium phosphate and autogenous bone
Collagen	β-TCP	Zou et al.	[78]	Mechanically mixed β-TCP powder with a collagen suspension cross-linked with glutaraldehyde
Collagen (porcine)	OCP	Shinji at al.	[82]	Collagen sponges combined with OCP
Collage (type I)	ACP, poorly crystalline carbonated apatite	Du et al.	[83,84]	Collagen sheets were immersed in solutions containing calcium and phosphate ions. By controlling the reaction conditions, different CaP phases were obtained
Collagen	Calcium phosphate	Pederson et al.	[85]	Liposomes loaded with calcium and phosphate ions that melt at body temperatures were introduced in a collagen matrix
Collagen (type I)	SiO_2 particles CaO-P_2O_5-SiO_2 bioactive glass	Eglin et al.	[86]	Sol–gel derived reinforcing phase, physically mixed with collagen suspensions
Alginate	HA	Lin et al.	[87]	Porous sponges with interconnected porosity of 85%

TABLE 8.1 (continued)
Composite Polymer-Ceramic Biomaterials for Bone Substitution

Polymer	Ceramic	Author	Reference	Description
Gelatin	α-TCP	Panzavolta et al.	[88]	Porous scaffolds. The ceramic phase partially hydrolyzes yielding OCP and poorly crystalline carbonated apatite
Gelatin	HA	Kim et al.	[89]	Freeze-dried sponges
PDLA	HA	Nazhat et al.	[90]	Hot-presses three phase material
PLLA	HA	Nehad et al.	[91]	Hot-pressed material
PDLLA	HA	Hasegawa et al.	[92]	Porous scaffold
PLA	α-TCP GB14N bioactive glass	Ignatius et al.	[93]	Particles of the reinforced phase dispersed in the matrix
PLA	$50P_2O_5$-40CaO-$5Na_2O$-$5Fe_2O_3$ bioactive glass	Ahmed et al.	[94]	Reinforcing phase consisting of heat-treated and untreated fibers
PCL	HA	Kim et al.	[95]	Composite including antibiotic factors, used as coating on porous HA scaffolds
PCL	HA	Causa et al.	[96]	Particles dispersed in polymer scaffold
PCL	HA	Guarino et al.	[97]	Polymer matrix reinforced with HA particles and PLA fibers
PLGC	β-TCP	Kikuchi et al.	[98]	Membrane made by heat-kneading method
pHEMA	HA	Song et al.	[17]	Hydrogel mineralized through a biomimetic process in the presence of urea
PU	HA	Khan et al.	[99]	HA nanoparticles prepared by the sol–gel method and chemically bonded on the polymer
PU	Bioglass	Bil et al.	[100]	PU foams coated with bioglass
PVA	HA	Wang et al.	[101]	Three-phase composite material. PVA/gelatin/HA ratio of 6:1:3
OPF	Poorly crystalline carbonated apatite	Leeuwenburgh et al.	[102]	A ceramic phase was synthesized directly in the hydrogel using a wet chemistry method
Poly-acrylamide	HA	Demitreas et al.	[103]	HA beads added in the monomer solution and polymerization in the presence of HA

CPC are mineral powders that are mixed with an aqueous solution to produce a paste that sets at room temperatures due to the precipitation and subsequent entanglement of calcium phosphate crystals [104]. These materials are biocompatible and osteoconductive but have slow degradation times. Ruhe et al. [105] incorporated rhBMP-2 into and on the surface of PLGA microspheres and added them in CPC. The release of the growth factor was slowed down by an interaction of rhBMP-2 with the cement. Fei et al. [106] also incorporated loaded PLGA particles with rhBMP-2. The growth factor was released in a controlled manner even after 28 days while the ALP and osteocalcin tests showed a dose response to the rhBMP-2, indicating that the activity of the growth factor was maintained.

TABLE 8.2

Composite Ceramic-Polymer Biomaterials for Bone Substitution

Ceramic	Polymer	Author	Reference	Description
HA	PCL	Kim et al.	[95]	Porous HA scaffold coated with a composite film of PCL with HA. The coating also included antibiotic agents
CPC	PLGA	Ruhe et al.	[105]	Commercially available cement (Calcibone®) reinforced with rhBMP-2 loaded microparticles
CPC	PLGA	Fei et al.	[106]	Loaded microspheres with rhBMP-2 were used as reinforcement in the cement
CPC	Collagen	Tamimi et al.	[107]	Acidic collagen solution was used as the liquid phase for the preparation of brushite CPC
CPC	Collagen	Mai et al.	[108]	Mineralized collagen fibrils in suspensions used as liquid phase for the preparation of the ceramic matrix
CPC	Alginate, PLGA	Qi et al.	[109]	Alginate hydrogel used as liquid phase for the preparation of CPC to make porous scaffolds. The pores of the scaffolds were filled with PLGA
CPC	Aramid Polyglactin	Xu et al.	[110]	CPC reinforced with fibers of variable length
HA+TCP	Chitosan	Zhang et al.	[111]	The ceramic porous substrate was prepared by using an organic mould that was burned out. After the immersion of the scaffold in a chitosan solution, chitosan sponges grew in the pores of the ceramic scaffold

Tamimi et al. [107] prepared a composite material consisting of a brushite CPC and collagen. In their study, they reported about the utilization of the solubility of collagen and the setting of the brushite cements at low pH. The collagen was dissolved in a solution containing citric acid and was used as the liquid phase for the preparation of the cement. The composite cements were tested in terms of mechanical properties and the combination of citric acid and collagen not only accelerated the precipitation of brushite crystals but also improved the cell adhesion capacity of the cement.

CPC intermixed with collagen was fabricated by Mai et al. [108]. The collagen fibers were mineralized before the addition into the cement using a wet-chemical precipitation method. The composite material was implanted in minipigs for a total of 18 months. After 12 months of implantation, the cement was completely replaced by spongeous bone and no left composite cement could be detected. The authors concluded that the degradation of the scaffold occurred simultaneously with the formation of new bone. As such, this material was considered to be a desirable material for clinical applications.

The group of Qi et al. [109] fabricated composite CPC with the addition of alginate. The alginate powder was dissolved in water in order to make a hydrogel that was used as the liquid phase for the cement synthesis. In order to control the pore size and the interconnectivity of the scaffolds, different liquid-to-powder ratios were used. After manufacturing the scaffold, its pores were filled with PLGA polymer by immersing the scaffolds in a PLGA polymer solution and imposing vacuum infiltration for 30 min. The addition of the PLGA polymer into the pores improved the mechanical properties of the alginate/CPC scaffolds significantly.

The effect of fiber reinforcement on CPC was studied by Xu et al. [110]. Two polymer-derived fibers (aramid and polyglactin) were tested among others, the latter being a copolymer of glycolide and lactide. The composite materials with the aramid fibers showed the highest tensile strength and the polyglactin/CPC showed the lowest. The fiber length in the composite seems to play an important role since an increase in ultimate strength was observed when longer fibers of the same polymer were used.

A very interesting approach is followed by Zhang et al. [111]. The group prepared a three-dimensional macroporous calcium phosphate scaffold with chitosan sponges. The CPC scaffold was prepared by dipping PU in a solution containing HA and TCP by burning the organic component (PU) out at 250°C. The result was a pure ceramic scaffold with interconnected pores of a mean size of 300 μm. The prepared scaffolds were then immersed in a chitosan solution under vacuum in order to achieve the penetration of the solution into the pores. The chitosan solution containing the ceramic scaffold was then transferred to −20°C to induce a solid–liquid phase separation and it was finally transferred to the freeze-dryer to result in a ceramic scaffold with chitosan sponges grown in its pores. The mechanical properties of the reinforced scaffold were improved in comparison with the porous scaffold without the polymer. Table 8.2 summarizes the aforementioned composite materials based on a continuous ceramic matrix.

8.8   CONCLUSIONS

In this chapter, several composite polymer/ceramic materials used for bone implants have been reviewed. Most of these materials have mechanical properties close to those of bone. The mechanical properties of these composites can be modified by a careful choice of the matrix and the dispersed phase. The mineral phase can provide nucleation sites for the crystallization of HA and subsequent bone formation.

Also, several classes of biodegradable composite materials for bone substitution were discussed. Compared to conventional single-component materials based on either ceramics or polymers, biodegradable polymer/ceramic composite materials display several advantages that are mainly related to the possibility to fine-tune the mechanical properties, degradation rate, and biological response by a careful selection of suitable composite components and appropriate composite preparation techniques. The ultimate goal is to tailor the functional properties of the composites in such way that their mechanical properties become similar to native bone tissue, while adjusting the degradation profile of the composites in order to match the rate of new bone formation. In that way, novel, temporary scaffolds can be created that instruct the physiological environment toward regeneration of diseased or damaged bone tissue.

REFERENCES

1. Kretlow, J.D. and Mikos, A.G., 2007. Review: Mineralization of synthetic polymer scaffolds for bone tissue engineering. *Tissue Engineering* 13(5): 927–938.
2. Moroni, L., Hamann, D., Paoluzzi, L. et al., 2008. Regenerating articular tissue by converging technologies. *PLoS ONE* 3(8): e3032.
3. Tadic, D. and Epple, M., 2004. A thorough physicochemical characterisation of 14 calcium phosphate-based bone substitution materials in comparison to natural bone. *Biomaterials* 25(6): 987–994.
4. Weiner, S. and Traub, W., 1992. Bone structure: From angstroms to microns. *The FASEB Journal* 6(3): 879–885.
5. Rho, J.-Y., Kuhn-Spearing, L., and Zioupos, P., 1998. Mechanical properties and the hierarchical structure of bone. *Medical Engineering and Physics* 20(2): 92–102.
6. Weiner, S., Traub, W., and Wagner, H.D., 1999. Lamellar bone: Structure-function relations. *Journal of Structural Biology* 126(3): 241–255.
7. Wong, S.-C., Baji, A., and Gent, A.N., 2008. Effect of specimen thickness on fracture toughness and adhesive properties of hydroxyapatite-filled polycaprolactone. *Composites Part A: Applied Science and Manufacturing* 39(4): 579–587.
8. Ritchie, R.O., Kinney, J.H., Kruzic, J.J. et al., 2005. A fracture mechanics and mechanistic approach to the failure of cortical bone. *Fatigue and Fracture of Engineering Materials and Structures* 28(4): 345–371.
9. Song, J., Malathong, V., and Bertozzi, C.R., 2005. Mineralization of synthetic polymer scaffolds: A bottom-up approach for the development of artificial bone. *Journal of the American Chemical Society* 127(10): 3366–3372.

10. Rezwan, K., Chen, Q.Z., Blaker, J.J. et al., 2006. Biodegradable and bioactive porous polymer/inorganic composite scaffolds for bone tissue engineering. *Biomaterials* 27(18): 3413–3431.

11. LeGeros, R., 2002. Properties of osteoconductive biomaterials: Calcium phosphates. *Clinical Orthopaedics and Related Research* 395: 81–98.

12. Zhou, Z., Liu, X., Liu, Q. et al., 2009. Evaluation of the potential cytotoxicity of metals associated with implanted biomaterials (I). *Preparative Biochemistry and Biotechnology* 39: 81–91.

13. Barrabés, M., Michiardi, A., Aparicio, C. et al., 2007. Oxidized nickel–titanium foams for bone reconstructions: Chemical and mechanical characterization. *Journal of Materials Science: Materials in Medicine* 18(11): 2123–2129.

14. Khan, Y., Yaszemski, M.J., Mikos, A.G. et al., 2008. Tissue engineering of bone: Material and matrix considerations. *Journal of Bone and Joint Surgery* 90(Suppl_1): 36–42.

15. Habraken, W.J.E.M., Wolke, J.G.C., and Jansen, J.A., 2007. Ceramic composites as matrices and scaffolds for drug delivery in tissue engineering. *Advanced Drug Delivery Reviews* 59(4–5): 234–248.

16. Bonfield, W., Wang, M., and Tanner, K.E., 1998. Interfaces in analogue biomaterials. *Acta Materialia* 46(7): 2509–2518.

17. Song, J., Saiz, E., and Bertozzi, C.R., 2003. A new approach to mineralization of biocompatible hydrogel scaffolds: An efficient process toward 3-dimensional bonelike composites. *Journal of the American Chemical Society* 125(5): 1236–1243.

18. Ho, E., Lowman, A., and Marcolongo, M., 2007. In situ apatite forming injectable hydrogel. *Journal of Biomedical Materials Research* 83A(1): 249–256.

19. Hutmacher, D.W., 2000. Scaffolds in tissue engineering bone and cartilage. *Biomaterials* 21(24): 2529–2543.

20. Schiller, C. and Epple, M., 2003. Carbonated calcium phosphates are suitable pH-stabilising fillers for biodegradable polyesters. *Biomaterials* 24(12): 2037–2043.

21. Kim, S.-S., Sun Park, M., Jeon, O. et al., 2006. Poly(lactide-co-glycolide)/hydroxyapatite composite scaffolds for bone tissue engineering. *Biomaterials* 27(8): 1399–1409.

22. Castner, D.G. and Ratner, B.D., 2002. Biomedical surface science: Foundations to frontiers. *Surface Science* 500(1–3): 28–60.

23. Nair, L. and Laurencin, C.T., 2006. Polymers as biomaterials for tissue engineering and controlled drug delivery. *Advances Biochemistry Engineering Biotechnology*, 102: 47–90.

24. Göpferich, A., 1996. Mechanisms of polymer degradation and erosion. *Biomaterials* 17(2): 103–114.

25. Seal, B.L., Otero, T.C., and Panitch, A., 2001. Polymeric biomaterials for tissue and organ regeneration. *Materials Science and Engineering: R: Reports* 34(4–5): 147–230.

26. Kyle, S., Aggeli, A., Ingham, E. et al., 2009. Production of self-assembling biomaterials for tissue engineering. *Trends in Biotechnology* 27(7): 423–433.

27. Cen, L., Liu, W., Cui, L. et al., 2008. Collagen tissue engineering: Development of novel biomaterials and applications. *Pediatric Research* 63(5): 492.

28. Kuijpers, A., Engbers, G., Krijgsveld, G. et al., 2000. Cross-linking and characterisation of gelatin matrices for biomedical applications *Journal of Biomaterial Science: Polymer Edition* 11: 225–243.

29. Campoccia, D., Doherty, P., Radice, M. et al., 1998. Semisynthetic resorbable materials from hyaluronan esterification. *Biomaterials* 19(23): 2101–2127.

30. Lloyd, L.L., Kennedy, J.F., Methacanon, P. et al., 1998. Carbohydrate polymers as wound management aids. *Carbohydrate Polymers* 37(3): 315–322.

31. Kuo, C.K. and Ma, P.X., 2001. Ionically crosslinked alginate hydrogels as scaffolds for tissue engineering: Part 1. Structure, gelation rate and mechanical properties. *Biomaterials* 22(6): 511–521.

32. Anna Gutowska, B.J.M.J., 2001. Injectable gels for tissue engineering. *The Anatomical Record* 263(4): 342–349.

33. Maurus, P.B. and Kaeding, C.C., 2004. Bioabsorbable implant material review. *Operative Techniques in Sports Medicine* 12(3): 158–160.

34. Heidemann, W., Jeschkeit, S., Ruffieux, K. et al., 2001. Degradation of poly($_{D,L}$)lactide implants with or without addition of calcium phosphates in vivo. *Biomaterials* 22(17): 2371–2381.

35. Middleton, J.C. and Tipton, A.J., 2000. Synthetic biodegradable polymers as orthopedic devices. *Biomaterials* 21(23): 2335–2346.

36. Ueda, H. and Tabata, Y., 2003. Polyhydroxyalkanonate derivatives in current clinical applications and trials. *Advanced Drug Delivery Reviews* 55(4): 501–518.

37. Ashammakhi, N., Suuronen, R., Tiainen, J. et al., 2003. Spotlight on naturally absorbable osteofixation devices. *Journal of Craniofacial Surgery* 14(2): 247–259.

38. Edlund, U. and Albertsson, A.C., 2003. Polyesters based on diacid monomers. *Advanced Drug Delivery Reviews* 55(4): 585–609.
39. Sinha, V.R., Bansal, K., Kaushik, R. et al., 2004. Poly-[epsilon]-caprolactone microspheres and nanospheres: An overview. *International Journal of Pharmaceutics* 278(1): 1–23.
40. Wu, X.S. and Wang, N., 2001. Synthesis, characterization, biodegradation, and drug delivery application of biodegradable lactic/glycolic acid polymers. Part II: Biodegradation. *Journal of Biomaterials Science, Polymer Edition* 12: 21–34.
41. Place, E.S., George, J.H., Williams, C.K. et al., 2009. Synthetic polymer scaffolds for tissue engineering. *Chemical Society Reviews* 38(4): 1139–1151.
42. Peppas, N.A., Keys, K.B., Torres-Lugo, M. et al., 1999. Poly(ethylene glycol)-containing hydrogels in drug delivery. *Journal of Controlled Release* 62(1–2): 81–87.
43. Jo, S., Shin, H., Shung, A.K. et al., 2001. Synthesis and characterization of oligo(poly(ethylene glycol) fumarate) macromer. *Macromolecules* 34(9): 2839–2844.
44. Shin, H., Temenoff, J.S., and Mikos, A.G., 2003. In vitro cytotoxicity of unsaturated oligo[poly(ethylene glycol)fumarate] macromers and their cross-linked hydrogels. *Biomacromolecules* 4(3): 552–560.
45. Temenoff, J.S., Shin, H., Conway, D.E. et al., 2003. In vitro cytotoxicity of redox radical initiators for cross-linking of oligo(poly(ethylene glycol) fumarate) macromers. *Biomacromolecules* 4(6): 1605–1613.
46. Kamitakahara, M., Ohtsuki, C., and Miyazaki, T., 2008. Review paper: Behavior of ceramic biomaterials derived from tricalcium phosphate in physiological condition. *Journal of Biomaterials Applications* 23(3): 197–212.
47. Dorozhkin, S., 2007. Calcium orthophosphates. *Journal of Materials Science* 42(4): 1061–1095.
48. Mirtchi, A.A., Lemaitre, J., and Terao, N., 1989. Calcium phosphate cements: Study of the [beta]-tricalcium phosphate—Monocalcium phosphate system. *Biomaterials* 10(7): 475–480.
49. Driessens, F.C.M., Boltong, M.G., Bermúdez, O. et al., 1994. Effective formulations for the preparation of calcium phosphate bone cements. *Journal of Materials Science: Materials in Medicine* 5(3): 164–170.
50. Yamamoto, H., Niwa, S., Hori, M. et al., 1998. Mechanical strength of calcium phosphate cement in vivo and in vitro. *Biomaterials* 19(17): 1587–1591.
51. Takagi, S., Chow, L.C., and Ishikawa, K., 1998. Formation of hydroxyapatite in new calcium phosphate cements. *Biomaterials* 19(17): 1593–1599.
52. Suzuki, O., Kamakura, S., Katagiri, T. et al., 2006. Bone formation enhanced by implanted octacalcium phosphate involving conversion into Ca-deficient hydroxyapatite. *Biomaterials* 27(13): 2671–2681.
53. Dekker, R.J., de Bruijn, J.D., Stigter, M. et al., 2005. Bone tissue engineering on amorphous carbonated apatite and crystalline octacalcium phosphate-coated titanium discs. *Biomaterials* 26(25): 5231–5239.
54. Kikawa, T., Kashimoto, O., Imaizumi, H. et al., 2009. Intramembranous bone tissue response to biodegradable octacalcium phosphate implant. *Acta Biomaterialia* 5(5): 1756–1766.
55. LeGeros, R.Z., 2001. Formation and transformation of calcium phosphates: Relevance to vascular calcification. *Zeitschrift für Kardiologie* 90(15): III116–III124.
56. Sinyaev, V.A., Shustikova, E.S., Levchenko, L.V. et al., 2001. Synthesis and dehydration of amorphous calcium phosphate. *Inorganic Materials* 37(6): 619–622.
57. Yoshikawa, H. and Myoui, A., 2005. Bone tissue engineering with porous hydroxyapatite ceramics. *Journal of Artificial Organs* 8(3): 131–136.
58. Hench, L.L., Splinter, R.J., Allen, W.C. et al., 1971. Bonding mechanisms at the interface of ceramic prosthetic materials. *Journal of Biomedical Materials Research* 5(6): 117–141.
59. Priya, S., Julian, R.J., Russell, S.P. et al., 2003. Bioactivity of gel-glass powders in the CaO-SiO$_2$ system: A comparison with ternary (CaO-P$_2$O$_5$-SiO$_2$) and quaternary glasses (SiO$_2$-CaO-P$_2$O$_5$-Na$_2$). *Journal of Biomedical Materials Research Part A* 66A(1): 110–119.
60. Priya, S., Julian, R.J., Sophie, V. et al., 2004. Binary CaO-SiO$_2$ gel-glasses for biomedical applications. *Bio-Medical Materials and Engineering* 14(4): 467–486.
61. Hench, L.L., 1997. Sol-gel materials for bioceramic applications. *Current Opinion in Solid State and Materials Science* 2(5): 604–610.
62. Jones, J.R., Ehrenfried, L.M., and Hench, L.L., 2006. Optimising bioactive glass scaffolds for bone tissue engineering. *Biomaterials* 27(7): 964–973.
63. Navarro, M., Michiardi, A., Castano, O. et al., 2008. Biomaterials in orthopaedics. *Journal of the Royal Society Interface* 5(27): 1137–1158.

64. Wang, M., 2003. Developing bioactive composite materials for tissue replacement. *Biomaterials* 24(13): 2133–2151.

65. Bonfield, W., 1988. Composites for bone replacement. *Journal of Biomedical Engineering* 10(6): 522–526.

66. Bonfield, W., 1988. Hydroxyapatite-reinforced polyethylene as an analogous material for bone replacement. *Annals of the New York Academy of Science* 523 (Bioceramics: Material Characteristics Versus In Vivo Behavior): 173–177.

67. Wahl, D.A. and Czermuszka, J.T., 2006. Collagen-hydroxyapatite composites for hard tissue repair. *European Cells and Materials Journal* 11: 43–56.

68. Yamauchi, K., Goda, T., Takeuchi, N. et al., 2004. Preparation of collagen/calcium phosphate multilayer sheet using enzymatic mineralization. *Biomaterials* 25(24): 5481–5489.

69. Leupold, J.A., Barfield, W.R., An, Y.H. et al., 2006. A comparison of ProOsteon, DBX, and collagraft in a rabbit model. *Journal of Biomedical Materials Research Part B: Applied Biomaterials* 79B(2): 292–297.

70. Wang, Y., Cui, F.Z., Hu, K. et al., 2008. Bone regeneration by using scaffold based on mineralized recombinant collagen. *Journal of Biomedical Materials Research Part B: Applied Biomaterials* 86B(1): 29–35.

71. Serre, C.M., Papillard, M., Chavassieux, P. et al., 1993. In vitro induction of a calcifying matrix by biomaterials constituted of collagen and/or hydroxyapatite: An ultrastructural comparison of three types of biomaterials. *Biomaterials* 14(2): 97–106.

72. Kikuchi, M., Itoh, S., Ichinose, S. et al., 2001. Self-organization mechanism in a bone-like hydroxyapatite/collagen nanocomposite synthesized in vitro and its biological reaction in vivo. *Biomaterials* 22(13): 1705–1711.

73. Kikuchi, M., Matsumoto, H.N., Yamada, T. et al., 2004. Glutaraldehyde cross-linked hydroxyapatite/collagen self-organized nanocomposites. *Biomaterials* 25(1): 63–69.

74. Sotome, S., Uemura, T., Kikuchi, M. et al., 2004. Synthesis and in vivo evaluation of a novel hydroxyapatite/collagen-alginate as a bone filler and a drug delivery carrier of bone morphogenetic protein. *Materials Science and Engineering: C* 24(3): 341–347.

75. Song, J.-H., Kim, H.-E., and Kim, H.-W., 2007. Collagen-apatite nanocomposite membranes for guided bone regeneration. *Journal of Biomedical Materials Research Part B: Applied Biomaterials* 83B(1): 248–257.

76. Wu, T.-J., Huang, H.-H., Lan, C.-W. et al., 2004. Studies on the microspheres comprised of reconstituted collagen and hydroxyapatite. *Biomaterials* 25(4): 651–658.

77. Joseph, E.Z., Sohrab, K., Robert, S. et al., 1992. Fibrillar collagen-biphasic calcium phosphate composite as a bone graft substitute for spinal fusion. *Journal of Orthopaedic Research* 10(4): 562–572.

78. Zou, C., Weng, W., Deng, X. et al., 2005. Preparation and characterization of porous [beta]-tricalcium phosphate/collagen composites with an integrated structure. *Biomaterials* 26(26): 5276–5284.

79. Jacqueline, A.S. and Melissa, K.C.B., 1999. Biocompatibility and degradation of collagen bone anchors in a rabbit model. *Journal of Biomedical Materials Research Part B: Applied Biomaterials* 48(3): 309–314.

80. Jui-Sheng, S., Feng-Huei, L., Yng-Jiin, W. et al., 2003. Collagen-hydroxyapatite/tricalcium phosphate microspheres as a delivery system for recombinant human transforming growth factor-β1. *Artificial Organs* 27(7): 605–612.

81. Vicente, V., Meseguer, L., Martinez, F. et al., 1996. Ultrastructural study of the osteointegration of bioceramics (whitlockite and composite β-TCP + collagen) in rabbit bone. *Ultrastructural Pathology* 20(2): 179–188.

82. Shinji, K., Kazuo, S., Takahiro, H. et al., 2007. The primacy of octacalcium phosphate collagen composites in bone regeneration. *Journal of Biomedical Materials Research Part A* 83A(3): 725–733.

83. Du, C., Cui, F.Z., Zhang, W. et al., 2000. Formation of calcium phosphate/collagen composites through mineralization of collagen matrix. *Journal of Biomedical Material Research* 50(4): 518–527.

84. Du, C., Cui, F.Z., Feng, Q.L. et al., 1998. Tissue response to nano-hydroxyapatite/collagen composite implants in marrow cavity. *Journal of Biomedical Materials Research* 42(4): 540–548.

85. Pederson, A.W., Ruberti, J.W., and Messersmith, P.B., 2003. Thermal assembly of a biomimetic mineral/collagen composite. *Biomaterials* 24(26): 4881–4890.

86. Eglin, D., Maalheem, S., Livage, J. et al., 2006. In vitro apatite forming ability of type I collagen hydrogels containing bioactive glass and silica sol-gel particles. *Journal of Materials Science: Materials in Medicine* 17(2): 161–167.

87. Lin, H.-R. and Yeh, Y.-J., 2004. Porous alginate/hydroxyapatite composite scaffolds for bone tissue engineering: Preparation, characterization, and in vitro studies. *Journal of Biomedical Materials Research Part B: Applied Biomaterials* 71B(1): 52–65.

88. Panzavolta, S., Fini, M., Nicoletti, A. et al., 2009. Porous composite scaffolds based on gelatin and partially hydrolyzed α-tricalcium phosphate. *Acta Biomaterialia* 5(2): 636–643.

89. Kim, H.-W., Knowles, J.C., and Kim, H.-E., 2005. Hydroxyapatite and gelatin composite foams processed via novel freeze-drying and crosslinking for use as temporary hard tissue scaffolds. *Journal of Biomedical Materials Research Part A* 72A(2): 136–145.

90. Nazhat, S.N., Kellomäki, M., Törmälä, P. et al., 2001. Dynamic mechanical characterization of biodegradable composites of hydroxyapatite and polylactides. *Journal of Biomedical Materials Research Part B: Applied Biomaterials* 58(4): 335–343.

91. Nenad, I., Edin, S., Jaroslava, B.-S. et al., 2004. Evaluation of hot-pressed hydroxyapatite/poly-$_L$-lactide composite biomaterial characteristics. *Journal of Biomedical Materials Research Part B: Applied Biomaterials* 71B(2): 284–294.

92. Hasegawa, S., Tamura, J., Neo, M. et al., 2005. In vivo evaluation of a porous hydroxyapatite/poly-$_{DL}$-lactide composite for use as a bone substitute. *Journal of Biomedical Materials Research Part A* 75A(3): 567–579.

93. Anita, A.I., Oliver, B., Peter, A. et al., 2001. In vivo investigations on composites made of resorbable ceramics and poly(lactide) used as bone graft substitutes. *Journal of Biomedical Materials Research Part B: Applied Biomaterials* 58(6): 701–709.

94. Ahmed, I.A., Cronin, P.S., Abou Neel, E.A. et al., 2009. Retention of mechanical properties and cytocompatibility of a phosphate-based glass fiber/polylactic acid composite. *Journal of Biomedical Materials Research Part B: Applied Biomaterials* 89B(1): 18–27.

95. Kim, H.-W., Knowles, J.C., and Kim, H.-E., Hydroxyapatite/poly(ε-caprolactone) composite coatings on hydroxyapatite porous bone scaffold for drug delivery. *Biomaterials* 25(7–8): 1279–1287.

96. Causa, F., Netti, P.A., Ambrosio, L. et al., 2006. Poly-ε-caprolactone/hydroxyapatite composites for bone regeneration: In vitro characterization and human osteoblast response. *Journal of Biomedical Materials Research Part A* 76A(1): 151–162.

97. Guarino, V. and Ambrosio, L., 2008. The synergic effect of polylactide fiber and calcium phosphate particle reinforcement in poly [epsilon]-caprolactone-based composite scaffolds. *Acta Biomaterialia* 4(6): 1778–1787.

98. Kikuchi, M., Koyama, Y., Yamada, T. et al., 2004. Development of guided bone regeneration membrane composed of [beta]-tricalcium phosphate and poly (-lactide-co-glycolide-co-[var epsilon]-caprolactone) composites. *Biomaterials* 25(28): 5979–5986.

99. Khan, A.S., Ahmed, Z., Edirisinghe, M.J. et al., 2008. Preparation and characterization of a novel bioactive restorative composite based on covalently coupled polyurethane-nanohydroxyapatite fibres. *Acta Biomaterialia* 4(5): 1275–1287.

100. Bil, M., Ryszkowska, J., Roether, J.A. et al., 2007. Bioactivity of polyurethane-based scaffolds coated with Bioglass. *Biomedical Materials* 2(2): 93–101.

101. Wang, M., Yubao, L., Wu, J. et al., 2008. In vitro and in vivo study to the biocompatibility and biodegradation of hydroxyapatite/poly(vinyl alcohol)/gelatin composite. *Journal of Biomedical Materials Research Part A* 85A(2): 418–426.

102. Leeuwenburgh, S.C., Jansen, J.A., and Mikos, A.G., 2007. Functionalization of oligo(poly(ethylene glycol)fumarate) hydrogels with finely dispersed calcium phosphate nanocrystals for bone-substituting purposes. *Journal of Biomaterials Science Polymer Edition* 18(12): 1547–1564.

103. Demirtaş, T.T., Karakeçili, A.G., and Gümüşderelioğlu, M., 2008. Hydroxyapatite containing superporous hydrogel composites: Synthesis and in-vitro characterization. *Journal of Materials Science: Materials in Medicine* 19(2): 729–735.

104. Kenny, S.M. and Buggy, M., 2003. Bone cements and fillers: A review. *Journal of Materials Science: Materials in Medicine* 14(11): 923–938.

105. Ruhé, P.Q., Boerman, O.C., Russel, F.G.M. et al., 2005. Controlled release of rhBMP-2 loaded poly(dl-lactic-co-glycolic acid)/calcium phosphate cement composites in vivo. *Journal of Controlled Release* 106(1–2): 162–171.

106. Fei, Z., Hu, Y., Wu, D. et al., 2008. Preparation and property of a novel bone graft composite consisting of rhBMP-2 loaded PLGA microspheres and calcium phosphate cement. *Journal of Materials Science: Materials in Medicine* 19(3): 1109–1116.

107. Tamimi, F., Kumarasami, B., Doillon, C. et al., 2008. Brushite-collagen composites for bone regeneration. *Acta Biomaterialia* 4(5): 1315–1321.

108. Mai, R., Reinstorf, A., Pilling, E. et al., 2008. Histologic study of incorporation and resorption of a bone cement-collagen composite: An in vivo study in the minipig. *Oral Surgery, Oral Medicine, Oral Pathology, Oral Radiology, and Endodontology* 105(3): e9–e14.

109. Qi, X., Ye, J., and Wang, Y., 2009. Alginate/poly (lactic-*co*-glycolic acid)/calcium phosphate cement scaffold with oriented pore structure for bone tissue engineering. *Journal of Biomedical Materials Research Part A* 89A(4): 980–987.

110. Xu, H.H.K., Eichmiller, F.C., and Giuseppetti, A.A., 2000. Reinforcement of a self-setting calcium phosphate cement with different fibers. *Journal of Biomedical Materials Research Part A* 52(1): 107–114.

111. Zhang, Y. and Zhang, M., 2002. Three-dimensional macroporous calcium phosphate bioceramics with nested chitosan sponges for load-bearing bone implants. *Journal of Biomedical Materials Research Part A* 61(1): 1–8.

9 Amphiphilic Systems as Biomaterials Based on Chitin, Chitosan, and Their Derivatives

Jacques Desbrieres

CONTENTS

9.1 INTRODUCTION

Chitin is the second most abundant polysaccharide in nature after cellulose. It is widely distributed in the animal and vegetal kingdom, constituting an important renewable resource. It was first isolated from fungi by Braconnot in 1811,[1] but the name chitin, from the Greek language, which means tunic or cover, was given by Odier, who in 1923 isolated it from the elytrum of the cock-chafer beetle by treatment with hot alkaline solutions.[2] Because of its insolubility in the vast majority of common solvents, chitin was considered an intractable polymer and for many years it remained mainly a laboratory curiosity. However, at present, chitin and its derivatives have become polymers of great interest in a large variety of areas of human activity. Chitin is generally represented as a linear polysaccharide composed of β $(1 \rightarrow 4)$ linked units of *N*-acetyl-2-amino-2-deoxy-D-glucose. A great structural similarity exists between chitin and cellulose; the difference between them consists in that the hydroxyl group of carbon C2 in cellulose is substituted by an acetamide group in chitin. Both biopolymers play similar roles, since they both act as structural support and defense materials in living organisms.

Chitin is the most abundant organic component of the skeletal structure of many classes comprising the group of invertebrates, such as arthropods, mollusks, and annelids. In animals, chitin occurs associated with other constituents, such as lipids, calcium carbonate, proteins, and pigments. It has been estimated that the crustacean chitin present in the sea amounts to 1560 million tons.[3] Chitin is also found as a major polymeric constituent of the cell wall of fungi and algae. Fungal chitin exhibits some advantages as compared with animal chitin, such as a greater uniformity in composition, a continuous availability in time, and the absence of inorganic salts in its matrix. However, in fungi, chitin is associated with other polysaccharides, such as cellulose, glucan, mannan, and polygalactosamine, which makes its isolation difficult.[4]

Chitosan is a linear polysaccharide obtained by extensive deacetylation of chitin. It is mainly composed of two kinds of β (1 → 4) linked structural units: 2-amino-2-deoxy-D-glucose and N-acetyl-2-amino-2-deoxy-D-glucose. However, since it is virtually impossible to completely deacetylate chitin, what is usually known as chitosan is a family of chitins with different degrees of acetylation (defined as the molar fraction of acetylated units) lower than 50%. The capacity of chitosan to dissolve in dilute aqueous solutions is the commonly accepted criterion to differentiate it from chitin. Chitosan is also present in significant quantities in some fungi, such as *Mucor rouxii* (30%) and *Choanephora cucurbitarum* (28%), although again associated with other polysaccharides.

Chitin and chitosan are structurally similar to heparin, chondroitin sulfate, or hyaluronic acid, which are biologically important mucopolysaccharides. By themselves they have a wide range of applications. Among them, they are remarkable biomaterials because of their numerous biological and immunological properties.[5] They show excellent biological properties such as biodegradation in the human body,[6,7] and immunological,[8,9] antibacterial,[10,11] and wound healing activity.[12–14] But the restricted solubilization restrained their use and it is necessary to consider their chemical or enzymatic modification to enlarge their domain of use. These polymers have two reactive groups suitable for this purpose, namely, primary (C6) and secondary (C3) hydroxyl groups in the case of chitin, whereas chitosan has additionally the amino (C2) group on each deacetylated unit. All these functions are susceptible to a variety of classical reactions that can be applied in a controlled fashion to obtain a vast array of novel materials based on the two polysaccharides which can also be modified by either cross-linking or graft copolymerization.

The presence of hydroxyl and highly reactive-free amino groups offers a great potential for further derivatization.[15,16] In a wider consideration, the application of chitin and chitin derivatives in many areas including food,[17–19] cosmetics,[20] biotechnology, and medicine is due to their promising colloidal properties such as the formation of complexes with surfactants (including fatty and amino acids, phospholipids), interaction with biomembranes, self-organization in ordered nano-structures at the interfaces or in solution, stabilization of emulsions or foams, formation of pH-, electrolyte-, or thermo-responsive physical gels. Moreover, the great interest in medical applications of chitosan and some of its derivatives is readily understood. The cationic character of chitosan is unique: it is the only artificial (derivative of a natural polymer) cationic polymer.

Several chitosan-based molecules have been shown to be non-toxic, biodegradable, and biocompatible with low immunogenicity.[17,21] These cationic polyelectrolytes are potential candidates for gene delivery[22–25] since they bind efficiently negatively charged *DNA* and protect it against nuclease degradation.[26–30] *DNA*-chitosan complexes have been applied for in vitro transfection assays and have been found useful for gene delivery.[31–33]

Numerous modifications of chitin and chitosan were described for biomedical applications[34] but we focused our review on the amphiphilic systems based on chitin, chitosan, and their derivatives.

9.2 AMPHIPHILIC POLYMERIC SYSTEMS

Amphiphilic polymeric systems combine in the macromolecular backbone or in grafted components a range of hydrophilic and hydrophobic groups. The former are typically based on hydroxyl, urethane, and carboxylic moieties, to name but a few, and the latter on alkyl or long fatty acid chains to cite the more prevalent examples. Amphiphilic polymers are, on one hand, able to self-assemble and as a consequence form mono- and polymolecular micelles or aggregates in aqueous dispersions[35–43] and, on the other hand, exhibit enhanced adsorption characteristics at interfaces. These results lead to sharp reductions in surface and interfacial tensions. In addition, aqueous dispersions tend to display enhanced viscosities due to intermolecular interactions between hydrophobic groups, thus leading to the formation of polymolecular associations and long-range structures. According to their structure, the aggregation mechanism will be different as the properties. But in any case the basic phenomenon will be the hydrophobic interactions between hydrophobic parts of the polymeric chains leading to junction zones. According to medium conditions, a physical gel, which is a loose tridimensional network, may be obtained. Hence, a large viscosity increase was observed. Self-assembly of such polymers allows also the elaboration of specific materials such as nanoparticles and gel-like aggregates. Among the many uses for polymers of this type are included environmentally friendly formulations or viscosity modifiers in tertiary oil recovery systems.[44–47] Moreover, such amphiphilic polymers may present hydrophobic interactions with other chemicals and especially coming from living systems. Hence although polysaccharide based systems have been considered for a range of applications[48–52] of particular interest, those allied to their use in biomedicine will be focused.

Polysaccharides are natural polymers that exhibit good biocompatibility and have been targeted for use in site-specific drug delivery, encapsulation, and rheological modifications due to their inherent stiffness. The amphiphilic characteristics of polysaccharides may be altered through modifying either hydrophilic or hydrophobic groups. As most polysaccharides are in fact polyelectrolytes, the hydrophilic character may be controlled by varying the charge density (or the number of ionic charges per saccharide unit), or when the ionic sites are weak acids or bases by changing the pH of the medium, which modifies the degree of ionization. The hydrophobic character may be adjusted, however, either by the chemical modification of the macromolecular chain structure through covalently bonding hydrophobic groups onto the hydrophilic backbone of the polysaccharide to give, for example, an alkylated polyelectrolyte, a so-called polysoap (see Figure 9.1a) or graft copolymerization,[53] or by the formation of non-stoichiometric dynamic associations between the polysaccharide and an oppositely charged molecule as a surfactant to give a surfactant-polyelectrolyte complex (SPEC), as exemplified by the modification of an O-carboxymethylchitin (Figure 9.1b) to the SPEC shown in Figure 9.1c or an amphiphilic polyelectrolyte leading to a polyelectrolyte complex (PEC) as shown in Figure 9.1d.

Thus, this allows adjusting the chemical structure and the architecture of polymeric chain to required properties. In these systems, hydrophobic associations through intra- or interpolymer interactions compete with electrostatic repulsions, which operate within the same polymer chain or between different chains. The preference of intra- or interpolymer hydrophobe association depends on the polymer concentration and the structural characteristics of the macromolecule, notably the content and the nature of the hydrophobic groups and the sequence distribution of electrolyte and hydrophobic monomer units along the polymer chain. These effects and the ionic strength of the system are factors that also are important[54,55] for the association efficiency and subsequently the dynamical behavior of solutions of hydrophobically modified polyelectrolytes.

Among the more discussed biomedical applications, encapsulation and controlled release of active compounds, on one hand, and stimuli-responsive polymeric gels,[56] on the other hand, are predominant. To be used in such domains polymeric systems need solubility control, specific controlled interactions, mechanical properties, fast response time in physiological environment, etc. The limitation of chitosan for the preparation of sustained release systems arises from its rapid

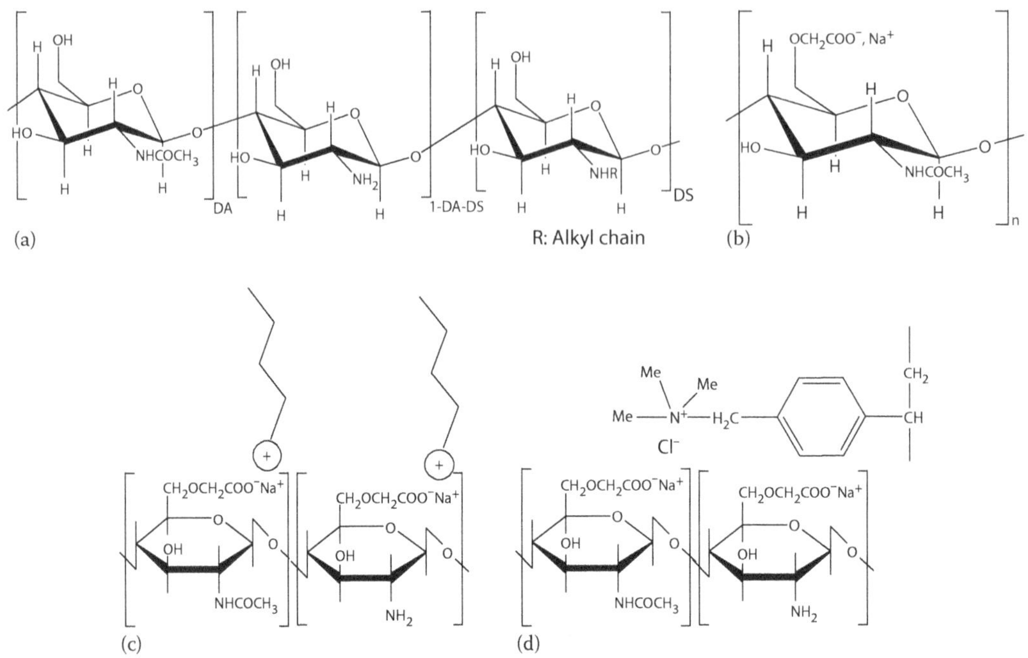

FIGURE 9.1 Chemical structures of chitin derivatives: (a) alkylchitosan, DA being the degree of acetylation and DS the degree of alkyl substitution, (b) *O*-carboxymethylchitin (CMCh), (c) SPEC formed from CMCh and cationic surfactant, and (d) PEC formed from CMCh and an amphiphilic cationic polyelectrolyte.

water absorption and high swelling degree in aqueous environments leading to fast drug release. The amphiphilicity is an important characteristic for the formation of self-assembled structures such as nanoparticles, potentially suited for drug delivery applications. The hydrophobic core of these nanoparticles could act as reservoirs or microcontainers for various active substances. Moreover, the small size of these particles leads them to be administrated via the intravenous injection for targeted drug delivery. For other applications such as DNA transfection, it was observed a small efficiency of chitosan assumed to be due to the low hydrophobicity of the complexes resulting from the condensation of DNAs by the chitosan chains. As a consequence it is important to be able to bring amphiphilic character to chitosan macromolecules. In order to overcome these problems, chitosan was modified with alkyl halides bearing hydrophobic alkyl chains. The utility of modifying chitosan with hydrophobic branches for controlling solubility properties has been demonstrated.[57,58] More recently, it was reported that chitosan derivatives with both hydrophobic groups (as long acyl groups) and hydrophilic groups (sulfate groups) could form micelles and solubilize hydrophobic compounds.[59] The possibility to modify water-soluble chitosan derivative (*N*-methylene phosphonic chitosan) allows to obtain an amphiphilic system in which the hydrophobic moiety counterbalances the electrostatic repulsion but it also gets more tensioactive properties.[58]

9.3 AMPHIPHILIC SYSTEMS BASED ON CHITIN AND CHITOSAN DERIVATIVES

Amphiphilic materials may exist under different types discussed from the chemical point of view. Moreover, they can be elaborated under several forms and among those specially devoted to biomedical applications is the hydrogel one.

9.3.1 ELABORATION

Chitin possesses one functional group, the hydroxyl one, whereas chitosan has two, the hydroxyl and the amino ones. This allows various chemical reactions on each of them and the capacity to perform specific reactions on the amino group, by contrast to cellulose or starch polymers.[60] Moreover, amphiphilic systems may be obtained from other strategies considering interactions that chitin or chitosan (and its derivatives) may build with other chemicals such as surfactants and other polyelectrolytes.

9.3.1.1 Amphiphilic Covalent Derivatives

The alkylated derivatives of chitosan and chitin were generally obtained by reductive amination adapting the procedure described by Yalpani.[61] This is a versatile and specific method for creating a covalent bond between a substrate and the amine function of the glucosamine unit. For considered chitosan it involves the reaction between the amine function of the unit and an aldehyde function. Generally the reaction of substitution of polymers is not homogeneous. Indeed, many reactions (e.g., on cellulose or starch) are performed in heterogeneous and the chemical modification can be observed only on the accessible sites of the macromolecular chain. Due to lack of accessibility, a block-wise substitution distribution is obtained and the compound is structurally inhomogeneous. A procedure was developed for performing homogeneous chemical reactions allowing a uniform distribution of substituants.[62] The reaction mechanism is as follows:

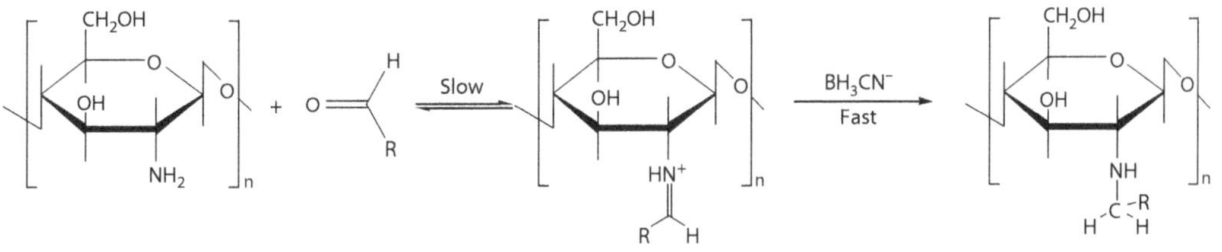

The reduction of imminium ion was performed using $NaCNBH_3$, which is more reactive and selective than usual reducing agents such as $NaBH_4$, selephenol (PhSeH) or pentacarbonyl ion in alcoholic potassium hydroxide. The advantage of such reducing agent is its stability in the acidic media.[63] Moreover, the reduction of imminium ion by BH_3CN^- anion is rapid at pH values of 6–7 and the reduction of aldehydes or ketones is negligible in this pH range (it becomes quick at pH values smaller than 3.5). Using the same process cyclodextrin-chitosan was synthesized using β-cyclodextrin monoaldehyde.[64] Cyclodextrins are cyclic oligosaccharides built from 6 to 8 D-glucose units and are formed during the enzymatic degradation of starch and related compounds.[65] The D-glucose units are covalently linked together by 1,4 linkages to form torus-like structures. The formed cavity, hydrophobic in nature, is capable of binding aromatic and other small organic molecules and therefore provides ideal binding sites.[66,67]

New techniques were developed to elaborate such amphiphilic derivatives. Graft copolymerization is a versatile method for providing functionality and regulating polymer properties for the resulting polymers.[68] Graft copolymerization may be initiated either by chemicals such as ammonium or potassium persulfate (*APS* or *KPS*) and ceric ammonium nitrate (*CAN*) or by γ-irradiation and enzymes. The grafting parameters such as grafting efficiency are greatly influenced by several parameters such as the type and concentration of initiator, monomer concentration, reaction time and time. Combining chitosan with poly(1,4-dioxan-2-one) (PPDO) would provide an amphiphilic copolymer. Chitosan-g-PPDO copolymer was synthesized in bulk by ring-opening polymerization of *p*-dioxanone initiated by the hydroxyl group or amino group of chitosan in the presence of stannous octoate $(SnOct_2)$[68] (Scheme 9.1).

The incorporation of PPDO into chitosan would lead to the controllability of physical properties by controlling the chemical structure of the graft copolymer. Similar architecture was obtained by grafting ε-caprolactone (ε-CL) oligomers onto the hydroxyl groups of chitosan via ring-opening polymerization.[69] In this reaction, methanesulfonic acid plays a dual solvent and catalyst role in the graft polymerization process. Due to protective protonation of amine group of chitosan in

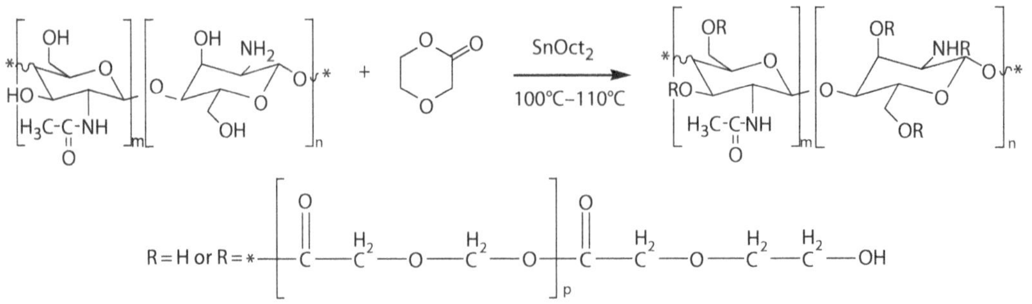

SCHEME 9.1 Grafting of chitosan with *p*-dioxanone.

acidic medium, grafting of ε-CL takes place mainly on hydroxyl groups of chitosan. The resultant chitosan-*g*-PCL showed improved solubility in a variety of organic solvents, and easy to be processed into nanofibers via electrospinning. Owing to the negative membrane potential of cells, it was promising that these amino-reserved chitosan-*g*-PCL copolymer fibers could generate a cationic surface to promote cell adhesion when used as tissue engineering scaffolds. Radiochemical techniques have recently caught the attention of researchers as an alternate process. Yu et al.[70] used γ-irradiation for grafting butylacrylate. They observed that grafting efficiency increases when the monomer concentration and total dose increase or when chitosan concentration and reaction temperature decrease. Compared with chitosan films, chitosans grafted with butylacrylate show larger hydrophobicity and best mechanical strength. Singh et al.[71] have grafted poly(acrylonitrile) onto chitosan using a microwave irradiation technique under homogeneous conditions. They obtained 70% efficiency within 90 s.

Recently, hydrophobically modified glycolchitosan was prepared by partially reacting the amino groups of glycolchitosan with 5β-cholanic acid by in situ activation using 1-ethyl-3-(3-dimethylaminopropyl)-carbodiimide hydrochloride (EDC) and N-hydroxysuccinimide (NHS; Scheme 9.2).[72] EDC catalyzes the formation of amide bonds between the carboxylic acid groups of organic acids and the amine groups of chitosan.

The enzymatic approach to the modification of chitin and chitosan is interesting owing to its specificity and environmental advantages compared with chemical modification. With respect to health and safety, enzymes offer the potential of eliminating the hazards associated with reactive agents. Chen[73] grafted hexyloxyphenol to chitosan using tyrosinase. On the basis of contact angle measurements, the heterogeneous modification of chitosan film was found to produce a hydrophobic surface due to the substituent. From the biochemically relevant quinines studied, it would be possible to prepare materials of medical interest; for instance, menadione, a synthetic naphtoquinone derivative having the physiological properties of vitamin K, is particularly prone to rapid reaction with chitosans, greatly modifying its spectral characteristics and increasing the surface hydrophobicity of treated chitosan films.[74]

Finally, another type of reaction was carried out to reduce the hydrophilic character of chitosan: Bordenave[75] reported an oxidation of the primary alcohol to carboxylic group of the chitosan units to produce a reactive site to further chemical modification such as hydrophobic chain grafting. The work focuses on the selective oxidation of the primary alcohols mediated by TEMPO. The oxidation was first conducted on the raw chitosan and degradation was observed. This phenomenon confirmed the necessity of a protection on the amino group.

9.3.1.2 Amphiphilic Block Copolymers

Poly(ethyleneglycol) (PEG) is amphiphilic and non-toxic. As chitosan is hydrophilic, mPEG-PEI-chitosan block copolymers were synthesized. Poly(ethylenimine) (PEI), a highly branched polyamine, is one of the most commonly used polycations for gene delivery. Methoxy PEG (mPEG) was used to increase the slow-releasing property and water-solubility when PEI was used to improve

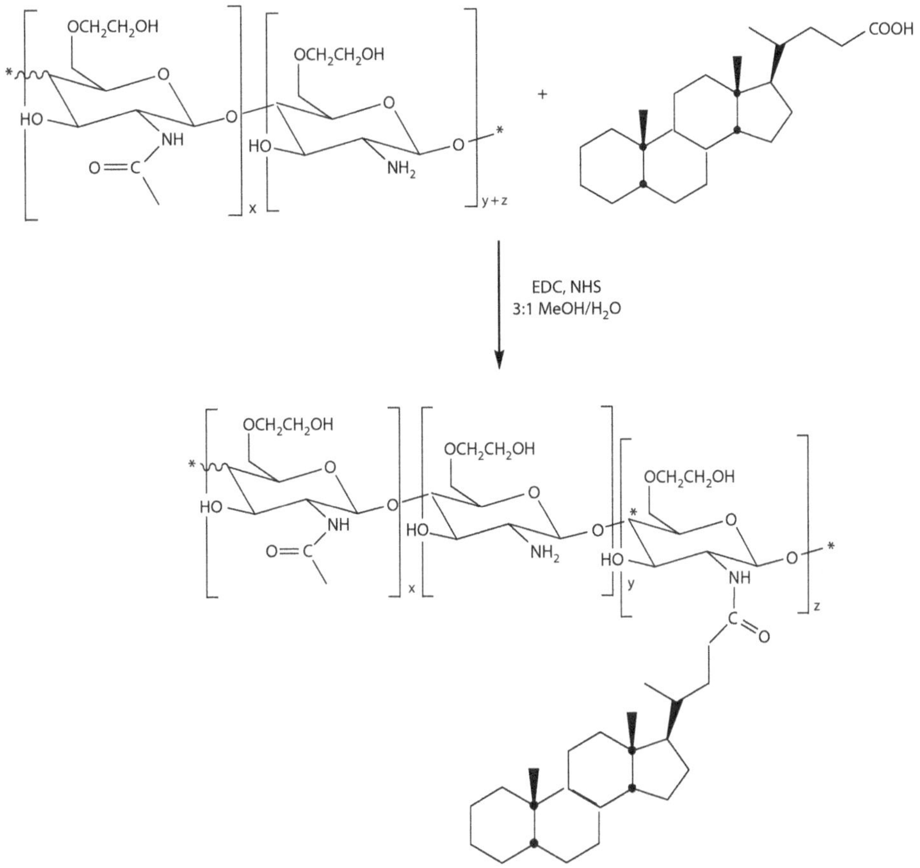

SCHEME 9.2   Reaction of glycolchitosan with 5β-cholanic acid.

transfection efficiency. To perform this synthesis, mPEG iodide was first prepared,[76] then synthesis of mPEG-PEI was carried out via the mPEG iodide attacking the amino groups of PEI.[77] Finally, the reaction of mPEG-PEI with *N*-phtaloyl chitosan iodide was performed in DMF.

9.3.1.3 Surfactant-Polyelectrolyte Complex or Polyelectrolyte Complex

PECs are formed by the reaction of a polyelectrolyte with an oppositely charged polyelectrolyte in an aqueous solution when SPECs are formed by the interaction of a polyelectrolyte with an oppositely charged surfactant. Essentially this is the result of electrostatic interactions between both ions (or polyions). They were applied in wide applications.[56] Yoshiwaza[78,79] prepared PEC films composed of a cationic chitosan and an anionic polyalkylenoxide/maleic acid (PAOMA) copolymer to be applied as drug carriers. PAOMA has an excellent biocompatibility and unique amphiphilic properties and PEC films were demonstrated to have pH- and temperature-sensitivity for equilibrium swelling.

The complexes between DNA and chitosan are now under investigation for elaboration of nanoparticles as gene carriers. It is necessary to have a minimal DP around 6–9 for stability of such complexes.[80] Their stability is reduced above a pH value of 7.4 (close to the physiological pH) and this is interpreted as one of the reasons for the observed high transfection activity of oligomer-based complex. The nanoparticles were prepared using a complex coacervation process[81] and the important parameters for this synthesis are concentrations of DNA and chitosan, molecular weights of polymers, temperature, and pH of solutions. Efficiency was improved by modifying the chitosan through hydrophobic modification.[82]

9.3.1.4 Hydrogels

Materials that can be elaborated from these polymeric systems are of two types according to the cross-linking process. Polymeric self-assembly materials (often called SA) are one of them and they were extensively studied concerning their autoassociative rheological or their micellar behavior. They form "hydrogels" using non-covalent cross-linking by exploiting either hydrogen bonding or hydrophobic interactions. These are biocompatible as gel formation does not require the use of organic solvents or chemical cross-linking reactions, which may be potentially deleterious to the active matter load. They are also called "physical gels."

The "chemical gels" (cross-linked by covalent bonds), the second type of hydrogels, can be obtained either by employing cross-linking agents such as epichlorohydrin or by grafting/cross-linking with acrylic monomers. Epichlorohydrin, as many other cross-linking agents, is known to be cytotoxic.[83] Moreover, short cross-linking bridges introduced by such a cross-linking agent between polymeric backbones would lead to a final product having a high rigidity in dry state. The increase of cross-linking bridge length can be accomplished by grafting/cross-linking of polysaccharide with monomers, among which acrylics occupy a significant position.[84,85] For this purpose, the polysaccharide should be at first modified with unsaturated binding functionalities. There are several advantages of this method, including introduction of longer and/or exhibiting temperature or pH-sensitive behavior bridges between polysaccharide chains. A less rigid structure, and an easier control of network density, as a function of number of double bonds introduced on the polysaccharide backbone, may thus be attained.

Among different kinds of hydrogels, "smart hydrogels" are of great interest because they can undergo a reversible discontinuous phase change in response to various external physicochemical factors. Among them, temperature- and pH-responsive hydrogels have been the most likely considered because these two factors can be applicable in vivo. Thermosensitive hydrogels are of great interest in therapeutic delivery systems and in tissue engineering as injectable depot systems. In a positive thermo-reversible system, polymer solutions having an upper critical solution temperature (UCST) shrink by cooling below this temperature. Polymer solutions having a lower critical solution temperature (LCST) contract by heating above LCST. These negative thermo-reversible hydrogels can be tuned to be liquid at room temperature (20°C–25°C) and to undergo gelation in contact with body fluids (36°C–37°C) due to an increase in temperature. Therefore, polymers having LCST below human body temperature have a potential for injectable applications. Different thermal setting gels have been described in the literature including, for example, the acrylic acid copolymers and the N-isopropylacrylamide (NIPAm).[86,87] Poly(N-isopropylacrylamide) (PNIPAm) is well known to have a thermally reversible property, exhibiting a LCST around 32°C in aqueous solution.[88] For biomedical applications, attention was paid to the development of stimuli-responsive polymeric gels with unique properties such as biodegradability and biocompatibility. They may be prepared by combining thermoresponsive polymers with natural-based polymeric components to form smart hydrogels.[89,90] When we have an assembly of two cross-linked polymers and at least one of which is synthesized and cross-linked in the presence of the other in the linear form, the system is termed as semi-Inter-Penetrating Network (semi-IPN). A number of polysaccharides were considered, including chitosan and alginate. Moreover due to the pH-sensitive character of chitosan, a combination of this polymer with a thermoresponsive material will produce dual-stimuli-responsive polymeric systems that can be used as delivery vehicles which respond to localized conditions of pH and temperature in the human body.

Graft copolymerization of NIPAm onto chitosan was carried out using usual initiator such as AIBN and CAN. Then, chitosan-g-PNIPAm particles were synthesized by a soapless emulsion copolymerization process[91] using APS or 2,2'-azobis-(2-methylpropionamidine) dihydrochloride (AIBA) as initiator. In this case, the swelling ratio of the copolymer decreased with increasing cross-linking density and pH value. Chemically cross-linked polymers are good candidates to improve wet strength. PNIPAm-containing chitosan/glutaraldehyde gels were elaborated.[92,93] Another example is glutaraldehyde cross-linked semi-*IPN* of chitosan and poly(acrylonitrile) (PAN).[94] Finally, full-*IPN* hydrogel was synthesized by the chemical combination of methylenebis(acrylamide) cross-linked PNIPAm network with a formaldehyde cross-linked chitosan network.[95] Hydrogels based on PNIPAm-grafted chitosan may be also obtained by applying γ-irradiation.[96] Kim et al. synthesized hydrogels based on grafting chitosan with epoxy-terminated poly(dimethylsiloxane) (PDMS) by using UV irradiation.[97]

9.3.2 PROPERTIES

9.3.2.1 "Self-Assembly" Properties

Macromolecules may aggregate due to secondary binding forces, solvation, steric factors, and interpenetration. As reflected by the reaction mechanisms in vivo, specific interactions between macromolecules are necessary for the display of a functionality regardless of the polymer chain structure. Amphiphilic polymers generate specific interactions called "hydrophobic" ones. They are not caused by direct cohesive forces between molecules but by the specific structure of water molecules. In order to minimize the contact surface area of hydrophobic groups with water, resulting in a decrease of the entropy, hydrophobic groups aggregate with each other, leading to a "self-assembly" phenomenon. The increase in the hydrophobicity of an amphiphilic polymer induces stronger intermolecular interaction and aggregation. In the chitosan-based aqueous system, there are electrostatic interaction, hydrophobic interaction, and hydrogen bonding, which will influence the solubility or the assembly behaviors of chitosan derivatives.[98] Hydrophobic modified chitosans prefer to form self-assembled aggregates in aqueous media. The main microstructural characteristic of such polymeric systems is their ability to give rise to weak intra- and intermolecular hydrophobic interactions in water solutions. Indeed the hydrophobic groups distributed along the macromolecular chain associate to minimize their exposure to water. As an example, hydrophilic polymers containing a few hydrophobic units can form temporary hydrophobic association networks in aqueous media at relatively low concentrations.[99] More generally different types of materials may be elaborated from amphiphilic polymers: micelles, gel-like or fiber-like aggregates, nanoparticles, etc.

In the semi-dilute or moderately concentrated regime, these intermolecular interactions are predominant and they form a temporary associated network which serves to reinforce the entangled network. Hence, a large viscosity increase was observed. The tendency for such polymeric systems to undergo association leads to various interesting and unusual linear and non-linear rheological behavior, which make them candidates as rheology modifiers in many industrial applications.[100] The preference of intra- or interpolymer hydrophobe association depends on the polymer concentration and the structural characteristics of the macromolecule, notably the content and the nature of the hydrophobic groups and the sequence distribution of electrolyte and hydrophobic monomer units along the polymer chain. These effects and the ionic strength of the system are factors that are also important[55] for the association efficiency and subsequently the dynamical behavior of solutions of hydrophobically modified polyelectrolytes. It is possible to adjust the chemical structure (content of hydrophobic and hydrophilic moieties) to reach desired properties.

Rheological curves (variation of the viscosity as a function of the shear rate) are determined for different polymer concentrations at different temperatures. At constant temperature and polymer concentration, the chitosan solution has a Newtonian behavior when the alkylchitosan solutions are more and more viscoelastic when the hydrophobicity (related with the length of the alkyl chain) was increased.[101] For the longest alkyl chain, a yield stress fluid was observed related to the formation of a gel as will be shown later and no Newtonian plateau was described in the used shear rate range. The higher the hydrophobic properties of the macromolecular chain, the larger the gap to the Newtonian behavior. Considering the viscosity at a low shear rate as a function of the polymer concentration (Figure 9.2), the viscosity rises with increasing polymer concentration. This increase is stronger as the hydrophobicity of the polymer increases (as the length of the alkyl chain or the

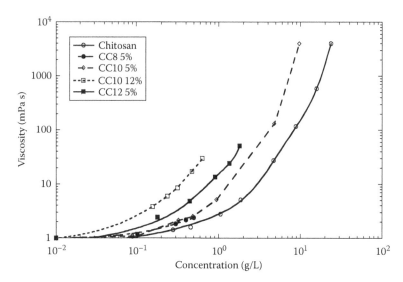

FIGURE 9.2 Role of polymer concentration on the zero-shear rate viscosity of chitosan and hydrophobic derivatives solutions (solvent AcOH 0.3M/AcONa 0.05 M, $T = 298$ K).

degree of substitution). Moreover, the concentration at which this increase is significant compared with the chitosan curve decreases when the hydrophobicity increases. This is due to the presence of interactions between polymer chains. The association effect on the viscosity becomes gradually more marked as the polymer concentration increases. This threshold concentration was sometimes called as the critical aggregation concentration (*cac*) and this viscosity increase was observed for polymer concentration lower than the overlap concentration. The viscosity increase is concomitant with the apparition of hydrophobic domains in the solution and it is shown that the junction points for the 3D-network formation are constituted of alkyl aggregates looking like a micellization.[101]

As the hydrophobic interactions are favored by temperature, it may be expected to obtain formulations presenting either thermal stability or a viscosity increase with temperature as already demonstrated for methylcellulose[102] or polyacrylate polymers grafted with poly(ethylene oxide) chains[103] as examples, these polymers belonging to the class of thermoassociative polymers. According to the structure of the polymer and the polymer concentration, temperature stable formulations or formulations presenting a viscosity increase with temperature may also be obtained.[104] Hence, it is possible to adjust the structure of the amphiphilic chitosan to the required rheological behavior. Application to cosmetic formulations was proved and solutions of alkylated carboxymethylchitin were used to stabilize such formulations. These are stable even when the temperature is increased and sensorial evaluation was found to be improved.[105]

In more concentrated media, self-assembly leads to aggregation. The morphologies of aggregates produced from amphiphilic polymer self-assembly can be diverse, such as liposomes, microsphere, vesicles, rods, lamellas, tubes, and large compound micelles.[106–110] They can be tailored by changing the hydrophobic/hydrophilic ratio or preparation conditions, since the self-assembly process has proven to be dominated by both thermodynamic and kinetic controls. The self-assembled polymer aggregates with unique structures and surface properties lead to a wide range of industrial applications, especially in pharmaceutical such as drug targeting devices, large quantity guest molecule encapsulation and gene therapy.[111–114] The effect of the descendible groups of chitosan derivatives on the aggregate morphology is far from well-understood. Chitosan-based self-aggregates were difficult widely apply to drug delivery systems because chitosan aggregates are insoluble in biological solution (pH ~ 7.4) and they are readily precipitated within a few days. To increase their stability in biological solution and decrease the cytotoxicity by acidic solution, where chitosan is soluble, water-soluble chitosan derivatives such as glycol chitosan were synthesized. It was then hydrophobically modified with deoxycholic acid[115] and nanoparticles were obtained from self-aggregation. The mean diameter of self-aggregates was controllable by changing the degree of substitution of deoxycholic acid in the range of 245–450 nm. They are stable for 10 days in physiological conditions and they provide the availability of nanoscale containers for hydrophobic drugs and genes. *N*-succinyl-chitosan can self-assemble regular nanosphere morphology in distilled water.[116] Within these nanospheres, there are hydrophobic domains formed and as they have non-toxic and cell-compatible properties, they present a great potential as a drug matrix in the controlled release delivery. Similarly hydrophobically modified glycolchitosan may form nanoparticles.[72,117] They have a globular shape with an average diameter of 350 nm. They present an enhanced distribution in the whole cells compared to the parent hydrophilic glycol chitosans. *N*-octyl-*O*-sulfate chitosan can therefore be used as a potential drug carrier.[118] To develop chitosan-based efficient gene carriers, highly purified chitosan oligosaccharides were chemically modified with deoxycholic acid. Owing to the amphiphilic character, the modified oligochitosans formed self-aggregated nanoparticles in aqueous medium.[119] Their physicochemical characterization revealed that the particle size of the nanoparticles was in the range of 200–240 nm and the *cac* was 0.012–0.046 g/L, depending on the degree of substitution.

N-maleoylchitosan was able to self-assemble, at concentrations higher than the *cac*, as fiber-like aggregates with an averaged diameter of around 2.5 μm and length of more than 100 μm.[120] These fiber-like aggregates have potential application in tissue engineering scaffold.

Amphiphilic copolymers consisting of hydrophilic and hydrophobic segments can finally form micelle structures with the hydrophobic inner core and the hydrophilic outer shell in aqueous media. The hydrophobic inner core is surrounded by a hydrophilic outer shell, which provides a stabilizing interface between the micelle core and the aqueous environment. Adjusting the structure of the amphiphilic copolymers, the size and morphology of the polymeric micelles can be easily controlled. In addition, polymeric micelles were also considerably more stable than surfactant micelles. Thus, these core-shell-type micelles may be used as drug delivery vehicles for poorly water-soluble drugs, especially when the micelles are made with suitable biodegradable polymers. By reacting DL-lactide (DLLA) and water-soluble chitosan in dimethylsulfoxide solution, an amphiphilic graft copolymer was synthesized[121] and micelles in water were obtained. According to the composition of the copolymer, the critical micelle concentration (*cmc*) was between 10^{-2} and 10^{-1} g/L. These values are lower than the critical micelle concentration of low molecular weight surfactants, indicating the stability of micelles from graft copolymers at dilute conditions. The increasing hydrophobicity by introduction of a large amount of hydrophobic groups further reduces the *cmc* values. The size of polymeric micelles was measured by dynamic light scattering and it varies between 150 and 200 nm. Finally, modifying sulfated chitosan by adding long alkyl chains to the amino groups at C-2 induces the creation of an amphiphilic molecule that forms micelles in water and therefore offers the potential to increase the solubility of poorly soluble drugs. With *N*-alkyl-*O*-sulfate chitosan (the length of the alkyl chain being between 8 and 12 C atoms) polymeric micelles of size around 100–400 nm were formed in water.[118]

9.3.2.2 Interfacial Properties

Interfacial or surface properties of such materials depend on the interactions that these may build with other media or molecules. As many of interesting media may be hydrophobic (or contain hydrophobic moieties), the amphiphilic character of used derivatives is of great importance. These interactions allow adsorption of hydrophobic active matter on surfaces, the formation of stable complexes, the ability of particles to migrate through membranes, etc.

Cationic polyelectrolyte chitosan (Ch) as well as anionic carboxymethylchitin (CMCh) has no tensioactive properties. This corresponds to the known property of weakly hydrophobic polyelectrolytes to manifest poor adsorption activity at both oil–water and air–water interfaces at relatively high degree of ionization.[122,123] But the surface activity of polyelectrolytes may be modified by preparing an amphiphilic chemical (as surfactants). As an example the alkyl carboxymethylchitin presents

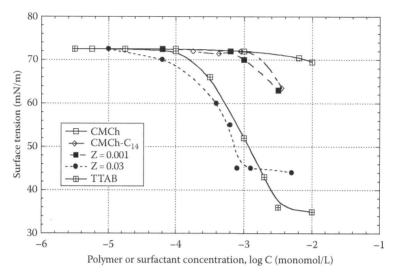

FIGURE 9.3 Surface tension isotherms of solution as a function of concentration of TTAB, chitin derivatives and SPECs (solvent: water, $T = 298\,\text{K}$).

some activity when the polymer concentration is larger than 10^{-3} monomol/L for C_{14} alkyl chain and a degree of substitution of 0.04 (CMCh-C_{14}, Figure 9.3). This critical concentration depends upon the length of the alkyl chain and the degree of alkyl substitution and this effect is larger with the hydrophobicity (e.g., longer alkyl chain for the same degree of alkylation).[124] This is not surprising taking into account that by increasing the hydrophobicity of the polysoaps their adsorption at the air–water interface is expected to be increased. Similar observations were done on (2-hydroxyl-3-dodecanoxyl)propyl-carboxymethylchitosans (HDP-CMCHS).[125] The introduction of hydrophobic substituents made the resulting derivatives become amphiphilic polymers, which decrease the surface tension. But different from the low molecule surfactants, the decrease of surface tension is slow and no clear transition points on the surface tension–concentration curve. Moreover, the addition of NaCl to the DHP-CMCHs solution can make the surface tension lower. It is because that the electrolyte can compress the electric double layer of the anionic carboxymethyl group on the hydrophilic backbone of the DHP-CMCh. So it is easy for the polymers to concentrate on the surface of the solution to reduce surface tension and to form aggregates by hydrophobic group association in solution as already observed from other amphiphilic derivatives.[102]

But the addition of a cationic surfactant (tetradecyltrimethylammonium bromide, TTAB, with a C_{14} alkyl chain as CMCh-C_{14}) able to interact with the highly charged anionic polymer causes a significant effect. With carboxymethylchitin, ion pairs are formed in the presence of TTAB by electrostatic interactions and tensioactive properties are affected (Figure 9.3). The effect is directly related to the stoichiometric ratio Z (defined as C_s/C_p, C_s being the concentration of surfactant molecules in mol/L and C_p, expressed in base-mol/L, the polymer concentration): the higher Z gives the lower surface tension at a given monomole concentration; in addition, it goes to a characteristic limit over a critical monomole concentration depending directly on Z. This quantity corresponds to a *cac* much lower than the critical micellar concentration (*cmc*) of TTAB (*cmc* ~ 4.10^{-3} mol/L). It was concluded that strong interactions were due to a mixed micelle formation including polyelectrolyte and surfactant molecules.[126–128] Comparing the isotherms $\sigma(C_p)$ for the alkylated CMCh sample with those of SPECs formed between nonalkylated CMCh and TTAB at different compositions Z of mixed solutions (Figure 9.3), one concludes that from the point of view of their effect on the surface activity the SPECs reveal to be formally much more effective than the alkylated CMCh. The SPECs formed in a mixed solution with the composition $Z \cong 0.001$ produce the same effect on the surface tension as the alkylated samples (polysoaps) with degree of alkylation $\tau = 0.04$ (for C_{14} chain), meaning that

40 times less of hydrophobic C_{14} alkyl chains is necessary to reach similar tensioactive properties. With the approximate same content of hydrophobic groups (Z = 0.03), the tensioactive properties are much more improved. This is probably related to the impossibility for the adsorbed alkyl side chains of the alkylated CMCh to form a dense packing at the interface because of the steric hindrance originating from the local stiffness of macromolecule backbone. The nonalkylated pieces of the polymeric backbone localized between the neighboring alkyl side chains "anchored" at the interface cannot form the "loops" if their average length l is not large enough. In contrast, the nonalkylated CMCh macromolecules form electrostatic dynamic associations with alkyl chains of the surfactant macromolecules, which are considered as more labile. As an expected result, more alkyl chains of bound surfactants could be anchored at the interface. This explains why the surface activity of the alkylated samples of CMCh reveals to be relatively low with regard to the nonalkylated CMCh which forms the dynamic complexes SPECs of the same expected content of alkyl chains of oppositely charged TTAB molecules. The very small value of the *cac*, compared to the *cmc* of the surfactant in pure solution, leads to the use of smaller quantities of surfactants to observe surface activity (Figure 9.3). This regime of very small addition of surfactants is the important domain for all the interfacial mechanisms in relation with the stabilization of emulsions and foams. With chitosan-SDS complexes, similar results were obtained.

Moreover, using the drop tensiometry technique,[129] the kinetics of formation of the film and the viscoelastic properties of the interfacial film may be studied. According to the hydrophobicity and the steric hindrance, the kinetics is not the same. Different observations were interpreted in terms of the formation of a gel-like structure in the adsorption layers of polysoaps and SPECs.[127,130] It has been found that the elastic modulus of the adsorbed layer increases with increasing alkyl chain content in the case of both polysoaps and SPECs, but unexpectedly, the kinetics of the adsorption and the elasticity of the SPEC adsorption layer are remarkably higher with regard to that of polysoaps. This is explained by higher diffusion mobility of alkyl groups of surfactants inside the SPEC with regard to that of polysoaps allowing the formation of more compact gel-like structure in the adsorption layers. It has been found that the conformation of the polysoap macroion and of SPEC in the bulk of solution has an impact on the kinetics of the structure formation in the adsorption layers.

Finally, wettability of surfaces was determined from static water contact angles on the films.

9.3.2.3 Interactions

Specific interactions of amphiphilic chitosan derivatives may lead to particular behavior and properties. As an example interactions between proteins (bovine serum albumin, BSA) and *N*-succinyl chitosan (NSCS) were characterized and it was demonstrated that chain entanglements are easily formed between BSA and NSCS and mainly driven by *H*-bond and hydrophobic interactions.[131] The NSCS entrapped BSA shows nanosphere morphology with a 100–200 nm diameter at high concentration of NSCS. The conformation of BSA entrapped in the matrix of NSCS does not change significantly, suggesting that NSCS is a potential matrix to load proteins or other bioactive drugs.

It may be mentioned that the association of chitosan with hydrophobic (or less hydrophilic) compounds such as fatty acid allows increase film resistance to water transmission.[132]

9.3.2.4 Swelling

Swelling of hydrogels in water and aqueous solutions is of great importance in relation with biomedical applications. It allows inclusion of active drugs and then their release considering diffusion process. The swelling behavior of a material refers to the case when water molecules enter the material and combine with the hydrophilic groups in the material molecules. Consequently, the content of hydrophilic groups and the intermolecular force are the two main factors that influence the material's swelling property. On one hand, the increase of the substitution degree lessens the intermolecular hydrogen bonds and reduces the compactness of the network. This will raise the swelling ration. On the other hand, the hydrophobicity of the network is enhanced upon increasing the amount of the hydrophobic groups. The increased hydrophobicity will result in a decrease of the swelling ratio.

When chitosan-*g*-PNIPAm hydrogels are more specially considered, below the *LCST*, the increase of either the chitosan/NIPAm weight ratio or the cross-linking agent decreased their swelling ratio. Indeed, both PNIPAm and chitosan have a hydrophilic behavior in this temperature range when for a temperature above *LCST*, PNIPAm was hydrophobic and difficult to swell in aqueous medium. So the hydrophilic behavior of chitosan dominated the swelling ratio of chitosan-*g*-PNIPAm. The increase of the chitosan/NIPAAm weight ratio increased the swelling ratio of the copolymer.[56]

9.4 BIOMEDICAL APPLICATIONS OF AMPHIPHILIC SYSTEMS

9.4.1 DRUG DELIVERY

Chitosan has interesting biopharmaceutical characteristics such as pH sensitivity, biocompatibility, and low toxicity.[133,134] Moreover, chitosan is metabolized by certain human enzymes, especially lysozyme, and is biodegradable.[133] As a consequence, the use of chitosan and its derivatives in drug delivery applications was extended. Moreover, in such applications it is also extremely important that chitosan be water-soluble and positively charged. This character enables it to interact with negatively charged polymers in an aqueous medium. But the limitation of chitosan for the preparation of sustained release systems arises for its rapid water absorption and high swelling degree in aqueous environments leading to fast drug release. In order to overcome these problems, chitosan was modified to increase its hydrophobicity. As mentioned earlier, self-assembly of hydrophobic polymers allows elaboration of materials with different morphologies.

Park et al.[135] elaborated glycol chitosan nanoparticles modified with hydrophobic bile acid analogs from self-assembly. They comprise the hydrophilic shell of glycol chitosan and hydrophobic multicores of bile acid analogs and they are promising therapeutic agents in the anticancer drug delivery system.[136,137] Varying the molecular weight of the polymer, the mean diameters of the nanoparticles (ranging from 230 to 310 nm when the molecular weight is between 20,000 and 250,000 g/mol) did not differ significantly. Furthermore, their in vitro colloidal stability and surfaces charges were similar. However, the higher molecular weight glycol chitosan particles circulate in the blood for longer periods of time and display higher tumor selectivity compared to those with lower molecular weight. This pharmacokinetic alteration in blood circulation may be associated with increased in vivo stability on nanoparticles, leading to high tumor selectivity. Particles elaborated from hydrophobically modified glycol chitosan with 5β-cholanic acid had a globular shape with an average diameter of 360 nm.[72] The internalization efficiency of *Cy5.5*-labeled nanoparticles (*Cy5.5*, a monofunctional cyanine as a carboxylic derivative being a fluorophore) in the *HeLa* cell cultures was higher than that of the glycolchitosan polymer analog or the control *Cy5.5*. The higher binding affinity of the hydrophobically modified glycol chitosan nanoparticles at the cellular membrane may account for their facile internalization. These nanoparticles possessing tunable physicochemical properties, low toxicity, and biocompatibility and exhibiting a fast cellular uptake through various routes are promising macromolecular carriers. *N*-succinyl chitosan was demonstrated to have potential applications for encapsulation of protein and other hydrophilic bioactive drugs.[131] Chitosan-*g*-PPDO (poly(*p*-dioxanone)) copolymer was synthesized by ring-opening polymerization of *p*-dioxanone.[68] The incorporation of PPDO into chitosan would lead to the controllability of physical properties by controlling the chemical structure of the graft copolymer. This copolymer was used as Ibuprofen carrier. The higher the degree of polymerization (DP) of PPDO grafts, the lower the cumulative release percentage, and it was observed that its release rate decreased compared with that of pure chitosan carrier. Those phenomena indicated that the release of ibuprofen was controlled by the diffusion and that the hydrophobicity of the carrier was crucial. The higher DP value made the copolymer more hydrophobic. The strong hydrophobicity of the carrier made water penetration difficult, and the drug release slowed. *N*-lauryl carboxymethylchitosan with both hydrophobic and hydrophilic groups was studied in connection with the delivery of taxol to cancerous tissues.[59] Chitosan-based systems bearing β-cyclodextrin cavities were considered as a matrix for controlled release.[138,139] Due to the presence of the hydrophobic cyclodextrin rings, these systems provide a slower release of the entrapped hydrophobic drugs.

Chitosan was modified with alkyl halides bearing different hydrophobic alkyl side chain[140] and delivery of vitamin B$_2$ was considered.[141] Liu et al.[142] examined the use of *N*-(aminoalkyl)chitosan microcapsules as a carrier for gene delivery. It was reported that *N*-(aminoalkyl)chitosan-modified poly(lactic acid) films (PDLLA) improved the biocompatibility of PDLLA.[143] Li et al.[144] prepared *N*-butylchitosan and it was demonstrated that the mechanical properties may be strengthened and the biocompatibility of materials with osteoblasts improved, leading this material to be a promising candidate for bone regeneration. Using the same chemical reaction (reductive amination), Taboada et al.[145] prepared new arylamine chitosan derivatives for trapping proteins.

Other systems were considered as non-covalent gels (or physical gels) based on chitosan; for example, the amphiphilic vesicle formed from palmitoyl glycol chitosan.[146] Varying the hydrophobicity of the gel material leads to variations of gel hydration, gel erosion, bioadhesion, and the release of entrapped macromolecules. Chitosan derivatives are ideal candidates materials for use in the fabrication of buccal drug delivery dosage forms. Indeed, polycations (such as chitosan) are reported to increase epithelial permeability by enhancing paracellular transport.[147,148] Systems combining chitosan and PNIPAm have drug release profiles which can be controlled by both pH and temperature,[149,150] constituting very promising materials.

Polyelectrolyte complexes (PEC) films prepared from chitosan and polyalkyleneoxide-maleic acid copolymer (PAOMA) were shown to change drug release rate in response to change in environmental pH.[78] The drug release was higher in pH 7.2 media compared to pH 3.8 because of the increase of repulsive force between carboxyl groups in PAOMA and anionic groups in model drugs (salicylic acid) and the increase of effective diffusion area. Temperature-sensitive drug release studies showed that the increase in temperature from 25°C to 50°C yielded the increase in the rate of drug release when using PAOMA with molecular weight of 20,000 and a cloud point equal to 35°C. This is attributed to the increase of release area due to the phase transition of PAOMA and the increase of repulsive forces between carboxyl groups in PAOMA and anionic groups in model drugs.

N-alkyl-*O*-sulfate chitosan derivatives carrying long alkyl chains ($n = 8$, 10, or 12) as hydrophobic moieties and sulfated groups as hydrophilic ones were used to elaborate polymeric micelles.[118,151] Taxol, a water-insoluble anticancer drug, was solubilized into the polymeric micelles by physical entrapment. The taxol concentration in the docyl derivative micellar solution was around 2 mg/mL, which is much higher than in water (<0.001 mg/mL). Another water-insoluble drug, Paclitaxel, was successfully tested.[151] As a consequence, *N*-octyl-*O*-sulfate chitosan may be used as a potential drug carrier. According to the nature of the drug to be released, it is obvious that the nature of the chitosan derivative can alter its profile release. By tailoring the properties of the polymer (hydrophilic/hydrophobic in our context), the desired drug release rate can be obtained. As an example, for bone implant and calcein release, a higher percentage of drug was released when the hydrophilic polymer *N*-octyl-sulfated chitosan was present in tablets compared with the tablets containing the hydrophobic polymer *N*-octyl chitosan.[152] Through grafting DL-lactide onto chitosan, leading to amphiphilic chitosan-polylactide graft copolymers, polymeric micelles were prepared.[121] As these micelles have a hydrophobic core, they are used as delivery carrier for the entrapment and controlled release of hydrophobic drugs.

Chitosan was hydrophobically modified by cholesterol to obtain amphiphilic cholesterol succinyl chitosan conjugates, which are used as the anchoring materials to coat on the liposome surface.[153] Indeed, chitosan coated liposomes (CCL) were hoped to be used as the carrier of anticancer drugs, but the chitosan layer formed by the electrostatic attractive interaction between the positively charged amino groups in chitosan molecules and the negatively charged surface of liposomes was not effective enough to significantly stabilize liposomes in vitro. To improve this stability hydrophobically modified chitosan derivatives (such as cholesterol succinyl or palmitoyl conjugates) were used as anchoring materials. Chitosan anchored liposomes (CAL) demonstrate larger sizes, higher zeta potential, better physical stability, and more sustained drug release behavior. Epoxypseudoisoeugenol-2-methyl butyrate (EPB) was chosen as a model anticancer drug to assess the potential of chitosan derivative anchored liposome as a novel carrier of anticancer drugs. Current results showed that cholesterol succinyl chitosan anchoring on the surface of EPB-loaded liposomes could significantly improve their physical stability and sustain the release of EPB in vitro.

Hydroxypropylchitosan and dihydroxypropylchitin[154] were considered for release of oxytetracycline and it may be adjusted according to the nature of the chitin derivative.

Sui et al. have synthesized amphiphilic derivatives of succinyl-chitosan ((2-hydroxypropyl-3-butoxy) propyl-succinyl chitosan).[155] It can concentrate on surface to decrease the surface tension and can associate with hydrophobic chains to form aggregates in solution. These derivatives may be used in controlled release of hydrophobic drug.

As already discussed SPECs may be obtained from ionic derivatives of chitin or chitosan and oppositely charged surfactant. The physical gels based on SPECs may be produced in the form of microbeads and capsules by the diffusion enhanced gelation.[156,157] When a droplet of an aqueous chitosan solution containing bioactive substance (as enzymes) falls into a receiving aqueous solution of oppositely charged surfactants (or polyelectrolytes), the gel-like wall of SPECs forms at the interface, whose thickness grows with time and finally the droplet transformed into a microcapsule. The gel-like wall of these capsules contains micelle-like aggregates with highly ordered nanostructure evidenced by small-angle X-ray scattering.[158] These capsules and microcapsules made by this method are polyfunctional: they may be loaded by enzymes in their aqueous core and contain in their gel-like walls the micelle-like aggregates able to be solubilized by the oil-soluble drugs. A preliminary study on the immobilization of enzymes (α-chymotrypsin, urease, etc.) in capsules made of chitosan-SDS complexes[159] has shown that these enzymes conserve their activity after encapsulation in spite of the possible contact with detergent molecules inside the gel.

9.4.2 Gene Transfection

Gene therapy is to introduce therapeutic genes into specific cells for the purpose of correcting the dysfunction of specific genes or over-expressing therapeutically useful proteins to take curative effects for human diseases or to provide immunologic protection. Potential clinical application of gene therapy depends greatly on the development of safe and efficient delivery systems, to which the most distinct drawback is the lack of effective gene ferrying vectors. Non-viral vectors have been attracting increasing research efforts owing to the consideration of safety and feasibility for manufacturing. Cationic polymers are among the most attractive non-viral gene carriers, which include PEI, polyallylamine, poly-L-lysine (PLL), and chitosan.[160] The advantage of chitosan-based vectors lies not only in avoiding cytotoxicity problems that are inherent in most synthetic polymeric vehicles but also in their unique capability for transcellular transport. The oligochitosans are water soluble in a wide range of pH and are still able to condense DNA. Nevertheless, recent reported data[161,162] have shown a rather limited transfection efficiency of oligochitosans in vitro as compared to cationic surfactants. These observations have been made also with quaternized forms of these oligomers where quaternary amino groups have been added in order to increase both their solubility and their electrostatic interaction with oppositely charged DNAs.[163] This rather low transfection efficiency is assumed to be due to low hydrophobicity of condensing agents, the oligochitosans, which produce the electrostatic condensation of DNA into the relatively bulky coil-like complexes. The higher transfection efficiency is attributed to increased entry into cells facilitated by hydrophobic interactions and easier unpacking of DNA from alkylated chitosans due to the weakening of electrostatic attractions between DNA and alkylated derivatives.

Indeed, the factors that are supposed to play an important role in the transfection efficiency of cationic surfactants are, primarily, the formation by these surfactants of micelle-like aggregates via hydrophobic interaction between the alkyl chains, and, secondly, the interaction of alkyl chains with the bilayer membranes of living cells by insertion of alkyl chains into the membrane hydrophobic core. The first factor, the hydrophobic attraction between the alkyl chains of cationic surfactants electrostatically bound to the oppositely charged DNA macroions, induces a transition (a collapse) from a coil-like conformation of DNA-surfactant complexes into smaller size and stable globule-like aggregates.[164] Smaller is the size of the complexes, higher is their transfection efficiency. Unlike cationic surfactants, cationic oligochitosans are giving only bulky coil-like electrostatically stabilized complexes with DNA macroions, with low transfection efficiency. The second factor, the hydrophobic interaction with the membrane bilayer, allows the absorption of the DNA-surfactant complexes onto the living cell surface by insertion of the alkyl chains of surfactants into the membrane core. This hydrophobic attachment is assumed to increase the ability of the DNA to penetrate into the cytoplasm. Unlike the surfactants, the oligochitosans whose electric charge is already compensated within the DNA-oligochitosan complexes are not able to interact with the negatively charged membranes of living cells even by electrostatic attraction.

Thus, covalent grafting of alkyl chains to oligochitosans is believed to increase, on one hand, their ability to form stable and small-size complexes with DNA and, on the other hand, the adhesion of DNA-oligochitosan complexes to the living cell surface. At the same time, the alkylation of oligochitosans is expected to decrease their solubility and their instability toward the hydrophobic aggregation in water. For example, as it has been reported earlier,[165,166] the alkylation of high molar mass chitosans decreases their solubility in aqueous media and consequently leads to the formation of complexes of larger size including ~50–100 DNA molecules per complex. Assuming that the solubility of the alkylated chitosans can be improved by the decrease of their molecular weight, a series of hydrophobically modified oligochitosans with different degree of covalent substitution of C12-chains in the form of *N-/2(3)-*(dodec-2-enyl)succinoyl groups were prepared[167] using a process already described by Hirano et al.[168]

Chitosan oligosaccharides were chemically modified with deoxycholic acid (DOCA).[119] They are able to form core-shell-type nanoparticles in aqueous medium. Compared to unmodified chitosan oligosaccharides, the hydrophobic derivative based nanoparticles showed elevated potential as gene carriers by efficient DNA condensation, and protection of the condensed DNA from DNase I attack. Based on a chitosan oligosaccharide of molecular weight in the range 1000–3000, more than 20 and 100 times enhanced gene transfection was observed by the particle mediated gene delivery than that of PLL in the absence and presence of FBS, respectively. Although the slightly lower pH value than physiological value for the efficient transfection may prevent systemic applications, hydrophobically modified chitosan oligosaccharides-mediated gene transfections can be a potential methodology for local gene deliveries such as mucosal and intestinal gene delivery. The self-aggregate DNA complex from deoxycholic acid-modified chitosan was shown to enhance the transfection efficiency over monkey kidney cells.[169] Self-aggregates can form charge complexes when mixed with plasmid DNA. These self-aggregate DNA complexes are considered to be useful for transfer of genes into mammalian cells in vitro and served as good delivery system composed of biodegradable polymeric systems. PEGylation of chitosan in order to increase its solubility, elongate the plasma circulation time, and prolong the gene transfer has been another proposed technique for the sustained DNA release. Microspheres physically combining PEG-grafted chitosan (PEG-*g*-Chi) with poly(lactide-*co*-glycolide) (PLGA) were formulated by Yun et al.[170] They reported that these microspheres were capable of sustained release of PEG-*g*-Chi/DNA for at least 9 weeks, and the rate of DNA release was not modulated by varying the amount of PEG-*g*-Chi.

The interaction of LMW-DDC-chitosans with model lipid membranes has been tested. The decrease of membrane fluidity (increase of membrane microviscosity) in the liquid crystalline phase induced by the interaction with LMW-DDC(18%)-chitosan molecules may be explained by the insertion of the dodecyl chains covalently bound to the glucosamine backbone into the membrane bilayer along to the fatty acid chains. Unlike the low molecular weight cationic detergents which exhibit some toxicity because of their high lateral mobility at the interface and their flip-flop ability between the two membrane leaflets, the LMW-DDC-chitosans are obviously much less prone to move laterally and across the bilayer, which probably explains their low toxicity.[82]

Cationic methoxy poly(ethyleneglycol) (mPEG)-polyethylenimine (PE)-chitosan was synthesized[77] and it showed a good DNA condensation capability, exhibiting negligible toxicity. The transfection of human embryonic kidney 293 (HEK 293) cells proved that mPEG-PEI-chitosan/VRaft-1 plasmid has little toxicity on the growth and gene expression of cells. Using this copolymer as transfection agent, the conversion of ω-6 to ω-3 fatty acids was successfully realized. Based on these superior properties, mPEG-PEI-chitosan should be a promising non-viral gene carrier for biomedical and biological applications.

Self-assembled nanoparticles were prepared using a hydrophobically modified glycol chitosan obtained by the reaction of the amino group with 5β-cholanic acid.[117] The modified chitosan spontaneously formed DNA nanoparticles by a hydrophobic interaction between hydrophobically modified glycolchitosan (HGC) and hydrophobized DNA. As the HGC content increased, the encapsulation efficiencies of DNA increased while the size of HGC nanoparticles decreased. Upon increasing HGC contents nanoparticles became less cytotoxic. The increased HGC content also facilitated endocytic uptakes of HGC nanoparticles by COS-1 cells. Moreover, the HGC nanoparticles showed increasing in vitro transfection efficiencies in the presence of serum. In vivo results also showed that the HGC particles have superior transfection efficiencies to naked DNA. Animal studies confirmed that HGC nanoparticles could be used as a potent gene delivery vehicle in vivo.

Low molecular weight chitosans grafted with N-/2(3)-(dodec-2-enyl)succinoyl groups with an average molecular weight of 5000 were considered as a new class of non-viral vectors for gene therapy.[82] These molecules were found to form micelles through hydrophobic interactions involving their tetradecenoyl chains and non-protonated glucosamine units. Interaction with large unilamellar vesicles taken as model membranes indicated that these chitosan derivatives interact mainly with vesicles mimicking the inner leaflet of biomembranes both through electrostatic and hydrophobic interactions. This preferential interaction may destabilize endosomal membranes and favor the DNA release into the cytoplasm in gene delivery applications. Moreover, since this interaction significantly decreased the membrane fluidity of these vesicles, the hydrophobically modified chitosans are thought to exhibit limited lateral mobility and flip-flop ability, and thus limited cytotoxicity.

Other derivatives such as N-methylenephosphonic chitosan (NMPCS) were synthesized and investigated as non-viral vector.[171] NMPCS/DNA or chitosan (CS)/DNA complexes were prepared using a complex coacervation method. MTT assay was employed to evaluate the cytotoxicity of the polymers and pGL3-control luciferase plasmid was utilized as a reporter gene to assess the transgenic efficacy of the polymers.[172] It was demonstrated that NMPCS was able to fully entrap the DNA at N/P ratio of 2:1, whereas chitosan entrapped the DNA completely at N/P ratio of 1:1. The expression of transgene mediated by NMPCS was much higher (more than 100-fold) than that mediated by chitosan, indicating that NMPCS was more efficacious gene ferrying vector than chitosan.

9.4.3 Tissue Engineering

Tissue engineering research is nowadays based on the seeding of cells onto porous biodegradable polymer matrixes. Chitosan and its derivatives have been reported as attractive candidates for scaffolding materials because they degrade as the new tissues are formed, eventually without inflammatory reactions or toxic degradation.[169,173] In tissue engineering applications, the cationic nature of chitosan is responsible for electrostatic interactions with anionic glycosylaminoglycans, proteoglycans, and other negatively charged molecules. Mixing biological properties of chitosan and mechanical properties of L-poly(lactic acid) (PLLA) may be of promising interest. Surface functionalization of biodegradable PLLA was performed by plasma coupling reaction of chitosan with PLLA.[174] The proliferation and morphology studies of ell lines (mouse fibroblasts and human hepatocytes) cultured on this surface showed that these cells hardly spread and tended to become round, but could proliferate at almost the same speed as cells cultured on glass surface. More recently, PLLA-chitosan hybrid scaffolds were proposed as tissue engineering scaffolds and simultaneously drug release carriers.[175] In this system, a chitosan porous structure, in which cells and tissues would mostly interact, is created within the pore structure of a stiffer PLLA scaffold.

N-maleoylchitosan was also considered for such applications.[121] The sample with a maleoyl substitution degree of 70% was synthesized from the reaction of maleic anhydride and the amino group of chitosan. With increasing concentration to around 0.5 mg/mL, which is 10 times the *cac*, these molecules self-assemble to fiber-like aggregates. The nanofibrous structure facilitates the attachment of human osteoblasts and chondrocytes. Therefore, these fiber-like aggregates might have potential applications in tissue engineering scaffold.

By grafting ε-caprolactone oligomers onto the hydroxyl groups of chitosan, chitosan-graft-polycaprolactone (PCL) copolymers were achieved.[69] The specimen synthesized from the feed molar ratio of glucosamine units in chitosan versus ε-caprolactone as 1:12 showed equivalent cytotoxicity to the neat chitosan and PCL against KB (human carcinoma) cell line, and the cell viability was almost close to 100%. In addition, chitosan-*g*-PCL exhibited good solubility in organic solvents, facilitating the formation of PCL/chitosan-*g*-PCL blend nanofibers via electrospinning with the use of DMF/CHCl$_3$ as solvent. Owing to the enhanced cellular attachment from cationic amino groups, it is promising that these copolymers are interesting substances for developing drug carriers and tissue engineering scaffolds.

9.4.4 CELLS AND PROTEINS ADHESION AND OTHER APPLICATIONS

Surface modification of biomaterials was significantly studied since it is the surface of such materials that first comes into contact with the biological surrounding. The change of surface property was found to affect adsorption of platelets, cells, and biomacromolecules such as proteins on the polymer surface. Numerous techniques were used to alter the chemical composition and thus the surface property of the materials (plasma treatment...). Grafting of polymer or oligomer species was considered and especially of monomethoxyethylene glycol oligomers on chitosan surface.[176] The hydrophobicity of the modified surface, determined by air–water contact angle, decreased when the ethyleneglycol derivatives were grafted on the film. The presence of such oligomers reduced the adsorption of proteins (albumin and lysozyme) on the films. PEG was recognized to suppress non-specific protein adsorption. Grafting was obtained from similar reaction than the elaboration of alkylated derivative (reductive amination). The choice of the chemical compound attached to the chitosan can affect its response to protein adsorption: hydrophobic (*N*-stearoyl chitosan) surface enhances protein adsorption when ethyleneglycol oligomer could reduce the protein adsorption.

A series of temperature-sensitive poly(NIPAm-*co*-CSA) hydrogels were synthesized by the copolymerization of acrylic acid derivatized chitosan (CSA) and *N*-isopropylacrylamide (NIPAm) in aqueous solution.[177] The investigation of L929 cell adhesion and detachment of poly(NIPAm-*co*-CSA) hydrogels indicated the cell adhesion and spreading was higher on the surface of these hydrogels than that of PNIPAm hydrogel at 37°C due to the incorporation of chitosan, which have excellent cell affinity. They showed more rapid detachment of cell sheet compared to PNIPAm hydrogel because of the highly hydrophilic and hygroscopic nature of chitosan chains when reducing the culture temperature from 37°C to 20°C.

The permeability of chitosan membranes grafted with hydroxymethylmethacrylate may be controlled by means of a plasma treatment for applications in dialysis domain.[178]

9.5 CONCLUDING REMARKS

Due to wide availability and their specific properties (biocompatibility, biodegradability, etc.) chitin, chitosan, and their amphiphilic-based systems have found applications in a large variety of areas of biomaterials. They are elaborated using various processes either from the chemical modification of the macromolecular chain or from specific interactions with surfactants or oppositely charged polyelectrolytes. Physicochemical properties (rheological, interfacial, and swelling) allow a great use of such systems as biomaterials. As a consequence, the domains of applications are wider and wider considering among them drug delivery, gene transfection, tissue engineering, and cell or protein adhesion. Further studies and development of such amphiphilic systems for new biomedical applications can be expected in the next century.

REFERENCES

1. Braconnot, H. Sur la nature des champignons, *Annales Chimie Physique*, 79 (1811): 265–304.
2. Odier, A. Mémoire sur la composition chimique des parties cornées des insectes, *Mémoires de la Société d'Histoire Naturelle*, 1 (1823): 29–42.
3. Cauchie, H.M. An attempt to estimate crustacean chitin production in the hydrosphere, in *Advances in Chitin science*, Domard, A., Roberts, G.A.F., Varum, K.M. (Eds.), Jacques Andre Publishers, Lyon, France, Vol. 2, 1998, pp. 32–39.
4. Peter, M.G. Chitin and chitosan in fungi, in *Polysaccharides II: Polysaccharides from Eukaryotes*, Vol. 6: Biopolymers, Steinbüchel, A. (Ed.), Wiley-VCH, Weinheim, Germany, 2002, pp. 123–157.
5. Kumar, M.N.V.R. A review of chitin and chitosan applications, *Reactive and Functional Polymers*, 46 (2000): 1–27.
6. Sashiwa, H., Saimoto, H., Shigemasa, Y., Ogawa, R., Tokura, S. Lysozyme susceptibility of partially deacetylated chitin, *International Journal of Biological Macromolecules*, 12 (1990): 295–296.
7. Shigemasa, Y., Saito, K., Sashiwa, H., Saimoto, H. Enzymatic degradation of chitins and partially deacetylated chitins, *International Journal of Biological Macromolecules*, 16 (1994): 43–49.
8. Nishimura, K., Nishimura, S., Nishi, N. Immunological activity of chitin and its derivatives, *Vaccine*, 2 (1984): 93–99.
9. Mori, T., Okumura, M., Matsuura, M., Ueno, K., Tokura, S., Okamoto, Y., Minami, S., Fujinaga, T. Effects of chitin and its derivatives on the proliferation and cytokine production of fibroblasts *in vitro*, *Biomaterials*, 18 (1997): 947–951.
10. Tokura, S., Ueno, K., Miyazaki, S., Nishi, N. Molecular weight dependent antimicrobial activity by chitosan, *Macromolecular Symposia*, 120 (1997): 1–9.
11. Tanigawa, T., Tanaka, Y., Sashiwa, H., Saimoto, H., Shigemasa, Y. Advances in chitin and chitosan, Brine, C.J., Sandford, P.A., Zikakis, J.P. (Eds.), in *Proceedings from the 5th International Conference on Chitin and Chitosan*, London, U.K., Elsevier, the Netherlands, 1992, pp. 206–215.
12. Okamoto, Y., Minami, S., Matsuhashi, A., Sashiwa, H., Saimoto, H., Shigesama, Y., Tanigawa, T., Tanaka, Y., Tokura, S. Polymeric N-acetyl-D-glucosamine (chitin) induces histrionic activation in dogs, *Journal of Veterinary Medical Science*, 55 (1993): 739–742.
13. Kweon, D.K., Song, S.B., Park, Y.Y. Preparation of water-soluble chitosan/heparin complex and its application as wound healing accelerator, *Biomaterials*, 24 (2003): 1595–1601.
14. Khor, E., Lim, L. Implantable applications of chitin and chitosan, *Biomaterials*, 24 (2003): 2339–2349.
15. Ramos, V.M., Rodriguez, N.M., Rodriguez, M.S., Heras, A., Agullo, E. Modified chitosan carrying phosphonic and alkyl groups, *Carbohydrate Polymers*, 51 (2003): 425–429.
16. Peter, M. Applications and environmental aspects of chitin and chitosan, *Journal of Macromolecular Science Part A Pure and Applied Chemistry*, A32 (1995): 629–640.
17. Muzzarelli, R.A.A. Chitosan-based dietary foods, *Carbohydrate Polymers*, 29 (1996): 309–316.
18. Fädlt, P., Bergenstahl, B., Claesson, P.M. Stabilization by chitosan of soybean oil emulsions coated with phospholipid and glycocholic acid, *Colloids and Surfaces A: Physicochemical and Engineering Aspects*, 71 (1993): 187–195.
19. Lindman, B., Carlsson, A., Gerdes, S., Karlström, G., Piculell, L., Thalberg, K., Zhang, K. *Food Colloids and Polymers: Stability and Mechanical Properties*, Dickinson, E., Walstra, P. (Ed.), Royal Society of Chemistry, Cambridge, U.K., 1993, pp. 113–125.
20. Baschong, W., Hüglin, D., Maier, T., Kulik, E. Influence of N-derivatization of chitosan to ist cosmetic activities, *SÖFW Journal*, 125 (1999): 22–24.
21. Muzzarelli, R.A.A. *Natural Chelating Polymers*: *Alginic Acid, Chitin and Chitosan,* International series of monographs in analytical chemistry, Belcher, R. (Ed.), Pergamon Press, Oxford, U.K., Vol. 144, 1973.
22. Romoren, K., Thu, B.J., Evensen, O. Immersion delivery of plasmid DNA. II. A study of the potentials of a chitosan based delivery system in rainbow trout (*Oncorhynchus mykiss*) fry, *Journal of Controlled Release*, 85 (2002): 215–225.

23. Borchard, G. Chitosans for gene delivery, *Advances in Drug Delivery Reviews*, 52 (2001): 145–150.

24. Rolland, A.P. From genes to gene medicines. Recent advances in nonviral gene delivery, *Critical Reviews in Therapeutic Drug Carrier Systems*, 15 (1998): 143–198.

25. Sato, T., Ishii, T., Okahata, Y. *In vitro* gene delivery mediated by chitosan. Effect of pH, serum and molecular mass of chitosan on the transfer efficiency, *Biomaterials*, 22 (2001): 2075–2080.

26. Venkatesh, S., Smith, T.J. Chitosan-membrane interactions and their probable role in chitosan-mediated transfection, *Biotechnology and Applied Biochemistry*, 27 (1998): 265–267.

27. Cui, Z., Mumper, R.J. Chitosan-based nanoparticles for topical genetic immunization, *Journal of Controlled Release*, 75 (2001): 409–419.

28. Illum, L., Jabbal-Gill, I., Hinchcliffe, M., Fisher, A.N., Davis, S.S. Chitosan as a novel nasal delivery system for vaccines, *Advanced Drug Delivery Reviews*, 51 (2001): 81–96.

29. Hejazi, R., Amiji, M. Chitosan-based gastrointestinal delivery systems, *Journal of Controlled Release*, 89 (2003): 151–165.

30. Fang, N., Chan, V., Mao, H.Q., Leong, K.W. Interactions of phospholipid bilayer with chitosan: Effect of molecular weight and pH, *Biomacromolecules*, 2 (2001): 1161–1168.

31. Mansouri, S., Lavigne, P., Corsi, K., Benderdour, M., Beaumont, E., Fernandes, J.C. Chitosan-DNA nanoparticles as non-viral vectors in gene therapy: Strategies to improve transfection efficacy, *European Journal of Pharmaceutics and Biopharmaceutics*, 57 (2004): 1–8.

32. Corsi, K., Chella, F., Yahia, L., Fernandes, J.C. Mesenchymal stem cells, MG63 and HEK293 transfection using chitosan-DNA nanoparticles, *Biomaterials*, 24 (2003): 1255–1264.

33. Erbacher, P., Zou, S., Bettinger, T., Steffan, A.M., Remy, J.S. Chitosan-based vector/DNA complexes for gene delivery: Biophysical characteristics and transfection ability, *Pharmaceutical Research*, 15 (1998): 1332–1339.

34. Sashiwa, H., Aiba, S.I. Chemically modified chitin and chitosan as biomaterials, *Progress in Polymer Science*, 29 (2004): 887–908.

35. Wang, K.T., Iliopoulos, I., Audebert, R. Viscometric behaviour of hydrophobically modified poly(sodium acrylate), *Polymer Bulletin*, 20 (1988): 577–582.

36. Bock, J., Pace, S.J., Schulz, D.N. Enhanced oil recovery with hydrophobically associating polymers containing N-vinyl-pyrrolidone functionality, U.S. Patent 4, 709, 759, 1987.

37. Hill, A., Candau, F., Selb, J. Properties of hydrophobically associating polyacrylamides: Influence of the method of synthesis, *Macromolecules*, 26 (1993): 4521–4532.

38. Kaczmarski, J.P., Glass, J.E. Synthesis and solution properties of hydrophobically-modified ethoxylated urethanes with variable oxyethylene spacer lengths, *Macromolecules*, 26 (1993): 5149–5152.

39. Wesslen, B., Wesslen, K.B. Preparation and properties of some water-soluble, comb-shaped, amphiphilic polymers, *Journal of Polymer Science Part A: Polymer Chemistry*, 27 (1989): 39153926.

40. Yahya, G.O., Hamad, E.Z. Solution behaviour of sodium maleate/1-alkene copolymers, *Polymer*, 36 (1995): 3705–3710.

41. Yahya, G.O., Asrof Ali, S.K., Al-Naafa, M.A., Hamad, E.Z. Preparation and viscosity behavior of hydrophobically modified poly(vinyl alcohol) (PVA), *Journal of Applied Polymer Science*, 57 (1995): 343–352.

42. Shulz, D.N., Kaladas, J.J., Maurer, J.J., Bock, H., Pace, S.J., Shulz, W.W. Copolymers of acrylamide and surfactant macromonomers: Synthesis and solution properties, *Polymer*, 28 (1987): 2110–2115.

43. Valint, P.L., Jr., Bock, J. Synthesis and characterization of hydrophobically associating block polymers, *Macromolecules*, 21 (1988): 175–179.

44. Chang, Y.H., Mc Cormick, C.L. Water-soluble copolymers. 49. Effect of the distribution of the hydrophobic cationic monomer dimethyldodecyl(2-acrylamidoethyl)ammonium bromide on the solution behavior of associating acrylamide copolymers, *Macromolecules*, 26 (1993): 6121–6126.

45. Shaw, K.G., Leipold, D.P. New cellulosic polymers for rheology control of latex paints, *Journal of Coatings Technology*, 57 (1985): 63–72.

46. Branham, K.D., Snowdey, H.S., Mc Cormick, C.L. Water-soluble copolymers. 64. Effects of pH and composition on associative properties of amphiphilic acrylamide/acrylic acid terpolymers, *Macromolecules*, 29 (1996): 254–262.

47. Taylor, K.C., Nasr-El-Din, H.A. Water-soluble hydrophobically associating polymers for improved oil recovery: A literature review, *Journal of Petroleum Science and Engineering*, 19 (1998): 265–280.

48. Durand, A., Marie, E., Rotureau, E., Leonard, M., Dellacherie, E. Amphiphilic polysaccharides: Useful tools for the preparation of nanoparticles with controlled surface characteristics, *Langmuir*, 20 (2004): 6956–6963.

49. Rotureau, E., Leonard, M., Dellacherie, E., Durand, A. Amphiphilic derivatives of dextran: Adsorption at air/water and oil/water interfaces, *Journal of Colloid and Interface Science*, 279 (2004): 68–77.

50. Langevin, D. Polyelectrolyte and surfactant mixed solutions. Behavior at surfaces and in thin films, *Advances in Colloid and Interface Science*, 89–90 (2001): 467–484.

51. Beltran, C.M., Guillot, S., Langevin, D. Stratification phenomena in thin liquid films containing polyelectrolytes and stabilized by ionic surfactants, *Macromolecules*, 36 (2003): 8506–8512.

52. Guillot, S., McLoughlin, D., Jain, N., Delsanti, M., Langevin, D. Polyelectrolyte-surfactant complexes at interfaces and in bulk, *Journal of Physics: Condensed Matter*, 15 (2003): S219–S224.

53. Alves, N.M., Mano, J.F., Chitosan derivatives obtained by chemical modifications for biomedical and environmental applications, *International Journal of Biological Macromolecules*, 43 (2008): 401–414.

54. Hara, M. *Polyelectrolytes*. Marcel Dekker, New York, 1993.

55. Magny, B., Iliopoulos, I., Audebert, R., Aggregation of hydrophobically modified polyelectrolytes in dilute solution: ionic strength effects in *Macromolecular Complexes in Chemistry and Biology*, Dubin, P., Bock, J., Davies, R.M., Schulz, D.N., Thies, C. (Edts), Springer, Berlin, Germany, 1994, pp. 51–62.

56. Prabaharan, M., Mano, J.F. Stimuli-responsive hydrogels based on polysaccharides incorporated with thermo-responsive polymers as novel biomaterials, *Macromolecular Bioscience*, 6 (2006): 991–1008.

57. Holme, K., Hall, L. Chitosan derivatives bearing C10-alkyl glycoside branches: A temperature-induced gelling polysaccharide, *Macromolecules*, 24 (1991): 3828–3833.

58. Nishimura, S., Miura, Y., Ren, L., Sato, M., Yamagashi, A., Nishi, N., Tokura, S., Kurita, K., Ishii, S. An efficient method for the syntheses of novel amphiphilic polysaccharides by regio- and thermoselective modifications of chitosan, *Chemistry Letters*, 22 (1993): 1623–1626.

59. Yoshioka, H., Nonaka, K., Fukuda, K., Kazama, S. Chitosan-derived polymer-surfactants and their micellar properties, *Bioscience Biotechnology and Biochemistry*, 59 (1995): 1901–1904.

60. Mourya, V.K., Inamdar, N.N., Chitosan-modifications and applications: Opportunities galore, *Reactive and Functional Polymers*, 68 (2008): 1013–1051.

61. Yalpani, M., Hall, L.D. Some chemical and analytical aspects of polysaccharide modifications. 3. Formation of branched-chain, soluble chitosan derivatives, *Macromolecules*, 17 (1984): 272–281.

62. Desbrieres, J., Martinez, C., Rinaudo, M. Hydrophobic derivatives of chitosan: Characterization and rheological behaviour, *International Journal of Biological Macromolecules*, 19 (1996): 21–28.

63. Lane, C.F. Sodium cyanoborohydride—A highly selective reducing agent for organic functional groups, *Synthesis*, 3 (1975): 135–146.

64. Auzely-Velty, R., Rinaudo, M. Chitosan derivatives bearing pendant cyclodextrin cavities: Synthesis and inclusion performance, *Macromolecules*, 34 (2001): 3574–2580.

65. Buschmann, H.J., Knittel, D., Schollmeyer, E. New textile applications of cyclodextrins, *Journal of Inclusion Phenomena*, 40 (2001): 169–172.

66. Lainé, V., Sarguet, C.A., Gadelle, A., Defaye, J., Perly, B., Pilard, D.F. Inclusion and solubilization properties of 6-S-glycosyl-6-thio derivatives of β-cyclodextrin, *Journal of Chemical Society, Perkin Transactions 2*, 2 (1995): 1479–1487.

67. Uekama, K. Recent aspects of pharmaceutical application of cyclodextrins, *Journal of Inclusion Phenomena*, 44 (2002): 3–7.

68. Liu, G.-Y., Zhai, Y.-L., Wang, X.-L., Wang, W.-T., Pan, Y.-B., Dong, X.-T., Wang, Y.-Z. Preparation, characterization, and in vitro release behavior of biodegradable chitosan-graft-poly(1,4-dioxan-2-one) copolymer, *Carbohydrate Polymers*, 74 (2008): 862–867.

69. Duan, K., Chen, H., Huang, J., Yu, J., Liu, S., Wang, D., Li, Y. One-step synthesis of amino-reserved chitosan-graft-polycaprolactone as a promising substance of biomaterial, *Carbohydrate Polymers*, 80 (2010): 499–504.

70. Yu, L., He, Y., Bin, L., Yue, F. Study of radiation-induced graft copolymerization of butyl acrylate onto chitosan in acetic acid aqueous solution, *Journal of Applied Polymer Science*, 90 (2003): 2855–2860.

71. Singh, V., Triparthi, D.N., Tiwari, A., Sanghi, R., Microwave promoted synthesis of chitosan-graft-poly(acrylonitrile), *Journal of Applied Polymer Science*, 95 (2005): 820–825.

72. Nam, H.Y., Kwon, S.M., Chung, H., Lee, S.-Y., Kwon, S.-H., Jeon, H., Kim et al. Cellular uptake mechanism and intracellular fate of hydrophobically modified glycol chitosan particles, *Journal of Controlled Release*, 135 (2009): 259–267.

73. Chen, T., Kumar, G., Harris, M.T., Smith, P.J., Payne, G.F. Enzymatic grafting of hexyloxyphenol onto chitosan to alter surface and rheological properties, *Biotechnology and Bioengineering*, 70 (2000): 564–573.

74. Muzzarelli, C., Muzzarelli, R.A.A. Reactivity of quinones towards chitosans, *Trends in Glycoscience and Glycotechnology*, 14 (2002): 223–229.

75. Bordenave, N., Grelier, S., Coma, V. Advances on selective C-6 oxidation of chitosan by TEMPO, *Biomacromolecules*, 9 (2008): 2377–2382.

76. Hu, Y., Jiang, H., Xu, C., Wang, Y., Zhu, K. Preparation and characterization of poly(ethylene glycol)-chitosan with water- and organosolubility, *Carbohydrate Polymers*, 61 (2005): 472–479.

77. Xu, Z., Wan, X., Zhang, W., Wang, Z., Peng, R., Tao, F., Cai, L., Li, Y., Jiang, Q., Gao, R. Synthesis and biodegradable polycationic methoxy poly(ethylene glycol)- polyethylenimine-chitosan and its potential as gene carrier, *Carbohydrate Polymers*, 78 (2009): 46–53.

78. Yoshiwaza, T., Shin-Ya, Y., Hong, K.J., Kajiuchi, T. pH- and temperature-sensitive release behaviors from polyelectrolyte complex films composed of chitosan and PAOMA copolymer, *European Journal of Pharmaceutics and Biopharmaceutics*, 59 (2005): 307–313.

79. Yoshiwaza, T., Shin-Ya, Y., Hong, K.J., Kajiuchi, T. pH- and temperature-sensitive permeation through polyelectrolyte complex films composed of chitosan and polyalkyleneoxide-maleic acid copolymer, *Journal of Membrane Science*, 241 (2004): 347–354.

80. Strand, S.P., Danielsen, S., Christensen, B.E., Varum, K.M. Influence of chitosan structure on the formation and stability of DNA-chitosan polyelectrolyte complexes, *Biomacromolecules*, 6 (2005): 3357–3366.

81. Mao, H.-Q., Roy, K., Troung-Le, V.L., Janes, K.A., Lin, K.Y., Wang, Y., August, J.T., Leong, K.W. Chitosan-DNA nanoparticles as gene carriers: Synthesis, characterization and transfection efficiency, *Journal of Controlled Release*, 70 (2001): 399–421.

82. Ercelen, S., Zhang, X., Duportail, G., Grandfils, C., Desbrieres, J., Karaeva, S., Tikhonov, V., Mely, Y., Babak, V. Physicochemical properties of low molecular weight alkylated chitosans: A new class of potential nonviral vectors for gene delivery, *Colloids and Surfaces B: Biointerfaces*, 51 (2006): 140–148.

83. Ginsberg, G.L., Pepelko, W.E., Goble, R.L., Hattis, D.B. Comparison of contact site cancer potency across dose routes: Case study with epichlorohydrin, *Risk Analysis*, 16 (1996): 667–681.

84. Athawale, V.D., Rathi, S.C. Graft polymerization: Starch as a model substrate, *Journal of Macromolecular Science—Reviews in Macromolecular Chemistry and Physics*, C39 (1999): 445.

85. Beck, R.H.F., Fitton, M.G., Kricheldorf, H.R. Chemical modification of polysaccharides in *Handbook of Polymer Synthesis, Part B*, Kricheldorf, H.R. (Ed.), Marcel Dekker, New York, Chapter 25, 1992, pp. 1517–1578.

86. Vernon, B., Kim, S.W., Bae, Y.H. Insulin release from islets of Langerhans entrapped in a poly(N-isopropylacrylamide-co-acrylic acid) polymer gel, *Journal of Biomaterials Science, Polymer Edition*, 10 (1999): 183–198.

87. Ebara, M., Aoyagi, T., Sakai, K., Okano, T. Introducing reactive carboxyl side chains retains phase transition temperature sensitivity in N-isopropylacrylamide copolymer gels, *Macromolecules*, 33 (2000): 8312–8316.

88. Schild, H.G. Poly(N-isopropylacrylamide): Experiment, theory and application, *Progress in Polymer Science*, 17 (1992): 163–249.

89. Bokias, G., Hourdet, D. Synthesis and characterization of positively charged amphiphilic water soluble polymers based on poly(N-isopropylacrylamide), *Polymer*, 42 (2001): 6329–6337.

90. Kubota, N., Tasumoto, N., Sano, T., Matsukawa, Y. Temperature-responsive properties of poly(acrylic acid-co-acrylamide)-graft-oligo(ethylene glycol) hydrogels, *Journal of Applied Polymer Science*, 80 (2001): 798–805.

91. Lee, C.F., Wen, C.J., Lin, C.L., Chiu, W.Y. Morphology and temperature responsiveness-swelling relationship of poly(N-isopropylamide-chitosan) copolymers and their application to drug release, *Journal of Polymer Science Part A: Polymer Chemistry*, 42 (2004): 3029–3037.

92. Wang, M., Qiang, J., Fang, Y., Hu, D., Cui, Y., Fu, X. Preparation and properties of chitosan-poly(N-isopropylacrylamide) semi-IPN hydrogels, *Journal of Polymer Science A: Polymer Chemistry*, 38 (2000): 478–481.

93. Goycoolea, F.M., Heras, A., Aranaz, I., Galed, G., Fernandez-Valle, M.E., Monal, W.A. Effect of chemical crosslinking on the swelling and shrinking properties of thermal and pH-responsive chitosan hydrogels, *Macromolecular Bioscience*, 3 (2003): 612–619.

94. Kim, S.J., Shin, S.R., Lee, S.M., Kim, I.Y., Kim, S.I. Water and temperature response of semi-IPN hydrogels composed of chitosan and polyacrylonitrile, *Journal of Applied Polymer Science*, 88 (2003): 2721–2724.

95. Wang, M., Fang, Y., Hu, D. Preparation and properties of chitosan-poly(N-isopropylacrylamide) full-IPN hydrogels, *Reactive and Functional Polymers*, 48 (2001): 215–221.

96. Cai, H., Zhang, Z.P., Sun, P.C., He, B.L., Zhu, X.X. Synthesis and characterization of thermo- and pH-sensitive hydrogels based on chitosan-grafted N-isopropylacrylamide via γ-radiation, *Radiation Physics and Chemistry*, 74 (2005): 26–30.

97. Kim, I.Y., Kim, S.J., Shin, M.S., Lee, Y.M., Shin, D.I., Kim, S.J. pH- and thermal characteristics of graft hydrogels based on chitosan and poly(dimethylsiloxane), *Journal of Applied Polymer Science*, 85 (2002): 2661–2666.

98. Zhu, A.P., Chan, M.B., Dai, S., Li, L. The aggregation behavior of O-carboxymethylchitosan in dilute aqueous solution, *Colloid and Surface B: Biointerfaces*, 43 (2005): 143–149.

99. Glass, J.E. *Polymers in Aqueous Media: Performance through Association*. Advances in Chemistry Series, American Chemical Society, Washington, DC, Vol. 223, 1989.

100. Hall, J.E., Hodgson, P., Krivanek, L., Malizia, P.J. Influence of rheology modifiers on the performance characteristics of latex paints, *Journal of Coatings Technology*, 58 (1986): 65–73.

101. Desbrieres, J. Autoassociative natural polymer derivatives: The alkylchitosans. rheological behaviour and temperature stability, *Polymer*, 45 (2004): 3285–3295.

102. Hirrien, M., Chevillard, C., Desbrieres, J., Axelos, M.A.V., Rinaudo, M. Thermogelation of methylcelluloses. New evidence for understanding the gelation mechanism, *Polymer*, 39 (1998): 6251–6259.

103. Hourdet, D., L'Alloret, F., Audebert, R. Reversible thermothickening of aqueous polymer solutions, *Polymer*, 35 (1994): 2624–2630.

104. Desbrieres, J. Contribution of chitin derivatives to the modification of physicochemical properties of formulations, *Polymer International*, 52 (2003): 494–499.

105. Desbrieres, J., Rinaudo, M., Klein, J.M., Mahler, B Dérivés du chitosane. Procédé pour sa préparation et composition cosmétiques contenant de tels dérivés, French Patent FR 2721933, 1994; European Patent EP 699433, 1995.

106. Marencic, A.P., Wu, M.W., Register, R.A., Chaikin, P.M. Orientational order in sphere-forming block copolymer thin films aligned under shear, *Macromolecules*, 40 (2007): 7299–7305.

107. Gohy, J.F., Lohmeijer, B.G.G., Alexeev, A., Wang, X.S., Manners, I., Winnik, M.A., Schubert, U.S. Cylindrical micelles from the aqueous self-assembly of an amphiphilic poly(ethylene oxide)-b-poly(ferrocenylsilane) (PEO-b-PFS) block copolymer with a metallo-supramolecular linker at the block junction, *Chemistry—A European Journal*, 10 (2004): 4315–4323.

108. Rodriguez-Hernandez, J., Lecommandoux, S. Reversible inside-out micellization of pH-responsive and water-soluble vesicles based on polypeptide diblock copolymers, *Journal of American Chemical Society*, 127 (2005): 2026–2027.

109. Pochan, D.J., Chen, Z.Y., Cui, H.G., Hales, K., Qi, K., Wooley, K.L. Toroidal triblock copolymer assemblies, *Science*, 306 (2004): 94–97.

110. Cameron, N.S., Corbierre, M.K., Eisenberg, A. 1998. E.W.R. Steacie award lecture asymmetric amphiphilic block copolymers solution: A morphological wonderland, *Canadian Journal of Chemistry*, 77 (1999): 1311–1326.

111. Discher, D.E., Eisenberg, A. Polymer vesicles, *Science*, 297 (2002): 967–973.

112. Yu, K., Bartels, C., Eiesnberg, A. Vesicles with hollow rods in the walls: A trapped intermediate morphology in the transition of vesicles to inverted hexagonally packed rods in dilute solutions of PS-b-PEO, *Macromolecules*, 31 (1998): 9399–9402.

113. Photos, P.J., Bacakova, L., Discher, B., Bates, F.S., Discher, D.E.J. Polymer vesicles *in vivo:* Correlations with PEG molecular weight, *Journal of Controlled Release*, 90 (2003): 323–334.

114. Mishima, K. Biodegradable particle formation for drug and gene delivery using supercritical fluid and dense gas, *Advanced Drug Delivery Reviews*, 60 (2008): 411–432.

115. Kim, K., Kwon, S., Park, J.H., Chung, H., Jeong, S.Y., Kwon, I.C. Physicochemical characterizations of self-assembled nanoparticles of glycol chitosan-deoxycholic acid conjugates, *Biomacromolecules*, 6 (2005): 1154–1158.

116. Aiping, Z., Tian, C., Lanhua, Y., Hao, W., Ping, L., Synthesis and characterization of N-succinyl-chitosan and its self-assembly of nanospheres, *Carbohydrate Polymers*, 66 (2006): 274–279.

117. Yoo, H.S., Lee, J.E., Chung, H., Kwon, I.C., Jeong, S.Y. Self-assembled nanoparticles containing hydrophobically modified glycol chitosan for gene delivery, *Journal of Controlled Release*, 103 (2005): 235–243.

118. Zhang, C., Ping, Q., Zhang, H., Shen, J. Preparation of N-alkyl-O-sulfate chitosan derivatives and micellar solubilization of taxol, *Carbohydrate Polymers*, 54 (2003): 137–141.

119. Chae, S.Y., Son, S., Lee, M., Jang, M.-K., Nah, J.-W. Deoxycholic acid-conjugated chitosan oligosaccharide nanoparticles for efficient gene carrier, *Journal of Controlled Release*, 109 (2005): 330–344.

120. Zhu, A., Lu, Y., Pan, Y., Dai, S., Wu, H. Self-assembly of N-maleoylchitosan in aqueous media, *Colloids and Surfaces B: Biointerfaces*, 76 (2010): 221–225.

121. Wu, Y., Zheng, Y., Yang, W., Wang, C., Hu, J., Fu, S. Synthesis and characterization of a novel amphiphilic chitosan-polylactids graft copolymer, *Carbohydrate Polymers*, 59 (2005): 165–171.

122. Ishimuro, Y., Ueberreiter K. The surface tension of poly(acrylic acid) in aqueous solution, *Colloid and Polymer Science*, 258 (1980): 928–931.

123. Babak, V.G. Thermodynamic and kinetic aspects of the stabilization of microscopic liquid films by the adsorbed layers of macromolecular surfactants, *Langmuir*, 3 (1987): 612–620.

124. Babak, V., Lukina, I., Vikhoreva, G., Desbrieres, J., Rinaudo, M. Interfacial properties of dynamic association between chitin derivatives and surfactants, *Colloids and Surfaces A: Physicochemical and Engineering Aspects*, 147 (1999): 139–148.

125. Sui, W., Song, G., Chen, G., Xu, G. Aggregate formation and surface activity property of an amphiphilic derivative of chitosan, *Colloids and Surfaces A: Physicochemical and Engineering Aspects*, 256 (2005): 29–33.

126. Babak, V.G., Desbrieres, J., Tikhonov, V.E. Dynamic surface tension and dilatational viscoelasticity of adsorption layers of a hydrophobically modified chitosan, *Colloids and Surfaces A: Physicochemical and Engineering Aspects*, 255 (2005): 119–130.

127. Babak, V.G., Desbrieres, J. Dynamic surface tension and dilational viscoelasticity of adsorption layers of alkylated chitosans and surfactant-chitosan complexes, *Colloid and Polymer Science*, 284 (2006): 745–754.

128. Desbrieres, J., Babak, V.G., Bousquet, C. Dynamic surface tension and viscoelastic properties of adsorption layers of amphiphilic chitosan derivatives systems, in *Advances in Chitin Science*, Domard, A., Guibal, E., Varum, K.M. (Eds.), Vol. 9, 2007, pp. 232–240.

129. Benjamins, J., Cagna, A., Lucassen-Reynderts, E.H. Viscoelastic properties of triacylglycerol/water interfaces covered by proteins, *Colloids and Surfaces A: Physicochemical and Engineering Aspects*, 114 (1996): 245–254.

130. Desbrieres, J., Babak, V.G. Interfacial properties of amphiphilic natural polymer systems based on derivatives of chitin, *Polymer International*, 55 (2006): 1177–1183.

131. Zhu, A.-P., Yuan, L.-H., Chen, T., Wu, H., Zhao, F. Interactions between N-succinyl-chitosan and bovine serum albumin, *Carbohydrate Polymers*, 69 (2007): 363–370.

132. Wong, D.W.S., Gastineau, F.A., Gregorski, K.S., Tillin, S.J., Pavlath, A.E. Chitosan-lipid films: Microstructure and surface energy, *Journal of Agricultural and Food Chemistry*, 40 (1992): 540–544.

133. Muzzarelli, R.A.A., Human enzymatic activities related to the therapeutic administration of chitin derivatives, *Cellular and Molecular Life Science*, 53 (1997): 131–140.

134. Bersch, P.C., Nies, B., Liebendörfer, A., Kreuler, J. *In vitro* evaluation of biocompatibility of different wound dressing materials, *Journal of Materials Science, Materials in Medicine*, 6 (1995): 201–205.

135. Park, K., Kim, J.-H., Nam, Y.S., Lee, S., Nam, H.Y., Kim, K., Park, J.H., Kim, I.-S., Choi, K., Kim, S.Y., Kwon, I.C. Effect of polymer molecular weight on the tumor targeting characteristics of self-assembled glycol chitosan nanoparticles, *Journal of Controlled Release*, 122 (2007): 305–314.

136. Kim, J.-H., Kim, Y.S., Kim, S., Park, J.H., Kim, K., Choi, K., Chung et al. Hydrophobically modified glycol chitosan nanoparticles as carriers for paclitaxel, *Journal of Controlled Release*, 111 (2006): 228–234.

137. Park, J.H., Kwon, S., Nam, J.O., Park, R.W., Chung, H., Seo, S.B., Kim, I.S., Kwon, I.C., Jeong, S.Y. Self-assembled nanoparticles based on glycol chitosan bearing 5β-cholanic acid for RGD peptide delivery, *Journal of Controlled Release*, 95 (2004): 579–588.

138. Prabaharan, M., Mano, J.F. Hydroxypropyl chitosan bearing β-cyclodextrin cavities: Synthesis and slow release of its inclusion complex with a model hydrophobic drug, *Macromolecular Science and Bioscience*, 5 (2005): 965–973.

139. Krauland, A.H., Alonso, M.J. Chitosan/cyclodextrin nanoparticles as macromolecular drug delivery system, *International Journal in Pharmaceutics*, 340 (2007): 134–142.

140. Li, F., Liu, W.G., Yao, K.D. Preparation of oxidized glucose-crosslinked N-alkylated chitosan membrane and in vitro studies of pH-sensitive drug delivery behaviour, *Biomaterials*, 23 (2002): 343–347.

141. Liu, W.G., Li, F. Yao, K.D. Oxidized glucose-crosslinked alkylated chitosan membrane for the delivery of Vitamin B2, in *Chitosan in Pharmacy and Chemistry*, Muzzarelli, R.A.A., Muzzarelli, C. (Eds.), Atec, Grottammare, Italy, 2002, pp. 93–99.

142. Liu, W.G., Sun, S.J., Zhang, X., Yao, K.D. Self-aggregation behavior of alkylated chitosan and its effect on the release of a hydrophobic drug, *Journal of Biomaterials Science Polymer Edition*, 14 (2003): 851–859.

143. Cai, K., Liu, W., Li, F., Yao, K., Yang, Z., Li, X., Xie, H. Modulation of osteoblast function using poly(D,L-lactic acid) surfaces modified with alkylation derivative of chitosan, *Biomaterials Science Polymer Edition*, 13 (2002): 53–66.

144. Li, J., Gong, Y., Zhao, N., Zhang, X. Preparation of N-butyl chitosan and study of its physical and biological properties, *Journal of Applied Polymer Science*, 98 (2005): 1016–1024.

145. Taboada, E., Cabrera, G., Cardenas, G. Synthesis and characterization of new arylamine chitosan derivatives, *Journal of Applied Polymer Science*, 91 (2004): 807–812.

146. Martin, L., Wilson, C.G., Koosha, F., Tetley, L., Gray, A.I., Senel, S., Uchegbu, I.F. The release of model macromolecules may be controlled by the hydrophobicity of palmitoyl glycol chitosan hydrogels, *Journal of Controlled Release*, 80 (2002): 87–100.

147. McEwan, G.T.A., Jepson, M.A., Hirst, B.H., Simmons, N.L. Polycation-induced enhancement of epithelial paracellular permeability is independent of tight junctional characteristics, *Biochima et Biophysica Acta—Biomembranes*, 1148 (1993): 51–60.

148. Artursson, P., Lindmark, T., Davis, S.S., Illum, L. Effect of chitosan on the permeability of monolayers of intestinal epithelial cells (Caco-2), *Pharmaceutical Research*, 11 (1994): 1358–1361.

149. Alvarez-Lorenzo, C., Concheiro, A., Dubovik, A.S., Grinberg, N.V., Burova, T.V., Grinberg, V.Y. Temperature-sensitive chitosan-poly(N-isopropylacrylamide) interpenetrated networks with enhanced loading capacity and controlled release properties, *Journal of Controlled Release*, 102 (2005): 629–641.

150. Bhattarai, N., Ramay, H.R., Gunn, J., Matsen, F.A., Zhang, M. PEG-grafted chitosan as an injectable thermosensitive hydrogel for sustained protein release, *Journal of Controlled Release*, 103 (2005): 609–624.

151. Zhang, C., Qu, G., Sun, Y., Yang, T., Yao, Z., Shen, W., Shen, Z., Ding, Q., Zhou, H., Ping, Q. Biological evaluation of N-octyl-O-sulfate chitosan as a new nano-carrier of intravenous drugs, *European Journal of Pharmaceutical Sciences*, 33 (2008): 415–423.

152. Green, S., Roldo, M., Douroumis, D, Bouropoulos, N., Lamprou, D., Fatouros, D.G. Chitosan derivatives alter release profiles of model compounds from calcium phosphate implants, *Carbohydrate Research*, 344 (2009): 901–907.

153. Wang, Y., Tu, S.L., Li, R.S., Yang, X.Y., Liu, L.R., Zhang, Q.Q. Cholesterol succinyl chitosan anchored liposomes: Preparation, characterization, physical stability, and drug release behavior, *Nanomedicine: Nanotechnology, Biology and Medicine,* 6 (2010): 471–477.

154. Sashiwa, H., Sumi, R., Saimoto, H., Okamoto, Y., Minami, S., Matsuhashi, A., Shigesama, Y. Evaluation of chitin and its derivatives as biomaterials, in *Chitin World*, Karnicki, Z.S., Wojtasz-Pajak, A., Brzeski, M.M., Bykowski, P.J. (Eds.), Wirtschaftsverlag NW, Bremerhaven, Germany, 1994, pp. 382–386.

155. Sui, W., Wang, Y., Dong, S., Chen, Y. Preparation and properties of an amphiphilic derivative of succinyl-chitosan, *Colloids and Surfaces A: Physicochemical and Engineering Aspects*, 316 (2008): 171–175.

156. Babak, V.G., Skotnikova, E.A. *Lipid and Surfactant Dispersed Systems. Fundamentals, Design, Formulation, Production*, AGPI, Moscow, Russia, Vol. 73, 1999.

157. Babak, V.G., Merkovich, E.A., Galbraich, L.S., Shtykova, E.V., Rinaudo, M. Kinetics of diffusionally induced gelation and ordered nanostructure formation in surfactant-polyelectrolyte complexes formed at water/water emulsion type interfaces, *Mendeleev Communications*, 3 (2000): 94.

158. Babak, V.G., Merkovich, E.A., Desbrieres, J., Rinaudo, M. Formation of an ordered nanostructure in surfactant-polyelectrolyte complexes formed by interfacial diffusion, *Polymer Bulletin*, 45 (2000): 77–81.

159. Babak, V.G., Rinaudo, M. Physico-chemical properties of chitin-surfactant complexes, in *Chitosan in Pharmacy and Chemistry*, Muzzarelli, R.A.A., Muzzarelli, C. (Eds.), Atec, Grottammare, Italy, 2002, pp. 277–284.

160. Lai, W.-F., Lin, M. C.-M. Nucleic acid delivery with chitosan and its derivatives, *Journal of Controlled Release*, 134 (2009): 158–168.

161. Jia, Z., Shen, D., Xu, W. Synthesis and antibacterial activities of quaternary ammonium salt of chitosan, *Carbohydrate Research*, 333 (2001): 1–6.

162. Sieval, A.B., Thanou, M., Kotze, A.F., Verhoef, J.C., Brussee, J., Junginger, H.E. Preparation and NMR characterization of highly substituted N-trimethyl chitosan chloride, *Carbohydrate Polymers*, 36 (1998): 157–165.

163. Thanou, M., Florea, B.I., Geldof, M., Junginger, H.E., Borchard, G. Quaternized chitosan oligomers as novel gene delivery vectors in epithelial cell lines, *Biomaterials*, 23 (2002), 153–159.

164. Mel'nikov, S.M., Sergeyev, V.G., Yoshikawa, K. Discrete coil—Globule transition of large DNA induced by cationic surfactant, *Journal of American Chemical Society*, 117 (1995): 2401–2408.

165. Liu, W.G., Yao, K.D., Liu, Q.G. Formation of a DNA/N-dodecylated chitosan complex and salt-induced gene delivery, *Journal of Applied Polymer Science*, 82 (2001): 3391–3395.

166. Liu, W.G., Zhang, X., Sun, S.J., Sun, G.J., Yao, K.D., Liang, D.C., Guo, G., Zhang, J.Y. N-alkylated chitosan as a potential nonviral vector for gene transfection, *Bioconjugate Chemistry*, 14 (2003): 782–789.

167. Tikhonov, V.E., Stepnova, E.A., Babak, V.G., Yamskov, I.A., Palma-Guerrero, J., Jansson, H.B., Lopez-Llorca, L.V. et al. Bactericidal and antifungal activities of a low molecular weight chitosan and its N-/2(3)-(dodec-2-enyl)succinoyl/-derivatives, *Carbohydrate Polymers*, 64 (2006): 66–72.

168. Hirano, S., Ohe, Y., Ono, H. Selective N-acylation of chitosan, *Carbohydrate Research*, 47 (1976): 315–320.
169. Kim, Y.H., Gim, S.H., Park, C.R., Lee, K.Y., Kim, T.W., Kwon, I.C., Chung, H., Jeong, S.Y. Structural characteristics of size-controlled self-aggregates of deoxycholic acid-modified chitosan and their application as a DNA delivery carrier, *Bioconjugate Chemistry*, 12 (2001): 932–938.
170. Yun, Y.H., Jiang, H., Chan, R., Chen, W. Sustained release of PEG-g-chitosan complexed DNA from poly(lactide-co-glycolide), *Journal of Biomaterials Science Polymer Edition*, 16 (2005): 1359–1378.
171. Zhu, D.W., Bo, J.G., Zhang, H.L., Liu, W.G., Leng, X.G., Song, C.X., Yin et al. Synthesis of N-methylene phosphonic chitosan (NMPCS) and its potential as gene carrier, *Chinese Chemical Letters* 18 (2007): 1407–1410.
172. Zhu, D., Yao, K., Bo, J., Zhang, H., Liu, L., Dong, X., Song, L., Leng, X. Hydrophilic/lipophilic N-methylene phosphonic chitosan as a promising non-viral vector for gene delivery, *Journal of Materials Science, Materials in Medicine*, 21 (2010): 223–229.
173. Tuzlakoglu, K., Alves, C.M., Mano, J.F., Reis, R.L. Production and characterization of chitosan fibers and 3-D fiber mesh scaffolds for tissue engineering applications, *Macromolecular Bioscience*, 4 (2004): 811–819.
174. Ding, Z., Chen, J., Gao, S., Chang, J., Zhang, J., Kang, E.T. Immobilization of chitosan onto poly-L-lactic acid film surface by plasma graft polymerization to control the morphology of fibroblast and liver cells, *Biomaterials*, 25 (2004): 1059–1067.
175. Prabaharan, M., Rodriguez-Perez, M.A., de Saja, J.A., Mano, J.F. Preparation and characterization of poly(L-lactic acid)-chitosan hybrid scaffolds with drug release capability, *Journal of Biomedical Biomaterials Research, Part B: Applied Biomaterials*, 81B (2007): 427–434.
176. Amornchai, W., Hoven, V.P., Tangpasuthadol, V. Surface modification of chitosan films-grafting ethylene glycol oligomer and its effect on protein adsorption, *Macromolecular Symposia*, 216 (2004): 99–107.
177. Wang, J., Chen, L., Zhao, Y., Guo, G., Zhang, R. Cell adhesion and accelerated detachment on the surface of temperature-sensitive chitosan and poly(N-isopropylacrylamide) hydrogels, *Journal of Materials, Materials in Medicine*, 20 (2009): 583–590.
178. Li, Y., Liu, L., Fang, Y. Plasma-induced grafting of hydroxyethyl methacrylate (HEMA) onto chitosan membranes by a swelling method, *Polymer International*, 52 (2003): 285–290.

10 Biomaterials of Natural Origin in Regenerative Medicine

Vijay Kumar Nandagiri, Valeria Chiono, Piergiorgio Gentile,
Franco Maria Montevecchi, and Gianluca Ciardelli

CONTENTS

10.1 INTRODUCTION

Biomaterials are nonviable materials used as medical devices intended to interact with biological systems. They display a combination of properties including chemical, mechanical, physical, and biological properties that render them suitable for safe, effective, and reliable use within a physiological environment. Biomaterials are developed to support damaged tissues, through suitable scaffolds for promoting cell differentiation and proliferation toward the formation of a new tissue (Williams 1987). The basic aim of scaffolds is to provide mechanical, morphological, and biological support for the damaged tissues and to recover them. To be successful, a scaffold should effectively repair the defect it covers without eliciting an adverse tissue reaction while maintaining mechanical and biological integrity for a desired amount of time from a few weeks to several years (Puppi et al. 2010). Natural-based polymers are or resemble the biological macromolecules, which the biological environment is prepared to recognize and deal with metabolically. The ECM is the optimized environment that nature has developed to maintain homeostasis and to direct tissue development (Mano et al. 2007). Owing to their similarity to the ECM macromolecules, natural polymers may serve as potential biomaterials. Natural polymers may also bypass the stimulation of chronic inflammation or immunological reactions and toxicities encountered by implants. The degradation of natural polymers is directed by naturally occurring enzymes and/or simple hydrolysis or oxidative processes which ensures that the implant degrades completely during normal metabolic processes. Furthermore, the rate of degradation of natural polymers can be modified by chemical cross-linking or other modifications to attain a predetermined biodegradation rate. So, with natural polymers, biomaterials can be designed such that they can function biologically at molecular level.

10.1.1 HISTORICAL BACKGROUND

The utilization of natural polymers as biomaterials is dated back to centuries. Ancient Egyptians used animal tendons as sutures. In the first century AD, in both Greece and India, physicians used natural biomaterials for surgery to repair mutilations from war. Susrutha, a medical encyclopedia of ancient India, describes a procedure similar to "autograft" which involved the transplantation of a tissue from one part of the body to another in the same individual. The method was used to obtain a prosthetic nose for patients whose nose had been cut off, by using a skin graft from the patient's cheek, and molded into the new prosthetic nose. Physicians succeeded to keep an open wound at the nose clean and viable, to detach a portion of patient's skin while maintaining the blood supply as to avoid necrosis and finally to reattach it to the patient so that it revascularized and the new nose sustained itself (Chari 2003).The Romans were able to perform simple surgical techniques such as repairing damaged ears around the first century BC. Due to religious reasons, they did not approve the dissection of both human beings and animals, thus their knowledge was based on the texts of their Greek predecessors (Santoni-Rugiu 2007). At around 1460, physicians performed rhinoplasty in which the skin from arm was used as prosthetic nose (Zimmerman and Veith 1993).

After two centuries, the first two cases of "xenograft" were performed in France and the Netherlands. In July 1667, Jean-Baptiste Denis, a French physician, transfused the blood of a sheep into a 15-year-old boy, who recovered. He also performed transfusion into a laborer, who also survived. But, in winter 1667, Denis performed several transfusions on Antoine Mauroy with calf's blood who on the third account died and the procedure was banned in France (Maluf 1954). A Dutch surgeon, Job van Meekeren, was the first to document the use of bone graft in 1668. He used a piece of canine skull to replace a skull defect in a Russian nobleman. The operation was successful but resulted in the excommunication of the patient by the church (Pacaccio 2005).

Whilst these attempts to advance medicine were opposed for political and religious reasons of that period, they set the stage for the introduction of prosthetics in regenerative medicine. Though early physicians lacked the sophistication and technology of modern surgeons, some of them were able to intuitively grasp concepts of biomimicry, necessary to successfully plan these operations. Of course, without knowledge of the immunogenicity of the implants they were using, many patients experienced complications or died soon after surgery. Implant technology progressed slowly until aseptic technique, anesthesia, and a basic understanding of cellular-level mechanisms matured. Understanding the history of the development of biomaterials can give a better perspective for the development of biomaterials (Coburn and Pandit 2007).

10.2 NATURAL POLYMERS AS BIOMATERIALS

Living organisms produce a wide variety of polymers such as polysaccharides, proteins, and polynucleotides which can function as biomaterials. These natural-based polymers offer the advantage of being similar to ECM allowing the design of biomimetic biomaterials, i.e., materials that function biologically at the molecular level. Natural ECM consists of nanofibers containing different chemical ligands that interact with cell surface receptors. The ECM is a complex mixture of matrix molecules, including the glycoproteins fibronectin, collagens, laminins, proteoglycans, and nonmatrix proteins including growth factors. Therefore, natural ECM provides cell signaling molecules to cells, including cell adhesion molecules and growth factors. A biomimetic substrate is able to emulate the functions of native ECM by mimicking its three-dimensionality, nanofibrous topography and plethora of chemical motifs. The use of natural polymers, which are component of the

ECM or similar to ECM macromolecules, allows the obtainment of substrates able to interact specifically with cell surface receptors (Albertsson and Karlsson 1996). Another interesting property of natural polymers is their ability to be degraded by naturally occurring enzymes guaranteeing that the implant will be finally metabolized by physiological mechanisms. Hence, it is possible to implant a natural-based scaffold which can deliver a specific function for a pre-determined period of time, following which it is expected to be degraded completely by normal metabolic processes. Moreover, the degradation rate of naturally occurring materials can be manipulated by introducing functionalities or cross-links which enable the regulation of the in vivo life-time of the implant (Elsie et al. 2009).

10.2.1 POLYSACCHARIDES AS BIOMATERIALS

Polysaccharides are crucial components of life matter. They display perfect biocompatibility and biodegradability, which are the basic characteristics for polymers used as biomaterials. Polysaccharides (called also glycans) consist of monosaccharides linked together by O-glycosidic linkages. Polysaccharides containing only one kind of monosaccharides are homopolysaccharides, or homoglycans, whereas those containing more than one kind of monosaccharides are heteropolysaccharides or heteroglycans. Polysaccharides differ not only in the nature of their component monosaccharides but also in the length of their chains and in the amount of chain branching. Although a given sugar residue has only one anomeric carbon and thus can form only one glycosidic linkage with hydroxyl groups on other molecules, each sugar residue carries several hydroxyls, one or more of which may be an acceptor of glycosyl substituents. This ability to form branched structures distinguishes polysaccharides from proteins and nucleic acids, which are linear polymers (Kobayashi and Uyama 2003).

Polysaccharides have several characteristics not found in other natural polymers such as water solubility, flow behavior, gelling potential, and/or surface and interfacial properties depending on the differences in the monosaccharide composition and linkage types (Barbosa et al. 2005). Polysaccharides perform different physiological functions and may offer a variety of potential applications as biomaterials. Three factors generate increased interest toward polysaccharide-based biomaterials: (1) the large and growing body of information pointing to the critical role of saccharide moieties in cell signaling schemes and particularly in the area of immune recognition; (2) the recent development of new powerful techniques with the potential for automated synthesis of biologically active oligosaccharides, able to allow decoding and exploiting the language of oligosaccharide signaling; (3) the explosion in tissue engineering research and the associated need for new materials with specific, controllable biological activity and biodegradability (Lakshmi and Cato 2007).

The presence of reactive functional groups along the polymer chains of polysaccharides facilitates their modification and allows the formation of hydrogels. Depending on the source of availability, polysaccharides can be further classified into plant polysaccharides (including algae), animal polysaccharides, and microbial polysaccharides. Different polysaccharides will be described in detail in the following with particular attention to their relevant applications in regenerative medicine.

10.2.1.1 Cellulose

Cellulose is a linear polysaccharide of repeated D-glucose units with β-1,4 linkages; it is an abundant, biodegradable, and biocompatible polymer and the principle constituent of plants and natural fibers, such as cotton and linen (Klemm et al. 2005). Plant cellulose (PC) forms the structural framework of plants and it is isolated in the form of microfibrils and possesses repeated D-glucose units linked by β-(1–4) glycosidic bonds which account for the high crystallinity of cellulose (usually in the range of 40%–60% for PC). In nature, the cellulose molecules assemble to form cellulose I-type crystal structures, with regularly packed, parallel-chain alignments along the longitudinal axis (Hieta et al. 1984). The regular intra- and intermolecular hydrogen bonds yield self-assembly and hierarchical organization of cellulose which corresponds to the crystal structure and polymorphism resulting in its insolubility in common solvents (Atalla and VanderHart 1984, Nishiyama et al. 2002, 2003). The biodegradation of cellulose leads to progressive decrease in molecular weight and mechanical strength and increase in water solubility. The presence of amorphous regions in cellulose can increase its biodegradability in tissue. In vivo biodegradability can be enhanced by lowering the crystallinity and intensifying the hydrophilicity of cellulose-based materials (Miyamoto et al. 1989). In order to enhance the water solubility and decrease the crystallinity of cellulose, various substitutions have been attempted along the anhydroglucose backbone. Introduction of hydrophobic groups such as methyl or hydroxypropyl moieties to the cellulose backbone yields the derivatives with hydrophilic-lipophilic balance values ranging from 10.0 to 11.25. Furthermore, introduction of these hydrophobic groups gives the polymer surface activity (as in polysoap) and unique hydration-dehydration characteristics and improves the water affinity of the cellulose (Sarkar and Walker 1995).

Cellulose can be derivatized in the form of ethers, esters, and acetals with varying solubility properties like methyl cellulose (MC), hydroxyl propyl cellulose (HPC), hydroxyl propyl methyl cellulose (HPMC), and carboxy methyl cellulose (CMC) which can be termed as cellulosics. Cellulose-based materials have been used in a number of biomedical applications. These include construction of cellulose membranes for hemodialysis (Wu 2002), construction of enzyme carriers for biosensors (Utpal et al. 2006) drug delivery (Malafaya et al. 2007), and preparation of scaffolds for the regeneration of various tissue types, such as bone (Barbosa et al. 2005), cartilage (Svensson et al. 2005, Muller et al. 2006), liver, skin, and blood vessels (Miyamoto et al. 1989, Seal et al. 2001). Precisely, due to their high diffusional permeability to most of the toxic metabolic solutes, cellulose membranes have been extensively investigated as hemodialysis membranes. Sevillano et al. (1990) designed hemodialysis membranes based on cellulose acetate (CA) and they observed that the membranes based on CA improved red blood cell function in hemodialysis patients. Artanareeswaran et al. (2004) demonstrated the influence of poly(ethylene glycol) (PEG) addition to CA membrane on pore size, permeate flux, and the protein rejection rates. Furthermore, Ani Idris et al. investigated the effect of incorporating PEGs with different molecular weights, such as PEG 200 kDa, PEG 400 kDa, and PEG 600 kDa on the dialysis membrane performance in terms of urea clearance and also on the morphology of membrane. They reported that the membrane performance in terms of solute clearance was significantly influenced by the amount of additives and their molecular weights (PEG molecular weight). For example, PEG with a low molecular weight produced superior dialysis membranes that can clear urea (Idris and Yet 2006).

Then, Entcheva et al. synthesized diffusion-controlling membranes and membrane carriers for enzyme immobilization in biosensors. They constructed biosensor membranes with glucose oxidase, covalently bound to acetylcellulose carriers activated with urea, and tested them to evaluate their application in the measurement of the glucose concentration in blood serum and plasma (Entcheva and Yotova 1994). Successively, Liang et al. prepared biodegradable cellulose/chitin blend porous-like membranes from blend solution of cellulose and chitin in 9.5 wt% NaOH/4.5 wt% thio urea aqueous solution coagulating with 5.0 wt% $(NH_4)_2SO_4$ and water as solvent. They studied the permeability, the partition coefficients, and the diffusion coefficients of three different drugs in these membranes. They concluded that the introduction of chitin influenced the morphology and enhanced the pore size and porosity of the membranes and thus the crystal structure of cellulose in the blend membranes was partly destroyed, resulting in a significantly different permeability. According to them, the diffusion of drugs through these membranes resulted by dual transport mechanism (pore mechanism and partition mechanism) with some hindrance of molecular diffusion via polymer obstruction (Liang et al. 2007).

Apart from these, cellulose derivatives have been extensively investigated for their applications such as dressings in the treatment of surgical incisions, burns, wounds, and various dermatological disorders (Peppas 1986). The dressings based on CMC fibers form a strong cohesive gel upon hydration and have been marketed under the trade name AQUACEL®. Cellulose derivatives can also undergo sol-gel transition in water at an optimum balance of substituting hydrophilic and hydrophobic moieties and can be employed for hydrogel fabrication (Walker et al. 2003).

Oxidized cellulose has been used for wound dressing in the form of a gauze which becomes gelatinous after contact with blood, thus adhering to the surrounding tissue and for osseous regeneration since decades under the trade name of Surgicel® (Johnson and Johnson) (Degenshein et al. 1963, Dias et al. 2003). Skoog demonstrated that Surgicel promotes osseous and soft tissue regeneration in maxillary reconstructive surgery, when used as superiosteal graft before flap closure (Skoog 1967). Owing to its biocompatibility, ability to integrate into surrounding tissue, and mechanical properties, cellulose is considered as a promising candidate for scaffolds in bone tissue engineering. To improve the osteoconduction of the membranes, Rhee et al. deposited a surface apatite layer on cellulose cloth with the aid of citric acid. Hydroxyapatite (HAp) formation was induced by soaking cellulose cloths in simulated body fluid (SBF) solutions (1.5× SBF) with ion concentrations 1.5 times that of citric acid (1.0 M citric acid) and they concluded that the formation of HAp could be due to the hydrogen bonding of citric acid to the cellulose cloth and its chelating ability toward calcium ions (Rhee and Tanaka 2000).

Moreover, regenerated cellulose (RC; cellulose regenerated by the viscose process, CRV®) sponges were demonstrated to promote bone regeneration and to have slow degradability (Martson et al. 1998). Regenerated cellulose hydrogels have also been implanted as an attachment material for the femoral component in hip prostheses, in place of the acrylic cement (Poustis et al. 1994).

Hofmann et al. attempted to coat nonwoven cellulose (regenerated, oxidized) fabrics with hydroxy carbonated apatite (HCA) using a procedure based on biomimetic method. They employed a high degree of supersaturation (5× SBF) to accelerate the biomimetic formation of bone-like apatite on the cellulose fabrics. Furthermore, the crystal growth was accelerated by successive soaking of the coated fabric into a second 5× SBF with reduced Mg^{2+} and HCO^{3-} concentrations. They found that the HCA layer thickness increased by increasing the soaking time in the second solution and concluded that the amount of CO_3^{2-} substituting PO_4^{3-} in the HAp lattice of the precipitates can be varied by changing the soaking time (Hofmann et al. 2006).

In a biomimetic approach (with the aim to mimic ECM functionalities), de Taillac et al. proposed the immobilization of bioactive molecules, like peptides bearing the RGD (Arg- Gly-Asp) sequence, onto cellulose. The ability of this peptide sequence to bind a variety of cells facilitates bone cell attachment and proliferation. In this approach, silane-derivatized spacer arms were first attached to the surface of cellulose as an intermediate for the covalent linkage of RGD-containing peptide (de Taillac et al. 2004). They reported that the biomimetic modification of cellulose surface enhanced the adhesion of human bone marrow stroma cells and proliferation of osteoprogenitor cells.

In other applications in tissue engineering, Entcheva et al. prepared scaffolds for growing functional cardiac cell constructs based on CA and RC to demonstrate the properties of CA and RC surfaces. They observed that CA and RC helped in promoting cardiac cell growth, enhancing cell connectivity (gap junctions) and electrical functionality (Entcheva et al. 2004). Luo et al. reported the preparation of cellulose/soya protein isolate (SPI) sponges by freeze drying process for neural tissue regeneration. They observed that the sponges were porous in structure and that the size of the pores increased and the thickness of the pore walls decreased as the SPI content of the sponges increased. The authors hypothesized that the incorporation of SPI into cellulose followed by freeze-drying process resulted in the formation of pores and thin pore walls in the composite sponges, which facilitated the migration of cells and tissue into the sponges, leading to gradual fusing with the implants. Thus, the new cellulose/SPI sponges exhibited good biocompatibility using L929 fibroblast cells and biodegradability (Luo et al. 2010).

Recently, Zaborowska et al. attempted to refine the method for preparing microporus bacterial cellulose scaffolds in order to attain a porosity of 300–500 μm and evaluated its suitability as a 3D scaffold for bone tissue engineering. They observed that the refined microporus cellulose-based scaffolds improved the osteoprogenitor cell seeding onto the scaffolding material resulting in the formation of cell clusters within the scaffolds and ultimately improved the bone regeneration (Zaborowska et al. 2010).

To summarize, cellulose has been attempted either directly or in modified form for various tissue engineering applications such as fabrication of dialysis membranes, surgical dressings, hydrogels, and as scaffolding material for bone, cartilage, and other tissue regeneration applications. Researchers also studied composite scaffolding materials based on cellulose either with addition of other polymeric materials or inorganic components.

10.2.1.2 Starch

Starch is a mixture of glycans that plants synthesize as their principal food reservoir. It is deposited in the chloroplasts of plant cells as insoluble granules composed of α-amylose (normally 20%–30%) and amylopectin (normally 70%–80%) (Morrison and Karkalas 1990). α-Amylose is a linear polymer of several thousands of glucose residues linked by α-(1–4) bonds. Amylopectin, consisting mainly of α-(1/4)-linked glucose residues, is a branched molecule with (1/6) branch points at every 24–30 glucose residues in average. Amylopectin molecules contain up to 106 glucose residues, making them some of the largest molecules in nature (Voet et al. 1999).

The physico-chemical and functional characteristics of starch and its uniqueness in various products vary with the biological origin of starch (Svegmark and Hermansson 1993). Starch is insoluble in cold water, but it is very hygroscopic and binds water reversibly. Heating starch solution results in rupturing of interior hydrogen bonding in the starch granule leading to the gelatinization of starch (Appelqvist and Debet 1997). Furthermore, the linear amylose molecules leach out of the granules into the solution which forms a suspension that contains a mixture of linear amylose molecules, swollen granules, and granule fragments which in turn result in the formation of a thick paste or gel, depending on the amount of water present (Whistler and Daniel 2005). Starch is degraded by amylase (EC number 3.2.1.1) and glucosidase (EC number 3.2.1.20) enzymes. The α $(1 \rightarrow 4)$ linkages are broken down by the enzyme amylase and the α $(1 \rightarrow 6)$ linkages are attacked by glucosidases. The degradation products of starch are low molecular weight starch chains, fructose, and maltose. Starch-based polymers and their derivatives exhibit a wide range of properties like ease in shaping them into thin films (Rindlava et al. 1997), fibers (Pavlov et al. 2004), or porous matrices (Gomes et al. 2001) which make them suitable for uses in several biomedical applications such as bone plates, screws, drug delivery carriers, and tissue engineering scaffolds (Marques et al. 2002). Starch-based materials were firstly proposed for bone-related applications in 1995 by Reis and Cunha (1995). These starch-based materials consist of blends of corn starch with ethylene vinyl alcohol (SEVA-C), cellulose acetate (SCA), poly (caprolactone) (SPCL), and poly (lactic acid) (SPLA). Up to now, several studies have shown the potential of these materials for several biomedical applications, namely, as bone substitutes (Sousa et al. 2000, 2003), drug delivery carriers (Malafaya et al. 2001, Elvira et al. 2002), partially degradable bone cements (Espigares et al. 2002, Boesel et al. 2004), and hydrogels (Pereira et al. 1998, Elvira et al. 2002).

Marques et al. (2002) demonstrated the in vitro cytocompatibility of two different blends of corn-starch SEVA-C and SCA and their respective composites with HAp, by cytotoxicity and cell adhesion tests using L929 mouse fibroblast cell line, and they reported that both blends are cytocompatible and can be adopted as promising material for bone replacement and/or tissue engineering applications. Then, Gomes et al. also attempted to fabricate three-dimensional porous architectures made of SEVA-C and SCA blends to obtain adequate mechanical properties. They also used hydroxyapatite as a reinforcement material for the polymeric scaffold. The scaffolds were fabricated by standard conventional injection molding process, in which a solid blowing agent based on carboxylic acids was used to generate the foaming of the bulk of the molded part in order to produce scaffolds with a compact skin and a porous core. They observed that scaffolds based on SCA have displayed better tensile properties than that of SEVA-C scaffolds and the stiffness of the materials was slightly increased (a maximum increase of 12% for SEVA-C and around 25% for SCA) until a certain value of added carboxylic acids as a blowing agent. They concluded that such blends are noncytotoxic and can be employed for various biomedical applications (Gomes et al. 2001).

Similarly, Salgado et al. have developed porous starch scaffolds based on a 50/50 (wt%) blend of corn starch/ethylene-vinyl alcohol by melt-based technology for bone tissue-engineering applications. The scaffolds showed 60% porosity degree with pore sizes between 200 and 900 μm and a reasonable degree of pore interconnectivity; furthermore, scaffolds were nontoxic and did not inhibit cell growth. These scaffolds showed a high compressive modulus of 117.50 ± 3.7 MPa and a compressive strength of 20.8 ± 2.4 MPa (Salgado et al. 2004).

Tuzlakoglu et al. designed nano- and microfiber combined scaffolds for bone tissue engineering based on starch and polycaprolactone (SPCL) by electrospinning method to mimic the biophysical structure of natural ECM. They tested the developed structures with two different cell types, a human osteoblast-like osteosarcoma SaOs-2 cell line and rat bone marrow stromal cells. They demonstrated that the nanofibers affected cell shape and cytoskeletal organization of the cells on the nano-/microfiber combined scaffolds. They concluded that the developed structures exhibited a great potential on the 3D organization and guidance of cells that is provided for engineering of 3D bone tissues (Tuzlakoglu et al. 2005).

Furthermore, Gomes et al. performed studies in dynamic culture conditions (by the use of a bioreactor) to demonstrate the expression of an array of bone growth factors by marrow stromal cells cultured on SPCL fiber-mesh scaffolds. Bone marrow cells cultured on starch-based scaffolds responded to flow perfusion conditions by augmenting the production of several bone-related growth factors, namely, TGF-β1, FGF-2, vascular endothelial growth factor (VEGF), and BMP-2 (Gomes et al. 2006), and Santosa et al. illustrated that SPCL fiber meshes supported the maintenance of extra cellular morphological structure and endothelial integrity and augmented the vascularization process. They suggested that SPCL fiber meshes are potential scaffolds for bone tissue engineering (Santosa et al. 2007).

Recently, Martins et al. developed a novel hierarchical multilayered starch-based scaffold by a combination of starch-polycaprolactone micro- and polycaprolactone nano-motifs, respectively produced by rapid prototyping and electrospinning techniques. Such multilayered scaffold displayed a combination of parallel aligned microfibers in a grid-like arrangement, inserted by a mesh-like structure with randomly distributed nanofibers. They concluded that such an integration of nanoscale fibers into 3D rapid prototyped scaffolds enhanced scaffold biological performance in bone tissue–engineering strategies (Martins et al. 2009). Azevedo and Reis (2009) reported compression molding method to tailor the degradation rates of starch by encapsulating α-amylase (a thermostable enzyme).

Starch can also be applied as a slow-release drug delivery system. Tuovinen et al. attempted to control the hydrophobicity and rate of enzymatic degradation of starch by acetyl substitution. They emphasized that the hydrophobicity of starch increases as a function of degree of substitution (Tuovinen et al. 2003). Furthermore, they prepared microparticles based on starch acetate having various DS from 2 to 6 and loaded with calcein to target retinal pigment epithelium (RPE) for ocular treatment. They observed that starch acetate microparticles are taken up by the RPE cells and the polymer can be degraded by the enzymes in these cells (Tuovinen et al. 2004).

Malafaya et al. developed starch microspheres as drug delivery carriers in tissue engineering applications. They loaded microparticles with a model nonsteroidal anti-inflammatory drug (Meclofenamic sodium salt) which is widely used in the treatment of arthritic inflammations and immobilized them into scaffolds or administrated separately with scaffolds. Furthermore, they speculated that by altering processing conditions such as pH and ionic strength, such a system can be potentially incorporated with different growth factors or even living cells and can be utilized for improving bone regeneration in tissue engineering (Malafaya et al. 2006). Silva et al. have synthesized composite microparticles composed of a starch-based material and a bioactive glass (BG 45S5®). These particles displayed a bioactive behavior by the formation of a calcium-phosphate layer resembling biological apatite at their surface upon immersion in SBF. The combination of bioactive behavior and the ability to encapsulate bioactive agents conferred these microparticles a great potential for controlled growth factor release in bone applications

(Silva et al. 2005). Starch-based biomaterials (SBB) surface properties can be changed by using oxygen-based radio frequency glow discharge (rfGD) treatment to modulate bone cells response to SBB. This Oxygen rfGD treatment uniformly functionalizes/activates the surface of SBB without affecting the bulk properties. RfGD-treated surfaces showed an increase in hydrophilicity and surface energy when compared to nonmodified SBB. The synergic effect of surface plasma treatment on the adhesion, growth, and morphology of bone-like cells was also reported (Alves et al. 2007). Duarte et al. described a method to prepare starch-based porous matrixes by using super critical fluid technology. They used supercritical immersion precipitation technique to prepare scaffolds of a polymeric blend of starch and poly (L-lactic acid) for tissue engineering purposes by using two organic solvents, dichloromethane and chloroform, and concluded chloroform as the most favorable solvent for the process as it dissolved the raw materials better and resulted in the formation of thicker and more porous matrix. They also revealed that pressure mostly affected the porosity, interconnectivity, and pore size of the matrix, whereas temperature had negligible affect (Duarte et al. 2009).

To conclude, starch has been used as a promising biomaterial for tissue regeneration applications either directly or in blended form with thermoplastic polymers to improve thermo-mechanical degradation, and make them less brittle and more easily processed. Blends of starch with ethylene vinyl alcohol, cellulose acetate, polycaprolactone, and poly(lactic acid) have been proposed as potential alternative biodegradable materials for a wide range of biomedical applications, including bone cements, hydrogels for controlled drug delivery, and bone substitutes. The structure and functional properties of these materials depend on blend components, material processing technique, incorporation and nature of additives, and reinforcement fillers. By incorporating α-amylases and cellulose enzymes into starch-based scaffolding systems, the degradation rates can be tailored.

10.2.1.3   Agar-Agarose

Agar is made of two components: agarose (a non-sulfated fraction) and agaropectin (a mixture of various sulfated molecules). The structure of agarose consists of a linear chain of β-D-galactopyranose residues linked through positions 1 and 3 (A units) and α-D/L-galactopyranose residues linked through positions 1 and 4 (B units) arranged in an alternating sequence (AB)$_n$. A unit may carry methyl ether groups on position 6, sulfate hemiester groups on positions 2, 4, or 6; some of A units may carry pyruvic acid, linked as a cyclic ketal bridging O-4 and O-6 (1-carboxyethylidene groups). B units can occur in either D or L form which carry methyl groups on position 2 or 4-oxymethyl-α-L-galactopyranosyl groups on position 6 or/and sulfate hemiester groups on position 2 or 6 or both (Painter 1983). This arrangement allows the chains to join together and adopt a double helix. The two chains are wrapped together so tightly that any gaps are closed, trapping any water inside the helix. Agarose forms thermally reversible gels (cold setting gels at approx. 38°C); however, the melting temperature is much higher, approximately 85°C; this gives agar gel a very large gelling/melting hysteresis (Izydorczyk et al. 2005). Degree of gelation depends on the methoxyl and sulfate content and it involves the formation of double helices followed by a phase separation to form a gel (Rinaudo 2008). Agarose gels have been attempted to prepare scaffolds for neural and cartilage tissue engineering (Kong and Mooney 2004); in particular, agarose-based hydrogels have been reported to be potential scaffolds for cartilage tissue engineering owing to their swelling nature and high water content (Elisseeff et al. 2005). For nerve tissue regeneration, agarose hydrogels have been reported to stimulate and maintain three-dimensional neurite extension from primary sensory ganglia in vitro (Luo and Shoichet 2004). However, a successful scaffold for neural tissue engineering is required to have suitable mechanical properties for neurite extension (Nisbet et al. 2008). Balgude et al. attempted to demonstrate the structure–function relationship between the mechanical stiffness of the agarose hydrogel and its ability to initiate and maintain maximal neurite extension. They explored agarose gels in the range of 0.75%–2.00% (wt/vol, increments of 0.25%), by dissolving powdered agarose in sterile saline and showed that pore size of agarose hydrogels increased with the increase in agarose concentration. Furthermore they explored the role of mechanical stiffness as a potential parameter that can be perturbed to enhance neurite extension in agarose hydrogel (Dillon et al. 1998, Balgude et al. 2001).

Moreover, agarose hydrogels have been used to provide a three-dimensional environment to investigate in vitro chondrocyte mechano-transduction pathways which involve (1) the encapsulation of cells within agarose hydrogel, (2) the application of cell-seeded hydrogel of defined levels of mechanical deformation or loading, and (3) the observation of the resulting changes in cellular structure and organization (Buckley et al. 2009).

Suspending chondrocytes in a three-dimensional matrix similar to their natural environment has been reported to permit the cells to retain their native phenotype and to produce their extracellular components. Benya et al. demonstrated that the culture of rabbit cartilage in agarose permitted restoration of the differentiated phenotype (Benya and Shaffer 1982). Chondrocytes cultured in agarose scaffolds have been reported to maintain their phenotype and synthesize normal levels of collagen II. Furthermore, chondrocyte-seeded agarose hydrogels have been shown to produce a functional extracellular matrix in free swelling culture (Mauck et al. 2002). Agarose hydrogels have been applied for cartilage tissue engineering applications to assess the elaboration of pericellular and extracellular matrix. Quinn et al. have demonstrated the matrix deposition around individual chondrocytes during de novo matrix synthesis in agarose suspension culture for a period of 1 month (long enough for cartilage-like material properties to begin to emerge). Cartilage regeneration was promoted by a differential recruitment, pericellular distributions of proteoglycan, and matrix protein deposition around the hydrogel (Quinn et al. 2002).

Mouw et al. (2005) compared the similarities and differences among scaffolding materials in proteoglycan accumulation and GAG composition. Maximum cell proliferation and extracellular matrix accumulation have been observed in agarose constructs in comparison with PGA, alginate, fibrin, and collagen. Furthermore, they reported that agarose constructs had the highest sGAG (sulfated GAG) to DNA ratio, while alginate and collagen I had the lowest levels (Mouw et al. 2005).

Even though agar is compatible with most proteins and other polysaccharides at near neutral pH, it has limited applications owing to its brittle mechanical properties (Freile-Pelegrin et al. 2007). In tissue engineering, the hierarchical organization of cells to promote the in vitro development of functional tissue has been benefited from the spatially controlled placement of cells in specific locations on porous, often heterogeneous, substrates. The use of printing technologies may facilitate the invasion of cells into the scaffold spatially and the generation of organized tissues. Agarose replica printing have been successfully used for generating patterns of cells with circular shapes and diameters of 200, 700, or 1000 μm on the surface of HAp scaffolds and glass slides (Stevens et al. 2005).

Agarose has been recently combined with ceramics as natural biodegradable binder to increase the biomaterial flexibility facilitating its placement into the bone defect. Roman et al. optimized the fabrication of biodegradable scaffolds composed of β-tricalcium phosphate (β-TCP) and agarose based on the thermal gelation technique. They succeeded in internalizing vancomycin during the scaffold preparation stage, without disturbing the scaffold consolidation process in order to improve the graft performance. They proposed these scaffolds as easily injectable scaffolds with reinforced hydrogel-like behavior (Roman et al. 2008).

Furthermore, Alcaide et al. studied the behavior of hydroxyapatite- β-TCP/agarose disks cultured with L929 fibroblasts and Saos-2 osteoblasts. They observed that both cell types adhered and proliferated on the biomaterial surface maintaining their characteristic morphology and that the biomaterial induced apoptosis in Saos-2 osteoblasts and a transitory stimulation of fibroblast mitochondrial activity. They concluded that the obtained scaffold can be potentially applied for bone substitution and repair (Alcaide et al. 2008).

Agar has also been used to encapsulate cells (such as islets of Langerhans, chondrocytes, Adherent Crandall-Reese feline kidney cells). Shinji et al. (2007) prepared spherical capsules of around 140 μm diameter by extruding aqueous agarose-gelatin conjugate and then encapsulated adherent crandall-reese feline kidney cells, reporting the enhancement of degree of proliferation and mitochondrial activity.

These studies demonstrate that agarose gels promote cell retention, proliferation, and chondrogenesis in vivo and in vitro. The intrinsic mechanical and cellular mechano-transduction properties of agarose hydrogels have been reported to be dependent on the agarose concentration. However, there is significant variation in the reported values for the compressive properties of agarose gel at any given concentration. Complete and concise characterization of the mechanical properties of these hydrogels at different concentrations is necessary to facilitate their use as a scaffold material in cartilage tissue engineering and to properly interpret the results of cellular mechano-transduction studies that utilize these gels.

10.2.1.4 Alginate

Alginate is a high-molecular mass polysaccharide extracted from various species of kelp. Alginates are also produced extracellularly by *Pseudomonas aeruginosa* and *Azotobacter vinilandii* (Johnson et al. 1997, Ertesvag and Valla 1998). The function of alginates in algae is primarily skeletal with the gel located in the cell wall and intercellular matrix conferring the strength and flexibility necessary to withstand the force of water in which the seaweed grows (Ertesvag et al. 1996). Alginates are composed of linear anionic block copolymers of heteropolysaccharides of β-D-mannuronic acid and α-L-guluronic acid residues, which can be arranged in different proportions and sequences along the polymer chain. Another important property of alginate gels is the occurrence of self sol-gel transition without any alteration of temperature. This sol-gel transition results in the formation of channel-like pores. The dimensions of these pores depend on concentration, nature, and conformation of the alginate.

The derivatives of alginates with monovalent metals (like sodium) are soluble in water and may form viscous solutions. Sodium alginate forms relatively stable hydrogels through ionotropic gelation in the presence of multivalent ions like Ca^{2+}. During calcium cross-linking, hydrogels do not undergo excessive swelling and subsequent shrinking; thus, they can maintain their form. The cross-linking process can be carried out under very mild conditions, at low temperature and in the absence of organic solvents and hydrogels of different shapes can be prepared. Alginate gels with uniform concentration can be obtained by cross-linking via internal setting method as described by Rassis et al. (2002) and Onsoyen (1992). By this method, a very regular gel network of alginate is formed through a slow release of calcium ions by using an inactive form of the cross-linking ion, either bound by a sequestering agent such as phosphate, citrate, or Ethylenediaminetetraacetic acid (EDTA) or as a very low solubility salt, as $CaSO_4$, or as a salt insoluble at neutral pH like $CaCO_3$, in association with a slowly hydrolyzing lactone, usually D-glucono-δ-lactone (GDL). Since GDL generates an acidic pH, the calcium ions are gradually released and captured by guluronic residues of alginate (Smidsrod and Draget 1996, Kuo and Ma 2001).

Owing to an outward flux of cross-linking calcium ions into the surrounding medium, alginate gels can gradually lose their mechanical stability in biological fluids. The instability is also favored due to the presence of high concentration of nongelling ions such as sodium and magnesium. To address this problem, Yeom and Lee (1998) have attempted to introduce stable covalent cross-links using bi-functional cross-linkers, such as glutaraldehyde. Stabilization can also be achieved by adding free calcium ions to the medium while maintaining Na:Ca ratio at 25:1 for high-guluronate alginates and 3:1 for low-guluronate alginates.

Alginate scaffolds with tube-like pores can be fabricated by using alginate/calcium phosphate composites without heating them to higher temperatures. Additionally, nanocrystalline hydroxyapatite could be synchronously precipitated during the sol-gel transition by adding phosphate ions into the alginate solution (Dittrich et al. 2006).

Gelinsky et al. have produced biphasic monolithic alginate scaffolds for the repair of joint osteochondral defects. The scaffolds contained both HAp phase and an alginate/hyaluronic acid composite hydrogel phase. Even if the scaffold consisted of two layers, the channel-like pores were running through the whole scaffold crossing the interface (Gelinsky et al. 2007).

Several therapeutic agents, including antibiotics, enzymes, growth factors, and DNA, have already been successfully incorporated in alginate gels, retaining a high percentage of biological activity (Rowley et al. 1999, Albarghouthi et al. 2000). Moreover, alginate hydrogels have been widely studied for cartilage and bone regeneration applications as scaffolds and vehicles for biologically active molecules or cells (Tortelli and Cancedda 2009).

10.2.1.5   Chitin and Chitosan

Chitin is the principal structural component of the exo-skeletons of invertebrates, such as crustaceans and insects, and it is also present in the cell walls of most fungi and many algae. It is a semicrystalline homopolymer of β-(1/4)-linked *N*-acetyl-D-glucosamine residues (GlcNAc) (Lezica and Quesada-Allue 1990). Depending on its source of availability, chitin occurs in two allomorphic forms, namely, α and β forms, α-Chitin being the most abundant. The allomorphic forms can be distinguished by IR spectroscopy and solid-state NMR and XRD (Rinaudo 2006). Owing to its biodegradability, nontoxicity, antimicrobial, and gel-forming properties, chitin has been widely employed in biomedical field (Kifune 1992, Shigemasa and Minami 1995, Hirota et al. 1996, Freier et al. 2005b). Chitin can be converted into soluble derivatives such as chitosan, carboxy methyl chitin, and glycol chitin (Kurita 2001). Among all the products of hydrophilic modifications of chitin, carboxymethylated-chitin (CMCH) attempted as a promising material for biomedical applications (Sashiwa and Aiba 2004). CMCH is an amphiprotic ether derivative, containing –COOH groups and –NH$_2$ groups in the molecule. CMCH is prepared by using monochloracetic acid in the presence of concentrated sodium hydroxide. CMCH exhibits high viscosity, large hydrodynamic volume, and film and gel-forming capacity which makes it an attractive option for biomedical applications. It was also observed that CMCH possesses wound healing and inhibition of keloid formation properties (Wongpanit et al. 2005).

Chitosan, a de-acetylated derivative of chitin, is a fiber-like substance and it is soluble in dilute acids. Chitosan is a high molecular, nontoxic, mainly biodegradable polymer similar to cellulose except in C-2 position. C-2 position of chitosan is occupied by chemically reactive free amino groups (–NH$_2$), whereas C-2 position of cellulose is occupied by hydroxyl (–OH) groups. Chitosan possesses positive ionic charges, which facilitates its chemical binding with negatively charged fats, lipids, cholesterol, metal ions, proteins, and macromolecules (Li et al. 1992). Cationic nature of chitosan is primarily responsible for electrostatic interactions with anionic glycosaminoglycans, proteoglycans, and other negatively charged molecules. Chitosan has free amine groups which are potentially reactive sites for many chemical reactions (Knaul et al. 1999).

The degree of deacetylation (DD) of chitosan varies between 56% and 99% depending on the crustacean species and the preparation methods. The number fraction (%) of *N*-acetyl-D-glucosamine GlcNAc residues in the polymer chain is designated by degree of acetylation and it influences chitosan physico-chemical properties such as solubility, reactivity, biodegradability, and cell response (Khor and Lim 2003 Freier et al. 2005a). Deacetylation also influences the biodegradability and immunological activity of chitosan (Tolaimate et al. 2000).

The degree of deacetylation can be determined by ninhydrin test (Curotto and Aros 1993), linear potentiometric titration, nuclear magnetic resonance spectroscopy, and infrared spectroscopy (Tan et al. 1998). Among these, the IR spectroscopy method is commonly used for the estimation of chitosan DD values (Rinaudo 2006). Chitosan can be chemically modified by attaching functional groups in order to favor its solubility in neutral or alkaline solutions and also to control some of its physical properties. The utilization of chitosan as a biomaterial can be drastically broadened by modifying it with hydrophilic and hydrophobic moieties. The hydrophilic modification includes PEG grafting, sulfonation, and carboxymethylation. Hydrophobic modifications have been attained by using palmitic acid, linolenic acid, linoleic acid, and deoxycholic acid (Kim et al. 2007). The chemical modification of chitosan also controls the interaction of chitosan with drugs/bioactive components, enhances the load capability, and tailors the release profiles (Prabaharan 2008).

Grafting of chitin and chitosan to synthetic polymers has also been performed as an interesting way to extend their use as biomaterials for tissue engineering application. The physico-chemical and biological properties of grafted co-polymers mainly depend on molecular structure, length, and number of side chains attached to it. Various grafting techniques have been exploited for chitosan such as radical or radiation-induced method, microwave irradiation, and dendronization. Graft copolymerization results in the formation of tailored hybrid materials of chitosan and synthetic polymers which are widely used as potential drug carriers (Carreira et al. 2010). For instance, grafting PEG onto chitosan has led to water-soluble chitosan derivatives, which have been proposed as carriers for anticancer drugs. The PEG-g-chitosans, prepared by using carbodiimide as initiator, have been found to aggregate in aqueous medium due to hydrogen bonding. When such derivatives have been prepared via reductive alkylation using PEG monoaldehydes, followed by further cross-linking, their water solubility has been found restricted and they have been found to gradually re-aggregate further, then becoming insoluble (Dal Pozzo 2000). Such a property is highly useful in wound dressings to prevent tissue adhesion in internal surgery. Ideally, such materials are insoluble when positioned in the surgical wound, but undergo progressive bioerosion leading to complete resorption. Based on choice of PEG substitution and cross-linking, derivatives having high swelling capacity and enhanced hydrophilicity can be obtained. Moreover, chitosan was largely used to prepare composite scaffolds for tissue regeneration. In particular, Zhang and Zhang (2001) have produced a strong macroporous, bioactive, and biodegradable composite based on chitosan and calcium phosphate. Two types of calcium phosphates, β-tricalcium phosphate and calcium phosphate inverted glass, have been incorporated into the composite. They have demonstrated that chitosan provides a structural skeleton fixing the scaffold form whereas the calcium phosphate bioactivity enhanced the attachment of osteoblasts. The mechanical strength of the scaffold depended on the weight ratio of chitosan with respect to the two types of calcium phosphates (Zhang and Zhang 2001).

On the other hand, several injectable materials based on chitosan and its derivatives have been attempted for bone regeneration (Kim et al. 2008). The mechanical characteristics of chitosan can be augmented by HAp and resulting materials can be applied as bone substitutes for bone repair and reconstruction (Huang et al. 2008). Another interesting application of hydroxyapatite–chitin–chitosan composites is the formation of self-hardening paste which can be employed as bone-filling material for guided tissue regeneration for the treatment of periodontal bony defects (Xu et al. 2008). In a recent study, Madhumathi et al. have reported chitin–HAp composite membranes by using an alternate soaking method. They employed $CaCl_2$ and Na_2HPO_4 solutions to prepare Chitin–Hap composite membranes within very short time based on the wet process of Hap. They observed the formation of apatite crystals on the surface of chitin membranes and revealed that the size and deposition of the crystals are directly proportional to the number of immersion cycles. They hypothesized that the presence of apatite layers on the surface of chitin membranes increased the biocompatibility and provided a suitable substratum to facilitate cell attachment, adhesion, and spreading (Madhumathi et al. 2009).

To conclude, many researchers have been attempting chitin and its derivatives, particularly chitosan, as promising materials for tissue regeneration applications. The advantage of chitin derivatives over other polysaccharides such as cellulose, starch, agar, etc., attributed to its chemical structure, which allows specific modifications without too many difficulties at the C-2 position, as described earlier. Furthermore, chitosan can be easily molded to various forms and its derivatives are biodegradable by lysozomal enzymes which make chitosan as an exciting and promising material for tissue regeneration applications. Table 10.1 represents some of the recent research publications (from 2009) which are based on chitin and/or chitosan and their modified forms.

10.2.1.6 Hyaluronan/Hyaluronic Acid/Hyaluronate

Hyaluronic acid (HA) is a GAG component of connective tissue, synovial fluid of vertebrates, and the vitreous humor of the eye (Drury and Mooney 2003). It also occurs as an extracellular polysaccharide in a variety of bacteria and it can be produced from *Streptococcus zooepidemicus* and *Streptococcus equi*. HA is composed of alternating disaccharide units (250–25,000) of D-glucuronic acid and N-acetyl-D-glucosamine with β(1 → 4) interglycosidic linkage. HA is an important component of ECM and displays a vital role in lubrication; water sorption; water-retention; and a number

TABLE 10.1
Scaffolds Based on Chitosan

Scaffold Type	Method of Preparation	Active Bio Molecule/Cell Types	Tissue Engineering Application	References
Collagen/hyaluronan/ chitosan composite sponges	EDCI crosslinking followed by freeze drying	—	Potential biomedical applications	Lin et al. (2009)
Hybrid chitosan-silica xerogels	Sol–gel process followed by freeze drying	TMOS, $CaCl_2$ and TEP as calcium and phosphorous oxide precursors	Bone regeneration	Lee et al. (2009)
Chitosan–nanoHAp composite scaffolds	Ultrasonic dispersion of Hap in chitosan solution followed by freeze drying	Nano HAp	Bone tissue engineering	Han and Misra (2009)
Chitosan–gelatin hybrid scaffolds	Cross linking with GTA solution followed by molding in SL apparatus and freeze drying	Hepatocytes	Hepatic tissue engineering	Jiankang et al. (2009)
CMCH/poly(vinyl alcohol) nanofibrous - HAp composite scaffolds	Electrospinning followed by cross linking with GTA followed by soaking in Hap suspension	HAp	Bone tissue engineering	Shalumon et al. (2009)
Alginate dialdehyde cross-linked chitosan/calcium polyphosphate composite scaffold	Freeze drying	CPP	Meniscus tissue engineering	Wang et al. (2010)
Chitosan-graft-β- cyclodextrin scaffolds	Grafting of CD on chitosan in presence of EDCI and NHS followed by freeze drying. Drug loaded scaffolds prepared by immersing in drug solutions in presence of GTA	Ketoprofen	Tissue engineering applications	Prabaharan and Jayakumar (2009)
Chitosan–gelatin/ nano-bioactive glass ceramic composite scaffolds	Mixing and GTA crosslinking followed by freeze drying	nBGC	Alveolar bone tissue engineering	Mathew et al. (2010)
Combined brushite- chitosan system	Immersing chitosan sponges in TCP cement mixture	β-TCP VEGF and PDGF	Bone regeneration	De La Riva et al. (2010)
Temperature- responsive chitosan hydrogels	Mixing CH-GP solution with cross-linking solution consisting filter-sterilized HEC	Sheep chondrocytes	Articular cartilage repair	Hao et al. (2010)
Chitosan collagen composite hydrogels	Mixing of appropriate volumes	hBMSC-	Bone tissue engineering	Wang and Stegemann (2010)

of cellular functions such as attachment, migration, and proliferation. In water solutions, HA occupies a volume approximately 1000 times higher than that in its dry state (Mano et al. 2007). Furthermore, its abundant repelling anionic groups bind cations and water molecules leading to a peculiar visco-elastic behavior of solutions (Gatj et al. 2005). Hyaluronate can be used as excellent biological absorber and lubricant. Moreover, HA enhances tissue formation and repairs, provides a protective matrix for reproductive cells, and serves as a regulator in the lymphatic system and acts as a lubricating fluid in joints.

Various composite materials based on HA with proteins like gelatin and collagen have been fabricated and successfully used for biomedical applications. Composites based on HA and gelatin have been obtained in the form of porous sponges from gelatin-hyaluronan solutions into 90% (vol/vol) acetone/water mixtures containing a small amount of a carbodiimide (EDCI) as crosslinking agent (Choi et al. 1999). This sponge-type biomaterial was constructed for wound dressings or scaffolds for tissue engineering. When impregnated with antibacterial silver sulfadiazine, this hyaluronan-gelatin sponge has been found to facilitate epidermal healing (Choi et al. 2001).

Composite materials of hyaluronan and collagen have been prepared by dissolution of the two components in aqueous acetic acid followed by cross-linking with glyoxal or a periodate-oxidized starch dialdehyde (Rehakova et al. 1996). These composites possess resistance to collagenase and stimulate fibroblast growth (Rehakova et al. 1996). Alternatively, composite materials consisting of hyaluronan and collagen have also been prepared using polyethylene oxide and hexamethylene diisocyanate by cross-linking the dried hyaluronan/collagen coagulates. In addition, composite materials of hydroxyapatite-collagen-hyaluronan have been prepared by adding hydroxyapatite particles to a hyaluronan solution followed by blending with an aqueous dispersion of collagen fibers. The final material, which consisted of 90% hydroxyapatite, 9.2% collagen, and 0.8% hyaluronan (wt/wt), was biocompatible and mechanically robust and was used as bone defect filler (Bakos et al. 1997). Alternatively, a porous collagen-hyaluronan matrix was prepared by crosslinking collagen with periodate-oxidized hyaluronan; the resulting material supported new bone formation and has potential uses for the delivery of growth factors or as an implantable cell-seeded matrix (Alini et al. 2003).

HA can be chemically modified to produce mechanically and chemically sturdy materials. It can be modified at three positions per repeat unit of its structure: (1) –OH in C-6 position of the N-acetyl glucosamine; (2) –COOH of the D-glucuronic acid, and (3) –NH position of the N-acetyl-D-glucosamine. Among these, the main reactions were performed on –OH groups by sulfation, esterification, or periodate oxidation. The sulfation of hydroxyl group in HA with a sulfur trioxide–pyridine complex in dimethyl formamide (DMF) produced $HyalS_x$, where x is the degree of sulfation per disaccharide ($x = 1–4$). The $HyalS_x$ compounds are promising components as blood-compatible materials for medical device coating. For instance, immobilized $HyalS_{3.5}$ onto plasma-processed polyethylene has been reported to produce blood-compatible, antithrombotic PE surfaces. Furthermore, Chen et al. (1997) have converted $HyalS_x$ to a photolabile azidophenylamino derivative and photoimmobilized onto a poly (ethylene terephthalate) film. They have described that $HyalS_x$-coated surfaces display significant reduction of cellular attachment, fouling, and bacterial growths compared with uncoated surfaces and the coating prevents the degradation by chondroitinase and hyaluronidase.

Many chemical modifications have been performed also on the carboxylic acid position of HA to increase its hydrophobicity. For example, Benedetti et al. have attempted to esterify HA by using its alkyl salt, tetra (n-butyl) ammonium salt, by alkylation with an alkyl halide in DMF solution (Benedetti et al. 1993). As the percentage of esterification increased, the hydrophobicity increased resulting in a water insoluble material, which has been named as HYAFF® (Fidia Advanced Biopolymers). These HA esters have been used in the production of membranes and fibers, lyophilized sponges, and spray-dried to microspheres. These polymers have shown good mechanical strength in a dry state and lower robustness in a wet state. The degree of esterification has been found to influence the size of hydrophobic patches leading to a polymer chain network which is more rigid and stable, and less susceptible to enzymatic degradation. Finally, hyaluronan benzyl ester materials have been used as meshes and sponges for growth of human fibroblasts and for culture of chondrocytes and bone marrow–derived mesenchymal cells for repair of cartilage and bone defects (Benedetti et al. 1993). Recently, Hirakura et al. reported a novel hybrid hydrogel composite material by using cross-linked HA as the biocompatible controlled release matrix and partially modified pullulan with cholesteryl group nanogels as the host molecule with molecular chaperone as a model protein to support neurite growth. They demonstrated that low substituted HA hydrogels degraded more quickly owing to higher recognition for enzymatic cleavage and bioactivity and they concluded that upon addition of bioactive components, HA hydrogels can serve as potential components to support neurite growth (Hirakura et al. 2010).

In another approach, Segura et al. have shown the construction of hydrogels based on cross-linked HA with poly (ethylene glycol) diglycidyl ether. They also have incorporated collagen into the cross-linked network in order to support cellular adhesion in vitro. This construct possessed low water content and low degradation rate and it could be topographically patterned during gelation. The authors have also functionalized HA and HA/collagen scaffolds by introducing biotin into the HA backbone before cross-linking reaction in order to introduce avidin-streptavidin and neutravidin molecules on the surface of hydrogel which can be widely employed for tissue engineering applications (Segura et al. 2005).

Pouyani et al. have hypothesized that hydrazides, which have pKa values for the corresponding conjugate acids range between 2 and 4, would have retained their nucleophilicity at pH 4.75 and would have coupled efficiently to carbodiimide-activated glucuronic acid residues of hyaluronan. They have referred the utilization of dihydrazide compounds, such as adipic dihydrazide (ADH), to modify the chemical structure of HA. This chemical modification has provided multiple pendant hydrazide groups which may be further derivatized with drugs, biochemical probes, and for cross-linking agents. These abundant surface hydrazide groups of ADH-modified hyaluronan can be loaded with drug molecules and then cross-linked into a hydrogel which facilitates confinement of the drug in the bulk of the hydrogel (Pouyani et al. 1994). Alternatively, a limited number of the hydrazide groups could be cross-linked and the remaining hydrazide groups on the resulting hydrogel could be loaded with a therapeutic agent. For example, Luo et al. have synthesized HA-Taxol® bioconjugates by the conjugation of HA-ADH and ester-activated Taxol. This conjugate can be labeled with a fluorescent dye allowing the monitoring of uptake by human breast, ovarian, and colon cancer cells and measurement of the corresponding selective toxicity of the polymeric prodrug (Luo et al. 2000).

Hahn et al. have reported different hydrogel preparation techniques based on HA for three different applications, such as the management of tissue adhesion formation and soft tissue augmentation. They have investigated the influence of functional groups involved and type of cross-linking covalent bonds and density on degradation of HA hydrogels. Initially they have grafted HA with adipic acid dihydrazide by cross-linking with bis(sulfosuccinimidyl) suberate (BS3) and then modified with methacrylic anhydride in the presence of dithiothreitol (DTT; Michael addition), Traut's reagent (iminothiolane) in order to produce methacrylated HA (HA-MA) and thiolated HA (HA-SH), respectively. The degradation rates were reported as HA-SH > HA-ADH > HA-MA. The degradation rate of HA hydrogels can be controlled by changing the functional group of HA derivatives and the cross-linking reagents (Hahn et al. 2007).

In addition, the gelling state of HA hydrogel was another factor in controlling the degradation rate. HA and HA-modified surfaces show potential medical applications. Important new products are already available in market. For example, Synvisc is a specific device for visco supplementation therapy in osteoarthritis which increases the viscoelastic properties of the synovial fluid and the intercellular matrix of the synovial tissue and capsule. Similarly, Hylans are hydrogels marketed by Biomatrix obtained by cross-linking hyaluronan-containing residual protein with formaldehyde in a basic solution. Hylan gels have greater elasticity and viscosity than soluble hylan materials, while still retaining the high biocompatibility of native hyaluronan. Hylans have been investigated in a number of medical applications. In the treatment of degenerative joint disease and rheumatoid arthritis, hylans have been found to protect cartilage and prevent further chondrocyte injury, although the effect is reversible and viscosity dependent.

10.2.2 PROTEINS AS BIOMATERIALS

Natural proteins are a class of materials with enormous different properties which perform very specific biochemical, mechanical, and structural properties. Amino acids are the building blocks of proteins and consist of a central carbon linked to an amine group, a carboxyl group, a hydrogen atom, and a side chain (R groups). R groups can be classified as nonpolar groups, uncharged polar groups, or charged polar groups, and their distribution along the protein backbone gives the proteins distinct characteristics. The amino acid sequences of polypeptides and proteins possess multiple functional groups, exact molecular weights, defined branch points, and predisposed three-dimensional structures conferring them unique and complex bioactive properties. Despite the tremendous research in synthesizing synthetic materials able to mimic the properties of natural proteins, these attempts were not successful. Protein-based polymers mimic many features of the extracellular matrix and potentially direct the physiological functions, such as cell migration, growth, and organization of cells during tissue regeneration and wound healing and the stabilization of encapsulated and transplanted cells. The major drawback of the natural–origin proteins is their possible batch variation. One interesting strategy to overcome this issue is the recombinant protein technology where the mono dispersity and precisely defined properties of proteins as well as the predictable placement of cross-linking groups (binding moieties at specific sites along the polypeptide chain) or their programmable degradation rates make them very attractive and useful for drug delivery and tissue engineering (Rodriguez-Cabello et al. 2005). In the following, the most important natural polymers and their application in biomedical field are discussed.

10.2.2.1 Collagen

Collagen is a major natural constituent of connective tissue, staging approximately 30% of all vertebrate body protein and a major structural protein of any organ. Collagen is synthesized by fibroblasts originated from pluripotential adventitial cells or reticulum cells. Collagen is the major protein component of the ECM and provides support to connective tissues such as skin, tendons, bones, cartilage, blood vessels, and ligaments. In its native environment, collagen interacts with cells in connective tissues and transduces essential signals for the regulation of cell anchorage, migration, proliferation, differentiation, and survival (Malafaya et al. 2007).

The basic unit of collagen is a polypeptide consisting of the repeating sequence of glycine, proline, and hydroxyproline. These three polypeptide chains twine around one another as in a three-stranded rope and combines with 12 others to form the left-handed triple helix structure in collagen. Each chain has an individual twist in the opposite directions. The principal feature that affects a helix formation is a high content of glycine and amino acid residues (Piez 1985).

The strands are held together primarily by hydrogen bonds between adjacent -CO and -NH groups, and also by covalent bonds. The basic collagen molecule is rod shaped with a length and a width of about 3000 and 15 Å, respectively, and has an approximate molecular weight of 300 kDa. The mechanical stability and other physical characteristics of collagen result from intermolecular cross-linking forming macromolecular fibers (Shoulders and Raines 2009).

The applications of collagen as a biomaterial was first reported by Ehrmann and Gey in 1956, who compared the growth of cells and tissue explants on collagen and on glass. They found that collagen gels augmented cellular growth. Since then, collagen substrates have been known to affect the morphology, the migration, the adhesion and, in certain cases, the differentiation of cells. Till date, more than 28 types of collagens have been identified, among which type I collagen is the most abundant. However, the most extensively investigated types are collagen type I, II, III, and IV. Type I collagen is present in tendons, ligaments, and bones and consists of striated fibers between

80 and 160 nm in diameter. Type II collagen consists of fibers that are less than 80 nm in diameter and it is present in cartilage and intervertebral disks. Type III forms reticular fibers in tissues and strengthens the walls of hollow structures such as arteries, intestine, and uterus. Type IV is a highly specialized form of collagen present as a loose fibrillar network in the basement membrane (Friess 1998). Type VI is microfibrillar collagen and type VII is anchoring fibril collagen (Samuel 1998). Type IX, XI, XII, and XIV collagens are fibril-associated collagens and possess small chains, which contain some nonhelical domains.

Biomaterials made of collagen offer several advantages as they are biocompatible and non-toxic and have well-documented structural, physical, chemical, and biological suitable properties. Additionally, collagen can be cross-linked to modulate its mechanical, degradation, and water uptake properties. Furthermore, it promotes blood coagulation, it can be formulated in a number of different forms, and it is compatible with many synthetic polymers. The disadvantages for the utilization of collagen as a biomaterial includes the high cost for purification (particularly for type I collagen) and the variability of isolated collagen in terms of cross-linking density, fiber size, and presence of impurities. Apart from this, the hydrophilicity of collagen may lead to swelling and more rapid release of incorporated bioactive components. Other drawbacks are variability in enzymatic degradation rate as compared with hydrolytic degradation. The degradation of collagen is influenced by many enzymes such as collagenase and gelatinase and many other nonspecific proteinases (Friess 1998). Moreover, handling of collagen degradation is difficult because several factors have an impact on degradability of collagen, for instance, the penetration of cells into the structure may cause contraction. Side effects of collagen include the risk of bovine sponge form encephalopathy (BSF) and the occurrence of undesired mineralization (Chvapil 1979, Todhunter et al. 1994, Brown and Timple 1995, Liu et al. 1995, Ramshaw et al. 1995, Rao 1995, Friess 1998, Exposito et al. 2002).

As collagen can form fibers with extra strength and stability through its self-aggregation and in vivo cross-linking, it is widely used for tissue engineering applications. But, once collagen is extracted and reprocessed from its original source, it has weaker mechanical properties, thermal instability, and it may undergo rapid proteolytic breakdown. To overcome these problems, collagen has been cross-linked by a variety of cross-linking agents (Lee et al. 2001). The modified cross-linked collagen degrades at a much slower rate than native collagen further enhancing its in vivo efficacy as a biomaterial. The degree of cross-linking of collagen can be estimated via free amino group analysis and resistance to enzymatic digestion. Glutaraldehyde (GTA) has been frequently used as the cross-linking reagent as it is easily available and inexpensive and can effectively cross-link collagenous tissues quite rapidly (Khor 1997). However, GTA is toxic if released into the host. As a consequence, natural, milder, efficient agents were attempted to strengthen collagen. Transglutaminases (EC 2.3.2.13) are a group of enzymes that can catalyze several post-translational reactions. The most important of these reactions results in the cross-linking of peptides or proteins to form multimers via ε- (γ-glutamyl)lysine linkage using the side chains of lysine and glutamine residues. Hence, transglutaminases can be adopted to cross-link collagen matrices to further increase their strength and also to incorporate cell adhesion factors within the gel matrix, resulting in an enhancement of cell proliferation. Collagen modified with transglutaminases demonstrated greater resistance to the total complement of cell-secreted proteases and improved resistance to cell-mediated degradation. Transglutaminase-cross-linked collagen may be more robust than native collagen, either through the presence of cross-links which may disturb the native conformation, or via disruption of the native fibrillar form during fibrillogenesis. Alternatively, due to the increased resistance of the cross-linked collagen to Matrix metallo proteinase (MMP) degradation, fibroblasts may elicit an enhanced MMP response in a futile attempt to increase the rate of collagen breakdown. The cross-linked collagen elicits different cellular response compared to untreated collagen. TG-treated collagen matrix is in contact with the cells longer due to its increased proteolytic resistance, thus providing the required integrin-mediated signal to the Human osteoblastic cells (HOB cells) necessary for differentiation (Yamane et al. 2004).

The other cross-linking methods include carbodiimide treatment, acylazide treatment, and physical methods like dry heat treatment and exposure to ultraviolet (Avery and Bailey 2008). Cross-linking with carbodiimides, especially 1-ethyl-3-(3-dimethylaminopropyl) carbodiimide (EDCI), involves the formation of amide bonds between carboxylic and amino groups on the collagen molecules without becoming part of the actual linkage. At pH 5, carbodiimides first couple to carboxylic groups to form *o*-isoacylurea structures. The resulting activated intermediate is attacked by a nucleophilic primary amino group to form an amide cross-link and the iso-ureaderivative of the applied carbodiimide is then eliminated and can be washed out (Grabarek and Gergely 1990). The susceptibility of carbodiimide-cross-linked collagen to enzymatic degradation can be controlled by varying the degree of cross-linking via the reaction conditions (Olde et al. 1996). Alternatively, acyl azide method involves the formation of an intermediate deriving from induction of amide bonds (Petite et al. 1990). The azide functionalities couple with amino groups to give amide cross-links with superior cytocompatibility (Petite et al. 1995). Physical methods in cross-linking of collagen include dehydrothermal treatment (DHT) and/or exposure to ultraviolet light at 254 nm. In case of DHT treatment, the degradation of the triple helices can be minimized through removal of water content via vacuum prior to heating. Even small amounts of residual moisture can cause breakdown of the helical structures and proteolysis (Yannas and Tobolksy 1967). The dehydration itself already induces amide formation and esterification between carboxy and free amino and hydroxyl groups, respectively. Cross-linking of collagen with UV irradiation is initiated by the formation of free radicals on aromatic amino acid residues (Weadock et al. 1995) which indicates a rather limited maximum degree of cross-linking due to the small number of tyrosine and phenylalanine residues in collagen (Weadock et al. 1996).

Owing to its physico-chemical properties, flexibility in physico-chemical modifications and versatile cross-linking properties, collagen has been applied for various tissue engineering applications including dermal, bone, and cartilage regeneration (Glowacki and Mizuno 2008). Collagen-based scaffolds containing other biopolymers or bio-ceramics have also been widely investigated. Particularly, in case of bone tissue engineering applications, the composites of osteoconductive hydroxyapatite and collagen have been extensively investigated as promising scaffolding materials where the addition of hydroxyapatite induces mechanical strength (Murphy et al. 2010). Table 10.2 represents some of the recent research activities which involve the use of collagen either in native form or in modified form as a scaffolding material for tissue regeneration applications.

10.2.2.2   Gelatin

Gelatin is a protein derived from collagen by thermal denaturation either with very dilute acid or alkaline water solution treatment (Zhou and Regenstein 2005). Depending on the pretreatment of collagen, gelatin is classified into two types: type A and type B. Type A gelatin is synthesized from pig skin by acid denaturation whereas type B gelatin is produced by alkaline denaturation of cattle hides and bones. Such treatments result in partial cleavage of collagen cross-links, resulting in the formation of warm water-soluble collagen, that is, gelatin (Schrieber and Gareis 2007, Seal et al. 2001). Collagen breakdown into gelatin is dependent on three factors: temperature, time, and pH. High temperatures and long periods of exposure to heat accelerate the process. Gelatin exhibits similar amino acid composition and to some extent similar backbone structure of collagen, but it also exhibits its own properties (Olsen et al. 2003). Gelatin is a heterogeneous mixture of single or multi-stranded polypeptides, each with extended left-handed proline helix conformations and containing between 50 and 1000 amino acids. Gelatin contains many glycine (almost one in three residues, arranged every third residue), proline, and 4-hydroxyproline residues. A typical structure of gelatin is -Ala-Gly-Pro-Arg-Gly-Glu-4Hyp-Gly-Pro- (Gomez et al. 2002). Recent studies revealed that the helical structure of gelatin is composed of a 7/1 helix (the three strands forming a 7/2 helix) with a 60 Å axial repeat, with tripeptides forming each unit (Okuyama et al. 2006). Each of the three strands in the triple helix requires about 21 residues to complete one turn and typically there would be between one and two turns per junction zone (Oakenfull and Scott 2003).

TABLE 10.2
Collagen Based Matrices/Scaffolds for Tissue Regeneration

Scaffold Type	Method of Preparation	Active Bio Molecule/Cell Types	Tissue Regeneration Application	References
Tailor-made porous collagen–GAG matrices	Covalent attachment of GAGs to collagen matrices performed using EDCI and NHS followed by lyophilization	Immobilized GAGs CS and HS	Regeneration of tissues like tendon and skin.	Pieper et al. (2000)
Cross-linked type I and type II collagen matrices	Crosslinking by using EDCI followed by lyophilization	Chondroitin sulfate	Regeneration of articular cartilage tissues	Buma et al. (2003)
Two-layered collagen–tricalcium phosphate cell-free construct	DHT crosslinking followed by lyophilization	Growth-factor mixture (GFM)	Treatment of deep osteochondral defects	Gotterbarm et al. (2006)
Thin collagen scaffolds	Cross-linking under a 254 nm UV lamp for 3.5 h and then vacuum-desiccated for 48 h with fresh Drierite	Cultured RPE cells for in vivo characteristics	Age-related macular degeneration	Lu et al. (2007)
pH-responsive alkali treated collagen hydrogels	Crosslinking with MAD	—	Tissue engineering and drug delivery systems	Saito et al. (2007)
Porous collagen-apatite nano-composite foams	Chemical crosslinking by using EDCI-NHS followed by freeze drying	Nano- HAp crystals	Bone regeneration	Pek et al. (2008)
Collagen biomimetic films	Obtained a film through forced evaporation of the solvent in air at 25°C by using glycerine as plasticizer	Calcium acetate and sodium silicate	Skin regeneration	Chirita (2008)
Biomimetic collagen/nano-hydroxyapatite scaffold	Compositional and structural gradient composite scaffold was formed by precipitation of nano-HAp crystallites within the collagen fibril matrix	Nano-HAp crystals	Bone regeneration	Liu (2008)
mPCL–TCP/collagen composites	Fused deposition modeling	rhBMP-2	Skull regeneration in a rat calvarial defect	Sawyer et al. (2009)
Collagen membranes	By grafting	Basic fibroblast growth factor	Bladder regeneration	Chen et al. (2010)
Porous collagen scaffolds with hyaluronic acid oligosaccharides	Homogenization followed by freezing for 24 h and freeze drying for 24 h followed by EDCI crosslinking.	Hyaluronic acid oligosaccharides	Improved revascularization process for rapid wound healing	Perng et al. (2009)
Three-dimensional porous HAp/collagen composite scaffolds	Crosslinking with mTGase followed by freeze drying for 36 h	Hydroxyapatite	Bone tissue regeneration	Ciardelli et al. (2009)

Gelatin is a natural derived polymeric product and it has been widely utilized for the fabrication of scaffolds owing to its peculiarities: (1) the presence of both acidic and basic functional groups in the gelatin macromolecules, which render superior chemical modification properties to gelatin; (2) its capacity to form a specific triple-stranded helical structure not observed in synthetic polymers (this structure is formed in solutions at low temperatures) (particularly, the rate of formation of a helical structure depends on many factors such as the presence of covalent cross-bonds, gelatin molecular weight, the presence of amino acids, and the gelatin concentration in the solution); and (3) its specific interaction with water which is different than that observed with synthetic hydrophilic polymers. This peculiarity governs the structural and physico-mechanical properties of gelatin in the solid state (de Wolf 2003, Okuyama et al. 2006).

At optimal concentrations, gelatin solutions are in the sol state at around 40°C and change into gels when they are cooled down to room temperature. This gelation normally occurs by physical cross-linking, which results in the formation of "junction zones" and finally a three-dimensional branched network. The sol-gel transformation is due to a conformational disorder-order transition of the gelatin chains which form thermo-reversible networks by associating helices in junction zones which are stabilized by hydrogen bonds (Bigi et al. 2000).

Throughout this dynamic process, bonds are continually breaking and reforming within the gel, due to weak polymer–polymer interactions. Thus, at elevated temperatures, gelatin macromolecules assume the conformation of a statistical coil. Under specific conditions (temperature, solvent, pH), gelatin macromolecules can display a flexibility sufficient to realize a wide variety of conformations. This makes it possible to vary gelatin characteristics depending on its molecular structure (Djabourov et al. 1988). Upon aging, the gel gradually assumes a more stable conformation. The detailed structure of the final product is highly dependent on the temperature at which the gel is set. Although the structure continues to change and the helix content of the gels continues to increase, the physical restrictions on the polymers prevent the gel from ever reaching equilibrium. Hence, the formation of gelatin gels is often described as a "frustrated renaturation process" (Djabourov et al. 1988).

Gelatin, similar to synthetic high polymers, displays a wide molecular weight distribution. Gelatin can form a large variety of super molecular structures; from the simplest globular structure, typical amorphous polymeric structure, to a well-developed fibrillar structure with various intermediate states (Young et al. 2005). Furthermore, gelatin displays various physico-mechanical properties. (1) First, gelatin behaves very differently depending on macromolecular conformations. In case of collagen-like helical conformation of gelatin macromolecules, gelatin samples exhibit properties suitable for service at ordinary temperatures and humidities, while in the coiled conformation conditions, gelatin behaves as a brittle impractical material. (2) In the dry state, gelatin behaves as a brittle material regardless of the molecular conformations. (3) Gelatin features a wide temperature range of the glassy state with the upper limit of 205°C–210°C. This characterizes gelatin as a material with a fairly wide range of heat resistance. Moreover (4) gelatin, as a hydrophilic biopolymer, specifically interacts with water and undergoes drastic changes of its physico-mechanical properties, depending on the moisture content (Lakshmi and Cato 2007). With respect to collagen, gelatin does not express antigenicity in physiological conditions, and it is much cheaper and easier to obtain in concentrated solutions. On the other hand, gelatin exhibits poor mechanical properties which limit its possible applications as a biomaterial. Moreover, gelatin is soluble in aqueous solution, so gelatin materials for long-term biomedical applications have to be cross-linked to improve both their thermal and the mechanical stability (Michon et al. 1997).

Several attempts have been made to modify the physico-mechanical properties of gelatin by chemical cross-linking. In chemical cross-linking methods, cross-linkers are used to bond functional groups of amino acids. Commonly used chemical cross-linkers include formaldehyde, glutaraldehyde, polyepoxy compounds, dimethyl suberimidate, carbodiimides, and acyl azide (Khor 1997), similar cross-linking agent described earlier (Section 10.2.2.1). Such cross-linking methods have been widely used in the manufacture of various gelatin-based biomaterials. However these

chemical cross-linking reagents are generally cytotoxic, impairing the biocompatibility of bioprostheses, resulting in an increasing demand for a cross-linking reagent that can form stable and biocompatible cross-linked products, without causing problems of cytotoxicity. Frequently, a low-toxic
and naturally occurring reagent, genipin, has been used as a cross-linking agent for fixing biological
tissue. Genipin and its related iridoid glucosides, extracted from the fruits of *Gardenia jasminoides
Ellis*, can react with amino acid or proteins to form stable cross-linked products (Sung et al. 1999).
The mechanism of the reaction of amino acids or proteins with genipin is still not well understood
at present. The mechanism proposed by Touyama's research group (Touyama et al. 1994) for the
formation of the genipin-methylamine monomer is through a nucleophilic attack by methylamine
on the olefinic carbon at C-3 in genipin, followed by the opening of the dihydropyran ring and an
attack by the secondary amino group on the resulting aldehyde group. The blue-pigment polymers
are presumably formed through the oxygen radical-induced polymerization and dehydrogenation of
several intermediary pigments.

Gelatin has been widely used in the pharmaceutical industry as well as in the biomedical fields to
make hard and soft capsules, microparticles, sealants for vascular prostheses, wound dressing, and
adsorbent pad for surgical use and for scaffold preparation for regenerative applications (Lakshmi
and Cato 2007). Table 10.3 depicts some of the published articles on gelatin-based scaffolding material either in native form or in cross-linked form. Apart from this, researchers have also prepared
hybrid scaffolding materials based on gelatin and other polymers. Table 10.4 represents some of the
recent publications which involve scaffolding materials based on blends between gelatin and other
natural polymers. Many researchers also fabricated scaffolds based on the combination of gelatin
and synthetic polymers such as PLGA, PLA, PCL, etc., to improve the biocompatibility of the substrates through bioartificial blending. Bioartificial polymeric materials were initially conceived to
produce new systems combining the characteristics of synthetic polymers, such as good mechanical
properties, good processability, low production and transformation costs, with the specific properties of biodegradability and compatibility with cells and tissues associated to biopolymers. In
addition to these studies, scientists also synthesized composite scaffolds based on native gelatin or
hybrid polymers with inorganic components such as hydroxyapatite and bioactive glass to attain
synergic bioactivity and biomechanical properties (Table 10.5).

10.2.2.3 Fibrin-Fibrinogen-Fibronectin

Fibrin is a nonglobular, fibrous, insoluble protein matrix which forms immediately at the site of skin
injury in order to prevent blood loss. Fibrin formation involves the enzymatic cleavage of fibrinopeptides A and B followed by noncovalent polymerization of thrombin (Mosesson et al. 2001).

Fibrinogen is a monomer of fibrin and is composed of three pairs of nonidentical polypeptide chains, which are arranged as one central domain (E domain) and two terminal domains
(D domains) linked together by disulfide bridging within the N-terminal E domain. The central
domain contains fibrinopeptide E and fibrinopeptides A and B whereas the two terminal domains
are based on fibrinopeptide D linked to the central domain by a coiled coil segment and sites that
participate in cross-linking. These domains contain constitutive binding sites that participate in
fibrinogen conversion to fibrin, fibrin assembly, cross-linking, and platelet interactions. Fibrinogen
and fibrin play an important, interdependent role in clot formation, fibrinolysis, cellular and matrix
interactions, inflammation, wound healing and neoplasia by fibrin formation and the cascade of its
complementary interactions between specific binding sites on fibrinogen and extrinsic molecules
including proenzymes, clotting factors, enzyme inhibitors, and cell receptors.

Fibronectin is a large extracellular glycoprotein (a disulfide-bonded dimer of 220–250 kDa subunits) and is a multifunctional component of the ECM. It induces cell attachment and serves as
a substrate for cell adhesion and migration events in important physiological processes such as
embryogenesis, wound healing, hemostasis, and thrombosis (Hynes 1990, Pankov and Yamada
2002). It is also involved in many cellular processes like tissue repair, embryogenesis, and
blood clotting.

TABLE 10.3
Scaffolds Based on Crosslinked Gelatin

Scaffold Type	Method of Preparation	Active Bio Molecule/Cell Types	Tissue Engineering Application	References
Gelatin-resorcinol formaldehyde-glutaraldehyde glue	Cross linking with formaldehyde (37%) and GTA (50%) (9:1 weight ratio)	Resorcinol, fibrin glue and antifibrinolytic agent (aprotinin)	Adhesive glue even in the presence of degenerative tissue	Kodama et al. (1997)
Gelatin as a sealant for vascular graft	Gelatin impregnation process by using partially succinylated mixture of USP ossein gelatin with polyester, 20% aqueous formaldehyde as cross linker	Rifampin and tobramycin (antibiotics)	Prevention of bacterial infection in vascular graft	Javerliat et al. (2007)
Gelatin films as wound dressing	Film formation by freezing at −4°C for 3 days and sterilization by Gamma irradiation 25 kGy	EGF	Wound healing	Akane and Toshikati (2005)
Gelatin sponges for wound dressing	GTA crosslinking and freeze drying for 24 h	Human recombinant EGF	Wound dressing	Ulubayram et al. (2001)
Gelatin gels	A citric acid derivative as a novel cross-linker	Divalent anions as cross linkers	Tissue adhesive	Aoki et al. (2004)
Aligned porous gelatin scaffolds with micro tubules orientation structure	GTA crosslinking, unidirectional freezing by using liquid nitrogen and freeze drying for 48 h	—	Aligned porous scaffolds for tissue engineering applications	Wu et al. (2010)
Gelatin scaffolds containing artificial skin	Salt leaching method by using NaCl followed by freeze drying. Crosslinking by immersing dried samples in acetone:water mixture (9:1 by volume) containing 30% EDCI by gelatin weight for 24 h	Artificial skin	Porous gelatin scaffolds for skin defects	Lee et al. (2005)
Porous gelatin cryogels	Cryogenic treatment and a subsequent freeze-drying	Human umbilical vein endothelial cells (HUVEC), osteoblasts (MG-63 and CAL 72), human foreskin fibroblasts, glial cells (U373-MG) and epithelial cells (HELA)	Cell delivery tool in tissue engineering	Van Vlierberghe et al. (2006)
Nanofibrous gelatin/apatite composite scaffolds	Combination of thermally induced phase separation and porogen leaching by using freeze-drying in a salt-ice bath for 4 days and then vacuum drying at room temperature for another 3 days	Paraffin spheres were used as a template to create macropores and MC3T3-E1 osteoblast cells	Bone tissue engineering	Liu et al. (2009b)

TABLE 10.4

Hybrid Polymeric Scaffolds Based on Gelatin and Other Natural Polymers

Scaffold Type	Method of Preparation	Active Bio Molecule/Cell Types	Application	References
Chitosan–gelatin hybrid polymer network	Freeze-drying technique	—	Tissue engineering	Mao et al. (2003)
Genipin cross linked chitosan-gelatin blends	Genipin cross linking followed by freeze drying	—	Tissue engineering	Chiono et al. (2008)
Chitosan-gelatin hybrid scaffolds with well-organized microstructures	Combination of rapid prototyping (RP), micro-replication and freeze–drying techniques	Hepatocytes isolated from male Sprague–Dawley rats	Hepatic tissue engineering	Jiankang et al. (2009)
Hydroxypropyl chitosan-gelatin scaffolds	Thermo-crosslinking by using 1,4-butanediol diglycidyl ether for 12 h followed by freezing at −50°C and drying for 12 h	—	Corneal stroma tissue engineering	Wang et al. (2009)
Glycidyl methacrylated dextran (Dex-GMA)/ gelatin scaffolds with microspheres loaded with BMPs	BMP-free or BMP-loaded Dex-GMA/PEG microspheres were prepared by polymerization of Dex-GMA in the discontinuous phase of an aqueous PEG solution and scaffolds prepared by radical crosslinking and low dose γ-irradiation	BMPs and human PDLCs	Periodontal regeneration	Fa-Ming et al. (2007)
Silk sericin gelatin 3-D scaffolds	Sericin/ gelatin 3-D scaffolds were fabricated by pre-freezing sericin/ gelatin mixture solution (s) at −20°C, followed by lyophilization	—	Tissue engineering and biomedical applications	Mandal et al. (2009)
Porous chitosan-gelatin scaffold containing plasmid DNA encoding transforming growth factor-β1	Freezing at −20°C for 1 h, −80°C for 2 h, and finally lyophilizing for 48 h	Plasmid DNA as expression vector consisting of the coding sequence of TGF-β1 and the cytomegalovirus enhancer inserted at the upstream (pCMV-TGF-β1, 6896 bp)	Cartilage defects regeneration	Guo et al. (2006)
Agarose-gelatin conjugate scaffolds	Coupling using 1,1-carbonyldiimidazole (CDI) at 25°C	—	A promising matrix for tissue engineering applications	Shinji et al. (2007)
Silk fibroin/gelatin multilayered films	Layer-by-layer (LbL) assembly of various blends	Trypan blue, FITC-inulin and FITC-BSA as model components for controlled release	Controlled release of a wide spectrum of bioactive molecules	Mandal et al. (2009)

TABLE 10.5

Composite Polymeric Scaffolds Based on Gelatin and Bioceramics

Scaffold Type	Method of Preparation	Active Bio Molecule/Cell Types	Tissue Engineering Application	References
Porous gelatin–starch composite scaffold with nanohydroxyapatite	Microwave vacuum drying	Nano hydroxy apatite crystals	Hard tissue (bone) engineering	Sundaram et al. (2008)
Composite chitosan-gelatin/ nanohydroxyapatite scaffolds	GTA crosslinking and freeze drying for 48 h	Nano-HAp crystals	Bone tissue engineering	Peter et al. (2010a)
Composite gelatin-nano hydroxyl apatite/fibrin scaffold loaded with rh BMP-2	Step 1: Cross-linking of gelatin with GTA in the presence of medical-grade nHAp granules to form hydrogel.			
	Step 2: Infusion of above hydrogel into a solution containing 100 μg rhBMP-2 and 50 mg fibronectin + 400 IU thrombin in 1 mL of $CaCl_2$ and freeze drying	Nano HAP crystals and rhBMP-2	Bone regeneration	Liu et al. (2009a)
Chitosan-gelatin/ nano-bioactive glass ceramic composite scaffold	GTA cross-linking and freeze drying for 48 h	Bioactive glass ceramic nanoparticles	Alveolar bone tissue engineering	Peter et al. (2010b)
Osteochondral scaffold of ceramic-gelatin assembly (CGA)	Composite ceramic part was prepared by pressure compression of TCP powder in the presence of amorphous calcium polyphosphate (CPP) powders. Genipin crosslinking and freeze drying for 48 h	Osteochondral cells	Articular cartilage repair	Lien et al. (2009)
3-dimensional gelatin/ montmorillonite–chitosan scaffolds	Glutaraldehyde crosslinking and freeze drying for 48 h	The rat stromal stem cells cells and montmorillonite	Tissue engineering applications	Zheng et al. (2007)
Composite gelatin films with hydroxyapatite and or bioactive glass	Genipin crosslinking followed by drying at room temperature for 48 h	Hydroxyapatite/bio active glass	Bone tissue engineering	Gentile et al. (2010)

Fibronectin exists in two main forms: (1) as an insoluble glycoprotein dimer that serves as a linker in the ECM (extracellular matrix), which is made by fibroblasts, chondrocytes, endothelial cells, macrophages, as well as certain epithelial cells and (2) as a soluble disulfide-linked dimer found in the plasma (plasma FN) which is synthesized by hepatocytes. Each monomer is composed of many repeats of three types (I, II, and III) of small domains (Baron et al. 1991).

Fibronectin can be incorporated onto the surface of numerous biomaterials to enhance the availability of its cell-binding site. The repeat domains of fibronectin reversibly unfold in a distinctive sequence under applied tension. Under certain mechanical conditions, cellular activity changes the tension applied to fibronectin molecules resulting in the formation of cryptic sites. These cryptic sites include additional sequences which bind integrin receptors on cells, other matrix molecules, and other fibronectin molecules, as well as antiadhesive sites. Thus, exposure of cryptic sites by mechanical unfolding may act as a trigger for a variety of cell-matrix interactions (Watanabe et al. 2000). On a scaffold with the surface having oriented fibronectin layer, human umbilical vein endothelial cells can spread significantly faster in a more spherical way (Calonder et al. 2005).

Fibrin-based materials are biocompatible and biodegradable, can be produced from human serum, and have been widely used in short-term conditions such as adhesion and sealing as they are soluble and unstable due to degradation by fibrinolysis. Furthermore, as they are autologous, the risks of toxic degradation or inflammatory reactions are negligible. The key aspect for the versatile utilization of fibrin as biomaterial relays on characteristic in vitro structural fiber network formation which exactly mimics that of in vivo plasma clots (Ye et al. 2000).

In the field of tissue engineering, fibrin has been attempted for a variety of tissue engineering applications such as to fabricate sealants in surgery and porous scaffold for bone regeneration (Eyrich et al. 2007). Biomaterials produced from fibrin display varying mechanical and structural properties depending on the initial fibrinogen concentrations, and hence, they are applied for various tissue regeneration applications. For instance, biomaterials produced by using higher concentrations of fibrinogen have been used as tissue sealants (Traver and Assimos 2006), whereas fibrin hydrogels, which have been produced at lower fibrinogen concentrations, have been studied as cardiovascular grafts (Tiwari et al. 2002, del Mel et al. 2008) and bioengineered cartilages. For use in tissue engineering applications, fibrin gels have been prepared by combining fibrinogen and thrombin solutions containing calcium ions. These preparations are not suitable in conditions where the scaffolds are needed to be intact for longer durations. One of the major challenges in using fibrin as a scaffold relays on its rapid degradation nature. However, the rate and extent of its degradation varies depending on cell type and culture conditions; in some cases, it is too rapid as compared to the rate at which ECM is produced by the cells. For instance, this happens in case of smooth muscle cells which rapidly degrade the fibrin, leading to the complete degradation of scaffold before significant ECM is produced to replace it (Ahmed et al. 2008). Rapid degradation of the fibrin matrix can be controlled by using serine protease inhibitor aprotinin. Tuan et al. have reported the successful inhibition of fibrin degradation by using aprotinin in concentrations of 500 KIU/mL (Tuan et al. 1996).

Herbert et al. have studied lysine analog, 1-aminocaproic acid (ACA), as an inhibitor to slow down fibrin degradation which competitively inhibits attachment of plasmin and plasminogen to fibrin (Herbert et al. 1996). Bach et al. utilized fibrin glue as a suspending vehicle for the delivery of urothelial cells in urethral reconstruction. They applied such suspension to a connective tissue capsule tube formed in vivo. The urothelial cells spread and formed an adherent and confluent cell layer within 2 weeks of implantation, at which time the fibrin clot had already been replaced with other connective tissue. Similar technique has also been applied to construct skin grafts and a composite neotrachea (Bach et al. 2001).

Han et al. have constructed bioengineered ocular surface by using fibrin as a matrix for corneal epithelial stem cells. They hypothesized that fibrin provides a favorable matrix environment for epithelial cell growth and differentiation during wound healing. In their work, fibrinogen and thrombin were isolated from blood plasma in the lab and contained a number of other plasma proteins. This demonstrated the possibility of using autologous sources for the fibrin gel. Human corneal stem cells suspended in the fibrin proliferated and exhibited markers of differentiation, expressing keratin 3, as well as keratin 19 in some cases. Keratin 3 is a marker of corneal-type epithelial differentiation, while keratin 19 is proposed as a marker for corneal stem cells. No functional tests were performed, but the tissue was found to be soft, pliable, and elastic (Han et al. 2002).

Fibrin has also been attempted for cardiac tissue engineering applications such as construction of bio-artificial arteries and tissue-engineered heart or venous valves based on the tendency of cells to remodel it and replace it with new ECM. Ye et al. attempted to construct heart valves by combining myofibroblasts with commercially available fibrinogen and thrombin to form a fibrin gel. However, the mechanical properties of the myofibroblast-seeded fibrin gel were not examined (Ye et al. 2000).

Neidert et al. enhanced the mechanical properties of fibrin by using growth factors such as TGF-β (transforming growth factor-β), plasmin, and insulin to increase the amount of ECM produced and observed that the fibrin constructs contained nearly 8 times the amount of collagen as those without the additives after 51 days of incubation. In addition, the ultimate tensile strength and modulus have also been improved by the addition of these growth factors. The results suggested that the tensile strength of fibrin with growth factors was improved 10-fold than that of the samples without growth factors (Neidert et al. 2002). Osathanon et al. have developed micro/nanoporous scaffolds based on fibrin and calcium phosphate with well-controlled porous architecture for bone regeneration. The scaffolds were fabricated by using poly (methyl methacrylate) (PMMA) sphere-templating and leaching technique (Osathanon et al. 2008).

Apart from this, significant efforts have also been made to develop fibrin as a drug delivery vehicle through both modification of the molecule and engagement of the native polymer system biology. For instance, functionalized polyethylene glycol was explored as a means to covalently bind the therapeutic proteins to fibrinogen backbone thus facilitating their delivery through fibrin polymer (Barker et al. 2001). Recently, Soon et al. (2010) introduced tetra peptide knob sequences of fibrin to other proteins to empower the binding capacities to them via knob-pocket interactions, thus improving their properties. They adopted integrin-binding fibronectin 9th and 10th type III repeats (FnIII9–10), which do not bind to fibrinogen as a model protein domain to demonstrate that knob–protein binds stably to fibrinogen via specific knob-pocket interactions.

To summarize, fibrin has been attempted to fabricate variety of scaffolds for tissue regeneration applications based on its inherent characteristics such as biocompatibility, biodegradability, and capability of entrapping cells directly. Furthermore, it has adequate mechanical properties and promotes cell growth and remodeling. Fibrin gels with large mesh size, low elastic modulus, and adequate binding sites for integrins and other cellular targets have been demonstrated to be suitable for tissue engineering applications. While there are some challenges to be addressed, fibrin has shown promise as a biopolymer scaffold in several tissue engineering applications including ocular implants, skin, blood vessels, and heart valves. The in vivo efficacy of fibrin gels cannot be assessed from its in vitro efficacy without considering the possible instability of gels at desired sites and the host immunologic response. Hence, researchers have attempted to derive fibrin from nonmammalian source such as farmed salmon to minimize the host immunological responses (Tavares-Dias and Oliveira 2009). More work is necessary to further develop these applications and examine new ones.

10.3 CONCLUDING REMARKS

Regenerative medicine is an alternative tool for reconstructive surgery and organ transplantation which relies on the induction of tissue regeneration through suitable scaffolding materials. At the current stage of research, tissue regeneration is still in its early development, and it will take a long time to become well established, although some approaches are nearly ready for clinical application. The basic mechanism of normal tissue regeneration involves the action of a variety of growth factors and bioactive components on cells so as to form their mutual networks. Time of release, site of release, and concentration of growth factors action are delicately regulated in the body.

Nonetheless, it will be impossible to replicate living systems only by the scientific knowledge and currently available technologies. If a key growth factor is supplied to the target site at the right time, for the appropriate period of time, and at the right concentration, the living body system will be naturally directed toward the process of tissue regeneration. Once the right direction is taken, it is highly possible that the intact biological system of the body will start to function, resulting in automatic achievement of tissue regeneration. Hence, successful tissue regeneration demands a collaborative research between material, pharmaceutical, biological, and clinical scientists. A great deal of research work is needed in order to obtain an increased number of commercially available and clinically successful natural-based systems for tissue regeneration. At the current scenario of research, several scaffolds based on natural origin polymers have been widely emerged as suitable tools for tissue regeneration applications. They display advantageous features in terms of intrinsic cellular interaction, degradability, biocompatibility, low cost and availability, and similarity with the extracellular matrix. However, these materials do also exhibit some disadvantages such as poor mechanical properties, difficulties in controlling the variability from batch to batch, mechanical properties or limited processability, which are the major obstacles for their widespread use. Therefore, it is necessary to increase the awareness about these natural polymers in order to enable the development of new approaches such as utilization of recombinant DNA technologies for producing more biocampatible polymers and better purification techniques to render them suitable material properties. To conclude, natural-based polymers or nature-inspired materials are attractive materials of choice for tissue regeneration applications. The purpose of this chapter can be accomplished if it provokes the reader's enthusiasm toward research field of tissue regeneration based on natural polymers and by using recent emerging suitable techniques to warranty complete functional tissue regeneration.

ABBREVIATIONS

ACA	1-aminocaproic acid
ADH	adipic dihydrazide
BMP-2	bone morphogenic protein-2
BSA	bovine serum albumin
BSF	bovine sponge form encephalopathy
CA	cellulose acetate
CD	cyclodextrin
CGA	ceramic-gelatin assembly
CH-GP	chitosan-β-sodium glycerophosphate
CMC	carboxy methyl cellulose
CMCH	carboxymethylated-chitin
CPP	calcium polyphosphate
DD	degree of deacetylation
Dex-GMA	glycidyl methacrylated-dextran
DHT	dehydrothermal treatment
DMF	dimethyl formamide
DNA	deoxyribonucleic acid
ECM	extracellular matrix
EDCI	1-ethyl-3-(3-dimethylaminopropyl) carbodiimide
EGF	epidermal growth factor
FGF-2	fibroblast growth factor
FITC	fluorescein isothiocyanate
FN	fibronectin
GAGs	glycosaminoglycans
GFM	growth-factor mixture

GlcNAc	*N*-acetyl-D-glucosamine
GTA	glutaraldehyde
HA	hyaluronic acid
Hap	hydroxyapatite
hBMSC	human bone marrow-derived stem cells
HEC	hydroxy ethyl cellulose
HELA	human epithelial cells
HOB	human osteoblastic cells
HPC	hydroxyl propyl cellulose
HPMC	hydroxyl propyl methyl cellulose
HUVEC	human umbilical vein endothelial cells
MAD	malic acid derivative
MC	methyl cellulose
MMP	matrix metallo proteinase
mPCL	methylated poly caprolactone
nBGC	bioactive glass ceramic nanoparticles
NHS	*N*-hydroxysuccinimide
PC	plant cellulose
pCMV	plasmid cyto megalo virus vector
PDGF	platelet derived growth factor
PDLC	periodontal ligament cells
PEG	poly(ethylene glycol)
PLA	poly(lactic acid)
PLGA	poly(lactic-co-glycolic acid)
RC	regenerated cellulose
rfGD	radio frequency glow discharge
RGD	(Arg-Gly-Asp)
RPE	retinal pigment epithelium
SBB	starch-based biomaterials
SBF	simulated body fluid
SCA	starch with cellulose acetate
SEVA-C	starch with ethylene vinyl alcohol
SL	stereo lithography
SPCL	starch with polycaprolactone
SPI	soya protein isolate
SPLA	starch with poly(lactic acid)
TCP	tri-calcium phosphate
TEP	triethyl phosphate
TG	transglutaminase
TGF	transforming growth factor
TMOS	tetramethylorthosilane
UV	ultra violet region
VEGF	vascular endothelial growth factor

REFERENCES

Ahmed, T.A., Dare, E.V., and Hincke, M. 2008. Fibrin: A versatile scaffold for tissue engineering applications. *Tissue Eng Part B Rev* 14:199–215.

Akane, T. and Toshikati, N. 2005. Acceleration of wound healing by gelatin film dressings with epidermal growth factor. *J Vet Med Sci* 67:909–913.

Albarghouthi, M., Fara, D.A., Saleem, M. et al. 2000. Immobilization of antibodies on alginate-chitosan beads. *Int J Pharm* 206:23–34.

Albertsson, A.C. and Karlsson, S. 1996. Macromolecular architecture-nature as a model for degradable polymers. *J Macromol Sci Pure Appl Chem* 33:1565–1570.

Alcaide, M., Serrano, M.C., Pagani, R. et al. 2008. L929 fibroblast and Saos-2 osteoblast response to hydroxyapatite-β-TCP/agarose biomaterial. *J Biomed Mater Res A* 89:539–549.

Alini, M., Li, W., Markovic, P. et al. 2003. The potential and limitations of a cell-seeded collagen/hyaluronan scaffold to engineer an intervertebral disc-like matrix. *Spine* 28:446–454.

Alves, C.M., Yang, Y., Carnes, D.L. et al. 2007. Modulating bone cells response onto starch-based biomaterials by surface plasma treatment and protein adsorption. *Biomaterials* 28: 307–315.

Aoki, H., Taguchi, T., Saito, H. et al. 2004. Rheological evaluation of gelatin gels prepared with a citric acid derivative as a novel cross-linker. *Mater Sci Eng C* 24:787–790.

Appelqvist, I.A.M. and Debet, M.R.M. 1997. Starch-biopolymer interactions—A review. *Food Rev Int* 13:163–224.

Artanareeswaran, G., Thanikaivelan, P., Srinivasn, K. et al. 2004. Synthesis, characterization and thermal studies on cellulose acetate membranes with additives. *Eur Polym J* 40:2153–2159.

Atalla, R.H. and VanderHart, D.L. 1984. Native cellulose: A composite of two distinct crystalline forms. *Science* 223:283–285.

Avery, N.C. and Bailey, A.J. 2008. Restraining cross-links responsible for the mechanical properties of collagen fibers: Natural and artificial. In *Collagen: Structure and Mechanics,* ed. P. Fratzl, pp. 81–110. Boston, MA: Springer.

Azevedo, H.S. and Reis, R.L. 2009. Encapsulation of α-amylase into starch-based biomaterials: An enzymatic approach to tailor their degradation rate. *Acta Biomater* 5:3021–3030.

Bach, A.D., Bannasch, H., Galla, T.J. et al. 2001. Fibrin glue as matrix for cultured autologous urothelial cells in urethral reconstruction. *Tissue Eng* 7:45–53.

Bakos, D., Soldan, M., and Hernández-Fuentes, I. 1997. Hydroxyapatite-collagen-hyaluronic acid composite. *Biomaterials* 20:191–195.

Balgude, A.P., Yu, X., Szymanski, A. et al. 2001. Agarose gel stiffness determines rate of DRG neurite extension in 3D cultures. *Biomaterials* 22:1077–1084.

Barbosa, M.A., Granja, P.L., Barrias, C.C. et al. 2005. Polysaccharides as scaffolds for bone regeneration. *ITBM-RBM* 26:212–217.

Barker, H.T., Fuller, G.M., Klinger, M.M. et al. 2001. Modification of fibrinogen with poly (ethylene glycol) and its effects on fibrin clot characteristics. *J Biomed Mater Res* 56:529–535.

Baron, M., Norman, D.G., and Campbell, I.D. 1991. Protein modules. *Trends Biochem Sci* 16:13–17.

Benedetti, L., Cortivo, R. Berti, T. et al. 1993. Biocompatibility and biodegradation of different hyaluronan derivatives (HYAFF) implanted in rats. *Biomaterials* 14:1154–1160.

Benya, P. D. and Shaffer, J. D. 1982. Dedifferentiated chondrocytes re-express the differentiated phenotype when cultured in agarose cells. *Cell* 30:215–224.

Bigi, A., Borghi, M., Cojazzi, G. et al., 2000. Structural and mechanical properties of crosslinked drawn gelatin films. *J Therm Anal Calorim* 61:451–459.

Boesel, L.F., Mano, J.F., and Reis, R.L. 2004. Optimization of the formulation and mechanical properties of starch based partially degradable bone cements. *J Mater Sci Mater Med* 15:73–83.

Brown, J.C. and Timple, R. 1995. The collagen superfamily. *Int Arc Allergy Immunol* 107:484–490.

Buckley, C.T., Thorpe, S.D., O'Brien, F.J. et al. 2009. The effect of concentration, thermal history and cell seeding density on the initial mechanical properties of agarose hydrogels. *J Mech Behav Biomed* 2:512–521.

Buma, P., Pieper, J.S., van Tienen, T. et al. 2003. Cross-linked type I and type II collagenous matrices for the repair of full-thickness articular cartilage defects—A study in rabbits. *Biomaterials* 24:3255–3263.

Calonder, C., Matthew, H.W., and Van Tassel, P.R. 2005. Adsorbed layers of oriented fibronectin: A strategy to control cell-surface interactions. *J Biomed Mater Res* 75A:316–323.

Carreira, A.S., Goncalves, F.A.M.M., Mendonça, P.V. et al. 2010. Temperature and pH responsive polymers based on chitosan: Applications and new graft copolymerization strategies based on living radical polymerization. *Carbohyd Polym* 80:618–630.

Chari, P.S. 2003. Susruta and our heritage. *Indian J Plast Surg* 36:4–13.

Chen, G., Ito, Y., Imanishi, Y. et al. 1997. Photoimmobilization of sulfated hyaluronic acid for antithrombogenicity. *Bioconjug Chem* 8:730–734.

Chen, W., Shi, C., Yi, S. et al. 2010. Bladder regeneration by collagen scaffolds with collagen binding human basic fibroblast growth factor. *J Urol* 183:2432–2439.

Chiono, V., Pulieri, E., Vozzi, G. et al. 2008. Genipin-crosslinked chitosan/gelatin blends for biomedical applications. *J Mater Sci Mater Med* 19:889–898.

Chirita, M. 2008. Mechanical properties of collagen biomimetic films formed in the presence of calcium, silica and chitosan. *J Bionic Eng* 5:149–158.

Choi, Y.S., Hong, S.R., Lee, Y.M. et al. 1999. Studies on gelatin-containing artificial skin: II Preparation and characterization of cross-linked gelatin-hyaluronate sponge. *J Biomed Mater Res* 48:631–639.

Choi, Y.S., Lee, S.B., Hong, S.R. et al. 2001. Studies on gelatin-based sponges. Part III: A comparative study of cross-linked gelatin/alginate, gelatin/hyaluronate and chitosan/hyaluronate sponges and their application as a wound dressing in full-thickness skin defect of rat. *J Mater Sci Mater Med* 12:67–73.

Chvapil, M. 1979. Industrial uses of collagen. In *Fibrous Proteins: Scientific, Industrial and Medical Aspects*, eds. D.A.D. Parry, and L.K. Creamer, pp. 247–629. London, U.K.: Academic Press.

Ciardelli, G., Gentile, P., Chiono, V. et al. 2009. Enzymatically crosslinked porous composite matrices for bone tissue regeneration. *J Biomed Mater Res A* 92A:137–151.

Coburn, J.C. and Pandit, A. 2007. Development of naturally-derived biomaterials and optimization of their biomechanical properties. In *Topics in Tissue Engineering*, eds. N. Ashammakhi, R. Reis, and E. Chiellini, Vol. 3, pp. 1–32 (Ebook).

Curotto, E. and Aros, F. 1993. Quantitative determination of chitosan and the percentage of free amino groups. *Anal Biochem* 211:240–241.

Dal Pozzo, A., Vanini, L., Fagnoni, M., Guerrini, A., De Benedittis, M., and Muzzarelli, R. A. A. 2000. Preparation and characterization of poly(ethylene glycol) crosslinked reacetylated chitosans. *Carbohydrate Polymers*, 42: 201–206.

De La Riva, B., Sànchez, E., Hernández, A. et al., 2010. Local controlled release of VEGF and PDGF from a combined brushite–chitosan system enhances bone regeneration. *J Control Release* 143:45–52.

Degenshein, G.A., Hurwitz, A., and Ribacoff, S. 1963. Experience with regenerated oxidized cellulose. *NY State J Med* 63:2639–2643.

Dias, G.J., Peplow, P.V., and Teixeira, F. 2003. Osseous regeneration in the presence of oxidized cellulose and collagen. *J Mater Sci Mater Med* 14:739–745.

Dillon, G.P., Yu, X., Sridharan, A. et al. 1998. The influence of physical structure and charge polarity on neurite extension in a 3D hydrogel scaffold. *J Biomater Sci Polym* 9:1049–1069.

Dittrich, R., Despang, F., Bernhardt, A. et al. 2006. Mineralized scaffolds for hard tissue engineering by ionotropic gelation of alginate. *Adv Sci Technol* 49:159–164.

Djabourov, M., Leblond, J., and Papon, P. 1988. Gelation of aqueous gelatin solutions. II. Rheology of the sol–gel transition. *J Phys France* 49:333–343.

Drury, J.L. and Mooney, D.J. 2003. Hydrogels for tissue engineering: Scaffold design variables and applications. *Biomaterials* 24:4337–4351.

Duarte, A.R.C., Mano, J.F., and Reis, R.L. 2009. Preparation of starch-based scaffolds for tissue engineering by supercritical immersion precipitation. *J Supercrit Fluids* 49:279–285.

Elisseeff, J., Puleo, C., Yang, F. et al. 2005. Advances in skeletal tissue engineering with hydrogels. *Orthod Craniofac Res* 8:150–161.

Elsie, S.P., Nicholas, D.E., and Molly, M.S. 2009. Complexity in biomaterials for tissue engineering. *Nat Mater* 8:457–470.

Elvira, C., Mano, J.F., San Roman, J. et al. 2002. Starch-based biodegradable hydrogels with potential biomedical applications as drug delivery systems. *Biomaterials* 23:1955–1966.

Entcheva, E., Bien, H., Yin, L. et al. 2004. Functional cardiac cell constructs on cellulose-based scaffolding. *Biomaterials* 25:5753–5762.

Entcheva, E.G. and Yotova, L.K. 1994. Analytical application of membranes with covalently bound glucose-oxidase. *Anal Chim Acta* 299:171–177.

Ertesvag, H. and Valla, S. 1998. Biosynthesis and applications of alginates. *Polym Degrad Stabil* 59:85–91.

Ertesvag, H., Valla, S., and Skjåk-Bræk, G. 1996. Genetics and biosynthesis of alginates. *Carbohyd Eur* 14:14–18.

Espigares, I., Elvira, C., and Reis, R.L. 2002. New partially degradable and bioactive acrylic bone cements based on starch blends and ceramic fillers. *Biomaterials* 23:1883–1895.

Exposito, J.Y., Cluzel, C., Garrone, R. et al. 2002. Evolution of collagens. *Anat Rec* 268:302–316.

Eyrich, D., Brandl, F., Appel, B. et al. 2007. Long-term stable fibrin gels for cartilage engineering. *Biomaterials* 28:55–65.

Fa-Ming, C., Yi-Min, Z., Rong, Z. et al. 2007. Periodontal regeneration using novel glycidyl methacrylated dextran (Dex-GMA)/gelatin scaffolds containing microspheres loaded with bone morphogenetic proteins. *J Control Release* 121:81–90.

Freier, T., Koh, H.S., Kazazian, K. et al. 2005a. Controlling cell adhesion and degradation of chitosan films by N-acetylation. *Biomaterials* 26:5872–5878.

Freier, T., Montenegro, R., Koh, H.S. et al. 2005b. Chitin-based tubes for tissue engineering in the nervous system. *Biomaterials* 26:4624–4632.

Freile-Pelegrin, Y., Madera-Santana, T., Robledo, D. et al. 2007. Degradation of agar films in a humid tropical climate: Thermal, mechanical, morphological and structural changes. *Polym Degrad Stabil* 92:244–252.

Friess, W. 1998. Collagen: Biomaterial for drug delivery. *Eur J Pharm Biopharm* 45:113–136.

Gatj, I., Popa, M., and Rinaudo, M. 2005. Role of the pH on hyaluronan behavior in aqueous solution. *Biomacromolecules* 6:61–67.

Gelinsky, M., Eckert, M., and Despang, F. 2007. Biphasic but monolithic scaffolds for the therapy of osteochondral defects. *Int J Mater Res* 8:749–755.

Gentile, P., Chiono, V., Ciardelli, G. et al. 2010. Composite films of gelatin and hydroxyapatite/bioactive glass for tissue-engineering applications. *J Biomater Sci Polym Ed* 21:1207–1226.

Glowacki, J. and Mizuno, S. 2008. Collagen scaffolds for tissue engineering. *Biopolymers* 89:338–344.

Gomes, M.E., Bossano, C.M., Johnston, C.M. et al. 2006. In vitro localization of bone growth factors in constructs of biodegradable scaffolds seeded with marrow stromal cells and cultured in a flow perfusion bioreactor. *Tissue Eng* 12:177–188.

Gomes, M.E., Ribeiro, A.S., Malafaya, P.B. et al. 2001. A new approach based on injection moulding to produce biodegradable starch-based polymeric scaffolds: Morphology, mechanical and degradation behaviour. *Biomaterials* 22:883–889.

Gomez, G.M.C., Turnay, J., and Monter, P. 2002. Structural and physical properties of gelatine extracted from different marine species: A comparative study. *Food Hydrocolloid* 16:25–34.

Gotterbarm, T., Richter, W., Jung, M. et al. 2006. An in vivo study of a growth-factor enhanced, cell free, two-layered collagen–tricalcium phosphate in deep osteochondral defects. *Biomaterials* 27:3387–3395.

Grabarek, Z. and Gergely, J. 1990. Zero-length crosslinking procedure with the use of active esters. *Anal Biochem* 18:131–135.

Guo, T., Zhao, J., Chang, J. et al. 2006. Porous chitosan-gelatin scaffold containing plasmid DNA encoding transforming growth factor-β1 for chondrocytes proliferation. *Biomaterials* 27:1095–1103.

Hahn, S.K., Park, J.K., Tomimatsu, T. et al. 2007. Synthesis and degradation test of hyaluronic acid hydrogels. *Int J Biol Macromol* 40:374–380.

Han, W.W. and Misra, R.D.K. 2009. Biomimetic chitosan–nanohydroxyapatite composite scaffolds for bone tissue engineering. *Acta Biomater* 5:1182–1197.

Han, B., Schwab, I., Madsen, T. et al. 2002. Fibrin-based bioengineered ocular surface with human corneal epithelial stem cells. *Cornea* 21:505–510.

Hao, T., Wen, N., Wang, H.B. et al. 2010. The support of matrix accumulation and the promotion of sheep articular cartilage defects repair in vivo by chitosan hydrogels. *Osteoarthr Cartil* 18:257–265.

Herbert, C.B., Bittner, G.D., and Hubbell, J.A. 1996. Effects of fibrinolysis on neurite growth from dorsal root ganglia cultured in two- and three-dimensional fibrin gels. *J Comp Neurol* 365:380–391.

Hieta, K., Kuga, S., and Usuda, M. 1984. Electron staining of reducing ends evidences a parallel-chain structure in valonia cellulose. *Biopolymers* 23:1807–1810.

Hirakura, T., Yasugi, K., Nemoto, T. et al. 2010. Hybrid hyaluronan hydrogel as a protein nano carrier: New system for sustained delivery if protein with a chaperone-like function. *J Control Release* 142:483–489.

Hirota, Y., Tanioka, S., Tanigawa, T. et al. 1996. Clinical applications of chitin and chitosan to human decubitus. In *Advances in Chitin Science,* eds. A. Domard, C. Jeuniaux, R. Muzzarelli et al., pp. 407–413. Lyon, France: Jacques Andre Ed.

Hofmann, I., Muller, L., Greil, P. et al. 2006. Calcium phosphate nucleation on cellulose fabrics. *Surf Coating Tech* 201:2392–2398.

Huang, Z.H., Dong, Y.S., Chu, C.L. et al. 2008. Electrochemistry assisted reacting deposition of hydroxyapatite in porous chitosan scaffolds. *Mater Lett* 62:3376–3378.

Hynes, R. O. 1990. *Fibronectins.* New York: Springer-Verlag.

Idris, A. and Yet, L.K. 2006. The effect of different molecular weight PEG additives on cellulose acetate asymmetric dialysis membrane performance. *J Membr Sci* 280:920–927.

Izydorczyk, M., Cui, S.W., and Wang, Q. 2005. *Food Carbohydrates: Chemistry, Physical Properties, and Applications.* Boca Raton, FL: Taylor & Francis Group.

Javerliat, I., Goëau-Brissonnière, O., Sivadon-Tardy, V. et al., 2007. Prevention of *Staphylococcus aureus* graft infection by a new gelatin-sealed vascular graft prebonded with antibiotics. *J Vasc Surg* 46:1026–1031.

Jiankang, H., Dichen, L., Yiaxiong, L. et al., 2009. Preparation of chitosan–gelatin hybrid scaffolds with well-organized microstructures for hepatic tissue engineering. *Acta Biomater* 5:453–461.

Johnson, F.A., Craig, D.Q.M., and Mercer, A.D. 1997. Characterization of the block structure and molecular weight of sodium alginates. *J Pharm Pharmacol* 49:639–643.

Khor, E. 1997. Methods for the treatment of collagenous tissues for bioprostheses. *Biomaterials* 18:95–105.

Khor, E. and Lim, L.Y. 2003. Implantable applications of chitin and chitosan. *Biomaterials* 24:2339–2349.

Kifune, K. 1992. Clinical application of chitin artificial skin. In *Advances in Chitin and Chitosan*, eds. C.J. Brine, P.A. Sandford, and J.P. Zikakis, pp. 9–15. London, U.K.: Elsevier.

Kim, T.H., Jiang, H.L. Jere, D. et al. 2007. Chemical modification of chitosan as a gene carrier in vitro and in vivo. *Prog Polym Sci* 32:726–753.

Kim, I.Y., Seo, S.J., Moon, H.S. et al. 2008. Chitosan and its derivatives for tissue engineering applications. *Biotechnol Adv* 26:1–21.

Klemm, D., Heublein, B., Fink, H.P. et al. 2005. Cellulose: Fascinating biopolymer and sustainable raw material. *Angew Chem Int Ed Engl* 44:3358–3393.

Knaul, J. Z., Hudson, S.M., and Creber, K.A.M. 1999. Crosslinking of chitosan fibers with dialdehydes: Proposal of a new reaction mechanism. *J Polym Sci B Polym Phys* 37: 1079–1094.

Kobayashi, S., and Uyama, H. 2003. Biomacromolecules and bio-related macromolecules. *Macromol Chem Phys* 204:235–256.

Kodama, K., Doi, O., Higashiyama, M. et al. 1997. Pneumostatic effect of gelatin-resorcinol formaldehyde-glutaraldehyde glue on thermal injury of the lung: An experimental study on rats. *Eur J Cardiothorac Surg* 11:333–336.

Kong, H. and Mooney, D.J. 2004. Polysaccharide-based hydrogels in tissue engineering. In *Polysaccharides*, 2nd edn, ed. S. Dumitriu, Chapter 36, pp. 817–837. New York: Marcel & Dekker.

Kuo, C.K. and Ma, P.X. 2001. Ionically crosslinked alginate hydrogels as scaffolds for tissue engineering: Part 1. Structure, gelation rate and mechanical properties. *Biomaterials* 22:511–521.

Kurita, K. 2001. Controlled functionalization of the polysaccharide chitin. *Prog Polym Sci* 26:1921–1971.

Lakshmi, S.N. and Cato, T.L. 2007. Biodegradable polymers as biomaterials. *Progr Polym Sci* 32:762–798.

Lee, S.B., Kim, Y.H., Chong, M.S. et al. 2005. Study of gelatin-containing artificial skin V: Fabrication of gelatin scaffolds using a salt-leaching method. *Biomaterials* 26:1961–1968.

Lee, E.J., Shin, D.S., Kim, H.E. et al. 2009. Membrane of hybrid chitosan–silica xerogel for guided bone regeneration. *Biomaterials* 30:743–750.

Lee, C.H., Singla, A., and Lee, Y. 2001. Biomedical applications of collagen. *Int J Pharm* 221:1–22.

Lezica, R.P. and Quesada-Allue, L. 1990. Chitin. In *Methods in Plant Biochemistry. Carbohydrates*, ed. P.M. Dey, pp. 443–481. London, U.K.: Academic Press.

Li, Q., Dunn, E.T., Grandmaison, E.W. et al. 1992. Applications and properties of chitosan. *J Bioact Compat Polym* 7:370–397.

Liang, S.M., Zhang, L.N., and Xu, J. 2007. Morphology and permeability of cellulose/chitin blend membranes. *J Membr Sci* 287:19–28.

Lien, S.-M., Chien, C.-H., and Huang, T.-J. 2009. A novel osteochondral scaffold of ceramic–gelatin assembly for articular cartilage repair. *Mater Sci Eng C* 29:315–321.

Liu, C.Z. 2008. Biomimetic synthesis of collagen/nano-hydroxyapatite scaffold for tissue engineering. *J Bionic Eng* 5:1–8.

Liu, Y., Lu, Y., Cui, G. et al. 2009a. Segmental bone regeneration using an rhBMP-2-loaded gelatin/nanohydroxyapatite/fibrin scaffold in a rabbit model. *Biomaterials* 30:6276–6285.

Liu, X., Smith, L.A., Hu, J. et al. 2009b. Biomimetic nanofibrous gelatin/apatite composite scaffolds for bone tissue engineering. *Biomaterials* 30:2252–2258.

Lin, Y.C., Tan, F.J., Marra, K.G. et al. 2009. Synthesis and characterization of collagen/hyaluronan/chitosan composite sponges for potential biomedical applications. *Acta Biomater* 5:2591–2600.

Liu, S.H., Yang, R.S., al-Shaikh, R. et al. 1995. Collagen in tendon, ligament and bone healing. *Clin Orthop Res* 318:265–278.

Lu, J.T., Lee, C.J., Bent, S.F. et al. 2007. Thin collagen film scaffolds for retinal epithelial cell culture. *Biomaterials* 28:1486–1494.

Luo, Y. and Shoichet, M.S. 2004. A photolabile hydrogel for guided three-dimensional cell growth and migration. *Nat Mater* 3:249–253.

Luo, L.H., Zhang, Y.F., Wang, X.M. et al. 2010. Preparation, characterization, and in vitro and in vivo evaluation of cellulose/soy protein isolate composite sponges. *J Biomater Appl* 24:503–526.

Luo, Y., Ziebell, M.R., and Prestwich, G.D. 2000. A hyaluronic acid-taxol antitumor bioconjugate targeted to cancer cells. *Biomacromolecules* 1:208–218.

Madhumathi, K., Nair, S.V., and Jayakumar, R. et al. 2009. Preparation and characterization of novel-chitin–hydroxyapatite composite membranes for tissue engineering applications. *Int J Biol Macromol* 44:1–5.

Malafaya, P.B., Elvira, C., Gallardo, A. et al. 2001. Porous starch-based drug delivery systems processed by a microwave route. *J Biomater Sci Polym Ed* 12:1227–1241.

Malafaya, P.B., Silva, G.A., and Reis, R.L. 2007. Natural–origin polymers as carriers and scaffolds for biomolecules and cell delivery in tissue engineering applications. *Adv Drug Deliv Rev* 59:207–233.

Malafaya, P., Stappers, F., and Reis, R.L. 2006. Starch-based microspheres produced by emulsion crosslinking with a potential media dependent responsive behaviour to be used as drug delivery carriers. *J Mater Sci Mater Med* 17:371–377.

Maluf, N.S. 1954. History of blood transfusion. *J Hist Med Allied Sci* 9:59–107.

Mandal, B.B., Mann, J.K., and Kundu, S.C. 2009. Silk fibroin/gelatin multilayered films as a model system for controlled drug release. *Eur J Pharm Sci* 37:160–171.

Mano, J.F., Silvia, G.A., Reis, R.L. et al. 2007. Natural origin biodegradable systems in tissue engineering and regenerative medicine: Present status and some moving trends. *J R Soc Interface* 4:999–1030.

Mao, J.S. Zhao, L.G., Yin Y.J. et al. 2003. Structure and properties of bilayer chitosan-gelatin scaffolds. *Biomaterials* 24:1067–1074.

Marques, A.P., Reis, R.L., and Hunt, J.A. 2002. The biocompatibility of novel starch-based polymers and composites: In vitro studies. *Biomaterials* 23:1471–1478.

Martins, A., Chung, S., and Pedro, A.J. et al. 2009. Hierarchical starch-based fibrous scaffold for bone tissue engineering applications. *J Tissue Eng Regen Med* 3:37–42.

Martson, M., Viljanto, J., and Hurme, T. et al. 1998. Biocompatibility of cellulose sponge with bone. *Eur Surg Res* 30:426–432.

Mathew, P., Binulal, N.S., Nair, S.V. et al. 2010. Novel biodegradable chitosan–gelatin/nano-bioactive glass ceramic composite scaffolds for alveolar bone tissue engineering. *Chem Eng J* 158:353–361.

Mauck, R.L., Seyhan, S.L., Ateshian, G.A. et al. 2002. Influence of seeding density and dynamic deformational loading on the developing structure/function relationships of chondrocyte-seeded agarose hydrogels. *Ann Biomed Eng* 30(8):1046–1056.

del Mel, A., Bolvin, C., Edirisinghe, M. et al. 2008. Development of cardiovascular bypass grafts: Endothelialization and applications of nanotechnology. *Expert Rev Cardiovasc Ther* 6:1259–1277.

Michon, C., Cuvelier, G., Relkin, P. et al. 1997. Influence of thermal history on the stability of gelatin gels. *Int J Biol Macromol* 20:259–264.

Miyamoto, T., Takahashi, S., Ito, H. et al. 1989. Tissue biocompatibility of cellulose and its derivatives. *J Biomed Mater Res* 23:125–133.

Morrison, W.R. and Karkalas, J. 1990. *Carbohydrates*. London, U.K.: Academic Press.

Mosesson, M.W., Siebenlist, K.R., and Meh, D.A. 2001. The structure and biological features of fibrinogen and fibrin. *Ann N Y Acad Sci* 936:11–30.

Mouw, J.K., Case, N.D., Guldberg, R.E. et al. 2005. Variations in matrix composition and GAG fine structure among scaffolds for cartilage tissue engineering. *Osteoarthr Cartil* 13:828–836.

Muller, F.A., Muller, L., Hofmann, I. et al. 2006. Cellulose-based scaffold materials for cartilage tissue engineering. *Biomaterials* 27:3955–3963.

Murphy, C.M., Haugh, M.G., and O'Brien, F.J. 2010. The effect of mean pore size on cell attachment, proliferation and migration in collagen–glycosaminoglycan scaffolds for bone tissue engineering. *Biomaterials* 31:461–466.

Neidert, M. R., Lee, E. S., Oegema, T.R. et al. 2002. Enhanced fibrin remodeling in vitro with TGF-β1, insulin and plasmin for improved tissue-equivalents. *Biomaterials* 23:3717–3731.

Nisbet, D. R., Crompton, K.E., Horne, M.K. et al. 2008. Neural tissue engineering of the CNS using hydrogels. *J Biomed Mater Res B Appl Biomater* 87B:251–263.

Nishiyama, P., Langan H., and Chanzy, H. 2002. Crystal structure and hydrogen-bonding system in cellulose I$_\beta$ from synchrotron X-ray and neutron fiber diffraction. *J Am Chem Soc* 124:9074–9082.

Nishiyama, Y., Sugiyama, J., Chanzy, H. et al. 2003. Crystal structure and hydrogen bonding system in cellulose I$_\alpha$ from synchrotron x-ray and neutron fiber diffraction. *J Am Chem Soc* 125:14300–14306.

Oakenfull, D. and Scott A. 2003. Gelatin gels in deuterium oxide. *Food Hydrocolloid* 17:207–210.

Okuyama, K., Wu, G., Jiravanichanun, N. et al. 2006. Helical twists of collagen model peptides. *Biopolymers* 84:421–432.

Olde, L.H.H.D., Dijkstra, P.J., van Luyn, M.J.A. et al. 1996. In vitro degradation of dermal sheep collagen crosslinked using a water-soluble carbodiimide. *Biomaterials* 17:679–684.

Olsen, D., Yang, C., Bodo, M. et al. 2003. Recombinant collagen and gelatin for drug delivery. *Adv Drug Deliv Rev* 55:1547–1567.

Onsoyen, E. 1992. Alginates. In *Thickening and Gelling Agents for Food,* ed. A. Imeson, pp. 1–24. London, U.K.: Blackie Academic & Professional.

Osathanon, T., Linnes, M.L. Rajachar, R.M. et al. 2008. Microporous nanofibrous fibrin-based scaffolds for bone tissue engineering. *Biomaterials* 29:4091–4099.

Pacaccio, D. 2005. Demineralized bone matrix: Basic science and clinical applications. *Clin Podiatr Med Surg* 22:599–606.

Painter, T.J. 1983. Algal polysaccharides. In *The Polysaccharides*, ed. G.O. Aspinall, pp. 195–285. New York: Academic Press.

Pankov, R. and Yamada, K.M. 2002. Fibronectin at a glance. *J Cell Sci* 115:861–863.

Pavlov, M.P., Mano, J.F., and Reis, R.L. 2004. Fibers and 3D mesh scaffolds from biodegradable starch-based blends: Production and characterization. *Macromol Biosci* 4:776–784.

Pek, Y.S., Shujun, G., Arshad, M.S.M. et al. 2008. Porous collagen-apatite nanocomposite foams as bone regeneration scaffolds. *Biomaterials* 29:4300–4305.

Peppas, N.A. 1986. *Hydrogel in Medicine and Pharmacy*. Boca Raton, FL: CRC Press.

Pereira, C.S., Cunha, A.M., Reis, R.L. et al. 1998. New starch-based thermoplastic hydrogels for use as bone cements or drug-delivery carriers. *J Mater Sci Mater Med* 9:825–833.

Perng, C.K., Wang, Y.J., Tsi, C.H. et al. 2009. In vivo angiogenesis effect of porous collagen scaffold with hyaluronic acid oligosaccharides. *J Surg Res* (doi.org/10.1016/j.jss. 2009. 09. 052).

Peter, M., Binulal, N.S., Nair, S.V. et al. 2010a. Novel biodegradable chitosan–gelatin/nano-bioactive glass ceramic composite scaffolds for alveolar bone tissue engineering. *Chem Eng J* 158:353–361.

Peter, M., Ganesh, N. Selvamurugan, N. et al. 2010b. Preparation and characterization of chitosan–gelatin/nanohydroxyapatite composite scaffolds for tissue engineering applications. *Carbohyd Polym* 80:687–694.

Petite, H., Duval, J.L., Frei, V. et al. 1995. Cytocompatibilty of calf pericardium treated by glutar- aldehyde and by the acyl azide methods in an organotypic culture model. *Biomaterials* 16:1003–1008.

Petite, H., Rault, I., Huc, A. et al. 1990. Use of the acyl azide method for crosslinking collagen-rich tissues such as pericardium. *J Biomed Mater Res* 24:179–187.

Pieper, J.S., Hafmans, T., Veerkamp, J.H. et al. 2000. Development of tailor-made collagen–glycosaminoglycan matrices: EDC/NHS crosslinking, and ultrastructural aspects. *Biomaterials* 21:581–593.

Piez, K.A. 1985. Collagen. In *Encyclopedia of Polymer Science and Engineering*, ed. J.I. Kroschwitz, pp. 699–627. New York: Wiley.

Poustis, J., Baquey, C., and Chauveaux, D. 1994. Mechanical properties of cellulose in orthopaedic devices and related environments. *Clin Mater* 16:119–124.

Pouyani, T., Harbison, G.S., and Prestwich, G.D. 1994. Novel hydrogels of hyaluronic acid: Synthesis, surface morphology, and solid-state NMR. *J Am Chem Soc* 116:7515–7522.

Prabaharan, M. 2008. Chitosan derivatives as promising materials for controlled drug delivery. *J Biomater Appl* 23:5–36.

Prabaharan, M. and Jayakumar R. 2009. Chitosan-graft-β-cyclodextrin scaffolds with controlled drug release capability for tissue engineering applications. *Int J Biol Macromol* 44:320–325.

Puppi, D., Chiellini, F., Piras, A.M. et al. 2010. Polymeric materials for bone and cartilage repair. *Prog Polym Sci* 35:403–440.

Quinn, T.M., Schmid, P., Hunziker, E.B. et al. 2002. Proteoglycan deposition around chondrocytes in agarose culture: Construction of a physical and biological interface for mechanotransduction in cartilage. *Biorheology* 39:27–37.

Ramshaw, J.A.M., Werkmeister, J.A., and Glattauer, V. 1995. Collagen based biomaterials. *Biotechnol Genet Eng Rev* 13:335–382.

Rao, K.P. 1995. Recent developments of collagen-based materials for medical applications and drug delivery. *J Biomater Sci* 7:623–645.

Rassis, D.K., Saguy, I.S., and Nussinovitch, A. 2002. Collapse, shrinkage and structural changes in dried alginate gels containing filles. *Food Hydrocolloid* 16:139–151.

Rehakova, M., Bakos, D., Vizarova, K. et al. 1996. Properties of collagen and hyaluronic acid composite materials and their modification by chemical cross linking. *J Biomed Mater Res* 30:369–372.

Reis, R.L. and Cunha, A.M. 1995. Characterization of two biodegradable polymers of potential application within the biomaterials field. *J Mater Sci Mater Med* 6:786–792.

Rhee, S.H. and Tanaka, J. 2000. Hydroxyapatite formation on cellulose cloth induced by citric acid. *J Mater Sci Mater Med* 11:449–452.

Rinaudo, M. 2006. Chitin and chitosan: Properties and applications. *Prog Polym Sci* 31:603–632.

Rinaudo, M. 2008. Main properties and current applications of some polysaccharides as biomaterials. *Polym Int* 57:397–430.

Rindlava, A., Hulleman, S.D.H., and Gatenholma, P. 1997. Formation of starch films with varying crystallinity. *Carbohyd Polym* 34:25–30.

Rodriguez-Cabello, J.C., Reguera, J., and Girotti, A. 2005. Developing functionality in elastin-like polymers by increasing their molecular complexity: The power of the genetic engineering approach. *Prog Polym Sci* 30:1119–1145.

Roman, J., Cabañas, M.V., and Vallet-Regi, M. 2008. An optimized β-tricalcium phosphate and agarose scaffold fabrication technique. *J Biomed Mater Res Part A* 84A:99–107.

Rowley, J.A., Madlambayan, G., and Mooney, D.J. 1999. Alginate hydrogels as synthetic extracellular matrix materials. *Biomaterials* 20:45–53.

Saito, H., Murabayashi, S., Mitamura, Y. et al. 2007. Characterization of alkali-treated collagen gels prepared by different crosslinkers. *J Mater Sci Mater Med* 19:1297–1305.

Salgado, A.J., Coutinho, O.P., and Reis, R.L. 2004. Novel starch-based scaffolds for bone tissue engineering: Cytotoxicity, cell culture, and protein expression. *Tissue Eng* 10:465–474.

Samuel, C.S., Coghlan, J.P., and Bateman, J.F. 1998. Effects of relaxin, pregnancy and parturition on collagen metabolism in the rat public symphysis. *J Endocrinol* 159:117–125.

Santoni-Rugiu, P. and Sykes, P.J. 2007. *A History of Plastic Surgery*. Chapter 7. Heidelberg: Springer.

Santosa, M.I., Sabine, F., and James, K. 2007. Response of micro- and macrovascular endothelial cells to starch-based fiber meshes for bone tissue engineering. *Biomaterials* 28:240–248.

Sarkar, N. and Walker, L.C. 1995. Hydration–dehydration properties of methylcellulose and hydroxyl propyl methyl cellulose. *Carbohyd Polym* 7:177–185.

Sashiwa, H. and Aiba, S. 2004. Chemically modified chitin and chitosan as biomaterials. *Prog Polym Sci* 29:887–808.

Sawyer, A.A., Song, S.J., Susanto, E. et al. 2009. The stimulation of healing within a rat calvarial defect by mPCL–TCP/collagen scaffolds loaded with rhBMP-2. *Biomaterials* 30:2479–2488.

Schrieber, R. and Gareis, H. 2007. *Gelatin Handbook*. Weinhem, Germany: Wiley-VCH GmbH & Co.

Seal, B.L., Otero, T.C., and Panitch, A. 2001.Polymeric biomaterials for tissue and organ regeneration. *Mater Sci Eng R* 34:147–230.

Segura, T., Anderson, B.C., Chung, P.H. et al. 2005. Crosslinked hyaluronic acid hydrogels: A strategy to functionalize and pattern. *Biomaterials* 26:359–371.

Sevillano, G., Rodriguez-Puyol, M., Martos, R. et al. 1990. Cellulose acetate membrane improves some aspects of red blood cell function in haemodialysis patients. *Nephrol Dial Transplant* 5:497–499.

Shalumon, K.T., Binulal, N.S., Selvamurugan, N. et al. 2009. Electrospinning of carboxy methylchitin/poly (vinyl alcohol) nano fibrous scaffolds for tissue engineering applications. *Carbohyd Polym* 77:863–869.

Shigemasa, Y. and Minami, S. 1995. Applications of chitin and chitosan for biomaterials. *Biotech Gen Eng Rev* 13:383–415.

Shinji, S., Hashimoto, I., and Kawakami, K. 2007. Agarose-gelatin conjugate for adherent cell-enclosing capsules. *Biotechnol Lett* 29:731–735.

Shoulders, M.D. and Raines, R.T. 2009. Collagen structure and stability. *Annu Rev Biochem* 78:929–958.

Silva, G.A., Pedro, A., Costa, F.J. et al. 2005. Soluble starch and composite starch Bioactive Glass 45S5 particles: Synthesis, bioactivity, and interaction with rat bone marrow cells. *Mater Sci Eng C* 25:237–246.

Skoog, T. 1967. The use of periosteum and Surgicel® for bone restoration in congenital clefts of the maxilla. *Scand J Plast Reconstr Surg* 1:113–130.

Smidsrod, O. and Draget, K.I. 1996. Chemistry and physical properties of alginates. *Carbohyd Eur* 14:6–13.

Soon, A.S.C., Stabenfeldt, S.E., Brown, W.E. et al. 2010. Engineering fibrin matrices: The engagement of polymerization pockets through fibrin knob technology for the delivery and retention of therapeutic proteins. *Biomaterials* 31:1944–1954.

Sousa, R.A., Kalay, G., Reis, R.L. et al. 2000. Injection molding of a starch/EVOH blend aimed as an alternative biomaterial for temporary applications. *J Appl Polym Sci* 77:1303–1315.

Sousa, R.A., Reis, R.L., Cunha, A.M. et al. 2003. Processing and properties of bone-analogue biodegradable and bioinert polymeric composites. *Comp Sci Tech* 63:389–402.

Stevens, M.M., Mayerb, M., Anderson, D.G. et al. 2005. Direct patterning of mammalian cells onto porous tissue engineering substrates using agarose stamps. *Biomaterials* 26:7636–7641.

Sundaram, J., Durance, T.D., and Wang, R. 2008. Porous scaffold of gelatin-starch with nano hydroxy apatite composite processed via novel microwave vacuum drying. *Acta Biomater* 4:932–942.

Sung, H.W., Huang, D.M., Chang, W.H. et al. 1999. Evaluation of gelatin hydrogel crosslinked with various crosslinking agents as bioadhesives: In vitro study. *J Biomed Mater Res* 46:520–530.

Svegmark, K. and Hermansson, A. M. 1993. Microstructure and rheological properties of composites of potato starch granules and amylose: A comparison of observed and predicted structures. *Food Struct* 12:181–193.

Svensson, A. Nicklasson, E., Harrah, T. et al. 2005. Bacterial cellulose as a potential scaffold for tissue engineering of cartilage. *Biomaterials* 26:419–431.

de Taillac, L.B., Porté-Durrieua, M.C., Labrugere, C. et al. 2004. Grafting of RGD peptides to cellulose to enhance human osteoprogenitor cells adhesion and proliferation. *Compos Sci Technol* 64:827–837.

Tan, S.C., Khor, E., Tan, T. K. et al. 1998. The degree of deacetylation of chitosan: Advocating the first derivative UV-spectrophotometry method of determination. *Talanta* 45:713–719.

Tavares-Dias, M. and Oliveira, S.R. 2009. A review of the blood coagulation system of fish. *Braz J Biosci* 7:205–244.

Tiwari, A., Salacinski, H.J., Punshon, G. et al. 2002. Development of a hybrid cardiovascular graft using a tissue engineering approach. *FASEB J* 16:791–796.

Todhunter, R.J., Wooton, J.A.M., Lust, G. et al. 1994. Structure of equine type I and type II collagens. *Am J Vet Res* 55:425–431.

Tolaimate, A., Desbrieres, J., Rhazi, M. et al. 2000. On the influence of deacetylation process on the physicochemical characteristics of chitosan from squid chitin. *Polymer* 41:2463–2469.

Tortelli, F. and Cancedda, F. 2009. Three-dimensional cultures of osteogenic and chondrogenic cells: A tissue engineering approach to mimic bone and cartilage in vitro. *Eur Cell Mater* 17:1–14.

Touyama, R., Inoue, K. Takeda, Y. et al. 1994. Studies on the blue pigments produced from genipin and methylamine. II. On the formation mechanisms of brownish-red intermediates leading to the blue pigment formation. *Chem Pharm Bull* 42:1571.

Traver, M.A. and Assimos, G. 2006. New generation tissue sealants and haemostatic agents: Innovative urological applications. *Rev Urol* 8:104–111.

Tuan, T.L., Song, A.K., Chang, S. et al. 1996. In vitro fibroplasia: Matrix contraction, cell growth, and collagen production of fibroblasts cultured in fibrin gels. *Exp Cell Res* 223:127–134.

Tuovinen, L., Peltonen, S., and Jarvinen, K. 2003. Drug release from starch acetate films. *J Control Release* 91:345–354.

Tuovinen, L, Ruhanen, L., Kinnarinen, T. et al. 2004. Starch acetate microparticles for drug delivery into retinal pigment epithelium-in vitro study. *J Control Release* 98:407–413.

Tuzlakoglu, K., Bolgen, N., Salgado, A.J. et al. 2005. Nano and micro-fiber combined scaffolds: A new architecture for bone tissue engineering. *J Mater Sci Mater Med* 16:1099–1104.

Ulubayram, K., Cakar, A.N., Korkusuz, P. et al. 2001. EGF containing gelatin-based wound dressings. *Biomaterials* 22:1345–1356.

Utpal, B., Pragya, S., Krishnamoorthy, K. et al. 2006. Photoreactive cellulose membranes—A novel matrix for covalent immobilization of biomolecules. *J Biotech* 126:220–229.

Van Vlierberghe, S., Cnudde, V., Jacobs, P.J.S. et al. 2006. Porous gelatin cryogels as cell delivery tool in tissue engineering. *J Control Release* 116: e95–e98.

Voet, D., Voet, J.G., and Pratt, C.W. 1999. *Fundamentals of Biochemistry*. New York: Wiley.

Walker, M., Hobot, J.A., Newman, G.R. et al., 2003. Scanning electron microscopic examination of bacterial immobilization in a carboxy methyl cellulose (AQUACEL) and alginate dressings. *Biomaterials* 24:883–890.

Wang, J., Fu, W., Zhang, D. et al. 2010. Evaluation of novel alginate dialdehyde cross-linkedchitosan/calcium polyphosphate composite scaffolds for meniscus tissue engineering. *Carbohyd Polym* 79:705–710.

Wang, S., Liu, W., Han, B. et al. 2009. Study on a hydroxypropyl chitosan–gelatin based scaffold for corneal stroma tissue engineering. *Appl Surf Sci* 255:8701–8705.

Wang, L. and Stegemann, J.P. 2010. Thermogelling chitosan and collagen composite hydrogels initiated with β-glycerophosphate for bone tissue engineering. *Biomaterials* 31:3976–3785.

Watanabe, K., Takahashi, H. Habu, Y. et al. 2000. Interaction with heparin and matrix metalloproteinase 2 cleavage expose a cryptic anti-adhesive site of fibronectin. *Biochem* 39:7138–7144.

Weadock, K.S., Miller, E.J., and Bellincampi, L.D. et al. 1995. Physical crosslinking of collagen fibers: Comparison of ultraviolet irradiation and dehydrothermal treatment. *J Biomed Mater Res* 29:1373–1379.

Weadock, K.S., Miller, E.J., Keuffel, E.L. et al. 1996. Effect of physical crosslinking methods on collagen-fiber durability in proteolytic solutions. *J Biomed Mater Res* 32:221–226.

Whistler, R.L. and Daniel, J.R. 2005. Starch. In *Kirk Othmer Encyclopedia of Chemical Technology*, ed. A. Seidel, Hoboken, NJ, pp. 699–719. John Wiley & Sons, Inc.

Williams, D.F. 1987. Definitions in biomaterials. In *Progress in Biomedical Engineering*, ed. D.F. Williams, Vol. 54, Amsterdam, the Netherlands: Elsevier.

de Wolf, F.A. 2003. Collagen and gelatin. In *Progress in Biotechnology*, eds. W.Y. Aalbersberg, R.J. Hamer, P. Jasperse et al., pp. 133–218. Amsterdam, the Netherlands: Elsevier Science B.V.

Wongpanit, P., Sanchavanakit, N., Pavasant, P. et al. 2005. Preparation and characterization of microwave-treated carboxymethyl chitin and carboxymethyl chitosan films for potential use in wound care application. *Macromol Biosci* 5:1001–1012.

Wu, X., Liu, Y., Li X. et al. 2010. Preparation of aligned porous gelatin scaffolds by unidirectional freeze-drying method. *Acta Biomater* 6:1167–1177.

Wu, J. and Yuan, Q. 2002. Gas permeability of a novel cellulose membrane. *Journal of Membrane Science* 204:185–194.

Xu, H.H.K., Weir, M.D., and Simon, C.G. 2008. Injectable and strong nano-apatite scaffolds for cell/growth factor delivery and bone regeneration. *Dent Mater* 24:1212–1222.

Yamane, T., Yamaguchi, N., Yoshida, Y. et al. 2004. Regulation of the extracellular matrix production and degradation of endothelial cells by shear stress. *Int Congr Ser* 1262:407–410.

Yannas, I.V. and Tobolksy, A.V. 1967. Crosslinking of gelatin by dehydration. *Nature* 215:509–510.

Ye, Q., Zünd, G., Benedikt, P. et al., 2000. Fibrin gel as a three dimensional matrix in cardiovascular tissue engineering. *Eur J Cardiothorac Surg* 17:587–591.

Yeom, C.K. and Lee, K.H. 1998. Characterization of sodium alginate membrane crosslinked with glutaraldehyde in pervaporation separation. *J Appl Polym Sci* 67:209–219.

Young, S., Wong, M., Tabata, Y. et al. 2005. Gelatin as a delivery vehicle for the controlled release of bioactive molecules. *J Control Release* 109:256–274.

Zaborowska, M., Bodin, A., Bäckdahl, H. et al. 2010. Microporous bacterial cellulose as a potential scaffold for bone regeneration. *Acta Biomater* 6:2540–2547.

Zhang, Y. and Zhang, M. 2001. Synthesis and characterization of macroporous chitosan/calcium phosphate composite scaffolds for tissue engineering. *J Biomed Mater Res* 55:304–312.

Zheng, J.P., Wang, C.Z., Wang, X.X. et al. 2007. Preparation of biomimetic three-dimensional gelatin/montmorillonite-chitosan scaffold for tissue engineering. *React Funct Polym* 67:780–788.

Zhou, P. and Regenstein, J.M. 2005. Effects of alkaline and acid pretreatments on Alaska Pollock skin gelatin extraction. *J Food Sci* 70:C392–C396.

Zimmerman, L.M. and Veith, I. 1993. *Great Ideas in the History of Surgery*. New York: Norman Publishers.

11 Natural Polymers as Components of Blends for Biomedical Applications

Alina Sionkowska

CONTENTS

11.1 INTRODUCTION

The biomedical sector requests more and more complex structures in attempts to fulfill the requirements of a multitude of different applications. For this reason, polymer scientists exploit existing compounds and also design new compounds with properties that can be considered as new materials for biomedical applications [1,2]. The applications of polymers are essential in surgery, for prosthetic systems, and in pharmacology, for drug formulation and controlled drug delivery. Polymeric biomaterials are usually used under complex, demanding conditions, and must fulfill stringent requirements relating to their chemical, biological, and physicomechanical properties. Therefore, there is an urgent need for new polymeric biomaterials that can satisfy the demanding performance requirements and standards required for these applications [3]. The development of innovative polymeric biomaterials can be an engine of innovation for new medical treatments and therapies. The existing knowledge clearly shows that the careful optimization of the chemical structure of polymers used in medical implants can result in better clinical outcomes for the implant recipients. Biodegradable polymers play a significant role in the preparation of polymeric materials for biomedical applications. They are preferred candidates for developing therapeutic devices such as temporary prostheses, three-dimensional porous structures as scaffolds for tissue engineering, and as controlled/sustained release drug delivery vehicles [4].

At the end of the twentieth century, a new trend in the development of polymeric materials was observed [5]. Many scientific laboratories started to work with polymer blends as potential new materials for medicine (biomaterials), the food industry (biodegradable and photodegradable food packaging materials), and for the electronic industry (conducting polymers). Especially, new materials for biomedical applications based on the blends of two polymers were studied. Cell-based transplantation, tissue engineering, and gene therapy are important therapeutic strategies for current and future regenerative medicine. One challenge is to present the target cells in a suitable matrix to allow the cells to survive wound contraction, tissue repair, and remodeling in certain tissues. Recently, functional biomaterial research has been directed toward the development of improved scaffolds and new drug delivery systems for regenerative medicine. During the last three decades, the increasing interest in new materials based on the blends of two or more natural polymers and synthetic polymers has been observed. Blends of two (or more) natural polymers and/or blends of synthetic and natural polymers can form a new class of materials with improved mechanical properties and biocompatibility compared with those of single components. In scientific literature, new words had been designed for the blends of natural polymers with synthetic ones. Materials based on such blends have been called bioartificial or biosynthetic polymeric materials [5–9]. In many scientific articles, one can find that this new vocabulary has been commonly used. Natural biomacromolecules, so-called biopolymers, serve as intrinsic templates for cell attachment and growth. In general, natural nonspecific polymers can offer a range of advantages when compared with synthetic polymer-based materials. Natural polymers are usually biocompatible, whereas synthetic polymers can contain a residue of initiators and other compounds/impurities that do not allow cell growth. Synthetic polymers possess good mechanical properties and thermal stability, much better than several naturally occurring polymers. There is also a limitation in the performance of several natural polymers in comparison to synthetic polymers. New polymeric materials in many cases should be biocompatible, and at the same time they should possess good thermal and mechanical properties. For this reason, blends of natural polymers and synthetic polymers have become promising materials for biomedical applications. That is why the new materials designed should contain both natural polymers and synthetic ones.*

New materials based on blends of natural and synthetic polymers are a rapidly growing research area [5–44]. Intensive research is currently underway in areas in which biopolymeric–polymeric materials can be used in the engineering of bone, tissue, and organs [5].

New materials based on the blends of natural polymers and man-made polymers can also be applied as packaging materials for drugs and medical devices. They can also be applied in the food packaging industry. The biodegradability of natural polymers is very useful in the packaging industry. Blends of synthetic and natural polymers are required in this field as bacteria start to degrade the natural polymer and therefore decrease the natural component making it easier to degrade the synthetic component. The blends of natural polymers and synthetic polymers can more readily undergo photodegradation than synthetic polymers. Natural polymers possess many chemical groups (chromophores) absorbing UV light from the sun. Free radicals produced in the natural polymers may initiate photochemical degradation in synthetic polymers [16–18]. Bacterial degradation of the blend is more efficient after partial photodegradation of synthetic polymers. On the other hand, UV light is used for sterilization of biomedical materials. Natural polymers are usually sensitive to the action of UV radiation, and they might be photochemically more stable in the blend with synthetic polymer [18]. Moreover, the thermal stability of the blend is usually different from that of the single component of the blend. Thermal and photochemical stability are very important during processing and sterilization by high temperature and UV irradiation [37].

Two components of the blend need to be combined into one versatile material. The components can be combined in solid state and/or dissolved in the same solvent. The problem is that many naturally occurring polymers are insoluble in common solvents. One has to find a soluble derivative of natural polymers or one has to hydrolyze the macromolecule to shorter chains. When a soluble biopolymer is achieved, then it is necessary to decide what kind of synthetic polymer to combine with the natural one. After preparing the mixture of two components in the same solvent, one has to study the interaction between the natural and synthetic polymers. One has to remember that the third component, solvent, is also present in the blend. Miscibility of the

---

* Reprinted from *Prog. Polym. Sci.*, 36, Sionkowska, A., Current research on the blends of natural and synthetic polymers as new biomaterials: review, 1254–1276, 2011. Copyright 2011, with permission from Elsevier.

components is an important aspect, and it determines the properties of the blend. The miscibility in polymer blends is assigned to specific interactions between polymeric components, which usually give rise to a negative free energy of mixing in spite of the high molecular weight of polymers. Several interactions can appear between the components of a polymeric blend. The most common interactions in the blends are hydrogen bonding (when polymers contain chemical groups capable of forming hydrogen bonds), ionic and dipole, pi-electrons, and charge-transfer complexes. Most polymers in the blend are immiscible with each other due to the absence of specific interactions. Sometimes, two polymers are partially miscible, only in solution, but they are immiscible in the solid state. Miscibility of two or more synthetic polymers has been studied by many research groups by viscometry, Fourier Transform Infrared spectroscopy (FTIR), and differential scanning calorimetry (DSC). The miscibility of two natural polymers as well as natural polymers with synthetic polymers is poorly understood. In the assessment of miscibility in solution, the main problem is to find a solvent suitable for both natural and synthetic polymers. The next problem is the hierarchical native structure of many natural polymers. It may happen that during mixing of biopolymers with synthetic polymers the natural one can change its native structure.

The most common natural polymers that can be used for the preparation of biomaterials based on the blends of synthetic polymers and natural ones are collagen, cellulose, chitosan, elastin, starch, and silk. Natural polymers, such as collagen and elastin, are usually insoluble in water and in organic solvents. Collagen extracted from young animals' tissues is soluble in dilute acetic acid. Chitosan is soluble in dilute acetic acid solution, but the concentration that can be reached is low and strongly depends on the molecular weight of this biopolymer. The solubility of collagen and chitosan in acetic acid may allow them to be blended with other water-soluble polymers. Silk, elastin, and keratin are highly insoluble natural polymers, and hence processing of these biopolymers is a problem. In many laboratories, elastin, keratin, and silk have been hydrolyzed to aid solubility.

The blends of collagen with other polymers have been studied widely by several research groups [5–7,12–14,16–62]. Collagen possesses several interesting properties that make it a good material for biomedical applications [63–83]. Chitosan also possesses several interesting properties to be applied as a potential biomedical material [82,84–102]. Chitosan can be blended with synthetic and/or other natural polymers [10,13–15,19,20,46,48,54,62,102–157]. Elastin as a natural polymer has also been considered for biomaterial preparation [69,158–166]. A very promising biopolymer to prepare biomaterials is silk [167–195].

11.2   MISCIBILITY AND COMPATIBILITY OF BLENDS

Natural polymers and synthetic polymers do not exist together as blends in nature, but the specific properties of each may be used to produce man-made blends that confer unique structural and mechanical properties. Two natural polymers sometimes exist together as a blend in nature, and in fact such blends possess unique structural and mechanical properties. To mimic nature and to produce new materials for biomedical applications, it is necessary to prepare polymeric blends in the laboratory under controlled conditions. If we use a relatively low-cost, low-pollution biopolymer with specific properties and an appropriate synthetic polymer, we could increase the potential for the development of a new generation of prosthetic implants.

An important aspect of the properties of a blend is the miscibility of its components. Miscibility in polymer blends is assigned to specific interactions between polymeric components, which usually give rise to a negative free energy of mixing in spite of the high molecular weight of polymers. The most common interactions in the blends are hydrogen bonding, ionic and dipole, π-electrons, and charge-transfer complexes. Most polymer blends are immiscible with each other due to the absence of specific interactions.

Several methods have been employed to assess the interactions between two components in the blend and then miscibility of two polymers [5–8,12,13,39,103,107].

In several laboratories, viscosity measurements have been employed to study the miscibility of two components. The main problem is that it is necessary to dissolve two polymers in the same solvent. In our laboratory, viscosity measurements have been used to assess the miscibility between components in the blend. For calculation, the following equations have been used [196]:

$$\frac{\eta_{sp}}{c} = [\eta] + bc, \tag{11.1}$$

where
  $[\eta]$ is the intrinsic viscosity
  b is the polymer–polymer interactions term at finite concentration related to the Huggins coefficient k_H by

$$b = k_H [\eta]^2. \tag{11.2}$$

Equation 11.1 can be adapted to a ternary system containing a solvent (component 1) and two polymers (components 2 and 3). For the ternary system, the total polymer–polymer interactions $(b_m c^2)$ are given by $c_2(c_2 b_{22} + c_3 b_{23})$ for component 2 and $c_3(c_2 b_{23} + c_3 b_{33})$ for component 3.

$$b_m c^2 = b_{22} c_2^2 + b_{33} c_3^2 + 2 b_{23} c_2 c_3, \tag{11.3}$$

where
  c_2 and c_3 are concentrations of components 2 and 3, respectively, in a mixed polymer solution
  c is the concentration of mixed polymer solution
  b_{22} and b_{33} are specific interaction coefficients of components 2 and 3 in a single polymer solution
  b_{23} represents the interaction between different types of polymer molecules in a mixed polymer solution

$$b_m = \frac{\left(b_{22} c_2^2 + b_{33} c_3^2 + 2 b_{23} c_2 c_3 \right)}{c^2}. \tag{11.4}$$

The intrinsic viscosity of mixed polymer solution

$$[\eta]_m = \left(\frac{\eta_{spm}}{c} \right)_{c \to 0} = [\eta]_2 \left(\frac{c_2}{c} \right)_{c \to 0} + [\eta]_3 \left(\frac{c_3}{c} \right)_{c \to 0}. \tag{11.5}$$

The intrinsic viscosity of a mixture of two polymers is the weight average of the intrinsic viscosity of the two polymers taken separately. For a ternary system the equation is

$$\frac{\eta_{spm}}{c} = [\eta]_2 \left(\frac{c_2}{c} \right) + [\eta]_3 \left(\frac{c_3}{c} \right) + \left[\frac{\left(b_{22} c_2^2 + b_{33} c_3^2 + 2 b_{23} c_2 c_3 \right)}{c^2} \right] c \tag{11.6}$$

$$\frac{\eta_{spm}}{c} = [\eta]_2 w_2 + [\eta]_3 w_3 + \left(b_{22} w_2^2 + b_{33} w_3^2 + 2 b_{23} w_2 w_3 \right) c, \tag{11.7}$$

where w designates the normalized weight fraction of polymer ($w_2 = c_2/c$, $w_3 = c_3/c$).
  The equation for ideal mixed polymer solution

$$\eta_{spm} = [\eta]_2 c_2 + [\eta]_3 c_3 + b_{22} c_2^2 + b_{33} c_3^2 + 2 \sqrt{b_{22} b_{33}} \, c_2 c_3. \tag{11.8}$$

Polymer mixtures might exhibit positive or negative deviations from the defined ideal behavior because of the existence or absence of interactions. The polymer–polymer miscibility can be determined by a parameter Δb.

$$\Delta b = b_{23} - \sqrt{b_{22}b_{33}},\qquad(11.9)$$

where b_{23} is obtained experimentally by substituting all the terms in Equation 11.4.

$$\Delta b > 0 - \text{miscibility}.\qquad(11.10)$$

In our research group, we used the viscosimetry method to investigate polymer–polymer interactions and for assessment of the miscibility between collagen and synthetic polymers as well as between chitosan and synthetic polymers.

11.3 NEW MATERIALS BASED ON BLENDS OF COLLAGEN WITH SYNTHETIC POLYMERS

Collagen is the main protein in an animal's body. This structural protein forms molecular lines that strengthen the tendons and vast, resilient sheets that support the skin and internal organs. Hard tissues such as bones and teeth are made of collagen with the addition of mineral crystals, mainly hydroxyapatites. There are 20 genetically distinct collagens in the collagen family. Collagen type I, type II, and type III are fibril-forming ones: type I (skin, tendon, and bone), type II (cartilage), and type III (skin and vasculature) [63–69]. In each collagen chain, there are 1000 amino acids, which form the sturdy structure by a repeated sequence of three amino acids. Every third amino acid is glycine, a small amino acid that fits perfectly inside the helix. In other positions in the chain, the following are located: proline and a modified version of proline, hydroxyproline. It is believed that hydroxyproline is responsible for collagen stability because of additional hydrogen bonds formed by the OH group. Hydrogen bonds play the main role in the stabilization of the collagen triple helix [64]. Each collagen molecule undergoes a strong molecular connection with neighboring collagen molecules by hydrogen bonding and other crosslinks [63–65,67–69]. Highly crosslinked collagen is usually insoluble in water [70]. Additional crosslinking of collagen can be made by exposure of collagen to UV irradiation [71].

Like most proteins, collagen loses all of its structure during heating. In the collagen molecule, atoms in the individual chains are held together with strong covalent bonds. However, the three chains are held in the triple-helical structure by weaker bonds (mainly hydrogen bonds). During heating, these bonds are broken and the three chains separate from one another and collapse into random coils [67,72,73]. In solution, the unfolding temperatures of collagen are within only a few degrees of the animal's body temperature, but when the molecules are aggregated to form fibers there is an increase in the transition temperature [74,75]. The temperature of thermal denaturation of collagen depends on water content, pH of environmental medium, and degree of crosslinking [76–80].

The next material that is widely used in the biomedical field is gelatin. Gelatin is usually obtained from collagen by exposing animal skins and bones to a controlled extraction process. Commercially, gelatin is most typically obtained from cattle hides and bones and pigskins. Based on their structural roles and compatibility within the body, collagen and gelatin are commonly used biomaterials in the medical, pharmaceutical, and cosmetic industries. Gelatin can also be obtained from the processing of tannery waste, but this requires further processing to serve as a commercially useful material. Gelatin is widely used in food preparation and also in many households in the form of cosmetics and toiletries, which utilize gelatin for its hypoallergenic and hydrating properties.

Collagen-based materials and gelatin-based materials are widely used in biomedical field and the cosmetics industry [5,10,19,21,27,41,43,49,51,54,56, 62,63,81,82]. The properties of collagen-based materials depend on the source of collagen and the method of preparation involving purification, fibril formation or casting, and subsequent crosslinking. Collagen has been identified as a mildly antigenic fibrous protein, and for this reason it has been considered suitable for employment as a biomaterial. Moreover, as a biological macromolecule, collagen has been found to have low immunogenicity, absorbability, biodegradation rate, and cell interaction. For these reasons, collagen has been identified as a candidate material to combine with water-soluble synthetic polymers to produce hybrid systems with various properties, compositions, and forms. The blending of collagen with synthetic polymers can be modulated by many factors. From the blends of collagen with other hydrophilic polymers, it is possible to get scaffolds for tissue engineering and good quality hydrogels. Hydrogels are three-dimensional polymeric networks held together by crosslinked covalent bonds and weak cohesive forces in the form of either hydrogen bonds or ionic bonds. Hydrogels are a broad class of hydrophilic polymeric materials that have the inherent ability to swell in water and other suitable solvents, capable of imbibing and retaining more than 10% of their weight in water within the gel structure. Attributes such as permeability to small molecules (such as tissue metabolites), a soft consistency, and a low interfacial tension between the gel and an aqueous solution are some of the important properties that have helped to generate interest in hydrogels as useful biomaterials. New materials based on the blends of collagen and synthetic polymer can be used as biodegradable polymeric scaffolds, which are widely used in medical applications. A basic advantage is the avoidance of surgery to remove the polymer from the body, when the implanted device is no longer needed. Highly porous structures of scaffolds maintain large areas for the cells to attach and proliferate. The blends of collagen and synthetic polymer can be used for preparation of drug delivery devices. Biodegradable polymers provide flexibility in the design of delivery systems of large molecular weight drugs such as peptides and proteins, which are not suitable for diffusion controlled release through non-biodegradable polymeric matrices. These systems release the bioactive substances as they degrade to the desired area with an effective dose, for a predetermined rate and duration. Drugs formulated in polymeric devices are released either by diffusion through the polymer barrier or by erosion of the polymer material, or by a combination of both. The polymers selected as delivery devices must meet several requirements like biocompatibility, drug compatibility, the possibility of their processing, and suitable biodegradation kinetics and mechanical properties. A wide variety of natural and synthetic biodegradable polymers have been investigated for drug targeting or prolonged drug release. However, only a few of them are biocompatible. Natural biodegradable polymers like bovine serum albumin (BSA), human serum albumin (HSA), collagen, gelatin, and hemoglobin have been studied for drug delivery [9,61,62]. Biodegradable polymeric systems can be designed in various shapes such as membranes, pellets, sponges, micro- or nanospheres, depending on the purpose of application. However, the use of natural polymers in the biomedical field is limited due to their high cost and questionable purity.

To summarize, the applications of collagen in tissue regeneration/engineering may include artificial skin, bone graft substitutes, dental implants, implants for incontinence, artificial tendons and blood vessels, corneal implants, and nerve regeneration [81,82].

Nevertheless, it should be emphasized that the use of animal-derived materials in the preparation of tissue-engineered products is generally viewed as a problem in the area of biomaterials. Although the extensive processing used provides a wide margin of safety, the preference for products that do not contain animal-derived materials is growing. For these reasons, we can expect that polymers and biomaterials derived from recombinant sources will become increasingly important in future products, potentially decreasing the importance of bovine collagen. Current tissue engineering approaches have employed the use of hybrid polymer systems composed of natural and synthetic macromolecules for various applications in biomaterials science. These hybrid polymeric materials are often used as hydrogels and other implantable specimens. These materials are often biocompatible and possess good physicochemical and mechanical properties.

New materials based on the blends of collagen and other polymers could result in a new generation of biodegradable polymeric materials that could be utilized as packaging material. However, packaging materials should be as cheap as possible. For this reason, mainly collagen wastes are considered and studied. To get soluble material, collagen wastes are usually hydrolyzed to short peptide. After enzymatic hydrolysis of collagen waste, it is possible to use collagen hydrolysate for different applications (fertilizer, adhesives, concrete mixtures, etc.) and also as additive to plastics increasing their sustainability. Synthetic polymers modified with collagen hydrolysate can form a sustainable plastic material workable to films for use in agriculture, forestry, packaging, etc. Collagen hydrolysate can be obtained using leather industry wastes, which are abundant [47].

The properties of new polymeric blends that have often been evaluated are the mechanical properties, like ultimate tensile strength (UTS), and percentage of elongation at break (EL). The barrier properties, like water vapor and oxygen permeability, are also important. Although most of the developed films show desirable oxygen permeability, their moisture barrier and mechanical properties are poor in comparison to the synthetic polymeric films. Inclusion of lipids in the film in emulsified form or following the technique of formation of a composite bilayer has been found to improve the moisture barrier property of the films. The next important factor for biopolymeric blends is their surface. Surface modifications that directly affect the biological function can be performed by producing specific surface functional groups, adsorption-adhesion of specific biomolecules and atoms, and by deposition of microparticles such as functional inorganic species, catalysts, and microcapsules to deliver specific reagents depending on environmental conditions. The surface can be also modified by UV irradiation [83]. The medical device and its packaging can be degraded irreversibly by the sterilization process. Most often, the earliest evidence of material degradation will appear in the surface layers of the product or its package. Surface plays a very important role in biology and medicine because most biological reactions occur at surfaces and interfaces. It is also important for most tissue engineering approaches aiming at the repair or regeneration of living tissues and the interactions of cells and polymeric biomaterial. UV irradiation is a good method of sterilization of polymeric biomedical materials, but it may also degrade them. The energy of UV light has been extensively applied for surface graft polymerization of polymers. The surface of a synthetic polymer can be grafted by natural polymers, and this procedure improves the biocompatibility of polymeric materials.

In our laboratory, water-soluble synthetic polymers have been blended with natural polymers like collagen, chitosan, and gelatin. We have used the following synthetic polymers: poly(vinylpyrrolidone) PVP, poly(vinylalcohol) PVA, poly(ethylene glycol) PEG, and poly(ethylene oxide) PEO. From the blends, we attempted to form films, hydrogels, and sponges. Collagen films can be obtained by solvent evaporation from collagen solution and by treating collagen gel solutions with different concentrations of glutaraldehyde. In our laboratory, the polymeric films were obtained by solvent evaporation from collagen solution in acetic acid [5,12,14,16–18,39].

Miscibility between two components has been studied by viscometry methods, Fourier transform infrared spectroscopy (FTIR) and differential scanning calorimetry (DSC) methods [39]. According to the classical Huggins equation (11.1), from the intercept and slope of the plots of η_{sp}/c versus c, $[\eta]$ and b_m (for solutions of individual polymers) are obtained. The value of b_{23} is obtained by substituting all the terms in Equation 11.4. Viscometric data for binary blends containing collagen and PVP show that collagen and PVP are miscible. Similar results have been obtained for collagen/PVA blends. The parameter $\Delta b > 0$ signifies the miscibility and is shown in Table 11.1. The intrinsic

TABLE 11.1
Δb Parameter for Binary Blends Containing Collagen and Synthetic Polymer

X_2 (Collagen)	X_3 (Synthetic)	Δb Coll/PVP	Δb Coll/PVA	Δb Coll/PEO	Δb Coll/PEG
0.00	1.00				
0.20	0.80	0.0867	1.9811	−0.2189	−0.2901
0.50	0.50	0.3942	0.0839	0.0765	−0.3811
0.80	0.20	1.2901	1.2691	1.4131	3.7500
1.00	0.00				

Source: Reprinted from *J. Mol. Liq.,* 145, Sionkowska, A., Skopińska, J., and Wiśniewski, M., Collagen synthetic polymer interactions in solution and thin films, 135–138, 2009. Copyright 2009, with permission from Elsevier.

TABLE 11.2

Amide Band Position of Collagen before and after Blending with Synthetic Polymers

Sample	Amide Band Position [cm⁻¹]			
	A	B	I	II
Collagen	3326	3094	1654	1557
Collagen/PVP blend	3330	3088	1658	1557
Collagen/PEO blend	3328	3082	1656	1557
Collagen/PVA blend	3325	—	1654	1559
Collagen/PEG blend	3228	3090	1656	1557

Source: Reprinted from *J. Mol. Liq.,* 145, Sionkowska, A., Skopińska, J., and Wiśniewski, M., Collagen synthetic polymer interactions in solution and thin films, 135–138, 2009. Copyright 2009, with permission from Elsevier.

viscosity measurements for the blends are different from the values from theoretical calculations (Table 11.1). The miscibility of these polymers in solid state has been confirmed by FTIR spectra of collagen and collagen/synthetic polymer blend. FTIR is a powerful technique to detect the intermolecular interactions between two polymers. The intermolecular interaction through hydrogen bonding can be characterized by FTIR, because the specific interaction affects the local electron density and the corresponding frequency shift can be observed. The position of amide band A in a blend with PVP is shifted to higher frequencies, while amide B to lower frequencies (Table 11.2). It suggests interaction between collagen and PVP by hydrogen bonds. Collagen, which is a hydrogen donor, forms a hydrogen bond with the carbonyl group from PVP. The pyrrolidone rings in PVP contain a proton-accepting carbonyl moiety, while collagen presents hydroxyl and amino groups as side groups. Therefore, a hydrogen-bonding interaction may take place between these two chemical moieties in a blend of collagen and PVP. The formation of hydrogen bonds between two different macromolecules competes with the formation of hydrogen bonds between molecules of the same polymer.

Viscometric data for binary blends containing collagen/PEO and collagen/PEG show that collagen and those polymers are immiscible, parameter $\Delta b < 0$ (Table 11.1) [39]. However, from the mixture it was possible to prepare films that can have a potential for future applications. FTIR spectra revealed some minor interactions between collagen and PEO and/or PEG in a solid state (thin film). The position of amide bands for collagen in blends was slightly different from that for pure collagen (Table 11.2).

FTIR spectra of collagen and collagen/PVA blends show that the positions of amide bands are at the same wavenumbers for all the blends and only slightly different than for pure collagen film (the example of FTIR spectra is shown in Figure 11.1, data are listed in Table 11.2). It suggests only a minor interaction between collagen and PVA in the solid state. It seems that hydrogen bond formation between collagen molecules competes with the formation of hydrogen bonds between molecules of PVA and collagen in the solid state.

In Figure 11.2, the biocompatibility of new polymeric materials based on the blends of collagen and synthetic polymers is compared with that of pure collagen film. The biocompatibility test of collagen–polymer blend sheets was carried out as follows: polymeric films were soaked in ethyl alcohol for 15 min to sterilize, then an appropriate medium was added and incubated until cells proliferate. When cells had spread adequately, the Resazurin assay test was performed [5,39]. After the Resazurin test, the fluorescence of each sample was measured with a spectrofluorometer. Figure 11.2 shows that fibroblasts grow with the fastest rate on collagen films. However, the biocompatibility of new polymeric materials based on the blends of collagen, PEG, and PVA seems to be sufficient for biomedical applications.

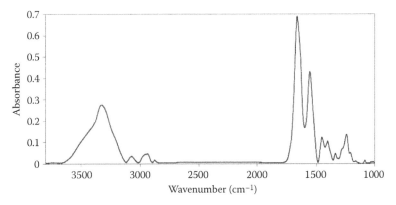

FIGURE 11.1 FTIR spectra of collagen before blending with synthetic polymers. (Reprinted from *J. Mol. Liq.,* 145, Sionkowska, A., Skopińska, J., and Wiśniewski, M., Collagen synthetic polymer interactions in solution and thin films, 135–138, 2009. Copyright 2009, with permission from Elsevier.)

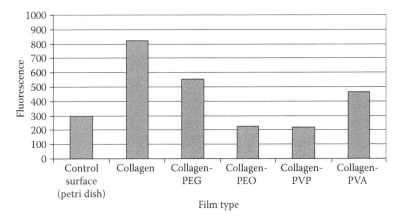

FIGURE 11.2 Fluorescence of Resazurine after 3.25 h reaction–fibroblast growth time of 5 days (gain = 60). (Reprinted from *J. Mol. Liq.,* 145, Sionkowska, A., Skopińska, J., and Wiśniewski, M., Collagen synthetic polymer interactions in solution and thin films, 135–138, 2009. Copyright 2009, with permission from Elsevier.)

The FTIR and viscosimetry results show that collagen macromolecules could form different types of hydrogen bonds with the macromolecule of synthetic polymer. The following hydrogen bonds can be formed:

- Between two hydroxyl groups (OH······OH)
- Between the hydrogen of the amide group and oxygen of the hydroxyl group (NH····OH)
- Between the hydrogen of heteroatoms and the carbonyl group of the amide (C=O····HO, C=O····HN)

These specific interactions between two polymers can modify their physicochemical properties. New polymeric materials based on the blends of collagen and synthetic polymer possess lower biocompatibility than pure collagen, but the biocompatibility can be further modified by UV irradiation. It is commonly known that living cells like to attach to polar chemical groups at the material surface [163]. During UV irradiation, photooxidation process takes place and new polar groups can be formed. The interaction of UV light with the blends is important, as the behavior of the blends needs to be understood under potentially harsh conditions. UV light may be used to refine the blend preparation process to produce blends of specific structural and chemical characteristics. After different doses of UV irradiation, the mechanical and surface properties of the film were investigated.

TABLE 11.3

Thermal Parameters of Collagen, Collagen/PVP, and Collagen/PVA Blends before and after UV Irradiation

Specimen	Time of Irradiation [h]	T_1 [°C]	Peak 1 ΔH_1 [J/g]	T_2 [°C]	Peak 2 ΔH_2 [J/g]	Glass Transition T_g [°C]
Collagen	0	80.3	222.3	212.8	3.6	—
Collagen	8	91.9	249.4	212.4	4.4	—
PVP	0	82.5	306.3	—	—	178.3
PVP	8	76.1	450.3	—	—	177.7
Collagen/PVP	0	87.1	119.5	206.8	2.4	—
Collagen/PVP	8	87.9	156.6	207.4	2.0	—
PVA	0	92.5	366.3	227.3	132.5	—
PVA	8	86.1	350.3	225.8	136.5	—
Collagen/PVA	0	85.1	129.5	216.8	122.4	—
Collagen/PVA	8	82.9	146.6	217.4	143.0	—

Source: Reprinted from Sionkowska, A., in: R.K. Bregg, ed., *Current Topics in Polymer Research*, Nova Science Publisher, New York, 2005. With permission.

During UV irradiation of polymers, excited molecules are formed in the first step and then, the secondary processes such as chain scission, crosslinking, and oxidation take place. For UV-irradiated blends, thermal properties were investigated. In the presence of thermally stable synthetic polymers, collagen is usually more stable and undergoes thermal degradation in higher temperature than collagen alone (Table 11.3).

Interesting materials for potential biomedical applications can also be obtained using the blends of collagen with chitosan. Such blends are considered in the next chapter.

11.4   NEW MATERIALS BASED ON BLENDS OF CHITOSAN WITH OTHER POLYMERS

Chitosan is a natural polymer derived from chitin. Chitin is a polysaccharide, a homopolymer composed of 2-acetamido-2-deoxy-β-D-glucopyranose units. Chitosan can be obtained from chitin by the deacetylation process. Some units in chitosan chains exist in the deacetylated form as 2-amino-2-deoxy-β-D-glucopyranose. Chitosan is a biodegradable natural polymer with great potential for pharmaceutical applications and cosmetic industry due to its biocompatibility, high charge density, nontoxicity, and mucoadhesion. It has been shown that it not only improves the dissolution of poorly soluble drugs but also exerts a significant effect on fat metabolism in the body [82,84–88]. Chitosan is widely applied in the biomedical field and also in waste water treatment [86]. It has been reported that chitosan is a potentially useful pharmaceutical material owing to its biocompatibility and low toxicity [85]. For this reason, it can be used in many applications in the formulations employed in drug delivery. Furthermore, the macromolecular chain of chitosan is stiff enough to stabilize a liquid crystalline phase in acetic acid solution where the persistence length is reported to be between 190 and 300 Å depending on the degree of acetylation [87]. Chitosan and its derivatives can be used to prepare nonviral gene delivery [88].

Chitosan itself is a very good material for biomedical applications, but after modification of properties of chitosan one can obtain versatile biomaterial for cell therapy, tissue engineering, and gene therapy [89–91].

Many research groups have studied chitosan-based materials and their applications in the field of orthopedic tissue engineering. Interesting characteristics that render chitosan suitable for this purpose are a minimal foreign body reaction, an intrinsic antibacterial nature, and the ability to be molded in various geometries and forms such as porous structures, suitable for cell ingrowth and osteoconduction. The cationic nature of chitosan allows it to complex DNA molecules making it an ideal candidate for gene delivery strategies [92–96]. Covalent and ionic crosslinking of chitosan leads to the formation of hydrogels, which can work as a drug delivery system under pH-controlled conditions [97]. Chitosan derivatives can also be applied in various tissue engineering applications, namely, skin, bone, cartilage, liver, nerve, and blood vessel [98–102].

In the scientific literature, one can find information about blending of chitosan with other natural polymers and synthetic polymers. There are several methods of preparing polymer blends of cellulose, chitin, and chitosan with natural and synthetic polymers [103]. It is possible to use the method of solid-phase blending of these polymers under conditions of high pressure and shear deformation.

Several synthetic polymers can be blended with chitosan. Water-soluble synthetic polymers, such as polyvinylalcohol (PVA), polyethyleneoxide (PEO), and polyvinylpyrrolidone (PVP), have been blended with chitosan to design new materials. Several research groups showed miscibility of chitosan and PVP and the possibility of using the blend material. Other studies showed that chitosan and PVA form an immiscible system in which interaction between macromolecules of PVA is stronger than that between PVA and chitosan [104,105]. Chitosan blends with hydrophilic polymers including polyvinylalcohol (PVA), polyethyleneoxide (PEO), and polyvinylpyrrolidone (PVP) were investigated as candidates for oral gingival delivery systems [15,104–106]. Moreover, chitosan/PVP blends can be used as a drug release system and to control the release profile of a poorly water-soluble drug. The mechanical and surface properties of chitosan, polyvinylpyrrolidone (PVP), and chitosan/PVP blends have been studied, and the influence of UV irradiation on these properties has been compared [107]. The results showed that the mechanical properties of the blends were greatly affected by UV irradiation with 254 nm wavelength, but the level of the changes of these properties was smaller in the blend than in pure chitosan and strongly dependent on the dose of irradiation and the composition of the blends (Figures 11.3 and 11.4). The surface alterations of chitosan-based blends were studied by contact angle measurements. The contact angle and the surface free energy of PVP/chitosan blends were altered by UV irradiation. The polarity of the surface of the blend was altered by UV irradiation due to the photooxidation process (Table 11.4) [107].

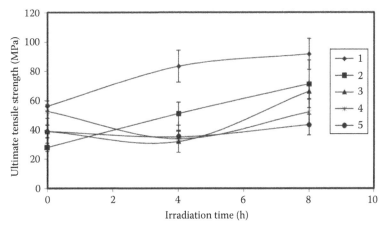

FIGURE 11.3 The influence of UV irradiation on ultimate tensile strength of chitosan (1) and chitosan/ PVP films (2—80:20, 3—60:40, 4—40:60, 5—20:80). Each value is reported with its standard deviation. (Reprinted from *Polym. Deg. Stab.,* 88, Sionkowska, A., Wisniewski, M., Skopinska, J., Vicini, S., and Marsano, E., The influence of UV irradiation on the mechanical properties of chitosan/poly(vinyl pyrrolidone) blends, 261–267, 2005. Copyright 2005, with permission from Elsevier.)

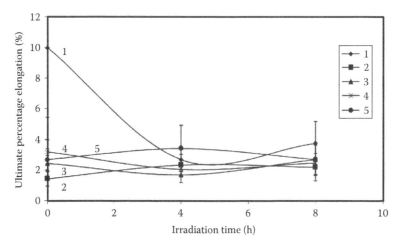

FIGURE 11.4 The influence of UV irradiation on ultimate percentage elongation of chitosan (1) and chitosan/PVP films (2—80:20, 3—60:40, 4—40:60, 5—20:80). Each value is reported with its standard deviation. (Reprinted from *Polym. Deg. Stab.*, 88, Sionkowska, A., Wisniewski, M., Skopinska, J., Vicini, S., and Marsano, E., The influence of UV irradiation on the mechanical properties of chitosan/poly(vinyl pyrrolidone) blends, 261–267, 2005. Copyright 2005, with permission from Elsevier.)

TABLE 11.4
Surface Energy Calculated by Owens-Wendt Method of the Chitosan, PVP, and Chitosan/PVP Blend after UV Irradiation

Sample	Irradiation Time [h]	Surface Energy (Total) [mN/m]	Disperse Part	Polar Part
Chitosan	0	27.2	21.8	5.4
	4	39.3	22.8	16.5
	8	50.0	20.0	30.0
PVP	0	38.7	31.7	7.0
	4	42.4	39.8	12.6
	8	49.4	26.5	22.9
Chitosan/PVP	0	27.7	22.4	5.3
80:20	4	38.7	21.4	17.3
	8	47.4	21.7	25.7
Chitosan/PVP	0	28.4	25.3	3.1
60:40	4	34.7	26.8	7.9
	8	43.1	23.4	19.7
Chitosan/PVP	0	28.6	26.1	2.5
50:50	4	34.3	26.6	7.7
	8	39.9	24.5	15.4
Chitosan/PVP	0	29.0	25.1	3.9
40:60	4	34.2	26.6	7.6
	8	43.0	22.3	20.7
Chitosan/PVP	0	31.8	24.2	7.6
20:80	4	35.3	30.6	4.6
	8	49.5	18.8	30.7

Source: Reprinted from *Polym. Deg. Stab.*, 88, Sionkowska, A., Wisniewski, M., Skopinska, J., Vicini, S., and Marsano, E., The influence of UV irradiation on the mechanical properties of chitosan/poly(vinyl pyrrolidone) blends, 261–267, 2005. Copyright 2005, with permission from Elsevier.

The interaction between chitosan and synthetic polymers is mainly due to hydrogen bonding. Another kind of highly crosslinked material can be obtained by chemical crosslinking using crosslinking agents. The most common crosslinking agent is glutaraldehyde. By crosslinking of chitosan/polyvinyl pyrrolidone blend by glutaraldehyde, a semi-interpenetrating polymer network can be obtained as potential controlled drug release system for antibiotic delivery [108]. Chitosan/PVP blends can be turned to hydrogels with porous structures using the freeze-drying method and/or electron beam irradiation at room temperature [109,110]. The blends of poly(vinyl alcohol) and chitosan can also be used for hydrogel preparation [111–113]. From the blends of PVA and chitosan in solution, nanofibers can be obtained using the electrospinning method. Nanofibrous microporous PVA/chitosan scaffolds can be fabricated using a cryogenic grinding method with subsequent compaction. Such multiscale porous structures may offer ideal matrices for tissue engineering applications [114]. The blend of poly(ethylene oxide) PEO and chitosan was used to develop novel drug delivery systems with pH-sensitive swelling and drug release properties for localized antibiotic delivery in the stomach [105,115,116]. The blend could be useful for localized delivery of antibiotics in the acidic environment of the gastric fluid.

Chitosan can be blended with nylon [117,118], polyacrylamide [119], and poly(lactic acid) [120]. Poly(lactic acid) is a biodegradable and resorbable polymer widely used in biomedical applications. Biodegradable film blends can be prepared (mainly by mixing the components in solution and film casting) [120]. Interesting membranes can also be obtained from the blend of chitosan and polyhedral oligomeric silsesquioxane derivatives [121], as well as poly(caprolactone) [122,123]. Poly(caprolactone) PCL, a synthetic polymer, can be used together with chitosan for preparation of several materials for medical applications. The blends of chitosan with PCL have been studied not only as a membrane but also as particles [122,123]. Such particles can form stable complexes with DNA. New materials based on chitosan can be made by mixing of three components. By mixing of chitosan and poly(caprolactone) with a third component (a whey-protein isolate), an increase in the water vapor resistivity of chitosan was observed [124].

New materials based on the blends of chitosan with synthetic polymers show interesting properties not only for biomedical applications [125–129]. Chitosan can also be blended with polymers that possess electrical properties, like conductivity [130–132]. The conductive materials based on the blends of chitosan with other polymers can be used in biomedical devices. Thermal properties of chitosan can be modified by blending it with thermally stable polymers [133].

Chitosan can be blended with several other naturally occurring polymers. The interaction of chitosan and collagen leads to the opportunity for preparation of new materials based on their blends [5,13,14,20,44,62,134–137]. Several research groups have shown that collagen and chitosan are miscible in both solution and solid state [13,134,135]. The blending of collagen with chitosan gives the possibility of producing new bespoke materials for potential biomedical applications [44,135]. The addition of chitosan does not denature collagen fibers [20]. Thermal properties of collagen/chitosan blends depend on the composition of the blend and are not significantly altered by UV irradiation (Table 11.5) [136]. However, the mechanical properties of the blends were greatly affected by the duration of UV irradiation (Figure 11.5) [136].

From collagen/chitosan blends, nanofibers have been prepared by electrospinning [137]. Intermolecular interactions varied with the chitosan content in electrospun fibers. New fibers were also produced by blending chitin with tropocollagen. The blended fiber was chemically modified at the fiber state by treatment with a series of carboxylic anhydrides and aldehydes to afford the corresponding *N*-modified fiber. Such fibers have improved blood compatibility [62]. Collagen and gelatin have been used as components of a binary blend with chitosan for biomedical applications [138–141]. Gelatin itself is a mixture of water-soluble proteins derived primarily from collagen. Usually, gelatin is obtained from collagen by exposing animal skins and bones to a controlled extraction process. Gelatin comes in various types. Based on their structural roles and compatibility within the body, collagen and gelatin

TABLE 11.5

Values of Thermal Parameters of Collagen and Its Blends with Chitosan Determined by Thermal Analysis in Nitrogen

Stage I: Weight Loss%

Time of Irradiation [h]	Collagen	Collagen/Chitosan 70/30	50/50	30/70	Chitosan
0	13	14	11.3	12.4	10.9
8	11.5	14.5	11.9	15.4	14.2

Stage II: Weight Loss%

Time of Irradiation [h]	Collagen	Collagen/Chitosan 70/30	50/50	30/70	Chitosan
0	67.9	67.5	54.8	59.6	57.7
8	55.1	57.6	50.1	63.9	62.9

Stage II: T_{max}

Time of Irradiation [h]	Collagen	Collagen/Chitosan 70/30	50/50	30/70	Chitosan
0	313	308	289	289	289
8	312	311	290	290	287

Source: Reprinted from *Polym. Deg. Stab.,* 91, Sionkowska, A., Wisniewski, M., Skopinska, J., Poggi, G.F., Marsano, E., Maxwell, C.A., and Wess, T.J., Thermal and mechanical properties of UV irradiated collagen/chitosan thin films, 3026–3032, 2006. Copyright 2006, with permission from Elsevier.

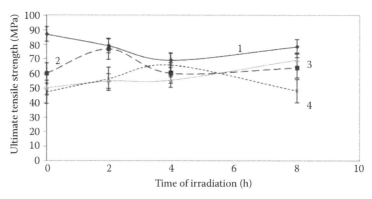

FIGURE 11.5   (See companion CD for color figure.) The influence of UV irradiation on ultimate tensile strength of collagen and collagen/chitosan films: (1) collagen; (2) 90/10 collagen/chitosan; (3) 70/30 collagen/chitosan; (4) 60/40 collagen/chitosan. Each value is the means of 20 determinations and is reported with its standard deviation. (Reprinted from *Polym. Deg. Stab.,* 91, Sionkowska, A., Wisniewski, M., Skopinska, J., Poggi, G.F., Marsano, E., Maxwell, C.A., and Wess, T.J., Thermal and mechanical properties of UV irradiated collagen/chitosan thin films, 3026–3032, 2006. Copyright 2006, with permission from Elsevier.)

are commonly used biomaterials in the medical, pharmaceutical, and cosmetic industries. It was found that several enzymes can catalyze the gel formation from gelatin and chitosan blends [138]. A novel chitosan-based membrane was made of hydroxypropyl chitosan, gelatin, and chondroitin sulfate [139]. It can potentially be used as a carrier for corneal endothelial cell transplantation. Glucose permeation results demonstrated that the blend membrane had higher glucose permeability than natural human cornea. The optical transparency of the membrane was as good as the natural human cornea [139]. Blends between chitosan and gelatin with various compositions were produced using genipin as crosslinking agent. Different amounts of genipin were used to crosslink the components of the blend, promoting the formation of amide and tertiary amine bonds between the macromolecules and the crosslinker. The mouse fibroblasts adhesion and proliferation on substrates depend on the blend composition and on the amount of crosslinker. Such materials are promising candidates for use in the field of nerve regeneration [140]. Blends based on chitosan and gelatin have gained much attention as scaffolds in various tissue engineering applications [141]. The results obtained by several research groups show significant influence of blending gelatin with chitosan on scaffold properties and cellular behavior.

Chitosan was blended with other proteins such as soy protein and silk fibroin [142–147]. From such blends, mainly biomaterials for applications in wound healing and skin tissue engineering scaffolding are considered. The formation of scaffolds from chitosan/fibroin blends from aqueous solution is possible but it depends on the pH value [144,145]. By electrospinning of the silk fibroin/chitosan blends with a chitosan content of up to 30%, the continuous fibrous structure of material can be obtained [147].

In scientific literature, one can also find reports that chitosan can be blended with starch [148,149], cellulose and its derivatives [150,151], alginates [102,152–154], pullulan [155], and dextran [156].

Cellulose can be used together with chitosan for biomedical applications. However, cellulose is a highly insoluble biopolymer, so it is not easy to get a chitosan/cellulose blend. The main problem is which solvent to choose to prepare a blend. Polysaccharide-based membranes of chitosan and cellulose blends were prepared using trifluoroacetic acid as a co-solvent [150]. The mechanical and dynamic mechanical thermal properties of the cellulose/chitosan blends appear to be dominated by cellulose and suggest that cellulose/chitosan blends are immiscible. It is believed that the intermolecular hydrogen bonding of cellulose breaks down to form cellulose–chitosan hydrogen bonding [150]. The chitosan/cellulose blend membranes may be suitable as a wound dressing with antibacterial properties. Chitosan can be blended with cellulose derivatives like hydroxypropyl cellulose, and the membranes can be prepared by the solution casting method followed by crosslinking with urea-formaldehyde–sulfuric acid mixture [151]. Material prepared in such a way possesses a high selectivity of moisture uptake.

Alginates are another class of biopolymers from marine sources and have been studied and utilized in the pharmaceutical and biotechnological fields as well as in product development for many years [152–154]. Alginates are usually extracted from brown algae. Alginates and chitosan have the common characteristic of forming gels, in addition to possessing viscosity-related properties. Alginates and chitosan can be utilized as hydrophilic drug carriers as coating material (they form film) and as matrix materials. Recently, these polysaccharides have received attention as potential rate-controlling excipients in drug release systems. Both chitosan and alginates can be used separately as materials for biomedical application as well as blends of these two biopolymers [152–154]. In addition to drug delivery systems, these polysaccharides have been used in encapsulation and immobilization of living cells. They can also be used as artificial organs consisting of hormone-producing cells. Several research groups have published articles regarding chitosan and alginate blends. The blend of alginate and carboxymethyl chitosan has been used for fiber preparation by spinning the mixture solution through a viscose-type spinneret into a coagulating bath containing aqueous $CaCl_2$ [152]. Improved miscibility between alginate and carboxymethyl chitosan due to the strong interaction from the intermolecular hydrogen bonds has been shown. Introduction of carboxymethyl chitosan in the blend fiber improved water-retention properties of the blend fiber compared with pure alginate fibers. From these alginate-chitosan blends, gel beads could be prepared based on Ca ions or dual crosslinking with various proportions of alginate and chitosan [153]. The homogeneous solution of alginate and chitosan was dripped into the solution of calcium chloride; the resultant Ca ion single crosslinked beads were dipped in the solution of sodium sulfate sequentially to prepare dual crosslinked beads. The dual crosslinkage effectively promoted the stability of beads under gastrointestinal tract conditions. The dual crosslinked beads have potential small intestine or colon site-specific drug delivery properties [153]. Based on the blends of chitosan and synthetic polymers, cationic and anionic microparticles can be prepared [102]. Such modified microparticles showed an enhanced model protein delivery into phagocytic cells.

Chitosan blended with pullulan is a potential moisture adsorbent and can be intended for wound dressing application [155]. The design and development of chitosan as a wound dressing can be improved by the addition of cornstarch or dextran and propylene glycol to obtain the films with optimal properties for wound management [156].

When chitosan is blended with egg phosphatidylcholine, advanced injectable drug delivery systems for pharmaceutical applications can be obtained [157]. The stability of such blends is due to the specific interactions between the molecules. Microdomains can be formed within the blends, which increase the stability.

11.5 MATERIALS BASED ON BLENDS OF ELASTIN AND OTHER POLYMERS

The next biopolymer that can be used for biomaterial preparation is elastin [69,158–160]. Elastin is composed of several amino acids, where the main amino acids are glycine, proline, and alanine. Elastin is a structural protein in mammals, where it is the main component of skin, blood vessels, such as the aorta, and tissues of the lung [69]. Elastin is a highly crosslinked material, and it is insoluble in water [158,159]. In biomedical applications, elastin can be used as raw material for preparation of several biomaterials, such as arterial prosthesis, dermal substitute, and hydrogels [161–163]. However, elastin is not a good material for blend preparation as it is highly insoluble. The only way to obtain soluble derivatives of elastin is hydrolysis of this biopolymer to smaller peptide and polypeptide [164–166]. In the scientific literature, there is a lack of publications regarding blends of elastin and elastin hydrolysates with synthetic polymers.

In our laboratory, we prepared new material based on the blend containing elastin hydrolysates and soluble collagen [163,166]. Elastin for hydrolysis was obtained from pig aorta. The molecular weight of the elastin hydrolysates was characterized by gel filtration using a superose gel column. The distribution of molecular weight of elastin hydrolysates was broad (Figure 11.6). As can be seen in Figure 11.6 (curve b), elastin hydrolysate was eluted as a broad peak, covering a molecular weight range from 180 to 10 kDa: about 70% of total elastin-derived peptides are in the range of 65–12 kDa. High molecular weight heterogeneity is observed compared with marker globular proteins (ovalbumin and myoglobin), which are eluted as well-separated peaks (Figure 11.6, curve a), superimposed. We tried to modify elastin hydrolysates with UV irradiation ($\lambda = 254$ nm). The short duration of UV irradiation (0.5 h) did not cause any larger differences in molecular weight distribution of elastin hydrolysates (Figure 11.6, curve c). After 2 h of UV irradiation, elastin hydrolysate was eluted as a broader peak than elastin before irradiation (Figure 11.6, curve d). These results showed high molecular weight heterogeneity of UV-irradiated elastin hydrolysates in comparison to nonirradiated elastin hydrolysates. The clear fraction of higher molecular weight than that for elastin before irradiation suggests that elastin hydrolysates undergo crosslinking reactions during UV irradiation.

The mixtures of elastin/collagen were prepared by mixing of appropriate volumes of elastin hydrolysates and collagen in acetic acid solution. From elastin/collagen mixture, thin films were obtained by solvent evaporation from solutions poured onto glass plates covered by polypropylene sheets. The obtained films were 0.015–0.030 mm thick. An example of elastin/collagen film is presented in Figure 11.7.

For elastin/collagen films, the surface properties were studied. Our results showed that the surface of a collagen/elastin film is enriched in the less polar component–collagen. Moreover, the surface of elastin/collagen films was modified by the exposure to UV light. To assess polarity of the surface, the contact angle can be measured and the surface free energy can be calculated. The surface free energy (γ_s) is one of the parameters that determine the quality of the material's surface

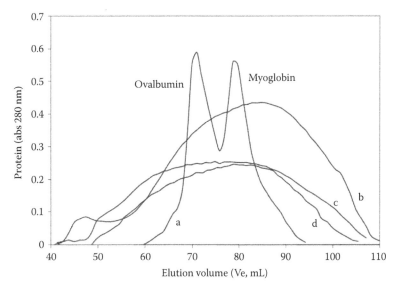

FIGURE 11.6 Gel filtration chromatograph of the standard protein mixture (a) (ovalbumin, 45 kDa and horse myoglobin, 17.8 kDa)—for comparative purposes and elastin hydrolysates before (b) and after 0.5 h (c), 2 h (d) irradiation. (Reprinted from *J. Photochem. Photobiol. B Biol.*, 86, Sionkowska, A., Skopińska, J., Wiśniewski, M., and Leżnicki, A., Spectroscopic studies into the influence of UV radiation on elastin hydrolysates in the presence of collagen, 186–191, 2007. Copyright 2007, with permission from Elsevier.)

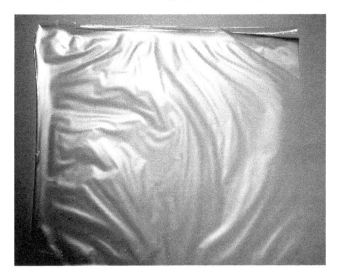

FIGURE 11.7 **(See companion CD for color figure.)** Image of thin film based on collagen and elastin hydrolysates. (Reprinted from *Appl. Surf. Sci.*, 225, Skopinska-Wisniewska, J., Sionkowska, A., Kaminska, A., Kaznica, A., Jachimiak, R., and Drewa T., Surface properties of collagen/elastin based biomaterials for tissue regeneration, 8286–8292, 2009. Copyright 2009, with permission from Elsevier.)

and its possible applications. It characterizes the disruption of intermolecular bonds that occur during the formation. The surface free energy can be calculated on the basis of the contact angle (Θ), which measures the noncovalent forces between a liquid and the first monolayer of material. Owens-Wendt method is one of the most commonly used calculating ways for polymeric materials. The method allows to estimate the dispersive (γ_s^d) and polar (γ_s) components of surface free energy, which provides more detailed information on the studied surfaces. In order to calculate the surface free energy of collagen, elastin, and its blends, two measuring liquids were used—glycerol (G) and

TABLE 11.6

Surface Free Energy (γ_s) of Elastin, Collagen, and Elastin/ Collagen Films before and after UV Irradiation (Calculated by Owens-Wendt Method)

Time of Irradiation [h]	Surface Free Energy			
	Elastin	Elastin-1:3	Collagen 5:95	Collagen
0	32.31	37.52	36.61	36.76
2	37.11	37.00	37.88	38.90
4	37.35	37.45	39.86	38.64
12	37.08	41.12	42.69	43.77
24	37.76	46.90	47.24	48.52

Source: Reprinted from *Appl. Surf. Sci.*, 225, Skopinska-Wisniewska, J., Sionkowska, A., Kaminska, A., Kaznica, A., Jachimiak, R., and Drewa T., Surface properties of collagen/elastin based biomaterials for tissue regeneration, 8286–8292, 2009. Copyright 2009, with permission from Elsevier.

TABLE 11.7

Polar Component of the Surface Free Energy γ_s^p of Elastin, Collagen, and Elastin/Collagen Films before and after UV Irradiation (Calculated by Owens-Wendt Method)

Time of Irradiation [h]	Polar Component of Surface Free Energy			
	Elastin	Elastin-1:3	Collagen 5:95	Collagen
0	8.45	5.81	4.93	3.77
2	11.87	6.42	7.67	6.27
4	15.59	7.18	8.87	8.09
12	11.77	11.83	12.83	13.60
24	13.27	18.00	18.00	18.93

Source: Reprinted from *Appl. Surf. Sci.*, 225, Skopinska-Wisniewska, J., Sionkowska, A., Kaminska, A., Kaznica, A., Jachimiak, R., and Drewa T., Surface properties of collagen/elastin based biomaterials for tissue regeneration, 8286–8292, 2009. Copyright 2009, with permission from Elsevier.

diiodomethane (D). The results of the surface free energy for films made of elastin, collagen, and the blend of these two biopolymers are presented in Tables 11.6 and 11.7.

The value of surface free energy is lower for nonirradiated elastin films than that for collagen film before irradiation. In addition, the surface free energy for the elastin/collagen blends is higher than that for elastin films. However, the differences between surface free energy for collagen film and the blend made of elastin and collagen are not significant. After UV treatment, we observed an increase in surface free energy of elastin/collagen blends. This change in the surface free energy may lead to the improvement of adhesion ability of the material. The values of polar component of surface free energy have been presented in Table 11.7. The results prove that elastin is more polar than collagen in spite of the presence of hydrophobic domains in the structure of the first one. It could be caused by relatively short length elastin hydrolysate chains and, as an effect, large content of polar terminal groups. The presence of collagen in the blended films causes the drop of the value γ_s^p. The polar component of the surface free energy for all kinds of studied samples increases after UV irradiation.

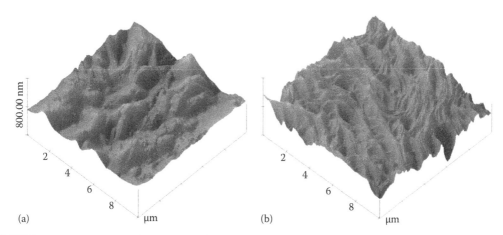

FIGURE 11.8 **(See companion CD for color figure.)** AFM images of: (a) nonirradiated film based on the blend of elastin/collagen (25/75) (b) the same film after 24 h of irradiation. (Reprinted from *Appl. Surf. Sci.*, 225, Skopinska-Wisniewska, J., Sionkowska, A., Kaminska, A., Kaznica, A., Jachimiak, R., and Drewa T., Surface properties of collagen/elastin based biomaterials for tissue regeneration, 8286–8292, 2009. Copyright 2009, with permission from Elsevier.)

Such results indicate that efficient photochemical reactions take place on the surface of biopolymeric films. The increase in polarity leads to improved interaction between the material's surface and cells. The higher content of elastin in biomaterial promotes cell adhesion and their viability on the surface. A suitable dose of UV light may improve the biocompatibility of the materials.

Atomic force microscopy (AFM) images have showed that the topography and roughness of the materials based on elastin/collagen blends were affected by UV irradiation (Figure 11.8).

11.6 SILK AND KERATIN BLENDS WITH OTHER POLYMERS

Silk fibroin produced by the domesticated *Bombyx mori* silkworm has been used as a raw material for decades. Natural fibroin is comprised of repeat sequences of alanine and glycine that readily form the β-sheet crystals responsible for the mechanical properties of silk. Natural (and also synthetic) silks' elasticity and strength make them important candidates for application in synthetic bone, ligament, and cartilage. Pure fibroin is biocompatible, can degrade slowly in vivo, supports attachment and proliferation of many cells, and supports osteoblastic differentiation. Silk fibroin can be processed into films and scaffolds to improve tissue regeneration in skin, nerve, bone, and cartilage. Blending the natural fibroin with the less expensive synthetic polymers is one approach to reduce the cost of materials. The surface of silk fibroin is usually improved when silk-based materials are used as blood-contacting materials, such as introducing heparin into them. In tissue engineering recently a big amount of silk is used.

Polymeric biomaterials based on the blends of silk with synthetic polymers and/or another natural polymer are widely studied [167–186]. From such blends, thin films can be produced, and also nanofiber can be prepared by the electrospinning method [170,172].

Several materials made from a chitin/silk fibroin blend were considered as a potential candidate for tissue engineering scaffolds with excellent cell attachment properties [178–180]. Silk fibroin was blended with chitosan [180], poly(vinyl alcohol) (PVA) [181–184], poly(ethylene glycol), and nylon 66 [182]. Silk fibroin after blending with poloxamer 407 macromer was considered as a matrix for wound dressing [185]. Using chitosan/silk fibroin blend films containing nanoparticles several nanocomposites can be obtained [186].

In our laboratory, we prepared several biomaterials based on the blends of chitosan and silk fibroin. Thin films were obtained from solution by solvent evaporation. For films made of the blend of chitosan and silk fibroin, several properties were measured. To assess the surface properties of

TABLE 11.8

The Surface Free Energy (γ_s) of Silk Fibroin, Chitosan, and Their Blends (Films) before and after UV Irradiation (Calculated by Owens-Wendt Method)

Time of UV Irradiation [h]	\(\gamma_s\) [mN/m]								
	SF	90/10	80/20	60/40	50/50	40/60	20/60	10/90	CTS
0	27.90	23.24	20.01	23.12	21.17	22.72	23.05	20.23	10.18
1	33.86	41.75	40.50	34.10	36.18	36.54	36.89	35.22	30.68
2	43.64	50.09	52.69	49.28	48.93	44.44	49.57	40.50	40.35
4	37.69	51.61	46.52	51.34	48.76	49.97	50.17	50.09	46.06
8	41.72	48.09	51.04	50.68	52.23	52.04	42.23	42.04	49.63
12	20.85	26.65	34.70	28.83	27.10	24.32	28.72	30.78	40.15
24	48.43	54.81	57.17	57.64	57.58	57.86	57.33	57.26	54.38

TABLE 11.9

The Polar Component of the Surface Free Energy γ_s^p of Surface Free Energy (γ_s) of Silk Fibroin, Chitosan, and Their Blends (Films) before and after UV Irradiation (Calculated by Owens-Wendt Method)

Time of UV [h]	γ_s^p [mN/m]								
	SF	90/10	80/20	60/40	50/50	40/60	20/60	10/90	CTS
0	2.87	1.97	1.83	2.28	3.07	3.80	4.15	4.12	0.26
1	7.52	11.08	10.14	4.89	4.07	3.34	5.66	4.20	0.50
2	17.35	19.50	22.04	20.78	18.95	17.32	21.08	9.93	39.72
4	10.68	19.89	11.71	14.36	11.59	12.27	12.38	11.49	18.59
8	22.32	13.48	17.02	16.49	17.06	17.08	7.01	8.14	21.15
12	10.50	8.91	7.09	21.98	22.14	13.77	16.73	9.00	16.74
24	16.55	20.53	22.77	25.27	28.15	26.48	24.88	25.11	18.69

thin films before and after UV treatment, the contact angle measurements were done and surface free energy was calculated. The results of the surface free energy for films made of chitosan, silk fibroin, and the blend of these two biopolymers are presented in Tables 11.8 and 11.9.

The value of surface free energy is lower for nonirradiated chitosan films than for silk fibroin film before irradiation. The surface free energy for the chitosan/silk fibroin blends is higher than for chitosan films. However, the differences between surface free energy for silk fibroin film and the blend made of silk fibroin and chitosan are not significant. After UV treatment, the increase in surface free energy of chitosan/silk fibroin blends was observed. The values of polar component of surface free energy have been presented in Table 11.9. The results prove that silk fibroin film is more polar than chitosan film. It could be caused by several hydrophilic functional groups in silk chain. The presence of chitosan in the blended films causes the drop in the value of the polar component of surface free energy (γ_s^p). The polar component of the surface free energy for chitosan/silk fibroin blends increased after UV irradiation. One can say that efficient photochemical reactions take place on the surface of biopolymeric films based on the blends of chitosan/silk fibroin. The increase in polarity can lead to improved interaction between the material's surface and cells. The polar component of surface free energy had the highest value for the blend with a composition 50% of chitosan and 50% of silk fibroin after 24 h of UV irradiation.

Structural changes on the surface of chitosan/silk fibroin films induced by UV irradiation were observed using AFM microscope. The microphotographs of nonirradiated and irradiated chitosan/silk

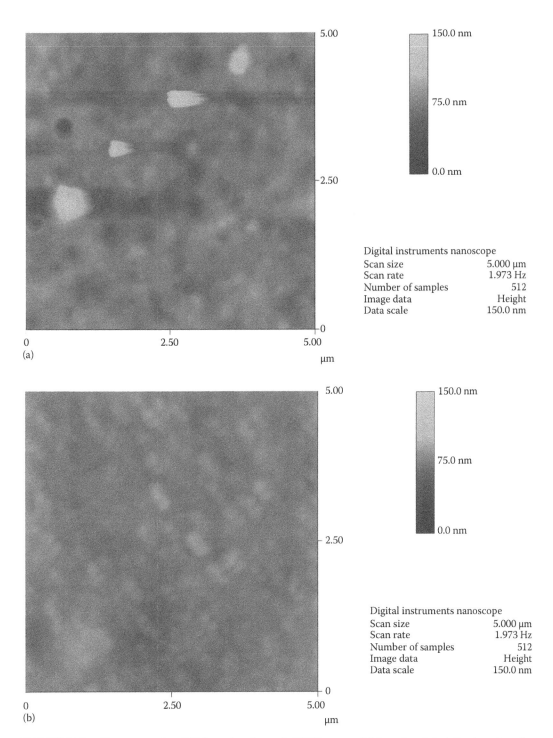

FIGURE 11.9 (See companion CD for color figure.) AFM images of (a) nonirradiated film based on the blend of chitosan/silkfibroin (50/50) (b) the same film after 2 h of irradiation.

TABLE 11.10

Surface Roughness of Chitosan Films, Silk Fibroin Films, and Films Based on the Blends of SF/CTS

Time of UV [h]	SF	50/50	CTS
		Rms [nm]	
0	22.9	13.4	4.8
2	19.2	9.1	3.4
8	3.9	4.2	2.8

TABLE 11.11

Mechanical Properties of the Blends of Chitosan/Silk Fibroin

Specimen	% Elongation	Young's Modulus [GPa]	Breaking Load [N]
Chitosan	1.15	0.54	5.88
CTS/SF (80/20)	0.97	1.11	15.48
CTS/SF (60/40)	1.14	2.68	15.69
CTS/SF (50/50)	0.48	2.01	5.04
CTS/SF (40/60)	0.75	2.14	13.52
CTS/SF (20/80)	0.07	0.58	0.11

fibroin films are shown in Figure 11.9. AFM images show differences in surface properties of the films before and after UV irradiation. The surface roughness is decreasing with increasing UV irradiation time. The atomic force microscope is often used to obtain a reasonable measure of surface roughness on the nanometer scale. Typically, AFM users rely on root mean square (*Rms*) roughness. The values of root-mean-square (*Rms*) surface roughness of films made of silk fibroin, chitosan, and their blends before and after UV irradiation are shown in Table 11.10. *Rms* value for silk fibroin film is much bigger than for pure chitosan film and also bigger than for chitosan/silk fibroin film. UV irradiation caused the decrease in surface roughness of silk fibroin films, chitosan films, and films made of chitosan/silk fibroin blend.

Mechanical properties of chitosan films and films based on the blends of chitosan and silk fibroin are shown in Table 11.11. Several films made of the blend of chitosan and silk fibroin with different composition were obtained after solvent evaporation from chitosan/silk fibroin solution. As one can see in Table 11.11, for chitosan/silk fibroin film with a composition of 60% of chitosan and 40% of silk fibroin, the mechanical properties, such as elongation at break, ultimate tensile strength, and Young's Modulus have the highest value. Thin silk fibroin films were very fragile, and it was not possible to measure mechanical properties for such films.

Keratin is a biopolymer, the major component of hair, feathers, nails, and horns of mammals, reptiles, and birds. This fibrous protein is composed of several repeating sequences of amino acids along its chain. In the biopolymer chain amino acid containing sulfur, cystine is present. Cystine in the keratin chain is able to form characteristic inter and intramolecular disulphide bonds, which determine the properties of keratin [187,188]. Keratin, similarly to elastin is a highly insoluble protein [189]. Keratin can be modified by UV radiation and chemical crosslinking [190–192]. In scientific literature, there are only a few papers regarding the blends of keratin with other polymers. There are reports regarding blends of keratin with poly(ethylene oxide) [193,194] and with chitosan [170,195].

11.7   NEW MATERIALS AND COMPOSITES BASED ON NATURAL POLYMERS, SYNTHETIC POLYMERS, AND THEIR BLENDS CONTAINING INORGANIC PARTICLES

Most natural polymers have a complex structure with interesting mechanical properties. These complex structures, which have risen from hundreds of million years of evolution, are inspiring scientists in the design of novel materials. Their defining characteristics, hierarchy, multifunctionality, and self-healing capability are often investigated [197]. Self-organization is also a fundamental feature of many biological materials and the manner by which the structures are assembled from the molecular level up. The basic building blocks are described, starting with the 20 amino acids and proceeding to polypeptides, polysaccharides, and polypeptide–saccharides. Biological materials very often contain inorganic nanoparticles. The "hard" phases are primarily strengthened by minerals, which nucleate and grow in a biomediated environment that determines the size, shape, and distribution of individual crystals. The most important mineral phases are hydroxyapatite, silica, and aragonite [197].

New materials based on the blends of two polymers or biopolymers that contain inorganic nanoparticles can be used as an implant in the context of both hard and soft tissues. Natural polymers can be used for preparation of such biomaterials for biomedical applications without any additives. Pure and blended collagen were widely studied [63–83], together with pure chitosan and its blends [15,20,44,62,82–156]. Nanostructured materials can be created by combining organic and inorganic compounds. It can be achieved by the intercalation of inorganic nanoparticles in a polymeric matrix. In order to develop novel bone substitutes, hydroxyapatite has been intensively investigated from the viewpoints of biocompatibility, bioactivity, and biodegradation. It is extremely difficult to connect nanoparticles with natural polymers, as many of them are insoluble. In many projects, insoluble natural polymers such as collagen, elastin, chitosan, keratin, and silk have been converted to soluble derivatives by hydrolysis (acidic, base, and enzymatic). These hydrolysates, after characterization, could be used to design and make new biomaterials for medical applications.

Regenerative medicine is currently using three strategies: transplantation of cells to form new tissue in the transplant site, implantation of bioartificial tissues constructed in vitro, and induction of regeneration in vivo from healthy tissues adjacent to an injury. New materials based on the blends of natural and synthetic polymers containing nanoparticles can be used as bioartificial implants in the context of both hard and soft tissues. The next several years will be critical for the maturation of nanomaterial science and its role in clinical medicine. Elucidation of the fundamental structure–function relationships in normal and diseased or damaged tissue through sound basic research will continue to be a necessary prerequisite. In addition, consideration of the different elements in the product development continuum from inspiration, product design, and bench studies, to the clinic is vital. Although natural polymers have separately been proposed as in vitro extracellular matrix materials, the influence of biopolymer–polymer composite matrices on cell morphology, differentiation, and function is not well studied. Recently, attention has been given to the applications of nanoparticles in combination with polymers and biopolymers. Biopolymer-based materials can be combined with a bonelike filler material (as bioactive glass, calcium phosphates, and other calcium salts). The properties of biomaterials that will be used in tissue regeneration as bone, scaffolds, and guiding structures are of key importance. In addition to gross properties, such as mechanical strength, biodegradability, biocompatibility, and lack of toxic or immune reactions, the biomaterial should exhibit the capacity to interact specifically with cell surfaces, in particular with cell-membrane receptors that are active in recognition, adhesion of cells to the extracellular matrix, and in mediation of signals to the cell. Not only should the biomaterial surfaces contain the respective recognizable structures, but they should also present these structures on a nanoscale, which would compare to the dimensions of adhesive plaques and clusters of receptor molecules at the surfaces of adhering cells. Properties of the biomaterial surface (such as topography and chemistry) control the type and magnitude of cellular and molecular events at the tissue–implant interface. Therefore, nanotechnologies represent a new approach to materials science. Several papers have been published so far in the area of biomaterials based on the blends of natural polymers and synthetic polymers containing inorganic nanoparticles [147,181,198–206]. We can expect that in the near future several new materials will be designed based on the blends of two or more natural polymers as well as with synthetic polymers containing nanoparticles. An economic impact can be expected, as materials based on the blends of natural polymers and synthetic polymers should be less expensive than materials made strictly from biopolymers.

11.8   CONCLUSIONS

Most natural materials are complex composites with outstanding mechanical properties. These complex structures, which have risen from many years of evolution, are inspiring materials scientists in the design of novel materials. Their defining characteristics, hierarchy, multifunctionality, and self-healing capability, are illustrated by several researchers. Self-organization is also a fundamental feature of many biological materials and the manner by which the structures are assembled from the molecular level up. The main biopolymers used in preparation of materials for biomedical applications are collagen, chitin, chitosan, keratin, silk, and elastin. These biopolymers can be blended with several synthetic polymers. Moreover, the blends can be strengthened by minerals, which nucleate and grow in a polymeric blend to produce the appropriate size, shape, and distribution of individual crystals, similar to hard tissue. The main goal of several research groups is the search for bioinspired materials and structures. Traditional approaches focus on design methodologies of biological materials using conventional synthetic materials. The new approach focuses on design methodologies of biological materials using blends of synthetic and natural polymers. The new frontiers reside in the synthesis of bioinspired materials through processes that are characteristic of biological systems; these involve nanoscale self-assembly of the components and the development of hierarchical structures. The development of new biomaterials for medical applications is one of the challenging tasks for many scientists today. Due to the mechanical, chemical, and biological properties of the natural-based composites, multifunctional biomaterials based on the blends of natural polymers with synthetic ones can be designed to be similar to those of normal human tissue. It is believed that it will present significant advantages over current materials such as human and animal grafts or artificial materials. In the near future, we can expect the development of adequate three-dimensional scaffolds from both natural and synthetic materials and also their blends. Such materials can subsequently be used as highways for cells in order to induce cell activity, being physiologically suitable for cell activity (adhesion, proliferation, differentiation) and having mechanical properties similar to those of the natural tissue. We can expect also the development of 3D culture technology by which functional tissue can be grown ex vivo in the three-dimensional scaffolds. The application of the blends can be very wide, including membranes for dialysis, wound dressings, artificial skin, and implantable devices for the delivery of biologically active substances and materials. New materials based on the blends have the potential to develop new material for packaging of medical devices and drugs. Our research and that of several research groups has shown that it is not easy to study the interaction between natural and synthetic polymers [207–218]. Very often, the properties of the blends do not meet the scientific and practical expectations. Nevertheless, substantial gains are expected both from a medical and an economic standpoint. We hope that in the near future there will be a huge increase in new products based on the blends of natural and synthetic polymers.

ACKNOWLEDGMENT

Financial support from the Ministry of Science (MNII, Poland) Grant No. N N507 3495325 is gratefully acknowledged.

REFERENCES

1. Vert M. Polymeric biomaterials: Strategies of the past vs. strategies of the future. *Prog. Polym. Sci.* 2007; 32: 755–761.
2. Howard MWT. Polymers for tissue engineering scaffolds. In: S Dumitriu, ed., *Polymeric Biomaterials*, 2nd edn., Marcel Dekker Inc., New York, 2002, p. 167.
3. Kohn J, Welsh WJ, and Knight D. A new approach to the rationale discovery of polymeric biomaterials. *Biomaterials* 2007; 28: 4171–4177.
4. Nair LS and Laurencin CT. Biodegradable polymers as biomaterials. *Prog. Polym. Sci.* 2007; 32: 762–798.
5. Sionkowska A. Current research on the blends of natural and synthetic polymers as new biomaterials: Review. *Prog. Polym. Sci.* 2011; 36: 1254–1276.
6. Cascone MG. Dynamic-mechanical properties of bioartificial polymeric materials. *Polym. Int.* 1997; 43: 55–69.
7. Giusti P, Lazzeri L, Petris S, Palla M, and Cascone MG. Collagen-based new bioartificial polymeric materials. *Biomaterials* 1994; 15: 1229–1233.
8. Giusti P, Lazzeri L, and Lelli L. Bioartificial polymeric materials: A new method to design biomaterials by using both biological and synthetic polymers. *TRIP* 1993; 1(9): 261–267.
9. Werkmeister JA, Edwards GA, Casagranda F, White JF, and Ramshaw JAM. Evaluation of a collagen-based biosynthetic materials for the repair of abdominal wall defects. *J. Biomed. Mater. Res.* 199; 39(3): 429–436.
10. Suh JKF and Matthew HWT. Application of chitosan-based polysaccharide biomaterials in cartilage tissue engineering: A review. *Biomaterials* 2000; 21(24): 2589–2598.
11. Leclerc E, Furukawa KS, Miyata F, Sakai Y, Ushida T, and Fujii T. Fabrication of microstructures in photosensitive biodegradable polymers for tissue engineering applications. *Biomaterials* 2004; 25(19): 4683–4690.
12. Sionkowska A. Interaction of collagen and poly(vinyl pyrrolidone) in blends. *Eur. Polym. J.* 2003; 39: 2135–2140.
13. Sionkowska A, Wisniewski M, Skopinska J, Kennedy CJ, and Wess TJ. Molecular interactions in collagen and chitosan blends. *Biomaterials* 2004; 25: 795–801.
14. Sionkowska A, Wisniewski M, Skopinska J, Kennedy CJ, and Wess TJ. The photochemical stability of collagen-chitosan blends. *J. Photochem. Photobiol. Part A; Chem.* 2004; 162: 545–554.
15. Marsano E, Vicini S, Skopińska J, and Sionkowska A. Chitosan and poly(vinyl pyrrolidone): Compatibility and miscibility of blends. *Macromol. Symp.* 2004; 218: 251–260.
16. Sionkowska A, Kaczmarek H, Kowalonek J, Wisniewski M, and Skopinska J. Surface state of UV irradiated collagen/PVP blends. *Surf. Sci.* 2004; 566–568: 608–612.
17. Sionkowska A, Wisniewski M, and Skopinska J. Photochemical stability of collagen/poly (vinyl alcohol) blends. *Polym. Degrad. Stab.* 2004; 83: 117–125.
18. Sionkowska A, Kaczmarek H, Wisniewski M, ElFeninat F, and Mantovani D. Ultraviolet irradiation of synthetic polymer/collagen blends: Preliminary results of atomic force microscopy. In: D. Mantovani, ed., *Advanced Materials for Biomedical Applications*, COM 2002, Quebec, Canada, p. 27.
19. Shanmugasundaram N, Ravichandran P, Neelakanta PR, Nalini R, Subrata P, and Rao KP. Collagen-Chitosan polymeric scaffolds for the in vitro culture of human epidermoid carcinoma cells. *Biomaterials* 2001; 22: 1943–1951.
20. Salome Machado AA, Martins VCA, and Plepis AMG. Thermal and rheological behaviour of collagen chitosan blends. *J. Therm. Anal. Calorim.* 2002; 67: 491–498.
21. Cascone MG, Di Pasquale G, La Rosa AD, Cristallini C, Barbani N, and Recca A. Blends of synthetic and natural polymers as drug delivery systems for growth hormon. *Biomaterials* 1995; 16(7): 569–574.
22. Yang JM, Su WY, Leu TL, and Yang MC. Evaluation of chitosan/PVA blended hydrogel membranes. *J. Memb. Sci.* 2004; 236(1–2): 39–51.
23. Da Cruz AGB, Góes JC, Figueiró SD, Feitosa JPA, and Ricardo NMPS. On the piezoelectricity of collagen/natural rubber blend films. *Eur. Polym. J.* 2003; 39(6): 1267–1272.
24. Coombes AGA, Verderio E, Shaw B, Li X, Griffin M, and Downes S. Biocomposites of non-crosslinked natural and synthetic polymers. *Biomaterials* 2002; 23(10): 2113–2118.
25. Piza MA, Constantino CJL, Venancio EC, and Mattoso LHC. Interaction mechanism of poly (o-ethoxyaniline) and collagen blends. *Polymer* 2003; 44(19): 5663–5670.
26. Yang Y, Porté MCh, Marmey P, El Haj AJ, and Amédée J. Covalent bonding of collagen on poly(L-lactic acid) by gamma irradiation. *Nucl. Instrum. Methods Phys. Res. Sec. B: Beam Interact. Mater. Atoms* 2003; 207(2): 165–174.

27. Lee PC, Huang LLH, Chen LW, Hsieh KH, and Tsai CL. Effect of forms of collagen linked to polyurethane on endothelial cell growth. *J. Biomed. Mater. Res.* 1996; 32(4): 645–653.
28. Goissis G, Piccirili L, Goes JC, Plepis AMD, and Das-Gupta DK. Anionic collagen: Polymer composites with improved dielectric and rheological properties. *Artif. Organs* 1998; 22(3): 203–209.
29. Lee SD, Hsiue GH, Chang PCT, and Kao CY. Plasma-induced grafted polymerization of acrylic acid and subsequent grafting of collagen onto polymer film as biomaterials. *Biomaterials* 1996; 17(16): 1599–1608.
30. Dufrene YF, Marchal TG, and Rouxhet PG. Influence of substratum surface properties on the organization of absorbed collagen films: In situ characterisation by atomic force microscopy. *Langmuir* 1999; 15(8): 2871–2878.
31. Naqvi A and Nahar P. Photochemical immobilization of proteins on microwave-synthesized photoreactive polymers. *Anal. Biochem.* 2004; 327(1): 68–73.
32. Cheng Z and Teoh SH. Surface modification of ultra thin poly(ε-caprolactone) films using acrylic acid and collagen. *Biomaterials* 2004; 25(11): 1991–2001.
33. Van Wachem PB, Hendriks M, Blaauw EH, Dijk F, Verhoeven MLPM, Cahalan PT, and van Luyn MJA. (Electron) microscopic observations on tissue integration of collagen-immobilized polyurethane. *Biomaterials* 2002; 23(6): 1401–1409.
34. Tyan YC, Liao JD, Klauser R, Wu ID, and Weng CC. Assessment and characterization of degradation effect for the varied degrees of ultra-violet radiation onto collagen-bonded polypropylene non-woven fabric surfaces. *Biomaterials* 2002; 23: 65–76.
35. De Cupere VM and Rouxhet PG. Collagen films adsorbed on native and oxidized poly(ethylene terephthalate): Morphology after drying. *Surf. Sci.* 2001; 491: 395–404.
36. Giusti P, Lazzeri L, Barbani N, Narducci P, Bonaretti A, Palla M, and Lelli L. Hydrogels of poly(vinyl alcohol) and collagen as new bioartificial materials: Physical and morphological study. *J. Mater. Sci.: Mater. Med.* 1993; 4(6): 538–542.
37. Barbani N, Lazzeri L, Bonaretti A, Seggiani M, Nerducci P, Polacco G, Pizzirani G, and Giusti P. Physical and biological stability of dehydro-thermally crosslinked collagen-poly(vinyl alcohol) blands. *J. Mater. Sci.: Mater. Med.* 1994; 5(12): 882–888.
38. Barbani N, Cascone MG, Giusti P, Lazzeri L, Polacco G, and Pizzirani G. Bioartificial materials based on collagen: 2. Mixtures of soluble collagen and poly(vinyl alcohol) cross-linked with gaseous glutaraldehyde. *J. Biomater. Sci. Polym. Ed.* 1995; 7(6): 471–484.
39. Sionkowska A, Skopińska J, and Wiśniewski M. Collagen synthetic polymer interactions in solution and thin films. *J. Mol. Liq.* 2009; 145: 135–138.
40. Cascone MG, Giusti P, Lazzeri L, Pollicino A, and Recca A. Surface characterisation of collagen-based bioartificial polymeric materials. *J. Biomater. Sci. Polym. Ed.* 1996; 7(10): 917–924.
41. Rao KP. Recent developments of collagen-based materials for medical applications and drug delivery systems. *J. Biomater. Sci. Polym. Ed.* 1995; 7(7): 623–631.
42. Shenoy V and Rosenblatt J. Diffusion of macromolecules in collagen and hyaluronic acid, rigid-rod-flexible polymer, composite matrices. *Macromolecules* 1995; 28(26): 8751–8756.
43. Barbani N, Lazzeri L, Cristallini C, Cascone MG, Polacco G, and Pizzirani G. Bioartificial materials based on blends of collagen and poly(acrylic acid). *J. Appl. Polym. Sci.* 1999; 72: 971–975.
44. Taravel MN and Domard A. Collagen and its interaction with chitosan. Some biological and mechanical properties. *Biomaterials* 1996; 17(4): 451–455.
45. Taniguchi T and Okamura K. New films produced from microfibrillated natural fibres. *Polym. Int.* 1998; 47: 291–294.
46. Chen XG, Wang Z, Liu WS, and Park HJ. The effect of carboxymethyl-chitosan on proliferation and collagen secretion of normal and keloid skin fibroblasts. *Biomaterials* 2002; 23(23): 4609–4614.
47. Alexy P, Bakoš D, Hanzelová S, Kukolíková L, and Kupec J. Poly(vinyl alcohol)–collagen hydrolysate thermoplastic blends: I. Experimental design optimisation and biodegradation behaviour. *Polym. Test.* 2003; 22(7): 801–809.
48. Mao JSh, Cui YL, Wang XH, Sun Y, Yin YJ, and Zhao HM. A preliminary study on chitosan and gelatin polyelectrolyte complex cytocompatibility by cell cycle and apoptosis analysis. *Biomaterials* 2004; 25(18): 3973–3981.
49. Quek ChH, Li J, Sun T, Chan MLH, Mao HQ, Gan LM, and Leong KW. Photo-crosslinkable microcapsules formed by polyelectrolyte copolymer and modified collagen for rat hepatocyte encapsulation. *Biomaterials* 2004; 25(17): 3531–3540.
50. Shan Y, Zhou Y, Cao Y, Xu Q, Ju Huangxian, and Wu Z. Preparation and infrared emissivity study of collagen-g-PMMA/In$_2$O$_3$ nanocomposite. *Mater. Lett.* 2004; 58(10): 1655–1660.

51. Lee SB, Kim YH, Chong MS, and Lee YM. Preparation and characteristics of hybrid scaffolds composed of β-chitin and collagen. *Biomaterials* 2004; 25(12): 2309–2317.

52. Wong Po Foo Ch and Kaplan DL. Genetic engineering of fibrous proteins: Spider dragline silk and collagen. *Adv. Drug Delivery Rev.* 2002; 4(8): 1131–1143.

53. Daamen WF, van Moerkerk HThB, Hafmans T, Buttafoco L, Poot AA, and Veerkamp JH. Preparation and evaluation of molecularly-defined collagen–elastin–glycosaminoglycan scaffolds for tissue engineering. *Biomaterials* 2003; 24(22): 4001–4009.

54. Ma L, Gao Ch, Mao Z, Zhou J, Shen J, Hu X, and Han Ch. Collagen/chitosan porous scaffolds with improved biostability for skin tissue engineering. *Biomaterials* 2003; 24(26): 4833–4841.

55. Dai NT, Williamson MR, Khammo N, Adams EF, and Coombes AGA. Composite cell support membranes based on collagen and polycaprolactone for tissue engineering of skin. *Biomaterials* 2004; 25(18): 4263–4271.

56. Lopes CMA and Felisberti MI. Mechanical behaviour and biocompatibility of poly(1-vinyl-2-pyrrolidone)-gelatin IPN hydrogels. *Biomaterials* 2003; 24: 1279–1284.

57. Dutoya S, Lefebvre F, Deminières C, Rouais F, Verna A, Kozluca A, and Le Bugle A. Unexpected original property of elastin derived proteins:spontaneous tight coupling with natural and synthetic polymers. *Biomaterials* 1998; 19 (1–3): 147–155.

58. Scotchford CA, Cascone MG, Downes S, and Giusti P. Osteoblast responses to collagen-PVA bioartificial polymers in vitro: The effects of cross-linking method and collagen content. *Biomaterials* 1998; 19: 1–11.

59. Sarti B and Scandola M. Viscoelastic and thermal properties of collagen/poly(vinyl alcohol) blends. *Biomaterials* 1995; 16: 785–792.

60. Nezu T and Winnik FM. Interaction of water-soluble collagen with poly(acrylic acid). *Biomaterials* 2000; 21: 415–419.

61. Thacharodi D and Rao KP. Collagen-chitosan composite membranes for controlled release of propranolol hydrochloride. *Int. J. Pharm.* 1995;120(1): 115–118.

62. Hirano S, Zhang M, Nakagawa M, and Miyata T. Wet spun chitosan–collagen fibers, their chemical N-modifications, and blood compatibility. *Biomaterials* 2000; 21(10): 997–1003.

63. Bailey AJ and Paul RG. Collagen - Is not so simple protein. *J. Soc. Leather Technol. Chem.* 1998; 82: 104–108.

64. Van der Rest R and Garrone M. Collagen family of proteins. *FASEB J.* 1991; 5: 2814–2823.

65. Ellis DO and McGavin S. The structure of collagen- on X-ray study. *J. Ultrastruct. Res.* 1970; 32: 191–211.

66. Prockop J and Fertala A. The collagen fibril: The almost crystalline structure. *J. Struct. Biol.* 1998; 122: 111–118.

67. Bella J, Eaton M, Brodsky B, and Berman HM. Crystal and molecular structure of a collagen-like peptide at 1.9 A resolution. *Science* 1994; 266: 75–81.

68. Piez KA. Molecular and aggregate structures of the collagen. In: KA Piez and AH Reddi, eds., *Extracellular Matrix Biochemistry*, Elsevier Science Publishing, New York, 1994, p. 1.

69. Silver FH. Connective tissue structure. In: *Biological Materials: Structure Mechanical Properties and Modelling of Soft Tissues*, New York University Press, New York, 1987, p. 7.

70. Miyahara T, Murai A, Tanaka T, Shiozawa S, and Kameyama M. Age-related differences in human skin collagen: Solubility in solvent, susceptibility to pepsin digestion, and the spectrum of the solubilized polymeric collagen molecules. *J. Gerontol.* 1982; 37: 651.

71. Sionkowska A, Kaminska A, Miles CA, and Bailey AJ. The effect of UV radiation on the structure and properties of collagen. *Polimery* 2001; 6: 379–389.

72. Rich A and Crick FHC. The molecular structure of collagen. *J. Mol. Biol.* 1961; 3: 483–506.

73. Fraser RDB, MacRea TP, and Suzuki E. Chain conformation in the collagen molecule. *J. Mol. Biol.* 1979; 129: 463–481.

74. Privalov PL. Stability of proteins which do not present a single co-operative system. *Adv. Protein Chem.* 1982; 25: 1–104.

75. Burjanadze TV. Thermodynamic substantiation of water-bridged collagen structure. *Biopolymers* 1992; 32: 941–949.

76. Flory PJ and Garrett RR. Phase transition in collagen and gelatin systems. *J. Am. Chem. Soc.* 1958; 80: 4836–4845.

77. Bigi A, Cojazzi, G, Roveri N, and Koch MHJ. Differential scanning calorimetry and X-ray diffraction study of tendon collagen thermal denaturation. *Int. J. Biol. Macromol.* 1987; 9: 363–367.

78. Luescher M, Ruegg M, and Scjindler P. Effect of hydration on thermal stability of tropocollagen and its effect of hydration on thermal stability of tropocollagen and its dependence on the presents of neutral salts. *Biopolymers* 1974; 13: 2489–2503.

79. Sionkowska A and Kamińska A. Thermal helix-coil transition in UV irradiated collagen from rat tail tendon. *Int. J. Biol. Macromol.* 1999; 24: 337–340.
80. Usha R and Ramasami T. The effects of urea and n-propanol on collagen denaturation: Using DSC, circular dichroism and viscosity. *Termochim. Acta* 2004; 409: 2001–206.
81. Lee CH, Singla A, and Lee Y. Biomedical applications of collagen. *Int. J. Pharm.* 2001; 221(1–2): 1–22.
82. Seal BL, Otero TC, and Panitch A. Polymeric biomaterials for tissue and organ regeneration. *Mater. Sci. Eng.: R. Reports* 2001; 34(4–5): 147–230.
83. Fischbach C, Tessmar J, Lucke A, Schnell E, Schmeer G, and Blunk T. Does UV irradiation affect polymer properties relevant to tissue engineering? *Surf. Sci.* 2001; 491(3): 333–345.
84. Struszczyk, MH. Chitin and Chitosan: Part II: Applications of chitosan. *Polimery* 2002; 47(6): 396–403.
85. Muzzarelli R, Baldassarre V, Conti F, Ferrara P, Biagini G, Gazzanelli G, and Vasi V. Biological activity of chitosan: Ultrastructural study. *Biomaterials* 1988; 9(3): 247–252.
86. Rinaudo M and Domard A. In: G Skiåk-Break, T Anthonsen, and P Sandford, eds., *Chitin and Chitosan*, Elsevier Applied Science, London, U.K., 1989, pp. 71–86.
87. Terbojevich M, Cosani A, Conio G, Marsano E, and Bianchi E. Chitosan: Chain rigidity and mesophase formation. *Carbohydr. Res.* 1991; 209: 251–260.
88. Gerrit B. Chitosans for gene delivery. *Adv. Drug Delivery Rev.* 2001; 52: 145–150.
89. Chunmeng S, Ying Z, Xinze R, Meng W, Yongping S, and Tianmin C. Therapeutic potential of chitosan and its derivatives in regenerative medicine. *J. Surg. Res.* 2006; 133: 185–192.
90. Berger J, Reist M, Mayer JM, Felt O, and Gurny R. Structure and interactions in chitosan hydrogels formed by complexation or aggregation for biomedical applications. *Eur. J. Pharm. Biopharm.* 2004; 57: 35–52.
91. Sevda S and McClure SJ. Potential applications of chitosan in veterinary medicine. *Adv. Drug Deliv. Rev.* 2004; 56: 1467–1480.
92. Di Martino A, Sittinger M, and Risbud MV. Chitosan: A versatile biopolymer for orthopaedic tissue-engineering. *Biomaterials* 2005; 26: 5983–5990.
93. Kim TH, Jiang H, Jere D, Park IK, Cho MH, and Nah JW. Chemical modification of chitosan as a gene carrier in vitro and in vivo. *Prog. Polym. Sci.* 2007; 32: 726–753.
94. Khor E and Lim LY. Implantable applications of chitin and chitosan. *Biomaterials* 2003; 24: 2339–2349.
95. Dodane V and Vilivalam VD. Pharmaceutical applications of chitosan. *Pharm. Sci. Technol. Today* 1998; 1: 246–253.
96. Sinha, VR, Singla AK, Wadhawan S, Kaushik R, Kumria R, Bansal K, and Dhawan S. Chitosan microspheres as a potential carrier for drugs. *Int. J. Pharm.* 2004; 274: 1–33.
97. Berger J, Reist M, Mayer JM, Felt O, Peppas NA, and Gurny R. Structure and interactions in covalently and ionically crosslinked chitosan hydrogels for biomedical applications. *Eur. J. Pharm. Biopharm.* 2004; 57: 19–34.
98. Kim IY, Seo SJ, Moon HS, Yoo MK, Park IY, Kim BC, and Cho CS. Chitosan and its derivatives for tissue engineering applications. *Biotech. Adv.* 2008; 26: 1–21.
99. Krajewska B. Application of chitin- and chitosan-based materials for enzyme immobilizations: A review. *Enzyme Microb. Technol.* 2004; 35: 126–139.
100. Ta HT, Dass CR, and Dunstan DE. Injectable chitosan hydrogels for localised cancer therapy. *J. Control. Release* 2008; 126: 205–216.
101. Grant J, Lee H, Soo P, Lim CJ, Piquette-Miller M, and Allen C. Influence of molecular organization and interactions on drug release for an injectable polymer-lipid blend. *Int. J. Pharma.* 2008; 360: 83–90.
102. Wischke C, Borchert HH, Zimmermann J, Siebenbrodt I, and Lorenzen DR. Stable cationic microparticles for enhanced model antigen delivery to dendritic cells. *J. Control. Release* 2006; 114: 359–368.
103. Rogovina SZ and Vikhoreva GA. Polysaccharide-based polymer blends: Methods of their production. *Glycoconj. J.* 2006; 23: 611–618.
104. Karavas E, Georgarakis E, and Bikiaris D. Adjusting drug release by using miscible polymer blends as effective drug carries. *J. Therm. Anal. Calorim.* 2006; 84: 125–133.
105. Khoo CGL, Frantzich S, Rosinski A, Sjöström M, Hoogstraate J. Oral gingival delivery systems from chitosan blends with hydrophilic polymers. *Eur. J. Pharm. Biopharm.* 2003; 55: 47–56.
106. Sakurai K, Maegawa T, and Takahashi T. Glass transition temperature of chitosan and miscibility of chitosan/poly(N-vinyl pyrrolidone) blends. *Polymer* 2000; 41: 7051–7056.
107. Sionkowska A, Wisniewski M, Skopinska J, Vicini S, and Marsano E. The influence of UV irradiation on the mechanical properties of chitosan/poly(vinyl pyrrolidone) blends. *Polym. Degrad. Stab.* 2005; 88: 261–267.
108. Risbud MV, Hardikar AA, Bhat SV, and Bhonde RR. pH-sensitive freeze-dried chitosan–polyvinyl pyrrolidone hydrogels as controlled release system for antibiotic delivery. *J. Control. Release* 2000; 68: 23–30.

109. Zhao L, Xu L, Mitomo H, and Yoshii F. Synthesis of pH-sensitive PVP/CM-chitosan hydrogels with improved surface property by irradiation. *Carbohydr. Polym.* 2006; 64: 473–480.

110. Smitha B, Sridhar S, and Khan AA. Chitosan–poly(vinyl pyrrolidone) blends as membranes for direct methanol fuel cell applications. *J. Power Sources* 2006; 159: 846–854.

111. Marsano E, Bianchi E, Vicini S, Compagnino L, Sionkowska A, Skopińska J, and Wiśniewski M. Stimuli responsive gels based on interpenetrating network of chitosan and poly(vinylpyrrolidone). *Polymer* 2005; 46: 1595–1600.

112. Don TM, King CF, Chiu WY, and Peng CA. Preparation and characterization of chitosan-g-poly(vinyl alcohol)/poly(vinyl alcohol) blends used for the evaluation of blood-contacting compatibility. *Carbohydr. Polym.* 2006; 63: 331–339.

113. Minoura N, Koyano T, Koshizaki N, Umehara H, Nagura M, and Kobayashi K. Preparation, properties, and cell attachment/growth behavior of PVA/chitosan-blended hydrogels. *Mater. Sci. Eng. C* 1998; 6: 275–280.

114. Sajeev US, Anoop AK, Menon D, and Nair S. Control of nanostructures in PVA, PVA/chitosan blends and PCL through electrospinning. *Bull. Mater. Sci.* 2008; 31: 343–351.

115. Patel VR and Amiji MM. Preparation and characterization of freeze-dried chitosan-poly(ethylene oxide) hydrogels for site-specific antibiotic delivery in the stomach. *Pharm. Res.* 1996; 13: 588–593.

116. Amiji MM. Permeability and blood compatibility properties of chitosan-poly(ethylene oxide) blend membranes for haemodialysis. *Biomaterials* 1995; 16: 593–599.

117. Kuo PC, Sahu D, and Yu Hsin H. Properties and biodegradability of chitosan/nylon 11 blending films. *Pol. Degrad. Stab.* 2006; 91: 3097–3102.

118. Shieh JJ and Huang RYM. Chitosan/*N*-methylol nylon 6 blend membranes for the pervaporation separation of ethanol–water mixtures. *J. Membr. Sci.* 1998; 148: 243–255.

119. Desai K and Kit K. Effect of spinning temperature and blend ratios on electrospun chitosan/poly(acrylamide) blends fibers. *Polymer* 2008; 49: 4046–4050.

120. Suyatma NE, Copinet A, Tighzert L, and Coma V. Mechanical and barrier properties of biodegradable films made from chitosan and poly (lactic acid) blends. *J. Polym. Environ.* 2004; 12: 1–6.

121. Strachota A, Tishchenko G, Matejka L, and Bleha M. Chitosan–oligo(silsesquioxane) blend membranes: Preparation, morphology, and diffusion permeability. *J. Inorg. Organometal. Polym.* 2001; 11: 165–182.

122. Haas J, Ravi K, Borchard G, Bakowsky U, and Lehr CM. Preparation and characterization of chitosan and trimethyl-chitosanmodified poly-(ε-caprolactone) nanoparticles as DNA carriers. *AAPS PharmSciTech* 2005; 6: E22–E30.

123. Sarasam A and Madihally SV. Characterization of chitosan–polycaprolactone blends for tissue engineering applications. *Biomaterials* 2005; 26: 5500–5508.

124. Olabarrieta I, Forsström D, Gedde UW, and Hedenqvist MS. Transport properties of chitosan and whey blended with poly(ε-caprolactone) assessed by standard permeability measurements and microcalorimetry. *Polymer* 2001; 42: 4401–4408.

125. Zeng M, Fang Z, and Xu C. Effect of compatibility on the structure of the microporous membrane prepared by selective dissolution of chitosan/synthetic polymer blend membrane. *J. Membr. Sci.* 2004; 230: 175–181.

126. Dufresne A, Cavaillé JY, Dupeyre D, and Garcia-Ramirez M. Morphology, phase continuity and mechanical behaviour of polyamide 6/chitosan blends. *Polymer* 1999; 40: 1657–1666.

127. Peesan M, Supaphol P, and Rujiravanit R. Preparation and characterization of hexanoyl chitosan/polylactide blend films. *Carbohydr. Polym.* 2005; 60: 343–350.

128. Wu H, Wan Y, Cao X, and Wu Q. Interlocked chitosan/poly(DL-lactide) blends. *Mater. Lett.* 2008; 62: 330–334.

129. Zhang X, Hua H, Shen X, and Yang Q. In vitro degradation and biocompatibility of poly(L-lactic acid)/chitosan fiber composites. *Polymer* 2007; 48: 1005–1011.

130. Thanpitcha T, Sirivat A, Jamieson AM, and Rujiravanit R. Preparation and characterization of polyaniline/chitosan blend film. *Carbohydr. Polym.* 2006; 64: 560–568.

131. Sang YN and Young ML. Pervaporation and properties of chitosan-poly(acrylic acid) complex membranes. *J. Membr. Sci.* 1997; 135: 161–171.

132. Abou-Aiad THM, Abd-El-Nour KN, Hakim IK, and Elsabee MZ. Dielectric and interaction behavior of chitosan/polyvinyl alcohol and chitosan/polyvinyl pyrrolidone blends with some antimicrobial activities. *Polymer* 2006; 47: 379–389.

133. Rao V and Johns J. Thermal behavior of chitosan/natural rubber latex blends TG and DSC analysis. *J. Therm. Anal. Calorim.* 2008; 92: 801–806.

134. Ye Y, Dan W, Zeng R, Lin H, Dan N, Guan L, and Mi Z. Miscibility studies on the blends of collagen/chitosan by dilute solution viscometry. *Eur. Polym. J.* 2007; 43: 2066–2071.

135. Sionkowska A. New materials based on the blends of collagen and other polymers. In: RK Bregg, ed., *Current Topics in Polymer Research*, Nova Science Publisher, New York, 2005, pp. 125–168.
136. Sionkowska A, Wisniewski M, Skopinska J, Poggi GF, Marsano E, Maxwell CA, and Wess TJ. Thermal and mechanical properties of UV irradiated collagen/chitosan thin films. *Polym. Degrad. Stab.* 2006; 91: 3026–3032.
137. Chen Z, Mo X, He C, and Wang H. Intermolecular interactions in electrospun collagen–chitosan complex nanofibers. *Carbohydr. Polym.* 2008; 72: 410–418.
138. Chen T, Embree HD, Brown EM, Taylor MM, and Payne GF. Enzyme-catalyzed gel formation of gelatin and chitosan: Potential for in situ applications. *Biomaterials* 2003; 24: 2831–2841.
139. Gao X, Liu W, Han B, Wei X, and Yang C. Preparation and properties of a chitosan-based carrier of corneal endothelial cells. *J. Mater. Sci.: Mater. Med.* 2008; 19: 3611–3619.
140. Chiono V, Pulieri E, Vozzi G, Ciardelli G, Ahluwalia A, and Giusti P. Genipin-crosslinked chitosan/gelatin blends for biomedical applications. *J. Mater. Sci.: Mater. Med.* 2008; 19: 889–898.
141. Huang Y, Onyeri S, Siewe M, Moshfeghian A, and Madihally SV. In vitro characterization of chitosan–gelatin scaffolds for tissue engineering. *Biomaterials* 2005; 26: 7616–7627.
142. Chen CH, Wang FY, Mao CF, Liao WT, and Hsieh CD. Studies of chitosan: II. Preparation and characterization of chitosan/poly(vinyl alcohol)/gelatin ternary blend films. *Int. J. Biol. Macromol.* 2008; 43: 37–42.
143. Silva SS, Goodfellow BJ, Benesch J, Rocha J, Mano JF, and Reis RL. Morphology and miscibility of chitosan/soy protein blended membranes. *Carbohydr. Polym.* 2007; 70: 25–30.
144. Sashina ES and Novoselov NP. Polyelectrolyte complexes of fibroin with chitosan. *Russ. J. Appl. Chem.* 2005; 78: 487–491.
145. Lu Q, Feng Q, Hu K, and Cui F. Preparation of three-dimensional fibroin/collagen scaffolds in various pH conditions. *J. Mater. Sci.: Mater. Med.* 2008; 19: 629–634.
146. Sashina ES, Janowska G, Zaborski M, and Vnuchkin AV. Compatibility of fibroin/chitosan and fibroin/cellulose blends studied by thermal analysis. *J. Therm. Anal. Calorim.* 2007; 89: 887–891.
147. Park WH, Jeong L, Yoo DI, and Hudson S. Effect of chitosan on morphology and conformation of electrospun silk fibroin nanofibers. *Polymer* 2004; 45: 7151–7157.
148. Zhai M, Zhao L, Yoshii F, and Kume T. Study on antibacterial starch/chitosan blend film formed under the action of irradiation. *Carbohydr. Polym.* 2004; 57: 83–88.
149. Wang Q, Zhang N, Hu X, Yang J, and Du Y. Chitosan/starch fibers and their properties for drug controlled release. *Eur. J. Pharm. Biopharm.* 2007; 66: 398–404.
150. Wu YB, Yu SH, Mi FL, Wu CW, Shyu SS, Peng CK, and Chao AC. Preparation and characterization on mechanical and antibacterial properties of chitsoan/cellulose blends. *Carbohydr. Polym.* 2004; 57: 435–440.
151. Veerapur RS, Gudasi KB, and Aminabhavi TM. Pervaporation dehydration of isopropanol using blend membranes of chitosan and hydroxypropyl cellulose. *J. Membr. Sci.* 2007; 304: 102–111.
152. Xu Y, Zhan C, Fan L, Wang L, and Zheng H. Preparation of dual crosslinked alginate–chitosan blend gel beads and in vitro controlled release in oral site-specific drug delivery system. *Int. J. Pharm.* 2007; 336: 329–337.
153. Fan L, Du Y, Zhang B, Yang J, Zhou J, and Kennedy JF. Preparation and properties of alginate/carboxymethyl chitosan blend fibers. *Carbohydr. Polym.* 2006; 65: 447–452.
154. Dornish M, Aarnold M, and Skaugrud Ø. Alginate and chitosan: Biodegradable biopolymers in drug delivery systems. *Eur. J. Pharm. Sci.* 1996; 4: S153.
155. Lazaridou A and Biliaderis CG. Thermophysical properties of chitosan, chitosan–starch and chitosan–pullulan films near the glass transition. *Carbohydr. Polym.* 2002; 48: 179–190.
156. Wittaya-areekul S and Prahsarn C. Development and in vitro evaluation of chitosan–polysaccharides composite wound dressings. *Int. J. Pharm.* 2006; 313: 123–128.
157. Croyle MA, Cheng X, Sandhu A, and Wilson JM. Development of novel formulations that enhance adenoviral-mediated gene expression in the lung in vitro and in vivo. *Mol. Ther.* 2001; 4: 22–28.
158. Samouillan V, Dandurand J, Lacabanne C, and Hornebeck W. Molecular mobility of elastin: Effect of molecular architecture. *Biomacromolecules* 2002; 3: 531–537.
159. Debelle L and Alix AJP. The structures of elastins and their function. *Biochimie* 1999; 81: 981–994.
160. Bonzon N, Carrat X, Daminiere C, Daculsi G, Lefebvre F, and Rabaud M. New artificial connective matrix made of fibrin monomers, elastin peptides and type I + III collagens: Structural study, biocompatibility and use as tympanic membranes in rabbit. *Biomaterials* 1995; 16: 881–885.
161. Klein B, Schiffer R, Hafemann B, Klosterhalfen B, and Zwadlo-Klarwasser G. Inflammatory response to a porcine membrane composed of fibrous collagen and elastin as dermal substitute. *J. Mater. Sci.: Mater. Med.* 2001; 12: 419–424.

162. Mithieux SM, Rasko JE, and Weiss AS. Synthetic elastin hydrogels derived from massive elastic assemblies of self-organized human protein monomers. *Biomaterials* 2004; 25: 4921–4927.

163. Skopinska-Wisniewska J, Sionkowska A, Kaminska A, Kaznica A, Jachimiak R, and Drewa T. Surface properties of collagen/elastin based biomaterials for tissue regeneration. *Appl. Surf. Sci.* 2009; 225: 8286–8292.

164. Reiersen H, Clarke AR, and Rees AR. Short elastin-like peptides exhibit the same temperature induced structural transitions as elastin polymer: Implications for protein engineering. *J. Mol. Biol.* 1998; 283: 255–264.

165. Sionkowska A, Skopinska J, Wisniewski M, Leznicki A, and Fisz JJ. Spectroscopic studies into the influence of UV radiation on elastin hydrolysates in water solution. *J. Photochem. Photobiol. B: Biol.* 2006; 85: 79–84.

166. Sionkowska A, Skopińska J, Wiśniewski M, and Leżnicki A. Spectroscopic studies into the influence of UV radiation on elastin hydrolysates in the presence of collagen. *J. Photochem. Photobiol. B Biol.* 2007; 86: 186–191.

167. Liu Y, Liu H, Qian J, Deng J, and Yu T. Two papers dealing with the use of a regenerated silk fibroin membrane in a biosensor. *Biosens. Bioelectron.* 1996; 11: x.

168. Qian J, Liu Y, Liu H, Yu T, and Deng J. Characteristics of regenerated silk fibroin membrane in its application to the immobilization of glucose oxidase and preparation of a p-benzoquinone mediating sensor for glucose. *Fresenius' J. Anal. Chem.* 1996; 354: 173–178.

169. Boschi A, Arosio C, Cucchi I, Bertini F, and Catellani M. Properties and performance of polypyrrole (PPy)-coated silk fibers. *Fibers Polym.* 2008; 9: 698–707.

170. Baek DH, Ki CS, Um IC, and Park YH. Metal ion adsorbability of electrospun wool keratose/silk fibroin blend nanofiber mats. *Fibers Polym.* 2007; 8: 271–277.

171. Chen H, Hu X, and Cebe P. Thermal properties and phase transitions in blends of Nylon-6 with silk fibroin. *J. Therm. Anal. Calorim.* 2008; 93: 201–206.

172. Liu H, Xu W, Zou H, Ke G, Li W, and Ouyang C. Feasibility of wet spinning of silk-inspired polyurethane elastic biofiber. *Mater. Lett.* 2008; 62: 1949–1952.

173. Marsano E, Corsini P, Canetti M, and Freddi G. Regenerated cellulose-silk fibroin blends fibers. *Int. J. Biol. Macromol.* 2008; 43: 106–114.

174. Hirano S, Nakahira T, Zhang M, Nakagawa M, Yoshikawa M, and Midorikawa T. Wet-spun blend biofibers of cellulose–silk fibroin and cellulose–chitin–silk fibroin. *Carbohydr. Polym.* 2002; 47: 121–124.

175. Yang G, Zhang L, and Liu Y. Structure and microporous formation of cellulose/silk fibroin blend membranes: I. Effect of coagulants. *J. Membr. Sci.* 2000; 177: 153–161.

176. Yang G, Zhang L, Cao X, and Liu Y. Structure and microporous formation of cellulose/silk fibroin blend membranes: Part II. Effect of post-treatment by alkali. *J. Membr. Sci.* 2002; 210: 379–387.

177. Lee KY and Ha WS. DSC studies on bound water in silk fibroin/S-carboxymethyl kerateine blend films. *Polymer* 1999; 40: 4131–4134.

178. Park KE, Jung SY, Lee SJ, Min BM, and Park WH. Biomimetic nanofibrous scaffolds: Preparation and characterization of chitin/silk fibroin blend nanofibers. *Int. J. Biol. Macromol.* 2006; 38: 165–173.

179. Yoo CR, Yeo IS, Park KE, Park JH, Lee S, Park WH, and Min BM. Effect of chitin/silk fibroin nanofibrous bicomponent structures on interaction with human epidermal keratinocytes. *Int. J. Biol. Macromol.* 2008; 42: 324–334.

180. Yongcheng L, Zhang X, Liu H, Tongying Y, and Jiaqi D. Immobilization of glucose oxidase onto the blend membrane of poly(vinyl alcohol) and regenerated silk fibroin: Morphology and application to glucose biosensor. *J. Biotechnol.* 1996; 46: 131–138.

181. Kweon HY, Um IC, and Park YH. Structural and thermal characteristics of *Antheraea pernyi* silk fibroin/chitosan blend film. *Polymer* 2001; 42: 6651–6656.

182. Liu YS, Zhengzhong Z, and Chen X. Thermal and crystalline behavior of silk fiborin/nylon 66 blend films. *Polymer* 2004; 45: 7705–7710.

183. Lee KH, Baek DH, Ki CS, and Park YH. Preparation and characterization of wet spun silk fibroin/poly(vinyl alcohol) blend filaments. *Int. J. Biol. Macromol.* 2007; 41: 168–172.

184. Li M, Lu S, Wu Z, Tan K, Minoura N, and Kuga S. Structure and properties of silk fibroin–poly(vinyl alcohol) gel. *Int. J. Biol. Macromol.* 2002; 30: 89–94.

185. Yoo MK, Kweon HY, Lee KG, Lee HC, and Cho CS. Preparation of semi-interpenetrating polymer networks composed of silk fibroin and poloxamer macromer. *Int. J. Biol. Macromol.* 2004; 34: 263–270.

186. Niamsa N, Srisuwan Y, Baimark Y, Phinyocheep P, and Kittipoom S. Preparation of nanocomposite chitosan/silk fibroin blend films containing nanopore structures. *Carbohydr. Polym.* 2009; 78: 60–65.

187. Salminem F and Rintala J. Anaerobic digestion of organic solid poultry slaughterhouse waste – A review. *Bioresour. Technol.* 2002; 83: 13–26.

188. Moncrieff RW. *Man Made Fibres*, Vol. 11, 6th edn, Butterworths Scientific, London, U.K., 1975, p. 231.
189. Aluigi A, Zoccola M, Vineis C, Tonin C, Ferrero F, and Canetti M. Study on the structure and properties of wool keratin regenerated from formic acid. *Int. J. Biol. Macromol.* 2007; 41: 266–273.
190. Millington KR and Church JS. The photodegradation of wool keratin II. Proposed mechanisms involving cystine. *J. Photochem. Photobiol. B: Biol.* 1997; 39: 204–212.
191. Smith GJ. New trends in photobiology (invited review) photodegradation of keratin and other structural proteins. *Photochem. Photobiol. B: Biol.* 1995; 27: 187–198.
192. Tanabe T, Okitsu N, and Yamauchi K. Fabrication and characterization of chemically crosslinked keratin films. *Mater. Sci. Eng. C* 2004; 24: 441–446.
193. Aluigi A, Vineis C, Varesano A, Tonin C, Ferrero F, Canetti M, Mazzuchetti G, Ferrero F, and Tonin C. Structure and properties of keratin/PEO blend nanofibres. *Eur. Polym. J.* 2008; 44: 2465–2475.
194. Tonin C, Aluigi A, Vineis C, Varesano A, Montarsolo A, and Ferrero F. Thermal and structural characterization of poly(ethylene-oxide)/keratin blend films. *J. Therm. Anal. Calorim.* 2007; 89: 601–608.
195. Tanabe T, Okitsu N, Tachibana A, and Yamauchi K. Preparation and characterization of keratin–chitosan composite film. *Biomaterials* 2002; 23: 817–825.
196. Pingping Z. A new criterion of polymer-polymer miscibility detected by viscometry. *Eur. Polym. J.* 1997; 33: 411–414.
197. Meyers M, Chen PY, Lin A, and Seki Y. Biological materials: Structure and mechanical properties. *Prog. Mater. Sci.* 2008; 53: 1–206.
198. Olszta MJ, Cheng X, Sang S, and Kumar R. Bone structure and formation: A new perspective. *Mater. Sci. Eng. R* 2007; 58: 77–116.
199. Almer JD and Stock SR. Micromechanical response of mineral and collagen phases in bone. *J. Struct. Biol.* 2007; 157: 365–370.
200. Chang MC, Ikoma T, Kikuchi M, and Tanaka J. The cross-linkage effect of hydroxyapatite/collagen nanocomposites on a self-organization phenomenon. *J. Mater. Sci.: Mater. Med.* 2002; 13: 993–997.
201. Gerber T, Holzhuter G, and Gotz W. Nanostructuring of biomaterials—A pathway to bone grafting substitute. *Eur. J. Trauma* 2006; 32: 132–140.
202. Chen G, Ushida T, and Tateishi T. Development of biodegradable porous scaffolds for tissue engineering. *Mater. Sci. Eng. C* 2001; 17: 63–69.
203. Yunoki S, Marukawa E, Ikoma T, Sotome S, Fan H, and Zhang X. Effect of collagen fibril formation on bioresorbability of hydroxyapatite/collagen composites. *J. Mater. Sci.: Mater. Med.* 2007; 18: 2179–2183.
204. Degirmenbasi N, Kalyon DM, and Birinci E. Biocomposites of nanohydroxyapatite with collagen and poly(vinyl alcohol). *Colloids Surf. B Biointerfaces* 2006; 48: 42–49.
205. Kweon H, Ha HC, Um IC, and Park YH. Physical properties of silk fibroin/chitosan blend films. *J. Appl. Polym. Sci.* 2001; 80: 928–934.
206. Wang L and Li C. Preparation and physicochemical properties of a novel hydroxyapatite/chitosan–silk fibroin composite. *Carbohydr. Polym.* 2007; 68: 740–745.
207. Sionkowska A. Photochemical stability of collagen/poly(ethylene oxide) blends. *J. Photochem. Photobiol. A Chem.* 2006; 177: 61–67.
208. Sionkowska A. The influence of UV light on collagen/poly(ethylene glycol) blends. *Polym. Degrad. Stab.* 2006; 91: 305–312.
209. Sionkowska A, Wiśniewski M, Kaczmarek H, Skopińska J, Chevallier P, Mantovani D, Lazare S, and Tokarev V. The influence of UV irradiation on surface composition of collagen/pvp blended films. *Appl. Surf. Sci.* 2006; 253: 1970–1977.
210. Sionkowska A, Wiśniewski M, Skopińska J, and Mantovani D. Effects of solar radiation on collagen based biomaterials. *Int. J. Photoenergy* 2006, article ID 29196, pp. 1–9.
211. Skopińska-Wiśniewska J, Sionkowska A, Joachimiak R, Kaźnica A, Drewa T, Bajer K, and Dzwonkowski J. The modification of new collagen-elastin hydrolysates biomaterials by ultraviolet irradiation. *Eng. Biomater.* 2007; 69–72: 67–69.
212. Drewa T, Joachimiak R, Kaźnica A, Wiśniewska-Skopińska J, and Sionkowska A. The comparison study on collagen scaffolds seeded with hair follicle epithelial cells and urothelials cells. *Tissue Eng.* 2007; 13: 1762–1763.
213. Rajan N, Couet F, Pennock W, Lagueux J, Sionkowska A, and Mantovani D. Low doses of UV radiation stimulate cell activity in collagen-based scaffolds. *Biotechnol. Prog.* 2008; 24(4): 884–889.

214. Sionkowska A, Kozlowska J, Planecka A, and Skopinska-Wisniewska J. Photochemical stability of poly(vinyl pyrrolidone) in the presence of collagen. *Polym. Degrad. Stab.* 2008; 93: 2127–2132.

215. Sionkowska A, Kozlowska J, Planecka A, and Skopinska-Wisniewska J. Collagen fibrils in UV irradiated poly(vinyl pyrrolidone) films. *Appl. Surf. Sci.* 2008; 255: 2030–2039.

216. Sionkowska A, Planecka A, Kozlowska J, and Skopinska-Wisniewska J. Collagen fibrils formation in poly(vinyl alcohol) and poly(vinyl pyrrolidone) films. *J. Mol. Liq.* 2009; 144: 71–74.

217. Sionkowska A, Planecka A, Kozlowska J, and Skopinska-Wisniewska J. Surface properties of UV-irradiated poly(vinyl alcohol) films containing small amount of collagen. *Appl. Surf. Sci.* 2009; 255: 4135–4139.

218. Sionkowska A, Planecka A, Kozlowska J, and Skopinska-Wisniewska J. Photochemical stability of poly(vinyl alcohol) in the presence of collagen. *Polym. Degrad. Stab.* 2009; 94: 383–388.

12 Metal–Polymer Composite Biomaterials

Takao Hanawa

CONTENTS

12.1 INTRODUCTION

The fast technological evolution of polymers has made it possible to apply these materials to medical devices over the last three decades. In particular, because of their excellent biocompatibility and biofunctions, polymers are expected to show excellent properties for use as biomaterials; in fact, many devices made from metals have been replaced by others made from polymers. In spite of this fact, about 80% of implant devices are still made from metals, and this percentage remains unchanged because of their high strength, toughness, and durability. Medical devices consisting of metals cannot be replaced with polymers at present.

The use of metals as raw materials has a long history, and it can be said that present materials science and engineering have been based on research into metals. However, metals are sometimes thought of as "unfavorite materials" for biomedical implants because of memories of the environmental and human damages caused by heavy metals. Since an improvement in the safety of metals for medical use is vital, strenuous efforts have been made to improve corrosion resistance and mechanical durability. In addition, metals are typically artificial materials and have no biofunctions, which makes them fairly unattractive as biomaterials.

On the other hand, polymers are widely used as biomaterials because of their high degree of flexibility, biocompatibility, and technologic properties.[1] Also, it is relatively easy to design biofunctional polymers based on biomimic technique, because biofunctional polymers exist in the human body as parts of biomolecules, cells, tissues, and organs. Therefore, some polymeric materials are widely known as biocompatible and biofunctional materials, as shown in other chapters in this book.

If these advantages of metals and polymers are mixed and disadvantages are eliminated by manufacturing metal–polymer composite materials, then humankind will obtain ideal materials with excellent mechanical properties and biofunctions. Two metal–polymer composite materials are feasible to design: one of them is a combination of bulk polymeric materials and bulk metallic materials, the other is immobilization of polymers to metal surfaces. In this chapter, these metal–polymer composite materials for biomedical use are explained and the corresponding researches are reviewed. To understand metallic materials against polymeric materials, an outline of metals is first presented. Then, surface properties of metals are described to know the reaction of metal surface with polymers that govern the manufacture of metal–polymer composites, followed by the main subject, metal–polymer composites.

12.2 PROPERTIES OF METALS AGAINST POLYMERS

Materials are categorized to metals, ceramics, and polymers (Figure 12.1). Metallic materials are generally multicrystal bodies consisting of metal bonds. For example, metal oxide, metal salt, metal complex, etc., contain metal elements; however, these are compounds consisting of ionic bonds or covalent bonds; these properties are completely different from those of metals consisting of metal bonds. Therefore, in the field of materials engineering, ceramics and metals are clearly distinguished, although they are categorized together in inorganic compounds. Each material has its own advantages and disadvantages, and the applications are determined according to their properties.

Metals have been utilized for dental restoration and bone fixation since 2500 years ago; they have a long history as biomaterials. Advantages of metals as biomaterials are listed as follows. These properties are caused by metal bonds:

1. Great strength
2. Great ductility, easy working
3. Fracture toughness
4. Elasticity and stiffness
5. Electroconductivity

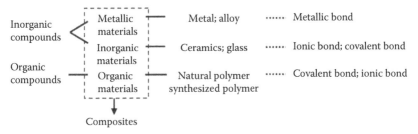

FIGURE 12.1 Category of materials.

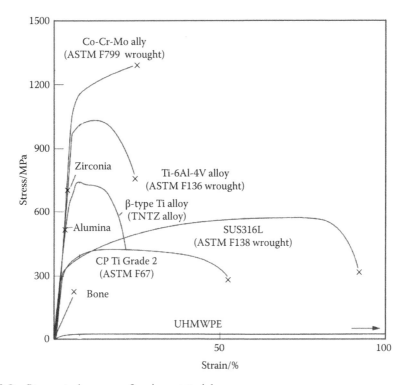

FIGURE 12.2 Stress–strain curves of various materials.

Metals and alloys are widely used as biomedical materials and are indispensable in the medical field. Advantages of metals compared with ceramics and polymers are great strength and difficulty to fracture. In particular, toughness, elasticity, rigidity, and electrical conductivity are essential properties for metals used in medical devices. Figure 12.2 shows stress–strain curves of typical metals, ceramics, and ultra high molecular weight polyethylene (UHMWPE). The strength and elongation of metals against ceramics and polymers are clearly understood.

Conventionally, metals are essential for orthopedic implants: Bone fixators, artificial joints, external fixators, etc., can substitute for the mechanical function of hard tissues in orthopedics. Stents and stent grafts are placed at angusty in blood vessels for dilatation. Therefore, elasticity or plasticity for expansion and rigidity for maintaining dilatation are required for the devices. In dentistry, metals are used for restorations, orthodontic wires, and dental implants.

TABLE 12.1
Comparison of Properties of Metals for Implant

Materials		Mechanical Property Tensile Strength	Workability Wear Resistance	Plasticity	Machinability	Corrosion Resistance Pitting
Stainless steel	Type316L	Good	Excellent → Poor	Excellent	Excellent	Excellent → Poor
Co-Cr alloy	Cast	Good	Good → Good	Poor	Poor	Good
	Annealed	Excellent	Good → Good	Good	Good	Fair
Ti; Ti alloy	CP Ti	Excellent → Poor	Excellent → Poor	Fair	Fair	Excellent
	Ti-6Al-4V	Excellent	Excellent → Poor	Fair	Excellent → Poor	Excellent

For mechanical reliability, metals must be used and cannot be replaced with ceramics or polymers. In noble metals and alloys, gold (Au) marker for the imaging of stent, platinum (Pt) for embolization wire, and Au alloys and silver (Ag) alloys for dental restoratives are utilized. In base metals and alloys, austenitic stainless steels, cobalt–chromium (Co-Cr) alloys, titanium (Ti), and Ti alloys, whose corrosion resistance is maintained with surface oxide film as passive film, are utilized for implant materials. On the other hand, wear resistance is required to decrease the generation of wear debris. Co-Cr-molybdenum (Mo) alloys have good wear resistance and are used for sliding parts of artificial joints. Comparison of various properties of metals used for implant is summarized in Table 12.1. The metals used for such devices are listed in Table 12.2. Safety to the human body is essential in biomaterials; no toxic material is used for biomaterials. Metals implanted in tissues do not show any toxicity without metal ion dissolution by corrosion and generation of wear debris by wear. Therefore, corrosion resistance is absolutely necessary for metals in biomedical use, resulting in the uses of noble or corrosion-resistant metals and alloys for medicine and dentistry.

A disadvantage of using metals as biomaterials is that they are typically artificial materials and have no biofunction. Therefore, additional properties are required for metals. Requests for metals for biomedical use are summarized in Table 12.3. To respond to these requests, new design of alloys and many techniques for surface modification of metals are attempted on a research stage and some of them are commercialized. Surface modification is necessary because biofunction cannot be added during manufacturing processes of metals such as melting, casting, forging, and heat treatment. Surface modification is a process that changes a material's surface composition, structure, and morphology, leaving the bulk mechanical properties intact. Reviews on surface modification of Ti have already been published on plasma spraying[2] and electrochemical treatments.[3]

The most important difference between a metal and polymer is cohesive and noncohesive bodies. In other words, the interface of those is a boundary between a rigid and condensed solid and a flexible and mobile molecule. Of course, metals mainly consist of metal bonds; polymers mainly consist of covalent bonds. Therefore, design and fabrication of metal–polymer composites are to decrease the mismatch of their properties at the interface. In addition, the mechanical and chemical properties of metallic materials are not identified, even though the compositions are determined. The phase of metallic materials is easily changed by working and heat treatment, and various factors influence the resultant structure. Factors governing the mechanical and chemical properties of metallic materials are summarized in Figure 12.3.

TABLE 12.2

Metals Used for Medical Devices

Clinical Division	Medical Device	Material
Orthopedic surgery	Spina fixation	316L stainless steel; Ti; Ti-6Al-4V; Ti-6Al-7Nb
	Bone fixation (bone plate, screw, wire, bone nail, mini-plate, etc.)	316L stainless steel; Co-Cr-W-Ni; Ti; Ti-6Al-4V; Ti-6Al-7Nb
	Artificial joint; bone head	316L stainless steel; Fe-Cr-Ni-Co; Co-Cr-Mo; Ti-6Al-4V; Ti-6Al-7Nb; Ti-15Mo-5Zr-3Al: Ti-6Al-2Nb-1Ta-0.8Mo
	Spina spacer	316L stainless steel; Ti-6Al-4; Ti-6Al-7Nb
Cardiovascular medicine and surgery	Implant-type artificial heart (housing)	Ti
	Pace maker (case)	Ti; Ti-6Al-4V
	(electric wire)	Ni-Co
	(electrode)	Ti; Pt-Ir
	(terminal)	Ti; 316L stainless steel; Pt
	Artificial valve (frame)	Ti-6Al-4V
	Stent	316L stainless steel; Co-Cr-Fe-Ni; Co-Ni-Cr-Mo; Co-Cr-Ni-W-Fe; Ti-Ni; Ta
	Guide wire	316L stainless steel; Ti-Ni; Co-Cr
	Embolization wire	Pt
	Clip	Ti-6Al-4V; 630 stainless steel; 631 stainless steel; Co-Cr-Ta-Ni; Co-Cr-Ni-Mo-Fe; Ti; Ti-6Al-4V
Otorhinology	Artificial inner ear (electrode)	Pt
	Artificial eardrum	316L stainless steel
Dentistry	Filling	Au foil; Ag-Sn(-Cu) amalgam
	Inlay, crown; bridge; clasp; denture base	Au-Cu-Ag; Au-Cu-Ag-Pt-Pd; Ag-Pd-Cu-Au; Ag-(Sn-In-Zn); Co-Cr-Mo; Co-Cr-Ni; Co-Cr-Ni-Cu; Ti; Ti-6Al-7Nb; 304 stainless steel; 316L stainless steel
	Thermosetting resin facing crown; porcelain-fused-to-metal	Au-Pt-Pd; Ni-Cr
	Solder	Au-Cu-Ag; Au-Pt-Pd; Au-Cu; Ag-Pd-Cu-Zn
	Dental implant	Ti; Ti-6Al-4V; Ti-6Al-7Nb; Au
	Orthodontic wire	316L stainless steel; Co-Cr-Fe-Ni; Ti-Ni; Ti-Mo
	Magnetic attachment	Sm-Co; Nd-Fe-B; Pt-Fe-Nb; 444 stainless steel; 447J1 stainless steel; 316L stainless steel
	Treatment device (bar, scaler, periodontal probe, dental tweezers, raspatory, etc.)	304 stainless steel
General surgery	Needle of syringe	304 stainless steel
	Scalpel	304 stainless steel; 316L stainless steel; Ti
	Catheter	304 stainless steel; 316L stainless steel; Co-Cr; Ti-Ni; Au; Pt-In
	Staple	630 stainless steel

TABLE 12.3
Requests of Metals for Medical Devices

Required Property	Target Medical Devices	Effect
Elastic modulus	Bone fixation; spinal fixation	Prevention of bone absorption by stress shielding
Superelasticity shape memory effect	Multipurpose	Improvement of mechanical compatibility
Wear resistance	Artificial joint	Prevention of generation of wear debris; improvement of durability
Biodegradability	Stent; artificial bone; bone fixation	Elimination of materials after healing; unnecessity of retrieval
Bone formation bone bonding	Stem and cup of artificial hip joint; dental implant	Fixation of devices in bone
Prevention of bone formation	Bone screw; bone nail	Prevention of assimilation
Adhesion of soft tissue	Dental implant; trans skin device; external fixation; pacemaker housing	Fixation in soft tissue; prevention of inflectional disease
Inhibition of platelet adhesion	Devices contacting blood	Prevention of thrombus
Inhibition of biofilm formation	All implant devices; treatment tools and apparatus	Prevention of infectious disease
Low magnetic susceptibility	All implant devices; treatment tools and apparatus	No artifact in MRI

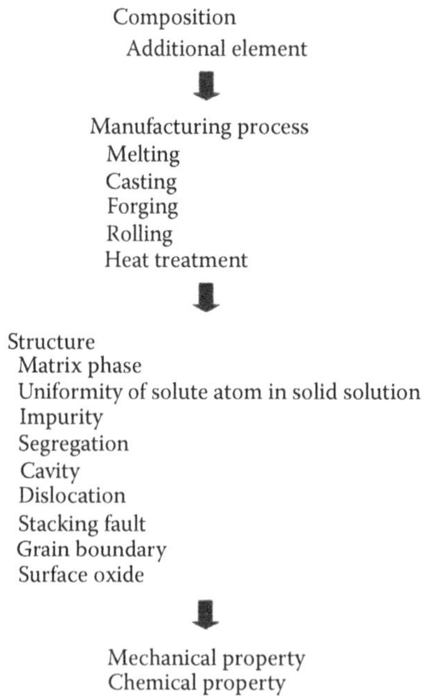

Composition
Additional element

Manufacturing process
Melting
Casting
Forging
Rolling
Heat treatment

Structure
Matrix phase
Uniformity of solute atom in solid solution
Impurity
Segregation
Cavity
Dislocation
Stacking fault
Grain boundary
Surface oxide

Mechanical property
Chemical property

FIGURE 12.3 Factors governing the mechanical and chemical properties of metallic materials through the manufacturing process.

12.3 SURFACE OF METALLIC BIOMATERIALS

12.3.1 IMPORTANCE OF METAL SURFACE

When a metallic material is implanted into a human body, immediate reaction occurs between its surface and the living tissues. In other words, immediate reaction at this initial stage straightaway determines and defines a metallic material's tissue compatibility. Since conventional metallic biomaterials are usually covered with metal oxides, surface oxide films on metallic materials play an important role not only against corrosion but also in tissue compatibility. To fabricate metal–polymer composites, knowledge of the material's surface composition is absolutely necessary. Adhesion of polymers and immobilization of molecules to metals are governed by the surface property of the substrate metals. Immobilization of molecules is a kind of surface modification of metals, because surface modification is a process that improves surface property by changing the composition and structure, while leaving the mechanical properties of the material intact. Surface properties of a metallic material may be controlled with surface modification techniques.

12.3.2 SURFACE OF METALS

Atoms in metallic materials located at the surface are considered partly reactive to the environment because atomic configuration terminates at the surface. The surface represents a property different from the inside of the material. Due to high surface energy, a single molecular layer forms easily on solid surface, where gas molecules are adsorbed at 1 Pa in 10^{-4} s. For example, in the presence of oxygen atoms, oxygen atoms and metal atoms chemically bond together to form an oxide layer. This phenomenon occurs even at the surface of Au—which is the most noble metal. Unlike polymers and ceramics, enrichment of component elements occurs easily at metal surfaces. This means that the surface composition of a metal is different from its inside composition in the order of nanometers. Therefore, the variant surface composition of a metallic material contributes significantly to defining the overall properties of the material.

A metal surface is usually covered with a surface oxide film. The surface oxide layer, on the other hand, is always covered with surface hydroxyl groups that are adsorbed by water, as shown in Figure 12.4. In particular, the surface oxide film and surface active hydroxyl groups are important to understand the surface reaction of metals.

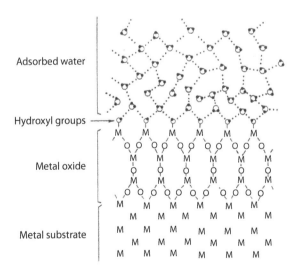

FIGURE 12.4 Schematic model of the structure of surface layer consisting of oxide layer, hydroxyl group layer, and adsorbed water layer on metals and alloys.

12.3.3 SURFACE OXIDE FILM

Except in reduction environments, the corrosion process always causes a reaction film to form on metallic materials. Passive film is one such reaction film, and it is particularly significant for corrosion protection. When solubility is extremely low and pores are absent, adhesion of film—which is formed in an aqueous solution—to the substrate will be strong. The film then becomes a corrosion-resistant or passive film. Passive film is about 1–5 nm thick and is transparent. Due to the tremendously fast rate at which it is formed, passive film readily becomes amorphous. For example, film on a Ti metal substrate was generated in 30 ms. This was estimated from the time transient of current of the Ti at 1 V versus a saturated calomel electrode (SCE) after exposing the metal surface, as shown in Figure 12.5. Since amorphous films hardly contain grain boundary and structural defects, they are corrosion resistant. However, corrosion resistance decreases with crystallization. Fortunately, passive films contain water molecules that promote and maintain amorphousness.

Metallic materials such as stainless steels, Co-Cr alloys, commercially pure Ti, and Ti alloys used for biomedical devices are covered by their characteristic passive films. These films self-repair when they are disrupted by some causes. Noble metals and alloys such as dental alloys are also covered with an oxide layer. While the oxide layer protects against corrosion, it is not chemically strong like the passive film.

12.3.3.1 Titanium

When Ti is polished in deionized water and analyzed using X-ray photoelectron spectroscopy (XPS), the Ti 2p spectrum obtained from the Ti gives four doublets according to valence: Ti^0, Ti^{2+}, Ti^{3+}, and Ti^{4+}. Published data[4] are used to determine the binding energy of each valence. Figure 12.6 shows an example of the decomposition of Ti 2p spectrum. A distinct Ti^0 peak at metallic state is observed, which accounts for a very thin surface oxide film at less than a few nanometers only. Besides, Ti^{4+} (TiO_2), Ti^{3+} (Ti_2O_3), and Ti^{2+} (TiO) are detected. Though Ti^{2+} oxide exists in the surface oxide film, Ti^{2+} formation is always thermodynamically less favorable than Ti^{3+} formation at the surface. On the other hand, Figure 12.7 shows O 1s spectrum obtained from polished Ti, which is decomposed to three peaks originating from oxide (O^{2-}), hydroxide or hydroxyl group (OH^-), and water (H_2O). The surface oxide film on Ti contains

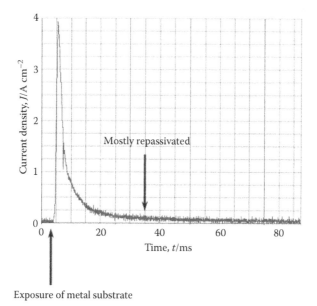

FIGURE 12.5 Time transient of anodic current of titanium in Hanks' solution under the 1 V charge versus SCE. Anodic current is generated with the dissolution and repassivation of titanium.

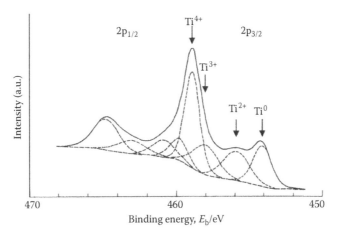

FIGURE 12.6 Decomposition of Ti 2p XPS spectrum obtained from titanium abraded and immersed for 300 s in water into eight peaks (2p3/2 and 2p1/2 electron peaks in four valences). Numbers with arrows are valence numbers.

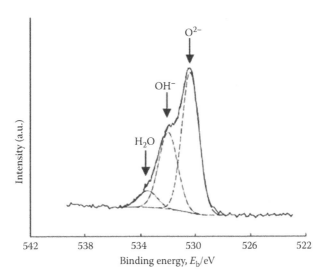

FIGURE 12.7 Typical O 1s spectrum obtained from polished Ti and its deconvolution into O^{2-}, OH^-, and H_2O components.

these chemical states. Therefore, the surface oxide film on Ti consists mainly of nonstoichiometric TiO_2 containing the hydroxyl group and water.

Since a considerable portion of oxidized titanium stays as Ti^{2+} and Ti^{3+} in the surface film, the oxidation process may proceed to the end just at the uppermost part of the surface film. As shown in Figure 12.8, the proportion of Ti^{4+} among titanium cations (Ti^{2+}, Ti^{3+}, and Ti^{4+}) in the film decreases with an increase in photoelectron take-off angle,[5] indicating that more Ti^{4+} exists near the outer layer in the film. Deducing from the take-off angle dependence of $[OH^-]/[O^{2-}]$ in Figure 12.8, oxygen atoms in the hydroxyl group are mainly located in the outer part of the surface film. This means that dehydration proceeds inside the surface film and only partly for Ti^{4+} oxide.

Thickness of the film is about 2 nm just after polishing and about 5 nm at 1 week after polishing (as shown in Figure 12.9). Note too that the thickness of the surface oxide film increases according to the logarithmic rule, which is common in the initial growth of oxide films of metallic materials.

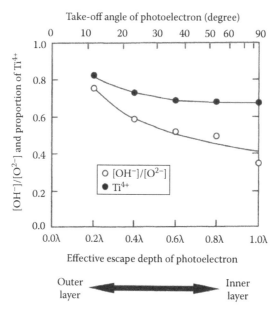

FIGURE 12.8 The ratio of the proportion of the concentration of OH^- to that of O^{2-}, $[OH^-]/[O^{2-}]$, and proportion of cationic fraction of Ti^{4+} among titanium species, in surface oxide film on Ti polished in water plotted against the average effective escape depth of photoelectrons for angle-resolved XPS measurements. Lambda (λ) is the average escape depth of O 1s and Ti $2p_{3/2}$ photoelectrons, and the effective escape depth is the escape depth times sin (take-off-angle). The values at small take-off angle of photoelectron or effective escape depth of photoelectron represent outer region information, and the larger ones represent inner region information.

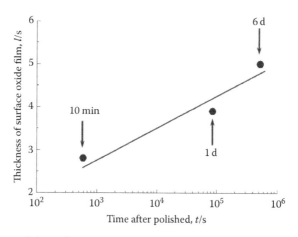

FIGURE 12.9 Thickness of the surface oxide film on Ti as a function of time after polishing.

12.3.3.2 Titanium Alloy

The film on Ti-6 aluminum (Al)-4 vanadium (V) alloy is almost the same as that on Ti containing a small amount of aluminum oxide.[6,7] In other words, the surface oxide film on Ti-6Al-4V is a TiO_2 containing small amounts of Al_2O_3, hydroxyl groups, and bound water. V contained in the alloy is not detected in the oxide film after the alloy is polished. The Ti-56 nickel (Ni) shape-memory alloy is covered by TiO_2-based oxide, with minimal amounts of Ni in both the oxide and metallic states.[6,7] The film on Ti-56Ni is a TiO_2 containing $Ni(OH)_2$, hydroxyl groups, and bound water.

In Ti-zirconium (Zr) alloy, the surface oxide film consists of titanium and zirconium oxides.[8] The relative concentration ratio of Ti to Zr in the film is almost the same as that in the alloy. The thickness of the film increased with increase in Zr content. The chemical state of Zr is more stable than that of Ti in the film.

12.3.3.3 Stainless Steel

Compositions of surface oxide films on stainless steels are well understood in the field of engineering. In an austenitic stainless steel, the surface oxide film consists of iron (Fe) and chromium containing a small amount of molybdenum. However, it does not contain nickel while in the air and in chloride solutions.[9,10]

On the other hand, surface oxide film on 316L steel polished mechanically in deionized water consists of oxide species of Fe, Cr, Ni, Mo, and manganese (Mn), and its thickness is about 3.6 nm.[11] The surface film contains a large amount of OH⁻—the oxide which is hydrated or oxyhydroxidized. The surface oxide film is also enriched with Fe, while the alloy substrate just under the film is enriched with Ni, Mo, and Mn.

12.3.3.4 Cobalt–Chromium Alloy

Surface oxide film of a Co-Cr-Mo alloy is characterized as containing oxides of cobalt and chromium without molybdenum.[12] On the other hand, the surface oxide film on another Co-Cr-Mo alloy polished mechanically in deionized water consists of oxide species of cobalt, chromium, and molybdenum, and its thickness is about 2.5 nm.[13] This surface film contains a large amount of OH⁻—the oxide which is hydrated or oxyhydroxidized. There are also more traces of Cr and Mo distributed at the inner layer of the film. In the surface oxide on Co-Cr-Ni Mo alloy for metallic stent known as MP35N, Cr is enriched and Co and Ni are depleted.[14]

12.3.3.5 Dental Precious Alloys

Au-copper (Cu)-Ag alloys and Ag-palladium (Pd)-Cu-Au alloys for dental restoration are covered by copper oxide and silver oxide.[15] An Ag-indium (In) alloy is covered by zinc oxide and indium oxide, and an Ag-Sn-zinc (Zn) alloy is covered by tin oxide and zinc oxide. While these oxides serve as a protection film against corrosion, they are not as chemically strong as the passive film.

12.3.4 Surface Active Hydroxyl Group

12.3.4.1 Formation

The surface of oxide reacts with moisture in air, and hydroxyl groups are rapidly formed. In the case of Ti, the surface oxide immediately reacts not only with water molecules in aqueous solutions but also with moisture in air and is covered by hydroxyl groups.[16,17] The surface oxide is always formed on conventional metallic biomaterials, and the surface of the surface oxide is active because of the same reason described earlier. Therefore, the oxide surface immediately reacts with water molecules and hydroxyl groups are formed, as shown in Figure 12.10a. The surface hydroxyl groups contain both terminal OH and bridge OH in equal amounts. Concentration of hydroxyl groups on the unit area of the surface is determined with various techniques.

12.3.4.2 Acidic and Basic Properties

Active surface hydroxyl groups dissociate in aqueous solutions and form electric charges, as shown in Figure 12.10b.[16–19] Positive or negative charge due to the dissociation is governed by the pH of the surrounding aqueous solution: Positive and negative charges are balanced and apparent charge is zero at a certain pH. This pH is the point of zero charge (pzc). The pzc is the unique value for an oxide and an indicator that the oxide shows acidic or basic property. For example, in the case of TiO_2, the pzc of rutile is 5.3 and that of anatase is 6.2.[16] In other words, anatase surface is acidic

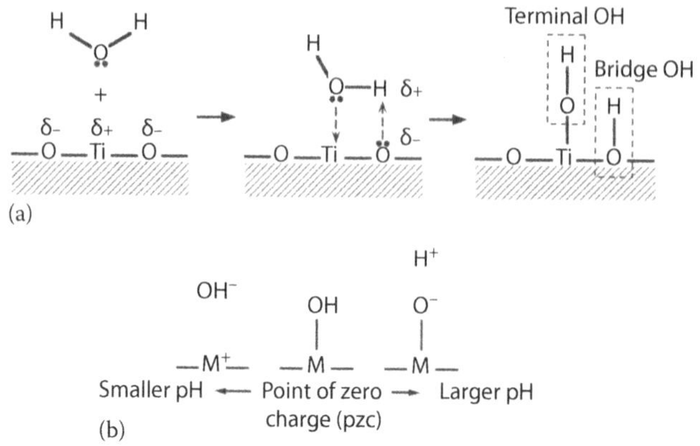

FIGURE 12.10 Formation process of hydroxyl group on titanium oxide (a) and dissociation of the hydroxyl group in aqueous solution and pzc (b).

at smaller pH and basic at larger pH than 6.2. Active surface hydroxyl groups and electric charges formed by the dissociation of the groups play important roles for the bonding with polymers and immobilization of molecules. Therefore, the concentration of surface hydroxyl group and pH are important factors for the bonding with polymeric materials and immobilization of molecules.

12.4   ADHESION OF POLYMERS TO METALS IN DENTISTRY

12.4.1   ADHESIVE REAGENTS

Metal restorations and prostheses such as inlay, crown, and bridge must be retained in a fixed position in a mouth. For this purpose, dental cements are used: glass ionomer cements, zinc phosphate cements, polyacrylate cements, resin-based cements, etc. In particular, resin-based cements generate chemical adhesion between tooth and metal. Cements based on resin composites are now used for cementation of crown and bridges and direct bonding of orthodontic brackets to enamel. Polymer-based filling, restorative, and luting materials are specified in ISO 4049.

Resin cements based on methyl methacrylate (MMA) have been available since 1952 for the use of cementation of inlays, crowns, etc. The resin composite for crown and resin was invented in the early 1970s. Cementation of alloy restorations is performed with self-cured composite cements. Two-paste systems are adopted for self-cured composite cements. One paste mainly consists of a diacrylate oligomer diluted with lower molecular weight dimethacrylate monomers. The other consists of silanated silica or glass. Peroxide amine is used for the initiator accelerator.

Although many efforts have been made to use industrial adhesives such as epoxy resins and cyanoacrylate as dental adhesives, adequate adhesion is not achieved under the extreme conditions of the oral cavity. Much effort has been expended to achieve adhesion between artificial compounds and tooth substances. In 1978, Takeyama et al.[20] synthesized a new dental adhesive that meets the requirements stated here. The dental adhesives contained 4-methacryloxyethyl-trimellitic anhydride (4-META) as an adhesive monomer with MMA. In 1983, Omura[21] synthesized an adhesive containing a dimethacrylate monomer bis-glycidyldimethacrylate (Bis-GMA) and phosphate monomers. 4-META is an ethyl anhydride of trimellitic acid and hydroxyethyl methacrylate and has both hydrophilic and hydrophobic groups, as shown in Figure 12.11. The 4-META cement is formulated with MMA monomer and acrylic resin filler and is catalyzed by tributylborane (TBB). Another adhesive resin cement is phosphonate cement supplied with a two-paste system, containing Bis-GMA resin and silanated quartz filler. Phosphonate molecule is very sensitive to oxygen, so a gel is provided to coat the margins of a restoration until setting has occurred.

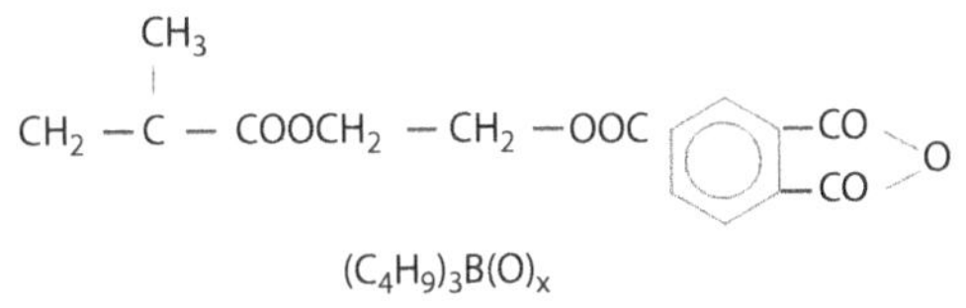

FIGURE 12.11 4-Methacryloxyethyl trimellitate anhydride (4-META) (top) and tri-butyl Borane (TBB) (bottom).

One of the main problems in dental adhesion techniques is that bonds between adhesives and metals are strong in a dry environment and weaken in a wet environment. For example, there are clinical reports of dental adhesion bridges breaking off from teeth after extended use, and this has led to a loss of confidence in dental adhesion techniques. Hence, a major problem awaiting solution in dentistry is the durability of bonding joints exposed to water. Researches on dental adhesives are well reviewed elsewhere.[22]

12.4.2 Bonding to Base Metal Alloys

The phosphate end of the phosphonate reacts with calcium of the tooth or with a metal oxide. The double-bonded ends of both 4-META and phosphonate cements react with other double bonds when available. Setting of resin cements results from self- or light-cured polymerization of carbon–carbon double bonds.

There are numerous studies that have been made on the adhesion to dental alloys; pioneering work by Tanaka on Ni-Cr alloys[23] and gold alloys;[24] studies on Ni-Cr alloys, Co-Cr alloys, and precious metal alloys.[25] There are several surface pretreatment methods for Ni-Cr alloys: immersion in concentrated HNO_3 solution after etching with HCl,[23] immersion in solutions containing an oxidizing agent after sandblasting, etching and passivating by an electrochemical method, and spraying of molten metal on the alloy surface. These methods increase bond strength by increasing the mechanical retention and chemical affinity between the adhesive and alloys. However, it is necessary to distinguish the contribution of these two effects, mechanical and chemical, to evaluate how the alloy surface structure improves the resin adhesion.

The adhesion of 4-META resin to Co-Cr and Ni-Cr alloys is examined by tensile test with and without thermally induced stress.[26] The resin bond to the as-polished surface of Co-Cr and Ni-Cr alloys is stronger than that to the oxidized surfaces. The thermal cycles cause clear differences in the surface states that affected the adhesion. The adhesive ability of 4-META resin to Ni-Cr alloy improves remarkably when the alloy surface is treated by concentrated HNO_3 solution. Adhesion to the alloy surfaces treated by concentrated HNO_3 is excellent, comparable with as-polished Co-Cr alloy, and is resistant to and protects against thermal cycling using liquid nitrogen.

The effect of a thick water layer adsorbed on the top of oxide surfaces on the bonding ability of 4-META resin is examined with a Co-Cr alloy and 18–8 stainless steel.[27] The alloys are heated 500°C in air and dehydrated by heating to 700°C at 1×10^{-4} Pa in silica glass tube. Dental adhesive resin containing 4-META is bonded to these alloy surfaces, and after thermal cycle treatment the bonding strengths and failure types for the dehydrated surfaces are compared with surfaces heated at 500°C. In both alloys, the specimens oxidized at 500°C in air show partial interface (alloy/resin) failure at the periphery. However, except for a few cases, dehydrated specimens display cohesive failure and the bonding strength is similar to that of the as polished Co-Cr alloy, showing excellent bonding ability. The bonding ability of the 4-META resin to the dehydrated oxide layer surface is excellent when adhesion procedures are performed in an atmosphere excluding water vapor. The cleaned metal surface obtained by hydrogen gas reaction shows excellent adhesive ability on Cr, Co, and 304 type stainless steel, showing a passive oxide film on the metal substrates.[28] Figure 12.12 shows the possible adhesion mechanism models of 4-META: Models (a), (b), and (c) are for the as-polished surface and models (d), (e), and (f) are for the oxidized surface.

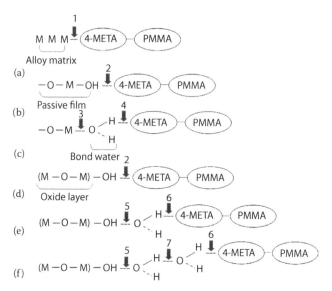

FIGURE 12.12 Suggested adhesion model of PMMA resin containing 4-META on as polished (a, b, and c) and oxidized (d, e, and f) surface: M is metal atom—is hydrogen bond. The chemical bonds that are similar in bonding strength are labeled by the same number at the bonding position.

12.4.3 DURABILITY IN WATER

Generally, joints of dental adhesive resin bonded to dental alloys weaken in a wet environment, although adhesion is strong in a dry environment. Durability of the adhesion in a wet environment, such as the oral cavity, is predominantly important in the bonding of dental adhesive materials to teeth and dental alloys.

With aluminum/epoxy resin,[29] aluminum/Li-polysulfone,[30] and mild-steel/4-META resin[31] in wet environments, water molecules enter the adhesion interface by diffusion through the adhesive resin rather than by passage along exposed bond lines. Polymethyl methacrylate (PMMA) film bonded by 4-META dental adhesive resin to mild steel is used as a specimen to study the mechanism of water permeation into the adhesive interface. Water enters the interface by diffusion through the resin rather than by passage along the interface. The water content penetrated to the interface is calculated from the solution to Fick's second equation.[31] After 3 days of immersion, no change is observed on the mild steel surface through the clear resin layer, but XPS analysis reveals the same chemical state as in the 2 week immersion specimen.[32] The hydrogen bonds appear to be destroyed with water penetrating through the resin layer. This was followed by corrosion, resulting in a complete destruction of the adhesion interface.

The intensity of thermal stress due to thermal shock, calculated by the three-dimensional finite element method, increases with increasing resin film thickness, indicating that the resin film separates without degradation due to water when the resin film is thick. Total interface failure occurs on specimens with resin thicker than 0.5 mm. The critical thermal shearing stress was calculated as 22.5 MPa. When the adhesion interface degrades, the adhesion is broken by lower shear stresses than the value for 0.25 mm layers, 16 MPa. Immersion time to reach equilibrium water content is 0.5 days for a 0.25 mm resin film.[33]

12.4.4 BONDING TO PRECIOUS ALLOYS

4-META adheres to hydroxyl groups on metal surface as explained already, and noble metals and precious alloys have a smaller number of hydroxyl groups on their surfaces. Therefore, the bonding strength of 4-META with noble metals and precious alloys is weaker than base metal alloys.

The resins bond independently to dental precious metal alloys because they have low chemical affinity for the precious metals. Several surface modification methods have been developed for improvement of adhesion to dental alloys: high-temperature oxidation, immersion in oxidizing agent, immersion in concentrated nitric acid,[23] anodizing, electropolishing with tin, SiOx coating,[34] and ion coating.[35] The principle of these techniques is to form oxide layer on precious metal alloys. However, these methods have drawbacks, such as complicated procedures, expensive equipment, and degradation of chemical agent.

The other method to obtain bonding of 4-META to precious metal alloy is to design new alloys containing In, Zn, and tin (Sn).[36] The water durability and bonding strength of 4-META resin to binary alloys of Au, Ag, Cu, or Pd containing In, Zn, or Sn are studied. With In in Au-based alloys, the In_2O_3 on the alloy surface plays an important role in the adhesion with 4-META. To obtain excellent adhesion, the element in an oxide with chemical affinity for 4-META must cover at least 50% of the alloy surface. The poor adhesive ability of 4-META resin to pure gold is considered to be caused by chemisorbed H_2O molecules on the surface.[37] The adhesion ability of binary alloy was improved by adding In, Zn, or Sn.

12.5 STENTS AND STENT GRAFTS

12.5.1 STENTS

Stents and stent grafts are tubes used for dilatation to counteract decreases in vessel or duct diameter and to maintain localized flow in stenotic blood vessels. Stents are also applied to vessels of the bile duct, esophagus, and other passages for dilation. Therefore, elasticity or plasticity for expansion and rigidity for the maintenance of dilatation and resistance to elastic recoil are required. The expandability and plasticity of a balloon-expandable stent and the elasticity of a self-expandable stent, as well as rigidity and resistance to the elastic recoil of blood vessels, are required for stent materials. Conventional metals cover these properties when proper metals are selected.

Stents are expandable tubes of metallic mesh that were developed to address the negative sequelae of balloon angioplasty. At the beginning of the twentieth century, glass tubes that became a prototype of stents were implanted into blood vessels of animals. In the next stage, the origin of percutaneous transluminal angioplasty (PTA), the expansion of blood vessels obtained by increasing the diameter of catheter tube, was attempted.[38] This trial failed because of migration and the development of thrombus. In 1985, Palmaz and his colleagues developed the first balloon-expanded stent,[39] and just 1 year later, Gianturco developed a balloon-expanded coil stent.[40] These stents were made of stainless steel. The first self-expandable stent, Wallstent®, made of a Co-Cr alloy (Elgiloy), was clinically applied in 1986.[41] Another popular self-expandable stent, SMART®, consists of a superelastic Ni-Ti alloy.[42] Since the 1990s, stents have been used in coronary arteries. Typical popular stents for this purpose are the Palmaz-Schatz® stent and the Gianturco-Roubin® stent. New designs and functions of stents have also been developed. Today, for example, the majority of patients undergoing percutaneous transluminal coronary angioplasty (PTCA) receive a stent. Since the mid-1990s, stents have also been applied to the treatment of cerebrovascular disease, and carotid stenting to prevent stroke, recoil, and restenosis has been attempted.[43–45] In addition, the first treatment for abdominal aortic aneurysms with stent grafts started in the 1990s.[46] Stent grafts are generally constructed of stainless steel or a Ni-Ti alloy and coated with compounds such as expanded polytetrafluoroethylene (ePTFE) that is a metal–polymer composite. Typical bare stent and stent graft are shown in Figure 12.13.

Metals are the main materials utilized for stents because of their mechanical properties and visibility by X-ray imaging. Metallic stents and stent grafts are now important devices in noninvasive medicine. New techniques for treatments using metallic stents are continuously being developed, and such development will continue into the foreseeable future. While pioneering ideas for

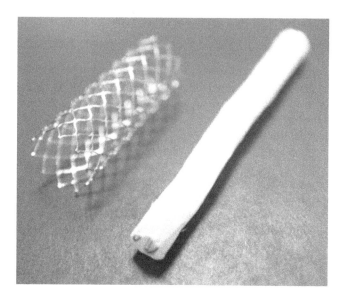

FIGURE 12.13 **(See companion CD for color figure.)** Examples of bare metallic stent and stent graft.

treatment are constantly being announced, the development of proper materials for stents is backward from the viewpoint of materials science.

12.5.2 Drug Eluting Stents

Initial complications associated with stenting generally involve subacute stent thrombosis, which occurs in 1%–3% of patients within 7–10 days. The major long-term complication is in-stent restenosis, which occurs within 6 months in 50% of patients. Although it is unknown whether the original cause of these problems is the metal from which the stents are made, the use of polymer-coated drug-eluting stents (DESs) is increasing.[47–49]

The most promising results have been attained with polymer-coated DESs[1] and DES is one of the metal–polymer composites. Two of the drugs currently in clinical trials are rapamycin and paclitaxel. These drugs are embedded in a polymer matrix (such as a copolymer of poly-*n*-butyl methacrylate and polyethylene-vinyl acetate or a gelatin–chondroitin sulfate coacervate film) that is coated onto the stent. The drug is related by diffusion and/or polymer degradation over varying periods of time that can be engineered by the specifics of the polymer-drug system. These coated stents have had excellent initial success, virtually eliminating restenosis over time periods of 2 years and longer and are felt to represent a major breakthrough in the treatment of coronary artery disease. Commercial DESs and their structures are summarized in Table 12.4.

In DES, chemicals such as immunosuppressants are coated on metal frames. Because direct immobilization of drugs to metals or coating of the polymers and ceramics impregnating the drugs is necessary to control their release rate, attempts have been made to coat the photosetting gelatin impregnating the drug.[50] To improve the biofunction of stents, a stent coated with a phospholipid-like 2-methacryloyloxyethyl phosphorylcholine (MPC) polymer to inhibit the adsorption of proteins and the adhesion of platelets has been developed.[51] For the same purpose, the immobilization of poly(ethylene glycol) (PEG) and compounding with medical polymers are effective.[52]

In DESs, all fractures occurred around areas of increased rigidity due to the overlapping of metals, which may have formed a fulcrum for metal deformation due to vessel movement.[53]

TABLE 12.4
Various Drug Eluting Stents

Manufacturer	Coating			Stent	
	Brand Name	Drug	Material	Name	Material
Johnson & Johnson	Cypher™	Sirolimus	Nonabsorbable polymer	Bx Velocity	316L stainless steel
Boston Scientific	Taxus™	Paclitaxel	Nonabsorbable polymer	Express 2	316L stainless steel
Medtronics	Endeavor™	Zotarolimus	PC coating	Driver	Co-Cr alloy
Abbot/Boston Scientific	Xience™ V	Everolimus	Acryl/fluorine polymer	ML-Vision	Co-Cr alloy
Biosensors/Termo	Nobori™	Biolimus	Absorbable polymer (PLA)	S-Stent	316L stainless steel

12.5.3 Stent Grafts

The other metal–polymer composite is stent graft. Stents for treatment of peripheral vascular disease are generally constructed of stainless steel or nitinol and may be coated with compounds such as ePTFE. Stent grafts are composed of a metallic frame covered by a fabric tube and combine the features of stents and vascular grafts; they can be deployed endovascularly. Stent grafts are used to treat aortic aneurysms, where the aortic wall has been weakened and threatens to rupture, as well as stenosis of other arterial sites. The graft portion, usually composed of polyester or ePTFE, can sit on either the luminal or abluminal (outside) aspect of the metallic stent and is intended to provide a mechanical barrier to prevent intravascular pressure from being transmitted to the weakened wall of the aneurysm, thus excluding the aneurysm from the flow of blood. These stents and stent grafts are developed in a similar manner to those in the coronary circulation, either as self-expanding units or covering an inflatable balloon. The stent is used for a given application unit or over an inflatable balloon. The stent used for a given application is selected by diameter, length, and geometry of the lesion and location of side braches or branch points.

Current synthetic vascular grafts are typically fabricated from poly(ethylene terephthalate) (Dacron) or ePTFE, with the Dacron graft being used for larger vessel applications and the ePTFE to bypass smaller vessels. These grafts can be made porous to enhance healing, but they are then impregnated with connective tissue proteins to aid clotting, reduce the blood loss through the pores of the graft upon implantation, and stimulate tissue ingrowth, and with antibiotics to reduce the risk of infection of the grafts that are not impregnated, which need to be preclotted with the patient's own blood for this same purpose.

In light of the complications associated with vascular grafts, current research has focused on improvement of synthetic vascular grafts and on alternatives such as tissue-engineered blood vessels. The purpose of obtaining the luminal surface is as follows: (1) prevent coagulation, (2) prevent platelet adhesion/aggregation, (3) promote fibrinolysis, (4) inhibit smooth muscle cell adhesion/proliferation, and (5) promote endothelial cell adhesion and proliferation. Endotherialization of the entire graft may also help prevent bacterial attachment to the graft and subsequent infection. Engineering an artery muscle and endotherial cells is a promising approach to solving the problem of developing an adequate small-diameter vascular graft.[54,55]

Severe pitting and crevice corrosion originating from crevice corrosion are observed in Ni-Ti alloys in stent grafts, as shown in Figure 12.14.[56] These problems may occur in Ni-Ti alloys and stainless steel at crevices between an artificial blood vessel and a metallic stent because of the electrochemical properties of the alloys. Corroded sites may then become the initiation site of a fracture. Alloys that are resistant to pitting and corrosion must be developed and used.

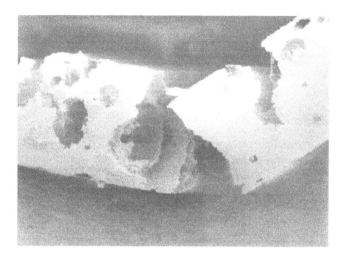

FIGURE 12.14   Severe crevice corrosion observed in Ni-Ti alloy used as stent graft for 6 months. (From Heintz, C. et al. 2001. Corroded Nitinol wires in explanted aortic endografts: An important mechanism of failure? *J Endovasc Ther* 8: 248–253.)

12.6 IMMOBILIZATION OF POLY(ETHYLENE GLYCOL)

12.6.1 POLY(ETHYLENE GLYCOL)

PEG is an oligomer or polymer of ethylene oxide, but historically PEG has tended to refer to oligomers and polymers with a molecular weight below 20,000. PEG has the structure shown in Figure 12.15. PEGylation is the act of covalently coupling a PEG structure to another larger molecule, for example, a therapeutic protein (which is then referred to as PEGylated). PEG is soluble in water, methanol, benzene, and dichloromethane and is insoluble in diethyl ether and hexane. It is coupled to hydrophobic molecules to produce nonionic surfactants. This property, combined with the availability of PEGs with a wide range of end functions, contributes to the wide use of PEGs in biomedical research: drug delivery, tissue engineering scaffolds, surface functionalization, and many other applications.[57]

12.6.2 CHEMICAL IMMOBILIZATION

The immobilization of biofunctional polymers on noble metals such as Au is usually conducted by using the bonding –SH or –SS– group; however, this technique can only be used for noble metals. The adhesion of platelets and adsorption of proteins, peptides, antibodies, and DNA are controlled by modifications of this technique. On the other hand, PEG is a biofunctional molecule on which adsorption of proteins is inhibited. Therefore, immobilization of PEG to metal surface is an important event to biofunctionalize the metal surface. Examples of immobilization of PEG to oxide surface are shown in Figure 12.16. A class of copolymers based on poly(L-lysine)-g-poly (ethylene glycol), PLL-g-PEG, has been found to spontaneously adsorb from aqueous solutions onto TiO_2, $Si_{0.4}Ti_{0.6}O_2$, and Nb_2O_5 to develop blood-contacting materials and biosensors.[52,58] In another case, TiO_2 and Au surfaces are functionalized by the attachment of PEG-poly(DL-lactic acid), PEG-PLA, and copolymeric micelles. The micelle layer can enhance the resistance to protein adsorption to the surfaces up

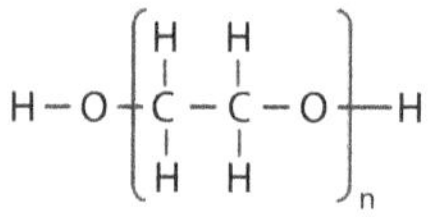

$$H-O\left[\begin{matrix} H & H \\ | & | \\ C & C \\ | & | \\ H & H \end{matrix} - O\right]_n H$$

FIGURE 12.15   Chemical structure of PEG.

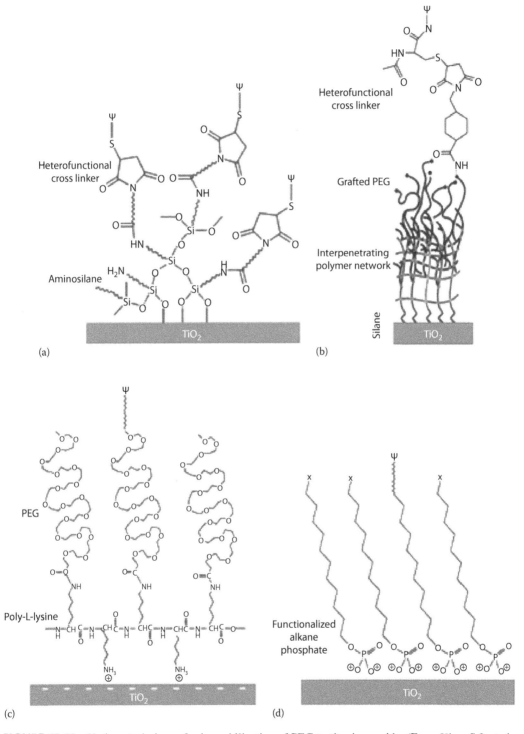

FIGURE 12.16 Various techniques for immobilization of PEG to titanium oxide. (From Xiao, S.J. et al., Biochemical modification of titanium surfaces, in: D.M. Brunrtte, P. Tenvall, M. Textor, and P. Thomsen, eds., *Titanium in Medicine*, Springer, Amsterdam, the Netherlands, p. 421, 2001.) (a) Silanization using APTES followed by covalent attachment of maleimide; (b) silanized surface followed by formation of a non-adhesive P(AAm*co*EG/AA); (c) polycationic poly(amino acid) grafted with poly(ethylene grycol) side chain; (d) self-assembled monolayers of long-chain alkanephosphate.

to 70%.[59] A surface of stainless steel was first modified by a silane coupling agent, SCA, (3-mercaptopropyl)trimethoxysilane. The silanized stainless steel, SCA-SS, surface was subsequently activated by argon plasma and then subjected to UV-induced graft polymerization of poly(ethylene glycol) methacrylate, PEGMA. The PEGMA graft-polymerized stainless-steel coupon, PEGMA-g-SCA-SS, with a high graft concentration and, thus, a high PEG content was found to be very effective to prevent the absorption of bovine serum albumin and γ-globulin.[60] These processes require several steps but are effective for immobilization; however, no promising technique for the immobilization of PEG to a metal surface has been so far developed. Photoreactive PEG is photoimmobilized on Ti.[61]

12.6.3 ELECTRODEPOSITION

Both terminals of PEG (MW: 1000) are terminated with $-NH_2$ ($NH_2-PEG-NH_2$), but only one terminal is terminated with $-NH_2$ (NH_2-PEG). The cathodic potential is charged to Ti from the open circuit potential to -0.5 V versus an SCE and maintained at this potential for 300 s. During charging, the terminated PEGs electrically migrate to and are deposited on the Ti cathode, as shown in Figure 12.17. Not only electrodeposition but also immersion leads to the immobilization of PEG onto a Ti surface. However, more terminated amines combine with Ti oxide as an NH-O bond by electrodeposition, while more amines randomly exist as NH_3^+ in the PEG molecule by immersion (Figure 12.18).[62,63] A scanning probe microscopic image is shown in Figure 12.19. The amounts of the PEG layer immobilized onto the metals are governed by the concentrations of the active hydroxyl groups on each surface oxide in the case of electrodeposition, which is governed by the relative permittivity of the surface oxide in the case of immersion.[64] The PEG-immobilized surface inhibits the adsorption of proteins and cells, as well as the adhesion of platelets[65] and bacteria[66] (Figure 12.20), indicating

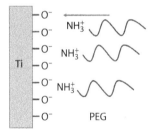

FIGURE 12.17   Attraction of PEG with positively charged terminal to cathodic Ti surface by electrodeposition.

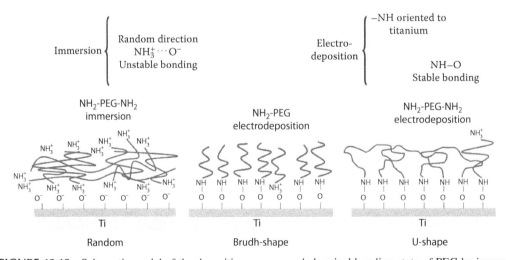

FIGURE 12.18   Schematic model of the deposition manner and chemical bonding state of PEG by immersion and electrodeposition.

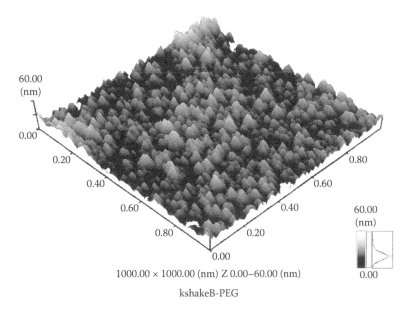

kshakeB-PEG

FIGURE 12.19 **(See companion CD for color figure.)** Scanning probe microscopic image of electrodeposited PEG to Ti surface.

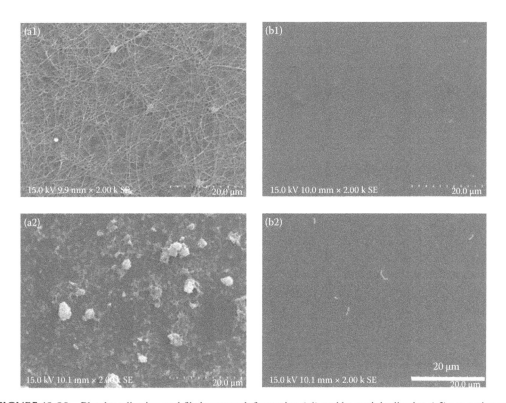

FIGURE 12.20 Platelet adhesion and fibrin network formation (a1) and bacterial adhesion (a2) are active on Ti, while they are inhibited on PEG-electrodeposited Ti surfaces (b1 and b2).

that this electrodeposition technique is useful for the biofunctionalization of metal surfaces. It is also useful for all electroconductive materials and materials with complex surface topography.

12.7   IMMOBILIZATION OF BIOMOLECULES

12.7.1   METAL–BIOMOLECULE INTERFACE

Organic coatings have been scarcely used on metallic implants. Nevertheless, it is expected that these materials will significantly improve their market share as improved biocompatibility becomes the decisive criterion for patients and surgeons in the future. This approach, which is based on the latest medical and cellular biological results, employs biopolymers (proteins) that have been immobilized on the surface of metallic implants. The intention is to reduce the region character of the implant for the body. This is accomplished by coating with substances that are normally found on the surface or in the vicinity of the tissue that has to be substituted by the implant. It has been found that these coatings act as local mediators of cell adhesion and, in consequence, as a stimulating factor for the growth and proliferation of the cells normally found around the substituted tissue. The tight attachment at the oxide-coated surface of the metallic implant and the conservation of the biological function of the proteins involved are prerequisites for obtaining these highly desirable properties.

Since the natural environment around the implant is aqueous while the surface of the implant is either bare or oxidized metal, specific demands are imposed on the coating to mediate successfully between these different structural entities. The purpose of these demands is to obtain the native conformation of all proteins and cells that are in contact with the coating and to avoid all forms of aggregation and other conformational changes that might lead to protein denaturalization or cell death.

One approach is the immobilization of biological molecules (growth factors, adhesive proteins) onto the implant surface in order to induce a specific cellular response and promote osseointegration. The application of large extracellular matrix proteins, however, can be unpractical due to their low chemical stability, solubility in biological fluids, and high cost. In addition, entire ECM molecules are usually of allogenetic or xenogenetic origin and thus associated with the risk of immune reaction and pathogen transfer. To immobilize biomolecules to metal surfaces, the following techniques are used: modification through silanized titania (thiol-directed immobilization, amino- and carboxyl-directed immobilization), modification through photochemistry, electrochemical techniques, and chemical modifications based on self-assembled monolayers.

The immobilization of biomolecules to a metallic surface can be achieved using self-assembled monolayers as cross-linkers. Self-assembled monolayers provide chemically and structurally well-defined surfaces that can often be manipulated using standard synthetic methodologies.[67] Thiol on self-assembled monolayers[68,69] and siloxane-anchored self-assembled monolayers[70] has been particularly well studied. A problem related to the application of immobilized biomolecules via silanization techniques is the hydrolysis of siloxane films when exposed to aqueous (physiological) conditions.[71] More recently, alkyl phosphate films that remain robust under physiological conditions[72] have been used to provide an ordered monolayer on tantalum oxide surfaces,[73,74] and alkylphosphonic acids have been used to coat the native oxide surfaces of metals and their alloys inducing iron,[75] steel,[76] and Ti.[77]

12.7.2   PEPTIDES

In a living tissue, the most important role played by the extracellular matrix has been highlighted to favor cell adhesion.[78] Studies have shown that interactions occur between cell membrane

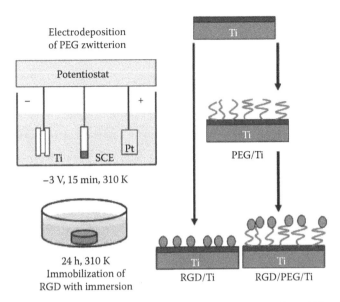

FIGURE 12.21 PEG twitter ion is electrodeposited to Ti firstly and RGD is immobilized on the PEG.

receptors and adhesion proteins (or synthetic peptides) derived from the bone matrix, such as type I collagen or fibronectin.[79] These proteins are characterized by an RGD (Arg-Gly-Asp) motif with special transmembrane connections between the actin cytoskeleton and the RGD motif, and the whole system can activate several intracellular signaling pathways modulating cell behavior (e.g., proliferation, apoptosis, shape, mobility, gene expression, and differentiation).[80] Due to the main role of the RGD sequence in cell adhesion, several research groups have developed biofunctionalized surfaces by immobilization of RGD peptides. Grafting RGD peptides has been performed on different biomaterials, such as Ti,[81–83] and has been shown to improve osteoconduction in vitro. Methodologies differ by the conformation of RGD (cyclic or linear) and by the technique used for peptide immobilization.[78,79,82–84] Since the graft of an RGD peptide is known to be efficient in bone reconstruction,[85] the challenge is to develop simple and cheap methods to favor cell anchorage on biomaterial surfaces.[83,84]

Self-assembled molecular monolayers bearing RGD moieties have been grafted to numerous surfaces, using either silanes,[86] phosphonates on oxidized surfaces,[83] or thiols on Au[84] but have analyzed some application problems for large-scale production. Phosphonates are known to adsorb on Ti. To be mechanically and physiologically stable, phosphonate layers have to be covalently bound to the material surface by using drastic conditions[77,87] (anhydrous organic solvents, high temperature), which are not compatible with biomolecule stability. Monolayers of RGD phosphonates have been achieved using a complex multistep process, which necessitates tethering a primer onto Ti surface, then a linker, and finally the peptide.[88] To immobilize RGD to the electrodeposited PEG on Ti, PEG with an $-NH_2$ group and a $-COOH$ group (NH$_2$–PEG–COOH) must be employed. One terminal group, $-NH_2$, is required to bind stably with a surface oxide on a metal. On the other hand, the other terminal group, $-COOH$, is useful to bond biofunctional molecules such as RGD, as shown in Figure 12.21.[89] This RGD/PEG/Ti surface accelerates calcification by MC3T3-E1 cell.[90] The calcification is the largest on the RGD/PEG/Ti surface (Figure 12.22).

Glycine (G)-arginine (R)-glycine (G)-asparaginic acid (D)-serine (S) sequence peptide, GRGDS peptide, is coated with chloride activation technique to enhance adhesion and migration of osteoblastic cells.[91] The expression levels of many genes in MC3T3-E1 cells are altered.

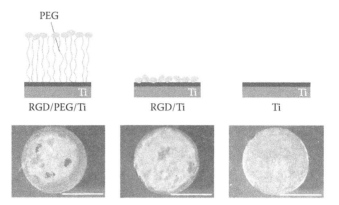

FIGURE 12.22   Calcification (dark regions) by MC3T3-El cells is more active on RGD/PEG/Ti specimen than on RGD/Ti and Ti. Scale bar represents 5 mm.

12.7.3   PROTEIN AND COLLAGEN

Among the relevant molecules involved in biochemical modification of bone-contacting surfaces, growth factor, such as bone morphogenetic protein-2 (BMP-2), is of primary interest. BMP-2 has been known to play an important role in bone healing processes and to enhance therapeutic efficiency. Ectopic bone formation by BMP-2 in animals has been well established following the first reports of BMP-2 by the Urist research group.[92–94] Synthetic receptor binding motif mimicking BMP-2 is covalently linked to Ti surfaces through a chemical conjunction process.[95] A complete and homogeneous peptide overlayers on the Ti surfaces; the content is further measured by gamma counting. Biological evaluations show that the biochemically modified Ti were active in terms of cell attachment behavior. Ti surfaces can enhance the rate of bone healing as compared with untreated Ti surface. Bone morphogenetic protein-4 (BMP-4) is immobilized on a Ti-6Al-4V alloy through lysozyme to improve the hard tissue response.[96] Proteins are silane coupled to the oxidized surfaces of the Co-Cr-Mo alloy, the Ti-6Al-4V alloy, Ti, and the Ni-Ti alloy to improve tissue compatibility.[97]

Biomolecules are also used to accelerate bone formation and soft tissue adhesion on a material. Type I collagen is immobilized by immersion in the collagen solution.[98] Type I collagen production increases with modification by ethane-1,1,2-triphosphonic acid and methylenediphosphonic acid grafted onto Ti.[99] Type I collagen is grafted through glutaraldehyde as a crosslinking agent.[100] For electrodeposition, it is found that an alternating current between −1 V and +1V versus SCE with 1 Hz is effective to immobilize type I collagen to Ti and durability in water is high.[101]

Fibronectin is immobilized directly on Ti using the tresyl chloride activation technique.[102] L-threonine and O-phospho-L-threonine are immobilized on acid-etched Ti surface.[103]

12.7.4   HYDROGEL AND GELATIN

Immobilization or coating of hydrogel to metal surface is currently attempting to add a drug delivery ability to orthopedic implant and stents or fluorescent sensing ability to microchips. Currently, synthetic polymeric hydrogels like poly(hydroxyethylmethacrylate) (pHEMA) and poly(hydroxyethylacrylate) (pHEA) are widely used as compliant materials particularly in the case of contact with blood or other biological fluids.[104] Despite hydrogel's good flexibility in the swollen state, hydrogels usually lack suitable mechanical properties, and this could greatly impair their use as coating materials for surgical procedure. Moreover, in case of inadequate adhesion between the hydrogel coating and the metal surface, a breakage at the coating–steel interface might occur.[105] A spray-coated method has been set up with the aim to control the coating of pHEMA onto the complex surface of a 316L steel stent for percutaneous coronary intervention (PCI).[106] The pHEMA coating evaluation of roughness wettablity together with its morphological and chemical stability after three cycles of expansion-crimping along with preliminary results after 6 months demonstrates the suitability of the coating for surgical implantation of stent.

An alternative promising synthetic route is represented by electrochemical polymerization, which leads to thin film coatings directly on the metal substrates with interesting applications either for corrosion protection or for the development of bioactive films.[107–110] As far as the orthopedic field is concerned, in recent years, many procedures based on surface modification have been suggested to improve the biocompatibility and biofunctionality of Ti-based implant.[111] 2-Hydroxy-ethyl-methacrylate (HEMA), a macromer poly(ethylene-glycol diacrylate)(PEGDE), and PEGDE copolymerized with acrylic acid were used to obtain hydrogels. A model protein and a model drug were entrapped in the hydrogel and released according to pH change.[112]

12.8 OTHER METAL–POLYMER COMPOSITES

12.8.1 BONDING OF POLYMERS WITH METAL THROUGH SILANE COUPLING AGENT

The interfacial chemical structure governing the bonding strength, especially at the nanometer level, is one of the most challenging aspects of the development of composite materials. The combination of a Ti alloy with a resin for crown facings has been attempted.[113] In particular, SCAs containing S-H groups and Si-O-CH$_3$ groups are comprehensively used to combine dental alloys with resins.[114] The S-H groups work as a bonding agent with polymers; the Si-O-CH$_3$ works as a bonding agent with metals. The mechanical properties and durability of composite resin increase with silanized filler.[115–118] However, in most studies about materials using SCAs in the field of dentistry, only the bonding strength is evaluated and discussed, and there are few reports that examine and discuss the chemical structures at the bonding interface and how they influence the bonding strength. Other studies on SCAs to combine polymers with metals have been performed in other fields. An aluminum–vegetable oil composite using a SCA has been developed.[119,120] Rubber-to-metal bonding by a SCA was investigated.[121] In addition, the surface modification of stainless steel by grafting of PEG using a SCA has been reported.[60] However, only the chemical structure is investigated in these studies. In other words, the relationship between the bonding strength and the interfacial chemical structure containing a SCA layer has not been studied.

The unequivocal relationship between the shear bonding strength and the chemical structure at the bonding interface of a Ti-segmentated polyurethane (SPU) composite through a SCA (γ-mercaptopropyl trimethoxysilane [γ-MPS]) is investigated.[122] Schematic models of the fractured region before and after the shear bonding test in the case of a thin γ-MPS layer and a thick γ-MPS layer are shown in Figure 12.23. On the other hand, the shear bond strength of the Ti/SPU interface increased with ultraviolet (UV) irradiation according to the increase in the cross-linkage in SPU. Platelet adhesion to Ti is inhibited by SPU, as shown in Figure 12.24. This technique is used for the creation of a new meta-based material with high strength, high toughness, and biofunction. UV irradiation to a Ti-SPU composite is clearly a factor governing the shear bond strength of the Ti/SPU interface.[123] In addition, active hydroxyl groups on the surface oxide film are clearly factors governing the shear bond strength.[124] After a good bonding between metal and polymer is produced, biofunctionalization techniques developed in the field of polymers could be applied to the composite materials.

12.8.2 POLYMERS CONDENSED IN POROUS TITANIUM

The Young's modulus of metallic materials is relatively larger than that of cortical bone: about 200 GPa in stainless steels and Co-Cr-Mo alloys, about 100 GPa in Ti and Ti alloys, and 15–20 GPa in cortical bone. When fractured bone is fixed with metallic bone fixator such as bone plate and

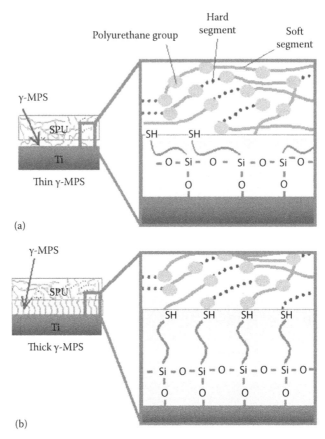

FIGURE 12.23 Schematic model of the fractured region before and after the shear bonding test in the case of a thin ā-MPS layer (a) and a thick γ-MPS layer (b).

screws and bone nail, during healing a load to fixation part is mainly received by metallic fixators because of the difference in their Young's modulus. This phenomenon is well known as the so-called "stress shielding" in orthopedics. This large Young's modulus generates other problems. When a metal is used as a metallic spacer in spinal fixation, the spacer is mounted in matrix bone. In the case of a dental implant, occlusal pressure is not absorbed by the implant and directly conducts to jaw bone.

To solve these problems, metals with low Young's modulus are required. Two approaches are feasible: the decrease in the Young's modulus of a metallic material itself and decrease in the apparent Young's modulus by forming a porous body. In the latter case, the pores are sometimes filled by polymers to control the apparent Young's modulus. UHMWPE[125] and PMMA[126] are attempted to fill the pores in porous Ti. Figure 12.25 shows porous Ti whose pores are filled by UHMWPE.

12.8.3 METALLIZATION OF POLYMERS

The properties of polymer surfaces with regard to their chemical, electrical, mechanical, and other properties are modified by metallization, and this technique could be applied to biomaterials.[127] In the metallization process, the metal atoms arrive as mobile individuals at a polymer surface. The metal layer formation on polymer will be influenced by the type and strength of metal–polymer interaction and the structure of the polymer. Metallization techniques are categorized by evaporation, sputtering, chemical vapor deposition, and electrochemical deposition.

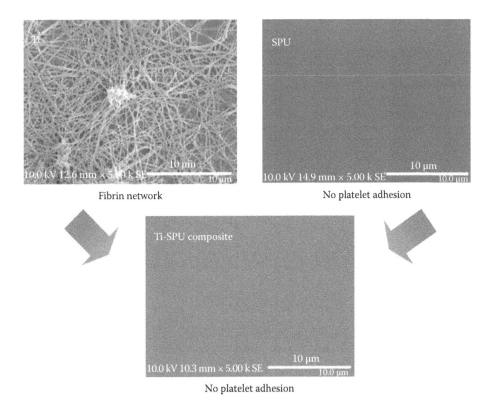

FIGURE 12.24 Platelet adhesion on Ti is inhibited by SPU layer.

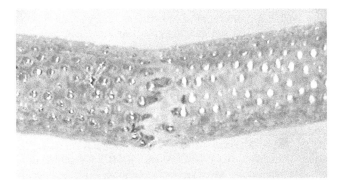

FIGURE 12.25 Porous Ti whose pores are filled by UHMWPE after bending test.

A number of conditions must be determined to accomplish metal penetration into polymers. The metal deposition rate must be low enough to prevent metal cluster formation at the polymer surface in an initial stage of the metallization process. The metal bonds in such clusters immobilize the atoms, and a strong interaction between the metal atom and polymer chain reduces the diffusion. The metal atoms must not react with the polymer. The polymer substrate temperature must be well above the glass transition of the polymer surface region. The polymer segment motions assist the metal atom transport deep into the material.

The diffusion front of a metal is not smooth, but it is rough and creates mechanical interlocking, which in turn enhances the mechanical strength between metal overlayer and polymer. The mechanical properties increase in the composite region compared with the pure polymer. Electric conductivity may be affected and the thermal expansion coefficients are matched to some extent by interphase.

12.9 CONCLUSIONS

Metallic materials are widely used in medicine for not only orthopedic and dental implants but also cardiovascular devices and other purposes. The metal surface may be biofunctionalized by various techniques such as dry and wet processes, the immobilization of biofunctional molecules, and the creation of metal–polymer composites. These techniques make it possible to apply metals to a scaffold in tissue engineering. Artificial materials such as metal–polymer composites will continue to be used as biomaterials in the future, because of their excellent biocompatibilities and biofunctions. Some metal–polymer composites are reviewed in other textbooks.[127–129]

REFERENCES

1. Ratner, B. D., Hoffman, A. S., Schoen, F. J., and J. E. Lemons. 2004. *Biomaterials Science: An Introduction to Materials in Medicine*. Amsterdam, the Netherlands: Elsevier.
2. Yang, Y., Kim, K. H., and J. Ong. 2005. A review on calcium phosphate coatings produced using a sputtering process—An alternative to plasma spraying. *Biomaterials* 26: 327–337.
3. Kim, K. H. and N. Ramaswany. 2009. Electrochemical surface modification of titanium in dentistry. *Dent Mater J* 28: 20–36.
4. Asami, K., Chen, S.C., Habazaki, H., and K. Hashimoto. 1993. The surface characterization of titanium and titanium-nickel alloys in sulfuric acid. *Corros Sci* 35: 43–49.
5. Hanawa, T., Asami, K., and K. Asaoka. 1998. Repassivation of titanium and surface oxide film regenerated in simulated bioliquid. *J Biomed Mater Res* 40: 530–538.
6. Hanawa T. and M. Ota. 1991. Calcium phosphate naturally formed on titanium in electrolyte solution. *Biomaterials* 12: 767–774.
7. Hanawa T. 1991. Titanium and its oxide film: a substrate for formation of apatite. In: J. E. Davies, ed., *The Bone-Biomaterial Interface*, pp. 49–61. Toronto, Ontario, Canada: University of Toronto Press.
8. Hanawa, T., Okuno, O., and H. Hamanaka. 1992. Compositional change in surface of Ti-Zr alloys in artificial bioliquid. *J Jpn Inst Met* 56: 1168–1173.
9. Bruesch, P., Muller, K., Atrens, A., and H. Neff. 1985. Corrosion of stainless-steels in chloride solution-an XPS investigation of passive films bruesch. *Appl Phys* 38: 1–18.
10. Jin, S. and A. Atrens. 1987. ESCA-studies of the structure and composition of the passive film formed on stainless steels by various immersion times in 0.1 M NaCl solution. *Appl Phys* A42: 149–165.
11. Hanawa, T., Hiromoto, S., Yamamoto, A., Kuroda, D., and K. Asami. 2002. XPS characterization of the surface oxide film of 316L stainless samples that were located in quasi-biological environments. *Mater Trans* 43: 3088–3092.
12. Smith, D. C., Pilliar, R. M., Metson, J. B., and N. S. McIntyre. 1991. Preparative procedures and surface spectroscopic studies. *J Biomed Mater Res* 25: 1069–1084.
13. Hanawa, T., Hiromoto, S., and K. Asami. 2001. Characterization of the surface oxide film of a Co-Cr-Mo alloy after being located in quasi-biological environments using XPS. *Appl Surf Sci* 183: 68–75.
14. Nagai, A., Tsutsumi, Y., Suzuki, Y., Katayama, K., Hanawa, T., and K. Yamashita. 2012. Characterization of air-formaed surface oxide film on a Co-Ni-Cr-Mo alloy (MP35N) and its change in Hanks' solution. *Appl Surf Sci* 258: 5490–5498.
15. Endo, K., Araki, Y., and H. Ohno. 1989. In vitro and in vivo corrosion of dental Ag-Pd-Cu alloys. In: T. Okabe and S. Takahashi, eds., *Transaction of International Congress on Dental Materials*, November 1–4, 1989, Honolulu, Hawaii, The Academy of Dental Materials and The Japanese Society for Dental Materials and Devices, pp. 226–227.
16. Parfitt, G. D. 1976. The surface of titanium dioxide. *Prog Surf Membr Sci* 11: 181–226.
17. Westall, J. and H. Hohl. 1980. A comparison of electrostatic models for the oxide/solution interface. *Adv Colloid Interface Sci* 12: 265–294.
18. Healy, T. W. and D. W. Fuerstenau. 1965. The oxide-water interface–Interreaction of the zero point of charge and the heat of immersion. *J Colloid Sci* 20: 376–386.
19. Boehm, H. P. 1971. Acidic and basic properties of hydroxylated metal oxide surfaces. *Discuss Faraday Soc* 52: 264–289.
20. Takeyama, M., Kashibuti, S., Nakabayashi, N., and E. Masuhara. 1978. Studies on dental self-curing resins(17)—Adhesion of PMMA with bovine enamel or dental alloys. *J Jpn Soc Dent Appar Mater* 19: 179–185.
21. Omura, I., Yamauchi, J., Nagase, Y., and F. Uemura. 1983. Jpn Published Unexamined Patent Application, 58-21607.

22. Ikemura, K. and T. Endo. 2010. A review of our development of dental adhesives –Effects of radical polymerization initiators and adhesive monomers on adhesion. *Dent Mater J* 29: 109–121.
23. Tanaka, T., Nagata, K., Takeyama, M., Atsuta, M., Nakabayashi, N., and E. Masuhara. 1981. 4-META opaque resin—A new resin strongly adhesive to nickel-chromium alloy. *J Dent Res* 60: 1697–1707.
24. Tanaka, T., Nagata, K., Takeyama, M., Nakabayashi, N., and E. Masuhara. 1980. Heat treatment of gold alloys to get adhesion with resin. *J Jpn Soc Dent Appar Mater* 21: 95–102.
25. Varga, J., Matsumura, H., Tabata, T., and E. Masuhara. 1985. Adhesive behavior of the alloys 'ALBABOND E' containing large percentage of Pd after various surface treatments. *Dent Mater J* 4: 181–190.
26. Ohno, H., Araki, Y., and M. Sagara. 1986. The adhesion mechanism of dental adhesive resin to the alloy—Relationship between Co-Cr alloy surface structure analyzed by ESCA and bonding strength of adhesive resin. *Dent Mater J* 5: 46–65.
27. Ohno, H., Araki, Y., Sagara, M., and Y. Yamane. 1986. The adhesion mechanism of dental adhesive resin to the alloy—Experimental evidence of the deterioration of bonding ability due to adsorbed water on the oxide layer. *Dent Mater J* 5: 211–216.
28. Ohno, H., Araki, Y., Endo, K., and K. Kawashima. 1989. The adhesion mechanism of dental adhesive resin to the alloy—Adhesive ability of dental adhesive resin to the clean metal surface obtained by hydrogen gas reduction method. *Dent Mater J* 8: 1–8.
29. Brewis, D. M., Comyn, J., and J. L. Tegg. 1980. The durability of some epoxide adhesive-bonded joints on exposure to moist warm air. *Int J Adhes Adhes* 1: 35–39.
30. Ko, C. U. and J. P. Wightman. 1988. Experimental analysis of moisture intrusion into the Al/Li-polysulfone interface. *J Adhes* 25: 23–29.
31. Ohno, H., Endo, K., Araki, Y., and S. Asakura. 1992. Destruction of metal-resin adhesion due to water penetrating through the resin. *J Mater Sci* 27: 5149–5153.
32. Ohno, H., Endo, K., Araki, Y., and Y. Asakura. 1993. ESCA study on the destruction mechanism of metal-resin adhesion due to water penetrating through the resin. *J Mater Sci* 28: 3764–3768.
33. Ohno, H., Araki, K., Endo, K., Yamane, Y., and I. Kawashima. 1996. Evaluation of water durability at adhesion interface by peeling test of resin film. *Dent Mater J* 15: 183–192.
34. Musil, R. 1987. Clinical verification of the Silicoater technique, results of three-years' experience. *Dent Lab* 35: 1709–1715.
35. Tanaka, T., Hirano, M., Kawahara, H., Matsumura, H., and M. Atsuta. 1988. A new io-coating surface treatment of alloys for dental adhesive resins. *J Dent Res* 67: 1376–1380.
36. Ohno, H., Araki, K., and K. Endo. 1992. A new method for promoting adhesion between precious metal alloys and dental adhesives. *J Dent Res* 71: 1326–1331.
37. Ohno, H., Endo, K., Yamane, Y., and I. Kawashima. XPS study on the weakest zone in the adhesion structure between resin containing 4-META and precious metal alloys treated with different surface modification methods. *Dent Mater J* 20: 330–337.
38. Dotter, C. T. and M. P. Judkins. 1964. Transluminal treatment of arteriosclerotic obstruction of a new technique and preliminary report of its application. *Circulation* 30: 654–670.
39. Palmaz, J. C., Sibbitt, R. R., Tio, F. O., Reuter, S. R., Peters, J. E., and F. Garcia. 1986. Expandable intra-luminal vascular graft. A feasibility study. *Surgery* 99: 199–205.
40. Roubin, G. S., Robinson, K. A., King, S. B., Gianturco, C., Black, A. J., Brown, J. E., Siegel, R. J., and J. S. Douglas. 1987. Early and late results of intracoronary arterial stenting after coronary angioplasty in dog. *Circulation* 76: 891–897.
41. Sigwart, U., Puel, J., Mirkovitch, V., Joffre, F., and L. Kappenberger. 1987. Intravascular stents to prevent occlusion and restenosis after trans-luminal angioplasty. *N Eng J Med* 316: 701–706.
42. Phatouros, C. C., Higashida, R. T., and A. M. Malek. 2000. Endovascular stenting for carotid artery stenosis: Preliminary experience using the shape-memory-alloy-recoverable-technology (SMART) stent. *Am J Neuroradiol* 21: 732–738.
43. Roubin, G. S., Yadav, S., Iyer, S. S., and J. Vitek. 1996. Carotid stent-supported angioplasty: A neurovascular intervention to prevent stroke. *Am J Cardiol* 78: 8–12.
44. Dietrich, E. B. 1996. Aortic endografting: Visions of things to come. *J Endovasc Surg* 3: R21–R23.
45. Wholey, M. H., Wholey, M., Bergeron, P., Diethrich, E. B., Henry, M., Laborde, J. C., Mathias, K. et al. 1998. Current global status of carotid artery stent placement. *Cathet Cardiovasc Diagn* 44: 1–6.
46. Richter, G. M., Palmaz, J. C., Allenberg, J.R., and G. W. Kauffmann. Percutaneous stent grafts for aortic-aneurysms–preliminary experience with a new procedure. *Radiology* 34: 511–518.
47. Fattori, R. and T. Piva. 2003. Drug-eluting stents in vascular intervention. *Lancet* 362: 247–249.
48. Sousa, J. E., Serruys, P. W., and M. A. Costa. 2003. New frontiers in cardiology: Drug-eluting stents – Pt. I. *Circulation* 107: 2274–2279.

49. Sousa, J. E., Serruys, P. W., and M. A. Costa. 2003. New frontiers in cardiology: Drug-eluting stents – Pt. II. *Circulation* 107: 2283–2289.

50. Nakayama, Y., Kim, J. Y., Nishi, S., Ueno, H., and T. Matsuda. 2001. Development of high-performance stent: Gelatinous photogel-coated stent that permits drug delivery and gene transfer. *J Biomed Mater Res* 57: 559–566.

51. Ishihara, K., Ueda, T., and N. Nakabayashi. 1990. Preparation of phospholipids polymers and their properties as hydrogel membrane. *Polym J* 30: 355–360.

52. Huang, N. P., Michel, R., Voros, J., Textor, M., Hofer, R., Rossi, A., Elbert, D. L., Hubbell, J. A., and N. D. Spencer. 2001. Poly(L-lysine)-g-poly(ethylene glycol) layers on metal oxide surfaces: Surface-analytical characterization and resistance to serum and fibrinogen adsorption. *Langmuir* 17: 489–498.

53. Sianos, G., Hofma, S., Lighthart, J. M. R., Saia, F., Hoye, A., Lemos, P. A., and P. W. Serruys. 2004. Stent fracture and restenosis in the drug-eluting stent era. *Catheter Cardiovasc Interv* 61: 111–116.

54. Consgny, P. M. 2000. Endotherial cell seeding on prosthetic surfaces. *J Long Term Eff Med Implants* 10: 79–95.

55. Seifalian, A. M., Tiwari, A., Hamilton, G., and H. J. Salacinski. 2001. Improving the clinical patency of prostetic vascular and coronary artery bypass grafts: The role of seeding and tissue engineering. *Artif Organs* 26: 489–495.

56. Heintz, C., Riepe, G., Birken, L., Kaiser, E., Chakfe, N., Morlock, M., Delling, G., and H. Imig. 2001. Corroded Nitinol wires in explanted aortic endografts. *J Endovasc Ther* 8: 248–253.

57. Mahato, R. I. 2005. *Biomaterials for Delivery and Targeting of Proteins and Nucleic Acids.* Boca Raton, FL: CRC Press (Ald. Cat. No. Z705102).

58. Kenausis, G. L., Vörös, J., Elbert, D. L., Huang, N., Hofer, R., Ruiz-Taylor, L., Textor, M., Hubbell, J.A., and N. D. Spencer. 2000. Poly(L-lysine)-g-poly(ethylene glycol) layers on metal oxide surfaces: Attachment mechanism and effects of polymer architecture on resistance to protein adsorption. *J Phys Chem* B104: 3298–3309.

59. Huang, N. P., Csucs, G., Emoto, K., Nagasaki, Y., Kataoka, K., Textor, M., and N. D. Spencer. 2002. Covalent attachment of novel poly(ethylene glycol)-poly(DL-lactic acid) copolymeric micelles to TiO_2 surfaces. *Langmuir* 18: 252–258.

60. Zhang, F., Kang, E. T., Neoh, K. G., Wang, P., and K. L. Tan. 2001. Surface modification of stainless steel by grafting of poly(ethylene glycol) for reduction in protein adsorption. *Biomaterials* 22: 1541–1548.

61. To, Y., Hasuda, H., Sakuragi, M., and S. Tsuzuki. 2007. Surface modification of plastic, glass and titanium by photoimmobilization of polyethylene glycol for antibiofouling. *Ast Biomater* 3: 1024–1032.

62. Tanaka, Y., Doi, H., Iwasaki, Y., Hiromoto, S., Yoneyama, T., Asami, K., Imai, H., and T. Hanawa. 2007. Electrodeposition of amine-terminated-poly(ethylene glycol) to Ti surface. *Mater Sci Eng* C27: 206–212.

63. Tanaka, Y., Doi, H., Kobayashi, E., Yoneyama, T., and T. Hanawa. 2007. Determination of the immobilization manner of amine-terminated poly(ethylene glycol) electrodeposited on a Ti surface with XPS and GD-OES. *Mater Trans* 48: 287–292.

64. Tanaka, Y., Saito, H., Tsutsumi, Y., Doi, H., Imai, H., and T. Hanawa. 2008. Active hydroxyl groups on surface oxide film of Ti, 316l stainless steel, and cobalt-chromium-molybdenum alloy and its effect on the immobilization of poly(ethylene glycol). *Mater Trans* 49: 805–811.

65. Tanaka, Y., Matsuo, Y., Komiya, T., Tsutsumi, Y., Doi, H., Yoneyama, T., and T. Hanawa. 2010. Characterization of the spatial immobilization manner of poly(ethylene glycol) to a titanium surface with immersion and electrodeposition and its effects on platelet adhesion. *J Biomed Mater Res* 92A: 350–358.

66. Tanaka, Y., Matin, K., Gyo, M., Okada, A., Tsutsumi, Y., Doi, H., Nomura, N., Tagami, J., and T. Hanawa. 2010. Effects of electrodeposited poly(ethylene glycol) on biofilm adherence to titanium. *J Biomed Mater Res* 95A: 1105–1113.

67. Balachander, N. and C. N. Sukenik. 1990. Monolayer transformation by nucleophilic substitution: Applications to the creation of new monolayer assemblies. *Langmuir* 6: 1621–1627.

68. Bain, C. D., Troughton, Y., Tao, Y. T., Evall, J., Whitesides, G. M., and R. G. Nuzzo. 1989. Formation of monolayer films by the spontaneous assembly of organic thiols from solution onto gold. *J Am Chem Soc* 111: 437–335.

69. Dubois, L. H. and R. G. Nuzzo. 1992. Synthesis, structure, and properties of model organic surfaces. *Ann Rev Phys Chem* 43: 437–463.

70. UlMan, A. 1996. Formation and structure of self-assembled monolayers. *Chem Rev* 96: 1533–1554.

71. Xiao, S. J., Textor, M., and N. D. Spencer. 1998. Covalent attachment of cell-adhesive, (Arg-Gly-Asp)-containing peptides to titanium surfaces. *Langmuir* 14: 5507–5516.

72. Gawalt, E. S., Avaltroni, M. J., Danahy, M. P., Silverman, B. M., Hanson, E. L., Midwood, K. S., Schwarzbauer, J. E., and J. Schwartz. 2003. Bonding organics to Ti alloys: Facilitating human osteoblast attachment and spreading on surgical implant materials. *Langmuir* 19: 200–204.

73. Brovelli, D., Hahner, G., Ruis, L., Hofer, R., Kraus, G., Waldner, A., Schlosser, J., Oroszlan, P., Ehart, M., and N. D. Spencer. 1999. Highly oriented, self-assembled alkanephosphate monolayers on tantalum(V) oxide surfaces. *Langmuir* 15: 4324–4327.

74. Textor, M., Ruiz, L., Hofer, R., Rossi, K., Feldman, K., Hahner, G., and N. D. Spencer. 2000. Structural chemistry of self-assembled monolayers of octadecylphosphoric acid on tantalum oxide surfaces. *Langmuir* 16: 3257–3271.

75. Fang, J. L., Wu, N. J., Wang, Z. W., and Y. Li. 1991. XPS, AES and Raman studies of an antitarnish film on tin. *Corrosion* 47: 169–173.

76. Van Alsten, J. G. 1999. Self-Assembled monolayers on engineering metals: structure, derivatization, and utility. *Langmuir* 15: 7605–7614.

77. Gawalt, E. S., Avaltroni, M. J., Koch, N., and J. Schwartz. 2001. Self-assembly and bonding of alkane-phosphonic acids on the native oxide surface of titanium. *Langmuir* 17: 5736–5738.

78. Verrier, S., Pallu, S., Bareille, R., Jonczyk, A., Meyer, J., Dard, M., and J. Amedee. 2002. Function of linear and cyclic RGD-containing peptides in osteoprogenitor cells adhesion process. *Biomaterials* 23: 585–596.

79. Reyes, C. D., Petrie, T. A., Burns, K. L., Schwartz, Z., and A. J. Garcia. 2007. Biomolecular surface coating to enhance orthopaedic tissue healing and integration. *Biomaterials* 28: 3228–3235.

80. Hynes, R. O. 2002. Integrins: bidirectional, allosteric signaling machines. *Cell* 110: 673–687.

81. Bagno, A., Piovan, A., Dettin, M., Chiarion, A., Brun, P., Gambaretto, R., Fontana, G., Di Bello, C., Palu, G., and I. Castagliuolo. 2007. Human osteoblast-like cell adhesion on titanium substrates covalently functionalized with synthetic peptides. *Bone* 40: 693–699.

82. Elmengaard, B., Bechtold, J. E., and K. Soballe. 2005. In vivo study of the effect of RGD treatment on bone ongrowth on press-fit titanium alloy implants. *Biomaterials* 26: 3521–3526.

83. Rammelt, S., Illert, T., Bierbaum, S., Scharnweber, D., Zwipp, H., and W. Schneiders. 2006. Coating of titanium implants with collagen. RGD peptide and chondroitin sulfate. *Biomaterials* 27: 5561–5571.

84. Auernheimer, J., Zukowski, D., Dahmen, C., Kantlehner, M., Enderle, A., Goodman, S. L., and H. Kessker. 2005. Titanium implant materials with improved biocompatibility through coating with phosphate-anchored cyclic RGD peptides. *ChemBiochem* 6: 2034–2040.

85. Ferris, D. M., Moodie, G. D., Dimond, P. M., Gioranni, C. W., Ehrlich, M. G., and R. F. Valentini. 1999. RGD-coated titanium implants stimulate increased bone formation in vivo. *Biomaterials* 20: 2323–2331.

86. Xiao, S. J., Textor, M., Spencer, N. D., Wieland, M., Keller, B., and H. Sigrist. 1997. Immobilization of the cell-adhesive peptide Arg-Gly-Asp-Cys (RGDC) on titanium surfaces by covalent chemical attachment. *J Mater Sci Mater Med* 8: 867–872.

87. Silverman, B. M., Wieghaus, K. A., and J. Schwartz. 2005. Comparative properties of siloxane vs phosphonate monolayers on a key titanium alloy. *Langmuir* 21: 225–228.

88. Schwartz, J., Avaltroni, M. J., Danahy, M. P., Silverman, B. M., Hanson, E. L., Schwarzbauer, J. E., Midwood, K. S., and E. S. Gawalt. 2003. Cell attachment and spreading on metal implant materials. *Mater Sci Eng* C23: 395–400.

89. Tanaka, Y., Saito, H., Tsutsumi, Y., Doi, H., Nomura, N., Imai, H., and T. Hanawa. 2009. Effect of pH on the interaction between zwitterion and titanium oxide. *J Colloid Interface Sci* 330: 138–143.

90. Oya, K., Tanaka, Y., Saito, H., Kurashima, K., Nogi, K., Tsutsumi, H., Tsutsumi, Y., Doi, H., Nomura, N., and T. Hanawa. 2009. Calcification by MC3T3-E1 cells on RGD peptide immobilized on titanium through electrodeposited PEG. *Biomaterials* 30: 1281–1286.

91. Yamanouchi, N., Pugdee, K., Chang, W. J., Lee, S. Y., Yoshinari, M., Hayakawa, T., and Y. Abiko. 2008. Gene expression monitoring in osteoblasts on titanium coated with fibronectin-derived peptide. *Dent Mater J* 27: 744–750.

92. Urist, M. R. 1965. Bone: Formation by autoinduction. *Science* 150: 893–899.

93. Lee, Y. M., Nam, S. H., Seol, Y. J., Kim, T. I., Lee, S. J., Ku, Y., Rhyu, I. C., Chung, C. P., Han, S. B., and S. M. Choi. 2003. Enhanced bone augmentation by controlled release of recombinant human bone morphogenetic protein-2 from bioabsorbable membranes. *J Periodontol* 74: 865–872.

94. Wikesjo, U. M., Lim, W. H., Thomson, R. C., Cook, A. D., Wozney, J. M., and W. R. Hardwick. 2003. Periodontal repair in dogs: Evaluation of a bioabsorbable space-providing macroporous membrane with recombinant human bone morphogenetic protein-2. *J Periodontol* 74: 635–647.

95. Seol, Y. J., Park, Y. J., Lee, S. C., Kim, K. H., Lee, J. Y., Kim, T. I., Lee, Y. M., Ku, Y., Rhyu, I. C., Han, S. B., and C. P. Chung. 2006. Enhanced osteogenic promotion around dental implants with synthetic binding motif mimicking bone morphogenetic protein (BMP)-2. *J Biomed Mater Res* 77A: 599–607.

96. Puleo, D. A., Kissling, R. A., and M. S. Sheu. 2002. A technique to immobilize bioactive proteins, including bone morphogenetic protein-4 (BMP-4), on titanium alloy. *Biomaterials* 23: 2079–2087.

97. Nanci, A., Wuest, J. D., Peru, L., Brunet, P., Sharma, V., Zalzal, S., and M. D. McKee. 1998. Chemical modification of titanium surfaces for covalent attachment of biological molecules. *J Biomed Mater Res* 40: 324–335.

98. Nagai, M., Hayakawa, T., Fukatsu, A., Yamamoto, M., Fukumoto, M., Nagahama, F., Mishima, H., Yoshinari, M., Nemoto, K., and T. Kato. 2002. In vitro study of collagen coating of titanium implants for initial cell attachment. *Dent Mater J* 21: 250–260.

99. Viornery, C., Guenther, H. L., Aronsson, B. O., Péchy, P., Descouts, P., and M. Grätzel. 2002. Osteoblast culture on polished titanium disks modified with phosphonic acids. *J Biomed Mater Res* 62: 149–155.

100. Chang, W. J., Qu, K. L., Lee, S. Y., Chen, J. Y., Abiko, Y., Lin, C. T., and H. M. Huang. 2008. Type I collagen grafting on titanium surfaces using low-temperature grow discharge. *Dent mater J* 27: 340–346.

101. Kamata, H., Suzuki, S., Tanaka, Y., Tsutsumi, Y., Doi, H., Nomura N., Hanawa, T., and K. Moriyama. 2011. Effects of pH, potential, and deposition time on the durability of collagen electrodeposited to titanium. *Mater Trans* 52: 81–89.

102. Pugdee, K., Shibata, Y., Yamamichi, N., Tsutsumi, H., Yoshinari, M., Abiko, Y., and T. Hayakawa. 2007. Gene expression of MC3T3-E1 cells on bibronectin-immobilized titanium using tresyl chloride activation technique. *Dent Mater J* 26: 647–655.

103. Abe, Y., Hiasa, K., Takeuchi, M., Yoshida, Y., Suzuki, K., and Y. Akagawa. 2005. New surface modification of titanium implant with phosphor-amino acid. *Dent mater J* 24: 536–540.

104. Cadotte, A. J. and T. B. DeMarse. 2005. Poly-HEMA as a drug delivery device for in vitro neutral network on micro-electrode arrays. *J Neural Eng* 2: 114–122.

105. Belkasm J. S., Munro, C. A., Shoichet, M. S., Johnston, M., and R. Midha. 2005. Long-term in vivo biochemical properties and biocompatibility of poly(2-hydroxyethyl methacrylate-co-methyl methacrylate) nerve conduits. *Biomaterials* 26: 1741–1749.

106. Indolfi, L., Causa, F., and P. A. Netti. 2009. Coating process and early stage adhesion evaluation of poly(2-hydroxy-ethyl-methacrylate) hydrogel coating of 316L steel surface for stent applications. *J Mater Sci: Mater Med* 20: 1541–1551.

107. Fenelon, A. M. and C. B. Breslin. 2003. The electropolymerization of pryrole at a CuNi electrode: Corrosion protection properties. *Corros Sci* 45: 2837–2850.

108. Mengoli, G. 1979. Feasibility of polymer film coatings through electroinitiated polymerization in aqueous medium. *Adv Polym Sci* 33: 1–31.

109. De Giglio, E., Guascito, M. R., Sabbatini, L., and G. Zambonin. 2001. Electropolymerization of pyrrole on titanium substrates for the future development of new biocompatible surfaces. *Biomaterials* 22: 2609–2616.

110. Rammelt, U., Nguyen, P. T., and W. Plieth. 2003. Corrosion protection by ultrathin films of conducting polymers. *Electrochem Acta* 48: 1257–1262.

111. De Giglio, E., Gennaro, I., Sabbatini, L., and G. Zambonin. 2001. Analytical characterization of collagen-and/or hydroxyapatite-modified polypyrrole films electrosynthesized on Ti-substrates for the development of new bioactive surfaces. *J Biomater Sci Polym Ed* 12: 63–76.

112. De Giglio, E., Cometa, S., Satriano, C., Sabbatini, L., and G. Zambonin. 2009. Electrosynthesis of hydrogel films on metal substrates for the development of coatings with tunable drug delivery performance. *J Biomed Mater Res* 88A: 1048–1057.

113. Taira, Y. and Y. Imai. 1995. Primer for bonding resin to metal. *Dent Mater* 11: 2–6.

114. Smith, N. A., Antoun, G. G., Ellis, A. B., and W. C. Crone. 2004. Improved adhesion between nickel-titanium shape memory alloy and polymer matrix via silane coupling agents. *Compos Part A-Apply S* 35: 1307–1312.

115. Abboud, M., Casaubieilh, L., Morval, F., Fontanille, M., and E. Duguet. 2000. PMMA-based composite materials with reactive ceramic fillers: IV. Radiopacifying particles embedded in PMMA beads for acrylic bone cements. *J Biomed Mater Res* 53: 728–736.

116. Yoshida, K., Tanagawa, M., and M. Atsuta. 2001. Effects of filler composition and surface treatment on the characteristics of opaque resin composites. *J Biomed Mater Res* 58: 525–530.

117. Kanie, T., Arikawa, H., Fujii, K., and K. Inoue. 2004. Physical and mechanical properties of PMMA resins containing γ-methacryloxypropyltrimethoxysilane. *J Oral Rehabil* 31: 161–171.

118. Ferracane, J. L., Berge, H. X., and J. R. Condon. 1998. In vitro aging of dental composites in water-effect of degree of conversion, filler volume, and filler matrix coupling. *J Biomed Mater Res* 42: 465–472.

119. Bexell, U., Olsson, M., Jhansson, M., Samuelsson, J., and P. E. Sundell. 2003. A tribological study of a novel pre-treatment with linseed oil bonded to mercaptosilane-treated aluminum. *Surf Coat Tech* 166: 141–152.

120. Bexell, U., Olsson, M., Sundell, P. E., Jhansson, M., Carlsson, P., and M. A. Hellsing 2004. ToF-SIMS study of linseed oil bonded to mercaptosilane-treated aluminum. *Appl Surf Sci* 231–232: 362–365.

121. Jayaseelan, S. K. and W. J. V. Ooji. 2001. Rubber-to-metal bonding by silanes. *J Adhes Sci Technol* 15: 967–991.

122. Sakamoto, H., Doi, H., Kobayashi, E., Yoneyama, T., Suzuki, Y., and T. Hanawa. 2007. Structure and strength at the bonding interface between a titanium-segmented polyurethane composite through 3-(trimethoxysilyl) propyl methacrylate for artificial organs. *J Biomed Mater Res* 82A: 52–61.

123. Sakamoto, H., Hirohashi, Y., Saito, H., Doi, H., Tsutsumi, Y., Suzuki, Y., Noda, K., and T. Hanawa. 2008. Effect of active hydroxyl groups on the interfacial bond strength of titanium with segmented polyurethane through γ-mercaptopropyl trimethoxysilane. *Dent Mater J* 27: 81–92.

124. Sakamoto, H., Hirohashi, Y., Doi, H., Tsutsumi, Y., Suzuki, Y., Noda, K., and T. Hanawa. 2008. Effect of UV irradiation on the shear bond strength of titanium with segmented polyurethane through γ-mercapto propyl trimethoxysilane. *Dent Mater J* 27: 124–132.

125. Nomura, N., Baba, Y., Kawamura, A., Fujinuma, S., Chiba, A., Masahashi, N., and S. Hanada. 2007. Mechanical properties of porous titanium compacts reinforced by UHMWPE. *Mater Sci Forum* 539–543: 1033–1037.

126. Nakai, M., Niinomi, M., Akahori, T., Yamanoi, H., Itsuno, S., Haraguchi, N., Itoh, Y., Ogasawara, T., Onishi, T., and T. Shindo. Effect of silane coupling treatment on mechanical properties of porous pure titanium filled with PMMA for biomedical applications. *J Jpn Inst Metal* 72: 839–845.

127. Possart, W. 1998. Adhesion of polymers. In: J. A. Hansen and H. J. Breme, eds., *Metals as Biomaterials*, pp. 197–218. New York: Wiley.

128. Worch, H. 1998. Special thin organic film. In: J. A. Hansen and H. J. Breme, eds., *Metals as Biomaterials*, pp. 177–196. New York: Wiley.

129. Xiao, S. J., Kenausis, G., and M. Textor. 2001. Biochemical modification of titanium surfaces. In: D. M. Brunrtte, P. Tenvall, M. Textor, and P. Thomsen, eds., *Titanium in Medicine*, pp. 417–455. Amsterdam, the Netherlands: Springer.

13 Evolution of Current and Future Concepts of Biocompatibility Testing

Menno L.W. Knetsch

CONTENTS

13.1 INTRODUCTION: BIOCOMPATIBILITY TESTING OF MEDICAL DEVICES

The clinical success of medical devices is ultimately determined by their biocompatibility. Medical devices have to possess all the required characteristics to ensure clinically acceptable biocompatibility. The definition of biocompatibility that is most commonly used these days was formulated by D. Williams some years ago as "The ability of a material to perform with an appropriate response in a specific application" [1,2]. This definition takes into account that a biomaterial does not only have a passive, supportive, or mechanical function but also has an important role in guiding the host response toward the biomedical implant. This means that the biofunctionality of the implant is an intricate characteristic of the device. It is unavoidable that the body reacts to a newly implanted material with a wide range of defense mechanisms. This response is known as the foreign body response [3]. Over the past decades, different strategies to obtain acceptable host response to medical implants have been applied. These range from trying to achieve perfect inertness to implants with highly bioactive or biofunctionalized surfaces to actively induce a desired host-tissue response. There has been success with many biomedical implants, but most of these do not carry bioactivity in or on the device [4]. Clinical practice demonstrates that design of a truly bioactive implant is difficult since the exact host-response cannot be predicted. The way in which the tissue will respond to the invading foreign material is dependent on many factors and differs between patients significantly. After a while, a fragile dynamic equilibrium is reached between the materials and the surrounding tissue. The implanted materials will perform well until a slight change in one of the interaction parameters, causing a series of events that will lead to the disturbance of the delicate equilibrium and may result in failure of the device. This also means that the variation between the treated patients makes prediction of the performance of a biomaterial virtually impossible. For instance when considering hip replacement, the quality of the bone in which the stem and cup are fixed are of critical importance for the clinical outcome of the device. One can easily imagine that patients suffering from osteoporosis may have a poorer clinical outcome than patients who do not suffer from this bone disease [5]. Another example that demonstrates the effect of, on first sight, minor details is the performance of mechanical heart valves. The thrombotic complications of metallic and polymeric parts of this medical device are suppressed by anticoagulants and/or antiplatelet drugs. A simple visit to the dentist, e.g., to remove a tooth, will require the reduction of this anticoagulant therapy for a short while. The risk of thrombosis will increase in this short period, although the materials do not change. There is a significant change in one of the parameters that determines the delicate equilibrium between the blood and the blood contacting surfaces of the mechanical valve. Also for patients that have received a coronary stent, patient compliance concerning the anticoagulation therapy has been shown to be the cause of late-stent thrombosis, leading to failure of the device and in worst cases death [6,7]. These examples show that reduction in the number of parameters determining the tissue-implant equilibrium can be beneficial for the clinical success of the medical device. In some cases this means that biofunctionalization of biomaterial surfaces is a viable strategy to increase the long-term performance. In the cases of the stents and heart valves, a coating that would inhibit/suppress surface-induced blood coagulation will not only decrease side effects, like prolonged bleeding, but also would prevent failure of the device because of poor patient compliance concerning the anticoagulation therapy. Therefore, the introduction of bioactivity in theory is a useful characteristic of any biomaterial. We should however not forget that next to the practical problems concerning bioactive biomaterials (sterilization, costs, limited shelf life, storage conditions, etc.), the intended biological response to the implant may differ considerably between patients. Even after thorough testing, the intended functionality of the implant may vary because of the subtle differences of the tissue environment that are encountered in different patients.

Of course, also the intended site of the implants determines the nature of the biomaterials used as well as the biofunctionalization. A hip implant will require totally different materials from a synthetic vascular prosthesis. The first will have to initiate bone formation close to and in direct contact with the surface. This may be achieved by the application of hydroxyapatite coatings [8,9], while a more sophisticated, but not more successful, method is the immobilization of specific growth factors that can stimulate bone growth locally [10,11]. For blood vessel prostheses, the biomaterials have to possess a certain compliance to avoid blood flow disturbances and the surface will have to accommodate vascular endothelial cells to obtain a biofunctionalized surface at the lumen of these devices. One can easily imagine that the nature of the bioactive compounds used for both applications is very different and also the need for the use of bioactive compounds differs greatly. Additionally, the expectation of patients and society in general determines the need for complicated and costly modifications of biomaterials. When patients are satisfied with a hip prosthesis that will function without noticeable problems for 15 years, a well-constructed inert metal implant will be the implant of choice. However, when younger patients require a much longer lifetime of the implant, the choice of implant may be very different, relying on the manipulation of natural processes improving the functionality of the hip prosthesis. For synthetic blood vessels and stents, the problem of biomaterial-induced thrombosis can be quite easily handled pharmacologically, by administering the right amounts of anticoagulant drugs. There is however a downside to this that the long-term use of such drugs can have some undesired side effects, like increased bleeding tendencies or interference of the drugs with other physiological processes, like the effect of coumarins on osteoporosis and vascular calcification [12].

All this means that for a (bio)material to be tested, the intended site of implantation strongly determines the characteristics of the materials used as well as the production parameters. This has to be taken into account when testing the biomaterials and medical devices before they can be applied clinically. The choice of the right method of biocompatibility and biofunctionality testing is therefore critical to obtain any useful information concerning the chance the implant will function satisfactory in patients.

Biocompatibility testing procedures have been developed to predict the performance of a biomedical implant. Most and for all, the biosafety of implants is thoroughly tested. The functionality of the biomedical implants is very hard to predict, if possible at all. The use of animal model systems is nowadays required before any implantation in human patients can be considered. In this chapter, these methods of testing are described extensively and also the value of the different biocompatibility testing methods is discussed.

13.1.1 History of Medical Devices and Biocompatibility Testing

Biomaterials have been used for thousands of years, and one might think that some form of testing has been in place during the same period of time. The discovery of a prosthetic tooth in an Egyptian mummy, dated at 5500 BC, provokes the question if the surgeons of that time would have put an experimental implant directly in a high-society person, or that they first would have tested this implant in slaves or humans belonging to a low caste [13]. This would be the ultimate way of testing (in humans), and one can easily imagine that ethical considerations as we know today would not have been an obstacle for the surgeons in the antiquities. The situation might be a little different in the case of the cranioplasty found in the Inca culture, approximately 1500 years ago [14]. It is still not clear why and under which circumstances the Inca surgeons would have decided to make holes in skulls. Because the Inca had a strong tradition in warfare, skull injuries would be frequently encountered, and closing the wounds the only option to save the warrior. Already then, and still today, wartime situations accelerate the application of new and untested treatments of severely wounded patients, who have nothing to lose and preferable need to return to the battlefield as soon as possible. The Inca surgeons closed skull holes with gold plates, and natural materials like shells, and in many cases the patients survived this treatment. Skulls were found in which the "biomaterial" was well incorporated in the bone of the patient. There is however no information about the quality of life of these patients, so the question if the biomaterial functioned satisfactory cannot be answered, but the lack of signs of severe infection or inflammation near the site of implantation does suggest successful treatment. The Inca were smart enough, or just discovered by chance, to use inert metals to close the skull holes, or to use materials that are close to the structure and composition of bone, like calcium-carbonate shells.

Some of the first detailed reports on the use of biomaterials concern cranioplasty, the repair of large cranial defects. At the start of the sixteenth century, Fallopius described the use of gold plates to replace fractured bone [14]. The first described bone graft for cranioplasty was reported by J.J. van Meekeren, a surgeon in Amsterdam, in 1668 [14]. He described the use of a canine bone graft in the treatment of a Russian nobleman, repairing his cranium successfully with a piece of dead dog's cranium, which was formed to fit the cranial defect exactly. The nobleman recovered but was excommunicated by the church because "the Christian head was too pure to be graced with the bone of a dog." When the nobleman asked for the removal of the piece of dog cranium, the surgeon was helpless because the bone was firmly fused to the patient's cranium. The lack of knowledge prompted the surgeons of those times to perform implantations of foreign body materials into their patients. Although extensive literature is not available, it is highly likely that many failures occurred during these implantations of such foreign materials in patients. Of course, some of the materials and methods were tested on animal models, but since there was no regulation and standardization concerning such experiments, the results were of limited use. Of course, severely toxic materials or just bad surgical methodology were eliminated before they could be used for human treatment.

Several developments in the nineteenth and twentieth centuries, like the discovery of microorganisms, sterilization procedures, and better hygiene conditions in general, improved the success rate of implantation procedures. However, it became more and more clear that some sort of regulation was required to ensure efficient and safe treatment of all patients. In the mean time, the fast increase in animal experimentation, which formed the basis of many of the developed implantation treatments, prompted the British government in 1876 to pass a law on animal cruelty (the Cruelty to Animals Act), fixing, among others, rules for the (more humane) treatment and experimental use of animals [15]. One could see this as the first regulatory action for biomedical experiments on animals. Up to that time, virtually all testing of materials to be applied in humans was done in live animals, under sometimes horrendous conditions. This is the so-called in vivo testing, in a living animal so that also functionality of the implant can be observed and studied. Later in this chapter, we extensively discuss the current in vivo testing procedures.

In the mean time, a number of scientists had been working for a long time with isolated organs, which were kept "alive" and functioning for some time outside of the body. This so-called ex vivo strategy bypassed all legislation concerning animal experimentation and was seen as an alternative for testing procedures [16]. Later in the twentieth century, it became possible to isolate cells of animal, and also human, origin and culture these for some time in the controlled environment of the laboratory. Cells from fast growing tumors were found to be able to grow indefinitely and soon became an easy tool for experimentation in vitro. The construction of immortalized cell lines by molecular genetic techniques, developed largely in the 1970s, gave rise to huge libraries of different cell lines [17]. It soon became clear that it was necessary to organize all these new cell lines into a central facility like the ATCC (American Type Culture Collection) and the ECACC (European Collection of Cell Cultures; now run by the British Health Protection Agency). The testing of biocompatibility soon embraced all new developments and was done in vivo, ex vivo, and in vitro. In the rest of this chapter, the three different forms of testing are extensively discussed, also in the context of different sorts of implants.

13.1.2 BIOCOMPATIBILITY REGULATIONS

During this time of great advance in biology and medicine, the first strict rules on the use of biomaterials in humans or in contact with human tissue came from the field of pharmacology in the form of a U.S. pharmacopeia regulation on the use of plastic cups for tissue, like e.g., blood, collection.

Without going into too much detail, both the FDA and European Union decided that a specific norm for the testing of biocompatibility of medical devices was necessary to tightly regulate and control the quality of safety of implanted medical devices and devices that contact human tissue [18]. First in the early 1980s, the American Association for Testing and Materials (ASTM) published the F748 standard and subsequent refinements over the years. In 1989, the International Standards Organization (ISO) installed Technical Committee (TC) 194 concerning the biological evaluation of medical devices. The resulting regulations and standards were finally published in the ISO-10993 norm that currently contains 20 different parts/chapters, all describing a different facet of biocompatibility testing. A summary of the ISO norm is given in Table 13.1, and the matrix on which the biological evaluation is based is in Table 13.2. The FDA Office of Device Evaluation (ODE) issued a slightly adapted guidance to the biomaterials reviews process, described in the ISO-10993 norm. This means that in general the ISO-10993 norm, and the therein described testing matrix, is the best guideline to decide which biocompatibility tests are required for testing of medical devices and implants. Also a flowchart, adapted from ISO-10993, describing the information and test required to make funded decisions about what steps to take in medical device testing is depicted in Figure 13.1.

It has become clear in the past, that the regulations or biocompatibility testing are ruled by the political situation and the public's request for safety and humane treatment of test animals. The risks associated with the implantation or use of medical devices have to be minimized as much as possible by well-planned testing procedures that utilize the most modern and reliable techniques available. Over the centuries, the increase in scientific knowledge, development of different attitudes toward the treatment of living beings, and changing reluctance to accept safety risks associated with medical procedures have led to a steady evolution of biocompatibility testing concepts.

TABLE 13.1
Different Chapters of the ISO-10993 Norm

ISO-10993-Part	Description
1	Guidance to biological evaluation and testing of medical products
2	Animal welfare requirements
3	Tests for genotoxicity, carcinogenicity, and reproductive toxicity
4	Selection for tests for interactions with blood
5	Tests for in vitro cytotoxicity
6	Tests for local effects after implantation
7	Ethylene oxide sterilization residuals
8	Selection of reference materials
9	Framework for identification and quantification of potential degradation products
10	Tests for irritation and delayed-type hypersensitivity
11	Test for systemic toxicity
12	Sample preparation and reference materials
13	Identification and quantification of degradation products from polymeric medical devices
14	Identification and quantification of degradation products from ceramics
15	Identification and quantification of degradation products from metals and alloys
16	Toxicokinetic study design for degradation products and leachables
17	Establishment of allowable limits for leachable substances
18	Chemical characterization of materials
19	Physicochemical, morphological, and topographical characterization of materials
20	Principles and methods for immunotoxicology testing of medical devices

The different parts describe separate aspects of biocompatibility testing of medical devices and products.

TABLE 13.2

Biocompatibility Testing Matrix Adapted from the ISO-10993 Norm

Device Category		Contact Duration	Biological Effect — Initial Evaluation Tests								Supplementary Tests			
Body Contact		S = Limited (≤24 h); M = Prolonged (24 h–30 Days); L = Permanent (>30 Days)	Cytotoxicity	Sensitization	Irritation or Intracutaneous Reactivity	Systemic Toxicity (Acute)	Subchronic Toxicity (Subacute Toxicity)	Genotoxicity	Implantation	Hemocompatibility	Chronic Toxicity	Carcinogenicity	Reproductive/ Developmental	Biodegradation
Surface devices	Skin	S	X	X	X									
		M	X	X	X									
		L	X	X	X									
	Mucosal membrane	S	X	X	X									
		M	X	X	X		•		•	•				
		L	X	X	X		•	X	X	•			•	
	Breached or compromised surfaces	S	X	X	X		•							
		M	X	X	X		•		•	•				
		L	X	X	X		•	X	X	•			•	
External communicating devices	Bloodstream indirect	S	X	X	X	X	X			X				
		M	X	X	X	X	X	•		X				
		L	X	X	•	X	X	X	•	X	X	X		X
	Tissue/bone/ dentine communicating	S	X	X	X	X	•							
		M	X	X	X	X	X	X	X					
		L	X	X	X	X	X	X	X				X	X
	Circulating blood	S	X	X	X	X	X		▲	X	X			
		M	X	X	X	X	X	X	X	X	X			
		L	X	X	X	X	X	X	X	X	X	X		X
Implant devices	Tissue/bone	S	X	X	X		•							
		M	X	X	X	X	X	X	X					
		L	X	X	X	X	X	X	X				X	X
	Blood	S	X	X	X	X	X		X	X				
		M	X	X	X	X	X	X	X	X				
		L	X	X	X	X	X	X	X	X	X	X		X

The X checked boxes indicate required tests for the medical device. The • indicates optional tests according to the ISO norm, while the ▲ indicates a potential test as described for devices used in extracorporal circulation. Boxes that are not checked indicate these tests are not required or recommended but can be performed voluntarily. For the contact duration S = short-, M = medium-, and L = long-period of contact.

For biocompatibility testing the material(s) or the medical device of interest and some form of living tissue or organism are required. The choice of the materials and isolated cells, tissue, or animal model that should be used is of critical importance for the significance of the results. Some of the experimentation procedures are standardized and are described, e.g., in the ISO-10993 norm. The norms and the value of standardization are obvious when considering extrapolation of results from in vitro, ex vivo, or in vivo testing of biomaterials and medical implants for their biocompatibility.

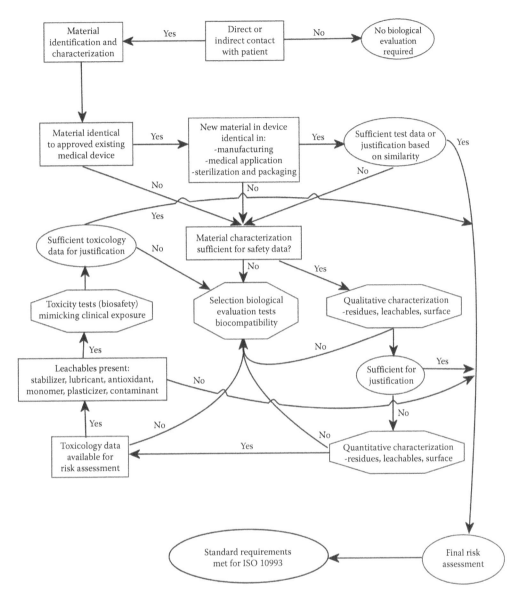

FIGURE 13.1 Flowchart describing decision making in biocompatibility testing. (Adapted from ISO-10993 Part 1; ISO norms can be purchased from www.iso.org)

Of course, the exact definition of biocompatibility declares if the results of such testing make the material or device acceptable for clinical application.

13.1.3 DEFINITION OF BIOCOMPATIBILITY

Biocompatibility is one of those words of which the exact meaning has constantly changed over time. The word *biocompatibility* is obviously composed of the words *bio* and *compatibility*. When analyzing these words, one could come to the definition for biocompatibility as "the capacity of a foreign, synthetic material to be introduced into the body of a (specified) individual without exciting a destructive reaction." This definition is deducted from the biological meaning of compatibility, which was derived from early blood-transfusion experiments in which it was noted that the recipient often demonstrated a violent reaction to the received blood, but this did not occur in monozygotic twins. This observation induced the development of "compatibility test" which should be performed before a transfusion to determine if the recipient had antibodies against the blood of the donor [18,19]. The other definite meaning of compatibility is mutual tolerance. This means that in case of a biological situation both the patient and the implant tolerate each other without any negative responses. After this look at the classical dictionaries, a Google search with the keywords "biocompatibility" and "definition" results in thousands of hits. The definition that comes up with the highest frequency is the definition that appears in the *Williams Dictionary of Biomaterials* [2]:

> The ability of a material to perform with an appropriate host response in a specific application.

Biocompatibility is a term that is still frequently used to describe the biosafety of a synthetic material that is destined to function in contact with tissues, organs, and cells of a living patient. Many scientists and students still feel that biocompatibility comprises the somewhat older definition:

> The capacity of blood, tissue, or an organ to be introduced into the body of a (specified) individual without exciting a destructive reaction.

Therefore, most of biocompatibility testing over the last 50 years has concentrated on testing for toxicity of the materials. Often this was done in vitro using immortalized cell lines, and still today many new biomaterials are presented, accompanied by data from in vitro cytotoxicity assays. This is of course invalid concerning the biocompatibility of the material or medical device, since functionality of materials or medical device is not tested. Therefore, the misconception that toxicity, at least the absence of toxicity, is what determines biocompatibility is still present in the biomaterials field. Especially David Williams has been instrumental in educating the biomaterial scientists in proper biocompatibility thinking. Recently, he also fine-tuned the definitions of biocompatibility, taking into account the upcoming field of tissue engineering. Additionally, the effects of long-term implants were not clearly defined in previous definitions, and so the slow and systemic effects were incorporated in the current biocompatibility definition, which now reads:

> Biocompatibility refers to the ability of a biomaterial to perform its desired function with respect to a medical therapy, without eliciting and undesirable local or systemic effects in the recipient of beneficiary of that therapy, but generating the most appropriate beneficial cellular or tissue response in that specific situation, and optimizing the clinically relevant performance of that therapy.

The above definition implies that a range of biological and physiological tests have to be performed before a biomaterial or medical device can be labeled as biocompatible. Not only in vitro toxicity measurements but also in vivo functionality and systemic toxicity determination have to be combined to make any funded statement about biocompatibility. In addition, the definition indicates that the current testing for biocompatibility will always be flawed by the use of some model system or model organism used for the initial testing. The simplification in testing and the proposed extrapolation from these simplified systems to the human situation make the decision on biocompatibility scientifically unreliable. It means that the definitive answer on the question if a biomaterial or medical device is biocompatible can only be gained from the clinical use in the patient.

13.2　BIOCOMPATIBILITY TESTING

Before a medical implant can be applied clinically, a series of tests have to be performed to ensure safe and efficient treatment of the patient with the medical device. The exact tests that have to be performed are laid down in a number of regulations of which the ISO-10993 norm is the most referred to norm (Table 13.2). However, the exact regulations for testing of biocompatibility differ somewhat between different countries, but we will not go into these minor differences here. It is good and useful to know that biocompatibility testing regulations differ slightly between for instance the United States and Europe. The tests that will have to be performed for a new biomaterial or medical device can be deduced from a flowchart that is given in Figure 13.1. This flowchart is an adaptation of the chart that has been published earlier in ISO-10993 part 1. One of the shortcomings of the chart is the absence of tissue-engineered products, but this class of bioengineered products is so different from the "classical" medical devices that a completely new set of tests is required for the biocompatibility testing of these products. One should not forget the fact that almost all tissue-engineered products contain some form of synthetic material, which is necessary as a scaffold to shape the engineered tissue as well as giving it the mechanical strength for the period of constructing the tissue in a biocontainer. Consequently, in the end also the scaffold or the remains of the scaffold are transferred into the patient's body. This means that, in principle, the scaffold materials should undergo the whole range of biocompatibility tests. It is however not completely clear what regulation will be applied in case of tissue-engineered products with scaffold still in. Next, ex vivo testing is performed for a range of biomaterials. One could think here about perfused organs or working with freshly obtained tissue, like cornea or pieces of skin. Another interesting option is to attach blood-contacting devices, in a flow setup, to extracorporal circulatory machines. In this way, a longer testing time can be achieved than with freshly obtained blood, which will deteriorate within hours after collection, thus limiting the window of testing blood-contacting devices ex vivo. The final step in testing for biocompatibility is of course implantation in an animal model, so-called in vivo testing. The way in which animal testing has been performed over the centuries strongly depended on the political situation. After the acceptance of the animal cruelty act (Great Britain in 1876), the use of test animals became more "humane." This is also one of the key mistakes that has been made in the early years of controlled animal testing, the fact that animals would make a reliable source of information about the performance, the biocompatibility, of a medical device or implant in human patients. The choice of the animal model as well as other parameters chosen in the experiment strongly influence the outcome of the test. The goal of implantation can range from relatively straightforward biosafety testing to functionality testing of the implant or device. Recently, there has been some discussion about the value of this animal testing since the extrapolation to the human situation has been shown to be very difficult and unreliable in many cases [20]. A well-known example is synthetic small-diameter vascular prostheses [21]. These will perform satisfactorily in animal models like dogs or pigs, but they perform with an extremely high failure rate in humans. Some of the considerations for in vivo testing are discussed later in this chapter.

13.3 IN VITRO BIOCOMPATIBILITY TESTING

In vitro biocompatibility tests are mainly concerned with safety of materials as well as adverse reactions of body fluids like blood. Immunological tests like irritations and long-term hypersensitization cannot be performed in vitro because no reliable alternatives for the complete immune system are available. In addition, experiments concerning the long-term systemic toxicity in a three-dimensional (3D) environment of tissue can currently not be reproduced in in vitro tests. Although there are now various bioreactor systems, in which it is possible to grow cells in a 3D environment on a carrier material, this is only useful for the testing of potential scaffold materials. But the further development of these 3D systems might in the near future lead to the "implantation" of a material in a small piece of 3D cell mass, followed by incubation for a chosen period of time. At the moment, only very small pieces of tissue can be produced in vitro, up to a couple of cubic centimeters. This of course is not of much use for material testing in vitro, but it makes for an outlook in which in vivo biosafety testing might be minimized, saving a large number of animals and also reducing the total costs of biocompatibility testing. Changes to the genetic information in the DNA can be studied relatively easily and reliably in simple laboratory tests. The underlying message is that in vitro testing will never suffice for medical devices. The real answer on the biocompatibility of a medical device is optimally studied in animal models and ultimately in human patients. Of course, in vitro tests are useful to filter out really "bad" materials that have very strong effects as toxins, mutagens, or carcinogens. However, one should remember that materials are tested in an environment that does not resemble the conditions they encounter after implantation. The presence of proteins, extracellular matrix proteins, enzymes, and cells of the immune system makes the environment of medical devices highly complex and the in vitro reaction of the materials or substances that leach from the materials unpredictable.

Therefore, the in vitro biocompatibility testing should be seen as the attempt to minimize the risk for the patient and increase biosafety. One can identify several levels of in vitro biocompatibility testing, namely:

1. Cytotoxicity—the effect of a synthetic material or the substances that leach from this material on stable cell lines in liquid culture media.
2. Cell growth—do cells attach and proliferate on the test material?
3. Functionality—does the material influence the "normal" cellular functions of differentiated cells?

These three main levels of in vitro testing are discussed in detail.

13.3.1 CONTROL MATERIALS

For biocompatibility testing, in general, the use of controls is vital for critical assessment of results. In biocompatibility testing, a comparison to positive (toxic) and negative (nontoxic) controls is necessary. The positive control demonstrates the sensitivity of the used system, and can also be used as a background measurement. The fact that cells, tissues, or animals will show a noncompatible response demonstrates that an appropriate model was chosen. Also for quantitative measurements, especially in vitro, the signal obtained with the positive control can be considered the background level of the assay used. The negative control can be compared to mock operations when assessing the efficiency and functionality of an implanted medical device or to the placebo in pharmacological clinical testing. A good negative control should have the same shape and physical properties as the tested material. Therefore, a negative control for a ceramic powder should also be a powder variant of standard material. A hydrophilic polymeric material should not be compared to poly(tetrafluoroethylene), which is highly hydrophobic. Finally, a degradable polymer should preferably be compared to a degradable polymer, or two different control materials should be used: one for the bulk, nondegraded state and one for the soluble degradation products.

A variety of materials is being used as control materials in (cyto)toxicity testing, but the ISO-10993 norm describes organo-tin stabilized poly(vinylchloride) as the preferred positive control, although many scientists use phenol or ethanol dilutions as such. The use of liquid materials is not preferred, because the prescribed extraction conditions are not mimicked in the positive control. As negative control materials, high- and low-molecular weight polyethylene are recommended [22]. These control materials represent the far ends of the biocompatibility scale. There is no reliable scale of materials that represent intermediate biocompatibility or toxicity. This leaves the field open to interpretation and subjective assessment of the results of biocompatibility tests. Several attempts on producing standardized materials with intermediate toxicity have been made, but the use of such a scale is unverified and remains to be determined. It is however clear that for in vitro biocompatibility testing a series of standard materials ranging from highly cytotoxic to noncytotoxic would improve objective evaluation of biocompatibility testing. Additionally, the interlaboratory variation between results and interpretation of results could be decreased. The biggest problem is how this range of standard materials correlates to in vivo biocompatibility performance. The choice of animal model and surgical procedures may significantly influence the biocompatibility and the performance of the materials in vivo. So the search for a measuring stick to create an objective biocompatibility score is still ongoing.

13.3.2 CYTOTOXICITY

The ISO-10993 part 5 norm describes the correct way to test the effect of biomaterials and medical devices on the viability of cells in vitro. The determination of cytotoxicity is one of the first tests that are performed after obtaining a new material that can be incorporated into a medical device. In addition, when a material is manufactured using a different procedure, when the manufacturing or sterilization protocol is changed, a simple start is to test the in vitro cytotoxicity. First, one should make a clear division in direct and indirect testing.

In direct testing, the living cells make direct physical contact with the test material, while in indirect testing the cells are merely used as a sensor to detect possible toxic substances released from the material. Also, indirect testing can be used to determine the toxicity of degradation products of (bio)-degradable biomaterials. The committee recognized that the preparation of the materials to be tested is of critical importance for the results of in vitro toxicity and compatibility testing. Therefore, separate parts of the ISO-10993 norm were created to specify conditions by which material extracts and degradation products should be produced. The ISO-10993 part 12 contains protocols describing the extraction conditions for obtaining leachables from biomaterials and medical implants. The extraction media, incubations times and temperatures, storage conditions, etc. are all specified. The ISO-10993 part 9 was included to specify identification and quantification of degradation products and describes how degradation products should be obtained. The correctly obtained material extracts or degradation products can then be used in standard cytotoxicity testing. This emphasis on carefully chosen conditions for sample preparation before in vitro biocompatibility testing is of critical importance to get to an objective and transparent testing procedure that mimics clinical conditions as good as possible. This should ensure results with optimal information concerning in vitro biocompatibility.

13.3.2.1 Indirect Testing

ISO-10993 is very clear in how indirect cytotoxicity testing should be performed and how the leachables from the test materials should be extracted (Figure 13.2). In the first possible cytotoxicity assay, a piece of material is extracted in for instance cell growth medium. This extract is then added to a semiconfluent cell layer and incubated. After the desired incubation period (>24 h), cells are analyzed for cell viability. Another possibility is that the material is placed on top of a filter (pores approximately 1–5 μm), which fits perfectly in the cell culture dish and keeps

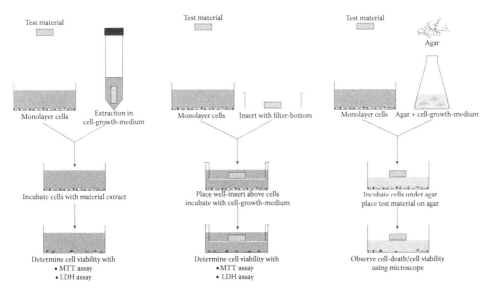

FIGURE 13.2 (See companion CD for color figure.) Overview of indirect-contact cytotoxicity testing. On the left is depicted cytotoxicity testing of material extracts that are separately isolated by incubating a sample of the material in an appropriate extraction medium. The middle part shows indirect incubation with the test material, the leachables can diffuse through the filter, and the cells will detect cytotoxic components. On the right, the material and cells are separated by a layer of soft agar through which the extractable compounds from the material can diffuse. A zone of cytotoxicity around the material may appear since diffusion is limited.

the test material a little distance above the cells. After incubation, the insert and test material are removed, and cell viability is determined. The last method shown in Figure 13.2 is an agar overlay assay. Cells are overlaid with a layer of (soft) agar, and the test material is placed on top of this agar. The leachables then diffuse though the agar and reach the cells. Here, the viability of cells is examined microscopically, and a zone of nonviable cells just beneath the test material might be observed. The extracts should be obtained in such a way that the extraction conditions exaggerate the conditions of clinical use. One should be careful as not to change the physical properties of the material during the testing, so fusion, melting, and alteration of the chemical structure should be avoided. The choice of the extraction medium also limits the extraction conditions. When choosing a cell culture medium with serum, the use of elevated temperatures above 50°C is not recommended to avoid denaturation and/or deactivation of important compounds from this serum. In such a case, a prolonged extraction at 37°C is most appropriate, for instance for more than 24 h and up to 120 h. When an accelerated extraction is required, higher extraction temperatures, 50°C or 70°C, can be used, but then only culture medium without serum or a buffered physiological salt solution are useful.

Although these conditions are described in the ISO norm, one has to be cautious about such extraction conditions since the presence of proteins in the extraction medium can greatly influence the extraction kinetics and the amount of substances leaching from the material. The next parameter important for the extraction of leachables is the material/medium ratio. How much of the material should be incubated in which volume of medium? There is no straightforward answer to this, since the shape and the physical structure of the test material are vital for the extraction efficiency. A solid piece of pMMA will have much slower extraction kinetics than a porous polyurethane piece. A rule of thumb describes the use of 6 cm^2/mL of extraction medium. In practice, this is sometimes impossible to calculate, as for porous materials, or materials that are powdered or broken up into nonuniform particles. Then for sheets of material, one could easily calculate the total surface area, but the thickness or bulk of the material will also greatly influence the total concentration of extractable substance in the extract. It is therefore clear that great care has to be taken during the extraction procedure. A guideline for the amount of material that has to be used for extraction of potential toxic chemicals, adapted from ISO-10993-12, is given in Table 13.3.

For the testing of cytotoxicity of extracts, the choice of cell line is not that critical. The cell in this case is only used as a sensor, albeit a living sensor, for toxins that influence the metabolism and integrity of the cell. The most commonly used cells are mouse fibroblasts, largely because these are easy to maintain in the laboratory at low cost. In addition, these fibroblasts are sensitive to a large array of chemical compounds, which is essential for the cell to be a useful toxin sensor.

TABLE 13.3
Amount or Surface Area to Volume Ratio Required for Proper Extraction of Test Materials according to the ISO-10993 Norm Part 12

Thickness of Material (mm)	Extraction Ratio ± 10% (Surface Area or Mass per Volume)	Examples of Test Materials
≤0.5	6 cm^2/mL	Synthetic polymers, ceramics, metals, sheets, composite films, tubings
>0.5	3 cm^2/mL	Synthetic polymers, ceramics, metals, composite films or tubings, preformed implants
≤1.0	3 cm^2/mL	Natural elastomers, latex, rubber
>1.0	1.25 cm^2/mL	Natural elastomers, latex, rubber
Nonregular	0.1–0.2 g/mL or 6 cm^2/mL	Microspheres, nanospheres, pellets, powders

13.3.2.2 Direct Contact Testing

For direct contact testing of biomaterials and medical devices or implants, a somewhat different approach must be taken. First of all, the fact that most used materials are not opaque makes them unfit for use in standard microscopic analysis. Also in direct contact testing, it is vital that the cells attach well to the material. This is important to realize because the lack of cells on a particular material can be caused by toxicity of the surface, inhibition of proliferation, or by lack of cell attachment. Direct contact testing should therefore be combined with testing of material extracts as described earlier. Also, the test material should first be extracted, so that no more toxic compounds leach from the material. Then the "clean" surface of the material can be tested, and the cells will demonstrate the effect of the surface on cell viability [23].

If this is not done properly, no real additional information is gained from performing such a direct-contact biocompatibility assay. The topography, charge, free energy, and the physicochemical composition of the surface all play a role in the interaction with cells and tissues. In addition, a pretreatment of the test material with proteins, i.e., medium including serum, can be of importance since protein adsorption may dramatically alter the interaction of cells with surfaces. In the in vivo situation, this also occurs, and thus it is advisable to mimic the in vivo situation as closely as possible. For instance, for a synthetic blood vessel the luminal surface of the device optimally will get covered with functional endothelial cells to prevent thrombotic complications. The "naked" surface of the device will never be encountered by the endothelial cells since plasma proteins will rapidly and irreversible adsorb to this surface. For in vitro testing of material endothelialization, plasma proteins should be allowed to adsorb to the surface prior to endothelial cell interaction. Only then will the situation as encountered after implantation be mimicked appropriately. It has been shown in several studies that the adsorption of proteins can dramatically change the behavior of cells on synthetic surfaces.

Direct contact testing can be performed in several ways, of which two are most popular (Figure 13.3). First, the test material can be placed on top of a (semi-)confluent cell layer, and

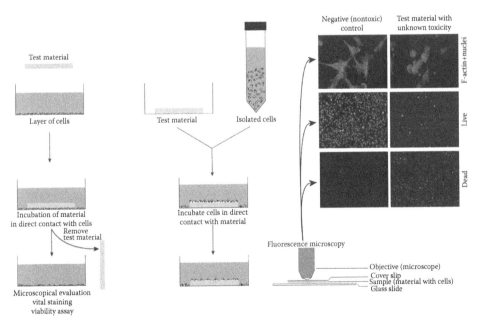

FIGURE 13.3 (See companion CD for color figure.) Direct-contact testing in vitro of test materials of medical devices. Materials are either placed on top of or act as a support for cells. After the appropriate incubation time, cells can be analyzed for cell viability, morphology, or cell function. Fluorescent techniques are often used because of the opaque nature of most synthetic materials. The cytoskeleton of attached cells can be visualized by staining with fluorescently labeled phalloidin (right top panel). Cells can also be analyzed for viability or cell death (right middle and bottom panels, respectively).

after incubation the viability and functionality of the cells directly underneath the material can be analyzed. Another, more elegant method is to place the cells directly on the test material. In this way, the cells will not obtain mechanical damage or suffer from nutrient or oxygen depletion due to limited diffusion underneath the piece of test material. Then, the analysis of the attached cells is problematic. In most cases, standard microscopic observation of the cells will not be possible due to the bad optical qualities of most medical devices and materials. However, the development of fluorescent probes and immuno-histological techniques overcomes this problem (Figure 13.3). Also, the cells can be isolated from the material and the differentiation and functionality can be tested by studying protein expression or cell-specific enzymatic activities.

Overall, a combination of direct and indirect contact testing will deliver the most complete and reliable information concerning cytotoxicity and support of cell-type specific function. This is the first step in biocompatibility testing, and these in vitro techniques will be developed further to improve their predictive power and extrapolation to the in vivo situation. One should however always be cautious with the extrapolation since the medical device is only one part of the patient's treatment, and the actual surgical procedure is another vital part. Often this is disregarded in biocompatibility testing, and devices are produced that are impossible to use properly in a clinical setting.

13.3.2.3 Cell Viability Assays

Cell viability is tested as a general marker for toxicity of a material or substance. The most widely used cell viability tests are based on conservation of mitochondrial integrity and activity. Loss of mitochondrial activity is one of the crucial steps during necrosis and apoptosis of cells. Therefore, chemicals like MTT (3-(4,5-Dimethyl-2-thiazolyl)-2,5-diphenyl-2H-tetrazolium bromide) and XTT (Sodium 3,3,-(Phenylamino)carbonyl]-3,4-Tetrazolium-Bis(4-methoxy-6-nitro) benzenesulfonic acid hydrate) are used as viability agents. These compounds are converted by the succinate dehydrogenase in the mitochondrion in an insoluble colored or fluorescent dye, respectively (Figure 13.4) [24].

Another variation on cell viability test is based on cell membrane integrity. The leaching of the enzyme lactate dehydrogenase (LDH) from treated cells can be used as a marker for cytotoxicity. The amount of LDH in the culture medium is then a measure for the (non-)viability of the cells.

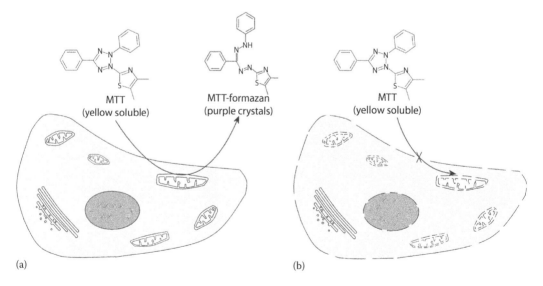

(a)   (b)

FIGURE 13.4 MTT is converted by viable cells (a) into a purple, water-insoluble compound that can easily be quantified by microscopy or spectrophotometry. Nonviable cells (b) will not convert MTT because mitochondrial activity has been lost.

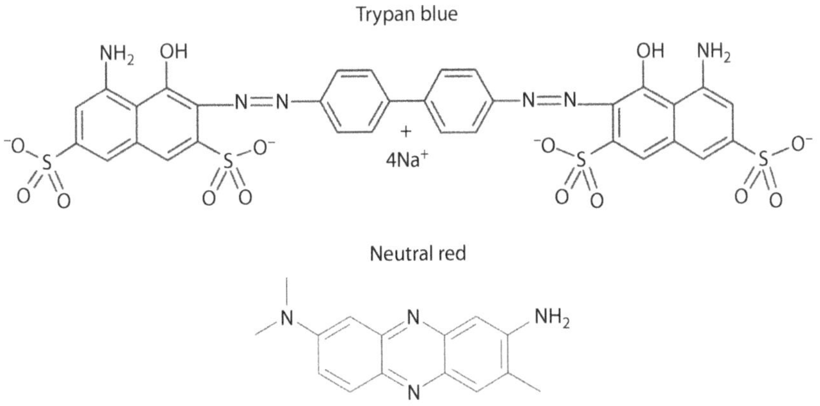

Trypan blue

Neutral red

FIGURE 13.5 Chemical structure of the viability stains trypan blue and neutral red. Trypan blue is excluded from viable cells and only stains necrotic cells, with compromised plasma membrane, blue. Neutral red is exclusively incorporated in viable cells.

A known disadvantage of this method is that many sera used for supplementing culture medium have a high basal level of LDH. The basal level of LDH activity will therefore be high and the sensitivity of the LDH-viability assay low.

The advantage of the MTT type dyes is that it is relatively easy to quantify cell viability of a large number of samples in a short period of time. Of course, the method critically depends on the use of suitable positive (toxic) and negative (nontoxic) controls. Also, MTT staining can be used for staining of tissue-engineered constructs and cells grown on top of medical devices in a 3D environment. Simple bright-field microscopy can then reveal viable cells, or by lysing the cells and dissolving the formed purple formazan, a measure for the number of viable cells can be obtained.

Other frequently used compounds to determine cell viability are the so-called viability stains or vital stains. The most widely used of these dyes are Neutral Red and Trypan Blue (Figure 13.5). Trypan Blue, which is negatively charged, cannot permeate through an intact plasma membrane of viable cells. When cell viability is lost and the plasma membrane gets compromised, the dye can enter the cell and the cell will take on the color of this blue dye. Trypan blue is often used in standard cell culture to determine the viability of cells. Cells that take up a blue color have a compromised cell membrane and should be considered as nonviable. Neutral red however is only taken up by viable cells and stored in lysosomes. Standard microscopic examination can then be used to determine the level of cell viability in a population of cells.

In the last few decades, a large number of more sophisticated fluorescent dyes were developed to assess cell viability [25]. These dyes can freely diffuse into the cell and are converted by intracellular enzymes into a fluorescent and membrane-impermeable compound inside the cell. Therefore, when cells are challenged with such a dye, only the fluorescent cells are viable in the population. The fluorescent dye Calcein AM is a compound that becomes fluorescent when hydrolyzed by intracellular esterases in intact, viable cells (Figure 13.6). The amount of observed fluorescence is a direct measure for the number of viable cells. Cells that have a compromised plasma membrane will not be able to retain the intracellular esterases as well as the formed polyanionic product of Calcein-AM hydrolysis.

Thus, nonviable cells will not be fluorescently labeled. The quantification of cell viability using these types of viability dyes is a bit more laborious since the fraction number of fluorescent cells has to be determined using flow cytometry. For this, the cells have to be released from the tissue culture plate and each sample has to be analyzed separately, which results in a rather time-consuming analysis. Also, a number of cells may get damaged during the procedure, influencing the measurement in this way, increasing the number of nonviable cells.

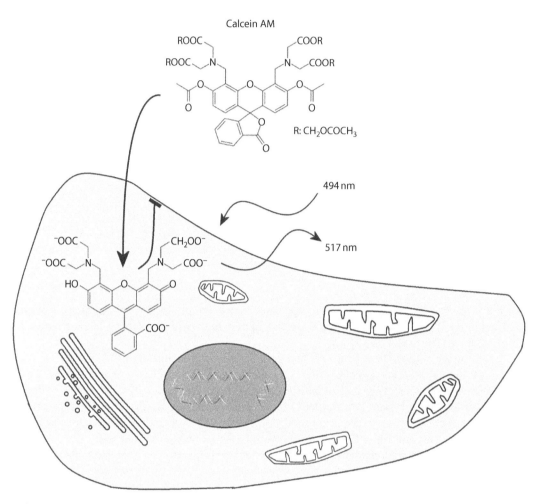

FIGURE 13.6 Calcein AM staining of viable cells results in uptake and hydrolysis of the dye. The result is an anionic fluorescent compound now trapped inside the cell. The excitation wavelength for detection is 494 nm and the emission 517 nm.

Another variation of these assays that are based on membrane integrity makes use of fluorescent dyes that are membrane impermeable. Only nonviable cells with a compromised plasma membrane will incorporate the dye and be visible under the appropriate wavelength of light. Since exclusively nonviable cells are stained by these dyes, they are often called "dead-dyes." Many of these dyes are nucleotide-binding dyes, like propidium-iodide or ethidium bromide analogues, that dramatically increase their fluorescence upon binding of DNA, which is normally unreachable inside the nucleus (Figure 13.3). With FACS analysis, the proportion of fluorescent cells is a direct measurement for the percentage of dead cells in a population.

Mostly the cell viability dyes are used as a rapid qualitative method to observe large differences in cell viability. Using two separate dyes, one viability dye like Calcein-AM, combined with a nonpermeable dead dye (e.g., propidium iodide or ethidium-homodimer 1), a rapid evaluation on the cell viability in a population of cells can be made using a standard fluorescence microscope setup (Figure 13.3).

All these viability dyes have in common the fact that they are most suited for qualitative analysis of a cell population. Quantitative analysis using these dyes is possible but can be time-consuming and can suffer from a high background level of dead cells due to the procedure. Therefore, for quantitative analysis, the metabolic dyes are more practical since an accurate measurement of cell viability for a large number of samples can be determined easily and rapidly.

13.3.2.4 Evaluation Cell Viability or Cytotoxicity

Determining the level of cell viability or cytotoxicity in a population of cells is rather straightforward, but the interpretation of the results is far from that. Where the ISO-10993 norm lacks detail is in the way in which the results of cytotoxicity assays should be interpreted. For qualitative assays, like microscopic examination of cells that have been in contact with toxic materials or solutes, a scale describing exact toxicity levels is lacking. The interpretation is now completely dependent on the subjective interpretation of the researcher analyzing the cells. The interpretation should be classified on the cytotoxicity scale as 0 = noncytotoxic, 1 = mildly cytotoxic, 2 = moderately cytotoxic, and 3 = severely cytotoxic. The problem of this qualitative evaluation is that there are no strict rules describing when a material is mildly cytotoxic or moderately cytotoxic. What percent of the cell population needs to be dead or nonviable for the material to be considered nonsafe and cytotoxic? This problem also arises for the quantitative assays. The results can be determined objectively, a percentage of cell viability of a standard control material. The question that remains is when does one consider a material toxic? Is a material that displays a cell viability of 75% of the negative control cytotoxic? What is missing here is a series of materials that have intermediate cytotoxicity (i.e., biocompatibility), from which is known how they behave both in vitro as well as in vivo concerning biosafety.

Obviously, the functionality of a material, which is an intricate part of biocompatibility, cannot be determined with these relatively simple toxicity tests. To conclude, in vitro cytotoxicity and viability testing is very important in determining the biosafety of a material or medical device, but the interpretation of the results is not fully standardized. This makes it very hard to compare the in vitro biocompatibility data between labs and even in the same lab between different investigators. There is a clear need for a series of standard materials that cover a range from severely cytotoxic to noncytotoxic, to standardize interpretation of in vitro biocompatibility and cytotoxicity tests.

13.3.3 Cell Proliferation In Vitro

Next to viability, the proliferation rate of cells can be of importance for the performance of implanted synthetic materials. Especially for materials that should allow regeneration of cells and tissues, cell growth is an important parameter. There are numerous ways to determine cell proliferation on top of synthetic materials. Most are based on the increase in DNA of biomass. The simplest method to determine cell proliferation is counting cells in a well-defined area or volume and compare the cell counts at different time intervals. The advantage is that except for a microscope, no equipment is required. When cells are cultured on a translucent material, the cell number can be directly determined in a defined field of vision and recorded over time. In case the cells are attached to an opaque material, the cells have to be released from this material and then counted. The disadvantage is that the efficiency of harvesting the cells prior to cell counting is crucial for the accuracy of the measurement. Therefore, this method is implemented as a crude indication of cell proliferation in general. Other classic cell proliferation assays were performed by addition of nucleotide analogues, both radioactive and nonradioactive. The incorporation of for example ^{3}H-thymidine over time is a direct measure for the amount of DNA synthesis and directly proportional to cell proliferation. The radioactive compound (^{3}H-thymidine) was slowly replaced by the nonradioactive 5-bromo-2' deoxyuridine (BrdU) because of practical considerations. The comparison of cells in contact with a standard material (toxic or nontoxic) demonstrates the rate of cell proliferation on the studied materials or implant. The setup of the experiment is the same as for cell viability experiments as shown in Figure 13.3. Another method to determine the increase in genetic material inside a cell population is based on the use of DNA-dependent fluorescent dyes. For these, cells must be lysed so that the dye can bind tightly to DNA and become fluorescent. The amount of fluorescence can be compared to a standard curve, linking fluorescence to cell number. One should of course beware of the difference between cell types and the standard curve should always be of the identical cell type (or mixture of cells) that is being studied. Another method to determine the number of cells undergoing cell division and thus proliferation is to study cell-cycle markers. One of these markers is the Ki-67 protein, which is only produced by cells undergoing cell division. The number of cells positive for this marker is a direct measure for cell proliferation in the population of cells. The more Ki-67 can be observed, the more cells are going through the cell cycle and cell division, so the higher the proliferation rate of this population of cells. Another method uses the activity of intracellular enzymes measured over a period of time to determine an increase in cell number. The assumption is that the number of, for instance, mitochondria per cell does not change over time and does not change by contact with a synthetic material or its soluble compounds. The most used substance here is the MTT dye that is frequently used for cell viability measurements. The disadvantage of the use of these metabolic dyes is that one does not actually determine the number of cells, but the relative number of living cells compared with a control or standard. For the answer if a synthetic material or a medical device influences the proliferation of a certain cell type, the results of all the described assays is satisfactory. One has to keep in mind that loss of viability and loss of proliferative potential are two different things. Cells may lose their ability to proliferate, to multiply by cell division, without losing their viability. Therefore, a combined approach determining the effect of biomaterials or their leachables on both viability and proliferation is preferred.

When screening a large library of different synthetic compounds, the cells used are just a sensor for the ability of the material to allow for cell proliferation. In that case, one may choose a cell line that is intrinsically fluorescent. This fluorescence can be achieved by incorporation of a fluorescent protein, i.e., green-fluorescent-protein (GFP). The gene encoding this protein is put under control of a constitutive promoter so that all the cells are brightly fluorescent. These cells are then incubated with the biomaterials to be tested, and the change in fluorescence over time is a direct measure for the proliferation rate. One can imagine that this method is easily applicable for large-scale and high-throughput screening of libraries of materials. Of course, one remark that has to be made is that the lack of fluorescent cells on a material can be caused by loss of viability, lack of proliferation, or the inability to adhere to the material. The problem with all the described proliferation assays is that the recorded increase in cell number is not only dependent on the rate of cell division but also on the attachment of the cells to their solid support. Cell growth is the addition of proliferation and attachment. The attachment of cells to a test surface can be easily determined. A high density of cells is incubated with the material for a short time so that no significant proliferation takes place. The number of attached cells can be determined as described here. In some studies, the force needed to detach cells from the surface is determined as a measure for the strength of attachment. The force needed to remove the cells is proportional to the attachment force of the cells for that specific material surface.

13.3.4 Cell Function In Vitro

The next level of in vitro biocompatibility testing is functionality of cells in contact with biomaterials (Figure 13.7). One can imagine that a material that is intended as a filler of gaps in bones should be nontoxic, support the growth and attachment of the bone-forming cells (osteoblasts), and should not interfere with the bone-forming characteristics of these cells. The state of differentiation of the cells should not be influenced by the materials, so no re- or de-differentiation should occur. This means that the production of functional proteins is a vital part of biocompatibility of materials in vitro. In the case of osteoblasts, the formation of calcified extracellular matrix is the important functional parameter. There are several different methods to determine whether the right set of genes gets expressed. The steady developments in the field of molecular genetics and biology are exploited to facilitate the biocompatibility studies. In the early days of gene-expression determinations, mRNA was isolated from cells in contact with biomaterials, and the expression of the gene of interest was determined using the northern blotting technique [26]. This method is rather time-consuming because there are a large number of steps involved before a signal is obtained. The development of polymerase chain reaction (PCR) in combination with the use of reverse transcriptase (RT) resulted in the RT-PCR technique. In the earlier version of this technique, a low amount of RNA was converted to DNA using RT and then used as a template in the subsequent PCR technique. By choosing the right primers for the reactions, a specific set of genes can be amplified. The produced DNA was analyzed on agarose gels followed by DNA staining. The quantification of this technique was rather inaccurate, and this problem was tackled

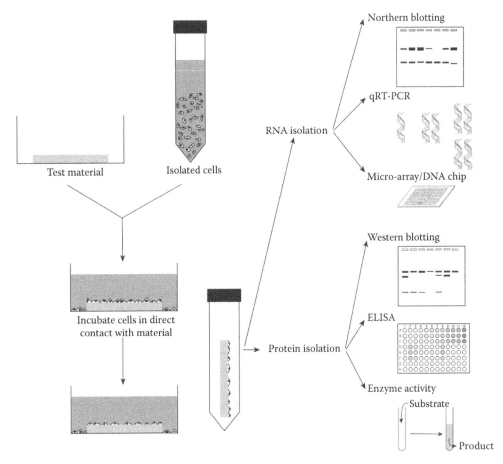

FIGURE 13.7   (See companion CD for color figure.) Scheme depicting possible test for cell function in vitro. Cells in direct contact with the test material can be analyzed for the presence of cell-type specific gene expression and protein production.

by the development of quantitative PCR (qRT-PCR). The incubation of fluorescent nucleotides during the PCR reaction, coupled to a sensitive fluorescent measurement of the produced DNA, is a real-time measure for the amount of mRNA present in the original sample. So comparing the cells to the appropriate controls delivers detailed information to the relative gene expression of cell-type specific genes. A more recent method to determine the level of gene expressions of cells is to make use of DNA microarrays. These libraries of DNA sequences, printed or synthesized on glass chips, are ideal for comparing mRNA profiles in a large variety of cell-types. In one reaction, a large number of different genes can be studied, so there is a massive amount of information available in one experiment. At this moment, the commercially available DNA arrays are rather expensive, but they can be reused. Also only one sample can be incubated with one array at the same time. The bioinformatics attached to DNA array analysis is improving all the time, and also prices for arrays are coming down. Therefore, it is reasonable to assume that most of the analyses of gene expression in biocompatibility testing will be performed by microarray in the near future. Different microarrays will become available for analyzing the transcriptional activity of a certain cell types, enabling a rapid estimation of the differentiation state of the cell.

The advantage of these molecular genetic methods is that they are fast and, in most cases, a large number of samples can be analyzed simultaneously. The information is also rather direct, and relatively small differences in gene-expression can be detected. The disadvantage is that the produced mRNA is not the functional component in the cell. The protein that is produced from the RNA is what determines the functionality of the cell. The change in mRNA is not always directly and proportionally reflected in the production of the encoded protein. From this point of view, the study of proteins to determine the differentiation state of the cells is much more direct and sensible.

The proteins produced by a cell can be analyzed by western blotting or ELISA (enzyme linked immunosorbent assay) methods that are based on the use of specific antibodies. For western blotting isolated proteins are separated by molecular weight and the protein of choice is visualized by the specific interaction with an antibody [27]. For ELISA, the protein of interest is caught out of a crude protein sample by a surface-attached antibody [28]. The immobilized protein can now be quantified with a second specific antibody and a standard curve. This analysis of the cell's protein make-up gives a direct representation of the differentiation and functionality of the cell. The development of proteomic sciences has greatly improved the determination of protein identities over the last decade. The production of reliable antibody-, protein- or peptide arrays is still not possible but is expected to be available over the next years. Proteins can then be identified on the basis of specific interaction with an antibody or peptide sequence on a glass chip, very similar to that described for the DNA array. For the moment, however, this technique is not fully developed, and the analysis of protein production by the classical techniques is preferred. In addition, the composition of deposited extracellular matrix is only possible by these methods. The production of the right proteins and enzymes is a direct measure of the functionality of the cells and therefore also for the biocompatibility of the tested material.

These described assays can be performed in an array setup, so that a large number of materials can be tested simultaneously. High-throughput screening will become more and more available when some remaining technical problems have been solved.

13.3.4.1 Evaluation of In Vitro Functional Biocompatibility Tests

One of the drawbacks of in vitro biocompatibility testing is to assess the relevance of the obtained results to the in vivo, the in patiento situation. It is of course important to show that a cell is remaining functional despite interaction with a foreign material. Also, the production of cell-type specific proteins and enzymes is a clear indicator for conserved functionality of cells. In the case of tissue engineering, often precursor or stem cells are used, which are stimulated by signaling molecules to adapt a certain phenotype. The analysis of the gene-expression profile can then confirm that the desired differentiation has indeed taken place and that a functional tissue is being formed. Because only a specific set of active genes defines the differentiation of a specific cell type, the number of genes that has to be studied is large. The set of studied genes must be large enough to exclude some intermediate differentiation state, in which the cell-type specific genes are expressed, along with genes specific for another, sometimes closely related, cell type. One can imagine that with the appropriate DNA micro arrays, this analysis will be possible. The increase in the expression of certain genes and decreased expression of others are then clear indicators of cellular function.

As stated before, the produced mRNA molecules are not the active compounds of the cell, but the proteins that are encoded by these messenger molecules. The analysis of large numbers of different proteins is a focus of the proteomic field, and the techniques are steadily developing. Therefore, it is feasible that in the near future a protein profile of a specific cell type can be analyzed in a chip format with similar speed as the DNA microarray. Until that time the somewhat slower protein analysis and identification techniques will have to do. A last possibility is histological analysis of cells and cell constructs that were grown in vitro. The overall morphology and the interaction with the produced extracellular matrix can all be assessed via microscopic evaluation of proper samples. The primary bottleneck however remains the lack of a 3D network of cells, a tissue that demonstrates proper homeostasis in vitro.

13.3.5 Genotoxicity

This form of testing is relatively modern, since 100 years ago nobody even considered toxic effects on the genetic material, which had only just been identified as being DNA. It was already known that some chemicals could cause phenotypic changes in *Drosophila* or plants, but the mechanism remained completely unknown for a long time. The developments in bacterial and eukaryote genetics, combined with cell culturing techniques, led to the development of several different genotoxicity assays. The publication of the structure of DNA stimulated the discovery of many chemical mutation mechanisms over the following decades. Subsequently, regulation was put in place to avoid human exposure to chemical mutagens [29]. The effect of biomaterials and their degradation products or leached compounds on the stability of the DNA is of great concern when determining biocompatibility. This specific form of toxicity is mostly observed after long-term incubations in vivo, but the consequences can be disastrous. Consider the use of asbestos in building projects during long periods of the twentieth century as insulating and heat-resistant material. Only decades later it turned out that inhaled fibers of asbestos remained present in the lungs, got encapsulated, and caused cancer long after the inhalation. The study of so-called genotoxicity has therefore gained importance over the past decades. Very important progress was made by the development of the Ames test in the 1970s by Bruce Ames and coworkers [30]. The test was designed for analysis of mutagenicity of chemical compounds and is still widely used today. Genotoxicity can be subdivided in different forms of damage to the DNA. Mutagenicity are changes to DNA that change the sequence in functional genes, resulting in an altered function of the gene product. These mutations mostly only affect small portions of the DNA and include frame shifts, base-pair mutations, and small insertions or altered splice sites. The most frequently used test to detect such DNA changes is the Ames test, which is described in Figure 13.8. The test is based on the use of bacterial auxotrophic markers. A histidine-dependent *Salmonella typhimurium* strain is treated with the test material, and the number of revertants is then assessed. A mutagenic substance will randomly evoke changes in the DNA of the bacteria. Some mutations will restore the defective gene in the histidine operon, and the resulting bacterium will now be autotrophic for histidine, so be able to grow on a

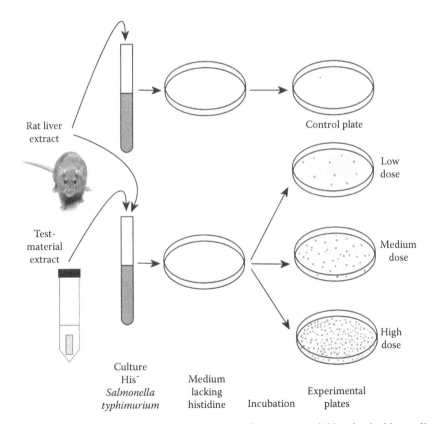

FIGURE 13.8 Scheme showing the Ames test. An extract of the test material is mixed with a rat liver extract and incubated with *Salmonella typhimurium* His⁻ strain. The number of revertants, histidine-independent growing bacterial colonies, is a measure for the mutagenic potential of the test material. Low, medium, and high doses of the extract are tested to improve reliability of the test and minimize false positives.

medium without added histidine. The number of bacteria that become histidine independent after treatment is a direct measure for the mutagenicity of a material.

One of the main criticisms on this assay is that it involves a prokaryote, and the DNA structure and DNA repair mechanisms are different from those found in eukaryotes. There is an alternative assay using mouse lymphoma cells, L5178Y. This method of determining genotoxicity is also based on the dependence of cells on certain vital nutrients in their medium. The mutations generated by incubation with a test material, in both metabolically active and inactive cells, result in clonal growth of cells under restricted medium conditions. The analysis of these mutant colonies can be directly linked to the level of mutagenicity of the test material.

Next to rather subtle changes to the DNA sequence, also larger DNA damage can occur, which destroys the integrity of chromosomes. The breaks in chromosomes, gaps, deletions, and exchanges can be relatively easily demonstrated in Chinese Hamster Ovary (CHO) cells after incubation with the test material. Staining of the CHO cells with a DNA stain and subsequent microscopic evaluation can reveal a large number of different defects caused by a mutagen.

Another assay, popular for the detection of chromosomal aberrations, is the mouse bone-marrow micronucleus test. From the bone marrow of mice, erythroblasts are isolated and these cells are treated with the test material. When the cells subsequently divide, only unaffected cells will give rise to two daughter cells with the entire chromosomal DNA in the nucleus. When however either the chromosomes or the mitotic apparatus gets damaged, small parts of chromosomes will be localized outside of the main nucleus in a so-called secondary nucleus, or the micronucleus. This micronucleus will remain present in the cells that will develop to form erythrocytes. These red blood cells normally do not contain a nucleus, but formed micronuclei will remain present and can be easily identified by staining for DNA. The number of erythrocytes with micronucleus is a direct measure for the mutagenicity of the test material or chemical.

A rather new method to determine genotoxicity is the alkaline comet assay, also known under the name single cell gel electrophoresis (SCGE). This is an ex vivo/in vivo assay that makes use of isolated single cells from tissue or peripheral blood. The assay is depicted in Figure 13.9. Isolated cells are

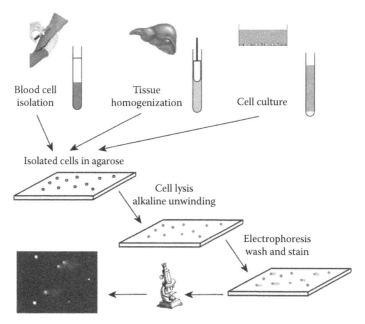

FIGURE 13.9 **(See companion CD for color figure.)** Alkaline comet assay for determining genotoxicity of materials. Isolated cells are incorporated in an agarose gel, lysed, and the DNA denatured. Damaged DNA will migrate away from the original site of the cell in an electric field, resulting in comet-like structures after staining of the DNA.

challenged with a potential genotoxic material and incorporated in an agarose layer. After alkaline (pH > 13) treatment, the DNA remains in the agarose, localized where the nucleus was. DNA-fragments will now migrate away from the site of the nucleus to the cathode in an electric field. The shape of the DNA, like a comet, determines if and how much of the DNA was damaged. The nondamaged DNA remains in the head of the comet, with the damaged DNA pieces forming the tail. The exact shape of the comet can also indicate the type of damage that was induced. The method is still under scrutiny, and the predictive value is under investigation [31]. Because large cohort studies in multilaboratory validation trials are still lacking, the reliability is still not clear. The assay is increasingly popular, but small differences in the protocols between different labs lead to different results [32]. Therefore, a definitive standardized protocol will be necessary for this assay to be useful in general for biocompatibility studies.

13.4 EX VIVO BIOCOMPATIBILITY TESTING

Because the direct extrapolation of in vitro data to the human, clinical situation is not feasible and certainly not accepted, further testing of biomaterials and medical devices is required. The in vitro testing lacks the 3D aspect in which cells in tissues normally exist. Furthermore, the interaction and communication between these cells is not present or very much inhibited in 2D cell cultures. Also in tissue-engineered constructs, the small piece of tissue lacks the communication between the cells since a constant flow of medium will dilute signaling molecules rapidly, so that no intercellular communication beyond direct cell-cell contact is unlikely. One option to circumvent this shortcoming is to use isolated tissues and/or organs for short-term biocompatibility testing. The problem, with this form of ex vivo testing, is the limited time the isolated tissue can be maintained outside of the body. An incubation of more than 2–3 days is not possible since the cells in the tissue will start deteriorating, leading to nonreliable results.

Perfusion of for instance kidneys or hearts can be used to test surgical patches or implants that deliver drugs to these organs. Another ex vivo approach is used in hemocompatibility testing, so the contact between blood and a biomaterial. For this, the material can be formed into a tube, or a patch of the material can be incorporated in a tube. This tube can then be attached to a vein of a test animal or subject, and the blood can be allowed to flow through the conduit for a specified time (Figure 13.10). The blood is then monitored during the experiment to detect changes in

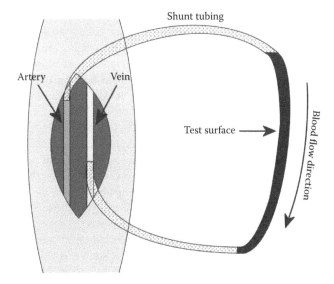

FIGURE 13.10 Ex vivo testing of a blood contacting material. The material is shaped or incorporated in a tube and attached to the circulation of an animal or human volunteer. The blood is analyzed for the generation of thrombin or other coagulation indicators. After the incubation period, the material is studied separately for thrombus formation on this surface.

blood composition. The appearance of for instance D-dimer indicates that a thrombus has formed inside the extracorporal conduit, and thus the materials could be thrombogenic, so have poor hemo-compatibility. After the incubation period, the tube can be analyzed for different parameters, among which the formation of thrombi on the surface is the most important. Later in this chapter, the determination of hemocompatibility is discussed separately as an example for a strategy of biocom-patibility testing.

Ex vivo testing with perfused organs is not frequently used because of the limited incuba-tions that are possible. In addition, it is not easy to test a series of biomaterials simultaneously, since the number of organs of good quality is limited. Furthermore, perfusion of tissues and organs is relatively expensive, making it an unpopular method. Finally, although the cells are now organized in a 3D matrix, extrapolation of the results to an in vivo situation, as encoun-tered in a patient, is not trivial. A careful prediction on the behavior of the biomaterial inside an organ is possible. However, the lack of blood circulation and cells of the immune system being transported to the site of implantation or contact makes this prediction limited. Therefore, the question is if ex vivo testing has a significant role to play in biocompatibility testing. Maybe in the future when tissue-engineered organs and tissues become readily available, this form of testing outside the human body may pose a relevant method of testing of medical devices in a 3D cellular environment. But up to that time ex vivo test results should be interpreted with extreme care.

13.5   IN VIVO BIOCOMPATIBILITY TESTING: INTRODUCTION

The use of animal model systems to test surgical procedures or pharmaceuticals dates back to the start of medical science. Most animal testing was based on vivisection, the anatomical study of animal physiology. In addition, surgeons might have practiced operating techniques on live ani-mals or human cadavers. However, this form of animal experimentation cannot be seen as in vivo testing of medical implants. Up to the end of the nineteenth century, there was no significant use of synthetic medical implanted devices. The Cruelty to Animals Act, which was introduced in Great Britain in the early nineteenth century, regulated the treatment of animals in general as well as the use of animals for scientific purposes. The Act of 1876 (fully called *An Act to amend the Law relating to Cruelty to Animals*) stipulated that researchers would be prosecuted for cruelty, unless they conformed to its provisions, which required that an experiment involving the infliction of pain upon animals to only be conducted when "the proposed experiments are absolutely necessary for the due instruction of the persons to save or prolong human life." Additionally, the act prescribed that the animal should only get anesthetics once and should be killed after the experiment was finished. These regulations appear obvious to us, but up to that time several scientists denied the perception of pain by animals. The steady evolution of medical science and synthetic chemistry resulted in the synthesis of medical implants in the middle of the twentieth century. It also became clear that the testing of such medical implants should include testing in a living organism, before clinical application could be warranted. However, for a long time the in vivo testing of medical devices was completely arbitrary and dependent on the choices made by the researchers. In addi-tion, the patient rights improved steadily, urging the companies that were marketing the medical devices to pay more attention to patient safety and functionality of the medical device or implant. The safety of medical devices as well as longevity and near perfect functionality became important for the marketing strategy of the devices. As a result, the researchers started to take more care in the strategy of in vivo testing. The choice of animal model system was long guided by convenience but is nowadays often directed by the suitability of the model to assess biocompatibility and func-tionality in vivo. Still, the extrapolation of in vivo test results to the clinical situation, i.e., to use in patients, remains a point of discussion. The limited predictive value remains a matter of concern, and therefore recently some researchers have openly discussed the value of and need for biocom-patibility testing in animal models.

13.5.1 SHORT HISTORY OF IN VIVO BIOCOMPATIBILITY TESTING

The evaluation of synthetic materials and implants with in vivo test methods is as old as the use of medical devices and implants. There is little evidence that the ancient Egyptian and Inca surgeons used to test their procedures on animals. From current understanding, it seems almost impossible to comprehend that also these surgeons did not test their implants and procedures in an animal model before treating patients. One should also not forget that in the past and also in some so-called "primitive" societies, the use of implants is widely distributed. The insertion of materials to pierce tissue or subcutaneous implants to create a desired look (cosmetic surgery) obeys the same rules as the implantation of a medical device today. There should be a desired response of the tissue, and the "patient" should not be harmed by the implant. This means that by simple trial and error, some materials were used and others were rejected because of unwanted effects or toxicity in the recipient. In principle, this is similar to the in vivo testing that is required today before clinical application of a medical device is considered. Of course, biocompatibility testing evolved together with improving biological and chemical techniques. Therefore, in the nineteenth century, strict rules were adopted for "humane" treatment of animals, and consequently more care and planning went into animal testing. Also the risks accepted for medical treatments of patients are under constant change. Electroshock therapy was widely used 50 years ago, and today it will be considered cruel by a vast majority of people. Heroine and cocaine were recognized drugs for treating a variety of disorders and were initially hailed as the compounds to combat morphine addiction. Changing opinions in society influence decision on the use of medical devices and how these are supposed to be tested prior to clinical use.

Often in the past, discovery of biocompatible materials was more or less coincidental and driven by current technical possibilities. The use of ceramic or glass for eye prostheses was driven by the possibility of producing these materials in the shape of eyes and coloring these to mimic the lost eye. The materials turned out to be inert enough not to cause damaging effects in continuous contact with the eye socket. The use of ceramic or glass eyes was thus the result of a more or less trial and error concept. Also the clinical use of poly(methylmethacrylate) (pMMA) was initiated after it had become clear that shrapnel from Second World War fighter planes (the windows of the cockpit) in the eyes of pilots remained intact without adverse effects for years. So the scientists concluded that pMMA was a biocompatible material although, in hindsight, only the biosafety of pMMA was established.

In vivo biocompatibility testing comprises the use of animal models for testing biocompatibility as well as implantation studies in human patients. Implantation studies are the final step in testing of a new medical device. The local and systemic response of cells and tissues is of critical importance for the efficiency of the treatment and the success of the medical device. The lack of toxicity can be predicted from the various in vitro cytotoxicity experiments. Genotoxicity can be studied by in vitro mutagenicity studies although a mouse-based assay has been developed, namely the comet assay (see Figure 13.9). So for a material to turn out toxic only in the latter stage of in vivo testing is rare. Of course, the completely assembled medical device, sterilized and packed, is generally tested only in vivo, because in vitro testing of a complete medical device is often not feasible. The disassembly of the device should be done and the parts then tested in vitro for toxicity, but this is seldom done properly. Wear within a medical device can reveal new particles or compounds that can only be studied realistically in vivo. Also the effect of different sterilization procedures on the medical device is often not tested properly in vitro. Part 7 of the ISO-10993 norm describes how the residues of ethylene oxide sterilization should be investigated, because these residues are highly toxic. Since the sterilization procedures for clinically used medical devices is different from those in laboratory testing, analyzing the effect of these procedures on the biosafety of the medical device is important.

13.5.2 IN VIVO BIOCOMPATIBILITY TESTING

There are some biocompatibility aspects that can only be observed in vivo, namely: (1) local toxicity, sensitization, and irritation; (2) systemic toxicity; (3) local effects of solid implants in a 3-D tissue;

(4) degradation or wear in living organism and effects of the degradation and wear particles; (5) functionality of the medical device in a living organism. Most of these biocompatibility aspects are discussed in the following sections.

13.5.2.1 Local and Systemic Toxicity

The most obvious test that should be performed in vivo is local and systemic toxicity, basically a biosafety determination. This is prompted by the requirement that the medical device should not harm the patient. This is the logical next step after in vitro cytotoxicity, which is a good predictor of toxicity of the material and medical device and chemicals released by these. Several levels of toxicity can be identified, namely sensitization, irritation, and direct toxic effects on neighboring and distant cells and tissues, called systemic toxicity. The exact tests that have to be performed are of course guided by the intended use of the medical device as depicted in Table 13.2. However, for all medical devices intended for clinical use, sensitization and irritation tests have to be performed.

13.5.2.2 Sensitization

Sensitization testing is rather straightforward, although in practice the test requires experience and is time-consuming. A film of the biomaterial, or extracts of the medical implant in saline solution or oil, is applied to the shaven skin on the back of a guinea pig. Guinea pigs are chosen because the skin resembles the human skin in sensitization reactions. In addition, the relatively low cost of working with these small rodents is of importance. The film of material or an extract-soaked patch is allowed to have contact with the skin for up to several weeks. The skin can be repeatedly challenged after recovery periods of approximately 2 weeks, to observe additive sensitization reactions. The skin reactions are analyzed, by determining the reddening and swelling of the skin, and are compared to those in negative controls.

In another experimental setup, the material extracts are first injected into the test animal along with an adjuvant, intended to enhance the immune response. Then after a couple of weeks, the skin of these animals is challenged with a topical patch soaked with the extract. Again after the incubation periods, the reddening and swelling of the skin are analyzed and compared to those of a negative control (e.g., the extraction medium).

In the past decades, the use of mice for a local lymph node assay (LLNA) has been introduced (Figure 13.11). LLNA is based on the proliferation of lymphocytes in the lymph nodes after contact with a test material. The increased incorporation of 3H-thymidine is a direct measure for lymphocyte proliferation and immune reaction [33,34]. The method has a number of advantages above the earlier described guinea pig assay in that less animals are needed for the assay and these animals suffer much less trauma. Also the assay has a quantitative outcome, which can be objectively determined, this in contrast to the subjective evaluation by a skilled technician for the guinea pig assay. Extensive testing of this method and comparison with the standard guinea pig assay also revealed much less interlaboratory variation. Therefore, the LLNA is the method of choice for large-scale sensitization and irritation testing as for instance under the new REACH (registration, evaluation, authorization and restriction of chemicals) program from the European Union [35].

13.5.2.3 Irritation

According to the ISO definition, irritation is a "localized inflammatory response to single, repeated, or continuous application of a test substance, without involvement of an immunological mechanism." The effect of chemicals on patients can be studied in different ways. The ISO-10993-10 norm proposes first a thorough search of the literature to find data on the test materials. Second, an in vitro cytotoxicity test as described earlier can also identify the presence of extractable chemicals that have unacceptable effects on cells in vitro. When no information on irritation can be obtained by these steps, in vivo experiments should follow. Finally, noninvasive clinical tests in human subjects are the definitive test, but they should only be performed after some initial form of irritation testing.

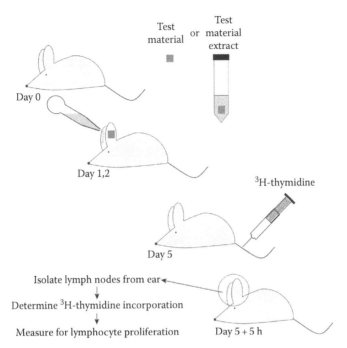

FIGURE 13.11   Local lymph node assay (LLNA). Mice are treated with a material or material extract for a defined period. After the incubation, 3H-thymicine is injected vial the tail vein, and after several hours lymph nodes from the ear are isolated and analyzed for lymphocyte proliferation.

The in vivo tests are very similar to standard allergy tests. The test material fluid extracts are injected into the skin of shaven albino rabbits. For medical devices, the use of saline and vegetable oil is sufficient to ensure the extraction of hydrophilic and hydrophobic compounds. A set volume of the extracts is injected and the injection sites are examined after 1, 2, or 3 days of incubation. The resulting reaction is scored and compared to adequate controls. Depending on the intended use of the medical device, additional tests are performed at the site of use. For this the ISO-10993-10 norm describes a number of tests for oral, penile, vaginal, rectal, and ocular irritation tests. The ocular irritation tests for instance are only performed for eye-contacting devices. The extract or a powder of the test material is directly put under the lower eyelid of rabbits. This irritation testing has been considered particularly cruel, also because it was mainly used in the cosmetic industry to test the safety of their products. Additionally, this method was rather laborious and not very reliable for predicting human skin and eye reactions. Therefore, a number of alternatives have been developed over the last 10–20 years. Most of these assays are however based on some form of cytotoxicity testing. The only advance in alternatives for ocular irritation testing is that not only cultured cells are used but also corneal tissues, isolated eyes, and pieces of epidermis. The common denominator is that all the assays evaluate adverse responses of the tissue or cells. The readout therefore is often loss of cell viability (as measured by MTT or viability dyes), loss of cell integrity (hemolysis, the disruption of red blood cells), or swelling of the tissue (isolated eyes and corneas). When the in vitro tests are performed carefully, the predictive power is as good as the tests on rabbits. At the end, none of the in vitro assays can fully replace the in vivo or clinical testing, but the large majority of the animal testing may be omitted in the near future [36].

13.5.2.4   Systemic Toxicity

The study of systemic toxicity is based on a rather simple idea, that when an exaggerated extract of a test material is injected into a test animal, observation of behavior and physiology can indicate acute or long-term toxicity. Acute effects are observed in a matter of days, while subchronic effects take 14–28 days (less than 10% of the expected life span of the animal). Chronic effects may take more than 10% of the animal's life span and are therefore only required for permanent implants. At the end of a systemic toxicity experiment, the organs can be examined and possible systemic toxic effects can be observed by an animal pathologist. Of course, the exact way in which the experiment is performed is vital for the reliability and significance of the obtained results. Is the extract of the material administered once or multiple times, at what doses, what intervals are chosen, how long are the animals allowed to recover before final examination, etc. These are all decisions that can seriously affect the outcome of systemic toxicity testing. It will therefore not come as a surprise that before extraction and test conditions were standardized, the variability of test results between different labs was enormous. Additionally, the result of the tests is dependent on the researchers' ability to observe aberrant behavior and or changes in the animals' physiology. In addition, the method of application of the extract or chemical is important, and several different ways have been suggested. Subcutaneous or intramuscular injection is the most obvious but also intraperitoneal (for oils) and intravenous (for saline extracts) are frequently used. Also application via inhalation has become more popular especially in the toxicity study of airborne nano- and microparticles.

In general, large numbers of rats or mice are used to perform these tests. In some cases, it might be preferable to use the implanted device in systemic toxicity tests, for instance for resorbable and degradable implants. Then the systemic toxicity of the degradation products is a vital part of biocompatibility testing.

13.5.2.5 Acute Systemic Toxicity

For acute systemic toxicity, an exaggerated extract of the medical device is prepared and applied to the animal. When toxic effects are observed with in 1 or 2 days, the material can be rejected because of acute systemic toxicity. This acute effect is not always caused directly by the materials comprising the medical device but can originate from contaminants, called pyrogens, that have attached to the device during its production. Rest of ethylene oxide, which is used for sterilization of the device, can cause rapid and dramatic toxicity. Also residues of bacteria that attach to the surface of the device, also known as endotoxins, can cause acute systemic toxicity. Therefore, methods were designed to determine the possible contamination of materials with such pyrogens. Often the pyrogens or endotoxins are introduced during manufacturing by the used machinery or the people working in the production line. The endotoxins are organic molecules of bacterial origin, like membrane components or proteoglycans that stick to the medical device. The classical assay to detect pyrogens makes use of rabbits that are injected with an extract of the medical device or test material. The reaction in body temperature of the rabbit is a measure for the presence of unacceptable levels of pyrogens. If most of the test rabbits demonstrate a rise in temperature of more than 0.5°C and the total rise in temperature over all affected rabbits in a group of 10 is more than 3.3°C, the test materials contained unacceptable levels of pyrogens. The assay detects both chemical and material-mediated pyrogenicity [37].

An alternative assay was developed that makes use of the lysate of Limulus Amoebocyte (the so-called LAL test) [37]. This test replaced the rabbit testing for materials and devices. An aqueous extract of the blood cells (amoebocytes) of the horseshoe crab *Limulus polyphemus* is used to detect the presence of membrane components of bacterial origin (endotoxins). The LAL test is the preferred way to detect bacterial endotoxins. The clotting of the limulus extract is strongly induced by endotoxins, and this clotting reaction can be detected. Some of the newer available LAL assays make use of recombinant forms of enzymes in the clotting cascade of the extract, and fluorescent substrates replace the formation of a solid clot as readout. The commercially available LAL test kits show a high sensitivity for endotoxins and demonstrated that it is very difficult to produce endotoxin-free medical devices.

Another alternative in vitro pyrogens test relies on the interaction of human blood with the pyrogens present on the test material or medical device. Upon interaction of the leukocytes in the blood with the pyrogens, interleukin-1 will be produced, and this can be detected using a specific ELISA.

The advantages of this assay are obvious since the blood of the targeted species (human) is used. Additionally, the assay is cheap and sensitive to all pyrogens, not exclusively for gram-negative endotoxins as is the case for the LAL test [38].

13.5.2.6 Subchronic and Chronic Systemic Toxicity

For long-term toxic effects, the methodology does not necessarily change, but the time span over which the animals are allowed to be exposed to the test material is longer. The rule of thumb is that for subchronic testing the incubation period does not exceed 10% of the animal's lifespan. Chronic systemic toxicity measurements take longer, in general > 10% of the animal's lifespan, and are therefore more expensive and only required for permanent implants. The testing for long-term systemic toxicity can be replaced by good chemical characterization of the test material. The identification and quantification of extractable chemicals from the test material should precede biological evaluation. If no detectable extractable chemicals can be identified, the necessity for systemic toxicity testing can be irrelevant. In addition, when there is extensive literature available on the toxicity of the exact chemicals that are identified in the extract, toxicity testing can be precluded (Figure 13.1).

The most used organism for systemic toxicity tests is the rat. This is because enough biological materials like blood and urine can be obtained from these animals to perform a range of testing. If not enough blood can be obtained, only limited analysis of the blood is possible and insufficient toxicity data will be the result. Therefore, rats and in earlier times also dogs, are the animal model of choice for systemic toxicity. After the desired time of exposure to the test material, the animal is sacrificed and a standard pathological examination is performed. This includes the examination of the organs, both macroscopically and microscopically. The organs are weighed and microscope samples are prepared according to standardized protocols. Furthermore, the clinical chemistry examination will yield valuable information concerning toxicity. The blood analysis can reveal changes in electrolyte balance, liver and kidney functions, lipid and carbohydrate metabolism, and many specific parameters that vary with the nature of the investigated medical device.

Of course, also systemic carcinogenicity and mutagenicity of the test material can be studied using the comet assay that was described earlier (Figure 13.9). For this, parts of organs are homogenized and the isolated cells are subsequently subjected to the alkaline comet assay [31]. The animals should also be carefully examined for the appearance of unexplained tumors. The long-term effects on stability of the DNA are of critical importance to assure safety of the material and medical implant.

13.5.2.7 Local Implant Effects

Part 6 of the ISO-10993 norm gives guidelines for examining the effects of implants on their direct environment, although no real border, where the local environment stops, for this is given. Mostly a test strip of a material (dimensions 1–3 × 1 × 10 mm [h × w × l]) is implanted in the paralumbar muscle of a rat, mouse, guinea pig, or rabbit. Simultaneously, a number of nontoxic control materials are implanted in the same animal. For short-term implants, the implants are assessed after 1, 4, and 12 weeks, while for long-term implants 12, 26, 52, and 78 weeks are specified by the ISO norm. After each incubation period, the implant is examined for the presence of a capsule and the implant is removed together with the surrounding tissue. The negative control materials will not have induced formation of a capsule around the implant. For further processing, often the implant is removed, although this is not preferred since the interface between the material and the tissue is actually the most interesting. The removal of the implant from the tissue will destroy this delicate interface layer of tissue and diminish the quality of the data. Of course, the embedding of the tissue with implant is critical, since a discrepancy between the mechanical properties of the embedding polymer and the test material can lead to problems during microscope sample preparation. The lack of necrotic cells close to the implant and the absence of inflammatory cells indicate a minimal response of the tissue toward the implant. High numbers of inflammatory cells can be accompanied by necrotic cells around an implant with inherent toxicity. The assay is rather straightforward, although there are some pitfalls for this method of testing. The choice of model organism can be of importance for the observed tissue effects [39]. In addition, the site of implantation should be considered carefully, since the subcutaneous implantation of a bone support device is not very logical. For simple biosafety experiments of medical implants, subcutaneous (on the back), or intramuscular (in the hind leg musculature) is sufficient. This is also because the animal will suffer the least discomfort and pain at these sites. In addition, because the animal has two virtually identical sides, there is always place for a control implant in the same animal.

In short, the implantation of a medical device can serve the purpose of examining the direct response of the local, contacting tissue. The preparation of the tissue sample after the specified time intervals is of critical importance to preserve the structures formed at the interface of the implant and tissue. In addition, the formation of a capsule that can be observed with a low-power microscope should be recorded, as it is a clear sign for the activation of a defense mechanism of the animal. The embedding and staining procedures should be carefully chosen so that no structures are damaged during the processing of the samples. The readout from the experiments is very direct, the presence of dead cells or large numbers of inflammatory cells indicates toxicity of the implant.

13.5.3 FUNCTIONALITY OF MEDICAL DEVICES IN VIVO: CHOICE OF MODEL

A very important part of biocompatibility testing is the evaluation of the function of the medical device. For instance, a heart valve needs to perform perfectly for the patient to benefit fully. Hip replacements have to perform for a long period of time without any complications as loosening and wear to prevent premature revision. Of course, most of the functions of the medical device can only be fully assessed in human patients. The use of model animal systems is complicated since the animals react often very differently to the implanted materials. Also the choice of animals and implantation procedure can be of great influence on the results.

Let us consider the following example. The replacement of occluded blood vessels is of great importance for a large group of human patients. Not only stenosis of coronary arteries but also of peripheral vessel (e.g., carotid arteries and femoral-popliteal artery) can often give rise to serious medical complications. With the sharp increase in diabetes mellitus type 2, the number of patients with peripheral artery disease is increasing rapidly. The diabetes foot is a well-known example of this. Therefore, the researchers of the biomedical engineering field have been looking for decades for synthetic solutions to this problem. The Holy Grail would be a fully synthetic vascular prosthesis that performs well over a long period of time. The prosthesis should regulate hemostasis as well as inhibit inflammation that can lead to failure of the device. The patients that would normally receive such a prosthesis are not healthy and mostly of higher age. It is therefore puzzling that the potential synthetic vascular prostheses are almost exclusively tested in young, healthy animals that have no underlying disease causing peripheral artery complications. Therefore, most of the synthetic vascular prostheses perform very well in the animal models. However, when applied in humans the efficiency of these implanted vascular grafts is disappointing [21]. The underlying problem is that in animals the inside of the prosthesis gets rapidly covered by a layer of vascular endothelial cells. This is possible because these cells are healthy and motile and proliferate well. In patients however the endothelial cells do not proliferate well and do not cover the luminal side of the prosthesis. Therefore, the regulation of hemostasis and inflammation is flawed and the implant will eventually fail. To really get a good testing procedure, older, preferably diseased animals should be used in biocompatibility testing, especially concerning functionality of the medical device. This is mostly not allowed by the medical and ethical commissions overseeing the animal testing procedures.

Another example is the testing of bulking agents used for treatment of stress urinary incontinence (SUI). These materials are injected next to the muscles of the pelvic floor to enhance stiffness of the soft tissue. This stiffening then allows for better closure of the urethra, and a decrease in SUI. The results of the SUI treatment are largely subjective since it is virtually impossible to get an objective readout (e.g., lower frequency of incontinence events and less urine leakage during an event) of the treatment. It is how the patients perceive their situation that indicates whether the treatment was successful and whether the patients feel that there is a decrease in urine leakage and an increase in quality of life. It is almost impossible to test such materials for their function in animal models. First of all, there are no real stress incontinent animal models available. Then it is very difficult to get an objective readout of the animals concerning the treatment. The animals cannot tell how they are doing or feeling. In such cases one feels that after well-performed biosafety experiments, there is no other option than to apply the medical implants in the clinic and carefully assess the outcome for each patient.

The choice of the right animal model is therefore of critical importance. Preferably, an animal as close to humans as possible should be used. The use of primates however is highly controversial and in most cases impossible, not only because of the extremely high costs but also because institutions that are allowed to work with primates are rare. The political pressure to avoid the use of primates in biomedical research is high, and it is virtually impossible to get approval for testing of medical devices in primates. The use of large mammals, pigs, horses, goats, dogs, etc., is limited because of the relatively high cost and limited numbers of animals. The time it takes to breed a new generation of these animals is high, and therefore a constant flow of test animals is not available, at least not in large numbers.

Additionally, the implantation procedure should be as similar as possible to the intended clinical application. The collaboration with surgeons that work in the clinic daily is of great value to improve the medical implant at this stage, since they can indicate exactly what can be improved to obtain the best and most efficient medical device. Only when the surgeon is happy with the implant and the most appropriate test model is chosen will the results of the experiments have any predictive value.

A last consideration is whether the use of clonal animals that are genetically all identical is appropriate [40,41]. In practice, the patients that await treatment are a genetically heterogeneous pool, which will demonstrate a wide range of reactions to the device. The problem arises when the identical genetic makeup of test animals (for instance a mouse strain or rat strain) has a direct influence on the tissue- or systemic response to the implant or device. The results are then biased and only valid for animals with that genetic makeup. In this way, good materials and devices could be eliminated and missed, or some devices are considered safe and good. The increase in variation between the test animals may improve the reliability of the obtained results. Recent studies have shown that this increase in heterogeneity does not significantly affect the reproducibility of the results [42]. Therefore, choosing a representative population of animals in preclinical studies would be an improvement over the current testing procedure. It seems logical to increase this heterogeneity, because in this way the animal group used resembles a patient population much more closely.

To conclude, the results of animal testing in general should be treated with extreme care since the animals are just different from the patients for whom the medical device was designed. The final, conclusive answer whether a medical device is biocompatible can only come from the clinical application of the device in real patients.

13.5.4 Future of In Vivo Testing

The rate of failure for medical implants in the clinic is relatively high and prompted a discussion during the *World Biomaterials Conference 2008* in Amsterdam with the provocative thesis that preclinical testing in animal models is a waste of time [20]. The predictive value of many animal models systems is limited. In addition, the in vitro use of cell lines and many ill-characterized primary cells is in most cases imperfect and inadequate to get sensible information concerning the biocompatibility of the medical device. At the end of the 2008 biomaterials conference, the consensus was that preclinical testing is not a complete waste of time, and the severely toxic and completely nonfunctional materials and devices can be filtered out quite well. However, biofunctionality testing remains a matter of concern, and researchers should take great care in the design of their animal experiments.

Another, maybe unexpected, reason for many unnecessary biocompatibility experiments is the reluctance to publish failed experiments or experiments describing toxic and noncompatible newly synthesized materials, composites, implants, and devices. One can imagine that by carefully studying the literature on identical materials or combination of materials, and how these performed in vitro and in vivo, we could eliminate many animal experiments, since the fact that these materials do not perform satisfactory has already been shown. When researchers and journals do not publish these data, other groups will have to perform the same, unnecessary experiments, coming to the same conclusion again.

Then there is the continuous pressure from the "animal rights and alternative experiments" lobby, practicing political pressure to minimize the use of animals for biomedical experimentation. Although the majority of the population in western countries is convinced of the necessity of animal testing of biomedical devices and procedures, the pressure to invent alternatives for animal testing may be useful to improve current in vivo experimentation or accelerate the development of proper alternatives. Already some alternatives have been published, which could replace some forms of animal testing.

One example we have already discussed is the use of LAL and human blood test for determination of pyrogen and endotoxin levels in medical devices (systemic toxicity). The assay has for a large part replaced the existing rabbit assay, lowering the number of required animals dramatically.

Another example is the development of in vitro cultured skin, which resembles natural skin ever more close. The recently published method for producing skin in the lab is a critical step in providing material for chemical and biomaterial testing in vitro on a living tissue that is identical to the "real" tissue [42]. In this way, the number of animals needed for testing can be reduced in the near future.

A last example is based on the use of cellular matrices, harvested from animals that are used as a scaffold for specialized cells. The vascular network and extracellular matrix are used for supporting and feeding specialized cells during a prolonged time [43]. This pseudo-tissue can then be used to test the effect of chemicals or material extract on for instance liver or kidney cells. This is important because especially in the liver, chemicals can be converted and subsequently show completely different toxicity profiles. So by performing testing with the help of such engineered tissue-like structures, unexpected toxicities can be minimized and thereby the number of unnecessary animal experiments can be decreased.

With the speed of evolution of biocompatibility testing increasing steadily, it is disappointing how few new biomaterials have reached clinical practice over the last few decades. The medical device companies seem to stick to the old and trusted materials that have been in use for a long time. The risk of introducing a new material is too large, and the requirements for risk assessment and compatibility testing sometimes work paralyzing for innovation. The constant improvement in biocompatibility testing, reducing the risk for the patient, will eventually break the deadlock and induce a greater use of more sophisticated, better performing, and safer materials for medical devices and implants.

13.6   IN SILICO TESTING

The development of increasingly larger databases, also in biomedical sciences, may present a good opportunity to develop in silico biocompatibility testing procedures. In silico design of novel drugs has been common practice for more than a decade now. However, in the field of biomedical engineering and biomaterials science, computer-aided design is underexplored. A potential biocompatibility database would consist of all experimental data obtained worldwide concerning in vitro, ex vivo, and in vivo biocompatibility testing on all biomaterials that are interesting for biomedical scientists. Of course, the completion of such a database can be considered a utopia, but the perceived advantages are enormous. First, and maybe most important, many unnecessary biocompatibility tests can be avoided, because these have been performed

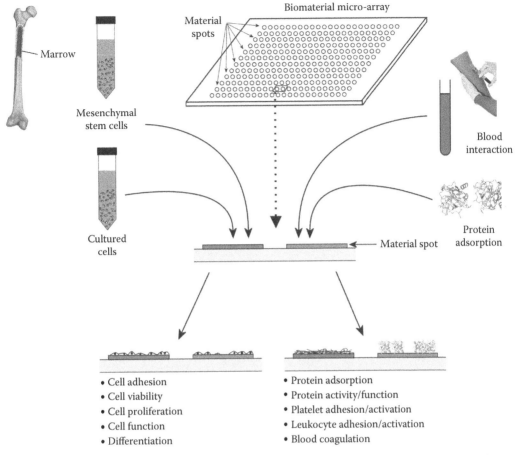

FIGURE 13.12   (See companion CD for color figure.) Scheme showing one possibility for high-throughput screening of biomaterials. Small spots of the materials are deposited on a glass slide. Cells are cultured on these spots and can be analyzed for viability, proliferation, or functionality. Alternatively, the materials can be challenged with proteins or body fluids like blood, and the subsequent interaction analyzed.

before, at some time in another laboratory. This would save not only time, effort, and funds but also would reduce the use of animals for in vivo testing.

An important development for such a biocompatibility database could be high-throughput screening, which is starting to gain ground in the biomaterials field [44,45]. In Figure 13.12, a possible scenario for high-throughput screening is depicted. A glass chip is covered with spots of: (1) chemically different biomaterials, (2) identical materials with different surface characteristics (roughness, topography, etc.), (3) different composites and blends of materials, (4) identical materials with varying biofunctionalization (e.g., protein adsorption of attachment).

The chip can then be submitted to a number of tests to assess the different aspects of biocompatibility. Interaction with a cell line that is genetically engineered to indicate different levels of cytotoxicity would be of great interest. Differentiation of stem cells into a desired differentiated cell lineage (for instance bone forming osteoblasts from mesenchymal stem cells) is another possibility. Cell adhesion and proliferation studies are also easy experiments that can be adapted for such high-throughput conditions. Interaction of the materials with blood, analyzing coagulation or immune response could prove a great tool in the design of blood-contacting materials. In fact, one can think of many different in vitro experiments to test some aspects of biocompatibility. These data should then be collected in an open-access database so that no unnecessary duplication of experiments is required. In the flowchart in Figure 13.1, one of the first steps that are taken by researchers working on a new medical device is to investigate whether biocompatibility and/or toxicity data are available on the materials they intend to use. So it is logical that the collection of all biocompatibility tests will help the development of medical devices and implants.

For in vivo biocompatibility experiments, it is a bit harder. There are so many different parameters influencing the final results, but a (central) database would be of use. If in vivo experiments on a biomaterial, composite, ceramic, or metal alloy have already been performed, the additional number of in vivo experiments may be reduced. The largest gain would be in saving funds and valuable time. The funds that are saved could then be used for new, valuable experiments. The gain in time might eventually even save the lives of patients, because an improved treatment comes to the clinic faster.

Of course, there are huge problems with such a database, and the chance of it ever seeing the light of day is slim. How are the results in such a database verified? What data are entered and what data are considered unwanted? The decision to include experimental data into a database means that other researchers have to be able to rely on these data. Because all data concerning biocompatibility are scattered over the scientific literature, the need for some structuring in these data would be of great use. Anyhow, the biocompatibility database is far, far away; however, some effort should be made to achieve this goal.

13.7 EXAMPLE: BIOCOMPATIBILITY TESTING OF A CENTRAL VENOUS CATHETER

Central venous catheters (CVCs) are percutaneous (through the skin) catheters that are frequently used in critical care patients. The catheters are mostly polymeric tubes that allow for monitoring of the patients as well as therapeutic treatment, i.e., the application of medication directly in the circulation. These catheters suffer from different problems among which the infection rate and material-induced thrombosis are most prominent. The development of new coatings for the central venous catheters has therefore attracted much attention of biomedical engineers and medical doctors. Here we take an existing polymeric catheter with a newly developed coating as an example to enlighten the process of biocompatibility testing. Because the coating is completely new, there are no existing biocompatibility or toxicity data available. Therefore, the new coating material needs to be carefully analyzed and biocompatibility tests according to the ISO10993 matrix have to be performed. From Table 13.2 we can deduce that the device is an external communicating device that is in contact with the circulating blood for a medium-long period of time (up to 30 days). The required tests therefore are cytotoxicity (in vitro), sensitization, irritation, acute and subchronic systemic toxicity, genotoxicity, implantation (local effects on tissue), and hemocompatibility. For the design of the experiments, careful planning and consideration of the intended use of the device determine how the experiments will be performed.

For the in vitro cytotoxicity, direct contact and indirect contact testing are warranted since the new material will be in direct contact with cells and tissue. The extract used for indirect testing is best made in medium containing serum, since this mimics the intended blood contact best. The choice of cell strains for the in vitro toxicity tests is relatively straightforward. Fibroblasts and possibly keratinocytes are the most obvious choices, although the use of endothelial cells could be considered since these cells cover the inside of blood vessels. Sensitization and irritation tests are very much standardized, and the only variation can be the way the material is presented to the animals. In this case, it would be best to produce strips of the tube-polymer with the new coating. These strips could then be applied to the skin of the animals. Extracts of the final CVCs can be produced in saline and vegetable oil and applied subcutaneously or intravenous to study the acute and subchronic systematic toxicity. Genotoxicity of the CVC can be easily analyzed with a standard AMES- or Comet test as described in Figures 13.8 and 13.9. Systemic genotoxicity assay is not required, since the amount of chemicals leaching from the thin coating of the CVC is extremely low. For the study of the local effect on tissue upon implantation, a clear choice can be made.

Small parts of the catheter can be implanted subcutaneously or intramuscularly in small rodents. Another option would be to produce a smaller version of the catheter and do the implantation study according to clinical use, so through the skin and into a vein. This is of course more work; one has to produce the miniature version of the catheter, and only relatively large animals (e.g., rabbits or dogs) are suited for such a study. This approach is closer to the clinical situation but the ISO norm does not prescribe such testing, so in this case subcutaneous implantation is for many scientists the preferred option. The number of hemocompatibility tests is large (ISO-10993-4). Many scientists study hemocompatibility by analyzing the interaction between the material and platelets from the blood. Also, often the adsorption of proteins from the blood-plasma is used as an indicator for blood compatibility. But when one carefully thinks about the function of CVCs, these tests are relatively meaningless. In the case of CVC, the actual problem encountered is formation of thrombus on the surface, which could lead to thromboemboli being released and vessel occlusion in other tissues. Therefore, the most relevant experiments that address this process are the generation of thrombin and subsequent thrombus formation at the material's surface. Which can be done in vitro. Small pieces of catheter are brought in contact with blood or platelet-rich plasma, and the thrombin generation is determined. Alternatively, the thrombogenicity of the CVC can be analyzed in an ex vivo approach (Figure 13.10), attaching a tube of the CVC material to the extracorporal circulation device of a patient. The readout will be the formation of thrombus at the surface and the generation of thrombin and clotting indicators (e.g., D-dimer) in the contacting blood. Of course, the interaction of the CVC surface with erythrocyte and leukocytes should be considered. Hemolysis of red blood cells should be determined because the device should not harm the patient. The interaction with white blood cells, however, is more difficult. The release of inflammatory markers in the blood can be determined, but the question remains what the actual consequences of this proinflammatory response for CVCs are. Activation of monocytes can result in faster blood coagulation. Because most CVCs are only used for several days, study of long-term immunological responses is not required for these devices.

The use of a good control material is of vital importance for the evaluation of the CVC. The basic CVC without the new coating can be used as a control, since this device is already approved for clinical use. The new coating should improve the performance of the CVC in the functional tests (hemocompatiblity and implantation) and should not perform worse in the biosafety tests (sensitization, irritation, cytotoxicity, genotoxicity).

REFERENCES

1. Williams, D. 2003. Revisiting the definition of biocompatibility. *Med Device Technol* 14:10–12.
2. Williams, D. J. 1999. *The Williams Dictionary of Biomaterials*. Liverpool, U.K.: Liverpool University Press.
3. Anderson, J. M., Rodriguez, A., and Chang, D. T. 2008. Foreign body reaction to biomaterials. *Semin Immunol* 20:86–100.
4. Williams, D. F. 2008. On the mechanisms of biocompatibility. *Biomaterials* 29:2941–2953.
5. Nixon, M., Taylor, G., Sheldon, P., Iqbal, S. J., and Harper, W. 2007. Does bone quality predict loosening of demented total hip replacements? *J Bone Joint Surg* 89-B:1303–1308.
6. Gurbel, P. A. and Tantry, U. S. 2007. Stent thrombosis: Role of compliance and nonresponsiveness to antiplatelet therapy. *Rev Cardiovasc Med* 8(Suppl):S19–S26.
7. McFadden, E. P., Stabile, E., Regar, E. et al. 2004. Late thrombosis in drug-eluting coronary stents after discontinuation of antiplatelet therapy. *Lancet* 364:1519–1521.
8. Dhert, W. J. 1994. Retrieval studies on calcium phosphate-coated implants. *Med Prog Technol* 20:143–154.
9. Geesink, R. G., deGroot, K., and Klein, C. P. 1987. Chemical implant fixation using hydroxyl-apatite coatings. The development of a human total hip prosthesis for chemical fixation using hydroxyl-apatite coatings on titanium substrates. *Clin Orthop Relat Res* 225:147–170.
10. Bessa, P. C., Casai, M., and Reis, R. L. 2008. Bone morphogenic proteins in tissue engineering: The road from laboratory to clinic, part II (BMP delivery). *J Tissue Eng Regen Med* 2:81–96.

11. Cowan, C. M., Soo, C., Ting, K., and Wu, B. 2005. Evolving concepts in bone tissue engineering. *Curr Topics Dev Biol* 66:239–285.

12. Cranenburg, E. C. M., Schurgers, L. J., and Vermeer, C. 2007. Vitamin K: Coagulation vitamin that became omnipotent. *Thromb Haemost* 98:120–125.

13. Irish, J. D. 2004. A 5,500 year old artificial human tooth from Egypt: A historical note. *Int J Oral Maxillofac Implants* 19:645–647.

14. Sanan, A. and Haines, S. J. 1997. Repairing holes in the head: A history of cranioplasty. *Neurosurgery* 40:588–603.

15. Balls, M. 1985. Scientific procedures on living animals: Proposals for reform of the 1876 cruelty to animals act. *Altern Lab Anim* 12:225–242.

16. Zimmer, H. G. 2001. Perfusion of isolated organs and the first heart-lung machine. *Can J Cardiol* 17:963–969.

17. Kaplan, J. and Hukku, B. 1998. Cell line characterization and authentication. *Methods Cell Biol* 57:203–216.

18. Braybrook, J. H. 1997. Biocompatibility assessment of medical devices and materials. Chichester, U.K.: Wiley.

19. Schneider, W. H. 1996. The history of research on blood group genetics: Initial discovery and diffusion. *Hist Philos Life Sci* 18:277–303.

20. Malda, J., Rouwkema, J., Leeuwenbergh, S., Dhert, W. J., and Kirkpatrick, C. J. 2008. Crossing frontiers in biomaterials and regenerative medicine. *Regen Med* 3:765–768.

21. Zilla, P., Bezuidenhout, D., and Human, P. 2007. Prosthetic vascular grafts: Wrong models, wrong questions and no healing. *Biomaterials* 28:5009–5027.

22. Belanger, M. C. and Marois, Y. 2001. Hemocompatibility, biocompatibility, inflammatory and in vivo studies of primary reference materials low-density polyethylene and polydimethylsiloxane: A review. *J Biomed Mater Res (Appl Biomater)* 58:467–477.

23. Knetsch, M. L. W., Olthof, N., and Koole, L. H. 2007. Polymers with tunable toxicity: A reference scale for cytotoxicity testing of biomaterial surfaces. *J Biomed Mater Res* 82:947–957.

24. Mosmann, T. 1983. Rapid colorimetric assay for cellular growth and survival: Application to proliferation and cytotoxicity assays. *J Immunol Methods* 65:55–63.

25. Johnson, I. 1998. Fluorescent probes for living cells. *Histochem J* 30:123–140.

26. Kroczek, R. A. 1993. Southern and northern analysis. *J Chromatogr* 618:133–145.

27. Fido, R. J., Tatham, A. S., and Shewry, P. R. 1995. Western blotting analysis. *Methods Mol Biol* 49:423–437.

28. Butler, J. E. 2000. Enzyme-linked immunosorbent assay. *J Immunoassay* 21:165–209.

29. Malling, H. V. 1970. Chemical mutagens as a possible genetic hazard in human populations. *Am Ind Hyg Assoc J* 31:657–666.

30. Ames, B. N., Lee, F. D., and Durston, W. E. 1973. An improved test system for the detection and classification of mutagens and carcinogens. *Proc Natl Acad Sci USA* 70:782–786.

31. Møller, P. 2006. The alkaline comet assay: Towards validation in biomonitoring of DNA damaging exposures. *Basic Clin Pharmacol Toxicol* 98:336–345.

32. Burlinson, B., Tice, R. R., Speit, G. et al. 2007. Fourth international workgroup on genotoxicity testing: Results in the in vivo comet assay workgroup. *Mutat Res* 627:31–35.

33. Gerberick, G. F., Ryan, C. A., Dearman, R. J., and Kimber, I. 2007. Local lymph node assay (LLNA) for detection of sensitization capacity of chemicals. *Methods* 41:54–60.

34. McGarry, H. F. 2007. The murine local lymph node assay: Regulatory and potency considerations under REACH. *Toxicology* 238:71–89.

35. Basketter, D. A., McFadden, J. F., Gerberick, F., Cockshott, A., and Kimber, I. 2009. Nothing is perfect, not even the local lymph node assay: A commentary and the implications for REACH. *Contact Dermat* 60(2):65–69.

36. Vinardell, M. P. and Mitjans, M. 2008. Alternative methods for eye and skin irritation tests: An overview. *J Pharm Sci* 97:46–59.

37. Hartung, T. 2002. Comparison and validation of novel pyrogens tests based on the human fever reaction. *Altern Lab Anim* 30(Suppl 2):49–51.

38. Fennrich, S., Fischer, M., Hartung, T., Lexa, P., Montag-Lessing, T., Sonntag, H. G., Weigandt, M., and Wendel, A. 1999. Detection of endotoxins and other pyrogens using human whole blood. *Dev Biol Stand* 101:131–139.

39. van Tienhoven, E. A. E., Korbee, D., Schipper, L., Verharen, H. W., and De Jong, W. H. 2006. In vitro and in vivo (cyto)toxicity assays using PVC and LDPE as model materials. *J Biomed Mater Res* 78:175–182.

40. Katsnelson, A. 2009. The scientists: Newsblog: Are lab standards harmful? http://www.the-scientist.com/blog/print/55552/ (accessed July 27, 2009).
41. Richter, S. H., Garner, J. P., and Würbel, H. 2009. Environmental standardization: Cure or cause of poor reproducibility in animal experiments? *Nat. Methods* 6:257–261.
42. El Ghalzbzouri, A., Commandeur, S., Rietveld, M. H., Mulder, A. A., and Willemze, R. 2008. Replacement of animal-derived collagen matrix by human fibroblast-derived dermal matrix for human skin equivalent products. *Biomaterials* 30:71–78.
43. Walles, T., Weimer, M., Linke, K., Michaelis, J., and Mertsching, H. 2007. The potential of bioartificial tissues in oncology research and treatment. *Onkologie* 30:388–394.
44. Yliperttula, M., Cung, B. G., Navaladi, A., Manchabi, A., and Urtti, A. 2008. High-throughput screening of cell responses to biomaterials. *Eur J Pharm Sci* 35:151–160.
45. Anderson, D. G., Levenberg, S., and Langer, R. 2004. Nanoliter-scale synthesis of arrayed biomaterials and application to human embryonic stem cells. *Nat. Biotechnol.* 22:863–866.

Dominique Chauvel-Lebret, Pascal Auroy, and Martine Bonnaure-Mallet

CONTENTS

14.1 OVERVIEW

14.1.1 INTRODUCTION

Elastomers, like plastics, are polymers. The term elastomer is currently used to designate all rubbers, i.e., natural or synthetic macromolecular materials with rubber-like elasticity. Medical applications for elastomers appeared soon after they began being produced on an industrial scale (Yoda 1998) following the discovery of vulcanization (crosslinking) in 1839 by the American Charles Goodyear. Over the past 15 years, the literature on biomedical applications for elastomers has grown considerably and reports have appeared regularly (Leeper and Wright 1983, McMillin 1994).

14.1.2 MISCELLANEOUS

14.1.2.1 Definition

While they are members of the broad family of polymers, elastomers behave differently than plastic materials or plastomers. Generally speaking, for materials to be considered elastomers they must be

- Flexible, i.e., have low rigidity (several megapascals)
- Highly deformable, i.e., are able to withstand strong deforming forces without rupturing and have an elongation at rupture over 200% while possessing relatively high tensile strength at ultimate elongation
- Elastic or resilient, i.e., are able to return to their original shape and size after the deforming force is removed and quantitatively release the energy used to deform them

(Table 14.1) (Pinchuk 1994).

Many so-called biomedical elastomers do not meet the second criterion. While it is difficult to set an elastic limit beyond which a polymer stops being a plastomer and becomes an elastomer, we will use the broadest medical definition and refer to all polymers with sufficient elasticity to return to their original shape after a substantial deformation as elastomers (Mardis et al. 1993). However, it will occasionally be more appropriate to talk about elastic polymers than true elastomers.

14.1.2.2 Classification of Elastomers

14.1.2.2.1 The Various Families of Elastomers

There are some 15 families of elastomers, some of which include 10–20 different grades. Modern polymerization processes increasingly allow the manufacture of customized products that address specific application problems and provide specific elastomeric properties.

The current classification system separates elastomers into four categories (the acronyms are based on the ISO 1629:1995 standard):

1. *General-use elastomers*, which include natural rubber (NR), synthetic polyisoprene (IR), styrene butadiene copolymers (SBR), and polybutadienes (BR)
2. *Special elastomers*, which include ethylene propylene and diene copolymers (EPM and EPDM), isobutylene and chloro- and bromo-isobutylene isoprene copolymers (IIR, BIIR, CIIR), nitrile butadiene copolymers (NBR), and polychloroprenes (CR)
3. *Very special elastomers* with high thermal and/or chemical resistance, including silicone elastomers (VMQ), fluoro-elastomers (FPM), chloro-polyethylenes and chloro-sulfonyl polyethylenes (CM, CSM), polyacrylates (ACM), ethylene vinyl acetate elastomers (EVA), ethylene methyl acrylate (EAM), hydrogenated nitrile elastomers (HNBR), and epichlorhydrin elastomers (CO, ECO, GECO)
4. *Thermoplastic elastomers*, which are in a separate category in that they do not need to be vulcanized since they can be used like any thermoplastic material

The first two families of elastomers account for 97% of total world consumption.

14.1.2.2.2 Very Special Elastomers

Polysiloxanes or silicone elastomers are the most abundantly produced very special elastomers. The backbone of these elastomers is not a chain of carbon atoms but an arrangement of silicon and oxygen atoms. Their most remarkable quality is their outstanding thermal performance. They retain their service properties from −85°C to 250°C. However, their mechanical properties are relatively weak. They are used for a broad range of applications in the electrical, electronic, and automotive industries. Their physiological inertness makes them well suited for use in the biomedical field.

TABLE 14.1

Some Commercially Available Elastomers Arranged by Tensile Strength

Elastomer	Trade Name	Supplier	Tensile Strength (PSI)	Tensile Strength (MPa)	Ultimate Elongation (%)
Butyl rubber (polyisobutylene)	Butyl	Polysar Inc., Akron, OH	700–1,500	4.8–10.3	200–250
Polyisoprene	Santoprene	Monsanto Co., Akron, OH	1,640	11.3	520
Silicone rubber	Silastic	Dow Corning, Midland, MI	500–1,500	3.4–10.3	200–1000
	Med	McGhan NuSil Corp., Carpenteria, CA	750–1,750	5.2–12.1	850
	SM	Sil-Med Corp., Taunton, MA	500–1,500	3.4–10.3	250–900
Silicone rubber copolymers	C-Flex	Concept polymer, Inc., Clearwater, FL	1,000–2,000	6.9–13.8	200–600
Polyvinylchloride (PVC)	Goodtouch	B.F. Goodrich, Cleveland, OH	1,700–1,850	11.7–12.7	430–460
	Tygon	Norton Perform. Plastic, Wayne, NJ	1,200–2,300	8.3–15.9	375–410
Ethylene vinyl acetate (EVA)	Ultrathene	U.S. Industrial Chem. Co., New York, NY	2,000	13.8	700
Acrylonitrile-butadiene	Krynac	Polysar Inc., Akron, OH	500–2,250	3.4–15.5	150–650
Polyhexene	Hexsyn®	Goodyear Tire CO., Akron, OH	1,500–2,400	10.3–16.5	250–350
Polychloroprene	Neoprene	E.I. DuPont, Wilmington, DE	1,000–2,500	6.9–17.2	200–600
Fluoro-elastomers	Fluorel	3M, St. Paul, MN	1,500–2,500	10.3–17.2	100–400
Styrene butadiene rubber (SBR)	Krayton	Shell Chem. Co., Houston, TX	750–2,650	5.2–18.3	550–850
	Plioflex	Goodyear Tire Co., Akron, OH	200–2,825	1.4–19.5	290–900
Ethylene vinyl acetate (EVA)	EVA	E.I. DuPont, Wilmington, DE	3,200	22.1	600–750
Natural rubber (cis-polyisoprene)	Natsyn	Goodyear Tire CO., Akron, OH	2,200–3,450	15.2–23.8	500–600
Polyester terephthalate glycolated	Hytrel®	E.I. DuPont, Wilmington, DE	4,000–6,800	27.6–46.9	350–700
Nylon (amorphous)	Pebax	Atochem Polymers, Philadelphia, PA	4,200–7,400	29.0–51.0	350–680
Thermoplastic polyether urethanes	Tecoflex	Thermedics, Inc., Woburn, MA	3,000–6,000	20.7–41.4	250–600
Thermoplastic polyester urethanes	Estane	B.F. Goodrich, Cleveland, OH	4,000–8,000	27.6–55.2	400–600
Thermoplastic polycarbonate ureth	Corethane®	Corvita Corp., Miami, FL	4,500–9,000	31.0–62.1	250–500
Chlorotrifluoroethylene	Aclar	Allied-Signal, Morristown, NJ	7,500–11,000	51.7–75.8	200–300
	Saran	Dow Chemical, Midland, MI	8,000–16,800	55.1–115.8	70–90

Source: Pinchuk, L., *J. Biomater. Sci. Polym. Ed.,* 6, 225, 1994.

Fluorocarbons are another family of high-performance elastomers containing 70% fluoride, which gives them outstanding resistance to heat and chemicals. They can be used at temperatures up to 250°C. Their cold resistance, however, is less than that of silicones (−30°C). Because of their passivity and insolubility, they are useful in the biomedical field.

Polyacrylates are specially designed to resist oils containing sulfur-based additives. They have high heat (180°C) and oil swell resistance. They are used as components of grafted and linear biomedical elastomers (Polyurethane (PU), Styrene Ethylbutylene Styrene (SEBS)).

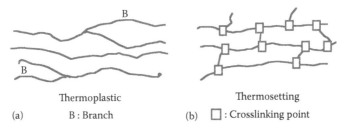

Thermoplastic   Thermosetting

(a)   B : Branch   (b)   ☐ : Crosslinking point

FIGURE 14.1 **(See companion CD for color figure.)** Schematic drawing of (a) thermoplastic and (b) thermosetting elastomer structure.

Chloro-polyethylenes and chloro-sulfonyl polyethylenes compete with polychloroprenes because of very similar chemical compositions. The high chemical saturation of the polymer chain gives them better heat resistance (120°C–130°C), and they are often used for wires and cables.

Hydrogenated nitrile, one of the most recent elastomers to arrive on the market, has hydrogenated double bonds, which provide it with better heat resistance (150°C) than conventional nitrile rubber as well as with excellent resistance to oils and ozone.

14.1.2.2.3 Thermoplastic Elastomers

Polymers are commonly defined by their thermal properties, i.e., thermoplastics and thermosetting materials. Thermoplastics are composed of independent, linear molecules (which may have side branches). They can be melted (e.g., polyethylene) and are processed by heating. Thermosetting materials are composed of networks of covalently linked chains. They are initially polymerized at room temperature or by heating and cannot be remelted once they have solidified (e.g., silicone elastomers). The reality is a little more complex since certain elastomers have intermediate structures and properties (Figure 14.1).

Many types of thermoplastic elastomers (TPEs) are being developed. They have certain advantages with respect to classic elastomers and are replacing them in some applications. Thermoplastic elastomers can be defined as materials that combine the flexibility or elasticity of rubber compounds with the thermoreversible versatility of thermoplastic polymers. Most require little or no formulation. If formulation is required, it can be performed by the producer or by an intermediary between the producer and the processor—the compounder. Thermoplastic elastomers are relatively simple to use and require fewer processing steps than thermosetting elastomers (no mixing or crosslinking), thus reducing costs. Processing time is also much shorter, reducing energy consumption. Processing methods are nonspecific and more numerous, being the same as those used for plastomers. Unlike classic elastomers, thermoplastics can be recycled. Manufacturing scraps and clippings can be reused with little or no change in their properties. Despite these numerous advantages, most of these elastomers do have drawbacks, including their temperature sensitivity. According to the Freedonia Group at the 11th International Thermoplastic Elastomers Conference in 2008, global TPE demand is expected to grow more than 6% per year and reach 3.7 million metric tons by 2011 for a market value of approximately U.S$12 billion. The use of TEPs in medical applications has risen from 3% to 5% in 10 years.

14.1.2.3 Elastomer Chemistry and Synthesis

14.1.2.3.1 Structure

Elastomers are composed of macromolecular chains that occasionally exceed 100,000 atoms in length, fold back on themselves when at rest, and have a very high degree of self-cohesion (bond energies of several hundred kJ/mole). On the other hand, cohesion between different chains is weak because of the small number of bonds and their relatively low energy (covalent, Van der Waals, polar bonds). Because of their ability to rotate around their carbon-carbon or silicon-silicon catenary backbones, the chains can twist into various shapes and are thus very flexible at their service temperature. When subjected to a deforming stress, these entangled macromolecular chains begin to unfold, sliding over each other, and separating along the direction of the stress, giving the material its elastomeric property. The greater the number of bonds between chains and the more chemically stable the bonds (crosslinking), the less likely the material is to deform under stress. What limits rotation of the catenary backbone and sliding is the formation of chemical bonds (generally covalent) between the chains, thereby leading to the formation of a stable, three-dimensional network. The greater the crosslinking, the more rigid the polymer. The formation of chemical bonds produces various active sites on the macromolecular chains (double bonds, hydrogen atoms, labile halogen atoms, etc.). From a thermodynamic perspective, when an elastomer is subjected to a tensile stress, it may stretch to several times its free length when the macromolecular chains unfold. The energy transferred to the elastomer makes it thermodynamically unstable. When the stress is removed, the macromolecules return to their initial, energetically stable, intertwined state, releasing most of the input energy.

14.1.2.3.2　Polymerization

True polymerization only involves the formation of macromolecules by the creation of multiple carbon–carbon bonds. However, in a larger sense, it simply means the formation of macromolecules. Monomers with accessible double bonds, i.e., bonds that are not obstructed by large substituents may form giant molecules. The double bonds can be rearranged by catalysts such as radiation, oxygen, and free radicals (like those produced by the dissociation of peroxides), giving a new radical that attacks another monomer, in a chain reaction. This is called *free radical polymerization*.

The catalyst may attack the main chain, producing side chains (Figure 14.2). However, in principle, bridges never form between two chains (see Crosslinking). Certain molecules with different functional groups can combine into a single molecule, e.g., acid and alcohol or aldehyde and amine, etc. Molecules with different functional groups can combine into macromolecules. Molecules with identical functional groups can also react with other molecules with antagonistic functional groups to produce macromolecules. Siloxanes are an example of this. They condense to form polysiloxanes (silicon elastomer).

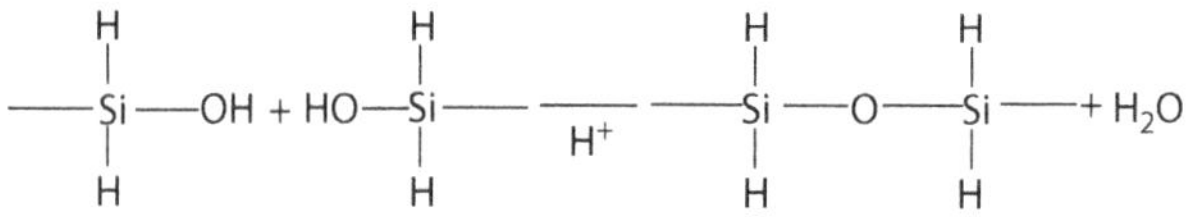

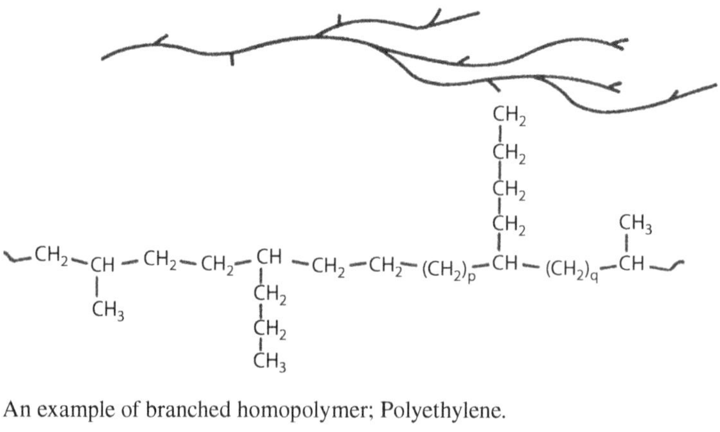

FIGURE 14.2　An example of branched homopolymer; Polyethylene.

Another example is carbonyl chloride (phosgene) and bisphenol A, which combine to produce polycarbonate.

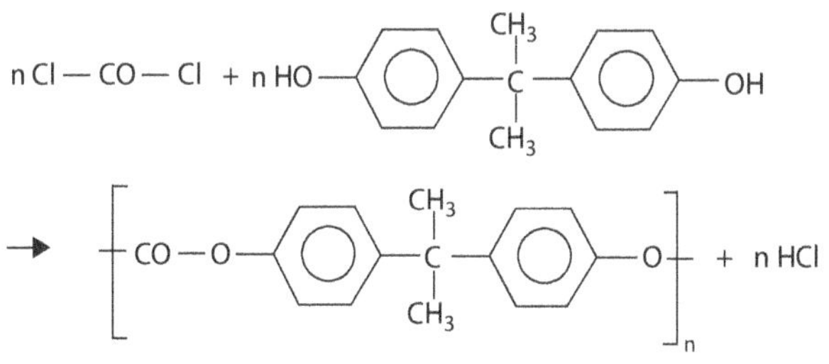

In the two examples given earlier, for each bond created, a small molecule (H_2O, HCl) is lost in a reaction called *polycondensation*. In other cases, as with the formation of urethanes

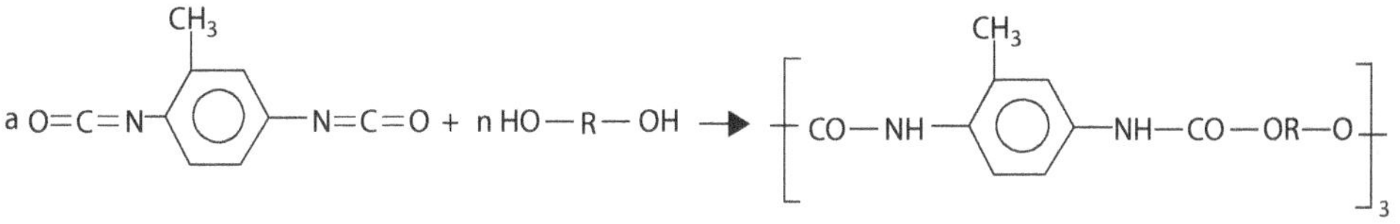

and other forms of polysiloxanes, no molecules are lost. This is referred to as *polyaddition*.

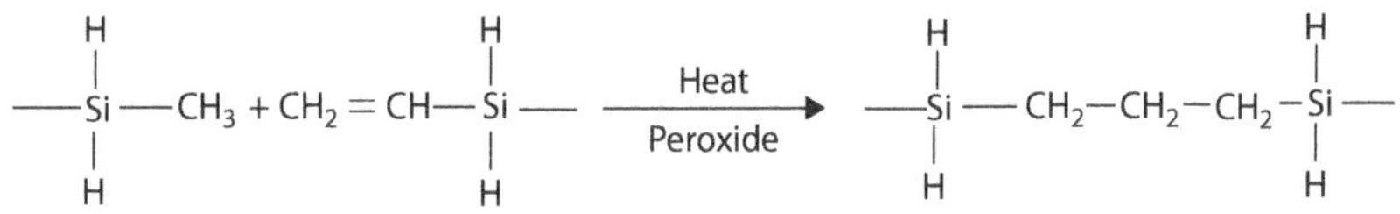

A wide variety of catalysts (specific to each reaction) are involved in free radical polymerization reactions (polyaddition or polycondensation), all of which lead to long polymers that concatenate either the same monomer (homopolymers) or different monomers (copolymers).

14.1.2.3.2.1 Homopolymerization The linear bonding of structurally identical small molecules (monomers) leads to the formation of homogenous polymers. Most commonly used biomedical elastomers are homopolymers (Figure 14.2). However, the possibility of mixing different monomers and/or small polymers is leading to the synthesis of new polymers with infinitely varied properties, which, because of their improved performance, are increasingly replacing homopolymers for specific applications.

14.1.2.3.2.2 Copolymerization Different types of monomers can be copolymerized or mixed with small reactive polymers to obtain alternating (-A-B-A-B-A-B-), sequential (-A-A-A-B-B-B-B-A-A-A-B-B-B-), or random (A-B-B-A-B-A-A-B-) copolymers with different block structures containing two, or occasionally more, identical monomers (Figure 14.3).

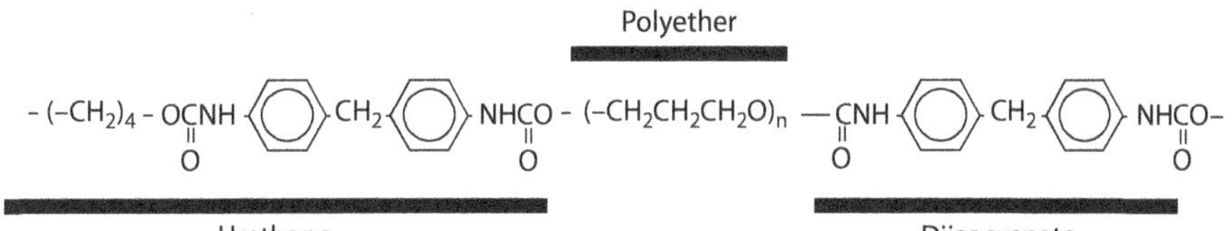

FIGURE 14.3 Different blocks of a terpolymer segmented polyetherurethane; Pellethane® (A-(Bn)-C).

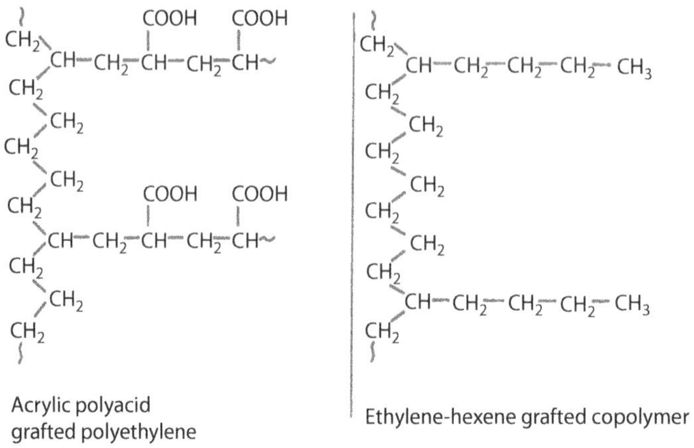

FIGURE 14.4 **(See companion CD for color figure.)** Monomers attached to existing chains to create branched or grafted copolymers.

Other types of monomers or small polymers can also be attached to existing chains to create branched or grafted copolymers (Figure 14.4). Alterations to the catenary or terminal structures of copolymers change their properties, creating new types of elastomers, some of which have potentially interesting biomedical applications. It would be more accurate to talk about bi and terpolymers when a reaction involves two or three types of monomers, but the term copolymer is used to describe all polymers containing two or more different types of monomers.

14.1.2.3.3 Crosslinking

When polymers have reactive sites within the chain (e.g., the double bonds of unsaturated polyesters) or at the terminal (e.g., the hydroxyl group of polyols), they can be reacted with a polyfunctional agent to create a network in a process referred to as crosslinking or vulcanization. Such agents are generally called catalysts, hardeners, or coupling agents, and the initial polymer is called a prepolymer (Figure 14.5).

Crosslinking can be performed by free radical polymerization using organic peroxides, which greatly increases the thermal resistance of the network by creating very strong carbon-carbon bridges. Other types of crosslinking agents can be used, especially for elastomers with specific reactive groups (carboxyls, halogens, etc.).

14.1.2.3.4 Formulation

Apart from a few rare exceptions, the properties of "green" elastomers do not allow them to be used as is. Various additives, all with well-defined functions, must be incorporated. This is what is called formulation. Despite a certain empirical approach, formulation requires a great deal of know-how to compensate for the occasionally antagonistic properties required by certain applications.

A standard formulation contains

- Elastomer(s): Specific properties, especially chemical properties, must address application specifications. A mixture of elastomers may be used.
- Filler(s): Fillers are used to provide superior mechanical properties, and also to improve handling and to lower production costs. Precipitated silica is the most common additive in biomedical elastomers. Chalks and kaolins are also occasionally used.
- Plasticizers: Plasticizers are used to make processing easier and improve low temperature flexibility. Heavy solvents, oils, and esters are generally used.
- Protective agents: These are used to protect the macromolecular chains from oxygen, ozone, and ultraviolet radiation. Amine and phenol derivatives are generally used.

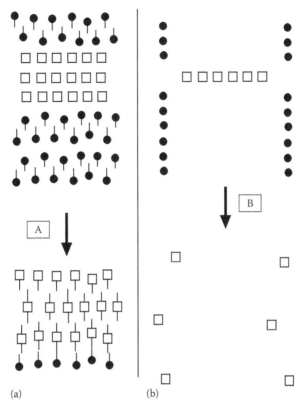

FIGURE 14.5 Crosslinking of a polymer. (a) Crosslinking of a polymer with reactive side groups. (b) Crosslinking of a polymer with reactive terminal groups.

- A crosslinker: These allow the formation of a stable, three-dimensional network without which the elastomer would be useless.
- Various other ingredients: Numerous other ingredients with very specific functions such as colorants, conditioning agents, swelling agents for manufacturing cellular elastomers, etc.

It is clear that the release of additives like plasticizers, antioxidants, initiators, catalysts, fillers, etc. by biomedical elastomers may have a significant impact on host tissues.

14.1.2.3.5 Elastomer Reinforcement

Reinforcement fillers are quite often added because elastomers have fairly low breaking strengths at ambient temperature. Adding more precipitated silica enhances breaking strength, but decreases flexibility. The amount of filler can be changed to generate variable stress-strain curves and produce elastomers with specific properties. Fillers like semi-reinforcing kaolins can be used to generate economical, clear mixtures. The same is true of chalks, which are considered as diluents. Elastomers are also occasionally reinforced by the addition of bulk fibers or tissues (glass, carbon, aramide) for very specialized aeronautics and medical applications for example.

Exogenous reinforcing materials can affect the biocompatibility of finished materials. However, natural networks composed of two or three interpenetrating polymers can have major advantages. Interpenetrating networks are based on at least two polymers chemically linked to form at least one network and are synthesized in the presence of or at the same time as the other constitutive polymers (Sperling and Hu 2002, Sperling and Misra 1997). The properties of interpenetrating networks depend on the kinetics of formation, network crosslinking, and polymer miscibility during synthesis.

However, numerous other factors play a major role, including the synthesis method itself, the composition of the polymers and their respective quantities, the density of potential crosslinking sites of each polymer, their glass transition temperature, and their crystallinity (Lipatov 1990, Xu et al. 2000).

Many classes of polymers can be used to synthesize interpenetrating networks, notably polyurethanes, which readily lend themselves to such a use. The mechanical properties of the synthesized materials are improved while there is little or no effect on their biocompatibility (Karabanova et al. 2006).

14.1.2.3.6   Surface Modification of Elastomeric Biomaterials

In recent years, progress in understanding the interactions between biomaterials and cells has revealed that the predominant factors in play are the physical and chemical properties of the surface of the biomaterials. The development of tissue-engineered biomaterials, that is, biomaterials capable of eliciting cell behavior (adhesion, migration, proliferation, differentiation, etc.) is testimony to this progress. The adhesion of cells to a biomaterial surface is a sine qua non for eliciting other cell behaviors by the biomaterial. The fact that cell adhesion is influenced by the surface properties of biomaterials (wettability, surface energy, roughness, topography, ionic interactions, electric charge, etc.) explains why so much effort is expended to modify surface properties to improve biological interactions. Of course, these modifications do not alter the bulk properties of the materials.

Surface modification techniques make it possible to change the morphology and physicochemical properties of the surface and to deposit biologically active molecules such as peptides and proteins on it. (Table 14.2) (Jiao and Cui 2007). Ma et al. recently published a complete review on this topic (Ma et al. 2007).

14.1.2.3.7   Elastomeric Biomaterials and Micro- and Nanotechnology

The emergence and rapid development of nano- and microtechnologies—i.e., the development and exploitation of structures and devices at nano- and micrometer levels—has opened vast new vistas in the life, health, biomaterials, and biomedical elastomer sciences. Nanotechnologies and nanomanufacturing can be used to optimize the polymer structure, formulations, strength, and surface properties of elastomeric biomaterials. Nanomanufacturing relies on the mastery of surface science, especially in biology. The nano- and microtopographies of surfaces are as important as physicochemical properties for biological interfaces (Kasemo 1998).

The creation of surfaces that copy the molecular organization of natural tissues at the nanometer level makes it possible for biomaterials to present the desired biological signal with much greater precision. The biomolecules used in what are called "precision immobilization strategies" to create biomimetic surfaces include proteins, polypeptides, nucleotides, lipids, and polysaccharides. The biomolecules immobilized on the surface of the material in one position, one conformation, and one specific density provide the surrounding tissue with one surface organized to better control cell interactions. (Castner and Ratner 2002). For example, the deposition of nanostructure films of amphiphilic macromolecules, that are macromolecules with both hydrophilic and hydrophobic domains, has been used to control cell adhesion (Hung et al. 2009b, Jiao and Cui 2007).

Nanoencapsulation of elastomeric biomaterials is another approach for optimizing surface properties through the deposition of nanofilms. Based on its physicochemical properties and thickness, a nanofilm deposited on the surface of a biomaterial will behave as a selective barrier regulating exchanges between the underlying biomaterial and the surrounding tissue. For example, encapsulating the silicone-hydrogel of contact lenses with a plasma-deposited nanofilm prevents the absorption of proteins by the hydrogel and the diffusion of constitutive macromolecules from the material while remaining permeable to small molecules such as water, carbon dioxide, and oxygen (Yasuda 2006).

Nanofibers are found naturally in living tissues. The fabrication of biomaterials with synthetic biomimetic nanofibers either incorporated in the biomaterial, exposed on their surfaces, or formed into networks has given rise to numerous biomedical applications, tissue engineering scaffolds, and drug delivery systems. Most of the polymer fibers, which have diameters ranging from a few nanometers to a few microns, are produced by electrospinning.

TABLE 14.2
Surface Modifications of Elastomeric Biomaterials

A. Surface modification techniques used to produce various surface properties of biomaterials

Type of Modification	Properties of Modified Surface
Morphological modifications	1. Coating containing pores to encourage tissue ingrowth (porosity and roughness) 2. Grooved surface to prevent epithelial down-growth on dental implants and direct bone formation along the particular region of an implant
Chendeal modifications	1. Glow discharge to increase surface free energy and tissue adhesion (composition and charge) 2. Crosslinked polymeric surface to decrease surface permeability and increase surface hardness 3. Plasma treatment with reactive gases to create new functional groups on polymer surface 4. Ion implantation to cause chemical and structural alterations to increase corrosion/wear resistance 5. Grafting macromolecules such as PEO to reduce protein adsorption and cell adhesion. Functional groups used to produce positively (amino) or negatively (acidic and sulfonate) charged surface
Biological modifications	1. Immobilizing molecules (ROD, heparin/heparin sulfate-binding peptides, proteins like fibronectin and grotl1 factors) on biomaterials to enhance cell adhesion and cell growth 2. Immobilizing phosphorylcholine on biomaterials to resist protein and cell adhesion for applications where the material comes into contact with blood

B. Surface modifications with biomacromolecules

Modification	Mechanism	Method
Chemical graft	Immobilizing bioactive peptides, proteins, and polysaccharides to improve cell adhesion and biocompatibility	Chemical graft, photochemical graft, plasma graft
Surface coating	Immobilizing bioactive macromolecules and growth factors to promote cell affinity	Physical adsorption
Entrapment	Immobilizing bioactive or water-soluble biomacromolecules to improve hydrophilicity and blood compatibility	Swelling in a mixture solution and entrapment in a nonsolvent to produce a stably modified surface
Electrostatic	Immobilizing charged, hydrophilic and bioactive macromolecules	Depositing polyanions and polycations alternately

Source: Jiao, Y.P. and Cui, F.Z., *Biomed. Mater.*, 2, R24, 2007.

This process is being used in protective clothing, air filtration, template nanofiber, and polymer nanotube applications. It uses an electrical charge to draw very fine fibers from a liquid. The process is noninvasive and does not require the use of coagulation chemistry or high temperatures to produce solid threads from solution. This makes the process particularly suited to the production of fibers using large and complex molecules of nano- and microfiber biomaterials. These fibers can be used, among other things, to create nanofiber tubes from biodegradable polyester to guide the regeneration of peripheral nerves, as well as bone, muscles, and endothelial tissues (Berkland et al. 2004, Bini et al. 2004, Gu and Ren 2005, Kim et al. 2003, Kwon et al. 2005, Luu et al. 2003, Mo et al. 2004, Wutticharoenmongkol et al. 2006a,b, Yang et al. 2005). The addition of metallic nanoparticles makes it possible to produce polymer–metal composites with better properties than elastomeric biomaterials. For example, adding gold nanoparticles to polyurethane induces a morphological transformation of the surface of the nanocomposite, providing it with long-lasting resistance to biodegradation (Chou et al. 2008, Hsu et al. 2006).

Many biomaterials never come in direct contact with living tissues as such, just with molecules, proteins, fluids, and tissue extracts of living organisms. This is the case for materials specifically developed for biological laboratory applications. Micro- and nanobiotechnologies have also given rise to new applications in this area. The infinitesimal machines that Richard Feynmann dreamed of (Feynmann 1992, 1993) are now a reality in a broad range of major applications in cellular biology, immunological testing, DNA sequencing, and protein crystallization (Balagadde et al. 2005, Fu et al. 2002, Hansen et al. 2002, Hong et al. 2004, Jiang et al. 2003, Kartalov and Quake 2004, Liu et al. 2002, 2003, Murakami et al. 2004, Sohn et al. 2000, Wu et al. 2004).

Many of these nanomachines are composed of polydimethylsiloxane elastomer, which is used in various nanolithography techniques (Kuncova-Kallio and Kallio 2006, Rogers and Nuzzo 2005, Wacaser et al. 2003, Whitesides et al. 2001). The moldability of silicone elastomers makes it possible to reproduce details less than 10 nm in size (Kartalov et al. 2006). In addition, they have no effect on molecules, proteins, cells, or tissues because of their chemical inertness and biocompatibility. They are thus often used to construct microfluidic channels that consist of a surface micromachined labyrinth with several inlets and outlets.

Soft polysiloxane elastomers have a promising future in micro- and nanotechnology biomedical applications and research (Kartalov et al. 2006, Quake and Scherer 2000).

14.1.3 SILICONE ELASTOMERS

14.1.3.1 Introduction

Silicones (a contraction of silicon ketone) make up a vast family of polymers with remarkable properties due to the presence of both silicon–oxygen and silicon–carbon bonds. Depending on the nature of the functional groups attached to the silicon atoms, silicone polymers may be resins, oils, or elastomers, making them well-suited for diverse applications in a wide variety of areas. Silicon elastomers are composed of very long polymers containing several thousand silicon atoms per molecule (Bischoff 1972). The first biomedical applications for silicone elastomers can be traced back to the 1950s, while industrial-scale manufacturing dates to 1965 (Jones 1988, Toub 1987). Today, silicone elastomers are the most widely used polymers in medical applications (Yoda 1998) because of the strong, very mobile bonds of their Si-O-Si (siloxane) catenary backbone, which provide elevated chemical inertness and exceptional flexibility. They are also very stable over time and at body temperature, show little tissue reactivity, and are highly resistant to chemical attack and heat, which allows them to be autoclaved. They also have exceptional mechanical properties such as high tear strength, outstanding elasticity, and high gas permeability, which make them suitable for many medical applications (contact lenses, special dressings, air/blood filter membranes, tissue expanders, artificial dermises, artificial esophaguses, intraocular lenses, artificial cornea, etc.) (Boone and Braley 1966, Ren et al. 2008, Vondracek 1981) (Table 14.3).

TABLE 14.3

Properties of Silicones for Use in Biomedical Applications

Good elastomeric and relatively uniform properties over a wide temperature range

Physiological indifference

Good low-temperature resistance and stable at high temperatures

Excellent resistance to biodegradation

Excellent resistance to oxidation and ultraviolet light

Moderate biocompatibility

Outstanding resistance to aging

Excellent dielectric behavior over a wide range of temperatures

14.1.3.2 Manufacture

Chlorosilanes, especially dichlorosilanes and trichlorosilanes (where R is an organic radical), are the basic building blocks of silicone elastomers.

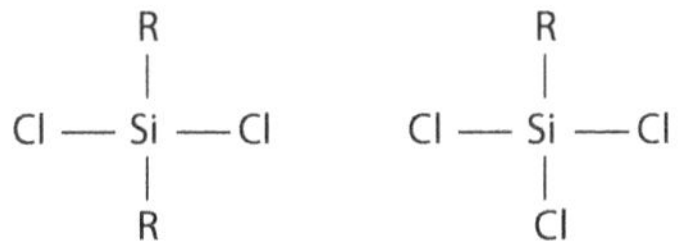

$$
\begin{array}{ccc}
& R & \quad\quad\quad R \\
& | & \quad\quad\quad | \\
Cl-&Si-Cl & \quad Cl-Si-Cl \\
& | & \quad\quad\quad | \\
& R & \quad\quad\quad Cl
\end{array}
$$

The main organic radicals used are methyl ($-CH_3$) and phenyl ($-C_6H_5$) groups. Others (ethyl, vinyl, etc.) are used for special application silicones. Chlorosilanes are complicated and costly to produce. They react with water to give silanols, which, being unstable, lose water molecules and condense into polymers—polysiloxanes.

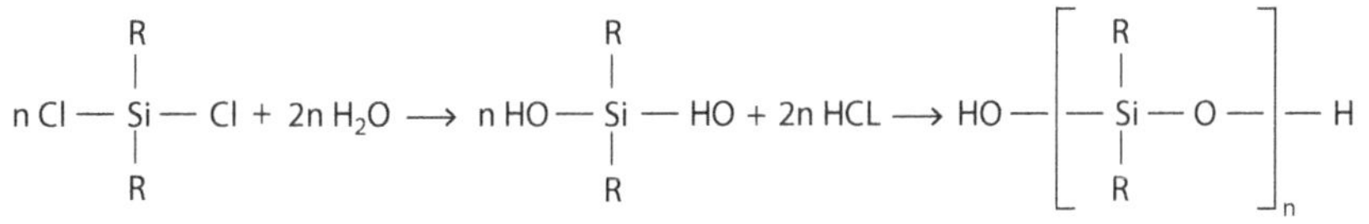

$$
\begin{array}{c}
R \\
| \\
n\,Cl-Si-Cl + 2n\,H_2O \longrightarrow n\,HO-Si-HO + 2n\,HCL \longrightarrow HO-\left[\begin{array}{c} R \\ | \\ Si-O \\ | \\ R \end{array}\right]_n-H \\
| \\
R
\end{array}
$$

Depending on the dichlorosilane-trichlorosilane ratio, different polymers are obtained. The greater the percentage of dichlorosilane, the more linear the polymer, and the greater the percent of trichlorosilane, the more branched the polymer. Depending on the reaction conditions, polymers with varying molecular masses can be produced. The terminal hydroxyl groups can react with other components or with each other in the presence of catalysts. The termination reaction produces the final silicone product.

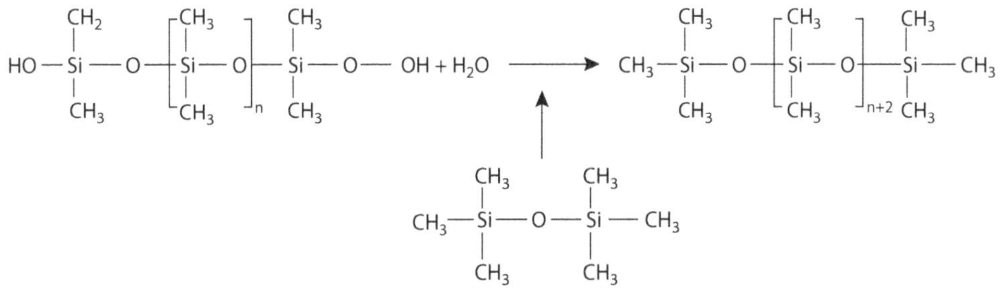

The properties of the finished product depend on the nature of the organic radicals grafted onto the catenary backbone, meaning that an infinite number of chemical structures—and properties—are possible.

The end products of this industrial process are polycondensates of basically linear or weakly branched polysiloxane molecules of varying molecular masses and consistencies with terminal reactive groups such as –Cl or –OH. These reactive polycondensates or prepolymers serve as the raw materials to produce the final silicone products by room temperature vulcanization (RTV) or high temperature vulcanization (HTV).

14.1.3.3 Homopolymerization of Silicone Elastomers

Most commonly used biomedical silicone elastomers are homopolymers. Increasingly, however, various polysiloxane prepolymers as well as other types of prepolymers with varying properties are being combined to synthesize copolymers for the biomedical market. Different polymerization processes can be used to produce a variety of silicone elastomers.

One-component RTV silicones result from a reaction between atmospheric moisture and a mixture of polycondensates and catalyst. Crosslinking thus occurs from the outside in, limiting the useful thickness.

$$
\text{—Si—OH} + \text{HO—Si—} \xrightarrow{\text{H+}} \text{—Si—O—Si—} + H_2O
$$

Such silicones are only used for adhesives, coatings, and sealants. Their biomedical utility arises from the absence of fillers and additives, which gives them excellent biocompatibility (Ashar et al. 1981).

Two-component RTV silicones result from a reaction between a polycondensate and a crosslinker added to initiate the reaction. They are used in the medical field for limited production runs of molded products.

• *Crosslinking can occur by polycondensation* of a relatively low molecular weight polydimethyl siloxane with a reactive hydroxyl group in the presence of tin octoate (i.e., condensation cure)

$$
\text{—Si—H} + \text{HO—Si—} \xrightarrow[\text{salt}]{\text{Metal}} \text{—Si—O—Si—} + H_2
$$

• *By polyaddition* of a relatively low molecular weight vinylsiloxane, a polysiloxane with a terminal hydrogen, and an organometallic catalyst like chloroplatinum acid (e.g., RTV 71556 Silbione from Rhodia, formerly Aventis or Rhône-Poulenc).

$$
\text{—Si—H} + CH_3\text{—CH—Si—} \xrightarrow{\text{Organometallic catalyst}} \text{—Si—}CH_3\text{—}CH_2\text{—Si—}
$$

HTV elastomers are produced by heat vulcanizing similar but higher molecular weight polycondensates premixed with crosslinkers (peroxides).

$$
\text{—Si—}CH_3 + CH_2\text{=CH—Si—} \xrightarrow[\text{Peroxide}]{\text{Heat}} \text{—Si—}CH_2\text{—}CH_2\text{—}CH_2\text{—Si—}
$$

They are used in small and large-scale industrial production runs by injection molding, extrusion, calendaring, etc. (e.g., MDX 4 4210 Silastic from Dow Corning).

14.1.3.4 Copolymerization of Silicone Elastomers

Silicone elastomer copolymers can be produced by addition or condensation. Condensation generally involves low to medium molecular weight prepolymers. For example, vinyl polymethylsiloxane is condensed with hydroxy polymethylsiloxane to create a new prepolymer—siloxyethanol ether—which can in turn be condensed with diisocyanate to produce silicone urethane copolymers whose blood compatibility and stiffness have been improved to the point where they can be used as blood catheters or blood pump coatings.

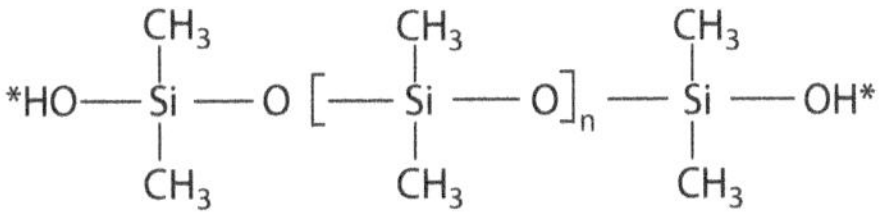

$$*HO-\underset{\underset{CH_3}{|}}{\overset{\overset{CH_3}{|}}{Si}}-O\left[-\underset{\underset{CH_3}{|}}{\overset{\overset{CH_3}{|}}{Si}}-O\right]_n-\underset{\underset{CH_3}{|}}{\overset{\overset{CH_3}{|}}{Si}}-OH*$$

Silanol terminated

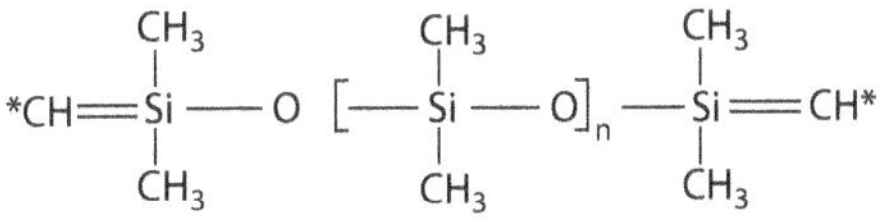

$$*CH=\underset{\underset{CH_3}{|}}{\overset{\overset{CH_3}{|}}{Si}}-O\left[-\underset{\underset{CH_3}{|}}{\overset{\overset{CH_3}{|}}{Si}}-O\right]_n-\underset{\underset{CH_3}{|}}{\overset{\overset{CH_3}{|}}{Si}}=CH*$$

Vinyl terminated

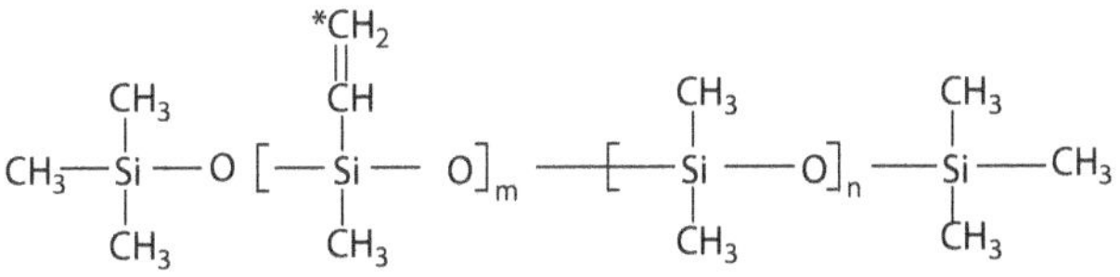

$$CH_3-\underset{\underset{CH_3}{|}}{\overset{\overset{CH_3}{|}}{Si}}-O\left[-\underset{\underset{CH_3}{|}}{\overset{\overset{\overset{*CH_2}{\|}}{CH}}{Si}}-O\right]_m-\left[-\underset{\underset{CH_3}{|}}{\overset{\overset{CH_3}{|}}{Si}}-O\right]_n-\underset{\underset{CH_3}{|}}{\overset{\overset{CH_3}{|}}{Si}}-CH_3$$

Vinyl terminated

$$CH_3-\underset{\underset{CH_3}{|}}{\overset{\overset{CH_3}{|}}{Si}}-O\left[-\underset{\underset{CH_3}{|}}{\overset{\overset{*H}{|}}{Si}}-O\right]_n-\underset{\underset{CH_3}{|}}{\overset{\overset{CH_3}{|}}{Si}}-CH_3$$

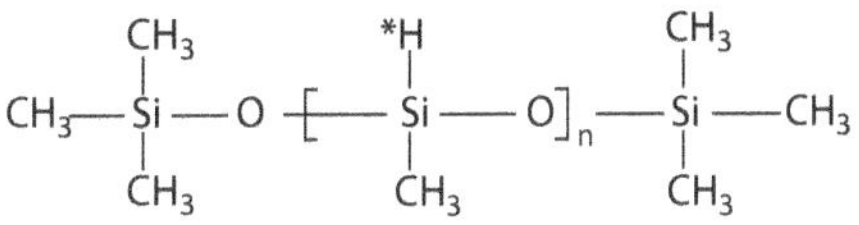

Polymethyl hydrosiloxane

Examples of different reactive silicone prepolymers (Whitford 1984)

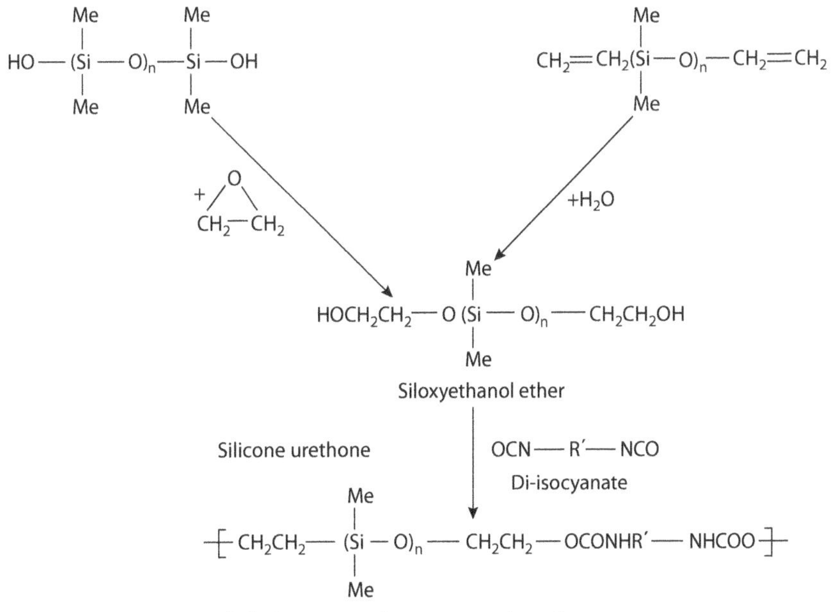

$$HO-\underset{\underset{Me}{|}}{\overset{\overset{Me}{|}}{(Si}}-O)_n-\underset{\underset{Me}{|}}{\overset{\overset{Me}{|}}{Si}}-OH$$

$$CH_2=CH_2(\underset{\underset{Me}{|}}{\overset{\overset{Me}{|}}{Si}}-O)_n-CH_2=CH_2$$

$$+\;\overset{O}{\overset{/\backslash}{CH_2-CH_2}}$$

$$+H_2O$$

$$HOCH_2CH_2-O\,(\underset{\underset{Me}{|}}{\overset{\overset{Me}{|}}{Si}}-O)_n-CH_2CH_2OH$$

Siloxyethanol ether

Silicone urethone

$$OCN-R'-NCO$$

Di-isocyanate

$$\left[CH_2CH_2-(\underset{\underset{Me}{|}}{\overset{\overset{Me}{|}}{Si}}-O)_n-CH_2CH_2-OCONHR'-NHCOO\right]$$

Polysiloxane copolymers — condensation

Copolymerization (Whitford 1984)

Condensation has also been used to produce silicone polycarbonate copolymers designed to transport ions. Their good mechanical resistance and high stability in aqueous environments make them useful for microelectrodes.

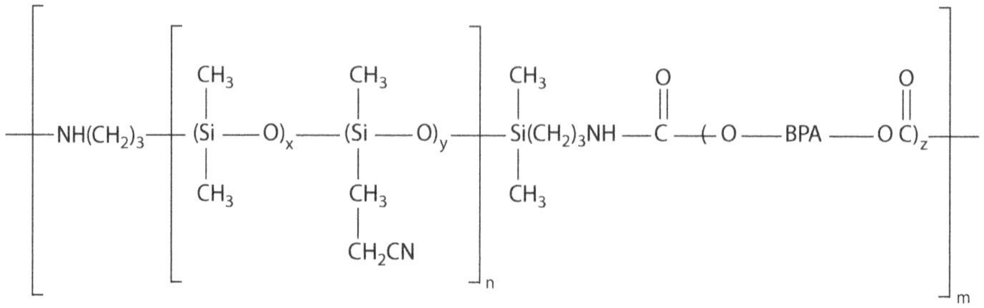

BPA = Bisphenol A

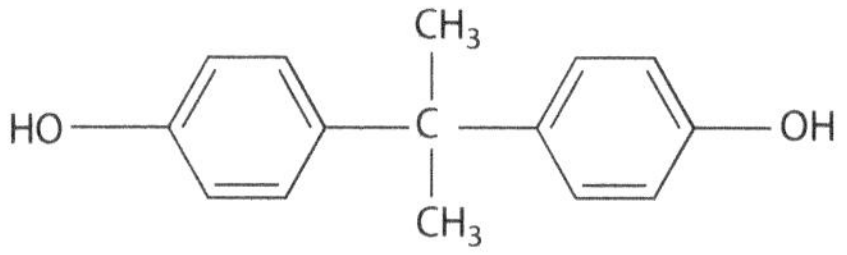

Silicone polycarbonate

(Whitford 1984)

Certain linear silicone elastomer copolymers are also prepared by addition curing. These are generally higher molecular weight polymers that can be processed or molded using industrial processes. Siloxane and methacrylate copolymers were developed to improve the properties of elastomer contact lenses (increasing their hardness, stiffness, transparency, wettability, and gas permeability) by combining the properties of the various monomers used to produce the copolymer.

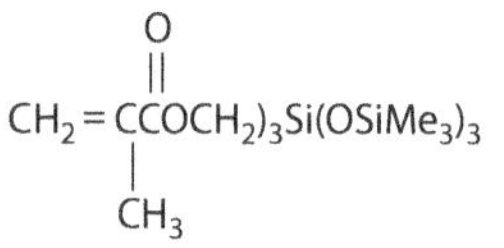

Methacrylaxypropyltris (trimethylsiloxy) silane

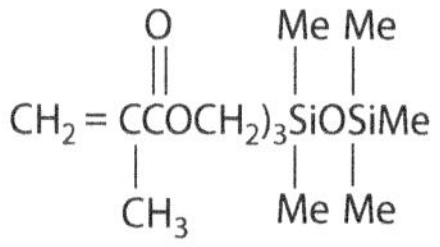

Methacryloxypropylpentamethyldisiloxane

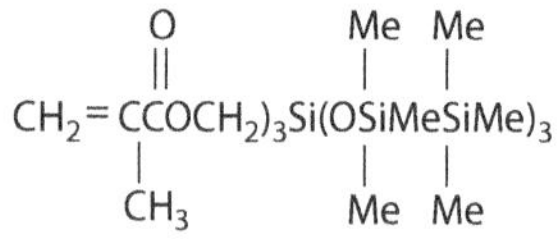

Methacryloxypropyltris (pentamethyldisiloxanyl)-silane

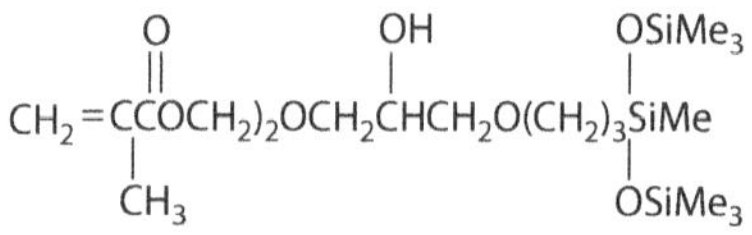

Various siloxane methacrylate copolymers (Whitford 1984)

Graft silicone elastomer copolymers can have two types of grafted side chains, one type on all the molecules in the elastomer and a second added by graft polymerization at the surface (see Surface Treatments). Silicone elastomers can be mixed relatively easily with numerous copolymers (Styrene butadiene styrene [SBS], SEBS, Polypropylene [PP], PU). Silicone elastomer copolymers are all very biocompatible because the addition of silicone gives them the elasticity that only the incorporation of plasticizers (phthalates) would otherwise provide. They are suitable for large-scale production runs by molding or injection because of the thermoplasticity provided by the SBS, SEBS, and PP copolymers. They also combine the surface properties of silicones with the intrinsic mechanical performance of the added copolymers. Polyurethanes are good examples of copolymers with greatly enhanced blood compatibility and mechanical strength. Polyurethane products are produced by mixing a vinylsiloxane polycondensate with a curing agent (crosslinker) followed by injection into an appropriate mold. The components crosslink to form a thermoplastic silicone–polyurethane network with remarkable physical and biomedical properties. These biomedical silicone elastomers can be further modified by the addition of other components before the final cure. These components may or may not be involved in the crosslinking, and they can have an impact on the physical and chemical properties and thus on the biocompatibility of the final elastomer products (Kossovsky 1995, Kossovsky and Freiman 1995).

14.1.3.5 Fillers

Biomedical silicone elastomers may be considered composite materials in the sense that they are composed of organomineral polymer matrixes containing fillers that are more or less linked to the network. Fillers are used to improve the mechanical performance of elastomers, notably to increase tear and tensile strength. They are generally mineral in nature, but may also be organic, like high molecular weight polyacrylic acid (Maturri et al. 1991). Amorphous silica is the most widely used filler and can be modified using a coupling agent that improves adherence to the polymer matrix (sizing). This is generally done by silanization because the most common coupling agents are silanes like $Y X_3Si(CH^2)nY$, where n = 0 to 3, X is the hydrolysable group, and Y is the organic group selected on the basis of the polymer matrix. Silica with reactive Si-H covalent bonds can be obtained by a reaction with methyldichlorosilane for example (Yoda 1998). Other more specific treatments have been reported like the attachment of antithrombogenic heparinoid substances to amorphous silica or nanohydroxyapatite fibers (Khan et al. 2008, Von Recum and LaBerge 1995, Von Recum and Van Kooten 1995). Other fibers (glass, aramide, carbon, etc.) can also be used to reinforce the structure of biomedical silicone elastomers. Sizing the fibers ensures good adherence between the reinforcement and the matrix (Auroy et al. 1996).

The fillers must never reduce material biocompatibility, either because they are not biocompatible themselves or are released in situ due to insufficient bonding to the polymer.

14.1.3.6 Surface Treatments

Changes to the basic silicone elastomer may be insufficient for special applications. The desire to improve the materials has led to a number of surface treatments to optimize the biocompatibility of surfaces in contact with living tissue, seal in undesirable residues or additives using a coating, and regulate excretion and/or absorption by the sealed elastomer using a selectively permeable surface (Colter et al. 1977). Silicone elastomers are hydrophobic and have high coefficients of friction, which in combination with their low surface energy (10–18 dynes/cm) ensures low wettability (Brown 1995, Mardis et al. 1993). While these biophysical characteristics can be advantageous (no tissue adherence to mammary implants, no bacterial adherence (Rodrigues et al. 2007), no sliding of urethral stents or blood catheters (Bambauer et al. 2004) due to surface friction caused by contact with living tissue), they can cause numerous biological and technical problems that need to be solved if they are to be used for other applications.

Various surface treatments can be performed using a variety of techniques:

- Plasma deposition of simple substances or composites. Heating a gas to a very high temperature using an electric arc (electric glow discharge) causes extreme atomic excitation resulting in partially or almost totally ionized gases, or plasmas. When the ions come into contact with a "cold" surface, they are deposited and become chemically bonded to the surface. Exposure of elastomers to such plasmas (He, Ar, etc.) allows the deposition and chemical bonding of monomers or polymers to the exposed surface in a process called plasma polymerization, or glow discharge polymerization to avoid confusion with blood plasma. The atomic, molecular, or macromolecular deposits obtained using this technique create a thin, smooth film several Å to several hundred Å thick. Most polymers or molecules with suitable biological or technical properties (polyethylene, polyvinylchloride, polytetrafluoroethylene, polycarbonate, polymethylmethacrylates, polysiloxanes, acrylic acid, allylamine, etc.) (Ren et al. 2008) can be applied to silicone elastomers.

  The plasma deposition of H2 +, N2+, O2+, and Ne+ ions breaks existing chemical bonds on the surface and creates new radicals such as =C=O using O+ plasma for example. The affinity of certain proteins (albumin, fibrin) for the surface of silicone elastomers modified in this way creates a layer of adsorbed protein that prevents the adherence of other blood constituents by steric repulsion, a well-known effect in colloidal chemistry (Park and Park 1996). The result is a powerful antithrombogenic effect.

  The deposition of ionic oxygen, argon, carbon dioxide, or ammonia makes the silicone elastomer surface more hydrophilic. However, the migration of free polymer to the elastomer surface affects the durability of such treatments (Everaert et al. 1995, Hsiue et al. 1996). The deposition of these ions also allows for polymer grafting on the elastomer surface.

- Surface polymer grafting is a more long-lasting solution, but more technically complicated. After argon plasma or solvent activation, the silicone elastomer is placed in a reactor containing a monomer solution (acrylic acid, hydroxyethyl methacrylate, bis amino polyethylene oxide, etc.). Heating promotes chemical bonding of the monomers to the elastomer and crosslinking among the grafted monomers, producing a smooth, continuous coating of grafted copolymer (polyacrylic or polyethylene for example) covering the elastomer surface. By alternating the chemical treatment of the various surfaces of an elastomer, a bifunctional, homobifunctional, or heterobifunctional material can be created (Figure 14.6). Such treatments have been used to improve the adherence and growth of

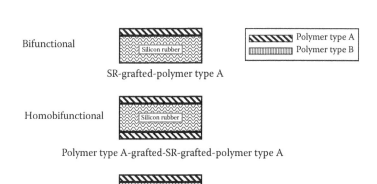

FIGURE 14.6   Schematic diagrams for grafted silicon rubber surfaces.

corneal epithelial cells by grafting 2-hydroxyethylmethacrylate to the outside surfaces of silicone elastomer corneas and to prevent adherence by grafting bis amino polyethylene oxide to the inside surfaces (Lee et al. 1996a,b).

- The denucleation of silicone elastomers helps remove the micro air bubbles in the surface irregularities of the elastomer. The goal is to remove air/blood interfaces to prevent complement activation and platelet aggregation. The procedure is relatively long but technically simple. It involves rinsing the elastomer in double distilled water for 12 h, soaking in 99% ethanol for 24 h, soaking in a buffer solution with vacuum degassing for 6 h, then gradually replacing the buffer solution with ethanol over a 6 h period. The elastomer is then stored in the buffer solution until used (Kalman et al. 1991). This treatment helps increase hemocompatibility without modifying the structure or composition of the silicone elastomer surface.

- Acylation or hydroxylation of the silicone elastomer surface increases its affinity for serum proteins (Tsai et al. 1991). Hydroxylation is performed by an oxymercuration/demercuration reaction or hydroboration, which transforms the polydimethylsiloxanes on the surface into hydroxymethylvinylsiloxane. Acylation of -OH groups is achieved through simple esterification of exposed carbons on the siloxane side chains (Tsai et al. 1991).

14.1.4 Polyurethane Elastomers

14.1.4.1 Introduction

Otto Bayer developed the first polyurethanes in the late 1930s. They now make up the largest family of thermoplastic and thermosetting plastics and range from very rigid to very flexible and from compact to cellular. They are used to produce varnishes, fibers, textiles, plastics, and elastomers. Polyurethane elastomers were studied in great detail in the 1950s and 1960s (Bonart 1968, Cooper and Toblosky 1966, Frisch and Saunders 1962). Pangman (Pangman 1958), Boretos (Boretos and Pierce 1967), and Yoda (Yoda 1998) proposed their use in biomedical applications (Pangman 1958). They have been widely used in the medical field for over 35 years now (Autian 1974, Boretos and Pierce 1967, Coury et al. 1984, Lelah and Cooper 1986, Stokes and Cobian 1982, Wilkes and Samuels 1973) to manufacture biomaterials that come into contact with blood, including catheters, valve protheses, artificial blood vessels, etc. They have also recently been used to fabricate microsphere vectors for drugs and bioactive substances, tissue-engineered scaffolds (membranes, biodegradable porous polymer structures, etc.) (Bouchemal et al. 2004, Haugen et al. 2006, Shukla et al. 2002). Cf lower 1 4 7 modified polyurethane elastomers (Table 14.4).

They are produced by polyaddition reactions between polyisocyanates and polyalcohols (polyols). Broadly speaking, all elastomers are isocyanate derivatives.

TABLE 14.4
Current Biomedical Applications of Polyurethanes

Blood bags, closures, fittings	Leaflet heart valves
Blood oxygenation tubing	Mechanical heart valve coatings
Breast prostheses	Orthopedic splints, bone adhesives
Cardiac assist pump bladders, tubing, coatings	Percutaneous shunts
Catheters	Reconstructive surgery materials
Dental cavity liners	Skin dressing and tapes
Endotracheal tubes	Surgical drapes
Heart pacemaker connectors, coatings	Suture materials
Hemodialysis tubing, membranes, connectors	Synthetic bile ducts
Lead insulators, fixation devices	Vascular grafts and patches

---

14.1.4.2 Manufacture

Polyurethane elastomers are formed by reactions between isocyanate and substances with mobile hydrogen ions, generally polyols, amines, and water.

$$R-N=C=O+H-A \rightarrow R-NH-\underset{\underset{A}{|}}{C}=O$$

Alcohols give urethanes:

$$R-N=CO+H-OR' \rightarrow R-NH-CO-OR'$$

Nontertiary amines give ureins or substituted ureas.
Water also give ureins together with carbon dioxide.

In most cases, the components, i.e., the raw materials, the catalysts, and the adjuvants, are mixed as liquids and then injected into molds or cavities. Aromatic polyisocyanates are the most frequently used raw materials. Toluenediisocyanate (TDI) is used as is or is dimerized to synthesize elastomers with high tear resistance. Diphenylmethanediisocyante (MDI) and naphthylenediisocyante (NDAI) are used to produce polyester elastomers.

Toluene diisocyanate (TDI)

Dimerized toluene diisocyanate (TDI)

Diphenylmethane 4-4'diisocyanate (MDI)

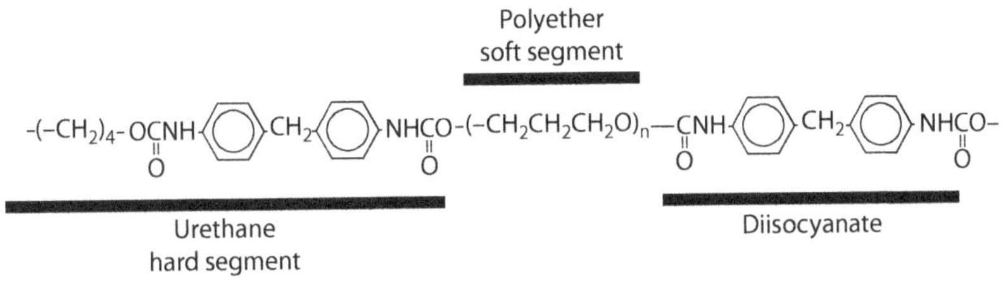

Polyether
soft segment

$-(-CH_2)_4- OCNH- \bigcirc -CH_2- \bigcirc -NHCO-(-CH_2CH_2CH_2O)_n-CNH- \bigcirc -CH_2- \bigcirc -NHCO-$

Urethane
hard segment

Diisocyanate

FIGURE 14.7   Polyetherurethane-segmented Pellethane®.

Naphthylene 1-5-diisocyanate (NDI)

Polyethers are the most frequently used polyols and result in polyetherurethanes. Polyesters are used less often, but the resulting polyesterurethanes are much more chemically stable.

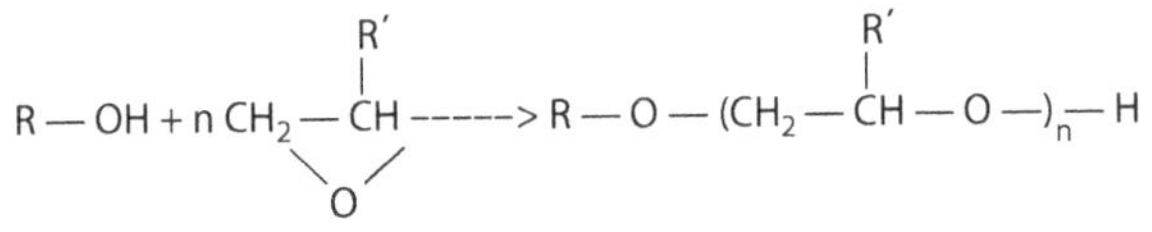

$$R-OH + n\, CH_2-\overset{R'}{\underset{O}{CH}} \xrightarrow{\hspace{1cm}} R-O-(CH_2-\overset{R'}{CH}-O-)_n-H$$

Formation of polyether-polyols

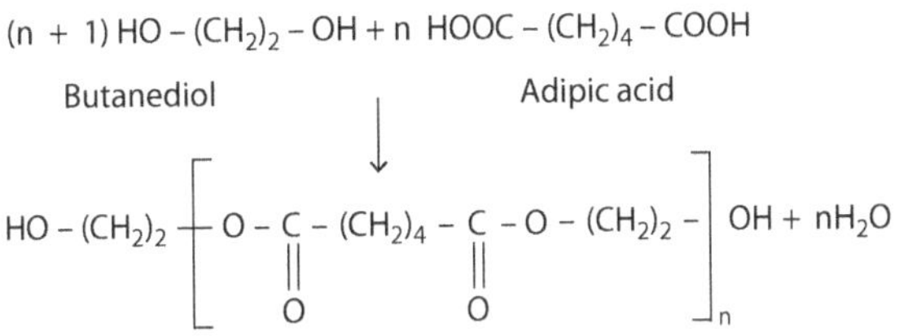

$$(n+1)\, HO-(CH_2)_2-OH + n\, HOOC-(CH_2)_4-COOH$$

Butanediol   Adipic acid

$$HO-(CH_2)_2 \left[O-\overset{O}{\underset{\parallel}{C}}-(CH_2)_4-\overset{O}{\underset{\parallel}{C}}-O-(CH_2)_2- \right]_n OH + nH_2O$$

Formation of polyester-polyols

Pellethane® is a poly(ether urethane) that has been used for many years in the medical field, and it is the result of a reaction between MDI and polytetramethyleneglycol (PTMG), which produces a soft segment ending in an isocyanate that reacts with the diol moiety of butanediol to extend the chain (Yoda 1998) (Figure 14.7).

The greater the number of reactive groups on the polyols used to synthesize the polyurethane, the more the end product is crosslinked and thus rigid. The polyols with the highest molecular mass and the least crosslinking give the most flexible elastomers. To improve the mechanical resistance of certain polyurethane elastomers, all or part of the polyol may be replaced by a reactive diamine (methylene dianailine or MDA, methylene bisorthochloroaniline or MOCA, ethylene dianiline or EDA), giving rise to segmented polyetherurethaneurea elastomers such as Biomer® and Mitrathane®, which are used in cardiovascular applications. Prepolymer processing (foaming) using expansion agents produces cellular elastomers: either water is reacted with isocyanate to produce a urein and carbon dioxide (chemical expansion) or low boiling point inert liquids are added and subsequently expelled by degassing during high temperature polymerization (physical expansion). Adjuvants are also required for crosslinking: catalysts (tertiary amines, organic tin salts), surfactants (silicones), fillers, reinforcements, pigments, etc. During manufacture (whatever the process) and storage, polyurethanes must be kept in a dry environment because their affinity to and reactivity with water can result in changes to their composition and/or create defects in the end product.

14.1.4.3   Mechanical Behavior of Polyurethane Elastomers

The hard segments crosslink among themselves within the mass of the elastomer to form agglomerations that act as fillers, improving the mechanical resistance of the material. The soft segments remain free and randomly arranged (Figures 14.8 and 14.9). Under a tensile stress

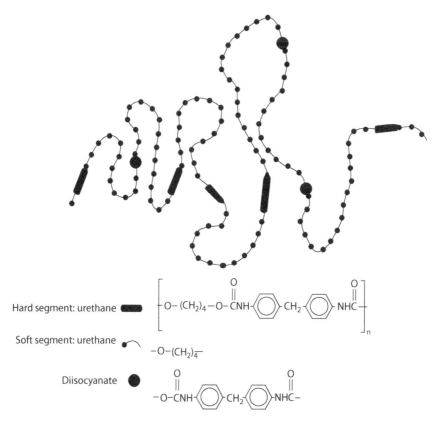

FIGURE 14.8 Linear polyetherurethane chain. (From Coury, A.J. et al., *J. Biomater. Appl.,* 3, 130, 1988.)

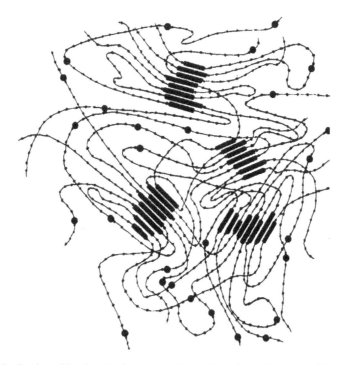

FIGURE 14.9 Distribution of hard and soft segments in a polyurethane elastomer. (From Coury, A.J. et al., *J. Biomater. Appl.,* 3, 130, 1988.)

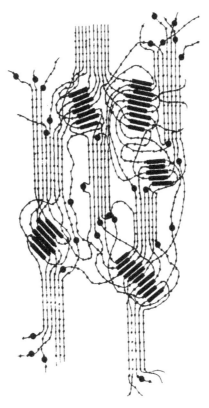

FIGURE 14.10 Crystallization of soft segments under a weak vertical stress. (From Coury, A.J. et al., *J. Biomater. Appl.*, 3, 130, 1988.)

of 150% (Bonart 1968), the polyether soft segments line up along the elongation axis, displacing the urethane hard segments so that they are more or less perpendicular to the vertical axis. Soft segments are brought into close proximity with each other by stress crystallization, a process that is completed when the elongation reaches approximately 250% (Figure 14.10). When the elongation increases even more (~500%), the chemical bonds crosslinking the hard segments break down and the now separate hard segments align along the vertical axis. At this point, the soft segments relax because of the stretching due to the release of the hard segments (Figure 14.11).

Segmented polyurethane elastomers thus have elastic behavior under low stress (deformation) conditions, which becomes plastic when the hard segment network breaks down. Deformed samples do not immediately return to their original shape following removal of the stress. However, rearrangement of the hard segments eventually results in a total or partial return to the initial shape, depending on the elastomer structure. The greater the concentration of hard segments, the more plastic the deformation. On the other hand, the lower the concentration of hard segments, the more elastic the behavior.

Chemical properties also depend on the soft/hard segment ratio. A large proportion of hard segments makes the material harder and provides better stress resistance but decreases elasticity and resistance to abrasion (Szycher and Reed 1992). The presence of long polyether or polyester soft segments provides superior break elongation but makes the material more sensitive to oxidation and degradation by biological fluids and thus to failure under repeated stress (Table 14.5). Other strategies for synthesizing segmented polyurethanes result in the formation of biodegradable

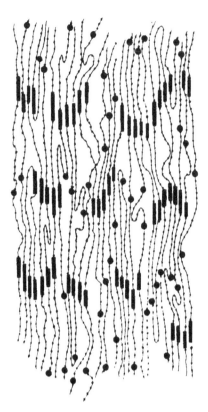

FIGURE 14.11   Reorientation of hard segments and relaxation of soft segments under a strong vertical stress. (From Coury, A.J. et al., *J. Biomater. Appl.*, 3, 130, 1988.)

polyurethanes that are extremely elastic (elongation at break 4700%) because they contain very few hard segments (Asplund et al. 2007).

14.1.4.4   Biodegradation of Polyurethane Elastomers

In vivo degradation of polyurethane elastomers occurs via four main mechanisms: calcification, macromolecular chain breakdown (Yoda 1998), hydrolysis, and environmental stress cracking. Calcification involves the deposition of calcium hydroxyapatite on the surfaces of implanted elastomers (Schoen et al. 1988). Macromolecular chain breakdown of polyurethane elastomers occurs via oxidation reactions with the soft segments and at the junctions between hard and soft segments (Takahara et al. 1991). Corrodable metallic residues favor oxidation by acting as catalysts in vivo (Pinchuk 1994, Stokes 1988). Hydrolysis depends on the susceptibility of the urea and urethane bonds and the concentration of hydrolytic enzymes at the cell/polymer interface. The extreme sensitivity of polyesterurethanes to hydrolysis limits their use in biomedical applications as opposed to polyetherurethanes, which are more resistant (Yoda 1998). However, polyurethane elastomers in general cannot be autoclaved because of the risk of hydrolysis.

Environmental stress cracking (ESC) occurs when mechanical stresses are combined with interactions with living tissue (Phillips et al. 1988) and can lead to the complete destruction of the implanted material. Three phenomena must coincide for ESC to occur: the *presence of enzymes* as a result of an inflammatory response at the interface (Anderson et al. 1992, Anderson and Miller 1984, Kao et al. 1994, Zhao et al. 1992), a *susceptible poly(ether urethane) elastomer*, and a *mechanical stress* on the elastomer (Phillips et al. 1988, Rambour 1973). ESC, which has never been observed

TABLE 14.5
Commercially Available Polyurethanes Used in Biomedical Applications

Type and Molecular Architecture	Processing	Trade Name	Supplier	
Segmented polyetherurethane				
$-(CH_2)_4-OCNH-\langle\bigcirc\rangle-CH_2-\langle\bigcirc\rangle-NHCO-(-CH_2CH_2CH_2CH_2O)_m-CNH-\langle\bigcirc\rangle-CH_2-\langle\bigcirc\rangle-NHCO-$	Extrusion Injection molding Solution casting	Pellethane™	Dow Chem. Co.	USA
$-(CH_2)_4-OCNH-\langle H \rangle-CH_2-\langle H \rangle-NHCO-(-CH_2CH_2CH_2CH_2O)_n-CNH-\langle H \rangle-CH_2-\langle H \rangle-NHCO-$	Extrusion Injection molding Solution casting	Tecoflex™	Thermedics Inc.	USA
Segmented polyetherurethaneurea				
$-NH-C-NH-\langle\bigcirc\rangle-CH_2-\langle\bigcirc\rangle-NHCO-(-CH_2CH_2CH_2CH_2O)_m-CNH-\langle\bigcirc\rangle-CH_2-\langle\bigcirc\rangle-NHCNHCH_2CH_2-$	Solution casting	Biomer™	Ethicon Co.	USA
$-NH-C-NH-\langle\bigcirc\rangle-CH_2-\langle\bigcirc\rangle-NHCO-(-CH_2CH_2CH_2CH_2O)_m-CNH-\langle\bigcirc\rangle-CH_2-\langle\bigcirc\rangle-NHCNHCH_2CH_2- \quad CH_3$	Solution casting	TM3™, TM5™	Toyobo Co.	Japan
Copolymer of segmented polyetherurethane and polydimethylsiloxane				
$-(-CH_2)_4-OC-N-\langle\bigcirc\rangle-CH_2-\langle\bigcirc\rangle-NHCO-(-CH_2CH_2CH_2CH_2O)_m-CNH-\langle\bigcirc\rangle-CH_2-\langle\bigcirc\rangle-NHCO-$ with $-(-Si-O-)_n-C-CH_3$ (CH$_3$ substituents)	Solution casting	Cardiothene51™	Kontron Inc. Nippon Zeon Co.	USA Japan

Source: Yoda, R., *J. Biomater. Sci. Polym. Ed.*, 9, 561, 1998.

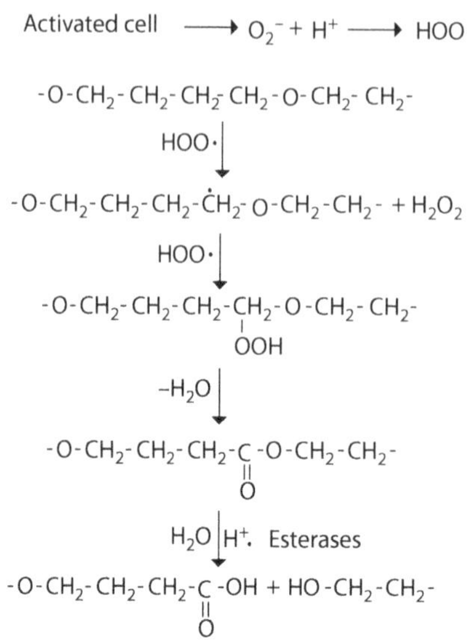

Activated cell $\longrightarrow$ O_2^- + H^+ $\longrightarrow$ HOO·

-O-CH$_2$-CH$_2$-CH$_2$-CH$_2$-O-CH$_2$-CH$_2$-

HOO· $\downarrow$

-O-CH$_2$-CH$_2$-CH$_2$-ĊH$_2$-O-CH$_2$-CH$_2$- + H$_2$O$_2$

HOO· $\downarrow$

-O-CH$_2$-CH$_2$-CH$_2$-CH$_2$-O-CH$_2$-CH$_2$-
|
OOH

$-$H$_2$O $\downarrow$

-O-CH$_2$-CH$_2$-CH$_2$-C-O-CH$_2$-CH$_2$-
‖
O

H$_2$O | H$^+$. Esterases

-O-CH$_2$-CH$_2$-CH$_2$-C-OH + HO-CH$_2$-CH$_2$-
‖
O

FIGURE 14.12 The degradation mechanism of a polyurethane at the carbon alpha to the ether group. (As proposed by Anderson, J.M. et al., Cell/polymer interactions in the biodegradation of polyurethanes, in *Biodegradable Polymers and Plastics*, eds. Vert, M., Feijen, J., Albertson, A., Scott, G., Chiellini, E., Royal Society of Chemistry, Cambridge, U.K., pp. 122–136, 1992.)

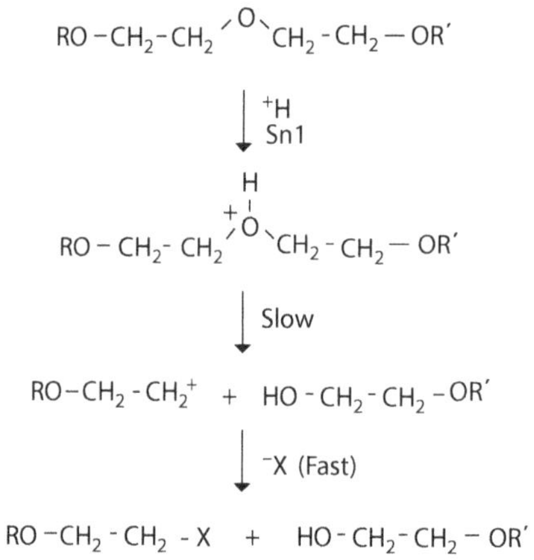

RO$-$CH$_2$-CH$_2$ $\diagup$O$\diagdown$ CH$_2$-CH$_2$$-$OR′

$\downarrow$ $^+$H Sn1

H
|
RO$-$CH$_2$-CH$_2$ $\overset{+}{\diagup}$O$\diagdown$ CH$_2$-CH$_2$$-$OR′

$\downarrow$ Slow

RO$-$CH$_2$-CH$_2^+$ + HO-CH$_2$-CH$_2$-OR′

$\downarrow$ $^-$X (Fast)

RO$-$CH$_2$-CH$_2$-X + HO-CH$_2$-CH$_2$-OR′

FIGURE 14.13 Cleavage of a polyether urethane by Sn1 type acid hydrolysis. (From Pinchuk, L., *J. Biomater. Sci. Polym. Ed.*, 6, 225, 1994.)

with polyurethanes in vitro, does not occur if a single factor is missing (Szycher and Reed 1992). ESC seems to be due to the oxidation of the methyl group on the soft segment polyether chains (Szycher and Reed 1992) (Figures 14.12 through 14.14).

Numerous studies have shown that soft segments are prone to oxidation after varying periods of exposure in vivo (Gogolewski 1991, Schubert et al. 1997, Zhao et al. 1990, 1991). It is now generally accepted that reactive oxygen species released by macrophages and foreign body giant cells are responsible for the biodegradation because they capture one proton on the alpha methylene of the polyether leading to either the rupture of the polymer or the creation of a new bond between two chains of neighboring polymer (Schubert et al. 1995, Zhao et al. 1991). The splitting of the polymer chains inexorably breaks down the biomaterial while, at the same time,

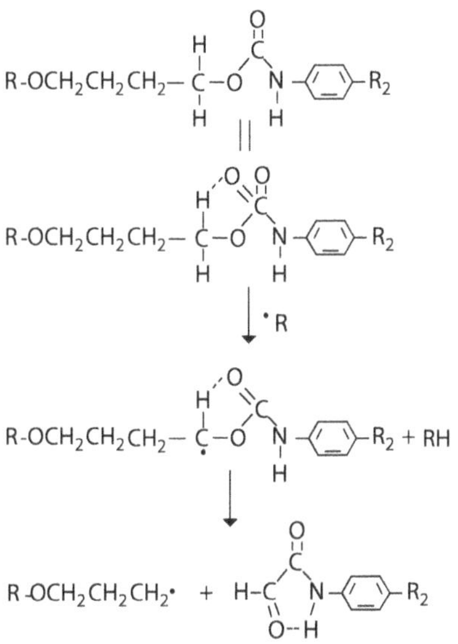

FIGURE 14.14 The degradation mechanism of a polyurethaneurea at the carbon alpha to the urethane group as proposed by Tyler and Ratner. (From Pinchuk, L., *J. Biomater. Sci. Polym. Ed.*, 6, 225, 1994.)

the increase in the number of bonds between the chains makes it more rigid, making it less and less elastomeric. At last, the biomaterial breaks down more and more and each piece becomes increasingly rigid.

Chemical properties are not the only factors at play. The hardness, mechanical resistance, and morphology or microstructure as well as the affinity for proteins also influence sensitivity to biodegradation (Zhang et al. 1994). The microporous surfaces of polyetherurethane vascular devices promote tissue colonization, neointima formation, and fibrotic capsule formation that in turn favor biodegradation by increasing the size of the biological interface with respect to the volume of the device (Marois et al. 1989, White 1988).

14.1.4.5 Biostable Polyurethane Elastomers

The impact of ESC on polyetherurethanes has stimulated research on polyurethane elastomers that do not contain ether groups and that remain stable when subjected to environmental stresses in vivo, i.e., biostable polyurethanes (Stokes 1983, Stokes et al. 1987, Szycher and Reed 1992). Coury et al. (Coury and Hobot 1991) have filed a patent for a polyurethane that contains no ether groups. An aliphatic complex replaces the soft segment polyethers with a very rigid biostable skeleton that makes the polymer much too hard for many applications. Numerous attempts have been made to modify or replace the soft segments in order to improve biostability. Soft segment polyethers have been replaced by polybutadienes and polymethylsiloxanes (Hergenrother et al. 1994, Lim et al. 1994, Takahara et al. 1991), polycarbonates (Capone 1992, Christenson et al. 2004b, Stokes et al. 1995), aliphatic hydrocarbons (Capone 1992), and sulfonated polyethylene oxide (Han et al. 2006). Polysiloxanes are attractive substitutes for polyethers because of their low toxicity, good thermal and oxidative stability, low coefficient of friction, and good hemocompatibility, and the fact that they do not affect the mechanical properties of the polyurethanes because their modulus of elasticity is similar to that of the soft segment they are replacing (Briganti et al. 2006, Martin et al. 2000, Simmons et al. 2004). Polycarbonate soft segments are also less sensitive to oxidation, although the hydrolysis of carbonate linkages in vivo is always

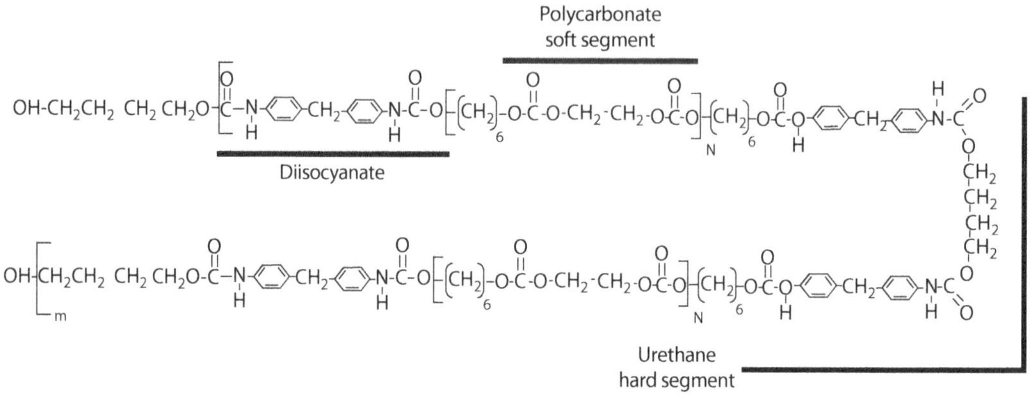

FIGURE 14.15 An example of a biostable polyurethane: Corethane® (polycarbonate urethane) produced by reacting poly(1,6-hexyl 1,2-ethyl carbonate)diol with MDI. (From Pinchuk, L., *J. Biomater. Sci. Polym. Ed.*, 6, 225, 1994.)

possible (Capone 1992, Coury et al. 1990, Pinchuk 1994, Pinchuk et al. 1993, Reed et al. 1994, Stokes et al. 1995, Ward et al. 1996) (Figure 14.15).

Polycarbonate urethanes are gradually supplanting polyurethanes because they are more resistant to oxidation. They were first commercialized by Corvita Corporation in 1992 and more recently, by the Polymer Technology Group as corethane, coremer, vascugraft, carbothane, chronoflex, bionate, etc. Like polyether urethane, they can be used in applications ranging from vascular to orthopedic. The addition of a triglyceride to 12-hydroxyoleic acid (the main component of castor oil) during the synthesis of polycarbonate urethanes produces significant improvements, including an affinity for albumin, antithrombogenic properties, and enhanced physical properties (Szelest-Lewandowska et al. 2003, 2007). The addition of carbon fibers or hydroxyapatite improves the physical and biological properties of polycarbonate urethane elastomers destined for orthopedic applications (Geary et al. 2008).

The complementary qualities of polyurethane and polyvinylsiloxane elastomers led to the idea of producing "alloys" in the hope of creating new polymers with enhanced resistance to biodegradation (Iwamoto et al. 1986, Nyilas and Ward 1977, Yoda 1998). This complex mixture of polyurethane and silicone (Cardiotane 51®) is difficult to process because it is neither soluble nor thermoplastic (Coury et al. 1988).

14.1.4.6 Shape Memory Polyurethane Elastomers

Shape memory elastomers are a class of shape memory polymers (SMPs) used in various medical applications. Light, easy to manufacture, and adaptable to a wide range of applications, most are shape memory polyurethanes. Most are rigid polymers, but some with the consistency of elastomers are used for specific applications such as stretchable stents, prosthetic valves, and self-tensioning sutures. They are generally biocompatible, nontoxic, and nonmutagenic. Patented in 1996 by Mitsubishi Heavy Industries, they are now distributed by its Diaplex subsidiary in the form of thermoset resin or thermoplastic (Cabanlit et al. 2007, Sokolowski et al. 2007).

14.1.4.7 Grafted Polyurethane Elastomers

The biomedical properties of polyurethane elastomers can be improved by encapsulation to isolate them from living tissue. Grafting polysiloxane molecules on the surface makes the material hydrophobic and protects the soft segments from oxidation and hydrolysis in vivo (Macocinschi et al. 2009, Mathur et al. 1997, Ward 1995, Ward et al. 1995). Grafting alkyl or polyethylene (oligoethylene monoalkyl(aryl)alcohol ether) groups to the sodium atoms of urethane hard segments or incorporating sulfate groups by nucleophilic molecular displacement of N-H urethane groups creates sulfonate polyurethane anionomers with chemical properties that limit biodegradation and thrombogenicity (Baumgartner et al. 1997, Grasel and Cooper 1989, Grasel et al. 1987, Han et al. 1998, Keogh et al. 1996, Lelah et al. 1985, Lim et al. 1993, Okkema et al. 1987). Researchers have also grafted proteins and other molecules on polyurethane elastomers in an attempt to improve hemocompatibility (albumin [Keogh and Eaton 1994]; heparin [Albanese et al. 1994]; adenosine diphosphate [Bakker et al. 1991]; fibronectin and gelatin [Hess et al. 1992]; amino acids [Wang et al. 2003]). The grafting processes, or impregnation in the case of gelatin, are complex and require numerous steps. Collagen grafted to carboxyl group-enriched polyurethanes via 1,2-bis(2,3-epoxypropoxy)ethane links allows better epithelial cell growth than raw polyurethanes (Lee et al. 1996a,b).

Seeding the lumen of microvessel vascular prostheses prepared by glow discharge allows the in vivo proliferation of a multicell layer close to but distinct from the elastomer wall. The cell layer is histologically similar to a native artery with antithrombogenic endothelial cells on the surface (Williams et al. 1992).

Plasma processing can be used to modify surfaces by depositing metallic silver (Boswald et al. 1995, Greil et al. 1999, Jansen et al. 1994, Li and Zhao 1995, Oloffs et al. 1994), fluorinated films (Pizzoferrato et al. 1995), nitrogen, and hexamethyldisiloxan polymers (Li and Zhao 1995) to, among other things, improve the antithrombogenic and antibacterial properties of cardiovascular implants.

14.1.4.8 Modified Polyurethane Elastomers

Solvent extraction (toluene, acetone, methanol, etc.) of small molecular weight surface polymers (which are different in composition from the bulk polyurethane) increases the molecular weight of the surface fraction and enriches it in hard segments, making it more polar and improving hemocompatibility (Grasel et al. 1988, Lelah and Cooper 1986, Lelah et al. 1986, Ratner 1983).

Since biogradation mainly affects the ether bonds of soft segments, which are very sensitive to oxidation, certain biomedical polyetherurethanes contain high concentrations of antioxidants (Schubert et al. 1996). Unfortunately, these elastomers (or their degradation products) may be toxic to host tissues (Szycher 1988). Researchers have recently begun examining natural antioxidants like vitamin E as possible alternatives to industrial additives.

Attempts to inhibit tissue responses have also been made. Adding hydroepiandrosterone (DHEA) to poly(etherurethane urea) reduces biodegradation by limiting macrophage activation. However, the effect is dose-dependent and since DHEA is very hydrosoluble, it rapidly diffuses away, making it useless for long-term protection (Collier et al. 1998, Szelest-Lewandowska et al. 2007).

Toyo Cloth Co. produces high molecular weight poly(ether/urethane/amide) from polytetramethylene oxide, MDI, ethylene glycol, and nylon 66. The addition of calcium chloride to this polyurethane-nylon polymer modifies its physical and biomedical properties (Kawashima and Sato 1997) by increasing water permeability, hydrophilicity, biocompatibility, thrombogenicity, and elasticity, and by decreasing breaking strength (tensile modulus). An affinity for calcium ions points to potential thrombogenic or antithrombogenic applications for heparinized versions of the elastomer (Albanese et al. 1994).

The addition of chitin during the synthesis of polyurethane improves its hydrophobicity and enhances its resistance to biodegradation in vivo (Zia et al. 2009a,b).

While polyurethanes can be encapsulated, they can also be used to coat medical devices like pacemakers and endovascular stents (De Scheerder et al. 1995a,b, Holmes et al. 1994, Schellhammer et al. 1999). Such coatings have various degrees of hemocompatibility (Rechavia et al. 1998) and occasionally shift because of a lack of adherence to vessel walls (Schellhammer et al. 1999).

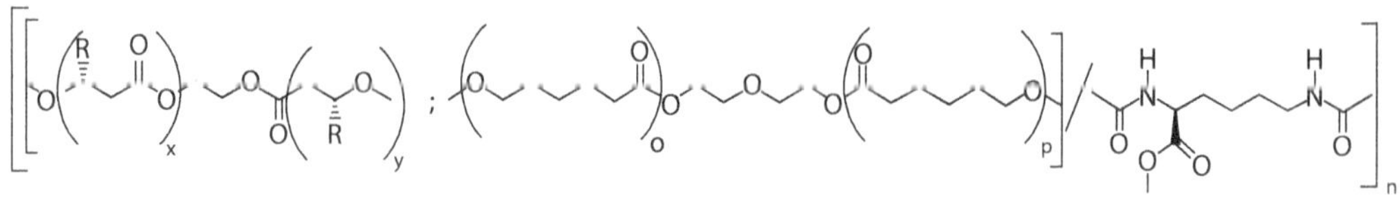

FIGURE 14.16 Example of a biodegradable polyurethane. (From Saad, B. et al., *J. Biomed. Mater. Res.*, 36, 65, 1997.)

14.1.4.9 Biodegradable Polyurethane Elastomers

The development of biodegradable polymers is a major advance in the research on biomaterials. Designed to gradually degrade in vivo over the implantation period, these polymers are used in numerous medical and surgical applications Biodegradable polyesterurethanes have been developed in recent years (Hirt et al. 1996, Saad et al. 1996, 1997). DegraPol/btc® is a polyesterurethane copolymer with polyhydroxybutyric acid hard segments and polycaprolactone soft segments (Figure 14.16).

The hard and soft segments can be replaced by a variety of substitutes that make the end product more or less biodegradable in vivo. By selecting the right substitute, it is possible to design cellular polyesterurethanes with varying degrees of rigidity and with degradation rates suitable for osteoblast colonization of bone implants (Saad et al. 1996). Polyesterurethane foams can also be used to guide soft tissue regeneration (van Minnen et al. 2005). They can lose half their mass within several months of implantation. They are not strictly speaking elastomers, but lengthening their soft segments may provide new mechanical properties and open the door to soft tissue applications.

The controlled degradation of biodegradable polyurethanes in vivo facilitates the growth of new tissue. Tissue engineering scaffolds are fabricated from biodegradable polyurethanes using various processes such as carbon dioxide injection foaming, electrospinning, wet spinning, thermal phase separation, and salt leaching.

These scaffolds support the growth of cells and tissues both in vitro and in vivo, and their progressive degradation can be controlled through their formulation and composition. In addition, the degradation products are not cytotoxic (Guelcher 2008).

14.1.5 Saturated Polyesters and Biopolyesters

These thermoplastic polymers have long been used in textiles (Dacron®, Tergal®) and plastic films. Polyethylene terephthalate (PET) holds the lion's share of the market.

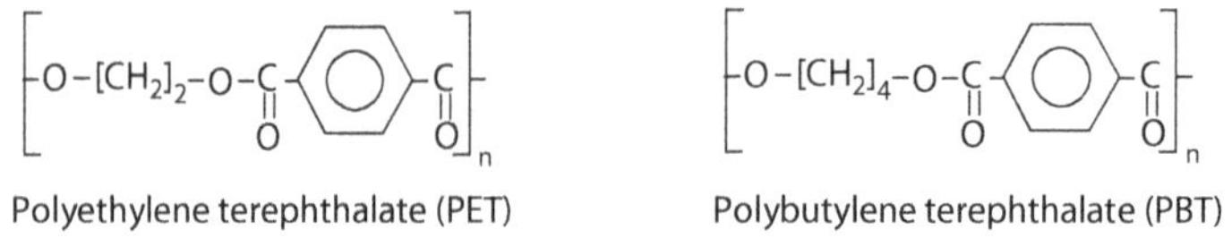

Polyethylene terephthalate (PET)

Polybutylene terephthalate (PBT)

PET was first used as a medical elastomer when DuPont introduced Hytrel® copolymer in 1972. PET, together with polybutylene terephthalate (PBT), which is used in injection molding, polycyclohexylenedimethylene terephthalate (PCT), and polyester-glycol copolymers (ethylenes and cyclohexane dimethanols) are excellent elastomers (polyesters and copolyesters) for biomedical applications. They are used in surgical field, bag, flask, bottle, and bandage-backing materials. PET and PBT are especially suitable for making copolymers with polycarbonate, and are also available with fiberglass reinforcements (~30%). PET and PBT, which theoretically can be esterified on their diacid or dialcohol groups, are in practice obtained by transesterification of dimethylterephthalate by ethylene glycol and butylene glycol, with the release of methanol.

The reaction is controlled by gradually increasing the temperature until the desired molecular mass is obtained. They are easy and cheap to produce on an industrial scale. They are sensitive to hydrolysis but are chemically resistant to a wide range of chemicals and solvents, are relatively nontoxic, contain no plasticizers, can be easily washed and sterilized (autoclave, gamma radiation), are very rigid, have the lowest coefficient of friction of any thermoplastic, and are very esthetic (smooth surface, very translucid). Thin films are permeable to water vapor and are very flexible and tear resistant (White 1991). When copolymerized with polycarbonate, they are very easily colonized by endothelial cells and may see use as vascular prostheses.

The first natural biopolyester—poly(hydroxyalkonoate) or PHA—was isolated by Lemoigne in 1925 (Chen et al. 2008). These water-insoluble polyesters are synthesized and used as storage materials by many bacterial species.

The general structure of PHAs (x = 1,2,3,4;
n =100–30000; R1, R2 = alkyl groups, Cl–C13)

An infinite variety of PHAs can be synthesized industrially using a variety of processes and substrates (Huijberts et al. 1992, Poirier et al. 1995). Biopolyesters produced by living organisms are interesting because they are easily degraded by living cells, tissues, and organisms and are a renewable biodegradable source of plastic polymers and elastomers (Barak et al. 1991, Imam et al. 1998).

Most PHAs are somewhat rigid but, since the 1990s, biodegradable elastomers have been fabricated using chemical reactions in the presence of sulfates or peroxides (Gagnon et al. 1994a,b) or by photopolymerization (Ashby et al. 1998, de Koning et al. 1991). These industrial polymers can be enzymatically hydrolyzed just like natural PHAs. They can be produced using various processes (injection, extrusion) and with different structures (granules, pasts) and show promise in biomedical applications (Duvernoy et al. 1995, Hocking and Marchessault 1994, Lafferty et al. 1988, Williams et al. 1999).

14.1.6 Polyvinyl Chlorides and Phthalate

Polyvinylchloride is a widely used polymer and PVC products can be produced with a broad range of properties (flexible, rigid, cellular). PVC has been widely used in the medical field for some

40 years (gloves, tubes, tubing, catheters, bags, etc.) because of its low cost and ease of processing. The (CH_2-CHCl) monomer is obtained from ethylene and chloride.

$$
\left[\begin{array}{cc} H & Cl \\ | & | \\ C & - C \\ | & | \\ H & H \end{array} \right]_n
$$

The high pressure polymerization of vinyl chloride monomer, which is a gas at room temperature and pressure, is activated by the addition of an initiator (peroxide) and controlled using temperature, mixing, and surfactants. The result is a white pulverulent polymer whose structure varies depending on the process (bulk, suspension, or emulsion polymerization). Vinyl chloride monomer lends itself to copolymerization, especially with vinyl chloroacetate (VCA). PVC/VCA copolymers generally contain 5%–15% vinyl chloroacetate. They can be processed at lower temperatures and pressures than PVC alone, more filler can be added, and they can be used to produce very transparent products. PVC is almost never used on its own. Small amounts of specific adjuvants such as heat stabilizers (to protect the resin during processing) and lubricants (to facilitate or make the processing possible) are often added. Fillers, pigments, reinforcing agents, light stabilizers, antioxidants, etc. may also be added. In addition to the adjuvants mentioned before, PVC elastomers also contain plasticizers (5–70 parts per 100 parts of resin). Since conventional plasticizers (phthalates) are incompatible with biomedical applications, PVC elastomers intended for biomedical applications are specially designed not to release toxic products (Fabre et al. 1998, Lindner et al. 1994). Phthalate-free PVC elastomers either contain plasticizers that are not released or are nontoxic (azelate, phosphate ester and polyester, citrate), or they are made without plasticizer by combining PVC with other elastomers to produce copolymers (polyurethanes, vinyl chloroacetate) or simple phase mixtures. The elastomer phase plasticizes the rigid PVC (polyadipates, polyesters, polyacrylonitrile-butadiene) (Branger et al. 1990).

PVCs contain plasticizers that have been known to be potentially toxic since the late 1970s (Fishbein 1984). Prolonged exposure to phthalates has been suspected since the early 1980s to have a profound, cumulative effect on the structure and function of living organisms (Ganning et al. 1984).

The most commonly used plasticizer, di(2-ethylhexyl)phthalate (DEHP), is toxic in animals, but its effects in man were in doubt until 2005 (Blass 1992, Hill et al. 2001, Hoenich et al. 2005, de Lemos et al. 2005).

According to the international organization Health Care Without Harm (2004), dioxins, polyvinylchloride, and di(2-ethylhexyl) phthalate are the three main toxins interfering with the goal to maintain a healthy environment. Exposure to these chemical products has been correlated with the appearance of cancers, cardiac and liver diseases, and reproduction and development disorders in animals and is strongly suspected to cause the same problems in man (Loff et al. 2008, Satoh et al. 2008, Tickner et al. 2001). Epidemiological studies have also shown an association between exposure to phthalates and the development of asthma and allergies in children (Jaakkola and Knight 2008).

Phthalates have been found in the saliva of children who place plastified PVC objects in their mouths. While the phthalate concentrations are low and the amounts ingested probably even lower than acceptable limits (Corea-Tellez et al. 2008), biomedical devices made of PVC elastomers such as mouth protectors, oral ortheses worn by patients with algo dysfunction of the manducator apparatus, oral respiratory devices worn by patients with sleep apnea syndrome, etc. are placed in the mouth much more frequently by adults than PVC toys by children. As such, other elastomers that do not contain plasticizers or DEHP-free elastomers should be used instead of plastified PVCs.

It is technically possible to replace the toxic plasticizers in PVCs with other substances that are reputed to be nontoxic and that do not leach from the PVC (Branger et al. 1990). This is crucial and urgent for medical and food packing materials and toys (Hansen 2008) and, a fortiori, elastomeric biomaterials. BASF began developing a new plasticizer, diisononyl phthalate or Hexamoll DINCH, in the late 1990s with an excellent toxicological profile. However, BASF only began using it in their PVC biomedical products in 2008 (Hansen 2008).

The surface properties of PVC elastomers can be altered by plasma treatments and hydrophilic monomer grafts (carboxylation, acrylic and metacrylic acid, vinyl pyrrolidone) (Singh and Agrawal 1992). Single-use antithrombogenic endovascular materials can be produced by grafting PU/SI copolymers on the surface of PVC tubing.

14.1.7 POLYOLEFINS

Polyolefins are produced by the polymerization of ethylene molecules where the hydrogen atoms are replaced by various hydrocarbon groups. Ethylene ($CH_2=CH_2$) can be used to produce polyethylene, polypropylene, etc. by simple substitution of the side group (H, CH_3, C_2H_5, C_4H_9, etc.).

Polyethylenes are produced by polymerizing gaseous ethylene under different conditions. Polyisobutylenes (PIB), for example, are polymerized in the presence of catalysts. Propylene/ethylene copolymers (often called polypropylene copolymers) are elastomeric in nature and have improved shock resistance properties (block copolymers). Random copolymers are especially well suited for manufacturing plastic films, hollow objects, plates, etc. Polymethylpentenes (PMPs) are obtained by high pressure polymerization in the presence of a catalyst. PMPs have good water and gas permeability and superior physicochemical properties, making them resistant to autoclaving and inert to most pharmaceuticals. Numerous compounds may be added to facilitate processing. PMPs must, however, be washed to eliminate potentially toxic residues. They are used for all sorts of packaging materials, as well as for flexible tubing, stoppers, implants, catheters, etc. The excellent resistance to flexural fatigue of polypropylene copolymers makes them more suitable than other elastomers for certain applications (extracorporeal pump diaphragms, synthetic finger joints made of Hexsyn®, for example). Grafting various monomers (vinyl pyrrolidone, ethyl methacrylate, acrylamide) on the surfaces of polypropylenes can provide excellent wettability while conserving their superior mechanical performance (Katbab et al. 1992, Mirzadeh et al. 1993). The recent production of polypropylene homopolymers containing alternating isotactic and atactic segments (relative configurations of successive asymmetric elements in the polymer chain) by a new mettalocene polymerization process opens the way to new medical applications for polyolefins (Dopico-Garcia et al. 2007, Hotta et al. 2006, Yoda 1998).

14.1.8 STYRENE COPOLYMERS

The styrene family contains a wide variety of polymers. Polystyrene is obtained by the polymerization of styrene produced from the alkylation of benzene by ethylene and the dehydrogenation of ethylbenzene. Liquid styrene can be polymerized in two ways:

1. Bulk polymerization: delicate method
2. Suspension polymerization: fine droplets (0.1–10 µm) of styrene monomer dispersed in an aqueous phase

Polystyrene

Styrene can be copolymerized with polybutadiene or ethylene/butylene to produce graft styrene-butadiene-styrene (SBS) and styrene-ethylene/butylene-styrene (SEBS) or Styrene-block-IsoButylene-block-Styrene) ("SIBS") which are widely used in medical applications. These elastomers not only have improved shock resistance but also do not require additives. They are very flexible and pure and do not release any residues in situ (Freij-Larsson et al. 1993). However, they are not very histocompatible and numerous attempts have been made to improve them. Polyhydroxyethylmethacrylate grafts improve blood compatibility, polysiloxane grafts impart the properties of silicone elastomers (Deisler 1987), while N+, F+, and Ar+ ions improve cell adhesion (Bacakova et al. 1996, El Fray et al. 2006).

Poly(styrene-*block*-isoButylene-*block*-styrene) (SIBS) is a biostable thermoplastic elastomer with physical properties that overlap silicone rubber and polyurethane. Research so far has shown that (1) SIBS does not activate thrombogenesis in the vascular system, (2) inflammatory reactions are not triggered around implants in the in the eye, skin, or vascular system, (3) it does not cause fibrosis or calcification in vivo, and (4) no degradation of the polymer has been observed to date in vivo. It appears that the inertness of SIBS is responsible for its remarkable stability (Pinchuk et al. 2008). SBS, SEBS, and SIBS elastomers can be sterilized by heat, vapor, gamma radiation, or ethylene oxide without losing their physical properties. They are used to manufacture catheters, stoppers, nonrigid containers, surgical fields, condoms, gloves, etc.

14.1.9 NATURAL RUBBERS

Natural rubbers are the most elastic and resistant of all biomedical elastomers, but are also the least hemocompatible due to the release of accelerator (dithiocarbamate) residues (Nakamura et al. 1990). Many attempts have been made to improve their blood compatibility. Methylmethacrylate grafts have shown great promise because they make natural rubber more hemocompatible than many silicone elastomers (Razzak et al. 1988). Another approach has been to vulcanize natural rubber by gamma radiation without additives. Natural rubbers polymerized in this way are very pure and demonstrate remarkably good histocompatibility (Balabanian et al. 2006). Untreated natural rubbers are mainly used to manufacture latex gloves while treated natural rubbers are used to produce catheters and tubing. Latex gloves have been shown to prevent cross-infections (Tucci et al. 1996). The well-balanced physical properties of natural rubbers (elasticity/tear resistance) make them ideal for this use. Hydrosoluble proteins are responsible for allergies associated with latex products. Simple washing is sufficient to completely remove the proteins and eliminate the risk of allergic reactions (Cormio et al. 1993).

14.1.10 HYDROGELS

Biomedical uses for hydrogels first appeared over 30 years ago (National Heart and Lung Institute 1971, Ratner et al. 1975). A number of polymers can be designed to act as hydrogels and retain water. Polyvinyl alcohol is the most widely used such polymer in ophthalmological applications (Yoda 1998). Polyvinylpyrrolidone is a relatively nontoxic, hydrosoluble polymer that can be impregnated with water or organic solvents once it has been crosslinked (Vijayasekaran et al. 1996). It can be grafted on the surfaces of silicone polymers (Ratner et al. 1975) and polyurethanes to improve thromboresistance (Chapiro 1983). Thermogelling polyurethane copolymers and newly synthesized tri-component multiblock poly(ether ester urethane)s have a release rate that can be controlled by the adjusting the composition of the poly(ether ester urethane)s or the concentration of the copolymer (Loh et al. 2006, 2007), which will make it possible to better control the quantities and concentration of bioactive molecules released by this type of hydrogel (Xinming et al. 2008). It is also suitable for use in soft contact lenses (Yoda 1998).

14.1.11 POLYPEPTIDE, COLLAGEN, POLY-GLYCEROL-SEBACATE, ELASTOMERS

Polypeptide elastomers of varying composition that mimic natural biological elastic polymers with different physical properties have been synthesized to and can be designed to be either biostable or biodegradable (Yoda 1998). They contain polypeptides (synthetic and/or natural) that can be copolymerized or not with artificial polymers (Ifkovits and Burdick 2007). Collagen is the most widely used substance for producing bulk or graft polypeptide elastomers. Elastin is also used to synthesize elastomeric biological tissues for numerous promising biomedical applications (Table 14.6).

TABLE 14.6
Polypeptide Elastomers for Medical Applications

Elastomer	*Composition*
Polypentapeptide	$(Val-Pro-Gly-Val-Gly)_n$
Polypeptide	Poly(ethylene-graft-γ-benzyl L-glutamate)
A-B-A type block copolymers	A: Poly(γ-benzyl L-glutamate) B: Polybutadiene
A-B-A type block copolymers	A: Poly(γ-ethyl L-glutamate) B: Polybutadiene
A-B-A type block copolymers	A: Poly(γ-methyl L- or D, L-glutamate) B: Polybutadiene
A-B-A type block copolymers	A: Poly (ε-N-benzyloxycarbonyl L-lysine) B: Polybutadiene
A-B-A type block copolymers	A: Poly(γ-benzyl L-glutamate) B: Polyisoprene
Collagen	Collagen-polyacrylates graft copolymer
Gelatin/chitin	Gelatin/carboxymethylchitin complex
Pseudo-poly(amino acids)	Poly(desaminotyrosyltyrosine hexylester carbonate)

Source: Yoda, R., *J. Biomater. Sci. Polym. Ed.*, 9, 561, 1998.

Poly(glycerol sebacate) (PGS) is a biodegradable elastomer synthesized from biocompatible monomers. It forms a covalently crosslinked three-dimensional network with hydroxyl groups attached to the backbone. The hydrogen bonding interactions between the hydroxyl groups contribute to the unique properties of this elastomer. In vitro and in vivo studies have shown that it has very good biocompatibility and is completely absorbed within several weeks of implantation (Wang et al. 2002). It can also be used to fabricate macroporous elastomer scaffold sponges with numerous micropores. The macropores are generated by glycerol vapor formed during curing. When the elastomer sponges are implanted, the micropores facilitate the colonization of the biomaterial by host cells.

The physicochemical properties and biodegradability of this biomaterial make it potentially very useful for nerve tissue engineering (Chen et al. 2008, Fidkowski et al. 2005, Sundback et al. 2005). It can also be modified to adhere strongly to tissues (Mahdavi et al. 2008). Since PGS can be prepared by photocuring, it can be used to encapsulate bioactive molecules that are destroyed by high temperature polymerization (Nijst et al. 2007).

14.1.12 POLYPHOSPHAZENES

Polyphosphazenes have long flexible chains with alternating phosphorus and nitrogen atoms.

Synthesis of poly (dichlorophosphazene) and poly (organophosphazenes)

(Yoda 1998)

Several hundred different polyphosphazene polymers have been produced simply by changing the radical on the phosphorus atom. The flexibility of the chain provides good elasticity and superior flexibility. Fluorinated polyphosphazenes are very hydrophobic, which means they are theoretically weakly thrombogenic. Given their mechanical properties, they are used in external devices that come into contact with blood (Heyde et al. 2007). Their good biocompatibility means that microporous polyphosphazene polymer matrices can be seeded with osteoblasts to serve as bone fillers or to induce bone regeneration. Blending etheric biodegradable polyphosphazene, poly[(50% ethyl glycinato)(50% methoxyethoxyethoxy)phosphazene] [PNEG(50)MEEP(50)] with poly(lactide-co-glycolide) [PLAGA] enhances the adherence of osteoblasts, pointing to a number of interesting orthopedic applications for this biodegradable polymer (Conconi et al. 2006, Deng et al. 2010, Nukavarapu et al. 2008). While they are also used to coat metallic vascular stents to improve biocompatibility, they can occasionally provoke strong tissue reactions (De Scheerder et al. 1995a,b, Deng et al. 2010) Certain polyphosphazenes with aminoester acid radicals are biodegradable (Crommen et al. 1993).

14.1.13 POLYAMIDES

Polyamides have an amide (–CO–NH–) bond that is the result of the condensation of an organic acid group on an amine. They make up a large family of textile fibers and widely used technical polymers. The first polyamide (Nylon 66), which was invented by Carothers, is a member of the group of polymers obtained by reactions between diacids ($HO-CO-(CH_2)b-CO-OH$) and diamines ($NH_2-(CH_2)a-NH_2$). Two other types, derived from polycondensation reactions between neighboring caprolactam and amino acid molecules ($-CH_2-COOH + H_2N-CH$ and $-CH_2-CO-NH-CH_2-+H_2O$), appeared a little later. Gradual heating to 250°C results in the elimination of the water and an increase in molecular mass. Diacid/diamine polyamides (e.g., Nylon 66) are produced by isolating and purifying a crystalline intermediate (nylon salt), which then undergoes polycondensation. A wide variety of copolymers can be produced in this way. Polyether block amides (PEBA) are a very large family of copolymers that have alternating sequences of crystalline polyamides and amorphous polyethers. These thermoplastic copolymers are intermediates between plastics and elastomers. Ferruti and Ranucci (1991) have produced numerous functional polyethylene glycol and polyamidoamine polymers that are used as grafts on a number of elastomeric biomaterials. Polyamide elastomers are used in a wide variety of applications: sterile packing material, cardiovascular surgery, and perfusion and transfusion materials. Nerve chambers composed of collagen and synthetic biodegradable polymers filled with polyamide filaments oriented along the tube axis or highly porous, insoluble analogs of the ECM are very effective at orienting the growth and regeneration of peripheral nerves in vitro (Yannas et al. 2007). When covalently linked, hairpin polyamides made up of N-methylpyrrole and N-methylimidazole can bind to DNA in a sequence-specific manner (Murty and Sugiyama 2004).

14.1.14 POLYACRYLICS

Polyacrylics are the result of the esterification of methacrylic acid by methanol. Methylmethacrylate is a clear liquid at 20°C. Polymerization can be achieved in two ways:

1. *Emulsion polymerization* is catalyzed by adding peroxide to an aqueous monomer emulsion. The result is a low molecular weight white power ready for molding.
2. *Molding polymerization* involves heating a monomer and catalyst in an oven between two glass plates, which act as the mold. Thermoplastic objects are rarely obtained directly by polymerization. The higher molecular weight (compared to emulsion polymerization) facilitates thermomolding by increasing the temperature range of the rubber platform.

$$
\left[\begin{array}{c} H \quad CO-OCH_3 \\ | \qquad | \\ C - C \\ | \qquad | \\ H \quad CH_3 \end{array}\right]_n
$$

Polymethylmethacrylate

PMMA is mainly known for its exceptional optical properties. This amorphous polymer is remarkably transparent (92% light transmission). A 3 m thick block of PMMA only absorbs 50% of incident light. This makes it ideal for use in ophthalmological devices (contact lenses, crystalline lenses). PMMA is also very biocompatible (Hollick et al. 1998, Jongebloed et al. 1994). The addition of plasticizers (phthalates) provides PMMAs with elastomeric properties, enabling them to be used for odontological applications (maxillofacial prostheses, mouthguards) (Chauvel-Lebret et al. 1999). They are also used in combination with other biomedical elastomers (polyurethanes, SEBS, etc.) to produce linear and graft copolymers. In addition, they are widely used to fabricate intraocular lenses with the capacity to release bioactive molecules (Kaur and Smitha 2002, McCulley 2003, Xiong et al. 2006).

They can only be sterilized using ethylene oxide or gamma radiation (Henry 1985) since they melt at low temperatures (80°C–100°C).

14.1.15 FLUORINATED ELASTOMERS

Fluorinated elastomers are a group of technical polymers with exceptional properties. Replacing the hydrogen with fluorine greatly improves heat and chemical resistance. Industrial fluorinated monomers are gases that liquefy at very low temperatures. They are polymerized at very high pressures using radical catalysts, generally by suspension or aqueous emulsion polymerization in order to control the strong exothermic reaction.

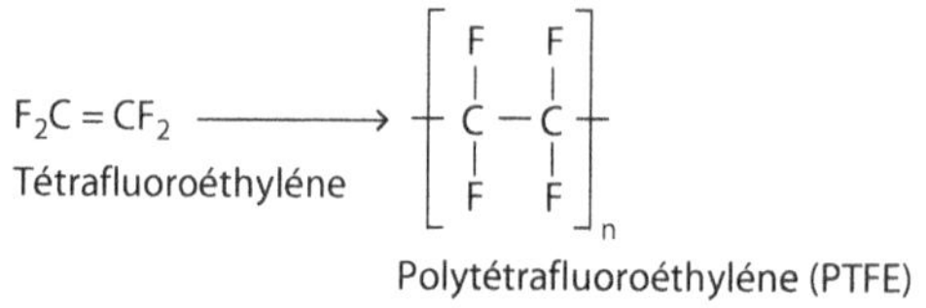

$$F_2C = CF_2 \longrightarrow \left[\begin{matrix} F & F \\ | & | \\ C & - C \\ | & | \\ F & F \end{matrix} \right]_n$$

Tétrafluoroéthyléne

Polytétrafluoroéthyléne (PTFE)

The very strong electronegativity of the PTFE fluoride ions protects the carbon chain and gives these elastomers their extreme chemical stability. They are inert and insoluble (Bischoff 1972, Wortman et al. 1983). They are naphthalene-plasticized resins and are extruded and made microporous for vascular graft prostheses (Teflon®, Impra®, Goretex®) (Huang et al. 1992, Sigot-Luizard et al. 1993). Their vascular biocompatibility is adequate but less than that of polyurethanes (Jeschke et al. 1999). They are used in orthopedic surgery to reduce friction between metallic and/or composite joint prostheses (Defrere and Franckart 1992) both alone and in combination with polyurethanes (Jahangir et al. 2002, Massa et al. 2005, Xie et al. 2008).

14.1.16 ELASTOMERIC MEDICAL DEVICES

See Tables 14.7 through 14.12 and Figure 14.17.

TABLE 14.7
Grouping of Medical Devices

Type	Environment	Duration	Application
I	Internal devices	Less than 30 days	Intravenous catheters
			Drainage tubes
		More than 30 days	Hip implants
			Pacemakers
			Artificial heart valves
II	External devices	Less than 30 days	Devices that contact the skin such as gloves, tapes dressings and orthopedic casts
		More than 30 days	Devices that contact the mucous membranes such as urinary catheters and intravaginal devices
III	Indirect devices (do not contact the body)		Hypodermic syringes
			Transfusion assemblies
			Dialysis components
IV	Non-patient contact devices (do not touch the body)		Dressing trays
			Packaging materials

Source: Yoda, R., *J. Biomater. Sci. Polym. Ed.*, 9, 561, 1998.

TABLE 14.8
Typical Research Concerning Small Diameter (Less than 6 mm I.D.) Vascular Grafts

Material	Important Design Feature
Polyetherurethane	Tube walls are porous and microfibrous three-dimensional mesh composed of polyurethane fibers of 1 μm diameter. Penetration depth of living tissue cell is limited to about 20–30 μm
Polyetherurethane	Graft with pore size 30–100 μm
Polyetherurethaneurea	It is of sufficient porosity to permit tissue ingrowth
	Its stress-strain curve is similar to that of the thoracic aorta
Polyetherurethaneurea/silicones	Microporous replamineform graft (20–30 μm pore size) composed of silicones was coated with commercial Biomer™
Polyetherurethaneurea	Fibrillar microporous hydrophobic graft made from commercial Mitrathane™
Polyetherurethane	Grafted with albumin and adenosine-cyclic monophosphate
	Surface has better blood compatibility
	Modified the surface further with urokinase
Degradable, aliphatic Polyesterurethane	It has well defined chemical and physical properties and controlled rates of degradation
	It induces the growth of functional neoartery which has cellular structure similar to that of the native artery
Expanded Teflon coated with silicones	Platelet accumulation was reduced by coating the graft surfaces with a smooth layer of silicones
Polyesterurethane (Vascugraft™)	Made from commercial polyesterurethane free from migrating additives such as catalysts and stabilizers
	Microfibrous, open pore structure
Polyurethane/poly (L-Lactide)	Microporous, compliant and biodegradable
	Function as a temporary scaffold for the generation of small-caliber arteries
Polyurethane/poly (L-Lactide)	Microporous and biodegradable graft consisted of two layers: the inner layer made from aliphatic polyetherurethane, the outer ply constructed by precipitating a physical mixture of polyesterurethane and poly (L-Lactide)
Polyetherurethane	Microporous graft prepared by an excimer laser ablation technique
	The pore size (100 μm) and the pore-to-pore distance (200 μm) are constant
Polyetherurethane	Made by a filament-winding technique comprising Lycra™ elastomeric fibers embedded in an elastomeric matrix of Pellethane™
Poly (carbonate urethane)/ pentapeptide	Glycine or fibronectin are covalently bound to polyurethane surface by succinyl dichloride coupling
Biodurable poly-(carbonate urethane) (ChronoFlex™)	Microporous grafts with excellent physical and mechanical characteristics. Self-sealing and maintenance of compliance
Polyurethane/poly (HEMA-block-styrene)	Polyurethane modified by coating with poly (HEMA- block-styrene) from solution
Copolymers of MPC	Copolymers of 2-methacryloyloxyethyl phosphorylcholine (MPC) with hydrophobic alkyl methacrylates
Polyurethane/ Endothelial cell	The hybrid artificial graft made of an endothelialized microporous polyurethane
Polyester/human vein	Vascular graft made of human vein and a highly porous polyester fabric vascular prosthesis

Source: Yoda, R., *J. Biomater. Sci. Polym. Ed.*, 9, 561, 1998.

TABLE 14.9
Elastomers Used for Making Artificial Hearts

Supplier or Investigator	Type of Blood Pump	Elastomer
Novacor's LVAD	Pusher-plate	Biomer™
Symbion	Diaphragm	Biomer
Aviomed	Tuberous	Angioflex™
Nimbus	Pusher-plate	Hexsyn™/gelatine
Thermedics LVAD	Tuberous	Tecoflex™
Nippon Zeon VAD	Sac	Cardiothane™
Totobo VAD	Diaphragm	TM-3™, TM-5™
Berlin University	Diaphragm	Pellethane™

Source: Yoda, R., *J. Biomater. Sci. Polym. Ed.*, 9, 561, 1998.

TABLE 14.10
Implant Applications for Medical Grade Silicone Elastomers

Treatment	Application
Plastic and reconstructive surgery	Reconstruction of nose, chin, ear armature, etc.
	Breast reconstruction
Ophthalmology	Correction of detached retina
	Prosthetric eye
	Repairing fracture of the floor of the orbit
Orthopedic surgery	Reconstruction of fingers, thumbs, wrists, elbows feet, tendons, temporomandibular joint, etc.
	Maxillofacial prosthesis
	Penile prosthesis.
Cardiovascular surgery	Ball in the ball-and-cage heart valve
	Coatings on pacemakers and leads-wires
	Construction in artificial hearts and heart assist devices

Source: Yoda, R., *J. Biomater. Sci. Polym. Ed.*, 9, 561, 1998.

TABLE 14.11
Elastomers Used in Transdermal Therapeutic Systems

Classification	Elastomer
Hydrophobic polymers	Poly(ethylene-co-vinyl acetate)
	Silicone rubber
	Polyurethane
	Poly(vinyl chloride)
	Collagen-based materials
	Ethylcellulose
	Cellulose acetate
Hydrogels	Poly(hydroxyethyl methacrylate)
	Crosslinked poly(vinyl alcohol)
	Crosslinked poly(vinylpyrrolidone)
	Polyacrylamide

Source: Yoda, R., *J. Biomater. Sci. Polym. Ed.*, 9, 561, 1998.

TABLE 14.12
Typical Materials for Contact Lenses

Classification	Polymer and Characteristics
Hydrogel soft lenses	Poly (hydroxyethyl methacrylate) (PHEMA)
	Generally crosslinked with ethylene glycol dimethacrylate, at a level of 26%–42% sufficient to give water content in equilibrium
	It may be copolymerized with other hydrophilic monomers, such as N-vinyl pyrrolidone, acrylamide and methacrylic acid
Elastic soft lenses	Polydimethylsiloxane and its copolymers
	They have good flexibility and very high oxygen permeability, but strong hydrophobicity results in considerable discomfort for the patient
	Surface treatment may improve wettability
	Perfluoropolyethers and its copolymers
	They have attractive combinations of flexibility and gas permeability
Rigid lenses	Poly(methyl methacrylate) (PMMA)
	It has good properties: very clear optically, good biocompatibility and good handling characteristics
	The main problem relates to poor oxygen permeability. There have been attempts to introduce arrays of very small holes ($15\,\mu m$) to facilitate oxygen transport
Gas-permeable rigid lenses	Cellulose acetate butyrate and others
	The butyrate is a highly oxygen-permeable and relatively rigid material
	The copolymerization of siloxany alkyl methacrylates of fluorosilicone methacrylates with methyl methacrylate provides a rigid methacrylate backbone with rubbery, highly permeable alkylsiloxane side groups

Source: Yoda, R., *J. Biomater. Sci. Polym. Ed.*, 9, 561, 1998.

14.2 METHODS TO EVALUATE THE BIOCOMPATIBILITY OF ELASTOMERS

14.2.1 DEFINITIONS

Most medical devices use synthetic biomaterials as their principal component. A biomaterial refers to any nonvital material intended to interact with biological systems within or on the human body. In general, elastomers, like other biomaterials, must be compatible and inert, must be nontoxic to living tissues, and must not destroy the cellular constituents of the body fluid with which they interface. A battle takes place between the organism and the implanted elastomer. Possible outcomes include acceptance, slow digestion of the implant, or rapid ejection accompanied by undesired secondary effects (Dumitriu and Dumitriu 1990). Medical devices must thus be designed to prevent rejection. This has led to the development of a new class of materials called biomaterials, which includes elastomers used in the medical field. The ability of biomaterials to fulfill their role in medical devices depends on their degree of biofunctionality and biocompatibility.

Biocompatibility is the ability of a material to perform it function in a host in a given situation (Williams 2003) when applied as intended. However, a "biocompatible" material may not be completely inert. Because of this, the host response is decisive (Schmalz and Arenholt-Bindslev 2009).

It is difficult to talk about the biocompatibility of elastomers since they are only in contact with physiological systems in the form of medical devices. In addition, the biocompatibility of elastomers depends on the type of device and its function. In theory, an evaluation of the biocompatibility of an elastomer should include

- The behavior of host tissue and the host itself in the presence of the biomaterial
- The behavior over time of the biomaterial in the host environment

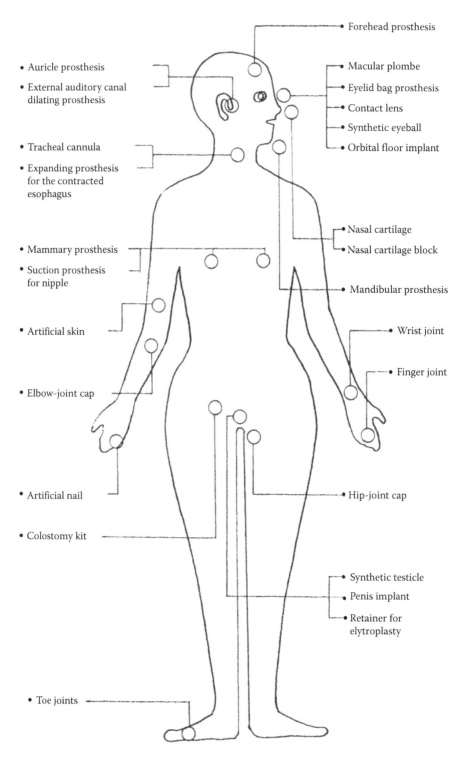

FIGURE 14.17　Medical device in the human body.

However, in practice, in vivo studies of all these parameters are difficult if not impossible to conduct, which is why the evaluation of biocompatibility begins with in vitro testing. Briefly, biomaterials have evolved through three generations:

1. First generation: technical performance (noncytotoxic = tolerance)
2. Second generation: biological performance (cytocompatibility)
3. Third generation: biological performance (biofunctionality)

Four trends currently dominate the evolution of biomaterials:

1. Giving materials a surface that mimics the biological properties required for the intended application, which requires fundamental knowledge of molecular/cellular reactions at the material/tissue interface
2. Taking implantation in young subjects into account, which implies long-term use and the need for immunological, mutagenic, and carcinogenic evaluations
3. Decreasing animal testing, which requires the development of alternative methods
4. Setting up marketing regulations, which includes the development of standard tests

14.2.2 GENERAL

Elastomers are very important in the manufacture of medical devices that may come into contact with host tissues for varying periods of time. Exposure times have been arbitrarily divided into three categories: limited exposure (<24h), prolonged exposure (24h to 30 days), and permanent exposure (>30 days). The tests used to evaluate biocompatibility depend on exposure time. It is thus easy to understand why elastomers used to make implants have to meet stricter standards than those used for dental impressions. This classification system is only a guide.

Various standards (mainly ISO 10993-(ISO 2003a)) apply to medical devices that are

1. In contact with surfaces like the skin or mucous membranes other than the skin
2. In direct or indirect contact with other tissues, including blood
3. Implantable devices

14.2.3 TESTING

14.2.3.1 Cytotoxicity

In vitro cytotoxicity testing evaluates lysis, growth inhibition, and other impacts on cells by medical devices, component materials, and leachates using morphological, biochemical, and metabolic criteria. These techniques have well-defined standards for material preparation, sterilization, cell lines, and the type of testing required. Whatever methods are used, controls must be included. Cell culture studies are usually the first step of the evaluation of the biocompatibility.

14.2.3.1.1 Material Preparation

The shape of the material used for biocompatibility testing varies greatly from author to author. Generally speaking, however, elastomers are tested as disks the size of multiwell plates used for cell cultures. Lehle et al. (2009) used 6mm diameter disks, Bal et al. (2009) used 14-mm-diameter disks, while Briganti et al. (2006) used 10×15mm pieces to evaluate polyurethanes. Other techniques involve producing blocks, which are then cut into thick slices using a microtome. Special attention must be paid to the cut surface, which must not only be identical in test products and controls but must also resemble the surface of the finished product. It is often preferable in the industrial research phase to test materials that correspond as closely as possible to the finished product. In the development and marketing phases, evaluations can be performed on randomly selected "parts" to avoid testing bias.

Many studies have been conducted on the cytotoxicity of finished elastomers. Extracts are prepared on the basis of recommended normative standards. A range of extraction protocols is available, including dynamic extraction, static extraction, and heat extraction.

The chemical substances extracted from elastomers are usually residues from accelerators (e.g., mercaptobenzothiazole and tetramethylthiuram disulfite), activators (e.g., zinc oxide), inert fillers (e.g., wax), plasticizers (e.g., oil), lubricants (e.g., stearic acid), and vulcanization products (e.g., zinc stearates and dithiocarbonate zinc salts). To minimize the amount of residue extracted from elastomers, they must be either cleaned or coated, or both.

14.2.3.1.2 Sterilization

The sterilization protocols for the various items used for cell cultures vary depending on the elastomer being evaluated. Most authors report using ethylene oxide (500–540 mg/L EO in 80% CO_2 at a relative humidity of over 35% for 3 h at 35°C–41°C). Others have used gamma radiation (2.5 Mrad for 5 h) (Pizzoferrato et al. 1995, Saad et al. 1996), dry heat or autoclaving (Bal et al. 2009), antiseptic solutions (Hunt et al. 1996, Menconi et al. 1992), antibiotics occasionally combined with UV radiation for 48 h (Menconi et al. 1992) or 30 min (Ozdemir et al. 2009), and immersion in 70% ethanol for 30 min followed by immersion in phosphate-buffered saline containing antibiotics (Morrison et al. 1995). In recent years, sterilization by exposure to gamma irradiation has increased in popularity with device manufacturers because it does not produce toxic residues. Most elastomers resist gamma radiation sterilization with minimal changes to their properties. All polymers are affected to some degree by irradiation, with crosslinking and molecular scission reported. While there have been reports that sterilization of polymers influences their performance in cell cultures with changes in cell attachment and morphology, certain authors do not mention the sterilization method used. There is an international standard regulating the management of residues following sterilization by ethylene oxide, glutaraldehyde, and formaldehyde (ISO 1995, 2002a), but there are no standards regulating sterilization procedures per se.

14.2.3.1.3 Cell Lines

Fibroblast cell lines (3T3, L929) are frequently used in cytotoxicity tests (Chen et al. 2008, Ozdemir et al. 2009, Zia et al. 2009a,b). Tests based on the medical end uses of materials are conducted with cell lines that are as similar as possible to the cells or tissues that will be in contact with the elastomers in vivo. For example, MG 63 osteoblast cell lines and RAW 264.7 macrophages (Yim et al. 2009) are used for testing materials to replace cartilage, and SV-HUC1 cells are used to test materials that will be in contact with the urinary tract (Pariente et al. 1998).

Primary cultures are more suitable for cytotoxicity testing. Various cells from different origins have been successfully used to evaluate the toxicity of elastomers. While primary cultures, especially those derived from human tissues, provide superior biocompatibility testing results because they are only one step removed from in vivo conditions, they do not allow interlaboratory comparisons to be made. Ideally, researchers should include both an established reference line and a primary culture in their studies, as was the case with Lehle et al. (Lehle et al. 2008).

Cells can be also grown in vitro in three dimensions, which allows better in vivo stimulation. A three-dimensional full-thickness engineered human oral mucosal model has recently been optimized (Moharamzadeh et al. 2008).

In dentistry, an ISO 7405 standard applies specifically to dental materials (semi-horizontal standard) (ISO 2008).

14.2.3.1.4 Tests

The crystal violet, neutral red, and trypan blue cell viability screening tool developed over 30 years ago for evaluating the toxicity of plastics is the most common test (Rosenbluth et al. 1965). Whatever the cell type, two methods can be used: direct tests where the cells are placed in direct contact with the material, and indirect tests where extracts of the medium used to culture the cells are placed in contact with a cell culture either by diffusion across a membrane or in an agar overlay (Johnson et al. 1983, Kirkpatrick and Mittermayer 1990). While the incubation period can vary from 24 h to 6 days, depending on the protocol, it also depends on the type of cells used. Generally speaking, the incubation period can be longer for established cell lines than for primary cultures. A tetrazolium-based colorimetric assay (MTT) is often used to evaluate polymer cytotoxicity or, more precisely, the metabolic activity of cells in contact with the material. The MTT assay is simple, reproducible, and does not use radioisotopes. Tests can be carried out using a fully automated microtiter plate system, making it possible to economically test polymers of limited availability. It has also been used to evaluate polyurethanes (Haugen et al. 2006, van Minnen et al. 2005), polyesterurethanes (Saad et al. 1997), elastomers (Chauvel-Lebret et al. 1999), and urinary stents (Pariente et al. 1998) The in vitro WST-1 cytotoxicity test based on the European ISO 10993-5 standard (ISO 1999a) is also used and, like the MTT test, involves the release of water soluble products from a tetrazolium salt by metabolically active cells. However, the results obtained using this test are different from those of the MTT test (Haugen et al. 2006).

In addition to cell viability and metabolic activity assays, cell morphology is also verified to complete the first-phase screening of elastomers. The impact on cell morphology is generally determined by scanning electron microscopy (Guidoin et al. 1992, Haugen et al. 2006, Lee et al. 1996a, Menconi et al. 1992). Samples are generally dehydrated by critical point drying and then metallicized with colloidal gold. Elastomers that have not been in contact with the cells are used as controls. While cell viability and morphology tests provide an initial screening for cytotoxicity, many researchers have been working to examine the biofunctionality of cells in contact with elastomers and elastomer extracts. Various tests have been used, including the evaluation of protein synthesis and mRNA expression. Fluorescence-activated cell sorting (FACS) and Western blotting are used to detect the impact of biomaterials on cell metabolism, especially on signaling pathways. These methods are used to evaluate radical oxygen species concentrations, apoptosis, DNA damage and repair, and synthesis of specific inflammation mediators (Hsu et al. 2008, Hung et al. 2009b, Lorenz et al. 2006).

The concept of biocompatibility is based on the interaction between a material and a biological environment, most often expressed as an inflammatory response. Implanted elastomers provoke an inflammatory response, which initiates the process of tissue repair and regeneration. Various in vitro tests have been developed to evaluate the inflammatory response and repair phases. The tests most frequently reported in the literature to determine the biocompatibility of polyurethanes involve measuring the concentrations of NO and TNFα, which are produced when macrophages and osteoblasts interact with polyesterurethanes (Saad et al. 1996, 1997), the concentrations of various types of fibronectin and collagen (Lee et al. 1996b, Saad et al. 1996, 1997, Sprague and Palmaz 2005), and the adsorption of proteins.

Techniques based on molecular approaches are being developed to characterize the interactions between cells, biological fluids, and biomaterials. Most of these techniques have become more accessible with the arrival of commercially available molecular biology kits. Gene expression using microarray test systems reveals information on the genes involved in the cellular stress response (Schweikl et al. 2008, Tare et al. 2009) and has been proposed for biocompatibility tests (Pernagallo et al. 2009). These tests obviously depend on the end use of the elastomer. With increasing information available on interactions between cells and biomaterials, new guiding principles for the in vitro evaluation of biocompatibility, including the use of relevant cells types and biological parameters and the need to use dynamic conditions will be necessary in the future (Owen et al. 2005).

14.2.3.1.5 Controls

Tests cannot be valid or objective unless internal controls are included. Tissue culture plastic is a commonly used negative control, although silicone rubber has also been used. Asbestos, phenol, and latex, which cause major disruptions to metabolic activity, are often used as positive or cytotoxic controls (Pariente et al. 1998). It should be noted that latex has been used as both a negative and positive control. Its cytotoxicity arises from the presence of accelerators and oxidants used during the vulcanization process (van Minnen et al. 2005, Oshima and Nakamura 1994). As indicated by Park and Park (Park and Park 1996), in the absence of clear criteria for evaluating biocompatibility, many materials have been mistakenly thought to be biocompatible and have been used as controls. Silicone rubber is a case in point. For a long time, silicone rubber was believed to be totally biocompatible. It is clearly necessary to reevaluate the biocompatibility of existing biomaterials and a fortiori, the negative controls. Positive controls are not reported in many studies, which undermines the validity of the results and proposed evaluation technique.

14.2.3.2 Sensitization Assays—Irritation or Intercutaneous Reactivity—Implantation

General Considerations: Even today, there are no totally satisfactory in vitro elastomer evaluation systems. While such assays provide an initial screening, in vivo testing on animals is required before implanting, injecting, or using elastomers in humans. A nonexhaustive search of the literature between 1990 and 2010 reveals that too many animals have been sacrificed just to study the implantation response of elastomers. It is vitally important that researchers meticulously plan their studies to extract the maximum amount of information from each animal.

14.2.3.2.1 Material Preparation

For solid implants, the physical properties (shape, hardness, surface finish) must be identical for each implant, except if one of the parameters is being singled out for study. The chemical and morphological aspects of elastomeric surfaces must be controlled (Chesmel and Black 1995). All contaminants must be removed from the implants, which must then be sterilized using the method that will be applied to the finished device. Careful attention must be paid to comparisons between in vivo responses and in vitro assays. For example, Rosengren et al. (2005) compared the in vivo tissue responses to in vitro cytotoxicity before and after implantation to evaluate the predictive value of cytotoxicity testing. They observed that when zinc diethyldithiocarbamate was added to polyurethane, fewer cytotoxins were released in vitro than in vivo conditions. They concluded that many clinically useful implant materials may be unnecessarily rejected due to the results of in vitro tests.

Tests on nonsolid materials such as powders, liquids, and particles (including mixtures) can be conducted in PTFE tubes. Some authors dispute the suitability of PTFE as a negative control, especially as regards surface properties. Cylindrical cages can be fabricated using stainless-steel wire mesh (Christenson et al. 2004a). At the very least, control tubes or cages must have the same physical properties as test tubes.

14.2.3.2.2 Animals

The choice of animal depends on the size of the implant, the test period, and known differences in biological responses of soft and hard tissues. Rats are the animals of choice for many studies. Implants are placed bilaterally in the back muscles. Two animals per material per implant period are used with a minimum of two implants per animal (Fabre et al. 1998, Hunt et al. 1996, van Minnen et al. 2005, Pomerantseva et al. 2009). When rabbits are used, 8–12 implants are placed in the back muscles (Briganti et al. 2006, Oloffs et al. 1994). Swine are also used to test biocompatibility and hemocompatibility (Richard et al. 2009). Dogs are used to study new materials for vasectomies and arteriovenous fistulas (Carroll et al. 1993, Schellhammer et al. 1999) and to test materials for temporary replacement organs. The number of animals used varies greatly depending on the type of research and the complexity of the procedure. Because of the differences, few studies can be compared.

14.2.3.2.3 Implantation Period

The implantation period varies greatly depending on the animals used and the material being evaluated. In the few chronic inflammation studies that have been reported, positive controls that could lead to a major inflammatory response within 12 weeks are missing. Generally speaking, observations are made over a 30 day (Fabre et al. 1998) or 12 week (Carroll et al. 1993, Huang et al. 1992, Hunt et al. 1996) period. The implantation period must be chosen to obtain a stable biological reaction, which depends on the nature and properties of the material and the damage caused by the surgery. A recent study published results of implantation after a 2 year period in dogs (Welsing et al. 2008).

14.2.3.2.4 Evaluation of Biological Reactivity

Evaluations of biological reactivity begin with an analysis of health of the animals (vigilance, nutrition, coat, weight, temperature, etc.). Macroscopic and histopathological reactions are evaluated over time. Histopathological examinations include the degree of inflammation, the number and type of inflammatory cells (leukocytes, PMNs, lymphocytes, plasmocytes, macrophages, polynuclear cells, etc.), necrosis, and the presence of material debris, granulomas, fat, and calcified tissue.

14.2.3.2.5 Irritation and Sensitization Assays

Irritation is a nonimmunological inflammatory reaction. Sensitization assays (contact allergies) and delayed hypersensitivity reactions involve the immune system. These assays are vital for understanding acute toxicity. Rabbits are often used for skin irritation assays with observations at 1, 24, 48, and 72 h. The test period may be prolonged in the event of persistent lesions to determine whether the lesions are reversible or permanent, but must in no case exceed 14 days. Less than 0.2 mL of extract is injected intradermally. Irritation is scored based on the level of edema and erythema. Generally speaking, 10 injections are made on the anterior and posterior portions of the trunk (including controls). Ocular and oral mucosal assays may also be conducted. To evaluate immune responses, tests developed for other purposes such as research on autoimmune disease animal models and conjunctive tissue pathologies are used. These tests are based on delayed hypersensitivity reactions induced by a given autoantibody (thyroglobulin), anti-collagen, etc. In the research reported by Naim et al. (1995), silicone gels were tested for toxicology, antigenic, and adjuvancy properties. In vitro skin irritation tests are being as alternatives to animal models.

14.2.3.2.6 Implantation

All elastomers act as foreign bodies when they are implanted, leading to an acute inflammatory response and the accumulation of phagocytes (Anderson et al. 2008). Elastomers are relatively inert, unreactive, and nontoxic. It is thus difficult to understand how they are detected by the immune system and how they provoke an inflammatory response. However, considering the mechanisms involved in such responses, a good starting point would be the initial interaction between the biomaterial and proteins in the surrounding host tissue. Immediately after implantation, hydrophobic polymers like polyurethanes, polyethylenes, polydimethylsiloxanes, and dacron become coated with host protein. Plasma and denatured interstitial fluid proteins rapidly colonize the implant, binding together to form a disorganized matrix on the implant surface. Since proteins rapidly adsorb to the surface of the biomaterial, host inflammatory cells and fibroblasts cannot come into direct contact with the implant. This protein coat determines future cell reactions to the implant and it is undoubtedly this initial phase that is key in determining biocompatibility. Albumin, immunoglobulin G, and fibrinogen are the main proteins coating the surfaces of implants. Conventional histological techniques involving hematoxylin, eosin, Von Gieson, and Schiff periodic acid stains are often used to study the inflammatory response. The inflammatory cells, which include macrophages, PMNs, and lymphocytes, are identified based on morphological criteria. The distribution of these cells around the implant provides a qualitative description of the response. A number of assays using computer assisted image analysis have been developed. Discriminating between different cells types has also greatly evolved with the arrival of specific monoclonal antibodies. Specific enzymatic assays and signal amplification kits (avidin, biotin, etc.) have considerably enhanced the ability to gauge implant inflammatory responses. In addition to these histochemical staining, immunocytochemical, and counting techniques, certain authors have published the results of in situ hybridization techniques for evaluating inflammatory protein production (TNFα) and detect the expression of TNFα mRNA. These techniques provide a clear picture of the intensity of the inflammatory response. The reaction of TNFα and other growth factors or enzymes must reflect a balance between the tendency to promote biocompatibility and the tendency to cause inflammatory damage and implant rejection (Hunt et al. 1996, Lobler et al. 1999).

A number of researchers have analyzed the quality and quantity of the inflammatory exudate using the in vivo cage implant system originally developed by Marchant et al. (1983). A number of biochemical tests (protein analyses, albumin concentration, extracellular enzymes) are possible. The exudate may also be analyzed by cytofluorometry for quantitative results (Fabre et al. 1998, Huang et al. 1992). Flow cytometry appears to be particularly well suited for studying exudate cell composition. The literature on exudate cell responses is relatively meager compared to that on tissue reactions. To our knowledge, there have been no reports correlating tissue inflammation (as measured by histomorphometric image analysis) and cellular response (as measured by flow cytometry).

14.2.3.3 Hemocompatibility and Blood Interaction Assays

14.2.3.3.1 General

Only materials in prolonged or permanent contact with blood pathways must be subjected to this type of testing, which can be divided into five categories depending on the primary process (thrombus, coagulation, platelet aggregation) or biological system involved (blood, immune). Various standards determine which tests will be used to evaluate hemocompatibility:

- Devices whose surfaces are not in direct contact with the blood
- Devices whose surfaces are in direct contact with the blood
- Implantable devices

In certain cases, it is difficult to determine whether biocompatibility and/or hemocompatibility tests should be performed. In general, experience has shown that materials that are hemocompatible will also be biocompatible. However, not all biocompatible materials are also hemocompatible. The situation is even more complex because there has been a tendency to designate materials as being biocompatible on the basis of toxicological tests while others have been designated as being hemocompatible on the basis of a single whole blood clotting test (Bruck 1980).

14.2.3.3.2 Tests

A given material may be adequate for one type of application, but not for others. Differences between arterial and venous blood, varying blood flow patterns, and the design and mechanical operation of a medical device are among the parameters that influence the performance of materials when they are in contact with blood. Most biomaterial researchers would like to be able to predict biological performance using simplified methodologies. However, no single procedure is adequate for this purpose. Tests that merely measure the amount of thrombus give only the final result of the blood coagulation sequence without any insight into the initial events. A rheology-based test allows the sequence of events to be measured over time. ISO 10993 guidelines for the biological evaluation of implantable medical devices (ISO 2002a,b) list a number of recommended in vitro and in vivo tests that can be used to study hemocompatibility. ISO recommends the hemolysis test as the standard screening method.

Many researchers use the acute ex vivo canine femoral anteriovenous series shunt technique, which allows the testing of a number of materials under similar physiological and hematological conditions for in vivo hemocompatibility studies (Chen et al. 1998, Peek et al. 2002). High levels of platelet and fibrinogen deposition are generally associated with a more thrombogenic polymeric surface. The morphology of the adherent platelets examined by SEM also provides valuable information. A measure of thrombogenicity is the tendency of a surface to cause adherent platelets to change shape and activate.

Hemocompatibility tests include evaluations of platelet adhesion, aggregation, activation, and release reactions under dynamic blood flow conditions together with measurements of protein adsorption and coagulation factor activation. The most common test for evaluating platelet adhesion uses fresh blood from healthy donors. The platelets are isolated and a standard microscopic or flow cytometric analysis is performed (Chen et al. 1998, Rechavia et al. 1998, Tarnok et al. 1999). Platelet counts can also be determined using the Petri disk model with flat sheet membranes. Complement activation and release reactions have also been explored. To evaluate the extent of in vitro complement activation by elastomers, the concentration of C3a des Arg has been measured according to the method of Wagner and Hugli (1984) using radioimmunoassay kits available from a number of manufacturers (Branger et al. 1990).

Tests evaluating contact activation, the intrinsic coagulation system, the Hageman factor (f XII) and dependent kinin formation and fibrinolysis pathways have been developed. A quantitative measurement of the activation of the contact system is required in studies examining the effect of artificial surfaces on blood. Van der Kamp and Van Oeveren (1994) proposed analyzing these cascades by using the kallikrein inhibitor aprotinin to calculate the activity of factor f XII and kallikrein.

One approach to improving the hemocompatibility of biomaterials has been to immobilize anticoagulants like heparin at the interface (Albanese et al. 1994, Hong et al. 1999). Tests including clotting assays (TT, reptilase time, and anti-XA activity), coagulation time, platelet counts, and Resonance Thrombography (RTG) have been used.

Previous studies have shown that blood-contacting materials activate blood monocytes. A highly sensitive technique has been proposed to detect changes at the mRNA level in circulating monocytes and to find suitable "gene markers" for assessing the hemocompatibility of biomaterials. Human blood can be recirculated in a modified Chandler Loop formed of test tubes. The recirculation of human blood in an in vitro model in combination with the immunomagnetic separation of monocytes and the Duplex RT-PCR method is a powerful tool for getting reliable results (Weber et al. 2001).

Also, with the development of nanomaterials, a new test has been developed to validate an in vitro method designed to analyze the potential of a variety of nanoparticles to damage red blood cells. Specifically, the study describes approaches to identify nanoparticle/erythrocyte interferences, when they occur and how to resolve them to get accurate results, i.e., to avoid false-positive or false-negative results (Dobrovolskaia et al. 2008).

14.2.3.4 Biodegradation Assays

These assays are used to evaluate both biodegradation and biostability. Degradation products are generated by the decomposition or chemical degradation of a material. Biodegradation is the degradation of a biological material involving a loss of integrity and/or performance during exposure to a physiological or simulated environment. Elastomer degradation is studied using toxicokinetic modeling. Many types of surgically implantable devices and drug delivery systems that only function for a relatively short time in vivo can be made from polymers that are eliminated from the body by hydrolytic degradation and subsequent metabolism after serving their intended purpose. Biomaterials made of biodegradable polymers are designed to degrade in vivo in a controlled manner over a predetermined period. The main in vivo degradation mechanism of polymers is hydrolytic degradation in which enzymes may play a role. In vitro cell culture and animal models may also be used (Piskin 1994, Singh et al. 1979).

So-called nonbiodegradable biomaterials may be slowly degraded over time by the organism. One example is the failure of PU medical devices like pacemaker leads, which are manufactured from poly(ether) polymers. PU failure consistently occurs 5–10 years after implantation. The mechanism of this process is still not well understood (Santerre et al. 2005). Although in vitro studies have been performed with single enzyme systems, which showed the release of products, the in vivo situation involves complex biosystems that act synergistically. Other assays using sophisticated technology such as attenuated total reflection–Fourier transform infrared spectroscopy (ATR-FTIR)

(Mathur et al. 1997), transmission Fourier transform infrared analysis (T–FTIR), and scanning electron microscopy (SEM) (de Brito Alves et al. 2003, Collier et al. 1998) have also been developed. These techniques can be used to evaluate inflammatory responses and throw light on the entire degradation process of a biomaterial.

No progress on the evaluation of degradation has appeared in the literature since 2000. An ISO framework (ISO 1999b) and standard (ISO 1998) are available.

14.2.3.5 Reproduction and Embryo Development—Assays

It is not known whether implant materials can have an impact on reproduction and embryo development. The few tests that have been conducted in this area involve the functional testing of biomaterials used for preparing sperm prior to artificial insemination, in vitro fertilization procedures, and embryo transfers. These assays can be conducted in vitro (e.g., spermatozoid cultures) or in vivo (e.g., sperm penetration tests using hamster ova and mouse embryo compatibility studies) (Hunter et al. 1988). To our knowledge, no standards have been established for these assays.

14.2.3.6 Genotoxicity

ISO 10993-3 (ISO 2003b) is a set of criteria based on the intended use of a product or device and establishes testing guidelines depending on the implantation period and the nature of the contact with the biomaterial. The criteria stipulate that any material or implantable device placed in contact with mucosal, bone, or oral tissue, when the contact exceeds 30 days, must be subjected to genotoxicity testing prior to commercialization. The tests are intended to detect genetic anomalies (mutations, chromosomal alterations) in cells (prokaryote, yeast, mammalian) placed in contact with the material. A review of the literature has revealed that four main techniques are used to evaluate genotoxicity: the sister chromatid exchange method, the micronucleus test, the Ames test, and the chromosomal aberration test (Nadeenko et al. 1997, Schendel et al. 1995). Other tests appear to evaluate genotoxicity using a micronucleus assay (Elespuru et al. 2009, Kleinsasser et al. 2006) and a high-throughput comet assay (Stang and Witte 2009). The high-throughput comet assay was developed to reduce processing time and increase sample-throughput of the assay (Tice et al. 2000). The number of published studies addressing the mutagenicity of elastomers is comparatively low.

14.2.3.7 Biocompatibility Testing in Humans

Clinical tests are better indicators of biocompatibility and also help improve biomaterials. While implantable devices are supposed to be biocompatible based on in vitro testing, in vivo clinical testing is the only way to be sure. Human tests are conducted using ethical research practices. These tests have been used in randomized clinical research protocols (Baumann et al. 2003, Hollick et al. 1998, 1999), medium to long-term epidemiological studies (Bambauer et al. 2004, Boldin et al. 2008, Johnell et al. 2005, Rubin and Yaremchuk 1997), and in one study with a 20 year follow-up (Donawa 2006a,b,c, Pritchett 2008).

14.2.3.8 Biocompatibility Testing—Microbial Colonization

There are no bacteriological standards or compulsory tests in the field of reproduction and embryo development. Elastomers used in implantable medical devices are by definition sterile (see Sterilization). Nevertheless, certain authors have included microbial (including yeast) adherence and colonization in their battery of tests (Gottenbos et al. 1999, Trevisani et al. 2005). Such testing is recommended for certain elastomers for which implantation failure may be caused by bacterial infections. The plasma proteins deposited on implant surfaces may mediate bacterial adherence, especially that of *Staphylococcus aureus*, a pathogen associated with recurrent infections. Bacterial adherence testing is conducted in radial flow chambers mounted on the motorized stage of a video microscopy system. Image processing software is used to perform automated data collection and image analysis (Baumgartner et al. 1997, Santos et al. 2008). Bacteria-induced infections in the presence of polymers have been studied in animals. It would be of interest to determine how bacteria adhere to polymer substrates and, in certain cases, to evaluate the efficacy of antibiotic/antiseptic treatments (Kockro et al. 2000). A better understanding of bacterial genomes should also lead to a better understanding of virulence factors produced by bacteria. These virulence factors may vary depending on the biomaterial as suggested by changes in bacterial phenotypes (unpublished data).

14.3 BIOCOMPATIBILITY OF DIFFERENT CLASSES OF ELASTOMERS

14.3.1 ELASTOMERS IN GENERAL

Biomaterials are materials designed for use with living tissues and/or biological fluids in order to evaluate, treat, modify, or replace a tissue, organ, or bodily function (Kalman et al. 1991). For materials to be biocompatible, they must not provoke allergic, inflammatory, or immunological reactions, they must be nonthrombogenic, nontoxic, and noncarcinogenic, and they must not damage surrounding tissues, plasma proteins, or enzymes (Kalman et al. 1991, Piskin 1994). However, in most cases, it is not a lack of response that is important but rather an appropriate host response to the specific application for which the biomaterial was designed. Biocompatibility is thus defined by the response of the biological system to the biomaterial, which is seen as a foreign body and which provokes a cascade of interrelated reactions both systematically as well as locally at the interface with the biomaterial (Piskin 1994).

Biocompatibility is directly dependent upon the magnitude and duration of the host inflammatory response (Kao et al. 1999, Roach et al. 2007). The body initiates a cascade of reactions against the foreign material (Brodbeck et al. 2005).

Biomaterials may thus lead to numerous undesirable side effects (clotting, infections, fibrosis, thrombosis, tissue necrosis, carcinogenesis, etc.). These effects can be caused by the degradation or mineralization of the biomaterial. In many cases, these adverse effects are associated with the rapid accumulation of large numbers of phagocytes (Anderson et al. 2008, Hu et al. 2001).

Over the long-term, the service life of the biomaterials depends on the mechanical and physico-chemical properties of the biomaterial on the micro and nano-scales (Roach et al. 2007).

14.3.2 LOCAL REACTIONS

When a biomaterial comes in contact with an intra or extravascular system, a complex sequence of events involving protein adsorption, monocyte/macrophage adhesion, and macrophage fusion to form foreign body giant cells (FBGC) takes place (Anderson et al. 2008, Schmidt and Kao 2007).

The initial responses are acute inflammation within a few days, followed by chronic inflammation (Anderson et al. 2008, Brodbeck et al. 2005, Schmidt and Kao 2007).

14.3.2.1 Protein Adsorption

Implanted biomaterials very quickly acquire a layer of host proteins well before the arrival of inflammatory cells (Hu et al. 2001, Roach et al. 2007). The first to adsorb are the more abundant small proteins, which are then replaced by larger proteins with a greater affinity for the surface of the biomaterial. The initial adsorption prevents direct interactions between cells and the implant material (Roach et al. 2007).

Protein cascades may be closely involved in the dynamic phenomenon of protein adsorption and desorption know as the Vroman effect (Anderson et al. 2008, Roach et al. 2007).

Protein adsorption is affected by surface chemistry and nanoscale topography. Protein deformation is induced by interactions with a surface and is affected by electrostatic forces, entropic effects, hydrophobic interactions, and conformational changes (Roach et al. 2007).

The presence of adsorbed proteins such as albumin, IgG, complement, vitronectin, gamma globulin, fibrinogen, and fibronectin modulate the host inflammatory response, cell activation, adhesion, and wound healing (Anderson et al. 2008, Lloyd 2002, Tang and Eaton 1995).

The primary interactions between cells and proteins occur via integrins. Integrins bind specifically to an arginine-glycine-aspartic acid (RGD) in adhesins such as fibronectin, vitronectin, laminin (Roach et al. 2007, Vasita et al. 2008).

The differences in integrin expression observed on different materials may account for the variations in cell attachment. Integrins are a large family of cell surface receptors that mediate binding to the extracellular matrix (ECM) and are also involved in intracellular signaling and a diverse range of cell functions (Anderson et al. 2008, Wilson et al. 2005).

14.3.2.2 Adherence and Activation of Inflammatory Cells

The type, concentration, and conformation of adsorbed proteins and thus the adhesion and survival of cells bound to the proteins depend on the surface properties of the biomaterial (Anderson et al. 2008).

Polymorphonuclear leukocytes, monocytes, and lymphocytes are the predominant cell involved in the inflammatory response and the development of foreign body reactions and fibrous capsule formation (Brodbeck et al. 2005).

Macrophages bound to the surface of a biomaterial fuse to form FBGC. IL4 induces macrophage fusion on biomaterial surfaces while IL3 induces the fusion of monocyte-derived macrophages (Anderson et al. 2008). All phagocytes contain lysosomes, intracellular organites that contain a great variety of proteolytic enzymes. When liberated in situ, these enzymes can destroy the ECM that supports connective and epithelial tissues. Lysosomes also release vasodilatating substances and activate the complement and kallicrein/kinin systems, coagulation, and fibrinolysis (Diebold et al. 1977, Kalman et al. 1991, Rakhorst et al. 1999, Van der Kamp and Van Oeveren 1994).

Platelets, which are very important in homeostasis, can act as inflammation mediators via substances in their intracytoplasmic granules (Trowbridge and Emling 1997).

14.3.2.3 Local Complications

Host inflammatory responses to contact with a biomaterial are normal reactions. In certain cases, local complications may occur, the most frequent being the formation of thromboses, fibrotic hyperplasia around the implant, and bacterial infections (Park and Park 1996, Sigler et al. 2005, Tang and Eaton 1995, Werner and Jacobasch 1999).

In humans and animal studies, protracted enhanced inflammation is associated with increased susceptibility to infections (Boelens et al. 2000).

While platelet aggregation may be a normal reaction when a biomaterial comes in contact with blood, it may become iatrogenic, with the formation of a thrombus and the risk of emboli (Park and Park 1996, Piskin 1994).

Thrombus formation is not only due to endothelial damage but also to a foreign body surface directly exposed to the blood stream. The consequences of a thrombus can be vessel occlusion or embolization (Sigler et al. 2005).

Factor XII activates the complement system. Activated fractions C3a and C5a cause platelet aggregation. C3b promotes cell adherence, increasing the risk of thrombosis (Kalman et al. 1991, Keogh et al. 1996).

Certain authors have reported that hemocompatibility is influenced by the type of proteins adsorbed to the surface of the implant. Albumin seems to prevent the adhesion of platelets, providing acceptable hemocompatibility by reducing the early proinflammatory and thrombogenic responses of inflammatory cells, while fibrinogen and gamma-globulin increase platelet adhesion, leading to bio-incompatibility (Asberg and Videm 2005, Casenave 1986, Dumitriu and Dumitriu 1990).

Platelets and neutrophils can bind to IgG adsorbed to a biomaterial surface. IgG-bound platelets induce neutrophils to release toxic reactive oxygen species (Asberg and Videm 2005).

Thrombogenicity is regulated by surface texture, electrostatic potential, and composition (Kannan et al. 2006).

Various techniques have been developed to diminish thrombogenicity, including coating the external surface of catheters with methacryloyloxyethyl phosphorylcholine-based copolymer (MPC) or silicone. These techniques improve hemocompatibility because the coatings have a low degree of thrombogenicity (Bambauer et al. 2004, Lloyd 2002, Nakabayashi and Williams 2003, Rose et al. 2005).

When vascular grafts are implanted, the response of endothelial cells is crucial because they play a role in maintaining the antithrombogenicity of the prosthesis. Endothelial cells (EC) have both anticoagulant and thrombostatic functions. Under physiological conditions, negatively charged EC prevent blood from clotting by releasing vascular relaxing factors, antiplatelet adhesion/aggregation factors, antithrombogenic factors, and fibrinolysis promoters. Under pathological or traumatic conditions, EC secrete repair proteins to maintain the integrity and continuity of blood vessels (Wang et al. 2003).

To achieve optimal integration of a vascular implant, the lumen of the prosthesis must be quickly and completely endothelialized, which prevents the formation of blot clots and vascular occlusion.

Various techniques have been developed to enhance EC adhesion and proliferation and optimize cell-polymer interfaces by giving the surface a positive charge and by immobilizing cell growth-promoting factors on it (fibronectin, laminin, Poly-D-lysine) (Wang et al. 2003).

The endothelial damage caused by the implantation activates the inflammatory and coagulation systems, which in turn activates the smooth muscle cells of the media. These cells then migrate into the intima leading to the formation of intimal hyperplasias (Sigler et al. 2005).

The formation of a fibrotic capsule is a normal reaction to the implantation of a foreign body. The process may become pathological if the fibroblasts proliferate too extensively and lead to fibrotic hyperplasia. The thickness of a fibrous capsule around an implant is part of the biological response to the degree of biocompatibility of the implant (Kaluzny et al. 2007). An intense host inflammatory response to a biomaterial stimulates phagocytes, especially macrophages and neutrophils, which are attracted in large number and produce cytokines, chemokines, and other mediators (Cabanlit et al. 2007, Miller and Anderson 1989, Park and Park 1996, Tang and Eaton 1995).

A lack of implant/tissue adherence and the space between the two surfaces causes a certain amount of friction that results in the accumulation of serous fluid and inflammatory cells. An intense inflammatory response at this interface can result in fibrotic hyperplasia (Von Recum and Van Kooten 1995). The contraction of the fibrotic capsule around the implant, which is more frequent with silicone elastomers than PUs, is a common complication with implanted biomaterials (Szycher and Siciliano 1991, Tang and Eaton 1995).

Bacterial infections are frequently associated with implanted biomaterials. Fibrinogen is known to mediate the adhesion of *Staphylococcus aureus*, the main pathogen involved in implant-associated infections. Fibrinogen can act as an opsonin and promote the aggregation of bacterial cells, especially staphylococci and streptococci (Trowbridge and Emling 1997). The access route for catheter-related infections remains a point of contention. Certain authors (Cooper and Hopkins 1985, Maki et al. 1977, 1991) have suggested that infections involving percutaneously inserted catheters are due to contamination of the external surfaces. Others have suggested that the contamination of the catheter hub can lead to intralumenal colonization and catheter-related septicemia (Boswald et al. 1995). Bacteria and fungi colonize the polymer surface, forming an adherent biofilm. The microorganisms embedded in these highly adherent films may, over time, secrete products that degrade certain biomaterials (Bakker et al. 1990a). In addition, biofilms make antibiotic treatments and host defense mechanisms less effective (Baumgartner et al. 1997, Jansen et al. 1994, Kossovsky 1995, Kossovsky and Freiman 1995).

The development of microdomain-structured surfaces (PUr SMA-coated catheters) is a promising technology for reducing bacterial colonization (Bambauer et al. 2004).

A novel silicone elastomer catheter grafted with a poly(vinylpyrrolidone) hydrogel (SEpvp) has been used to show that the adherence of *Staphylococcus epidermidis* on SEpvp in vitro is significantly lower than on conventional silicone elastomer catheters (Boelens et al. 2000).

14.3.3 GENERAL REACTIONS

14.3.3.1 Degradation

Degradation is a complex mechanism that affects all elastomeric biomaterials and depends on the chemical, physical, and structural properties of the implant and the inflammatory response of the host (Pinchuk 1994). Degradation occurs at the molecular, macromolecular, microscopic, and macroscopic levels (Maki et al. 1991). All implanted materials provoke an inflammatory response from the host, which attempts to attack, destroy, and reject the foreign body. Leukocytes, especially macrophages and FBGCs, are involved in this mechanism, which leads to the degradation of the elastomer (Szycher and Reed 1992, Tang and Eaton 1995).

Macrophages appear to play a predominant role in the degradation of elastomers. They are involved in the pinocytosis and phagocytosis of cell and tissue debris as well as polymer contaminants on the surfaces of biomaterials. Phagolysosomes digest the particles they have ingested and the degradation products are then released into the extracellular environment together with lysosomal enzymes, peroxides, and hydrogen ions (Collier et al. 1998, Marchant et al. 1984b, Szycher and Reed 1992, Tang and Eaton 1995). Lysosomal enzymes like enzymes—cholesterol esterase (CE) and phospholipase A2 (PLA) have often been implicated in the degradation process despite the difficulties in analyzing this phenomenon in vivo (Labow et al. 1997, Marchant et al. 1984a). The release of lysosomal enzymes, peroxides, and hydrogen ions, which are very concentrated at the cell/polymer surface interface, leads to significant degradation at this site. The degradation is faster for porous elastomers than dense elastomers. Stress and tension combined with the chemical properties of the elastomer are also involved in the degradation process (Collier et al. 1998). Degradation leads to surface cracking and a loss of molecular weight, which in turn results in a loss of tensile strength (Pinchuk 1994).

14.3.3.2 Carcinogenic Reactions

The carcinogenic potential of materials implanted in rodents has been recognized since 1950. This process is called the Oppenheimer effect. Many substances have been shown to induce foreign body carcinogenesis at the implantation site (Pinchuk 1994). In 1975, Brand et al. (Brand et al. 1975) described the various steps in foreign body tumorigenesis:

- Proliferation and tissue infiltration
- Progressive formation of a fibrotic tissue capsule
- Quiescence of the tissue reaction
- Direct contact of clonal preneoplastic cells on the foreign body surface
- Maturation of preneoplastic cells
- Sarcomatous proliferation (Brand et al. 1975)

Tumorigenesis is controlled by a number of factors. Foreign body carcinogenesis is influenced by physical features such as shape, size, hardness, and porosity (Nakamura et al. 1992, Pinchuk 1994, Stokes and Cobian 1982). Implant size is critical. Tumorigenesis does not occur unless the implant is of sufficient size. The bigger the implant, the greater the probability that preneoplastic clones will attach to the foreign body surface (Brand et al. 1975). Other researchers have reported that chemical properties play an important role in tumorigenesis (Nakamura et al. 1992). Numerous studies have examined the incidence of tumor formation as a function of substrate and, in particular, PUs and silicones. PUs appear more carcinogenic than silicones. Differences between the two elastomers in terms of cell adhesion and thus cell proliferation are important factors in the carcinogenic process (Nakamura et al. 1992, Tsuchiya et al. 1995a,b). The higher thermal sensitivity and water absorption of PUs with respect to silicones also plays an important role in this process (Stokes and Cobian 1982). In addition, a degradation product of PU (MDA or 4,4′ methylene dianiline) has been implicated in the development of cancer in rats (Brinton and Brown 1997, Nakamura et al. 1992,

Pinchuk 1994, Tang and Eaton 1995). Heat (autoclave) and gamma radiation sterilization have also been implicated in the formation of this toxic, carcinogenic compound (Yoda 1998). The mechanism remains very controversial. On the other hand, a 1995 study produced no evidence that sterilization leads to the formation of MDA (Shintani 1995). The tumorigenesis mechanism is not clear despite numerous studies on the carcinogenic potential of biomaterials. Nakamura (Nakamura et al. 1992) has tried to provide an explanation by suggesting a two-step mechanism: initiation followed by promotion. The initiation step depends on active oxygen species produced by cells like macrophages and foreign body giant cells or on initiators in the biomaterial (Tsuchiya et al. 1995a). The role of surface properties in the promotion of tumors has been examined by gap junction intercellular communication inhibition studies using collagen immobilized on polyethylene film. Studies have shown that the inhibition of gap junction intercellular communication reduces the tumorigenic potential of polyethylene by decreasing tumor promotion activity (Nakaoka et al. 1997).

14.3.4 SILICONES

Silicone elastomer is one of the most commonly used biomaterials in medical devices such as voice protheses, artificial dermises, artificial phages, tissue expanders, intraocular lens, and artificial cornea (Ai et al. 2003, Ren et al. 2008).

While it is accepted that silicone elastomers have good hemocompatibility, low toxicity, and long-term stability, the biocompatibility of these elastomers remains an open question. Undesirable effects from silicone implants have been reported (Spiller et al. 2007). Iatrogenic effects are caused by the formation of a hypertrophic fibrotic capsule around the implant, which later leads to the contraction of the capsule, causing severe pain, local tissue damage, and implant function impairment (Backovic et al. 2007). A comparison of silicone elastomer with methacrylate hydrogel, both of which are nonabsorbable, and with absorbable crosslinked Na-hyaluronate, showed that crosslinked Na-hyaluronate induces the least inflammation and a thicker fibrous capsule around the implant than silicone (Kaluzny et al. 2007). Silicone has occasionally been associated with severe foreign body reactions and with autoimmune disorders (Caballero et al. 2003). An analysis of epithelial cell adhesion to various elastomers gave similar results for both silicones and PUs. However, cell growth was poorer on silicone elastomers than on PUs. These results tend to minimize the role of cell adhesion to the substrate.

Cytotoxicity is likely related to the release of toxic substances (Bakker et al. 1988). Scanning electron microscopic analyses of the behavior of cells grown in contact with silicone elastomers have shown that the cells grow in aggregates with little spreading and produce numerous adherence structures linking them to one another, but few binding them to the substrate (Chauvel-Lebret et al. 1999). The properties of silicone elastomers described before affect cell adherence to the substrate and thus cell growth but have no effect on viability. Various techniques have been used to enhance cell behavior and growth on silicone. Microwave plasma surface modification of silicone with allylamine causes an increase in cell growth and viability with respect to unmodified silicone (Ren and Mahon 2007). Modifying silicone by physical entrapment of a synthetic hydrophilic lipid increases cells growth and enhances biocompatibility (Hodgins et al. 2007).

In vivo studies on the biocompatibility of elastomer implants have shown that a moderate inflammatory response occurs with the appearance of low numbers of FGBC (Kalicharan et al. 1991, Wolfaardt et al. 1992, Wortman et al. 1983).

Hemocompatibility studies on silicones have shown that albumin and fibrinogen strongly adsorb to these elastomers and that platelets also adhere in large quantities. Wettability can be calculated using contact angle measurements. There is a very large decrease in the contact angle of silicone elastomers over time, changing them from hydrophobic to hydrophilic biomaterials. According to the authors of one study (Rakhorst et al. 1999), variations in wettability have a greater impact on hemocompatibility than initial wettability (Van der Kamp and Van Oeveren 1994, Vignon 1995). A study on these systems has shown that silicones in contact with blood preferentially activate the kallicrein/kinin system. This system is more active on hydrophobic surfaces, which leads to vasodilatation, platelet aggregation, and an increase in permeability (Van der Kamp and Van Oeveren 1994). Various treatments have been attempted to improve the hemocompatibility of silicones. The elimination of micro air bubbles on the surface of the elastomer by denucleation reduces complement activation and platelet aggregation (Kalman et al. 1991). The affinity of albumin, which acts as an activation inhibitor of various biological reactions, can be increased by a number of different processes. This reduces inflammatory cell migration, greatly improves hemocompatibility, decreases complement activation, and prolongs contact time in vivo (Tsai et al. 1990, 1991). The presence of fibers in polymethylvinylsiloxane does not modify hemocompatibility (Fountain et al. 1979).

Chronic inflammation and myofibroblast proliferation often results in dense collagenous fibrosis shortly after implantation (Backovic et al. 2007). Another study showed that the FBGC have greater difficulty phagocytosing silicone particles than Teflon particles. When FBGC do not phagocytose silicone particles, they develop into large multinucleated cells surrounding the silicone particle, and a fibrous layer covers both the granulomatous lesion and the implant (Caballero et al. 2003). Monocytes and macrophages secrete more functional IL-1ß when grown on a film of fibrinogen and fibronectin than on a surface without a protein film. It has also been shown that the substrate itself plays a role in the production of IL-1ß. In the presence of an identical protein film on different substrates, monocytes/macrophages produce more functional IL-1ß when grown on silicone than on Teflon, Dacron, polyethylene, or Biomer (Bonfield and Anderson 1993). The nature of the substrate and the adsorbed protein film also plays a role in molecular conformation (Kossovsky and Freiman 1995). Backovic et al. (2007) showed that silicone promotes the adhesion of altered self proteins, which may trigger an autoimmune response. The cells of the ECM are particularly susceptible to modification when they adsorb to silicone. The proteins in the hydrophobic domain act like crystallization foci for other ECM proteins and are recognized as foreign by the immune system. Still other proteins adsorbed to the silicone surface are involved in the innate (complement component, C reactive protein) and immune (IgG, alpha 1 microglobulin) response. The process of protein modification is slow end allows metabolic and immune reactions but prevents the acute foreign body reaction (Backovic et al. 2007).

14.3.5 POLYURETHANES

PUs have a number of medical uses because of their acceptable level of biocompatibility, their high mechanical resistance, and their elastomeric properties (Spiller et al. 2007, Yoda 1998). PUs, together with silicones, are among the few elastomers than can be implanted for long periods in man (Pinchuk 1994). They are materials of choice for soft tissue and cardiovascular applications, especially for catheters (Brown 1995, Jeschke et al. 1999, Szycher and Reed 1992). PUs have also been cited as the best biomaterial for applications requiring a combination of hardness, durability, biocompatibility, and biostability (Coury et al. 1987). They are divided into two main families: polyetherurethanes and polyesterurethanes. Since the ester bonds of the polyesterurethanes are subject to hydrolysis (vascugraft), these elastomers are used less frequently in medical applications. Another class of PUs includes poly(ether urethane-urea) in which the diol group is replaced by a diamine.

Numerous studies have compared the biocompatibilities of various PUs. In vitro biocompatibility analyses comparing PU elastomers (polyether and polyester) and copolymers (polyether-polyester) have shown that epithelial cells have the same growth patterns and explant morphologies as cells grown on control substrates. Comparable results have obtained with the artificial aging technique (Bakker et al. 1988). Endothelial cells proliferative more strongly on PUs within 1 week despite slower spontaneous endothelialization than on other polymers like PTFE (Bschorer et al. 1994). The biocompatibilities of a polyetherurethane and a porous copolymer (polyether-polyester) have been studied following implantation in rats. The degradation of these PUs, which is less rapid than other elastomers, does not result in the release of toxic products. The proliferation of fibroblasts and the growth of fibrous and bone tissue are signs of acceptable implant fixation despite the presence of macrophages and foreign body giant cells (Bakker et al. 1990b).

A number of biocompatibility studies have been conducted to evaluate the role of wettability and implant structure (porosity) on cell adhesion, endothelial cell proliferation, inflammation, and cicatrization. The studies evaluated a polyesterurethane (Vascugraft) with interconnected pores and two polyetherurethaneureas (Mitrathane), one hydrophilic with closed pores and the other hydrophobic with open pores. Following implantation in rats, the histological analysis revealed a slight cell reaction to the polyether and a moderate reaction to the polyester. The fibrotic capsule surrounding the polyester implant pointed to good cicatrization (Huang et al. 1992). Cell adhesion to the polyester was weak, slightly stronger to the hydrophobic polyether, and equivalent to the control substrate for the hydrophilic polyether. Cell proliferation followed the same pattern, although a uniform monolayer of endothelial cells formed on the polyester while multiple layers formed on the hydrophilic polyether (Guidoin et al. 1992, Huang et al. 1992). Similar results were reported in a study on hydrophilic and hydrophobic polyetherurethaneureas. The formation of cell monolayers on a hydrophobic substrate is facilitated more by the presence of interconnecting pores than wettability (Sigot-Luizard et al. 1993). A porous structure promotes fibroblast proliferation and the production of new collagen on polyester. Tissue growth is a function of material porosity (Huang et al. 1992, Pollock et al. 1981). If there are no interconnecting pores, a thin but linear fibrotic capsule is deposited around the implant, but fibrotic tissue does not form within the implant (Ives et al. 1984, Martz et al. 1987). Open pores should allow the development of fibrotic tissue within the implant. A porous structure thus prevents deformation and reduces the risk of infection and contraction of the capsule around the implant (Huang et al. 1992, Szycher and Siciliano 1991, Tang and Eaton 1995). These reports show the importance of the physical properties of substrates such as porosity and surface wettability in elastomer biocompatibility. A recent study reported a novel processing method for injection-molded polyether urethane scaffolds. NaCl was used as a porogen to generate an open cell culture. This study showed an increase in fibroblast viability and uniform proliferation on the scaffold surface (Haugen et al. 2006).

Polyetherurethanes (PUs) can be aromatic or aliphatic. When aliphatic PUs are produced using HMDI (hydrogenated methylene diisocyanate) rather than MDI (methylene diisocyanate), they no longer release 4.4' methylene dianiline as a decomposition product, which has been shown to be carcinogenic in rats. In addition, these aliphatic PUs are not affected by ultraviolet radiation (Yoda 1998). As for hemocompatibility, platelet adherence is similar for both aromatic and aliphatic PUs, although more fibrinogen is deposited on aliphatic PUs (Lelah et al. 1986). Aliphatic PUs are more difficult to synthesize and have slightly weaker mechanical properties than aromatic PUs (Lelah et al. 1986, Yoda 1998).

Aliphatic polyether-coated stents (Tecoflex) and uncoated stents display the same level of endothelialization and neointimal formation. On the other hand, the inflammatory tissue response associated with aliphatic PU-coated stents may be an indication of low biocompatibility (Rechavia et al. 1998). A new type of aliphatic PU (UTA) has been shown to increase the speed of cell adhesion, making it equal to that of hydrophilic polyetherurethaneurea. After 1 week, both the UTA and hydrophilic PU surfaces were covered by multiple cell layers while the hydrophobic PU was only covered by a monolayer (Sigot-Luizard et al. 1993). Extraction cytotoxicity analyses have revealed that this family of elastomers does not release toxic contaminants and has no clear superiority in terms of biocompatibility over the other elastomers tested (polyetherurethaneurea and PTFE: Goretex, Impra) (Bischoff 1972). The author of this study also pointed to the role of biomaterial hydrophobicity in cell adhesion.

The rate of degradation of a biomaterial is affected by morphological variations in the surface of the material in direct contact with living tissue, indicating that there is a correlation between the surface properties of the biomaterial, cell behavior, and degradation (Wang et al. 2000). The degradation of PUs has been studied in great detail and involves three main mechanisms—mineralization, oxidation, and environmental stress cracking (ESC).

Mineralization or calcification (Marchant et al. 1984a) involves the incorporation of crystalline inorganic substances on the surface of the implanted material. This mechanism often results in hardening, distortion, and mechanical degradation of the elastomer (Coury et al. 1987). This mechanism most often comes into play when elastomers are in contact with the bloodstream or are in zones subject to dynamic flexing. Microdefects in the material can catch small particles of cellular debris, which in turn can act as nuclei for the formation and growth of calcium-phosphate-rich crystals (Coury et al. 1987, Szycher and Reed 1992, Yoda 1998).

Oxidation is induced by free radicals that break bonds, opening the elastomer up to attack by oxygen molecules. This can be influenced by factors such as the flexibility of the medical device, the type of soft segment, and the ability to interact with ions (Pinchuk 1994).

Environmental stress cracking is mediated by surface oxidation, stress, polyether soft segment chemistry, molecular morphology, and the presence of monocyte-derived macrophages, FBGC, and unknown biological elements (UBE) (Santerre et al. 2005). The degradation process may be better described by the term environmental biodegradation (EB), which designates a unique environment with multiple parameters, stress, and degradation of the polymer in a biological medium (Santerre et al. 2005). Szycher and Reed (1992) have mentioned the notion of BI-ESC in vitro. Cell and humoral-mediated inflammatory responses may also be involved in the surface degradation of PUs. ESC first occurs in the direction of the stress and then takes on a three-dimensional aspect that may extend to the very heart of the material, leading to total failure (Szycher and Reed 1992). Oxidative degradation in vivo is implicated in the degradation of PU, but the role of enzymatic hydrolysis remains controversial. Several authors think that hydrolytic enzymes, especially cholesterol esterase, are involved in the biodegradation of PU (Santerre et al. 2005). Others think that hydrolytic enzymatic degradation in vivo is negligible (Christenson et al. 2004b, 2006b). The problem with degradation has been minimized with the introduction of biostable PUs. Coury (Coury and Hobot 1991) have synthesized PUs with no ether functions. In vitro, the absence of an ether bond removes a point of attack for enzymes and chemical oxidants, thus lowering the risk of degradation (Pinchuk 1994, Szycher and Reed 1992). Szycher and Reed (1992) detected no degradation, cracking, or failure in biostable PUs after 12 weeks of implantation.

Modified PU elastomers (polycarbonate urethanes (PCU) have also been produced using polycarbonate glycol as a soft segment. While polyether urethanes undergo significant degradation after 6 months of implantation, the surface of PCUs remains smooth and crack-free (Pinchuk 1994). PCUs are more biostable than polyether urethane (PEU). PCUs display better cell attachment and proliferation in vitro, cause less platelet activation, and induce a lower inflammatory response (Hsu et al. 2004). Christenson compared the biodegradation in vitro of PCUs and PEUs, and showed that monocytes adhere to, differentiate on, and fuse to form FBGCs on these two PUs. The reactive oxygen released by macrophages and FBGCs initiates the degradation of PEUs. PCUs are also susceptible to biodegradation by products released by adherent cells (Christenson et al. 2004b). PCU degradation is caused by oxidation. The biodegradation rate of PCUs is significantly less than for PEUs, with PCUs being resistant to surface cracking. The degradation appears to be confined to a very thin surface layer (Christenson et al. 2004a, 2006a). The even greater biostability of polycarbonate PUs has been attributed to their carbonate bonds (Mathur et al. 1997). When used as vascular prostheses, polycarbonate PUs become endothelialized much more quickly than PTFEs. In addition, chronic proliferation is lower than with PTFEs, decreasing the risk of hyperplasia and occlusion (Jeschke et al. 1999).

New elastomers such as polycarbonate PUs and PUs endcapped with polydimethylsiloxane have made their appearance. Studies have recently been published comparing the biocompatibility and biostability of these new PUs with classic polyetherurethanes and PTFEs. They show that all PUs share the same biological reactions (acute and chronic inflammation). The hydrophobicity of polydimethylsiloxane leads to a lower accumulation of FBGCs on the surface. Polydimethylsiloxane endcapped PUs are less subject to biodegradation because the endcap provides protection against the oxygenated free radicals released by macrophages and FBGCs. Few studies have examined the biocompatibility of poly(ether)urethane modified by the addition of polydimethylsiloxane to produce PCU-PDMS. Briganti showed that these news elastomers have no cytotoxic effects on human endothelial cells and mouse fibroblasts (Briganti et al. 2006). Another study showed that PEU-PDMS are more hemocompatible (Spiller et al. 2007). PU elastomers are synthesized with varying proportions of poly(hexamethylene)oxide and polydimethylsiloxane(PDMS) macrodiols. The formulation with 80% PDMS produced the best result in terms of flexibility, strength, and biostability (Martin et al. 2000). A study with silicone-modified PEU and PCU showed that silicone modification did not fully inhibit the oxidative degradation of PEU and PCU soft segments but the rate of chain scission seemed to be slower than for the control PU soft segments (Christenson et al. 2005).

The biocompatibility of elastomers can be improved by modifying the surface or internal properties of polymers. However, these physical, chemical, and biological modifications must not affect their bulk properties. To quote Han et al., "practically every physical and chemical material property has been suggested as being important in biocompatibility for areas such as thrombus, calcification and infection." (Han et al. 1998).

Other techniques have been developed to improve the biocompatibility of PU elastomers. The incorporation of glycerophosphocholine as a chain extender in poly(tetramethylene oxide)-based PU significantly decreases bacterial adhesion and protein adsorption. A new MDI-based poly(ether)urethane with glycerophosphorylcholine (GPC) and glycerophosphorylserine(GPS) were synthesized and their biocompatibilities compared. This study showed that the elastomers were not cytotoxic while the GPC elastomer displayed a low level of interaction with blood (D'Arrigo et al. 2007). The incorporation of dehydroepiandrosterone (DHEA) in polyetherurethaneurea decreases macrophage adhesion and FBGC formation for up to 7 days. After 7 days, the biocompatibility of all the PUs (with and without DHEA) was similar (Collier et al. 1998). The incorporation of vitamin E (tocopherol) generates a very stable PU, something that is not observed with DHEA (Collier et al. 1998). Calcium chloride added to poly(ether-urethane-amide) improves biocompatibility, hydrophilicity, and plasticity. However, it also increases thrombogenicity (Kawashima and Sato 1997). Grafting collagen onto PU increases endothelial cell growth, which is dependent on the morphology and diameter of the collagen fibers. However, the collagen may also activate and aggregate platelets, making the material thrombogenic (Lee et al. 1996a). Coating prostheses (PU + collagen) with a confluent monolayer of endothelial cells should resolve this problem. All these studies clearly demonstrate the complexity of the biocompatibility of elastomers in general. PU has been modified with incorporation of chitin and modified chitin. These materials are nontoxic, biodegradable, antibacterial, and have good biocompatibility (Zia et al. 2009a,b).

Manufacturing processes, composition, and the surface properties of PU all play a role in PU/blood interactions (Lelah et al. 1986). The roles of the various components of PU have been studied. The nature of the soft segment makes PU more or less thrombogenic. Polyethylene oxide is more thrombogenic than polypropylene oxide or polytetramethylene oxide, and retains more platelets (Lelah et al. 1986). The nature of the hard segments is also important in the blood response. Aliphatic PUs do not have the same properties as aromatic PUs. They both attract platelets to an equal degree, but aliphatic PUs evoke an increased fibrinogen response (Bschorer et al. 1994). The quantity of soft and hard segments influences hemocompatibility. PUs with large proportions of hard segments are subject to rapid cell adhesion and sustained cell proliferation (Lelah et al. 1986). The lower the proportion of hard segments, the weaker the blood interactions (Chen et al. 1998). The molecular mass of the hard segments is important when elastomer devices must be in contact with blood. PTMO 1000 is more hemocompatible than PTMO 2000 (Chen et al. 1998). This shows that both the concentration and the conformation of the hard segments play a role in hemocompatibility (Lelah et al. 1986). Two main types of chain extender can be used—diols (polyether urethanes) and diamines (polyetherurethane urea). Diamines provide enhanced thromboresistance and hydrophilicity (Lelah et al. 1986).

Solvent casting affects the blood response by changing the orientation and/or conformation of the diisocyanate and the components of the chain extender (Bschorer et al. 1994). The biocompatibility, durability, and adhesive properties of aliphatic PUs increase in line with the decrease in the amount of plasticizer (Lindner et al. 1994).

To get around the problem of thromboresistance, researchers have tried adding and/or replacing components. Sulfonated PUs (PEO SO₃) produced by various techniques exhibit a lower level of platelet adhesion and platelet factor IV release from the time they first come in contact with blood to up to 12–24 h later (Han et al. 1998, McCoy et al. 1990). Hemocompatibility parallels bacterial repellence—the more hemocompatible a polymer, the more resistant it is to infections (Han et al. 1998).

The hemocompatibility of sulfonated PUs modified by a surface derivation technique using 2-acrylamido-2-methyl-propane-sulfonic-acid (AMPS) varies depending on the species, with in vivo and in vitro studies providing very contradictory results (Keogh et al. 1996). While very hemocompatible in vitro, sulfonated PUs attract large numbers of neutrophils when implanted in mice, increase macrophage recruitment to unexpected levels, and lead to the formation of thromboses when implanted in dogs (Keogh et al. 1996).

In a recent study, a sulfonated-polyethylene oxide-grafted PU (PU-PEO-SO3) was used to coated heart valves and vascular grafts. Less calcium was deposited on the PU-PEO-SO3 while platelet adhesion and thrombus formation appeared to be significantly lower on PU-PEO-SO3 (Han et al. 2006).

Applying a layer of heparin to PUPA [PU + poly(amido-amine) + HMDI] provides excellent antithrombogenic properties, increasing hemocompatibility (Albanese et al. 1994).

PUs that incorporate gold or silver nanoparticles have been developed. Nanocomposites have been prepared from water-borne polyether–type PU and water-borne polyester-type PU containing different amounts of gold nanoparticles. PU-Au enhances cell proliferation, and reduces platelet and monocyte activation, bacterial adhesion, and oxidative degradation (Hsu et al. 2006, 2008). The gold or silver particles improve the biostability of water-borne PU by inhibiting oxidation and crosslinking polyether soft segments (Chou et al. 2008).

To reduce the risk of infections associated with medical devices, a number of metallic ions known for their bactericidal effect, especially silver, have been investigated. Numerous studies have examined the biocompatibility, infection rates, and thrombogenicity of silver-impregnated and silver-coated PUs and silicones (Boswald et al. 1995, 1999, Greil et al. 1999, Jansen et al. 1994, Oloffs et al. 1994). Coating PU, silicone, and Dacron catheters with silver ions prevents microbial colonization both in vitro and in vivo (Boswald et al. 1995, Boswald et al. 1999, Jansen et al. 1994, Oloffs et al. 1994, Zhao and Stevens 1998).

14.3.6 Biodegradable Polyurethanes

The chemical properties of elastomers have also been modified to produce biodegradable products. Studies on biodegradable polyesterurethanes have shown that these elastomers have good cell compatibility and that cell/substrate interactions do not lead to the release of toxic substances or the activation of macrophages. Relatively strong adhesion and acceptable macrophage and osteoblast growth occurs. Fibroblasts keep their phenotype for 12 days. The main problem seems to the actual degradation of the elastomer, which results in the production of crystalline poly(r-3-hydroxybutyric acid) or PHB-P particles. Macrophages, and to a lesser degree osteoblasts, may internalize the particles by phagocytosis. These particles may be toxic for macrophages and osteoblasts at concentrations of 200 pg PHB-P/cell and 400 pg PHB-P/cell, respectively (Saad et al. 1996, 1997).

A few studies have analyzed the biocompatibility of biodegradable PU. Three types of biodegradable PU membranes were synthesized—smooth membrane, shrunken, and particulate. Cell adhesion was the best and the cells were better spread out and flattened on the particulate membrane. The smooth membrane was degraded the fastest, while the particulate membrane had not changed significantly 8 weeks post-implantation (Wang et al. 2000). A few studies have shown that biodegradable PUs display no signs of cytotoxicity in vitro and in vivo and do not provoke a toxic tissue response (Asplund et al. 2007, van Minnen et al. 2005). These elastomers support the ingrowth of cells and tissues in vivo and are degraded in a controlled fashion to noncytotoxic decomposition products (Guelcher 2008).

14.3.7 POLYESTERS

Polyethylene Terephthalate: Dacron

Dacron, which has been used since 1940 as a bone and cartilage substitute, is considered to be biocompatible (Schellhammer et al. 1999, Yoda 1998). It is used in numerous medical applications, especially in vascular surgery and for stent coatings (Schellhammer et al. 1999). However, chronic, systemic inflammation has been reported following the implantation of a Dacron device. Lymphocytes appeared after 30 days and the size of giant cells increased over time (Blum et al. 1996, Cavallini et al. 1987, Feldman et al. 1983, Granstroem et al. 1986, Link et al. 1996, Lodi et al. 1988, Maturri et al. 1991, Sato et al. 1986). Their tendency to provoke intimal hyperplasia and platelet stimulation—and thus their thrombogenicity—limits their clinical use to blood vessels greater than 6 mm in diameter (Yoda 1998). Polytetraphthalate-coated stents are preferred to PU stents, which are much more subject to dislocation (Schellhammer et al. 1999) The polyesters, such as poly(L-Lactide) PLLA have been reported to have an excellent biodegradability, biocompatibility and non toxicity (Jiao and Cui 2007). When a bioadsorbable (polylactide)acid PLA was compared with silicone sponge, the PLA displayed good biocompatibility, but the fibrous tissue encapsulation was thicker and there were more inflammatory cells than on the silicone sponge (Lansman et al. 2005).

Few polyesters have been modified to enhance the biocompatibility such as surface modification (Jiao and Cui 2007). A film of poly(3 hydroxybutyrate) modified by microwave ammonia treatment resulted in a hydrophilic surface with good long-term stability (Nitschke et al. 2002). A new class of biodegradable elastomeric APS polymers has been developed by copolymerization of poly(esteramide) elastomer and a polyol and a diacid. The cells interacted favorably with the elastomer, and displayed good morphology and metabolic activity. The reduced fibrous capsule thickness and the limited macrophage recruitment suggest that this elastomer would be well accepted in vivo (Bettinger et al. 2008).

14.3.8 POLYVINYLCHLORIDE: PVC

PVC is increasingly being used for medical applications, with an annual increase of 5.4%. It has been used in the healthcare field since 1940 because of its broad biocompatibility, ease of manufacture, and low cost (Yoda 1998). A study in 1991 reported chronic inflammation associated with PVC (Jayabalan et al. 1991). Two days after implantation, the inflammatory exudate around the PVC contained mainly neutrophils, while the exudate around the PU and silicone contained a mixture of neutrophils and monocytes (Fabre et al. 1998).

Additives have often been implicated in harmful host responses. Additives include plasticizers, antioxidants, pigments, and UV stabilizers. Plasticizers in particular can release substances that are incompatible with biological applications (Yoda 1998). A study in 1994 (Lindner et al. 1994) on the biocompatibility of and cell response to PVC showed that there is a clear correlation between the amount of plasticizer and the host inflammatory response. Various processes have been used in an attempt to make PVCs more biocompatible. The most commonly used plasticizer is phthalic ester, but it can be replaced by trioctyl trimellitate, azelate, or phosphate ester, which are nontoxic and are not released (Hansen 2007, Yoda 1998). It has been shown that the loss of adenoviral vector infectivity by effect of DEHP can be prevented by preflushing PVC tubes with albumin or methanol (Bilbao et al. 2004).

The effect of plasticizers on hemocompatibility has been studied using PVC blood tubing manufactured using various plasticizers (phthalate alone, trimellitate or TD 360, and phthalate coextruded with PU). The concentration of C3a anaphylatoxin and protein morphology were examined by electron microscopy. Phthalate-free PVC is more biocompatible while PU-coated PVC releases less C3a anaphylatoxin, although its hemocompatibility did not change (Branger et al. 1990).

PVC surfaces have been coated with three different substances—heparin, phosphatidylcholine (DMPC), and phosphatidylethanolamine (DMPE). The most biocompatible material was heparin-coated PVC, followed by DMPE and DMPC (Moreira Pda et al. 2004).

14.3.9 POLYTETRAFLUOROETHYLENE: PTFE

Polytetrafluoroethylene (PFTE) is used for many medical applications, especially for peripheral vascular surgery (Bschorer et al. 1994). in vitro extraction and direct contact analyses have revealed no cytotoxic contaminants (Sigot-Luizard et al. 1993). In vivo, the inflammatory response resorbs within 9 weeks, indicating acceptable biocompatibility (Huang et al. 1992, Sigot-Luizard et al. 1993). Fibroblasts respond to vascular PFTE prostheses by forming a thick external fibrotic capsule and a thin layer coating the pores (Guidoin et al. 1992, Sigot-Luizard et al. 1993). Despite an initial quick, spontaneous endothelialization process, complete endothelialization of the lumen of a PTFE prosthesis can take up to 6 months because of the rough surface, which, compared to PUs, makes endothelial cell proliferation more difficult. Chronic proliferation is, however, significant. After 6 months, the cells are still dividing, increasing neointimal formation and thus the risk of hyperplasia and thrombosis (Bschorer et al. 1994, Jeschke et al. 1999). Surface texture also influences cell migration. Cell migration, which can prevent thrombosis and occlusion, is slow on PTFE (Jeschke et al. 1999).

Following a 500 min contact between human plasma and a PTFE prosthesis, variations in contact angle were minimal and there was little fibrinogen adsorption, while platelet adhesion was elevated (Rakhorst et al. 1999). Blood contact with hydrophobic materials preferentially activates the kallicrein/kinin system, leading to vasodilatation, platelet aggregation, and an increase in vascular permeability (Van der Kamp and Van Oeveren 1994).

14.3.10 HYDROGELS

Hydrogels have been used in a wide variety of medical applications such as contact lenses, prosthetic devices, and drug delivery systems for over 30 years (Yoda 1998). General properties include swelling in water, high mobility of surface chains, low interfacial tension, low surface friction, and appreciable elasticity, making them reasonably biocompatible. Their hydrophilicity and low surface tension decreases protein adsorption and cell adhesion, which makes suitable for use in contact lenses. The low surface friction reduces fibroblast stimulation and prevents the formation of a fibrotic capsule (Park and Park 1996). The high elasticity and low mechanical resistance means that they have to be combined with other more solid elastomers to produce hydrogels with excellent mechanical and biological properties.

Aqueous solutions of poly(ethylene glycol) dimethacrylate (PEGDM) and poly(ethylene glycol) urethane dimethacrylate (PEGUDM) have been photopolymerized to yield hydrogels. The biocompatibility studies showed that the cells remained completely viable in both types of hydrogels after 2 weeks (Lin-Gibson et al. 2004).

14.3.11 POLYETHYLENES

A 1983 study evaluated the cell response to polyethylene following implantation in the peritoneal cavities of mice (Wortman et al. 1983). The rough surface of the elastomer becomes covered with numerous macrophages and FBGCs. The tissue reaction to polypropylene oxide includes an acute inflammatory response involving numerous macrophages and lymphocytes. Polyethylenes are degraded within 5 months of implantation. The resulting host reactions indicate that the degradation products are toxic. The study by Bakker et al. (1990a) showed that polypropylene oxide cannot be used for alloplastic tympanic membranes. Hemocompatibility studies have shown strong albumin adsorption and weak platelet adhesion to high-density polyethylene. In addition, analyses of the kallicrein/kinin system and factor XII have shown that both systems are activated to the same extent, indicating acceptable hemocompatibility (Rakhorst et al. 1999, Van der Kamp and Van Oeveren 1994).

14.3.12 Natural Rubbers

While natural rubbers are used in surgical gloves, urinary catheters, and tubing, their low hemocompatibility means they cannot be used in many other applications. Methylmethacrylate grafts can improve hemocompatibility (Razzak et al. 1988, 1989). Natural rubbers also cause cytotoxic and allergic reactions of various etiologies (Cormio et al. 1993). The proteins in natural rubbers cause Type I IgE-induced allergic reactions. It is possible to remove these proteins by simple washing. Gamma radiation vulcanization can be used to produce nitrosamine-free, low toxicity elastomers (Yoda 1998). The cytotoxicity of natural rubbers is due to dithiocarbamates (Nakamura et al. 1990). Various improvements mean that natural rubbers could soon see more frequent use in medical applications.

In a recent study, granules of natural latex were implanted in the dental alveoli of rats after extraction. This study shows that the latex was biocompatible, with the granules integrating into the alveolar bone, accelerating bone formation, and playing an important role in the healing process (Balabanian et al. 2006).

14.3.13 Styrene Copolymers

Polystyrene block- isoButhylene-Block-styrene (SIBS) is a thermoplastic elastomer and is oxidatively, hydrolytically and enzymatically stable and is biostable with a relatively low foreign body reaction. SIBS does not activate platelets in the vascular system. In this study, the authors have observed a few of polymorphonuclearleukocytes around SIBS implants, no encapsulation, no calcification and no degradation of SIBS implants. The excellent biocompatibility is due to the inertness of SIBS and lack of cleavable moieties (Pinchuk et al. 2008).

ASIBS with 30% of polystyrene has been studied and compared with silicone. The results show that SIBS 30 has similar biocompatibility and superior dynamic properties in comparison with silicone (El Fray et al. 2006).

14.3.14 Poly(Glycerol-CO-Sebacate)

Poly(glycerol-co-sebacate) (PGS) is an elastomer with excellent biocompatibility, the inflammatory and fibrotic responses to PGS implants are minimal in subcutaneous and intramuscular implantation sites. The mechanisms of PGS degradation are most likely controlled by an enzymatic surface erosion process and the hydrolytic cleavage may have played a minor role in PGS degradation (Pomerantseva et al. 2009).

Another study has synthesized three PGS at 110°C, 120°C, and 130°C and the degradation has been studied. The PGS have a wide range of degradability from being degradable in 2 weeks to being nearly inert (Chen et al. 2008).

A photocurable polymer has been developed: the poly(glycerol-co-sebacate) acrylate PGSA. In this study, PGSA has a good in vitro biocompatibility by sufficient cell adherence and subsequent proliferation into a confluent cell monolayer (Nijst et al. 2007).

A porous scaffold prepared from PGSA could be used to support the growth and differentiation of encapsulated cells. The biocompatibility is similar between the porous PGSA and no porous PGSA, but porous PGSA promotes tissue ingrowth as compared with nonporous PGSA (Gerecht et al. 2007).

ACKNOWLEDGMENTS

The authors thank Céline Allaire (Université de Rennes 1) for editorial assistance and for the preparation of the manuscript. English revision was made by Gene Bourgeau (Canada).

REFERENCES

Ai, H., H. Meng, I. Ichinose et al. 2003. Biocompatibility of layer-by-layer self-assembled nanofilm on silicone rubber for neurons. *J Neurosci Methods* 128:1–8.

Albanese, A., R. Barbucci, J. Belleville et al. 1994. In vitro biocompatibility evaluation of a heparinizable material (PUPA), based on polyurethane and poly(amido-amine) components. *Biomaterials* 15:129–136.

Anderson, J. M., A. Hiltner, Q. H. Zhao et al. 1992. Cell/polymer interactions in the biodegradation of polyurethanes. In *Biodegradable Polymers and Plastics*, eds. Vert, M., Feijen, J., Albertson, A., Scott, G., Chiellini, E., pp. 122–136. Cambridge, U.K.: Royal Society of Chemistry.

Anderson, J. M. and K. M. Miller. 1984. Biomaterial biocompatibility and the macrophage. *Biomaterials* 5:5–10.

Anderson, J. M., A. Rodriguez, and D. T. Chang. 2008. Foreign body reaction to biomaterials. *Semin Immunol* 20:86–100.

Asberg, A. E. and V. Videm. 2005. Activation of neutrophil granulocytes in an in vitro model of a cardiopulmonary bypass. *Artif Organs* 29:927–936.

Ashar, B., R. S. Ward, Jr., and L. R. Turcotte. 1981. Development of a silica-free silicone system for medical applications. *J Biomed Mater Res* 15:663–672.

Ashby, R. D., A. M. Cromwick, and T. A. Foglia. 1998. Radiation crosslinking of a bacterial medium-chain-length poly(hydroxyalkanoate) elastomer from tallow. *Int J Biol Macromol* 23:61–72.

Asplund, J. O., T. Bowden, T. Mathisen, and J. Hilborn. 2007. Synthesis of highly elastic biodegradable poly(urethane urea). *Biomacromolecules* 8:905–911.

Auroy, P., P. Duchatelard, N. E. Zmantar, and M. Hennequin. 1996. Hardness and shock absorption of silicone rubber for mouth guards. *J Prosthet Dent* 75:463–471.

Autian, J. 1974. Biological models for the testing of the toxicity of biomaterials. In *Polymers in Medicine and Surgery*, ed. Kronenthal, R., pp. 181–203. New York: Plenum Press.

Bacakova, L., V. Svorcik, V. Rybka et al. 1996. Adhesion and proliferation of cultured human aortic smooth muscle cells on polystyrene implanted with N+, F+ and Ar+ ions: Correlation with polymer surface polarity and carbonization. *Biomaterials* 17:1121–1126.

Backovic, A., H. L. Huang, B. Del Frari, H. Piza, L. A. Huber, and G. Wick. 2007. Identification and dynamics of proteins adhering to the surface of medical silicones in vivo and in vitro. *J Proteome Res* 6:376–381.

Bakker, D., C. A. Van Blitterswijk, W. T. Daems, and J. J. Grote. 1988. Biocompatibility of six elastomers *in vitro*. *J Biomed Mater Res* 22:423–439.

Bakker, D., C. A. van Blitterswijk, S. C. Hesseling, W. T. Daems, W. Kuijpers, and J. J. Grote. 1990a. The behavior of alloplastic tympanic membranes in *Staphylococcus aureus*-induced middle ear infection. I. Quantitative biocompatibility evaluation. *J Biomed Mater Res* 24:669–688.

Bakker, D., C. A. van Blitterswijk, S. C. Hesseling, H. K. Koerten, W. Kuijpers, and J. J. Grote. 1990b. Biocompatibility of a polyether urethane, polypropylene oxide, and a polyether polyester copolymer. A qualitative and quantitative study of three alloplastic tympanic membrane materials in the rat middle ear. *J Biomed Mater Res* 24:489–515.

Bakker, W. W., B. van der Lei, P. Nieuwenhuis, P. Robinson, and H. L. Bartels. 1991. Reduced thrombogenicity of artificial materials by coating with ADPase. *Biomaterials* 12:603–606.

Bal, B. T., H. Yilmaz, C. Aydin, S. Karakoca, and S. Yilmaz. 2009. In vitro cytotoxicity of maxillofacial silicone elastomers: Effect of accelerated aging. *J Biomed Mater Res B Appl Biomater* 89:122–126.

Balabanian, C. A., J. Coutinho-Netto, T. L. Lamano-Carvalho, S. A. Lacerda, and L. G. Brentegani. 2006. Biocompatibility of natural latex implanted into dental alveolus of rats. *J Oral Sci* 48:201–205.

Balagadde, F. K., L. You, C. L. Hansen, F. H. Arnold, and S. R. Quake. 2005. Long-term monitoring of bacteria undergoing programmed population control in a microchemostat. *Science* 309:137–140.

Bambauer, R., R. Latza, S. Bambauer, and E. Tobin. 2004. Large bore catheters with surface treatments versus untreated catheters for vascular access in hemodialysis. *Artif Organs* 28:604–610.

Barak, P., Y. Coquet, T. R. Halbach, and J. A. E. Molina. 1991. Biodegradability of polyhydroxybutyrate (co-hydroxyvalerate) and starch-incorporated polyethylene plastic films in soils. *J Environ Qual* 20:173–179.

Baumann, M., O. Witzke, R. Dietrich et al. 2003. Prolonged catheter survival in intermittent hemodialysis using a less thrombogenic micropatterned polymer modification. *ASAIO J* 49:708–712.

Baumgartner, J. N., C. Z. Yang, and S. L. Cooper. 1997. Physical property analysis and bacterial adhesion on a series of phosphonated polyurethanes. *Biomaterials* 18:831–837.

Berkland, C., D. W. Pack, and K. K. Kim. 2004. Controlling surface nano-structure using flow-limited field-injection electrostatic spraying (FFESS) of poly(D,L-lactide-co-glycolide). *Biomaterials* 25:5649–5658.

Bettinger, C. J., J. P. Bruggeman, J. T. Borenstein, and R. S. Langer. 2008. Amino alcohol-based degradable poly(ester amide) elastomers. *Biomaterials* 29:2315–2325.

Bilbao, R., D. P. Reay, B. M. Koppanati, and P. R. Clemens. 2004. Biocompatibility of adenoviral vectors in poly(vinyl chloride) tubing catheters with presence or absence of plasticizer di-2-ethylhexyl phthalate. *J Biomed Mater Res A* 69:91–96.

Bini, T. B., S. J. Gao, T. C. Tan et al. 2004. Electrospun poly(L-lactide-co-glycolide) biodegradable polymer nanofibre tubes for peripheral nerve regeneration. *Nanotechnology* 15:1459–1464.

Bischoff, F. 1972. Organic polymer biocompatibility and toxicology. *Clin Chem* 18:869–894.

Blass, C. R. 1992. PVC as a biomedical polymer—Plasticizer and stabilizer toxicity. *Med Device Technol* 3:32–40.

Blum, U., M. Langer, G. Spillner et al. 1996. Abdominal aortic aneurysms: Preliminary technical and clinical results with transfemoral placement of endovascular self-expanding stent-grafts. *Radiology* 198:25–31.

Boelens, J. J., S. A. Zaat, J. Meeldijk, and J. Dankert. 2000. Subcutaneous abscess formation around catheters induced by viable and nonviable *Staphylococcus epidermidis* as well as by small amounts of bacterial cell wall components. *J Biomed Mater Res* 50:546–556.

Boldin, I., A. Klein, E. M. Haller-Schober, and J. Horwath-Winter. 2008. Long-term follow-up of punctal and proximal canalicular stenoses after silicone punctal plug treatment in dry eye patients. *Am J Ophthalmol* 146:968–972.

Bonart, R. 1968. X-ray investigations concerning the physical structure of crosslinking in segmented urethane elastomers. *J Macromol Sci Phys* B7:115–138.

Bonfield, T. L. and J. M. Anderson. 1993. Functional versus quantitative comparison of IL 1 beta from monocytes/macrophages on biomedical polymers. *J Biomed Mater Res* 27:1195–1199.

Boone, J. L. and S. A. Braley. 1966. Resistance of silicone rubbers to body fluids. *Rubber Chem Technol* 39:1293–1297.

Boretos, J. W. and W. S. Pierce. 1967. Segmented polyurethane: A new elastomer for biomedical applications. *Science* 158:1481–1482.

Boswald, M., M. Girisch, J. Greil et al. 1995. Antimicrobial activity and biocompatibility of polyurethane and silicone catheters containing low concentrations of silver: A new perspective in prevention of polymer-associated foreign-body-infections. *Zentralbl Bakteriol* 283:187–200.

Boswald, M., K. Mende, W. Bernschneider et al. 1999. Biocompatibility testing of a new silver-impregnated catheter *in vivo*. *Infection* 27:S38–S42.

Bouchemal, K., S. Briancon, E. Perrier, H. Fessi, I. Bonnet, and N. Zydowicz. 2004. Synthesis and characterization of polyurethane and poly(ether urethane) nanocapsules using a new technique of interfacial polycondensation combined to spontaneous emulsification. *Int J Pharm* 269:89–100.

Brand, K. G., L. C. Buoen, K. H. Johnson, and I. Brand. 1975. Etiological factors, stages, and the role of the foreign body in foreign body tumorigenesis. *Cancer Res* 35:279–286.

Branger, B., M. Garreau, G. Baudin, and J. C. Gris. 1990. Biocompatibility of blood tubings. *Int J Artif Organs* 13:697–703.

Briganti, E., P. Losi, A. Raffi, M. Scoccianti, A. Munao, and G. Soldani. 2006. Silicone based polyurethane materials: A promising biocompatible elastomeric formulation for cardiovascular applications. *J Mater Sci Mater Med* 17:259–266.

Brinton, L. A. and S. L. Brown. 1997. Breast implants and cancer. *J Natl Cancer Inst* 89:1341–1349.

de Brito Alves, S., A. A. de Queiroz, and O. Z. Higa. 2003. Digital image processing for biocompatibility studies of clinical implant materials. *Artif Organs* 27:444–446.

Brodbeck, W. G., M. Macewan, E. Colton, H. Meyerson, and J. M. Anderson. 2005. Lymphocytes and the foreign body response: Lymphocyte enhancement of macrophage adhesion and fusion. *J Biomed Mater Res A* 74:222–229.

Brown, J. M. 1995. Polyurethane and silicone: Myths and misconceptions. *J Intraven Nurs* 18:120–122.

Bruck, S. D. 1980. Problems and artefacts in the evaluation of polymeric materials for medical uses. *Biomaterials* 1:103–107.

Bschorer, R., G. B. Koveker, G. Gehrke, M. Jeschke, and V. Hermanutz. 1994. Experimental improvement of microvascular allografts with the new material polyurethane and microvessel endothelial cell seeding. *Int J Oral Maxillofac Surg* 23:389–392.

Caballero, M., M. Bernal-Sprekelsen, C. Calvo, X. Farre, L. Quinto, and L. Alos. 2003. Polydimethylsiloxane versus polytetrafluoroethylene for vocal fold medialization: Histologic evaluation in a rabbit model. *J Biomed Mater Res B Appl Biomater* 67:666–674.

Cabanlit, M., D. Maitland, T. Wilson et al. 2007. Polyurethane shape-memory polymers demonstrate functional biocompatibility *in vitro*. *Macromol Biosci* 7:48–55.

Capone, C. D. 1992. Biostability of a non-ether polyurethane. *J Biomater Appl* 7:108–129.

Carroll, J. C., S. D. Schwaitzberg, A. A. Ucci, Jr., R. M. Schlesinger, D. Lauritzen, and G. R. Sant. 1993. New matrix material for potential use in "reversible" vasectomy. Preliminary animal biocompatibility studies. *Urology* 41:34–37.

Casenave, J. P. 1986. Interaction of platelets with surfaces. In *Blood Surface Interactions, Biological Principles Underlying Hemocompatibility with Artificial Surfaces*, eds. Casenave, J. P., Davies, J. A., Kazatchkine, M. D.,Van Aken, W. G., pp. 89–105. Amsterdam, the Netherlands: Elsevier.

Castner, D. G. and B. D. Ratner. 2002. Biomedical surface science: Foundations to frontiers. *Surf Sci* 500:28–60.

Cavallini, G., M. Lanfredi, M. Lodi, M. Govoni, and M. Pampolini. 1987. Detection and measurement of a cellular immune-reactivity towards polyester and polytetrafluoroethylene grafts. Leukocyte adherence inhibition test. *Acta Chir Scand* 153:179–184.

Chapiro, A. 1983. Radiation grafting of hydrogels to improve the thromboresistance of polymers. *Eur Polym J* 19:859.

Chauvel-Lebret, D. J., P. Pellen-Mussi, P. Auroy, and M. Bonnaure-Mallet. 1999. Evaluation of the in vitro biocompatibility of various elastomers. *Biomaterials* 20:291–299.

Chen, Q. Z., A. Bismarck, U. Hansen et al. 2008. Characterisation of a soft elastomer poly(glycerol sebacate) designed to match the mechanical properties of myocardial tissue. *Biomaterials* 29:47–57.

Chen, J. H., J. Wei, C. Y. Chang, R. F. Laiw, and Y. D. Lee. 1998. Studies on segmented polyetherurethane for biomedical application: Effects of composition and hard-segment content on biocompatibility. *J Biomed Mater Res* 41:633–648.

Chesmel, K. D. and J. Black. 1995. Cellular responses to chemical and morphologic aspects of biomaterial surfaces. I. A novel in vitro model system. *J Biomed Mater Res* 29:1089–1099.

Chou, C. W., S. H. Hsu, and P. H. Wang. 2008. Biostability and biocompatibility of poly(ether)urethane containing gold or silver nanoparticles in a porcine model. *J Biomed Mater Res A* 84:785–794.

Christenson, E. M., J. M. Anderson, and A. Hiltner. 2004a. Oxidative mechanisms of poly(carbonate urethane) and poly(ether urethane) biodegradation: In vivo and in vitro correlations. *J Biomed Mater Res A* 70:245–255.

Christenson, E. M., J. M. Anderson, and A. Hiltner. 2006a. Antioxidant inhibition of poly(carbonate urethane) in vivo biodegradation. *J Biomed Mater Res A* 76:480–490.

Christenson, E. M., M. Dadsetan, and A. Hiltner. 2005. Biostability and macrophage-mediated foreign body reaction of silicone-modified polyurethanes. *J Biomed Mater Res A* 74:141–155.

Christenson, E. M., M. Dadsetan, M. Wiggins, J. M. Anderson, and A. Hiltner. 2004b. Poly(carbonate urethane) and poly(ether urethane) biodegradation: In vivo studies. *J Biomed Mater Res A* 69:407–416.

Christenson, E. M., S. Patel, J. M. Anderson, and A. Hiltner. 2006b. Enzymatic degradation of poly(ether urethane) and poly(carbonate urethane) by cholesterol esterase. *Biomaterials* 27:3920–3926.

Collier, T., J. Tan, M. Shive, S. Hasan, A. Hiltner, and J. Anderson. 1998. Biocompatibility of poly(etherurethane urea) containing dehydroepiandrosterone. *J Biomed Mater Res* 41:192–201.

Colter, K. D., M. Shen, and A. T. Bell. 1977. Reduction of progesterone release rate through silicone membranes by plasma polymerization. *Biomater Med Devices Artif Organs* 5:13–24.

Conconi, M. T., S. Lora, A. M. Menti, P. Carampin, and P. P. Parnigotto. 2006. In vitro evaluation of poly[bis(ethyl alanato)phosphazene] as a scaffold for bone tissue engineering. *Tissue Eng* 12:811–819.

Cooper, G. L. and C. C. Hopkins. 1985. Rapid diagnosis of intravascular catheter-associated infection by direct Gram staining of catheter segments. *New Eng J Med* 312:1142–1147.

Cooper, S. and A. Toblosky. 1966. Properties of linear elastomeric polyurethanes. *J Appl Polym Sci* 10:1837–1844.

Corea-Tellez, K. S., P. Bustamante-Montes, M. Garcia-Fabila, M. A. Hernandez-Valero, and F. Vazquez-Moreno. 2008. Estimated risks of water and saliva contamination by phthalate diffusion from plasticized polyvinyl chloride. *J Environ Health* 71:34–39, 45.

Cormio, L., K. Turjanmaa, M. Talja, L. C. Andersson, and M. Ruutu. 1993. Toxicity and immediate allergenicity of latex gloves. *Clin Exp Allergy* 23:618–623.

Coury, A., K. Cobian, P. Cahalan, and A. Jevne. 1984. Biomedical uses of polyurethanes. In *Advances in Urethane Science and Technology*, eds. Frisch, K., Klempner, D., pp. 133–139. Lancaster, PA: Technomic Publishing Co, Inc.

Coury, A. J. and C. N. Hobot. 1991. Method for producing polyurethanes form poly-(hydroxyalkyl urethane). U.S. Patent 5, 001, 210.

Coury, A. J., C. M. Hobot, P. C. Slaikeu, K. B. Stokes, and P. T. Cahalan. 1990. A new family of implantable biostable polyurethanes. *Transactions of the 16th Annual Meeting for the Society for Biomaterials*, Charleston, SC, p. 158.

Coury, A. J., P. C. Slaikeu, P. T. Cahalan, K. B. Stokes, and C. M. Hobot. 1988. Factors and interactions affecting the performance of polyurethane elastomers in medical devices. *J Biomater Appl* 3:130–179.

Coury, A. J., K. B. Stokes, P. T. Cahalan, and P. C. Slaikeu. 1987. Biostability considerations for implantable polyurethanes. *Life Support Syst* 5:25–39.

Crommen, J., J. Vandorpe, and E. Schacht. 1993. Degradable poly-phazenes for biomedical applications. *J Control Release* 24:167–180.

D'Arrigo, P., C. Giordano, P. Macchi, L. Malpezzi, G. Pedrocchi-Fantoni, and S. Servi. 2007. Synthesis, platelet adhesion and cytotoxicity studies of new glycerophosphoryl-containing polyurethanes. *Int J Artif Organs* 30:133–143.

De Scheerder, I. K., K. L. Wilczek, E. V. Verbeken et al. 1995a. Biocompatibility of biodegradable and nonbiodegradable polymer-coated stents implanted in porcine peripheral arteries. *Cardiovasc Intervent Radiol* 18:227–232.

De Scheerder, I. K., K. L. Wilczek, E. V. Verbeken et al. 1995b. Biocompatibility of polymer-coated oversized metallic stents implanted in normal porcine coronary arteries. *Atherosclerosis* 114:105–114.

Defrere, J. and A. Franckart. 1992. Teflon/polyurethane arthroplasty of the knee: The first 2 years preliminary clinical experience in a new concept of artificial resurfacing of full thickness cartilage lesions of the knee. *Acta Chir Belg* 92:217–227.

Deisler, P. F., Jr. 1987. New silicone modified TPEs for medical applications. *Rubber World* 196:24–29.

Deng, M., L. S. Nair, S. P. Nukavarapu et al. 2010. Biomimetic, bioactive etheric polyphosphazene-poly(lactide-co-glycolide) blends for bone tissue engineering. *J Biomed Mater Res A* 92:114–125.

Diebold, J., J. P. Camilleri, M. Reynes, and P. Callard. 1977. *Anatomie Pathologique Générale*. Paris, France: Baillière.

Dobrovolskaia, M. A., J. D. Clogston, B. W. Neun, J. B. Hall, A. K. Patri, and S. E. McNeil. 2008. Method for analysis of nanoparticle hemolytic properties *in vitro*. *Nano Lett* 8:2180–2187.

Donawa, M. 2006a. New efforts to harmonise clinical evaluation. *Med Device Technol* 17:28, 30, 32.

Donawa, M. 2006b. Proposed amendments to the medical devices Directives. *Med Device Technol* 17:22–25.

Donawa, M. 2006c. Managing clinical data for worldwide acceptance. *Med Device Technol* 17:26–28.

Dopico-Garcia, M. S., J. M. Lopez-Vilarino, and M. V. Gonzalez-Rodriguez. 2007. Antioxidant content of and migration from commercial polyethylene, polypropylene, and polyvinyl chloride packages. *J Agric Food Chem* 55:3225–3231.

Dumitriu, S. and D. Dumitriu. 1990. Biocompatibility of polymers. In *Polymeric Biomaterials*, ed. Dumitriu, S., pp. 100–158. New York: Dekker Inc.

Duvernoy, O., T. Malm, J. Ramstrom, and S. Bowald. 1995. A biodegradable patch used as a pericardial substitute after cardiac surgery: 6- and 24-month evaluation with CT. *Thorac Cardiovasc Surg* 43:271–274.

El Fray, M., P. Prowans, J. E. Puskas, and V. Altstadt. 2006. Biocompatibility and fatigue properties of polystyrene-polyisobutylene-polystyrene, an emerging thermoplastic elastomeric biomaterial. *Biomacromolecules* 7:844–850.

Elespuru, R., R. Agarwal, A. Atrakchi et al. 2009. Current and future application of genetic toxicity assays: The role and value of in vitro mammalian assays. *Toxicol Sci* 109:172–179.

Everaert, E. P., H. C. Van Der Mei, J. De Vries, and H. J. Busscher. 1995. Hydrophobic recovery of repeatedly plasma-treated silicone rubber. I. Storage in air. *J Adhes Sci Technol* 9:1263.

Fabre, T., J. Bertrand-Barat, G. Freyburger et al. 1998. Quantification of the inflammatory response in exudates to three polymers implanted *in vivo*. *J Biomed Mater Res* 39:637–641.

Feldman, D. S., S. M. Hultman, R. S. Colaizzo, and A. F. von Recum. 1983. Electron microscope investigation of soft tissue ingrowth into Dacron velour with dogs. *Biomaterials* 4:105–111.

Feynmann, R. F. 1992. There's plenty of room at the bottom [data storage]. *J Microelectromech Syst* 1:60–66.

Feynmann, R. F. 1993. Infinitesimal machinery. *J Microelectromech Syst* 2:4–14.

Fidkowski, C., M. R. Kaazempur-Mofrad, J. Borenstein, J. P. Vacanti, R. Langer, and Y. Wang. 2005. Endothelialized microvasculature based on a biodegradable elastomer. *Tissue Eng* 11:302–309.

Fishbein, L. 1984. Toxicity of the components of poly(vinylchloride) polymers additives. *Prog Clin Biol Res* 141:113–136.

Fountain, S. W., J. Duffin, C. A. Ward, H. Osada, B. A. Martin, and J. D. Cooper. 1979. Biocompatibility of standard and silica-free silicone rubber membrane oxygenators. *Am J Physiol* 236:H371–H375.

Freij-Larsson, C., M. Kober, B. Wesslen, E. Willquist, and P. Tengvall. 1993. Effects of a polymeric additive in a biomedical poly(ether urethaneurea). *J Appl Polym Sci* 49:815–821.

Frisch, K. and J. Saunders. 1962. *Polyurethanes: Chemistry and Technology*. New-York: Interscience Publishers.

Fu, A. Y., H. P. Chou, C. Spence, F. H. Arnold, and S. R. Quake. 2002. An integrated microfabricated cell sorter. *Anal Chem* 74:2451–2457.

Gagnon, K. D., R. W. Lenz, R. J. Farris, and R. C. Fuller. 1994a. Chemical modification of bacterial elastomers: 1 Peroxide crosslinking. *Polymer* 35:4358.

Gagnon, K. D., R. W. Lenz, R. J. Farris, and R. C. Fuller. 1994b. Chemical modification of bacterial elastomers: 2 sulfur vulcanization. *Polymer* 35:4368.

Ganning, A. E., U. Brunk, and G. Dallner. 1984. Phthalate esters and their effect on the liver. *Hepatology* 4:541–547.

Geary, C., C. Birkinshaw, and E. Jones. 2008. Characterisation of Bionate polycarbonate polyurethanes for orthopaedic applications. *J Mater Sci Mater Med* 19:3355–3363.

Gerecht, S., S. A. Townsend, H. Pressler et al. 2007. A porous photocurable elastomer for cell encapsulation and culture. *Biomaterials* 28:4826–4835.

Gogolewski, S. 1991. in vitro and in vivo molecular stability of medical polyurethanes: A review. *Trends Polym Sci* 1:47–61.

Gottenbos, B., H. C. Van Der Mei, H. J. Busscher, D. W. Grijpma, and J. Feijen. 1999. Initial adhesion and surface growth of *Pseudomonas aeruginosa* on negatively and positively charged poly(methacrylates). *J Mater Sci Mater Med* 10:853–855.

Granstroem, L., L. Backam, and S. E. Dahlgren. 1986. Tissue reaction to polypropylene and polyester in obese patients. *Biomaterials* 7:455–458.

Grasel, T. G. and S. L. Cooper. 1989. Properties and biological interactions of polyurethane anionomers: Effect of sulfonate incorporation. *J Biomed Mater Res* 23:311–338.

Grasel, T. G., D. C. Lee, A. Z. Okkema, T. J. Slowinski, and S. L. Cooper. 1988. Extraction of polyure-thane block copolymers: Effects on bulk and surface properties and biocompatibility. *Biomaterials* 9:383–392.

Grasel, T. G., J. A. Pierce, and S. L. Cooper. 1987. Effects of alkyl grafting on surface properties and blood compatibility of polyurethane block copolymers. *J Biomed Mater Res* 21:815–842.

Greil, J., T. Spies, M. Boswald et al. 1999. Analysis of the acute cytotoxicity of the Erlanger silver catheter. *Infection* 27 Suppl 1:S34–S37.

Gu, S. Y. and J. Ren. 2005. Process optimization and empirical modeling for electrospun poly(D, L-lactide) fibers using response surface methodology. *Macromol Mater Eng* 290:1097–1105.

Guelcher, S. A. 2008. Biodegradable polyurethanes: Synthesis and applications in regenerative medicine. *Tissue Eng Part B Rev* 14:3–17.

Guidoin, R., M. Sigot, M. King, and M. F. Sigot-Luizard. 1992. Biocompatibility of the Vascugraft: Evaluation of a novel polyester urethane vascular substitute by an organotypic culture technique. *Biomaterials* 13:281–288.

Han, D. K., K. D. Park, and Y. H. Kim. 1998. Sulfonated poly(ethylene oxide)-grafted polyurethane copolymer for biomedical applications. *J Biomater Sci Polym Ed* 9:163–174.

Han, D. K., K. Park, K. D. Park, K. D. Ahn, and Y. H. Kim. 2006. In vivo biocompatibility of sulfonated PEO-grafted polyurethanes for polymer heart valve and vascular graft. *Artif Organs* 30:955–959.

Hansen, O. G. 2007. Phthalate labelling of medical devices. *Med Device Technol* 18:10–12.

Hansen, O. G. 2008. New developments in PVC. *Med Device Technol* 19:17–19.

Hansen, C. L., E. Skordalakes, J. M. Berger, and S. R. Quake. 2002. A robust and scalable microfluidic meter-ing method that allows protein crystal growth by free interface diffusion. *Proc Natl Acad Sci USA* 99:16531–16536.

Haugen, H., J. Aigner, M. Brunner, and E. Wintermantel. 2006. A novel processing method for injection-molded polyether-urethane scaffolds. Part 2: Cellular interactions. *J Biomed Mater Res B Appl Biomater* 77:73–78.

Henry, T. J. 1985. *Guidelines for the Preclinical Safety Evaluation of Materials Used in Medical Devices*. Washington, DC: Health Industry Manufacturers Association.

Hergenrother, R. W., X. H. Yu, and S. L. Cooper. 1994. Blood-contacting properties of polydimethylsiloxane polyurea-urethanes. *Biomaterials* 15:635–640.

Hess, F., R. Jerusalem, O. Reijnders et al. 1992. Seeding of enzymatically derived and subcultivated canine endothelial cells on fibrous polyurethane vascular prostheses. *Biomaterials* 13:657–663.

Heyde, M., M. Moens, L. Van Vaeck, K. M. Shakesheff, M. C. Davies, and E. H. Schacht. 2007. Synthesis and characterization of novel poly[(organo)phosphazenes] with cell-adhesive side groups. *Biomacromolecules* 8:1436–1445.

Hill, S. S., B. R. Shaw, and A. H. Wu. 2001. The clinical effects of plasticizers, antioxidants, and other contaminants in medical polyvinylchloride tubing during respiratory and non-respiratory exposure. *Clin Chim Acta* 304:1–8.

Hirt, T. D., P. Neuenschwander, and U. W. Suter. 1996. Telechelic diols from poly(r)-3hydroxybutyric acid and poly(r)-3-hydroxybutyric acid-co-(r)-3-hydroxyvaleric acid. *Macromol Chem Phys* 197:1609–1614.

Hocking, P. J. and R. H. Marchessault. 1994. *Chemistry and Technology of Biodegradable Polymers*. London, U.K.: Chapman and Hall.

Hodgins, D., J. M. Wasikiewicz, M. F. Grahn et al. 2007. Biocompatible materials developments for new medical implants. *Med Device Technol* 18:30, 32–35.

Hoenich, N. A., R. Levin, and C. Pearce. 2005. Clinical waste generation from renal units: Implications and solutions. *Semin Dial* 18:396–400.

Hollick, E. J., D. J. Spalton, and P. G. Ursell. 1999. Surface cytologic features on intraocular lenses: Can increased biocompatibility have disadvantages? *Arch Ophthalmol* 117:872–878.

Hollick, E. J., D. J. Spalton, P. G. Ursell, and M. V. Pande. 1998. Biocompatibility of poly(methyl methacrylate), silicone, and AcrySof intraocular lenses: Randomized comparison of the cellular reaction on the anterior lens surface. *J Cataract Refract Surg* 24:361–366.

Holmes, D. R., A. R. Camrud, M. A. Jorgenson, W. D. Edwards, and R. S. Schwartz. 1994. Polymeric stenting in the porcine coronary artery model: Differential outcome of exogenous fibrin sleeves versus polyurethane-coated stents. *J Am Coll Cardiol* 24:525–531.

Hong, J., K. Nilsson Ekdahl, H. Reynolds, R. Larsson, and B. Nilsson. 1999. A new in vitro model to study interaction between whole blood and biomaterials. Studies of platelet and coagulation activation and the effect of aspirin. *Biomaterials* 20:603–611.

Hong, J. W., V. Studer, G. Hang, W. F. Anderson, and S. R. Quake. 2004. A nanoliter-scale nucleic acid processor with parallel architecture. *Nat Biotechnol* 22:435–439.

Hotta, A., E. Cochran, J. Ruokolainen et al. 2006. Semicrystalline thermoplastic elastomeric polyolefins: Advances through catalyst development and macromolecular design. *Proc Natl Acad Sci USA* 103:15327–15332.

Hsiue, G. H., S. D. Lee, and P. C. Chang. 1996. Surface modification of silicone rubber membrane by plasma induced graft copolymerization as artificial cornea. *Artif Organs* 20:1196–1207.

Hsu, S. H., Y. C. Kao, and Z. C. Lin. 2004. Enhanced biocompatibility in biostable poly(carbonate)urethane. *Macromol Biosci* 4:464–470.

Hsu, S. H., C. M. Tang, and H. J. Tseng. 2006. Biocompatibility of poly(ether)urethane-gold nanocomposites. *J Biomed Mater Res A* 79:759–770.

Hsu, S. H., C. M. Tang, and H. J. Tseng. 2008. Biostability and biocompatibility of poly(ester urethane)-gold nanocomposites. *Acta Biomater* 4:1797–1808.

Hu, W. J., J. W. Eaton, T. P. Ugarova, and L. Tang. 2001. Molecular basis of biomaterial-mediated foreign body reactions. *Blood* 98:1231–1238.

Huang, B., Y. Marois, R. Roy, M. Julien, and R. Guidoin. 1992. Cellular reaction to the Vascugraft polyesterurethane vascular prosthesis: In vivo studies in rats. *Biomaterials* 13:209–216.

Huijberts, G. N., G. Eggink, P. de Waard, G. W. Huisman, and B. Witholt. 1992. *Pseudomonas putida* KT2442 cultivated on glucose accumulates poly(3-hydroxyalkanoates) consisting of saturated and unsaturated monomers. *Appl Environ Microbiol* 58:536–544.

Hung, W. C., M. D. Shau, H. C. Kao, M. F. Shih, and J. Y. Cherng. 2009b. The synthesis of cationic polyurethanes to study the effect of amines and structures on their DNA transfection potential. *J Control Release* 133:68–76.

Hunt, J. A., B. F. Flanagan, P. J. McLaughlin, I. Strickland, and D. F. Williams. 1996. Effect of biomaterial surface charge on the inflammatory response: Evaluation of cellular infiltration and TNF alpha production. *J Biomed Mater Res* 31:139–144.

Hunter, S. K., J. R. Scott, D. Hull, and R. L. Urry. 1988. The gamete and embryo compatibility of various synthetic polymers. *Fertil Steril* 50:110–116.

Ifkovits, J. L. and J. A. Burdick. 2007. Review: Photopolymerizable and degradable biomaterials for tissue engineering applications. *Tissue Eng* 13:2369–2385.

Imam, S. H., L. Chen, S. H. Gordon, R. L. Shogren, D. Weisleder, and R. V. Greene. 1998. Biodegradation of injection molded starch-poly (3-hydroxybutyrate-*co*-3-hydroxyvalerate) blends in a natural compost environment. *J Polym Environ* 6:91–98.

ISO. 1995. ISO 10993-7: Biological evaluation of medical devices. Part 7: Ethylene oxide sterilisation residuals.

ISO. 1998. ISO 10993-13: Biological evaluation of medical devices. Part 13: Identification and quantification of degradation products from medical polymeric medical devices.

ISO. 1999a. ISO 10993-5: Biological evaluation of medical devices. Part 5: Tests for in vitro cytotoxicity.

ISO. 1999b. ISO 10993-9: Biological evaluation of medical devices. Part 9: Framework for identification and quantification of potential degradation products.

ISO. 2002a. ISO 10993-4: Biological evaluation of medical devices. Part 4: Selection of tests for interaction with blood.

ISO. 2002b. ISO 10993-17: Biological evaluation of medical devices. Part 7: Establishment of allowable limits for leachable substances.

ISO. 2003a. ISO 10993-1: Biological evaluation of medical devices. Part 1: Evaluation and testing.

ISO. 2003b. ISO 10993-3: Biological evaluation of medical devices. Part 3: Test for genotoxicity, carcinogenicity and reproductive toxicity.

ISO. 2008. ISO 7405: Preclinical evaluation of biocompatibility of biomedical devices used in dentistry - Test methods for dental materials.

Ives, C. L., J. L. Zamora, S. G. Eskin et al. 1984. In vivo investigation of a new elastomeric vascular graft (Mitrathane). *Trans Am Soc Artif Intern Organs* 30:587–590.

Iwamoto, R., K. Ohta, T. Matsuda, and K. Imachi. 1986. Quantitative surface analysis of Cardiothane 51 by FT-IR-ATR spectroscopy. *J Biomed Mater Res* 20:507–520.

Jaakkola, J. J. and T. L. Knight. 2008. The role of exposure to phthalates from polyvinyl chloride products in the development of asthma and allergies: A systematic review and meta-analysis. *Environ Health Perspect* 116:845–853.

Jahangir, A. R., W. G. McClung, R. M. Cornelius, C. B. McCloskey, J. L. Brash, and J. P. Santerre. 2002. Fluorinated surface-modifying macromolecules: Modulating adhesive protein and platelet interactions on a polyether-urethane. *J Biomed Mater Res* 60:135–147.

Jansen, B., M. Rinck, P. Wolbring, A. Strohmeier, and T. Jahns. 1994. In vitro evaluation of the antimicrobial efficacy and biocompatibility of a silver-coated central venous catheter. *J Biomater Appl* 9:55–70.

Jayabalan, M., N. S. Kumar, K. Rathinam, and T. V. Kumari. 1991. In vivo biocompatibility of an aliphatic crosslinked polyurethane in rabbit. *J Biomed Mater Res* 25:1431–1432.

Jeschke, M. G., V. Hermanutz, S. E. Wolf, and G. B. Koveker. 1999. Polyurethane vascular prostheses decreases neointimal formation compared with expanded polytetrafluoroethylene. *J Vasc Surg* 29:168–176.

Jiang, X., J. M. Ng, A. D. Stroock, S. K. Dertinger, and G. M. Whitesides. 2003. A miniaturized, parallel, serially diluted immunoassay for analyzing multiple antigens. *J Am Chem Soc* 125:5294–5295.

Jiao, Y. P. and F. Z. Cui. 2007. Surface modification of polyester biomaterials for tissue engineering. *Biomed Mater* 2:R24–R37.

Johnell, M., R. Larsson, and A. Siegbahn. 2005. The influence of different heparin surface concentrations and antithrombin-binding capacity on inflammation and coagulation. *Biomaterials* 26:1731–1739.

Johnson, H. J., S. J. Northup, P. A. Seagraves, P. J. Garvin, and R. F. Wallin. 1983. Biocompatibility test procedures for materials evaluation *in vitro*. I. Comparative test system sensitivity. *J Biomed Mater Res* 17:571–586.

Jones, D. P. 1988. High quality silicones still dominate biomedical market after three decades. *Elastomerics* 120:12–16.

Jongebloed, W. L., G. van der Veen, D. Kalicharan, M. V. van Andel, G. Bartman, and J. G. Worst. 1994. New material for low-cost intraocular lenses. *Biomaterials* 15:766–773.

Kalicharan, D., W. L. Jongebloed, G. van der Veen, L. I. Los, and J. G. Worst. 1991. Cell-ingrowth in a silicone plombe. Interactions between biomaterial and scleral tissue after 8 years in situ: A SEM and TEM investigation. *Doc Ophtalmol* 78:307–315.

Kalman, P. G., C. A. Ward, N. B. McKeown, D. McCullough, and A. D. Romaschin. 1991. Improved biocompatibility of silicone rubber by removal of surface entrapped air nuclei. *J Biomed Mater Res* 25:199–211.

Kaluzny, J. J., W. Jozwicki, and H. Wisniewska. 2007. Histological biocompatibility of new, non-absorbable glaucoma deep sclerectomy implant. *J Biomed Mater Res B Appl Biomater* 81:403–409.

Kannan, R. Y., H. J. Salacinski, D. S. Vara, M. Odlyha, and A. M. Seifalian. 2006. Review paper: Principles and applications of surface analytical techniques at the vascular interface. *J Biomater Appl* 21:5–32.

Kao, W. J., A. Hiltner, J. M. Anderson, and G. A. Lodoen. 1994. Theoretical analysis of in vivo macrophage adhesion and foreign body giant cell formation on strained poly(etherurethane urea) elastomers. *J Biomed Mater Res* 28:819–829.

Kao, W. J., J. A. Hubbell, and J. M. Anderson. 1999. Protein-mediated macrophage adhesion and activation on biomaterials: A model for modulating cell behavior. *J Mater Sci Mater Med* 10:601–605.

Karabanova, L. V., A. W. Lloyd, S. V. Mikhalovsky et al. 2006. Polyurethane/poly(hydroxyethyl methacrylate) semi-interpenetrating polymer networks for biomedical applications. *J Mater Sci Mater Med* 17:1283–1296.

Kartalov, E. P., W. F. Anderson, and A. Scherer. 2006. The analytical approach to polydimethylsiloxane microfluidic technology and its biological applications. *J Nanosci Nanotechnol* 6:2265–2277.

Kartalov, E. P. and S. R. Quake. 2004. Microfluidic device reads up to four consecutive base pairs in DNA sequencing-by-synthesis. *Nucleic Acids Res* 32:2873–2879.

Kasemo, B. 1998. Biological surface science. *Curr Opin Solid State Mater* 3:451–459.

Katbab, A. A., R. P. Burford, and J. L. Garnett. 1992. Radiation graft modification of EPDM rubber. *Int J Radiat Appl Instrum C Radiat Phys Chem* 39:293–302.

Kaur, I. P. and R. Smitha. 2002. Penetration enhancers and ocular bioadhesives: Two new avenues for ophthalmic drug delivery. *Drug Dev Ind Pharm* 28:353–369.

Kawashima, K. and H. Sato. 1997. Calcium effect on the membrane preparation of segmented poly(ether/urethane/amide) (PEUN) as a biomedical material. *J Biomater Sci Polym Ed* 8:467–480.

Keogh, J. R. and J. W. Eaton. 1994. Albumin binding surfaces for biomaterials. *J Lab Clin Med* 124:537–545.

Keogh, J. R., M. F. Wolf, M. E. Overend, L. Tang, and J. W. Eaton. 1996. Biocompatibility of sulphonated polyurethane surfaces. *Biomaterials* 17:1987–1994.

Khan, A. S., Z. Ahmed, M. J. Edirisinghe, F. S. Wong, and I. U. Rehman. 2008. Preparation and characterization of a novel bioactive restorative composite based on covalently coupled polyurethane-nanohydroxyapatite fibres. *Acta Biomater* 4:1275–1287.

Kim, K., M. Yu, X. Zong et al. 2003. Control of degradation rate and hydrophilicity in electrospun non-woven poly(D,L-lactide) nanofiber scaffolds for biomedical applications. *Biomaterials* 24:4977–4985.

Kirkpatrick, C. J. and C. Mittermayer. 1990. Theoretical and practical aspects of testing potential biomaterials *in vitro*. *J Mater Sci Mater Med* 1:9–13.

Kleinsasser, N. H., K. Schmid, A. W. Sassen et al. 2006. Cytotoxic and genotoxic effects of resin monomers in human salivary gland tissue and lymphocytes as assessed by the single cell microgel electrophoresis (Comet) assay. *Biomaterials* 27:1762–1770.

Kockro, R. A., J. A. Hampl, B. Jansen et al. 2000. Use of scanning electron microscopy to investigate the prophylactic efficacy of rifampin-impregnated CSF shunt catheters. *J Med Microbiol* 49:441–450.

de Koning, G. J. M., H. M. M. van Bilsen, P. J. Lemstra et al. 1991. A biodegradable rubber by crosslinking poly(hydroxyalkanoate) from *Pseudomonas oleovorans*. *Polymer* 35:2090.

Kossovsky, N. 1995. Can the silicone controversy be resolved with rational certainty? *J Biomater Sci Polym Ed* 7:97–100.

Kossovsky, N. and C. J. Freiman. 1995. Physicochemical and immunological basis of silicone pathophysiology. *J Biomater Sci Polym Ed* 7:101–113.

Kuncova-Kallio, J. and P. J. Kallio. 2006. PDMS and its suitability for analytical microfluidic devices. *Conf Proc IEEE Eng Med Biol Soc* 1:2486–2489.

Kwon, I. K., S. Kidoaki, and T. Matsuda. 2005. Electrospun nano- to microfiber fabrics made of biodegradable copolyesters: Structural characteristics, mechanical properties and cell adhesion potential. *Biomaterials* 26:3929–3939.

Labow, R. S., J. P. Santerre, and G. Waghray. 1997. The effect of phospholipids on the biodegradation of polyurethanes by lysosomal enzymes. *J Biomater Sci Polym Ed* 8:779–795.

Lafferty, R. M., B. Korsatko, and W. Korsatko. 1988. *Biotechnology*. Weinheim, Germany: VCH Verlagsgesellschaft.

Lansman, S., P. Paakko, J. Ryhanen et al. 2005. Histologic analysis of bioabsorbable scleral buckling implants: An experimental study on rabbits. *Retina* 25:1032–1038.

Lee, P. C., L. L. Huang, L. W. Chen, K. H. Hsieh, and C. L. Tsai. 1996a. Effect of forms of collagen linked to polyurethane on endothelial cell growth. *J Biomed Mater Res* 32:645–653.

Lee, S. D., G. H. Hsiue, C. Y. Kao, and P. C. Chang. 1996b. Artificial cornea: Surface modification of silicone rubber membrane by graft polymerization of pHEMA via glow discharge. *Biomaterials* 17:587–595.

Leeper, H. M. and R. M. Wright. 1983. Elastomers in medicine. *Rubber Chem Technol* 56:523–556.

Lehle, K., M. Stock, T. Schmid, S. Schopka, R. H. Straub, and C. Schmid. 2009. Cell-type specific evaluation of biocompatibility of commercially available polyurethanes. *J Biomed Mater Res* 90:312–318.

Lelah, M. D. and S. L. Cooper. 1986. *Polyurethanes in Medicine*. Boca Raton, FL: CRC Press.

Lelah, M. D., T. G. Grasel, J. A. Pierce, and S. L. Cooper. 1986. Ex vivo interactions and surface property relationships of polyetherurethanes. *J Biomed Mater Res* 20:433–468.

Lelah, M. D., J. A. Pierce, L. K. Lambrecht, and S. L. Cooper. 1985. Polyetherurethane ionomers: Surface property/*ex vivo* blood compatibility relationships. *J Colloid Interface Sci* 104:422–439.

de Lemos, M. L., L. Hamata, and T. Vu. 2005. Leaching of diethylhexyl phthalate from polyvinyl chloride materials into etoposide intravenous solutions. *J Oncol Pharm Pract* 11:155–157.

Li, D. and J. Zhao. 1995. Surface biomedical effects of plasma on polyetherurethane. *J Adhes Sci Technol* 9:1249–1261.

Lim, F., C. Z. Yang, and S. L. Cooper. 1994. Synthesis, characterization and ex vivo evaluation of polydimethylsiloxane polyurea-urethanes. *Biomaterials* 15:408–416.

Lim, F., X. H. Yu, and S. L. Cooper. 1993. Effects of oligoethylene oxide monoalkyl(aryl) alcohol ether grafting on the surface properties and blood compatibility of a polyurethane. *Biomaterials* 14:537–545.

Lin-Gibson, S., S. Bencherif, J. A. Cooper et al. 2004. Synthesis and characterization of PEG dimethacrylates and their hydrogels. *Biomacromolecules* 5:1280–1287.

Lindner, E., V. V. Cosofret, S. Ufer et al. 1994. Ion-selective membranes with low plasticizer content: Electroanalytical characterization and biocompatibility studies. *J Biomed Mater Res* 28:591–601.

Link, J., B. Feyerabend, M. Grabener et al. 1996. Dacron-covered stent-grafts for the percutaneous treatment of carotid aneurysms: Effectiveness and biocompatibility-experimental study in swine. *Radiology* 200:397–401.

Lipatov, Y. S. 1990. Peculiarities of self-organization in the production of interpenetrating polymer networks. *J Macromol Sci Rev Macromol Chem Phys* C30:209–231.

Liu, J., M. Enzelberger, and S. Quake. 2002. A nanoliter rotary device for polymerase chain reaction. *Electrophoresis* 23:1531–1536.

Liu, J., C. Hansen, and S. R. Quake. 2003. Solving the "world-to-chip" interface problem with a microfluidic matrix. *Anal Chem* 75:4718–4723.

Lloyd, A. W. 2002. Interfacial bioengineering to enhance surface biocompatibility. *Med Device Technol* 13:18–21.

Lobler, M., M. Sass, P. Michel, U. T. Hopt, C. Kunze, and K. P. Schmitz. 1999. Differential gene expression after implantation of biomaterials into rat gastrointestine. *J Mater Sci Mater Med* 10:797–799.

Lodi, M., G. Cavallini, A. Susa, and M. Lanfredi. 1988. Biomaterials and immune system: Cellular reactivity towards PTFE and Dacron vascular substitutes pointed out by the leukocyte adherence inhibition (LAI) test. *Int Angiol* 7:344–348.

Loff, S., T. Hannmann, U. Subotic, F. M. Reinecke, H. Wischmann, and J. Brade. 2008. Extraction of diethylhexylphthalate by home total parenteral nutrition from polyvinyl chloride infusion lines commonly used in the home. *J Pediatr Gastroenterol Nutr* 47:81–86.

Loh, X. J., S. H. Goh, and J. Li. 2007. Hydrolytic degradation and protein release studies of thermogelling polyurethane copolymers consisting of poly[(R)-3-hydroxybutyrate], poly(ethylene glycol), and poly(propylene glycol). *Biomaterials* 28:4113–4123.

Loh, X. J., K. K. Tan, X. Li, and J. Li. 2006. The in vitro hydrolysis of poly(ester urethane)s consisting of poly[(R)-3-hydroxybutyrate] and poly(ethylene glycol). *Biomaterials* 27:1841–1850.

Lorenz, M. R., V. Holzapfel, A. Musyanovych et al. 2006. Uptake of functionalized, fluorescent-labeled polymeric particles in different cell lines and stem cells. *Biomaterials* 27:2820–2828.

Luu, Y. K., K. Kim, B. S. Hsiao, B. Chu, and M. Hadjiargyrou. 2003. Development of a nanostructured DNA delivery scaffold via electrospinning of PLGA and PLA-PEG block copolymers. *J Control Release* 89:341–353.

Ma, Z., Z. Mao, and C. Gao. 2007. Surface modification and property analysis of biomedical polymers used for tissue engineering. *Colloids Surf B Biointerfaces* 60:137–157.

Macocinschi, D., D. Filip, M. Butnaru, and C. D. Dimitriu. 2009. Surface characterization of biopolyurethanes based on cellulose derivatives. *J Mater Sci Mater Med* 20:775–783.

Mahdavi, A., L. Ferreira, C. Sundback et al. 2008. A biodegradable and biocompatible gecko-inspired tissue adhesive. *Proc Natl Acad Sci USA* 105:2307–2312.

Maki, D. G., C. E. Weise, and H. W. Sarafin. 1977. A semiquantitative culture method for identifying intravenous-catheter-related infection. *New Engl J Med* 296:1305–1309.

Maki, D. G., S. J. Wheeler, and S. M. Stolz. 1991. Study of a novel antiseptic-coated central venous catheter. *Crit Care Med* 19:99.

Marchant, R. E., J. M. Anderson, K. Phua, and A. Hiltner. 1984a. in vivo biocompatibility studies. II. Biomer: Preliminary cell adhesion and surface characterization studies. *J Biomed Mater Res* 18:309–315.

Marchant, R., A. Hiltner, C. Hamlin, A. Rabinovitch, R. Slobodkin, and J. M. Anderson. 1983. In vivo biocompatibility studies. I. The cage implant system and a biodegradable hydrogel. *J Biomed Mater Res* 17:301–325.

Marchant, R. E., K. M. Miller, and J. M. Anderson. 1984b. In vivo biocompatibility studies. V. in vivo leukocyte interactions with Biomer. *J Biomed Mater Res* 18:1169–1190.

Mardis, H. K., R. M. Kroeger, J. J. Morton, and J. M. Donovan. 1993. Comparative evaluation of materials used for internal ureteral stents. *J Endourol* 7:105–115.

Marois, Y., R. Guidoin, D. Boyer et al. 1989. In vivo evaluation of hydrophobic and fibrillar microporous poly-etherurethane urea graft. *Biomaterials* 10:521–531.

Martin, D. J., L. A. Warren, P. A. Gunatillake, S. J. McCarthy, G. F. Meijs, and K. Schindhelm. 2000. Polydimethylsiloxane/polyether-mixed macrodiol-based polyurethane elastomers: Biostability. *Biomaterials* 21:1021–1029.

Martz, H., R. Paynter, J. C. Forest, A. Downs, and R. Guidoin. 1987. Microporous hydrophilic polyurethane vascular grafts as substitutes in the abdominal aorta of dogs. *Biomaterials* 8:3–11.

Massa, T. M., M. L. Yang, J. Y. Ho, J. L. Brash, and J. P. Santerre. 2005. Fibrinogen surface distribution correlates to platelet adhesion pattern on fluorinated surface-modified polyetherurethane. *Biomaterials* 26:7367–7376.

Mathur, A. B., T. O. Collier, W. J. Kao et al. 1997. In vivo biocompatibility and biostability of modified poly-urethanes. *J Biomed Mater Res* 36:246–257.

Maturri, L., A. Azzolini, G. L. Campiglio, and E. Tardito. 1991. Are synthetic prostheses really inert? Preliminary results of a study on the biocompatibility of Dacron vascular prostheses and Silicone skin expanders. *Int Surg* 76:115–118.

McCoy, T. J., H. D. Wabers, and S. L. Cooper. 1990. Series shunt evaluation of polyurethane vascular graft materials in chronically AV-shunted canines. *J Biomed Mater Res* 24:107–129.

McCulley, J. P. 2003. Biocompatibility of intraocular lenses. *Eye Contact Lens* 29:155–163.

McMillin, C. R. 1994. Elastomers for biomedical application. *Rubber Chem Technol* 67:417.

Menconi, M. J., T. Owen, K. A. Dasse, G. Stein, and J. B. Lian. 1992. Molecular approaches to the characteriza-tion of cell and blood/biomaterial interactions. *J Card Surg* 7:177–187.

Miller, K. M. and J. M. Anderson. 1989. In vitro stimulation of fibroblast activity by factors generated from human monocytes activated by biomedical polymers. *J Biomed Mater Res* 23:911–930.

van Minnen, B., M. B. van Leeuwen, B. Stegenga et al. 2005. Short-term in vitro and in vivo biocompatibility of a biodegradable polyurethane foam based on 1,4-butanediisocyanate. *J Mater Sci Mater Med* 16:221–227.

Mirzadeh, H., A. A. Katbab, and R. P. Burford. 1993. CO-pulsed laser induced surface grafting of acrylamide onto ethylene-propylene rubber. *Int J Radiat Appl Instrum C Radiat Phys Chem* 42:53–56.

Mo, X. M., C. Y. Xu, M. Kotaki, and S. Ramakrishna. 2004. Electrospun P(LLA-CL) nanofiber: A biomi-metic extracellular matrix for smooth muscle cell and endothelial cell proliferation. *Biomaterials* 25:1883–1890.

Moharamzadeh, K., I. M. Brook, R. Van Noort, A. M. Scutt, K. G. Smith, and M. H. Thornhill. 2008. Development, optimization and characterization of a full-thickness tissue engineered human oral muco-sal model for biological assessment of dental biomaterials. *J Mater Sci Mater Med* 19:1793–1801.

Moreira Pda, L., P. R. Marreco, A. M. Moraes, M. L. Wada, and S. C. Genari. 2004. Analysis of cellular mor-phology, adhesion, and proliferation on uncoated and differently coated PVC tubes used in extracorpo-real circulation (ECC). *J Biomed Mater Res B Appl Biomater* 69:38–45.

Morrison, C., R. Macnair, C. MacDonald, A. Wykman, I. Goldie, and M. H. Grant. 1995. In vitro biocompatibil-ity testing of polymers for orthopaedic implants using cultured fibroblasts and osteoblasts. *Biomaterials* 16:987–992.

Murakami, Y., T. Endo, S. Yamamura, N. Nagatani, Y. Takamura, and E. Tamiya. 2004. On-chip micro-flow polysty-rene bead-based immunoassay for quantitative detection of tacrolimus (FK506). *Anal Biochem* 334:111–116.

Murty, M. S. and H. Sugiyama. 2004. Biology of N-methylpyrrole-N-methylimidazole hairpin polyamide. *Biol Pharm Bull* 27:468–474.

Nadeenko, V. G., I. R. Gol'dina, Z. D'Iachenko O, and L. V. Pestova. 1997. Comparative informative value of chromosome aberrations and sister chromatid exchanges in the evaluation of metals in the environment. *Gig Sanit* 3:10–13.

Naim, J. O., R. J. Lanzafame, and C. J. van Oss. 1995. The effect of silicone-gel on the immune response. *J Biomater Sci Polym Ed* 7:123–132.

Nakabayashi, N. and D. F. Williams. 2003. Preparation of non-thrombogenic materials using 2-methacryloy-loxyethyl phosphorylcholine. *Biomaterials* 24:2431–2435.

Nakamura, A., Y. Ikarashi, T. Tsuchiya et al. 1990. Correlations among chemical constituents, cytotoxicities and tissue responses: In the case of natural rubber latex materials. *Biomaterials* 11:92–94.

Nakamura, A., Y. Kawasaki, K. Takada et al. 1992. Difference in tumor incidence and other tissue responses to polyetherurethanes and polydimethylsiloxane in long-term subcutaneous implantation into rats. *J Biomed Mater Res* 26:631–650.

Nakaoka, R., T. Tsuchiya, K. Kato, Y. Ikada, and A. Nakamura. 1997. Studies on tumor-promoting activity of polyethylene: Inhibitory activity of metabolic cooperation on polyethylene surfaces is markedly decreased by surface modification with collagen but not with RGDS peptide. *J Biomed Mater Res* 35:391–397.

National Heart and Lung Institute. 1971. Annual Report of the Medical Devices Applications Program of the National Heart and Lung Institute, Bethesda, MD, U.S. Department of Health, Education, and Welfare, Public Health Service, National Institutes of Health.

Nijst, C. L., J. P. Bruggeman, J. M. Karp et al. 2007. Synthesis and characterization of photocurable elastomers from poly(glycerol-co-sebacate). *Biomacromolecules* 8:3067–3073.

Nitschke, M., G. Schmack, A. Janke, F. Simon, D. Pleul, and C. Werner. 2002. Low pressure plasma treatment of poly(3-hydroxybutyrate): Toward tailored polymer surfaces for tissue engineering scaffolds. *J Biomed Mater Res* 59:632–638.

Nukavarapu, S. P., S. G. Kumbar, J. L. Brown et al. 2008. Polyphosphazene/nano-hydroxyapatite composite microsphere scaffolds for bone tissue engineering. *Biomacromolecules* 9:1818–1825.

Nyilas, E. and R. S. Ward, Jr. 1977. Development of blood-compatible elastomers. V. Surface structure and blood compatibility of avcothane elastomers. *J Biomed Mater Res* 11:69–84.

Okkema, A. Z., T. A. Giroux, T. G. Grasel, and S. L. Cooper. 1987. Ionic polyurethanes: Surface and blood contacting properties. *MRS Symposium on Biomedical Materials and Devices*, Boston, MA.

Oloffs, A., C. Grosse-Siestrup, S. Bisson, M. Rinck, R. Rudolph, and U. Gross. 1994. Biocompatibility of silver-coated polyurethane catheters and silver-coated Dacron material. *Biomaterials* 15:753–758.

Oshima, H. and M. Nakamura. 1994. A study on reference standard for cytotoxicity assay of biomaterials. *Biomed Mater Eng* 4:327–332.

Owen, G. R., D. O. Meredith, I. ap Gwynn, and R. G. Richards. 2005. Focal adhesion quantification - A new assay of material biocompatibility? Review. *Eur Cell Mater* 9:85–96; discussion 85–96.

Ozdemir, K. G., H. Yilmaz, and S. Yilmaz. 2009. In vitro evaluation of cytotoxicity of soft lining materials on L929 cells by MTT assay. *J Biomed Mater Res B Appl Biomater* 90:82–86.

Pangman, W. J. 1958. U.S. Patent N° 2, 842, 775.

Pariente, J. L., L. Bordenave, R. Bareille et al. 1998. First use of cultured human urothelial cells for biocompatibility assessment: Application to urinary catheters. *J Biomed Mater Res* 40:31–39.

Park, H. and K. Park. 1996. Biocompatibility issues of implantable drug delivery systems. *Pharm Res* 13:1770–1776.

Peek, G. J., R. Scott, H. M. Killer, and R. K. Firmin. 2002. An in vitro method for comparing biocompatibility of materials for extracorporeal circulation. *Perfusion* 17:125–132.

Pernagallo, S., J. J. Diaz-Mochon, and M. Bradley. 2009. A cooperative polymer-DNA microarray approach to biomaterial investigation. *Lab Chip* 9:397–403.

Phillips, R. E., M. C. Smith, and R. J. Thoma. 1988. Biomedical applications of polyurethanes: Implications of failure mechanisms. *J Biomater Appl* 3:207–227.

Pinchuk, L. 1994. A review of the biostability and carcinogenicity of polyurethanes in medicine and the new generation of 'biostable' polyurethanes. *J Biomater Sci Polym Ed* 6:225–267.

Pinchuk, L., Y. P. Kato, M. L. Eckstein, G. J. Wilson, and D. C. MacGregor. 1993. Polycarbonate urethanes as elastomeric materials for long-term implant applications. *Transactions of the 16th Annual Meeting for the Society for Biomaterials*, Charleston, SC. p. 22.

Pinchuk, L., G. J. Wilson, J. J. Barry, R. T. Schoephoerster, J. M. Parel, and J. P. Kennedy. 2008. Medical applications of poly(styrene-block-isobutylene-block-styrene) ("SIBS"). *Biomaterials* 29:448–460.

Piskin, E. 1994. Review of biodegradable polymers as biomaterials. *J Biomater Sci Polymer Ed* 6:775–795.

Pizzoferrato, A., C. R. Arciola, E. Cenni, G. Ciapetti, and S. Sassi. 1995. In vitro biocompatibility of a polyurethane catheter after deposition of fluorinated film. *Biomaterials* 16:361–367.

Poirier, Y., C. Nawrath, and C. Somerville. 1995. Production of polyhydroxyalkanoates, a family of biodegradable plastics and elastomers, in bacteria and plants. *Biotechnology* 13:142–150.

Pollock, E., E. J. Andrews, D. Lentz, and K. Sheikh. 1981. Tissue ingrowth and porosity of biomer. *Trans Am Soc Artif Intern Organs* 27:405–409.

Pomerantseva, I., N. Krebs, A. Hart, C. M. Neville, A. Y. Huang, and C. A. Sundback. 2009. Degradation behavior of poly(glycerol sebacate). *J Biomed Mater Res* 91:1038–1047.

Pritchett, J. W. 2008. Curved-stem hip resurfacing: Minimum 20-year followup. *Clin Orthop Relat Res* 466:1177–1185.

Quake, S. R. and A. Scherer. 2000. From micro- to nanofabrication with soft materials. *Science* 290:1536–1540.

Rakhorst, G., H. C. Van der Mei, W. Van Oeveren, H. T. Spijker, and H. J. Busscher. 1999. Time-related contact angle measurements with human plasma on biomaterial surfaces. *Int J Artif Organs* 22:35–39.

Rambour, R. P. 1973. A review of crazing and fracture in thermoplastics. *J Polym Sci Macromol Rev* 7:1–154.

Ratner, B. D. 1983. Surface characterization of biomaterials by electron spectroscopy for chemical analysis. *Ann Biomed Eng* 11:313–336.

Ratner, B. D., T. Horbett, A. S. Hoffman, and S. D. Hauschka. 1975. Cell adhesion to polymeric materials: Implications with respect to biocompatibility. *J Biomed Mater Res* 9:407–422.

Razzak, M. T., K. Otsuhata, Y. Tabata, F. Ohashi, and A. Takeuchi. 1988. Modification of natural rubber tubes for biomaterials. I. radiation induced grafting of N,N dimethyl acrylamide onto natural rubber tubes. *J Appl Polym Sci* 36:645.

Razzak, M. T., K. Otsuhata, Y. Tabata, F. Ohashi, and A. Takeuchi. 1989. Modification of natural rubber tubes for biomaterials. II. Radiation induced grafting of N,N dimethylaminoethylacrylate (DMAEA) onto natural rubber (NR) tubes. *J Appl Polym Sci* 38:829.

Rechavia, E., F. Litvack, M. C. Fishbien, M. Nakamura, and N. Eigler. 1998. Biocompatibility of polyurethane-coated stents: Tissue and vascular aspects. *Cathet Cardiovasc Diagn* 45:202–207.

Reed, A. M., J. Potter, and M. Szycher. 1994. A solution grade biostable polyurethane elastomer: ChronoFlex AR. *J Biomater Appl* 8:210–236.

Ren, Y. and D. Mahon. 2007. Evaluation of microwave irradiation for analysis of carbonyl sulfide, carbon disulfide, cyanogen, ethyl formate, methyl bromide, sulfuryl fluoride, propylene oxide, and phosphine in hay. *J Agric Food Chem* 55:32–37.

Ren, T. B., T. Weigel, T. Groth, and A. Lendlein. 2008. Microwave plasma surface modification of silicone elastomer with allylamine for improvement of biocompatibility. *J Biomed Mater Res A* 86:209–219.

Richard, R., M. Schwarz, K. Chan, N. Teigen, and M. Boden. 2009. Controlled delivery of paclitaxel from stent coatings using novel styrene maleic anhydride copolymer formulations. *J Biomed Mater Res A* 90:522–532.

Roach, P., D. Eglin, K. Rohde, and C. C. Perry. 2007. Modern biomaterials: A review - Bulk properties and implications of surface modifications. *J Mater Sci Mater Med* 18:1263–1277.

Rodrigues, L., I. M. Banat, J. Teixeira, and R. Oliveira. 2007. Strategies for the prevention of microbial biofilm formation on silicone rubber voice prostheses. *J Biomed Mater Res B Appl Biomater* 81:358–370.

Rogers, J. A. and R. G. Nuzzo. 2005. Recent progress in soft lithography. *Mater Today* 8:50–56.

Rose, S. F., S. Okere, G. W. Hanlon, A. W. Lloyd, and A. L. Lewis. 2005. Bacterial adhesion to phosphorylcholine-based polymers with varying cationic charge and the effect of heparin pre-adsorption. *J Mater Sci Mater Med* 16:1003–1015.

Rosenbluth, S. A., G. R. Weddington, W. L. Guess, and J. Autian. 1965. Tissue culture method for screening toxicity of plastic materials to be used in medical practice. *J Pharm Sci* 54:156–159.

Rosengren, A., L. Faxius, N. Tanaka, M. Watanabe, and L. M. Bjursten. 2005. Comparison of implantation and cytotoxicity testing for initially toxic biomaterials. *J Biomed Mater Res A* 75:115–122.

Rubin, J. P. and M. J. Yaremchuk. 1997. Complications and toxicities of implantable biomaterials used in facial reconstructive and aesthetic surgery: A comprehensive review of the literature. *Plast Reconstr Surg* 100:1336–1353.

Saad, B., T. D. Hirt, M. Welti, G. K. Uhlschmid, P. Neuenschwander, and U. W. Suter. 1997. Development of degradable polyesterurethanes for medical applications: In vitro and in vivo evaluations. *J Biomed Mater Res* 36:65–74.

Saad, B., S. Matter, G. Ciardelli et al. 1996. Interactions of osteoblasts and macrophages with biodegradable and highly porous polyesterurethane foam and its degradation products. *J Biomed Mater Res* 32:355–366.

Santerre, J. P., K. Woodhouse, G. Laroche, and R. S. Labow. 2005. Understanding the biodegradation of polyurethanes: From classical implants to tissue engineering materials. *Biomaterials* 26:7457–7470.

Santos, L., D. Rodrigues, M. Lira et al. 2008. Bacterial adhesion to worn silicone hydrogel contact lenses. *Optom Vis Sci* 85:520–525.

Sato, O., Y. Tada, and A. Takagi. 1986. The biologic fate of Dacron double velour vascular prostheses: A clinicopathological study. *Japan J Surg* 19:301–311.

Satoh, K., R. Nonaka, K. Ohyama, F. Nagai, A. Ogata, and M. Iida. 2008. Endocrine disruptive effects of chemicals eluted from nitrile-butadiene rubber gloves using reporter gene assay systems. *Biol Pharm Bull* 31:375–379.

Schellhammer, F., M. Walter, A. Berlis, H. G. Bloss, E. Wellens, and M. Schumacher. 1999. Polyethylene terephthalate and polyurethane coatings for endovascular stents: Preliminary results in canine experimental arteriovenous fistulas. *Radiology* 211:169–175.

Schendel, K. U., L. Erdinger, G. Komposch, and H. G. Sonntag. 1995. Neon-colored plastics for orthodontic appliances. Biocompatibility studies. *Fortschr Kieferorthop* 56:41–48.

Schmalz, G. and D. Arenholt-Bindslev. 2009. Basic aspects. In *Biocompatibility of Dental Materials*, eds. Schmalz, G., Arenholt-Bindslev, D., pp. 1–12. Berlin, Germany: Springer-Verlag.

Schmidt, D. R. and W. J. Kao. 2007. The interrelated role of fibronectin and interleukin-1 in biomaterial-modulated macrophage function. *Biomaterials* 28:371–382.

Schoen, F. J., H. Harasaki, K. M. Kim, H. C. Anderson, and R. J. Levy. 1988. Biomaterial-associated calcification: Pathology, mechanisms, and strategies for prevention. *J Biomed Mater Res* 22:11–36.

Schubert, M. A., M. J. Wiggins, J. M. Anderson, and A. Hiltner. 1997. Role of oxygen in biodegradation of poly(etherurethane urea) elastomers. *J Biomed Mater Res* 34:519–530.

Schubert, M. A., M. J. Wiggins, K. M. DeFife, A. Hiltner, and J. M. Anderson. 1996. Vitamin E as an antioxidant for poly(etherurethane urea): In vivo studies. Student Research Award in the Doctoral Degree Candidate Category, *Fifth World Biomaterials Congress (22nd Annual Meeting of the Society for Biomaterials)*, Toronto, Canada, May 29–June 2, 1996. *J Biomed Mater Res* 32:493–504.

Schubert, M. A., M. J. Wiggins, M. P. Schaefer, A. Hiltner, and J. M. Anderson. 1995. Oxidative biodegradation mechanisms of biaxially strained poly(etherurethane urea) elastomers. *J Biomed Mater Res* 29:337–347.

Schweikl, H., K. A. Hiller, A. Eckhardt et al. 2008. Differential gene expression involved in oxidative stress response caused by triethylene glycol dimethacrylate. *Biomaterials* 29:1377–1387.

Shintani, H. 1995. Formation and elution of toxic compounds from sterilized medical products: Methylenedianiline formation in polyurethane. *J Biomater Appl* 10:23–58.

Shukla, P. G., B. Kalidhass, A. Shah, and D. V. Palaskar. 2002. Preparation and characterization of microcapsules of water-soluble pesticide monocrotophos using polyurethane as carrier material. *J Microencapsul* 19:293–304.

Sigler, M., T. Paul, and R. G. Grabitz. 2005. Biocompatibility screening in cardiovascular implants. *Z Kardiol* 94:383–391.

Sigot-Luizard, M. F., M. Sigot, R. Guidoin et al. 1993. A novel microporous polyurethane blood conduit: Biocompatibility assessment of the UTA arterial prosthesis by an organo-typic culture technique. *J Invest Surg* 6:251–271.

Simmons, A., J. Hyvarinen, R. A. Odell et al. 2004. Long-term in vivo biostability of poly(dimethylsiloxane)/poly(hexamethylene oxide) mixed macrodiol-based polyurethane elastomers. *Biomaterials* 25:4887–4900.

Singh, J. and K. K. Agrawal. 1992. Modification of poly(vinyl chloride) for biocompatibility improvement and biomedical application. *Polymer-Plastics Technology and Engineering* 31:203–212.

Singh, M., A. R. Ray, P. Vasudevan, K. Verma, and S. K. Guha. 1979. Potential biosoluble carriers: Biocompatibility and biodegradability of oxidized cellulose. *Biomater Med Devices Artif Organs* 7:495–512.

Sohn, L. L., O. A. Saleh, G. R. Facer, A. J. Beavis, R. S. Allan, and D. A. Notterman. 2000. Capacitance cytometry: Measuring biological cells one by one. *Proc Natl Acad Sci USA* 97:10687–10690.

Sokolowski, W., A. Metcalfe, S. Hayashi, L. Yahia, and J. Raymond. 2007. Medical applications of shape memory polymers. *Biomed Mater* 2:S23–S27.

Sperling, L. H. and R. Hu. 2002. In *Polymer Blends Handbook*, ed. Utracki, L. A., Vol. 56. Dordrecht, the Netherlands: Kluwer.

Sperling, L. H. and V. Misra. 1997. In *IPNs around the World: Science and Engineering*, eds. Kim, S. C., Sperling, L. H., Vol. 16. New York: Wiley.

Spiller, D., P. Losi, E. Briganti et al. 2007. PDMS content affects in vitro hemocompatibility of synthetic vascular grafts. *J Mater Sci Mater Med* 18:1097–1104.

Sprague, E. A. and J. C. Palmaz. 2005. A model system to assess key vascular responses to biomaterials. *J Endovasc Ther* 12:594–604.

Stang, A. and I. Witte. 2009. Performance of the comet assay in a high-throughput version. *Mutat Res* 675:5–10.

Stokes, K. B. 1983. The biostability of various polyether polyurethanes under stress. Medtronic, Inc., Minneapolis, MN.

Stokes, K. B. 1988. Polyether polyurethanes: Biostable or not? *J Biomater Appl* 3:228–259.

Stokes, K. and K. Cobian. 1982. Polyether polyurethanes for implantable pacemaker leads. *Biomaterials* 3:225–231.

Stokes, K., R. McVenes, and J. M. Anderson. 1995. Polyurethane elastomer biostability. *J Biomater Appl* 9:321–354.

Stokes, K. B., P. Urbanski, and R. Cobian. 1987. New test methods for the evaluation of stress cracking and metal catalysed oxidation in implanted polymers. In *Polyurethanes in Biomedical Engineering II*, ed. Planck, H., pp. 109–127. Amsterdam, the Netherlands: Elsevier.

Sundback, C. A., J. Y. Shyu, Y. Wang et al. 2005. Biocompatibility analysis of poly(glycerol sebacate) as a nerve guide material. *Biomaterials* 26:5454–5464.

Szelest-Lewandowska, A., B. Masiulanis, A. Klocke, and B. Glasmacher. 2003. Synthesis, physical properties and preliminary investigation of hemocompatibility of polyurethanes from aliphatic resources with castor oil participation. *J Biomater Appl* 17:221–236.

Szelest-Lewandowska, A., B. Masiulanis, M. Szymonowicz, S. Pielka, and D. Paluch. 2007. Modified polycarbonate urethane: Synthesis, properties and biological investigation *in vitro*. *J Biomed Mater Res A* 82:509–520.

Szycher, M. 1988. Biostability of polyurethane elastomers: A critical review. *J Biomater Appl* 3:297–402.

Szycher, M. and A. M. Reed. 1992. Biostable polyurethane elastomers. *Med Device Technol* 3:42–51.

Szycher, M. and A. A. Siciliano. 1991. Polyurethane-covered mammary prosthesis: A nine year follow-up assessment. *J Biomater Appl* 5:282–322.

Takahara, A., A. J. Coury, R. W. Hergenrother, and S. L. Cooper. 1991. Effect of soft segment chemistry on the biostability of segmented polyurethanes. I. In vitro oxidation. *J Biomed Mater Res* 25:341–356.

Tang, L. and J. W. Eaton. 1995. Inflammatory responses to biomaterials. *Am J Clin Pathol* 103:466–471.

Tare, R. S., F. Khan, G. Tourniaire, S. M. Morgan, M. Bradley, and R. O. Oreffo. 2009. A microarray approach to the identification of polyurethanes for the isolation of human skeletal progenitor cells and augmentation of skeletal cell growth. *Biomaterials* 30:1045–1055.

Tarnok, A., A. Mahnke, M. Muller, and R. J. Zotz. 1999. Rapid in vitro biocompatibility assay of endovascular stents by flow cytometry using platelet activation and platelet-leukocyte aggregation. *Cytometry* 38:30–39.

Tice, R. R., E. Agurell, D. Anderson et al. 2000. Single cell gel/comet assay: Guidelines for in vitro and in vivo genetic toxicology testing. *Environ Mol Mutagen* 35:206–221.

Tickner, J. A., T. Schettler, T. Guidotti, M. McCally, and M. Rossi. 2001. Health risks posed by use of Di-2-ethylhexyl phthalate (DEHP) in PVC medical devices: A critical review. *Am J Ind Med* 39:100–111.

Toub, M. R. 1987. Technical innovations enhance commercial value of silicone rubber. *Elastomerics* 119:20–22.

Trevisani, L., S. Sartori, M. R. Rossi et al. 2005. Degradation of polyurethane gastrostomy devices: What is the role of fungal colonization? *Dig Dis Sci* 50:463–469.

Trowbridge, H. O. and R. C. Emling. 1997. *Inflammation. A Review of the Process*. Carol Stream, IL: Quintessence Books.

Tsai, C. C., M. L. Dollar, A. Constantinescu, P. V. Kulkarni, and R. C. Eberhart. 1991. Performance evaluation of hydroxylated and acylated silicone rubber coatings. *Transactions of the American Society for Artificial Internal Organs Meeting* 37(3):M192–193.

Tsai, C. C., H. H. Huo, P. Kulkarni, and R. C. Eberhart. 1990. Biocompatible coating with high albumin affinity. *Transactions of the American Society for Artificial Internal Organs Meeting*, pp. 307–310.

Tsuchiya, T., K. Fukuhara, H. Hata et al. 1995a. Studies on the tumor-promoting activity of additives in biomaterials: Inhibition of metabolic cooperation by phenolic antioxidants involved in rubber materials. *J Biomed Mater Res* 29:121–126.

Tsuchiya, T., H. Hata, and A. Nakamura. 1995b. Studies on the tumor-promoting activity of biomaterials: Inhibition of metabolic cooperation by polyetherurethane and silicone. *J Biomed Mater Res* 29:113–119.

Tucci, M. G., M. Mattioli Belmonte, E. Toschi et al. 1996. Structural features of latex gloves in dental practice. *Biomaterials* 17:517–522.

Van der Kamp, K. W. and W. Van Oeveren. 1994. Factor XII fragment and kallikrein generation in plasma during incubation with biomaterials. *J Biomed Mater Res* 28:349–352.

Vasita, R., I. K. Shanmugam, and D. S. Katt. 2008. Improved biomaterials for tissue engineering applications: Surface modification of polymers. *Curr Top Med Chem* 8:341–353.

Vignon, D. 1995. Physiologie de l'hémostase. *Encyclopédie Médico Chirurgicale Stomatologie–Odontologie* 1:1–8. Elsevier Masson.

Vijayasekaran, S., T. V. Chirila, Y. Hong et al. 1996. Poly(1-vinyl-2-pyrrolidinone) hydrogels as vitreous substitutes: Histopathological evaluation in the animal eye. *J Biomater Sci Polym Ed* 7:685–696.

Von Recum, A. F. and M. LaBerge. 1995. Educational goals for biomaterials science and engineering: Prospective view. *J Biomater Appl* 6:137–144.

Von Recum, A. F. and T. G. Van Kooten. 1995. The influence of micro-topography on cellular response and the implications for silicone implants. *J Biomater Sci Polym Ed* 7:181–198.

Vondracek, P. 1981. Some aspects of the medical application of silicone rubber. *Int Polymer Sci Technol* 8:16.

Wacaser, B. A., M. J. Maughan, I. A. Mowat, T. L. Niederhauser, M. R. Linford, and R. C. Davis. 2003. Chemomechanical surface patterning and functionalization of silicon surfaces using an atomic force microscope. *Appl Phys Lett* 82:808–810.

Wagner, J. L. and T. E. Hugli. 1984. Radioimmunoassay for anaphylatoxins: A sensitive method for determining complement activation products in biological fluids. *Anal Biochem* 136:75–88.

Wang, Y., G. A. Ameer, B. J. Sheppard, and R. Langer. 2002. A tough biodegradable elastomer. *Nat Biotechnol* 20:602–606.

Wang, D. A., L. X. Feng, J. Ji, Y. H. Sun, X. X. Zheng, and J. H. Elisseeff. 2003. Novel human endothelial cell-engineered polyurethane biomaterials for cardiovascular biomedical applications. *J Biomed Mater Res A* 65:498–510.

Wang, J. H., C. H. Yao, W. Y. Chuang, and T. H. Young. 2000. Development of biodegradable polyesteru-rethane membranes with different surface morphologies for the culture of osteoblasts. *J Biomed Mater Res* 51:761–770.

Ward, R. S. 1995. Surface modification prior to surface formation: Control of polymer surface properties via bulk composition. *Med Plastics Biomater* 2:34–41.

Ward, R. S., K. A. White, R. S. Gill, and F. Lim. 1996. The effect of phase separation and endgroup chemistry on the in vivo biostability of polyurethanes. *Transactions of the American Society for Artificial Internal Organs Meeting*, Washington, DC. p. 17.

Ward, R. S., K. A. White, R. S. Gill, and C. A. Wolcott. 1995. Development of biostable thermoplastic polyure-thanes with oligomeric polydimethylsiloxane end groups. *Transactions of the 21st Meeting of the Society for Biomaterials* (March 18–22, 1995), San Francisco, CA.

Weber, N., H. P. Wendel, and G. Ziemer. 2001. Gene monitoring of surface-activated monocytes in circulating whole blood using duplex RT-PCR. *J Biomed Mater Res* 56:1–8.

Welsing, R. T., T. G. van Tienen, N. Ramrattan et al. 2008. Effect on tissue differentiation and articular car-tilage degradation of a polymer meniscus implant: A 2-year follow-up study in dogs. *Am J Sports Med* 36:1978–1989.

Werner, C. and H. J. Jacobasch. 1999. Surface characterization of polymers for medical devices. *Int J Artif Organs* 22:160–176.

White, L. 1991. Clean TPEs find medical uses. *Eur Rubber J* 173:26–29.

White, R. A. 1988. The effect of porosity on the variability of the neointima. An histological investigation on implanted synthetic vascular prostheses. *Trans Am Soc Artif Intern Organs*. pp. 95–100.

Whitesides, G. M., E. Ostuni, S. Takayama, X. Jiang, and D. E. Ingber. 2001. Soft lithography in biology and biochemistry. *Annu Rev Biomed Eng* 3:335–373.

Whitford, M. J. 1984. The chemistry of silicone materials for biomedical devices and contact lenses. *Biomaterials* 5:298–300.

Wilkes, G. L. and S. L. Samuels. 1973. Porous segmented polyurethanes—Possible candidates as biomaterials. *J Biomed Mater Res* 7:541–544.

Williams, D. F. 2003. Biomaterials and tissue engineering in reconstructive surgery. *Sadhana* 28:563–574.

Williams, S. K., T. Carter, P. K. Park, D. G. Rose, T. Schneider, and B. E. Jarrell. 1992. Formation of a multi-layer cellular lining on a polyurethane vascular graft following endothelial cell sodding. *J Biomed Mater Res* 26:103–117.

Williams, S. F., D. P. Martin, D. M. Horowitz, and O. P. Peoples. 1999. PHA applications: Addressing the price performance issue: I. Tissue engineering. *Int J Biol Macromol* 25:111–121.

Wilson, C. J., R. E. Clegg, D. I. Leavesley, and M. J. Pearcy. 2005. Mediation of biomaterial-cell interactions by adsorbed proteins: A review. *Tissue Eng* 11:1–18.

Wolfaardt, J. F., P. Cleaton-Jones, J. Lownie, and G. Ackermann. 1992. Biocompatibility testing of a silicone maxillofacial prosthetic elastomer: Soft tissue study in primates. *J Prosthet Dent* 68:331–338.

Wortman, R. S., K. Merritt, and S. A. Brown. 1983. The use of the mouse peritoneal cavity for screening for biocompatibility of polymers. *Biomater Med Devices Artif Organs* 11:103–114.

Wu, H., A. Wheeler, and R. N. Zare. 2004. Chemical cytometry on a picoliter-scale integrated microfluidic chip. *Proc Natl Acad Sci USA* 101:12809–12813.

Wutticharoenmongkol, P., N. Sanchavanakit, P. Pavasant, and P. Supaphol. 2006a. Novel bone scaffolds of electrospun polycaprolactone fibers filled with nanoparticles. *J Nanosci Nanotechnol* 6:514–522.

Wutticharoenmongkol, P., N. Sanchavanakit, P. Pavasant, and P. Supaphol. 2006b. Preparation and character-ization of novel bone scaffolds based on electrospun polycaprolactone fibers filled with nanoparticles. *Macromol Biosci* 6:70–77.

Xie, X., H. Tan, J. Li, and Y. Zhong. 2008. Synthesis and characterization of fluorocarbon chain end-capped poly(carbonate urethane)s as biomaterials: A novel bilayered surface structure. *J Biomed Mater Res A* 84:30–43.

Xinming, L., C. Yingde, A. W. Lloyd et al. 2008. Polymeric hydrogels for novel contact lens-based ophthalmic drug delivery systems: A review. *Cont Lens Anterior Eye* 31:57–64.

Xiong, X. Y., K. C. Tam, and L. H. Gan. 2006. Polymeric nanostructures for drug delivery applications based on Pluronic copolymer systems. *J Nanosci Nanotechnol* 6:2638–2650.

Xu, H., H. Toghiani, and C. U. Pittman, Jr. 2000. Modeling domain mixing in semi-interpenetrating polymer networks composed of poly(vinyl chloride) and 5% to 15% of crosslinked thermosets. *Polym Eng Sci* 40:2027–2036.

Yang, F., R. Murugan, S. Wang, and S. Ramakrishna. 2005. Electrospinning of nano/micro scale poly(L-lactic acid) aligned fibers and their potential in neural tissue engineering. *Biomaterials* 26:2603–2610.

Yannas, I. V., M. Zhang, and M. H. Spilker. 2007. Standardized criterion to analyze and directly compare various materials and models for peripheral nerve regeneration. *J Biomater Sci Polym Ed* 18:943–966.

Yasuda, H. 2006. Biocompatibility of nanofilm-encapsulated silicone and silicone-hydrogel contact lenses. *Macromol Biosci* 6:121–138.

Yim, E. S., B. Zhao, D. Myung et al. 2009. Biocompatibility of poly(ethylene glycol)/poly(acrylic acid) interpenetrating polymer network hydrogel particles in RAW 264.7 macrophage and MG-63 osteoblast cell lines. *J Biomed Mater Res A* 91:894–902.

Yoda, R. 1998. Elastomers for biomedical applications. *J Biomater Sci Polym Ed* 9:561–626.

Zhang, Z., M. W. King, R. Guidoin et al. 1994. Morphological, physical and chemical evaluation of the Vascugraft arterial prosthesis: Comparison of a novel polyurethane device with other microporous structures. *Biomaterials* 15:483–501.

Zhao, Q., M. P. Agger, M. Fitzpatrick et al. 1990. Cellular interactions with biomaterials: In vivo cracking of pre-stressed Pellethane 2363–80A. *J Biomed Mater Res* 24:621–637.

Zhao, O. H., J. M. Anderson, A. Hiltner, G. A. Lodoen, and C. R. Payet. 1992. Theoretical analysis on cell size distribution and kinetics of foreign-body giant cell formation in vivo on polyurethane elastomers. *J Biomed Mater Res* 26:1019–1038.

Zhao, G. and S. E. Stevens, Jr. 1998. Multiple parameters for the comprehensive evaluation of the susceptibility of *Escherichia coli* to the silver ion. *Biomaterials* 11:27–32.

Zhao, Q., N. Topham, J. M. Anderson, A. Hiltner, G. Lodoen, and C. R. Payet. 1991. Foreign-body giant cells and polyurethane biostability: In vivo correlation of cell adhesion and surface cracking. *J Biomed Mater Res* 25:177–183.

Zia, K. M., M. Barikani, M. Zuber, I. A. Bhatti, and M. Barmar. 2009a. Surface characteristics of polyurethane elastomers based on chitin/1,4-butane diol blends. *Int J Biol Macromol* 44:182–185.

Zia, K. M., M. Zuber, I. A. Bhatti, M. Barikani, and M. A. Sheikh. 2009b. Evaluation of biocompatibility and mechanical behavior of polyurethane elastomers based on chitin/1,4-butane diol blends. *Int J Biol Macromol* 44:18–22.

15 Preparation and Applications of Modulated Surface Energy Biomaterials

*Blanca Vázquez, Luis M. Rodríguez-Lorenzo,
Gema Rodríguez-Crespo, Juan Parra,
Mar Fernández, and Julio San Román*

CONTENTS

15.1   INTRODUCTION

Surface properties of materials are the key parameters in the interaction between an implanted device and the host tissue (Chen et al. 2008; Xu and Siedlecki 2009). Hydrophobicity or hydrophilicity of a material is usually expressed as surface wettability and determines protein adsorption, platelet adhesion, and cell and bacterial adhesion (Xu and Siedlecki 2007) and also their subsequent structural and biological activity (Hylton et al. 2005; Wu et al. 2005b). Hydrophobic surfaces, namely poorly wettable, tend to absorb proteins in a greater proportion than hydrophilic surfaces, namely highly wettable. In hydrophilic surfaces, repulsive solvation forces arising from strongly bound water occur (Israelachvili and Wennerstrom 1996; Noh and Vogler 2006). A stark transition between protein adherent and protein nonadherent materials has been found to happen on water contact angles 60°–65° (Xu and Siedlecki 2007), which is consistent with the appearance of hydrophobic interactions (Yoon et al. 1997). However, surface wettability seems not to be the only parameter affecting the adhesion of proteins. The phenomenum of protein adhesion is rather a competition between the hydrophobic effect and the dehydration of the surface. The former one expels proteins from solution but proteins will only be adsorbed on a surface when the energy necessary for dehydrating the surface is lower, i.e., $\Delta G_{0 \, (Phobic \, effect)} > \Delta G_{0 \, (dehydration)}$ (Noh and Vogler 2006). The hydrophobic effect is the energy that expels protein from solution to recover hydrogen bonds among water molecules otherwise separated by proteins.

The desirable biological response to a surface depends largely on the specific applications and remains as an important target area of research in the preparation of biocompatible polymeric materials. It is necessary to remember at this point that biocompatibility is defined as the ability of a material/device to perform in a specific environment (Williams 2008). And it is still an open issue, whether highly energetic surfaces, i.e., highly wettable or low energetic surfaces, i.e., poorly wettable, are more or less adequate for each specific application (Navarro et al. 2006).

The design of blood contacting devices must involve the engineering of surfaces to avoid protein adhesion. Platelet adhesion is the cause of thrombus formation and severe associated vascular problems (Rodrigues et al. 2006). Fibrinogen, which is one of the most abundant protein plasmas in blood, is a ligand for the platelet integrin receptor RIIb_3 (GPIIb/IIIa), and the interaction between this ligand and receptor is responsible for immobilization, activation, and aggregation of platelets. As an example, several platelet adhesion and activation studies with and without preadsorbed fibrinogen, albumin, and plasma have been performed based on the effect of surface wettability of self-assembled monolayers (SAMs) containing different ratios of methyl- and hydroxyl-terminated alkanethiols (Martins et al. 2003a; Rodrigues et al. 2006; Goncalves et al. 2009).

There are also numerous reports of adsorption on hydrophilic surfaces, either ionic or hydrogels (Vogler 1998; Navarro et al. 2006). On hydrogels, complications introduced by water-hydrated polymers must be considered since it becomes very difficult to distinguish between absorption and adsorption. Some adsorption phenomena on hydrophilic surfaces have been reported which cannot be easily dismissed. Adsorption on hydrophilic surfaces must be explained assuming that adsorbents are not physicochemically bonded to the surface but rather associated through a hydration layer or "bound" to the water layer.

Mammalian cell attachment has been described as a different process from protein adsorption and it occurs efficiently on hydrophilic surfaces but inefficiently on hydrophobic ones. Thus, the hydrophobic/hydrophilic balance in the biological response to materials, often disputed in biomaterials science, should be studied from the perspective of water structure and reactivity at surfaces (Vogler 1998).

Attachment and adhesion are different phenomena. Adhesion is related to the force necessary to separate adherents and is controlled by short-range forces that result from the formation of covalent, ionic, hydrogen, and charge-transfer bonds. Thus, adhesion occurs only when cellular surfaces (or some extension of the cellular surface) touch the substratum and contact at atomic level can be made. Attachment can be a physical phenomenon where cells are held close to the surface by an attractive potential where attractive dispersion forces overcome electrostatic repulsion. Attachment happens before adhesion. Once cells are attached, they can proliferate and spread on a surface.

The results obtained from studies carried out with culture medium supplemented with 10% fetal calf serum (FCS) suggested that the complex mixture of proteins present in FCS presented a higher affinity for hydrophilic surfaces.

Cell cultures performed with MG63 osteoblast-like human cells on materials with different surface wettability have shown that the initial attachment of the cells is better for the poly(lactic acid) (PLA)/phosphate glass (G5) composite, the most hydrophilic of the tested materials, than for the other two materials. PLA, the most hydrophobic of the tested materials, is the substrate with the lowest amount of attached cells. Besides, proliferation and differentiation assays have suggested that the most hydrophilic surface triggered the differentiation process earlier than the hydrophobic surfaces (Navarro et al. 2006).

For biosensing applications, the surface energy must be tailored to bind targeted proteins while rejecting all other proteins. Experiments have been carried out to monitor protein adsorption as a function of the hydrophilic or hydrophobic nature of the terminal functional groups in ethylene glycol (EG) (Clare et al. 2005). Because the hydroxyl group of EG can, in theory, be oxidized to a charged carboxylate group, the hydroxyl-terminated EG could lead to an increased nonspecific adsorption of positively charged proteins (Ostuni et al. 2001). Replacing the terminal hydroxyl with a methoxy group to form Me-EG3-ene could therefore provide additional resistance to protein adsorption, especially under oxidizing conditions. However, some reports have shown that methyl-terminated EG3 (Me-EG3) and hydroxyl-terminated EG3 are comparably effective at reducing the amount of nonspecifically adsorbed protein. Some proteins such as fibrinogen possess both hydrophilicity and hydrophobic domains and they can be adsorbed onto both hydrophilic and hydrophobic surfaces. The elongated structure of fibrinogen produces orientation-dependent changes, and these physical packing forces may dominate the adsorption dynamics, thus weakening the effect of surface termination (Figure 15.1).

The problem with many synthetic materials is that they absorb a complex nonspecific layer of proteins (Castner and Ratner 2002). Nonspecific protein layers are not used in nature. Nature uses specific proteins in fixed conformations and orientations so that they optimally deliver signals. The host may react to this nonphysiological proteinaceous layer as to a foreign invader that must be walled off. Thus, surfaces of biomaterials must be designed to control the conformation and orientation of the proteins they adsorb.

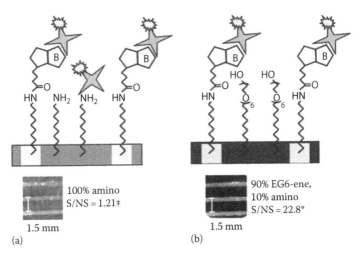

FIGURE 15.1 Scheme of avidin nonspecifically adsorbed onto (a) 100% amino-terminated and (b) 90% EG6/10% amino-terminated monolayer on silicon. (From Clare, T.L. et al., *Langmuir*, 21, 6344, 2005.)

TABLE 15.1
Summary of Strategies for Surface Modification

Surface strategy	Comment
PEG[a]	Effective but dependent on chain density at the surface; damaged by oxidants
PEG-like surfaces by plasma deposition	Applicable for the treatment of many substrates and geometries; highly effective
PEG oligomers in SAMs	Highly effective; applicable for precision molecularly engineered surfaces; durability to elevated temperature is low
PEG-containing surfactants adsorbed to the surface	A simple method for achieving nonfouling surfaces; durability may be low and high surface densities are hard to reach
PEG blocks in other polymers coated on the surface	May provide a relatively low density of surface PEG groups
Hydrogen bond acceptors	Possibly, this principle imparts nonfouling properties to PEG surfaces; this is leading to new discoveries of surface functional groups for nonfouling surfaces
Protein layer	A preadsorbed protein layer resists further adsorption of proteins; this approach, long used by biologists, is easy to implement but of low durability
Hydrogels	PEG is in this class; many other hydrogels have shown nonfouling behavior

[a] Also called poly(ethylene oxide) (PEO).

Many strategies to inhibit nonspecific protein adsorption, i.e., nonfouling surfaces, have been developed as expressed in Table 15.1 taken from Castner and Ratner (2002).

Cells relate to surface through proteins, thus, cells require surfaces with specific adhesive properties (Falconnet et al. 2006). Cells require an adhesive surface to exert forces and consequently spread. The ability to constrain the spreading to a specific cell–surface contact area has been shown to dramatically affect cellular development. Mechanical compliance of cell-adhering substrates can also substantially affect the cellular response and development. The ability to spatially and temporally control the chemistry, the pattern geometry, and the local substrate stiffness will continue to provide new insights into the fundamental aspects of cell–surface interactions and ligands (peptides, proteins, etc.) (Andersson et al. 2003; Barber et al. 2003; Falconnet et al. 2006). Thus, the ability to suppress nonspecific interactions between the surface and the protein-containing media is crucial in order to generate unbiased experimental outcomes.

Advances in surface chemistry have made possible the synthesis of numerous so-called nonfouling surfaces that significantly reduce or eliminate the nonspecific adsorption of proteins and other biomolecules from biological fluids, e.g., through the incorporation of molecules such as carbohydrates, agarose, and mannitol as well as albumin (Luk et al. 2000). The most widely used system is poly(ethylene glycol) or PEG with the monomeric repeat unit $[-CH_2-CH_2-O-]-$ (also known as poly(ethylene oxide) or PEO). The factors governing protein resistance of a PEG-graft copolymer (PLL-g-PEG) has been investigated, with quantitative information provided on the interfacial architecture of PEG chains and their influence on protein resistance (Daria et al. 2002; Dalsin et al. 2003). Many different PEG surface-immobilization strategies have been successfully applied. One example is the use of triblock copolymers PEG–PPO–PEG that assemble spontaneously on hydrophobic surfaces through hydrophobic–hydrophobic interactions. This polymer class is also known as poloxamers or Pluronics® and has been extensively studied (Daria et al. 2002; De Silva et al. 2004).

One major issue that biopolymers must deal with is the possibility of bacterial adhesion, which causes biomaterial-centered infection, and, as a consequence, the lack of successful tissue integration (Gristina 1987). Modifications of biomaterial surfaces at the atomic level will allow the programming of cell-to-substratum events, thus diminishing infection. Bacterial adhesion is the consequence of a competition between binding proteins of the host tissue and binding proteins of the bacteria.

Several approaches can be found on literature to overcome this shortcoming, the first one is to modify the surface of the biomaterial to enhance tissue compatibility or integration (Gristina 1987) and the second one is to modify the material to directly inhibit bacterial adhesion. This second approach can distinguish between those materials with intrinsic antimicrobial properties, such as those functionalized with amine or ammonium groups (Kenawy et al. 2005; Kenawy et al. 2006) and those that immobilize water-soluble or emulsifiable disinfectants onto macromolecular surfaces such as acrylic derivates of eugenol on hydrophilic 2-hydroxyethyl methacrylate (Rojo et al. 2008).

Finally, cellular adhesion and spreading, which is the ultimate purpose of surface modification on biomaterials, must deal with differences in behavior from in vitro to in vivo experiments. RGD peptide sequence is found in a wide variety of ECM proteins and implicated in a number of ligand–receptor interactions (Reyes and Garcia 2003), making it exceptionally useful for promoting general cell adhesion. However, the RGD tripeptide sequence exhibits limited specificity for specific integrins and allows minimal control over cellular responses resulting in the loss of full biological activity. This has been attributed to the absence of additional functional domains present in the native, three-dimensional structure of the ECM molecule (Reyes and Garcia 2003). Modification of surfaces with peptide sequences containing a three-dimensional structure has shown better effectiveness on in vivo experiments. The glycine-phenylalanine-hydroxyproline-glycine-glutamate-arginine (GFOGER) sequence that can be found on the $\alpha_1(I)$ chain of type I collagen is recognized by the integrin $\alpha_2\beta_1$. Integrin recognition is entirely dependent on the triple-helical conformation of the ligand similar to that of native collagen (Reyes and Garcia 2004). Modifications with the GFPGER sequence are lately being proposed as a robust strategy for enhancing cell–material interactions and encouraging biospecific cell adhesion that may provide with good results on in vivo experiments (Reyes and Garcia 2003, 2004).

Thus, there are several proposals on what surface properties and what surface modifications are the most appropriate to tailor biological interaction of biomaterials to specific applications. The purpose of this chapter is to review methods that can be found in the current literature to modify biomaterial surfaces, surface characterization methods, and the interactions of host tissues with different sorts of surfaces.

15.2 SURFACE MODIFICATION METHODS

Chemical, physical, and physicochemical modifications of biomaterial surfaces comprise a number of ways to control cellular functions on a biomaterial surface. Surface characteristics such as hydrophilicity, surface charge density, surface micro-morphology, free energy, and specific chemical groups can affect and regulate a wide variety of biological functions. Several chemical modification techniques have been developed to improve wetting, adhesion, and printing of polymer surfaces by introducing a variety of polar groups and specific functional groups.

15.2.1 PHYSICAL METHODS

Dry surface treatment techniques, such as corona discharges (Zhu et al. 2006c), plasma (Kwang et al. 2009), or ultraviolet (UV) radiation, are among the most common methods used to modify the physical–chemical exterior properties of biopolymers.

The principal effects of these treatment techniques are summarized in the following (Sham et al. 2009):

1. *Cleaning*: through conversion of contaminants into gaseous, volatile products
2. *Ablation*: etching out the weak layer on the microscopic scale or selective leaching of low–molecular weight polymer chains
3. *Surface strengthening*: further cross-linking or branching the near surface polymer molecules
4. Modification of chemical structures of surface through reactions of free radicals with surrounding gases

15.2.1.1 UV/Ozone Ablation

UV light surface modification is an easy process that does not require vacuum, avoiding complicated machinery, or chemical reagents that avoid the production of residual polluting by-products.

UV light can modify polymers through two different mechanisms. It can directly affect the bonds inside the polymer by local excitations, which introduce photocrosslinking and/or photodegradation. Or it can form a strongly oxidant ozone layer, containing some atomic oxygen and oxygen radicals as well, on the air layer outside the polymer (Øiseth et al. 2004). The ozone attacks and chemically modifies the top layers of the polymer, which are oxidized and/or etched primarily through the following reaction:

$$O_3 + Polymer \rightarrow O_2 + Oxidized\ polymer$$

The extent of these photo-induced reactions depends on the mobility of the polymer chains, the UV absorption coefficient, the photon penetration depth into the polymer, the glass transition temperature, the diffusion coefficients of gases in the polymer core, and the wavelength (Andrady et al. 2003). Polyethylene is mostly activated at wavelengths around 330 nm. However, a very strong absorption below 160 nm that originates from the dissociative excitation of the C–C and C–H bonds can be observed (Øiseth et al. 2004). Effects of UV light and ozone have been studied for various polymer surfaces, including polyethylene (PE) (Øiseth et al. 2004; Poulsson et al. 2009), polyethylenetherephthate (PET) (Teare et al. 2000), polypropylene (PP) (Macmanus et al. 1999), polyetheretherketone (PEEK) (Romero-Sánchez et al. 2001), poly(vinylmethyl siloxane) (PVMS) (Efimenko et al. 2005), polydimethylsiloxane (PDMS) (Ye et al. 2006), polystyrene (PS) (Mitchell et al. 2005), as well as polyesters and mix of fabric surfaces (Michael et al. 2004).

Using PDMS as an example, the atomic oxygen originating from ozone splitting reacts with the PDMS surface forming radicals that will oxidize and remove organic parts of the polymer in the form of CO_2, H_2O, and other volatile compounds. UV-light of 355 nm may break Si–CH_3 bonds since the energy of that radiation is 3.5 eV, which is close to the bond energy in Si–CH_3 (3.2 eV), but it will not affect the backbone of the polymer since Si–O bond energy is 4.6 eV. The surface of the PDMS will then have an SiOx structure that will be covered with –OH groups. Thus, it becomes more hydrophilic than when it is covered with the usual –CH_3 groups of a normal PDMS surface (Sham et al. 2009).

15.2.1.2 Radio Frequency Plasma Treatment

Plasma technology is used to alter the surface properties of polymers without affecting bulk properties. Physicochemical properties such as surface free energy, hydrophilicity, and surface morphology can be modified to make the surface of the biomaterial adequate for cell–material interactions. Plasma treatment can improve wettability, oxidize the surface, and enhance cell growth and adhesion. The effects of plasma on a polymer surface can be classified as surface modification, grafting, and film deposition. Plasma surface modification can be achieved using gases such as oxygen, nitrogen, argon, and helium. The main effects are removal of chemical contaminants (Kim et al. 2005) and/or weakly bound polymer layers but mainly, introduction of chemical groups such as hydroxyl, carbonyl, carboxylic, amino, or peroxyl groups on the surface that permit covalent bonding (Sanchis et al. 2006). Oxidation, nitration, hydrolization, and amination processes induced by plasma are used to improve the surface energy and reactivity. In plasma-induced grafting, free radical formation using inert gas plasma is followed by the introduction of unsaturated monomers into the reaction chamber that react with the free radicals to yield a grafted polymer (Choi et al. 2004). Plasma polymerization process is used to produce thin films with unique physicochemical properties (Cheng et al. 2004). Methane, ethylene, propylene, fluorocarbon monomers, and organosilicon compounds can be polymerized by this method. The plasma-polymerized thin films are generally homogeneous, highly cross-linked, and strongly bound to the surface.

Radio frequency (RF) plasma treatment of polymeric biomaterials has been used to improve biocompatibility. The effects of nitrogen and oxygen plasma treatment on surface modification of polyethyleneterephthalate (PET) have been studied (Vesel et al. 2008) and used to produce protein-resistant properties (Vandencasteele et al. 2008). The surface of PET polymer was modified by this method to achieve improved attachment of fucoidan, with antithrombogenic properties (Junkar et al. 2009). Helium plasma treatment has been applied on polyurethane film coated on glass substrate to enhance endothelial cell growth and adhesion (De et al. 2004). The critical surface tension (γ_C) of polyurethane (PU) film increased by 2 dynes/cm due to helium plasma treatment (De et al. 2005). Griesser et al. studied the attachment and growth of human endothelial cells and fibroblasts on polymer surfaces fabricated by the polymerization of volatile amine and amide compounds in a low-pressure gas plasma and by the treatment of various surfaces in ammonia plasmas, which served to increase the nitrogen content of the surface layers (Griesser et al. 1994). In other studies, they demonstrated that the covalent surface immobilization of polysaccharides, proteins, and synthetic oligopeptides can be achieved via nanometres thick, interfacial bonding layers deposited by gas plasma methods (Griesser et al. 2002). The adhesion of endothelial cells on plasma-treated PS varied with plasma treatment time, increasing with longer periods of time (Van Kooten et al. 2004). Surface energy increased after oxygen plasma treatment of fiber-reinforced PET composite and enhanced cell adhesion (Cioffi et al. 2003). Poly(lactide-co-glycolide) (PLGA) films were treated by oxygen plasma. Under specific conditions, cells stretched very well and the ability to endure the shear stress was improved greatly. Results showed that improved cell adhesion was attributed to the combination of surface chemistry and surface morphology of PLGA during plasma etching (Wan et al. 2004). Wan et al. (2003) modified poly(L-lactic acid) (PLLA) films by plasma treatment using O_2, N_2, Ar, and NH_3 gas atmospheres under different conditions. Cell retention on the NH_3 plasma-modified PLLA surface was much higher suggesting that a high polar component and N-containing groups may play a role in enhancing cell resistance to shear stress. Ultrathin antibacterial polyammonium coating given on plasma-treated polyethylene (PE) surface required much smaller quantities of the antibacterial agent than other conventional methods to produce the same effects (Thome et al. 2003). PE coated with triclosan and bronopol was treated with oxygen plasma followed by argon and hydrogen plasma, hydrogen plasma resulting more efficient to improve antibacterial properties (Zhang et al. 2006c). Carbon fiber-reinforced polyether ether ketone (PEEK) for joint endoprostheses and fracture fixation plates was treated by oxygen plasma and N_2/O_2 plasma to get better surface activation for joining and coating processes (Ha et al. 1997). A poly(ε-caprolactone) (PCL) surface was modified by an O_2 plasma surface treatment to form oxygen-containing functional groups to promote the precipitation of a bone-like apatite layer in vitro (Oyane et al. 2005). Film deposition was performed on porous PU matrix loaded with ciprofloxacin. The antibiotic released at a controlled rate when the film was coated with poly(n-butyl methacrylate) by RF plasma deposition (Kwok et al. 1999). To prevent biofilm formation in dental implants, adsorption of histatin 5 was produced onto poly(methyl methacrylate) (PMMA) surfaces after modification of the polymer surface using a cold plasma technique. Adsorption increased due to the increased surface hydrophilicity and the formation of the carboxyl groups (Yoshinari et al. 2006). Improvement of corneal epithelial cell growth on different substrates for artificial corneas such as poly(vinyl alcohol), silicone rubber, PS, and polycarbonate was analyzed and it was found that plasma-induced graft copolymerized poly(hydroxyethyl methacrylate) (pHEMA) on silicone rubber provided the best growth rate (George and Pitt 2002). Plasma processes have been used to obtain immobilization of biomolecules on various substrates. Insulin and heparin were immobilized on PET films grafted previously with acrylic acid using oxygen plasma to improve blood compatibility (Kim et al. 2000) and to enhance growth of smooth muscle cells (Gupta et al. 2002). Chen et al. (2004c) investigated the allylamine polymerization using pulsed RF plasma as an adhesion layer immobilizing DNA probe for DNA hybridization. Patterning of amine groups was achieved by selectively depositing plasma-polymerized ethylenediamine films on specific areas of a glass surface. These results demonstrated that the formation of amine groups on glass slides by plasma-enhanced chemical vapor deposition resulted an excellent tool for DNA array technology (Jung et al. 2006).

15.2.1.3 Corona Discharges

Corona discharge was first commercially used in the 1960s to modify the surface of different polymeric materials such as PP, PS, and polyethylene-terephthalate (Romero-Sánchez et al. 2001). A corona is a process that is developed from an electrode with a high potential in a neutral fluid that results in the ionization of the fluid. When the ionized fluid finds its way to an oppositely charged object, a spark or continuous arc of plasma is produced. A setup for corona discharge is shown in Figure 15.2.

When a corona discharge is applied to a polymeric surface, relatively stable free radicals are formed that produce changes in surface properties. Some examples of corona discharge uses follow. It has been used for changing the surface energy on rubber (Romero-Sánchez et al. 2001). The procedure removes low-molecular weight moieties on the surface, increases the amount of silica, and incorporates oxygen that is associated to Si as SiO_2. This is attributed to the fact that the energy of part of photons from corona discharge is larger than the binding energy of Si-O-Si (8.3 eV) and Si-CH_3 (4.5 eV) bonds. Hydrophilic OH groups can be formed as by-products instead of the hydrophobic CH_3 on the surface of rubber, which allows the hydrophobicity of polymeric materials to be reduced (Zhu et al. 2006c).

Corona discharges have been also applied to produce covalently linked hydroxyapatite (OHAp) on silicona. Hydrophobic surfaces are unable to nucleate OHAp due to a low Lewis acid–base surface tension (Wu et al. 1997). A corona-discharge treatment has been used for graft polymerization of acrylic acid onto silicone sheets. A composite with an amino-modified OHAp was then possible to prepare (Furuzono et al. 2001).

15.2.1.4 Ion Implantation

In ion implantation processes, ions are accelerated with energies in the range 20–200 keV and directed to the surface of a polymer. Penetration of these ions is of several hundred nanometers and can modify the surface of the materials without affecting bulk properties (Cui et al. 2008). Properties that can be tailored following this procedure include hardness, wear resistance, coagulability, wettability, and biocompatibility of the material (Cui and Luo 1997). Two different mechanisms have been found to explain the character of chemical processes in ion implantation processes. The first one is electronic excitation, also named as inelastic collision and the second one is nuclear collisions, also named as elastic collisions (Sviridov 2002). The electronic excitation is responsible for the formation of mobile radical species, e.g., H^+, whereas nuclear collisions produce massive ruptures of chemical bonds as a consequence of atom displacements. High-energy light ions deposit their energy by electronic

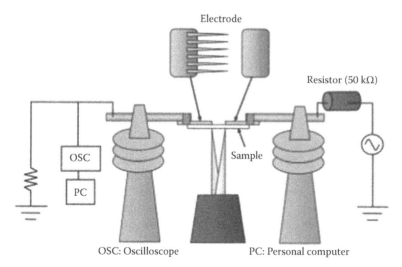

FIGURE 15.2 **(See companion CD for color figure.)** A corona discharge setup. (Taken from Zhu, Y. et al., *Polym. Degrad. Stab.*, 91, 1448, 2006c.)

excitations while the main mechanism for heavy ions is nuclear collision (Sviridov 2002). Thus, stopping ions involves the deposition of high energy within the tracks followed by the ions and the formation of radicals, secondary ions, and electrons. Ion implantation of partially halogenated aliphatic polymers causes radical-induced dehydrogenation and carbonization on the surfaces (Rao et al. 1993). The effects of ion implantation can be modulated by changing the energy of the ions used. An irradiation of Sb + (150 keV) on polyamide-6 results in an important change of polarity, whereas N+ irradiation on poly(ethylene therephthalate) (PETP) only changes polarity by 30% (Sviridov 2002). Concerning applications in the medical field, a few works have reported the use of this procedure to improve the blood biocompatibility and anti-thrombogenicity of PDMS surfaces (Dupas-Bruzek et al. 2009).

15.2.2 CHEMICAL METHODS

15.2.2.1 Wet Chemical Methods

This technique involves the treatment of the biomaterial surface with liquid chemical reagents to generate reactive functional groups. It is adequate for porous three-dimensional substrates due to the high penetration of the liquids inside the substrate (Liu and Ma 2004); however, this method can produce a range of oxygen-containing functional groups, and the degree of surface functionalization may not be repeatable depending on molecular weight, crystallinity, or tacticity of the polymer. Chromic acid oxidation has been applied to polyethylene (PE) films to introduce carboxylic acid groups (Goddard and Hotchkiss 2008). Partial oxidation of sodium alginate with potassium permanganate ($KMnO_4$) has been attempted to improve degradability of alginate for biomedical applications (Lu et al. 2009). Acid-catalyzed hydrolysis and aminolysis have been used to yield carboxyl and amine-terminated PMMA surfaces (Brown et al. 2006). George and co-workers (Zhu et al. 2004b) introduced free amino groups onto polyurethane (PU) scaffolds which increased surface energy and provided a convenient way to further immobilize bioactive species and to enhance the cell–PU interaction in the endothelium regeneration process. Electrospun nanofibers have been chemically functionalized after surface modification by wet chemical method and other different routes including plasma treatment, surface graft polymerization, and co-electrospinning of surface active agents and polymers (Yoo et al. 2009). Dalton et al. (1998) studied the effect of hydrophilicity on colonization by human bone-derived (HBD) cells, comparing untreated PS and a sulfuric acid-treated PS surface for mechanisms of cell migration. The chemically treated PS surface contained various oxidative groups and a minor amount of sulfonate groups and HBD cells migration was dependent on the presence of vitronectin (Vn) and it was higher on the hydrophilic acid-treated surface. Alferiev et al. (2006) investigated a chemical approach consisting of bulk carboxylation of polyurethanes via bromoalkylation to enable surface heparinization for thromboresistance to improve hemocompatibility of vascular prostheses. Some investigators have reported that surface modification of biodegradable polymers in alkaline solution could be used to generate a hydrophilic and rough surface for cell attachment (Nam et al. 1999). The treatment of poly(lactic-co-glycolic acid) (PLGA) with Na-OH solution enhanced vascular smooth muscle cell adhesion and proliferation but decreased those of endothelial cells (Miller et al. 2004). Aminolysis of PLLA membranes improved the cytocompatibility of PLLA using human umbilical endothelial cells culture (Zhu et al. 2004). Two chemical methods, base hydrolysis and aminolysis, were used to introduce carboxylic acid or primary and secondary amine groups onto PLGA surface and provide satisfactory results, creating a truly biomimetic scaffold for tissue engineering (Croll et al. 2004). Surface thiolation of poly(methacrylic acid)-based microparticles was carried out by coupling L-cysteine with a water-soluble carbodiimide. This method is proposed as an interesting strategy to improve paracellular permeability of hydrophilic macromolecules (Sajeesh et al. 2010). Biological ligands (biotin and peptides) were immobilized onto hydrolyzed poly(glycolic acid) (PGA) sutures through an amide bond between the amine (ligand) and the carboxylic groups of the modified surface. This strategy can be applied to other polyesters to tailor their properties for biomedical applications (Lee et al. 2003). Modification of poly(ethylene terphathalate) fabric with glycerol polyglycidyl ether has been achieved by means of supercritical carbon dioxide using a series of immobilization processes to finish with the incorporation of natural agents such as sericin, collagen, or chitosan (Ma et al. 2010).

15.2.2.2 Grafting Polymerization

Grafting polymerization is used to produce surface modification when the bioactive compound loses activity if it is directly linked to the hydrophobic polymer surface. Acrylic acid has been UV graft polymerized to PCL films pretreated with Argon plasma to generate carboxylic acid groups for immobilization of collagen (Cheng and Teoh 2004). Surface modification of a PLLA with collagen has been carried out by covalent bonding to polymer via gamma irradiation grafting with poly(acrylic acid) as a coupling agent (Yang et al. 2003). Vinyl acetate, acrylic acid, and acrylamide were grafted onto PLLA surface films separately or in combinations, using a UV-induced photo-polymerization process. Combinations of acrylamide with vinyl acetate or acrylic acid produced the hydrophilicity of the modified surface (Janorkar et al. 2006). Kubies and co-workers studied the influence of different hydrophilic groups, hydroxyl, carboxyl, or amide, introduced onto PLLA membrane surface via photocopolymerization of the corresponding monomers, on the cytocompatibility for chondrocytes. Results showed that whereas for hydroxyl and amide groups cytocompatibility was greatly improved, for carboxyl groups it was even worse although all modified surfaces possessed a similar hydrophilicity (Kubies et al. 2003). Polysulfone (PSF) membranes were treated with ozone to introduce peroxides, and then grafted with either acrylic acid or chitosan, followed by the immobilization of heparin to improve the hydrophilicity and blood compatibility of PSF membrane (Yang and Lin 2003). Surface of nylon membranes was modified by graft copolymerization of glycidyl methacrylate using a redox initiation system to promote further antibody immobilization (Jackeray et al. 2010). PEG was covalently immobilized to a layer of branched poly(ethylenimine) (PEI) to control bacteria adhesion (Kingshott et al. 2003). Silk fibroin surfaces were PEGylated by reaction with cyanuric chloride-activated PEG and proposed as useful antiadhesion and antothrombogenic biomaterials (Vepari et al. 2010). Different functional monomers, like hydroxyethyl methacrylate, acrylic acid, N-vinyl pyrrolidone, and glycidyl methacrylate, have been grafted onto the surface of films, using simultaneous photo-grafting and cold plasma-grafting techniques, to alter the surface properties of elastomeric substrates, such as hydrophilicity and biocompatibility. Hydroxyethyl methacrylate and acrylic acid showed exceptionally high cell compatibility in terms of cell adhesion and proliferation (Desai et al. 2003). Synthetic phospholipid analogs have been photografted onto PP hemodialysis membranes to reduce fouling and thrombus formation and improve biocompatibility (Xu et al. 2004). Graft polymerization of a phospholipid polymer 2-methacryloyloxyethyl phosphorylcholine (MPC) onto the polyethylene surface has been investigated by Moro et al. (2004) to reduce friction factors frequently involved in periprosthetic osteolysis—bone loss in the vicinity of a prosthesis. PU membranes were modified by grafting the zwitterion of sulfobetaine monomer using a novel three-step procedure to enhance hemocompatibility with good results (Huang and Xu 2010; Figure 15.3).

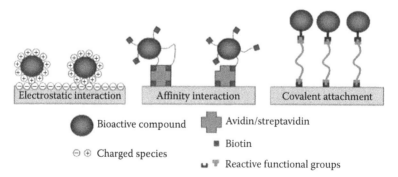

FIGURE 15.3 Mechanism of immobilization. (Taken from Goddard, J.M. and Hotchkiss, J.H., *Prog. Polym. Sci.*, 32, 698, 2007.)

15.2.2.3 Polyelectrolyte Complex

A polyelectrolyte complex (PEC) is a compound made from an electrically neutralized molecular complex of polyanions and polycations. The effects of PEC films composed of polysaccharides on cell behavior have been studied and it was reported that PEC can stimulate differentiation of osteoblasts and periodontal ligament fibroblasts (Hamano et al. 2002; Nagahata et al. 2005). Chitosan covalently immobilized onto PLA film surface was made to react with heparin (Hp) solution to form a PEC on the PLA surface. PLA surface modified by chitosan/heparin complex inhibited platelet adhesion and activation and showed enhanced cell adhesion (Zhu et al. 2002). PECs were formed through the mixing of aqueous solutions of naturally occurring sulfated fucoidans with hydroxyethyl cellulose derivatives containing cationic and hydrophobic substituents. Formation of PECs was determined by the type and arrangement of substituents and the degree of charging, as well as the composition of polysaccharide molecules (Shchipunov et al. 2003). Based on this principle, a way of surface modification has been focused on the PEC formation on the biomaterial surface to attain different goals. Polyelectrolyte films built up by the alternated adsorption of cationic and anionic polyelectrolyte layers constitute a novel and promising technique to modify surfaces in a controlled way (Ladam et al. 2002). Poly-(DL-lactide) substrates were activated by PEI to obtain stable positively charged surface and polyelectrolytes such as alginate and poly-(L-lysine) were alternatively deposited onto the activated substrates. Results indicated that the multilayer-modified substrates promoted the chondrocyte cytocompatibility (Zhu et al. 2004c). The surface modification with PEC multilayer of hyaluronic acid (HA) and poly(L-lysine) (PLL) resulted in drastically reduced peripheral blood mononuclear cell attachment according to the lactate dehydrogenase assay for cell counting. This nano-scale control of material deposition may be successfully applied for surface modification of various biomaterials (Hahn and Hoffman 2005). Poly(lysine)/poly(glutamic acid) (PLL/PGA) multilayer films ending by several PLL/PGA-g-PEG bilayers, PGA-g-PEG corresponding to PGA grafted by PEG, can be used to drastically reduce both protein adsorption and bacterial adhesion. This kind of anti-adhesive films represents a new and very simple method to coat any type of biomaterial for protection against bacterial adhesion and therefore limiting its pathological consequences (Boulmedais et al. 2004). Modification of intraocular lens of poly(methyl methacrylate) using layer-by-layer deposition of poly(sodium 4-styrenesulfonate) and PEI has been proposed to facilitate anti-bacterial drug adsorption (e.g., ampicillin) onto the surface as a way to protect eye and the device from infections (Manju and Kunnatheeri 2010). The adhesion of L929 cells to PCL nanofibers was successfully improved via coating with polyelectrolyte multilayer thin films of poly(diallyldimethylammonium chloride) and poly(sodium 4-styrene sulfonate), which enhanced the potential of this material as a scaffold in tissue engineering applications (Dubas et al. 2009). A new type of polymeric hybrid coating consisting of poly(acrylic acid) and poly(acrylamide) was created by layer-by-layer deposition of polyelectrolyte multilayers onto nano-patterned polymer brushes derived from a surface-reactive rod-coil block copolymer, PS-block-poly[3-(triethoxysilyl)-propylisocyanate], offering a new organic hybrid coating with novel surface properties (Yang et al. 2008). Layer-by-layer polyelectrolyte films offer extensive potentials to enhance surface properties of vascular biomaterials. Three films composed of poly(sodium 4-styrenesulfonate)/poly(allylamine hydrochloride), PLL/hyaluronan, and PLL/poly(L-glutamic acid) were built onto isolated PET filaments, thread, and vascular prostheses, demonstrating improved mechanical resistance (Rinckenbach et al. 2008). The layer-by-layer technique applied to the coating of an oral prostheses gave successful results and it was proposed for the coating of other oral prosthesis devices (Etienne et al. 2006).

15.2.2.4 Click Chemistry

The "click" concept introduced by Sharpless and co-workers in 2001 is undeniably one of the most promising trends in synthesis and functionalization chemistry (Kolb et al. 2001). Among the different chemical reactions that can be considered of the "click" type, the copper(I)-catalyzed azide-alkyne cycloaddition (CuAAC) is generally regarded as the example of click chemistry (Rostovtsev et al. 2002).

CuAAC has been used for building, crosslinking, or functionalizing polymeric micelles, nanogels, or liposomes (Lutz 2007). Wooley et al. first reported the advantages of click CuAAC for functionalizing block copolymer micelles of PS core and poly(acrylic acid) shell (Joralemon et al. 2005; O'Reilly et al. 2006). The in situ shell-functionalization of micelles composed of a cholesterol-based hydrophobic core and a thermoresponsive PEG-based polymer shell has been reported (Lutz et al. 2007). Functionalization of lipid vesicles have been made using different alkyne-functional surfactants, assuring that, by this method, only the outer surface of the vesicles was modified (Cavalli et al. 2006; Hassane et al. 2006). An artificial oxygen carrier was constructed by conjugating hemoglobin molecules to biodegradable micelles using the click chemistry. First, a series of triblock copolymers (PEG-PMPC-PLA) in which the middle block contains pendant propargyl groups were self-assembled into core-shell micelles in aqueous solution, and then, micelles were conjugated to azidized hemoglobin molecules via click reaction between the propargyl and azido groups (Shi et al. 2009). N-azidated chitosan was prepared using click chemistry by four different methods (Kulbokaite et al. 2009). In other studies thrombin inhibitors were fixed on poly(ethylene terephthalate) (PET) and poly(butylene terephthalate) (PBT) by wet chemistry treatment (activation of hydroxyl chain-ends) and photochemistry (nitrene insertion by photoactivation of aromatic azide) (Salvagnini and Marchand-Brynaert 2006). UV grafting of a carbohydrate-bearing photoreactive azide group has been performed on poly(ethylene terphthalate) film (Roger et al. 2010).

15.2.2.5 Biochemical Modification

The attachment of biomolecules to polymer surfaces could be an important tool for controlling cell–biomaterial interaction. Covalent immobilization of various integrin-binding peptides as well as entire proteins has been shown to mediate the adhesion of many types of cells. Polymeric surfaces must have specific functionality to be covalently attached to a bioactive compound. Common functional groups in bioconjugation chemistry include thiols, aldehydes, carboxylic acids, hydroxyls, and primary amines. Crosslinking agents are commonly used to link the bioactive compound directly to the functionalized surface or through a spacer of several angstroms (Hermanson 1996). Electrically conductive polypyrrole (PPY) was surface functionalized with HA and sulfated hyaluronic acid (SHA) by the use of a cross-linker having the appropriate functional groups to improve its surface biocompatibility (Cen et al. 2004). Gelatin and chitosan have been immobilized onto biodegradable poly(lactide-co-glycolide) acid (PLGA) producing a significant increase in cell adhesion and spreading (Zhu et al. 2006b). Surface functionalization of biodegradable PLLA was achieved by plasma coupling reaction of chitosan. The results suggest that the new substrate can be used to control the morphology of cells and has potential applications in tissue engineering (Ding et al. 2004). The conjugation of heparin with a biodegradable PLA has been achieved to improve blood compatibility and related biocompatibility and suggested for implantable medical devices and tissue engineering (Jee et al. 2004). Since Arg-Gly-Asp (RGD) sequences have been found to promote cell adhesion in 1984, numerous polymers have been functionalized with RGD peptides to achieve biomimetic materials (Niu et al. 2005). RGD has been covalently attached to amine-functionalized PLA using a glutaraldehyde linkage and the results indicated that modifying the surface of PLA scaffolds with RGD, bone cell attachment, and differentiation were enhanced, suggesting that this strategy may improve bone tissue regeneration more efficiently in wound models (Hu et al. 2003). BisGMA/TEGDMA thermosets and composites have been surface modified to expose carboxylic groups on the surface and the activated material was further modified by deposition of a lactose-modified chitosan decorated with RGD peptide driven by strong electrostatic interactions. The modified thermoset was carried out in vitro very similar to clinically used roughened titanium (Travan et al. 2010). Biochemical modification of fibrous scaffolds by interfacial PEC was achieved by immobilization of RGD chemically conjugated to a thiol-active maleimidylated form of the scaffold (Tai et al. 2010). Surface immobilization of fibronectin-derived PHSRN peptide has been attained by spontaneous adsorption on thin polysiloxane thin films having different surface energy. The effect of the immobilization of this sequence, which induces cell invasion and accelerates wound healing, on fibroblasts was different depending on the hydrophilicity of the surface (Satriano et al. 2010).

On the other hand, PEG has been used for conjugation of proteins since 1970 (Davis 2002). Nowadays, this technique is well established and it is called PEGylation (Milton Harris and Chess 2003). PEG possesses interesting properties for conjugation. It is a highly water-soluble polymer that extends to give a hydrodynamic radius greater than that of a globular protein of equivalent molecular weight and it can act as a hydrophilic spacer between the polymeric surface and the bioactive compound. Acrylated antibodies were photografted on polymer surfaces previously treated with PEG to facilitate rapid and sensitive antigenic detection (Sebra et al. 2005). A dodecapeptide (H12) has been conjugated to the end of PEG chains on the surface of a phospholipid vesicle with an average diameter of 220 nm to prepare H12-PEG-vesicles (Okamura et al. 2005). Covalent immobilization of heparin onto electrically conductive polypyrrole (PPY) film was attained after graft copolymerization with poly(ethylene glycol) methacrylate (PEGMA) and activation with cyanuric chloride. After surface modification, the film improved wettability while retaining significant electrical conductivity and with immobilized heparin, platelet adhesion and platelet activation on PPY film were significantly suppressed (Li et al. 2003). Peptides incorporating the RGD and PHSRN integrin-binding sequences were immobilized onto a polyethyleneglycol-based polymer substrate to study the effects of fibronectin in supporting macrophage fusion to form foreign body giant cells (Kao and Hubbell 1998). Chemoselectivity of click chemistry is another route presently used to obtain polymer bioconjugates. It has been demonstrated that sequence-defined oligopeptides can be efficiently linked to synthetic macromolecules using CuAAC ligation (Dirks et al. 2005; Parrish et al. 2005). Click cycloaddition of short peptides such as RGD with well-defined polymers synthesized by atom transfer radical polymerization (ATRP) has been investigated (Lutz et al. 2006). Peptide bioconjugation via huisgen cycloadditions were explored on linear or dendritic synthetic macromolecules with carbohydrates (Wu et al. 2005a; Fernandez-Megia et al. 2006). Other more complex biological entities such as proteins (Reynhout et al. 2007) or even viruses, bacteria, or cells (Deiters et al. 2003; Link et al. 2004) have also been conjugated to polymers or low-molecular weight functional groups using azide-alkyne click chemistry.

15.3 METHODS OF ANALYSIS OF POLYMER SURFACES

Surface characterization of biomaterials is a central issue in biomaterials research since surface properties can have an enormous effect on the success or failure of a biomaterial device. This characterization can be carried out with different analytical tools whose selection should be based on the nature of the surface, the specificity required, and the resources available. A general classification of these techniques divides them into the following three categories:

- Spectroscopic methods
- Nonspectroscopic methods
- Microscopic techniques

While nonspectroscopic and microscopic techniques describe in a precise manner the changes in surface chemistry, equipment is costly and analysis can be time-consuming. Therefore, spectroscopic analytical techniques are more useful for laboratory routine due to its simplicity and the speed with which results can be obtained.

15.3.1 Spectroscopic Methods

15.3.1.1 X-Ray Photoelectron Spectroscopy

X-ray photoelectron spectroscopy (XPS), also called electron spectroscopy for chemical analysis (ESCA), is a powerful technique to characterize the chemical composition of polymer surfaces, where a depth of typically 7 nm is probed. Therefore, the information provides only affects to some surface atomic layers of the material.

The principle of XPS is based on the emission of photoelectrons from the surface in response to irradiation by a beam of monochromatic x-rays. The kinetic energy (K_E) of these emitted photoelectrons is unique for the different elements and is also sensitive to the chemical state of the atoms; therefore, the comparison of K_E obtained to known values allows the user to identify the elements present on the surface and its oxidation state (McArthur 2006). The resulting spectrum is a plot of intensity (arbitrary units) versus binding energy (eV), the intensity of the ejected photoelectrons is directly related to the material surface atomic distribution and, therefore, can be used to quantify atomic compositions and stoichiometric ratios.

XPS can be used to identify specific functional groups present in complex polymers by using derivatization reactions that provide a more accurate identification and determination of functional groups by selective labeling. For example, trifluoroacetic acid anhydride (TFAA) is often used to determine hydroxyl groups, while aldehydes like pentafluorobenzaldehyde or trifluoromethylbenzaldehyde are reported as derivatization agents for quantification of amines (Pippig et al. 2009; Strola S 2010).

In biomaterials field, XPS has been widely used to determine elemental composition of solid surfaces. It has been applied to different surfaces and surface modifications in studies of adsorption and retention of chemicals such as antibiotics, for the detection of immobilized proteins, understanding the chemistry of the structure, formation and stability of plasma-treated surfaces, as well as for the characterization of the steps formation of thin coatings (Sabbatini and Zambonin 1996; McArthur 2006; Hook et al. 2010).

15.3.1.2 Fourier Transform Infrared Spectroscopy

Fourier transform infrared spectroscopy (FTIR) is a spectroscopic technique based on the use of infrared beam to analyze the chemical functionalities present in a molecule. When an infrared beam (IR) hits a sample, chemical bonds stretch, contract, and bend causing it to absorb IR radiation in a defined wavelength number (Goddard and Hotchkiss 2007). This technique provides information about the molecular structure, inter- and intramolecular interactions, crystallinity, conformation, and orientation of molecules.

Infrared spectroscopy can also be used in attenuated total reflection mode (ATR-FTIR), which is widely used in biomaterial surface analysis. In this mode, the incident IR beam first passes through an internal reflection element made from inorganic crystals like germanium, ZnSe, or diamond. Total reflection occurs at the interface between the sample and this internal reflection element, but instead of reflecting totally, a fraction of the incident IR enters into the sample in a form of electromagnetic wave with an exponentially decreasing magnitude and with a penetration depth from several hundred nanometers to more than 1 μm. This kind of electromagnetic wave is called the evanescent wave, which has a sampling depth depending on the wavelength and the refraction index of both the reflection element and the sample.

Advancement in FTIR characterization of materials has been achieved with the development of FTIR microscopy (Bhargava and Levin 2001). This technique has significant advantages compared to many other imaging methods for the characterization of biomedical materials, because it relies on the characteristic absorbance of corresponding molecular vibrations in the sample. Therefore, it does not require the use of added dyes or labeling methods for visualization of different chemical components in the sample (Kazarian and Chan 2006).

An IR microscope is similar to an optical microscope, being the optical paths for both the visible and the IR radiation identical through the sample plane. The user can focus on specific features of the specimen, use the apertures to delineate the sample region, and collect the single beam IR spectrum for the sample. As a microscopic analysis, this technique has a resolution limit dictated by diffraction effects. When two samples are less than 50 μm apart, they cannot be clearly resolved over the mid-IR range; thus, pure spectra from each sample cannot be obtained. This means that when the spectrum of one sample is obtained, a certain amount of the energy from the second sample is being detected. Thus, pure spectra from each sample cannot be obtained (Sahlin and Peppas 1997). This technique has been extensively used in analyzing polymer/polymer and polymer/solute diffusion (Sahlin and Peppas 1997; Miller-Chou and Koenig 2002; Gupper et al. 2004; Gupper and Kazarian 2005; González-Benito and Koenig 2006; Peter 2006), drug release and tablet dissolution or water sorption in pharmaceutical formulations (Chan and Kazarian 2004a,b; Kazarian et al. 2005). Kazarian and Chand (2003, 2006) have used ATR-FTIR imaging to investigate the in situ dissolution of polymer/drug formulations in water based on PEG and different model drugs.

15.3.1.3 Secondary Ion Mass Spectrometry

Secondary ion mass spectrometry (SIMS) is a mass spectroscopy-based technique for analyzing elemental and chemical compositions of the outermost molecular or atomic layer of a solid surface (Stamm 2008). This technique provides a high surface specificity, since only several atomic layers on the sample surface are analyzed. The material surface is bombarded with a beam of energetic ions, usually argon or gallium. When the ions hit the surface, atoms, clusters, and molecules will be removed from the material surface in a process called sputtering. A fraction of the sputtered particles will be ionized to produce secondary ions, which are separated as a function of the ratio of mass per electric charge in a mass spectrometer.

SIMS can be developed in static or dynamic mode (Ma et al. 2007). Static SIMS uses a low-energy primary ion beam to scan the sample surface in a static surface analysis. In this mode, only a monolayer is removed from the surface, it allows, e.g., the determination of surface contaminants. In dynamic SIMS, a high-energy primary ion beam is applied in short times eroding away the material surface continuously. Real-time signal is recorded simultaneously. The signal is plotted against the depth of the sample to obtain a high-resolution depth (several nanometers) profiling the chemical compositions/structure from the surface into the bulk of the sample.

Time-of-flight SIMS (ToF-SIMS) is a very efficient method for characterizing the elemental (including H and isotopes) and molecular composition of the top surface of biomaterials (Belu et al. 2003). In this case, the secondary ions are accelerated to a constant kinetic energy and then allowed to pass in a field-free environment before they are collected on an intensity-sensitive detector. The time of flight of the ions is correlated with individual masses; the positively charged secondary ion fragments are characteristic of chemical structures on the material surface and can provide significant information for the surface.

In the last 20 years, SIMS has progressed from a novel research methodology to an analytical tool for determination of the surface chemistry of polymers in biomedical applications. SIMS spectra can provide characteristic fingerprints for biomaterial surfaces. This technique has also been used to obtain direct proof of covalent binding of molecules to a surface or to correlate surface chemistry with cell growth, as well as to evaluate surface homogenicity by two-dimensional chemical maps (Chilkoti et al. 1991; Perez-Luna et al. 1997; Leonard et al. 1998; Kingshott et al. 2003; Mouhib et al. 2010).

15.3.2 NONSPECTROSCOPIC METHODS

15.3.2.1 Contact Angle

Measurement of the contact angle of a liquid test droplet on a solid surface is a useful technique that reveals surface energetic information inaccessible by surface spectroscopy. Although this method represents perhaps one of the earliest methods used to investigate surface structure, it still yields very useful information. Basically, a drop of fluid is placed on the biomaterial surface of interest and an equilibrium position is achieved. The contact angle is determined from the tangent associated with the drop/polymer surface. For immobile surfaces, contact angles are believed to be sensitive to the outermost 3–10 Å of a surface (Merrett et al. 2002).

Measurements can be carried out by several different methods such as the Wilhelmy plate method, the sessile drop method or the captive bubble method. In biomaterials bibliography, contact angles are typically measured using a goniometer by the sessile drop method using water as test fluid, as well as other liquids with different polarities such as diiodomethane or glycerol. Water contact angle gives information on the hydrophilic/hydrophobic balance of the surface by measuring how much a droplet of water spreads on it. The lower the contact angle, the more hydrophilic the surface is. Hydrophilicity of materials is an important parameter to predict the interaction of biomaterials with the surrounding tissues, and therefore to evaluate their biocompatibility (Hernández et al. 2007; Servoli et al. 2009; Xiong 2010).

In addition, contact angle measurements can be used to study the influence of surface topography on wettability (Öner and McCarthy 2000). For this purpose, dynamic wettability has to be addressed by measuring the hysteresis, i.e., the difference between the increase (advancing contact angle) and decrease (receding contact angle) in the test droplet volume. Contact angle hysteresis has also been reported to occur with surfaces that are chemically heterogeneous or contaminated with surface-active reagents, as well as for surfaces that reorient upon exposure to different fluids (Mathieu et al. 2003).

15.3.2.2 Surface Plasmon Resonance

Surface plasmon resonance (SPR) is a surface-sensitive technique able to monitor any dynamic process at the biomaterial surface, such as adsorption or degradation, in real time and without the need to label the adsorbate or complex sample preparation. It is an optical technique that has been widely used in biomaterials field to investigate biological interactions like protein adsorption process and is capable of real-time in situ analysis to define protein adsorption and desorption rate (Green et al. 1999; Green et al. 2000; Vicente 2010).

SPR is based on the excitation of surface plasmons. A surface plasmon is a longitudinal charge density wave that propagates along the interface of two media, one is a metal and the other is a dielectric. The metal used must exhibit free electron behavior, silver and gold being the most commonly used. Most SPR instruments are designed based on the Kretchman configuration (Figure 15.4), which is based on the phenomenon of total internal reflection. This occurs when light traveling through an optically dense medium, e.g., glass, reaches an interface between this medium and reflects back into the dense medium. Although the incident light is internally reflected, a component of this light, the evanescent wave or field, penetrates the interface into the less dense medium to a distance of one wavelength. In SPR, a monochromatic p-polarized light source is used and the interface between the two optically dense media is coated with a thin metal film of thickness less than one wavelength of light. The evanescent wave of the incoming light is able to couple with the free oscillating electrons (plasmons) in the metal film at a specific angle of incidence, and thus the surface plasmon is

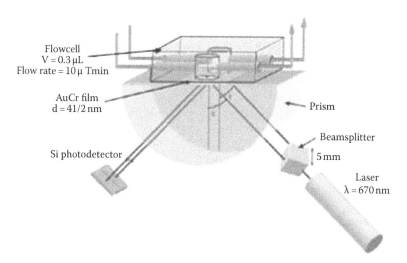

FIGURE 15.4 The Kretchmann configuration for SPR. (From Aguilar, M.R. et al., *Macromol. Biosci.*, 4(7), 631, 2004.)

resonantly excited. This causes a loss of energy from the incident light to the metal film, resulting in a reduction in the intensity of the reflected light which can be detected by a two-dimensional array of photodiodes or charge-coupled detectors (CCD). Therefore, if the refractive index above the metal film changes by the adsorption of a protein layer, a change in the angle of incidence required for exciting a surface plasmon will occur. By monitoring the angle at which resonance occurs during the adsorption process with respect to time, an SPR adsorption profile can be obtained.

SPR has been extensively used to follow interactions between polymers and biomolecules, especially proteins (Li et al. 2005; He et al. 2007; Servoli et al. 2008, 2009). Aguilar et al. (2004) investigated the formation of PECs between chitosan and synthetic polymers derived from 2-acrylamido-2-methylpropane sulfonic acid (AMPS), as well as the absorption of albumin and fibrinogen on these complex surfaces.

15.3.2.3 Zeta Potential

When a charged solid surface is in contact with a liquid phase, an electrical potential develops at the interface. A double layer is established, one of them between surface ionizable groups and tightly bound liquid phase ions of opposite charge forming the fixed layer and the other with loosely bound liquid phase ions of opposite charge forming the mobile layer. The zeta potential is the change in potential across this double layer (Goddard and Hotchkiss 2007). In general, zeta potential is measured indirectly using one of the three means: (1) by measuring electro-osmotic mobility, (2) by measuring streaming current or streaming potential generated by pressure-driven flow through a conduit, or (3) by measuring response of a small spherical particle in an applied E-field (Kirby and Hasselbrink 2004).

Zeta potential can be used to determine surface isoelectric points and quantify a change in surface ionisable groups (Richey et al. 2000). It is also a useful tool to predict the stability of colloidal suspensions or emulsions, and therefore, in biomaterials field, it is widely used for the investigation of physicochemical stability of micro- and nanoparticles for drug delivery (Mosqueira et al. 2000; Heurtault et al. 2003; Hoffart et al. 2003; DeVolder 2010).

15.3.2.4 Quartz Crystal Microbalance

A quartz crystal microbalance (QCM) is a high-resolution mass-sensing technique, which measures mass by measuring the change in frequency of a piezoelectric quartz crystal, when it is disturbed by the addition of tiny masses such as proteins or any other tiny objects intended to be used. This technique possesses a wide detector range. At the low mass end, it can detect a monolayer surface covered by small molecules or polymer films. At the upper end, it is capable of detecting much larger masses bound to the surface, such as complex arrays of biopolymers and biomacromolecules or even whole cells (Marx 2003).

The instrument consists of a thin quartz disk with electrodes placed on it. The application of an external electrical potential to a piezoelectric material produces internal mechanical stress. As the QCM is piezoelectric, an oscillating electric field applied across the device induces an acoustic wave. This wave propagates through the crystal and meets minimum impedance when the thickness of the device is a multiple of a half wavelength of the acoustic wave. The deposition of a thin film on the crystal surface decreases the frequency in proportion to the mass of the film. A resonant oscillation is achieved by including the crystal into an oscillation circuit where the electric and the mechanical oscillations are near to the fundamental frequency of the crystal. Changes in resonant frequency are simply related to the mass accumulated on the crystal. Thus, for detection of analytes in air, the frequency is simply related to the change in mass (O'Sullivan and Guilbault 1999).

QCM is becoming a good alternative analytical method in a great deal of applications such as biosensors, analysis of biomolecular interactions, study of bacterial adhesion at specific interfaces, study of polymer film-biomolecule or cell-substrate interactions, immuno-sensors, and an extensive use in fluids and polymer characterization, and electrochemical applications between others (Hook et al. 2002; Arnau 2008; Olsson et al. 2010).

A special variation of QCM technique is the so-called quartz crystal microbalance with dissipation monitoring (QCM-D). The main advantage of QCM-D compared with conventional QCM is that the former estimates changes in both the mass and the viscoelastic constant for the adsorbed layer through measurements of frequency and dissipation (Dixon 2008; Huang 2010; Martins et al. 2010).

15.3.3 MICROSCOPIC METHODS

15.3.3.1 Atomic Force Microscopy

With atomic force microscopy (AFM), three-dimensional images of a surface are created by scanning a tip attached to a flexible cantilever across the sample surface and monitoring the minute forces of interaction between the sample surface and probe. This technique allows the measurement of surface morphology at atomic scale. The change in surface height is then measured by the location of the reflected laser beam in the quadrant photo-detector, and a topographical map is generated from which surface roughness can be calculated (Alessandrini and Facci 2005). Surface roughness is an important parameter in biomedical materials since it can affect cell adhesion, specially important for blood-contacting materials, in which an increase of surface roughness can compromise hemocompatibility since turbulent flow can promote hemolysis (Aksoy et al. 2008).

The AFM can operate in different modes (Jandt 2001), as shown in Figure 15.5. In the *constant force mode*, also called *contact mode*, the force between the tip and the sample surface is kept

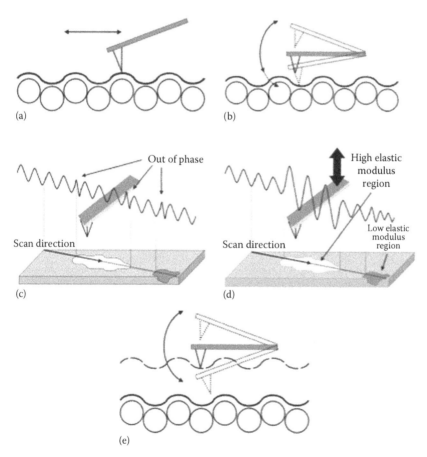

FIGURE 15.5 Schematic representation of the most common operating modes in AFM. (a) Nondynamic mode: contract mode, (b) dynamic mode: tapping mode, (c) phase imaging, (d) force modulation, (e) dynamic mode: noncontact mode. (From Jandt, K.D. et al., *Surf. Sci.*, 491(3), 303, 2001.)

constant via an electronic feedback loop. An image of the sample's surface topography is created from the cantilever deflection required to maintain a constant force over the surface. In this mode, the tip is in permanent physical contact with the sample; therefore it may not always be appropriate for biomaterials applications.

The *tapping mode* uses an oscillating tip (frequencies of 50–500 kHz in air and approximately 10 kHz in fluids) that lightly taps the sample surface with amplitude typically ranging from 20 to 100 nm. A split photodiode detector monitors the deflection of the laser light reflected from the tip of the cantilever. A feedback loop maintains constant the oscillation amplitude or frequency as the scanner moves laterally across the surface. This mode maintains the high-resolution capabilities of contact mode but it is not destructive since there are no lateral frictional forces applied to the sample. For this reason, it has been proved to be very successful for high-resolution studies of polymeric biomaterials allowing the characterization of nanometer-scale features not visible by other microscopic techniques (Ortega-Vinuesa et al. 1998; Bowen et al. 2000; Willemsen et al. 2000).

Phase imaging mode can be used to map the surface composition of a sample. In this mode, the cantilever vibrates at or near its resonance frequency and lightly taps the sample surface. A phase image shows the phase difference between oscillation of the piezoelectric crystal that drives the cantilever and oscillation of the cantilever itself as it interacts with the surface. AFM in phase imaging mode can be performed at the same time as topographic imaging with tapping mode in a single scan. The main advantage of this mode is that the tip–sample interactions depend not only on the sample's topography but also on different sample characteristics, e.g., sample hardness, elasticity, or adhesion.

In *noncontact mode*, attractive forces are measured. The scanning tip is oscillated perpendicular to, and just above the sample surface with amplitude typically less than 10 nm. This mode may work well with hydrophobic polymers but it is not recommended for hydrophilic polymers or samples in a liquid environment due to the lower resolution.

AFM can be also developed using *coated tips*. In this mode, tips are modified with functional groups or even biomolecules to achieve very high sensitivity in the detection of surface properties and molecular recognition. This technique has been applied to study specific interactions of biomaterials with biological systems (Schneider et al. 1999; Flores et al. 2006; Weder et al. 2010). It can also be applied for the production of chemical affinity maps. In this sense, the introduction of carbon nanotube tips, which can be specifically functionalized even with single molecules, has remarkably increased the performances of AFM, giving the potential for even greater affinity map resolution (Alessandrini and Facci 2005).

15.3.3.2 Scanning Electron Microscopy and Transmission Electron Microscopy

Scanning electron microscopy (SEM) and transmission electron microscopy (TEM) make use of a primary beam of electrons to bombard the sample in a vacuum environment, resulting in a secondary electron emission and x-rays (Merrett et al. 2002). The intensity of the secondary electrons is detected to generate a high-resolution three-dimensional surface image. X-rays can be detected to develop elemental analysis.

SEM sample preparation involves fixation (if cells or tissue are present), followed by drying, attachment to a metallic stub and coating with a metal prior to data collection. The metallic coating (20–30 nm) is usually applied by sputter coating, gold, platinum, or gold/platinum alloy being the most common conductive metals used. The drying and metal coating of the sample might alter surface morphology in polymeric materials, particularly those that could undergo changes in hydrated environment. In this sense, the introduction of environmental scanning electron microscopy (ESEM) has represented a new perspective. ESEM does not operate at high vacuum and allows imagery of the material sample in wet mode without further sample preparation (Muscariello et al. 2005). This instrument has a unique secondary electron detector capable of forming high-resolution images at pressures in the range of 0.1–20 Torr. At these relatively high pressures, nonconductive specimens are observed without metallic coating or any special preparation.

In addition to imaging the surface morphology of polymeric biomaterials, SEM can be combined with other analysis methods such as energy dispersive x-ray analysis (EDX) to determine elemental composition and IR and Raman spectroscopy to monitor surface modification procedures. EDX results are typically obtained from a sampling depth on the order of micrometers; thus, they are more representative of the bulk material rather than the surface.

TEM sample preparation involves fixation (if proteins, cells or tissue are present), processing, embedding, and sectioning. Embedding media can include methacrylates, polyester, and acrylic resins. Specimens are typically sectioned using a microtome and need to be very thin since electrons with an accelerating voltage of 100 kV will not penetrate specimens more than 1 μm thick. Good resolution and clarity of detail can normally be obtained with sample thickness in the order of 50–90 nm. It should be noted that embedding and sectioning used in the preparation of some polymeric materials may alter the polymeric material itself or the quality of the image obtained due to factors such as drying, thickness variations, wrinkling, or compression.

SEM and ESEM have been extensively used for studying both surface morphology and cellular response to biomaterials (Rodríguez-Lorenzo et al. 2006, 2009; Kirsebom et al. 2007; López Donaire et al. 2009). Moreover, TEM techniques are more useful for the investigation of the interface between tissue and biomaterials (Porter 2006; Lamers et al. 2010).

15.4 INTERACTIONS WITH BIOLOGICAL ENVIRONMENT AND APPLICATIONS

The biocompatibility study includes the analysis of the interaction between materials and biological systems in a wide range of applications and molecular processes, taking into account both the host response against the material as well as the evolution of the material properties after insertion into a biological environment (Tirrell et al. 2002). Due to the fact that the interaction between the material and the environment is developed through the interface by means of "bio-recognition processes" (Elbert and Hubbell 1996), it will consequently depend heavily on the properties of the material surface. When a material is in contact with a biological environment, its structure, chemical composition, and topography will determine the adsorption of proteins, the interaction with the cells and, ultimately, the response of the host.

The material incorporation into a biological system triggers an immediate and nonspecific adsorption of proteins on its surface, independent of its nature. This phenomenon will determine the subsequent biological reactions (platelet adhesion, activation of the foreign body reaction, etc.) so that the design of biocompatible surfaces requires a deep knowledge of the interaction between proteins and biomaterial surfaces (Ratner and Bryan 2004). Recently, the in vitro and in vivo studies on the interaction between biomaterial surfaces and biological systems have shown a remarkable progress which has led to the development of surfaces resistance to nonspecific protein adsorption as well as surfaces with selectivity toward specific proteins by immobilization of biomolecules or biomimetic ligands (Anderson et al. 2004; Levenberg et al. 2004; Goldberg et al. 2007; Vasita et al. 2008). Currently, the main lines of research in the field of polymeric biomaterials are focused on studying in depth the biological mechanisms involved in the interaction between the polymer surface and proteins that influence cellular response, on developing polymer surfaces for the repulsion of proteins; and developing polymeric bioactive surfaces by selective adsorption of biomolecules.

15.4.1 CELL–POLYMER INTERACTION DETERMINES BIOCOMPATIBILITY

Proteins are natural biopolymers that result from the polymerization of at least 20 L-α-amino acids and have very complex physical and chemical characteristics. Proteins are involved in most cellular functions leading to many forms of cell–substrate specific interactions that arise from the presence of one or more active domains available to react with specific substrates. When a polymer comes into contact with a biological system, cells in this system tend to identify the polymer as a foreign body, a fact that triggers an immune response and inflammation (Anderson 2001; Anderson et al. 2008). However, if the polymer includes any ligand capable of being specifically recognized by the proteins of host cells, a specific and selective response can be obtained.

There are two basic kinds of molecules involved in establishing such interactions, extracellular proteins and cell membrane proteins. Extracellular proteins play an important role in cell adhesion, propagation, and growth regulation and are divided into extracellular matrix (ECM) components and plasma proteins (Schwartz and Ginsberg 2002). The ECM components are directly or indirectly attached to the cells. In the organic component of the ECM, there are the two most important types of extracellular macromolecules: fibrous proteins and proteoglycans. The plasma proteins circulate in blood plasma and are primarily responsible for the regulation of cell proliferation, differentiation, and immune response (Chen et al. 2008). The proteins embedded in cell membranes are classified, according to their functionality, in transporters, receptors, adhesion molecules, and enzymes. They are responsible for substrate transfer and signal transmission across the membrane, connection and recognition among cells, interaction with the extracellular matrix, immune response to foreign objects, and matrix formation for cell embedding (Barclay 2003; Wierzbicka-Patynowski and Schwarzbauer 2003; Ehrhardt et al. 2004; Arnaout et al. 2005; Leckband and Prakasam 2006). These proteins, usually modified with carbohydrates and lipids, directly interact with the matrix substrates or signal like growth factors via its extracellular domain. Growth factors, together with ECM proteins are responsible for the regulation of cell proliferation, differentiation, and adhesion (Yamada and Even-Ram 2002). Adhesion between cells is accomplished by connections through ECM proteins and specific receptors anchored on cell surface. The connections of ECM proteins and these receptors trigger a specific focal adhesion formation and signal transduction. These focal adhesion phenomena entrench the union of the cell with the surrounding microenvironment. The signal transduction to adjacent cells is essential in regulating cell migration, proliferation, and differentiation (Schwartz and Ginsberg 2002).

The main biomolecules involved in cell adhesion are proteins. This fact has motivated the development of various research projects that aim to focus on the study of cell–surface interactions by immobilizing adhesive proteins derived from the ECM onto polymer surfaces, as fibronectin (FN), collagen, laminins, or vitronectin (Linnola et al. 2000; Vasita et al. 2008). The key advantage of such modifications lies in that ECM proteins are natural ligands of integrins so that their presence will not cause direct harmful effects. However, its application can also lead to several problems: the orientation of immobilized ECM proteins is difficult to control, their conformation can be changed, ECM proteins are easily degraded by enzymes so that the long-term applications may be impeded, and the ECM proteins can induce immune responses in the host (Chen et al. 2008).

The mechanisms employed in the FN immobilization on polymer surfaces have been developed by both physical adsorption and covalent conjugation. UV ozone has been used to activate hydrophobic materials derived from poly(dimethylsiloxane), obtaining hydrophilic surfaces with physically adsorbed FN that promote cell adhesion and spreading significantly (Toworfe et al. 2004). Koenig and co-workers analyzed the influence of the FN adsorption in different systems of varied chemical composition, using human umbilical vein endothelial cells (HUVECs) cultures, concluding that the adsorption of FN on the surface induced an increase in cell adhesion and spreading dependent on the polarity of these systems (Koenig et al. 2003). With increasing surface hydrophobicity, they observed a decrease in both cell adhesion and effective growth in these cultures, probably because the surface of such formulations induced the denaturalization of the FN. The latter result has been corroborated in other studies, such as that developed by Qu et al., who obtained a higher growth of HUVECs and rabbit aorta smooth muscle cells (SMCs) on the surface of a copolymer of 3-hydroxybutyrate and 3-hydroxyhexanoate treated with ammonia plasma (hydrophilic variant), compared with the same copolymer untreated (hydrophobic variant), although the amount of adsorbed FN was greater in the latter. This result incises again on the importance of the maintenance of the FN proper conformation to preserve unchanged its bioactivity (Qu et al. 2005).

There are also several examples of successful immobilization of such proteins on the surface of polymers by covalent conjugation, among them the functionalization of the poly(L-lactide-co-caprolactone) surface with both FN and collagen, using the bridge of amino group linkages introduced on the surface of these systems through aminolyzing with 1,6-hexanediamine and glutaraldehyde as coupling agents (Zhu et al. 2006a) or the development of FN-modified Pluronic conjugates, whose surfaces have shown a specific bioactivity to the dorsal root ganglia neurite outgrowth of postnatal rats (Zhang et al. 2005). Covalent conjugates have also been developed with other proteins, such as growth factors. Different studies have produced systems with enhanced bioactivity through the covalent binding of epidermal growth factor (EGF) (Gobin and West 2003; Klenkler et al. 2005), vascular endothelial growth factor (Zisch et al. 2001), transforming growth factor-β (TGF-β) (Mann et al. 2001), or nerve growth factor (Chen et al. 2004b), using PEG spacers or not.

With regard to the result obtained with the two methods described earlier, it is noteworthy that several systems with proteins covalently conjugated showed a performance advantage over systems with physically adsorbed proteins on its surface (Biran et al. 2001; Webb et al. 2001; Klee et al. 2003), although other researchers have suggested that the latter have a more favorable behavior for certain cell lines (Zhang et al. 2007).

With the aim of contributing to the inactivation of cellular recognition processes arising from conformational changes that affect proteins, specific polypeptides were covalently anchored to the surface of different polymers with diverse functional domains involved in molecular adhesion. Fibronectin has several cell membrane integrin-binding domains (RGD sequence, in repeat III10 module, PHSRN, in repeat III9 module, LDV y REDV from spliced V region, IDAPS y KLDAPT, in repeats III14 y III5 modules, y EDGIHEL, in spliced Eda segment) and heparin-binding domains (PRRARV sequence, in III13 module, and WQPPRARI, in III14 module) (Wierzbicka-Patynowski and Schwarzbauer 2003; Chen et al. 2008). Other peptide sequences are involved in cell adhesion processes through ligand–acceptor interaction, as are the sequences KKQRFRHRNRKG, from vitronectin; CDPGYIGSR, YIGSR, IKVAV, RNIAEIIKDI, YFQRYLI and PDSGR, from laminin; DGEA, from collagen I; VAPG, from elastin; and FHRRIKA, from bone sialoprotein (Tirrell et al. 2002; Chen et al. 2008).

Arg–Gly–Asp (RGD) sequence is present in several ECM proteins, a fact which has motivated numerous studies to develop biomimetic surfaces using this sequence. There are other advantageous characteristics in the use of this sequence, such as its low immunogenic activity, the ability to synthesize it with a controlled conformation, its high stability against sterilization methods, high temperatures or pH variations, or its low price compared with the ECM proteins. Some of the disadvantages associated with their use are the risk of enzymatic degradation and the lower binding activity obtained for specific subtypes of integrins compared with that shown by ECM proteins.

The methodologies used to incorporate the RGD sequence are varied, the first one being simple adsorption (McConachie et al. 1999; Dettin et al. 2002). This method does not significantly improve cell adhesion due to structural problems, such as the entrapment of peptides into depolymers and undesirable peptide density. Other methods are spacing (Rezania and Healy 1999), blending (Quirk et al. 2001), copolymerization (Shin et al. 2002), and chemical or physical treatment (Marchand-Brynaert et al. 1999; Carlisle et al. 2000). One of the techniques most commonly used in biomedical applications is the chemical conjugation. By applying this type of technique, RGD or RGD-containing peptides have been immobilized onto polymer surfaces using specific reactions or chemical functional groups. Systems that improve the adhesion and differentiation of osteogenic precursor cells were obtained using RGD peptide conjugated with PLA surface via the reaction of –CHO and –NH$_2$ (Hu et al. 2003), as well as conjugates with RGD-containing peptides that also promote the cellular functions (Ebara et al. 2004; Smith et al. 2005). The RGD sequence can also be immobilized by copolymerization techniques, obtaining copolymers such as poly[N-p-vinyl-benzyl-D-maltonamide-co-6-(p-vinylbenzamido)-hexanoic acid-GRGDS], in which the inclusion of the GRGDS peptide improves growth and specific functions of hepatocytes and phechromocy-toma cells (PC12), albumin secretion, and urea synthesis in hepatocytes (Park 2002), and dopamine secretion by PC12 cells (Park et al. 2004). Another example is the RGD-grafted poly-D-lysine-graft-(polyethylene glycol), adsorbed onto substrates by electrostatic interaction between positive charges in poly-D-lysine and negative charges in the substrate, which inhibit nonspecific protein adsorption in an aqueous medium and allow the adhesion and proliferation of human dermal fibroblasts (VandeVondele et al. 2003).

During the cell adhesion process, four overlapping steps occur: (1) cell attachment, when a specific integrin–ligand binding between the cell and the surface is established, (2) cell spreading, (3) actin–cytoskeleton organization, and (4) focal adhesion formation (Lebaron and Athanasiou 2000). One of the key points to achieve a solid union between cells and polymeric surface is the correct formation of focal adhesion complex. The main participants in this complex formation are integrins along with several transmembrane, membrane, and cytosolic molecules, such as tetraspanins, growth factor receptors, syndecans, lipids, tensin, vinculin, talin, paxillin, and focal adhesion kinase (Pande 2000; Petit and Thiery 2000; Geiger and Bershadsky 2001; Zamir and Geiger 2001). Integrins connect the extracellular components with intracellular proteins and are mainly responsible for signal transduction that regulates important processes such as cell proliferation, differentiation, migration, or apoptosis. Integrins consist of two transmembrane subunits (α and β subunits) associated by noncovalent bonds. The human genome can code for different types of subunits with specificity for certain ligands, resulting in 24 different heterodimers with different specificities (van der Flier and Sonnenberg 2001). Some of these integrins exhibit a high specificity, while others have the ability to bind to a large number of ligands, such as $\alpha v \beta 3$ integrin which binds to vitronectin, fibronectin, von Willebrand factor, osteopontin, tenascin, bone sialoprotein, and thrombospondin. Moreover, certain ECM molecules such as fibronectin are ligands for several types of integrins (Plow et al. 2000; van der Flier and Sonnenberg 2001). The transmembrane signaling by integrin mechanisms and binding mechanisms to different ligands are very complex and involve numerous conformational change (Gottschalk et al. 2002; Gallant et al. 2005). Integrins have a major role in the application of RGD-modified polymers, since at least half of the heterodimers existing in humans recognize this peptide as its ligand (van der Flier and Sonnenberg 2001). There are numerous examples in which the development and application of RGD-modified polymers have improved adhesion, proliferation, and/or differentiation mediated by integrins in different cell lines: osteogenic active cells (Rezania and Healy 1999; Itoh et al. 2002; Barber et al. 2003; Behravesh and Mikos 2003; Hu et al. 2003; Huang et al. 2003; Zreiqat et al. 2003; Shin et al. 2004), fibroblast (Koo et al. 2002; Gümüşderelioğlu and Karakeçili 2003; VandeVondele et al. 2003; Rajagopalan et al. 2004), hepatocytes (Bhadriraju and Hansen 2000; Park 2002), myoblast (Rowley and Mooney 2002; Smith et al. 2005), endothelial cells (Ebara et al. 2004; Sagnella et al. 2005; Chen et al. 2006; Sheardown et al. 2007), marrow stromal cells (Mapili et al. 2005), and neuron-like cells (Park et al. 2004). Moreover, other studies have shown that a more "complete" cellular response (e.g., cell attachment, spreading, focal contact formation and organized cytoskeletal assembly) is obtained through application of systems with two types of peptide sequences, cell-binding (RGD domain) and heparin-binding sequences of FN or bone sialoprotein (Tirrell et al. 2002).

15.4.2 Passivating Polymer Surfaces for Protein Repelling

One of the main objectives in the development of materials with biomedical applications, crucial for those that will be in contact with blood vessels, is the reduction of foreign body immune response by the inhibition of the nonspecific protein adsorption that occurs when the material comes in contact with the biological environment (Hench and Polak 2002; Rodriguez et al. 2009) (Figure 15.6).

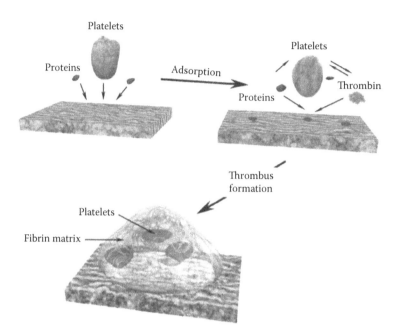

FIGURE 15.6 (See companion CD for color figure.) An example where the nonspecific adsorption of proteins to a surface can lead to serious problems is the thrombus formation. (Taken from Tirrell, M. et al., *Surf. Sci.*, 500, 61, 2002.)

Within the basic variables that regulate the adsorption of proteins, the host response is necessary to emphasize the hidrophilic/hydrophobic balance of the material*** surface (Ma et al. 2007) that can be defined using a specific parameter, pure water adhesion tension ($\tau°$) (Vogler 1998). The surfaces identified as hydrophilic have a value of $\tau° > 30\,\text{dyn/cm}$, while for hydrophobic, it is $\tau° > 30\,\text{dyn/cm}$. Hydrophobic surfaces favor nonspecific adsorption from aqueous solution thermodynamically and can contribute to denaturalize the native conformation of proteins involved in recognition and cell adhesion processes, and hence may affect their bioactivity. Moreover, hydrophilic materials tend to inhibit protein adsorption. Although the results obtained in this field do not always provide matching conclusions, it seems clear that an improvement in the surface hydrophilicity of the hydrophobic materials is necessary to adsorb an appropriate amount of proteins while maintaining its natural conformation and, hence, to obtain a positive cellular response (Vogler 1998; Tziampazis et al. 2000).

Along with the physical techniques developed to modulate the hydrophilicity/hydrophobicity ratio of synthetic polymers, the main research trends are focused on improving the resistance of materials against nonspecific adsorption of proteins by introducing chemical modifications on their surfaces (Shen et al. 2003; Roach et al. 2005, 2007; Ladd et al. 2008; Zhang et al. 2008). The number and type of functional groups resistant to protein adsorption has been determined by combining SAMs and SPR spectroscopy. This knowledge has allowed establishing the basic principles to be considered in the design of inert groups. These groups must be polar, hydrogen-bond acceptors, not hydrogen-bond donors and electrically neutral groups (Chapman et al. 2000; Ostuni et al. 2001), although not all inert surfaces possess all these features. One example is SAMs bearing mannitol groups with a high number of hydrogen-bond donors that are inert to protein absorption (Luk et al. 2000). According to these principles and the great diversity of interactions between proteins and material surfaces, different methodologies have been developed to obtain protein-repelling surfaces (Higuchi et al. 2002; Zhao et al. 2003; Zhou et al. 2005). In order to obtain inert surfaces, several polymeric layers grafted to a substrate have been developed (Gupta et al. 2002; Higuchi et al. 2002; Martins et al. 2003b; Vermette and Meagher 2003; Combellas et al. 2004; Wan et al. 2005; Joshi and Sinha 2006). The interaction between proteins and tethered polymer chains is very complex and depends on the properties of both proteins and polymer surface. Despite the high complexity of this interaction, three generic modes of protein adsorption in such systems have been described: primary adsorption, when the particle size is less than the distance between grafted chains, a fact which causes the diffusion of particles inside the brush, and their subsequent adsorption on the substrate surface; secondary adsorption, when the particles cannot physically penetrate into the brush, a fact which causes its adsorption to the brush-solvent interface; and ternary adsorption, when the particles can diffuse into the brush, where they are retained (Halperin 1999; Currie et al. 2003).

Two of the most commonly used polymers in the preparation of modified surfaces are PEG and PEO, which share the same repeating unit but are derived from different monomers via different polymerization methods (Chen et al. 2008). The modification of surfaces with these polymers is very interesting not only because they are efficient inhibitors of protein adsorption but also because they provide easy ether linkages with proteins. With regard to resistance to protein adsorption on PEG-modified surfaces, it is necessary to emphasize the role of different thermodynamic factors (Hoffman and Ratner 2004), the steric repulsion, and the hydration or water structuring layer formation (Morra 2000; Vermette and Meagher 2003) along with structural factors, as the length of the PEG chain, the density of surface coverage or the chains organization on the material surface (Currie et al. 2003). Ideally, the inhibition of protein adsorption requires a high density and a long chain. In the event in which PEG chains of different lengths are incorporated on the same support, it is possible to obtain surfaces with high efficiency in specific repulsion for certain proteins and not for others, depending on their size (Lazos et al. 2005). It is also possible to control the hard charged proteins adsorption, alternating the charge of PEG-modified surfaces (Pasche et al. 2005). Moreover, it is possible to modulate the bioactivity and conformational changes of adsorbed proteins, which will allow monitoring of cellular interaction and biocompatibility by varying the PEG concentration. For example, it was found that both the amount of adsorbed FN and its specific activity decreased with increasing PEG concentration. The decrease in the activity of this protein results from the globular conformation, more compact, which appears when the PEG concentration on the material surface is high, and hinders the participation of the RGD domain in the process of adhesion (Thom et al. 2000; Tziampazis et al. 2000). There are numerous examples that illustrate the influence of polymer surfaces modified with PEO or PEG, by physical adsorption techniques, by establishing covalent bonds, or by surface grafting of PEO/PEG-containing molecules. Among these examples, there are systems based on polyurethane (Cornelius et al. 2002; Kim et al. 2003; Archambault and Brash 2004a,b) and silicone (Chen et al. 2004a, 2005a,b) polymers with anticoagulant activity, that inhibit platelet adhesion or protein adsorption; PS systems (Byun et al. 2004; Bosker et al. 2005; Lazos et al. 2005), which inhibit specifically the adsorption of certain proteins depending on their size; or poly(methylmetacrylate) (Liu et al. 2004; Bi et al. 2006), polylactide- (Otsuka et al. 2000) and polyaniline- (Li and Ruckenstein 2004) based polymers that also significantly reduce platelet adhesion and adsorption of certain proteins.

Another type of surface modification aimed at obtaining biomimetic systems with nonthrombogenic characteristics is carried out using phosphorylcholine (PC)-based polymers. The development of this approach arises after discovering that the lipids that form part of the outer membrane of erythrocytes possess antithrombogenic character, unlike those found in the inner membrane, resulting from the presence of PC moieties (Zwaal et al. 1977). This fact has motivated the synthesis of different PC-containing monomers, such as MPC (Ishihara et al. 1990), which has a zwitterionic PC head group and whose polymerization has given rise to polymers with different properties, e.g., the ability to control the behavior of adsorbed proteins (Feng et al. 2004, 2005; Iwata et al. 2004). Different mechanisms have been proposed to explain protein adsorption resistance of these systems (Lewis 2000). One of them attributed this property to the formation of an organized bilayer of plasma lipid, which would act as a protective shield for the material surface

(Heiden et al. 1998). Other works consider that the large hydration layer formed by interaction of numerous free water molecules with PC groups on the material surface is responsible for the biomimetic character of these systems by not allowing significant changes in protein conformation (Sheng et al. 1995; Ishihara et al. 1998). It has also been shown that the flexibility and mobility of the PC group plays an important role in the reduction of protein–surface interactions (Parker et al. 2005). Different MPC-modified polymers based on polyurethane were obtained through blending, semi-IPN, coating, chemical or photochemical graft, photoinduced graft, ozone-induced graft, or swelling-deswelling methods, systems (Korematsu et al. 2002; Morimoto et al. 2002, 2004). These systems significantly reduce the adhesion of FN, reduce or suppress platelet adhesion, or inhibit the fibroblasts adhesion. Other systems based on polyethylene (Iwasaki et al. 2002; Ishihara et al. 2004), PSF (Hasegawa et al. 2002; Ishihara et al. 2002), and PDMS (Xu et al. 2003; Goda et al. 2006; Seo et al. 2008) also reduce platelet adhesion and total adsorption of proteins, or modulate specifically the adsorption of certain proteins.

Like the PC-containing polymers, sulfobetaine- and carboxybetaine-based polymers have cationic and anionic groups on the same monomer residue; this, coupled with the fact that both methacrylic derivatives of sulfobenaine (SBMA) and carboxybetaine (CBMA) are more stable and easier to obtain than MPC makes them suitable candidates for the development of biomedical systems with reduced protein adsorption (Wang et al. 2003; Zhang et al. 2006a). The most important factors affecting the effectiveness of sulfobetaine and carboxybetaine surfaces as inhibitors of protein adsorption are the modification method, the elaboration of surface packing, and the surface density. Sulfobetaine polymers have been incorporated on polyurethane, poly(ether urethane), and segmented poly(ether urethane) by graft polymerization and ozone-induced polymerization (Yuan et al. 2003a, 2004a; Yuan et al. 2003b, 2004b), silicone (Yuan 2004c), and cellulose (Zhang et al. 2003), giving systems with a high biocompatibility with surfaces that inhibit adsorption protein and/or platelet adhesion. With regard to the carboxybetaine methacrylic derivative, poly(carboxybetaine methacrylate) (PCBMA), it is noteworthy that this polymer not only exhibits a high efficacy to inhibit protein adsorption but also has several functional groups available for immobilizing biomolecules, such as fibrinogen, lysozyme, or human chorionic gonadotropin (hCG) (Zhang et al. 2006b). This dual functionality makes such systems of great interest for various biomedical applications (diagnostics, biomedical/tissue engineering, drug delivery).

15.4.3 BIOACTIVITY OF POLYMER BIOCONJUGATE SURFACES

To achieve biomaterial–biological environment interactions, one of the most innovative strategies is based on incorporating bioactive molecules on the biomaterial surfaces, to obtain hybrid macromolecules with unprecedented properties (Lutz and Zarafshani 2008; Lutz and Börner 2008; Börner 2009). The possibility of incorporating a "biological functionality" to biomaterials by combining synthetic materials and biological structures at different hierarchical levels of organization (monomeric, oligomeric, polymeric, supramolecular or highly organized structures) is a powerful and flexible tool that multiplies the applications of the new systems developed, through the regulation of protein–protein, protein–nucleic acid and protein–carbohydrates interactions. Although it is possible to modify biologically the surface of most biomaterials, synthetic and natural polymers are especially useful for this purpose because their physical or chemical combination with different biological moieties to obtain polymer bioconjugates is relatively simple (Goddard and Hotchkiss 2007; Hermanson 2008). The polymer surface may contain numerous reactive groups or can be easily changed by adding reactive groups that can be used to link biomolecules covalently. In addition, the possibility of varying composition and molecular structure through the synthesis of homopolymers and random, alternating, block and grafting copolymers and the versatility of polymers to be processed in many forms (films, membranes, tubes, hollow fibers, fabrics, particles, capsules, porous systems, etc.), further multiplies the possibilities. Biological molecules used in the preparation of these bio-hybrids and bio-inspired systems can be found in living organisms (lipids, nucleic acids, peptides or carbohydrates), but they can also be produced synthetically (de Graaf et al. 2009). There are different techniques to obtain man-made copies of natural molecules such as oligonucleotides or peptides with automated synthesizers. Although the combination of polymers with peptides and proteins has been widely practiced in the field of pharmaceutical chemistry during the past decades (Langer and Tirrell 2004), the recent and explosive development of biotechnology and nanotechnology has enlarged the applications of polymer bioconjugates in materials science, e.g., as biosensors, artificial enzymes, biometrics, light-harvesting systems or photonics, and nano-electronic devices (Klok 2005; Whitesides 2005).

Most of the polymer bioconjugates are obtained by applying covalent chemistry techniques, although it is notable that there are different alternatives to obtain bioconjugates by applying non-covalent methods, among which are electrostatic interactions and ligand–receptor pairing (as in biotin–avidin) (Niemeyer 2001). The biotin–(strept) avidin system is based on the recognition of the low-molecular weight biotin by larger glycoproteins such as avidin or streptavidin. The fact that the noncovalent bond established in these systems is the strongest reported (up to 250 pN) justifies the application of this system in the biotechnology field (DNA and proteins microarrays, immunological assays, or bio-separation devices), and the increased effort applied to the development of new biofunctionalized systems by anchoring of avidin-modified biological entities (Goddard and Hotchkiss 2007). There are other possibilities to prepare polymer bioconjugates such as cofactor reconstitution routes for modifying enzymes and transport proteins (Boerakker et al. 2002; Reynhout et al. 2007) and the supramolecular recognition of sugar moieties and corresponding lectins (You et al. 2002). On the other hand, less specific noncovalent interactions can be used for connecting synthetic polymers and biological objects. The ionic complexation of oppositely charged entities is a powerful alternative to conjugate biological structures that exhibit charged surfaces. For example, the charged surfaces of proteins, enzymes, or carbohydrates may be modified with appropriate polyelectrolytes, in order to implement them for the development of bioactive polymer bioconjugates by complexation (Dufresne and Leroux 2004; Jaturanpinyo et al. 2004). Moreover, the negatively charged backbone of DNA can be complexed with various synthetic polycations or block-copolymers containing polycationic segments, resulting compact biohybrid particles known as polyplexes, which have been extensively studied in the field of nonviral gene delivery (Lutz and Börner 2008).

Considering the nature of the biomolecules used in the preparation of polymer conjugates, three major types of polymer conjugation can be distinguished: (1) peptide bioconjugation, (2) bioconjugation with nucleobases, oligonucleotides, and nucleic acids, and (3) bioconjugation with carbohydrates. Moreover, the possibility of modifying highly organized biosystems also deserves attention.

Depending on the complexity of the peptide sequence incorporated into the polymer chain, peptide bioconjugates can be classified into three categories: (1) amino acid building blocks, (2) oligo and polymeric peptides, and (3) high-molecular weight proteins (Lutz and Börner 2008). The incorporation of single amino acids in synthetic polymers is the lowest level of complexity. This modification can modulate certain properties of polymer conjugates, such as the hydrophobic/hydrophilic balance (Geng et al. 2006). The increase in the length of the sequence improves the chemical diversity and the number and complexity of biological functions in which these systems may be involved. Thus, the control of more complex properties, such as selective catalytic activity, highly specific recognition or directional functions, and the combination of multiple functions can only be achieved through the preparation of polymer systems with native or high-molecular weight proteins (Klok 2005; Maskarinec and Tirrell 2005). Although the development of this latter type of polymers is significantly more expensive, and their interaction with the biological environment is less predictable, remarkable progress was attained in the field of protein engineering and biomaterials design in recent years (Langer and Tirrell 2004; Maskarinec and Tirrell 2005; Smeenk et al. 2005).

Another group of biomolecules used in polymer bioconjugates are the nucleobases, oligo-nucleotides, and nucleic acids, due to both structural and functional reasons. These biomolecules provide exceptional properties for organizing supramolecular synthetic nanomaterials and comprise a valuable tool for the development of biotechnology devices among which are biosensors, diagnostic kits, or bioseparation devices (Sessler and Jayawickramarajah 2005; Sivakova and Rowan 2005; He and He 2009). Polysaccharides and carbohydrates are biomolecules involved in many biological functions such as storage and transport of energy (starch, glycogen), generation of structure and support (cellulose, chitin, chitosan), control of osmotic pressure in connective tissue (proteoglycans), control of lubricity and cushion in the extracellular matrix (glycosaminoglycans), and control of blood coagulation (heparin) (Yoshida 2001). The cell membrane has carbohydrates, glycoproteins, and glycolipids, involved in biological processes as cell growth phenomena, inflammatory processes, or different specific molecular recognition processes that modulate cell adhesion and recognition mechanisms for toxins and viruses (Dwek 1996; Yoshida et al. 1999). Given the great diversity of biological functions in which carbohydrates are involved, and the plasticity of its chemical structure, the development of carbohydrate-containing synthetic polymers has grown rapidly in order to implement the biofunctionality of these natural macromolecules in biomedical, pharmaceutical, and cosmetic applications (Ladmiral et al. 2004). Glycoconjugates combining a synthetic polymer backbone and pendent mono- or oligo-saccharide segment in the side chain of each monomer repeat unit have been developed with the aim of incorporating the lubricating, cushioning, or anti-inflammatory properties of native carbohydrates (Yoshida 2001; Ladmiral et al. 2004). Maltoheptose polymeric systems containing units such as side-chains based on poly(methyl methacrylate), which posses anti-HIV activity and prevent blood coagulation (Yoshida et al. 1999), and polymers containing statistical distributions of D-glucaric residues, which inhibit the activity of β-D-glucuronidase (Hashimoto et al. 2006), are also described. Likewise, systems that mimic the native properties of proteoglycans have been developed by grafting sugar-containing methacrylates from each repeat unit of a poly(methyl acrylate) backbone (Muthukrishnan et al. 2005). Among the polymer–carbohydrate conjugates, the systems with heparin need to be highlighted. Heparin is a sulfated polysaccharide that participates in various physiological and pathophysiological processes resulting from its interaction with heparin-binding proteins. The result of the immobilization of heparin on the material surface is characterized by an improvement in the in vitro and in vivo biocompatibility (Tanzi 2005) derived from the catalytic effects of heparin on antithrombin III and the accompanying forces of the thrombin complexes (Sánchez et al. 1997), the heparin-induced reduction in the selective adsorption of other plasma proteins, and the contribution to the maintenance of the native structural state of these proteins (Wendel et al. 1999; Wendel and Ziemer 1999).

Finally, it is noteworthy to mention that those biological entities with an extremely high morphological and functional complexity such as viruses, bacteria, and cells are difficult to be explored. Their complexity hinders the design and characterization of bioconjugated systems, since it is difficult to predict how they will interact with biological systems. However, several research lines are aimed at the functionalization of polymers using these types of highly organized biological structures (Raja et al. 2003; Croyle et al. 2004; Prescher et al. 2004; Scott and Chen 2004; Lee et al. 2005; Lee et al. 2006) by applying in most cases PEO or PEG coatings. The main property that this modification provides to the system is the reduction of the immunogenic response, because the polymeric coating inhibits the adsorption of proteins that triggers the immune response cascade, when viruses, bacteria, or exogenous cells penetrate into the host organism. Among the works that illustrate this application, modified retroviruses or adenoviruses with a modified immunogenic activity (Raja et al. 2003; Croyle et al. 2004; Lee et al. 2005) are used as delivery vectors in gene therapy (Verma and Somia 1997) after replacing their genetic material by therapeutic genes. Living cells have also been used with this type of coating, in order to reduce the immune response in blood transfusion or tissue transplantation (Scott and Chen 2004; Lee et al. 2006).

REFERENCES

Aguilar, M. R., A. Gallardo et al. (2004). Modulation of proteins adsorption onto the surface of chitosan complexed with anionic copolymers. Real time analysis by surface plasmon resonance. *Macromolecular Bioscience* 4(7): 631–638.

Aksoy, A. E., V. Hasirci et al. (2008). Surface modification of polyurethanes with covalent immobilization of heparin. *Macromolecular Symposia* 269(1): 145–153.

Alessandrini, A. and P. Facci (2005). AFM: A versatile tool in biophysics. *Measurement Science and Technology* 16(6): R65–R92.

Alferiev, I. S., J. M. Connolly et al. (2006). Surface heparinization of polyurethane via bromoalkylation of hard segment nitrogens. *Biomacromolecules* 7(1): 317–322.

Anderson, J. M. (2001). Biological responses to materials. *Annual Review of Materials Science* 31: 81–110.

Anderson, D. G., J. A. Burdick et al. (2004). MATERIALS SCIENCE: Smart Biomaterials. *Science* 305(5692): 1923–1924.

Anderson, J. M., A. Rodriguez et al. (2008). Foreign body reaction to biomaterials. *Seminars in Immunology* 20(2): 86–100.

Andersson, A. S., K. Glasmastar et al. (2003). Cell adhesion on supported lipid bilayers. *Journal of Biomedical Materials Research Part A* 64A(4): 622–629.

Andrady, A. L., H. S. Hamid et al. (2003). Effects of climate change and UV-B on materials. *Photochemical and Photobiological Sciences* 2(1): 68–72.

Archambault, J. G. and J. L. Brash (2004a). Protein repellent polyurethane-urea surfaces by chemical grafting of hydroxyl-terminated poly(ethylene oxide): Effects of protein size and charge. *Colloids and Surfaces B: Biointerfaces* 33(2): 111–120.

Archambault, J. G. and J. L. Brash (2004b). Protein resistant polyurethane surfaces by chemical grafting of PEO: Amino-terminated PEO as grafting reagent. *Colloids and Surfaces B: Biointerfaces* 39(1–2): 9–16.

Arnaout, M. A., B. Mahalingam et al. (2005). Integrin structure, allostery, and bidirectional signaling. *Annual Review of Cell and Developmental Biology* 21(1): 381–410.

Arnau, A. (2008). A review of interface electronic systems for AT-cut quartz crystal microbalance applications in liquids. *Sensors* 8(1): 370–411.

Barber, T. A., S. L. Golledge et al. (2003). Peptide-modified p(AAm-co-EG/AAc) IPNs grafted to bulk titanium modulate osteoblast behavior in vitro. *Journal of Biomedical Materials Research Part A* 64A(1): 38–47.

Barclay, A. N. (2003). Membrane proteins with immunoglobulin-like domains—A master superfamily of interaction molecules. *Seminars in Immunology* 15(4): 215–223.

Behravesh, E. and A. G. Mikos (2003). Three-dimensional culture of differentiating marrow stromal osteoblasts in biomimetic poly(propylene fumarate-*co*-ethylene glycol)-based macroporous hydrogels. *Journal of Biomedical Materials Research Part A* 66A(3): 698–706.

Belu, A. M., D. J. Graham et al. (2003). Time-of-flight secondary ion mass spectrometry: Techniques and applications for the characterization of biomaterial surfaces. *Biomaterials* 24(21): 3635–3653.

Bhadriraju, K. and L. K. Hansen (2000). Hepatocyte adhesion, growth and differentiated function on RGD-containing proteins. *Biomaterials* 21(3): 267–272.

Bhargava, R. and I. W. Levin (2001). Fourier transform infrared imaging: Theory and practice. *Analytical Chemistry* 73(21): 5157–5167.

Bi, H., S. Meng et al. (2006). Deposition of PEG onto PMMA microchannel surface to minimize nonspecific adsorption. *Lab on a Chip* 6(6): 769–775.

Biran, R., K. Webb et al. (2001). Surfactant-immobilized fibronectin enhances bioactivity and regulates sensory neurite outgrowth. *Journal of Biomedical Materials Research* 55(1): 1–12.

Boerakker, M. J., J. M. Hannink et al. (2002). Giant amphiphiles by cofactor reconstitution13. *Angewandte Chemie International Edition* 41(22): 4239–4241.

Börner, H. G. (2009). Strategies exploiting functions and self-assembly properties of bioconjugates for polymer and materials sciences. *Progress in Polymer Science* 34(9): 811–851.

Bosker, W. T. E., P. A. Iakovlev et al. (2005). BSA adsorption on bimodal PEO brushes. *Journal of Colloid and Interface Science* 286(2): 496–503.

Boulmedais, F., B. Frisch et al. (2004). Polyelectrolyte multilayer films with pegylated polypeptides as a new type of anti-microbial protection for biomaterials. *Biomaterials* 25(11): 2003–2011.

Bowen, W. R., R. W. Lovitt et al. (2000). Application of atomic force microscopy to the study of micromechanical properties of biological materials. *Biotechnology Letters* 22(11): 893–903.

Brown, L., T. Koerner et al. (2006). Fabrication and characterization of poly(methylmethacrylate) microfluidic devices bonded using surface modifications and solvents. *Lab on a Chip—Miniaturisation for Chemistry and Biology* 6(1): 66–73.

Byun, J.-W., J.-U. Kim et al. (2004). Surface-grafted polystyrene beads with comb-like poly(ethylene glycol) chains: Preparation and biological application. *Macromolecular Bioscience* 4(5): 512–519.

Carlisle, E. S., M. R. Mariappan et al. (2000). Enhancing hepatocyte adhesion by pulsed plasma deposition and polyethylene glycol coupling. *Tissue Engineering* 6(1): 45–52.

Castner, D. G. and B. D. Ratner (2002). Biomedical surface science: Foundations to frontiers. *Surface Science* 500(1–3): 28–60.

Cavalli, S., A. R. Tipton et al. (2006). The chemical modification of liposome surfaces via a copper-mediated [3 + 2] azide-alkyne cycloaddition monitored by a colorimetric assay. *Chemical Communications* (30): 3193–3195.

Cen, L., K. G. Neoh et al. (2004). Assessment of in vitro bioactivity of hyaluronic acid and sulfated hyaluronic acid functionalized electroactive polymer. *Biomacromolecules* 5(6): 2238–2246.

Chan, K. L. and S. G. Kazarian (2004a). FTIR spectroscopic imaging of dissolution of a solid dispersion of nifedipine in poly(ethylene glycol). *Molecular Pharmaceutics* 1(4): 331–335.

Chan, K. L. A. and S. G. Kazarian (2004b). Visualisation of the heterogeneous water sorption in a pharmaceutical formulation under controlled humidity via FT-IR imaging. *Vibrational Spectroscopy* 35(1–2): 45–49.

Chapman, R. G., E. Ostuni et al. (2000). Surveying for surfaces that resist the adsorption of proteins. *Journal of the American Chemical Society* 122(34): 8303–8304.

Chen, H., M. A. Brook et al. (2004a). Silicone elastomers for reduced protein adsorption. *Biomaterials* 25(12): 2273–2282.

Chen, H., M. A. Brook et al. (2005a). Surface properties of PEO-silicone composites: Reducing protein adsorption. *Journal of Biomaterials Science, Polymer Edition* 16: 531–548.

Chen, H., M. A. Brook et al. (2006). Generic bioaffinity silicone surfaces. *Bioconjugate Chemistry* 17(1): 21–28.

Chen, P. R., M. H. Chen et al. (2004b). Biocompatibility of NGF-grafted GTG membranes for peripheral nerve repair using cultured Schwann cells. *Biomaterials* 25: 5667–5673.

Chen, Q., R. Förch et al. (2004c). Characterization of pulsed plasma polymerization allylamine as an adhesion layer for DNA adsorption/hybridization. *Chemistry of Materials* 16(4): 614–620.

Chen, H., L. Yuan et al. (2008). Biocompatible polymer materials: Role of protein-surface interactions. *Progress in Polymer Science* 33(11): 1059–1087.

Chen, H., Z. Zhang et al. (2005b). Protein repellant silicone surfaces by covalent immobilization of poly(ethylene oxide). *Biomaterials* 26(15): 2391–2399.

Cheng, T. S., H. T. Lin et al. (2004). Surface fluorination of polyethylene terephthalate films with RF plasma. *Materials Letters* 58(5): 650–653.

Cheng, Z. and S. H. Teoh (2004). Surface modification of ultra thin poly(ε-caprolactone) films using acrylic acid and collagen. *Biomaterials* 25(11): 1991–2001.

Chilkoti, A., B. D. Ratner et al. (1991). Plasma-deposited polymeric films prepared from carbonyl-containing volatile precursors: XPS chemical derivatization and static SIMS surface characterization. *Chemistry of Materials* 3(1): 51–61.

Choi, H. S., Y. Kim Young-Sun et al. (2004). Plasma-induced graft co-polymerization of acrylic acid onto the polyurethane surface. *Surface and Coatings Technology* 182(1): 55–64.

Cioffi, M. O. H., H. J. C. Voorwald et al. (2003). Surface energy increase of oxygen-plasma-treated PET. *Materials Characterization* 50(2–3): 209–215.

Clare, T. L., B. H. Clare et al. (2005). Functional monolayers for improved resistance to protein adsorption: Oligo(ethylene glycol)-modified silicon and diamond surfaces. *Langmuir* 21(14): 6344–6355.

Combellas, C., A. Fuchs et al. (2004). Surface modification of halogenated polymers. 6. Graft copolymerization of poly(tetrafluoroethylene) surfaces by polyacrylic acid. *Polymer* 45(14): 4669–4675.

Cornelius, R. M., J. G. Archambault et al. (2002). Adsorption of proteins from infant and adult plasma to biomaterial surfaces. *Journal of Biomedical Materials Research* 60(4): 622–632.

Croll, T. I., A. J. O'Connor et al. (2004). Controllable surface modification of poly(lactic-co-glycolic acid) (PLGA) by hydrolysis or aminolysis I: Physical, chemical, and theoretical aspects. *Biomacromolecules* 5(2): 463–473.

Croyle, M. A., S. M. Callahan et al. (2004). PEGylation of a vesicular stomatitis virus G pseudotyped lentivirus vector prevents inactivation in serum. *Journal of Virology* 78(2): 912–921.

Cui, F. Z., Y. P. Jiao et al. (2008). Functionalization of polymer surface for nerve repair. *Journal of Photopolymer Science and Technology* 21(2): 231–244.

Cui, F. Z. and Z. S. Luo (1999). Biomaterials modification by ion-beam processing. *Surface and Coating Technology* 112(1–3): 278–285.

Currie, E. P. K., W. Norde et al. (2003). Tethered polymer chains: Surface chemistry and their impact on colloidal and surface properties. *Advances in Colloid and Interface Science* 100–102: 205–265.

Dalsin, J. L., B. H. Hu et al. (2003). Mussel adhesive protein mimetic polymers for the preparation of nonfouling surfaces. *Journal of the American Chemical Society* 125(14): 4253–4258.

Dalton, B. A., C. D. McFarland et al. (1998). Polymer surface chemistry and bone cell migration. *Journal of Biomaterials Science, Polymer Edition* 9(8): 781–799.

Daria, V. R., P. J. Rodrigo et al. (2002). Dynamic formation of optically trapped microstructure arrays for biosensor applications. Presented at *Conference on Biomedical Applications of Micro-and-Nano Engineering*, Melbourne, Victoria, Australia, Published in the proceedings of International Society for Optical Engineering, Bellingham, WA, pp. 41–48.

Davis, F. F. (2002). The origin of pegnology. *Advanced Drug Delivery Reviews* 54(4): 457–458.

De, S., R. Sharma et al. (2004). Enhancement of blood compatibility of implants by helium plasma treatment. 39th IEEE/IAS Annual Meeting (IEEE Industry Applications Society), Seattle, Washington, October, 2004.

De, S., R. Sharma et al. (2005). Plasma treatment of polyurethane coating for improving endothelial cell growth and adhesion. *Journal of Biomaterials Science, Polymer Edition* 16(8): 973–989.

De Silva, M. N., R. Desai et al. (2004). Micro-patterning of animal cells on PDMS substrates in the presence of serum without use of adhesion inhibitors. *Biomedical Microdevices* 6(3): 219–222.

Deiters, A., T. A. Cropp et al. (2003). Adding amino acids with novel reactivity to the genetic code of Saccharomyces cerevisiae. *Journal of the American Chemical Society* 125(39): 11782–11783.

Desai, S., D. Bodas et al. (2003). Tailor-made functional surfaces: Potential elastomeric biomaterials I. *Journal of Biomaterials Science, Polymer Edition* 14(12): 1323–1338.

Dettin, M., M. T. Conconi et al. (2002). Novel osteoblast-adhesive peptides for dental/orthopedic biomaterials. *Journal of Biomedical Materials Research* 60(3): 466–471.

DeVolder R. J., K. H.-J. (2010). Three dimensionally flocculated proangiogenic microgels for neovascularization. *Biomaterials* 31(25): 6494–6501.

Ding, Z., J. Chen et al. (2004). Immobilization of chitosan onto poly-L-lactic acid film surface by plasma graft polymerization to control the morphology of fibroblast and liver cells. *Biomaterials* 25(6): 1059–1067.

Dirks, A. J., S. S. Van Berkel et al. (2005). Preparation of biohybrid amphiphiles via the copper catalysed Huisgen [3 + 2] dipolar cycloaddition reaction. *Chemical Communications* (33): 4172–4174.

Dixon, M. C. (2008). Quartz crystal microbalance with dissipation monitoring: Enabling real-time characterization of biological materials and their interactions. *Journal of Biomolecular Techniques: JBT* 19(3): 151–158.

Dubas, S. T., P. Kittitheeranun et al. (2009). Coating of polyelectrolyte multilayer thin films on nanofibrous scaffolds to improve cell adhesion. *Journal of Applied Polymer Science* 114(3): 1574–1579.

Dufresne, M.-H. and J.-C. Leroux (2004). Study of the micellization behavior of different order amino block copolymers with heparin. *Pharmaceutical Research* 21(1): 160–169.

Dupas-Bruzek, C., O. Robbe et al. (2009). Transformation of medical grade silicone rubber under Nd:YAG and excimer laser irradiation: First step towards a new miniaturized nerve electrode fabrication process. *Applied Surface Science* 255(21): 8715–8721.

Dwek, R. A. (1996). Glycobiology: Toward understanding the function of sugars. *Chemical Reviews* 96(2): 683–720.

Ebara, M., M. Yamato et al. (2004). Immobilization of cell-adhesive peptides to temperature-responsive surfaces facilitates both serum-free cell adhesion and noninvasive cell harvest. *Tissue Engineering* 10(7–8): 1125–1135.

Efimenko, K., J. A. Crowe et al. (2005). Rapid formation of soft hydrophilic silicone elastomer surfaces. *Polymer* 46(22): 9329–9341.

Ehrhardt, C., C. Kneuer, et al. (2004). Selectins—an emerging target for drug delivery. *Advanced Drug Delivery Reviews* 56(4): 527–549.

Elbert, D. L. and J. A. Hubbell (1996). Surface treatments of polymers for biocompatibility. *Annual Review of Materials Science* 26(1): 365–394.

Etienne, O., C. Picart et al. (2006). Polyelectrolyte multilayer film coating and stability at the surfaces of oral prosthesis base polymers: An in vitro and in vivo study. *Journal of Dental Research* 85(1): 44–48.

Falconnet, D., G. Csucs et al. (2006). Surface engineering approaches to micropattern surfaces for cell-based assays. *Biomaterials* 27(16): 3044–3063.

Feng, W., J. Brash et al. (2004). Atom-transfer radical grafting polymerization of 2-methacryloyloxyethyl phosphorylcholine from silicon wafer surfaces. *Journal of Polymer Science Part A: Polymer Chemistry* 42(12): 2931–2942.

Feng, W., S. Zhu et al. (2005). Adsorption of fibrinogen and lysozyme on silicon grafted with poly(2-methacryloyloxyethyl phosphorylcholine) via surface-initiated atom transfer radical polymerization. *Langmuir* 21(13): 5980–5987.

Fernandez-Megia, E., J. Correa et al. (2006). A click approach to unprotected glycodendrimers. *Macromolecules* 39(6): 2113–2120.

van der Flier, A. and A. Sonnenberg (2001). Function and interactions of integrins. *Cell and Tissue Research* 305(3): 285–298.

Flores, S. M., A. Shaporenko et al. (2006). Control of surface properties of self-assembled monolayers by tuning the degree of molecular asymmetry. *Surface Science* 600(14): 2847–2856.

Furuzono, T., K. Sonoda et al. (2001). A hydroxyapatite coating covalently linked onto a silicone implant material. *Journal of Biomedical Materials Research* 56(1): 9–16.

Gallant, N. D., K. E. Michael et al. (2005). Cell adhesion strengthening: Contributions of adhesive area, integrin binding, and focal adhesion assembly. *Molecular Biology of the Cell* 16(9): 4329–4340.

Geiger, B. and A. Bershadsky (2001). Assembly and mechanosensory function of focal contacts. *Current Opinion in Cell Biology* 13(5): 584–592.

Geng, Y., D. E. Discher et al. (2006). Grafting short peptides onto polybutadiene-*block*-poly(ethylene oxide): A platform for self-assembling hybrid amphiphiles13. *Angewandte Chemie International Edition* 45(45): 7578–7581.

George, A. and W. G. Pitt (2002). Comparison of corneal epithelial cellular growth on synthetic cornea materials. *Biomaterials* 23(5): 1369–1373.

Gobin, A. S. and J. L. West (2003). Effects of epidermal growth factor on fibroblast migration through biomimetic hydrogels. *Biotechnology Progress* 19(6): 1781–1785.

Goda, T., T. Konno et al. (2006). Biomimetic phosphorylcholine polymer grafting from polydimethylsiloxane surface using photo-induced polymerization. *Biomaterials* 27(30): 5151–5160.

Goddard, J. M. and J. H. Hotchkiss (2007). Polymer surface modification for the attachment of bioactive compounds. *Progress in Polymer Science* 32(7): 698–725.

Goddard, J. M. and J. H. Hotchkiss (2008). Tailored functionalization of low-density polyethylene surfaces. *Journal of Applied Polymer Science* 108(5): 2940–2949.

Goldberg, M., R. Langer et al. (2007). Nanostructured materials for applications in drug delivery and tissue engineering. *Journal of Biomaterials Science, Polymer Edition* 18(3): 241–268.

Goncalves, I. C., M. C. L. Martins et al. (2009). Protein adsorption and clotting time of pHEMA hydrogels modified with C18 ligands to adsorb albumin selectively and reversibly. *Biomaterials* 30(29): 5541–5551.

González-Benito, J. and J. L. Koenig (2006). Nature of PMMA dissolution process by mixtures of acetonitrile/alcohol (poor solvent/nonsolvent) monitored by FTIR-imaging. *Polymer* 47(9): 3065–3072.

Gottschalk, K. E., P. D. Adams et al. (2002). Transmembrane signal transduction of the $\alpha_{IIb}\beta_3$ integrin. *Protein Science* 11(7): 1800–1812.

de Graaf, A. J., M. Kooijman et al. (2009). Nonnatural amino acids for site-specific protein conjugation. *Bioconjugate Chemistry* 20(7): 1281–1295.

Green, R. J., M. C. Davies et al. (1999). Competitive protein adsorption as observed by surface plasmon resonance. *Biomaterials* 20(4): 385–391.

Green, R. J., R. A. Frazier et al. (2000). Surface plasmon resonance analysis of dynamic biological interactions with biomaterials. *Biomaterials* 21(18): 1823–1835.

Griesser, H. J., R. C. Chatelier et al. (1994). Growth of human cells on plasma polymers: Putative role of amine and amide groups. *Journal of Biomaterials Science. Polymer Edition* 5(6): 531–554.

Griesser, H. J., P. G. Hartley et al. (2002). Interfacial properties and protein resistance of nano-scale polysaccharide coatings. *Smart Materials and Structures* 11(5): 652–661.

Griffiths, P. R., J. A. de Haseth. (2006). *Fourier Transform Infrared Spectrometry*, pp. 303–320. Hoboken, NJ: J. W. sons.

Gristina, A. G. (1987). Biomaterial-centered infection: Microbial adhesion versus tissue integration. *Science* 237(4822): 1588–1595.

Gümüşderelioğlu, M. and A. G. Karakeçili (2003). Uses of thermoresponsive and RGD/insulin-modified poly(vinyl ether)-based hydrogels in cell cultures. *Journal of Biomaterials Science, Polymer Edition* 14: 199–211.

Gupper, A., K. L. A. Chan et al. (2004). FT-IR imaging of solvent-induced crystallization in polymers. *Macromolecules* 37(17): 6498–6503.

Gupper, A. and S. G. Kazarian (2005). Study of solvent diffusion and solvent-induced crystallization in syndiotactic polystyrene using FT-IR spectroscopy and imaging. *Macromolecules* 38(6): 2327–2332.

Gupta, B., C. Plummer et al. (2002). Plasma-induced graft polymerization of acrylic acid onto poly(ethylene terephthalate) films: Characterization and human smooth muscle cell growth on grafted films. *Biomaterials* 23(3): 863–871.

Ha, S. W., R. Hauert et al. (1997). Surface analysis of chemically-etched and plasma-treated polyetheretherketone (PEEK) for biomedical applications. *Surface and Coatings Technology* 96(2–3): 293–299.

Hahn, S. K. and A. S. Hoffman (2005). Preparation and characterization of biocompatible polyelectrolyte complex multilayer of hyaluronic acid and poly-L-lysine. *International Journal of Biological Macromolecules* 37(5): 227–231.

Halperin, A. (1999). Polymer brushes that resist adsorption of model proteins: Design parameters. *Langmuir* 15(7): 2525–2533.

Hamano, T., D. Chiba et al. (2002). Evaluation of a polyelectrolyte complex (PEC) composed of chitin derivatives and chitosan, which promotes the rat calvarial osteoblast differentiation. *Polymers for Advanced Technologies* 13(1): 46–53.

Hasegawa, T., Y. Iwasaki et al. (2002). Preparation of blood-compatible hollow fibers from a polymer alloy composed of polysulfone and 2-methacryloyloxyethyl phosphorylcholine polymer. *Journal of Biomedical Materials Research* 63(3): 333–341.

Hashimoto, K., H. Saito et al. (2006). Glycopolymeric inhibitors of beta-glucuronidase. III. Configurational effects of hydroxy groups in pendant glyco-units in polymers upon inhibition of beta-glucuronidase. *Journal of Polymer Science Part A: Polymer Chemistry* 44(16): 4895–4903.

Hassane, F. S., B. Frisch et al. (2006). Targeted liposomes: Convenient coupling of ligands to preformed vesicles using "click chemistry". *Bioconjugate Chemistry* 17(3): 849–854.

He, P. and L. He (2009). Synthesis of surface-anchored DNA-polymer bioconjugates using reversible addition-fragmentation chain transfer polymerization. *Biomacromolecules* 10(7): 1804–1809.

He, J., X. Lü et al. (2007). Study of protein adsorption on biomaterial surfaces by SPR. *Key Engineering Materials* 342–343: 825–828.

Heiden, A. P. v. d., G. M. Willems et al. (1998). Adsorption of proteins onto poly(ether urethane) with a phosphorylcholine moiety and influence of preadsorbed phospholipid. *Journal of Biomedical Materials Research* 40(2): 195–203.

Hench, L. L. and J. M. Polak (2002). Third-generation biomedical materials. *Science* 295(5557): 1014–1017.

Hermanson, G. T., Ed. (1996). *Bioconjugate Techniques*. New York, Academic Press.

Hermanson, G. T. (2008). *Bioconjugate Techniques*. Oxford, U.K.: Elsevier's Science & Tecnology.

Hernández, L., B. Vázquez et al. (2007). Acrylic bone cements with bismuth salicylate: Behavior in simulated physiological conditions. *Journal of Biomedical Materials Research Part A* 80(2): 321–332.

Heurtault, B., P. Saulnier et al. (2003). Physico-chemical stability of colloidal lipid particles. *Biomaterials* 24(23): 4283–4300.

Higuchi, A., K. Shirano et al. (2002). Chemically modified polysulfone hollow fibers with vinylpyrrolidone having improved blood compatibility. *Biomaterials* 23(13): 2659–2666.

Hoffart, V., Ubrich, N., Lamprecht, A., Bachelier, K., Vigneron, C., Lecompte, T., Hoffman, M., and P. Maincent (2003). Microencapsulation of low molecular weight heparin into polymeric particles designed with biodegradable and nonbiodegradable polycationic polymers. *Drug Delivery* 10: 1–7.

Hoffman, A. S. and B. D. Ratner (2004). Nonfouling surfaces. In: B. D. Ratner, A. S. Hoffman, F. J. Schoen and J. E. Lemons, Eds., *Biomaterials Science: An Introduction to Materials in Medicine*, pp. 197–201. Oxford, U.K,: Elsevier's Science & Tecnology.

Hook A. L., A. D., Langer R., Williams P., Davies M. C., and M. R. Alexander (2010). High throughput methods applied in biomaterial development and discovery. *Biomaterials* 31(2): 187–198.

Hook, F. F., J. Vörös et al. (2002). A comparative study of protein adsorption on titanium oxide surfaces using in situ ellipsometry, optical waveguide lightmode spectroscopy, and quartz crystal microbalance/dissipation. *Colloids and Surfaces B: Biointerfaces* 24(2): 155–170.

Hu, Y., S. R. Winn et al. (2003). Porous polymer scaffolds surface-modified with arginine-glycine-aspartic acid enhance bone cell attachment and differentiation in vitro. *Journal of Biomedical Materials Research Part A* 64(3): 583–590.

Huang, C.-J., T. P.-Y., and Y.-C. Chang (2010). Effects of extracellular matrix protein functionalized fluid membrane on cell adhesion and matrix remodelling.. *Biomaterials* 31(27): 7183–7195.

Huang, J. and W. Xu (2010). Zwitterionic monomer graft copolymerization onto polyurethane surface through a PEG spacer. *Applied Surface Science* 256(12): 3921–3927.

Huang, H., Y. Zhao et al. (2003). Enhanced osteoblast functions on RGD immobilized surface. *Journal of Oral Implantology* 29(2): 73–79.

Hylton, D. M., S. W. Shalaby et al. (2005). Direct correlation between adsorption-induced changes in protein structure and platelet adhesion. *Journal of Biomedical Materials Research Part A* 73A(3): 349–358.

Ishihara, K., T. Hasegawa et al. (2002). Protein adsorption-resistant hollow fibers for blood purification. *Artificial organs* 26(12): 1014–1019.

Ishihara, K., D. Nishiuchi et al. (2004). Polyethylene/phospholipid polymer alloy as an alternative to poly(vinylchloride)-based materials. *Biomaterials* 25(6): 1115–1122.

Ishihara, K., H. Nomura et al. (1998). Why do phospholipid polymers reduce protein adsorption? *Journal of Biomedical Materials Research* 39(2): 323–330.

Ishihara, K., T. Ueda et al. (1990). Preparation of phospholipid polymers and their properties as polymer hydrogel membranes. *Polymer Journal* 22(5): 355–360.

Israelachvili, J. and H. Wennerstrom (1996). Role of hydration and water structure in biological and colloidal interactions. *Nature* 379(6562): 219–225.

Itoh, D., S. Yoneda et al. (2002). Enhancement of osteogenesis on hydroxyapatite surface coated with synthetic peptide (EEEEEEEPRGDT) *in vitro*. *Journal of Biomedical Materials Research* 62(2): 292–298.

Iwasaki, Y., S. Uchiyama et al. (2002). A nonthrombogenic gas-permeable membrane composed of a phospholipid polymer skin film adhered to a polyethylene porous membrane. *Biomaterials* 23(16): 3421–3427.

Iwata, R., P. Suk-In et al. (2004). Control of nanobiointerfaces generated from well-defined biomimetic polymer brushes for protein and cell manipulations. *Biomacromolecules* 5(6): 2308–2314.

Jackeray, R., S. Jain et al. (2010). Surface modification of nylon membrane by glycidyl methacrylate graft copolymerization for antibody immobilization. *Journal of Applied Polymer Science* 116(3): 1700–1709.

Jandt, K. D. (2001). Atomic force microscopy of biomaterials surfaces and interfaces. *Surface Science* 491(3): 303–332.

Janorkar, A. V., S. E. Proulx et al. (2006). Surface-confined photopolymerization of single- and mixed-monomer systems to tailor the wettability of poly(L-lactide) film. *Journal of Polymer Science Part A: Polymer Chemistry* 44(22): 6534–6543.

Jaturanpinyo, M., A. Harada et al. (2004). Preparation of bionanoreactor based on core-shell structured polyion complex micelles entrapping trypsin in the core cross-linked with glutaraldehyde. *Bioconjugate Chemistry* 15(2): 344–348.

Jee, K. S., H. D. Park et al. (2004). Heparin conjugated polylactide as a blood compatible material. *Biomacromolecules* 5(5): 1877–1881.

Joralemon, M. J., R. K. O'Reilly et al. (2005). Shell click-crosslinked (SCC) nanoparticles: A new methodology for synthesis and orthogonal functionalization. *Journal of the American Chemical Society* 127(48): 16892–16899.

Joshi, J. M. and V. K. Sinha (2006). Graft copolymerization of 2-hydroxyethylmethacrylate onto carboxymethyl chitosan using CAN as an initiator. *Polymer* 47(6): 2198–2204.

Jung, D., S. Yeo et al. (2006). Formation of amine groups by plasma enhanced chemical vapor deposition and its application to DNA array technology. *Surface and Coatings Technology* 200(9): 2886–2891.

Junkar, I., A. Vesel et al. (2009). Influence of oxygen and nitrogen plasma treatment on polyethylene terephthalate (PET) polymers. *Vacuum* 84(1): 83–85.

Kao, W. J. and J. A. Hubbell (1998). Murine macrophage behavior on peptide-grafted polyethyleneglycol-containing networks. *Biotechnology and Bioengineering* 59(1): 2–9.

Kazarian, S. G. and K. L. A Chan (2003). "Chemical photography" of drug release. *Macromolecules* 36(26): 9866–9872.

Kazarian, S. G. and K. L. A. Chan (2006). Applications of ATR-FTIR spectroscopic imaging to biomedical samples. *Biochimica et Biophysica Acta—Biomembranes* 1758(7): 858–867.

Kazarian, S. G., K. W. T. Kong et al. (2005). Spectroscopic imaging applied to drug release. *Food and Bioproducts Processing* 83(2 C): 127–135.

Kenawy, E. R., F. I. Abdel-Hay et al. (2005). Biologically active polymers: Modification and anti-microbial activity of chitosan derivatives. *Journal of Bioactive and Compatible Polymers* 20(1): 95–111.

Kenawy, E. R., F. Imam Abdel-Hay et al. (2006). Synthesis and antimicrobial activity of some polymers derived from modified amino polyacrylamide by reacting it with benzoate esters and benzaldehyde derivatives. *Journal of Applied Polymer Science* 99(5): 2428–2437.

Kim, Y. H., D. K. Han et al. (2003). Enhanced blood compatibility of polymers grafted by sulfonated PEO via a negative cilia concept. *Biomaterials* 24(13): 2213–2223.

Kim, Y. J., I. K. Kang et al. (2000). Surface characterization and in vitro blood compatibility of poly(ethylene terephthalate) immobilized with insulin and/or heparin using plasma glow discharge. *Biomaterials* 21(2): 121–130.

Kim, D. K., Y. K. Park et al. (2005). Removal efficiency of organic contaminants on Si wafer surfaces by the N2O ECR plasma technique. *Materials Chemistry and Physics* 91(2–3): 490–493.

Kingshott, P., J. Wei et al. (2003). Covalent attachment of poly(ethylene glycol) to surfaces, critical for reducing bacterial adhesion. *Langmuir* 19(17): 6912–6921.

Kirby, B. J. and E. F. Hasselbrink Jr (2004). Zeta potential of microfluidic substrates: 1. Theory, experimental techniques, and effects on separations. *Electrophoresis* 25(2): 187–202.

Kirsebom, H., M. R. Aguilar et al. (2007). Macroporous scaffolds based on chitosan and bioactive molecules. *Journal of Bioactive and Compatible Polymers* 22(6): 621–636.

Klee, D., Z. Ademovic et al. (2003). Surface modification of poly(vinylidenefluoride) to improve the osteoblast adhesion. *Biomaterials* 24(21): 3663–3670.

Klenkler, B. J., M. Griffith et al. (2005). EGF-grafted PDMS surfaces in artificial cornea applications. *Biomaterials* 26(35): 7286–7296.

Klok, H.-A. (2005). Biological-synthetic hybrid block copolymers: Combining the best from two worlds. *Journal of Polymer Science Part A: Polymer Chemistry* 43(1): 1–17.

Koenig, A. L., V. Gambillara et al. (2003). Correlating fibronectin adsorption with endothelial cell adhesion and signaling on polymer substrates. *Journal of Biomedical Materials Research Part A* 64A(1): 20–37.

Kolb, H. C., M. G. Finn et al. (2001). Click chemistry: Diverse chemical function from a few good reactions. *Angewandte Chemie—International Edition* 40(11): 2005–2021.

Koo, L. Y., D. J. Irvine et al. (2002). Co-regulation of cell adhesion by nanoscale RGD organization and mechanical stimulus. *Journal of Cell Science* 115(7): 1423–1433.

Korematsu, A., Y. Takemoto et al. (2002). Synthesis, characterization and platelet adhesion of segmented polyurethanes grafted phospholipid analogous vinyl monomer on surface. *Biomaterials* 23(1): 263–271.

Kubies, D., L. Machová et al. (2003). Functionalized surfaces of polylactide modified by Langmuir-Blodgett films of amphiphilic block copolymers. *Journal of Materials Science: Materials in Medicine* 14(2): 143–149.

Kulbokaite, R., G. Ciuta et al. (2009). N-PEG'ylation of chitosan via "click chemistry" reactions. *Reactive and Functional Polymers* 69(10): 771–778.

Kwang, H. L., H. K. Gu et al. (2009). Hydrophilic electrospun polyurethane nanofiber matrices for hMSC culture in a microfluidic cell chip. *Journal of Biomedical Materials Research Part A* 90(2): 619–628.

Kwok, C. S., T. A. Horbett et al. (1999). Design of infection-resistant antibiotic-releasing polymersII. Controlled release of antibiotics through a plasma-deposited thin film barrier. *Journal of Controlled Release* 62(3): 301–311.

Ladam, G., P. Schaaf et al. (2002). Protein adsorption onto auto-assembled polyelectrolyte films. *Biomolecular Engineering* 19(2–6): 273–280.

Ladd, J., Z. Zhang et al. (2008). Zwitterionic polymers exhibiting high resistance to nonspecific protein adsorption from human serum and plasma. *Biomacromolecules* 9(5): 1357–1361.

Ladmiral, V., E. Melia et al. (2004). Synthetic glycopolymers: An overview. *European Polymer Journal* 40(3): 431–449.

Lamers, E., X. Frank Walboomers et al. (2010). The influence of nanoscale grooved substrates on osteoblast behavior and extracellular matrix deposition. *Biomaterials* 31(12): 3307–3316.

Langer, R. and D. A. Tirrell (2004). Designing materials for biology and medicine. *Nature* 428(6982): 487–492.

Lazos, D., S. Franzka et al. (2005). Size-selective protein adsorption to polystyrene surfaces by self-assembled grafted poly(ethylene glycols) with varied chain lengths. *Langmuir* 21(19): 8774–8784.

Lebaron, R. G. and K. A. Athanasiou (2000). Extracellular matrix cell adhesion peptides: Functional applications in orthopedic materials. *Tissue Engineering* 6(2): 85–103.

Leckband, D. and A. Prakasam (2006). Mechanism and dynamics of cadherin adhesion. *Annual Review of Biomedical Engineering* 8(1): 259–287.

Lee, G. K., N. Maheshri et al. (2005). PEG conjugation moderately protects adeno-associated viral vectors against antibody neutralization. *Biotechnology and Bioengineering* 92(1): 24–34.

Lee, D. Y., S. J. Park et al. (2006). A new strategy toward improving immunoprotection in cell therapy for diabetes mellitus: Long-functioning PEGylated islets in vivo. *Tissue Engineering* 12(3): 615–623.

Lee, K. B., K. R. Yoon et al. (2003). Surface modification of poly(glycolic acid) (PGA) for biomedical applications. *Journal of Pharmaceutical Sciences* 92(5): 933–937.

Leonard, D., Y. Chevolot et al. (1998). ToF-SIMS and XPS study of photoactivatable reagents designed for surface glycoengineering: Part 2. N-[m-(3-(trifluoromethyl)diazirine-a-yl)phenyl]-4-(-3-thio(-1-D-galactopyrannosyl)-maleimidyl)butyramide (MAD-GaI) on diamond. *Surface and Interface Analysis* 26(11): 793–799.

Levenberg, S., R. Langer et al. (2004). Advances in tissue engineering. *Current Topics in Developmental Biology*, Vol. 61, pp. 113–134, San Diego, CA, Academic Press..

Lewis, A. L. (2000). Phosphorylcholine-based polymers and their use in the prevention of biofouling. *Colloids and Surfaces B: Biointerfaces* 18(3–4): 261–275.

Li, L., S. Chen et al. (2005). Protein adsorption on oligo(ethylene glycol)-terminated alkanethiolate self-assembled monolayers: The molecular basis for nonfouling behavior. *Journal of Physical Chemistry B* 109(7): 2934–2941.

Li, Y., K. G. Neoh et al. (2003). Physicochemical and blood compatibility characterization of polypyrrole surface functionalized with heparin. *Biotechnology and Bioengineering* 84(3): 305–313.

Li, Z. F. and E. Ruckenstein (2004). Grafting of poly(ethylene oxide) to the surface of polyaniline films through a chlorosulfonation method and the biocompatibility of the modified films. *Journal of Colloid and Interface Science* 269(1): 62–71.

Link, A. J., M. K. S. Vink et al. (2004). Presentation and detection of azide functionality in bacterial cell surface proteins. *Journal of the American Chemical Society* 126(34): 10598–10602.

Linnola, R. J., L. Werner et al. (2000). Adhesion of fibronectin, vitronectin, laminin, and collagen type IV to intraocular lens materials in pseudophakic human autopsy eyes: Part 1: Histological sections. *Journal of Cataract & Refractive Surgery* 26(12): 1792–1806.

Liu, X. and P. Ma (2004). Polymeric scaffolds for bone tissue engineering. *Annals Biomedical Engineering* 32: 477–486.

Liu, J., T. Pan et al. (2004). Surface-modified poly(methyl methacrylate) capillary electrophoresis microchips for protein and peptide analysis. *Analytical Chemistry* 76(23): 6948–6955.

López Donaire, M. L., J. Parra-Cáceres et al. (2009). Polymeric drugs based on bioactive glycosides for the treatment of brain tumours. *Biomaterials* 30(8): 1613–1626.

Lu, L., P. Zhang et al. (2009). Study on partially oxidized sodium alginate with potassium permanganate as the oxidant. *Journal of Applied Polymer Science* 113(6): 3585–3589.

Luk, Y. Y., M. Kato et al. (2000). Self-assembled monolayers of alkanethiolates presenting mannitol groups are inert to protein adsorption and cell attachment. *Langmuir* 16(24): 9604–9608.

Lutz, J. F. (2007). 1,3-Dipolar cycloadditions of azides and alkynes: A universal ligation tool in polymer and materials science. *Angewandte Chemie, International Edition* 46(7): 1018–1025.

Lutz, J. F., H. G. Börner et al. (2006). Combining ATRP and "click" chemistry: A promising platform toward functional biocompatible polymers and polymer bioconjugates. *Macromolecules* 39(19): 6376–6383.

Lutz, J. F. and H. G. Börner (2008). Modern trends in polymer bioconjugates design. *Progress in Polymer Science* 33: 1–39.

Lutz, J. F., S. Pfeifer et al. (2007). In situ functionalization of thermoresponsive polymeric micelles using the "click" cycloaddition of azides and alkynes. *QSAR and Combinatorial Science* 26(11–12): 1151–1158.

Lutz, J.-F. and Z. Zarafshani (2008). Efficient construction of therapeutics, bioconjugates, biomaterials and bioactive surfaces using azide-alkyne "click" chemistry. *Advanced Drug Delivery Reviews* 60(9): 958–970.

Ma, Z., Z. Mao et al. (2007). Surface modification and property analysis of biomedical polymers used for tissue engineering. *Colloids and Surfaces B: Biointerfaces* 60(2): 137–157.

Ma, W. X., C. Zhao et al. (2010). A novel method of modifying poly(ethylene terephthalate) fabric using supercritical carbon dioxide. *Journal of Applied Polymer Science* 117(4): 1897–1907.

Macmanus, L. F., M. J. Walzak et al. (1999). Study of ultraviolet light and ozone surface modification of polypropylene. *Journal of Polymer Science Part A: Polymer Chemistry* 37(14): 2489–2501.

Manju, S. and S. Kunnatheeri (2010). Layer-by-Layer modification of poly(methyl methacrylate) intra ocular lens: Drug delivery applications. *Pharmaceutical Development and Technology* 15(4): 379–385.

Mann, B. K., R. H. Schmedlen et al. (2001). Tethered-TGF-[beta] increases extracellular matrix production of vascular smooth muscle cells. *Biomaterials* 22(5): 439–444.

Mapili, G., Y. Lu et al. (2005). Laser-layered microfabrication of spatially patterned functionalized tissue-engineering scaffolds. *Journal of Biomedical Materials Research Part B: Applied Biomaterials* 75(2): 414.

Marchand-Brynaert, J., E. Detrait et al. (1999). Biological evaluation of RGD peptidomimetics, designed for the covalent derivatization of cell culture substrata, as potential promotors of cellular adhesion. *Biomaterials* 20(19): 1773–1782.

Martins, G. V., J. F. Mano et al. (2010). Nanostructured self-assembled films containing chitosan fabricated at neutral pH. *Carbohydrate Polymers* 80(2): 570–573.

Martins, M. C. L., B. D. Ratner et al. (2003a). Protein adsorption on mixtures of hydroxyl- and methylterminated alkanethiols self-assembled monolayers. *Journal of Biomedical Materials Research Part A* 67A(1): 158–171.

Martins, M. C. L., D. Wang et al. (2003b). Albumin and fibrinogen adsorption on PU-PHEMA surfaces. *Biomaterials* 24(12): 2067–2076.

Marx, K. A. (2003). Quartz crystal microbalance: A useful tool for studying thin polymer films and complex biomolecular systems at the solution—Surface interface. *Biomacromolecules* 4(5): 1099–1120.

Maskarinec, S. A. and D. A. Tirrell (2005). Protein engineering approaches to biomaterials design. *Current Opinion in Biotechnology* 16(4): 422–426.

Mathieu, H. J., Y. Chevolot et al. (2003). Engineering and characterization of polymer surfaces for biomedical applications. *Advances in Polymer Science* 162: 1–34.

McArthur, S. L. (2006). Applications of XPS in bioengineering. *Surface and Interface Analysis* 38(11): 1380–1385.

McConachie, A., D. Newman et al. (1999). The effect on bioadhesive polymers either freely in solution or covalently attached to a support on human macrophages. *Biomedical Sciences Instrumentation* 35: 45–50.

Merrett, K., R. M. Cornelius et al. (2002). Surface analysis methods for characterizing polymeric biomaterials. *Journal of Biomaterials Science, Polymer Edition* 13(6): 593–621.

Michael, M. N., N. A. El-Zaher et al. (2004). Investigation into surface modification of some polymeric fabrics by UV/ozone treatment. *Polymer—Plastics Technology and Engineering* 43(4): 1041–1052.

Miller, D. C., A. Thapa et al. (2004). Endothelial and vascular smooth muscle cell function on poly(lactic-co-glycolic acid) with nano-structured surface features. *Biomaterials* 25(1): 53–61.

Miller-Chou, B. A. and J. L. Koenig (2002). FT-IR imaging of polymer dissolution by solvent mixtures. 3. Entangled polymer chains with solvents. *Macromolecules* 35(2): 440–444.

Milton Harris, J. and R. B. Chess (2003). Effect of pegylation on pharmaceuticals. *Nature Reviews Drug Discovery* 2(3): 214–221.

Mitchell, S. A., A. H. C. Poulsson et al. (2005). Orientation and confinement of cells on chemically patterned polystyrene surfaces. *Colloids and Surfaces B: Biointerfaces* 46(2): 108–116.

Morimoto, N., Y. Iwasaki et al. (2002). Physical properties and blood compatibility of surface-modified segmented polyurethane by semi-interpenetrating polymer networks with a phospholipid polymer. *Biomaterials* 23(24): 4881–4887.

Morimoto, N., A. Watanabe et al. (2004). Nano-scale surface modification of a segmented polyurethane with a phospholipid polymer. *Biomaterials* 25(23): 5353–5361.

Moro, T., Y. Takatori et al. (2004). Surface grafting of artificial joints with a biocompatible polymer for preventing periprosthetic osteolysis. *Nature Materials* 3(11): 829–836.

Morra, M. (2000). On the molecular basis of fouling resistance. *Journal of Biomaterials Science, Polymer Edition* 11: 547–569.

Mosqueira, V. C. F., P. Legrand et al. (2000). Poly(D,L-lactide) nanocapsules prepared by a solvent displacement process: Influence of the composition on physicochemical and structural properties. *Journal of Pharmaceutical Sciences* 89(5): 614–626.

Mouhib, T., D. A., Poleunis, C., Henry, M., and P. Bertrand (2010). C60 SIMS depth profiling of bovine serum albumin protein-coating films: A conformational study. *Surface and Interface Analysis* 42(6–7): 641–644.

Muscariello, L., F. Rosso et al. (2005). A critical overview of ESEM applications in the biological field. *Journal of Cellular Physiology* 205(3): 328–334.

Muthukrishnan, S., M. Zhang et al. (2005). Molecular sugar sticks: Cylindrical glycopolymer brushes. *Macromolecules* 38(19): 7926–7934.

Nagahata, M., R. Nakaoka et al. (2005). The response of normal human osteoblasts to anionic polysaccharide polyelectrolyte complexes. *Biomaterials* 26(25): 5138–5144.

Nam, Y. S., J. J. Yoon et al. (1999). Adhesion behaviours of hepatocytes cultured onto biodegradable polymer surface modified by alkali hydrolysis process. *Journal of Biomaterials Science, Polymer Edition* 10(11): 1145–1158.

Navarro, M., C. Aparicio et al. (2006). Development of a biodegradable composite scaffold for bone tissue engineering: Physicochemical, topographical, mechanical, degradation, and biological properties. *Ordered Polymeric Nanostructures at Surfaces* 200: 209–231.

Niemeyer, C. M. (2001). Nanoparticles, proteins, and nucleic acids: Biotechnology meets materials science. *Angewandte Chemie International Edition* 40(22): 4128–4158.

Niu, X., Y. Wang et al. (2005). Arg-Gly-Asp (RGD) modified biomimetic polymeric materials. *Journal of Materials Science and Technology* 21(4): 571–576.

Noh, H. and E. A. Vogler (2006). Volumetric interpretation of protein adsorption: Mass and energy balance for albumin adsorption to particulate adsorbents with incrementally increasing hydrophilicity. *Biomaterials* 27(34): 5801–5812.

Øiseth, S. K., A. Krozer et al. (2004). Ultraviolet light treatment of thin high-density polyethylene films monitored with a quartz crystal microbalance. *Journal of Applied Polymer Science* 92(5): 2833–2839.

Okamura, Y., I. Maekawa et al. (2005). Hemostatic effects of phospholipid vesicles carrying fibrinogen γ-chain dodecapeptide in vitro and in vivo. *Bioconjugate Chemistry* 16(6): 1589–1596.

Olsson, A. L. J. V. D. M. H., Busscher, H. J., and P. K. Sharma (2010). Novel analysis of bacterium-substratum bond maturation measured using a quartz crystal microbalance. *Langmuir* 26(13): 11113–11117.

Öner, D. and T. J. McCarthy (2000). Ultrahydrophobic surfaces. Effects of topography length scales on wettability. *Langmuir* 16(20): 7777–7782.

O'Reilly, R. K., M. J. Joralemon et al. (2006). Facile syntheses of surface-functionalized micelles and shell cross-linked nanoparticles. *Journal of Polymer Science Part A: Polymer Chemistry* 44(17): 5203–5217.

Ortega-Vinuesa, J. L., P. Tengvall et al. (1998). Molecular packing of HSA, IgG, and fibrinogen adsorbed on silicon by AFM imaging. *Thin Solid Films* 324(1–2): 257–273.

O'Sullivan, C. K. and G. G. Guilbault (1999). Commercial quartz crystal microbalances—Theory and applications. *Biosensors and Bioelectronics* 14(8–9): 663–670.

Ostuni, E., R. G. Chapman et al. (2001). A survey of structure-property relationships of surfaces that resist the adsorption of protein. *Langmuir* 17(18): 5605–5620.

Otsuka, H., Y. Nagasaki et al. (2000). Surface characterization of functionalized polylactide through the coating with heterobifunctional poly(ethylene glycol)/polylactide block copolymers. *Biomacromolecules* 1(1): 39–48.

Oyane, A., M. Uchida et al. (2005). Simple surface modification of poly(ε-caprolactone) to induce its apatite-forming ability. *Journal of Biomedical Materials Research Part A* 75(1): 138–145.

Pande, G. (2000). The role of membrane lipids in regulation of integrin functions. *Current Opinion in Cell Biology* 12(5): 569–574.

Park, K.-H. (2002). Arg-Gly-Asp (RGD) sequence conjugated in a synthetic copolymer bearing a sugar moiety for improved culture of parenchymal cells (hepatocytes). *Biotechnology Letters* 24(17): 1401–1406.

Park, K.-H., K. Na et al. (2004). Immobilization of Arg-Gly-Asp (RGD) sequence in sugar containing copolymer for culturing of pheochromocytoma (PC12) cells. *Journal of Bioscience and Bioengineering* 97(3): 207–211.

Parker, A. P., P. A. Reynolds et al. (2005). Investigation into potential mechanisms promoting biocompatibility of polymeric biomaterials containing the phosphorylcholine moiety: A physicochemical and biological study. *Colloids and Surfaces B: Biointerfaces* 46(4): 204–217.

Parrish, B., R. B. Breitenkamp et al. (2005). PEG- and peptide-grafted aliphatic polyesters by click chemistry. *Journal of the American Chemical Society* 127(20): 7404–7410.

Pasche, S., J. Vörös et al. (2005). Effects of ionic strength and surface charge on protein adsorption at PEGylated surfaces. *The Journal of Physical Chemistry. B* 109(37): 17545–17552.

Perez-Luna, V. H., K. A. Hooper et al. (1997). Surface characterization of tyrosine-derived polycarbonates. *Journal of Applied Polymer Science* 63(11): 1467–1479.

Petit, V. and J.-P. Thiery (2000). Focal adhesions: Structure and dynamics. *Biology of the Cell* 92(7): 477–494.

Pippig, F., S. Sarghini et al. (2009). TFAA chemical derivatization and XPS. Analysis of OH and NHx polymers. *Surface and Interface Analysis* 41(5): 421–429.

Plow, E. F., T. A. Haas et al. (2000). Ligand binding to integrins. *Journal of Biological Chemistry* 275(29): 21785–21788.

Porter, A. E. (2006). Nanoscale characterization of the interface between bone and hydroxyapatite implants and the effect of silicon on bone apposition. *Micron* 37(8): 681–688.

Poulsson, A. H. C., S. A. Mitchell et al. (2009). Attachment of human primary osteoblast cells to modified polyethylene surfaces. *Langmuir* 25(6): 3718–3727.

Prescher, J. A., D. H. Dube et al. (2004). Chemical remodelling of cell surfaces in living animals. *Nature* 430(7002): 873–877.

Qu, X.-H., Q. Wu et al. (2005). Enhanced vascular-related cellular affinity on surface modified copolyesters of 3-hydroxybutyrate and 3-hydroxyhexanoate (PHBHHx). *Biomaterials* 26(34): 6991–7001.

Quirk, R. A., W. C. Chan et al. (2001). Poly(-lysine)-GRGDS as a biomimetic surface modifier for poly(lactic acid). *Biomaterials* 22(8): 865–872.

Raja, K. S., Q. Wang et al. (2003). Hybrid virus-polymer materials. 1. Synthesis and properties of PEG-decorated cowpea mosaic virus. *Biomacromolecules* 4(3): 472–476.

Rajagopalan, P., W. A. Marganski et al. (2004). Direct comparison of the spread area, contractility, and migration of balb/c 3T3 fibroblasts adhered to fibronectin- and RGD-modified substrata. *Biophysical Journal* 87(4): 2818–2827.

Rao, G. R., Z. L. Wang et al. (1993). Microstructural effects on surface mechanical-properties of ion-implanted polymers. *Journal of Materials Research* 8(4): 927–933.

Ratner, B. D. and S. J. Bryan (2004). Biomaterials: Where we have been and where we are going. *Annual Review of Biomedical Engineering* 6: 41–75.

Reyes, C. D. and A. J. Garcia (2003). Engineering integrin-specific surfaces with a triple-helical collagen-mimetic peptide. *Journal of Biomedical Materials Research Part A* 65A(4): 511–523.

Reyes, C. D. and A. J. Garcia (2004). Alpha(2)beta(1) integrin-specific collagen-mimetic surfaces supporting osteoblastic differentiation. *Journal of Biomedical Materials Research Part A* 69A(4): 591–600.

Reynhout, I. C., J. J. L. M. Cornelissen et al. (2007). Self-assembled architectures from biohybrid triblock copolymers. *Journal of the American Chemical Society* 129(8): 2327–2332.

Rezania, A. and K. E. Healy (1999). Biomimetic peptide surfaces that regulate adhesion, spreading, cyto-skeletal organization, and mineralization of the matrix deposited by osteoblast-like cells. *Biotechnology Progress* 15(1): 19–32.

Richey, T., H. Iwata et al. (2000). Surface modification of polyethylene balloon catheters for local drug delivery. *Biomaterials* 21(10): 1057–1065.

Rinckenbach, S., J. Hemmerlé et al. (2008). Characterization of polyelectrolyte multilayer films on polyethylene terephtalate vascular prostheses under mechanical stretching. *Journal of Biomedical Materials Research Part A* 84(3): 576–588.

Roach, P., D. Eglin et al. (2007). Modern biomaterials: A review—Bulk properties and implications of surface modifications. *Journal of Materials Science: Materials in Medicine* 18(7): 1263–1277.

Roach, P., D. Farrar et al. (2005). Interpretation of protein adsorption: Surface-induced conformational changes. *Journal of the American Chemical Society* 127(22): 8168–8173.

Rodrigues, S. N., I. C. Gonçalves et al. (2006). Fibrinogen adsorption, platelet adhesion and activation on mixed hydroxyl-/methyl-terminated self-assembled monolayers. *Biomaterials* 27(31): 5357–5367.

Rodriguez Emmenegger, C., E. Brynda et al. (2009). Interaction of blood plasma with antifouling surfaces. *Langmuir* 25(11): 6328–6333.

Rodríguez-Lorenzo, L. M., R. García-Carrodeguas et al. (2006). Wollastonite-poly(ethylmethacrylate-co-vinylpyrrolydone) nanostructured materials: Mechanical properties and biocompatibility. *Key Engineering Materials* 309–311 II: 1149–1152.

Rodríguez-Lorenzo, L. M., R. García-Carrodeguas et al. (2009). Synthesis, characterization, bioactivity and biocompatibility of nanostructured materials based on the wollastonite-poly(ethylmethacrylate-co-vinylpyrrolidone) system. *Journal of Biomedical Materials Research Part A* 88(1): 53–64.

Roger, P., L. Renaudie et al. (2010). Surface characterizations of poly(ethylene terephthalate) film modified by a carbohydrate-bearing photoreactive azide group. *European Polymer Journal* 46(7): 1594–1603.

Rojo, L., J. M. Barcenilla et al. (2008). Intrinsically antibacterial materials based on polymeric derivatives of eugenol for biomedical applications. *Biomacromolecules* 9(9): 2530–2535.

Romero-Sánchez, M. D., M. M. Pastor-Blas et al. (2001). Surface modifications of a vulcanized rubber using corona discharge and ultraviolet radiation treatments. *Journal of Materials Science* 36(24): 5789–5799.

Rostovtsev, V. V., L. G. Green et al. (2002). A stepwise huisgen cycloaddition process: Copper(I)-catalyzed regioselective "ligation" of azides and terminal alkynes. *Angewandte Chemie—International Edition* 41(14): 2596–2599.

Rowley, J. A. and D. J. Mooney (2002). Alginate type and RGD density control myoblast phenotype. *Journal of Biomedical Materials Research* 60(2): 217–223.

Sabbatini, L. and P. G. Zambonin (1996). XPS and SIMS surface chemical analysis of some important classes of polymeric biomaterials. *Journal of Electron Spectroscopy and Related Phenomena* 81(3): 285–301.

Sagnella, S., E. Anderson et al. (2005). Human endothelial cell interaction with biomimetic surfactant polymers containing peptide ligands from the heparin binding domain of fibronectin. *Tissue Engineering* 11(1–2): 226–236.

Sahlin, J. J. and N. A. Peppas (1997). Near-field FTIR imaging: A technique for enhancing spatial resolution in FTIR microscopy. *Journal of Applied Polymer Science* 63(1): 103–110.

Sajeesh, S., K. Bouchemal et al. (2010). Surface-functionalized polymethacrylic acid based hydrogel micropar-ticles for oral drug delivery. *European Journal of Pharmaceutics and Biopharmaceutics* 74(2): 209–218.

Salvagnini, C. and J. Marchand-Brynaert (2006). Immobilization of thrombin inhibitors on polyesters surface: An original approach towards materials blood compatibilization. *Advanced Materials Forum Iii Parts 1 and 2* 514–516: 961–965.

Sánchez, J., G. Elgue et al. (1997). Inhibition of the plasma contact activation system of immobilized heparin: Relation to surface density of functional antithrombin binding sites. *Journal of Biomedical Materials Research* 37(1): 37–42.

Sanchis, M. R., V. Blanes et al. (2006). Surface modification of low density polyethylene (LDPE) film by low pressure O2 plasma treatment. *European Polymer Journal* 42(7): 1558–1568.

Satriano, C., G. M. L. Messina et al. (2010). Surface immobilization of fibronectin-derived PHSRN peptide on functionalized polymer films—Effects on fibroblast spreading. *Journal of Colloid and Interface Science* 341(2): 232–239.

Schneider, S. W., M. E. Egan et al. (1999). Continuous detection of extracellular ATP on living cells by using atomic force microscopy. *Proceedings of the National Academy of Sciences of the USA* 96(21): 12180–12185.

Schwartz, M. A. and M. H. Ginsberg (2002). Networks and crosstalk: Integrin signalling spreads. *Nature Cell Biology* 4(4): E65–E68.

Scott, M. D. and A. M. Chen (2004). Beyond the red cell: Pegylation of other blood cells and tissues. *Transfusion Clinique et Biologique* 11(1): 40–46.

Sebra, R. P., K. S. Masters et al. (2005). Surface grafted antibodies: Controlled architecture permits enhanced antigen detection. *Langmuir* 21(24): 10907–10911.

Seo, J.-H., R. Matsuno et al. (2008). Surface tethering of phosphorylcholine groups onto poly(dimethylsiloxane) through swelling-deswelling methods with phospholipids moiety containing ABA-type block copolymers. *Biomaterials* 29(10): 1367–1376.

Servoli, E., D. Maniglio et al. (2008). Quantitative analysis of protein adsorption via atomic force microscopy and surface plasmon resonance. *Macromolecular Bioscience* 8(12): 1126–1134.

Servoli, E., D. Maniglio et al. (2009). Comparative methods for the evaluation of protein adsorption. *Macromolecular Bioscience* 9(7): 661–670.

Sessler, J. L. and J. Jayawickramarajah (2005). Functionalized base-pairs: Versatile scaffolds for self-assembly. *Chemical Communications* (15): 1939–1949.

Sham, M. L., J. Li et al. (2009). Cleaning and functionalization of polymer surfaces and nanoscale carbon fillers by uv/ozone treatment: A review. *Journal of Composite Materials* 43(14): 1537–1564.

Shchipunov, Y. A., O. G. Mukhaneva et al. (2003). Polyelectrolyte complexes of naturally occurring fucoidans with cationically and hydrophobically modified hydroxyethyl cellulose. *Polymer Science Series A* 45(3): 295–303.

Sheardown, H., M. A. Brook et al. (2007). Cell Interactions with PDMS surfaces modified with cell adhesion peptides using a generic method. *Materials Science Forum* 539–543(Part 1): 704–709.

Shen, M., M. S. Wagner et al. (2003). Multivariate surface analysis of plasma-deposited tetraglyme for reduction of protein adsorption and monocyte adhesion. *Langmuir* 19(5): 1692–1699.

Sheng, Q., K. Schulten et al. (1995). Molecular dynamic simulation of immobilized artificial membranes. *Journal of Physical Chemistry* 99(27): 11018–11027.

Shi, Q., Y. Huang et al. (2009). Hemoglobin conjugated micelles based on triblock biodegradable polymers as artificial oxygen carriers. *Biomaterials* 30(28): 5077–5085.

Shin, H., S. Jo et al. (2002). Modulation of marrow stromal osteoblast adhesion on biomimetic oligo[poly(ethylene glycol) fumarate] hydrogels modified with Arg-Gly-Asp peptides and a poly(ethylene glycol) spacer. *Journal of Biomedical Materials Research* 61(2): 169–179.

Shin, H., K. Zygourakis et al. (2004). Attachment, proliferation, and migration of marrow stromal osteoblasts cultured on biomimetic hydrogels modified with an osteopontin-derived peptide. *Biomaterials* 25(5): 895–906.

Sivakova, S. and S. J. Rowan (2005). Nucleobases as supramolecular motifs. *Chemical Society Reviews* 34(1): 9–21.

Smeenk, J. M., M. B. J. Otten et al. (2005). Controlled assembly of macromolecular beta-sheet fibrils13. *Angewandte Chemie International Edition* 44(13): 1968–1971.

Smith, E., J. Yang et al. (2005). RGD-grafted thermoreversible polymers to facilitate attachment of BMP-2 responsive C2C12 cells. *Biomaterials* 26(35): 7329–7338.

Stamm, M. (2008). *Polymer Surfaces and Interfaces. Characterization, Modification and Applications*, pp. 1–16. Springer-Verlag, Berlin, Germany.

Strola, S. C. G., Gilliand, D., Valsesia, A., Lisboa, P., and F. Rossi (2010). Comparison of surface activation processes for protein immobilization on plasma-polymerized acrylic acid films. *Surface and Interface Analysis* 42(6–7): 1311–1315.

Sviridov, D. V. (2002). Chemical aspects of implantation of high-energy ions into polymeric materials. *Russian Chemical Reviews* 71(4): 315–327.

Tai, B. C. U., A. C. A. Wan et al. (2010). Modified polyelectrolyte complex fibrous scaffold as a matrix for 3D cell culture. *Biomaterials* 31(23): 5927–5935.

Tanzi, M. C. (2005). Bioactive technologies for hemocompatibility. *Expert Review of Medical Devices* 2(4): 473–492.

Teare, D. O. H., C. Ton-That et al. (2000). Surface characterization and ageing of ultraviolet-ozone-treated polymers using atomic force microscopy and x-ray photoelectron spectroscopy. *Surface and Interface Analysis* 29(4): 276–283.

Thom, V. H., G. Altankov et al. (2000). Optimizing cell-surface interactions by photografting of poly(ethylene glycol). *Langmuir* 16(6): 2756–2765.

Thome, J., A. Holländer et al. (2003). Ultrathin antibacterial polyammonium coatings on polymer surfaces. *Surface and Coatings Technology* 174–175: 584–587.

Tirrell, M., E. Kokkoli et al. (2002). The role of surface science in bioengineered materials. *Surface Science* 500(1–3): 61–83.

Toworfe, G. K., R. J. Composto et al. (2004). Fibronectin adsorption on surface-activated poly(dimethylsiloxane) and its effect on cellular function. *Journal of Biomedical Materials Research Part A* 71A(3): 449–461.

Travan, A., I. Donati et al. (2010). Surface modification and polysaccharide deposition on BisGMA/TEGDMA thermoset. *Biomacromolecules* 11(3): 583–592.

Tziampazis, E., J. Kohn et al. (2000). PEG-variant biomaterials as selectively adhesive protein templates: Model surfaces for controlled cell adhesion and migration. *Biomaterials* 21(5): 511–520.

Van Kooten, T. G., H. T. Spijker et al. (2004). Plasma-treated polystyrene surfaces: Model surfaces for studying cell-biomaterial interactions. *Biomaterials* 25(10): 1735–1747.

Vandencasteele, N., B. Nisol et al. (2008). Plasma-modified PTFE for biological applications: Correlation between protein-resistant properties and surface characteristics. *Plasma Processes and Polymers* 5(7): 661–671.

VandeVondele, S., J. Vörös et al. (2003). RGD-grafted poly-l-lysine-*graft*-(polyethylene glycol) copolymers block non-specific protein adsorption while promoting cell adhesion. *Biotechnology and Bioengineering* 82(7): 784–790.

Vasita, R., K. Shanmugam et al. (2008). Improved biomaterials for tissue engineering applications: Surface modification of polymers. *Current Topics in Medicinal Chemistry* 8(4): 341–353.

Vepari, C., D. Matheson et al. (2010). Surface modification of silk fibroin with poly(ethylene glycol) for antiadhesion and antithrombotic applications. *Journal of Biomedical Materials Research Part A* 93(2): 595–606.

Verma, I. M. and N. Somia (1997). Gene therapy—Promises, problems and prospects. *Nature* 389(6648): 239–242.

Vermette, P. and L. Meagher (2003). Interactions of phospholipid- and poly(ethylene glycol)-modified surfaces with biological systems: Relation to physico-chemical properties and mechanisms. *Colloids and Surfaces B: Biointerfaces* 28(2–3): 153–198.

Vesel, A., I. Junkar et al. (2008). Surface modification of polyester by oxygen- And nitrogen-plasma treatment. *Surface and Interface Analysis* 40(11): 1444–1453.

Vicente, T., Mota, J. P. B., Peixoto, C., Alves, P. M., and M. J. T. Carrondo (2010). Modeling protein binding and elution over a chromatographic surface probed by surface plasmon resonance. *Journal of Chromatography A* 1217(13): 2032–2041.

Vogler, E. A. (1998). Structure and reactivity of water at biomaterial surfaces. *Advances in Colloid and Interface Science* 74(1–3): 69–117.

Wan, Y., X. Qu et al. (2004). Characterization of surface property of poly(lactide-co-glycolide) after oxygen plasma treatment. *Biomaterials* 25(19): 4777–4783.

Wan, L.-S., Z.-K. Xu et al. (2005). Copolymerization of acrylonitrile with N-vinyl-2-pyrrolidone to improve the hemocompatibility of polyacrylonitrile. *Polymer* 46(18): 7715–7723.

Wan, Y., J. Yang et al. (2003). Cell adhesion on gaseous plasma modified poly-(L-lactide) surface under shear stress field. *Biomaterials* 24(21): 3757–3764.

Wang, D.-a., C. G. Williams et al. (2003). Synthesis and characterization of a novel degradable phosphate-containing hydrogel. *Biomaterials* 24(22): 3969–3980.

Webb, K., K. D. Caldwell et al. (2001). A novel surfactant-based immobilization method for varying substrate-bound fibronectin. *Journal of Biomedical Materials Research* 54(4): 509–518.

Weder, G. G.-G. O., Matthey, N., Montagne, F., Heinzelmann, H., Vörös, J., and M. Liley (2010). The quantification of single cell adhesion on functionalized surfaces for cell sheet engineering. *Biomaterials* 31(25): 6436–6443.

Wendel, H. P., N. Weber et al. (1999). Increased adsorption of high molecular weight kininogen to heparin-coated artificial surfaces and correlation to hemocompatibility. *Immunopharmacology* 43(2–3): 149–153.

Wendel, H. P. and G. Ziemer (1999). Coating-techniques to improve the hemocompatibility of artificial devices used for extracorporeal circulation. *European Journal of Cardio-Thoracic Surgery* 16(3): 342–350.

Whitesides, George M. (2005). Nanoscience, nanotechnology, and chemistry13. *Small* 1(2): 172–179.

Wierzbicka-Patynowski, I. and J. E. Schwarzbauer (2003). The ins and outs of fibronectin matrix assembly. *Journal of Cell Science* 116(16): 3269–3276.

Willemsen, O. H., M. M. E. Snel et al. (2000). Biomolecular interactions measured by atomic force microscopy. *Biophysical Journal* 79(6): 3267–3281.

Williams, D. F. (2008). On the mechanisms of biocompatibility. *Ratner Symposium 2006*, Maui, HI. *Biomaterials* 29(20): 2941–2953.

Wu, P., M. Malkoch et al. (2005a). Multivalent, bifunctional dendrimers prepared by click chemistry. *Chemical Communications* (46): 5775–5777.

Wu, Y. G., F. I. Simonovsky et al. (2005b). The role of adsorbed fibrinogen in platelet adhesion to polyurethane surfaces: A comparison of surface hydrophobicity, protein adsorption, monoclonal antibody binding, and platelet adhesion. *Journal of Biomedical Materials Research Part A* 74A(4): 722–738.

Wu, W. J., H. Z. Zhuang et al. (1997). Heterogeneous nucleation of calcium phosphates on solid surfaces in aqueous solution. *Journal of Biomedical Materials Research* 35(1): 93–99.

Xiong, X. L. Q., Lu, J.-W., Guo, Z.-X., and J. Yu (2010). Poly(lactic acid)/soluble eggshell membrane protein blend films: Preparation and characterization. *Journal of Applied Polymer Science* 117(4): 1955–1959.

Xu, Z. K., Q. W. Dai et al. (2004). Covalent attachment of phospholipid analogous polymers to modify a polymeric membrane surface: A novel approach. *Langmuir* 20(4): 1481–1488.

Xu, L.-C. and C. A. Siedlecki (2007). Effects of surface wettability and contact time on protein adhesion to biomaterial surfaces. *Biomaterials* 28(22): 3273–3283.

Xu, L. C. and C. A. Siedlecki (2009). Atomic force microscopy studies of the initial interactions between fibrinogen and surfaces. *Langmuir* 25(6): 3675–3681.

Xu, J., Y. Yuan et al. (2003). Ozone-induced grafting phosphorylcholine polymer onto silicone film grafting 2-methacryloyloxyethyl phosphorylcholine onto silicone film to improve hemocompatibility. *Colloids and Surfaces B: Biointerfaces* 30(3): 215–223.

Yamada, K. M. and S. Even-Ram (2002). Integrin regulation of growth factor receptors. *Nature Cell Biology* 4(4): E75–E76.

Yang, S. Y., D. Y. Kim et al. (2008). Stimuli-responsive hybrid coatings of polyelectrolyte multilayers and nanopatterned polymer brushes. *Macromolecular Rapid Communications* 29(9): 729–736.

Yang, M. C. and W. C. Lin (2003). Protein adsorption and platelet adhesion of polysulfone membrane immobilized with chitosan and heparin conjugate. *Polymers for Advanced Technologies* 14(2): 103–113.

Yang, Y., M. C. Porté et al. (2003). Covalent bonding of collagen on poly(L-lactic acid) by gamma irradiation. *Nuclear Instruments and Methods in Physics Research, Section B: Beam Interactions with Materials and Atoms* 207(2): 165–174.

Ye, H., Z. Gu et al. (2006). Kinetics of ultraviolet and plasma surface modification of poly(dimethylsiloxane) probed by sum frequency vibrational spectroscopy. *Langmuir* 22(4): 1863–1868.

Yoo, H. S., T. G. Kim et al. (2009) Surface-functionalized electrospun nanofibers for tissue engineering and drug delivery. *Advanced Drug Delivery Reviews* 61: 1033–1042.

Yoon, R. H., D. H. Flinn et al. (1997). Hydrophobic interactions between dissimilar surfaces. *Journal of Colloid and Interface Science* 185(2): 363–370.

Yoshida, T. (2001). Synthesis of polysaccharides having specific biological activities. *Progress in Polymer Science* 26(3): 379–441.

Yoshida, T., T. Akasaka et al. (1999). Synthesis of polymethacrylate derivatives having sulfated maltoheptaose side chains with anti-HIV activities. *Journal of Polymer Science Part A: Polymer Chemistry* 37(6): 789–800.

Yoshinari, M., T. Kato et al. (2006). Adsorption behavior of antimicrobial peptide histatin 5 on PMMA. *Journal of Biomedical Materials Research Part B: Applied Biomaterials* 77(1): 47–54.

You, L.-C., F.-Z. Lu et al. (2002). Glucose-sensitive aggregates formed by poly(ethylene oxide)-block-poly(2-glucosyl-oxyethyl acrylate) with concanavalin A in dilute aqueous Medium. *Macromolecules* 36(1): 1–4.

Yuan, Y., F. Ai et al. (2004a). Polyurethane vascular catheter surface grafted with zwitterionic sulfobetaine monomer activated by ozone. *Colloids and Surfaces B: Biointerfaces* 35(1): 1–5.

Yuan, J., L. Chen et al. (2004b). Chemical graft polymerization of sulfobetaine monomer on polyurethane surface for reduction in platelet adhesion. *Colloids and Surfaces B: Biointerfaces* 39(1–2): 87–94.

Yuan, J., C. Mao et al. (2003a). Chemical grafting of sulfobetaine onto poly(ether urethane) surface for improving blood compatibility. *Polymer International* 52(12): 1869–1875.

Yuan, Y., J. Zhang et al. (2003b). Surface modification of SPEU films by ozone induced graft copolymerization to improve hemocompatibility. *Colloids and Surfaces B: Biointerfaces* 29(4): 247–256.

Yuan, Y., X. Zang et al. (2004c). Grafting sulfobetaine monomer onto silicone surface to improve haemocompatibility. *Polymer International* 53(1): 121–126.

Zamir, E. and B. Geiger (2001). Molecular complexity and dynamics of cell-matrix adhesions. *Journal of Cell Science* 114(20): 3583–3590.

Zhang, Y., C. Chai et al. (2007). Fibronectin immobilized by covalent conjugation or physical adsorption shows different bioactivity on aminated-PET. *Materials Science and Engineering: C* 27(2): 213–219.

Zhang, Z., T. Chao et al. (2006a). Superlow fouling sulfobetaine and carboxybetaine polymers on glass slides. *Langmuir* 22(24): 10072–10077.

Zhang, Z., S. Chen et al. (2006b). Dual-functional biomimetic materials: Nonfouling poly(carboxybetaine) with active functional groups for protein immobilization. *Biomacromolecules* 7(12): 3311–3315.

Zhang, W., P. K. Chu et al. (2006c). Antibacterial properties of plasma-modified and triclosan or bronopol coated polyethylene. *Polymer* 47(3): 931–936.

Zhang, Z., R. Yoo et al. (2005). Neurite outgrowth on well-characterized surfaces: Preparation and characterization of chemically and spatially controlled fibronectin and RGD substrates with good bioactivity. *Biomaterials* 26(1): 47–61.

Zhang, J., J. Yuan et al. (2003). Chemical modification of cellulose membranes with sulfo ammonium zwitterionic vinyl monomer to improve hemocompatibility. *Colloids and Surfaces B: Biointerfaces* 30(3): 249–257.

Zhang, Z., M. Zhang et al. (2008). Blood compatibility of surfaces with superlow protein adsorption. *Biomaterials* 29(32): 4285–4291.

Zhao, C., X. Liu et al. (2003). Blood compatible aspects of DNA-modified polysulfone membrane—Protein adsorption and platelet adhesion. *Biomaterials* 24(21): 3747–3755.

Zhou, J., J. Yuan et al. (2005). Platelet adhesion and protein adsorption on silicone rubber surface by ozone-induced grafted polymerization with carboxybetaine monomer. *Colloids and Surfaces B: Biointerfaces* 41(1): 55–62.

Zhu, Y., K. S. Chian et al. (2006a). Protein bonding on biodegradable poly(L-lactide-co-caprolactone) membrane for esophageal tissue engineering. *Biomaterials* 27(1): 68–78.

Zhu, A. P., N. Fang et al. (2006b). Adhesion contact dynamics of 3T3 fibroblasts on poly(lactide-co-glycolide acid) surface modified by photochemical immobilization of biomacromolecules. *Biomaterials* 27(12): 2566–2576.

Zhu, Y., C. Gao et al. (2004a). Endothelial cell functions in vitro cultured on poly(L-lactic acid) membranes modified with different methods. *Journal of Biomedical Materials Research Part A* 69(3): 436–443.

Zhu, Y., C. Gao et al. (2004b). Endothelium regeneration on luminal surface of polyurethane vascular scaffold modified with diamine and covalently grafted with gelatin. *Biomaterials* 25(3): 423–430.

Zhu, H., J. Ji et al. (2004c). Construction of multilayer coating onto poly-(DL-lactide) to promote cytocompatibility. *Biomaterials* 25(1): 109–117.

Zhu, Y., M. Otsubo et al. (2006c). Loss and recovery in hydrophobicity of silicone rubber exposed to corona discharge. *Polymer Degradation and Stability* 91(7): 1448–1454.

Zhu, A., M. Zhang et al. (2002). Covalent immobilization of chitosan/heparin complex with a photosensitive hetero-bifunctional crosslinking reagent on PLA surface. *Biomaterials* 23(23): 4657–4665.

Zisch, A. H., U. Schenk et al. (2001). Covalently conjugated VEGF-fibrin matrices for endothelialization. *Journal of Controlled Release* 72(1–3): 101–113.

Zreiqat, H., F. A. Akin et al. (2003). Differentiation of human bone-derived cells grown on GRGDSP-peptide bound titanium surfaces. *Journal of Biomedical Materials Research Part A* 64A(1): 105–113.

Zwaal, R. F. A., P. Comfurius et al. (1977). Membrane asymmetry and blood coagulation. *Nature* 268(5618): 358–360.

16 Electrospinning for Regenerative Medicine

Toby D. Brown, Cedryck Vaquette,
Dietmar W. Hutmacher, and Paul D. Dalton

CONTENTS

16.1 INTRODUCTION

Regenerative medicine is an interdisciplinary field combining knowledge of cell biology, materials science, and medicine with the aim to create constructs capable of healing or replacing lost and damaged tissue due to sickness or accidents, eliminating the need for donated tissue. For example, the effective healing of critical size bone defects, as a consequence of injury or disease, is an increasing problem in an aging population. The majority of current treatment strategies involve autografts, where bone is extracted from another part of the patient's body, usually from the iliac crest or the fibula. Although this approach is often successful, there are issues associated with infections and hematomas due to the creation of a secondary injury at the donor site, while the amount of bone available is relatively scarce. Another approach involves using allografts; however, donated bone introduces the risk of transferring diseases between the donor and host, as well as increases the risk of infections and donor site rejection. Tissue engineering (TE) offers the possibility to avoid these problems by incorporating three-dimensional (3D) constructs, or scaffolds, with osteoinductive and osteoconductive properties to replace the auto/allograft.

Scaffold-based TE concepts often involve the combination of viable cells, biomolecules, and a structural scaffold combined into a tissue-engineered construct (TEC) to promote repair or regeneration [1]. The TEC is intended to support cell migration, growth, and differentiation and guide tissue development and organization into a mature and healthy state. Preferably, the scaffold should be absorbed by the surrounding tissue at the same rate as new tissue is formed so that by the time it breaks down the newly formed tissue will take over mechanical loads. Because this field is still in its infancy, the requirements for an ideal scaffold are complex and still not clear and may vary depending on specific tissue types and applications but involve consideration of factors including architecture, structural mechanics, surface properties, degradation properties and by-products, and the changes of these factors with time. However, it is widely accepted that scaffolds in TECs have minimum essential requirements. They should

1. Provide sufficient initial mechanical strength and stiffness to maintain a degree of mechanical function of the diseased or damaged tissue
2. Maintain sufficient structural integrity during the tissue growth and remodeling process in order to achieve stable biomechanical conditions and vascularization

3. Have a scaffold architecture that allows for initial cell attachment and subsequent migration into and through the matrix, mass transfer of nutrients and metabolites, and provide sufficient space for development and later remodeling of organized tissues
4. Have degradation and resorption kinetics based on the relationships between required mechanical properties, mass loss, and tissue development
5. Possess optimized surface properties for the attachment and migration of cell types depending on the desired tissue. Furthermore, the gross size and shape may be customized for a particular application and individual patient [2]

In TE, cellular regeneration requires scaffold support systems to guide the growth of cells and stimulate the development of native extracellular matrix (ECM). Cell proliferation and adhesion are dependent on the specifications of the scaffolds, especially upon the porosity and interconnectivity. Scaffolds for TE are obtained through a variety of fabrication techniques, described elsewhere, each with advantages and disadvantages [3]. These include rapid prototyping, 3D printing; selective laser sintering; extrusion/ direct writing; inkjet or organ printing; solid free-form fabrication; and stereo-lithography. However, these techniques have limits to the degree of miniaturization possible. The latest trend in TE has been to use solution electrospun scaffolds, which offer fibers with diameters comparable to those of the fibrils typically found in tissue ECM, rather than other methods for in vitro studies. Electrospinning is a technique that fabricates continuous nano- and microscale filaments with potential for regenerative medicine, in particular for growth factor delivery and TE. Even though the production of filamentous material for such biomedical applications has been used for many decades, electrospinning was only recently proposed for TE (in 2002) [4,5] and growth factor delivery (2005) [6,7]. Electrospun materials (also described as meshes) have widely been proposed as TE scaffolds, despite the clear evidence that many of these materials do not form a cell-invasive structure, which is a major criterion for a TE scaffold. Exceptions to this include specific strategies that introduce porosity into the electrospun mesh, such as co-electrospraying [8,9], electrospinning onto ice crystals [10,11], and the formation of bimodal scaffolds, which deliberately contain differently scaled fibers [12,13]. In comparison, growth factor delivery from these meshes does not necessarily require cell penetration and has been used in surgical paradigms. In this review, we focus on the approaches to make and modify electrospun scaffolds, rendering the material suitable for applications in regenerative medicine.

16.2   FUNDAMENTALS OF ELECTROSPINNING

There are numerous configuration permutations for electrospinning, but a schematic of the most common electrospinning design—using a single collector—is shown in Figure 16.1. In electrospinning, an electrical potential difference is generated between a liquid and a collector, which is

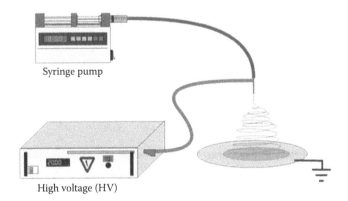

Syringe pump

High voltage (HV)

FIGURE 16.1   (See companion CD for color figure.) Schematic of a typical electrospinning setup.

achieved by applying high voltage to either the liquid [5,14], collector [13,15], or both [16]. A spinneret (small orifice or flat-tipped needle) may be used; however, nonspinneret "nozzle-less" systems have also been adopted [17,18]. Typically, as shown in Figure 16.1, a (positive or negative) voltage is applied to a polymer solution or melt that is pumped through a spinneret facing a collector at a different electrical potential. As the polymer droplet emerging from the spinneret becomes charged, electrostatic repulsion counteracts the surface tension and the droplet is stretched. Upon reaching a critical voltage, the surface tension of the polymeric fluid is overcome by the charge buildup, and the droplet elongates into a (Taylor) cone, where a continuous jet is ejected and drawn by the electrostatic force toward the collector. The potential difference between the spinneret and collector typically ranges from 10 to 20 kV, although voltages as low as 0.5 kV [19] and as high as 130 kV [20] have been reported.

The ejected electrospinning polymeric jet, which normally consists of a stable region and an unstable zone (usually called bending or whipping zone), has been subject to significant mathematical analysis. If there are sufficient molecular entanglements in the polymer, the jet does not break up into droplets due to Raleigh instabilities. Initially, the electrified jet is stable and travels directly toward the collector. The surface charge density of the liquid (which caused the ejection of the jet from the droplet) again increases with decreasing distance to the collector and "twists" the jet, resulting in bending instabilities. In this zone, the instabilities in the jet's path cause it to spiral or "whip," where it rapidly accelerates laterally to the flight path in a conical envelope, leading to further stretching. As the electrified jet passes through the air, the solvent evaporates or the molten polymer cools, depending on whether electrospinning of a polymer solution or melt is occurring, and the jet undergoes magnitudes (up to 5 orders) in diameter reduction by the time it reaches the collector. Electrospinning can be performed with the spinneret(s) and collector(s) configured so that the path of the jet is either vertical or horizontal, since the effects of aerodynamics and gravity are minimal compared to those of the electrostatic forces involved [21]. The path of the polymer solution jet, in particular, is extremely dynamic, and the fiber collection is rapid. Such high speeds, plus long spiraled traveling distances, make accurately controlled deposition of the electrospun fiber technically challenging. Electrospinning commonly results in a nonwoven mat of fibers, although other collection techniques shown here are expanding the morphological nature of the scaffolds.

A variation of electrospinning is melt electrospinning [22], which avoids issues such as toxicity and solvent removal associated with solution electrospinning. Melt electrospinning behaves differently to solution electrospinning, in that the initial stable region of the jet is relatively much longer and thus there is less lateral movement of the jet [23] resulting in less reduction in the diameter of the jet. However, the larger micron scale fibers that typify melt electrospinning may offer the ability to fabricate thicker structures with greater pore size and porosity, promoting more cell invasiveness. Advances in how melt electrospun fibers can be collected may provide patterned aligned fibrous substrates useful for building complex 3D structures that are of interest in applications such as TE [22,24].

16.3 POLYMERS USED FOR ELECTROSPINNING IN REGENERATIVE MEDICINE

There are literally hundreds of different polymers that have been electrospun, and extensive lists of these polymers can be found in other articles [25–28]. Tables 16.1 through 16.3 provide a selection of polymers chosen for electrospinning applications in regenerative medicine. Polymers can be categorized into two distinct classes—those derived from biological sources and those that are synthetically produced.

16.3.1 BIOLOGICALLY DERIVED POLYMERS

It is understandable that when trying to replace ECM fibrils within the body, the most obvious polymers to use are those that constitute the natural ECM. Such biologically derived polymers (Table 16.1) provide biological signals to cells and are typically degradable materials.

TABLE 16.1
Selected Biologically Derived Polymers

Polymer	Reference	Polymer	Reference
Collagen	[5,29,32]	Hyaluronic acid	[33,34]
Gelatin	[35,36]	Cellulose	[37]
Silk	[31,38]	Laminin	[39,40]
Chitosan	[41]	Fibronectin	[42,43]
Fibrinogen	[44,45]	Chitosan/agarose	[46]

TABLE 16.2
Selected Synthetic Polymers

Polymer	Reference	Polymer	Reference
PLGA	[13,48,49]	PLLA	[50,51]
PCL	[52]	PLLA-CL	[53,54]
PS	[55,56]	PEG-PCL	[57,58]
PU	[8,59–61]	PEO	[62,63]
PNIPAM	[64,65]	sP(EO-stat-PO)-PLGA	[66]

TABLE 16.3
Selected Hybrid Electrospun Systems

Polymer	Reference	Polymer	Reference
Collagen/PCL	[67]	Silk/PEO	[68]
Gelatin/PLGA	[69]	HAp/PLLA	[70]
Chitosan/PVA	[71]	Hap/Collagen/PVA	[72]

However, the process of dissolving the polymer into a solvent can create problems, particularly denaturation. For example, collagen can be denatured to gelatin during electrospinning, due to both the solvent used (often 1,1,1,3,3,3-hexa-fluoro-2-propanol [HFIP]) and the high voltage applied [29]. Silk fibroin is also used; however, it requires treatment in order to dissolve into a solution capable of being electrospun. A second issue with silk is also a generic issue with the use of aqueous solutions: a degree of post-treatment is needed after generating the electrospun fibers so that they do not redissolve upon placement in aqueous media [27,30,31]. Biologically derived polymers have also been blended with synthetic polymers to improve the processing ability of the fiber.

Numerous studies that blend different biologically derived polymers, such as chitosan and agarose [46], gelatin and hyaluronic acid, are reported [47]. In addition to the selected articles listed in Table 16.1, in-depth reviews on the electrospinning of biologically derived polymers are available [27,28].

16.3.2 SYNTHETIC POLYMERS

The number of different synthetic polymers electrospun (Table 16.2) is definitely greater than that of those derived from biological sources. Synthesized polymers have widely been electrospun due to the benefits of reproducibility and the ability to tailor important material properties such as molecular weight, which appears to facilitate improved electrospinning than when using biologically derived polymers. There are many specialized reviews on various polymers, with poly(caprolactone) (PCL) and polyethylene glycol (PEG) being two of the most studied polymers for regenerative medicine.

The lack of biological signals provided by synthetic polymers (to trigger and control cellular responses) requires the adsorption of proteins onto the surface, or a surface-engineered strategy: one common approach is to blend biologically derived polymers into a synthetic polymer solution in order to incorporate protein into the fibers. However, denaturation of the protein is still an issue. Furthermore, the two different polymers often phase separate into separate domains because of their intrinsic properties. For example, collagen (hydrophilic) and PCL (hydrophobic) are immiscible and will phase segregate. Table 16.3 provides a list of some commonly blended "hybrid" polymers.

16.4 ELECTROSPINNING PARAMETERS

It is widely appreciated that three broad factors influence the electrospinning process, and thus the morphology of the resulting fibers: the properties of the polymeric fluid used; the instrument conditions (e.g., voltage and flow rate); and the configuration of the instrument. The most widely reported characteristic of electrospun material is the diameter of the fibers. The diameter is known to increase with increasing polymer concentration and also decrease with voltage applied. While the diameter is an important aspect and is discussed later, the morphology and structure of the electrospun materials significantly have impacts on other important properties such as pore interconnectivity and cell invasiveness. Due to the large number of parameters that can be altered, and the interdependency of most of these parameters, studying the effects of the processing conditions on scaffold morphology can become excessively difficult. However, through careful experimental design it is possible to deduce definitive relationships.

16.4.1 POLYMERIC PARAMETERS

When electrospinning polymeric solutions, there are a number of parameters that can be altered to generate different types of structures or different fiber diameters.

16.4.1.1 Solution Viscosity

One of the most important electrospinning parameters is the viscosity of the polymer solution: i.e., the degree of macromolecular entanglements in the polymer that prevents breakup of the liquid jet into droplets. One way by which viscosity can be controlled is by adjusting the concentration of the polymer solution. Chain entanglement can be determined by the molecular weight used and also depends on the polymer type (linear or branched). There is a strong relationship between polymer concentration and fiber diameter, which is well described for a large number of polymers in a large number of studies. Usually highly concentrated solutions produce relatively thick and bead-free fibers. Mckee et al. [73] extensively studied the electrospinnability and resultant fiber morphology of polymer solutions with respect to their rheological properties. The dependence of polymer viscosity on solution concentration can be described by identifying four distinct concentration regimes: the dilute, semidilute unentangled, semidilute entangled, and the concentrated. The boundary between the semidilute un/entangled regimes is named entanglement concentration C_e and corresponds to the point at which there is significant overlap of the polymer chain to constrain the chain motion causing entanglement and couplings [73]. McKee observed that for a concentration below C_e only polymer beads were obtained. Whereas for concentrations between C_e and 2–2.5 times C_e, beaded fibers form, for concentrations above 2.5 times C_e, uniform bead-free fibers occur. This work showed that there exists a minimum polymer concentration required for solution electrospinning. A fiber diameter prediction model was proposed where the diameter universally scales with the normalized concentration $(C/C_e)^{2.6}$, where C is the concentration of the polymer solution and C_e the entanglement concentration [73].

Typically, electrospinning of natural polymers in aqueous solutions is challenging due to a number of factors including polymer chain rigidity (which reduces the degree of polymer chain entanglements) and high surface tension in the solution. Increasing the entanglements has been shown to be an efficient way to facilitate the aqueous electrospinning of alginate, a natural polymer [74]: in this strategy better electrospinnability was achieved by using glycerol, a strong polar solvent, which is believed to disrupt the strong intramolecular structure of the polymer. Glycerol also forms new hydrogen bonds with the polymer chains and improves their flexibility. As a result, the viscosity of the solution usually increases due to increased entanglements, whereas surface tension decreases, enhancing the electrospinnability of these polymers. A similar effect was observed when electrospinning hyaluronic acid in a mixture of water, formic acid, and dimethylformamide [75]. Here again the rigid inner structure of the natural polymer was disrupted by the formic acid, which changed the chain conformation from alpha helix to a coil conformation improving chain flexibility, increasing entanglements and therefore electrospinnability.

16.4.1.2 Surface Tension

During the electrospinning process, a jet forms once the electrostatic forces overcome the surface tension in the droplet emerging from the spinneret. Therefore, the surface tension is a critical parameter and in polymer solutions it can be controlled by blending solvents [76–78], adding a surfactant or salt [79], or varying the temperature [80]. It is generally assumed that high surface tension solutions are difficult to electrospin and require higher voltage for the initiation of the jet. Because it tends to minimize the surface area per mass of a fluid, resulting in a spherical shape, surface tension is also thought to participate in the formation of beads on fibers when electrospinning relatively low or moderately viscous solutions. Generally, reduced surface tension produces bead-free smooth fibers [79].

16.4.1.3 Solvent Conductivity

The forces that tend to elongate the electrospinning jet are created by a buildup of charge within and on the surface of the polymer droplet. For polymer solutions, the net charge carried by the jet is directly bound to the conductivity and the mobility of these charges in the solution. Generally, greater stretching of the jet, and thus thinner fibers, is to be expected when the net charge of the jet is increased, which is facilitated by an increased conductivity. Furthermore, in most cases increasing the conductivity of the solution enhances the bending instability (because it enhances the charge repulsion). Therefore, the jet lengthens and is subject to more whipping, causing more stretching and leading to further reduction in fiber diameter [81].The significance of this electrospinning parameter is demonstrated by the fact that changes in the conductivity of the same solvent can lead to discrepancies in fiber diameter and scaffold morphology from one experiment to another [55]. For the same conditions and using the same solvent but from different suppliers, Uyar et al. [55] obtained both beaded and smooth fibers. They correlated these varying fiber morphologies to the differences in the solution conductivities, which varied from 0.7 to 7 µS/cm. The highest conductivity produced the smallest fibers.

The solvent or combination of solvents that are part of electrospun fiber preparation affects the fiber structure, which in turn affects cell behavior [82]. Solvent mixtures have widely been used to alter, generally to increase, the conductivity of the electrospun polymer solution. This approach is valid as long as the viscosity of the solution does not vary too greatly; otherwise, conclusive observations on the effects of conductivity may be difficult. In some cases, adding a highly conductive solvent without a significant variation in the viscosity of the solution is possible [76]. When dimethylformamide (DMF) is added to methylene chloride, the viscosity of a 13 wt% PCL solution does not change, but the conductivity increases up to 10-fold, responsible for a sharp decrease in fiber diameter [76]. The stability of the process may also benefit from the addition of a highly conductive solvent. For example, chloroform or dichloromethane solutions can have an instable Taylor cone, and consequent spinneret clogging can require the operator to clean it frequently. However, upon addition of DMF to the solution the process is stabilized and can be run for several hours without the need to clean the spinneret.

Another way to enhance the conductivity of a polymer solution is to add salt. A substantial amount of literature on this subject has generally shown that thinner fibers are produced when the salt concentration is increased [81,83,84]. It has also been observed that increasing the salt concentration tends to reduce bead formation [85]. Interestingly, while comparing several types of salt, Zong et al. [83] observed that ions with smaller atomic radii produce thinner fibers. This effect was attributed to the higher charge density of smaller ions, providing higher mobility in the solution and therefore producing stronger elongational forces.

Although an increase in polymer solution conductivity generally leads to the production of thinner electrospun fibers, the opposite trend has also been reported [80]. This may be due to additional mass flow created by higher elongational forces.

16.4.2 Instrument Parameters

16.4.2.1 Flow Rate

The effect of mass delivery, or flow rate, on the nature of electrospun scaffolds has been studied extensively [24,83,86–88]. It is widely accepted that increasing the flow rate produces larger fibers: as more material is extruded through the spinneret per unit time, the volume of the Taylor cone increases for the same collection distance. This leads not only to larger diameter fibers but also to the deposition of wet fibers, which may fuse together due to a greater amount of residual solvent. Therefore, more fusion at the points of contact between solution electrospun fibers is generally expected for higher feed rates. This can significantly affect the mechanical properties of the resultant meshes (high fusion density results in tougher scaffolds) and affect the interconnectivity and therefore the cell invasiveness of the construct. For given conditions, increasing the flow rate can induce instability into the Taylor cone. This usually happens when the amount of polymer solution exceeds the amount of polymer that can be drawn by the electrostatic forces. As a result, breaking up of the jet can be observed, and fibers with large beads can also be formed [89]. Therefore, increasing the voltage may be necessary to maintain a stable jet when the flow rate is increased.

16.4.2.2 Applied Voltage

Since an electrical field is inherently required for electrospinning to be possible, the applied voltage plays a major role in the electrospinning process. In most cases, a direct-current (DC) apparatus is utilized to produce the high voltage. However, alternative-current (AC) setups are also possible. The main advantage of using AC voltage resides in the ability to electrospin or electrospray onto nonconductive collectors [90]. Using AC voltage, the alternating polarity induces a neutralizing effect, leading to the reduction of net charge on the jet. Consequently, compared to using DC voltage, the electrostatic forces that produce jet instability and thus the whipping instability are significantly reduced. As a result, fibers obtained using AC voltage are generally larger than those fabricated using DC voltage [90].

Several phenomenological observations can be made when the DC voltage is increased during electrospinning. Firstly, the shape of the Taylor cone will change from a droplet to a spindle. Several studies agree that there is a reduction in the volume of the Taylor cone with increasing voltage (for a given flow rate) [83,91]. As the voltage is increased, the jet acceleration is greater and more volume is drawn from the Taylor cone; therefore, the cone volume and also the jet diameter is reduced [83]. Secondly, increasing the voltage may cause the formation of multiple jets from the Taylor cone [92]: higher voltage favors the formation of thin secondary jets, which may broaden the fiber diameter distribution [81] or produce bimodal fiber distributions [93]. Usually, when secondary jets are not formed the fiber diameter distribution appears to narrow [94–96]. When the voltage is further increased, the Taylor cone can even recede into the spinneret. Thirdly, when the voltage becomes too high corona discharge can be observed, perturbing the deposition [97].

It is generally accepted that an increase in electrospinning voltage produces a decrease in fiber diameter [86,94,95]: as the jet is increasingly accelerated it is further stretched. Katti et al. observed that after an initial sharp decrease in fiber diameter, a constant diameter is obtained when the voltage is further increased [95]. However, several other studies report no change or increase in the fiber diameter at higher voltages. Therefore, there is no definitive consensus on the effect of the applied voltage; which appears to be dependent on the solution used.

At each voltage extreme (too low or too high voltage), bead formation has frequently been reported. However, once beads appear during fiber deposition, the higher the voltage the higher the bead density [86,91,98]. Lee et al. observed that the morphology of beads from a low viscosity solution could also be controlled by the applied voltage, where they adopted a more pronounced spindle shape at higher voltages [99].

16.4.2.3   Spinneret Diameter

The diameter of the spinneret through which the polymer is delivered during electrospinning plays a significant role in controlling fiber diameter as it mostly determines the initial size of the Taylor cone. Several studies have shown that smaller spinneret diameters produce smaller fibers [94,95]. A smaller diameter orifice also increases the surface tension of the emerging droplet and thus the critical voltage required to initiate the Taylor cone. It is also believed to reduce bead formation and clogging of the spinneret [94], while a smaller Taylor cone provides less surface area in contact with the air for solvent evaporation. However, it can prove difficult to extrude highly viscous solutions through small diameter orifices, and the buildup of back pressure in syringe-based systems may affect the feed rate and therefore consistent fiber deposition.

16.4.2.4   Temperature

The control over the solution/melt temperature can be obtained by developing a jacket-type heat exchanger in which water [80,100,101] or silicone oil [102] is circulated. Other technical solutions can be utilized such as glass fiber heating tape or infrared emitter heating systems [103]. Increasing the temperature generally decreases the fiber diameter since the viscosity of the solution is greatly reduced [92]. Wang et al. systematically studied the effect of elevated temperature over the properties of polyacrylonitrile solution in DMF and on the morphology of the resulting fibers [102]. This study found, as expected, that the viscosity of the solution decreased, but also that the entanglement density was not significantly affected by the elevated temperature. In addition, the surface tension of the solution was found to decrease with temperature, whereas the opposite trend was observed for the conductivity, which increased with temperature. The later phenomenon was attributed to the higher mobility of the molecules caused by the increased temperature. The reduction of the viscosity, surface tension, and the increase in the conductivity lead to smaller fiber diameters.

Besides reducing the fiber diameter, elevated temperatures have other implications for the process of electrospinning. Demir et al. observed improved uniformity in fiber diameter distribution when electrospinning a polyurethane solution at 70°C compared with room temperature [92]. Wang et al. also reported the increase in the Taylor cone volume, which resulted in an increase in the voltage required to electrospin the polymer solution [102]; In Wang's and the work by Givens et al. [103], a high electrospinning temperature was responsible for a slight decrease in polymer crystallinity. Electrospinning at elevated temperatures can also be used to dissolve polymers that are insoluble at room temperature: linear low density polyethylene (LLDPE) was successfully electrospun at 105°C–110°C using p-xylene as a solvent [103]. This strategy can also be used to improve the incorporation efficacy of poorly soluble drugs into fibrous polymer matrices, because the solubility of most compounds is improved at high temperatures [104].

16.4.2.5   Collection Distance

Varying the collection distance, also known as "tip-target distance" or "working distance," strongly impacts the strength of the electrical field and the time of flight of the polymer jet.

Although it is generally accepted that a collection distance that is too short will create adverse effects, such as excessive fiber fusion caused by the low solvent evaporation, there is no global consensus on the effects of collection distance on fiber morphology. Several studies have shown very small or insignificant changes in fiber diameter and morphology when increasing the collection distance [81,105,106], while others have observed significant changes in the fiber diameter [107]. However, the collection distance directly affects the volume of the Taylor cone and plays a significant role in the initial diameter of the jet [105]: small distances increase the field strength and tend to reduce the volume of the Taylor cone. Smaller distances may also lead to greater acceleration of the jet, but not necessarily greater stretching, where induced bead formation has been reported [86]. Longer distances reduce the field strength and have also been associated with the formation of beads or increased bead diameter [108]. Collection distance appears to have less of an effect on fiber diameter and morphology than deposition area. As the collector is moved further away from the spinneret, the base of the cone formed by whipping becomes larger, resulting in the deposition of the same quantity of material over a larger surface area. Therefore, thinner scaffolds are obtained for the same deposition time. When the working distance is increased, the time of flight is also increased. Consequently, this allows more solvent to evaporate and reduces the fusion density between the fibers. This can have a significant impact upon the mechanical properties of the scaffolds, as more fiber fusion increases the tensile strength of the electrospun mat. Therefore, varying the collection distance could also be a means to control the macroscopic mechanical properties of the resultant scaffolds. Although the deposition of melt electrospun fibers is relatively more focused compared to solution electrospinning, a similar effect is experienced where longer collection distances result in less focus of the deposited fibers [24].

16.4.2.6 Humidity

Humidity is not always controlled during the electrospinning process; however, it seems to play a significant role in the morphology of the fibers obtained. There are several studies that demonstrated the possibility of tailoring the surface morphology of the fiber by controlling the ambient humidity of the electrospinning chamber. It has been shown that increasing the relative humidity leads to the formation of spherical "pores" on the fiber surface, and it is generally accepted that the higher the relative humidity, the larger the pore diameter until the pores fuse together [109,110]. The formation of such pores is attributed to the condensation of water molecules onto the fibers during deposition. As a consequence, the specific surface area is greatly increased by the presence of these pores on the fibers [110]. Relative humidity can have significant impact not only on the superficial pattern of a single fiber but also on the gross architecture of an entire scaffold. When electrospinning an aqueous-based solution with various humidity contents, it has been observed that larger diameter fibers are formed at low humidity percentages, due to the higher evaporation rate of the water in the "dry" environment, resulting in less drawing of the jet by the applied voltage. Whereas the fiber diameter was observed to decrease when the relative humidity was increased, up to a point where beads were formed [111].

16.4.3 Instrument Configuration

This section focuses on the utilization of different configurations to generate discrete structures within an electrospun matrix. The variation of instruments described in the literature is far greater than presented in this chapter. However, the focus here is on the ability to discretely position and place fibers in specific locations and patterns. Central to many fibrous systems are woven materials, and therefore creating woven structures or threads using electrospinning is an interesting and relevant component of TE. Initially, static electrospinning instrument configurations are discussed, followed by dynamic collection systems that result in structured electrospun substrates.

16.4.3.1 Single Collection Systems

Using a single, solid collector is the most commonly used electrospinning configuration and results in a nonwoven mesh that is non-cell invasive. It is usually the arrangement that most polymeric materials are initially electrospun onto, since the nonwoven mesh is readily handled and manipulated. In this configuration, the electric field between the spinneret and the collector is uniform or symmetrical. Although in most cases electrospun fibers are deposited onto a solid collector (Figure 16.2a and b), common liquids such as water can be used for fiber collection [112,113]. A straightforward method of collecting a continuous yarn from electrospun fibers by first depositing them onto a liquid medium (as shown in Figure 16.2c) was not demonstrated until 2005 by Smit et al. and Khil et al. [114,115], even though Formhals had patented various continuous fiber yarn setups in the 1930s [116,117]. This was an important technical breakthrough, because of the potential for continuous yarns of electrospun nanofibers to be woven into textiles, with applications including protective clothing, high performance fabrics, composites, and TE [118]. The method proposed by Smit and colleagues [114] is worthwhile for further discussion. It involved the deposition of

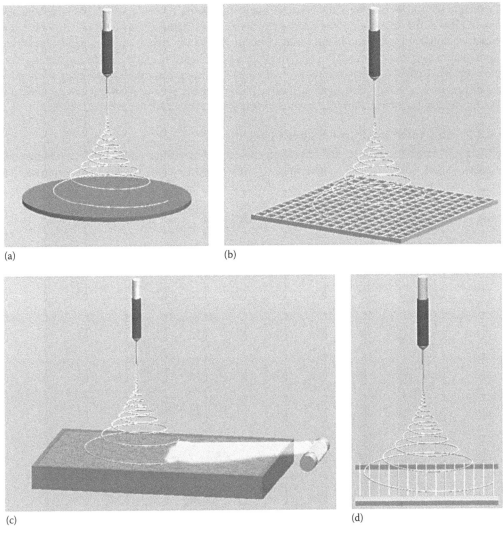

(a)   (b)

(c)   (d)

FIGURE 16.2 (See companion CD for color figure.) Various collection systems for electrospinning. (a) Single collector; (b) Patterned collector; (c) Collection onto water with rotating mandrel; and (d) Collection across two parallel collectors.

randomly oriented electrospun fibers onto the surface of a water bath, which were then drawn to the edge and collected on a rotating mandrel. This caused elongation of the mesh and alignment of the fibers as they were drawn over the water. Surface tension caused the fibrous mesh to collapse into a yarn when it was lifted off the surface of the water onto the rotating collector. As the resultant yarn was rolled onto the mandrel at a rate of 3 m/min, more fibers were deposited on the water surface to feed the drawing process. Electrospun poly(vinyl acetate), poly(vinylidene fluoride), and polyacrylonitrile nanofiber yarns were fabricated using this method. Khil et al. employed a similar concept but rotated the mandrel at 30 m/min and used a mixture of water and methanol [115]. However, no comment was made on the difference between yarn collected on water or water/methanol. According to Teo and Ramakrishna [118], the high surface tension provided by water may be favorable to other liquids with lower surface tension because the electrospun fibers may become submerged, creating a higher drag force as they are drawn to the rotating collector. More studies are required to determine the influence of the liquid bath properties on the yarn.

16.4.3.2   Dual Collector Systems

An example of a dual collector system is schematically illustrated in Figure 16.2d while a photograph in Figure 16.3, shows highly aligned fibers across the separation between two rings placed in parallel, equidistant from the spinneret were deposited [119]. An important advantage of this type of approach, which is also termed the "gap method of alignment," is that the perimeters of the electrode rings where the fibers deposit allow precise and consistent controlled positioning of the fiber bundle (Figure 16.4) [120]. Such a gap method of alignment has been used to manufacture oriented fibers for in vitro use, particularly for applications in the nervous system [16,67,121–123].

16.4.3.3   Structured Electroconductive Collectors

Methods to create patterned 2D and 3D electrospun structures using static electroconductive patterned collectors have also been developed. However, the random formation of honeycombed

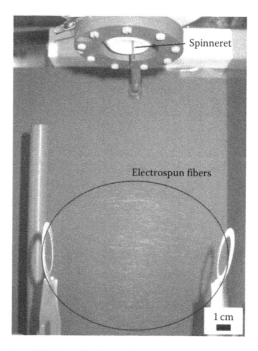

FIGURE 16.3   (See companion CD for color figure.) Oriented electrospun PCL fibers collected using the gap method of alignment. The collectors are spaced 10 cm apart, and the oriented fibers that collect in between are generated in approximately 30 s. (Previously unpublished photograph.)

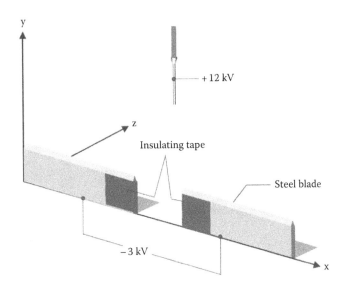

FIGURE 16.4 Experimental setup for precise deposition of electrospun fibers over two known points. (Reproduced from Teo, W.E. and Ramakrishna, S., *Nanotechnology*, 16, 1878, 2005.)

[63,124] and dimpled structures [125] using this approach has been observed, due to the buildup of electrostatic charges on the deposited fibers and/or the collector, which prevent new deposition directly onto the collector. Furthermore, it is currently reported that the fibers forming 3D structures are very loosely packed and can be easily compressed: rendering them unsuitable for applications where structural integrity is required [125]. Thus, the conditions required to form 3D structures using this method are still unclear.

Patterned electro-conductive collectors, either in the form of ridges and indentations or grids with pores in between, have been used to form patterned 2D solution electrospun meshes. Ramakrishna et al. reported the use of a collector with conducting ridges and nonconducting indentations where the electrospun fibers were randomly deposited according to time; however, the resultant mesh resembled the pattern of the collector [125]. Aligned fibers formed on the conducting ridges and randomly deposited fibers on the nonconducting regions. Zhang and Chang also demonstrated that protrusions/patterns in the collectors are important parameters that may greatly affect the structures of electrospun meshes (Figure 16.5) [126,127]. The density of uniaxially aligned fibers deposited between the protrusions on a collector was shown to be dependent on the distance between two protrusions, decreasing with increasing distance between the protrusions. Woven structures were then generated through time-dependent control of the arrangement of a patterned series of electroconductive protrusions: one pair of thin parallel-aligned protrusions, insulated from one another, was mechanically raised higher than the other pairs of protrusions. Fibers were drawn to deposit on the higher pair due to stronger electric attraction forming a parallel-aligned fiber bundle between the protrusions. After a certain time, the first pair was lowered and another pair oriented at 90° was raised and allowed to collect for a similar period. The newly deposited fibers were also parallel but oriented perpendicular to the initial fiber bundle. The process was repeated with additionally located protrusions at 90° to create woven fibrous architectures [127].

16.4.3.4 Rotating Collectors

Several researchers have shown that it is possible to obtain aligned fibers by using a rotating collector/mandrel [5,128,129]. Matthews et al. demonstrated the effect of the rotating speed of a mandrel on the degree of solution electrospun collagen fiber alignment [5]. At a speed of

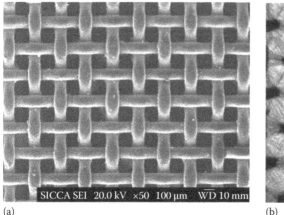

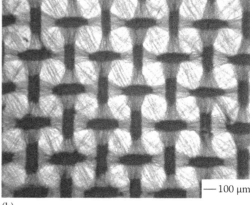

(a)   (b)

FIGURE 16.5 (a) Patterned electro-conductive collector and (b) resulting patterned electrospun mesh. (Reproduced from Zhang, D.M. and Chang, J., *Adv. Mater.*, 19, 3664, 2007.)

less than 500 rpm, a random mix of collagen fibers was collected. However, when the rotating speed of the mandrel was increased to 4500 rpm (approximately 1.4 m/s at the surface of the mandrel), the collagen fibers showed significant alignment along the axis of rotation. At such a high rotational speed, the velocity of the electrospinning jet may be slower than the linear velocity of the rotating mandrel. The reported average velocity range of the electrospinning jet is from 2 m/s [130,131] to 186 m/s [114]. A separate study by Zussman et al. using a rotating disk collector demonstrated that at high enough rotational speeds, necking of the electrospun fibers occurs [132]. Therefore, the mandrel rotational speed directly influences fiber orientation as well as the material properties of the engineered fibrous structure. In cases where disoriented fibers are collected using a rotating mandrel at sufficient speed, residual charge accumulation on the deposited fibers may interfere with the alignment of incoming fibers. According to Teo and Ramakrishna, reducing this residual charge accumulation on the rotating mandrel as well as reducing the chaotic path of the jet are possible ways to achieve greater fiber alignment, in solution electrospinning applications [120]. Kessick et al. used an AC high-voltage supply instead of the typically used DC supply when solution electrospinning polyethylene oxide (PEO) [90] onto a rotating mandrel to improve fiber alignment. Using an AC potential to charge the solution may have a dual function: the electrospinning jet may consist of short segments of alternating polarity, which significantly reduce its chaotic path and allows the fibers to be wound onto the rotating mandrel with greater ease and alignment. Then, given the presence of both positive and negative charges on the surface of the rotating mandrel, there may be a neutralizing effect over the area of the fibers, thus minimizing residual charge accumulation. Another simple approach to reduce charge accumulation is to apply the opposite polarity to the collector than to the spinneret, instead of grounding the collector [16].

Although the methods incorporating a rotating device described so far have all formed 2D aligned fiber meshes, 3D conduits can be fabricated by depositing fibers onto smaller diameter (<5 mm) collectors. For the construction of vascular grafts, Stitzel et al. electrospun a conduit 120 mm long and 1 mm thick using a polymer solution containing a mixture of collagen type I, elastin, and poly(D,L-lactide-co-glycolide) on a circular mandrel of diameter 4.75 mm [133]. However, in this case the rotational speed was only 500 rpm to facilitate homogeneous deposition of random fibers. Longitudinal and circumferential mechanical testing showed no significant difference between the two directions. Another advantage offered by the electrospinning process when forming small diameter conduits for vascular grafts is the ease of spinning different materials to form layered composites [134]. This way, materials such as collagen, which encourages cell proliferation and attachment but is mechanically weak, can be electrospun as an inner layer, while artificial materials such as polymers with superior mechanical properties can form an outer layer.

16.4.3.5 Combining Electric Field Manipulation with a Rotating Collector

The combination of electric field manipulation and a dynamic collector has been used to achieve greatly ordered fiber assemblies. Theron et al. used a knife-edged rotating disk collector to take advantage of the fiber alignment created by rotary motion as well as the convergence of electric field lines toward the edge of the disk [135]. To collect the fibers, glass coverslips or any nonconducting substrate can be attached around the edge of the discs used in TE studies [136]. Crossed fiber arrays were obtained by Bhattarai et al. using a similar concept to the rotating disk collector, but instead winding copper wire as an electrode on an insulated cylinder and then rotating it at an optimum speed of ~2000 rpm [137]. The size of the fiber bundle was controllable by varying the diameter of the collector. Instead of using static electrodes placed in parallel, Katta et al. used a rotating wire drum [138], whereas Sundaray et al. used a sharp pin as an electrode on a rotating collector to focus deposition of the electrospinning jet and introduced lateral movement to the collector to obtain cross-bar patterned fibers [131]. Instead of situating an electrode directly below the spinneret to focus electrospun fiber deposition, Teo et al. incorporated a negatively charged knife-edged electrode placed at an oblique angle to the spinneret [120]. The electrospinning jet was observed to travel at the same angle as that of the diagonal electric field lines from the tip of the spinneret to the knife edge. Diagonally aligned fibers were then collected on a rotating tube placed in the path of the electrospinning jet. This enabled the fabrication of laminar composites with aligned fibers in different orientations. While the combination of dynamic mechanical devices and electrical field manipulation leads to parallel arrays and grids of electrospun fibers with improved alignment, these methods do not provide sufficient control over the pitch width or the ability to create complex patterns such as circles [139].

16.4.3.6 Electrospinning Writing

Currently, a limited number of writing approaches using electrospinning are available. Buckling phenomena have been studied where electrically charged solution jets depositing onto collectors moving laterally at a constant velocity produced buckling patterns with limited controllability [140]. While in a different study, the bending and buckling of a molten polymer jet was shown to be close to the collector [141]. There are processes that take advantage of the initially stable region of the solution electrospinning jet, by significantly reducing the spinneret-to-collector distance to below that at which bending instabilities develop along the jet. This results in relatively focused fiber deposition. Then, matching the speed of a translating collector to that of the electrospinning jet enables the ability to write with a continuous filament. Scanning tip electrospinning [128] and near-field electrospinning [19,139,142] processes have demonstrated relatively predictable control over the deposition of submicron fibers using an automated x–y stage as the translating collector. Alternatively, a polymer melt jet will remain stable over relatively large distances and is well directed toward the collector. Dalton and colleagues utilized this phenomenon in combination with a translating collector, demonstrating the accurate patterning and drawing of a cell adhesive scaffold with a PEG-b-PCL/PCL blended melt across the surface of a microscope slide [24]. Recently Brown et al. demonstrated a direct writing process using melt electrospun pure PCL, where complex patterns could be drawn with a continuous filament. Layers of melt electrospun fibers were then consistently located on top of each other, to build complex 3D structures up to 1 mm in thickness (Figure 16.6). This process also enabled the fabrication of porous TE scaffolds, where control over the location of fiber placement was demonstrated with micron scale resolution [143].

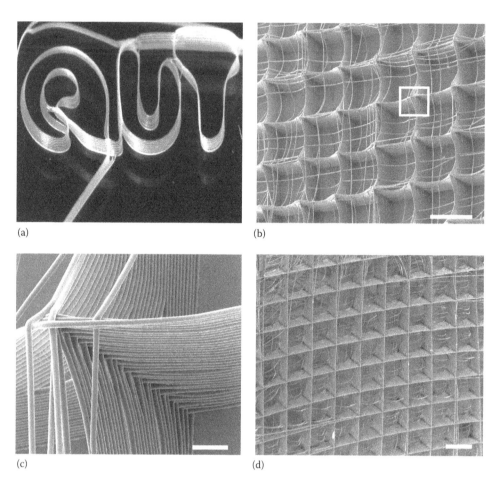

FIGURE 16.6 Electrospinning writing of a polymer melt. Photograph of Queensland University of Technology logo "QUT" written on a microscope slide (a), showing stacking of fibers upon each other. SEM image of stacked and interwoven fibers until the stacking becomes affected by either the height of the sample or the uneven z-direction height, the latter caused by differences due to the interweaving of the fibers (b and magnified in c). The underneath of this box-like structure is shown in (d). Scale bars for b, c, and d are 1 mm, 100 μm, and 1 mm, respectively. Unpublished images from the same study as reported in Ref. [143].

16.5 CELL-INVASIVE ELECTROSPUN MATERIALS

Solution electrospinning offers relatively consistent control over fiber diameters, generally in the submicron range. However, due to the nature of the process, it is characterized by disorderly fiber deposition, which has limited its full potential. Despite this, numerous techniques have been developed and are still evolving, which have improved the orderly placement of fibers, including mechanical devices such as rotating collectors, liquid deposition methods, manipulation of the electric field, combining electric field manipulation with a rotating collector, static patterned collectors, and translating collectors. These approaches are important as they progress electrospinning toward true TE scaffold manufacture. Presented here is a selection of techniques developed to produce cell-invasive scaffolds.

16.5.1 HYDROSPINNING

In this approach, water is used as a collection medium, and a layer of electrospun fibers forms on the surface. The film of fibers is transferred to a microscope slide and the process repeated, generating multiple layers [113]. The water is then removed with a vacuum and the resulting scaffolds contain considerable volume, are both highly porous, and cell invasive. In this same study, by Levenberg and colleagues, myoblasts were introduced into the process and successfully cultured with an equal density throughout the electrospun scaffold.

16.5.2   Electrospinning onto Ice Crystals

In another study aimed at increasing the porosity of electrospun scaffolds, Simonet et al. electrospun fibers directly onto ice crystals, as they grew off the surface of a rotating drum. The rotating drum was hollow and filled with solid CO_2, causing water vapor in the surrounding air to crystallize on the collector [10]. The ice/fiber mixture was lyophilized, and the remaining scaffold was impressively large, with interconnected pores. Extending on this process, the porous material was mechanically pulled and transformed into a cotton wool-like material, which was then used as a TE scaffold in this form [11].

16.5.3   Simultaneous Electrospinning and Electrospraying

A similar concept involving building a conductive substrate up from the surface of a collector was devised by researchers trying to manufacture TE constructs. However, instead of using ice crystals, a layer of hydrogel was "grown" from the collector surface: gelatin containing smooth muscle cells was electrosprayed concurrently while electrospinning polyurethane [8]. Similarly, Ekaputra et al. electrosprayed a thiolated hyaluronic acid solution while electrospinning PCL/collagen onto a collector [9]. Human fetal osteoblasts were then observed to migrate up to $200\,\mu m$ into the scaffold in a mixture of electrospun fibers and gel. In both these instances, a rotating collector was used and the two spinnerets were oriented at $180°$ to each other.

16.5.4   Bimodal Electrospun Scaffolds

It is recognized that ultrafine electrospun fibers do not readily create a cell-invasive scaffold, because they are too tightly packed and thus do not provide adequate pore size and porosity. However, combining these fibers with a larger structure may allow the benefits of both "nano" and "micro" structures. This approach leads to the generation of scaffolds with high surface area due to the smaller fibers (promoting cell attachment and assisting in higher seeding of the scaffold), as well as suitable volume and pore sizes due to the microfibers. Gentsch et al. showed that solution electrospinning at very low flow rates induced both small and large diameter electrospun materials from a single spinneret. In this instance, PCL was used in conjunction with Chinese hamster ovary (CHO-K1) cells for infiltration experiments with improvements using a bimodal approach. Alternatively, simultaneous melt electrospinning (which produced micron-diameter fibers of $28.0 \pm 2.6\,\mu m$) and solution electrospinning (which produced submicron diameter fibers of $530 \pm 240\,nm$) of PLGA were performed onto a rotating mandrel, to form a thick cell-invasive bimodal scaffold [13].

16.5.5   Combining Conventional Technologies with Electrospinning

Electrospun fibers can be mixed with conventionally fabricated fibers to create a scaffold similar in principle to a bimodal electrospun scaffold, where the ultrafine electrospun fibers are mechanically supported by the larger structures. In one case, this was achieved by melt spinning poly(L-lactide-co-epsilon-caprolactone) (PLCL) into a cylindrical structure for vascular TE [144]. In another notable case, Kim et al. reported 3D extrusion combined with electrospinning to significantly increase porosity (Figure 16.7a). In a different study, an automated scaffold assembly system was used to construct the scaffold: a fused deposition modeling (FDM) process was interrupted between each layer to deposit

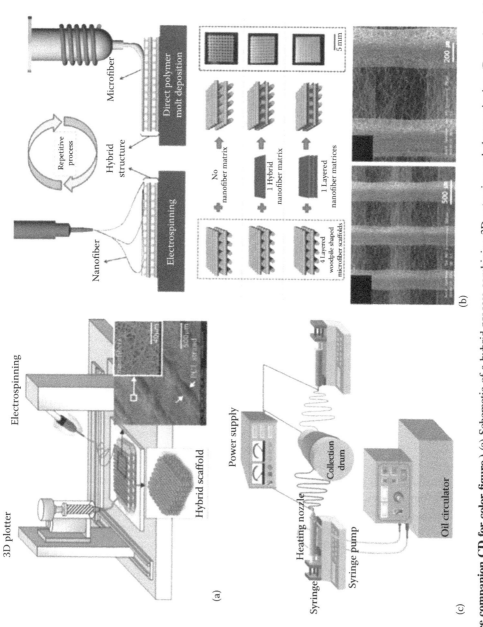

FIGURE 16.7 (See companion CD for color figure.) (a) Schematic of a hybrid process combining 3D extrusion and electrospinning. (Reproduced from Kim, G. et al., *Macromol. Rapid Commun.*, 29, 1577, 2008.) (b) Combining Electrospinning with a melt extrusion technique to create an electrospun carrier. (Reproduced from Park, S.H. et al., *Acta Biomater.*, 4, 1198, 2008.) (c) Combining melt electrospinning and solution electrospinning to create a bimodal scaffold. (From Kim, S.J. et al., *Polymer* 51, 1320, 2010.)

an electrospun PCL/collagen fiber mesh (Figure 16.7b) [145]. The resultant constructs contained large pores, while the electrospun fibers provided suitable structures for cell adhesion, effectively increasing the surface area available for penetrating cells to attach. Following this approach, a tubular construct for vascular TE was manufactured, incorporating nanostructured electrospun substrates and FDM fibers for sufficient mechanical strength [146]. Finally, a bimodal scaffold was successfully formed by simultaneously solution electrospinning (low diameter fibers) and melt electrospinning (large fibers) onto a rotating mandrel, using PLGA for both types of electrospinning (Figure 16.7c).

16.5.6 ELECTROSPINNING SMALL DIAMETER MICROFIBERS

There is a reported correlation between fiber diameter and pore size in electrospun meshes: a minimum fiber diameter (2–3 μm) is required for cell penetration (pore size 10–15 μm) [147]. Thus, the lack of pore size and interconnectivity resulting from the chaotic deposition of electrospun nanofibers creates a barrier to cell invasion. Interestingly, there are only a handful of studies on cellular interactions on meshes with fiber diameters between 2 and 10 μm, meaning that fibrous scaffolds with both high surface area *and* cell invasiveness are underinvestigated for TE. When such scaffolds are seeded, cells penetrate 1200 μm into the scaffold under flow and 300 μm under static conditions [147].

16.6 GROWTH FACTOR DELIVERY FROM ELECTROSPUN MATERIALS

Electrospun fibers can be used to deliver growth factors and other proteins to their surrounding environment. The growth factor can be included by simple addition to the polymer solution to be electrospun or incorporated using a variety of other means [148,149]. It is important in this case that the bioactivity of the growth factor is retained, since dissolution within a solvent can denature proteins, thus altering their bioactivity. Table 16.4 provides a summary of different growth factors incorporated within electrospun fibers.

As shown in Table 16.4, the delivery of growth factors can be achieved through the release of soluble factors or by tethering the molecules to the surface of the electrospun fibers. Drug delivery in the latter approach requires the sustained availability of the growth factor, in a bioactive and natural form. Therefore, the surface conjugation of biomolecules to electrospun fibers is an important area of research and includes single and multistep conjugation approaches for tethering bioactive molecules.

TABLE 16.4
Selected Examples of Growth Factors Delivered from Electrospun Fibers

Growth Factor	Polymer	Comments	References
BMP-2	Silk fibroin		[150]
BMP-2	PLGA/Hap		[151,152]
BMP-2	PLGA/Hap	BMP-2 plasmid delivered	[153]
BDNF	PCL	Surface functionalized	[154]
EGF	PLLA	GF both adsorbed and heparin bound	[155]
FGF-2	PLGA	Two approaches used—normal and coaxial	[156]
FGF-2	Gelatin or collagen	Introduced with perlecan binding domain	[157]
FGF-2	PLLA	Via surface bound heparin	[155,158]
GDNF	PCLEEP		[159]
IGF-1	Polyurethane	Delivered via PLGA microspheres in fibers	[160]
NGF	PLCL	Emulsion electrospinning	[161]
NGF	PCLEEP	NGF co-delivered with BSA	[7]
PDGF	Dextran/PLCL	Coaxial	[162]
VEGF	Gelatin/PCL	Coaxial—VEGF surface bound to heparin	[163]

16.7 SURFACE MODIFICATION OF ELECTROSPUN FIBERS

Biofunctionalization of the interface between a polymer and a biological entity is an important means to direct cell behavior and/or prevent bacterial infection [164]. The surface of the polymer fiber also influences the properties of the electrospun fiber mesh, such as wettability. Broadly, there are two methods possible to modify the electrospun fiber surface: postprocessing of the fibers, or more recently biofunctionalization was performed in a single electrospinning step [122,165].

16.7.1 POSTPROCESSING APPROACHES

The treatment of electrospun fibers after their manufacture is a common route to biofunctionalization. It is simple, because traditional conjugation chemistry such as 1-ethyl-3-(3-dimethylaminopropyl) carbodiimide hydrochloride (EDC) or N-hydroxysulfosuccinimide (sulfo-NHS) can be adopted. One problem with this approach, however, is that chemical removal and washing in different media is required, and the repetitive transition to different liquids may damage delicate parts of an electrospun scaffold.

16.7.1.1 Adsorbing Proteins onto Electrospun Fibers

Proteins can also adsorb onto surfaces, particularly hydrophobic surfaces. This is a simple method of introducing biological signals onto electrospun fibers but needs to be compared with including the protein as part of the electrospinning solution. For example, Ramakrishna and colleagues adsorbed laminin onto electrospun fibers. However, when laminin was included within the electrospinning solution, neurite growth was significantly higher compared to electrospun fibers with adsorbed laminin [40]. Some drugs adsorb onto electrospun fibers through simple physical adsorption using electrostatic interactions, hydrogen bonding, and hydrophobic interactions [166,167].

16.7.1.2 Chemical Binding Bioactive Functionalities

The direct conjugation of bioactive molecules onto electrospun fibers is an important approach to functionalization. Through various chemistries, brain-derived neurotrophic factor (BDNF) [154], heparin [158], Cibacron blue F3GA (CB), and bovine serum albumin (BSA) [168] have been covalently attached onto electrospun fibers. By incorporating heparin onto the surface of the fibers, growth factors such as FGF-2 and EGF-1 were bound [155,158]. This set the platform for many different growth factors to be bound using heparin binding sites. Combining adsorption and chemical conjugation, DNA was electrostatically incorporated onto electrospun fibers that have matrix metalloproteinase (MMP) cleavable functionalities [169].

16.7.2 PROTEIN RESISTANT BIOFUNCTIONALIZATION

There has been some focused work on generating protein resistant electrospun fibers with bioactive surface functionalities. Using amphiphilic block copolymer polymers with the correct hydrophilicity and chemistry, the resulting fibers were hydrophilic [58]. When these amphiphilic polymers were end modified with arginine-glycine-aspartic acid (RGD), biotin, or were thiolated, then they were protein resistant, hydrophilic, and functionalized [57,58]. Interestingly, only when mixed solvents are used (chloroform/methanol instead of chloroform only) there is strong resistance to BSA or streptavidin adsorption. Thiolated electrospun fibers in particular are able to then chemically bind with numerous bioactive molecules [170]. A more simplified approach to that described here has been devised by the same group that uses an amphiphilic and isocyanate star-shaped macromer, a reactive additive to a PCL or PLGA solution, to create both biofunctionalized and hydrophilic fibers [122,165]. Protein resistant fibers were produced using solutions containing only 7% star-shaped macromer relative to the PLGA or PCL. Functionalization was achieved by adding a bioactive peptide to the macromer/polymer solution prior to electrospinning with RGDS peptide and RDES modified electrospun substrates. The functionalized and hydrophilic fibers exhibited distinct differences in cell adhesion

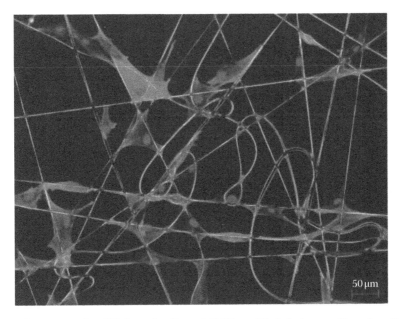

FIGURE 16.8   (See companion CD for color figure.) RGD-modified electrospun fibers deposited onto reactive substrates that prevent adhesion of cell—fibroblasts can only adhere onto the fibers. (Reproduced from Grafahrend, D. et al., *Nat. Mater.*, 10, 67, 2011.)

when investigated in vitro for neural applications [122]. When such electrospun fibers are placed upon reactive substrates that prevent cell adhesion, cells can only adhere to the fibers (Figure 16.8).

16.8   IN VITRO STUDIES OF ELECTROSPUN MATERIALS

As previously described, typical electrospinning does not result in cell-invasive scaffolds. Despite this, there are numerous excellent reviews on the interactions between electrospun fibers and cells in vitro that address aspects of cell biology [26,171] as well as demonstrate the diversity of polymeric, morphological, and structural properties. In general, electrospun fibers are excellent substrates for cells: except for a few examples, when the fibers are hydrophilic and protein resistant [57,58,165,170], cells adhere readily onto electrospun material. The cell shape, however, changes with electrospun fiber morphology: random meshes allow cell spreading similar to that seen on flat substrates, whereas cells tend to self-orient onto aligned fibers, particularly if there is no other substrate for them to adhere to Ref. [123]. In studies where "dilute" quantities of fibers were collected onto reactive substrates, the cells were only permitted to adhere to the fibers, inducing a very oriented structure. Interestingly, phenotypical differences have been observed between cells cultured on oriented fibers rather than random meshes.

16.8.1   Fibroblasts

Fibroblasts are the most abundant cells in connective tissue, and for various reasons, they are by far the most investigated cell type with electrospun material. Fibroblasts excrete ECM, including collagen, and are an important part of the wound-healing process. In addition to collagen, they produce glycosaminoglycans, reticular and elastic fibers, and glycoproteins. Electrospun fibers have often been compared to collagen, which can be fibrillar in structure. Since collagen is widely present in such connective tissue, it is appropriate that electrospun fibers and fibroblasts are so widely studied. For example, there are studies where the integration of fibroblasts into an electrospun scaffold is a crucial component of the final application such as in ligament of skin tissue engineering. Table 16.5 provides a list of selected references where fibroblasts were cultured with electrospun fibers.

TABLE 16.5
Selected In Vitro Studies with Fibroblasts

Polymer	Comments	References
PLGA	Cells adhered and spread according to fiber direction	[4]
PLGA	Bimodal d = 530 nm and 28 μm. Higher cell attachment/spreading in the nano/microfiber than the microfibrous scaffolds without nanofibers	[13]
PCL/PEG-b-PCL blends	Electrospun fibers were drawn onto substrates for discrete cell adhesion	[24]
Fibrinogen	Neonatal rat cardiac fibroblasts readily migrated into and remodel electrospun fibrinogen scaffolds with deposition of native collagen	[44]
PEG-b-PCL	Fibers were both protein resistant and surface functionalized with RGDS	[57,58]
PCL/PEG-b-PCL blends	Direct in vitro electrospinning with melt electrospinning	[100]
PCL/sP(EO-stat-PO)-RGDS	Single step functionalization—fibers were both protein resistant and surface functionalized with RGDS	[165]
Thiolated PEG-PDLLA	Adherence and proliferation on RGDC-functionalized fibers with reduced protein adsorption	[170]
Collagen	The aligned scaffold exhibited lower rabbit conjunctiva fibroblast adhesion but higher cell proliferation	[179]
Polyurethane	Rotating cylindrical collector aligned fibers. After mechanical strain, ligament fibroblasts had a spindle shape on aligned fibers and secreted more ECM than random ones	[180]
Polyamide	Cells rearranged their actin cytoskeleton to a more in vivo–like morphology	[181]
Collagen and PCL	Supported attachment and proliferation	[182]
Chitosan/silk fibroin	Supported attachment and proliferation	[183]
Ag-containing PVA	The Ag ions and Ag nanoparticles are cytotoxic to cells	[184]
PLLA/silk fibroin/gelatin	After 12 DIV, the scaffold supported attachment, spreading, and growth	[185]
Collagen and hyaluronate	The ratio of foreskin fibroblasts expression of TIMP1 to MMP1 was lower than collagen fibers	[186]
PDLA/PLLA (50:50)	After 5 DIV, the fibroblasts adhered, migrated, and proliferated	[187]
PLGA and PLGA/MWCNTs	MWCNT incorporation promoted attachment, spreading, and proliferation	[188]
PLGA/collagen	Effective as wound-healing accelerators in early stage wound healing	[189]
PEG/PPG/PCL	Supported excellent cell adhesion, comparable to pure PCL	[190]
PCL and PCL/PEO	Proliferated fastest with pores greater than 6 pm. Conformation changed as the pores grew from 12 to 23 pm; cells aligned along single fibers instead of attaching to multiple fibers	[191]
PHB/chitosan	Promoted attachment/proliferation	[192]
pDNA PEI/HA in PCL/PEG core shell	Complexes of pDNA with PEI-HA released from fiber mesh scaffolds could successfully transfect cells and induce expression of enhanced green fluorescent protein (EGFP)	[178]
PLGA/chitosan and core-shell PLGA/chitosan	PLGA/chitosan and core-shell PLGA/chitosan showed better adhesion/ viability than PLGA	[193]
PCL and ECM proteins	Presence of collagens, tropoelastin, fibronectin, and glycosaminoglycans in scaffolds	[194]
PDLLA/PEG	Dermal fibroblasts interacted and integrated with the fibers containing 20% and 30% PEG	[195]
Plasma silk fibroin	O_2-treated fibers showed higher cellular activities for fibroblasts	[196]
Silk fibroin	L929 fibroblasts showed good cell adhesion after 1, 3, and 7 DIV with confluence at 7 DIV	[197]
Polyurethane/MWNT	MWNT incorporation exhibited highest cell adhesion/proliferation and migration/aggregation	[198]
PVA	Cells grew significantly after 7 DIV but were round-shaped	[199]

TABLE 16.5 (continued)
Selected In Vitro Studies with Fibroblasts

Polymer	Comments	References
Collagen Type I	Qualitative analysis of rabbit corneal fibroblasts morphology and protein expression suggests fibers suitable for engineering a corneal tissue replacement	[200]
Bombyx mori silk fibroin	Attachment/growth with no difference between fibers electrospun with different solvents	[201]
Polyurethane elastomer	Human fibroblasts adhere and migrate, proliferate, and produce components of an ECM	[202]
PLGA	Human skin fibroblasts spread and showed growth. Collagen type III gene expression was upregulated on matrices with fiber diameters in the range of 350–1100 nm	[172]
Hexanoyl chitosan	Fibrous scaffolds supported attachment/proliferation	[174]
Polyethyleneimine	Normal human fibroblasts attached/spread on fibers	[203]
PCL	PCL was electrospun using an auxiliary electrode and chemical blowing agent. Good adhesion with the blown web relative to a normal electrospun mat	[204]

Because wound closure is an important part of fibroblast function, these cells have been used in wound-healing applications. Experiments with electrospun materials have focused on fibroblast adhesion, proliferation, and migration: the majority of electrospinning publications involving fibroblasts are superficial, in that the cells help to determine adhesion and toxicity toward the fiber, rather than as part of a focused wound-healing experiment. However, there are some studies providing in depth analysis of fibroblast function, including upregulation of important genes, such as collagen III production, which is optimal on fibers with diameters between 350 nm and 1.1 μm [172].

In many instances, fibroblasts are used as a proof of cytotoxicity [173] and biocompatibility [174]. These cells can also be used to ascertain whether a strategy of electrospinning produces feasible fiber morphology [175], if surface modification of a polymer was successful [176], or if a drug delivery system is viable [177]. These issues are important, as cells may be the most sensitive measure of changes in bioactivity of a surface, since instrumentation techniques have detection limitations. For example, fibroblasts have successfully been transfected with a nonviral vector incorporated within electrospun fibers so that they express green fluorescent protein [178]. In another study, fibroblasts were used as part of the collection system, to demonstrate that melt electrospun fibers could be deposited directly onto cells without a loss in vitality [100]. Similarly, recent work on making protein resistant and functionalized fibers involved fibroblasts to determine the success of the surface modification approach [57,58,122,165,170]. Although many of the aforementioned electrospinning investigations could be undertaken with almost any other cell type, fibroblasts are inexpensive, readily accessible, and relatively simple to work with. Suspended fibers have also been used with in vitro studies, with a minimum separation distance between fibers required for fibroblasts to bridge the gap (Figure 16.9) [345].

16.8.2 Epithelial Cells

Simple (one cell thick) or stratified epithelium covers and lines connective tissue (containing fibroblasts) surfaces in the body—from skin, cornea, to the inside of blood vessels, and intestine—thus protecting the underlying tissues. Epithelium also functions to secrete enzymes, hormones, as well as ECM and is separated from connective tissue by a basement membrane.

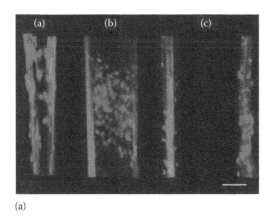

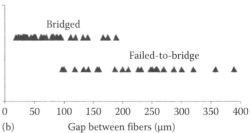

(a)   (b)   Gap between fibers (μm)

FIGURE 16.9 **(See companion CD for color figure.)** (a) Human dermal fibroblasts cultured on suspended aligned fibers at different distances apart. (b) The effect of a gap between suspended fibers and whether cells are able to bridge the gap. (Reproduced from Sun, T. et al., *Biotechnology and Bioengineering*, 97, 1318–1328, 2007.)

TABLE 16.6
In Vitro Studies with Epithelial Cells

Polymer	Comments	References
P(LLA-CL) 75:25	Adhered and proliferated well	[94]
Polyamide	T47D breast epithelial cells underwent morphogenesis to form multicellular spheroids	[181]
Chitosan and alginate	Layer-by-layer scaffolds have biocompatibility with Beas-2B human bronchial epithelial cells	[205]
PDLA	Both protein adsorption and porcine esophageal epithelial cells attachment	[206]
PHBV; PHBV/collagen, and PHBV/gelatin	Co-culture of hair follicular cells with epithelial outer root sheath/dermal sheath cells 3–5 DIV showed PHBV promoted wound closure and reepithelization more than PHBV/collagen	[207]
Fibronectin-g-PLLCL	Promotes epithelium regeneration, using esophageal epithelial cells	[208]
PCL and PVA co-spinning	Human prostate epithelial cells attached/proliferated more with PVA electrospun PCL mats	[209]

It is substituting the basement membrane that attracts researchers to electrospun fibers. These specialized epithelial cells have been used with electrospun fibers to create layered cellular structures. However, there has been less emphasis on creating cell-invasive substrates, since epithelium should sit on top of electrospun meshes and form a layered construct. Where electrospun fibers fit into epithelium regenerative medicine strategies is not as well defined as compared to other cell types. Despite this, epithelia are crucial to wound-healing strategies, and they will be increasingly used in electrospun fiber research. Table 16.6 lists a selection of epithelial studies using electrospun fibers.

16.8.3 ENDOTHELIAL CELLS

Endothelial cells are specialized epithelial cells that line the interior surface of blood vessels to create a barrier between circulating blood and the rest of the vessel wall. Endothelia are of wide interest to electrospinning researchers, which may be due to the structure they comprise: although the most widely used electrospinning collection configuration is a single flat collector, the second most

TABLE 16.7
In Vitro Studies with Endothelial Cells

Polymer	Comments	References
PLLCL (70:30)	Human coronary artery endothelial cells were evenly distributed and well spread from 1 to 10 DIV and maintained expression of PECAM-1	[210]
(P(LLA-CL))	Both random and aligned fibers preserved human coronary artery endothelial cell phenotype; expression of PECAM-1, fibronectin, and collagen type IV	[211]
Polyurethane	Small-diameter grafts with microfibers and microgrooves endothelial cells formed confluent monolayers. The cells expressed cadherin	[212]
PCL and CNTs	Good cerebro-microvascular endothelial cell viability	[213]
Collagen/chitosan	Endothelial cells proliferated well	[214]
PCL/collagen	Endothelial cells showed enhanced cellular orientation and focal adhesion	[215]
Coaxial cellulose acetate and chitosan	As mechanical stiffness increased, endothelial cell growth/adhesion and migration promoted. Fibers did not induce platelet aggregation or activation	[216]
Silk fibroin	Iliac endothelial cells attached and proliferated in comparison with cast SF films	[217]
Silk fibroin/PLLCL	Silk fibroin/PLLCL blended fibers increased Iliac endothelial cell growth compared with PLLCL	[218]
Polyurethane/collagen	Iliac endothelial cells migrated inside the scaffold within 24 h of culture	[219]
Span-80/PLLACL coaxial core/shell	No differences in iliac endothelium cell growth on the fiber types	[220]
HMA:MMA:MAA terpolymer	Cytocompatible allowing human blood outgrowth, endothelial growth, maintained phenotype	[221]
Polyphosphazene	After 16 DIV, neuromicrovascular endothelial cells formed a monolayer on the whole surface	[222]
P(LLA-CL 70:30)	Collagen-coated fibers enhanced spreading, viability, and attachment of human coronary artery endothelial cells and preserved phenotype	[223]
Gelatin-grafted-PET	Gelatin grafting improved the spreading/proliferation of the ECs and preserved phenotype	[224]
PLLA	Endothelial cell function was enhanced on smooth solvent cast rather than on the rough electrospun fibers. Electrospun substrates favored vascular smooth muscle cell behavior	[225]

commonly used is a single, rotating collector to generate tubular structures. Since blood vessels are such an important tissue, numerous researchers have looked at incorporating such cells and fibers together to create a functional epithelium on electrospun tubes, as shown in Table 16.7. Desirable mechanical properties of the electrospun tubular mesh are a critical requirement. Electrospun tubes of poly(urethane) and PLLCL in particular have been used, due to their elastic nature and strength [210–212]. In addition, nano and microfibers have been combined to obtain nanoscale structures with sufficient mechanical strength.

Endothelial cells derived from umbilical cords (human umbilical vascular endothelium cells; HUVECs), with the potential to function in regenerative medicine, are an important and increasingly researched cell type, as illustrated in Table 16.8. HUVECS are popular because they are derived from an unrequired tissue available at birth, and there are few ethical issues with using this form of human tissue [226].

16.8.4 SMOOTH MUSCLE CELLS

Blood vessels also consist of nonstriated muscle, and smooth muscle cells (SMCs) are an important cell type in this tissue. SMCs are also found in other tissues, including the aorta, gastrointestinal tract, lymphatic vessels, uterus, urinary bladder, the iris, and the ciliary muscle.

TABLE 16.8

In Vitro Studies with Human Umbilical Vascular Endothelium Cells

Polymer	Comments	References
HMA:MMA:MAA terpolymer	Cytocompatible fibers allowed growth, maintained phenotype	[221]
Methacrylic terpolymer	Fiber morphology controlled proliferation, metabolic activity, and morphology of cells	[227]
MWCNT/PU	Fibers activated Rac and Cdc42, while CNT regulated the activation of Rho	[228]
PCL	D = 0.8 or 3.6 μm. The larger fibers had higher vitality and improved interaction	[229]
Silk/gelatin	Gelatin improved the mechanical and biological properties	[230]
PLLCL	VE-cadherin expression not detected, but a high degree of tissue integrity was achieved	[231]
PLCL (50:50)	Adhered well and proliferated on small diameter fiber fabrics (0.3 and 1.2 μm; reduced adhesion, cell spreading and no signs of proliferation on the 7.0 μm diameter fibers	[232]

Table 16.9 shows a list of electrospun material cultured with SMCs, where emphasis is placed on seeding cells on tubular constructs mostly for vascular tissue engineering [233,234]. Cell seeding onto 3D constructs creates many challenges and the more anisotropic the scaffold, the more complex and difficult cell seeding becomes. This step is crucial in the in vitro culture of cellular implants since homogenous cell distribution appears necessary for successful TE strategies,

TABLE 16.9

In Vitro Studies with Smooth Muscle Cells

Polymer	Comments	References
[P(LLA-CL)] (75:25)	Rotating disk collector aligned fibers. SMCs attached and migrated along aligned fibers; the distribution of cytoskeleton proteins were parallel to the fibers	[136]
Type I and II collagen	Mandrel rotated at approximately 500 rpm. Fibers possessed the typical 67 nm banding pattern observed in native collagen	[5]
PU	Microintegrated cells into a biodegradable elastomer fiber matrix	[8]
[P(LLA-CL)] (75:25)	Adhered and proliferated well	[94]
PCL	Larger fiber diameter improved SMC infiltration into the bilayered vascular scaffolds	[215]
PLGA/COL	Co-culturing of SMCs with ECs, under a pulsatile perfusion system, leads to the enhancement of vascular EC development, as well as the retention of the differentiated cell phenotype	[233]
PU	Cells integrated into a strong, compliant biodegradable tubular matrix	[234]
Collagen-coated PCL	Proliferation on collagen-coated PCL fibers was not different than that on TCP	[236]
PCL	Contains heparin	[237]
Collagen I, II, and elastin	A three-layered vascular construct, integrating fibroblasts and endothelial cells	[235]
PLLA + PLLA/PVP	Morphology/proliferation improved with decreasing of the weight ratio of PLLA/PVP. In platelet adhesion decreased with increment of PVP content in the films	[238]
COL-coated PCL	Same proliferation rate on polycaprolactone (PCL) nanofibrous matrices coated with collagen or tissue culture plates. PCL coated with collagen showed more migration	[239]

which explains why there is such an emphasis placed on developing cell seeding techniques in this area. Multilayered vascular constructs have been prepared, demonstrating how complex tissue can be generated using the electrospinning technique [235]. Boland et al. fabricated a three-layered collagen/elastin vascular construct, which was built with separate media and adventitia seeded with SMCs and fibroblasts for the structural component, while the lumen was seeded with endothelial cells. Stankus et al. also addressed the cell-invasive issue associated with using submicron fibers by simultaneously electrospinning fibers and electrospraying cells in gelatin, to form tubular constructs, which were then included into a bioreactor [8]. The fibers and cells were laid down simultaneously, rather than constructing a scaffold and seeding them in a second step.

16.8.5 CARDIOMYOCYTES, CARDIAC STEM CELLS, AND CARDIAC CELL LINES

An excellent article on contractile cells was provided in 2004 by the Vacanti laboratory, where cardiomyocytes were grown on a mesh of PCL, which had been collected and suspended on a metallic ring [240]. This thin sheet began contractile movements, and the metallic wire ring acted as a passive mechanical support for the cells. Important cardiac tissue markers such as myosin, connexin 43, and cardiac troponin I were expressed after 14 DIV (days in vitro) [241]. Interestingly, electrospun fibers were also embedded into collagen sponges, with an increase in cardiac stem cell attachment on the collagen scaffolds with fibers embedded. Table 16.10 provides a list of in vitro studies with cardiomyocytes, cardiac stem cells, and cardiac cell lines.

16.8.6 MESENCHYMAL STEM CELLS

Mesenchymal stem cells (MSCs) are multipotent stem cells that can differentiate into different cell types, including chondrocytes, osteoblasts, and adipocytes. They are found in bone marrow and can be extracted using a biopsy, with MSCs often but not only sourced from humans. Table 16.11 provides a list of different publications where MCSs were cultured on electrospun materials and demonstrates that this is one of the most investigated cell types with electrospun materials due to their ability to differentiate. With MSCs, the lineage of the cultured cell is of particular interest. Some researchers use differentiation media with electrospun fibers to induce a particular type of cell line with success [246,247]. Therefore, electrospun fibers do not interfere with conditions that are designed to differentiate MSCs into particular lineages. Alternatively, electrospun scaffolds can be used as a drug delivery platform to induce stem cell differentiation into the targeted cell lineage.

TABLE 16.10
In Vitro Studies with Cardiomyocytes, Cardiac Stem Cells, and Cardiac Cell Lines

Polymer	Cells	Comments	References
PCL	Cardiomyocytes	A 10 μm thick mesh is suspended across a wire ring. Myosin, connexin 43, and cardiac troponin I expressed after 14 DIV	[240,241]
PLLA	Cardiomyocytes	Cells with fiber-guided filipodialike protrusions and developed into sarcomeres	[242]
PGA	Cardiac stem cells	Fibers embedded in collagen sponge and more cells attached to the collagen sponge incorporating fibers compared to without. The attachment and proliferation were enhanced by incubation in a bioreactor perfusion system	[243]
Polyurethane	Cardiomyocytes	Cells were elongated, but morphological changes induced by material macrostructure did not directly correlate to functional differences	[244]
PPDL	H9c2	Fibers were not cytotoxic and supported cell proliferation	[245]

TABLE 16.11
In Vitro Studies with Mesenchymal Stem Cells

Polymer	Comments	Reference
PLGA/TCP	Cotton wool-like fibers from collection on ice. Osteocalcin showed osteogenic differentiation	[11]
HAp/TCP/PCL	The solvent or combination for fiber preparation affects properties and cell behavior	[82]
PLLA/PCL	Combined electrospinning with FDM to make a tubular construct	[146]
Silk/PEO/HAp	HAp particles improved bone formation. BMP-2 and HAp resulted in the highest calcium deposition and upregulation of BMP-2 transcript levels	[150]
PLGA/HAp	Encapsulated DNA/chitosan nanoparticles have higher cell attachment, viability, and transfection	[153]
PLGA/HAp	Encapsulation of HAp could enhance cell attachment to scaffolds and lower cytotoxicity	[152]
PLGA	Under hypoxia/nutrient starvation conditions, the IGF-1 loaded scaffolds improved MSC survival	[160]
PLCL and PLCL/coll	Neural differentiation was induced with soluble factors	[246]
PCL	Chondrogenic differentiation was enhanced after incubation in chrondrogenic media	[247]
PCL	Delayed initial attachment and proliferation on meshes, but enhanced mineralization at a later time point	[248]
PLGA and PCL	Supported attachment, proliferation, and differentiation	[249]
PCL	Differentiated to adipogenic, chondrogenic, or osteogenic lineages by specific differentiation media	[250]
PCL	Differentiated to chondrocytic phenotype—a zonal morphology is described	[251]
PCL	Matrix mineralization and collagen type I throughout after 4 weeks	[252]
Collagen	Combined electrospinning with SFF; Enhanced initial cell attachment and cell compactness between pores of the scaffold relative to SFF 3D collagen scaffold only	[253]
HAp/PLGA	Increased ALP activity and expression of osteogenic genes w calcium mineralization of MSCs	[254]
Gelatin/siloxane	Resulted in apatite formation and stimulated proliferation and differentiation	[255]
PCL	Increased osteocalcin and osteopontin 3 weeks of culture	[256]
PCL	Multimodal distribution of fiber size allowed increased cell seeding	[257]
Collagen or PLLA	Early adhesive behavior is affected by texture of surface. Collagen and collagen coated PLLA fiber had similar outcomes	[258]
PCL	Thicker scaffolds provide a better substrate for cell proliferation	[259]
PLCL	MSCs expressed CD44 and CD105 but did not express CD34 or CD14	[260]
P(3HB-co-4HB)	Porcine aortic heart valves were decellularized, coated with electrospun fibers	[261]
Silk fibroin	Fiber alignment affected morphology and orientation	[262]
Chitin/PVA	Fibers supports cell adhesion/attachment and proliferation	[263]
PLA or PLA/DBP	Mineralization with osteogenic supplements was greater PLA after 14 DIV but same level 21 DIV	[264]
PCL or PCL/COL	Multilayer construct, supported cell attachment similar tissue culture polysterene plate	[265]

From a practical viewpoint, it was shown that insulin growth factor (IGF-1) loaded electrospun scaffolds improved MSC survival under hypoxia or nutrient reduced conditions [160]. With such a complex and heterogeneous environment generated during stem cell culture, all conditions and parameters need to be reported in order to allow comparison between studies and to draw definitive conclusions.

16.8.7 OSTEOBLASTS

Ideally, bone TE should provide a substitute for the autologous bone graft, and there is a lot of research to develop a load-bearing material that can integrate with bone defects. Since electrospun fibers do not generate sufficient compressive strength, their use is more to support bone formation in other surgical paradigms, including an artificial periosteum and as a structure to contain hematomas between bone fractures [266]. Mineralization is a crucial process in bone formation, and many of the in vitro studies with osteoblasts and electrospun fibers measure the alkaline phosphatase activity (ALP), which indicates deposition of mineralized matrix (early bone formation). A distinctive feature of electrospun fibers within bone TE is the inclusion of inorganic substances within the fibers to improve the biological outcome when using osteoblasts by rendering scaffolds more osteoinductive. Such materials include hydroxyapatite (HAp), silica, titanium dioxide (TiO_2), and bioactive glass. The cells used for bone TE can also be MSCs, as they are the stem cell precursor for osteoblasts. Another issue in this type of TE strategy is the degree of cell invasiveness of the electrospun scaffolds, which has been addressed by Ekaputra et al. [9]. These authors simultaneously electrospun and electrosprayed different materials, which made the fibrous mesh less dense. For example, when a hyaluronic acid solution is electrosprayed with electrospun PCL onto the same collector, there was increased osteoblast penetration into the scaffold.

A further step in the development of osteoinductive scaffolds is the deposition of a biomimetic calcium phosphate coating on the fibers. This appears to enhance the ALP activity and the osteopontin expression of a MC3T3-E1 mouse osteoblastic cell line under osteoblastic induction [267]. In a different study, the same cell line displayed similar behavior onto biomimetically coated electrospun PCL. The expressions of ALP, RunX2, collagen I, and osteocalcin were upregulated confirming the beneficial effect of a calcium phosphate coating on osteoblast differentiation [268].

As seen in Table 16.12, there is significant interest in the interactions between osteoblasts and electrospun fibers. This partial list contains only selected journal articles, and the actual publication list is much greater. Jang et al.'s work is an excellent review on electrospinning for bone regeneration [284].

16.8.8 CHONDROCYTES

There is a clear medical need for transplantable cartilage tissue, as cartilage has a limited capacity to repair itself. Cartilage TE research emphasizes regeneration, rather than repair, and an ideal cartilage scaffold should possess similar mechanical properties to the natural tissue. The ECM, which contributes the most to the mechanical properties, is comprised of collagen fibrils (mainly Type II) and proteoglycans and is distributed into zones with distinct mechanical properties [285,286].

A promising approach to cartilage repair is 3D printing of hydrogels [287,288]. As previously described, this approach can be combined with electrospinning to form a different class of "bimodal scaffold," where the scaffold consists of structures with distinctly different scales. In this instance, the extruded hydrogels form struts approximately 100 to 200 μm in diameter, while electrospun fibers are incorporated within the larger struts (Figure 16.7b) [289]. The smaller diameter fibers aid in increased cell seeding efficiency and are included in the printing process by electrospinning separate layers during the printing process. Table 16.13 summarizes in vitro studies with chondrocytes and electrospun fibers.

16.8.9 NEURONS AND GLIA

Electrospun fibers have been investigated for the guidance of neurons and glia, which naturally exist in highly oriented and structured tissues, particularly in the peripheral nerve and the white matter of the central nervous system (CNS). The research with such cells can be divided into two main areas of interest—in vitro guidance systems and tubular structures for containing damaged nerves.

TABLE 16.12

In Vitro Studies Performed with Osteoblasts

Polymer	Cell Type	Comments	References
PCL/collagen and PEO; PCL/collagen/Heprasil	Fetal osteoblasts	All three scaffold types supported attachment/proliferation; better penetration was with mPCL/Collagen fibers co-electrosprayed with Heprasil	[9]
PCL/b-TCP	MC3T3-E1 cells	Viscoelastic and biomechanical properties increased with DIV	[269]
Gelatin and HAp	Fetal osteoblasts	Combining of electrospinning of gelatin and electrospraying of HA increased ALP activity and enhanced mineralization	[270]
PCL-PEG-PCL and HAp	Osteoblasts	Scaffolds could provide a suitable environment for attachment, and MTT revealed the scaffolds had good biocompatibility and nontoxicity	[271]
HA/TiO$_2$	MC3T3-E1 cells	Collagen treated HA/TiO$_2$ fibers possess better cell adhesion and significantly higher proliferation and differentiation than untreated fibrous mats	[272]
PCL-silica	Pre-osteoblast cells	Good proliferation and differentiation of cells	[273]
PLA/bioactive glass	Osteoblasts	Bioactive glass phase enhanced the in vitro apatite formation and cells adhered	[274]
PLLA	Osteoblasts	After mechanical expansion of the microfibrous mats, the scaffolds proliferated osteoblasts that actively penetrated the inside of the 3D scaffold	[275]
p(TMC-co-CL)-block-p(p-dioxanone)	MC3T3-E1	Cells proliferated 1.2 times faster at 4 DIV, 1.5 times faster at 7 DIV after seeding; At 28 DIV alkaline phosphatase was four times more than control	[276]
RGD-mod PLCL	Osteoblasts	Adhesion and proliferation was greater on RGD modified fibers meshes up to 7 DIV. ALP activity and calcium content on the RGD-AAc-PLCL meshes were approximately 7.5 and 6.7 times higher than those on the other meshes, respectively. Expression of Cbfa1, ALP, and Osteocalcin (OCN), was upregulated (at least 5–9.7 times greater) on the RGD-AAc-PLCL meshes	[277]
PHBV and cHAp	SaOS-2	Both PHBV and CHA/PHBV supported the proliferation; The alkaline phosphatase was higher than the CHA/PHBV after 14 DIV	[278]
PCL/collagen	Pig bone marrow mesenchymal cells	Osteogenic differentiation markers were more pronounced on PCL/Col fibrous meshes	[279]
PLCL/gelatin/HAp	Fetal osteoblasts	Electrospinning PCL/gelatin and electrospraying of HA nanoparticles made a rough surface with better proliferation, mineralization, and alkaline phosphatase activity	[280]
Silica	MC3T3-E1 cells	ALP activity and expressions of type I collagen and osteocalcin were higher for cells cultured on nonwoven silica gel fabrics than on TCP	[281]
PLLA and PLLA/HAp	Osteoblasts	n-HA deposition improved seeding efficacy at 20 min time point	[282]
PLA/bioactive glass		As bioglass concentration increased (from 5% to 25%), the bioactivity improved. Cells attached, grew well, and secreted collagen. ALP expression was higher than pure PLA	[283]

TABLE 16.13

In Vitro Studies with Chondrocytes

Polymer	Comments	References
PCL/collagen	Favorable conditions for cell adhesion/proliferation, attributed to nanotopography, chemical composition, and an enlarged surface for cell attachment and growth	[145]
Chitosan	Fibers allowed chondrocytes to proliferate and produce glycosaminoglycan	[290]
PCL and 3D prototyping	This approach is feasible for fabricating 3D scaffolds for chondrocytes	[289]
PCL	Fetal bovine chondrocytes upregulated collagen type IIB expression after 3 weeks	[291]
HA/collagen	Chondrocyte adhesion/proliferation were enhanced maintained chondroblastic morphology	[292]
3D printed PEOT or BPT and electrospun PCL	Combined scaffolds enhanced articular chondrocyte entrapment compared to 3DF scaffolds and higher GAG/DNA ratio after 28 DIV. Spread morphology was observed on 3DF scaffolds, suggesting a direct influence of fiber dimensions on cell differentiation	[293]
PCL on PLA microfibers	Increased cellular infiltration with higher porosity	[294]
PHBV	Rabbit auricular chondrocytes attached more efficiently to electrospun fibers (30% after 2 h) than cast films (19%)	[295]
Collagen type II	Scaffolds produce a suitable environment for chondrocyte growth	[296]

Fibers have an inherent guiding property on neuron processes when their diameter is below approximately 250 μm [297,298]. Further reducing the fiber diameter down to submicron levels, as seen in electrospinning, still maintains this guidance effect [67,121,123]. Since nervous tissue is often bundled into organized arrays, the orientation of electrospun fibers is a particularly relevant aspect of the technique. As mentioned previously, electrospun fibers can be aligned by collecting them onto a rapidly rotating mandrel. The oriented electrospun sheet can then be flattened with minimal curvature if the mandrel is of sufficiently large diameter [299]. Alternatively, it is possible to collect electrospun fibers across the gap between two collectors, previously outlined in Section 16.4.3. This has been performed using parallel bars, sharp blades, and metallic rings. When short collection times are used in this approach, the fibers that are suspended across the gap remain "individual," in the sense that they do not adhere to other fibers. However, performing cell culture on such individual fibers can prove difficult, as changes in culture media are difficult to perform.

Recent work on chemically reactive substrates has permitted the study of a single cell on a single oriented fiber, quantifying the effects of changes in surface chemistry by measuring cell elongation and process extension [67,121,123]. Fibers are transferred to a substrate by passing a prepared cover slip through an oriented suspension of fibers. The reactive substrate binds the fibers, but then continues to react with water, so that it is ultimately resistant to both protein and cell adhesion. In this configuration, neurons, oligodendrocytes, Schwann cells, astrocytes, and dorsal root ganglia have all been cultured, and changes in the fiber composition (and reactive substrate chemistry) provide distinct differences in migration and adhesion of the cell. Table 16.14 summarizes these experiments and other investigations where oriented electrospun fibers are cultured with neural derived cells. Co-culture of DRGs on astrocytes/electrospun fibers induced directional growth along the fibers with increased neurites in the presence of astrocytes (Figure 16.10).

The other main regenerative medicine approach involving neurons and glia is to manufacture tubular structures to enclose nervous tissue, either the spinal cord or more commonly, the peripheral nerve. Here, tubular electrospun materials have been implanted within animal models, described further in Section 16.9. Such tubular materials must resist compressive forces and be flexible enough to avoid significant mechanical mismatch with the nerve [310].

TABLE 16.14
In Vitro Studies with Neurons and Glia

Polymer	Cell Type	Comments	References
Laminin/PLLA	PC-12	Functionalized fibers increased axonal extensions; blended laminin and PLLA is a simple method to modify fibers	[40]
PCL/collagen	Schwann Cells; OECs; DRG explants	Cells aligned on oriented fibers; collagen/PCL blends provided better process growth	[67]
PCL/collagen	Dissociated Schwann Cells; DRGs	Schwann cell processes were longer on collagen containing PCL fibers. When substrate is made adhesive, fibers reduced their guidance	[121]
PCL/sP(EO-stat-PO)-RGDS	DRG	Resistance to protein adsorption with functionalization	[122]
PCL/collagen	U373; hNP-ACs; SH-SY5Y	Glia aligned on the oriented electrospun substrates	[123]
PLGA	C17.2 nerve stem cells	Cells attached and differentiated along the direction of the fibers	[300]
PCL-gelatin	Schwann Cells	Aligned and random PCL/gelatin are better substrates compared with PCL	[301]
Polysialic acid	Schwann Cells	Good viability and directed cell proliferation along the fibers	[302]
Matrigel™	DRG	Supported attachment, elongation of neurites, and migration of Schwann cells similar to electrospun collagen type I fibers	[303]
PLLA	DRG; Schwann Cells	Guided growth along the aligned fibers. Schwann cells had bipolar phenotype, average neurite length was not different from fiber densities	[304]
PPy coated PLLA/PLGA	DRG	Electrical stimulation increased maximum neurite length by 83% and 47%, respectively, for random and aligned samples	[305]
PCL	PC-12	Aligned fibers had similar growth irrespective of diameter. Neurites on aligned fibers were longer than on randomly oriented fibers	[306]
Plasma PCL and PCL/COL	Schwann Cells	17% increased proliferation on plasma modified PCL compared with PCL/collagen. Schwann cells had bipolar elongations	[307]
PCL/chitosan and PCL	Schwann Cells	48% more cell proliferation on PCL/chitosan after 8 DIV	[308]
Polydioxanone	Astrocytes; DRG	Astrocytes had directional growth along the fibers. Increased neurites when co-cultured with DRGs	[309]

16.8.10　Ligament Cells

Ligaments and tendons are soft connective tissues that respectively attach bones to bones or bones to muscles. Their major function is usually to transmit movement and to ensure joint stability. The most predominant proteins in these tissues are collagen type I and III [311]. Ligaments and tendons display a specific structure made of crimped aligned collagen fibers in which cells are embedded, providing high tensile strength and elasticity. Electrospun nano- or microfibers have been mainly used to develop scaffolds for the rotator cuff [312], periodontal ligament [313–315], and also for orthopedic applications [316–319] (such as regeneration of anterior cruciate ligament or tendon). As the structure of the rotator cuff and the anterior cruciate ligament are similar, the scaffolds developed for their regeneration are

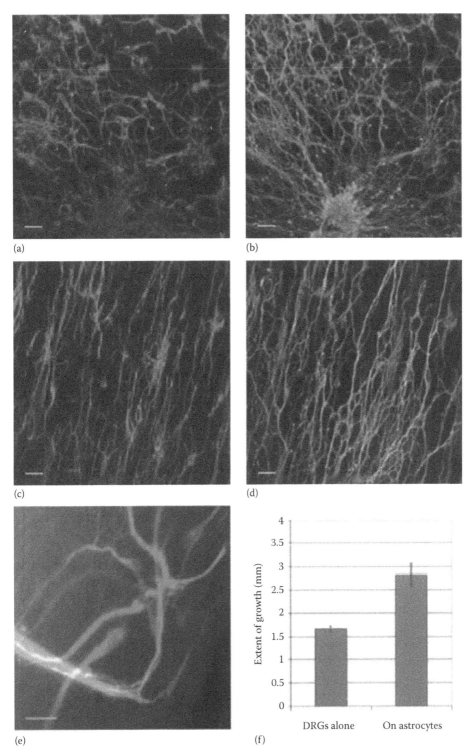

FIGURE 16.10 **(See companion CD for color figure.)** Images showing DRGs grown on astrocytes and how the directionality of growth is affected by the astrocytes. (a) and (b) are random electrospun meshes while the cells in (c) and (d) are grown on oriented fibers. (e) shows how neurite processes change direction when encountering an astrocyte. (f) is a graph showing that astrocytes increase neurite growth. Scale bars: a–d, 100 μm; e, 30 μm. (Reproduced from Chow, W.N. et al., *Neuron Glia Biol.*, 3, 119, 2007.)

also alike. In these regenerative strategies, the electrospun fibers were utilized as a cell culture support providing a topographic guide to the cells. There is evidence showing that cell orientation occurs onto aligned electrospun fibers mimicking the morphology of the native tissue. Even though the guidance provided by the aligned fibers appears to be cell specific, it is generally assumed that collagen type I and type III are upregulated as a result of the cell alignment [180,320]. It also seems that aligned fibers enhance the deposition of orientated ECM [319,321].

As discussed previously, there are a number of means to fabricate scaffolds with aligned fibers. It is also possible to obtain self-crimped aligned electrospun fibers (as long as the polymer used is not too rigid), in order to further mimic the morphology of the native tissue [322]. When aligning fibers onto a rotating mandrel, residual stress is created in the electrospun fibers. Once the fibers are removed from their collection support, spontaneous fiber relaxation occurs, resulting in fiber shrinking producing crimped structures. The crimping process can be controlled by adjusting the temperature at which the fibers are left to relax. This crimped structure has a significant effect on the mechanical properties at low strain (<6% strain) as it results in a J-shape stress–strain curve similar to that observed for native tissue. The particular shape of the tensile curve is explained by the recruitment of the collagen fibers during stretching; as the fibers are being aligned no stress increase is observed (corresponding to the initial toe-region in the J-curve); once the fibers are aligned they take over the load applied resulting in a rapid stress increase (known as the linear part of the curve).

Although it is possible to provide topographic cell guidance and to obtain scaffolds with similar stretching mechanisms to ligaments or tendons, one major drawback of the use of aligned electrospun scaffolds comes from the weak mechanical performance of these structures. There is a striking difference between the ultimate tensile force of a human anterior cruciate ligament (around 2500 N [323]) and an electrospun membrane (several tenths of N). Therefore, several groups have developed composite scaffolds in an attempt to overcome this limitation [316–319]. These structures are made of a mechanically resistant knitted construct onto which nano- or microfibers are electrospun. Here again it is also possible to fully mimic the morphology of the ligament and to provide a topographic cell guide by aligning the electrospun fibers onto the knitted structure [319]. These structures displayed a higher ultimate strength of up to 250 N (compared to several tenths of Newtons for electrospun scaffolds only), but this is still far from the ultimate strength of a human ligament. Therefore, research must be pursued in this direction, that is, to increase the mechanical properties of the scaffolds in order to develop functional tissue-engineered constructs.

There are two different strategies for periodontal applications. The first strategy concerns the so-called tissue-guided regeneration in which a biodegradable membrane is used to isolate the damaged region from the surrounding connective tissue. In this strategy, the membrane has to be biocompatible but cell-occlusive in order to prevent gingival cell infiltration and to allow the remaining periodontal cell to proliferate onto the root surface. Therefore, electrospun membranes can be considered good candidates due to their intrinsic small pore size that prevents cell infiltration. Gelatin with several levels of crosslinking [314] as well as synthetic biodegradable materials [324] have been used in the past, and they showed promising outcomes in terms of barrier property and stability. Another strategy involved the design of a graded electrospun composite in which hydroxyapatite nanoparticles were incorporated in the bulk of the fibers to enhance osteo-integration while maintaining cell occlusiveness [315]. The opposite side of the graded membrane in contact with the root surface can also be loaded with drugs to combat periodontal pathogens and/or favor periodontal fibroblast proliferation. The second strategy using the principles of TE utilizes a cellular scaffold to regenerate the damaged regions [313]. Here, electrospun scaffolds can be used as cell culture supports enabling the fabrication of a multilayered cellular construct upregulating the expression of osteopontin and sialoprotein. Other structures for periodontal tissue regeneration consisted of a complex arrangement of parallel and cross-aligned electrospun fibers, which were drawn onto a rotating mandrel [325]. These structures displayed higher tensile strength than randomly orientated fibers and also enhanced the proliferation and migration of rat periodontal ligament fibroblasts. Cell elongation onto the aligned fibers was also observed to promote a higher level of tissue organization in these scaffolds (Table 16.15).

TABLE 16.15

In Vitro Studies with Ligament Cells

Polymer	Cell Type	Comments	References
Polyurethane	Human ligament fibroblast	Aligned nanofibers had spindle-shaped, oriented cells; more collagen was synthesized on aligned fibers	[180]
PLGA (85/15)	Human rotator cuff fibroblast	Aligned nanofibers upregulated collagen type I from 3 DIV and type III after 1 DIV, integrin α2 expression was higher for the aligned fibers	[312]
PLGA	Periodontal ligament cells	Osteogenic differentiation and osteopontin, osteocalcin, and bone sialoprotein marker expression. At 3, 6, 9, and 12 DIV cell membrane layers deposited and multilayered structures established	[313]
Gelatin	Periodontal ligament cells	Cells cultured on the membrane in vitro exhibited good attachment, growth, and proliferation	[314]
PLGA	Bone marrow stromal cells	Electrospun onto PLGA knitted scaffold. Higher expression of collagen I, decorin, and biglycan genes	[316]
PLGA (65:35)	Porcine bone marrow stromal cells	Nanofiber coated knitted scaffolds could sustain cell growth and proliferation better than woven scaffolds and promoted easy cell seeding	[317]
PLGA(85:15);	Rabbit bone marrow stromal cells	More resistant scaffolds when cell seeded after 21 DIV	[318]
PLCL	Rat bone MSCs	Collagen type III started to form a fibrous network 14 DIV	[319]
Poly(ester urethane)	Bone marrow stromal cells	Spindle-shaped morphology increased with fiber diameter and degree of alignment. Expression of collagen 1 alpha 1, decorin, and tenomodulin was greatest on the smallest fibers	[320]
PLCL (80/20)	Bovine ligament fibroblasts	Immunohistochemistry staining revealed ligament ECM markers collagen type I, collagen type III, and decorin organizing and accumulating along the fibers	[321]
PLCL (80/20)	Bovine ligament fibroblasts	Deposition of ECM onto the fibers	[322]
PLLA/MWNTs/HA	Periodontal ligament cells	PLLA/MWNTs/HA enhanced adhesion/proliferation and inhibited the adhesion and proliferation of gingival epithelial cells	[324]
PLGA (50:50)	Rat periodontal ligament cells	Cells orientation along the fibers, higher level of tissue organization than random scaffold	[325]

16.8.11 KERATINOCYTES

For effective skin reconstruction, both the dermis and the epidermis need to be reconstructed. This requires the successful culture of keratinocytes. Such culture on a Petri dish presents challenges including the transfer of the cells. Therefore, thin membranes acting as mechanical supports and reinforcements are used during cell culture to aid in handling the cells, thus facilitating their transplantation. Transfer materials such as Laserskin, for example, assist in growing and expanding keratinocytes ready for implantation [326]. In this way, electrospun fibers can be seen as a potential transfer material when fabricated as a sheet and used for in vitro culture. Table 16.16 provides a list of studies where electrospun materials are used as a substrate for keratinocyte culture.

16.8.12 OTHER CELLS

Due to the number of specialized cells, tissue, and organs in the body, it is not possible to list all of the different areas where electrospun fibers are used in regenerative medicine approaches. However, Table 16.17 provides a selection of these different cell types and electrospun fibers.

TABLE 16.16
In Vitro Studies with Keratinocytes

Polymer	Comments	References
PLGA	Bimodal d = 530 nm and 28 μm. Higher cell attachment/spreading in the nano/microfiber than the microfibrous scaffolds without nanofibers	[13]
Silk fibroin	Electrospun material was more preferable than SF film and SF microfibers	[30]
Ag-containing PVA	The Ag ions and Ag nanoparticles are cytotoxic to cells	[184]
Silk fibroin	O_2-treated fibers showed higher cellular activities for normal human epidermal keratinocytes	[196]

TABLE 16.17
In Vitro Studies with Other Cell Types

Polymer	Cell Type	Comments	References
PCL	CHO-K1	Bimodal d = 400 nm and 6 μm; or d = 4 μm. Increased penetration of cells into scaffolds	[12]
Polyamide	Rat kidney cells	Cells rearranged their actin cytoskeleton to a more in vivo–like morphology	[181]
PCL	Human amniotic fluid stem cell	Kinetics of osteogenic differentiation were observed between hMSCs and hAFS cells, with the hAFS cells displaying a delayed alkaline phosphatase peak, but elevated mineral deposition, compared with hMSCs	[248]
PCL	CHO	Proliferation was influenced by the thickness of the scaffold (0.1 and 0.6 mm); scaffolds with 0.6 mm thickness were a better substrate for proliferation	[259]
PCL and PCL/gelatin	Adipose-derived stem cells	Good attachment, although migration was more with PCL/gelatin	[327]
PCL and PCL/HAp	Embryonic stem cells	Cells proliferated at the same rate as cells growing on TCP and maintained pluripotency markers	[328]

16.9 IN VIVO RESULTS OF ELECTROSPUN MATERIALS

There are understandably a greater number of in vitro experiments using electrospun polymers than there are in vivo studies, due to the increased time and effort to perform animal studies. To date, in vivo experiments with electrospun fibers are all generally positive, with the electrospun material demonstrating potential for the particular application.

16.9.1 Biocompatibility and In Vivo Cell Penetration

Electrospun scaffolds do not usually produce any major adverse reaction once implanted. Several studies have demonstrated relatively good biocompatibility with nano- and microfibers [329–331]. Cao et al. compared the foreign body response of PCL randomly orientated nanofibers, aligned nanofibers, and films in a subcutaneous rat model. They found that films and the two types of fibrous scaffold displayed good biocompatibility as shown by a very early resolution of acute and chronic inflammation. Two weeks post implantation no, or minimal, inflammation was observed. It was also found that aligned nanofibers induced a less severe healing response when compared to random nanofibers. Although no definitive conclusion could be made, it was hypothesized that the aligned nanofibers induced a more moderate inflammation due to better tissue integration as higher cell infiltration was found in this type of scaffold.

As opposed to most in vitro studies, implanting electrospun materials generally leads to full [265] or nearly full colonization of the scaffolds [332]. However, increasing the pore size, using for example the sacrificial fiber removal technique, produced better cell infiltration and tissue integration [332].

16.9.2 NEURAL TISSUE ENGINEERING

In neural TE, most approaches involve attempts at mimicking the natural ECM by fabricating pores, ledges, and fibers so as to present the growing axons with appropriate cues and a permissive environment. Perhaps the most numerous studies involve nerve guides, where cylinders of electrospun material are generated on a rotating mandrel and used to bridge a gap defect in the peripheral nerve. Nerve guidance channels (NGCs) are one of the most frequently implanted electrospun material constructs; however, the in vivo results are not particularly compelling so far, when compared with previous nerve guides. For peripheral nerve injuries, a critical gap length for different species dictates what maximum distance regenerates 50% of the nerves [333]. For NGCs from electrospun fibers, regeneration does not appear to extend beyond what would be expected for traditional biomaterials. However, electrospinning is still developing and there are approaches to manufacture electrospun tubes that may provide an advantage over traditional techniques. For example, the Bellamkonda laboratory has been filling NGCs with electrospun meshes, showing encouraging outcomes for repair (Figure 16.11) [334].

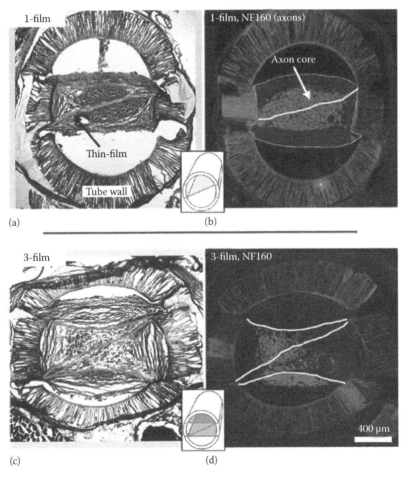

FIGURE 16.11 **(See companion CD for color figure.)** Nerve guides implanted within a peripheral nerve regeneration model, containing single (a&b) or three (c&d) sheets of electrospun meshes. (Reproduced from Clements, I.P. et al., *Biomaterials*, 30, 3834, 2009.)

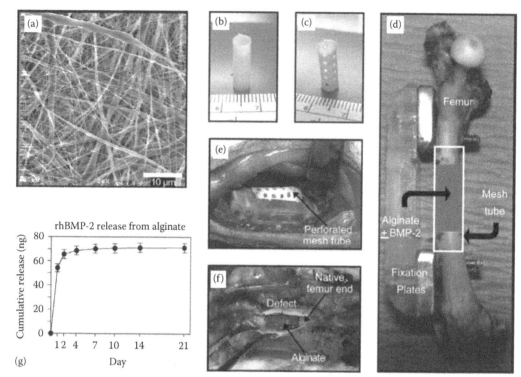

FIGURE 16.12 **(See companion CD for color figure.)** (a) SEM image of electrospun PCL. (b) Hollow electrospun tubular implants without perforations and (c) with perforations. (d) Scheme of implant in segmental bone defect. (e) Defect after placement of a perforated mesh tube with alginate seen inside the tube through the perforations. (f) One week postsurgery and alginate was present inside the defect, with hematoma present at the bone ends. (g) Alginate release kinetics of the rhBMP-2 during the first 4 days.

16.9.3 Bone Tissue Engineering

Electrospun scaffolds were also used in bone TE although for non-load bearing implantation due to the poor intrinsic mechanical properties of the fibrous mats. Generally, electrospun scaffolds do not induce spontaneous and significant bone formation unless surface treated with osteo-inductive materials [335,336] or seeded with osteo-inducted cells [337], or loaded with growth factors such as BMP-2 [338]. However, when one or several of these conditions are used bone formation can occur in vivo even in a nonbony environment such as subcutaneous implantation [335,336]. A recent study on perforated electrospun tubes between bone defects shows some fascinating outcomes, in that the perforated holes were crucial for attaining good bone formation (Figure 16.12).

16.9.4 Cartilage Tissue Engineering

TE using electrospun scaffolds offers the potential to develop treatments for defective articular cartilage, where isolated articular chondrocytes or MSCs are seeded onto a scaffold before implantation into a damaged joint. Here, the scaffold architecture is critical, as it should simulate the ECM of cartilage in order to promote cellular adhesion, proliferation, differentiation, and migration, whilst also providing resistance to tensile, compressive, and shear stresses. Such a strategy was used to regenerate a cartilage defect in a mini-pig model. Li et al. observed that MSCs performed better than chondrocytes seeded onto PCL nanofiber constructs. Fibrous scaffolds seeded with MSCs resulted in the formation of hyaline cartilage–like tissue at 6 months

TABLE 16.18
In Vivo Studies with Electrospun Materials

Polymer	Organ	Comments	Reference
PCL	Peripheral nerve	Preliminary data on scaffold made using gap method of alignment	[16]
Poly(urethane)	Wound dressing	Rate of epithelialization is increased and dermis is well organized if wounds are covered with electrospun membrane	[59]
P(LLA-CL) (70:30)	Inferior superficial epigastric vein	Tubular scaffolds kept structural integrity for 7 weeks	[210]
PCL/COL	Subcutaneous	Good integration/neovascularization	[265]
PCL-silica	Bone	Biocompatible and bioresorbable	[273]
PLLA/MWNTs/HA	Periodontal ligament cells	Cell/fibers were implanted into the leg muscle pouch	[324]
P(AN-co-MA)	Peripheral nerve	Sheet-like structures in a conventional nerve guide	[334]
PCL	Bone	Woven bone-like appearance, with mineralized matrix	[337]
Chitosan	Peripheral nerve	Similar to autograft	[341]
PLGA/PCL	Peripheral nerve	Modest regeneration over 10 mm gap after 16 weeks	[342]
Bilayered chitosan tube	Peripheral nerve	YIGSR modified tubes regenerated similar to isograft after 5 and 10 weeks	[343]
PCL	Abdominal aortic substitute	Good patency, endothelialization, and cell ingrowth	[344]

postimplantation, whereas the chondrocytes produced a fibrocartilage tissue [339]. It is also possible to surface treat the nanofibers with cationized gelatin to enhance their tissue integration while maintaining or even promoting the secretion of cartilage ECM such as collagen II [340] (Table 16.18).

16.9.5 VASCULAR TISSUE ENGINEERING

Alternatively, suitable vascular graft substitutes must be able to withstand pulsation and the high pressure and flow rate of the blood stream. Electrospinning offers the potential to control the composition, mechanical properties, and structure of a graft while making it theoretically possible to match the compliance of the synthetic scaffold to the native artery. Topographically aligned submicron fibers have similar circumferential orientations to the cells and fibrils found in the medial layer of the native artery. Because of these similarities between electrospun scaffolds and natural ECM, and because of the large variety of materials that can be used, cell viability is often superior to other scaffold designs. One crucial point in developing vascular grafts is the stability of the implants. Any blockage of the inner part of the cylinder due to thrombosis and/or collapse of the electrospun materials should be avoided. PCL-collagen tubular scaffolds have been demonstrated to be very stable after 4 weeks postimplantation under normal hemodynamic conditions in a rabbit aortoiliac by-pass model. This type of scaffold maintained structural integrity and could resist the development of aneurisms [330].

16.10 SUMMARY

With the sudden interest in electrospinning and the desire to control the process to fabricate different assemblies, various setups have been reported. The use of both rotating and translating devices, the manipulation of the electric field profile, and liquid and patterned stationary collectors have proven to be successful in the fabrication of various fibrous assemblies. However, most of the work has focused on solution electrospinning. Here, since increased deposition of fibers will often result in an accumulation of charges, it remains to be seen whether these methods are able to obtain thick, patterned fibrous meshes. The conditions required to form 3D structures are still unclear: currently, the fibers forming reported 3D structures are very loosely packed and easily compressed, rendering them unsuitable for applications where structural integrity is required. Further advances in how scaffolds can be collected using melt electrospinning, coupled with improvements in yield, may provide methods to fabricate aligned fibrous substrates with increased porosity: useful for building complex structures that are of interest in TE applications. By far the most studied of the two types of electrospinning, solution electrospun substrates are interesting in many applications for regenerative medicine. More importantly, the diversity of materials employed is excellent, and polymers from those both biologically derived and synthetically generated (and blends of both) can be readily used. The ability to generate electrospun scaffolds with a variety of morphologies is also important, with highly oriented, random, and patterned meshes already demonstrated. In this approach, surface segregation of polymeric additives is interesting, since it provides a simple route to both biofunctionalize a fiber and render it protein resistant. Undoubtedly, electrospinning will remain dominated by the use of solutions for the foreseeable future, and advances in controlling the morphology and utility of polymeric scaffold fabrication can advance electrospinning even further. At this point, it is pertinent to consider that electrospinning for TE was proposed less than a decade ago and that the next decade will provide further advances and breakthroughs in the process. We anticipate that electrospinning will remain an important aspect of regenerative medicine research and become clinically important as the field builds on the foundation of the last 10 years.

REFERENCES

1. D. W. Hutmacher, T. Woodfield, P. D. Dalton, and J. Lewis, in: C. Van Blitterswijk et al., Eds., Scaffold design and fabrication. *Tissue Engineering* (Academic Press, New York, 2008), pp. 403–450.
2. D. W. Hutmacher and A. K. Ekaputra, in: P. K. Chu and X. Liu, Eds., Design and fabrication principles of electrospinning of scaffolds. *Biomaterials Fabrication and Processing Handbook* (Taylor & Francis Group, Boca Raton, FL, 2008), pp. 115–139.
3. P. D. Dalton, T. Woodfield, and D. W. Hutmacher, Snapshot: Polymer scaffolds for tissue engineering (vol. 30, p. 701, 2009). *Biomaterials* 30, 2420 (Apr 2009).
4. W. J. Li, C. T. Laurencin, E. J. Caterson, R. S. Tuan, and F. K. Ko, Electrospun nanofibrous structure: A novel scaffold for tissue engineering. *Journal of Biomedical Materials Research* 60, 613 (June 2002).
5. J. A. Matthews, G. E. Wnek, D. G. Simpson, and G. L. Bowlin, Electrospinning of collagen nanofibers. *Biomacromolecules* 3, 232 (Mar–Apr 2002).
6. C. L. Casper, N. Yamaguchi, K. L. Kiick, and J. F. Rabolt, Functionalizing electrospun fibers with biologically relevant macromolecules. *Biomacromolecules* 6, 1998 (Jul–Aug 2005).
7. S. Y. Chew, J. Wen, E. K. F. Yim, and K. W. Leong, Sustained release of proteins from electrospun biodegradable fibers. *Biomacromolecules* 6, 2017 (Jul–Aug 2005).
8. J. J. Stankus, J. J. Guan, K. Fujimoto, and W. R. Wagner, Microintegrating smooth muscle cells into a biodegradable, elastomeric fiber matrix. *Biomaterials* 27, 735 (Feb 2006).
9. A. K. Ekaputra, G. D. Prestwich, S. M. Cool, and D. W. Hutmacher, Combining electrospun scaffolds with electrosprayed hydrogels leads to three-dimensional cellularization of hybrid constructs. *Biomacromolecules* 9, 2097 (Aug 2008).
10. M. Simonet, O. D. Schneider, P. Neuenschwander, and W. J. Stark, Ultraporous 3D polymer meshes by low-temperature electrospinning: Use of ice crystals as a removable void template. *Polymer Engineering and Science* 47, 2020 (Dec 2007).
11. O. D. Schneider et al., Cotton wool-like nanocomposite biomaterials prepared by electrospinning: In vitro bioactivity and osteogenic differentiation of human mesenchymal stem cells. *Journal of Biomedical Materials Research Part B-Applied Biomaterials* 84B, 350 (Feb 2008).
12. R. Gentsch, B. Boysen, A. Lankenau, and H. G. Borner, Single-step electrospinning of bimodal fiber meshes for ease of cellular infiltration. *Macromolecular Rapid Communications* 31, 59 (Jan 2010).

13. S. J. Kim, D. H. Jang, W. H. Park, and B. M. Min, Fabrication and characterization of 3-dimensional PLGA nanofiber/microfiber composite scaffolds. *Polymer* 51, 1320 (Mar 2010).
14. Y. M. Shin, M. M. Hohman, M. P. Brenner, and G. C. Rutledge, Electrospinning: A whipping fluid jet generates submicron polymer fibers. *Applied Physics Letters* 78, 1149 (Feb 2001).
15. R. J. Deng et al., Melt electrospinning of low-density polyethylene having a low-melt flow index. *Journal of Applied Polymer Science* 114, 166 (Oct 2009).
16. B. S. Jha et al., Two pole air gap electrospinning: Fabrication of highly aligned, three-dimensional scaffolds for nerve reconstruction. *Acta Biomaterialia* 7, 203 (Jan 2011).
17. N. Shimada, H. Tsutsumi, K. Nakane, T. Ogihara, and N. Ogata, Poly(ethylene-co-vinyl alcohol) and nylon 6/12 nanofibers produced by melt electrospinning system equipped with a line-like laser beam melting device. *Journal of Applied Polymer Science* 116, 2998 (Jun 2010).
18. A. Zajicova et al., Treatment of ocular surface injuries by limbal and mesenchymal stem cells growing on nanofiber scaffolds. *Cell Transplant* 19, 1281 (Jun 2011).
19. D. H. Sun, C. Chang, S. Li, and L. Lin, Near-field electrospinning. *Nano Letters* 6, 839 (Apr 2006).
20. S. N. Malakhov et al., Method of manufacturing nonwovens by electrospinning from polymer melts. *Fibre Chemistry* 41, 355 (Nov 2009).
21. D. H. Reneker, A. L. Yarin, H. Fong, and S. Koombhongse, Bending instability of electrically charged liquid jets of polymer solutions in electrospinning. *Journal of Applied Physics* 87, 4531 (May 2000).
22. D. W. Hutmacher and P. D. Dalton, Melt electrospinning. *Chemistry-An Asian Journal* 6, 44 (Jan 2011).
23. P. D. Dalton, D. Grafahrend, K. Klinkhammer, D. Klee, and M. Moller, Electrospinning of polymer melts: Phenomenological observations. *Polymer* 48, 6823 (Nov 2007).
24. P. D. Dalton, N. T. Joergensen, J. Groll, and M. Moeller, Patterned melt electrospun substrates for tissue engineering. *Biomedical Materials* 3, 34109 (Sep 2008).
25. P. Zahedi, I. Rezaeian, S. O. Ranaei-Siadat, S. H. Jafari, and P. Supaphol, A review on wound dressings with an emphasis on electrospun nanofibrous polymeric bandages. *Polymers for Advanced Technologies* 21, 77 (Feb 2010).
26. Y. K. Wang, T. Yong, and S. Ramakrishna, Nanofibres and their influence on cells for tissue regeneration. *Australian Journal of Chemistry* 58, 704 (2005).
27. M. J. Beglou and A. K. Haghi, Electrospun biodegdadable and biocompatible natural nanofibers: A detailed review. *Cellulose Chemistry and Technology* 42, 441 (Oct–Dec 2008).
28. J. D. Schiffman and C. L. Schauer, A review: Electrospinning of biopolymer nanofibers and their applications. *Polymer Reviews* 48, 317 (2008).
29. D. I. Zeugolis et al., Electro-spinning of pure collagen nano-fibres—Just an expensive way to make gelatin? *Biomaterials* 29, 2293 (May 2008).
30. B. M. Min et al., Formation of silk fibroin matrices with different texture and its cellular response to normal human keratinocytes. *International Journal of Biological Macromolecules* 34, 281 (Oct 2004).
31. X. H. Zhang, M. R. Reagan, and D. L. Kaplan, Electrospun silk biomaterial scaffolds for regenerative medicine. *Advanced Drug Delivery Reviews* 61, 988 (Oct 2009).
32. K. S. Rho et al., Electrospinning of collagen nanofibers: Effects on the behavior of normal human keratinocytes and early-stage wound healing. *Biomaterials* 27, 1452 (Mar 2006).
33. S. S. Xu et al., Chemical crosslinking and biophysical properties of electrospun hyaluronic acid based ultra-thin fibrous membranes. *Polymer* 50, 3762 (Jul 2009).
34. Y. Ji et al., Electrospun three-dimensional hyaluronic acid nanofibrous scaffolds. *Biomaterials* 27, 3782 (Jul 2006).
35. S. A. Sell, M. J. McClure, K. Garg, P. S. Wolfe, and G. L. Bowlin, Electrospinning of collagen/biopolymers for regenerative medicine and cardiovascular tissue engineering. *Advanced Drug Delivery Reviews* 61, 1007 (Oct 2009).
36. H. C. Chen, W. C. Jao, and M. C. Yang, Characterization of gelatin nanofibers electrospun using ethanol/formic acid/water as a solvent. *Polymers for Advanced Technologies* 20, 98 (Feb 2009).
37. M. W. Frey, Electrospinning cellulose and cellulose derivatives. *Polymer Reviews* 48, 378 (2008).
38. H. M. Powell and S. T. Boyce, Fiber density of electrospun gelatin scaffolds regulates morphogenesis of dermal-epidermal skin substitutes. *Journal of Biomedical Materials Research Part A* 84A, 1078 (Mar 2008).
39. R. A. Neal et al., Laminin nanofiber meshes that mimic morphological properties and bioactivity of basement membranes. *Tissue Engineering Part C-Methods* 15, 11 (Mar 2009).
40. H. S. Koh, T. Yong, C. K. Chan, and S. Ramakrishna, Enhancement of neurite outgrowth using nanostructured scaffolds coupled with laminin. *Biomaterials* 29, 3574 (Sep 2008).

41. R. Jayakumar, M. Prabaharan, S. V. Nair, and H. Tamura, Novel chitin and chitosan nanofibers in biomedical applications. *Biotechnology Advances* 28, 142 (Jan–Feb 2010).

42. L. Nivison-Smith, J. Rnjak, and A. S. Weiss, Synthetic human elastin microfibers: Stable cross-linked tropoelastin and cell interactive constructs for tissue engineering applications. *Acta Biomaterialia* 6, 354 (Feb 2010).

43. Y. Ner, J. A. Stuart, G. Whited, and G. A. Sotzing, Electrospinning nanoribbons of a bioengineered silk-elastin-like protein (SELP) from water. *Polymer* 50, 5828 (Nov 2009).

44. M. C. McManus, E. D. Boland, D. G. Simpson, C. P. Barnes, and G. L. Bowlin, Electrospun fibrinogen: Feasibility as a tissue engineering scaffold in a rat cell culture model. *Journal of Biomedical Materials Research Part A* 81A, 299 (May 2007).

45. G. E. Wnek, M. E. Carr, D. G. Simpson, and G. L. Bowlin, Electrospinning of nanofiber fibrinogen structures. *Nano Letters* 3, 213 (Feb 2003).

46. S. H. Teng, P. Wang, and H. E. Kim, Blend fibers of chitosan-agarose by electrospinning. *Materials Letters* 63, 2510 (Nov 2009).

47. J. X. Li et al., Electrospinning of hyaluronic acid (HA) and HA/gelatin blends. *Macromolecular Rapid Communications* 27, 114 (Jan 2006).

48. D. Puppi, A. M. Piras, N. Detta, D. Dinucci, and F. Chiellini, Poly(lactic-co-glycolic acid) electrospun fibrous meshes for the controlled release of retinoic acid. *Acta Biomaterialia* 6, 1258 (Apr 2010).

49. L. Zhao et al., Preparation and cytocompatibility of PLGA scaffolds with controllable fiber morphology and diameter using electrospinning method. *Journal of Biomedical Materials Research Part B-Applied Biomaterials* 87B, 26 (Oct 2008).

50. J. M. Corey et al., The design of electrospun PLLA nanofiber scaffolds compatible with serum-free growth of primary motor and sensory neurons. *Acta Biomaterialia* 4, 863 (Jul 2008).

51. R. Inai, M. Kotaki, and S. Ramakrishna, Structure and properties of electrospun PLLA single nanofibres. *Nanotechnology* 16, 208 (Feb 2005).

52. G. H. Kim, Electrospun PCL nanofibers with anisotropic mechanical properties as a biomedical scaffold. *Biomedical Materials* 3, 25010 (Jun 2008).

53. G. Y. Liao, K. F. Jiang, S. B. Jiang, and H. Xia, Synthesis and characterization of biodegradable poly(epsilon-caprolactone)-b-poly(L-lactide) and study on their electrospun scaffolds. *Journal of Macromolecular Science Part A-Pure and Applied Chemistry* 47, 1116 (2010).

54. J. Zeng, X. S. Chen, Q. Z. Liang, X. L. Xu, and X. B. Jing, Enzymatic degradation of poly(L-lactide) and poly (epsilon-caprolactone) electrospun fibers. *Macromolecular Bioscience* 4, 1118 (Dec 2004).

55. T. Uyar and F. Besenbacher, Electrospinning of uniform polystyrene fibers: The effect of solvent conductivity. *Polymer* 49, 5336 (Nov 2008).

56. G. T. Kim et al., The morphology of electrospun polystyrene fibers. *Korean Journal of Chemical Engineering* 22, 147 (Jan 2005).

57. D. Grafahrend et al., Control of protein adsorption on functionalized electrospun fibers. *Biotechnology and Bioengineering* 101, 609 (Oct 2008).

58. D. Grafahrend et al., Biofunctionalized poly(ethylene glycol)-block-poly(epsilon-caprolactone) nanofibers for tissue engineering. *Journal of Materials Science-Materials in Medicine* 19, 1479 (Apr 2008).

59. M. S. Khil, D. I. Cha, H. Y. Kim, I. S. Kim, and N. Bhattarai, Electrospun nanofibrous polyurethane membrane as wound dressing. *Journal of Biomedical Materials Research Part B-Applied Biomaterials* 67B, 675 (Nov 2003).

60. A. Pedicini and R. J. Farris, Mechanical behavior of electrospun polyurethane. *Polymer* 44, 6857 (Oct 2003).

61. S. A. Riboldi, M. Sampaolesi, P. Neuenschwander, G. Cossu, and S. Mantero, Electrospun degradable polyesterurethane membranes: Potential scaffolds for skeletal muscle tissue engineering. *Biomaterials* 26, 4606 (Aug 2005).

62. K. Arayanarakul, N. Choktaweesap, D. Aht-ong, C. Meechaisue, and P. Supaphol, Effects of poly(ethylene glycol), inorganic salt, sodium dodecyl sulfate, and solvent system on electrospinning of poly(ethylene oxide). *Macromolecular Materials and Engineering* 291, 581 (Jun 2006).

63. J. M. Deitzel, J. D. Kleinmeyer, J. K. Hirvonen, and N. C. B. Tan, Controlled deposition of electrospun poly(ethylene oxide) fibers. *Polymer* 42, 8163 (Sep 2001).

64. H. Okuzaki, K. Kobayashi, and H. Yan, Non-woven fabric of poly(N-isopropylacrylamide) nanofibers fabricated by electrospinning. *Synthetic Metals* 159, 2273 (Nov 2009).

65. D. N. Rockwood, D. B. Chase, R. E. Akins, and J. F. Rabolt, Characterization of electrospun poly(N-isopropyl acrylamide) fibers. *Polymer* 49, 4025 (Aug 2008).

66. D. Grafahrend et al., Degradable polyester scaffolds with controlled surface chemistry combining minimal protein adsorption with specific bioactivation. *Nature Materials* 10, 67 (Jan 2011).

67. E. Schnell et al., Guidance of glial cell. migration and axonal growth on electrospun nanofibers of poly-epsilon-caprolactone and a collagen/poly-epsilon-caprolactone blend. *Biomaterials* 28, 3012 (Jul 2007).

68. H. J. Jin, S. V. Fridrikh, G. C. Rutledge, and D. L. Kaplan, Electrospinning *Bombyx mori* silk with poly(ethylene oxide). *Biomacromolecules* 3, 1233 (Nov–Dec 2002).

69. Z. X. Meng et al., Electrospinning of PLGA/gelatin randomly-oriented and aligned nanofibers as potential scaffold in tissue engineering. *Materials Science and Engineering C-Materials for Biological Applications* 30, 1204 (Oct 2010).

70. X. L. Deng, G. Sui, M. L. Zhao, G. Q. Chen, and X. P. Yang, Poly(L-lactic acid)/hydroxyapatite hybrid nanofibrous scaffolds prepared by electrospinning. *Journal of Biomaterials Science-Polymer Edition* 18, 117 (2007).

71. B. Duan et al., A nanofibrous composite membrane of PLGA-chitosan/PVA prepared by electrospinning. *European Polymer Journal* 42, 2013 (Sep, 2006).

72. A. S. Asran, S. Henning, and G. H. Michler, Polyvinyl alcohol-collagen-hydroxyapatite biocomposite nanofibrous scaffold: Mimicking the key features of natural bone at the nanoscale level. *Polymer* 51, 868 (Feb 2010).

73. M. G. McKee, G. L. Wilkes, R. H. Colby, and T. E. Long, Correlations of solution rheology with electrospun fiber formation of linear and branched polyesters. *Macromolecules* 37, 1760 (Mar 2004).

74. H. R. Nie et al., Effects of chain conformation and entanglement on the electrospinning of pure alginate. *Biomacromolecules* 9, 1362 (May 2008).

75. Y. Liu et al., Effects of solution properties and electric field on the electrospinning of hyaluronic acid. *Carbohydrate Polymers* 83, 1011 (2011).

76. K. H. Lee, H. Y. Kim, M. S. Khil, Y. M. Ra, and D. R. Lee, Characterization of nano-structured poly(epsilon-caprolactone) nonwoven mats via electrospinning. *Polymer* 44, 1287 (Feb 2003).

77. K. H. Lee, H. Y. Kim, Y. J. Ryu, K. W. Kim, and S. W. Choi, Mechanical behavior of electrospun fiber mats of poly(vinyl chloride)/polyurethane polyblends. *Journal of Polymer Science Part B-Polymer Physics* 41, 1256 (Jun 2003).

78. H. Fong, W. D. Liu, C. S. Wang, and R. A. Vaia, Generation of electrospun fibers of nylon 6 and nylon 6-montmorillonite nanocomposite. *Polymer* 43, 775 (Feb 2002).

79. W. W. Zuo et al., Experimental study on relationship between jet instability and formation of beaded fibers during electrospinning. *Polymer Engineering and Science* 45, 704 (May 2005).

80. C. Mit-uppatham, M. Nithitanakul, and P. Supaphol, Ultratine electrospun polyamide-6 fibers: Effect of solution conditions on morphology and average fiber diameter. *Macromolecular Chemistry and Physics* 205, 2327 (Nov 2004).

81. C. X. Zhang, X. Y. Yuan, L. L. Wu, Y. Han, and J. Sheng, Study on morphology of electrospun poly(vinyl alcohol) mats. *European Polymer Journal* 41, 423 (Mar 2005).

82. A. Patlolla, G. Collins, and T. L. Arinzeh, Solvent-dependent properties of electrospun fibrous composites for bone tissue regeneration. *Acta Biomaterialia* 6, 90 (Jan 2010).

83. X. H. Zong et al., Structure and process relationship of electrospun bioabsorbable nanofiber membranes. *Polymer* 43, 4403 (Jul 2002).

84. Y. You, S. J. Lee, B. M. Min, and W. H. Park, Effect of solution properties on nanofibrous structure of electrospun poly(lactic-co-glycolic acid). *Journal of Applied Polymer Science* 99, 1214 (Feb 2006).

85. Z. Jun, H. Q. Hou, A. Schaper, J. H. Wendorff, and A. Greiner, Poly-L-lactide nanofibers by electrospinning - Influence of solution viscosity and electrical conductivity on fiber diameter and fiber morphology. *E-Polymers*, Paper 9 (Mar 2003).

86. S. Megelski, J. S. Stephens, D. B. Chase, and J. F. Rabolt, Micro- and nanostructured surface morphology on electrospun polymer fibers. *Macromolecules* 35, 8456 (Oct 2002).

87. Y. Z. Zhang et al., Coaxial electrospinning of (fluorescein isothiocyanate-conjugated bovine serum albumin)-encapsulated poly(epsilon-caprolactone) nanofibers for sustained release. *Biomacromolecules* 7, 1049 (Apr 2006).

88. Z. G. Chen, B. Wei, X. M. Mo, and F. Z. Cui, Diameter control of electrospun chitosan-collagen fibers. *Journal of Polymer Science Part B-Polymer Physics* 47, 1949 (Oct 2009).

89. Y. Z. Zhang, Z. M. Huang, X. J. Xu, C. T. Lim, and S. Ramakrishna, Preparation of core-shell structured PCL-r-gelatin Bi-component nanofibers by coaxial electrospinning. *Chemistry of Materials* 16, 3406 (Sep 2004).

90. R. Kessick, J. Fenn, and G. Tepper, The use of AC potentials in electrospraying and electrospinning processes. *Polymer* 45, 2981 (Apr 2004).

91. J. M. Deitzel, J. Kleinmeyer, D. Harris, and N. C. B. Tan, The effect of processing variables on the morphology of electrospun nanofibers and textiles. *Polymer* 42, 261 (Jan 2001).
92. M. M. Demir, I. Yilgor, E. Yilgor, and B. Erman, Electrospinning of polyurethane fibers. *Polymer* 43, 3303 (May 2002).
93. J. Ayutsede et al., Regeneration of Bombyx mori silk by electrospinning. Part 3: Characterization of electrospun nonwoven mat. *Polymer* 46, 1625 (Feb 2005).
94. X. M. Mo, C. Y. Xu, M. Kotaki, and S. Ramakrishna, Electrospun P(LLA-CL) nanofiber: A biomimetic extracellular matrix for smooth muscle cell and endothelial cell proliferation. *Biomaterials* 25, 1883 (May 2004).
95. D. S. Katti, K. W. Robinson, F. K. Ko, and C. T. Laurencin, Bioresorbable nanofiber-based systems for wound healing and drug delivery: Optimization of fabrication parameters. *Journal of Biomedical Materials Research Part B-Applied Biomaterials* 70B, 286 (Aug 2004).
96. F.-L. Zhou, R.-H. Gong, and I. Porat, Three-jet electrospinning using a flat spinneret. *Journal of Materials Science* 44, 5501 (2009).
97. C. Zhang, X. Yuan, L. Wu, Y. Han, and J. Sheng, Study on morphology of electrospun poly(vinyl alcohol) mats. *European Polymer Journal* 41, 423 (2005).
98. N. Bolgen, Y. Z. Menceloglu, K. Acatay, I. Vargel, and E. Piskin, In vitro and in vivo degradation of non-woven materials made of poly(epsilon-caprolactone) nanofibers prepared by electrospinning under different conditions. *Journal of Biomaterials Science-Polymer Edition* 16, 1537 (2005).
99. K. H. Lee, H. Y. Kim, H. J. Bang, Y. H. Jung, and S. G. Lee, The change of bead morphology formed on electrospun polystyrene fibers. *Polymer* 44, 4029 (Jun 2003).
100. P. D. Dalton, K. Klinkhammer, J. Salber, D. Klee, and M. Moller, Direct in vitro electrospinning with polymer melts. *Biomacromolecules* 7, 686 (Mar 2006).
101. N. Detta et al., Melt electrospinning of polycaprolactone and its blends with poly(ethylene glycol). *Polymer International* 59, 1558 (Nov 2010).
102. C. Wang et al., Electrospinning of polyacrylonitrile solutions at elevated temperatures. *Macromolecules* 40, 7973 (Oct 2007).
103. S. R. Givens, K. H. Gardner, J. F. Rabolt, and D. B. Chase, High-temperature electrospinning of polyethylene microfibers from solution. *Macromolecules* 40, 608 (Feb 2007).
104. D. G. Yu et al., Multicomponent amorphous nanofibers electrospun from hot aqueous solutions of a poorly soluble drug. *Pharmaceutical Research* 27, 2466 (Nov 2010).
105. C. Wang, C. H. Hsu, and J. H. Lin, Scaling laws in electrospinning of polystyrene solutions. *Macromolecules* 39, 7662 (Oct 2006).
106. J. P. Chen, K. H. Ho, Y. P. Chiang, and K. W. Wu, Fabrication of electrospun poly(methyl methacrylate) nanofibrous membranes by statistical approach for application in enzyme immobilization. *Journal of Membrane Science* 340, 9 (Sep 2009).
107. C. J. Thompson, G. G. Chase, A. L. Yarin, and D. H. Reneker, Effects of parameters on nanofiber diameter determined from electrospinning model. *Polymer* 48, 6913 (Nov 2007).
108. T. Jarusuwannapoom et al., Effect of solvents on electro-spinnability of polystyrene solutions and morphological appearance of resulting electrospun polystyrene fibers. *European Polymer Journal* 41, 409 (Mar 2005).
109. C. L. Casper, J. S. Stephens, N. G. Tassi, D. B. Chase, and J. F. Rabolt, Controlling surface morphology of electrospun polystyrene fibers: Effect of humidity and molecular weight in the electrospinning process. *Macromolecules* 37, 573 (Jan 2004).
110. J.-Y. Park and I.-H. Lee, Relative humidity effect on the preparation of porous electrospun polystyrene fibers. *Journal of Nanoscience and Nanotechnology* 10, 3473 (2010).
111. S. De Vrieze et al., The effect of temperature and humidity on electrospinning. *Journal of Materials Science* 44, 1357 (Mar 2009).
112. G. Srinivasan and D. H. Reneker, Structure and morphology of small-diameter electrospun aramid fibers. *Polymer International* 36, 195 (Feb 1995).
113. R. Tzezana, E. Zussman, and S. Levenberg, A layered ultra-porous scaffold for tissue engineering, created via a hydrospinning method. *Tissue Engineering Part C-Methods* 14, 281 (Dec 2008).
114. E. Smit, U. Buttner, and R. D. Sanderson, Continuous yarns from electrospun fibers. *Polymer* 46, 2419 (Mar 2005).
115. M. S. Khil, S. R. Bhattarai, H. Y. Kim, S. Z. Kim, and K. H. Lee, Novel fabricated matrix via electrospinning for tissue engineering. *Journal of Biomedical Materials Research Part B-Applied Biomaterials* 72B, 117 (Jan 2005).
116. A. Formhals, U. S. P. Office, Ed. (1934).
117. A. Formhals, U. S. P. Office, Ed. (1944).

118. W. E. Teo and S. Ramakrishna, A review on electrospinning design and nanofibre assemblies. *Nanotechnology* 17, R89 (Jul 2006).

119. P. D. Dalton, D. Klee, and M. Moller, Electrospinning with dual collection rings. *Polymer* 46, 611 (Jan 2005).

120. W. E. Teo and S. Ramakrishna, Electrospun fibre bundle made of aligned nanofibres over two fixed points. *Nanotechnology* 16, 1878 (Sep 2005).

121. K. Klinkhammer et al., Deposition of electrospun fibers on reactive substrates for in vitro investigations. *Tissue Engineering Part C-Methods* 15, 77 (Mar 2009).

122. K. Klinkhammer et al., Functionalization of electrospun fibers of poly(epsilon-caprolactone) with star shaped NCO-poly(ethylene glycol)-stat-poly(propylene glycol) for neuronal cell guidance. *Journal of Materials Science-Materials in Medicine* 21, 2637 (Sep 2010).

123. J. Gerardo-Nava et al., Human neural cell interactions with orientated electrospun nanofibers in vitro. *Nanomedicine* 4, 11 (Jan 2009).

124. J. M. Deitzel, J. Kleinmeyer, D. Harris, and N. C. B. Tan, The effect of processing variables on the morphology of electrospun nanofibers and textiles. *Polymer* 42, 261 (2001).

125. S. Ramakrishna, K. Fujihara, W. E. Teo, T. C. Lim, and Z. Ma, *An Introduction to Electrospinning and Nanofibers* (World Scientific Publishing, Singapore, 2005).

126. D. M. Zhang and J. Chang, Electrospinning of three-dimensional nanofibrous tubes with controllable architectures. *Nano Letters* 8, 3283 (Oct 2008).

127. D. M. Zhang and J. Chang, Patterning of electrospun fibers using electroconductive templates. *Advanced Materials* 19, 3664 (Nov 5, 2007).

128. J. Kameoka et al., A scanning tip electrospinning source for deposition of oriented nanofibres. *Nanotechnology* 14, 1124 (Oct 2003).

129. A. Subramanian, D. Vu, G. F. Larsen, and H. Y. Lin, Preparation and evaluation of the electrospun chitosan/PEO fibers for potential applications in cartilage tissue engineering. *Journal of Biomaterials Science-Polymer Edition* 16, 861 (2005).

130. T. A. Kowalewski, S. Barral, and T. Kowalczyk, Modeling electrospinning of nanofibers. *Iutam Symposium on Modelling Nanomaterials and Nanosystems* 13, 279 (2009).

131. B. Sundaray et al., Electrospinning of continuous aligned polymer fibers. *Applied Physics Letters* 84, 1222 (Feb 2004).

132. E. Zussman, A. Theron, and A. L. Yarin, Formation of nanofiber crossbars in electrospinning. *Applied Physics Letters* 82, 973 (Feb 2003).

133. J. Stitzel et al., Controlled fabrication of a biological vascular substitute. *Biomaterials* 27, 1088 (Oct 2006).

134. S. Kidoaki, I. K. Kwon, and T. Matsuda, Mesoscopic spatial designs of nano- and microfiber meshes for tissue-engineering matrix and scaffold based on newly devised multilayering and mixing electrospinning techniques. *Biomaterials* 26, 37 (Jan 2005).

135. A. Theron, E. Zussman, and A. L. Yarin, Electrostatic field-assisted alignment of electrospun nanofibres. *Nanotechnology* 12, 384 (Sep 2001).

136. C. Y. Xu, R. Inai, M. Kotaki, and S. Ramakrishna, Aligned biodegradable nanofibrous structure: A potential scaffold for blood vessel engineering. *Biomaterials* 25, 877 (Feb 2004).

137. N. Bhattarai, D. Edmondson, O. Veiseh, F. A. Matsen, and M. Q. Zhang, Electrospun chitosan-based nanofibers and their cellular compatibility. *Biomaterials* 26, 6176 (Nov 2005).

138. P. Katta, M. Alessandro, R. D. Ramsier, and G. G. Chase, Continuous electrospinning of aligned polymer nanofibers onto a wire drum collector. *Nano Letters* 4, 2215 (Nov 2004).

139. C. Chang, K. Limkrailassiri, and L. W. Lin, Continuous near-field electrospinning for large area deposition of orderly nanofiber patterns. *Applied Physics Letters* 93, 123111 (Sep 2008).

140. T. Han, D. H. Reneker, and A. L. Yarin, Buckling of jets in electrospinning. *Polymer* 48, 6064 (Sep 2007).

141. H. J. Zhou, T. B. Green, and Y. L. Joo, The thermal effects on electrospinning of polylactic acid melts. *Polymer* 47, 7497 (Oct 2006).

142. C. Hellmann et al., High precision deposition electrospinning of nanofibers and nanofiber nonwovens. *Polymer* 50, 1197 (Feb 2009).

143. T. D. Brown, P. D. Dalton, and D. W. Hutmacher, Direct writing by way of melt electrospinning. *Advanced Materials* 23, 5651 (Dec 2011).

144. S. Chung, A. K. Moghe, G. A. Montero, S. H. Kim, and M. W. King, Nanofibrous scaffolds electrospun from elastomeric biodegradable poly(L-lactide-co-epsilon-caprolactone) copolymer. *Biomedical Materials* 4, 15019 (Feb 2009).

145. S. H. Park, T. G. Kim, H. C. Kim, D. Y. Yang, and T. G. Park, Development of dual scale scaffolds via direct polymer melt deposition and electrospinning for applications in tissue regeneration. *Acta Biomaterialia* 4, 1198 (Sep 2008).

146. M. Centola et al., Combining electrospinning and fused deposition modeling for the fabrication of a hybrid vascular graft. *Biofabrication* 2, 14102 (Mar 2010).

147. Q. P. Pham, U. Sharma, and A. G. Mikos, Electrospun poly(epsilon-caprolactone) microfiber and multi-layer nanofiber/microfiber scaffolds: Characterization of scaffolds and measurement of cellular infiltration. *Biomacromolecules* 7, 2796 (Oct 2006).

148. N. Ashammakhi, A. Ndreu, L. Nikkola, I. Wimpenny, and Y. Yang, Advancing tissue engineering by using electrospun nanofibers. *Regenerative Medicine* 3, 547 (Jul 2008).

149. M. Goldberg, R. Langer, and X. Q. Jia, Nanostructured materials for applications in drug delivery and tissue engineering. *Journal of Biomaterials Science-Polymer Edition* 18, 241 (Mar 2007).

150. C. M. Li, C. Vepari, H. J. Jin, H. J. Kim, and D. L. Kaplan, Electrospun silk-BMP-2 scaffolds for bone tissue engineering. *Biomaterials* 27, 3115 (Jun 2006).

151. Y. C. Fu, H. Nie, M. L. Ho, C. K. Wang, and C. H. Wang, Optimized bone regeneration based on sustained release from three-dimensional fibrous PLGA/HAp composite scaffolds loaded with BMP-2. *Biotechnology and Bioengineering* 99, 996 (Mar 2008).

152. H. Nie, B. W. Soh, Y. C. Fu, and C. H. Wang, Three-dimensional fibrous PLGA/HAp composite scaffold for BMP-2 delivery. *Biotechnology and Bioengineering* 99, 223 (Jan 2008).

153. H. M. Nie and C. H. Wang, Fabrication and characterization of PLGA/HAp scaffolds for delivery of BMP-2 plasmid composite DNA. *Journal of Controlled Release* 120, 111 (Jul 2007).

154. M. K. Horne, D. R. Nisbet, J. S. Forsythe, and C. L. Parish, Three-dimensional nanofibrous scaffolds incorporating immobilized BDNF promote proliferation and differentiation of cortical neural stem cells. *Stem Cells Dev* 19, 843 (Jun 2010).

155. H. J. Lam, S. Patel, A. J. Wang, J. Chu, and S. Li, In vitro regulation of neural differentiation and axon growth by growth factors and bioactive nanofibers. *Tissue Engineering Part A* 16, 2641 (Aug 2010).

156. S. Sahoo, L. T. Ang, J. C. H. Goh, and S. L. Toh, Growth factor delivery through electrospun nanofibers in scaffolds for tissue engineering applications. *Journal of Biomedical Materials Research Part A* 93A, 1539 (Jun 2010).

157. C. L. Casper, W. D. Yang, M. C. Farach-Carson, and J. F. Rabolt, Coating electrospun collagen and gelatin fibers with perlecan domain I for increased growth factor binding. *Biomacromolecules* 8, 1116 (Apr 2007).

158. S. Patel et al., Bioactive nanofibers: Synergistic effects of nanotopography and chemical signaling on cell guidance. *Nano Letters* 7, 2122 (Jul 2007).

159. S. Y. Chew, R. F. Mi, A. Hoke, and K. W. Leong, Aligned protein-polymer composite fibers enhance nerve regeneration: A potential tissue-engineering platform. *Advanced Functional Materials* 17, 1288 (May 2007).

160. F. Wang, Z. Q. Li, K. Tamama, C. K. Sen, and J. J. Guan, Fabrication and characterization of prosurvival growth factor releasing, anisotropic scaffolds for enhanced mesenchymal stem cell survival/growth and orientation. *Biomacromolecules* 10, 2609 (Sep 2009).

161. X. Q. Li et al., Encapsulation of proteins in poly(L-lactide-co-caprolactone) fibers by emulsion electrospinning. *Colloids and Surfaces B-Biointerfaces* 75, 418 (Feb 2010).

162. H. Li et al., Controlled release of PDGF-bb by coaxial electrospun dextran/poly(L-lactide-co-epsilon-caprolactone) fibers with an ultrafine core/shell structure. *Journal of Biomaterials Science-Polymer Edition* 21, 803 (2010).

163. Y. Lu, H. L. Jiang, K. H. Tu, and L. Q. Wang, Mild immobilization of diverse macromolecular bioactive agents onto multifunctional fibrous membranes prepared by coaxial electrospinning. *Acta Biomaterialia* 5, 1562 (Jun 2009).

164. N. Bolgen, I. Vargel, P. Korkusuz, Y. Z. Menceloglu, and E. Piskin, In vivo performance of antibiotic embedded electrospun PCL membranes for prevention of abdominal adhesions. *Journal of Biomedical Materials Research Part B-Applied Biomaterials* 81B, 530 (May 2007).

165. D. Grafahrend, K.-H. Heffels, M. Beer, P. Gasteier, M. Möller, G. Boehm, P. D. Dalton, and J. Groll, Degradable polyester scaffolds with controlled surface chemistry combining minimal protein adsorption with specific bioactivation. *Nature Materials* 10, 67–73 (2010).

166. H. S. Yoo, T. G. Kim, and T. G. Park, Surface-functionalized electrospun nanofibers for tissue engineering and drug delivery. *Advanced Drug Delivery Reviews* 61, 1033 (Oct 2009).

167. M. Yoshida, R. Langer, A. Lendlein, and J. Lahann, From advanced biomedical coatings to multifunctionalized biomaterials. *Polymer Reviews* 46, 347 (Oct–Dec 2006).

168. Z. W. Ma, K. Masaya, and S. Ramakrishna, Immobilization of Cibacron blue F3GA on electrospun polysulphone ultra-fine fiber surfaces towards developing an affinity membrane for albumin adsorption. *Journal of Membrane Science* 282, 237 (Oct 2006).

169. H. S. Kim and H. S. Yoo, MMPs-responsive release of DNA from electrospun nanofibrous matrix for local gene therapy: In vitro and in vivo evaluation. *Journal of Controlled Release* 145, 264 (Aug 2010).

170. R. Losel, D. Grafahrend, M. Moller, and D. Klee, Bioresorbable electrospun fibers for immobilization of thiol-containing compounds. *Macromolecular Bioscience* 10, 1177 (Oct 2010).

171. J. B. Chiu et al., Electrospun nanofibrous scaffolds for biomedical applications. *Journal of Biomedical Nanotechnology* 1, 115 (Jun 2005).

172. S. G. Kumbar, S. P. Nukavarapu, R. James, L. S. Nair, and C. T. Laurencin, Electrospun poly(lactic acid-co-glycolic acid) scaffolds for skin tissue engineering. *Biomaterials* 29, 4100 (Oct 2008).

173. P. Wutticharoenmongkol, N. Sanchavanakit, P. Pavasant, and P. Supaphol, Preparation and characterization of novel bone scaffolds based on electrospun polycaprolactone fibers filled with nanoparticles. *Macromolecular Bioscience* 6, 70 (Jan 2006).

174. A. Neamnark, N. Sanchavanakit, P. Pavasant, R. Rujiravanit, and P. Supaphol, In vitro biocompatibility of electrospun hexanoyl chitosan fibrous scaffolds towards human keratinocytes and fibroblasts. *European Polymer Journal* 44, 2060 (Jul 2008).

175. C. A. Bashur, L. A. Dahlgren, and A. S. Goldstein, Effect of fiber diameter and orientation on fibroblast morphology and proliferation on electrospun poly(D,L-lactic-co-glycolic acid) meshes. *Biomaterials* 27, 5681 (Nov 2006).

176. K. Park, Y. M. Ju, J. S. Son, K. D. Ahn, and D. K. Han, Surface modification of biodegradable electrospun nanofiber scaffolds and their interaction with fibroblasts. *Journal of Biomaterials Science-Polymer Edition* 18, 369 (2007).

177. R. A. Thakur, C. A. Florek, J. Kohn, and B. B. Michniak, Electrospun nanofibrous polymeric scaffold with targeted drug release profiles for potential application as wound dressing. *International Journal of Pharmaceutics* 364, 87 (Nov 19, 2008).

178. A. Saraf, L. S. Baggett, R. M. Raphael, F. K. Kasper, and A. G. Mikos, Regulated non-viral gene delivery from coaxial electrospun fiber mesh scaffolds. *Journal of Controlled Release* 143, 95 (Apr 2010).

179. S. P. Zhong et al., An aligned nanofibrous collagen scaffold by electrospinning and its effects on in vitro fibroblast culture. *Journal of Biomedical Materials Research Part A* 79A, 456 (Dec 2006).

180. C. H. Lee et al., Nanofiber alignment and direction of mechanical strain affect the ECM production of human ACL fibroblast. *Biomaterials* 26, 1261 (Apr 2005).

181. M. Schindler et al., A synthetic nanofibrillar matrix promotes in vivo-like organization and morphogenesis for cells in culture. *Biomaterials* 26, 5624 (Oct 2005).

182. J. Venugopal and S. Ramakrishna, Biocompatible nanofiber matrices for the engineering of a dermal substitute for skin regeneration. *Tissue Engineering* 11, 847 (2005).

183. Z. X. Cai et al., Fabrication of chitosan/silk fibroin composite nanofibers for wound-dressing applications. *International Journal of Molecular Sciences* 11, 3529 (Sep 2010).

184. J. Y. Chun et al., Epidermal cellular response to poly(vinyl alcohol) nanofibers containing silver nanoparticles. *Colloids and Surfaces B-Biointerfaces* 78, 334 (Jul 2010).

185. Y. Gui-Bo et al., Study of the electrospun PLA/silk fibroin-gelatin composite nanofibrous scaffold for tissue engineering. *Journal of Biomedical Materials Research Part A* 93A, 158 (Apr 2010).

186. F. Y. Hsu, Y. S. Hung, H. M. Liou, and C. H. Shen, Electrospun hyaluronate-collagen nanofibrous matrix and the effects of varying the concentration of hyaluronate on the characteristics of foreskin fibroblast cells. *Acta Biomaterialia* 6, 2140 (Jun 2010).

187. P. J. Kluger et al., Electrospun poly(d/l-lactide-co-l-lactide) hybrid matrix: A novel scaffold material for soft tissue engineering. *Journal of Materials Science-Materials in Medicine* 21, 2665 (Sep 2010).

188. F. J. Liu et al., Effect of the porous microstructures of poly(lactic-co-glycolic acid)/carbon nanotube composites on the growth of fibroblast cells. *Soft Materials* 8, 239 (2010).

189. S. J. Liu et al., Electrospun PLGA/collagen nanofibrous membrane as early-stage wound dressing. *Journal of Membrane Science* 355, 53 (Jun 2010).

190. X. J. Loh, P. Peh, S. Liao, C. Sng, and J. Li, Controlled drug release from biodegradable thermoresponsive physical hydrogel nanofibers. *Journal of Controlled Release* 143, 175 (Apr 2010).

191. J. L. Lowery, N. Datta, and G. C. Rutledge, Effect of fiber diameter, pore size and seeding method on growth of human dermal fibroblasts in electrospun poly(epsilon-caprolactone) fibrous mats. *Biomaterials* 31, 491 (Jan 2010).

192. G. P. Ma et al., Organic-soluble chitosan/polyhydroxybutyrate ultrafine fibers as skin regeneration prepared by electrospinning. *Journal of Applied Polymer Science* 118, 3619 (Dec 2010).

193. L. L. Wu et al., Composite fibrous membranes of PLGA and chitosan prepared by coelectrospinning and coaxial electrospinning. *Journal of Biomedical Materials Research Part A* 92A, 563 (Feb 2010).

194. K. Schenke-Layland et al., The use of three-dimensional nanostructures to instruct cells to produce extracellular matrix for regenerative medicine strategies. *Biomaterials* 30, 4665 (Sep 2009).

195. W. G. Cui, X. L. Zhu, Y. Yang, X. H. Li, and Y. Jin, Evaluation of electrospun fibrous scaffolds of poly(DL-lactide) and poly(ethylene glycol) for skin tissue engineering. *Materials Science and Engineering C-Materials for Biological Applications* 29, 1869 (Aug 2009).

196. L. Jeong et al., Plasma-treated silk fibroin nanofibers for skin regeneration. *International Journal of Biological Macromolecules* 44, 222 (Apr 2009).

197. B. Marelli, A. Alessandrino, S. Fare, M. C. Tanzi, and G. Freddi, Electrospun silk fibroin tubular matrixes for small vessel bypass grafting. *Materials Technology* 24, 52 (Mar 2009).

198. J. Meng et al., Enhancement of nanofibrous scaffold of multiwalled carbon nanotubes/polyurethane composite to the fibroblasts growth and biosynthesis. *Journal of Biomedical Materials Research Part A* 88A, 105 (Jan 2009).

199. Y. H. Nien et al., Fabrication and cell affinity of poly(vinyl alcohol) nanofibers via electrospinning. *Journal of Medical and Biological Engineering* 29, 98 (2009).

200. L. S. Wray and E. J. Orwin, Recreating the microenvironment of the native cornea for tissue engineering applications. *Tissue Engineering Part A* 15, 1463 (Jul 2009).

201. F. Zhang, B. Q. Zuo, and L. Bai, Study on the structure of SF fiber mats electrospun with HFIP and FA and cells behavior. *Journal of Materials Science* 44, 5682 (Oct 2009).

202. E. Borg et al., Electrospinning of degradable elastomeric nanofibers with various morphology and their interaction with human fibroblasts. *Journal of Applied Polymer Science* 108, 491 (Apr 2008).

203. N. Khanam, C. Mikoryak, R. K. Draper, and K. J. Balkus, Electrospun linear polyethyleneimine scaffolds for cell growth. *Acta Biomaterialia* 3, 1050 (Nov 2007).

204. G. Kim and W. Kim, Highly porous 3D nanofiber scaffold using an electrospinning technique. *Journal of Biomedical Materials Research Part B-Applied Biomaterials* 81B, 104 (Apr 2007).

205. H. B. Deng et al., Layer-by-layer structured polysaccharides film-coated cellulose nanofibrous mats for cell culture. *Carbohydrate Polymers* 80, 474 (Apr 2010).

206. M. F. Leong, K. S. Chian, P. S. Mhaisalkar, W. F. Ong, and B. D. Ratner, Effect of electrospun poly(D,L-lactide) fibrous scaffold with nanoporous surface on attachment of porcine esophageal epithelial cells and protein adsorption. *Journal of Biomedical Materials Research Part A* 89A, 1040 (June, 2009).

207. I. Han et al., Effect of poly(3-hydroxybutyrate-co-3-hydroxyvalerate) nanofiber matrices cocultured with hair follicular epithelial and dermal cells for biological wound dressing. *Artificial Organs* 31, 801 (Nov 2007).

208. Y. B. Zhu, M. F. Leong, W. F. Ong, M. B. Chan-Park, and K. S. Chian, Esophageal epithelium regeneration on fibronectin grafted poly(L-lactide-co-caprolactone) (PLLC) nanofiber scaffold. *Biomaterials* 28, 861 (Feb 2007).

209. C. H. Kim, M. S. Khil, H. Y. Kim, H. U. Lee, and K. Y. Jahng, An improved hydrophilicity via electrospinning for enhanced cell attachment and proliferation. *Journal of Biomedical Materials Research Part B-Applied Biomaterials* 78B, 283 (Aug 2006).

210. W. He et al., Tubular nanofiber scaffolds for tissue engineered small-diameter vascular grafts. *Journal of Biomedical Materials Research Part A* 90A, 205 (Jul 2009).

211. W. He et al., Biodegradable polymer nanofiber mesh to maintain functions of endothelial cells. *Tissue Engineering* 12, 2457 (Sep 2006).

212. P. Uttayarat et al., Micropatterning of three-dimensional electrospun polyurethane vascular grafts. *Acta Biomaterialia* 6, 4229 (Nov 2010).

213. A. Bianco et al., Microstructure and cytocompatibility of electrospun nanocomposites based on poly(epsilon-caprolactone) and carbon nanostructures. *International Journal of Artificial Organs* 33, 271 (May 2010).

214. Z. G. Chen, P. W. Wang, B. Wei, X. M. Mo, and F. Z. Cui, Electrospun collagen-chitosan nanofiber: A biomimetic extracellular matrix for endothelial cell and smooth muscle cell. *Acta Biomaterialia* 6, 372 (Feb 2010).

215. Y. M. Ju, J. S. Choi, A. Atala, J. J. Yoo, and S. J. Lee, Bilayered scaffold for engineering cellularized blood vessels. *Biomaterials* 31, 4313 (May 2010).

216. D. A. Rubenstein et al., In vitro biocompatibility of sheath-core cellulose-acetate-based electrospun scaffolds towards endothelial cells and platelets. *Journal of Biomaterials Science-Polymer Edition* 21, 1713 (2010).

217. K. H. Zhang, X. M. Mo, C. Huang, C. L. He, and H. S. Wang, Electrospun scaffolds from silk fibroin and their cellular compatibility. *Journal of Biomedical Materials Research Part A* 93A, 976 (Jun 2010).

218. K. H. Zhang et al., Fabrication of silk fibroin blended P(LLA-CL) nanofibrous scaffolds for tissue engineering. *Journal of Biomedical Materials Research Part A* 93A, 984 (Jun 2010).

219. R. Chen, L. J. Qiu, Q. F. Ke, C. L. He, and X. M. Mo, Electrospinning thermoplastic polyurethane-contained collagen nanofibers for tissue-engineering applications. *Journal of Biomaterials Science-Polymer Edition* 20, 1513 (2009).

220. X. Q. Li et al., Sorbitan monooleate and poly(L-lactide-co-epsilon-caprolactone) electrospun nanofibers for endothelial cell interactions. *Journal of Biomedical Materials Research Part A* 91A, 878 (Dec 2009).

221. A. N. Veleva et al., Interactions between endothelial cells and electrospun methacrylic terpolymer fibers for engineered vascular replacements. *Journal of Biomedical Materials Research Part A* 91A, 1131 (Dec 2009).

222. P. Carampin et al., Electrospun polyphosphazene nanofibers for in vitro rat endothelial cells proliferation. *Journal of Biomedical Materials Research Part A* 80A, 661 (Mar 2007).

223. W. He, Z. W. Ma, T. Yong, W. E. Teo, and S. Ramakrishna, Fabrication of collagen-coated biodegradable polymer nanofiber mesh and its potential for endothelial cells growth. *Biomaterials* 26, 7606 (Dec 2005).

224. Z. W. Ma, M. Kotaki, T. Yong, W. He, and S. Ramakrishna, Surface engineering of electrospun polyethylene terephthalate (PET) nanofibers towards development of a new material for blood vessel engineering. *Biomaterials* 26, 2527 (May 2005).

225. C. Y. Xu, F. Yang, S. Wang, and S. Ramakrishna, In vitro study of human vascular endothelial cell function on materials with various surface roughness. *Journal of Biomedical Materials Research Part A* 71A, 154 (Oct 2004).

226. S. Welin, in: C. Van Blitterswijk et al., Eds., Ethical issues in tissue engineering. *Tissue Engineering* (Academic Press, New York, 2008), pp. 685–703.

227. D. E. Heath, J. J. Lannutti, and S. L. Cooper, Electrospun scaffold topography affects endothelial cell proliferation, metabolic activity, and morphology. *Journal of Biomedical Materials Research Part A* 94A, 1195 (Sep 2010).

228. J. Meng et al., Electrospun aligned nanofibrous composite of MWCNT/polyurethane to enhance vascular endothelium cells proliferation and function. *Journal of Biomedical Materials Research Part A* 95A, 312 (Oct 2010).

229. C. Del Gaudio, A. Bianco, M. Folin, S. Baiguera, and M. Grigioni, Structural characterization and cell response evaluation of electrospun PCL membranes: Micrometric versus submicrometric fibers. *Journal of Biomedical Materials Research Part A* 89A, 1028 (Jun 2009).

230. G. Yin et al., Study on the properties of the electrospun silk fibroin/gelatin blend nanofibers for scaffolds. *Journal of Applied Polymer Science* 111, 1471 (Feb 2009).

231. H. Inoguchi, T. Tanaka, Y. Maehara, and T. Matsuda, The effect of gradually graded shear stress on the morphological integrity of a huvec-seeded compliant small-diameter vascular graft. *Biomaterials* 28, 486 (Jan 2007).

232. I. K. Kwon, S. Kidoaki, and T. Matsuda, Electrospun nano- to microfiber fabrics made of biodegradable copolyesters: Structural characteristics, mechanical properties and cell adhesion potential. *Biomaterials* 26, 3929 (Jun 2005).

233. S. I. Jeong et al., Tissue-engineered vascular grafts composed of marine collagen and PLGA fibers using pulsatile perfusion bioreactors. *Biomaterials* 28, 1115 (Feb 2007).

234. J. J. Stankus et al., Fabrication of cell microintegrated blood vessel constructs through electrohydrodynamic atomization. *Biomaterials* 28, 2738 (Jun 2007).

235. E. D. Boland et al., Electrospinning collagen and elastin: Preliminary vascular tissue engineering. *Frontiers in Bioscience* 9, 1422 (May 2004).

236. J. Venugopal, L. L. Ma, T. Yong, and S. Ramakrishna, In vitro study of smooth muscle cells on polycaprolactone and collagen nanofibrous matrices. *Cell Biology International* 29, 861 (2005).

237. E. Luong-Van et al., Controlled release of heparin from poly(epsilon-caprolactone) electrospun fibers. *Biomaterials* 27, 2042 (Mar 2006).

238. F. Xu et al., Improvement of cytocompatibility of electrospinning PLLA microfibers by blending PVP. *Journal of Materials Science-Materials in Medicine* 20, 1331 (Jun 2009).

239. J. Venugopal, L. L. Ma, T. Yong, and S. Ramakrishna, In vitro study of smooth muscle cells on polycaprolactone and collagen nanofibrous matrices. *Cell Biology International* 29, 861 (Oct 2005).

240. M. Shin, O. Ishii, T. Sueda, and J. P. Vacanti, Contractile cardiac grafts using a novel nanofibrous mesh. *Biomaterials* 25, 3717 (Aug 2004).

241. O. Ishii, M. Shin, T. Sueda, and J. P. Vacanti, In vitro tissue engineering of a cardiac graft using a degradable scaffold with an extracellular matrix–like topography. *The Journal of Thoracic and Cardiovascular Surgery* 130, 1358 (2005).

242. X. H. Zong et al., Electrospun fine-textured scaffolds for heart tissue constructs. *Biomaterials* 26, 5330 (Sep 2005).

243. H. Hosseinkhani, M. Hosseinkhani, S. Hattori, R. Matsuoka, and N. Kawaguchi, Micro and nano-scale in vitro 3D culture system for cardiac stem cells. *Journal of Biomedical Materials Research Part A* 94A, 1 (Jul 2010).

244. J. D. Fromstein et al., Seeding bioreactor-produced embryonic stem cell-derived cardiomyocytes on different porous, degradable, polyurethane scaffolds reveals the effect of scaffold architecture on cell morphology. *Tissue Engineering Part A* 14, 369 (Mar 2008).

245. M. L. Focarete et al., Electrospun scaffolds of a polyhydroxyalkanoate consisting of omega-hydroxylpentadecanoate repeat units: Fabrication and in vitro biocompatibility studies. *Journal of Biomaterials Science-Polymer Edition* 21, 1283 (2010).

246. M. P. Prabhakaran, J. R. Venugopal, and S. Ramakrishna, Mesenchymal stem cell differentiation to neuronal cells on electrospun nanofibrous substrates for nerve tissue engineering. *Biomaterials* 30, 4996 (Oct 2009).

247. J. K. Wise, A. L. Yarin, C. M. Megaridis, and M. Cho, Chondrogenic differentiation of human mesenchymal stem cells on oriented nanofibrous scaffolds: Engineering the superficial zone of articular cartilage. *Tissue Engineering Part A* 15, 913 (Apr 2009).

248. Y. M. Kolambkar, A. Peister, A. K. Ekaputra, D. W. Hutmacher, and R. E. Guldberg, Colonization and osteogenic differentiation of different stem cell sources on electrospun nanofiber meshes. *Tissue Engineering Part A* 16, 3219 (Oct 2010).

249. R. S. Tuan, G. Boland, and R. Tuli, Adult mesenchymal stem cells and cell-based tissue engineering. *Arthritis Research and Therapy* 5, 32 (2003).

250. W. J. Li, R. Tuli, X. X. Huang, P. Laquerriere, and R. S. Tuan, Multilineage differentiation of human mesenchymal stem cells in a three-dimensional nanofibrous scaffold. *Biomaterials* 26, 5158 (Sep 2005).

251. W. J. Li et al., A three-dimensional nanofibrous scaffold for cartilage tissue engineering using human mesenchymal stem cells. *Biomaterials* 26, 599 (Feb 2005).

252. H. Yoshimoto, Y. M. Shin, H. Terai, and J. P. Vacanti, A biodegradable nanofiber scaffold by electrospinning and its potential for bone tissue engineering. *Biomaterials* 24, 2077 (May 2003).

253. S. Ahn, Y. H. Koh, and G. Kim, A three-dimensional hierarchical collagen scaffold fabricated by a combined solid freeform fabrication (SFF) and electrospinning process to enhance mesenchymal stem cell (MSC) proliferation. *Journal of Micromechanics and Microengineering* 20, 129901 (Jun 2010).

254. J. H. Lee, N. G. Rim, H. S. Jung, and H. Shin, Control of osteogenic differentiation and mineralization of human mesenchymal stem cells on composite nanofibers containing poly lactic-co-(glycolic acid) and hydroxyapatite. *Macromolecular Bioscience* 10, 173 (Feb 2010).

255. L. Ren et al., Fabrication of gelatin-siloxane fibrous mats via sol-gel and electrospinning procedure and its application for bone tissue engineering. *Materials Science and Engineering C-Materials for Biological Applications* 30, 437 (Apr 2010).

256. T. T. Ruckh, K. Kumar, M. J. Kipper, and K. C. Popat, Osteogenic differentiation of bone marrow stromal cells on poly(epsilon-caprolactone) nanofiber scaffolds. *Acta Biomaterialia* 6, 2949 (Aug 2010).

257. S. Soliman et al., Multiscale three-dimensional scaffolds for soft tissue engineering via multimodal electrospinning. *Acta Biomaterialia* 6, 1227 (Apr 2010).

258. C. K. Chan et al., Early adhesive behavior of bone-marrow-derived mesenchymal stem cells on collagen electrospun fibers. *Biomedical Materials* 4, 35006 (Jun 2009).

259. L. Ghasemi-Mobarakeh et al., The thickness of electrospun poly (epsilon-caprolactone) nanofibrous scaffolds influences cell proliferation. *International Journal of Artificial Organs* 32, 150 (Mar 2009).

260. C. M. Li et al., Preliminary investigation of seeding mesenchymal stem cells on biodegradable scaffolds for vascular tissue engineering in vitro. *Asaio Journal* 55, 614 (Nov–Dec 2009).

261. H. Hong et al., Fabrication of a novel hybrid scaffold for tissue engineered heart valve. *Journal of Huazhong University of Science and Technology-Medical Sciences* 29, 599 (Oct 2009).

262. A. J. Meinel et al., Optimization strategies for electrospun silk fibroin tissue engineering scaffolds. *Biomaterials* 30, 3058 (Jun 2009).

263. K. T. Shalumon et al., Electrospinning of carboxymethyl chitin/poly(vinyl alcohol) nanofibrous scaffolds for tissue engineering applications. *Carbohydrate Polymers* 77, 863 (Jul 2009).

264. E. K. Ko et al., In vitro osteogenic differentiation of human mesenchymal stem cells and in vivo bone formation in composite nanofiber meshes. *Tissue Engineering Part A* 14, 2105 (Dec 2008).

265. S. Srouji, T. Kizhner, E. Suss-Tobi, E. Livne, and E. Zussman, 3-D Nanofibrous electrospun multilayered construct is an alternative ECM mimicking scaffold. *Journal of Materials Science-Materials in Medicine* 19, 1249 (Mar 2008).

266. Y. M. Kolambkar et al., An alginate-based hybrid system for growth factor delivery in the functional repair of large bone defects. *Biomaterials* 32, 65 (Jan 2011).

267. B. Mavis, T. T. Demirtas, M. Gumusderelioglu, G. Gunduz, and U. Colak, Synthesis, characterization and osteoblastic activity of polycaprolactone nanofibers coated with biomimetic calcium phosphate. *Acta Biomaterialia* 5, 3098 (Oct 2009).

268. H. S. Yu, J. H. Jang, T. I. Kim, H. H. Lee, and H. W. Kim, Apatite-mineralized polycaprolactone nanofibrous web as a bone tissue regeneration substrate. *Journal of Biomedical Materials Research Part A* 88A, 747 (Mar 2009).

269. C. Erisken, D. M. Kalyon, and H. J. Wang, Viscoelastic and biomechanical properties of osteochondral tissue constructs generated from graded polycaprolactone and beta-tricalcium phosphate composites. *Journal of Biomechanical Engineering-Transactions of the Asme* 132, 91013 (Sep 2010).

270. L. Francis et al., Simultaneous electrospin-electrosprayed biocomposite nanofibrous scaffolds for bone tissue regeneration. *Acta Biomaterialia* 6, 4100 (Oct 2010).

271. S. Z. Fu et al., Preparation and characterization of nano-hydroxyapatite/poly(epsilon-caprolactone)-poly(ethylene glycol)-poly(epsilon-caprolactone) composite fibers for tissue engineering. *Journal of Physical Chemistry C* 114, 18372 (Nov 2010).

272. H. M. Kim et al., Composite nanofiber mats consisting of hydroxyapatite and titania for biomedical applications. *Journal of Biomedical Materials Research Part B-Applied Biomaterials* 94B, 380 (Aug 2010).

273. E. J. Lee et al., Nanostructured poly(epsilon-caprolactone)-silica xerogel fibrous membrane for guided bone regeneration. *Acta Biomaterialia* 6, 3557 (Sep 2010).

274. K. T. Noh, H. Y. Lee, U. S. Shin, and H. W. Kim, Composite nanofiber of bioactive glass nanofiller incorporated poly(lactic acid) for bone regeneration. *Materials Letters* 64, 802 (Apr 2010).

275. I. K. Shim et al., Novel three-dimensional scaffolds of poly((L)-lactic acid) microfibers using electrospinning and mechanical expansion: Fabrication and bone regeneration. *Journal of Biomedical Materials Research Part B-Applied Biomaterials* 95B, 150 (Oct 2010).

276. T. J. Shin, S. Y. Park, H. J. Kim, H. J. Lee, and J. H. Youk, Development of 3-D poly(trimethylenecarbonate-co-epsilon-caprolactone)-block-poly(p-dioxano ne) scaffold for bone regeneration with high porosity using a wet electrospinning method. *Biotechnology Letters* 32, 877 (Jun 2010).

277. Y. M. Shin, H. Shin, and Y. M. Lim, Surface modification of electrospun poly(L-lactide-co-epsilon-caprolactone) fibrous meshes with a RGD peptide for the control of adhesion, proliferation and differentiation of the preosteoblastic cells. *Macromolecular Research* 18, 472 (May 2010).

278. H. W. Tong, M. Wang, Z. Y. Li, and W. W. Lu, Electrospinning, characterization and in vitro biological evaluation of nanocomposite fibers containing carbonated hydroxyapatite nanoparticles. *Biomedical Materials* 5, 54111 (Oct 2010).

279. A. K. Ekaputra, Y. F. Zhou, S. M. Cool, and D. W. Hutmacher, Composite electrospun scaffolds for engineering tubular bone grafts. *Tissue Engineering Part A* 15, 3779 (Dec 2009).

280. D. Gupta, J. Venugopal, S. Mitra, V. R. G. Dev, and S. Ramakrishna, Nanostructured biocomposite substrates by electrospinning and electrospraying for the mineralization of osteoblasts. *Biomaterials* 30, 2085 (Apr 2009).

281. Y. M. Kang, K. H. Kim, Y. J. Seol, and S. H. Rhee, Evaluations of osteogenic and osteoconductive properties of a non-woven silica gel fabric made by the electrospinning method. *Acta Biomaterialia* 5, 462 (Jan 2009).

282. M. Ngiam et al., Fabrication of mineralized polymeric nanofibrous composites for bone graft materials. *Tissue Engineering Part A* 15, 535 (Mar 2009).

283. H. W. Kim, H. H. Lee, and G. S. Chun, Bioactivity and osteoblast responses of novel biomedical nanocomposites of bioactive glass nanofiber filled poly(lactic acid). *Journal of Biomedical Materials Research Part A* 85A, 651 (Jun 2008).

284. J. H. Jang, O. Castano, and H. W. Kim, Electrospun materials as potential platforms for bone tissue engineering. *Advanced Drug Delivery Reviews* 61, 1065 (Oct 2009).

285. A. R. Poole et al., Composition and structure of articular cartilage—A template for tissue repair. *Clinical Orthopaedics and Related Research* 1, S26 (Oct 2001).

286. E. B. Hunziker, Articular cartilage repair: Basic science and clinical progress. A review of the current status and prospects. *Osteoarthr Cartilage* 10, 432 (Jun 2002).

287. T. J. Klein, J. Malda, R. L. Sah, and D. W. Hutmacher, Tissue engineering of articular cartilage with biomimetic zones. *Tissue Engineering Part B-Reviews* 15, 143 (Jun 2009).

288. T. J. Klein et al., Strategies for zonal cartilage repair using hydrogels. *Macromolecular Bioscience* 9, 1049 (Nov 10, 2009).

289. G. Kim, J. Son, S. Park, and W. Kim, Hybrid process for fabricating 3D hierarchical scaffolds combining rapid prototyping and electrospinning. *Macromolecular Rapid Communications* 29, 1577 (Oct 2008).

290. I. K. Shim et al., Chitosan nano-/microfibrous double-layered membrane with rolled-up three-dimensional structures for chondrocyte cultivation. *Journal of Biomedical Materials Research Part A* 90A, 595 (Aug 2009).

291. W. J. Li, K. G. Danielson, P. G. Alexander, and R. S. Tuan, Biological response of chondrocytes cultured in three-dimensional nanofibrous poly(epsilon-caprolactone) scaffolds. *Journal of Biomedical Materials Research Part A* 67A, 1105 (Dec 2003).

292. T. G. Kim, H. J. Chung, and T. G. Park, Macroporous and nanofibrous hyaluronic acid/collagen hybrid scaffold fabricated by concurrent electrospinning and deposition/leaching of salt particles. *Acta Biomaterialia* 4, 1611 (Nov 2008).

293. L. Moroni, R. Schotel, D. Hamann, J. R. de Wijn, and C. A. van Blitterswijk, 3D fiber-deposited electrospun integrated scaffolds enhance cartilage tissue formation. *Advanced Functional Materials* 18, 53 (Jan 2008).

294. A. Thorvaldsson, H. Stenhamre, P. Gatenholm, and P. Walkenstrom, Electrospinning of highly porous scaffolds for cartilage regeneration. *Biomacromolecules* 9, 1044 (Mar 2008).

295. I. S. Lee, O. H. Kwon, W. Meng, and I. K. Kang, Nanofabrication of microbial polyester by electrospinning promotes cell attachment. *Macromolecular Research* 12, 374 (Aug 2004).

296. K. J. Shields, M. J. Beckman, G. L. Bowlin, and J. S. Wayne, Mechanical properties and cellular proliferation of electrospun collagen type II. *Tissue Engineering* 10, 1510 (Sep 2004).

297. R. M. Smeal and P. A. Tresco, The influence of substrate curvature on neurite outgrowth is cell type dependent. *Experimental Neurology* 213, 281 (Oct 2008).

298. R. M. Smeal, R. Rabbitt, R. Biran, and P. A. Tresco, Substrate curvature influences the direction of nerve outgrowth. *Annals of Biomedical Engineering* 33, 376 (Jan 2005).

299. F. Yang, R. Murugan, S. Wang, and S. Ramakrishna, Electrospinning of nano/micro scale poly(L-lactic acid) aligned fibers and their potential in neural tissue engineering. *Biomaterials* 26, 2603 (May 2005).

300. T. B. Bini, S. J. Gao, S. Wang, and S. Ramakrishna, Poly(l-lactide-co-glycolide) biodegradable microfibers and electrospun nanofibers for nerve tissue engineering: An in vitro study. *Journal of Materials Science* 41, 6453 (Oct 2006).

301. D. Gupta et al., Aligned and random nanofibrous substrate for the in vitro culture of Schwann cells for neural tissue engineering. *Acta Biomaterialia* 5, 2560 (Sep 2009).

302. U. Assmann et al., Fiber scaffolds of polysialic acid via electrospinning for peripheral nerve regeneration. *Journal of Materials Science-Materials in Medicine* 21, 2115 (Jul 2010).

303. R. C. de Guzman, J. A. Loeb, and P. J. VandeVord, Electrospinning of matrigel to deposit a basal lamina-like nanofiber surface. *Journal of Biomaterials Science-Polymer Edition* 21, 1081 (2010).

304. H. B. Wang et al., Creation of highly aligned electrospun poly-L-lactic acid fibers for nerve regeneration applications. *Journal of Neural Engineering* 6, 16001 (Feb 2009).

305. J. W. Xie et al., Conductive core-sheath nanofibers and their potential application in neural tissue engineering. *Advanced Functional Materials* 19, 2312 (Jul 2009).

306. L. Yao, N. O'Brien, A. Windebank, and A. Pandit, Orienting neurite growth in electrospun fibrous neural conduits. *Journal of Biomedical Materials Research Part B-Applied Biomaterials* 90B, 483 (Aug 2009).

307. M. P. Prabhakaran, J. Venugopal, C. K. Chan, and S. Ramakrishna, Surface modified electrospun nanofibrous scaffolds for nerve tissue engineering. *Nanotechnology* 19, 455102 (Nov 2008).

308. M. P. Prabhakaran et al., Electrospun biocomposite nanofibrous scaffolds for neural tissue engineering. *Tissue Engineering Part A* 14, 1787 (Nov 2008).

309. W. N. Chow, D. G. Simpson, J. W. Bigbee, and R. J. Colello, Evaluating neuronal and glial growth on electrospun polarized matrices: Bridging the gap in percussive spinal cord injuries. *Neuron Glia Biology* 3, 119 (2007).

310. P. D. Dalton, L. Flynn, and M. S. Shoichet, Manufacture of poly(2-hydroxyethyl methacrylate-co-methyl methacrylate) hydrogel tubes for use as nerve guidance channels. *Biomaterials* 23, 3843 (Sep 2002).

311. D. E. Birk and R. Mayne, Localization of collagen types I, III and V during tendon development. Changes in collagen types I and III are correlated with changes in fibril diameter. *European Journal of Cell Biology* 72, 352 (1997).

312. K. L. Moffat et al., Novel nanofiber-based scaffold for rotator cuff repair and augmentation. *Tissue Engineering Part A* 15, 115 (2009).

313. B. Inanc, Y. E. Arslan, S. Seker, A. E. Elcin, and Y. M. Elcin, Periodontal ligament cellular structures engineered with electrospun poly(DL-lactide-co-glycolide) nanofibrous membrane scaffolds. *Journal of Biomedical Materials Research Part A* 90A, 186 (Jul 2009).

314. S. Zhang et al., Gelatin nanofibrous membrane fabricated by electrospinning of aqueous gelatin solution for guided tissue regeneration. *Journal of Biomedical Materials Research Part A* 90A, 671 (Sep 2009).

315. M. C. Bottino, V. Thomas, and G. M. Janowski, A novel spatially designed and functionally graded electrospun membrane for periodontal regeneration. *Acta Biomaterialia* 7, 216 (2011).

316. S. Sahoo, H. Ouyang, J. C. H. Goh, T. E. Tay, and S. L. Toh, Characterization of a novel polymeric scaffold for potential application in tendon/ligament tissue engineering. *Tissue Engineering* 12, 91 (Jan 2006).

317. S. Sahoo, J. C.-H. Goh, and S. L. Toh, Development of hybrid polymer scaffolds for potential applications in ligament and tendon tissue engineering. *Biomedical Materials* 2, 169 (2007).

318. S. Sahoo, S. L. Toh, and J. C.-H. Goh, PLGA nanofiber-coated silk microfibrous scaffold for connective tissue engineering. *Journal Biomedical Material Research Part B: Applied Biomaterials* 95B, 19 (2010).

319. C. Vaquette et al., Aligned poly(L-lactic-co-e-caprolactone) electrospun microfibers and knitted structure: A novel composite scaffold for ligament tissue engineering. *Journal of Biomedical Materials Research Part A* 94A, 1270 (Sep 2010).

320. C. A. Bashur, R. D. Shaffer, L. A. Dahlgren, S. A. Guelcher, and A. S. Goldstein, Effect of fiber diameter and alignment of electrospun polyurethane meshes on mesenchymal progenitor cells. *Tissue Engineering Part A* 15, 2435 (Sep 2009).

321. J. W. S. Hayami, D. C. Surrao, S. D. Waldman, and B. G. Amsden, Design and characterization of a biodegradable composite scaffold for ligament tissue engineering. *Journal of Biomedical Materials Research Part A* 92A, 1407 (Mar 2010).

322. D. C. Surrao, J. W. S. Hayami, S. D. Waldman, and B. G. Amsden, Self-crimping, biodegradable, electrospun polymer microfibers. *Biomacromolecules* 11, 3624 (2010).

323. S. L. Woo, J. M. Hollis, D. J. Adams, R. M. Lyon, and S. Takai, Tensile properties of the human femur-anterior cruciate ligament complex. The effects of specimen age and orientation. *American Journal of Sport Medicine* 19, 217 (1991).

324. F. Mei et al., Improved biological characteristics of poly(L-lactic acid) electrospun membrane by incorporation of multiwalled carbon nanotubes/hydroxyapatite nanoparticles. *Biomacromolecules* 8, 3729 (Dec 2007).

325. S. H. Shang, F. Yang, X. R. Cheng, X. F. Walboomers, and J. A. Jansen, The effect of electrospun fibre alignment on the behaviour of rat periodontal ligament cells. *European Cells and Materials* 19, 180 (Jan–Jun 2010).

326. P. K. Lam et al., Development and evaluation of a new composite Laserskin graft. *The Journal of Trauma* 47, 918 (Nov 1999).

327. S. Heydarkhan-Hagvall et al., Three-dimensional electrospun ECM-based hybrid scaffolds for cardiovascular tissue engineering. *Biomaterials* 29, 2907 (Jul 2008).

328. A. Bianco et al., Electrospun poly(epsilon-caprolactone)/Ca-deficient hydroxyapatite nanohybrids: Microstructure, mechanical properties and cell response by murine embryonic stem cells. *Materials Science and Engineering C-Materials for Biological Applications* 29, 2063 (Aug 2009).

329. D. R. Nisbet, A. E. Rodda, M. K. Horne, J. S. Forsythe, and D. I. Finkelstein, Neurite infiltration and cellular response to electrospun polycaprolactone scaffolds implanted into the brain. *Biomaterials* 30, 4573 (Sep 2009).

330. B. W. Tillman et al., The in vivo stability of electrospun polycaprolactone-collagen scaffolds in vascular reconstruction. *Biomaterials* 30, 583 (Feb 2009).

331. H. Q. Cao, K. McHugh, S. Y. Chew, and J. M. Anderson, The topographical effect of electrospun nanofibrous scaffolds on the in vivo and in vitro foreign body reaction. *Journal of Biomedical Materials Research Part A* 93A, 1151 (Jun 2010).

332. J. L. Ifkovits, K. Wu, R. L. Mauck, and J. A. Burdick, The influence of fibrous elastomer structure and porosity on matrix organization. *Plos One* 5, e15717 (2010).

333. P. D. Dalton, A. R. Harvey, M. Oudega, and G. W. Plant, in: C. Van Blitterswijk et al., Eds., Tissue engineering of the nervous system. *Tissue Engineering* (Academic Press, New York, 2008), pp. 611–647.

334. I. P. Clements et al., Thin-film enhanced nerve guidance channels for peripheral nerve repair. *Biomaterials* 30, 3834 (Aug 2009).

335. E. Seyedjafari, M. Soleimani, N. Ghaemi, and I. Shabani, Nanohydroxyapatite-coated electrospun poly(L-lactide) nanofibers enhance osteogenic differentiation of stem cells and induce ectopic bone formation. *Biomacromolecules* 11, 3118 (Nov 2010).

336. A. Nandakumar, L. Yang, P. Habibovic, and C. van Blitterswijk, Calcium phosphate coated electrospun fiber matrices as scaffolds for bone tissue engineering. *Langmuir* 26, 7380 (May 18, 2010).

337. M. Shin, H. Yoshimoto, and J. P. Vacanti, In vivo bone tissue engineering using mesenchymal stem cells on a novel electrospun nanofibrous scaffold. *Tissue Engineering* 10, 33 (2004).

338. S. Srouji et al., A model for tissue engineering applications: Femoral critical size defect in immunodeficient mice. *Tissue Eng Part C Methods* 17, 597 (May 2011).

339. W. J. Li et al., Evaluation of articular cartilage repair using biodegradable nanofibrous scaffolds in a swine model: A pilot study. *Journal of Tissue Engineering and Regenerative Medicine* 3, 1 (Jan 2009).

340. J. P. Chen and C. H. Su, Surface modification of electrospun PLLA nanofibers by plasma treatment and cationized gelatin immobilization for cartilage tissue engineering. *Acta Biomaterialia* 7, 234 (Jan 2011).

341. W. Wang et al., Effects of Schwann cell alignment along the oriented electrospun chitosan nanofibers on nerve regeneration. *Journal of Biomedical Materials Research Part A* 91A, 994 (Dec 2009).

342. S. Panseri et al., Electrospun micro- and nanofiber tubes for functional nervous regeneration in sciatic nerve transections. *BMC Biotechnology* 8, 39 (Apr 2008).

343. W. Wang et al., Enhanced nerve regeneration through a bilayered chitosan tube: The effect of introduction of glycine spacer into the CYIGSK sequence. *Journal of Biomedical Materials Research Part A* 85A, 919 (Jun 2008).

344. B. Nottelet et al., Factorial design optimization and in vivo feasibility of poly(epsilon-caprolactone)-micro- and narofiber-based small diameter vascular grafts. *Journal of Biomedical Materials Research Part A* 89A, 865 (Jun 2009).

345. T. Sun, D. Norton, R. J. Mckean, J. W. Haycock, A. J. Ryan, and S. MacNeil, Development of a 3D cell culture system for investigating cell interactions with electrospun fibers, *Biotechnology and Bioengineering* 97, 1318–1328 (2007).

17 Polymeric Nanoparticles for Targeted Delivery of Bioactive Agents and Drugs

Cesare Errico, Alberto Dessy, Anna Maria Piras, and Federica Chiellini

CONTENTS

17.1 INTRODUCTION

The interdisciplinary field of nanobiotechnology, which combines chemistry, biology, engineering, and medicine, is revolutionizing the development of drug-delivery systems (DDS) and devices. Novel materials and formulations enable the site-specific targeting and controlled release of traditional pharmaceuticals, recombinant proteins, vaccines, and nucleic acids (Goldberg et al. 2007); safeguard drugs from spoilage as caused by attacks of enzymes, and enhance the penetration of the active agent in the diseased tissues improving its bioavailability, with increase in efficacy and toxicity reduction (Veronese and Caliceti 2002).

The field of drug delivery has attracted the attention of not only academic researchers but also pharmaceutical industries because it offers a strategic tool to expand current drug markets; new delivery technologies could repackage classic drugs, thus offering a competitive edge after the expiration of patents and preclude competition from generics (Parveen et al. 2011). According to a recent estimate, the development of a new drug for human use involves around \$800 million and 10–12 years of research inputs. However, 9 out of 10 drugs fail in their clinical study phase causing huge loss to the investigating organization. For DDS development, an U.S. estimate shows the development cost to be around \$40 million and a period between 3 months and 3 years. Thus the burden on company's exchequer is quite less with good chances of ensured returns (Mandal and Mandal 2010).

Similarly, the development of nanotechnology products may play an important role in adding a new armamentarium of therapeutics to the pipelines of pharmaceutical companies (Farokhzad and Langer 2009) and is likely to have a significant impact on the drug-delivery sector (Parveen et al. 2011). Depending on the final role of the developed DDS, its dimension and shape play a fundamental role, but an engineered nanoscale system is preferred for a defined interaction with the biological environment at the molecular level (Wagner et al. 2006).

Nanoparticles can be correctly envisioned as the future of drug-delivery technology as they have the potential to become useful therapeutic and diagnostic tools in the near future, with many potential applications in clinical medicine and research (Parveen et al. 2011).

Nanoparticles are frequently defined as solid, colloidal particles with size in the range of 10–1000 nm. The term "nanoparticles" is a collective term given for any type of polymer nanoparticles, but specifically, it can be distinguished in nanospheres and nanocapsules. Nanospheres are matrix particles, i.e., particles whose entire mass is solid, and bioactive molecules may be adsorbed at the sphere surface or encapsulated within the particle bulk. In general, they are spherical, but "nanospheres" with a nonspherical shape and those that are uniformly distributed are also described in the literature. Nanocapsules are vesicular systems, acting as a kind of reservoir, in which the entrapped substances are confined to a cavity consisting of a liquid core (either oil or water) surrounded by a solid material shell (Rao and Geckeler 2011).

The relative interest of using nanoparticulate rather than microparticulate systems in pharmaceutical applications is in their potential to increase the absorption rate, improve bioavailability, and enable target drug delivery and intravenous delivery systems.

Nanoparticles are taken up by cells more efficiently than larger micromolecules and, therefore, could be used as effective transport and delivery systems. Drugs can either be integrated in the matrix of the particle or attached to the particle surface (Suri et al. 2007) and their release kinetics (Goldberg et al. 2007) can be tuned by surface or bulk erosion, diffusion through the matrix, swelling followed by diffusion, or in response to the local environment inputs (Peer et al. 2007).

For a rapid and effective clinical translation, the developed nanocarriers should be made from a material that is biocompatible, well characterized, and easily functionalized; exhibit high differential uptake efficiency in the target cells over normal cells (or tissue), either soluble or colloidal under aqueous conditions for increased effectiveness; and have an extended circulating half-life, a low rate of aggregation, and a long shelf life (Peer et al. 2007). Currently, both polymers and lipids are typically used as drug delivery vectors. However, polymers are the most commonly explored materials for constructing nanoparticle-based drug carriers (Figure 17.1).

Polymeric nanoparticles can be made from synthetic polymers, including among the other biocompatible poly(lactic acid) (PLA) and poly(lactic co-glycolic acid), or from natural polymers such as chitosan and collagen and may be used to encapsulate drugs without major chemical modification (Peer et al. 2007). The role of polymers in DDS covers multiple aspects, from the enhancement of the physical–chemical stability of the drug to the regulation of drug release profile and targeting (Vasir et al. 2005). They have highly tunable physical–chemical characteristics, and in some cases, polymeric materials can be further processed or functionalized to more coherent systems. Accordingly, they very likely represent the best-suited class of materials for the modern drug delivery technology (Kashyap et al. 2004).

Several different types of nanoparticles have successfully made their way into preclinical studies in animals, clinic trials in patients, or even successful commercial products used in routine

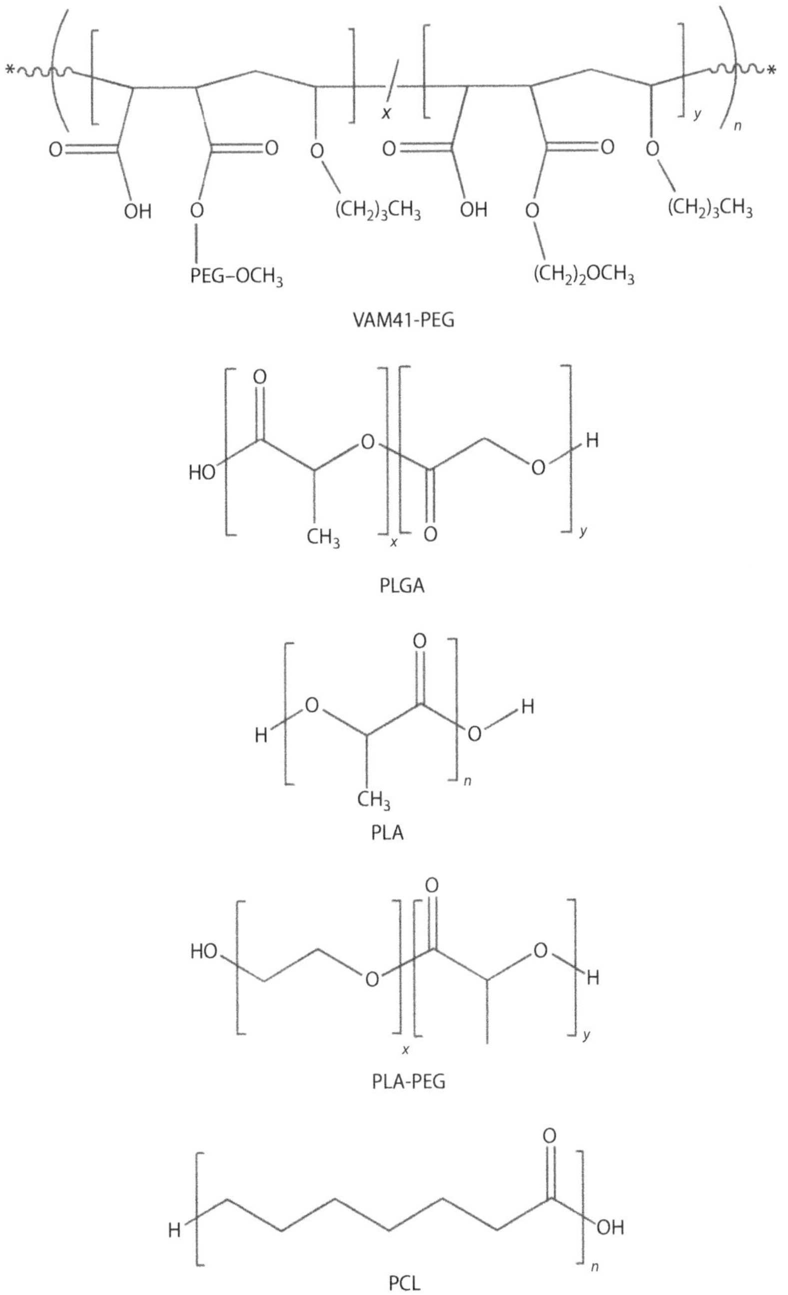

FIGURE 17.1 List of commonly used polymer as matrices for drug-loaded nanoparticle preparation.

clinical practice (Wang et al. 2009). Nanoscale albumin-based drug carriers have recently reached clinic exploitation. Examples include Abraxane®, a nanoparticle of albumin and paclitaxel, and Albuferon-α®, a conjugate of albumin and interferon-α (Hoffman 2008). A few drugs have been marketed as nanoparticulate systems such as Amphotericin B, some antineoplastics, and others including proteins, peptides, and macromolecules have been investigated in animal and clinical models. The promises of nanoparticles as DDS are only limited by the choice of suitable biocompatible polymers (Mandal and Mandal 2010). Compared with the total pharmaceutical and medical device market, nanostructured medical formulations currently constitute a tiny niche. However, great strength lies in the versatility of nanomedicine: Nanotechnology has the potential to add innovative functionality and bioefficacy to many pharmaceutical products and medical devices (Wagner et al. 2006).

17.2 TECHNIQUES FOR NANOPARTICLE PREPARATIONS

For the preparation of nanoparticles using biodegradable or biostable, biocompatible polymers, it is important to choose an appropriate encapsulation process which meets the following requirements: chemical stability and biological activity of the incorporated drug, high encapsulation efficiency, reasonable size range, and reproducible release profile.

There are a number of techniques available for the preparation of drug-loaded nanoparticulate systems such as the *emulsion solvent evaporation/extraction method*, *spray drying*, *phase separation–coacervation*, *interfacial deposition*, and in situ *polymerization*. Each method has its own advantages and disadvantages. The choice of a particular technique depends on polymer and drug features, site of action, and therapy regimes (Errico et al. 2009b).

The methods currently used to produce nanoparticles can be divided into three main groups:

1. Physicochemical methods, e.g., the creation of nanoparticles using preformed polymers and inducing their precipitation by emulsification–solvent evaporation, diffusion, or reverse salting-out
2. In situ chemical synthesis methods of macromolecules, giving rise for instance to bulk polymerization or interfacial polycondensation reactions
3. Mechanical methods, e.g., use of high-energy devices like high-pressure homogenizers, sonifiers, or high-energy wet milling apparatuses [Li 2010]

The techniques that follow concise descriptions concerning the preparation of nanoparticles from preformed polymers are reported.

17.2.1 EMULSION–SOLVENT EVAPORATION/EXTRACTION METHODS

17.2.1.1 Single Emulsion Method

The *solvent evaporation technique* is generally used for the preparation of microspheres based on polyesters. However, microspheres prepared by the typical technique have a relatively large particle size of 5–100 μm, do not show a very narrow size distribution, and the solvent used may not be completely removed. It is a very popular method because it is easy and it mainly allows efficient encapsulation of numerous compounds of lipophilic nature (Desgouilles et al. 2003).

This method has been primarily used to encapsulate hydrophobic drugs through oil-in-water (o/w) emulsification process (Park et al. 2005). The hydrophobic drug is dissolved or dispersed in an organic solvent into the polymer solution, and the resulting mixture, after emulsification by high-speed homogenization or sonication, is added into an aqueous solution to make an o/w emulsion with the aid of amphiphilic macromolecules, which are known as emulsifier/stabilizer/additives (Si-Feng 2004). The solvent in the emulsion is removed by either evaporation at elevated temperatures or extraction in a large amount of water, resulting in the formation of compact particles (Figure 17.2).

The solvent evaporation method has been used extensively to prepare PLA and PLGA micro- and nanoparticles containing many different drugs (Fonseca et al. 2002; Kim and Martin 2006). Several variables have been identified that can influence the properties of the nanoparticles, including drug solubility, internal morphology, solvent type, diffusion rate, temperature, polymer composition, medium viscosity, and drug loading (Panyam et al. 2004). This method, however, is only available for the hydrophobic drugs because the hydrophilic drugs may diffuse out or may partition out from the dispersed oil phase into the aqueous phase, leading to poor encapsulation efficiencies

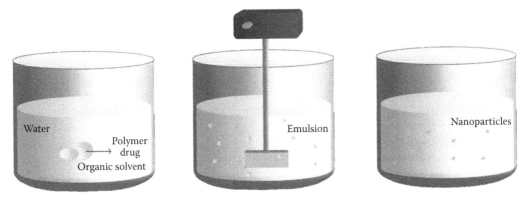

FIGURE 17.2 Nanoparticles preparation: oil in water emulsion technique.

(Park et al. 2005). Many types of drugs with different physical and chemical properties have been formulated into polymeric systems, including anticancer drugs, narcotic agents, local anesthetics, steroids, and fertility control agents (O'Donnell and McGinity 1997).

17.2.1.2   Double Emulsion Method

In double emulsion formulations, microparticles with different morphologies are generated due to the presence of the aqueous-phase microdroplets inside the emulsion droplet. During the solvent elimination, these microdroplets generally coalesce under the pressure of the precipitating polymer. Depending mainly on the polymer concentration and emulsification energies, the final micropar-ticles will be a mixture of honeycomb, capsule, or plain structures. During the shrinkage due to the incompressibility of the inner microdroplets, the precipitating polymer wall around them may break forming holes through which the encapsulated substance partly leaks. Through these holes, the encap-sulated substance is further partitioned with the external aqueous phase during solvent evaporation and contributes to the initial burst release during the therapeutical administration (Rosca et al. 2004).

Several water-soluble drugs have been encapsulated by water-in-oil-in-water (w/o/w) methods. The aqueous solution of the water-soluble drug is emulsified with polymer-dissolved organic solu-tion to form the water-in-oil (w/o) emulsion. The emulsification is carried out using either high-speed homogenizers or sonicators. This primary emulsion is then transferred into an excess amount of water containing an emulsifier under vigorous stirring, thus forming a final w/o/w emulsion. In the subsequent procedure, the solvent is removed by either evaporation or extraction process.

One advantage of this method is bound to the possibility of encapsulating hydrophilic drugs in an aqueous phase with a high encapsulation efficiency. For this reason, the w/o/w emulsion system has been used widely for the development of protein delivery systems. The characteristics of the particles prepared by the double emulsion method are dependent upon the properties of the poly-mer (such as composition and molecular weight), the ratio of polymer to drug, the concentration and nature of the emulsifier, temperature, and the stirring/agitation speed during the emulsifica-tion process (Park et al. 2005). Polymers such as poly(lactic acid), poly(lactic acid)/poly(ethylene glycol), poly(D,L-lactic acid-co-glycolic acid), and poly(ethylene oxide)/poly(lactic acid) are the most utilized in the described technical methodology (Rao and Geckeler 2011).

17.2.2   Miniemulsion Methods

In recent years, miniemulsions have been studied and successfully applied in cosmetics and thera-peutics for skin treatments. They can be defined as heterophase systems consisting of thermo-dynamically stable nanodroplets (possessing an average diameter between 50 and 500 nm) in a continuous phase. Miniemulsions can be basically divided into o/w and w/o systems (W/O); the aforementioned systems are commonly studied both in the development of miniemulsion polymerization processes and for the production of emulsion-based DDS for the time-controlled administration of a variety of drug and bioactive agents. These comprise antitumor agents, peptide drugs, sympatholytics, local anaesthetics, steroids, anxiolytics, anti-infective drugs, vitamins, anti-inflammatory drugs, and dermatological products. The use of miniemulsions in the development of skin care products presents many advantages related to the system stability against sedimentation and to the use of skin-friendly ingredients. Dead Sea Minerals (DSM)-loaded polymeric nanoparticles can be prepared by means of a w/o combined miniemulsion/solvent evaporation process. In the treatment of skin diseases, the beneficial effects of Dead Sea Water (DSW) are well known. Therapeutic baths in the Dead Sea are commonly used as a treatment both for psoriasis, atopic dermatitis, and UV-damaged skin. DSW salt composition is represented by a high concentration of different minerals such as magnesium, calcium, bromide, sodium, potassium, zinc, and strontium. Ahava, an Israeli cosmetics company manufactures skin care products made of mud and mineral-based compounds from the DSW.

The first step consists in the development of a stable nanoemulsion. Nanoparticles were prepared by dissolving an amphiphilic ionomeric polymer such as 2-methoxyethyl hemiester of poly(maleic anhydryde-*alt*-butylvinylether) grafted with 5% of methoxy-PEG2000 (VAM41-PEG) and DSM in a water/ethanol mixture while nonionic surfactants are dissolved in paraffin oil. The two phases are mixed by means of ultrasounds to obtain a miniemulsion. An oily nanoparticles suspension can then be obtained by removing ethanol and water under vacuum (Dessy et al. 2011a).

17.2.3 PHASE SEPARATION

This method involves phase separation of a polymer solution by adding an organic nonsolvent. Drugs are first dispersed or dissolved in a polymer solution. To this solution an organic nonsolvent (e.g., silicon oil) is added under continuous stirring, by which the polymer solvent is gradually extracted and soft coacervate droplets containing the drug are generated. The rate of nonsolvent addition affects the extraction rate of the solvent, the size of the particles, and encapsulation efficiency of the drug. The commonly used nonsolvents include silicone oil, such as dimethyl vegetable oil, light liquid paraffin, and low-molecular-weight polybutadiene. The coacervate phase is then hardened by exposing it into an excess amount of another nonsolvent such as hexane, heptane, and diethyl ether. The final characteristics of the colloidal product are affected by the molecular weight of the selected polymer, viscosity of the nonsolvent, and polymer concentration. The main disadvantage of this method is represented by the high possibility of ending with large aggregates. Infact extremely sticky coacervate droplets frequently adhere to each other before complete phase separation.

This technique is promising for preparation of protein-loaded microcapsules. For example, conventional methods of preparing microparticles involve extensive exposure of proteins to the interface between aqueous and organic phases, to hydrophobic polymer matrix, and to acidic/basic microenviroments resulting from degradation of the polymer. These unfavorable interactions are reported to induce conformational changes of proteins. On the contrary, the interfacial phase separation technique is shown to minimize these sources of protein inactivation (Yeo et al. 2004) thus enabling a satisfactorily maximum therapeutical efficacy of the prepared systems.

17.2.4 SOLVENT DISPACEMENT

17.2.4.1 Nanoprecipitation Method

The nanoprecipitation method, also called solvent displacement, was developed by Fessi et al. It is one of the easiest preparation procedures of nanospheres. Additionally due to its simplicity, this procedure is reproducible, fast, and economic and it uses preformed polymers as starting materials rather than monomers (Vauthier and Bouchemal 2008).

The polymers commonly used are biodegradable polyesters, especially poly(ε-caprolactone) (PCL), polylactide (PLA), and poly(lactide-co-glycolide) (PLGA). Eudragit, poly(butyl methacylate-co-(2-dimethylaminoethyl) methacrylate-co-methyl methacrylate), can also be used as many other polymers such as polyalkylcyanoacrylate (PACA). Natural polymers such as allylic starch and dextran ester were also used, though synthetic polymers have higher purity and better reproducibility than chemically modified natural polymers. Some of the used polymers are characterized by the presence in the macromolecular structure of poly(ethylene glycol) (PEG) fragments as grafts or as blocks inherent in the chain backbone. The presence of PEG in the polymer matrix prevents the opsonization of the nanoderived formulations with consequent abatement of the decrease nanoparticle recognition by the reticular endothelial system (Rao and Geckeler 2011).

The nanoparticle formation is instantaneous and the entire procedure is carried out in only one step. The polymer and the drug are dissolved together and precipitated into a nonsolvent, miscible with the former one. Nanoprecipitation occurs by a rapid desolvation of the polymer when the polymer solution is added to the nonsolvent. Indeed, as soon as the polymer-containing solvent has diffused into the dispersing medium, the polymer precipitates, involving immediate drug entrapment. The rapid nanoparticle formation is governed by the so-called Marangoni effect (Scriven and Sternling 1960), which is due to interfacial turbulence taking place at the interface of the solvent and the nonsolvent and results from complex and cumulated phenomena such as flow, diffusion, and surface tension variations. Nanoprecipitation often enables the production of small nanoparticles (100–300 nm) with narrow unimodal distribution and a wide range of preformed polymers can be used, such as PLGA, cellulose derivatives, or poly caprolactones.

This method does not require extended shearing/stirring rates, sonication, or very high temperatures, and is characterized by the absence of oily-aqueous interfaces, all conditions that might damage the structure of very sensitive bioactive agents and drug as proteins. Moreover, surfactants are not always needed and unacceptable toxic organic solvents are generally excluded from this procedure.

However, the original nanoprecipitation method suffers because of some limitations correlated to the nature of the drug to be loaded. This technique is mostly suitable for compounds having a hydrophobic nature such as indomethacin, which is soluble in ethanol or acetone, but displays very limited solubility in water. Recently, formulation and process modifications were investigated to improve the versatility of the nanoprecipitation technique, particularly with respect to the encapsulation of hydrophilic drugs (e.g., proteins and poly saccharides) (Bilati et al. 2005).

The nanoprecipitation method has been used, however, extensively to prepare nanoparticles containing many different drugs (Fonseca et al. 2002; Dong and Feng 2004).

17.2.4.2 Coprecipitation Method

The coprecipitation method is an original and straightforward procedure recently developed (Cowdall et al. 1999a,b) that appears advantageous for the loading of protein drugs into a polymer matrix. In this case, the polymer is dissolved in a water-miscible organic solvent and added dropwise to an aqueous solution containing the selected loading water soluble drug (proteins or oligonucleotides) and appropriate stabilizers (protective colloids). During the coprecipitation process, the polymeric material gives rise to microphase separation because of its low water solubility and the concurrent interaction with macromolecular drug leads to nanoparticle formation (Figure 17.3).

This methodology does not entail the use of chlorinated solvents and it does not require vigorous shear mixing, preventing appreciable denaturation of the entrapped bioactive macromolecular drug that typically occurs with other techniques. It is well established, in fact, that exposure of bioactive drug such as proteins to denaturation factors including organic solvent and sonication during the encapsulation process can reduce the biological activity of proteic active agents (Park et al. 1995; Zambaux et al. 1999; Pawar et al. 2004). Albumin, λ-interferon, trypsin, urokinas, and hemoglobin (Hb) have been loaded into bioerodible polymeric nanoparticles by following the coprecipitation technique (Chiellini et al. 2006, 2007; Piras et al. 2006, 2007; Dinucci et al. 2007).

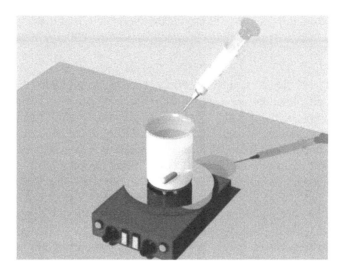

FIGURE 17.3 Nanoparticles preparation: coprecipitation technique.

In particular the feasibility of loading Hb into polymeric nanoparticles by means of a nonaggressive technique, like the coprecipitation, can lead to potential applications in the biomedical field. The encapsulation of Hb into nanostructures has in fact attracted the attention of the scientific community over the past decades, due to the perspective of developing artificial systems that can substitute blood transfusions in restoring oxygen homeostatic concentration. At present, blood transfusions are in fact mostly applied for this purpose, but the need of the right type of blood and its short shelf life are still serious problems to be overcome. In this case, the use of autologous transfusions prevents the need of cross-matching and, although autologous transfusions are considered the safest, they are not always feasible because they may cause perioperative anemia and are more expensive than allogenic transfusions (Pape 2007). In the meanwhile, the direct administration of Hb is not feasible due to its short half life, abnormal high oxygen affinity, and the incoming of serious sides effects, namely malaise, abdominal pain, hemoglobinuria, and renal toxicity (Ness and Cushing 2007). The potential benefits of developing artificial blood substitutes can thus be related to the universal compatibility, the on-need availability, freedom from disease transmission, and long-term storage potential. In the past decades, many strategies have been investigated in order to obtain artificial oxygen carriers that could solve the aforementioned donor transfusion associated drawbacks. Many studies regarding the employment of fluorocarbons (Clark and Gollan 1966; Riess 2001; Castro and Briceno 2010) or chemically modified Hb are reported in the literature (Li et al. 2006; Asanuma et al. 2007; Hu et al. 2007); anyway both approaches have revealed the presence of severe side effects and altered oxygen carrier ability once injected in vivo (Sharma et al. 2011). The encapsulation of Hb into structures able of mimicking red blood cell membranes (such as polymeric nanoparticles, liposomes, and solid lipid nanoparticles) represents an alternative strategy that has shown favorable results in terms of biocompatibility and maintenance of protein physicochemical properties. Different polymeric matrices have been tested for the production of Hb-loaded polymeric nanoparticles by means of the coprecipitation technique; among these the employment of VAM41-PEG polymer has shown favorable results. As reported in literature (Dessy et al. 2011b), stealth nanoparticles possessing suitable features to be used as injectable systems were obtained by applying the *coprecipitation* technique under controlled atmosphere and temperature (Figure 17.4).

Moreover, the introduction of different reducing agents in the VAM41-PEG Hb-loaded nanoparticles formulation system, aimed at minimizing protein oxidative phenomena, did not alter nanoparticles features in terms of dimension, shape, and protein loading, highlighting the extreme versatility of the *coprecipitation* technique (Dessy et al. 2011b).

FIGURE 17.4 VAM41-PEG Hb-loaded polymeric nanoparticles: (a) SEM micrograph and (b) water suspension.

17.2.5 DIALYSIS METHOD

Some disadvantages of the conventional methods include the difficulties and necessities of removal of solvent and surfactant residues, low particle yields, excessive time-consuming steps for preparation, and the necessity to use a high concentration of surfactant for the preparation of small spherical particles and polymeric micelles. The dialysis method is a simple and effective preparation method for small and narrow-sized distributed nanoparticles mostly using either block or graft copolymers and other amphiphilic materials. Polymer, drug, and surfactants are dissolved in the same organic solvent and placed inside a dialysis tube with proper molecular weight cut-off. Dialysis is performed against a nonsolvent, miscible with the former one. The displacement of the solvent inside the membrane is followed by the progressive aggregation of polymer, drug, and surfactants due to a loss of solubility and by the formation of homogeneous suspensions of micro- nanoparticles (Errico et al. 2009) (Figure 17.5).

Trials of preparing nanoparticles in the absence of surfactants were performed. The dialysis method was applied to PLGA starting from different solvent solutions and avoiding the use of surfactants. The solubility of the particle components in the initial solvents may affect the physicochemical properties such as particle size and drug contents, whereas the absence of surfactant led to some drawbacks such as instability of the re-dispersed freeze-dried PLGA nanoparticles and low drug-loading efficiency (Jeong et al. 2001). PLA/Tween 80 copolymers were synthesized, aiming at incorporating a widely used emulsifier for biomedical nanoparticles preparation directly in the polymer backbone. Nanoparticles loaded with anticancer drugs were obtained by using the dialysis method and without the use of additional surfactants/emulsifiers (Zhang and Feng 2006).

Dialysis method was applied in the preparation of PLGA and PHB nanoparticles encapsulating retinoic acid (RA) and some processing conditions varied in order to observe their effect on nanoparticle characteristics. The use of a binary solvent solution composed of a mix of polymer and drug solvent and nonsolvent proved to be effective for preparing nanoparticles with small size and a narrow distribution of diameters. The small diameter of PHB nanoparticles prepared by this method was quite exceptional when compared with those of PHB microsphere prepared by other techniques commonly used such as solvent evaporation (Errico et al. 2009).

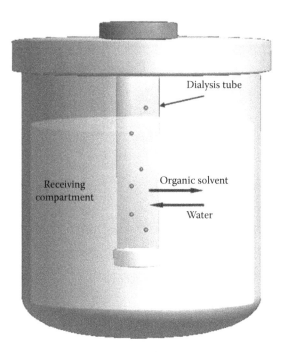

FIGURE 17.5   Nanoparticles preparation: dialysis technique.

17.2.6   Self-Assembling

The concept of self-assembly has been extensively used to prepare organic nanoparticles. For example, liposomes and nanoscaled vesicles that were prepared using self-assembly of phospholipids can serve as powerful nanocarriers for drug and gene delivery. Self-assembled amphiphilic copolymer building blocks spontaneously form nanoparticles, which can be used for drug delivery and molecular imaging (Wang et al. 2009).

Such association depends on many factors including coulombic interactions, hydrophobicity of the polymer–molecule pair, and the conformational features of the polymer. One special class of such systems is represented by the complexes formed from macromolecular monomers of opposite charges. The solution behavior of these complexes strongly depends on their composition. Electroneutral complexes that contain equivalent amounts of polyion units and monomers are water insoluble.

Nonstoichiometric complexes containing an excess of one of the components are generally soluble in water. Since these complexes are capable of forming aggregates of nanometer size, they have been termed as polyion complex (PIC) micelles or block ionomer complexes (BICs) (Gupta et al. 2006).

Micellar DDS enjoy several advantages over other particulate configurations. For example, they can significantly enhance the water solubility of hydrophobic drugs for improved bioavailability. They usually exhibit low critical micelle concentration (CMC), rendering the drug-loaded micelles stable in the bloodstream to achieve fairly long circulation time (Liu et al. 2003).

Typical studies were focused on polystyrene-b-poly(ethylene oxide), polystyrene-b-poly(acrylic acid), polystyrene-b-poly(methacrylic acid), etc. The traditional micelles are unable to sense a signal and respond by changing their structures. Incorporating external control over structure and physical properties offers the possibility of constructing systems capable of both tunable transformation and controlled transmission of energy or information. For this purpose, thermosensitive (Liu et al. 2003; Wei et al. 2006), pH-sensitive (Hruby et al. 2005; Hsiue et al. 2006), and pH- and temperature-sensitive micelles (Determan et al. 2005) had been engineered.

17.2.7 RAPID EXPANSION OF SUPERCRITICAL FLUID SOLUTION

Supercritical fluids and dense gases have unique properties that make them suitable for various applications such as extractions, reactions, and particle formation processes. In recent years, considerable interest has been expressed in the field of supercritical fluid and dense gas technologies in an attempt to find alternative and improved methods of pharmaceutical processing where biodegradable compounds are used frequently (Mishima 2008).

A supercritical fluid is loosely defined as a solvent at a temperature above the critical temperature, at which the fluid remains a single phase regardless of pressure. However, for practical purposes, such as high density for solubility considerations, fluids of interest in materials processing are maintained typically at near-critical temperatures.

Among the most important properties of a supercritical fluid are the low and tunable densities, which can be easily varied from gaslike to liquidlike through a simple change in pressure at constant temperature, and the unusual solvation effects at densities near the critical density (often discussed in terms of solute–solvent and solute–solute clusterings). The production of polymeric nanoparticles through "Rapid Expansion of a Supercritical Solution" into either air (RESS) or Liquid Solvent (RESOLV) may conceptually be divided into two somewhat related processes. One is the initial formation of nanoparticles in the rapid expansion, and the other is the stabilization of the suspended nanoparticles. Evidently, the protection of initially formed polymeric nanoparticles represents a different set of technical challenges, which are largely independent of the rapid expansion process itself, especially if the protection agent is added immediately after the expansion. The good point is that many methods aimed at stabilizing nanoparticle suspensions are already available in the literature, some of which have shown promise in use with RESOLV.

Supercritical fluid processing techniques can play a significant role in the particle formation and production. In particular, the RESOLV technique offers a unique way to prepare clean and narrowly distributed polymeric nanoparticles. Since the nanoparticles obtained in RESOLV are suspended, they may be protected from agglomeration by using existing stabilization methods and agents. These stable nanoparticle suspensions may find many interesting and important applications, as already discussed in the literature (Sun et al. 2005; Meziani et al. 2006). Furthermore, a new method called Supercritical Fluid Extraction of Emulsions (SFEE) has been investigated. The method combines the advantages of traditional emulsion-based techniques, namely control of particle size and surface properties, with the advantages of continuous supercritical fluid extraction process, such as efficient scale-up, higher product purity, and shorter processing times (Chattopadhyay et al. 2006).

17.2.8 SPRAY DRYING

Spray drying technology is widely known and used to transform liquids (solutions, emulsions, suspension, slurries, pastes, or even melts) into solid powders. A traditional *spray dryer* is generally used to transform liquid substances into powders rapidly and efficiently. The speed of the process and the consequently short drying time enable the drying of even temperature-sensitive products without degradation. This spray drying process is particularly used to improve product conservation in dried solid form (Li et al. 2010).

Compared to other conventional methods, spray drying offers several advantages. It shows good reproducibility, involves relatively mild conditions, allows for control of the particle size, and is less dependent on the solubility of the drug and the polymer. Generally, the polymer is dissolved in volatile solvents and the drug is dispersed or dissolved in the polymer solution.

Solutions or dispersions are sprayed against a stream of cold air ($-60°C$; top-spraying) using a two-fluid pneumatic nozzle with heating facilities. The frozen droplets formed by this spray-freezing step are dried during the following atmospheric freezedrying in the cold desiccated air stream by sublimation. A filter holds the fine product back in the drying chamber, while the water vapor is removed by the circulating air in the cooling systems, where the humidity condenses on the refrigerated surfaces (Leuenberger 2002). Recently this method has been used to prepare dry powder aerosol particles (Azarmi et al. 2006), a powder formulation for controlled delivery of Paclitaxel (Mu et al. 2005), and powders for aerosol delivery to the lung (Sham et al. 2004).

In an attempt to minimize aggregation of the microparticles, a double-nozzle spray-drying technique was developed. While the polymer/drug solution is sprayed from one nozzle, aqueous mannitol solution is simultaneously sprayed in order to coat the particulate with an anti-adherent agent. The results indicate that the coating of the microspheres with mannitol reduces the extent of aggregation and augments the yield of the product (Takada et al. 1995). When a protein drug is encapsulated by means of spray drying, a loss of its biological activity may occur especially when aqueous phase systems are involved (Elversson and Millqvist-Fureby 2005). The atomization of a w/o emulsion or a cryogenic nonaqueous processes can be used as an alternative to spray drying (De Rosa et al. 2005). In the latter technique, the liquid droplets of the polymer/drug solution are produced through the spraying nozzle, collected in liquid nitrogen containing frozen ethanol, and hardened by placing them at −80°C where the solvent extraction occurs (Johnson et al. 1996).

In a recent study, Li et al. presented the Nano Spray Dryer B-90, a revolutionary new sprayer developed by Büchi, use of which can lower the size of the produced dried particles by an order of magnitude attaining submicron sizes. The strength of the BüchiNanoSprayDryerB-90 lies in its vibration mesh spray technology, creating tiny droplets (before evaporation) in a size range of a smaller order of magnitude than in classical spray dryers. This is a revolution in spray drying technology, making it possible to produce powders in the submicron-size range with very narrow distributions and high formulation yields. The Büchi Nano Spray Dryer B-90 appeared to provide very satisfactory results for the formulation of submicron particles, with relatively high yields (70%–90%) for small sample amounts. Five representative polymeric wall materials (Arabic gum, whey protein, polyvinyl alcohol, modified starch, and maltodextrin) were spray dried and the resulting size distributions were shown to be mainly below the 1-μm scale, attaining sizes as low as ~350 nm, which is a very noteworthy result for spray drying technology (Li et al. 2010).

17.2.9 COLLOIDAL COATING

Although a number of different methods exist for the preparation of nanoparticles, it is highly desirable to provide alternative methods, aimed at producing polymeric particles that do not entail the utilization of organic solvents, which are undesirable for their potential toxicity. Such methods are expected to receive a particular attention in biotechnological field. An organic solvent-free method has been developed for the preparation of nanoparticles based on a copolymer of maleic anhydride and butyl vinyl ether (VAM41) nanoparticles loaded with RA. This method is easy to perform, effective for the preparation of relatively small nanoparticles with narrow size distribution, and operator friendly since toxic organic solvents need not be handled.

The novel nanoparticle preparation method was conceived by taking into account the characteristics of the drug, polymer, and stabilizing agent (HSA) and the behavior of their different mixture in response to variation of water pHs. At pHs below 7, RA self-associates into a micellar-like structure while it is soluble at pHs above 8 in its deprotonated form. Alike, VAM41, due to the presence of the carboxylic ionizable group, is soluble in water at pHs above 8 and is insoluble at pHs below 7. A basic solution of the drug, polymer, and stabilizing agent was prepared and the subsequent acidification of the solution provoked the polymer to absorb and coat uniformly the surface of the forming micellar suspension of RA, leading to the formation of small and narrow-size distribution nanoparticles.

Somehow this process resembles the practice of *tablet coating* in which tablets are loaded in large rotating pans and coated with synthetic polymers or polysaccharides. In the case of tablet coating, it is the evaporation of the solvent that triggers the coating process, while in this case, it is the decrease of pH (Errico et al. 2009).

17.3 TARGETING APPROACHES OF POLYMERIC NANOPARTICLES

The biophysical-chemical properties of the vehicle, such as size, charge, surface hydrophilicity, and the nature and density of the ligands on their surface, can all impact the circulating half-life of the particles as well as their biodistribution (Farokhzad and Langer 2009).

The introduction of three key technologies stimulated the immense activity and clinical success of nanotherapeutics from the late 1980s to the present. The first was the concept of "PEGylation," which refers to polyethylene glycol-conjugated drugs or drug carriers (Veronese and Pasut 2005). The second is the concept of "active targeting" of the drug conjugate by conjugating cell membrane receptor antibodies, peptides, or small molecule cell ligands to the polymer carrier (Timko et al. 2011) (Figure 17.6).

The third was the discovery of the "enhanced permeation and retention (EPR) effect", wherein nanoscale carriers are entrapped within solid tumors due to leaky vasculature of the fast-growing tumor. This is called "passive" targeting as contrasted with active targeting (Hoffman 2008), whereas the former two were considered a kind of active targeting (Figure 17.7).

More recently, surface properties of nontargeted drug-delivery vehicles such as ordered striations of functional groups as well as their shape and size have also been shown to enhance particle uptake (Farokhzad and Langer 2009).

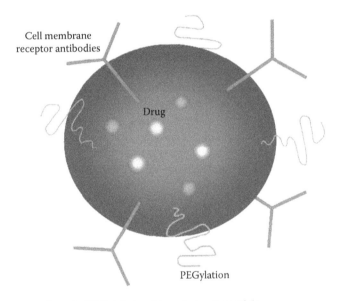

FIGURE 17.6 Representation of a PEGylated and targeted nanoparticle.

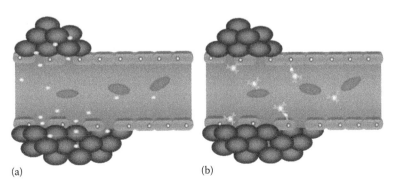

(a) (b)

FIGURE 17.7 Schematic representation of passive (a) and active (b) targeting.

To reach the site of action, the carrier has to cross many biological barriers such as other organs, cells, and intracellular compartments, where it can be inactivated or can express undesirable effects. The striking advantages of drug targeting can be summarized as (1) simplification of drug administration protocols, (2) reduction of the drug quantity required to achieve a therapeutic effect with abatement of therapy costs, and (3) sharp increase of the drug concentration in the required sites without negative effects on nontarget compartments.

17.3.1 Passive Targeting

Passively targeting nanocarriers first reached clinical trials in the mid-1980s, and the first products, based on liposomes and polymer–protein conjugates, were marketed in the mid-1990s (Peer et al. 2007). Passive targeting is generally based on the relation between the size of the drug carrier and the tissue characteristics, such as their permeability.

A typical example of passive targeting based on particle density, shape, and size is the targeted delivery of drugs to specific lung tissues. Before releasing the pharmacological agent, inhaled particles need to reach and deposit in the site of action. The deposit of particles in the respiratory tract is governed by impaction (inertial deposition), sedimentation (gravitational deposition), brownian diffusion, interception, and electrostatic precipitation. In mammals, the upper airways (nose, mouth, larynx, and pharynx) and the branching anatomy of the tracheobronchial tree act as a series of filters for inhaled particles. Thus, aerosol particles bigger than 100 μm generally do not enter the respiratory tract and are trapped in the naso/oropharynx. Particles bigger than 10 μm will not penetrate the tracheo-bronchial tree; they must be less than 5 μm to reach the alveolar space and bigger than 0.5 μm to not be exhaled without deposition (Crowder et al. 2002). The carrier particles should meet special requirements such as an appropriate mass median aerodynamic diameter (MMAD) and a suitable fine particle fraction (FPF). Direct delivery of drug-loaded nanoparticles to the lungs combines the concepts of localized delivery with the advantages of using nanoparticles in lung cancer therapy. The recent developments of inhalable biodegradable nanoparticles, large porous particles, and liposomal dry powders make inhalation a feasible alternative approach to deliver macromolecules such as insulin and for the treatment of lung-specific diseases such as tuberculosis. One issue with nanoparticle pulmonary delivery is that their size is not suitable for deep lung deposition. In fact, a carrier system such as lactose microparticles is required for deep lung delivery (Roa et al. 2011).

Concerning the intravenous administration of nanoparticles, passive targeting to body compartments may be achieved by extravasation in the diseased tissue or organ (Palmer et al. 1984). The enhanced therapeutic potencies are mainly due to the EPR effect that is attributed to high vascularization and enhanced permeability of tumor blood vessels combined with limited lymphatic clearance of macromolecules from the tumor environment (Min et al. 2008).

The ability of vascular endothelium to increase its permeability was first noticed in tumors and in hypoxic areas of infarcted myocardium (Palmer et al. 1984).

In tumors, the defective vascular architecture is due to the rapid vascularization necessary to serve fast-growing cancers, coupled with poor lymphatic drainage (Brannon-Peppas and Blanchette 2004).

In such areas with increased vascular permeability, even relatively large particles ranging from 10 to 500 nm in size can extravasate, accumulate inside the interstitial space, and release the loaded active agent. Because the "cut-off" size of the permeabilized vasculature can vary from case to case, the size of the DDS may be used to control the efficacy of such spontaneous or "passive" drug delivery.

However, to exploit this targeting method, drug carriers should circulate in blood long enough to provide acceptable accumulation of the bioactive molecule in the area of interest.

Particles with 0.3- to 7-μm diameter have short circulation half-lives due to their fast removal by the natural defense mechanisms of the body. After opsonization, these particles are seized by the Mononuclear Phagocytic System (MPS), also known as RES. Therefore, in order to take advantage of the EPR effect, the particle surfaces need to be modified to be "invisible" to opsonization (Maeda et al. 2009).

In a recent study, Min et al. (2008) used a hydrophobically modified glycol chitosan-based (HGC) nanoparticle as a Camptothecin (CPT) delivery carrier because it has been previously confirmed that such nanoparticles are promising therapeutic DDS for hydrophobic drugs such as doxorubicin, Paclitaxel, and therapeutic peptide drugs. CPT, a natural plant alkaloid extracted from *Camptotheca acuminate*, has been shown to be a potent anticancer drug targeting intracellular topoisomerase I. CPT and CPT analogs are increasingly explored in clinical treatment and show great utility in the treatment of various cancer forms including primary and metastatic colon carcinoma, small-cell lung carcinoma, and ovarian, breast, pancreatic, and stomach cancers. The achieved results showed that compared to free CPT, CPT-HGC nanoparticles showed a longer circulation time in the blood, a better targeting of drug to tumor tissue, and a substantial enhancement of antitumor activity (Min et al. 2008).

In a site of inflammation, one of the cardinal features of the inflammatory reaction is the increase in vascular permeability to solutes and macromolecules (Rossler et al. 1995). This phenomenon determines the extravasation of the nanoparticles at the diseased site (Coester et al. 2000). As an example, during multiple scleroses (MS), the permeability of the blood brain barrier (BBB) is increased and the passage of nanoparticulate systems could be facilitated. PACA nanoparticles have been investigated in experimental allergic encephalomyelitis (EAE), which serves as an animal model for MS, in order to determine their capacity to reach the central nervous system (CNS). Their concentration has been found to increase in the CNS, particularly in the white substance. Long-term circulating PACA nanoparticles with PEG moieties were found in the CNS in greater amounts than non-PEGylated PACA nanoparticles (Gupta et al. 2004).

Seki et al. developed and studied the biodistribution of lipoprotein-like particles, Lipid Nano-Sphere (LNS®), incorporating dexamethasone palmitate (DMP). LNS® easily and selectively passed through the leaky capillary wall at the inflamed sites by passive diffusion. Because of the lower uptake of these particles by the liver, LNS® showed good recovery from the liver and prolonged plasma half-life of DMP after intravenous injection. High efficiency in the targeting of DMP into inflammation sites and high anti-inflammatory efficacy were observed (Seki et al. 2004).

17.3.2 DIRECT APPLICATION OF DRUGS

The simplest method of driving a drug to a specific region within the body consists of locally applying the drug to the selected area. Even if this strategy cannot properly be assimilated to targeted delivery, it is at the borderline between systemic delivery and targeted delivery. Indeed, typical applications reach the goal to limit the area within which the drug can diffuse, thus avoiding most systemic side effects. When a drug is dispensed systemically, most of the product is wasted because of its diffusion through healthy organs, which can suffer damage, and only a small amount of the drug can reach the target tissue or organ. Direct application allows for better control over the amount of administered drug that reaches the area of interest; increasing the local concentration while reducing the drug diffusion within the rest of the body.

The successful examples of this approach include the intra-articular administration of hormonal drugs in the therapy of arthritis (Williams et al. 1996), intracoronary infusion of thrombolytic enzymes in the therapy of thrombus-induced myocardial infarction (Khaw et al. 2007) and ocular formulations (Vandervoort and Ludwig 2007). Nanoparticles also can be delivered to distant target sites by localized delivery using a catheter-based approach with a minimal invasive procedure (Panyam and Labhasetwar 2003), which allows the drug to be retained at the site action for a long period of time. However, the applicability of such straightforward approaches is rather limited.

BBB, which is formed by the tight junctions within the capillary endothelium of the brain, constitutes a formidable barrier to the CNS delivery of therapeutic agents. Although selective transport mechanisms are present in the BBB such as diffusion, carrier-mediated transport, receptor-mediated, adsorptive, and fluid-phase endocytosis, the transport of therapeutic agents via systemic administration to the brain is limited. To overcome these challenges, the direct injection of drugs into the brain has been investigated by means of microinjection techniques. This approach could be effective in targeted delivery to a certain area of the brain; however, it would not be as effective if the whole brain is addressed, due to the poor diffusion of the therapeutic agents through the brain tissue. To improve diffusion of the injected drug, some of the strategies include conjugation of the therapeutic proteins with nontoxic neuronal binding domain of tetanus toxin (Tetanus Toxin fragment C-TTC). Proteins linked to TTC show not only a better retention, but also a superior distribution in the extracellular space on direct injection into brain (Francis et al. 2004).

In recent years, a novel approach to direct application of drugs has been studied. The active agent or the nanocarrier can be loaded in implants, such as cardiac valves and vascular stents, and be released from there. The loading of a nanoparticulate carrier instead of the direct loading of the drug would provide a fine control over drug release kinetics (Finkelstein et al. 2003).

Concerning growth factors (GF) delivery, it has been shown that their direct injection into the site to be regenerated is generally not effective because they rapidly diffuse from the injected site and are enzymatically digested or deactivated (Muzzarelli and Muzzarelli 2005). Therefore, there are demands for either extremely higher doses and/or frequent injections. For clinically beneficial outcome, these GF require a delivery system to guide tissue regeneration and prevent the rapid dispersal of the factors from the site (Douglas and Tabrizian 2005). First, it is essential to identify the key GF or factors for a particular tissue application. The mode of factor delivery must target the desired cell population and minimize signal propagation to nontarget tissues and cells. Moreover, the controlled delivery of GF requires a relatively long-term maintenance of biologic activity within the time. Finally, the release profile of the GF from the system should be controlled temporally and spatially, making these systems similar to an artificial Extracellular Matrix (ECM) (Agnihotri et al. 2004).

17.3.3 Physical Targeting

The concept behind physical targeting consists of the targeted delivery of drugs mediated by physical alterations of the area of interest. This technique can be applied following one of the two main approaches. The first one relies on the intrinsic properties of the injured area; indeed, inflamed or neoplastic areas differ from normal tissues, often showing higher temperature or lower pH. In such conditions, it is possible to employ drug carriers that are able to survive in normal tissues, but subjected to degradation at lower pH or higher temperatures; in this case, drugs are only released in the injured area, without undesired systemic effects. The second approach consists of the application of an external stimulus, like heat or a magnetic field, to degrade the carrier and release the drug. When the application is strictly localized, the drug accumulates only inside the area of interest.

The latter is already employed in medicine for contrast agents in magnetic resonance-based diagnostics and it is now widely studied for magnetic drug targeting processes (Chithrani et al. 2009; Hoeffel et al. 2009). In fact, by surrounding iron oxide nanoparticles in supermagnetic conditions with a biocompatible polymer able to absorb a drug, it is possible to direct them onto a specific body target, by means of an external localized magnetic field (Zhang et al. 2009). DDS based on nanoparticles mainly endowed with magnetic properties resulted to be also very interesting for the gene therapy (Pan et al. 2007; McBain et al. 2008). Furthermore, the presence of magnetic particles is useful for the treatment of tumors, by creating a synergistic effect of drug release and local hypertermia therapies. Local heating of the tumoral cells between 37°C and 45°C typically 42.5°C induces cellular apoptosis. Over 45°C, thermoablation is meant to occur (Müller 2009).

17.3.4 Active Targeting

The use of targeting moieties is an active strategy that relies on specific interactions at target sites and such interactions include antibody–antigen and ligand–receptor interactions. To maximize specificity, a surface marker (antigen or receptor) should be overexpressed on target cells relative to normal cells (Peer et al. 2007).

The "magic bullet" initially theorized by Paul Ehrlich (1906) has evolved in a three-parts system composed of a therapeutic agent a carrier for the drug, and a targeting moiety combined together. The targeting moiety has to be specific for the area of interest, thus making the distribution of the drug independent from the EPR effect. This targeting strategy provides evident advantages such as high specificity for the injured area and reduced side effects, but it can show its best performances in cancer therapy, especially for the cure of delocalized tumors or tumors in their early stages of development, when the vasculature is still immature (Solaro et al. 2010). The identification and characteristics of the targeting moiety are important for circulation time, cellular uptake, affinity, and extravasation (Byrne et al. 2008).

Several cellular receptors or integral membrane proteins have been investigated as target for tumor-specific drug delivery because they are overexpressed in tumor tissues, compared to normal tissues. Folate receptor (FR) is a well-known tumor marker that binds vitamin folate and folate–drug conjugates with a high affinity, and transferrin receptor is a carrier protein for Transferrin that imports iron by internalizing the transferrin–iron complex through receptor-mediated endocytosis. Glycans, which are expressed on the extracellular side of the plasma membrane, also play an important role as targeting moieties. Cancer cells often express different glycans compared with their normal counterparts (Cho et al. 2008).

Despite many of these still standing challenges, targeted nanocarriers represent a promising strategy for future development of targeted delivery therapeutics.

Today, over 200 delivery systems based on antibodies or their fragments are in preclinical and clinical trials (Peer et al. 2007).

Monoclonal antibodies (mAb) can be attached to particles by physical adsorption (Kubiak et al. 1988) or by covalent linkages (Cirstoiu-Hapca et al. 2007). Antibody fragments containing only the variable region of the antibody are now more commonly used for active targeting of therapeutics because they retain the specificity of their target, while lacking the constant "Fc effector region" that could result in complement activation or undesirable interaction with other cells (Chapman 2002). Furthermore, animal-derived antibodies are easily recognized as foreign elements and may cause strong immune responses. Chimeric antibodies, combining human constant regions and mouse variable regions, have shown to reduce but not completely prevent immune reactions (Stacy 2005). Humanized antibodies, which contain only the binding regions of the mouse antibodies fused with a human antibody, have shown reduction in immunogenicity but sometimes at the expense of affinity for the target (Stacy 2005). Antibodies which are 100% human have been developed in transgenic animals (Van Dijk and van de Winkel 2001) and through phage display techniques (Goletz et al. 2002). The advantages of using small-molecular weight fragments of mAb are the lower steric hindrance and the possibility of grafting a larger number of targeting moieties on carrier surfaces, thus leading to improved pharmacokinetics profiles.

The final concern in the incorporation of antibodies or antibody fragments (Fab) to DDS is the method of conjugation. Abs or Fabs fragments are normally conjugated either directly to nanoparticle surfaces or through linker molecules such as PEG. Conjugation of the antibody can be random or site specific. Random conjugation is commonly carried out by carbodiimide-mediated chemistry, which creates stable amide bonds between carboxylic acid groups in the nanocarrier and primary amine groups, including lysines and the N-terminus amine, of the antibody (Chapman 2002). Due to the lack of specificity of this conjugation route, antibodies can be bound to the nanocarrier in a number of ways, some of which can even block access to the binding site of the antibody (Lee et al. 1999).

Lectin-carbohydrate is another classical example of active drug targeting. Lectins are proteins of nonimmunological origin capable of recognizing and binding to glycoproteins expressed on cell surface or capable of circulating. The interaction of lectins with certain carbohydrates is very specific. This interaction is as specific as the enzyme–substrate or antigen–antibody interactions. Lectins may bind free sugar or sugar residues of polysaccharides, glycoproteins, or glycolipids. One major property of lectins is their specific saccharide-binding sites. Some lectins are composed of subunits with different binding sites. This ligand–carbohydrate interaction can be made use of in the development of nanoparticles containing carbohydrate moieties that are directed to certain lectins (direct lectin targeting) as well as incorporating lectins into nanoparticles that are directed to cell surface carbohydrates (reverse lectin targeting).

Presently, a powerful targeting strategy is represented by the use of aptamers. Aptamers are RNA or DNA molecules that fold by intramolecular interaction into unique three-dimensional conformations for target recognition. They can be selected by a process referred to as SELEX (Systematic Evolution of Ligands by Exponential Enrichment) from a pool of DNA or RNA by repetitive binding of the target molecules (Ellington and Szostak 1990; Turek and Gold 1990). Aptamers possess numerous advantageous characteristics, including small size, lack of immunogenicity, and ease of synthesis, all of which rival those of other molecular probes such as antibodies (Brody and Gold 2000). While most aptamers reported so far have been selected for single targets, such as protein, drug, or amino acid, whole living cells have been used as targets for the selection of a panel of aptamers (cell-SELEX) for specific cell recognition (Shangguan et al. 2006; Tang et al. 2007). Aptamers are also amenable to a wide variety of chemical modifications to insert radioscopic or fluorescent reporters and affinity tags for molecular recognition or to create nuclease-resistant aptamers. It is also possible to chemically modify aptamers to facilitate covalent conjugation to nanomaterials, e.g., with $5'$ or $3'$ amino or thiol groups.

Aptamers that were conjugated to nanoparticles resulted in increased targeting and more efficient therapeutics, as well as more selective diagnostics. For example, nanoparticle–aptamer conjugates could target the prostate-specific membrane antigen (Farokhzad 2004), a transmembrane protein that is upregulated in prostate cancer (Lupold 2002). This was the first report of an aptamer that enabled the targeted delivery of drug-encapsulated nanoparticles. In addition, aptamers have been used as vehicles to deliver anthracycline chemotherapeutics by their own, through intercalation within the aptamer double-stranded regions (Bagalkot et al. 2006). This concept has enabled the engineering of polymeric nanoparticle systems to co-deliver two drugs in a temporally distinct manner (Zhang 2007).

Localized diseases such as inflammation not only have leaky vasculature but are also characterized by an overexpression of some epitopes or receptors (Coester et al. 2000) and an accumulation of macrophages (Gupta et al. 2004) that can be used as target moieties. Therefore, nanomedicines can also be actively targeted to these sites. In this view, macrophage-targeted therapies can be applied for atherosclerosis (Gupta et al. 2004) and rheumatoid arthritis (RA) (Furusawa et al. 2004).

Paclitaxel (PCX) is an antitumor agent with a unique mechanism of action promoting the assembly of microtubules from tubule dimers and prevents them from depolarizing. This leads to the loss of normal microtubule dynamics necessary for cell division and other vital processes and consequently causes cell death. The ability of this drug to stabilize microtubules makes it significantly effective against various types of solid tumors including breast cancer, advanced ovarian carcinoma, lung cancer, head and neck carcinomas, and acute leukemias. However, its clinical application is mainly limited by its narrow therapeutic index and very poor solubility in water or in other pharmaceutically acceptable solvents (Bilensoy et al. 2008). Zhang et al. formulated PCX-loaded nanoparticles that exposed folate moieties on the surface for targeting the cancer cells rich of the FRs. They showed that the nanoparticle formulations significantly promoted targeted delivery of the drug to the corresponding cancer cells, substantially enhancing its therapeutic effects (Zhang and Feng 2006).

17.4   CONCLUSIONS

Polymeric nanoparticles are rapidly progressing and are being implemented to improve the therapeutic value of various water-soluble/insoluble medicinal drugs and bioactive molecules, by solving several limitations of conventional DDS such as nonspecific biodistribution and targeting, poor oral bioavailability, and low therapeutic indices.

The present chapter is focused on the technical aspects of nanoparticles preparation and provides an overview of the exploitable targeting approaches that make them suitable for the treatment of a multitude of diseases. The flexibility to modify and adapt nanoparticles to meet the needs of pathological conditions or as a diagnostic tool (targeted molecular imaging) is just one of the important characteristics of this technology that appears to be promising for far-reaching consequences in drug-delivery practice.

REFERENCES

Agnihotri, S.A., Mallikarjuna, N.N., and Aminabhavi, T.M. 2004. Recent advances on chitosan-based micro- and nanoparticles in drug delivery. *J. Control. Release* 100: 5–28.

Asanuma, H., Nakai, K., Sanada, S., Minamino, T., Takashima, S., Ogita, H., Fujita, M., Hirata, A. et al. 2007. S-nitrosylated and pegylated hemoglobin, a newly developed artificial oxygen carrier, exerts cardioprotection against ischemic hearts. *J. Mol. Cell. Cardiol.* 42(5): 924–930.

Azarmi, S., Tao, X., Chen, H., Wang, Z., Finlay, W.H., Loebenberg, R., and Roa, W.H. 2006. Formulation and cytotoxicity of doxorubicin nanoparticles carried by dry powder aerosol particles. *Int. J. Pharm.* 319(1–2): 155–161.

Bagalkot, V., Farokhzad, O.C., Langer, R., Jon, S. 2006. An aptamer-doxorubicin physical conjugate as a novel targeted drug-delivery platform. *Angew. Chem. Int. Ed. Engl.* 45: 8149–8152.

Bilati, U., Allemann, E., and Doelker, E. 2005. Development of a nanoprecipitation method intended for the entrapment of hydrophilic drugs into nanoparticles. *Eur. J. Pharm. Sci.* 24: 67–75.

Bilensoy, E., Gurkaynak, O., Ertan, M., Murat, A., and Atilla, H. 2008. Development of nonsurfactant cyclodextrin nanoparticles loaded with anticancer drug paclitaxel. *J. Pharm. Sci.* 97: 19.

Brannon-Peppas, L. and Blanchette, J.O. 2004. Nanoparticle and targeted systems for cancer therapy. *Adv. Drug Deliv. Rev.* 56: 1649–1659.

Brody, E.N. and Gold, L.J. 2000. Aptamers as therapeutic and diagnostic agents. *Biotechnol.* 74: 5–13.

Byrne, J.D., Betancourt, T., and Brannon-Peppas, L. 2008. Active targeting schemes for nanoparticle systems in cancer therapeutics. *Adv. Drug Deliv. Rev.* 60: 1615–1626.

Castro, C.I. and Briceno, J.C. 2010. Perfluorocarbon-based oxygen carriers: Review of products and trials. *Artif. Organs* 34(8): 622–634.

Chapman, A.P. 2002. PEGylated antibodies and antibody fragments for improved therapy: A review. *Adv. Drug Deliv. Rev.* 54(4): 531–545.

Chattopadhyay, P., Huff, R., and Shekunov, B.Y. 2006. Drug encapsulation using supercritical fluid extraction of emulsions. *J. Pharm. Sci.* 95(3): 667–679.

Chiellini, F., Bartoli, C., Dinucci, D., Piras, A.M., Anderson, R., and Croucher, T. 2007. Bioeliminable polymeric nanoparticles for proteic drug delivery. *Int. J. Pharm.* 343(1–2): 90–97.

Chiellini, E., Chiellini, F., and Solaro, R. 2006. Bioerodible polymeric nanoparticles for targeted delivery of proteic drugs. *J. Nanosci. Nanotechnol.* 6(9/10): 3040–3047.

Chiellini, F., Piras, A.M., Fiumi, C., Anderson, R., Muckova, M., Bartoli, C., Dinucci, D. et al. 2006. Bioerodible/ bioeliminable nanoparticles as versatile vectors for the targeted and controlled release of protein drugs, bioactive principles and as oxygen carriers. *20th European Conference on Biomaterials.* September 27– October 1. Cité des Congrès, Nantes, France.

Chithrani, B.D., Stewart, J., Allen, C., and Jaffray, D.A. 2009. Intracellular uptake, transport, and processing of nanostructures in cancer cells. *Nanomedicine* 5(2): 118–127.

Cho, K., Wang, X., Nie, S., Chen, Z., and Shin, D.M. 2008. Therapeutic nanoparticles for drug delivery in cancer. *Clin. Cancer Res.* 14(5): 1310–1316.

Cirstoiu-Hapca, A., Bossy-Nobs, L., Buchegger, F., and Gurny, F. 2007. Differential tumor cell targeting of anti-HER2 (Herceptin®) and anti-CD20 (Mabthera®) coupled nanoparticles. *Int. J. Pharm.* 331(2): 190–196.

Clark, L.C. and Gollan, F. 1966. Survival of mammals breathing organic liquids equilibrated with oxygen at atmospheric pressure. *Science* 152(3730): 1755–1756.

Coester, C.J., Langer, K., van Briesen, H., and Kreuter, J. 2000. Gelatin nanoparticles by two step desolvation, a new preparation method, surface modifications and cell uptake. *J. Microencapsul.* 17: 187–193.

Cowdall, J., Davies, J., Roberts, M., Carlsson, A., Solaro, R., Chiellini, E., Chiellini, F. et al. 1999a. Microparticles based on hybrid polymeric materials for controlled release of biologically active molecules. A process for preparing the same and their uses for in vivo and in vitro therapy, prophylaxis and diagnostics. *PCT Int. Appl.* WO9902131.

Cowdall, J., Davies, J., Roberts, M., Carlsson, A., Solaro, R., Chiellini, E., Chiellini, F. et al. 1999b. Microparticles for controlled delivery of biologically active molecules. *PCT Int. Appl.* WO9902135.

Crowder, T.M., Rosati, J.A., Schroeter, J.D., Hickey, A.J., and Martonen, T.B. 2002. Fundamental effects of particle morphology on lung delivery: Predictions of stokes' law and the particular relevance to dry powder inhaler formulation and development. *Pharm. Res.* 19(3): 239–245.

De Rosa, G., Larobina, D., La Rotonda, M., Musto, P., Quaglia, F., and Ungaro, F. 2005. How cyclodextrin incorporation affects the properties of protein-loaded PLGA-based microspheres: The case of insulin/hydroxypropyl-beta-cyclodextrin system. *J. Control. Release* 102(1): 71–83.

Desgouilles, S., Vauthier, C., Bazile, D., Vacus, J., Grossiord, J.L., Veillard, M., and Couvreur, P. 2003. The design of nanoparticles obtained by solvent evaporation: A comprehensive study. *Langmuir* 19: 9504–9510.

Dessy, A., Kubowicz, S., Alderighi, M., Bartoli, C., Piras, A.M., Schmid, R., and Chiellini, F. 2011a. Dead Sea Minerals loaded polymeric nanoparticles. *Colloids Surf. B Biointerfaces* 87(2): 236–242.

Dessy, A., Piras, A.M., Schirò, G., Levantino, M., Cupane, A., and Chiellini, F. 2011b. Hemoglobin loaded polymeric nanoparticles: Preparation and characterizations. *Eur. J. Pharm. Sci.* 43: 57–64.

Determan, M.D., Cox, J.P., Seifert, S., Thiyagarajan, P., and Mallapragada, S.K. 2005. Synthesis and characterization of temperature and pH-responsive pentablock copolymers. *Polymer* 46: 6933–6946.

Dinucci, D., Bartoli, C., Piras, A.M., and Chiellini, F. 2007. Intracellular fate investigation of bioeliminable polymers and relative nanoformulates by confocal laser scanning microscopy. *7th International Symposium on Frontiers in Biomedical Polymers*. June 24–27. Ghent, Belgium.

Dong, Y. and Feng, S.S. 2004. Methoxy poly(ethylene glycol)-poly(lactide) (MPEG-PLA) nanoparticles for controlled delivery of anticancer drugs. *Biomaterials* 25: 2843–2849.

Douglas, K.L. and Tabrizian, M. 2005. Effect of experimental parameters on the formation of alginate-chitosan nanoparticles and evaluation of their potential application as DNA carrier. *J. Biomater. Sci. Polym. Ed.* 16(1): 43–56.

Ehrlich, P. 1906. In: *Collected Studies on Immunity*, John Wiley, New York, p. 442.

Ellington, A.D. and Szostak, J.W. 1990. In vitro selection of RNA molecules that bind specific ligands. *Nature* 346: 818–822.

Elversson, J. and Millqvist-Fureby, A. 2005. Aqueous two-phase systems as a formulation concept for spray-dried protein. *Int. J. Pharm.* 294(1–2): 73–87.

Errico, C., Bartoli, C., Chiellini, F., and Chiellini, E. 2009a. Poly(hydroxyalkanoates)-based polymeric nanoparticles for drug delivery. *J. Biomed. Biotechnol.*: 1–10.

Errico, C., Gazzarri, M., and Chiellini, F. 2009b. A novel method for the preparation of retinoic acid loaded nanoparticles. *Int. J. Mol. Sci.* 10: 2336–2347.

Farokhzad, O.C. 2004. Nanoparticle–aptamer bioconjugates: A new approach for targeting prostate cancer cells. *Cancer Res.* 64: 7668–7672.

Farokhzad, O.C. and Langer, R. 2009. Impact of nanotechnology on drug delivery. *ACS Nano* 3(1): 16–20.

Finkelstein, A., McClean, D., KarKaname Takizawa, S., Varghese, K., Baek, N., Park, K. et al. 2003. Local drug delivery via a coronary stent with programmable. *Release Pharmacokinetics Circ.* 107: 777.

Fonseca, C., Simoes, S., and Gaspar, R. 2002. Paclitaxel-loaded PLGA nanoparticles: Preparation, physicochemical characterization and in vitro antitumoral activity. *J. Control. Release* 83: 273–286.

Francis, J.W., Bastia, E., Matthews, C.C., Parks, D.A., Schwarzschild, M.A., Brown, R.H., and Fishman, P.S. 2004. Tetanus toxin fragment C as a vector to enhance the delivery of proteins to the CNS. *Brain Res.* 1011: 7–13.

Furusawa, K., Terao, K., Nagasawa, N., Yoshii, F., Kubota, K., and Dobashi, T. 2004. Nanometer-sized gelatin particles prepared by means of gamma-ray irradiation. *Colloid Polym. Sci.* 283(2): 229–233.

Goldberg, M., Langer, R., and Jiaj, X. 2007. Nanostructured materials for applications in drug delivery and tissue engineering. *Biomater. Sci. Polym. Ed.* 18(3): 241–268.

Goletz, S.P., Christensen, A., Kristensen, P., Blohm, D., Tomlinson, I., Winter, G., and Karsten, U. 2002. Selection of large diversities of antiidiotypic antibody fragments by phage display. *J. Mol. Biol.* 315(5): 1087–1097.

Gupta, K., Ganguli, M., Pasha, S., and Maiti, S. 2006. Nanoparticle formation from poly(acrylic acid) and oppositely charged peptides. *Biophys. Chem.* 119: 303–306.

Gupta, A.K., Gupta, M., Yarwood, S.J., and Curtis, A.S. 2004. Effect of cellular uptake of gelatin nanoparticles on adhesion, morphology and cytoskeleton organisation of human fibroblasts. *J. Control. Release* 95(2): 197–207.

Hoeffel, C., Mulé, S., Romaniuk, B., Ladam-Marcus, V., Bouché, O., and Marcus, C. 2009. Advances in radiological imaging of gastrointestinal tumors. *Crit. Rev. Oncol. Hematol.* 69(2): 153–167.

Hoffman, A.S. 2008. The origins and evolution of "controlled" drug delivery systems. *J. Control. Release* 132: 153–163.

Hruby′, M., Konàk, C., and Ulbrich, K. 2005. Polymeric micellar pH-sensitive drug delivery system for doxorubicin. *J. Control. Release* 103: 137–148.

Hsiue, G.H., Wang, C.H., Lo, C.L., Wang, C.H., Li, J.P., and Yang, J.L. 2006. Environmental-sensitive micelles based on poly(2-ethyl-2-oxazoline)-b-poly(l-lactide) diblock copolymer for application in drug delivery. *Int. J. Pharm.* 317: 69–75.

Hu, T., Manjula, B.N., Li, D., Brenowitz, M., and Acharya, S.A. 2007. Influence of intramolecular cross-links on the molecular, structural and functional properties of PEGylated haemoglobin. *Biochem. J.* 402(1): 143–151.

Jeong, Y.I., Cho, C.S., Kim, S.H., Ko, K.S., Kim, S.I., Shim, Y.H. et al. 2001. Preparation of poly (DL-lactide-coglycolide) nanoparticles without surfactant. *J. Appl. Polym. Sci.* 80: 2228–2236.

Johnson, O.L., Cleland, J.L., Lee, H.J., Charnis, M., Duenas, E., Jaworowicz, W. et al. 1996. A month-long effect from a single injection of microencapsulated human growth hormone. *Nat. Med.* 2: 795–799.

Kashyap, N., Modi, S., Jain, J.P., Bala, I., Hariharan, S., Bharadwaj, R. et al. 2004. Polymers for advanced drug delivery. *CRIPS* 5: 7–12.

Khaw, B., Jose, D.S., and Wiiliam, H.C. 2007. Cytoskeletal-antigen specific immunoliposome targeted in vivo preservation of myocardial viability. *J. Control. Release* 120: 35–40.

Kim, D.H. and Martin, D.C. 2006. Sustained release of dexamethasone from hydrophilic matrices using PLGA nanoparticles for neural drug delivery. *Biomaterials* 27: 3031–3037.

Kubiak, C., Manil, L., and Couvreur, P. 1988. Sorbtive properties of antibodies onto cyanoacrylic nanoparticles. *Int. J. Pharm.* 41: 181–187.

Lee, L.S., Conover, C., Shi, C., Whitlow, M., and Filpula, D. 1999. Prolonged circulating lives of single-chain Fv proteins conjugated with polyethylene glycon: A comparison of conjugating chemistries and compounds. *Bioconjug. Chem.* 10(6): 973–981.

Leuenberger, H. 2002. Spray freeze-drying—The process of choice for low water soluble drugs? *J. Nanopart. Res.* 4: 111–119.

Li, D., Manjula, B.N., and Acharya, A.S. 2006. Extension arm facilitated PEGylation of hemoglobin: Correlation of the properties with the extent of PEGylation. *Protein J.* 25(4): 263–274.

Li, X., Anton, X., Arpagaus, C., Belleteix, F., and Vandamme, T.F. 2010. Nanoparticles by spray drying using innovative new technology: The Büchi nano spray dryer B-90. *J. Control. Release* 147: 304–310.

Liu, X.M., Yang, Y.Y., and Leong, K.W. 2003. Thermally responsive polymeric micellar nanoparticles self-assembled from cholesteryl end-capped random poly(N-isopropylacrylamide-co-N,N-imethylacryl-amide): Synthesis, temperature-sensitivity, and morphologies. *J. Colloid. Interface Sci.* 266: 295–293.

Lupold, S.E. 2002. Identification and characterization of nuclease-stabilized RNA molecules that bind human prostate cancer cells via the prostate-specific membrane antigen. *Cancer Res.* 62: 4029–4033.

Maeda, H., Bharate, G.Y., and Daruwalla, J. 2009. Polymeric drugs for efficient tumor-targeted drug delivery based on EPR-effect. *Eur. J. Pharm. Biopharm.* 71(3): 409–419.

Mandal, S.C. and Mandal, M. 2010. Current status and future prospects of new drug delivery system. *Pharma. Times* 42(4): 13–16.

McBain, S.C., Yiu, H.P., and Dobson, J. 2008. Magnetic nanoparticles for drug and gene delivery. *Int. J. Nanomed.* 3: 169–180.

Meziani, M.J., Pathak, P., Desai, T., and Sun, Y.P. 2006. Supercritical fluid processing of nanoscale particles from biodegradable and biocompatible polymers. *Ind. Eng. Chem. Res.* 45(10): 3420–3424.

Min, H.K., Park, K., Kim, Y.S. et al. 2008. Hydrophobically modified glycol chitosan nanoparticles-encapsulated camptothecin enhance the drug stability and tumor targeting in cancer therapy. *J. Control. Release* 127: 208–218.

Mishima, K. 2008. Biodegradable particle formation for drug and gene delivery using supercritical fluid and dense gas. *Adv. Drug Deliv. Rev.* 60: 411–432.

Mu, L., Teo, M.M., Ning, H.Z., Tan, C.S., and Feng, S.S. 2005. Novel powder formulations for controlled delivery of poorly soluble anticancer drug: Application and investigation of TPGS and PEG in spray-dried particulate system. *J. Control. Release* 103: 565–575.

Müller, S. 2009. Magnetic fluid hyperthermia therapy for malignant brain tumours—An ethical discussion. *Nanomedicine* 5(4): 387–393.

Muzzarelli, R.A.A. and Muzzarelli, C. 2005. Chitosan chemistry: Relevance to the biomedical sciences. *Adv. Polym. Sci.* 186: 151–209.

Ness, P.M. and Cushing, M.M. 2007. Oxygen therapeutics pursuit of an alternative to the donor red blood cell. *Arch. Pathol. Lab. Med.* 131(5): 734–741.

O'Donnell, P. and McGinity, J. 1997. Preparation of microspheres by the solvent evaporation technique. *Adv. Drug Deliv. Rev.* 28: 25–42.

Palmer, T.N., Caldercourt, M.A., and Kingaby, R.O. 1984. Liposome drug delivery in chronic ischemia. *Biochem. Soc. Trans.* 12: 344–345.

Pan, B., Cui, D., Sheng, Y., Ozkan, C., Gao, F., He, R., Li, Q., Xu, P., and Huang, T. 2007. Dendrimer-modified magnetic nanoparticles enhance efficiency of gene delivery system. *Cancer Res.* 67: 8156–8163.

Panyam, J. and Labhasetwar, V. 2003. Biodegradable nanoparticles for drug and gene delivery to cells and tissue. *Adv. Drug Deliv. Rev.* 55: 329–347.

Panyam, J., Williams, D., Dash, A., Leslie-Pelecky, D., and Labhasetwar, V. 2004. Solid-state solubility influences encapsulation and release of hydrophobic drugs from PLGA/PLA nanoparticles. *J. Pharm. Sci.* 93(7): 1804–1814.

Pape, A. 2007. Alternatives to allogeneic blood transfusions. *Best Pract. Res. Clin. Anaesthesiol.* 21(2): 221–239.

Park, T.G., Lu, W., and Crotts, G. 1995. Importance of in vitro experimental conditions on protein release kinetics, stability and polymer degradation in protein encapsulated poly(D,L–lactic acid–*co*–glycolic acid) microspheres. *J. Control. Release* 33: 211–222.

Park, J.H., Ye, M., and Park, K. 2005. Biodegradable Polymers for microencapsulation of drugs. *Molecules* 10: 146–161.

Parveen, S., Misra, R., and Sahoo, S.K. 2011. Nanoparticles: A boon to drug delivery, therapeutics, diagnostics and imaging. *Nanomedicine*: 1–20.

Pawar, R., Ben-Ari, A., and Domb, A.J. 2004. Protein and peptide parental controlled delivery. *Expert Opin. Biol. Ther.* 4(8): 1203–1212.

Peer, D., Kar, J.M., Hong, S.P., Farokhzad, O.C., Margalit, M., and Langer, R. 2007. Nanocarriers as an emerging platform for cancer therapy. *Nat. Nanotechnol.* 2: 751–760.

Piras, A.M., Chiellini, F., Fiumi, C., Bartoli, C., and Chiellini, E. 2007. A new biocompatible nanoparticle delivery system for targeted release of fibrinolytic drugs. *Int. J. Pharm.* 357(1–2): 260–271.

Piras, A.M., Nikkola, L., Chiellini, F., Ashammakhi, N., and Chiellini, E. 2006. Bioerodible electrospun fibers for the controlled release of conventional and protein drugs. *20th European Conference on Biomaterials.* September 27–October 1. Nantes, France.

Rao, J.P. and Geckeler, K.E. 2011. Polymer nanoparticles: Preparation techniques and size-control parameters. *Prog. Polym. Sci.* 36: 887–913.

Riess, J.G. 2001. Oxygen carriers ('blood substitutes') Raison d'etre, chemistry, and some physiology. *Chem. Rev.* 101(9): 2797–2894.

Roa, W.H., Azarmi, S., Al-Hallak, M.H.D., Finlay, W.H., Magliocco, A.M., and Löbenberg, R. 2011. Inhalable nanoparticles, a non-invasive approach to treat lung cancer in a mouse model. *J. Control. Release* 150: 49–55.

Rosca, I.D., Watari, F., and Uo, M. 2004. Microparticle formation and its mechanism in single and double emulsion solvent evaporation. *J. Control. Release* 99(2): 271–280.

Rossler, B., Kreuter, J., and Scherer, D. 1995. Collagen microparticles: Preparation and properties. *J. Microencapsul.* 12: 49–57.

Scriven, L.E. and Sternling, C.V. 1960. The marangoni effects. *Nature* 187(4733): 186–188.

Seki, J., Sonoke, S., Saheki, A., Fukui, H., Sasaki, H., and Mayumi, T. 2004. A nanometer lipid emulsion, lipid nano-sphere (LNS(R)), as a parenteral drug carrier for passive drug targeting. *Int. J. Pharm.* 273(1–2): 75–83.

Sham, J.O.-H., Zhang, Y., Finlay, W.H., Roa, W.H., and Lobenberg, R. 2004. Formulation and characterization of spray-dried powders containing nanoparticles for aerosol delivery to the lung. *Int. J. Pharm.* 269(2): 457–467.

Shangguan, D., Li, Y., Tang, Z., Cao, Z.C., Chen, H.W., Mallikaratchy, P., Sefah, K., Yang, C.J., and Tan, W. 2006. Aptamers evolved from live cells as effective molecular probes for cancer study. *Proc. Natl. Acad. Sci. USA* 103: 11838–11843.

Sharma, A., Arora, A., Grewal, P., Dhillon, V.K., and Kumar, V. 2011. Recent innovations in delivery of artificial blood substitute: A review. *Int. J. App. Pharm.* 3(2): 1–5.

Si-Feng, S. 2004. Nanoparticles of biodegradable polymers for new-concept chemotherapy. *Expert Rev. Med. Devices* 1(1): 115–125.

Solaro, R., Chiellini, F., and Battisti, A. 2010. Targeted delivery of protein drugs by nanocarriers. *Materials* 3: 1928–1980.

Stacy, K.M. 2005. Therapeutic MAbs: Saving lives and making millions. *The Scientist*: 17–19.

Sun, Y.P., Meziani, M.J., Pathak, P., and Qu, L. 2005. Polymeric nanoparticles from rapid expansion of super-critical fluid solution. *Chem. Eur. J.* 11: 1366–1373.

Suri, S.S., Fenniri, H., and Singh, B. 2007. Nanotechnology-based drug delivery systems. *J. Occup. Med. Toxicol.* 2:16.

Takada, S., Uda, Y., Toguchi, H., and Ogawa, Y. 1995. Application of a spray drying technique in the production of TRH-containing injectable sustained-release microparticles of biodegradable polymers. *PDA J. Pharm. Sci. Technol.* 49(4): 180–184.

Tang, Z.W., Shangguan, D., Wang, K.M., Shi, H., Sefah, K, Mallikratchy, P., Chen, H.W., Li, Y., and Tan, W.H. 2007. Selection of aptamers for molecular recognition and characterization of cancer cells. *Anal. Chem.* 79: 4900–4907.

Timko, B.P., Whitehead, K., Gao, W., Kohane, D.S., Farokhzad, O., Anderson, D., and Langer, R. 2011. Advances in drug delivery. *Ann. Rev. Mater. Res.* 41: 1–20.

Turek, C. and Gold, L. 1990. Systematic evolution of ligands by exponential enrichment—RNA ligands to bacteriophage-T4 DNA-polymerase. *Science* 249: 505–510.

Van Dijk, M.A. and van de Winkel, J.G. 2001. Human antibodies as next generation therapeutics. *Curr. Opin. Chem. Biol.* 5(4): 368–374.

Vandervoort, J. and Ludwig, A. 2007. Ocular drug delivery: nanomedicine applications. *Nanomedicine* 2(11): 21.

Vasir, J.K., Reddy, M.K., and Labhasetwar, V.D. 2005. Nanosystems in drug targeting: Opportunities and challenges. *Curr. Nanosci.* 1: 47–64.

Vauthier, C. and Bouchemal, K. 2008. Methods for the preparation and manufacture of polymeric nanoparticles. *Pharm. Res.* 26(5): 1025–1058.

Veronese, F.M. and Caliceti, P. 2002. Drug delivery systems. In: R. Barbucci, ed., *Integrated Biomaterials Science*, pp. 833–873. Academic/Plenum Publishers, New York.

Veronese, F.M. and Pasut, G. 2005. PEGylation, successful approach to drug delivery. *Drug Discov. Today* 10(21): 1451–1458.

Wagner, V., Dullaart, A., Bock, A.K., and Zweck, A. 2006. The emerging nanomedicine landscape. *Nat. Biotechnol.* 24(10): 1211–1217.

Wang, S., Su, H., Chen, K., Armijo, A., Lin, W., Wang, Y. et al. 2009. A supramolecular approach for preparation of size-controlled nanoparticles. *Angew. Chem.* 121(24): 4408–4412.

Wei, H., Zhang, X.Z., Zhou, Y., Cheng, S.X., and Zhuo, R.X. 2006. Self-assembled thermoresponsive micelles of poly(N-isopropylacrylamide-b-methyl methacrylate). *Biomaterials* 27: 2028–2034.

Williams, A.S., Camilleri, J.P., Goodfellow, R.M., and Williams, B.D. 1996. A single intra-articular injection of liposomally conjugated methotrex- ate suppresses joint inflammation in rat antigen-induced arthritis. *Br. J. Rheumatol.* 35: 719–724.

Yeo, Y., Chen, A.U., Basaran, O.A., and Park, K. 2004. Solvent exchange method: A novel microencapsulation technique using dual microdispensers. *Pharm. Res.* 21(8): 1419–1427.

Zambaux, M.F., Bonneaux, F., Gref, R., Dellacherie, E., and Vigneron, C. 1999. Preparation and characterization of protein C-loaded PLA nanoparticles. *J. Control. Release* 60(2–3): 179–188.

Zhang, L. 2007. Co-delivery of hydrophobic and hydrophilic drugs from nanoparticle–aptamer bioconjugates. *Chem. Med. Chem.* 2: 1268–1271.

Zhang, T., Brown, J., Oakley, R.J., and Faul, C.F.J. 2009. Towards functional nanostructures: Ionic self-assembly of polyoxometalates and surfactants. *Curr. Opin. Colloid Interface Sci.* 14(2): 62–70.

Zhang, Z. and Feng, S.S. 2006. In vitro investigation on poly(lactide)-Tween 80 copolymer nanoparticles fabricated by dialysis method for chemotherapy. *Biomacromolecules* 7: 1139–1146.

18 Polymeric Materials Obtained through Biocatalysis

Florin Dan Irimie, Csaba Paizs, and Monica Ioana Tosa

CONTENTS

18.1 INTRODUCTION

Modern life would be inconceivable without synthetic or artificial polymers. By catalytically assisted processes, polymers of a large diversity are produced in impressive amounts; their fabrication consumes both energy and material resources. Therefore, any measure that leads to lowering the cost and increasing the quality of the polymer product, or reduces environmental impact and the costs related to it, can make the difference between the producing companies. During the last two decades we assist to an explosion of papers that deal with the topic of enzyme utilization in polymerization reactions.

The use of enzymes is legitimated by the advantages they possess. Enzymes are proteins that catalytically assist in an integrated manner, all the biochemical reactions in a living organism (metabolic reactions). Being produced and constantly improved by nature over millions of years of evolution, they possess in relation to natural substrates performances that are unattainable in the world of chemocatalysts.

Enzymes have all the properties of catalysts: they assist the transformation of substrate(s) by providing alternative mechanistic pathways with more reduced activation energies; their quantities are not modified during reactions, and therefore, they can participate in several catalytic cycles.

Moreover, as compared to chemocatalysts, enzymes have in dedicated instances the advantages of outstanding selectivity and activity, of mild reaction conditions regarding temperature, pressure, ionic strength and, last but not least, an almost perfect compatibility with the environment.

A high enzymatic activity can be achieved when the protein provides alternative reaction routes; an optimum combination of proximity effect, general and specific acid-base catalysis with covalent and ionic catalysis will lower the activation energy of the reaction. Thus more molecules would have access to transformation in the same period; in other words the transformation rate would increase.

The high, but not absolute, selectivity enables enzymes to assist transformation of non-natural substrates. They can be used in pure state or as enzyme preparations with different degrees of purity. To increase their usability enzymes can be immobilized on various supports, entrapped, or used in reticulated form.

Enzymes can assist polymer production: (i) in vivo, when biopolymers are obtained by transforming a precursor or an environmental component, in a metabolic process, with the participation of several different enzymes; (ii) in vitro, when a polymerization (polycondensation) reaction is assisted by a single enzyme, present in a more or less purified state, and (iii) enzymatic modification of already existing polymers.

At industrial scale all these procedures are applicable. Industrial in vivo obtaining of polymers is achieved by fermentative processes conducted with intact cells. Application of such a process requires a complex metabolic approach that exceeds the topic we discuss here.

A modified definition of that found in literature [1] can be enzymatic polymerization is an in vitro polymerization process of a non-natural substrate, catalyzed by a more or less purified enzymatic preparation, *via* a nonmetabolic process.

18.2 ENZYMATIC SYNTHESIS OF OLIGO- AND POLYSACCHARIDES

There are serious difficulties in providing stereo- and regioselectivity to successive glycosylation for polymer synthesis by manipulating the leaving groups and the activators, or the selective protection and de-protection. These difficulties, together with the challenging aspects of chemical achievement of a sugar oligomer or polymer, involve a great work expense with mostly modest outcomes.

Therefore, the need for improvement focused the attention of researchers on enzymatic assistance of polymerization process. The target enzymes studied for polymerization (polyglycosylation) process were glycosidases (glycosid hydrolases) and glycosyl transferases. The so-called glycosynthases, more adapted for in vitro synthesis of multimeric structures, have been created by genetic manipulations during these investigations.

18.2.1 Glycosyltransferases-Assisted Synthesis of Oligosaccharides

The idea of using enzymes for the synthesis of polysaccharides led to the question: Which are the most suitable enzymes that could assist the formation of certain polysaccharides, in conditions of regio- and stereoselectivity, with acceptable rate and molecular mass dispersion?

Seemingly, the most useful type of enzymes for the synthesis of oligo- and polysaccharides are *glycosyltransferases*. In biosynthetic pathways they can ensure the gradual generation of glycosidic bonds between a growing, accepting oligo/polysaccharide chain and a carbohydrate component originating from an activated donor. In vivo, one glycosyltransferase can transfer processively the sugar to form homopolysaccharides, or in combination with other glycosyltransferases, to form heteropolysaccharides. In vivo, glycosyltransferases are responsible for the synthesis of most cell glycoconjugates. An exhaustive classification of glycosyltransferases can be found in CAZy (Carbohydrate-Active enZymes) database. Recent syntheses, of high quality, on in vivo utilization of transferases, including glycosyltransferases, to obtain oligo- and polysaccharides are those of Visser [2] and van de Vilst [3].

Glycosyltransferases are type II membrane-bound glycoproteins consisting of a short *N*-terminal cytoplasmic domain, a transmembrane domain with a "stem" or "neck" region which is easily cleavable by proteases, and the *C*-terminal catalytic domain [4].

The canonical model at hand is cellulose synthase (UDP forming), E.C. 2.4.1.12 [5] (Scheme 18.1) [1] responsible for cellulose synthesis. A similar enzyme is cellulose synthase (GDP forming), E.C. 2.4.1.29 [6].

Glycogen and starch are similarly produced by glycogen (starch) synthase (E.C. 2.4.1.11.)

Attempts to use these enzymes at a reasonable scale to produce artificial/synthetic cellulose or amylose have been abandoned due to the unavailability of a stable enzyme and the prohibitive price of the substrate.

Glycosyltransferases are a subclass of transferases (E.C. 2.4.x.x.) which transfer carbohydrates from their conjugates with nucleoside monophosphates (CMP NeuAc), diphosphates (UDP-Glc, UDP-Man, UDP-GlcNAc), lipids (dolichol phosphate oligosaccharides), or unsubstituted phosphates. Glycosyltransferases that use sugar nucleotides as carbohydrate donors are known as Leloir glycosyltransferases, whereas all the others are non-Leloir glycosyltransferases. Most of the glycosyltransferases responsible for the biosynthesis of mammalian glycoproteins and glycolipids are Leloir glycosyltransferases. The products of glycosyltransferase reactions are conjugates of the transferred carbohydrates with (oligo-, poly-)saccharides, lipids, proteins, or nucleic acids.

The glycosyltransferases transfer the sugar with either retention or inversion of configuration at the anomeric center (Scheme 18.2).

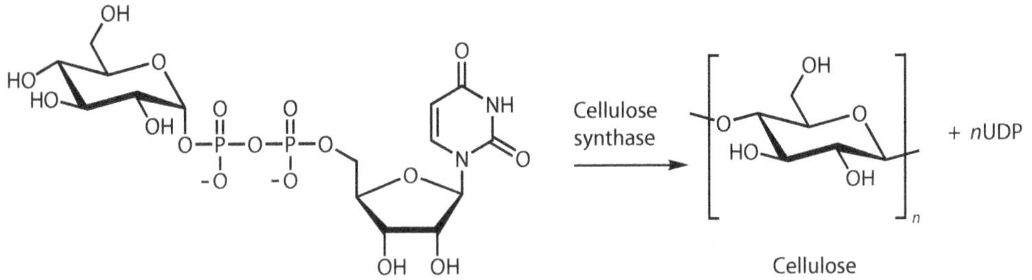

SCHEME 18.1   Synthesis of cellulose chain assisted by cellulose synthase. The monomer is a very expensive UDP-α-D-glucose.

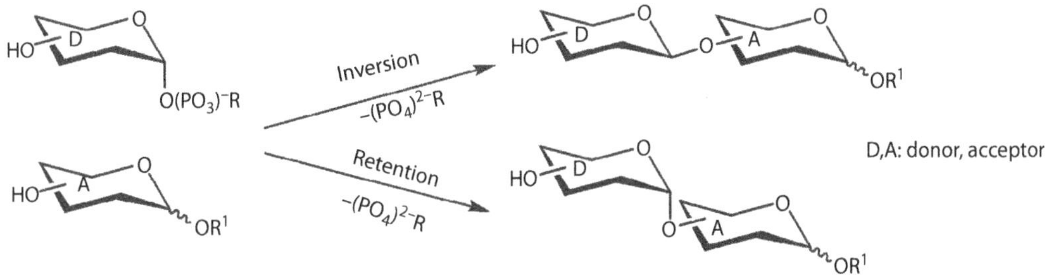

SCHEME 18.2   Formation of glycosidic bond between two adjacent sugars can be done with retention or inversion of anomeric carbon of donor sugar.

18.2.1.1 Leloir Glycosyltransferases

The use of Leloir glycosyltransferases encountered two major drawbacks: limited enzyme availability, the high costs of donors, and, related to this, the necessity of their regeneration.

Nevertheless, for the preparation of complex and highly pure oligosaccharides, methods based on the application of glycosyltransferases are currently recognized as being the most effective [2]. The methods are extremely useful particularly for monoglycosylation of a sugar or oligosaccharide. By performing the reactions with various donors and enzymes, oligosaccharide structures of high complexity can be obtained.

For example, a crude β-1,3–N-Acetylglucosaminyltransferase (β3GnT) prepared from bovine serum by precipitation with 25%–30% saturated ammonium sulfate (Scheme 18.3) was used to obtain certain oligosaccharide sequences which are present in human milk [7]. The synthesis of these compounds includes the transfer of N-acetylglucosamine (GlcNac). These sequences containing N-acetylglucosamine are useful for investigation of their immunomodulating, probiotic, and inhibitor in human infections properties.

Another modality of attaching a galactose residue by an α(1–3) bond was the use of a bifunctional fusion enzyme, obtained by cloning, which contains both uridine-5′-diphospho-galactose 4-epimerase and galactosyltransferase (GalT). GalE allows in situ generation of the expensive UDP-Gal, from the cheaper UDP-Glc and galactose (Scheme 18.4) [8].

Many experimental investigations have been patented [9–12].

UDP formed in the transglycosylation reaction is a powerful inhibitor for transglycosidase and therefore it should be removed. By introducing in the reaction environment a calf intestinal phosphatase, an enzyme that degrades UDP, a significant increase of the conversion rate has been noticed [13].

So, if enzyme availability, including its high price, could be resolved by recombinant DNA techniques, the price contribution of nucleotidic conjugates to the product value could be reduced by using in situ regeneration systems [14].

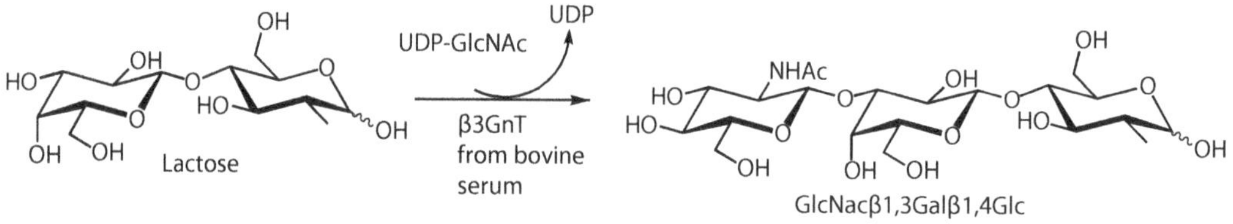

SCHEME 18.3 The transfer of GlcNAc from UDP-GlcNAc assisted by β1,3–N-acetylglucosaminyltransferase, (β3GnT).

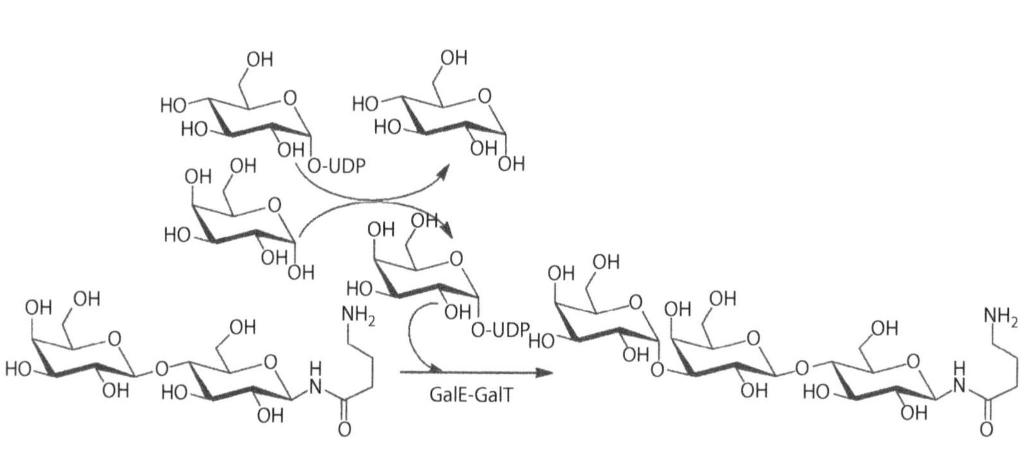

SCHEME 18.4 Use a bifunctional fusion enzyme for Gal transfer and "in situ" formation of UDP-Gal from inexpensive UDP-Glc.

Moreover, to obtain the desired structures with financially advantageous conditions, a multi enzymatic procedure should be used and also substrate engineering is often necessary. An approach to these techniques can be found in an excellent review by Homann and Seibel [15].

18.2.1.2 Non-Leloir Glycosyltransferases

Besides Leloir glycosyltransferases which ensure the biosynthesis in vivo of the oligo- and polyglucidic structures, as stated earlier, the researcher and the engineer now have at their hands glycosyltransferases functioning in vivo as catabolic, depolymerizing enzymes (phosphorylases, sucrases).

The enzyme that normally catalyzes the direct path can be used for the assistance of reverse reaction by handling the factors that influence the equilibrium: raising the substrate concentration or reducing the concentrations of products by eliminating them from the system. In this way, enzymes that are commonly depolymerizing can be used to obtain large amounts of polymers.

18.2.1.3 Phosphorylases

Among non-Leloir glycosyltransferases, α-glucan phosphorylases are most used in the attempts to obtain oligo- or glycopolymers. These enzymes that catalyze phosphorolysis at the nonreducing ends of α-1,4-glucans with retention of configuration at the anomeric carbon constitute a large family of very similar enzymes, of which glycogen phosphorylase, or simply phosphorylase, is best investigated [16]. The overall reaction catalyzed by phosphorylase is

$$(1,4\text{-}\alpha\text{-D-glucosyl})_n + \text{inorganic phosphate (Pi)} \ \Delta \ (1,4\text{-}\alpha\text{-D-glucosyl})_{n+1} + \text{glucose-1-phosphate}$$

The enzyme functions only in the presence of pyridoxal phosphate. The role of this one in ensuring the stereoselectivity of the phosphorolytic attack is still questionable [16]. The mechanism proposed by Helmreich remains likely: the breaking of the glycosidic bond and the forming of the new one take place on the same side of the plan of the oxocarbenium intermediary, the retention of configuration being achieved at C1 [17]. Scheme 18.5 illustrates the reversibility of the

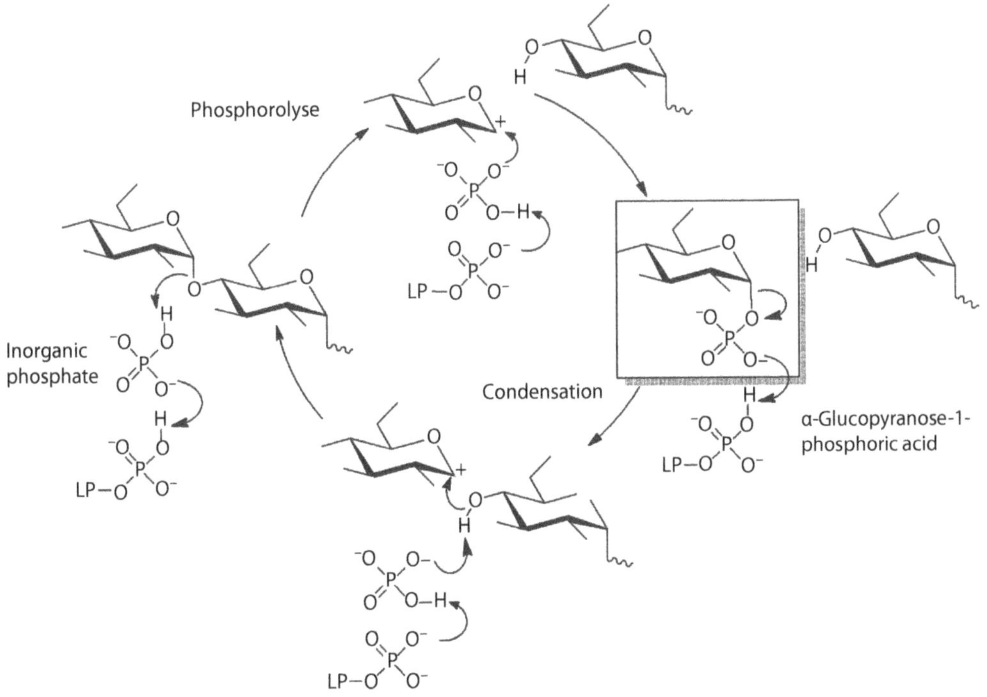

SCHEME 18.5   The phosphorylase-assisted phosphorolysis and condensation.

catalyzed reaction: it shows both the direct sense—phosphorolysis and the inverse one, of utility in polymerization—the condensation.

The problem concerning the production of synthetic amylose by condensation of glucose-1-phosphate is that the process cannot flow directly from glucose-1-phosphate and needs the presence of a pre-formed chain or primer (a poly- or oligomaltosaccharide).

The smallest primer able to trigger the polymerization was maltotriose (Weibull and Tisselius, cited by Vilst [3]).

In situ generation of glucose-1-phosphate was investigated and solved by the team of Kurichi, using an auxiliary enzymatic equipment, consisting of sucrose phosphorylase, or cellobiose phosphorylase [18–20], and also using auxiliary substrates: sucrose or cellobiose, respectively, and inorganic phosphate.

The dependence on primers of the glycogen phosphorylase-assisted polycondensation opens the way for making hybrid structures, starting from modified primers, recognized by the enzymatic equipment that performs the polycondensation.

A synthesis of the most recent results can be found in the paper of Vlist and Loos [3].

By the enzymatic polycondensation of alkylated maltose primers, structures with modified hydrophilicity and interesting properties were produced. Attaching of primers to appropriate surfaces allows the growing of polymers on these surfaces—a technique named "grafting from." The initial surface takes a "brush" appearance with hydrophilic hairs and can acquire new properties, such as biocompatibility, wettability, and corrosion resistance.

The stands initially used were silica or silicone surfaces. After an activation of the surfaces with (γ-aminopropyl)trimetoxysylane or chlorodimethylsilane, the primers were attached chemically. Using glucose-1-phosphate as monomer, the growing of primers was assisted by potato phosphorylase. The consumption of monomer was monitored by spectrometric determination of inorganic phosphate, and it was correlated with the average length of the amylose polymers. Due to the chiral nature of glucose units, the modified surfaces have a good chiral discrimination, when employed as stationary phases in the chiral chromatography.

The investigations done, since 1954 (Bailey cited in Vlist and Loos [3]), have shown that the mechanism of the reaction is of a "multichain" type, which means that after each step of monomer binding, enzyme dissociates from the elongating chain, whereas a "single-chain"-type mechanism ensures the continuous growth of the primer length without dissociation. Moreover, it has been noticed that the length of the primer influences the distribution of amylose chain dimension. Starting from four monosaccharide units, the lengths of amylose chains show a Poisson-type distribution, compatible with a "multichain"-type mechanism: the attachment of monomers is random, but the chains grow with a constant rate. When maltotriose was used as a primer, a bimodal distribution was noticed, which can be explained by a more difficult, and consequently slower attachment of the first monomers on the initial, short chain, followed by the attachment of the next monomers at a constant, 400 times higher rate [3].

The known helicoidal structure of amylose allows the formation of inclusion compounds, even during the polymerization reaction. Through phosphorylase assisted by polycondensation, a α-D-glucose 1-phosphate (Glc-1-P), initiated by maltoheptose, in the presence of polyTHF [poly(tetrahydrofuran]—a hydrophobic polymer, the polycondensation proceeded with the formation of the amylose-polyTHF inclusion complex (Scheme 18.6). By degree of polymerization and length measurements, it was shown that one unit of polyTHF is included by two amylose units, with a degree of polymerization of 75–90 for each. This inclusion complex could not be formed by mixing amylose and polyTHF having the same length [1]. The helicoidal structure wrapped around the hydrophobic polymer resembles vines of a plant growing around a rod. Therefore, this polymerization system was named "vine twining polymerization [21]." Besides polyTHF, other hydrophobic polymers have been used: poly(oxetane), poly(ε-caprolactame), and poly(γ-valerolactone). During polycondensation, the growing polymer chooses the hydrophobic polymer from a mixture of two resembling polymers. For example, if amylose synthesis is done in the presence of polyTHF and poly(oxetane), the polymer that will participate in the inclusion complex will be polyTHF and not poly(oxetane). Similarly, from a mixture of poly(ε-caprolactame) and poly(γ-valerolactone), the last one will be chosen [22].

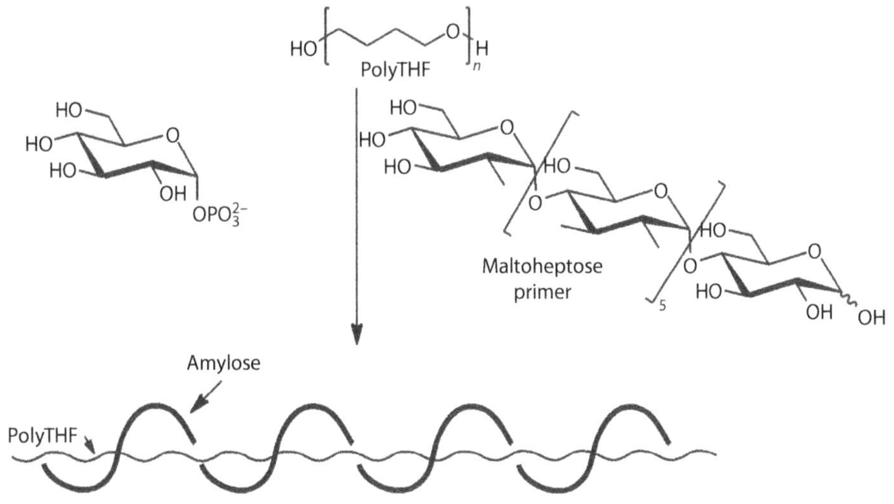

SCHEME 18.6 Poly(THF) template polycondensation of glucose-6-phosphate using maltoheptose primer.

Moreover, when the synthesis of amylose is performed in the presence of polyTHF with different molecular weights, the inclusion complex partner is selected from a specific range of degree of polymerization [23].

Besides phosphorylase, branching enzymes (E.C. 2.4.1.18) could be successfully used in vitro. The glycogen branching enzyme belongs to the transferase family and is able to transfer short, α-(1−4) linked, oligosaccharides from the nonreducing end of starch to an α-(1−6) position in intra- or intermolecular manner.

The group of Katja Loos reported the use of an enzymatic tandem reaction in which the potato phosphorylase and *Deinococcus geothermalis* glycogen branching enzyme (Dg GBE) catalyze the synthesis of branched polyglucans from glucose-1-phosphate (G-1-P) (Scheme 18.7). The resulting branched polyglucans have a high degree of branching of 11%. The reaction was primer dependent.

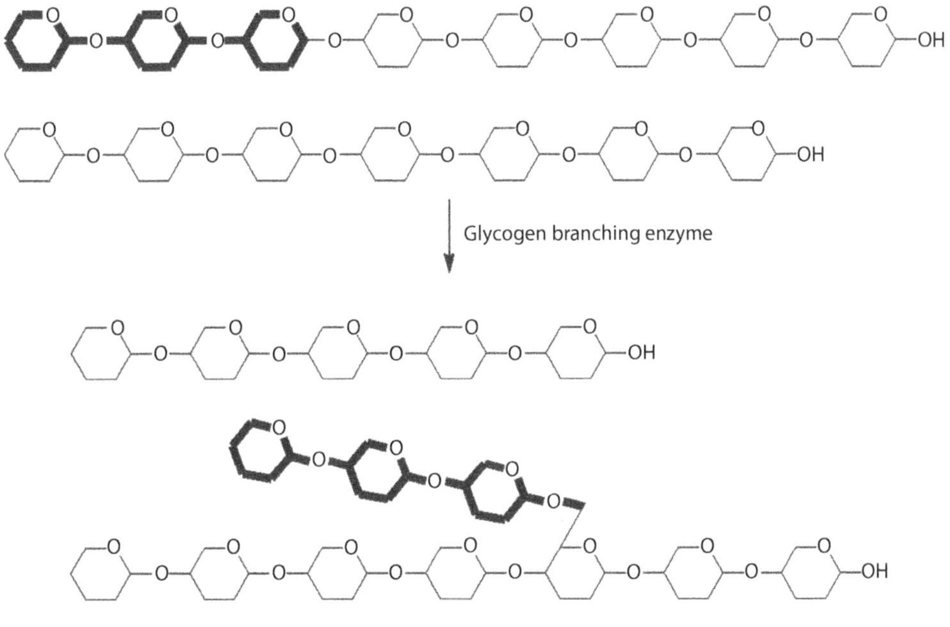

SCHEME 18.7 Using branching enzymes for introducing 1-6-α-glycosidic bonds in amylose chain.

The fact is important since it allows synthesizing by modification of the primer at the reducing side, a class of amylose-based materials, including star-shaped structures, block copolymers, like amylose block-polystyrene, PEG-block-amylose, L-glutamic acid-block-amylose, comb-type copolymers like polysiloxane-graft-amylose, and amylose-functionalized silica particles [24].

18.2.2 GLYCOSIDASES (GLYCOSIDE HYDROLASES)-ASSISTED SYNTHESIS OF OLIGO- AND POLYSACCHARIDES

In normal conditions, these enzymes hydrolytically split the glycosidic bonds of oligo- and polysaccharides, thus producing lower weight sugars. Through their characteristic degradation mechanism exercised over polysaccharides, glycosidases contribute to degradation of cellulose biomass, and participate in body defense through lysis of the bacterial peptidoglycan wall (lysozyme), to expression of pathogenicity (neuraminidase), etc. In suitable conditions, glycosidases can catalyze the opposite reaction, i.e., condensation reaction, in which water is no more a reactant, but a product, i.e., condensation reaction. These conditions can be achieved either through thermodynamic control, which forces the equilibrium toward the condensation reaction (increasing the monomer concentration, using an organic co-solvent, to reduce water activity) or by using activated donors in a kinetic control. More than that, the condensation reaction is facilitated because the solubility of the obtained polymer is lower than that of substrates and it will leave the system by precipitation.

The benefits of using glycosidases include their stability, availability, the large number of sources, with a diversity of substrate selectivity, and last but not least, the fact that glycosyl donors are much cheaper compounds as compared to those of glycosyltransferases [25].

As glycosyltransferases, some glycosidases can function with configurational retention at the anomeric carbon atom involved in the hydrolyzed/formed glycosidic bond, or with configurational inversion.

Most glycosidases used for synthetic purposes are retaining glycosidases [26]. The mechanism of this reaction, as established by Koshland in 1953, is based on the existence of two reaction centers [27] against which the anomeric carbon and oxygen atom of the glycosidic bond that will be cleaved or formed are positioned. This mechanism is confirmed today by numerous experimental data [28] (Scheme 18.8).

The two centers are represented by carboxyl group (the acid-base center) and carboxylate (the nucleophile center), respectively. The reaction undergoes with the retention of configuration and involves two steps: (1) the carboxylate attacks the anomeric carbon, which weakens the C–O bond, and produces a partial shift of the pyranose oxygen electron pair. In addition, the glycosidic oxygen comes near carboxyl group proton. The first in sequence transition state, with oxocarbenium ion-like structure, evolves to the covalent intermediate glycosyl–enzyme. The anomeric carbon of the newly formed glycosyl bond has a changed configuration as compared to the substrate, as an effect of the nucleophile attack on the opposite side of the leaving group. (2) This intermediate can be further transformed by a nucleophilic attack of oxygen from a water molecule or from a glycosidic bond of an acceptor oligosaccharide structure. Similar to the first step, the nucleophilic attack will be done on the opposite side of the leaving group (anomeric carbon–carboxylic oxygen bond). The substitution reaction is facilitated by the attack of the conjugated base (the carboxylate ion) which will take the proton attached to the nucleophile. Transition state 2 has a similar structure to the first one and it will evolve toward the reaction product. The latest one depends on the nucleophile that attacks the glycosyl–enzyme intermediate. When the nucleophile is water, a hydrolysis reaction takes place, and when the nucleophile is a sugar compound, the reaction is a transglycosylation. A relative reduction of the water content is done by using a co-solvent. Water cannot be totally removed, because a minimal amount is needed to ensure the conformational mobility of the enzyme.

Thermodynamic forcing of the direction of hydrolysis/transglycosylation reaction through water concentration is limited to the synthesis of some simple disaccharides and alkyl glycosides.

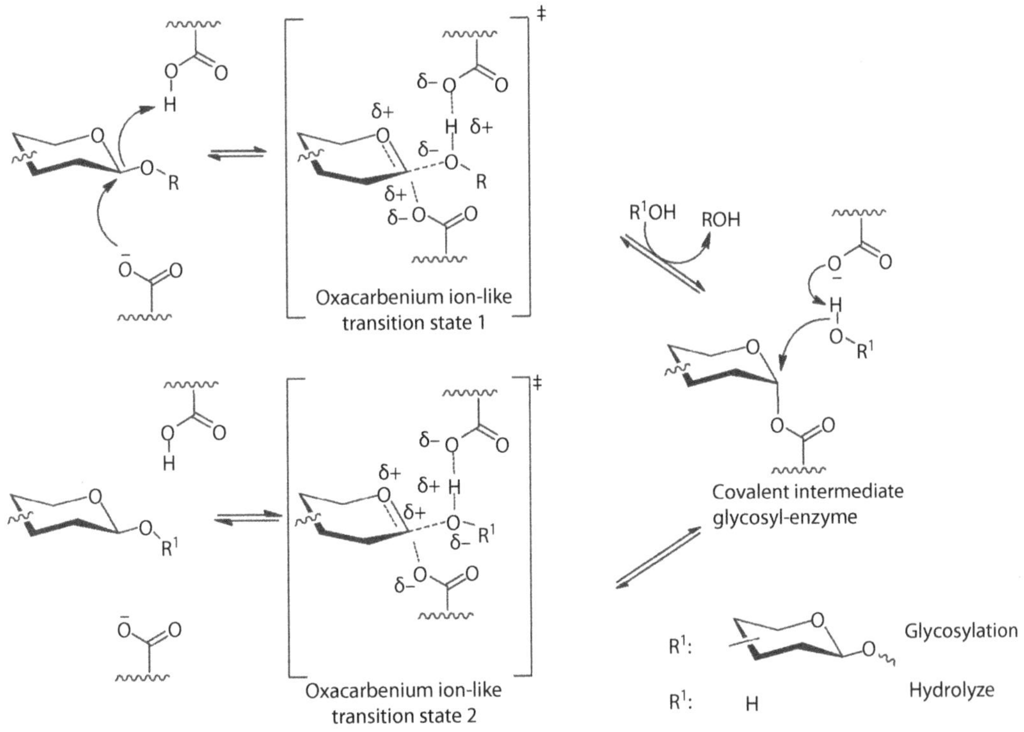

SCHEME 18.8 Biocatalytic mechanism of retaining glycosidases.

Therefore, the strategies of polysaccharide synthesis are based either on kinetic control of polymerization reaction or on the use of engineered glycosidases. In the first case, activated donor sugar (fluorosugars or nitrophenyl esters) is used and an additional thermodynamic control is provided by a reduced water concentration. In the second case, by dissolution of the nucleophilic site, the enzymatic action is reduced to the control of a single nucleophilic step (the attack of the acceptor sugar), which, as will be shown, will lead to polymers with a higher molecular mass, and obtained with a superior yield.

18.2.2.1 Polymerization of the Activated Substrates with Glycosidases

There are commonly two types of activated monomers that can be used to obtain multimeric structures: sugar fluoride–type monomer used for synthesis of cellulose and its derivatives via "polycondensation" and sugar oxazoline–type monomer used in synthesis of chitin and related polymers, all of these polymers having a 2-*N*-acetyl group in the core used as donor.

When sugar fluoride–type monomer is used, the classic example is the polycondensation of di- or trisaccharide glycosyl fluoride donor units, as substrates for self-polycondensation by known glycosidases such as cellulases, xylanases, amylases, and β-glucanases.

The use of fluorine, attached to the anomeric carbon of the di- or triglucidic monomer (cellobiose, cellotriose, or derivatives), in the same configuration as of the oxygen of the newly formed glycosidic bond, is motivated by the following: (1) cellobiose is the smallest monomer unit recognized as acceptor by cellulase, (2) the dimension of fluorine (covalent radius 0.64 Å) is very close to that of oxygen (covalent radius 0.68 Å), which enables it to accommodate in the catalytic site of the enzyme, (3) the fluoride anion is a better leaving group than oxygen because of both its electronegativity and its capability to form hydrogen bonds in the presence of suitable partners, and (4) among sugar haloids, fluoro derivatives are the most stable [1].

A particular mention must be done for synthetic cellulose. This was the first polymer obtained through enzymatic synthesis by Kobayashi [29] starting from cellobiosyl-fluoride in acetonitrile/acetate buffer,

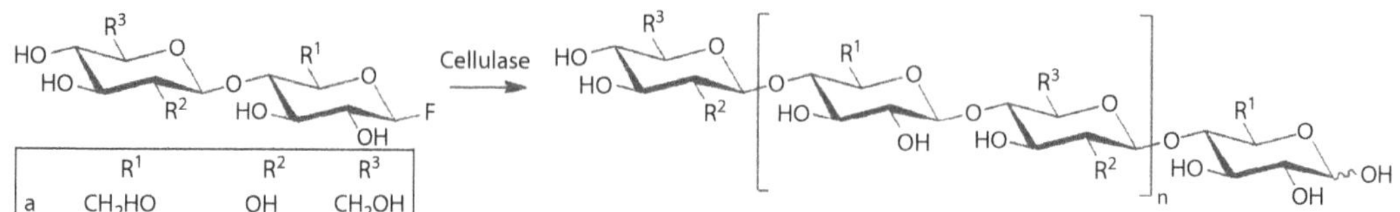

	R^1	R^2	R^3
a	CH$_2$HO	OH	CH$_2$OH
b	CH$_2$OCH$_3$	OH	CH$_2$OH
c	CH$_2$OH	NHAc	CH$_2$OH
d	H	OH	H

SCHEME 18.9   Cellulose derivatives obtained by cellulose or cellulose like enzymes from fluorodiglucides.

with a cellulase from *Trichoderma viride*. The degree of polymerization was 22, superior to the limit of 10 which is the demarcation between oligo- and polysaccharides.

Further investigations of Kobayashi team demonstrated that the structure of cellulose obtained by polymerization with this cellulase depends on enzyme source and its purity [25].

Besides cellulose, some of its unnatural derivatives have been synthesized, starting from adequately derivatized monomers (Scheme 18.9).

A related example, an alternating 6-*O*-methylated cellulose derivative has been produced (Scheme 18.9b). A hybrid cellulose–chitin polysaccharide was also synthesized in the presence of a cellulase from *T. Viride* as well as with (Scheme 18.9c) a chitinase from *Bacillus* sp. (see the following). The polymerization degree of the hybrid is only 8. Starting from a β-D-xylobiosyl fluoride as self-polycondensation monomer, a synthetic xylan has been obtained (Scheme 18.9d). The main degree of polymerization grade was 23, with a perfect stereo- and regioselectivity [30].

Besides enzymatic polycondensation of cellobiosyl fluoride and its derivatives, the polycondensation of α-maltosyl fluoride in the presence of α-amylases of different sources has also been investigated. A preference of the enzyme for the self-condensation reaction was noticed in which the fluoromaltose donor attached to the elongating chain, as compared to the normal hydrolysis reaction (Scheme 18.10). Unfortunately, the maximum polymerization degree was only 6 (12 glucosyl units) [31].

The activation of the monomer by introducing an oxazolinic structure at C1 can be a valuable method used to obtain chitin, chondroitin, chondroitin sulfate, and keratan-type carbohydrate polymers. These polymers contain a *N*-acetyl group in position C2, adjacent to the anomeric carbon involved in the formation or breakdown of glycosidic bond. Therefore, the utilization of oxazolinyl-monomer provides it a structural similarity with the intermediate of hydrolysis reaction. This intermediate possesses a strongly positivated C1 atom, induced by the topology of the two carboxyl-carboxylate residues of the catalytic site and by the susceptibility to protonation of the oxazolinic cycle. Therefore, the intermediate will be attacked from the opposite side of the C1-oxazolinic oxygen bond by the nucleophilic water. (Reaction I-B in Scheme 18.11, which shows the mechanisms of hydrolysis/polymerization reactions catalyzed by chitinase).

If the oxazolinic monomer is in excess and the water amount is reduced, the nucleophile which will open the oxazolinic ring will be the hydroxylic oxygen attached to C4 of the nonreducing end of the acceptor (Reaction II-B in Scheme 18.11). In the condition of an activation similar to that of water molecule, the nucleophilic attack takes place with the same enantiopreference. The result is the formation of a new glycosidic bond. The repetitive process is, in fact, a ring-opening polymerization.

Certain data were published concerning the polymerization of oxazoline-activated monomers with chitinase from *Bacillus* sp [32]. (Scheme 18.12).

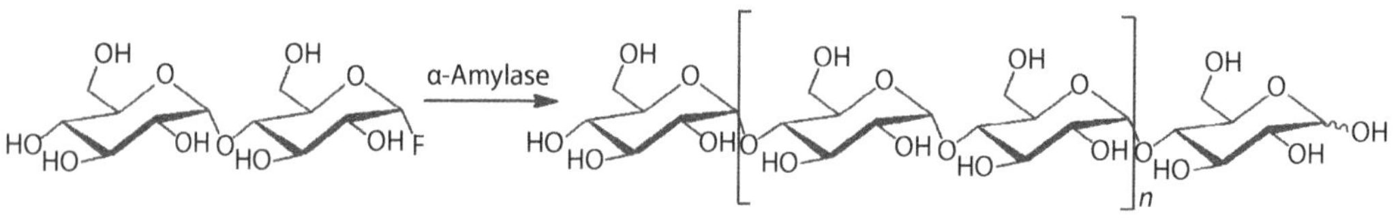

SCHEME 18.10   Polycondensation of maltose units assisted by α-amylase.

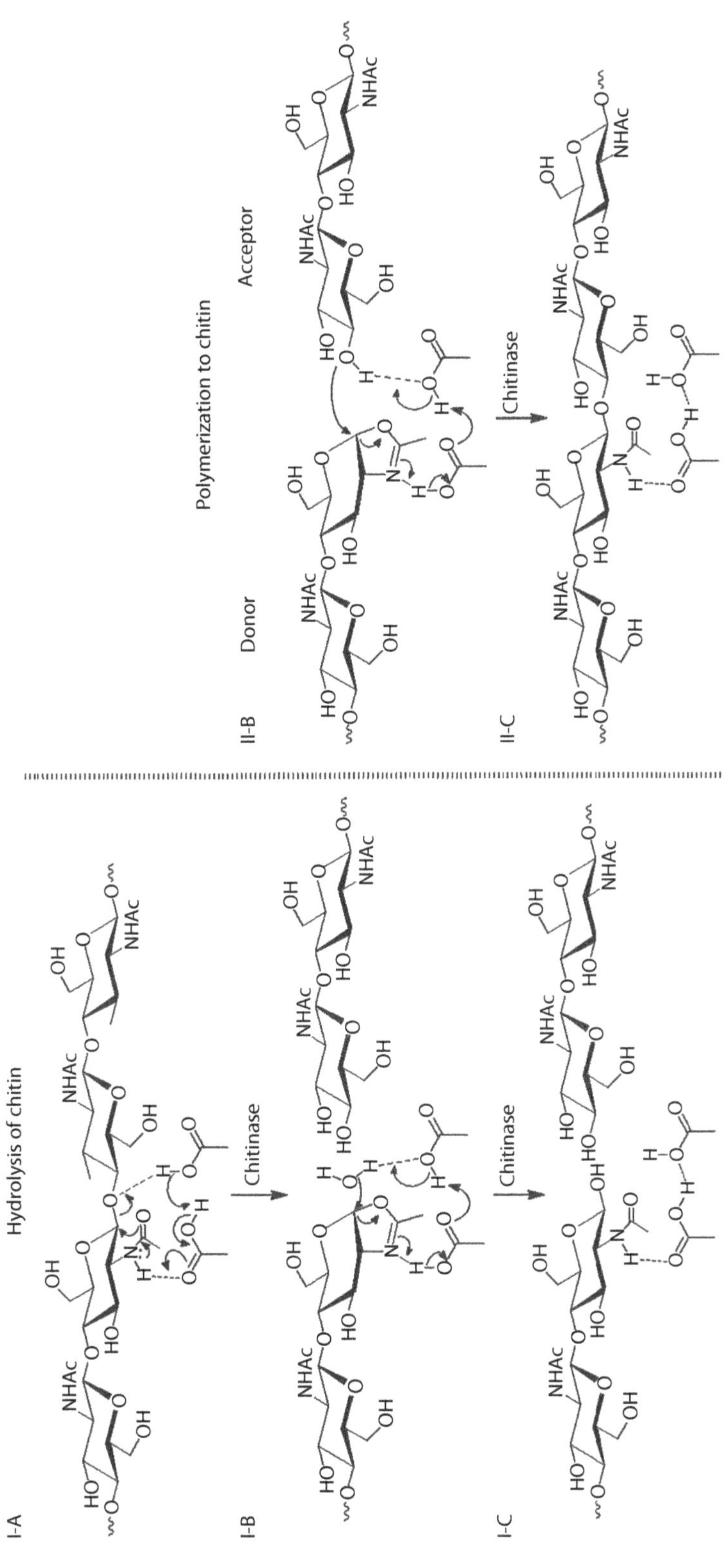

SCHEME 18.11 Mechanisms of hydrolysis/polymerization reactions catalyzed by chitinase.

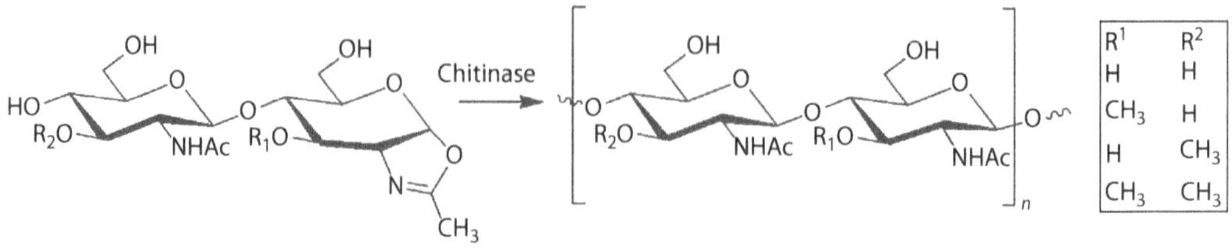

SCHEME 18.12 Polymerization of oxazoline-activated monomers with chitinase from *Bacillus* sp.

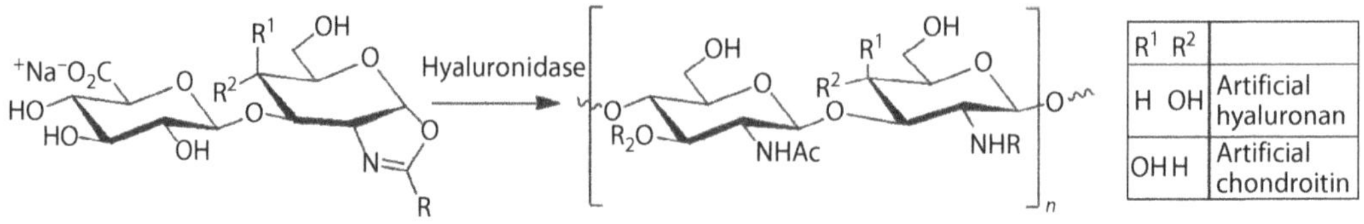

SCHEME 18.13 Hyaluronidase-assisted polymerization of oxazoline activated diglucides monomers.

Besides chitinase, hyaluronidase, an extremely promising catalyst for the synthesis of glycosaminoglycans, was another enzyme used in obtaining polysaccharides starting from oxazoline-activated monomers.

Ovine testicular hyaluronidase was found to be more efficient than bovine testicular hyaluronidase and allowed the formation with 40%–53% yield of hyaluronan-type polysaccharides with $(8–13.3) \times 10^4$ molecular weight values. The same enzyme was used for the synthesis of chondroitin-type polysaccharides, giving a 19%–50% yield with molecular weight values of $(2.1–4.6) \times 10^3$ (Scheme 18.13) [33].

18.2.2.2 Transglycosidases

These enzymes of bacterial origin are very helpful in the synthesis of polysaccharides because they can directly act on unactivated substrates, using the energy of the broken bonds to re-form new glycosidic bonds [34]. All these enzymes work by the same retaining mechanism as retaining glycosidases.

The name of transglycosidases is questionable, as far as glycosyltransferases make the same sugar transfer; however, the latest enzyme prefers the substrates that are activated by phosphorylation. Glucansucrases are a special class of transglycosidases which use sucrose as glucose donors in oligomerization reaction, releasing D-fructose.

Glucan sucrases are transglycosidases classified in the glycoside-hydrolase (GH) family 70 according to the CAZy system. These enzymes can produce homopolyglucans as dextran, containing above 50% α-1,6 linkages by dextransucrase (E.C. 2.4.1.5), mutan, containing above 50% α-1,3 linkages by mutansucrase (E.C. 2.4.1.5), reuteran, containing above 50% α-1,4 linkages by reuteransucrase (E.C. 2.4.1.5), and alternan, containing alternating α-1,3 and α-1,6 linkages by alternansucrase (E.C. 2.4.1.140).

Amylosucrase (E.C. 2.4.1.4), another transglycosidase classified in GH family 13, assists the transfer of glucose moiety to a growing chain of amylase [35]:

$$\text{Sucrose} + (1,4\text{-}\alpha\text{-}D\text{-glucosyl})_n \Delta \text{ } D\text{-fructose} + (1,4\text{-}\alpha\text{-}D\text{-glucosyl})_{n+1}$$

18.2.2.3 Glycosynthases

The difficulties encountered when glycosyltransferases and glycosidases were used for the synthesis of oligo- and polysaccharides have been partly overcome with the occurrence of glycosynthases. Glycosynthases are precise molecular instruments for making specifically linked oligosaccharides [36].

These enzymes derived from glycosidases have been engineered to catalyze the formation of glycosidic bond without catalyzing the hydrolysis reaction. This modification has been done by replacing the amino acid that bears the nucleophilic carboxylate in the catalytic site (Glu, Asp) with a less voluminous amino acid that bears a non-nucleophilic residue (Gly, Ala) (Scheme 18.14). By removing the

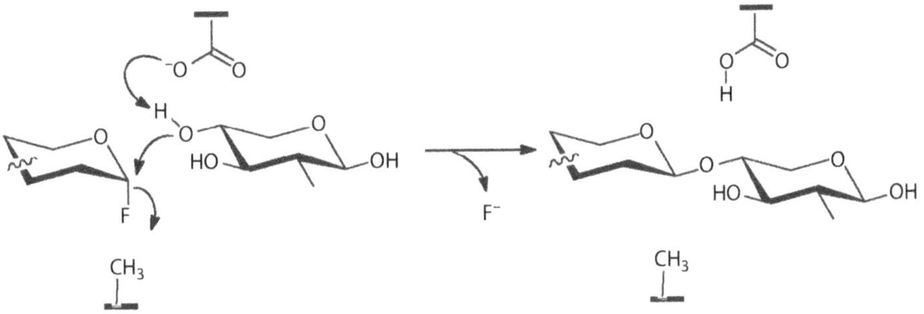

SCHEME 18.14 Condensation of two sugars assisted by a synthase.

nucleophilic center, the hydrolytic activity of the mutated enzyme decreases with five orders of magnitude, being virtually abolished. In this case the covalent intermediate glycosyl–enzyme intermediate cannot form. It can be mimed by a substrate having attached, at the anomeric carbon, a substitute with good fugacity, and also less voluminous, that could accommodate in the free space remained in the catalytic site by the replacement of a methylene carboxylate or carboxylate with a methyl or hydrogen. It is obvious that the position of this substitute must be opposite to the forming glycosidic bond. The most used substitute is fluorine, reasons for its suitability being discussed in a reference paper [37]. The intact acid-base site in its acid form will facilitate the discharge of the donor leaving group, or in its conjugated base form (carboxylate) will generate the nucleophile that forms the glycosidic bond [38]. The functioning of the mutant as a condensation enzyme (synthase) will be more efficient than using glycosidases for transglycosylation, because this reaction will no more compete with the hydrolysis reaction.

Glycosynthases are obtained from both exoglycosidases and endoglycosidases. Similar to wild-type enzyme, exoglycosynthases derived from exoglycosidases show moderate substrate- and regioselectivity, similar to those of the wild-type enzymes. They can catalyze the forming of a large variety of glycosidic bonds only in short-chain oligosaccharides (di-, tri- and tetrasaccharides) as major products. On the other hand, endoglycosynthases derived from endoglycosidases have high regioselectivity; they can catalyze the forming of a single type of glycosidic bond. Unlike exoglycosynthases, they can use longer glycosyl donors which can form polysaccharides by self-condensation or transfer to suitable acceptors [39].

The first types of used glycosynthases were derived from retaining glycosidases. These enzymes keep the acceptor-related regioselectivity of original glycosidases, but use donors with a configuration of glycosidic carbon opposed to the one of the condensation product. The mutant E358A of β-glycosidase from *Agrobacterium* sp. (Abg) is inactive in hydrolysis, but performs, with a high yield, glycosylation of acceptors bearing carbohydrate units, including self-glycosylation. It forms β-(1,4) and β-(1,3) bonds, respectively, in the case of acceptors bearing xylose units, using as donors both α-glucosyl fluoride and α-galactosyl fluoride. However, these donors cannot glycosylate acceptors bearing a hydroxyl in the position 4 axial [40].

Beside sugar polycondensation, glycosynthases are able to mediate not only the glycosylation of sugars, but also glycosylation of sugars with noncarbohydrate acceptors (aglicons). For example, Abg glycosynthase catalyzed the formation of 4-methylumbelliferil cello-trioside and cello-tetraoside, used as chromogenic substrates for cellulases. The same enzyme catalyzed the synthesis of 2,4-dinitrophenyl 2-deoxy-2-fluoro-β-cellobioside and -cellotrioside, which are mechanism-based inactivators for cellulases [41].

Some papers report glycosynthases derived from *Thermus thermophilus* β-glycosidase, which are able to catalyze with a yield higher than 90% the formation of β-1,3 glycosidic bonds [42]. Besides mutant enzymes from *Agrobacterium* with xylosynthase activity, starting from β-xylanases from *Cellulomonas fimi* and *Geobacillus stearothermophilus,* other engineered enzymes have also been created. Mutant β-glucuronidases produced by *Thermotoga maritima* [43] and *Escherichia coli* [44] are to transfer both glucuronyl and glucuronyl moieties.

The synthesis of polysaccharides with a high polymerization degree became possible through endoglycosynthases, which possess a wider binding site than that of exoglycosynthases.

During the last decade a great progress was achieved concerning the endoglycosynthases-mediated preparation of oligo- and polymers. Usually self-condensation has been used, namely glycosyl fluoride serving as donor and acceptor. Numerous mutants have been employed, such as E197 of cellulase Cel7B from *Humicola insolens*, E231 of $\beta(1\rightarrow3)$ glucanase from barley, and E134A of $(1\rightarrow3,1\rightarrow4)$-$\beta$-D-glucanase from *Bacillus licheniformis*.

Each of these mutants accepts several types of substituted or nonsubstituted glycosyl fluorides as donors.

18.2.2.3.1 β-(1→4)-Glucans (Synthetic Cellulose–Type Glycans)

The mutant E197 of cellulase Cel7B from *H. insolens* through β-$(1\rightarrow4)$ bonds can catalyze the attachment of α-cellobiosyl- or lactosyl fluoride to different donors. Moreover, the mutant accepts donors functionalized at C6', with amino, bromo, azido groups, or even glycosyl substitution (R^1 substituent in Scheme 18.15), which allows it to be used in a large variety of applications to produce cellodextrins or diversely functionalized cellodextrinic structures [45]. Although the transfer yields of lactobiosyl fluoride on different mono- or diglucidic acceptors were relatively high (51%–100%), only trimers and tetramers were obtained, according to the acceptors used, because galactose at the nonreducing end is not an acceptor.

Using the same glycosynthase, but with substituted α-cellobiosyl fluorides as donors, polymers with a molecular weight of 5000 ± 1000 have been obtained; this is the case of polymer with R^1:NH_2 (Scheme 18.15) which means a polymerization degree of about 16 ± 3 [45].

18.2.2.3.2 β-(1→3)-Glucans (Synthetic Laminarin–Type Glucans)

The mutant E231 of $\beta(1\rightarrow3)$ glucanase from barley is capable of catalyzing the self-condensation of fluorine-activated monomer glucidic units, which will be linked through $\beta(1\rightarrow3)$ bonds (Scheme 18.16). The resulting polymer is similar to recrystallized curdlan, and has a degree of polymerization in the range of 28–44, with an average of 30 [46]. The mutant can assist the self-condensation of 3-thio-α-laminaribiosyl fluoride, producing oligosaccharides in which monomers are linked through alternating *O*- and *S*-glycosidic bonds, respectively.

18.2.2.3.3 β-(1→3)-β-(1→4) Alternating Mixed Glucans

An interesting use of glycosynthases in the synthesis of sugar polymers refers to the self-condensation of α-laminaribiosyl fluoride catalyzed by the mutant E134A of $(1\rightarrow3,1\rightarrow4)$-$\beta$-D-glucanase from

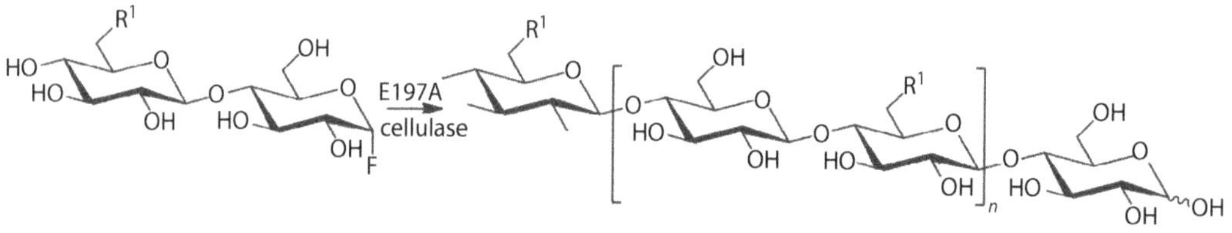

SCHEME 18.15 Polycondensation reaction catalyzed by the E197A cellulase (glycosynthase) from *H. insolens*.

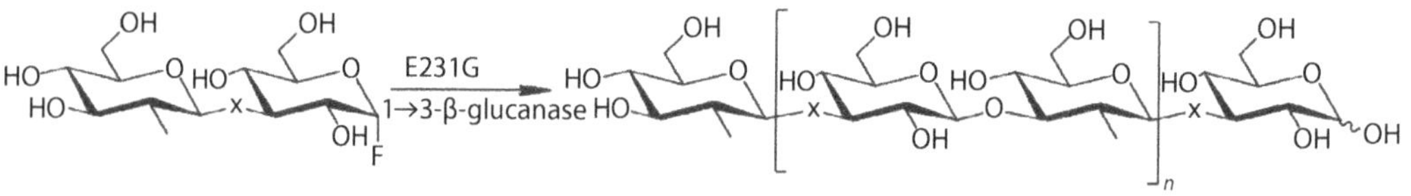

SCHEME 18.16 Polycondensation reaction catalyzed by the E231Gβ $(1 \rightarrow 3)$-β-D-glucanase (glycosynthase) from barley.

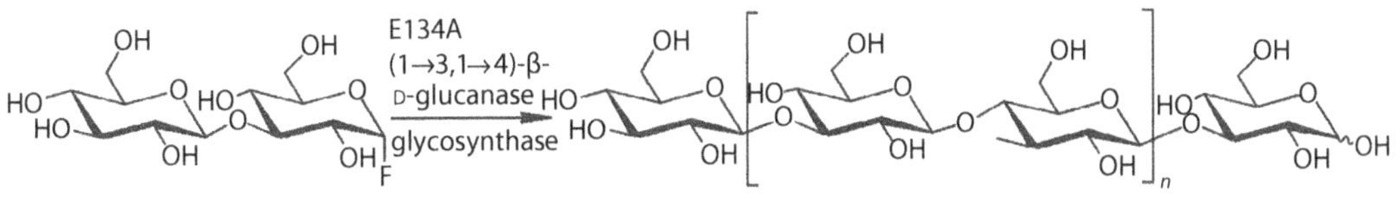

SCHEME 18.17 Polycondensation reaction catalyzed by the E134A $(1 \to 3,1 \to 4)$-β-D-glucanase (glycosynthase) from *B. licheniformis*.

B. licheniformis, which leads to the generation of a mixture of oligosaccharides, the most abundant and having a degree of polymerization of 12 [47] (Scheme 18.17).

The *H. insolens* GH7 glycosynthase, HiCel7B E197S, is capable of synthesizing nongalactosylated, XXXG-based homoxyloglucan up to Mw 60,000 [G = Glcβ(1→4); X = Xylα(1→6)Glcβ(1→4); L = Galβ(1→2)Xylα(1→6)Glcβ(1→4)], which is among the largest products so far obtained with glycosynthase technology [48].

18.3 ENZYMATIC SYNTHESIS OF POLYESTERS

In the living world polyesters are the fourth class of biopolymers after nucleic acids (DNA and RNA), proteins (polypeptides), and polysaccharides. They can be found as structures of omega hydroxy acids and their derivatives, which are interlinked *via* ester bonds, forming an indeterminate size polyester-type polymer. They are main components of the plant cuticle which covers all aerial surfaces of plants. Polyhydroxyalkanoates are another class of natural polyesters found in both plants and bacteria. It seems that they serve as energy and carbon stores. The interest for polyhydroxyalkanoates is rising, because they possess the general properties of synthetic polyesters, and in addition have the advantages of biodegradability and biocompatibility. These qualities make them useful in a wide range of applications, including the medical ones [49].

Another example of a natural polyester is shellack produced by female lac bug, *Laccifer* (*Tachardia*) *lacca*, and stored on tree barks, in India and Thailand, from where it is scraped.

Generally, polyesters are high tonnage synthetic products with broad utilization, starting with textile industry and finishing with the well-known PETs (PolyEthyleneTerephthalate). Today, when more and more environmental restrictions are imposed on chemical synthesis, the enzymatic alternative becomes increasingly attractive, not only due to the general advantages of enzyme utilization but also due to the high diversity of existing lipases, the most used enzymes, in both the formation and the breaking of esteric bonds [50].

However, utilization of enzymes in polymerization processes has certain peculiarities, as compared to organic synthesis biotransformations. Temperature introduces a number of restrictions which need to be optimized. In chemocatalysis, polycondensation temperatures can reach values of 180°C–280°C. These high values are required by the catalyst, and also by the need to maintain a reduced viscosity of the environment in which a polymer with high melting point is formed. The latter condition is necessary for the efficient mass, impulse, and heat transfer. On the other hand, the high temperature can induce secondary reactions in thermolabile polymers, as water release in the case of diols, opening of oxiranic cycles, uncontrolled branching in unsaturated polymers, etc. It is evident that polymer thermolability increases with their complexity. Besides temperature, the metals existing in the structure of chemocatalysts bring additional problems concerning the environmental impact. Moreover, these metals affect biocompatibility, as it is difficult to totally eliminate them from the product destinated to be used in the organism.

Not least, as compared to chemocatalysts, enzymes have the advantage of chemo-, regio-, and stereoselectivity. A summary of the advantages offered by enzymes in obtaining polyesters through polycondensation is presented in a recent, excellent review [51].

Lipases are present in all organisms, but the majority of commercial ones come from microorganisms. In vivo, lipases (E.C. 3.1.1.3.) split esteric bonds in triacylglycerols, converting them to diacylglycerols, monoacylglycerols, glycerol, and fatty acids.

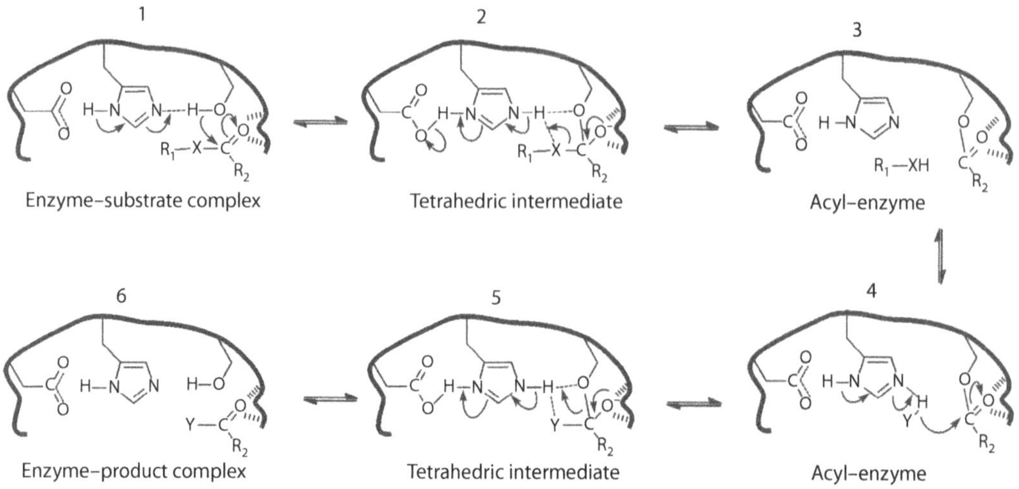

SCHEME 18.18 Catalytic mechanism of lipases.

The mechanism by which lipases assist the breaking of the esteric or amidic bond is shown in Scheme 18.18 and is based on the existence in the catalytic site of the catalytic triad consisting of a carboxylate (derived from Asp/Glu), an imidazole residue derived from His, and a hydroxymethylene residue derived from a Ser unit. The topological sequence of these residues ensures an increased nucleophilicity to the serine oxygen.

The reaction occurs in two main steps. In the first one, a covalent bond is formed between the carbonylic carbon atom of esteric/amidic bond to be hydrolyzed and the serine from the catalytic site. The acyl–enzyme intermediate is formed through a negatively charged transition state at the carbonylic oxygen atom. The role of general base played by the catalytic His facilitates the generation of this transition state. This role is potentiated by the Asp/Glu residue which stabilizes the partial positive charge that occurs at the imidazolic nucleus of His. The serinic oxygen becomes strongly nucleophilic and therefore can generate a tetrahedric transition state at the carbonylic carbon. This transition state is stabilized by hydrogen bonds which attach the oxyanion of the carbonylic carbon (intermediate 2—Scheme 18.18). The hydrogen bonds form in the so-called oxyanionic pocket. Breaking of the esteric/amidic bond leads to the formation of acyl–enzyme (intermediate 3).

In the second step, under the influence of the same general base but in the presence of water or of another nucleophile Y–H (Y=HO, RO, HOO, RNH), a nucleophilic attack will occur at the same C1 (state 4). In the second transition state, the oxyanion is also stabilized by hydrogen bonds present in the same oxyanion pocket (state 5). The enzyme is regenerated by breaking the esteric bond between serine and the fragment R_2CO- (state 6).

18.3.1 STEREOSELECTIVITY OF LIPASES

The chiral center may belong to the acyldonor (R^2) or can be attached to the nucleophile or to the acylated constituent (X or Y).

Lipase enantioselectivity toward the acyldonor is low to moderate [52], while enantioselectivity toward secondary alcohols, acylated amines, or amines in train to be acylated is high.

Lipases show an enantiopreference for the chiral center of the substrate to be transformed, located in the component that is acylated. The enantiopreference is preserved, whatever the nature of the substituent at the chiral center. Moreover, the so-called rule of Kazlauskas has been formulated on a statistical basis, applicable to secondary alcohols and to their O-Acyl derivatives. According to this rule, in most cases the enantiomer R will be transformed faster than enantiomer S (Scheme 18.19).

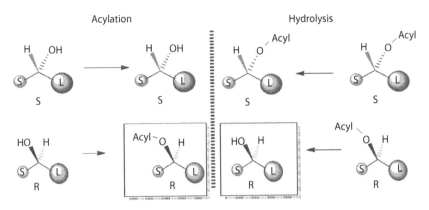

SCHEME 18.19 Enantiopreference *R* of lipases, illustrated on a chiral secondary alcohol.

Structurally, this enantiopreference is reflected in the topology of lipases' active site. The attachment of the acylated, or in train to be acylated, component to the enzyme is done at the stereo-complementary chiral center with two pockets: a small pocket that can easily accommodate a small (S) substituent, usually a methyl or at most ethyl group, and a large pocket that can accommodate a large substituent (L). In this way, the attachment of the other enantiomer is less firm, and a difference in the binding energy between the two stereoisomers occurs. This energy difference will be expressed as a preference for the complementary enantiomer.

It should be noted that enantiopreference in both steps is a manifest of acyl–enzyme intermediate formation and the transacylation step, when the enzyme is deacylated.

Activity, selectivity, availability, and lack of cofactor have made the lipases to be some of the most used enzymes in the biotransformation of organic substrates [53–55]. In polymer chemistry, the valorization of the enzymes' enantioselective potential leads to the obtaining of stereoregulated polyesters, with the most exquisite applications.

Lipases can be used in (poly)condensation by providing conditions in which the system can evolve in the opposite direction (high alcohol and acid/acyl derivative concentration, as low as possible concentration of water).

Practically, the lipase-assisted obtaining of polyesters can be achieved in two different ways:

1. Polycondensation
   a. Polycondensation of hydroxyacids
   b. Polycondensation of diacids/their derivatives with diols/their acylated derivatives
2. Polymerization (with opening of the lactone ring)

The most common ways of lipase-assisted forming of the esteric bond in polycondensation reaction are

1. The direct reaction of the acid and alcohol:

$$RCO_2H + R^1OH \; \Delta \; RCO_2R^1 + H_2O$$

2. Transesterification, which implies a direct interaction of the ester with the alcohol (alcoholysis), with the acid (acidolysis), or with another ester (interesterification):

$$RCO_2R^1 + R^2OH \quad \Delta \quad RCO_2R^2 + R^1OH \quad \text{(alcoholysis)}$$
$$RCO_2R^1 + R^2CO_2H \quad \Delta \quad R^2CO_2R^1 + RCO_2H \quad \text{(acidolysis)}$$
$$RCO_2R^1 + R^2CO_2R^3 \quad \Delta \quad RCO_2R^3 + R^2CO_2R^1 \quad \text{(interesterification)}$$

18.3.2 Polycondensation Synthesis of Polyesters

18.3.2.1 Polycondensation of Hydroxyacids

Because it is a dehydration reaction, the polycondensation of hydroxyacids must be conducted in conditions which do not favor the hydrolysis of the formed product, implying a reduced water concentration. Hydrophobic organic solvents, such as benzene, toluene, and diphenylether, are preferred. These solvents remove water from the reaction, solubilize the polymeric product, and force the shift of equilibrium toward the product. However, in hydrophobic solvents, the enzyme may suffer a partial denaturation. This is produced by transforming the intermolecular hydrogen bonds in intramolecular ones. Intermolecular hydrogen bonds are established between hydrophilic components of the enzyme and the existing water molecules in the natural environment of the enzyme. In a hydrophobic solvent, the hydrophilic groups on the surface of the enzyme will interact, reducing the conformational mobility of the enzyme, and consequently reducing its activity. A solution to maintain the enzyme activity in a reasonable range is the attachment of polyethylene glycol (PEG) chains on its proteic structure. These chains interact with the hydrophobic solvent and provide to enzyme the necessary mobility to interact with the substrate. In case of monomers and polymerization products that are in a liquid state at the working temperature (60°C–80°C), the reaction can be done in a solvent-free environment [56,57].

Moreover, water can be removed from the system by using low-pressure, molecular sieves or temperatures as high as enzymes can tolerate.

The polymerization of a hydroxyacid was first mentioned in 1985. 10-hydroxydecanoic acid was polymerized in benzene using a PEGylated lipase [58], with a degree of polymerization higher than 5. An oligomer of glycolic acid was also obtained with a PEGylated esterase [58].

The degree of polymerization of polyesters obtained from hydroxyacids is generally low and depends on carbon chain length. It was noticed that for the lipase from *Candida antarctica*, 10-hydroxydecanoic acid was close to optimum length, as monomers with shorter and longer carbon chains (4-hydroxybutyric acid, DL-2-hydroxybutyric acid, glycolic acid, and 16-hydroxyhexadecanoic acid) failed to polymerize.

The degree of polymerization can be increased by raising the enzyme/monomer mass ratio.

An interesting result is the polycondensation of ricinoleic acid ((R,Z)-12-hydroxyoctadec-9-enoic acid) in presence of an immobilized lipase (from *C. antarctica* or *Burkholderia cepacia*) and a synthetic zeolite (Scheme 18.20). The working temperature ranges from 60°C to 80°C, the time was 168 h, and MS 4A was used as zeolite [59]. In the conditions described in the patent, the average molecular mass was not less than 20,000.

In the same license the alternate method of polycondensation by transesterification, starting from the methyl- or ethyl ester, is described. The authors demonstrated that the polymer obtained by crosslinking has elastomer properties comparable with those of synthetic rubbers.

Another interesting example of polycondensation, which emphasized the regioselective character of the enzyme, was the obtaining of cholic acid oligomers (Scheme 18.21). Their average mass, of only 920, was modest, but the esteric bound was achieved only with the hydroxy in position 3 [60].

The chemocatalysis of the polycondensation reaction can be made more performant using carboxyl-activated substrates through electron-withdrawing groups (acid chlorides, trifluoroethyloxy). The enzymatic catalysis does not need such an expensive activation.

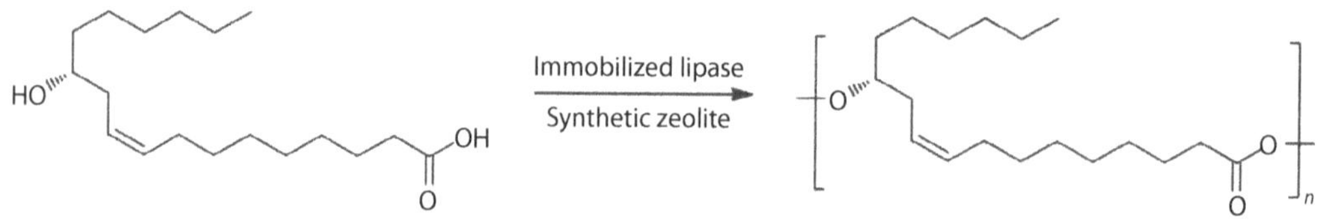

SCHEME 18.20 Polycondensation of ricinoleic acid.

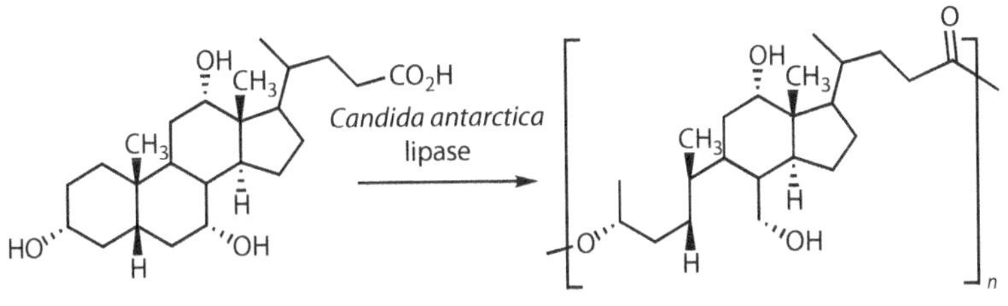

SCHEME 18.21　Polycondensation of cholic acid.

Polycondensation of diacids and of their diol derivatives offer more diverse possibilities. Instead of free dicarboxylic acids, the use of dicarboxylic acid esters, is largely employed, because they are liquid at the working temperature. Besides, in case of vinyl esters, during transesterification, usually alcoholysis, the released vinyl alcohol tautomerizes to acetaldehyde which, due to its volatility, leaves the system and forces the shift of the equilibrium toward the condensation reaction. The average molecular mass and the reaction time depend on the working temperature, on the presence or absence of the solvent and on the solvent's nature.

The structure of monomers is an important influence factor. Lipases usually accept monomers with a relatively long carbon chain, because the substrate is attached through the hydrophobic portion, and the reaction occurs in the hydrophilic area of the substrate.

The formation of acyl–enzyme intermediate is a direct condensation reaction between the carboxyl group of the free acid and the serine hydroxyl group, or a transesterification (alcoholysis) between the substrate ester and the serine hydroxyl. The second step is almost always an alcoholysis of acyl–enzyme bond and the alcohol substrate (monomer).

All the earlier-mentioned parameters deeply depend on the nature of the enzyme. There is no such thing as an universal enzyme suitable for all reactions. Therefore, trials remain at hand; if the enzyme's structure is known, docking calculations can be made, or evolutionary means for adapting its structure and functionality to the reaction's needs can be ultimately used.

In the enzyme-assisted synthesis of polyesters, starting from diesters of dicarboxylic acids and diols, linear polyesters are obtained together with oligoesteric macrocycles. The concentration of the latter ones in the product mixture depends on the structure of monomers and the initial concentration of the reactive functions [61].

18.3.2.2　Polycondensation of Carboxylic Acids/Esters with Diols

The easiest reaction is simple dehydration. The first paper on polymerization by condensation (polycondensation) in enzymatic catalysis of α,ω-dicarboxylic acids with α,ω-glicols reports the identification, among esterification products, of α,ω-dicarboxylic acids (a = 4,5,6,8,10,11,12) with 1,2 ethanediol b = 2 and 1,3 propanediol b = 3 (Scheme 18.22) [62].

Oligomers: a trimer (m = 1), a pentamer (m = 2), and a heptamer (m = 3), respectively, have been identified only for the couple 1,13-tridecanoic acid (a = 11) and 1,3-propanediol (b = 3). The reactions have been conducted at 30°C for 16h. The reaction mixture has not been based on equimolarity of the two types of components.

Usually (poly)condensation reactions are run in a nonaqueous environment, on the reason of shifting the equilibrium toward water and also (poly)ester generation. However, it has been noticed that certain polycondensations can run in water with good yields. Moreover, it has been

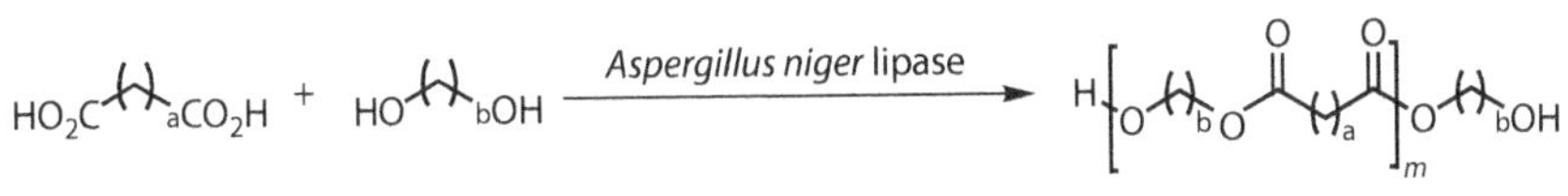

SCHEME 18.22　General polycondensation between of a α,ω-dicarboxylic acid and 1,2-ethanediol or 1,3-propane diol.

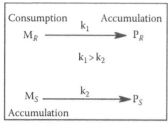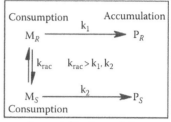

| Kinetic resolution | Dynamic kinetic resolution |

SCHEME 18.23 Kinetic resolution versus dynamic kinetic resolution.

stated that in the polycondensation of α,ω-dicarboxylic acids with α,ω-glicols, the length of the carbon chain influences the yield; the best results were obtained with compounds having similar hydrophobicity [63].

18.3.2.2.1 Enantioselective Polycondensation

As it has been shown, one advantage of enzymes in general, and of lipases in particular, is their stereoselectivity. This property can be valorized in obtaining polymers with high enantiopurity. Enantioselective polymerization is based on the higher incorporation rate of a certain stereoisomer, say (M_R), as compared to its optic antipode (M_S). The process is developed in organic chemistry biotransformations and is known as (enzymatic) kinetic resolution (Scheme 18.23, Kinetic resolution); its application in obtaining chiral polyesters is similar. Despite the advantage of obtaining a highly enantiopure polymer, the major disadvantage of the process is a maximum 50% conversion of the racemic mixture; the untransformed enantiomer (M_S in our example) may have a use difficult to identify. More than that, its separation greatly complicates the whole process.

The solution is to find a system that could racemize the monomer, at a rate higher than its incorporation into the polymer. This system is known as (enzymatic) dynamic kinetic resolution and it is applied in enzymatic biotransformations of organic substrates. In the absence of racemization system, the preferred enantiomer (M_R) is consumed, and the relative concentration of the enantiomer (M_S) increases. When the racemization system is present, the enantiomer (M_S), which is in a higher concentration, is racemized to (M_R) and (M_S); the (M_R) enantiomer is further incorporated into the product, while (M_S) enantiomer is transformed to (M_R) and so on (Scheme 18.23, dynamic kinetic resolution).

The most suitable example is the obtaining of a high enantiopurity polyester, starting from a racemic mixture of 1,1′-(1,4-phenylene)diethanol and dimethyl adipate (Scheme 18.24) [64].

Racemization was achieved with the catalyst $[RuCl_2Cymene]_2$. The reaction lasted for 4 days, the molecular mass reached 3000–4000, and the optical rotation of the mixture increased from −0.6° to +128°.

18.3.3 RING-OPENING POLYMERIZATION OF CYCLIC MONOMERS

Aliphatic polyesters prepared by ring-opening polymerization of lactones are now used worldwide as bioresorbable devices in surgery (orthopedic devices, sutures, stents, tissue engineering, and adhesion barriers) and in pharmacology (controlled drug delivery) [65].

SCHEME 18.24 Dynamic kinetic resolution used for synthesis of a high enantipure polymer.

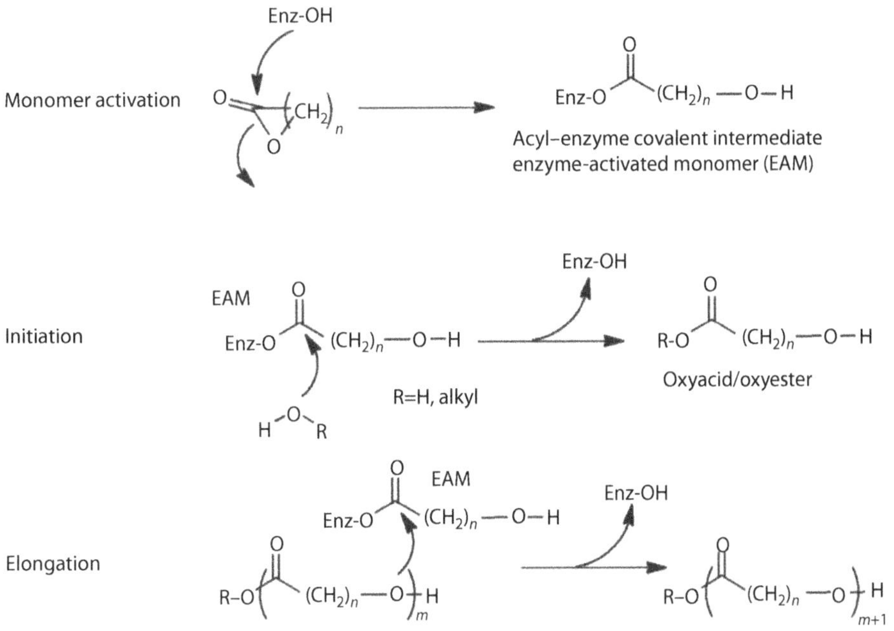

SCHEME 18.25 The mechanism of lactone polymerization by ring opening.

The first references on the use of cyclic monomers in enzymatic polymerization by ring opening issued in 1993. Knani et al. have described the polymerization of ε-caprolactone in *n*-hexane, in the presence of crude *porcine pancreatic lipase* (PPL) with a degree of polymerization (DP) of 35 [66]; practically in the same time, Uyama and Kobayashi reported the polymerization of δ-valerolactone and ε-caprolactone using lipases from *Pseudomonas fluorescens*, *Candida cilindracea*, and porcine pancreatic lipase. They worked in the absence of solvent (bulk polymerization), obtaining after 10 days, at 60°C a polymer having a Mn = 7000, and a conversion of the monomer of 92% [67,68].

The mechanism of the polymerization reaction, as described by Kobayashi [69], is essentially depicted in Scheme 18.25.

The main step of the process is the reaction of the enzyme with the lactone, which leads to monomer activation, by attaching to the enzyme. The enzyme is positioned with the serine hydroxyl close to the lactone carbonyl. This configuration actually represents the complex enzyme–substrate, in which the two components are not attached by covalent bonds. In this enzyme–substrate complex, the attack of serine hydroxyl on the carbonylic carbon opens the lactone ring, with the formation of the covalent intermediate acyl–enzyme (monomer-enzyme or enzyme-activated monomer—EAM).

Once formed, the acyl–enzyme intermediate will release the oxyacid, through the nucleophilic attack of water or of an aliphatic alcohol present in the reaction environment, or the oxyester respectively, which resulted from the initial lactone. This is the initiation step that generates the structure from which elongation begins. For the monomer 12-dodecanolide (DDL) it has been shown, from the similarity of its conversion-time curves, in the presence and absence of the alcohol that releases the enzyme in the initiation step that this nucleophile does not influence the consumption rate of the monomer. Consequently, the release step is much faster than the formation of EAM intermediate, this latter one being the rate limiting step.

In the elongation step, the oxyacid or oxyester will interact with the activated monomer (EAM) and attach it releasing the enzyme, which will be able to activate another monomer molecule.

This "monomer-activated mechanism" found in case of lipases is different from the conventional anionic mechanism of ring-opening polymerization, which is an "active-chain mechanism."

In chemocatalysis, lactone reactivity increases from macrolides to medium ring-sized mono-mers and finally to small-size compounds. The reactivity of lactones in the polymerization reaction directly correlates with the ring-strain.

In enzymatic catalysis of ring-opening polymerization, the minimum reactivity is found for five-, six-, and seven-membered lactones, which do not have a sufficiently high ring-strain to energetically facilitate the ring opening, nor an extended hydrophobic area, to provide an efficient contact with the enzyme, for the interaction of serine in the active site with the ester group. Lactones with smaller rings are more reactive because of the tension accumulated in the ring, and those with higher dimen-sions are more reactive than the five-, six-, and seven-membered ones, because of the increased hydrophobic contact with the lipase. So the polymerization of ε-caprolactones is a challenge.

When monomer concentration lowers and oxoesteric oligomer concentration rises, the enzyme will attach such an oligomer, obviously at the free or esterified carboxyl group. Its detachment can be performed with the participation of an OH group of an oligomer—when another, linear oligomer results, or with the participation of the OH group of the molecule attached to the enzyme, when a cyclic oligomer will be formed (Scheme 18.26).

The use of organic solvents leads to the lowering of monomer concentration and consequently to an increased probability of obtaining cyclic oligomers [70].

Ring-opening polymerization was investigated on lactones with a different number of members, unsubstituted and substituted. Several very good quality reviews synthesize this area of research.

The general features of ring-opening polymerization show that enzymatic polymerization, simi-lar to chemical polymerization, conducts to polymers with higher molecular weights and higher conversions, as compared to polymerization through condensation of oxyacids [71].

Investigations showed that lipases acted as efficient catalysts for the ring-opening polymerization of simple 4–17-membered unsubstituted lactones, which demonstrates that enzymes are an increas-ingly competitive alternative to chemocatalysts [72].

Enzymatic ring-opening polymerization can be performed in bulk or in organic solvents. When a low molecular mass alcohol is not present in the environment, the obtained polymer will always have at its ends a carboxyl group and a hydroxyl group, respectively, which means that the nucleophile necessary for monomer detachment from the enzyme is water (R=H in Scheme 18.26). Therefore, the polymers obtained by ring-opening polymerization belong to the class of telechelics.

The most studied lactone was ε-caprolactone. Although it has a lower reactivity, as compared to four-membered lactones and to macrolides, using of lipases conducted to good results. When using Porcine Pancreas Lipase in the polymerization of ε-caprolactone to heptanes, a polymer with an average molecular mass $Mn = 2300$ was obtained in 4 days at 65°C. Lipases from other sources have been also tested. Among them, *C. antarctica* lipase B (CALB) used as a commercial product immo-bilized on acrylic resin (Novozym 435) proved to be the most efficient. The performance parameters were clearly superior to other tested lipases. Thus, in a bulk polymerization, the percentage ratio

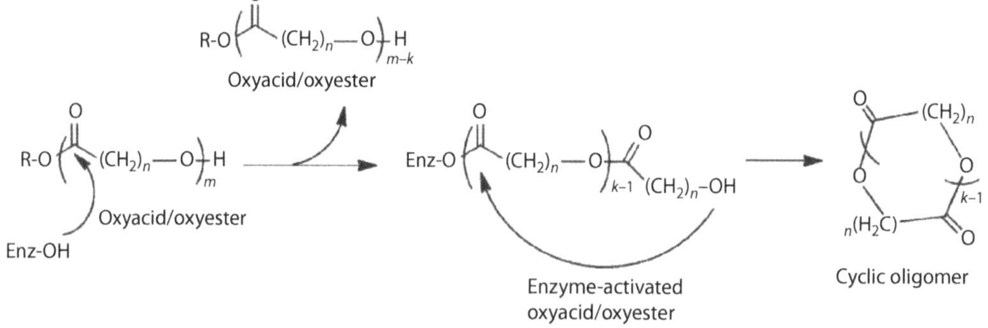

SCHEME 18.26   Formation of cyclic oligomer in polycondensation of oxyacids/oxyesters.

enzyme/monomer lowered from 20%–50% to 1%, and the reaction time was reduced to 10h, at a working temperature of 60°C [72].

Polymerization of methyl-substituted ε-caprolactones has also been studied. If α-Me-ε-caprolactone and γ-Me-ε-caprolactone had approximately the same polymerizability as unsubstituted ε-caprolactone, ω-Me-ε-caprolactone showed a significantly lower polymerizability than unsubstituted ε-caprolactone [71]. The reason seems to be the crowding of the nucleophilic center (the terminal OH, with a crucial role in the propagation step).

Four-membered lactones unsubstituted or substituted have been easily polymerized due to the ring tensions (Scheme 18.27).

A description of the polymerization level of these lactones is done by Matsumura [71].

The significance of β-butyrolactone polymerization must be noticed, which conducts to poly(3-hydroxybutirates), polymers similar to those existing in bacteria [73]. The polymerization products contain, besides linear polyesters with OH and COOH ends, cyclic oligo-/polyesters and crotonate-type polyesters, which underwent elimination at the level of the last hydroxyl (Scheme 18.28).

Polymalate is also a natural product. It has been first isolated from cells of *Penicillium cyclopium* [74]. Polymalate is an attractive polymer for the biomedical field because of its water soluble, biodegradable structure [75]. Its degradation in vivo leads to the nontoxic malic acid.

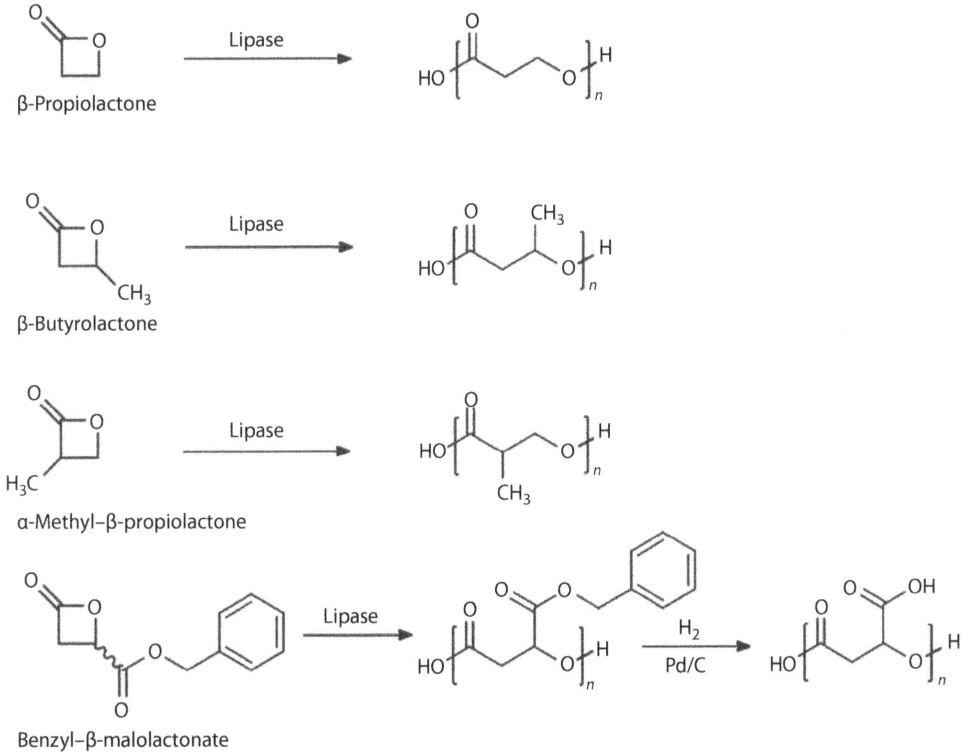

SCHEME 18.27 Polymerization of four-membered lactones unsubstituted or substituted.

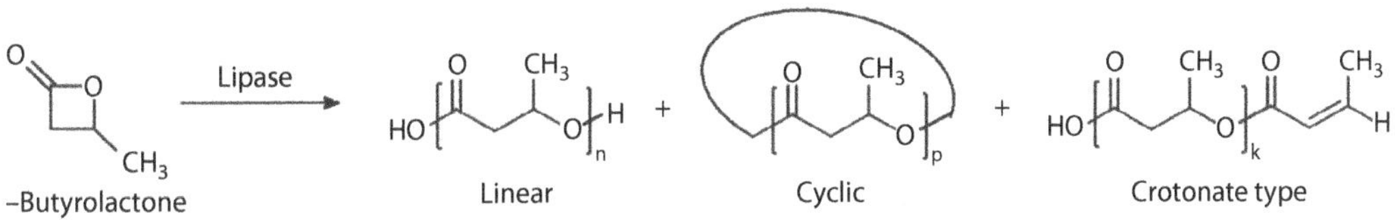

SCHEME 18.28 β-Butyrolactone polymerization products.

The presence of carboxyl side-chain groups attached to a stereogenic center allows, through conjugation with other structures, the fine tuning of physical and physiological properties of the polymer.

It is obtained by ring-opening polymerization of benzyl–β-malolactonate. The product, poly(benzyl malate) with a Mw higher than 7000, is submitted to a quantitative debenzylation by catalytic reduction, to generate poly(β-malic acid). The sodium salt of the polyacid is formed by neutralization [76].

From the five-membered lactones series, γ-butyrolactone has been polymerized. Chemical polymerization of this compound is not possible; the reaction was conducted with lipase from pork pancreas or from *Pseudomonas* sp. [73].

From the six-membered lactones series several structures have been polymerized, such as δ-valerolactone, α-methyl valerolactone [77], and derivatives as 1,4-dioxan-2-one [78]. The latter is of interest due to its biocompatibility, good flexibility, and tensile strength. Another type of six-membered lactone is lactide (the di-ester of lactic acid), which conducted by polymerization with lipase to a polymer with high molecular weight, but with a low yield [71].

Higher-membered lactone lipase polymerizations are detailed in literature [1,72].

18.3.3.1 Enantioselectivity in Ring-Opening Polymerization of Lactones

Unsubstituted lactones are nonchiral. The introduction of a substituent generates an asymmetric carbon atom. Enzymes in general, and lipases in particular, are enantioselective catalysts, which show preference for one of the two enantiomers in an equimolar or nonequimolar mixture. Enzymatic polymerization with lipases is a selective process, but can be influenced by numerous factors, and their interaction may produce hardly predictable effects. The essential characteristic of ring-opening polymerization of substituted lactones is that a monomer structure bearing an asymmetric center contacts an enzyme at least twice. The first contact takes place at the ring opening of lactone and formation of enzymea-activated monomer—EAM (Scheme 18.25). Then the opened structure, released by the enzyme, will contact the enzyme once again, in order to retrieve the activated monomer from the enzyme through the nucleophilic attack with the OH group. Each contact will represent an enantiomeric selection; therefore the overall selectivity of the lipase should be high.

The first systematic studies on ring-opening polymerization of substituted lactones have been done by Kobayashi and coworkers, using *P. fluorescens* and *C. antarctica* lipases. They have clearly revealed the enantiopreference of the enzyme [77,79]. The type of enantiopreference, (R) or (S), and its magnitude calculated as E (enantiomeric ratio), defined as the ratio of the specific constants, $E = (k_{cat}/K_M)_R/(k_{cat}/K_M)_S$, depend on the position of the substituents, the size of the substituents, and the size of the ring.

The effect of substituent position was evaluated in an experiment on methyl-substituted ε-caprolactone [80]. Methyl-substituted substrates in positions 3–6 were polymerized in 1,2,3,4-trimethylbenzene at 45°C. Substituents in positions 3, 4, and 6 induced an (S) selectivity, while the substituent in position 5 was transformed with (R) selectivity. The recorded E values ranged from 93 ± 27 for 4-methyl ε-caprolactone to 13 ± 4 for 3-methyl ε-caprolactone; due to an extremely low conversion, E for 6-methyl ε-caprolactone was not recorded, only its (S) selectivity was established.

The size of substituent also influences the polymerization rate and selectivity. Thus, Novozym 435-catalyzed ring-opening polymerizations of 4-substituted ε-caprolactones showed that the polymerization rate decreases by a factor of 2 upon the introduction of a Me substituent at the 4-position. Moreover, 4-EtCL polymerizes five times slower than 4-MeCl and 4-PrCL is even 70 times slower. The decrease in polymerization rate is accompanied by a strong decrease in enantioselectivity: while the E-ratio of 4-MeCL polymerization is 16.9, the E-ratios of 4-EtCL and 4-PrCL are 7.1 and 2.0, respectively. Interestingly, Novozym 435 displays S-selectivity for 4-MeCL and 4-EtCL in the polymerization reaction, but the enantioselectivity is changed to the (R)-enantiomer in the case of the 4-PrCL [80].

The influence of lactone ring size on its lipase-catalyzed ring opening by transesterification showed that enzymes have a pronounced tendency for transforming (S) stereoisomers of rings up

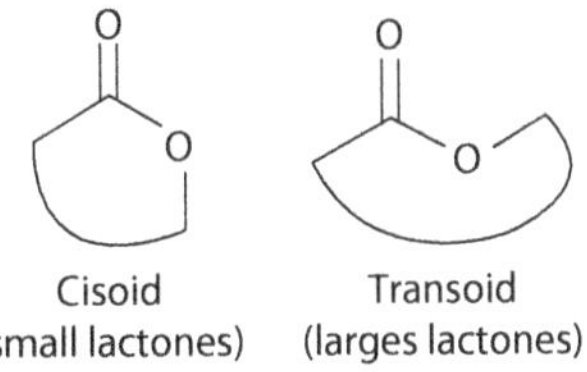

Cisoid
(small lactones)

Transoid
(larges lactones)

FIGURE 18.1 Cisoid and transoid lactone rings.

to seven members; for rings with more than seven members, the tendency is directed toward (R) stereoisomers [81]. (S)-enantiopreference is unexpected, in relation with statistical data that allowed validation of Kazlauskas's rule. This switch was explained by conformational change adopted by large, less tense rings, from transoid structures to cisoid conformation specific for small rings [82] (Figure 18.1). The enantiopreference switch is accompanied by an increase of the transformation rate from small (cisoid) rings to large (transoid) rings. The rate increase is about one order of magnitude. The rate of alkaline hydrolysis changes in the opposite direction, when compared to ring-opening polymerization rate. Small rings will be hydrolyzed more rapidly than large rings. The decrease of hydrolysis rate with the increase of ring size is more than four orders of magnitude.

18.3.3.2 End-Functionalized Polyesters

Ring-opening polymerization can be used for the synthesis of polyester macromonomers using functionalized initiators and/or terminators.

Initiation of the polymerization reaction (release of the structure from which elongation begins) can be done, besides water, with an alcohol (R=alkyl, Scheme 18.25), thiol, or amine [83]. In this case, one end of the polymer will be a hydroxyl and the other end will be an ester, thiolester, or amine, as appropriate. The alcoholic end group can be modified by introducing in the environment of a "terminator" that contains a carboxyl group capable to acylate, in the presence of lipase, the terminal OH group of the polyester. By modifying the end groups of the oligomer/polymer, end-functionalized polyesters can be obtained, useful in block-copolymerization processes or in further derivatization.

18.3.3.2.1 Initiator Method

This method allows the introduction of desired groups using alcohols that contain them. For example, the introduction of vinyl group can be done by using hydroxy (met)acrylates or ω-alkenyl/alynyl-1-alcohol as initiators [84] (Scheme 18.29).

When using alcohols that contain esteric bonds as initiators (see the HEMA example), a problem occurs with the transesterification reaction. This reaction takes place because lipases interact, with

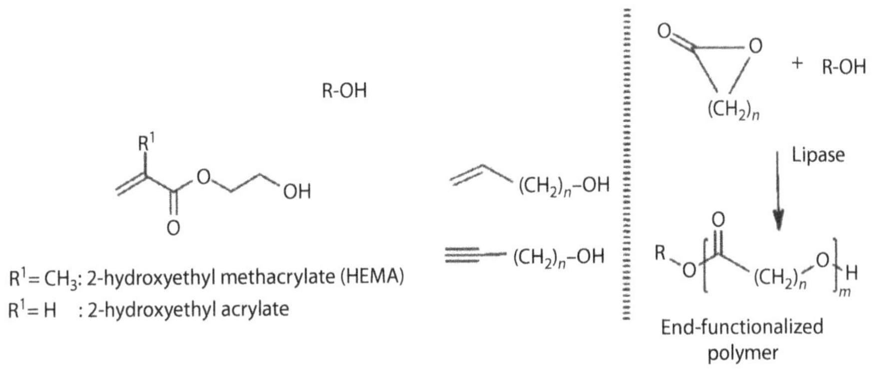

R¹ = CH₃: 2-hydroxyethyl methacrylate (HEMA)
R¹ = H : 2-hydroxyethyl acrylate

End-functionalized polymer

SCHEME 18.29 Obtaining end-functionalized polymers.

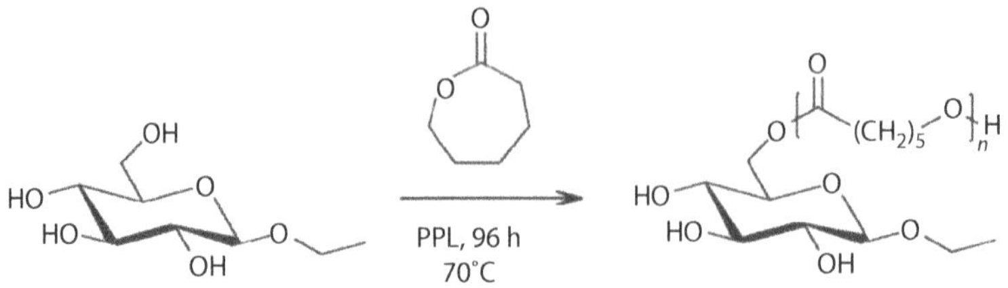

SCHEME 18.30 Using ethyl glucoside as initiator in ring-opening polymerization of ε-caprolactone.

a higher or smaller selectivity, with esters leading to acidolysis, alcoholysis, or interesterification. Therefore, 1,2-ethanediol or methacrylated hydroxyl end-groups [85] can be found, besides HEMA, among terminal structures attached to the carboxylic end of the polyester.

An interesting example initiated by Gross was the use of ethyl glucoside [86] as an alcoholic component. Ethyl glucoside in the presence of porcine pancreatic lipase (PPL), in the reaction with ε-caprolactone and in the absence of solvent, initiated a polymerization process that, after 96 h at 70°C, resulted in a polymer with Mn = 2200, in which the carboxylic end was esterified with ethyl glucoside in position 6 (Scheme 18.30).

18.3.3.2.2 Terminator Method

The OH end of the obtained polyesters can be acylated, using the acylation potential of lipases, with structures that bear reactive groups. Thus, the quantitative acylation of the terminal OH was achieved by using lipases derived from *Pseudomonas* family. Methacryl- and ω-alkenyl-type polyester macromonomers were synthesized by the polymerization of 12-dodecanolide in the presence of vinyl methacrylate or vinyl 10-undecenoate. Vinyl methacrylate is an useful acylating agent, as through the alcoholysis produced during the formation of acyl–enzyme intermediate, the acetaldehyde that ensures the irreversibility of the process is released. This process could be extended to single-step synthesis of polyester telechelics. The polymerization in the presence of divinyl sebacate produced a telechelic polymer having a carboxylic acid group at each end. By using a fatty acid ethyl ester or an acetic acid alkyl ester as additive, on the other hand, the quantitative introduction of the corresponding acyl group was not achieved [87].

18.3.3.3 Polycarbonates from Cyclic Carbonates

Cyclic carbonates are actually carbonic acid esters. By their ring-opening polymerization, polycarbonates are generated. These macromolecular structures are biodegradable, because of their biocompatibility properties, and have a high applicative potential, especially in the biomedical area, as a result of their reduced toxicity.

The first cyclic carbonate submitted to ring-opening enzymatic polymerization was trimethylene carbonate [88]. Enzymatic polymerization was first approached, almost simultaneously, by the groups of Kobayashi et al., Matsumura et al., [89] and Gross et al., [90] which communicated the results on lipase-catalyzed polymerization of trimethylenecarbonate (Scheme 18.31).

The enzymes used were Porcine Pancreatic Lipase and Novozym 435 (Lipase B from *C. antarctica*). The highest molecular weight of poly(trimethylenecarbonate) has been obtained by Matsumura et al. (Mw = 156,000, Mw/Mn = 3.8). In the paper of Gross, it has been found that

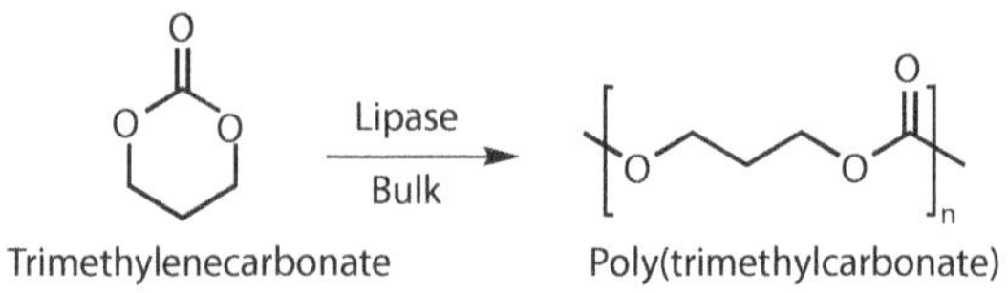

Trimethylenecarbonate   Poly(trimethylcarbonate)

SCHEME 18.31 Polymerization of trimethylene carbonate.

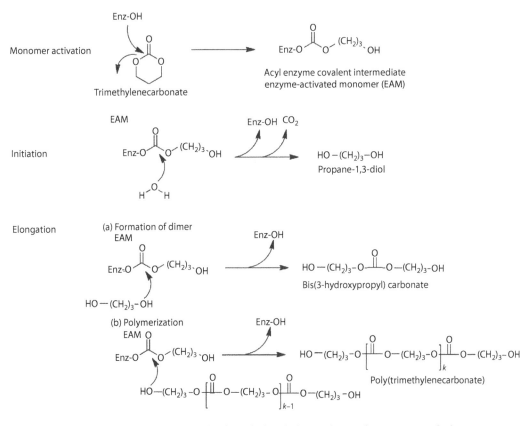

SCHEME 18.32 Mechanism of polymerization of trimethylenecarbonate in enzyme catalysis.

the polymer obtained by Novozym 435 [90] catalysis has both ends terminated in $-CH_2-OH$. The author explains this by a mechanism that involves a decarboxylation step (Scheme 18.32).

Although molecular masses did not exceed 24,400, monomer conversions were almost complete. Trimethylcarbonate polymerization was resumed in 2008 with Porcine Pancreatic Lipase immobilized on silica particles; good conversions were obtained and also high molecular weight [91].

An interesting example is the synthesis of an enantiomerically pure functional polycarbonate from a novel seven-membered cyclic carbonate monomer derived from naturally occurring ring L-tartaric acid. The monomer was synthesized in three steps and was polymerized at 80°C. An optically active polycarbonate with Mn = 15,500 g/mol was obtained by catalysis with lipase Novozyme-435. Deprotection of the ketal group was achieved with minimal polymer chain cleavage (Mn = 10,000 g/mol) [92] (Scheme 18.33).

Other examples of cyclic carbonates which were polymerized by ring-opening enzymatic catalysis can be found in the literature (Scheme 18.34) [93].

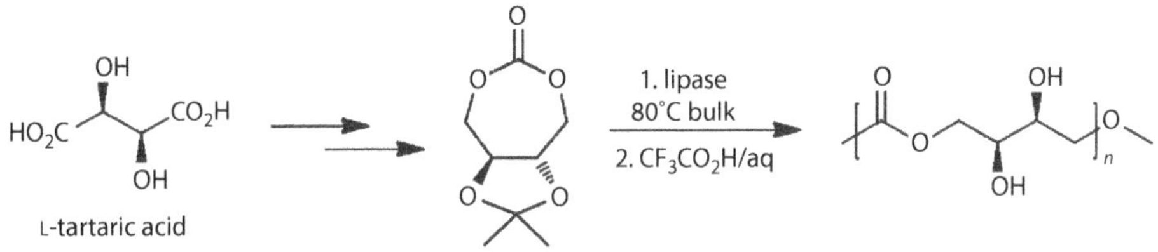

SCHEME 18.33 Polymerization of cyclic carbonate derived from a L-tartaric in lipase catalysis.

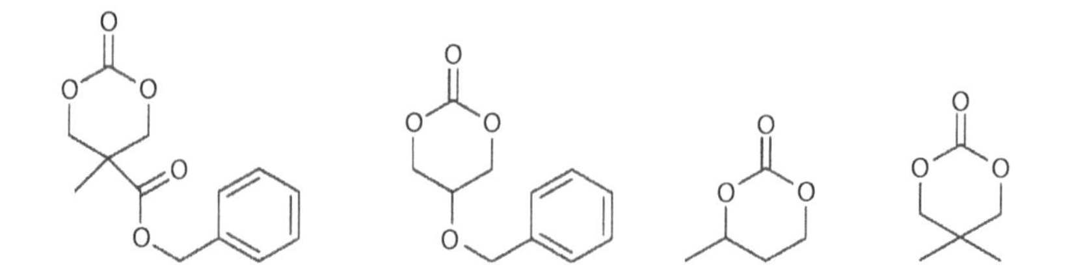

5-Methyl-5-benzoyloxycarbonyl-   5-(Benzyloxy)-1,3-   4-Methyl-1,3-   5,5-Dimethyl-1,3-
      1,3-dioxane-2-one   dioxan-2-one   dioxan-2-one   dioxan-2-one

SCHEME 18.34   Cyclic carbonates submitted to ring-opening polymerization in hydrolases assistance.

18.4   ENZYMATIC SYNTHESIS OF PHENOLIC POLYMERS

Phenolic compounds are widely spread in the living world, as, for example, the ubiquitary tyrosine. In plants, phenolic structures are well expressed. They are present in tree bark, in the content of vacuole, as well as in intercellular spaces. Flavonoids, anthocyans, and resveratrol are several poly-phenolic structures with known antioxidant properties, used in cancer and circulatory disorders prevention, as antimicrobial and anti-inflammatory agents. As a general rule, the term polyphenol refers to the existence of more than two OH groups attached to an aromatic structure. Polymers produced starting from phenolic monomers can also be called polyphenols. To avoid confusion, we stick to the convention proposed by Ritter [94] to use the term "polyphenol" for nonpolymer compounds, and the term "phenolic polymer" for macromolecular phenolic compounds. Phenolic polymers can be produced in the desired amounts by combining radicals resulted from phenolic structures. Radicals are generated with the help of enzymes able to mediate a monoelectronic transfer from a phenol to an oxidizing agent that can be a peroxide (even hydrogen peroxide) or oxygen. The most significant results have been obtained with peroxidases and laccases.

18.4.1   PHENOLIC POLYMERS BY PEROXIDASE-ASSISTED POLYMERIZATION

Peroxidases (E.C. 1.11.1.7.) are a large family of heme-containing enzymes widely spread in living organisms. Most of the peroxidases studied so far are heme enzymes with ferric protoporphyrin IX (protoheme) as a prosthetic group (Figure 18.2). However, the active centers of some peroxidases

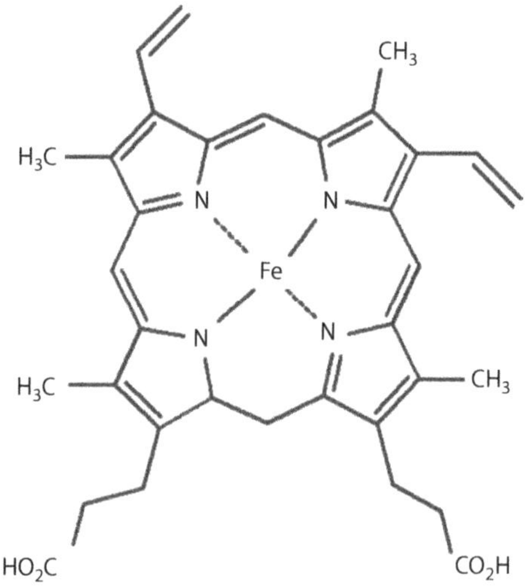

FIGURE 18.2   Ferric protoporphyrin IX as prosthetic group of peroxidases.

also contain selenium (glutathione peroxidase), vanadium (bromoperoxidase) most commonly found in marine environments, manganese (manganese peroxidase), and flavin as a nonmetal cofactor (flavoperoxidase) in case of some bacterial peroxidases [95].

These enzymes perform monoelectronic oxidation of a substrate using as an oxidative cosubstrate, usually hydrogen peroxide. But in some cases an organic peroxide (alkyl or benzyl). They have a variety of functions in the organism. Most of them are involved in detoxification of hydrogen peroxide in antimicrobial activity, in biosynthesis of plant hormones, in cytochrome c metabolism, and in degradation of lignin [94].

Peroxidases transform electron-rich substrates in radicals, allowing their further transformation by recombinations, or by radical transfer with nonradical molecules, in particular cases, to polymers.

The need to replace certain chemical transformations with enzymatic ones, the biodegradation of phenols, as easily oxidizing compounds, from industrial wastes became a consistent object of study.

One of the first systematic studies on phenol polymerization dates back to 1987 [96]. Its authors showed that in an appropriate environment (solvent, buffer) these compounds, in the presence of peroxidase, can be transformed into polymers. On the one hand, phenol polymerization with peroxidase has found applications in replacing the production of phenol-formaldehyde-type resins like resitol, resole, and novolak, avoiding in the manufacturing process the use of the highly toxic formaldehyde, and on the other hand provides a reliable alternative for the treatment of wastewaters. In the first case, the obtained polymers have similar properties to those of commercial phenol-formaldehyde polymers. The tentative to impose the oxidative polymerization with peroxidase in the treatment of wastewaters with high phenol content is based on the insolubility of the obtained polymer.

The mechanism of electron transfer put to work by the peroxidase engine is still in debate. Based on spectral data, a model of the process is described in detail by Montellano [97].

A simplified model, which captures the essence of electron and atom transfer, is presented in Scheme 18.35. According to this model, the oxidation of the phenolic substrate takes place in three steps and leads to the formation of a phenoxy radical. Through recombinations or radical transfer steps, this radical will be able to form polymers.

In the first stage, the resting state, the central ion having the state Fe(III) will react with H_2O_2 (or with another peroxide) generating the iron-oxo derivative known as compound I and releasing water. This compound has two additional oxidizing equivalents as compared to the resting state (Fe(III)), also having a ferilic structure Fe(IV)=O and a structure of π cation radical. Compound I is able to

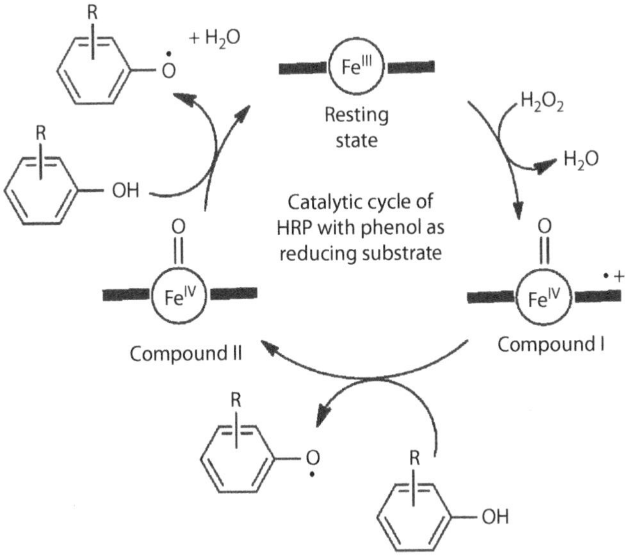

SCHEME 18.35 A simplified catalytic cycle of peroxidase.

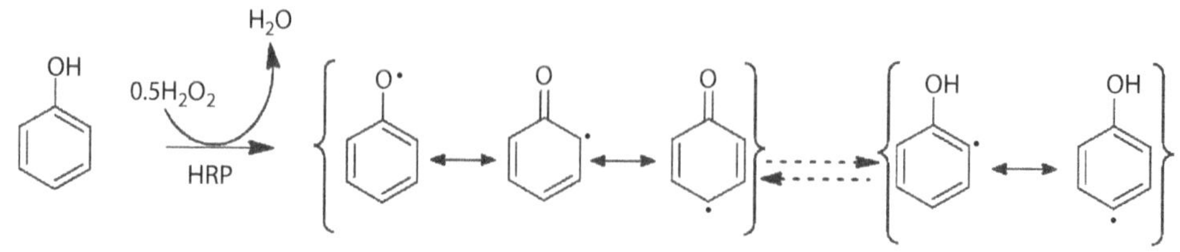

SCHEME 18.36   Radical species generated by peroxidase from phenol.

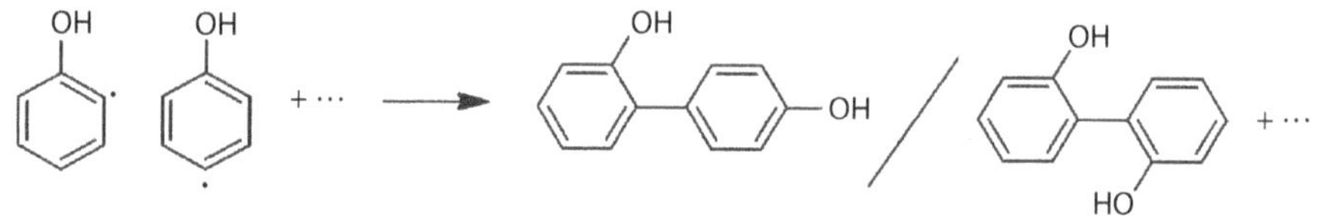

SCHEME 18.37   Radical coupling reaction(s).

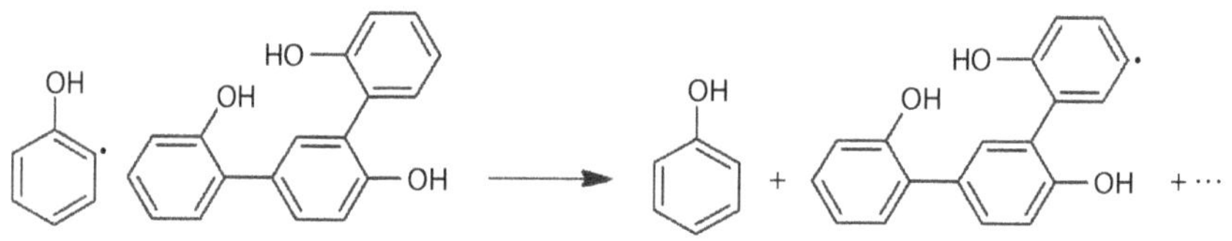

SCHEME 18.38   Radical transfer reaction(s).

extract an electron from the phenolic substrate, which will be transformed into a phenoxy radical. The extraction is assisted by a base which captures the proton from the same phenolic hydroxyl (Ar–OH=Ar–O• +H+ + 1e−). By this transfer compound I is reduced to compound II.

The resting state is regenerated by a new monoelectronic reduction of compound II, while a new phenoxy radical is generated. The ferryl oxygen, together with two protons derived eventually from phenol, will generate a new water molecule.

The resting state is regenerated by a new monoelectronic reduction of compound II, with the generation of a new phenoxy radical. The ferryl oxygen, together with two protons derived eventually from phenol, will generate a new water molecule.

Due to the conjugation of the unpaired electron with the π-electrons of the aromatic ring, phenolic radicals will possess to reaction centers (Scheme 18.36).

The formed radicals, which initially are in high amounts, will further undergo coupling reactions (Scheme 18.37). In these reactions, the concentration of radical species lowers rapidly because through reaction of two radical species will result a nonradical species. Monoelectronic transfer reactions take place in parallel, but also sequentially, with a dynamics depending on the environment, but especially on the structure of phenolic components (Scheme 18.38). In these reactions, from small radicals which are continuously generated, oligomeric radicals are formed, which recombine with other radical species, thus increasing the chain. The process continues until radical species are depleted.

When a radical species cannot participate in a recombination or an electron transfer reaction, it will be oxidized to a ketone structure.

Phenol oxidation at room temperature, in an aqueous environment, precipitates polymers composed of phenylene and oxyphenylene units (Scheme 18.39). From the point of view of oxidative polymerization, phenol is a multifunctional monomer. The radicalic mechanism, which involves highly reactive species with similar affinity to their reaction partners, explains the difficulty in controlling the regio- and chemoselectivity of the process, for the obtaining of a stereoregulated polymer. In the case of phenol, a certain control of regioselectivity can be obtained by changing the solvent. Obviously, regioselectivity is measured (expressed) as phenylene/oxyphenylene unit ratio (Scheme 18.39). Thus,

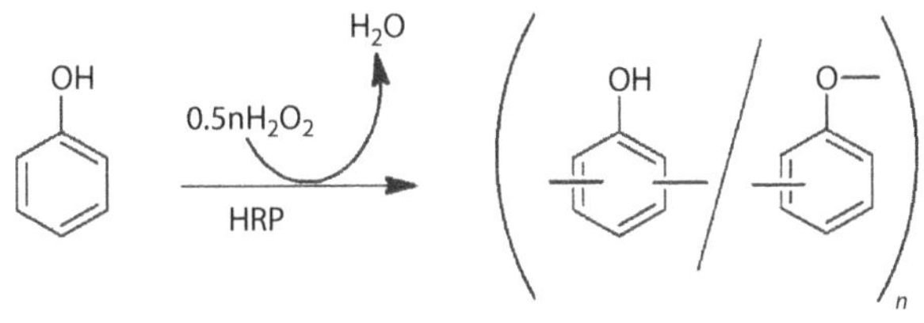

SCHEME 18.39 Phenolic monomer units in the polymer obtained using peroxidase.

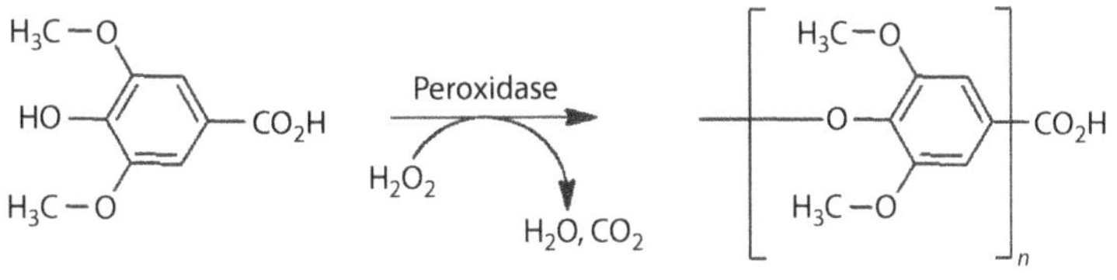

SCHEME 18.40 Peroxidase-assisted polymerization of syringic acid.

a polymer with a molecular mass of 2100–6000, soluble in DMF, was obtained by polymerization in aqueous methanol. A modification of the share of phenylene unit between the limits of 32% and 66% has succeeded, by changing the solvent composition [98].

The presence of the substituents on the phenol ring increases the solvent-mediated control on the regioselectivity of enzymatic polymerization [99]. This control was investigated for the 4-substituted phenols. Besides phenol, *p*-phenylphenol [95], *meta*-cresol [100], 4,4′-Dihydroxydiphenylether [101], A and F bisphenols, have also been polymerized.

An interesting example is offered by syringic acid polymerization, which undergoes by decarboxylation and is catalyzed by HRP or by soja beans peroxidase. The molecular weight of the obtained polymer was of approximately 1.3×10^4 (Scheme 18.40) [102].

The working environment for enzymatic oxidative polymerization must provide, on the one hand, solubilization of phenolic monomers and oligomers, and an efficient buffering on the other hand, taking into consideration the existence of a proton transfer, associated to electron traffic between the substrate and the enzyme. Polar solvents such as methanol, ethanol, acetone, *N,N′*-dimethylformamide, 1,4-dioxane, and methyl formate have been proved to be the most efficient organic solvents.

Utilization of modified cyclodextrins, such as 2,6-di-*O*-methyl-β-cyclodextrin, eliminates the need for organic solvents because these structures, when are used in catalytic amounts, include the substrate in a hydrophilic coat, allowing the enzyme to interact with the exposed substrate area. Thus, by phenol polymerization in the presence of cyclodextrin in buffer only, a polymer with the weight in the range 1000–1400 was obtained with a high yield [103]. The polymer was soluble in acetone and DMF.

Phosphate buffer and the organic buffer *N*-(2-hydroxyethyl)piperazine-*N′*-(2-ethanesulphonic acid) (HEPES) were most often used as they maintain the pH in the neutral domain.

The reaction kinetics depends on the structure of phenolic monomers. Phenols and aromatic amines with electron-withdrawing groups could hardly be polymerized by HRP catalysis, but phenols and aromatic amines with electron-donating groups could easily be polymerized. The reaction rate of either the *para*-substituted or *meta*-substituted substrate was higher than that of *ortho*-substituted substrate. When *ortho*- and *para* position related to the hydroxylic group of phenols are occupied by an electron-repulsive group, the reaction rate increased [104].

One solution to avoid the introduction of hydrogen peroxide or other peroxides is to use peroxidase together with glucose oxidase. In this way, hydrogen peroxide is generated in situ using glucose as a co-substrate. This one will be oxidized to gluconolactone (Scheme 18.41) simultaneously with the generation of hydrogen peroxide that will be used by peroxidase [105].

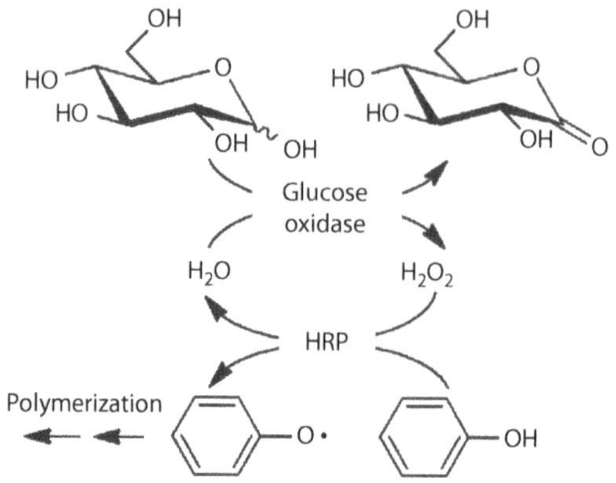

SCHEME 18.41 The combined action of two enzymes for in situ generation of hydrogen peroxide.

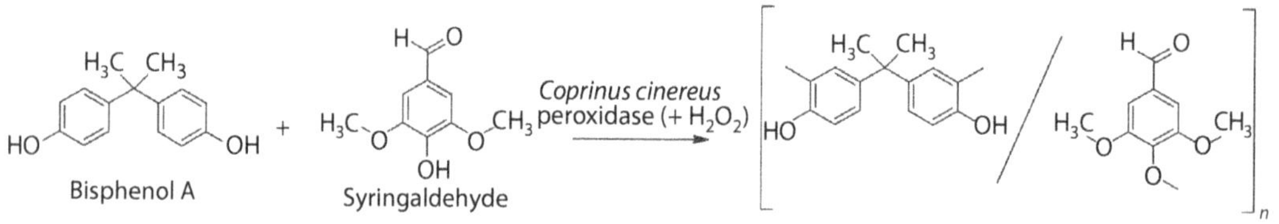

SCHEME 18.42 Copolymerization of syringaldehyde and bisphenol A.

An interesting case is that of syringaldehyde. This compound is abundant in plants, especially as glycosides. In contrast to the earlier-presented syringic acid, which can be polymerized by decarboxylation with the participation of peroxidase, syringaldehyde seems unable to polymerize, because positions 2, 4, and 6 related to the hydroxylic group are firmly blocked. In a recent paper it has been demonstrated that although the products of radicalic enzymatic polymerization of syringaldehyde with peroxidase from *Coprinus cinereus* (CiP) have a low molecular mass, this monomer may be included, by a radicalic mechanism, into a co-polymer with bisphenol A [106].

The incorporation into the copolymer can be explained by the attachment of syringaldehyde to the ends of the polymer, its grafting on bisphenol A units, and also by transformation of syringaldehyde, through an oxidative decarboxylation process, in 2,6-dimethoxybenzoquinone, followed by a reduction in the reaction environment, to 1,4-dihydroxy-2,6-dimethoxy-benzene. This diphenolic product can be incorporated into the main chain.

The yield of its incorporation into the polymer, the number average molecular weight (Mn), weight average molecular weight (Mw), and also its polydispersity increase with the lowering of syringaldehyde content in the initial mixture. The presence of syringaldehyde in the copolymer has been proven by FT-IR spectrometry and also by ^{13}C NMR. Incorporation of syringaldehyde with bisphenol A leads to a copolymer with an increased thermal stability (Scheme 18.42) [102].

18.4.2 PHENOLIC POLYMERS BY LACCASE-ASSISTED POLYMERIZATION

Laccases (E.C. 1.10.3.2.) are monophenoloxidases that oxidize a phenol group to a radical and transfers hydrogen via an enzyme-bound cupric ion to a molecular oxygen. Hence, in contrast to peroxidases, laccase uses Cu instead of Fe and the electron acceptor is oxygen instead of hydrogen peroxide. Laccases are widely distributed in plants, fungi, and bacteria. If in plants their main biological role is achieved together with peroxidases and consists in participation to lignification process, in pathogenic fungi laccase participates *inter alia* to inactivation of certain phytoalexinic and taninic toxins.

Laccase or, if we refer to the multitude of sources from which they come, laccases have a historical determination as being the first applicative example of enzyme use in polymerization [107].

This refers to the fact that, because of a high laccase content, the latex from Japanese/Chinese lacquer tree (*Toxicodendron vernicifluum* or formerly *Rhus vernicifera*) hardened in presence of air [108]. Through this report, laccase also became one of the first enzymes described in the literature.

The common substrates for laccases are phenols and aliphatic or aromatic amines. The electron transfer process takes place through a cluster of four cupric ions which form the catalytic core of the enzyme. Two of these ions directly interact. This interaction confers to the enzymes an intense blue color [109]. By completing a catalytic cycle which involves the four cupric ions, from one molecule of oxygen are obtained four water molecules, respectively four radicals, which can interact according to the model shown in peroxidases, through combination and electronic transfer. Laccase-mediated electron transfer from the substrate is not always possible, either because of the large dimensions of the substrate which does not have access to the active site of the enzyme, or because of the elevated redox potential of the substrate [110]. Using the model of nature, human factor can introduce mediators. These are substances able to participate in redox reactions. They will be inserted into the original electron chain. The role of mediators is not only to increase the efficiency of substrate use, but also to widen the range of substrates (Scheme 18.43) [111].

Laccases normally oxidize substrates with a redox potential inferior to the source potential ($E°$ in the range +475 to +790 mV according to the source of laccase). Using mediators, laccases can oxidize molecules with a superior redox potential [111]. A couple of mediators are shown in Figure 18.3. Although more than 100 types of mediators are known, the most commonly used are still 1-hydroxybenzotriazole (HBT) and 2,2′-azino-bis-(3-ethylbenzothiazoline-6-sulfonic acid) (ABTS).

There are relatively few examples of polymers obtained by action of laccases:

Polymerization of syringic acid (4-hydroxy-3,5-dimethoxybenzoic acid) produces a polymer (poly(oxy-2,6-dimethoxy-1,4-phenylene) similar to the one formed by polymerization of 2,6-dimethyphenol. It is important to note that laccase-assisted polymerization, just as the HRP-assisted one, occurs through decarboxylation. Polymers with masses up to 1.8×10^4, slightly higher than in peroxidase-assisted polymerization have been obtained in the presence of laccases from *Pycnoporus coccineus* and *Myceliophthore* [102].

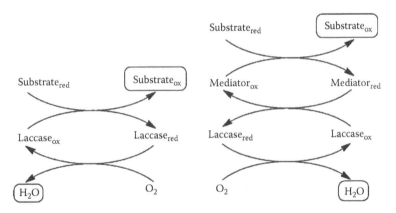

SCHEME 18.43 Insertion of mediators in the redox chain of substrates oxidation by laccase.

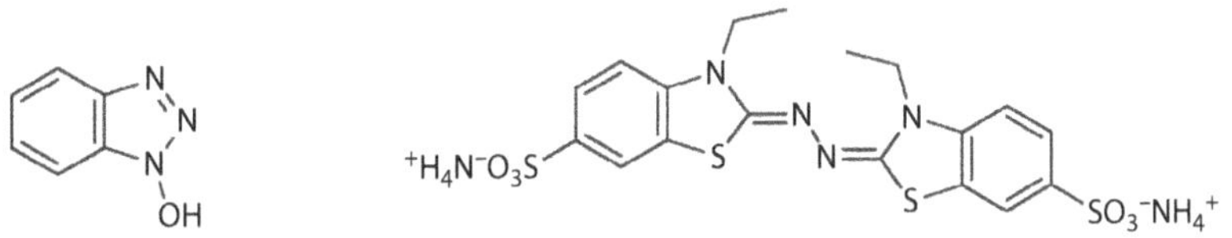

1*H*-benzo[*d*][1,2,3]triazol-1-ol (HBT)   2,2′-Azino-bis-(3-ethylbenzothiazoline-6-sulfonic acid) (ABTS)

FIGURE 18.3 Two laccase mediators.

Polymerization of 4-hydroxy-benzoic acid assisted by peroxidases from horse radish, soybean, and *Coprinus cinerius*, or by laccases from *P. coccineus*, *Myceliophtore*, and *Pyricularia oryzae* did not succeed, as mentioned by Ikeda [102] in the earlier study. On the contrary, it was found that in the presence of laccase from *Trametes villosa*, expressed in an *Aspergillus* host, 4-hydroxy-benzoic acid can polymerize. A polymer with weight-average molecular mass (Mw) of 17,550 was obtained after 4h, and of 17,800 after 24h, at 45°C. The ability of laccase to bind 4-hydroxy-benzoic acid on lignin structures has been also highlighted [112]. Because of a more reduced steric hindrance, 4-hydroxy-benzoic acid is grafted easily on lignin components in the presence of laccase, as compared to vanillic or syringic acids.

18.5 ENZYMATIC POLYMERIZATION OF VINYLIC MONOMERS

The first mention on the possibility of enzymatic polymerization of acrylic monomers was issued in 1951 and is due to Parravano [113]. He describes the possibility to initiate the polymerization of aqueous methyl methacrylate with biocatalytic systems such as *E. coli* cells and xanthine oxidase, in the absence of oxygen. Until the 1990s, this topic was not significantly approached. In 1992, a paper of Derango and co-workers reported the polymerization of certain acrylic monomers (acrylamide, hydroxyethyl metacrylate, acrylic acid, and methyl acrylate in the presence of xanthine oxidase, horseradish peroxidase, chloroperoxidase, and alcohol oxidase) [114]. They motivated the use of these enzymatic equipments by the fact known at that time that acrylic monomers do not polymerize through an ionic mechanism, but through a radicalic one. Therefore, the used enzymes were selected on the basis of their ability to produce radicals. Although this paper does not specify the working conditions and the characteristics of the obtained polymers, we can state that it is the first modern presentation of enzymatically assisted polymerization of vinylic monomers. Since then, there was a constant increase of interest for enzymatic polymerization of vinylic monomers. The most investigated were acrylic and styrenic monomers [1].

Oxidative polymerization of aromatic compounds by peroxidase or laccase systems was described earlier. These enzymes develop, by specific mechanisms, in presence of hydrogen peroxide or molecular oxygen, respectively, radical species which can conduct to macromolecular products by recombination with a reduced specificity.

Further studies showed that in the polymerization of acrylic substrates, peroxidases are the most useful, followed by laccases, even if the first ones have the disadvantage of using hydrogen peroxide as an oxidizing agent to obtain the initiator in free radical state.

Several factors can influence the parameters of polymerization reaction. Among these are the nature, the presence, and the concentration of initiator.

It has been emphasized that utilization of a species that can generate free radicals (the initiator, or the initiator's analogue in chemical polymerization) in the enzyme–substrate polymerization system greatly improves the yield of the process. Thus, enzymatic polymerization differs from the chemical one only by the way radicals are generated (formation of the active initiator). Propagation and termination steps are similar.

In principle, polymerization of a vinylic substrate can also take place without an initiator; in this case the source of radicals would be one of the nondedicated components in the environment. This statement is valid for laccases, but not for peroxidases, which need the presence of the ternary system: substrate, initiator source (usually a β-diketone), and the oxidant. However, for laccases, the yield of reaction increases in the presence of a dedicated initiator. Thus, polymerization of acrylamide without an initiator was carried out in water, at relatively low temperature (50°C), with laccase from *P. coccineus*. The result was a polymer of high molecular weight. In the presence of 2,4-pentanedione, laccase efficiently mediated the vinyl polymerization at room temperature [115].

β-diketones can be used as radicalic initiators because of the liability of the proton (hydrogen atom) located between the two keto groups. The taking over of this atom (proton + electron) produces a mesomeric stabilized radical (Scheme 18.44).

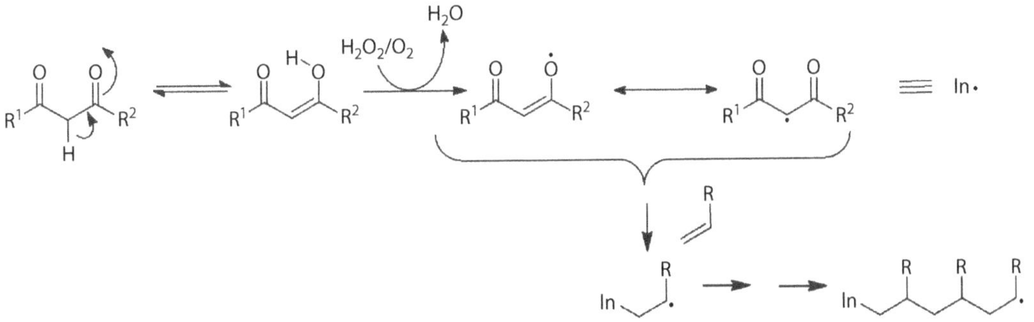

SCHEME 18.44 Mechanism of generation, stabilization, and initiation of polymerization by radicals derived from β-diketonic initiators, in polymerization with peroxidase.

Acrylamide polymerization in presence of horse radish peroxidase at room temperature, conducted to polymers with Mn in the range 0.15–0.46×10^6 g/mol, and a dispersion Mw/Mm = 2.0–2.4. The yield was in the range of 72%–95%; these values lowered, with a decrease in hydrogen peroxide concentration under 1.0/0.66 (mol × L^{-1}/mol × L^{-1}) [acrylamide]$_0$/[H$_2$O$_2$]$_0$. The absence from the reaction environment of hydrogen peroxide, of enzyme or of 2,4-pentanedione, was independently tested. In all cases when one of the three components was missing, the reaction did not occur, proving that polymerization does not take place in absence of the oxidant, the reducer (initiator source), and the enzyme [116].

Again in case of acrylamide polymerization, but in the presence of laccase from *P. coccineus*, and without diketone, at 50°C–80°C, a polymer was obtained with a degree of polymerization similar to the one in the previous example: Mn ≈ 1×10^6 g/mol and Mw/Mm ≈ 2.0 [116].

In a study dedicated to styrene polymerization catalyzed by horseradish peroxidase at ambient temperature, molecular weight and yield of polystyrene were influenced by solvent, concentration of hydrogen peroxide, and initiator (β-diketones, coumarin). THF:H$_2$O (v/v) and hydrogen peroxide (0.082 mol/L) provided maximum yield of polymer (21.2% weight conversion of styrene to polystyrene) with 2,4-pentanedione as initiator. 1,3-Cyclopentanedione and dibenzoylmethane as initiators resulted in higher yield of polymer (60%) and a higher molecular weight (Mn = 96,504, polydispersity = 2.16), respectively. This enzymatic strategy was also used for the synthesis of polymers from styrene derivatives, 4-methylstyrene and 2-vinylnaphthalene, the latter resulting in a >90% yield of polymer [117]. The presence of the initiators in the polymer chains was reported, which is in complete agreement with the mechanism shown in Scheme 18.44. Several types of β-diketones as well as cumarin have been used as initiators in enzymatic polymerization experiments (Figure 18.4).

The molecular weight of polymer depends on enzyme concentration. When the enzyme concentration is higher, the average molar weight of the obtained polymer is smaller [118]. It is obvious that

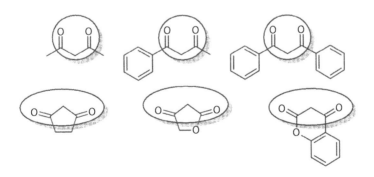

FIGURE 18.4 Types of initiators used in styrene polymerization.

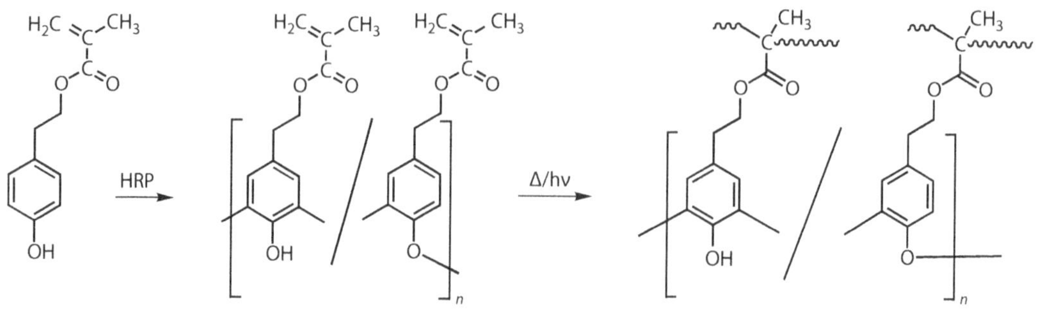

SCHEME 18.45 Chemoselective polymerization of a phenol derivative having a methacryl group by peroxidase catalyst.

high enzyme concentrations simultaneously initiate more polymerization chains; thus the average molar mass will be smaller, even if monomer conversion is greater.

The influence of hydrogen peroxide concentration on polymerization reaction parameters is insignificant, within the limits of a small inferior concentration, necessary for initiator formation by the existing (small) amount of enzyme, and a high concentration, from which oxidation processes of the enzyme may take place [118].

Methyl metacrylate polymerization can be carried out at ambient temperatures in the presence of low concentrations of hydrogen peroxide and 2,4-pentanedione in a mixture of water and a water-miscible solvent [119]. Polymers of MMA formed were highly stereoregular with predominantly syndiotactic sequences (syn-dyad fractions from 0.82 to 0.87). Analyses of the chloroform-soluble fraction of syndio-PMMA products by GPC showed that they have number-average molecular weights, Mn, that ranged from 7,500 to 75,000. By using 25% v/v cosolvents: dioxane tetrahydrofuran, acetone, and dimethylformamide, 85%, 45%, 7%, and 2% product yields, respectively, resulted after 24 h. Increasing the proportion of dioxane to water from 1:3 to 1:1 and 3:1 resulted in a decrease in polymer yield from 45% to 38% and 7%, respectively. An increase in the enzyme concentration from 70 to 80 and 90 mg/mL resulted in increased reaction rate. By adjustment of the molar ratio of 2,4-pentanedione to hydrogen peroxide between 1.30:1.0 and 1.45:1.0, the product yields and Mn values were increased. On the basis of the catalytic properties of HRP and studies herein, we believe that the keto—enoxy radicals from 2,4-pentanedione are the first radical species generated. Then, initiation may take place through this radical or by the radical transfer to another molecule.

Based on the aforementioned experiments, it can be concluded that peroxidases can initiate oxidative polymerization of aromatic rings and also polymerization of (meth)acrylic motifs. The polymerization reaction of acrylic monomers is much slower than that of phenol-type substituted aromatic structures. When both moieties coexist in the same molecule, due to the difference in reaction rate, peroxidases will selectively polymerize the aromatic structures (chemoselectivity). Completion of the reaction, in the sense of also achieving a polymerization of the vinylic structure, can be done chemically by thermic or photochemical classical initiation; in this case structures with two chains will be obtained, as for monomers of 4-hydroxyphenethyl methacrylate type (Scheme 18.45) [120].

18.6 CONCLUSIONS

This review, with no claim of completeness, presents the state of the art in obtaining polymers by enzymatic biotransformation. Enzymatic obtaining of polymers is certainly an area with great prospects. The research orientations required in the future are optimization of the reaction conditions, to produce polymers with desired properties; identification of new enzymes or biocatalysts for specific structural purposes, and improvement, by evolutionary techniques, of existing enzymes performances.

All these can be achieved by joint efforts whose motivation is highlighted by the utility of polymer product.

REFERENCES

1. Kobayashi, S. and A. Makino, Enzymatic polymer synthesis: An opportunity for green polymer chemistry. *Chemical Reviews*, 2009. **109**(11): 5288–5353.
2. Weijers, C.A.G.M., M.C.R. Franssen, and G.M. Visser, Glycosyltransferase-catalyzed synthesis of bioactive oligosaccharides. *Biotechnology Advances*, 2008. **26**(5): 436–456.
3. van der Vlist, J. and K. Loos, Transferases in polymer chemistry, in *Enzymatic Polymerization*, A.R.A. Palmans and A. Heise, Eds. 2011, Springer, Berlin, Germany, pp. 21–54.
4. Öhrlein, R., Glycosyltransferase-catalyzed synthesis of non-natural oligosaccharides, in *Biocatalysis: From Discovery to Application*, W.-D. Fessner, Ed. 1999, Springer Verlag, Berlin, Germany, pp. 227–254.
5. Glaser, L., The synthesis of cellulose in cell-free extracts of *Acetobacter xylinum. Journal of Biological Chemistry*, 1958. **232**(2): 627–636.
6. Flowers, H.M. et al., Biosynthesis of cellulose in vitro from guanosine diphosphate d-glucose with enzymic preparations from *Phaseolus aureus* and *Lupinus albus. Journal of Biological Chemistry*, 1969. **244**(18): 4969–4974.
7. Murata, T. and T. Usui, Enzymatic synthesis of oligosaccharides and neoglycoconjugates. *Bioscience, Biotechnology, and Biochemistry*, 2006. **70**(5): 1049–1059.
8. Zhang, Y. et al., Glycopolymers: The future antiadhesion drugs, in *Polymer Biocatalysis and Biomaterials II*, H.N. Cheng and R.A. Gross, Eds. 2008, American Chemical Society, Washington, DC, pp. 342–361.
9. Deguchi, K.S., H. Genzou, I. Masahito, N. Hiroaki, and N. Shinichiro, Oligosaccharide synthesizer. 2006. United States Patent 7070988.
10. Johnson, K.F.H., D.J. Bezila, D.E. Taylor, J. Simala-grant, and D. Rasko, Synthesis of oligosaccharides, glycolipids, and glycoproteins using bacterial glycosyltransferases. 2009, Neose Technologies, Inc., Horsham, PA, Governors of the University of Alberta, Edmonton, Alberta, Canada. United States Patent 7524655.
11. Nilsson, I.K.G.L., Method for synthesis of oligosaccharides. 1993, Procur AB, Lund, SE. United States Patent 5246840.
12. Roth, S.G., Apparatus for glycosyltransferase-catalyzed saccharide synthesis. 2003, The Trustees of the University of Pennsylvania, Philadelphia, PA. United States Patent 6544778.
13. Boons, G.-J. and K. Hale, *Organic Synthesis with Carbohydrates*. 2000, Sheffield Academic Press, Sheffield, U.K., Vol. xi, 336pp.
14. Wong, C.H. et al., Enzymes in organic synthesis: Application to the problems of carbohydrate recognition (part 2). *Angewandte Chemie International Edition in English*, 1995. **34**(5): 521–546.
15. Homann, A. and J. Seibel, Towards tailor-made oligosaccharides—Chemo-enzymatic approaches by enzyme and substrate engineering. *Applied Microbiology and Biotechnology*, 2009. **83**(2): 209–216.
16. Frey, P.A. and A.D. Hegeman, *Enzymatic Reaction Mechanisms*. 2007, Oxford University Press, Oxford, U.K., Vol. xviii, 831pp.
17. Palm, D. et al., The role of pyridoxal 5′-phosphate in glycogen phosphorylase catalysis. *Biochemistry*, 1990. **29**(5): 1099–1107.
18. Ohdan, K. et al., Phosphorylase coupling as a tool to convert cellobiose into amylose. *Journal of Biotechnology*, 2007. **127**(3): 496–502.
19. Fujii, K. et al., Bioengineering and application of novel glucose polymers. *Biocatalysis and Biotransformation*, 2003. **21**(4–5): 167–172(6).
20. Yanase, M., T. Takaha, and T. Kuriki, α-Glucan phosphorylase and its use in carbohydrate engineering. *Journal of the Science of Food and Agriculture*, 2006. **86**(11): 1631–1635.
21. Kadokawa, J.-I. and S. Kobayashi, Polymer synthesis by enzymatic catalysis. *Current Opinion in Chemical Biology*, 2010. **14**(2): 145–153.
22. Kaneko, Y., K. Beppu, and J.-I. Kadokawa, Amylose selectively includes a specific range of molecular weights in poly(tetrahydrofuran)s in vine-twining polymerization. *Polymer Journal*, 2009. **41**(9): 792–796.
23. Kaneko, Y. et al., Selectivity and priority on inclusion of amylose toward guest polyethers and polyesters in vine-twining polymerization. *Polymer Journal*, 2009. **41**(4): 279–286.
24. van der Vlist, J. et al., Synthesis of branched polyglucans by the tandem action of potato phosphorylase and *Deinococcus geothermalis* glycogen branching enzyme. *Macromolecular Rapid Communications*, 2008. **29**(15): 1293–1297.
25. Faijes, M. and A. Planas, In vitro synthesis of artificial polysaccharides by glycosidases and glycosynthases. *Carbohydrate Research*, 2007. **342**(12–13): 1581–1594.

26. Plou, F.J., A.G.D. Segura, and A. Ballesteros, Application of glycosidases and transglycosidases in the synthesis of oligosaccharides, in *Industrial Enzymes*, J. Polaina and A.P. MacCabe, Eds. 2007, Springer, Dordrecht, the Netherlands, pp. 141–157.

27. Koshland, D.E., Stereochemistry and the mechanism of enzymatic reactions. *Biological Reviews*, 1953. **28**(4): 416–436.

28. Vocadlo, D.J. and G.J. Davies, Mechanistic insights into glycosidase chemistry. *Current Opinion in Chemical Biology*, 2008. **12**(5): 539–555.

29. Kobayashi, S. et al., Novel method for polysaccharide synthesis using an enzyme: The first in vitro synthesis of cellulose via a nonbiosynthetic path utilizing cellulase as catalyst. *Journal of the American Chemical Society*, 1991. **113**(8): 3079–3084.

30. Kobayashi, S., Challenge of synthetic cellulose. *Journal of Polymer Science Part A: Polymer Chemistry*, 2005. **43**(4): 693–710.

31. Okada, G., D.S. Genghof, and E.J. Hehre, The predominantly nonhydrolytic action of alpha amylases on [alpha]-maltosyl fluoride. *Carbohydrate Research*, 1979. **71**(1): 287–298.

32. Kobayashi, S. and M. Ohmae, Enzymatic polymerization to polysaccharides. *Enzyme-Catalyzed Synthesis of Polymers*, 2006. **194**: 159–210.

33. Linhardt, R.J. and M. Weïwer, Synthesis of glycosaminoglycans and their oligosaccharides in *Comprehensive Glycoscience, Volume 1*, J. Kamerling et al., Eds. 2007, Elsevier science, Oxford, U.K., pp. 713–745.

34. Plou, F. et al., Glucosyltransferases acting on starch or sucrose for the synthesis of oligosaccharides. *Canadian Journal of Chemistry*, 2002. **80**(6): 743–752.

35. Monsan, P., M. Remaud-Siméon, and I. André, Transglucosidases as efficient tools for oligosaccharide and glucoconjugate synthesis. *Current Opinion in Microbiology*, 2010. **13**(3): 293–300.

36. Vasur, J. et al., Synthesis of cyclic β-glucan using laminarinase 16A glycosynthase mutant from the basidiomycete *Phanerochaete chrysosporium*. *Journal of the American Chemical Society*, 2010. **132**(5): 1724–1730.

37. Williams, S.J. and S.G. Withers, Glycosyl fluorides in enzymatic reactions. *Carbohydrate Research*, 2000. **327**(1–2): 27–46.

38. Perugino, G. et al., Oligosaccharide synthesis by glycosynthases. *Trends in Biotechnology*, 2004. **22**(1): 31–37.

39. Hommalai, G. et al., Enzymatic synthesis of cello-oligosaccharides by rice BGlu1 β-glucosidase glycosynthase mutants. *Glycobiology*, 2007. **17**(7): 744–753.

40. Shaikh, F. and S. Withers, Teaching old enzymes new tricks: Engineering and evolution of glycosidases and glycosyl transferases for improved glycoside synthesis. *Biochemistry and Cell Biology*, 2008. **86**(2): 169–177.

41. Mackenzie, L.F. et al., Glycosynthases: Mutant glycosidases for oligosaccharide synthesis. *Journal of the American Chemical Society*, 1998. **120**(22): 5583–5584.

42. Drone, J. et al., *Thermus thermophilus* glycosynthases for the efficient synthesis of galactosyl and glucosyl-(1→3)-glycosides. *European Journal of Organic Chemistry*, 2005. **2005**(10): 1977–1983.

43. Müllegger, J. et al., Thermostable glycosynthases and thioglycoligases derived from thermotoga maritima β-glucuronidase. *ChemBioChem*, 2006. **7**(7): 1028–1030.

44. Wilkinson, S.M. et al., *Escherichia coli* glucuronylsynthase: An engineered enzyme for the synthesis of β-glucuronides. *Organic Letters*, 2008. **10**(8): 1585–1588.

45. Fort, S. et al., Highly efficient synthesis of β(1 → 4)-oligo- and -polysaccharides using a mutant cellulase. *Journal of the American Chemical Society*, 2000. **122**(23): 5429–5437.

46. Hrmova, M. et al., Mutated barley (1,3)-β-d-glucan endohydrolases synthesize crystalline (1,3)-β-d-glucans. *Journal of Biological Chemistry*, 2002. **277**(33): 30102–30111.

47. Faijes, M. et al., In vitro synthesis of a crystalline (1→3, 1→4)-beta-D-glucan by a mutated (1→3, 1→4)-beta-D-glucanase from *Bacillus*. *Biochemical Journal*, 2004. **380**(Pt 3): 635.

48. Gullfot, F. et al., Functional characterization of xyloglucan glycosynthases from GH7, GH12, and GH16 scaffolds. *Biomacromolecules*, 2009. **10**(7): 1782–1788.

49. Volova, T., *Polyhydroxyalkanoates—Plastic Materials of the 21st Century: Production, Properties, Applications*. 2004, Nova Science Pub Inc., Hauppauge, New York.

50. Guisan, J., *Immobilization of Enzymes and Cells*. 2006, Humana Press Inc., Totowa, NJ.

51. Gross, R.A., M. Ganesh, and W. Lu, Enzyme-catalysis breathes new life into polyester condensation polymerizations. *Trends in Biotechnology*, 2010. **28**(8): 435–443.

52. Heise, A. and A. Palmans, Hydrolases in polymer chemistry: Chemoenzymatic approaches to polymeric materials, in *Enzymatic Polymerisation*, A.R.A. Palmans and A. Heise, Eds. 2011, Springer, Berlin, Germany, pp. 79–113.

53. Bencze, L.C. et al., CaL-B a highly selective biocatalyst for the kinetic resolution of furylbenzthiazole-2-yl-ethanols and acetates. *Tetrahedron: Asymmetry*, 2010. **21**(16): 1999–2004.

54. Bencze, L.C. et al., Synthesis of enantiomerically enriched (R)- and (S)-benzofuranyl- and benzo[b]thio-phenyl-1,2-ethanediols via enantiopure cyanohydrins as intermediates. *Tetrahedron: Asymmetry*, 2010. **21**(4): 443–450.

55. Brem, J. et al., Lipase-catalyzed kinetic resolution of racemic 1-(10-alkyl-10H-phenothiazin-3-yl)etha-nols and their butanoates. *Tetrahedron: Asymmetry*, 2010. **21**(16): 1993–1998.

56. Mahapatro, A., A. Kumar, and R.A. Gross, Mild, solvent-free ω-hydroxy acid polycondensations cata-lyzed by Candida antarctica lipase B. *Biomacromolecules*, 2004. **5**(1): 62–68.

57. Mahapatro, A. et al., Solvent-free adipic acid/1,8-octanediol condensation polymerizations catalyzed by Candida antarctica lipase B. *Macromolecules*, 2003. **37**(1): 35–40.

58. Ajima, A. et al., Polymerization of 10-hydroxydecanoic acid in benzene with polyethylene glycol-modi-fied lipase. *Biotechnology Letters*, 1985. **7**(5): 303–306.

59. Ebata, H. and S. Matsumura, Polyricinoleate composition and process for producing the same. 2007. United States Patent application 0270550 (2009).

60. Noll, O. and H. Ritter, Enzymes in polymer chemistry, 9. Polymerizable oligoesters from cholic acid via lipase catalyzed condensation reactions. *Macromolecular Rapid Communications*, 1996. **17**(8): 553–557.

61. Berkane, C. et al., Lipase-catalyzed polyester synthesis in organic medium. Study of ring–chain equilib-rium. *Macromolecules*, 1997. **30**(25): 7729–7734.

62. Okumura, S., M. Iwai, and Y. Tominaga, Synthesis of ester oligomer by *Aspergillus niger* lipase. *Agricultural and Biological Chemistry*, 1984. **48**(11): 2805–2808.

63. Kobayashi, S., Lipase-catalyzed polyester synthesis—A green polymer chemistry. *Proceedings of the Japan Academy, Series B*, 2010. **86**(4): 338–365.

64. Hilker, I. et al., Chiral polyesters by dynamic kinetic resolution polymerization. *Angewandte Chemie International Edition*, 2006. **45**(13): 2130–2132.

65. Albertsson, A.-C. and I.K. Varma, Recent developments in ring opening polymerization of lactones for biomedical applications. *Biomacromolecules*, 2003. **4**(6): 1466–1486.

66. Knani, D., A. Gutman, and D. Kohn, Enzymatic polyesterification in organic media. Enzyme-catalyzed syn-thesis of linear polyesters. I. Condensation polymerization of linear hydroxyesters. II. Ring-opening polym-erization of ε-caprolactone. *Journal of Polymer Science Part A: Polymer Chemistry*, 1993. **31**(5): 1221–1232.

67. Varma, I.K. et al., Enzyme catalyzed synthesis of polyesters. *Progress in Polymer Science*, 2005. **30**(10): 949–981.

68. Uyama, H. and S. Kobayashi, Enzymic ring-opening polymerization of lactones catalyzed by lipase. *Chemistry Letters*, 1993. **7**: 1149–1150.

69. Kobayashi, S., Enzymatic ring-opening polymerization of lactones by lipase catalyst: Mechanistic aspects. *Macromolecular Symposia*, 2006. **240**(1): 178–185.

70. Córdova, A. et al., Lipase-catalysed formation of macrocycles by ring-opening polymerisation of [epsilon]-caprolactone. *Polymer*, 1998. **39**(25): 6519–6524.

71. Matsumura, S., Enzymatic synthesis of polyesters via ring-opening polymerization, in *Enzyme-Catalyzed Synthesis of Polymers*, S. Kobayashi, H. Ritter, and D. Kaplan, Eds. 2006, Springer, Berlin, Germany, pp. 95–132.

72. Kobayashi, S., Recent developments in lipase-catalyzed synthesis of polyesters. *Macromolecular Rapid Communications*, 2009. **30**(4–5): 237–266.

73. Nobes, G.A.R., R.J. Kazlauskas, and R.H. Marchessault, Lipase-catalyzed ring-opening polymerization of lactones: A novel route to poly(hydroxyalkanoate)s. *Macromolecules*, 1996. **29**(14): 4829–4833.

74. Shimada, K. et al., Poly-(L)-malic acid: A new protease inhibitor from *Penicillium cyclopium*. *Biochemical and Biophysical Research Communications*, 1969. **35**(5): 619.

75. Liu, S. and A. Steinbüchel, Investigation of poly(β-L-malic acid) production by strains of *Aureobasidium pullulans*. *Applied Microbiology and Biotechnology*, 1996. **46**(3): 273–278.

76. Matsumura, S. and S. Yoshikawa, Biodegradable poly(carboxylic acid) design in *Agricultural and Synthetic Polymers, Biodegradability and Utilization*, J.E. Glass and G. Swift, Eds. 1990, ACS Symposium Series, American Chemical Society, Washington, DC.

77. Küllmer, K. et al., Lipase-catalyzed ring-opening polymerization of α-methyl-δ-valerolactone and α-methyl-ε-caprolactone. *Macromolecular Rapid Communications*, 1998. **19**(2): 127–130.

78. Nishida, H. et al., Synthesis of metal-free poly(1,4-dioxan-2-one) by enzyme-catalyzed ring-opening polymerization. *Journal of Polymer Science Part A: Polymer Chemistry*, 2000. **38**(9): 1560–1567.

79. Kikuchi, H., H. Uyama, and S. Kobayashi, Lipase-catalyzed enantioselective copolymerization of substi-tuted lactones to optically active polyesters. *Macromolecules*, 2000. **33**(24): 8971–8975.

80. Peeters, J.W. et al., Lipase-catalyzed ring-opening polymerizations of 4-substituted ε-caprolactones: Mechanistic considerations. *Macromolecules*, 2005. **38**(13): 5587–5592.

81. van Buijtenen, J. et al., Switching from S- to R-selectivity in the Candida antarctica lipase B-catalyzed ring-opening of ω-methylated lactones: Tuning polymerizations by ring size. *Journal of the American Chemical Society*, 2007. **129**(23): 7393–7398.

82. Veld, M.A.J. et al., Lactone size dependent reactivity in Candida antarctica lipase B: A molecular dynamics and docking study. *ChemBioChem*, 2009. **10**(8): 1330–1334.

83. Geus, M.D. et al., Insights into the initiation process of enzymatic ring-opening polymerization from monofunctional alcohols using liquid chromatography under critical conditions. *Biomacromolecules*, 2008. **9**(2): 752–757.

84. Uyama, H., S. Suda, and S. Kobayashi, Enzymatic synthesis of terminal-functionalized polyesters by initiator method. *Acta Polymerica*, 1998. **49**(12): 700–703.

85. Takwa, M. et al., Lipase catalyzed HEMA initiated ring-opening polymerization: In situ formation of mixed polyester methacrylates by transesterification. *Biomacromolecules*, 2008. **9**(2): 704–710.

86. Bisht, K.S. et al., Ethyl glucoside as a multifunctional initiator for enzyme-catalyzed regioselective lactone ring-opening polymerization. *Journal of the American Chemical Society*, 1998. **120**(7): 1363–1367.

87. Uyama, H., H. Kikuchi, and S. Kobayashi, Single-step acylation of polyester terminals by enzymatic ring-opening polymerization of 12-dodecanolide in the presence of acyclic vinyl esters. *Bulletin of the Chemical Society of Japan*, 1997. **70**(7): 1691–1695.

88. Kobayashi, S., H. Kikuchi, and H. Uyama, Lipase-catalyzed ring-opening polymerization of 1,3-dioxan-2-one. *Macromolecular Rapid communications*, 1997. **18**(7): 575–579.

89. Matsumura, S., K. Tsukada, and K. Toshima, Enzyme-catalyzed ring-opening polymerization of 1,3-dioxan-2-one to poly(trimethylene carbonate). *Macromolecules*, 1997. **30**(10): 3122–3124.

90. Bisht, K.S. et al., Lipase-catalyzed ring-opening polymerization of trimethylene carbonate[†]. *Macromolecules*, 1997. **30**(25): 7735–7742.

91. He, F., Immobilized porcine pancreas lipase for polymer synthesis, in *Polymer Biocatalysis and Biomaterials II*, H.N. Cheng1 and R.A. Gross, Eds. 2008, American Chemical Society, Washington, DC, pp. 144–154.

92. Wu, R., T.F. Al-Azemi, and K.S. Bisht, Functionalized polycarbonate derived from tartaric acid: Enzymatic ring-opening polymerization of a seven-membered cyclic carbonate. *Biomacromolecules*, 2008. **9**(10): 2921–2928.

93. Albertsson, A.-C. and R.K. Srivastava, Recent developments in enzyme-catalyzed ring-opening polymerization. *Advanced Drug Delivery Reviews*, 2008. **60**(9): 1077–1093.

94. Reihmann, M. and H. Ritter, Synthesis of phenol polymers using peroxidases, in *Enzyme-Catalyzed Synthesis of Polymers*, S. Kobayashi, H. Ritter, and D. Kaplan, Eds. 2006, Springer, Berlin, Germany, pp. 1–49.

95. Valderrama, B., M. Ayala, and R. Vazquez-Duhalt, Suicide inactivation of peroxidases and the challenge of engineering more robust enzymes. *Chemistry & Biology*, 2002. **9**(5): 555–565.

96. Dordick, J.S., M.A. Marletta, and A.M. Klibanov, Polymerization of phenols catalyzed by peroxidase in nonaqueous media. *Biotechnology and Bioengineering*, 1987. **30**(1): 31–36.

97. Montellano, P.R.O.D., Catalytic mechanisms on heme peroxidases, in *Biocatalysis Based on Heme Peroxidases*, E. Torres and M. Ayala, Eds. 2010, Springer-Verlag, Berlin, Germany, pp. 79–105.

98. Kobayashi, S. et al., Regio- and chemo-selective polymerization of phenols catalyzed by oxidoreductase enzyme and its model complexes. *Macromolecular Symposia*, 2001. **175**(1): 1–10.

99. Mita, N. et al., Precise structure control of enzymatically synthesized polyphenols. *Bulletin of the Chemical Society of Japan*, 2004. **77**(8): 1523–1527.

100. Ayyagari, M., J. Akkara, and D. Kaplan, Solvent-enzyme-polymer interactions in the molecular-weight control of poly(m-cresol) synthesized in nonaqueous media in *Enzymes in Polymer Synthesis*, A.C. Society, Ed. 1998, ACS Symposium Series, American Chemical Society, Washington, DC.

101. Fukuoka, T. et al., Peroxidase-catalyzed oxidative polymerization of 4,4′-dihydroxydiphenyl ether. Formation of α,ω-hydroxyoligo(1,4-phenylene oxide) through an unusual reaction pathway. *Macromolecules*, 2000. **33**(24): 9152–9155.

102. Ikeda, R. et al., Enzymatic oxidative polymerization of 4-hydroxybenzoic acid derivatives to poly(phenylene oxide)s. *Polymer International*, 1998. **47**(3): 295–301.

103. Mita, N. et al., Enzymatic oxidative polymerization of phenol in an aqueous solution in the presence of a catalytic amount of cyclodextrin. *Macromolecular Bioscience*, 2002. **2**(3): 127–130.

104. Xu, Y.P., G.L. Huang, and Y.T. Yu, Kinetics of phenolic polymerization catalyzed by peroxidase in organic media. *Biotechnology and Bioengineering*, 1995. **47**(1): 117–119.

105. Uyama, H., H. Kurioka, and S. Kobayashi, Novel bienzymatic catalysis system for oxidative polymerization of phenols. *Polymer Journal*, 1997. **29**(2): 190–192.

106. An, E.S. et al., Peroxidase-catalyzed copolymerization of syringaldehyde and bisphenol A. *Enzyme and Microbial Technology*, 2010. **46**(3–4): 287–291.

107. Cañas, A.I. and S. Camarero, Laccases and their natural mediators: Biotechnological tools for sustainable eco-friendly processes. *Biotechnology Advances*, 2010. **28**(6): 694–705.

108. Yoshida, H., Chemistry of lacquer (urshi), part 1. *Journal of Chemical Society*, 1883. **43**: 472–486.

109. Piontek, K., M. Antorini, and T. Choinowski, Crystal structure of a laccase from the fungus Trametes versicolor at 1.90-Å resolution containing a full complement of coppers. *Journal of Biological Chemistry*, 2002. **277**(40): 37663.

110. Riva, S., Laccases: Blue enzymes for green chemistry. *Trends in Biotechnology*, 2006. **24**(5): 219–226.

111. Kunamneni, A. et al., Engineering and applications of fungal laccases for organic synthesis. *Microbial Cell Factories*, 2008. **7**(1): 32.

112. Chandra, R., C. Felby, and A. Ragauskas, Improving laccase-facilitated grafting of 4-hydroxybenzoic acid to high-kappa kraft pulps. *Journal of Wood Chemistry and Technology*, 2005. **24**(1): 69–81.

113. Parravano, G., Chain reactions induced by enzymic systems. *Journal of the American Chemical Society*, 1951. **73**(1): 183–184.

114. Derango, R. et al., Enzyme-mediated polymerization of acrylic monomers. *Biotechnology Techniques*, 1992. **6**(6): 523–526.

115. Ikeda, R. et al., Laccase-catalyzed polymerization of acrylamide. *Macromolecular Rapid Communications*, 1998. **19**(8): 423–425.

116. Emery, O. et al., Free-radical polymerization of acrylamide by horseradish peroxidase-mediated initiation. *Journal of Polymer Science Part A: Polymer Chemistry*, 1997. **35**(15): 3331–3333.

117. Singh, A., D. Ma, and D.L. Kaplan, Enzyme-mediated free radical polymerization of styrene. *Biomacromolecules*, 2000. **1**(4): 592–596.

118. Durand, A. et al., Enzyme-mediated radical initiation of acrylamide polymerization: Main characteristics of molecular weight control. *Polymer*, 2001. **42**(13): 5515–5521.

119. Kalra, B. and R.A. Gross, Horseradish peroxidase mediated free radical polymerization of methyl methacrylate. *Biomacromolecules*, 2000. **1**(3): 501–505.

120. Uyama, H. et al., Chemoselective polymerization of a phenol derivative having a methacryl group by peroxidase catalyst. *Macromolecules*, 1998. **31**(2): 554–556.

19 Polymer-Based Colloidal Aggregates as a New Class of Drug Delivery Systems

Cesare Cametti

CONTENTS

19.1 INTRODUCTION

Drug delivery occurs when an appropriate carrier, whether natural or synthetic, is combined in a suitable manner with a drug or other active bioagents in such a way that the drug itself is transferred to the target (biological structures, both tissues or biological cells) and then released in a predesigned and controlled manner.

The idea of pairing up a bioactive agent (the drug) with a transporter vehicle was borne in the early 1970s and forcefully advanced by Ringsdorf [1] who attracted attention on how drug conjugation technique requires two equally contributing elements, i.e., the polymer which acts as a carrier and transport vehicle and the drug that is inert during the transport process and exerts its bioactivity, under environmental conditions, only in the target tissue. In practical terms, the technique comprises the bioreversible binding of a bioactive agent, for example, an antitumor drug, to a water-soluble sub-micrometer-molecular carrier designed and synthesized in conformance with pharmaceutical requirements. This strategy has been proved to be particularly advantageous in the treatment of solid tumors, because polymer conjugates tend to accumulate thanks to the enhanced intra-tumoral vascular permeability, allowing for substantial leakage of the polymeric molecules into the tumor tissue [2].

This methodology presents several advantages compared with more traditional methods, namely the possibility to concentrate the drug in the region of interest, leaving practically unaffected the surrounding regions, the maintenance of the drug level within the desired range, the optimization of the drug treatment to which it follows an increase patient compliance. Moreover, the great flexibility of this technique, due to the variety of the structures the polymers are able to produce, allows modulating the drug release in a large extent in a controlled way, eliminating or reducing the potential risk of both the under- and overdosing.

So far, viral vectors are the most efficient and widely used gene delivery systems. However, their use in clinical trials is limited by safety problems. For this reason, large effort has been devoted in developing nonviral vectors employing nucleic acid-based drugs, nanoscopic or mesoscopic structures based on lipid self-assembly and, more recently, on polymeric aggregates of different structural complexity.

Until recently, structures based on lipid self-assembling have assumed a primary importance. When phospholipids are placed in an aqueous medium, the hydrophilic interaction of the lipid head groups with water results in the formation of multilamellar and/or unilamellar vesicles. These structures (lipoplexes), which resemble a biological membrane, form spontaneously as a result of electrostatic interactions between positively charged groups of the polycation and the negatively charged phosphate groups of DNA. Because of their entrapping ability, liposomes are considered as drug carrier structures. Lipoplexes introduce anionic polymers (DNA) inside cells in vitro and in vivo by means of adsorptive endocytosis and pore formation and/or fusion [3]. These structures promise to be a viable alternative to the viral vectors, because of absence of risk of infections, the low immunogenicity, as well as the ease of manufacturing and their broad applicability.

Many factors are known to affect the efficiency of the gene transfer by means of lipoplex carriers. Among these parameters, there are the composition of the liposomes, the cationic lipid-to-DNA ratio, and even more there is the transfection protocol, and various efforts have been made in the direction of a rationalization of this very complex phenomenology. However, there are also some drawbacks, such as the lack of colloidal stability, the relatively low transfection efficiency, the relatively low duration of expression and, most importantly, the nonspecific interactions with many cells and tissues.

The past few years have witnessed important advances in polymer-based controlled release drug delivery systems and significant efforts have been made [4–6]. Up to earlier years, these methods are based on natural polymers such as carbohydrates or proteic compounds, giving rise to adverse features including immunogenicity and backbone toxicity, moderate biodegradation and poor stability. More recently, synthetic polymers imposed themselves in drug conjugation for different reasons, which encompass synthetic versatility, easy control of physical and chemical behavior and, above all, minimization in synthetic toxicity and immunogenic properties.

The ideal drug delivery system should be biocompatible, capable of achieving high drug loading, and biodegradable. Most of these advantages are based on the nano- or meso-scale size of these delivery systems.

This approach is based on the premise that polymers, whose physicochemical characteristics must be assigned, can be used to formulate biodegradable polymer particle delivery compositions, for time release of bioactive agents in a consistent and reliable manner. The trend has been toward utilizing biodegradable polymer excipientes which allow sustained release of therapeutic agents while eliminating the necessity of removing the polymer after the drug is spent [7].

Polymer-based approaches to the delivery of DNA have advanced notably over the past few years. The incorporation of multiple functional components within a single carrier, for example, combining tumor targeting, stimulated release of drugs and delivery of imaging agents has been proved essential to address the challenges of tumor heterogeneity in order to obtain treatment of cancer, realizing a nanomedicine platform that can diagnose, deliver, and monitor cancer pathologies [8].

Future directions in controlled drug delivery are oriented toward responsive systems where drugs, from implantable devices, are released in response to changes in the biological environment, i.e., the system is able to deliver, or cease to deliver, drugs according to these changes. Successful developments of these novel formulations are based on responsive polymers with specifically designed microscopic structural and chemical features.

In the following, we will review most of the recent progresses obtained in polymer-based drug delivery systems, focusing on preparation, characterization, and potential applications of the most important types of polymeric aggregates as drug carriers, showing how polymers can be used in a smart fashion leading to multiple responses at the desired point of action. Over recent years, the synthesis of new polymeric materials has become more sophisticated and is presently in continuing evolution. There are a vast number of papers available on this topic and therefore only a selection of examples will be reviewed and discussed.

19.2 POLYMERS IN CONTROLLED DRUG DELIVERY

19.2.1 Basic Mechanisms of the Controlled Release

The basic mechanisms according to which the active agent can be released from the delivery system can be classified into three different classes, i.e., diffusion, degradation, and swelling, followed by diffusion. Here, we will briefly comment about each one of them.

Diffusion, which is the most common mechanism that occurs, takes place when the drug passes from the polymer matrix to the external environment. The rate of the drug release normally decreases upon time since the drug has a progressively longer distance to travel and therefore requires a longer diffusion time to release.

The swelling increases the aqueous solvent content within the formulation, accompanied by an increase of the polymer mesh size. This effect enables the drug to diffuse through the swollen network into the external environment. Most of the materials employed in swelling-controlled release systems are based on hydrogels, which can adsorb a great deal of fluid (on the order of 60%–90% of the total mass). As stated earlier, in these systems, the drug release is a consequence of the polymer swelling. Because in the most cases the swelling is controlled by pH and pH-sensitive polymers swell at high pH and collapse at low pH, the drug delivery occurs only upon an increase of pH in the environment. This peculiar feature is exploited, for example, in systems such as oral delivery, in which the drug is not released at low pH values, as in the stomach, but rather in the upper small intestine, at higher pH values.

Finally, biodegradable systems are based on materials that degrade within the body as a result of natural biological processes. This class of drug release has the advantage that it is unnecessary to remove the drug delivery system after the release of the active agent.

Most biodegradable polymers degrade as a result of hydrolysis of the polymer chains into progressively smaller compounds. For example, in the case of polylactides, polyglycolides and their copolymers, the polymers break down to lactic acid and glycolic acid, which, entering into the Kreb's cycle, will be further broken into carbon dioxide and water and finally excreted through normal processes. The degradation may take place through bulk hydrolysis or may occur only at the surface of the polymer matrix, as, for example, in the case of polyanhydrides and polyorthoesters, resulting in a release that is proportional to the surface area of the drug delivery system.

The most commonly used colloidal particle-based delivery systems have been reviewed by Kostarelos [9]. They are summarized in Figure 19.1, together with their typical average dimensions, ranging from 1 nm, in the case of polymer-drug complexes, to some micrometers, in the case of hydrogels. This wide size interval can be conveniently exploited in the different formulations.

19.2.2 Polymer Aggregates

Various types of particles, such as polymer vesicles, microcapsules, microspheres, micelles, or colloidal dispersions, have been designed, according to the particular purpose to be attained. The possible structures, resulting from the polymer aggregation, have a different hierarchy and act in

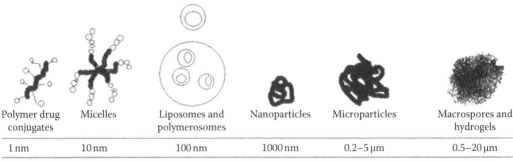

Polymer drug conjugates	Micelles	Liposomes and polymerosomes	Nanoparticles	Microparticles	Macrospores and hydrogels
1 nm	10 nm	100 nm	1000 nm	0.2–5 μm	0.5–20 μm

Typical size

FIGURE 19.1 A sketch of the different polymeric structures involved in drug delivery systems, ordered by increasing size. They vary from polymer-drug conjugates employing water-soluble polymers, micelles based on amphiphilic di- (and tri-) block copolymers, liposomes, and polymerosomes based on phospholipid vesicles up to micro- and macro-particles and hydrogels. The typical size of the aggregates ranges from 1 nm, in the case of the smallest structure, up to 10–20 μm, in the case of hydrogels.

a different way. Selected examples will be described and we will review key advances of some of these polymeric structures and discuss their potential in drug delivery.

19.2.3 POLYMERIC MICELLES

Polymeric micelles are characterized by a supramolecular core-shell structure and were first proposed as drug carriers by Bader et al. [10] in 1984. These structures have been the object of a growing interest as carriers for water-insoluble drugs and for their ability of avoiding scavenging by the phagocyte system and offering considerable advantage for cancer diagnosis and therapy [11].

These kinds of micelles are generally formed in an aqueous environment by self-assembling of biocompatible amphiphilic block copolymers having an A-B diblock structure, where the hydrophobic chains form the exterior shell (a coronal layer) and the hydrophilic chains form a semisolid core. A sketch is shown in Figure 19.2. The term "block" denotes a linear polymer in which the end of one segment is covalently linked to the end of the other adjacent segment to give the diblock A-B or multiblock (A-B)$_n$ copolymers.

This special arrangement makes polymeric micelles suitable as long-circulating drug nanocontainers. Being loaded in the hydrophobic core, the drug is well protected from inactivation under the effect of biological surrounding.

Recently, block copolymer micelles have been drawing significant attention as promising carriers in drug delivery systems [12]. Micelles entrapping drugs exhibited several important positive features such as high colloidal stability, reduced interaction with biological components, prolonged circulation in the blood, and a relatively low toxicity, which are all particularly advantageous for in vivo applications. One of the unique features that have made polymeric micelles an attractive

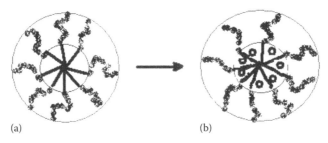

(a)   (b)

FIGURE 19.2 Schematic representation of a polymeric micelle formed by amphiphilic diblock copolymers (a) and the drug-encapsulated structure drug carrier, (b).

method for the tumor-targeted delivery of chemotherapeutic agents is the flexibility of the polymer organization, which permits the simultaneous chemical modification of the micellar core and the shell structure without one affecting the other.

Insoluble drugs can be incorporated into the micelles by chemical conjugation or by physical entrapment through, for example, dialysis or emulsion techniques. In the first case, a covalent bond between a specific group of the drug molecules and the hydrophobic chain of the polymer is required. In the second case, the physical entrapment may occur in two different ways. After the solubilization of the drug and the polymer (its hydrophobic part) in a solvent in which they are both soluble, an extensive dialysis provokes a continuous replacement of the solvent with one that is suitable for the hydrophilic part of the polymer, in most cases, water. As a consequence, the hydrophobic portions of the polymers associate into a micellar core, incorporating the drug too.

On the other hand, the emulsion method consists in preparing an aqueous solution of the copolymer to which drug in a water-insoluble solvent is added. The mixture forms an oil-in-water emulsion and the following solvent evaporation favors the micelle-drug complexes.

The hydrophobic core generally consists of biodegradable polymers such as poly(β-benzyl-L-aspartate) [PBLA], poly(LD-lactic acid) [PLA], poly(ϵ-caprolactone) [PCL], and many others.

The shell, which is responsible for the micelle stabilization in an aqueous solvent, consists of hydrophilic, biocompatible, but not biodegradable polymers such as poly(ethylene oxide) [PEO] and poly(ethylene glycol) [PEG], just to name a few.

The size of the micelles is determined in a large extent by the balance between opposite forces, i.e., the interfacial energy and the conformational entropy of the polymer strands. Moreover, an important factor that influences the size of the polymeric micelles is the length of the charged block copolymer. For example, in the case of poly(ethylene glycol)-b-poly(L-lysine) [PEG-b-PLL] diblock copolymer, micelles have a hydrodynamic radius of 24 and 37 nm as the degree of polymerization of poly(L-lysine) [PPL] increases from 18 to 78, the molecular weight of poly(ethylene glycol) [PEG] being equal (12 kD) [13]. These aggregates generally present a considerably small size distribution characterized by a low polydispersity index (in the most case less than 0.1) and by a unimodal size distribution.

The final geometrical arrangement these micelles assume is, in some ways, conditioned by the nature of the specific compound entrapped. While micelles entrapping oligonucleotides, oligopeptides, or proteins have a spherical or spheroidal shape, cationic micelles entrapping plasmid DNA present a large variety of topologies, the most frequent of them being the toroidal shape. These different topologies can affect the transfection efficiency and these aspects are the subject of intense investigation, especially as far as the DNA condensation is concerned.

The physicochemical parameters that characterize the behavior of a polymeric micelle are those typical of a charged colloidal particle in an aqueous medium. Among these, the ζ-potential controls in a large extent the colloidal stability of the aggregate and modulates the interaction of the micelle with the biological entities. Most of the properties of colloidal systems are determined by the electrical charge on the particle surface and hence by the surface potential. Moreover, ζ-potential and the size R of the aggregates are generally strongly correlated. A typical example of this connection, taken from Ref. [14], is shown in Figure 19.3, where the dependence of ζ-potential and the size R on the concentration of the agent that induces aggregation is reported. Even if these data refer to a system composed by polymeric particles stuck together by an oppositely charged polyion, and therefore a system that differs from the one consisting of polymeric micelles, as those we are dealing with, nevertheless, the basic behavior holds.

Polymeric micelles have found practical applications in a variety of pharmacological fields, even if, up to today, the most promising applications have been made as anticancer drug delivery. Compared to conventional small-molecule-based therapy, polymeric micellar systems have several potential advantages including higher payload capacity, higher circulation time, and a generally reduced toxicity to healthy tissues.

Typical examples are polymeric micelles formed by Adriamycin [ADR, an anthracycline anticancer drug] bound to the poly(aspartic acid) chain of a diblock copolymer formed by poly(ethylene glycol)

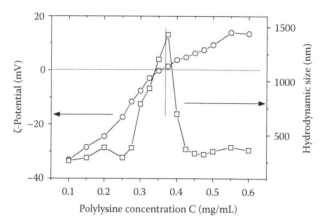

FIGURE 19.3 ζ-potential (on the left) and hydrodynamic size (on the right) of poly(lactic acid) particles aggregated together to form a cluster, as a function of ε-polylysine concentration in the aqueous solution. The maximum in the aggregate size corresponds to a ζ-potential close to zero. The behavior of the two parameters evidences two typical phenomena in the aggregation of colloidal particles governed exclusively by electrostatic interactions, i.e., the charge inversion (documented by the change of sign in the ζ-potential close to the isoelectric point) and the reentrant condensation (documented by the nonmonotonous increase of the aggregate size). (Data taken and redrawn from Zuzzi, S. et al., *Langmuir*, 25, 5910, 2009.)

and poly(aspartic acid) [PEG-P[Asp(ADR)]] [15]. These complexes form a micellar structure of size on the order of 50 nm, with a narrow size distribution and a very good stability in a phosphate-buffered saline solution. The system works, for example, against mouse leukemia, demonstrating a reduced toxicity as compared to an administration of free Adriamycin.

This polymeric micelle has been proved to be efficient also against solid tumors [16] and moreover it was observed in blood at much higher concentrations with a longer half-life than ADR, after the intravenous injection. An example of the efficiency of these systems in reducing the tumor growth after the inoculation is shown in Figure 19.4. The stabilization of ADR in blood, by binding

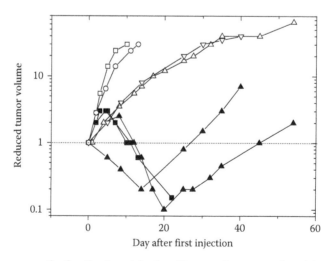

FIGURE 19.4 Tumor growth after the drug injection. Tumor volumes are plotted in ratios to the volume at the first drug injection. Control: (o, □, Δ, ∇) 0.9% NaCl solution in a volume of 0.1 mL/10 g body weight. (■): C26 tumor, 150 mg/kg PEG-P[Asp(ADR)]; (●): M5076 tumor, 150 mg/kg PEG-P[Asp(ADR)]; (▲): C38 tumor, 150 mg/kg PEG-P[Asp(ADR)]; (▼): MX-1 tumor, 150 mg/kg PEG-P[Asp(ADR)]. (Data redrawn from Yokoyama, M. et al., *Cancer Res.*, 51, 3229, 1991.)

to the block copolymer, was attributed to the micellar structure of the carrier which possesses the hydrated outer layer composed by poly(ethylene glycol) chains.

Polymeric micellar systems allow the use of a large variety of existing drugs otherwise deemed too hydrophobic to be inherently water insoluble. For example, the low water solubility of paclitaxel (PTX, an anticancer agent which inhibits the microtubule growth) can be increased from 0.0015 to 2 mg/mL (more that three order of magnitude) when encapsulated within the hydrophobic core of a polymeric micelle [17]. Several polymeric systems are currently in phase I/II clinical trials for delivery of doxorubicin [DOX] and paclitaxel [PTX].

As mentioned earlier, the hydrophilic shell with its brush-like coating inhibits or reduces the phagocytic clearance by the reticuloendothelial system [18].

While in other nanotherapeutic systems, such as polymer-drug conjugates or dendrimers, it has proved to be necessary the covalent conjugation of the drug molecules to the polymer (the carrier), polymeric micelles offer the advantage to need a noncovalent encapsulation (compatible with hydrophobic drugs). This fact extends the use of this approach to a large class of drugs since the presence of functionalizable chemical groups on the drug molecules necessary to the covalent conjugation is not necessary.

An aspect that makes polymeric micelles an emerging and powerful platform for therapeutic applications is the possibility to incorporate multiple drug components within a single micelle. Furthermore, polymeric micelles can be employed in cancer diagnosis, for example, by incorporation of gadolinium-based contrast agents at their surface for use in NMR imaging. The presence of gadolinium complexes on the surface of polymeric micelles modifies the relaxation time T_1, increasing the sensitivity of detection [19,20].

A typical example that summarizes in a clear way the multifunctional use of a polymeric micelle is reported by Nasongka et al. [21]. These authors employ a micelle that incorporates different functions due to the presence of targeting ligand and of an MRI-visible agent and of an anticancer drug, realizing micelles 45 nm in size containing doxorubicin [DOX] and a cluster of superparamagnetic iron oxide [SPIO] particles into the core of a poly(ethylene glycol)-poly(lactic acid) [PEG-PLA] micelle with a cyclic(Arg-Gly-Asp-D-Phe-Lys) [cRGD] ligand on the micelle surface. The peptide targets the $\alpha_v\beta_3$ integrin overexpressed on the surface of angiogenic tumor vessels.

Figure 19.5 shows the antitumor efficacy of cRGD-DOX-SPIO micelles against subcutaneous A549 tumor xenografts in nude mice [data redrawn from Ref. [21]]. As can be seen, a notable tumor volume reduction is observed.

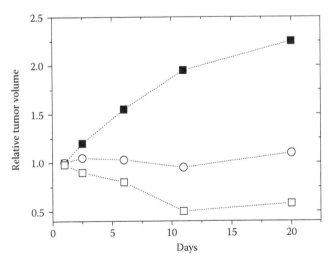

FIGURE 19.5 Antitumor efficacy data for PBS control (■), DOX-SPIO micelles with 0 (o) and 16% (□) surface density of cRGD in subcutaneous A549 tumor xenografts in nude mice. (Data redrawn from Nasongka, N. et al., *Nano Lett.*, 6, 2427, 2006.)

A similar approach was developed by Xiong et al. [22], who investigated multi-functional polymeric micelles based on RGD decorated poly(ethylene oxide)-b-poly(ester) diblock copolymer with potential for selective delivery of DOX to metastatic tumor cells. In order to provide a selective intracellular delivery of DOX in tumor over healthy cells, these authors decorated the exterior micellar shell with internalizing antagonist peptide for $\alpha_v\beta_3$ integrin. DOX loading was carried out through physical encapsulation or chemical conjugation. Mechanism of DOX release is different. In the case of physical encapsulation, DOX release is dependent on the rate of drug diffusion from the micellar core and the core degradation. In the case of DOX conjugation, the drug release is controlled essentially by the core degradation.

Polymeric micelles can be considered as stimulus-responsive drug release systems and several strategies have been explored including pH-, temperature-, and ultrasound-stimulated release. We will go into details of these aspects in the next section which is handling general aspects of stimuli-responsive polymers.

While polymeric micelles characterized by a hydrophilic shell have been extensively studied, fewer studies have been reported on micelles in organic solvents (reverse micelles), where the role of the core and the shell is exchanged (a polar hydrophilic core and a hydrophobic shell). In the last decade, reverse micelles have been prepared from dendrimers [23] or hyperbranched polymers [24]. The amphiphilic polymers were obtained from the hydrophobic modification of poly(glycerol) of poly(amidoamine), and the hydrophobic shell-forming segments consisted of covalently linked hydrocarbon chains resulting in micelles soluble in organic solvents. To these structures, a potential as drug delivery systems has been attributed and, because of these properties, they are mentioned here.

Recently, the water-soluble poly(glycerol methacrylate) backbones were modified through the esterification of pendant hydroxyl functions with acyl chlorides and the resulting compounds were shown to self-assemble into reverse micelles (20–60 nm in size) in organic solvents [25]. These structures, besides having applications in organic chemistry (extraction of anionic dyes from water), have a large use in pharmacological technology. For example, Jones et al. [25] reported that the subcutaneous administration of loaded reverse micelles to rats prolonged the effect of vasopressin for more than a factor of four.

Most of the properties suitable for drug delivery depend on the architecture of the hydrophobic tail of the amphiphilic copolymer employed in the polymeric micelle formation. For applications in drug delivery, the shape of the micelle, that is ultimately governed by the self-assembly of the hydrophobic core, influences in a rather strong way its area of application. Micelles with different sizes would locate at different tissues and organs with a consequent different ability to stabilize their circulation time in blood.

Recently, the effect of four different tail architecture on the self-assembly of amphiphilic copolymers in dilute solutions was investigated through Brownian dynamics simulations by Cheng and Cao [26], who considered linear, dendritic, starlike, and branched copolymers. Their main results reveal that linear tail copolymers begin to associate at a very low critical micellar concentration [cmc] forming spheroidal polymeric micelles with a large size and a narrow size distribution. In the case of branched polymers, the cmc is inversely proportional to the branching parameter and the size of the aggregates increases exponentially with the branching parameter, presenting a higher thermal stability. This study offers the possibility of providing important information to design the self-assembling systems to prepare polymeric micelles for applications in drug delivery.

Up to today, there are several poly(amino acid) micellar carriers under clinical evaluation. Following the recent review by Matsumura [27], the first example at the stage of phase I clinical trial we are dealing with, concerns a polymeric micelle incorporating paclitaxel [PTX], one of the most useful anticancer agent helpful in various cancers, including ovarian, breast, and lung cancers. Micelles were self-assembled from diblock copolymers consisting of poly(ethylene glycol) and poly(aspartate), the PTX being incorporated by physical entrapment owing to hydrophobic interactions with poly(aspartate) chains. The resulting formulation, named NK105, had a weight average diameter of about 85 nm. These micelles present a marked antitumor activity against colon cancers as compared to administration of free PTX. This is a clear example of how the drug, in nanostructure, increases its efficiency. Moreover, in this specific case, tumors disappeared after the first dosing to mice treated with NK105 (with an equivalent dose of 100 mg/kg) up to 40 days after the drug injection. In his review, Matsumura [27], besides antitumor activity, considers the neurotoxicity and the radiosensitizing effect of NK105 in comparison to free PTX effect. As far as the last property is concerned, the radiosensitization is considered to be important in combined treatments based on PTX-based chemotherapy and radiotherapy, as demonstrated by clinical studies [28].

The second example deals with cisplatin-incorporating polymeric micelles [29], prepared starting from poly(ethylene glycol) [PEG], which constitutes the outer shell of the micelle, and the coordinate complex of poly(glutamic acid) [P(Glu)] and cisplatin [CDDP], which constitutes the inner core of the micelle. The resulting aggregate [NC-6004] has an average size of 30 nm, with a narrow size distribution. Cisplatin [CDDP] is a chemotherapy drug for cancer including lung and gastrointestinal cancers. The driving force for the complex formation is the ligand substitution of platinum(II) atom from chloride to carboxylate in the side chain of P(Glu). In this case too, even if the CDDP and NC-6004 administration at the same dose level showed no significant difference in the tumor growth rate, the nephrotoxicity and the neurotoxicity are clearly in favor of the nanostructured delivery system.

The third and last example taken from the review by Matsumura [27] and which is worth mentioning here is the use of a camptothecin [CPT] analog, irinotecan hydrochloride [CPT-11], which is converted to 7-ethyl-10-hydroxy-CPT [SN-38] by carboxylesterases. SN-38 is a potent cytotoxic agent against various cancer cells. The polymeric micelles were built up by self-assembling of copolymers PEG-P(Glu)(SN-38) [16]. The resulting aggregate [NK012] is about 20 nm in size, within a relatively narrow range.

Recently, a micellar-based carrier for honokiol delivery in cancer chemotherapy has been proposed by Wei et al. [30]. Honokiol is an active component derived from Chinese traditional herb magnolia able to inhibit growth and induce apoptosis in different cancer cell lines (human ovarian tumor SKOV3 and COC1 cells [31], lung cancer CH27 cells [32], and leukemia HL-60 cells [33]). In this case, honokiol-loaded polymeric micelles were prepared by self-assembling of triblock copolymer from biodegradable poly(ε-caprolactone)-poly(ethylene glycol) polymers. This system, resulting in particles of 60 nm in size with a relatively low surface charge density (ζ-potential of about −0.5 mV), offers the advantage of a slow honokiol release over a prolonged time up to several weeks, maintaining a comparable cytoxicity with free honokiol.

Micelles formed by self-assembly of copolymers specifically designed for protein drug delivery have been proposed by Yiang et al. [32], who presented the synthesis of a series of hyper-branched poly(amine ester)-co-poly(DL-lactide) copolymers. As we have stated earlier, poly(DL-lactic acid) is widely used in drug delivery and tissue engineering because of its nontoxic biocompatible and bioabsorbable properties. However, poly(DL-lactide) deactivates proteins when it forms protein-loaded nanoparticles. The combining of poly(DL-lactide) with hyper-branched poly(amine ester)s overcomes this difficulty. The system has been proved to be effective in the interaction with bovine serum albumin [BSA]. BSA-loaded micelles showed a marked encapsulation efficiency and, overall, the structural stability of BSA was maintained over the whole release process.

19.3 POLYMEROSOMES

Polymerosomes (also referred to as polymeric vesicles) represent a class of vesicles ranging in size from 50 nm to 1–2 μm made up of amphiphilic synthetic block copolymers which form the vesicle membrane. The term polymerosomes was coined in 1999 by Discher et al. [34] to indicate a new structure for encapsulation with an increased stability and reduced permeability when compared with more traditional liposomes and with more efficiency in manipulating the physicochemical characteristic of the membrane. Polymerosomes are similar to liposomes formed from naturally occurring lipids, but, compared with these, they exhibit increased stability and reduced permeability.

Polymerosomes are generally composed of diblock copolymeric amphiphiles, where one block is hydrophilic and the other block is hydrophobic. In aqueous solution, the diblocks form an arrangement where the hydrophobic blocks tend to associate with each other in order to minimize the direct exposure to water, whereas the more hydrophilic blocks form the two interfaces of a typical bilayer membrane.

These structures have rapidly attracted growing interest thanks to their intriguing aggregation phenomena, their virus-mimicking size and, above all, to their potential applications in medicine and biotechnology [35].

Polymerosomes have a fluid-filled core with walls that consist of entangled polymer chains separating the core from the external medium. Compared to liposomes, polymerosomes have many different properties that render these structures particularly attractive. They mainly concern with the elasticity, permeability, and mechanical stability associated with the membrane thickness which can be controlled by the molecular weight of the hydrophobic block of the polymers. Moreover, due to the high molecular weight of the polymers as compared to lipids, the membrane of polymerosomes are generally thicker, stronger and are more stable than conventional liposomes. Finally, polymerosomes can encapsulate hydrophilic molecules within the aqueous interior and can integrate hydrophobic molecules within the membrane.

The use of diblock copolymers to generate, in general, polymer particles and, in particular, polymerosomes is widely recognized as something very frequently done. Polymerosomes can be spontaneously formed by precipitating the block copolymers by adding a poor solvent for one of the diblocks [36]. Analogous to the standard method to form lipid vesicles from dried phospholipid films, polymerosomes can also be formed by rehydration of a dried film of the copolymers [37] followed by sonication or extrusion, generating unilamellar micrometer-sized vesicles, as well as monodisperse sub-micro-meter-sized vesicles [38]. More recently, Lorenceau et al. [39] have described a new method to create highly uniform polymerosomes using a one-step process where the inner and the outer fluids are maintained as completely separated streams. In this way, a highly efficient encapsulation is assured.

There are a number of polymers available for in vivo applications, as controlled release drug delivery systems. They include both hydrophilic and hydrophobic block copolymers. To the ones of the former class belong poly(ethylene glycol) [PEG/PEO] [40], poly(2-methyleoxazoline) [41], while to the second class belong poly(dimethyl siloxane) [PDMS] [41], poly(caprolactone) [PCL] [42], poly(lactide) [40], poly(methyl methacrylate) [PMMA] [43].

Recently, Rameez et al. [42] describe the preparation of polymerosomes made by biodegradable and biocompatible amphiphilic diblock copolymers composed of poly(ethylene oxide) [PEO] and poly(caprolactone) [PCL] or poly(ethylene oxide) [PEO] and poly(lactide) [PLA]. In the diblock copolymers, PEO acts as the hydrophilic block and either PLC or PLA can function as the hydrophobic blocks. The important feature of this formulation is that PEO, PCL, and PLA are biocompatible polymers and the last two polymers are biodegradable. Polymerosomes based on amphiphilic block copolymers are interesting vesicles for drug delivery (and also for diagnostic purpose), because, by selecting different block copolymers, the membrane properties, the stability, and the rate of degradation can be conveniently tuned. For example, PEO-PCL and PEO-PLA polymerosomes have been demonstrated to be able to encapsulate hemoglobin and they are promising candidates for hemoglobin-based oxygen carriers.

Figure 19.6 shows a comparison of the hemoglobin encapsulation efficiency of the two types of polymerosomes (PEO-PCL and PEO-PLA, respectively) (data redrawn from Ref. [42]). As the molecular weight of the PEO block of the copolymer increases, the hemoglobin encapsulation efficiency decreases. This is a simple consequence of the fact that, with the increase of the molecular weight, the length of the outer leaflet of the polymerosomes becomes larger and less volume in the core will be available for hemoglobin encapsulation. Moreover, the hemoglobin-polymer weight ratio can be appropriately modified in order to make the oxygen carrying capacity of these polymerosomes comparable to the one of blood.

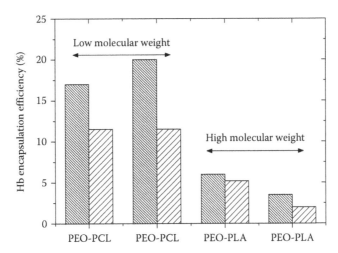

FIGURE 19.6 Comparison of hemoglobin encapsulation efficiency in polymersosomes composed of poly(ethylene oxide-b-caprolactone) [PEO-PCL] and of poly(ethylene oxide-b-lactic acid) [PEO-PLA]. The polymers have different molecular weights, in the range 10–15 kDa (referred to as high molecular weight) and in the range 1–3 kDa (referred to as low molecular weight). Darker histograms refer to bovine hemoglobin and lighter histograms refer to human hemoglobin. (Redrawn and adapted from Rameez, S. et al., *Bioconj. Chem.*, 19, 1025, 2008.)

Together with polymerosomes, we mention a new multicomponent structure, known as vesosome, consisting of drug-loaded liposomes encapsulated with another bilayer. Even if vesosomes are currently too large for optimal intravenous use, they represent a promising and interesting drug carrier, as far as the retention and stability are concerned, compared with traditional liposomes.

19.3.1 Polymer Vesicles

One of the major problems with the aggregation of lipidic molecules into closed lipid bilayers (liposomes) is a reduced stability that, as far as the drug delivery context is concerned, results in a rapid clearance of the encapsulated drug, for example, after the intravascular administration. This disadvantage can be avoided, and the stability increased, by surface grafting of hydrophilic polymers [44] or by polymerization of reactive lipid molecules in the vesicular aggregates [45].

Another approach involves the layer-by-layer deposition of polyelectrolytes on the surface of charged nanoparticles followed by the subsequent dissolution of the templating particles [46]. Finally, a different approach consists in the self-assembly of amphiphilic diblock copolymers into micelles, in a selective cross-linking of their hydrophilic shell and the subsequent degradation of the hydrophobic core [47].

As stated earlier, block copolymers have been studied extensively because of their interesting properties and their self-assembling behavior in aqueous solution. Recently, attention has been focused on a specific class of hybrid block copolymers, employing a well-defined peptide sequence combined with synthetic polymers, resulting in novel aggregates for drug delivery. Just to mention some of them, we remember here copolymers of β-sheet folding peptides prepared by Sogah and coworkers [48], consisting of alanine (or alanyl-glycine) repeats and poly(ethylene glycol) or the diblock copolymers of polybutadiene and poly(glutamic acid) prepared by Klok et al. [49] that form vesicles whose size could be altered by variation of the pH.

More recently, a well-defined ABA triblock copolymer based on the atom transfer radical polymerization of methyl methacrylate was mentioned by Ayres et al. [43], resulting into aggregates that were a mixture of polymerosomes and large compound micelles, ranging from 400 nm to 10 μm in size. As noted by these authors, their method for the preparation of peptide-containing triblock copolymers leads to a wide range of fascinating structures in the nanometer size range whose characteristics can be appropriately tuned by varying the peptide, the length of the polymer, or even the kind of the monomer that is polymerized.

Finally, the synthesis of a triblock copolymer composed by poly(2-methyleoxazoline) [PMOXA], poly(dimethylsiloxane) [PDMS], PMOXA-PDMS-PMOXA, carrying polymerizable methacrylate groups at both chain ends has been reported by Nardin et al. [41]. This polymer forms vesicular structures in dilute aqueous solution which can be polymerized to hallow nanospheres.

19.3.1.1 Microcapsules

A microcapsule can be considered as a microsphere containing many small cavities inside. When a drug is included in one of these cavities, before being released, it should permeate through several regions of polymer, sustaining a release over a long period of time. For example, in the case of polypeptide microcapsules [50] built up by poly[L-Lys(Z)]-block-poly(Sar) [PKZ-b-PS] and poly[L-Glu(OMe)-block-poly(Sar) [PMG-b-PS], the release of fluorescein-labeled dextran [FITC-dextran] entrapped within the microcapsule during the stage of preparation occurs over period of time on the order of ten days (Figure 19.7).

One of the more intriguing properties of polypeptide microcapsules is that these structures are capable of modulating the drug release in response to external signals due to the easiness of a chemical modification at the site of the functional groups of the amino acid components. If we stick to the examples described by Kimura et al. [50], a pH-responsive microcapsule can be prepared by the connection of poly(L-Glu) to the side chain amino groups in partially deprotected poly[L-Lys(Z)]-b-poly(Sar), [PKZ-b-PS]. In this case, the carboxyl groups are located in the graft chains connected to the positively charged main chain. The influence of pH in the release of fluorescein-labeled dextran [FITC-dextran] encapsulated in the microcapsule is shown in Figure 19.8.

Figure 19.9 shows the release of FITC-dextran from microcapsules when the temperature goes up and down between 25°C and 40°C (data taken and redrawn from ref. [51]). When the temperature was raised to 40°C, the pNIPAAM gel in the small cavities shrank, producing a pathway where dextran could diffuse more or less freely.

19.3.1.2 Microspheres

Some of the most widely studied materials for drug delivery are aliphatic polyester-based polymers which include poly(lactic acid) and its copolymers. Poly(lactic acid) is a biodegradable and

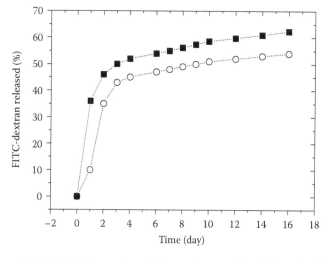

FIGURE 19.7   Amount of FITC-dextran released from microcapsules built up by PZK-b-PS (o) and by PGM-b-PS (■). (Redrawn and adapted from Kimura, S. et al., *Polym. Adv. Technol.*, 12, 85, 2001.)

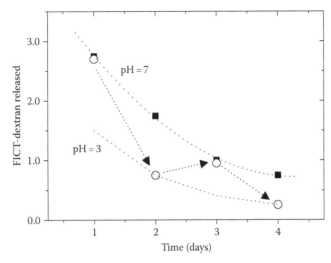

FIGURE 19.8 Release of FITC-dextran from poly(L-Glu)-grafted-PKZ-b-PS microspheres for different values of pH, as a function of time. (■): pH = 7; (○): pH has been changed from 7.0 to 3.0 and then after brought back to the value of pH = 7.0. (Redrawn and adapted from Kimura, S. et al., *Polym. Adv. Technol.*, 12, 85, 2001.)

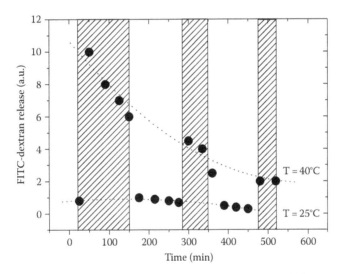

FIGURE 19.9 Release of FITC-dextran from pNIPAAM-loaded microcapsule in response to temperature jump between 25°C and 40°C. (Redrawn and adapted from Kidchob, T. et al., *J. Chem. Soc. Perkin Trans.*, 2, 2195, 1997.)

thermoplastic polymer (that has been approved by the Food and Drug Administration [USA]) for use in human subjects. As biocompatible polymer, it has been employed as suture material, interbody cages, scaffolds for tissue engineering and for biodegradable drug delivery systems [52–54]. This poly(lactic acid) has been one of the most widely used biodegradable polymers for drug delivery systems and porous scaffolds [55,56].

Another classic biodegradable polymer able to form nanospheres and microspheres is poly(DL-Lactide-co-Glycolide) [PLGA] polymer. Together with poly(DL lactic acid), PLGA has been one of the most widely used polymers because of its excellent tissue compatibility and safety for use in humans [57].

Nanoparticles, because of their surface properties, have been proved to guarantee a prolonged in vivo circulation and a sustained drug release.

Hyaluronic acid-coated poly(butyl cyanoacrylate) nanoparticles (of about 300 nm in size) were synthesized by He et al. [58]. These particles represent an effective and safe vehicle for hydrophobic anticancer drugs such as paclitaxel [PTX], a model anticancer drug. In this system, whereas the PTX encapsulation was obtained with an efficiency up to 90%, the in vitro drug release profile clearly shows that the presence of hyaluronic acid slows down the drug release and reduces the initial burst in the first 10 h. Moreover, the particles have a very good biocompatibility, the hemolysis ratio being well less than 2%.

Recently, an increasing interest as drug carrier in clinical setting has been turned to albumin [59]. Albumin is an acidic protein stable in the pH range of 4–9, that preferentially uptakes in tumor and inflamed tissues. These properties, besides ready availability, biodegradability, and lack of toxicity, make it a very interesting candidate for drug delivery. A recent review that gives account of albumin in different drug delivery systems has been made by Kratz [59]. Here, we will report some results dealing with micro- and nanoalbumin particles. Albumin microspheres, which are usually in the range 1–100 μm, can carry therapeutic or diagnostic agents. However, the size is the main factor for the biodistribution of the microspheres. Whereas small (1–3 μm) microspheres are taken up by the reticuloendothelial system, accumulating in the liver and spleen, as well as in solid tumors, larger microspheres (>15 μm) accumulate in capillary bed of the lungs.

Recently, albumin nanoparticles ideal for the encapsulation of lipophilic drugs in the size range comparable with the size of small liposomes (100–200 nm) have been developed by American Bioscience, Inc. As an example, albumin nanoparticles with paclitaxel [nab-paclitaxel], approximate diameter of 130 nm, have been extensively investigated preclinically and clinically, showing superior antitumor efficacy over paclitaxel in a number of human tumor models [60].

Figure 19.10 shows the antitumor activity of nab-paclitaxel (30 mg/kg/d) and paclitaxel (13.4 mg/ kg/d) in the MX-1 xenograft model.

Among polymeric particles, we mention hydrogel nanoparticles that have gained considerable attention as a promising drug delivery system, combining the characteristics of a hydrogel (hydrophilicity and large water content) with the one of the nanoparticles (size on the order of 50–100 nm). In the recent years, several polymeric hydrogel nanoparticles have been prepared and characterized, based on both natural and synthetic polymers [61].

The most common monomers employed in hydrogels for pharmaceutical applications include Hydroxyethyl methacrylate [HEMA], Ethylene glycol dimethacrylate [EGDMA], N-isopropyl Aam

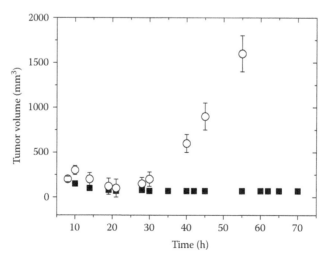

FIGURE 19.10 Antitumor activity of nab-paclitaxel (30 mg/kg/d) (■) and paclitaxel (13.4 mg/kg/d) (○) in the mamma carcinoma MX-1 model (at equitoxic dose). (Data redrawn and adapted from Desai, N. et al., *Clin. Cancer Res.*, 12, 1317, 2006.)

[NIPAAm], Ethylene glycol [EG], and many others. Hydrogels must have an acceptable biodegradability and biocompatibility that can be assured by the different improved crosslinking methods.

Hydrogel-based systems can be divided into two categories, according to the manner they provide the drugs, i.e., time-controlled systems and stimuli-responsive systems. These latter systems will be considered in detail in the next section. As far as the former one is concerned, the drug release is controlled by diffusion, according to Fick's first law of diffusion (with constant or variable diffusion constants), depending on the mesh size within the matrix of the gel. Typical mesh size in biomedical hydrogels range from 5 to 100 nm [62,63], the drug diffusion being not considerably inhibited in the swollen state.

Hydrogel particles of nanoscopic size (nanogel), merging the features and the characteristics that hydrogel and nanoparticles separately possess, have been the subject of considerable amount of effort in drug delivery approaches. They summarize the hydrophilicity, flexibility, high water content, biocompatibility typical of the hydrogels with the advantages of nanoparticles which allow long life in the circulation system and the possibility of targeting the desired site. Detailed classification and applications of different hydrogel polymeric materials have been reviewed by Coviello et al. [64], Lin et al. [65] and Peppas ct al. [66].

19.4 STIMULI-RESPONSIVE POLYMERS

An ideal drug delivery system should respond to physiological requirements and modify the drug release profile accordingly. Much research has been directed toward single- or dual-stimuli-responsive hydrogel for drug delivery.

Stimuli-responsive polymeric drug delivery is a method of increased importance in which drugs can be delivered to the body that is able to respond to physiological requirements and modify the drug-release profile accordingly [67]. The main characteristic of these systems is their ability to undergo changes from a hydrophilic to a hydrophobic state triggered by small changes in the environment. Moreover, these changes are reversible in the sense that the system is able to return to its initial state as the external stimulus is removed. Compared to the more conventional delivery methods, where the drug concentration results in a peak followed by a rapid fall, these systems allow maintaining the desired drug concentration within the single administration.

The external stimuli able to induce a response in the polymer structure are extremely various, ranging from electrical, magnetic, physicochemical, thermal, or mechanical stimuli. A summary of them, together with the associated release mechanisms, has been reviewed by Soppimath et al. [68]. Here, we will consider only the most important of them and will report and discuss some of the more typical examples.

19.4.1 THERMO-RESPONSIVE POLYMERS

The characteristic of this class of polymers, which are the most commonly studied class of stimuli-responsive polymers, is the presence of a critical solution temperature at which a phase transition occurs. In other words, if the polymer becomes insoluble upon heating, hydrophobic interactions prevail and therefore hydrogels made of these polymers shrink as the temperature increases. These hydrogels are made of polymer chains that possess a mixture of hydrophilic and hydrophobic segments.

The most widely synthetic thermo-responsive polymer is poly(N-isopropyl acrylamide) [pNIPAAM] which exhibits a coil-globule transition at its lower critical solution temperature [LCST]. Its biocompatibility and the occurring of the LCST at 32.3°C makes pNIPAM a very interesting material for controlled release applications. Moreover, the LCST is largely independent of the molecular weight and concentration and its phase transition temperature can be easily tuned around an appropriate temperature ($\simeq 40°C$) by shifting the hydrophilic-hydrophobic balance by means of copolymerization achieved by the introduction of a hydrophilic co-monomer [69]. Hydrophilic copolymers decrease the LCST whereas hydrophobic copolymers have the opposite effect.

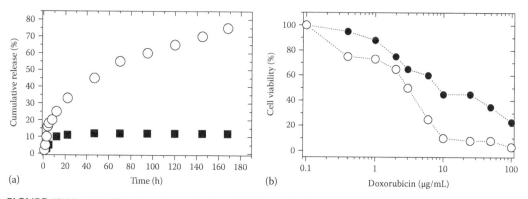

FIGURE 19.11   (a) DOX release behavior from DOX-mixed micelles under acidic (pH = 5.0, (○)) and neutral (pH = 7.4, (■)) conditions. (b) Cell viability of HeLa cells treated with various concentrations of free DOX for 1 day (●) and 3 days (○) incubation. (Data redrawn from Lo, C.-L. et al., *Biomaterials*, 30, 3961, 2009.)

Polymers that exhibit a temperature phase transition have been the subject of many investigations [70], leading to many intriguing applications such as artificial muscle, nanoactuators, and motor systems [71,72].

Recently, two other polymers attracted attention. They are poly(N,N'-diethyl acrylamide) [pDEAAM] that has lower critical solution temperature in the range of 25°C–35°C [73] and poly(N-(1)-1-hydroxymethyl propyl methacrylamide) [HMpMAAM], which is a novel thermo-responsive polymer designed for optical activity [74].

Mixed micelles formed from critical micelle concentration and temperature-sensitive di-block copolymers for biomedical applications have been widely investigated. However, different impediments prevent their use. Among these, there is the lack of biocompatibility of the most part of temperature sensitive copolymers, the uncontrolled mechanism of drug release, because of their phase transition temperature, and, over all, the poor micellar stability.

A new class of mixed micelles composed of temperature-sensitive diblock copolymers and Cmc diblock copolymers has been recently proposed by Lo et al. [75], which exhibit an improved stability under various physiological conditions and can be applied to intravenous drug delivery of different anticancer drugs. In this particular case, these authors employed methoxy poly(ethylene glycol)-block-poly(DL-Lactide) [mPEG-b-PLA], a biocompatible Cmc diblock polymer, and methoxy poly(ethylene glycol)-block-poly(N,n-propylacrylamide-co-vinylimidazole) [mPEG-b-P(NnPAAm-co-Vim)], a temperature-sensitive copolymer.

The drug release from mixed micelles incorporating doxorubicin [DOX] is shown, as an example, in Figure 19.11, together with the cell viability of human cervical epithelioid carcinoma cells treated with DOX-mixed micelles for one (or three) day incubation.

This new micellar system has various remarkable characteristics important for practical applications such as smaller and uniform particle size, marked stability, and a rapid response to stimuli, besides an easy and economic method of preparation.

19.4.2   ELECTRO-RESPONSIVE AND MAGNETICALLY RESPONSIVE POLYMERS

Polyelectrolytes, which are polymers with a high concentration of ionizable groups along the backbone chain, are sensitive to an external electric field. Under this influence, electro-responsive hydrogels generally shrink or swell, allowing the release of a pre-loaded drug under controlled conditions.

Magnetic targeting is based on the attraction of magnetic particles to an external magnetic field and a translational force will be exerted on the drug-encapsulated particle complex (magnetic drug targeting [76]).

19.4.3 Light-Responsive Polymers

A typical example of a light-response polymer is a polymer network where a leuco derivative molecule has been inserted in order to produce a UV light–responsive hydrogel [77]. Leuco derivatives are normally neutral but dissociate into ion pairs under an ultraviolet irradiation. This causes a swell due to an increase in osmotic pressure due to the increased ion concentration. Notably, hydrogels shrink when the light is removed.

An interesting and versatile system investigated by Sumaru et al. [78] is based on pNIPAM and acrylamide-functionalized spirobenzopyran that is able to inter-convert between neutral and zwitterionic form upon irradiation. This photochemically switchable polymer undergoes four distinct states consisting of a colorless spiro form, a colored merocyanine, and of the corresponding protonated merocyanine and spiropyran, which exhibit different solubilities with temperature. This behavior, which is extremely photosensitive during UV irradiation, is caused by the hydrophilic character of merocyanin and the hydrophobic character of spiropyran.

19.5 INTERNAL STIMULI-RESPONSIVE SYSTEMS

19.5.1 pH-Responsive Polymers

The physiological pH varies according to the various tissues and cellular compartment. Its value is strongly acid in the stomach (pH = 1.0–3.0), shifting to basic values (pH = 5–8) in the intestine or close to neutral in the blood (pH = 7.35–7.45). This property can be favorably addressed in pH-responsive polymers. Basically, weak acids and bases linked to a polymer chain may change their ionization upon variation of pH, with a consequent conformational change of the whole polymer chain or a swelling behavior in the case of a hydrogel. The pH-responsive swelling and the collapsing behavior has been used to induce release of caffein [79], indomethacin [80], or lysozyme [81].

A change in pH is used in drug release in cancer targeting, because tumor tissue has generally an extracellular pH slightly lower than normal tissue (pH = 6.5–7.2 compared with pH = 7.4 in normal condition) [82].

The pH-sensitive micelles appear to be a promising candidate for photosensitizer delivery for photodynamic therapy. As a matter of fact, it is advantageous in therapy of this kind if a drug delivery system is used that releases the drug (in this case the photosensitizer) selectively at the target site. In this context, pH-triggered photosensitizer release using polymeric micelle formulation, has been suggested by Leroux el al. [83] and by the group of Kataoka [84]. Their main results have been recently summarized by van Nostrum [85]. For example, the latter author employed a polycationic porphyrin dendrimer mixed with a polyanionic poly(ethylene glycol)-poly(aspartic acid) diblock copolymer forming high-stable micelles [PIC micelles [86]], with an average diameter of 56 nm. These micelles are internalized in Lewis lung carcinoma cells with no marked toxicity exhibiting a high photodynamic efficiency. These systems offer many opportunities to be further optimized, for example, using degradable systems in order to prevent accumulation and long-term toxic effects.

Stimuli-responsive polymers may respond as well to applied stimuli originated from internal physicochemical changes occurring within various body tissues. A typical example is a pH-responsive polymer designed for oral delivery which experiences marked pH changes along different gastrointestinal tracts, passing from pH $\simeq$ 1.0–3.0 in the stomach to pH $\simeq$ 7.0–7.5 in the colon [87].

This class of polymers present acidic or basic groups that allow the accepting or the release of protons in response to a change in pH. Polymers undergo a conformational change with a consequent change in the swelling behavior of the hydrogel when ionizable groups are linked to the polymer structure.

To this class of polymers, poly(acrylamide) [PAAM], poly(acrylic acid) [PAA], poly(methacrylic acid) [PMMA], and many others belong. On the exposure to aqueous media of appropriate pH, pendant groups ionize and produce a charge distribution on the polymer network causing swelling or deswelling of the hydrogel originated by electrostatic repulsive forces. This process ultimately controls the drug delivery [88].

pNIPAM has been extensively studied for oral delivery of calcitonin and insulin. These substances are immobilized in the polymeric beads which are stable passing through the stomach. Conversely, in the alkaline intestin, the beads disintegrate with the consequent drug release. This approach was followed by Serres et al. [89] for the intestinal delivery of calcitonin and by Kim et al. [90] for the delivery of insulin. In these cases, the thermal stimulus from the pNIPAM is not needed for the delivery, but it is appropriately exploited for the preparation of the loaded beads.

For drug delivery, a response to a local environmental stimulus at the disease site is, in principle, an ideal way to direct a therapeutic drug toward the target. At least from a theoretical point of view, factors able to control a polymer response are extremely wide, including biological factors such as over-expressed intracellular enzymes, cell surfaces receptors, inflammation signaling molecules, or physicochemical factors such as hyperthermia. This is the reason that has led to increased interest in polymers in biomedical and biological research, giving rise to a branch known as bioresponsive chemistry in drug delivery.

As an example of this approach, we can mention the method proposed by Shabat and co-workers [91], who used a "self-immolating" system, i.e., macromolecules able to break down by a cascade process following biochemical reactions and triggered by biochemical stimuli. In the earlier-stated example, the first process in realizing the target molecule employs a nucleophilic intramolecular cyclization to generate a stable cyclic species and the second employs a quinine methide rearrangement followed by a self-elimination to release the therapeutic agent.

When a controlled drug release profile is required, for example, in diseases such as diabetes, reversible responsive polymers have been implemented. Diabetes is one disease that received great attention because of the potential for therapies using controlled drug release.

Polymers that respond to changes in local glucose concentration by releasing insulin have been studied extensively investigated by Miyata et al. [92]. Acidic copolymer hydrogels in which glucose oxide has been immobilized represent a significative example. In the initial state, because of the extended polycarboxylates, the system is impermeable to macromolecular transport (e.g., to insulin) but permeable to sugar, resulting in the production of gluconic acid. Because of the protonation of the gel, caused by pendant glucose oxidase, following the ingress of glucose, a contraction of the polymer matrix occurs, allowing insulin through.

The "molecular gates" systems, shown in Figure 19.12, consist of an insulin-containing reservoir with a polymeric membrane in which glucose-oxidase has been immobilized. The gel expands at

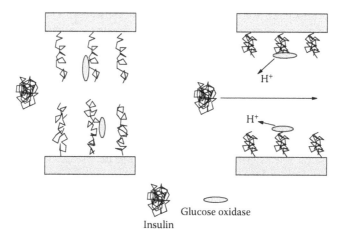

FIGURE 19.12 Sketch of a glucose-responsive polymer for "gating" a flow of insulin. Left-hand side: chain-extended polycarboxylates are impermeable to macromolecular transport (insulin), but permeable to glucose. Right-hand side: Pendant glucose oxidase generates protons following ingress of glucose, which in turn causes polymer protonation and chain collapse. The membrane opens and allows insulin through. (Redrawn and adapted from Alexander, C. and Shakesheff, K.M., *Adv. Mater.*, 18, 3321, 2006.)

high pH values, closing the gate, and shrinks at low pH values (due to the interaction of glucose with immobilized glucose-oxidase), opening the gates. Control of the insulin delivery depends on the size of the gates and on the rate of the gates' opening and closing [93].

There is more. Since conversion of glucose to gluconic acid needs the presence of oxygen, the effectiveness of the system could be limited by the oxygen availability in the hydrogel matrix. This limitation can be at least partially overcome by incorporation of a further enzyme, catalase, into the polymer matrix, with the charge of reducing hydrogen peroxide generated during the glucose oxidation.

19.6 CONCLUDING REMARKS

Polymers, that represent an exceptional versatile class of materials, have led to the development of several novel drug delivery systems. The new strategy, derived from the amalgamation of polymer science and pharmaceutical science, is based on the symbiotic union between two partners, the polymer, that acts as a carrier to move the drug to the target, and the drug itself, that, in virtue of the polymer interaction, is preserved during the transport and exerts its bioactivity only in the target tissue.

Within this context, polymeric aggregates are promising candidates for a wide range of applications in the field of drug delivery systems. This strategy, based on polymer aggregates of different physicochemical characteristics and started more than a quarter of a century ago, with the advent of polymer-drug conjugates, is destined to get on in the delivery of multiple agents to target several pathways to overcome the challenge of tumor heterogeneity.

Incorporation of a drug in a polymeric structure provides a lot of opportunities for optimizing polymer-based drug delivery systems, such as enhancing drug loading efficiency, biocompatibility, a controlled release of the drug from the polymeric matrix, and the minimization of the side-effects.

Future advances in polymer science will be based on modifying the physical and chemical properties of the polymer in order to reach a wide acceptability and distinct advantages such as the possibility of tuning the polymer aggregate size, high solubility, simple sterilization and a wide biocompatibility and, finally, a controlled release of drugs. It is expected that continuous investigations on polymer aggregates with various structures will lead to an optimal formulation for drug delivery systems, opening a new stage in the chemotherapy development to achieve a personalized therapy of cancer.

REFERENCES

1. H. Ringsdorf, Structure and properties of pharmacologically active polymers, *J. Polym. Sci. Polym. Symp.*, 51, 1975, 135–153.
2. E.W. Neuse, Synthetic polymers as drug delivery vehicles in medicine, *Metal Based Drugs*, 2008, 1–19; doi: 10.1155/2008/469531.
3. D.D. Lasic, *Liposomes in Gene Delivery*, Boca Raton, FL, CRC Press, 1997.
4. I. Brigger, C. Dubernet, P. Couvreur, Nanoparticles in cancer therapy and diagnosis, *Adv. Drug. Deliv. Rev.*, 54, 2002, 631–651.
5. K. Cho, X. Wang, Z.C. Chen, D.M. Shin, Therapeutic nanoparticles for drug delivery in cancer, *Clin. Cancer Res.*, 14, 2008, 1310–1316.
6. M. Nahar, T. Dutta, S. Murugesan, A. Asthana, D. Mishra, V. Rajkumar, M. Tare, S. Saraf, N.K. Jain, Functional polymeric nanoparticles: an efficient and promising tool for active delivery of bioactives, *Crit. Rev. Ther. Drug. Carr. Syst.*, 23, 2006, 259–318.
7. O. Pillai, R. Panchagnula, Polymers in drug delivery, *Curr. Opin. Chem. Biol.*, 5, 2001, 447–451.
8. B. Sumer, J. Gao, Theranostic nanomedicine for cancer, *Nanomedicine*, 3, 2008, 137–140.
9. K. Kostarelos, Rational design and engineering of delivery systems for therapeutics: biomedical exercises in colloid and surface science, *Adv. Colloid Interface Sci.*, 106, 2003, 147–168.
10. H. Bader, H. Ringsdorf, B. Schmidt, Water soluble polymers in medicine, *Angew. Makromol. Chem.*, 123/124, 1984, 457–485.
11. E. Blanco, C.W. Kessinger, B.D. Sumer, J. Gao, Multifunctional micellar nanomedicine for cancer therapy, *Exp. Biol. Med.*, 234, 2009, 123–131.

12. K. Kataoka, A. Harada, Y. Nagasaki, Block copolymer micelles for drug delivery: design characterization and biological significance, *Adv. Drug Deliver Rev.*, 47, 2001, 113–131.

13. A. Harada, H. Togawa, K. Kataoka, Physico-chemical properties and nuclease resistance of antisense-oligonucleotides entrapped in the core of polyion complex micelles composed of poly(ethylene glycol) - poly(L-lysine) block copolymers, *Eur. J. Pharm. Sci.*, 13, 2001, 35–42.

14. S. Zuzzi, C. Cametti, G. Onori, S. Sennato, S. Tacchi, Polyion-induced cluster formation in different colloidal polyparticle aqueous suspensions, *Langmuir*, 25, 2009, 5910–5917.

15. M. Yokoyama, M. Miyauchi, N. Yamada, T. Okano, Y. Sakurai, K. Kataoka, Characterization and antcancer activity of the micelle-forming polymeric anticancer drug adriamycin-conjugated poly(ethylene glycol)-poly(aspartic acid) block copolymer, *Cancer Res.*, 50, 1990, 1693–1700.

16. M. Yokoyama, T. Okano, Y. Sakurai, H. Ekimoto, C. Shibazaki, K. Kataoka, Toxicity and antitumor activity against solid tumors of micelle- forming polymeric anticancer drug and its extremely long circulation in blood, *Cancer Res.*, 51, 1991, 3229–3236.

17. V.P. Torchilin, A.N. Lukyanov, Z. Gao, B. Papahadiopoulos -Stenberg, Immunomicelles: targeted pharmaceutical carriers for poorly soluble drugs, *Proc. Natl. Acad. Sci., USA*, 100, 2003, 6039–6044.

18. V.P. Torchilin, PEG-based micelles as carriers of contrast agents for different imaging modalities, *Adv. Drug Deliv. Rev.*, 54, 2002, 235–252.

19. E. Nakamura, K. Makino, T. Okano, T. Yamamoto, M. Yokoyama, A polymeric micelle MRI contrast agent with changeable relaxivity, *J. Control. Release*, 114, 2006, 325–333.

20. G. Zhang, R. Zhang, X. Wen, L. Li, C. Li, Micelles based on biodegradable poly(L-glutamic acid)-b-polylactide with paramagnetic Gd ions chelated to the shell layer as a potenntial nanoscale MRI-visible delyvery system, *Biomacromolecules*, 9, 2008, 36–42.

21. N. Nasongka, E. Bey, J. Ren, H. Ai, C. Khemtong, J.S. Guthi, S.F. Chin, A.D. Sherry, D.A. Boothman, J. Gao, Multifunctional polymeric micelles as cancer- targeted MRI-ultrasensitive drug delivery system, *Nano Lett.*, 6, 2006, 2427–2430.

22. X.-B. Xiong, A. Mahmud, H. Uludag, A. Lavasanifar, Multifuctional polymeric micelles for enhanced intracellular delivery of Doxorubicin to metastatic cancer cells, *Pharm. Res.*, 25, 2008, 2555–2566.

23. Y. Sayed-Sweet, D.M. Hedstrand, R. Spinder, D.A. Tomalia, Hydrophobically modified poly(amidoamine) (PAMAM) dendrimers:their properties at the air-water interface and use as nanoscopic container molecules, *J. Mater. Chem.*, 7, 1997, 1199–1205.

24. Y. Chen, Z. Shen, H. Frey, J. Perez-Prieto, S.-E. Stiriba, Synergistic assembly of hyperbranched polyethylenimine and fatty acids leading to unusual supramolecular nanocapsules, *Chem. Commun.*, 48, 2005, 755–757.

25. M.C. Jones, H. Gao, J.C. Leroux, Reverse polymeric micelles for pharmaceutical applications, *J. Control. Release*, 132, 2008, 208–215.

26. L. Cheng, D. Cao, Effect of tail architecture on self-assembly of amphiphiles for polymeric micelles, *Langmuir*, 25, 2009, 2749–2756.

27. Y. Matsumura, Poly(amino acid) micelle nanocarriers in preclinical and clinical studies, *Adv. Drug Deliv. Rev.*, 60, 2008, 899–914.

28. H. Chou, W. Arkeley, H. Safran, S. Graziano, C. Chung, T. Williams, B. Cole, T. Kennedy, Multiinstutional phase II trial of paclitaxel, carboplatin and concurrent radiation therapy for locally advanced non-small-cell lung cancer, *J. Clin. Oncol.*, 16, 1998, 3316–3322.

29. H. Uchino, Y. Matsumura, T. Negishi, F. Koizumi, T. Hayashi, T. Honda, Cisplatin-incorporating polymeric micelles (NC-6004) can reduce nephrotoxicity and neurotoxicity of cisplatin in rats, *Br. J. Cancer*, 93, 2005, 678–687.

30. X.W. Wei, C.Y.Gong, S. Shi, S.Z. Fu, K. Men, S. Zeng, X. L. Zeng et al. Self-assembled Honokiol-loaded micelles based on poly(ε-caprolactone)-poly(ethylene glycol)-poly(ε-caprolactone) copolymer, *Int. J. Pharm.*, 369, 2009, 170–175.

31. H. Liu, C. Zang, A. Emde, M.D. Planas-Silva, M. Roshe, A. Kuhnl, C.O. Schulz, E. Elstner, K. Possinger, J. Eucker, Anti-tumor effect of Honokiol alone and in combination with other anticancer agents in breast cancer, *Eur. J. Pharm.*, 591, 2008, 43–51.

32. S.E. Yang, M.T. Hsieh, T.H. Tsai, S.L. Hsu, Down modulation of Bcl-XL, release of cytochrome c and sequential activation opases during Honokiol-induced apoptosis in human squamous lung cancer CH27 cells, *Biochem. Pharmacol.*, 63, 2002, 1641–1651.

33. W.F. Fong, K.W. Tse Anfernee, K.H. Poon, C. Wang, Magnolon and Honokiol enhance HL-60 human leukemia cell differentiation induced by 1.25-dihydroxyvitamin D3 and retinoic acid, *Int. J. Biochem. Cell. Biol.*, 37, 2005, 427–441.

34. B.M. Discher, Y.Y. Won, D.S. Ege, J.C. Lee, F.S. Bates, D.E. Discher, D.A. Hammer, Polymersomes: tough vesicles made from diblock copolymers, *Science*, 284, 1999, 1143–1146.

35. F. Meng, Z. Zhong, J. Feijen, Stimuli-responsive polymerosomes for programmed drug delivery, *Biomacromolecules*, 10, 2009, 197–209.
36. H. Shen, A. Eisenberg, Morphological phase diagram for a ternary system of block copolymer PS310-b-PAA52/H$_2$O, *J. Phys. Chem. B*, 103, 1999, 9473–9487.
37. M. Antonietti, S. Foster, Vesicles and liposomes. A self assembly principle beyond lipids, *Adv. Mater.*, 15, 2003, 1323–1333.
38. J.C.M. Lee, H. Bermudez, B.M. Discher, M.A.Won, Y.-Y. Bates, D.E. Discher, Preparation, stability and in vitro performance of vesicles made with diblock copolymers, *Biotechnol. Bioeng.*, 73, 2001, 135–145.
39. E. Lorenceau, A.S. Utada, D.R. Link, G. Cristobal, M. Joanicott, D.A. Weitz, Generation of polymerosomes from double emulsions, *Langmuir*, 21, 2005, 9138–9186.
40. F. Ahmed, E.D. Discher, Self-poroting polymerosomes of PEG-PLA and PEG-PCL: hydrolysis-triggered controlled release vesicles, *J. Control. Release*, 96, 2004, 37–53.
41. C. Nardin, T. Hirt, J. Leukel, W. Meier, Polymerized ABA triblock copolymer vesicles, *Langmuir*, 16, 2000, 1035–1041.
42. S. Rameez, H. Alosta, A.F. Palmer, Biocampatible and biodegradable polymerosomes encapsulated hemoglobin: a potential oxygen carrier, *Bioconj. Chem.*, 19, 2008, 1025–1032.
43. L. Ayres, P. Hans, J. Adams, D.W.P.M. Loewik, J.C.M. van Hest, Peptide-polymer vesicles prepared by atom transfer radical polymerization, *J. Polym. Sci. Part A Polym. Chem.*, 43, 2005, 6355–6366.
44. J. Ding, G. Liu, Water soluble hollow nanospheres as potential drug carriers, *J. Phys. Chem. B*, 102, 1998, 6107–6113.
45. H. Ringsdorf, B. Schlarb, J. Venzmer, Molekulare architektur und funktion von polymeren orientierten systemen, *J. Angew. Chem.*, 100, 1988, 117–121.
46. E. Donath, G.B. Sukhorukov, F. Caruso, S.A. Davis, H. Möhwald, Novel hollow polymer shells by colloid-templated assembly of polyelectrolytes, *Angew. Chem. Int. Ed. Engl.*, 37, 1998, 2202–2205.
47. H. Huang, E.E. Remsen, T. Kowalewski, K.L. Wooley, Nanocages derived from shell cross-linked micelle templates, *J. Am. Chem. Soc.*, 121, 1999, 3805–3806.
48. O. Rathore, D.Y. Sogah, Self-assembly of beta-sheets into nanostructures by poly(alanine) segments incorporated in multiblock copolymers inspired by spider silk, *J. Am. Chem. Soc.*, 123, 2001, 5231–5239.
49. F. Checot, S. Lecommandoux, Y. Gnanou, H.A. Klok, Water-soluble stimuli-responsive vesicles from peptide-based diblock copolymers, *Angew. Chem. Int. Ed.*, 41, 2001, 1339–1343.
50. S. Kimura, T. Kidchob, Y. Imanishi, Controlled release from amphiphilic polymer aggregates, *Polym. Adv. Technol.*, 12, 2001, 85–95.
51. T. Kidchob, S. Kimura, Y. Imanishi, Thermoresponsive release from poly(L-lactic acid) microcapsules containing poly(N-isopropylacrylamide) gel, *J. Chem. Soc. Perkin Trans.*, 2, 1997, 2195–2199.
52. M. van Dijk, T.H. Smith, F.M. Arnoe, E.H. Burger, P.I. Wuisman, The use of poly-L-lactic acid in lumbar interbody cages: design and biomechanical evaluation in vitro, *Eur. Spine J.*, 12, 2003, 34–42.
53. A.G. Anderopoulos, E.C. Hatzi, M. Doxastakis, Controlled release systems based on poly(lactic acid). An in vitro and in vivo study, *J. Mater. Sci. Mater. Med.*, 11, 2000, 393–397.
54. Y.M. Lin, A.R. Boccaccini, J.M. Polak, A.E. Bishop, V. Maquet, Biocompatibility of poly-dl-lactic acid (pdlla) for lung tissue engineering, *J. Biomater. Appl.*, 21, 2006, 109–118.
55. R.Langer, J.P. Vacanti, Tissue engineering, *Science*, 260, 1993, 920–926.
56. J. Klompmaker, H.W.B. Jansen, R.P.H. Verth, J.H. deGroot, A.J. Nijenhuius, J.A. Pennings, Porous polymer implant for repair of meniscal lesions: a preliminary study in dogs, *Biomaterials*, 12, 1991, 810–816.
57. C. Berkland, M.J. Kipper, B. Barasimhan, K. Kim, D.W. Pack, Microsphere size, precipitation kinetics and drug distribution control drug release from biodegradable polyanhydride microspheres, *J. Control. Release*, 94, 2004, 129–141.
58. M. He, Z. Zhao, L. Yin, C. Tang, C. Yin, Hyaluronic acid coated poly(butyl cyanoacrylate) nanoparticles as anticancer drug carriers, *Int. J. Pharm.*, 373, 2009, 165–173.
59. F. Kratz, Albumin as a drug carrier: design of prodrugs, drug conjugates and nanoparticles, *J. Control. Release*, 132, 2008, 171–183.
60. N. Desai, V. Trieu, Z. Yao, L. Louie, S. Ci, A. Yang, C. Tao, T. De, B. Beals, D. Dykes, P. Noker, R. Yao, E. Labao, M. Hawkins, P. Soon-Shiong, Increased antitumor activity, intratumor paclitaxel concentration and endothelial cell transport of cremophor-free, albumin-bound paclitaxel, ABI-007, compared with cremophor-based paclitaxel, *Clin. Cancer Res.*, 12, 2006, 1317–1324.
61. M. Hamidi, A. Azadi, P. Rafiei, Hydrogel nanoparticles in drug delivery, *Adv. Drug Deliv. Rev.*, 60, 2008, 1638–1649.
62. M.N. Mason, A.T. Metters, C.N. Bowman, K.S.Anseth, Predicting controlled release behavior of degradable PLA-b-PEG-b-PLA hydrogels, *Macromolecules*, 34, 2001, 4630–4635.

63. G.M. Cruise, D.S. Scharp, J.A. Hubbell, Characterization of permeability and network structure of interfacially photopolymerized poly(ethylene glycol) diacrylate hydrogels, *Biomaterials*, 19, 1998, 1287–1294.

64. T. Coviello, P. Matricardi, C. Marianecci, F. Alhaique, Polysaccharide hydrogels for modified release formulations, *J. Control. Release*, 119, 2007, 5–24.

65. C.C. Lin, A.T. Metters, Hydrogels in controlled release formulation: network, design and mathematical modeling, *Adv. Drug. Deliv. Rev.*, 58, 2006, 1379–1408.

66. N.A. Peppas, P. Bures, W. Leobandung, H. Ichikawa, Hydrogel in pharmaceutical formulation, *Eur. J. Pharm. Biopharm.*, 50, 2000, 27–46.

67. P. Bawa, V. Pillay, Y.E. Choonara, C. du Toit, Stimuli-responsive polymers and their applications in drug delivery, *Biomed. Mater.*, 4, 2009, 022001–022015.

68. K.S. Soppimath, T.M. Aminabhavi, A.M. Dave, S.G. Kumbar, W.E. Rudzinski, Stimulus-responsive "smart" hydrogels as novel drug delivery systems, *Drug. Dev. Ind. Pharm.*, 28, 2002, 957–974.

69. L.H. Gan, D.G. Roshan, X.J. Loh, Y.Y. Gan, New stimuli responsive copolymers and N- acryloyl -N'-alkyl piperazine and methyl methacrylate and their hydrogels, *Polymer*, 42, 2001, 65–69.

70. E.S. Gil, S.A. Hudson, Stimuli-responsive polymers and their bioconjugates, *Prog. Polym. Sci.*, 29, 2004, 1173–1222.

71. T. Farhan, O. Azzaroni, W.T.S. Huck, AFM study of cationically charged polymer brushes: switching between soft and hard matter, *Soft Matter*, 1, 2005, 66–68.

72. S.S. Pennadam, K. Firman, C. Alexander, D.G. Gorecki, Protein-polymer nano-machines. Towards synthetic control of biological processes, *J. Nanobiotechnol.*, 2, 2004, 8–22.

73. Y. Qiu, K. Park, Environment-sensitive hydrogels for drug delivery, *Adv. Drug. Deliv. Rev.*, 53, 2001, 321–339.

74. T. Aoki, M. Muramatsu, A. Nishina, K. Sanui, N. Ogata, Thermosensityvity of optically active hydrogels constructed with N-(1)-(hydroxymethyl) propylmethacrylamide, *Macromol. Biosci.*, 4, 2004, 943–949.

75. C.-L. Lo, S.-J. Lin, H.C. Tsai, W.H. Chan, C.H. Tsai, C.D.H. Cheng, G.H. Hsiue, Mixed micelle systems formed from critical micellar concentration and temperature-sensitive diblock copolymers for doxorubicin delivery, *Biomaterials*, 30, 2009, 3961–3970.

76. E.E. Carpenter, Iron nanoparticles as potential magnetic carriers, *J. Magn. Mater.*, 225, 2001, 17–20.

77. A. Mamada, T. Tanaka, D. Kungwachakun, M. Irie, Photo-induced phase transition of gels, *Macromolecules*, 23, 1990, 1517–1519.

78. K. Sumaru, M. Kameda, T. Kanamori, T. Shinbo, Characteristic phase transition of aqueous solutions of poly(N-isopropylacrylamide) functionalized with sphirobenzopyran, *Macromolecules*, 37, 2004, 4949–4955.

79. K. Nakamae, T. Nizuka, T. Miyata, M. Furukawa, T. Nishino, K. Kato, T. Inoue, A.S. Hoffman, Lysozyme loading and release from hydrogels carrying pendant phosphate groups, *J. Biomater. Sci. Polym. Ed.*, 9, 1997, 43–53.

80. L.C. Dong, A.S. Hoffman, A novel approach for preparation of pH-sensitive hydrogels for enteric drug delivery, *J. Control. Release*, 15, 1991, 141–152.

81. K. Nakamae, T. Nizuka, T. Miyata, T. Uragami, A.S. Hoffman, Y. Kanzaki, *Advanced Biomaterials in Biomedical Engineering and Drug Delivery Systems*, N. Ogata, S.W.Kim, J.Feijen, T.Okano Eds., Springer-Verlag, Tokyo, Japan 1996.

82. C. Alexander, K.M. Shakesheff, Responsive polymers at the biology/materials science interface, *Adv. Mater.*, 18, 2006, 3321–3328.

83. D. Le Garrec, J. Taillefer, J.E. van Lier, V. Lenaerts, J.-C. Leroux, Optimizing pH-responsive polymeric micelles for drug delivery in a cancer photodynamic therapy model, *J. Drug Target.*, 10, 2002, 420–437.

84. G.-D. Zang, N. Nishiyama, A. Harada, D.-L. Jiang, T. Aida, K. Kataoka, pH-sensitive assembly of high-harvesting dendrimer zinc porphyrin bearing peripheral groups of primary amine poly(ethylene glycol)-b-poly(aspartic acid) in aqueous solution, *Macromolecules*, 36, 2003, 1304–1309.

85. C.F. van Nostrum, Polymeric micelles to deliver photosentisizers for photodynamic therapy, *Adv. Drug Deliv. Rev.*, 56, 2004, 9–16.

86. Y.Kakizawa, K. Kataoka, Block copolymer micelles for delivery of gene and related compounds, *Adv. Drug Deliv. Rev.*, 54, 2002, 203–222.

87. B. Schmaljohann, Thermo- and pH-responsive polymers in drug delivery, *Adv. Drug Deliv. Res.*, 58, 2006, 1655–1670.

88. P. Gupta, K. Vermani, S. Garg, Hydrogels: from controlled release to pH-responsive drug delivery, *Drug Discov. Today*, 7, 2002, 569–579.

89. A. Serres, M. Baudys, S.W. Kim, Temperature and pH-sensitive polymers for human calcitonin delivery, *Pharm. Res.*, 13, 1996, 196–201.
90. Y.H. Kim, Y.H. Bae, S.W. Kim, pH-temperature sensitive polymers for macromolecular drug loading and release, *J. Control. Release*, 28, 1994, 143–152.
91. K. Hada, M. Porkov, M. Shamis, R.A. Lerner, C.F. Barbas, D. Shabat, Single triggered trimeric prodrugs, *Angew. Chem. Int. Ed.*, 44, 2005, 716–720.
92. T. Miyata, J. Jikihara, K. Nakamae, A.S. Hoffman, Preparation of reversible glucose-responsive hydrogels by covalent immobilization of lecitin in polymeric networks having pendant glucose, *J. Biomater. Sci. Polym. Ed.*, 15, 2004, 1085–1098.
93. C.M. Dorski, F.J. Doyle, N.A. Peppas, Preparation and characterization of glucose sensitive P(MAA-g-EG) hydrogels, *Polym. Mater. Sci. Eng. Proceed.*, 76, 1997, 281–282.

20 Photoresponsive Polymers for Control of Cell Bioassay Systems

Kimio Sumaru, Shinji Sugiura, Toshiyuki Takagi, and Toshiyuki Kanamori

CONTENTS

20.1 INTRODUCTION

Corresponding to recent increase in need for innovative technologies in terms of tools for biological research and medical diagnosis, cell chips have attracted lots of attention of many researchers and pharmaceutical industry as new biochips following DNA chips, protein chips, and glyco chips (Singhvi et al. 1994; Ito et al. 1997; Ziauddin and Sabatini 2001; Anderson and van den Berg 2004; Suh et al. 2004; Hui and Bhatia 2007). A number of critical problems arising in the integration of bioassays on the basis of conventional cell-screening methods have been expected to be cleared feasibly by using the cells cultured in array on a chip in a cost-effective manner. However, unlike existing biochips to array molecules such as DNA, proteins, and sugar chains (Fodor et al. 1991; Chee et al. 1996; Kiessling and Cairo 2003), cell chips are made of living cells and need certain conditions to keep cells alive until the end of bioassays, imposing severe restriction in preparation, storage, and distribution of the cell chips. Therefore, compatibility to the on-site preparation is also considered in designing the construction and the operation of cell chips.

In order to implement high-throughput screening, integrated microfluidic systems need to be designed to generate myriad bioassay conditions; mediums or drug solutions should be dispensed to each culture chamber according to each different schedule.

Further, it is required to induce the cultured cells to express and maintain their high-order functions by bringing the environment of the cell chips close to that in vivo, since the cell bioassay is based on functions of cells in internal organs or body tissues (Fukuda and Nakazawa 2005; Nakazawa et al. 2006; Sakai and Nakazawa 2007; Tamura et al. 2008). Therefore, the methods to construct elaborated co-culture systems using multiple cell types are a key to prepare biomimetic cell chips that mimic in vivo conditions in addition to the development of microfluidic portions of cell chips.

Thus, the development of practically applicable cell chips needs to break down many obstacles, which are hard to get over with existing technologies. In these circumstances, many researchers are developing a variety of new technology components, which could make a breakthrough. Among them, we proposed a scheme to manipulate cell adhesion on culture substrate and fluid flow in microfluidic systems freely by means of computer-controlled patterned light irradiation. This scheme is to manipulate living cells and fluid flows individually in a parallel manner taking full advantage of light irradiation as a means of signaling: the applicability to the object locally and instantaneously in a noncontact manner. In this chapter, we will briefly review the research on photoresponsive biopolymers as technology components to construct and control integrated cell bioassay system.

20.2   WET AND SOFT SYSTEM MANIPULATED BY LIGHT IRRADIATION

Not to point out, light has been playing quite an important role in biological research. Optical microscopes have been widely used as the most general means to observe cells and tissues since the dawn of biology. Confocal scanning laser microscopes which enable us to observe the 3-dimensional structure of cells and tissues, as well as fluorescent microscopes, have spread rapidly since 1990 in the research field of biology, and have become common tools for most researchers nowadays. Further, accompanied with the rapid progress in optoelectronics, the single molecule analysis (Kitamura et al. 1999) and super-resolving microscopy (Fujita et al. 2007; Kawata et al. 2008) have been developed and applied in mainly biological research.

As seen earlier, the use of light has been common especially in a passive scheme such as a detection of the light from the objects in the research field of biology. On the other hand, a few technologies utilizing light as a means to control the objects in an active manner have been developed. Above all, "optical tweezer," which is a method to capture a small object by applying light focused with an objective lens of high power (Ashkin 1970, 1992), has been applied to manipulate suspended cells (Ashkin and Dziedzic 1989a,b). More recently, pulsed laser irradiation with high energy was introduced as a tool to kill living cells or to make a hole in cellular membrane (Hosokawa et al. 2004; Iwanaga et al. 2006; Smith et al. 2006), and several products based on this method have already been released (Szaniszlo et al. 2006).

In addition to these existing technologies to let the light affect the objects directly, other strategies to manipulate small objects such as cultured cells and fluid flows in microfluidic system via the photoresponsive materials have been investigated (Figure 20.1), and new class of photoresponsive materials, which are wet and soft, have been developed so far to construct the systems implementing such strategies. In practice, several research groups including us have developed culture substrates, whose cell adhesion is controlled by light irradiation, and the hydrogels, which exhibit volume change in response to light irradiation. Mediated by these materials, the light irradiation applied as an input signal is transduced to the change in cell adhesion of culture substrate, or to change in the volume and the shape of a hydrogel.

On the other hand, an apparatus to irradiate light in arbitrary patterned area in the microscope field of view at arbitrary timing is necessary in order to control the properties of photoresponsive materials locally and instantaneously in a noncontact manner. To meet the requirement, a computer-controlled micropattern irradiation system (Kikuchi et al. 2009) was developed by the cooperation of our research group and Engineering System Co. (Matsumoto, Japan). A schematic illustration is shown in Figure 20.2. A projection unit and an image sensor unit, built in the optical system of an inverted biological microscope, are both controlled by a computer software, and the users can specify the area, where the light should be irradiated, just by dragging the pointer on the monitored microscope field of view. Also, light irradiation and microscopic observation, which are associated with each other, can be easily automated due to their high compatibility to information technology.

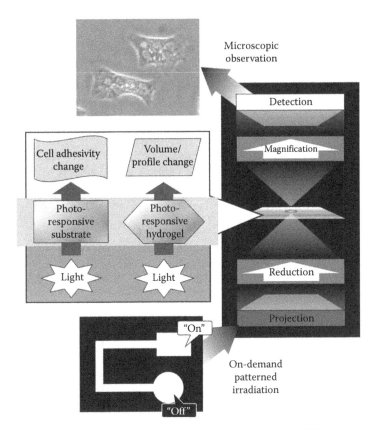

FIGURE 20.1 Scheme to manipulate small objects based on the conversion of light to material property.

20.3 PHOTORESPONSIVE POLYMERS

Organic photochromic materials whose chemical structures and properties are varied by light are promising as a means to transduce an input signal of light irradiation to the surface property or the morphology of material. Among them, spiropyrans, azobenzenes, and diarylethenes are known to exhibit reversible changes (Figure 20.3). Research on their application has been carried out very actively, and some of them have already been used practically in, for example, optical storage media.

In terms of the attempts to provide polymer materials with photoresponsive property by introducing photochromic groups, Goodman and Falxa reported in 1967 that the conformational transition was induced by light irradiation for an azobenzene-functionalized polypeptide in trifluoroacetic acid (Goodman and Falxa 1967), and many related researches were carried out subsequently (Pieroni et al. 1980, 1985; Fissi et al. 1996). Then, Ishihara et al. reported that the surface property such as water wettability and dye adsorption of azobenzene-functionalized poly(2-hydroxyethyl methacrylate) was changed by light irradiation (Ishihara et al. 1982). Later, in the first report on the photoresponsive solution property observed in aqueous system, Kungwatchakun and Irie described that the lower critical solution temperature (LCST, transition temperature between miscible and immiscible conditions) was largely shifted by light irradiation for an aqueous solution of poly(N-isopropylacrylamide)(pNIPAAm) functionalized with azobenzene (Kungwatchakun and Irie 1988). Later, on pNIPAAm functionalized with azobenzene in aqueous system, several researches have improved the photoresponsive properties (Desponds and Freitag 2003; Akiyama and Tamaoki 2004).

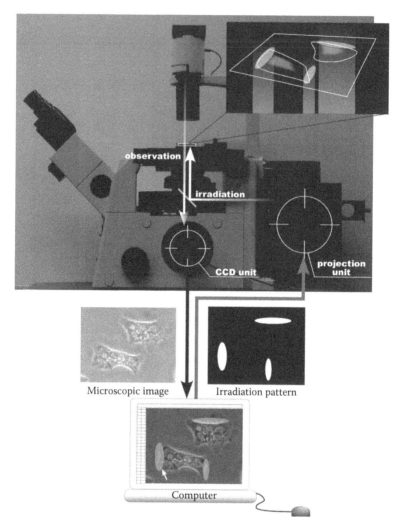

FIGURE 20.2 Schematic illustration showing construction of computer-controlled micropatterned irradiation system.

On the other hand, Kröger et al. found that the LCST of aqueous solution of poly(N,N-dimethylacrylamide) functionalized with azobenzene was shifted by at most 20°C in response to light irradiation (Kröger et al. 1994). Afterward, Shimoboji et al. modified an enzyme with this photoresponsive copolymer and demonstrated that the enzymatic activity could be controlled by light irradiation (Shimoboji et al. 2002). As aqueous systems of photoresponsive polymers other than azobenzene-functionalized ones, there were several reports on aqueous solutions of polymer functionalized with leuco chromophores (Irie and Hosoda 1985).

Also, there were some reports on photoresponsive polymers functionalized with spiropyrans which were among the most studied organic photochromic materials along with azobenzenes and leuco dyes since 1970s (Smets et al. 1978; Irie et al. 1979, 1985; Menju et al. 1981; Pieroni et al. 1992). However, all these reports were on the polymer property in organic solvents, and research on aqueous system had hardly been carried out until 2000s.

Later, we found that pNIPAAm functionalized with spiropyran chromophore (pSpNIPAAm) exhibited drastic photoresponsive dehydration in acidic aqueous solution of pH 4 or less (Sumaru et al. 2004a,b). The polymer solution which had been transparent and yellowish in the dark turned colorless in quick response to blue light irradiation due to the structural change (isomerization) of the

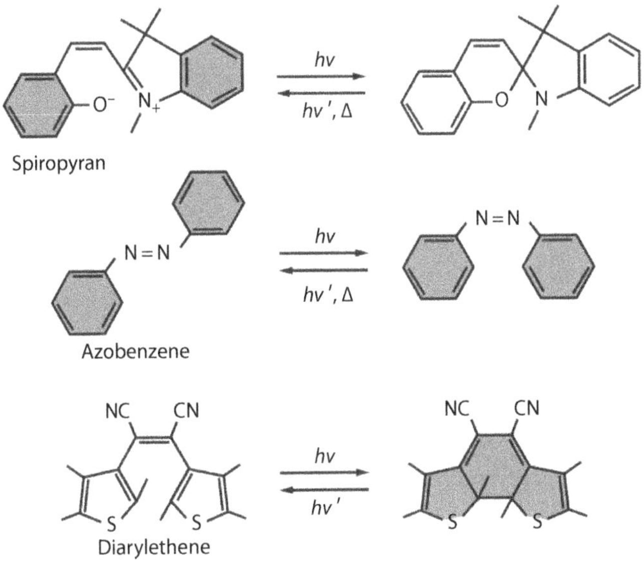

Spiropyran

Azobenzene

Diarylethene

FIGURE 20.3 Structural change of organic photochromic materials through photoisomerization.

chromophore. In addition to the change in color, the change in the physical properties of the chromophore accompanied by the photoisomerization is so substantial as to influence much on the hydration of pNIPAAm main chain; drastic dehydration was triggered by the isomerization of the chromophore introduced at a small functionalization ratio of only one monomer mol%. As a result, drastic turbidity increase followed by the precipitation was observed in the polymer solution after light irradiation at temperature from 30°C to 37°C (Figure 20.4). After stopping the irradiation, however, the solution returned to its original transparent state in one or a few hours, and this reversible change was repeatable for at least 10 times without any detectable degradation.

Thereafter, our detailed analysis revealed that the spiropyran chromophores introduced at side chain of pNIPAAm were in a protonated open-ring form, which exhibited yellow color due to strong absorbance at around 420 nm. These positively charged chromophores contributed to stabilize the

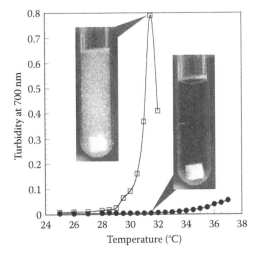

FIGURE 20.4 Influence of the light irradiation and temperature change on the turbidity of 0.10 wt% pSpNIPAAm aqueous solution containing 0.20 mM HCl. Open squares: under irradiation, closed circles: in the dark.

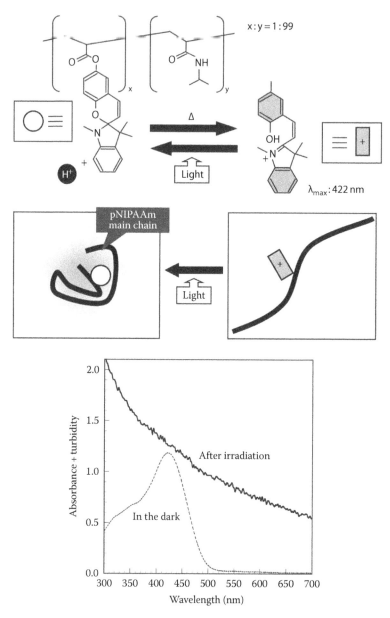

FIGURE 20.5 Schematic illustration showing the photo-induced structural change and the following dehydration of pSpNIPAAm.

hydration of pNIPAAm main chain even at temperatures higher than LCST of pNIPAAm aqueous solution. Irradiated with blue light, on the other hand, most of the chromophores isomerized effectively to a nonionic and hydrophobic form through the photo-induced ring closure and proton dissociation, destabilizing the hydration of main chain (Figure 20.5).

As a spiropyran-functionalized water-soluble polymer other than the pSpNIPAAm described earlier, we examined dextran functionalized with nitrospiropyran (NSp-Dex) (Edahiro et al. 2006). An aqueous mixture solution of dextran and poly(ethylene glycol) (PEG) at concentrations in a certain range was known to separate into two aqueous phases at temperatures higher than a certain critical temperature, while the mixture solution was uniform at the lower temperature (Tjerneld et al. 1985; Andersson and Hahn-Hagerdal 1990). By functionalizing dextran with nitrospiropyran composing

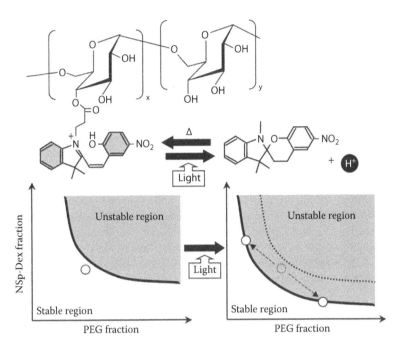

FIGURE 20.6 Schematic illustration showing the photo-induced structural change of NSp-Dex and the mechanism of photo-induced phase separation of aqueous mixture solution of NSp-Dex and PEG.

such a mixture solution, phase separation was induced by light irradiation under the condition of constant temperature (Figure 20.6). Since the physicochemical stability of such a solution is quite sensitive to the balance between the hydrophilicities of the two polymer components, drastic phase separation was triggered by light even though the functionalization ratio was only 0.3 mol% of the total number of hydroxyl groups in dextran. Phase separation of aqueous mixture solutions, which had been examined for the application to the less-invasive separation of proteins and living cells (Albertsson 1958; Guan et al. 1996; Edahiro et al. 2005a), was expected to gain wider application due to the property controllable by light as well as temperature change.

20.4 DEVELOPMENT OF PHOTORESPONSIVE HYDROGELS

Many researchers have studied so far not only for aqueous polymer solutions, but also for photoresponsive hydrogels. In the pioneering research on photoresponsive hydrogels, Irie et al. reported that the volume of the hydrogel functionalized with leuco chromophore increased largely by UV irradiation (Irie and Kunwatchakun 1986; Mamada et al. 1990). Although the photoresponsive swelling was so slow as to take more than an hour, it was shown experimentally for the first time that the signal of light irradiation could be transduced to the change in the volume of material. Later, several research groups reported independently the photoresponsive ion permeability of leuco-functionalized hydrogel membranes (Kimura et al. 1995; Kono et al. 1995; Kodzwa et al. 1999). The leuco chromophore was advantageous in modifying the hydration of hydrogel since it is ionized in response to light irradiation. However, its photoreaction requires high-energy UV light of the wavelength less than 300 nm, and therefore its degradation was critical due to the side reaction.

Without using photochromic materials, Suzuki and Tanaka demonstrated that the pNIPAAm hydrogel colored with chlorophyll shrank in response to the intensive irradiation with visible light (Suzuki and Tanaka 1990). This photoresponsive shrinking was due to the temperature rise induced through the conversion of light absorbed by chlorophyll to heat ("heat-mode"). Because of inevitable influence of the thermal diffusion, however, such a thermal-mode scheme is not valid in application to the manipulation in micrometer scale, where the use of light seemed mostly suitable. Although Garcia et al. reported later that the hydrodynamic radius of the polyarylamine microgel containing spiropyran in water decreased by light irradiation, its variation was at most 20% and the temperature range to obtain the photoresponse was limited (Garcia et al. 2007).

In these circumstances, we prepared the hydrogel composed of the pSpNIPAAm described earlier, and observed that the hydrogel, which had been swollen in the acidic aqueous system, shrank by 70% in quick response to blue light irradiation, and returned gradually to the former swollen state after stopping the irradiation (Figure 20.7) (Sumaru et al. 2006; Szilagyi et al. 2007). This drastic and quick photoresponsive shrinking was observed in the wide temperature range between 20°C and 30°C, and was repeatable for many times.

In order to examine the applicability of the pSpNIPAAm hydrogel to mass transfer control, we functionalized a porous membrane with the hydrogel, and evaluated the influence of light irradiation on its permeability for acidic aqueous solution (Sumaru et al. 2006). As a result, the permeability of the membrane was doubled by light at 30°C, maintained the increased value under irradiation, and returned gradually to the former low value over a few hours. On the other hand, we prepared a rod composed of the photoresponsive hydrogel and observed that the rod in an aqueous solution, when irradiated with blue light from one side, bent toward the light source.

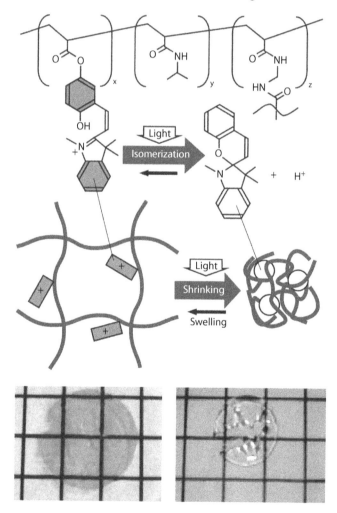

FIGURE 20.7 Photo-induced shrinking of pSpNIPAAm hydrogel.

20.5 MICROMANIPULATION OF PHOTORESPONSIVE HYDROGELS

As described at the beginning, light is an advantageous means of signaling due to its local and instantaneous applicability to small object. In order to demonstrate this feature, we composed a microfluidic system having photoresponsive microvalves using the pSPNIPAAm hydrogel (Sugiura et al. 2007). Three photoresponsive microvalves were fabricated by incorporating pieces of the hydrogels at prescribed positions in microfluidic channel with 500 μm width and 100 μm depth, through in situ photo-polymerization. In order to visualize the fluid flow through the microvalves, dilute acidic aqueous solution of indigocalmin was introduced into a microchannel with constant applied pressure. Before irradiation, all the hydrogels were in swollen state, stopping the fluid flow in these microvalves stably. In response to the local irradiation of blue light with 436 nm wavelength at 20 mW/cm^2 intensity, a hydrogel in the right-side microvalve shrank to allow the fluid flow toward downstream. The openings of the center and the left-side microvalves were also achieved independently by the following local irradiation of blue light in series (Figure 20.8). Further, simultaneous opening of multiple microvalves by temperature rise was demonstrated utilizing the temperature-dependence of pSpNIPAAm hydrogel.

In clear contrast to the methods based on the conversion of light to heat (Sershen et al. 2005; Chen et al. 2008), the photoresponsive valve openings working in pure "photon mode" would be insusceptible to thermal interference between adjacent microvalves even in such a case that many of the highly integrated microvalves were manipulated at the same time. And unlike the methods based on the thermal melting (Moriguchi et al. 2002; Hattori et al. 2004; Park et al. 2007; Hua et al. 2008) and wettability change (Caprioli et al. 2007) of materials caused by light irradiation, fluid flow could be controlled reversibly.

By using a computer-controlled micropattern irradiation system described previously, we can control the photoresponsive hydrogel with resolutions down to a micrometer scale. So we tried on-demand control of the microrelief profile on the surface of photoresponsive hydrogel sheet by modulating its local thickness by micropatterned light irradiation (Szilagyi et al. 2007). First, we prepared a pSpNIPAAm hydrogel sheet of initial thickness of ~250 μm on an acrylated glass surface so that the hydrogel sheet was attached covalently to the surface of the flat glass plate. In order to characterize the photoresponse of the hydrogel sheet prior to the experiment of microrelief formation, we measured the change in its thickness induced by blue light irradiation. As a result, the thickness in the irradiated area decreased by half shortly after irradiation for a few seconds, and the shrunk state was retained for about 10 min. Then the thickness returned gradually to the former swollen state over 2.5 h (Figure 20.9).

Then, we applied the micropatterned irradiation onto the pSpNIPAAm hydrogel sheet by using a computer-controlled micropattern irradiation system. As a result, a very sharp microrelief was etched on the surface of hydrogel sheet right after micropatterned irradiation for a second. The resultant fine microrelief was retained stably after stopping irradiation, indicating determinately that the shrinking of the hydrogel was not due to the thermally induced process through light-heat conversion ("heat mode" process) but due to pure "photon mode" process through the photoisomerization of chromophores. The microrelief was erased by leaving the sheet in the dark for a while, and new microrelief was formed by other micropatterned irradiation. Further, as a demonstration of 3-dimensional control of the microrelief profile, a microrelief with four-step height was formed by the stepwise micropatterned irradiation (Figure 20.10).

In order to apply such a controllability of surface profile to a new microfluidic system with high flexibility of design, we developed a universal microfluidic system composed of the photoresponsive hydrogel in which we could instantly form a microchannel along arbitrary path just by micropatterned light irradiation (Sugiura et al. 2009). The working principle is shown in Figure 20.11. The system was constructed by mounting another glass plate with multiple inlet and outlet ports, on the photoresponsive hydrogel sheet attached to a glass plate. As a result of light irradiation along an

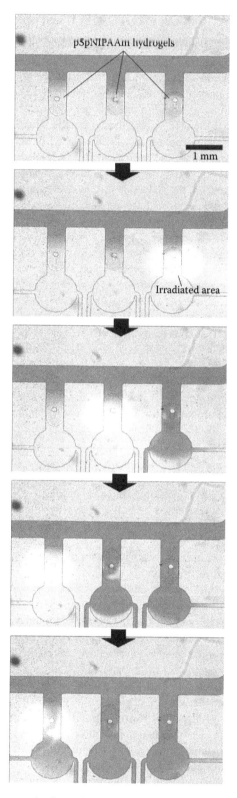

FIGURE 20.8 Independent control of multiple photoresponsive microvalves by means of local light irradatiation.

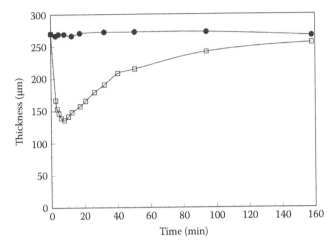

FIGURE 20.9 Change in the thickness of a pSpNIPAAm hydrogel sheet after blue light irradiation.

arbitrary path from one of the inlet ports to one of the outlet ports, the hydrogel in the irradiated area shrank, and a microchannel was formed to lead a fluid supplied from the inlet port to the outlet port.

Figure 20.12 shows the experimental result of the practical demonstration using the constructed universal microfluidic system. After micropatterned irradiation, a microchannel was formed in the hydrogel sheet along the irradiated pattern, and the fluorescent latex suspension supplied from an inlet port started to flow through the newly formed microchannels within a few minutes. A straight microchannel, a bent microchannel, a serpentine microchannel, and branched microchannels were repeatedly formed by micropatterned light irradiation. Note that it took several minutes to 10 min to fill up the long microchannels. In this scheme of microfluidic control based on the conversion of light to the volume change of the photoresponsive hydrogel, microchannels with an arbitrary width and depth along arbitrary path could be formed at an arbitrary timing. This result demonstrated that the universal microfluidic system based on this scheme would provide microfluidics with a powerful method of mass transfer control.

Further, combining the universal microfluidic system described earlier and conventional fixed microchannel, we attempted the construction of new integrated photoresponsive microvalve array (Sugiura et al. 2009). The system was constructed by stacking a polydimethylsiloxane microchannel chip and a glass plate with mechanically fabricated through-holes, on a photoresponsive hydrogel sheet attached to a glass plate (Figure 20.13). As a result of micropatterned light irradiation, the hydrogel in the irradiated area adjacent to a pair of through-holes shrank, inducing the opening of the microvalve. Figure 20.14 shows the construction of microvalve array and the experimental result of the practical demonstration. In the microvalve array, 10 photoresponsive microvalves were integrated within 1 cm². Each microvalve connected a 1.2 mm-wide inlet microchannel to 400 μm-wide outlet microchannel via two 400 μm through holes and the photoresponsive hydrogel sheet. The individual microvalves were successfully controlled through the opening of a single microvalve by local light irradiation while they had stopped stably the flow of fluid applied at the pressure of 5 kPa before the irradiation. Parallel control of multiple microvalves was also successfully demonstrated as two microvalves were simultaneously opened within several minutes by micropatterned irradiation.

In the photoresponsive microvalves fabricated by incorporating pieces of the hydrogels at prescribed positions in microfluidic channel as shown in Figure 20.8, reproducible polymerization with high precision was required to achieve both the stable stopping under unirradiated condition and the prompt opening upon irradiation (Sugiura et al. 2007). On the other hand, such a requirement was easily cleared in the construction stacking a flat glass plate on a photoreponsive hydrogel sheet with uniform thickness attached to another flat glass plate.

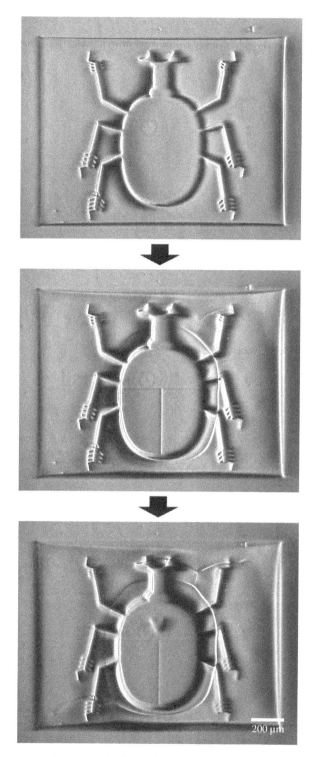

FIGURE 20.10 The process of a microrelief with 4-step light by stepwise micropatterned irradiation on pSpNIPAAm hydrogel sheet.

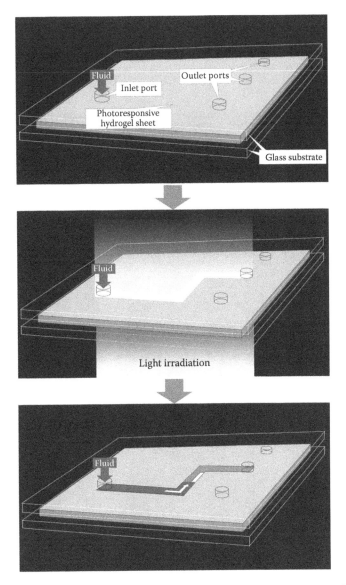

FIGURE 20.11 Schematic illustration showing a working principle of a universal microfluidic system composed of photoresponsive hydrogel sheet.

20.6 PHOTO-CONTROL OF CELL ADHESION

To implement the integrated cell bioassay system, which was mentioned at the beginning, a technology to control the cell culture system precisely is required in addition to the photo-controlled manipulation of microfluidics. Considering the advantage of light control as mentioned earlier, it is important to develop the cell culture substrate that transduces light irradiation to the change in its cell adhesion property. As attempts to control cell adhesion property of culture substrates by light irradiation, patterned culture substrates were developed by applying conventional photo-resist process from 1980s to 1990s (Pierschbacher and Ruoslahti 1984). Later, Nakayama et al. developed a culture surface whose cell adhesion property was increased by UV irradiation with the wavelength of 290–400 nm (Nakayama et al. 2003). Since these methods require photo-resist process or the irradiation of high-energy UV, light irradiation must be done in prior to cell seeding on the substrate.

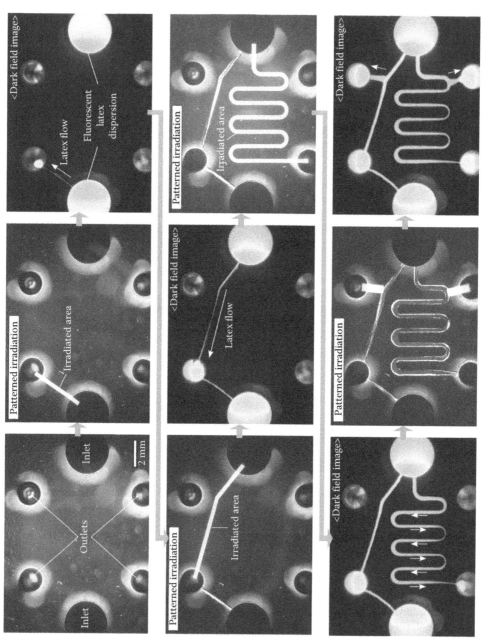

FIGURE 20.12 On-demand formation of microchannels with arbitrary pathways in the universal microfluidic system by micropatterned light irradiation. (From Ref. [77]. With permission.)

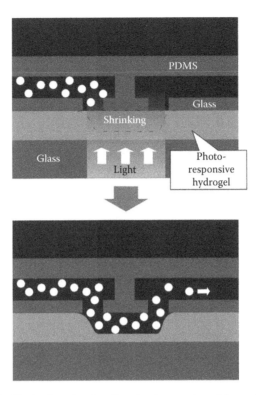

FIGURE 20.13   Schematic illustration showing the structure and working principle of microvalve array composed of photoresponsive hydrogel sheet.

In such situations, we proposed a new scheme of cell manipulation through in situ control of cell adhesion property of the culture substrate by light irradiation after cell seeding (Figure 20.15). First, living cells are seeded and cultured on a culture substrate whose surface property can be changed by mild light irradiation. These cells are observed and classified, for example, by the shape and/or fluorescence markers, then the light is irradiated in the area where the classified cells are. As a result, cell adhesion property of the culture substrate changes (increases in the case shown in Figure 20.15) only in the irradiated area, and subsequent washing of the substrate surface selectively removes the cells whose adhesion to the substrate surface is relatively weak. Implementation of such a technology will enable us to separate or purify the cultured cells having a certain specific property in the adherent state without exfoliating from the substrate.

We synthesized a number of photoresponsive polymers functionalized with spiropyrans and azobenzenes, prepared cell culture substrates, and examined the influence of light irradiation on their cell adhesion property. After a number of experiments which turned negatively, we observed that the cell adhesion on the surface of nitrospiropyran-functionalized pNIPAAm (pNSpNIPAAm) increased significantly in response to the light irradiation in the presence of living cells under a specific condition (Edahiro et al. 2005b, 2008). CHO-K1 cells were seeded and cultured for 24 h on a photoresponsive culture substrate functionalized with pNSpNIPAAm, and confluent cells were irradiated with the patterned UV light of 365 nm wavelength at 30 mW/cm^2 for 5 min. After subsequent cooling at 10°C for 20 min, the surface was washed uniformly with a phosphate buffer solution cooled at 10°C. As a result, it was observed that the cells were removed from the unirradiated area effectively, while most cells remained in the irradiated area. After cell pattern was successfully formed, cells were stained with CMTPX, a fluorescence dye staining only living cells. Fluorescence was observed from most of the cells, indicating that a series of the process including light irradiation and cooling did not cause critical damages on the cell viability (Figure 20.16).

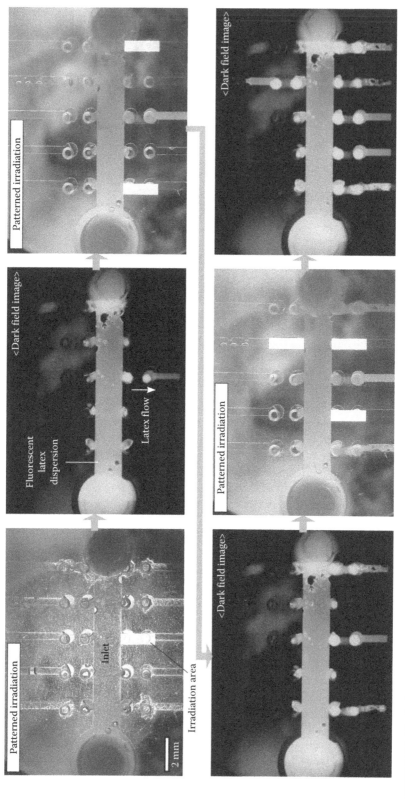

FIGURE 20.14 Independent and parallel flow control in photoresponsive microvalve array composed of pSpNIPAAm hydrogel sheet. (From Ref. [77]. With permission.)

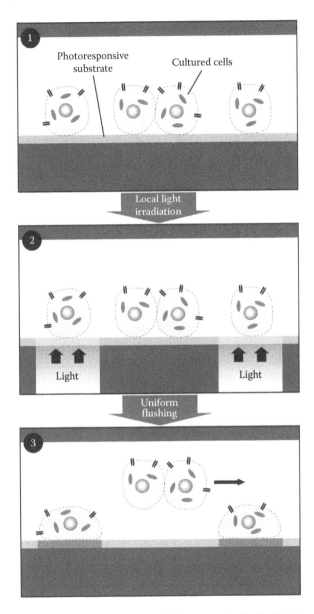

FIGURE 20.15 Scheme of cell manipulation through in situ control of cell adhesion property of the culture substrate by light irradiation after cell seeding.

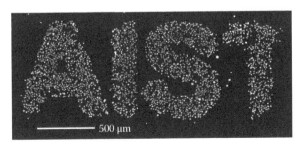

FIGURE 20.16 Fluorescence microscopic image of CHO-K1 cells patterned by in situ photo-control of cell adhesion and stained with CMTPX.

This photoinduced increase of cell adhesion was applicable only to the cells in a certain adherent condition, and this unique anchoring effect could not be produced with suspended cells and firmly adhering cells cultured for a long time. In spite of this practical problem, the result was important as the first report on in situ control of cell adhesion by light irradiation. Further, we also demonstrated that the different kinds of cells were immobilized in certain areas in a closed microchannel by using this method (Tada et al. 2006b).

20.7 ELABORATE CELL PATTERNING ON PHOTORESPONSIVE CULTURE SUBSTRATE

As seen in Figure 20.16, cell patterning could be easily achieved even in the culture environment by using this photoinduced increase of cell adhesion mentioned earlier. However, the photoinduced anchoring was effective only in temporary cell adhesion increase, and this method did not restrict the growth or movement of the cells. So, the patterned cells continued to grow and spread beyond the irradiated area in the further incubation. To keep cells within fine patterns, making confined cell adhesive areas in the nonadhering surface by light irradiation seemed reasonable to explore.

Nakanishi et al. developed a new class of photoresponsive culture substrate based on "caged" (protected by a photocleavable protecting group) technology (Nakanishi et al. 2004, 2007, 2008). They prepared a substrate by coating adhesion inhibitors such as bovine serum albumin (BSA) or block copolymer of ethylene glycol and propylene glycol (Pluronic) on a caged glass surface to provide the property of cell adhesion inhibition. By irradiation with UV light of 365 nm wavelength, which is compatible with the optical systems of widely used biological microscopes, the adhesion inhibitors were released via the "uncaging" mechanism, allowing cell adhesive moieties such as fibronectin to be adsorbed on. In clear contrast to the conventional photo-patterned substrates, which require the photo-resist process or the irradiation of high-energy UV as mentioned earlier, the cell adhesion property of the substrate can be changed by light irradiation, even being in cell culture environment. By using the substrate, they prepared various cellular patterns at single-cell resolution. Further, they demonstrated that the single cells, which had localized within independent confined areas respectively, were able to migrate along the paths appended afterward by the subsequent irradiation.

On the other hand, we developed a substrate, whose surface attribute could be switched from being cell-inhibitory to being cell-adhesive just by light irradiation without any treatment before cell introduction (Tada et al. 2006a; Kikuchi et al. 2009). Figure 20.17 shows the mechanism of the photoresponsive change in cell adhesion property and a scheme to construct an elaborately patterned co-culture system using the substrate. We prepared the substrate by immobilizing PEG on the surface as the cell adhesion inhibitor via poly(methyl methacrylate) functionalized with nitrospiropyran (pNSpMMA). In response to irradiation with UV light of 365 nm wavelength, pNSpMMA became hydrophilic through photoisomerization of nitrospiropyran and released the PEG from the surface, and then cell adhesive area appeared instantly. When the suspension of adherent cells was introduced onto the surface, the cells adhered only to the previously irradiated area and grew within the area in the subsequent incubation. After removing the floating cells above unirradiated area, highly contrasted cellular pattern appeared, and the constructed cell pattern was maintained stably. The cell adhesive area was surrounded by the barrier of PEG, which prevented the adhering cells effectively from further spreading. Since the cell adhesive area could be appended just by light irradiation even in the culture environment, we could construct the elaborately co-culture system, which was composed of multiple kinds of living cells arranged in arbitrary patterns by repeating patterned irradiation, introducing and removing cells on the photoresponsive culture substrate in a stepwise manner.

In the earlier-mentioned method based on caged technology, application of cell adhesive moieties such as fibronectin was necessary after light irradiation to obtain sufficient cell adhesion

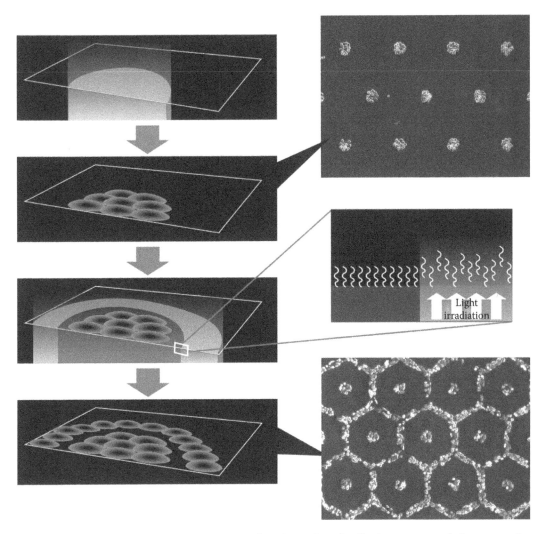

FIGURE 20.17 Mechanism of the photoresponsive change in cell adhesion property of photoresponsive culture substrate and a scheme to construct an elaborately patterned co-culture system via stepwise cell patterning on the substrate.

in the irradiated area (Nakanishi et al. 2004, 2007). In contrast, the bare surface of pNSpMMA which appears after photoinduced removal of PEG from our photoresponsive substrate was sufficiently cell adhesive, and most anchorage-dependent cells could adhere. As a result, highly contrasted cellular pattern was available just by light irradiation without further treatment. This feature is advantageous especially in stepwise fabrication of elaborate cellular pattern. PEG used here as a cell adhesion inhibitor is known to be basically harmless and is used in a large amount in cell fusion (Wojcieszyn et al. 1983; Khetani and Bhatia 2008). Additionally, the amount of PEG immobilized on the pNSpMMA layer was quite small, and the PEG released into medium by light irradiation was removed from the culture environment by subsequent medium replacement.

Here we show several examples of practical cell patterning implemented by the photoresponsive culture substrates and a computer-controlled micropatterned irradiation system. Figure 20.18 shows the appearance of a simple cellular pattern composed of MDCK cells. The MDCK cells were seeded on a surface of a photoresponsive culture substrate which had been irradiated with a stripe pattern of 80 μm line width. After incubation for 3 h, the cells floating

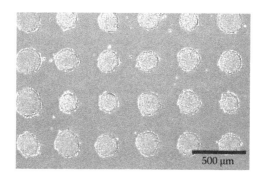

FIGURE 20.18   The microscopic image of MDCK cells patterned on a photoresponsive culture substrate.

above unirradiated area were removed subsequently. Figure 20.18 shows the appearance of the cells after additional incubation for half a day, and all the irradiated area was filled with the grown MDCK cells. The constructed cellular pattern was maintained stably even after 3 days. As a result, it was demonstrated that sufficient cell adhesion was obtained on the substrate surface just by light irradiation in the culture environment while potent inhibition was stably maintained in unirradiated area.

Meanwhile, the regulation of the culture environment by this cell-patterning method has much influence on the bioactivity of cultured cells. The HepG2 cells were seeded on a surface of a photoresponsive culture substrate which had been irradiated with a polka-dot pattern. HepG2 is a cell line derived from human hepatoma which is widely used in the research concerning the function of human liver. After incubation for 9 h, the cells floating above unirradiated area were removed. The patterned cells continued to grow and propagated within the irradiated area confined with a wall of PEG, forming evenly spaced uniform organoids in 5 days (Figure 20.19). Then balb/3T3 fibroblasts were introduced into the space among HepG2 organoids to form patterned co-culture system. Analysis on liver-specific gene expressions proved that the hepatic function of HepG2 such as cytochrome P450 activities and albumin production improved remarkably when cultured in such a patterned co-culture system compared with the conventional monoculture systems (Kikuchi et al. 2009). This experimental result suggested strongly that the patterned co-culture would increase the reliability of bioassay using liver-derived cells culture in vitro, which had been used as a means of high-throughput and low-cost drug screening (Norde and Gage 2004; Otsuka et al. 2004; Fukuda et al. 2006; O'Brien et al. 2006).

FIGURE 20.19   Microscopic image of HepG2 organoid array fabricated by using photoresponsive culture substrate.

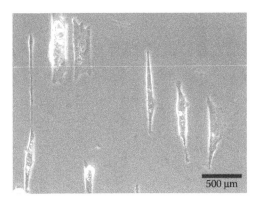

FIGURE 20.20 Microscopic image of balb/3T3 fibroblasts oriented by using photoresponsive culture substrate.

Figure 20.20 shows the appearance of balb/3T3 fibroblast cultured on a photoresponsive culture substrate which had been irradiated with a fine stripe pattern of 4 μm-wide lines and 12 μm-wide spaces. Adhering cells extended along the direction of stripe pattern, showing that the photoresponsive culture substrate enables to control through patterned light irradiation not only the distribution of adhering cells, but also the cell orientation.

Although the cell adhesion ability varies widely with the type of cells, the cell adhesion property of the photoresponsive culture substrate is easily tunable by varying the amount and molecular weight of PEG immobilized on the pNSpMMA layer (Tada et al. 2006a). By using properly tuned substrate, most of anchorage-dependent cells can be patterned in this way. We have confirmed that this cell-patterning method is applicable to all the cell types we examined such as CHO-K1, HeLa, MEF, Hek293, HOS, HUVEC, and NIH/3T3, as well as MDCK, HepG2, and balb/3T3 mentioned earlier. In addition, we can irradiate light in arbitrary patterns according to monitored microscope field of view with very high reproducibility by using a computer-controlled micropatterned irradiation system we developed. Taking these advantages, we demonstrated to fabricate more elaborate culture system than the previously mentioned system composed of HepG2 and balb/3T3, by repeating cell patterning process in a stepwise manner as shown schematically in Figure 20.17. Figure 20.21 shows the result of stepwise fabrication of an elaborate cellular pattern composed of MDCK cells. As patterned irradiation and cell seeding were repeated, cellular patterns were added on the substrate sequentially. The cell patterning process was repeated three times taking a day in total. The cellular pattern fabricated in former patterning process was maintained stably during the latter process including the removal of floating cells. The MDCK cells introduced in each step were stained differently in advance, and we confirmed that the cells in the resultant pattern are introduced successfully into the prescribed areas as we intended.

20.8 SUMMARY AND FUTURE OUTLOOK

In this chapter, we briefly reviewed the researches on photoresponsive biomaterials as technology components to construct and control integrated cell bioassay system by means of light irradiation. In particular, we focused on two technology components: the method to fabricate elaborated co-culture systems by light and photo-controlled manipulation of microfluidics. Combination of these technologies is expected to realize an integrated cell bioassay system with high reliability, in which all the process including cell culture and cell bioassay is thoroughly automated via computer-controlled micropatterned light irradiation. The schematic illustration of such system is shown in Figure 20.22. The use of light is advantageous to control high-throughput cell bioassay since non-contact regulation of closed and sterile systems and parallel control of massive objects are feasible. Since the photoresponsive hydrogels we have developed so far can be driven only under acidic

FIGURE 20.21 Stepwise fabrication of an elaborate cellular pattern composed of MDCK cells on photoresponsive culture substrate.

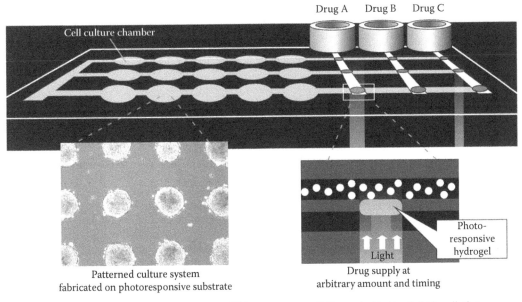

Drug A   Drug B   Drug C

Cell culture chamber

Photo-responsive hydrogel

Light

Patterned culture system
fabricated on photoresponsive substrate

Drug supply at
arbitrary amount and timing

FIGURE 20.22 Construction of integrated cell bioassay system fully controlled by light irradiation.

conditions, we cannot apply these hydrogels to the components of cell bioassay system as they are now. To solve this problem, we are now studying the properties of various copolymers functionalized with structurally modified photochromic materials to achieve precise photo-control in the culture environment.

In recent years, there is a tendency to create new technologies by gathering and combining technology components developed in various different research fields. Also the photo-manipulation methods for integrated cell bioassay system introduced in this chapter have been accomplished only by gathering and combining the various technology components such as synthesis and characterization of photoresponsive polymer materials, culture and analysis of cells, and construction and control of optoelectronic systems. The cell bioassay, which is biologically accurate and fully controllable by computer, will provide the drug screening, which has already been highly automated, with higher reliability and throughput. In these days, the rapid progress in optoelectronics is launching various new devices, and we expect that the photo-control technology for cell culture and bioassay will contribute to the progress in the research field of biology.

ACKNOWLEDGMENTS

The authors gratefully acknowledge Dr. Jun-ichi Edahiro, Dr. Andras Szilagyi, Dr. Koji Hattori, Ms. Kyoko Kikuchi, Mr. Yuichi Tada, and Ms. Yuki Ooshima for their cooperation in our research on the development and the application of photoresponsive polymer materials. The authors also acknowledge financial supports by Industrial Technology Research Grant Program in 2002 and 2005 from the New Energy Development Organization (NEDO), the Creation and Support Program for Start-ups from Universities in 2005 from Japan Science and Technology Agency (JST), and KAKENHI (20350110).

REFERENCES

Akiyama, H. and Tamaoki, N. 2004. Polymers derived from *N*-isopropylacrylamide and azobenzene-containing acrylamides: Photoresponsive affinity to water. *J. Polym. Sci. A Polym. Chem.* 42: 5200.
Albertsson, P.-A. 1958. Partition of proteins in liquid polymer–polymer two-phase systems. *Nature* 182: 709.
Andersson, H. and van den Berg, A. 2004. *Lab-on-Chips for Cellomics,* Kluwer Academic Publishers, Dordrecht, the Netherlands.

Andersson, E. and Hahn-Hagerdal, B. 1990. Bioconversion in aqueous two-phase systems. *Enzyme Microb. Technol.* 12: 242.

Ashkin, A. 1970. Acceleration and trapping of particles by radiation pressure. *Phys. Rev. Lett.* 24: 156.

Ashkin, A. 1992. Forces of a single-beam gradient laser trap on a dielectric sphere in the ray optics regime. *Biophys. J.* 61: 569.

Ashkin, A. and Dziedzic, J. M. 1989a. Internal cell manipulation using infrared laser traps. *Proc. Natl. Acad. Sci. USA* 86: 7914.

Ashkin, A. and Dziedzic, J. M. 1989b. Optical trapping and manipulation of single living cells using infra-red laser beams. *Ber. Bunsenges. Phys. Chem.* 93: 254.

Caprioli, L., Mele, E., Angilè, F. E. et al. 2007. Photocontrolled wettability changes in polymer microchannels doped with photochromic molecules. *Appl. Phys. Lett.* 91: 113113.

Chee, M., Yang, R., Hubbel, E. et al. 1996. Accessing genetic information with high-density DNA arrays. *Science* 274: 610.

Chen, G., Svec, F., and Knapp, D. R. 2008. Light-actuated high pressure-resisting microvalve for on-chip flow control based on thermo-responsive nanostructured polymer. *Lab Chip* 8: 1198.

Desponds, A. and Freitag, R. 2003. Synthesis and characterization of photoresponsive *N*-isopropylacrylamide cotelomers. *Langmuir* 19: 6261.

Edahiro, J., Sumaru, K., Tada, Y., Ohi, K., Takagi, T., Kameda, M., Shinbo, T., Kanamori, T., and Yoshimi, Y. 2005a. In-situ control of cell adhesion using photoresponsive culture surface. *Biomacromolecules* 6: 970.

Edahiro, J., Sumaru, K., Takagi, T., Kanamori, T., and Shinbo, T. 2006. Photoresponse of an aqueous two-phase system composed of photochromic dextran. *Langmuir* 22: 5224.

Edahiro, J., Sumaru, K., Takagi, T., Shinbo, T., Kanamori, T., and Sudoh, M. 2008. Analysis of photo-induced hydration of a photochromic poly(N-isopropylacrylamide)—Spiropyran copolymer thin layer by quartz crystal microbalance. *Eur. Polym. J.* 44: 300.

Edahiro, J., Yamada, M., Seike, S., Kakigi, Y., Miyanaga, K., Nakamura, M., Kanamori, T., and Seki, M. 2005b. Separation of cultured strawberry cells producing anthocyanins in aqueous two-phase system. *J. Biosci. Bioeng.* 100: 449.

Fissi, A., Pieroni, O., Balestreri, E., and Amato, C. 1996. Photoresponsive polypeptides. photomodulation of the macromolecular structure in poly(*N*- ((phenylazophenyl)sulfonyl)-l-lysine). *Macromolecules* 29: 4680.

Fodor, S. P., Read, J. L., Pirrung, M. C., Stryer, L., Lu, A. T., and Solas, D. 1991. Light-directed, spatially addressable parallel chemical synthesis. *Science* 251: 767.

Fujita, K., Kobayashi, M., Kawano, S., Yamanaka, M., and Kawata, S. 2007. High-resolution confocal microscopy by saturated excitation of fluorescence. *Phys. Rev. Lett.* 99: 228105.

Fukuda, J. and Nakazawa, K. 2005. Orderly arrangement of hepatocyte spheroids on a microfabricated chip. *Tissue Eng.* 11: 1254.

Fukuda, J., Sakai, Y., and Nakazawa, K. 2006. Novel hepatocyte culture system developed using microfabrication and collagen/PEG microcantact printing. *Biomaterials* 27: 1061.

Garcia, A., Marquez, M., Cai, T. et al. 2007. Photo-, thermally, and pH-responsive microgels. *Langmuir* 23: 224.

Goodman, M. and Falxa, M. L. 1967. Conformational aspects of polypeptide structure. XXIII. Photoisomerization of azoaromatic polypeptides. *J. Am. Chem. Soc.* 89: 3863.

Guan, Y., Lilley, T. H., Treffry, T. E., Zhou, C.-L., and Wilkinson, P. B. 1996. Use of aqueous two-phase systems in the purification of human interferon-alphal from recombinant *E. coli*. *Enzyme Microb. Technol.* 19: 446.

Hattori, A., Moriguchi, H., Ishiwata, S., and Yasuda, K. 2004. A 1480-nm/1064-nm dual wavelength photothermal eching system for non-contact three-dimensional microstructure generation into agar microculture chip. *Sens. Actuat. B Chem.* 100: 455.

Hosokawa, Y., Takabayashi, J., Shukunami, C., Hiraki, Y., and Masuhara, H. 2004. Nondestructive isolation of single cultured animal cells by femtosecond laser- induced shockwave. *Appl. Phys. A* 79: 795.

Hua, Z., Pal, R., Srivannavit, O., Burns, M. A., and Gulari, E. 2008. A light writable microfluidic flash memory: Optically addressed actuator array with latched operation for microfluidic applications. *Lab Chip* 8: 488.

Hui, E. E. and Bhatia, S. N. 2007. Microscale control of cell contact and spacing via three-component surface patterning. *Langmuir* 23: 4103.

Irie, M. and Hosoda, M. 1985. Photoresponsive polymers. Reversible solution viscosity change of poly (*N,N*-dimethylacrylamide) with pendant triphenylmethane leucohydroxide residues in methanol. *Makromol. Chem. Rapid Commun.* 6: 533.

Irie, M., Iwayanagi, T., and Taniguchi, Y. 1985. Photoresponsive polymers. 7. Reversible solubility change of polystyrene having pendant spirobenzopyran groups and its application to photoresists. *Macromolecules* 18: 2418.

Irie, M. and Kunwatchakun, D. 1986. Photoresponsive polymers. 8. Reversible photostimulated dilation of polyacrylamide gels having triphenylmethane leuco derivatives. *Macromolecules* 19: 2476.

Irie, M., Menju, A., and Hayashi, K. 1979. Photoresponsive polymers. Reversible solution viscosity change of poly (methyl methacrylate) having spirobenzopyran side groups. *Macromolecules* 12: 1176.

Ishihara, K., Okazaki, A., Negishi, N. et al. 1982. Photo-induced change in wettability and binding ability of azoaromatic polymers. *J. Appl. Polym. Sci.* 27: 239.

Ito, Y., Chen, G., Guan, Y., and Imanishi, Y. 1997. Patterned immobilization of thermoresponsive polymer. *Langmuir* 13: 2756.

Iwanaga, S., Smith, N., Fujita, K., Kawata, S., and Nakamura, O. 2006. Slow Ca^{2+} wave stimulation using low repetition rate femtosecond pulsed irradiation. *Opt. Express* 14: 717.

Kawata, S., Ono, A., and Verma, P. 2008. Subwavelength colour imaging with a metallic nanolens *Nat. Photon.* 2: 438.

Khetani, S. R. and Bhatia, S. N. 2008. Microscale culture of human liver cells for drug development. *Nat. Biotechnol.* 26: 120.

Kiessling, L. L. and Cairo, C.W. 2003. Hitting the sweet spot. *Nat. Biotechnol.* 20: 234.

Kikuchi, K., Sumaru, K., Edahiro, J. et al. 2009. Stepwise assembly of micropatterned co-cultures using photoresponsive culture surfaces and its application to hepatic tissue arrays. *Biotechnol. Bioeng.* 1003: 552.

Kimura, K., Kaneshige, M., and Yokoyama, M. 1995. Cation complexation, photochromism, and photoresponsive ion-conducting behavior of crowned malachite green leuconitrile. *Chem. Mater.* 7: 945.

Kitamura, K., Tokunaga, M., Iwane, A. H., and Yanagida, T. 1999. A single myosin head moves along an actin filament with regular steps of 5.3 nanometres. *Nature* 397: 129.

Kodzwa, M. G., Staben, M. E., and Rethwisch, D. G. 1999. Photoresponsive control of ion -exchange in leuco-hydroxide containing hydrogel membranes. *J. Membr. Sci.* 158: 85.

Kono, K., Nishihara, Y., and Takagishi, T. 1995. Photoresponsive permeability of polyelectrolyte complex capsule membrane containing triphenylmethane leucohydroxide residues. *J. Appl. Poly. Sci.* 56: 707.

Kröger, R., Menzel, H., and Hallensleben, M. L. 1994. Light controlled solubility change of polymers: Copolymers of *N,N*-dimethylacrylamide and 4-phenylazophenyl acrylate. *Macromol. Chem. Phys.* 195: 2291.

Kungwatchakun, D. and Irie, M. 1988. Photoresponsive polymers. Photocontrol of phase separation temperature of aqueous solutions of Poly(*N*-isopropylacrylamide) with pendant azobenzene groups. *Makromol. Chem. Rapid. Commun.* 9: 243.

Mamada, A., Tanaka, T., Kungwatchakun, D., and Irie, M. 1990. Photoinduced phase transition of gels. *Macromolecules* 23: 1517.

Menju, A., Hayashi, K., and Irie, M. 1981. Photoresponsive polymers. 3. Reversible solution viscosity change of poly (methacrylic acid) having spirobenzopyran pendant groups in methanol. *Macromolecules* 14: 755.

Moriguchi, H., Wakamoto, Y., Sugio, Y., Takahashi, K., Inoue, I., and Yasuda, K. 2002. An agar-microchamber cell-cultivation system: Flexible change of microchamber shapes during cultivation by photo-thermal etching. *Lab Chip* 2: 125.

Nakanishi, J., Kikuchi, Y., Inoue, S., Yamaguchi, K., Takarada, T., and Maeda. M. 2007. Spatiotemporal control of migration of single cells on a photoactivatable cell microarray. *J. Am. Chem. Soc.* 129: 6694.

Nakanishi, J., Kikuchi, Y., Takarada, T., Nakayama, H., Yamaguchi, K., and Maeda, M. 2004. Photoactivation of a substrate for cell adhesion under standard fluorescence microscopes. *J. Am. Chem. Soc.* 126: 16314.

Nakanishi, J., Takarada, T., Yamaguchi, K., and Maeda, M. 2008. Recent advances in cell micropatterning techniques for bioanalytical and biomedical sciences. *Anal. Sci.* 24: 67.

Nakayama, Y., Furumoto, A., Kidoaki, S., and Matsuda, T. 2003. Photocontrol of cell adhesion and proliferation by a photoinduced cationic polymer surface. *Photochem. Photobiol.* 77: 480.

Nakazawa, K., Izumi, Y., Fukuda, J., and Yasuda, T. 2006. Hepatocyte spheroid culture on a poly-dimethylsiloxane chip having microcavities. *J. Biomater. Sci. Polym. Ed.* 17: 859.

Norde, W. and Gage, D. 2004. Interaction of bovine serum albumin and human blood plasma with PEO-tethered surfaces: Influence of PEO chain length, grafting density, and temperature. *Langmuir* 20: 4162.

O'Brien, P. J., Irwin, W., Diaz, D. et al. 2006. High concordance of drug-induced human hepatotoxicity with in vitro cytotoxicity measured in a novel cell- based model using high content screening. *Arch. Toxicol.* 80: 580.

Otsuka, H., Hirano, A., Nagasaki, Y., Okano, T., Horiike, Y., and Kataoka, K. 2004. Two-dimensional multiarray formation of hepatocyte spheroids on a microfabricated PEG-brush surface. *Chem. Bio. Chem.* 5: 850.

Park, J.-M., Cho, Y.-K., Lee, B.-S., Lee, J.-G., and Ko, C. 2007. Multifunctional microvalves control by optical illumination on nanoheaters and its application in centrifugal microfluidic devices. *Lab Chip* 7: 557.

Pieroni, O., Fissi, A., Houben, J. L., and Ciardelli, F. 1985. Photoinduced aggregation changes in photochromic polypeptides. *J. Am. Chem. Soc.* 107: 2990.

Pieroni, O., Fissi, A., Viegi, A., Fabbri, D., and Ciardelli, F. 1992. Modulation of the chain conformational of spiropyran-containing poly(L-lysine) by the combined action of visible light and solvent. *J. Am. Chem. Soc.* 114: 2734.

Pieroni, O., Houben, J. L., Fissi, A., Costantino, P., and Ciardelli, F. 1980. Reversible conformational changes induced by light in poly(L-glutamic acid) with photochromic side chains. *J. Am. Chem. Soc.* 102: 5913.

Pierschbacher, M. D. and Ruoslahti, E. 1984. Cell attachment activity of fibronectin can be duplicated by small synthetic fragments of the molecule. *Nature* 309: 30.

Sakai, Y. and Nakazawa, K. 2007. Technique for the control of spheroid diameter using microfabricated chips. *Acta Biomater.* 3: 1033.

Sershen, S. R., Mensing, G. A., Ng, M., Halas, N. J., Beebe, D. J., and West, J. L. 2005. Independent optical control of microfluidic valves formed from optomechanically responsive nanocomposite hydrogels. *Adv. Mater.* 17: 1366.

Shimoboji, T., Larenas, E., Fowler, T., Kulkarni, S., Hoffman, A. S., and Stayton, P. S. 2002. Photoresponsive polymer–enzyme switches. *Proc. Natl. Acad. Sci. USA* 99: 16592.

Singhvi, R., Kumar, A., Lopez, P., Stephanopoulos, G. N., Wang, D. I., Whitesides, G. M., and Ingber, D. E. 1994. Engineering cell shape and function. *Science* 264: 696.

Smets, G., Braeken, J., and Irie, M. 1978. Photomechanical effects in photochromic systems. *Pure Appl. Chem.* 50: 845.

Smith, N. I., Iwanaga, S., Beppu, T., Fujita, K., Nakamura, O., and Kawata, S. 2006. Photostimulation of two types of Ca^{2+} waves in rat pheochromocytoma PC12 cells by ultrashort pulsed near-infrared laser irradiation. *Laser Phys. Lett.* 3: 154.

Sugiura, S., Sumaru, K., Ohi, K., Hiroki, K., Takagi, T., and Kanamori, T. 2007. Photoresponsive polymer gel microvalves controlled by local light irradiation. *Sens. Actuat. A: Phys.* 140: 176.

Sugiura, S., Szilagyi, A., Sumaru, K. et al. 2009. On-demand microfluidic control by micropatterned light irradiation of a photoresponsive hydrogel sheet. *Lab Chip* 9: 196.

Suh, K. Y., Seong, J., Khademhosseini, A., Laibainis, P. E., and Langer, R. 2004. A simple soft lithographic route to fabrication of poly(ethyleneglycol) microstructures for protein and cell patterning. *Biomaterials* 25: 557.

Sumaru, K., Kameda, M., Kanamori, T., and Shinbo, T. 2004a. Characteristic phase transition of aqueous solution of poly(N-isopropylacrylamide) functionalized with spirobenzopyran. *Macromolecules* 37: 4949.

Sumaru, K., Kameda, M., Kanamori, T., and Shinbo, T. 2004b. Reversible and efficient proton dissociation of spirobenzopyran-functionalized poly(N-isopropylacrylamide) in aqueous solution triggered by light irradiation and temporary temperature rise. *Macromolecules* 37: 7854.

Sumaru, K., Ohi, K., Takagi, T., Kanamori, T., and Shinbo, T. 2006. Photo-responsive properties of poly(N-isopropylacrylamide) hydrogel partly modified with spirobenzopyran. *Langmuir* 22: 4353.

Suzuki, A. and Tanaka, T. 1990. Phase transition in polymer gels induced by visible light. *Nature* 346: 345.

Szaniszlo, P., Rose, W. A., Wang, N. et al. 2006. Scanning cytometry with a LEAP: Laser-enabled analysis and processing of live cells *in situ. Cytometry* 69A: 641.

Szilagyi, A., Sumaru, K., Sugiura, S. et al. 2007. Rewritable microrelief formation on photoresponsive hydrogel layers. *Chem. Mater.* 19: 2730.

Tada, Y., Edahiro, J., Sumaru, K. et al. 2006b. Fabrication of a flow-type micro cell chip based on photo-induced cell capturing. *Proceedings of the μTAS 2006* 1: 966.

Tada, Y., Sumaru, K., Kameda, M. et al. 2006a. Development of a photoresponsive cell culture surfaces: Regional enhancement of living cell adhesion induced by local light irradiation. *J. Appl. Polym. Sci.* 100: 495.

Tamura, T., Sakai, Y., and Nakazawa, K. 2008. Two-dimensional microarray of HepG2 spheroids using collagen/polyethylene glycol micropatterned chip. *J. Mater. Sci. Mater. Med.* 19: 2071.

Tjerneld, F., Persson, I., Albertsson, P.–A., and Hahn-Hagerdal, B. 1985. Enzymatic hydrolysis of cellulose in aqueous two-phase systems. I. Partition of cellulases from *Trichoderma reesei. Biotechnol. Bioeng.* 27: 1036.

Wojcieszyn, J. W., Schlegel, R. A., Lumley-Sapanski, K., and Jacobson, K. A. 1983. Studies on the mechanism of polyethylene glycol-mediated cell fusion using fluorescent membrane and cytoplasmic probes. *J. Cell Biol.* 96: 151.

Ziauddin, J. and Sabatini, D. M. 2001. Microarrays of cells expressing defined cDNAs. *Nature* 411: 107.

21 Lignin in Biological Systems

Valentin I. Popa

CONTENTS

21.1 INTRODUCTION

In the polymer industry, utilization of plant and animal compounds is the key to sustainable development. Carbohydrates have already been used significantly in the food, medical, and cosmetic industries. Plant materials containing cellulose, hemicelluloses, and lignin are the largest organic resources but with the exception of cellulose, they are not very well utilized. Hemicelluloses are significantly underutilized. Lignin, production of which is over 20 million tons per year worldwide, is mostly burnt as fuel, although it is one of the most useful natural resources (Ranby 2001).

Lignins are three-dimensional polymers consisting of phenylpropane or C_9 units. There are three types of structural units that are available in a wide range of proportions in the woody tissue of different plant species: *p*-hydroxyphenylpropane, guiacyclopropane, and syringylpropane units. The proportion of these three C_9 units in lignin available in the woody tissues of plants depends on the nature of plant species. In general, gymnosperm (softwood) lignins consist mostly of guiacyclopropane units and angiosperm (hardwood) lignins are mainly composed of guiacyl- and syringylpropane units with varying ratios depending on the wood species. By contrast, grass (Gramineae) lignins are composed of *p*-hydroxyphenyl-, guiacyl-, and syringylpropane units and consist of lignin cores and peripheral units (Le Digabel and Averous 2006). The lignin cores are essentially lignins of the guiacyl-syringyl type, while the peripheral units consist of p-hydroxycinnamic acid and ferulic acid groups that are bonded by ester linkages mostly to hydroxyl groups at C-γ in the C_9 units of lignin cores (Figure 21.1).

It has been shown that lignin is a versatile molecule that possesses multiple properties such as antioxidant (i.e., radical scavenger), UV-absorption antifungal, and antibiotic activity. It has been suggested that lignin can be applied for stabilization of food and feed, due to their antioxidant, antifungal, and antiparasitic properties. Also, anticarcinogenic, apoptosis-inducing antibiotic, and anti-HIV activities of lignin have been reported.

However, very little attention has been paid to the biological activities and mechanism of action of lignins or lignin-containing/derived substances. To enhance the industrial use of lignins, there is

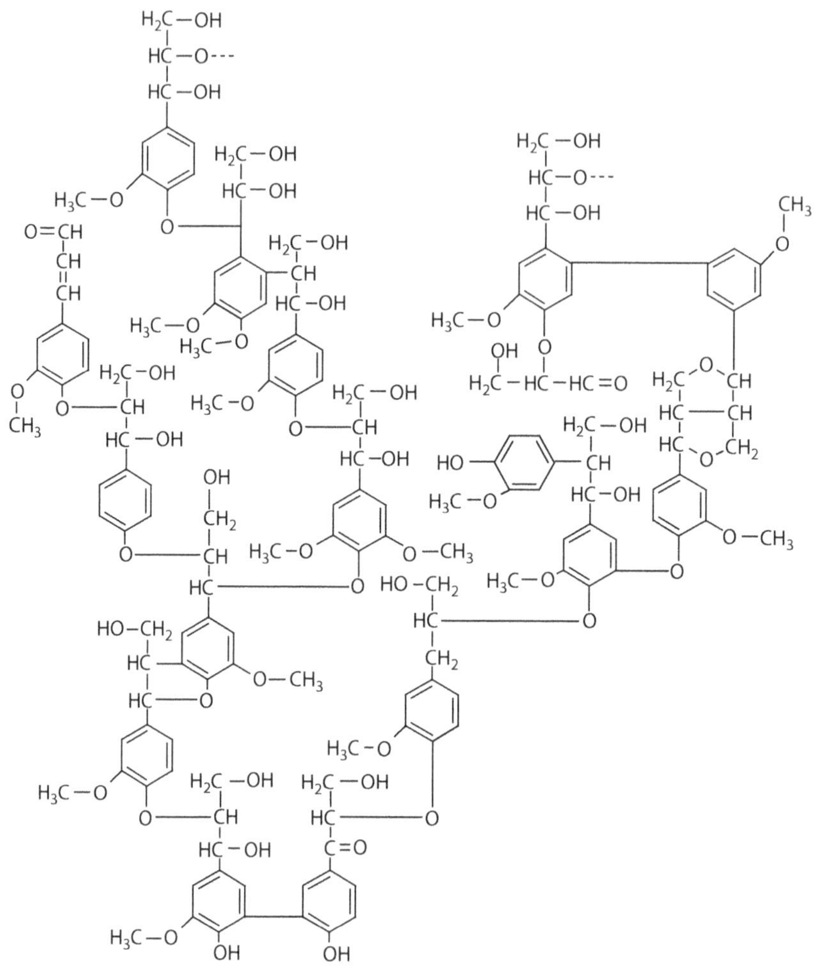

FIGURE 21.1 Schematic representation of lignin molecules from wheat straw.

a need for a continuous supply of lignin products with constant quality as related to purity, chemical composition, functional properties, and their reproducibility (Popa 2007).

Sometimes, the use of lignin has not yet been successfully achieved, because lignin molecules lack stereoregularity and also the repeating units in the molecule are heterogeneous and complex. In addition, lignocellulose components cannot be separated by simple extraction processes such as the pulping process, which make lignin molecules even more heterogeneous. That is why, lignophenols, as lignin derivatives are characterized by high content of phenolic function, high stability, and less heterogeneity compared with conventional lignin preparation. It was assumed that lignophenols derived from native lignin have different phenolic functionalities to exhibit various biological activities (Akao et al. 2004). Moreover, the functionality of lignophenols is able to be easily modulated by various modifications (Figure 21.2).

Lignophenols, such as lignocatechol, lignoresorcinol, and lignopyrogallol, which have many phenolic hydroxyl groups, are highly hydrophilic. On the other hand, lignophenols combined with monohydric phenols such as cresol are water insoluble. When water-insoluble lignophenols were carboxymethylated (CM) and/or alkaline treated (AT), they became highly hydrophilic and water soluble. Especially, CM-lignophenols showed a high binding affinity for proteins because of the carboxyl groups that were introduced.

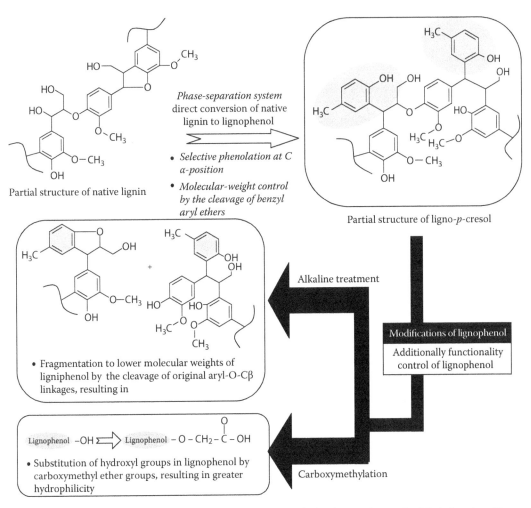

FIGURE 21.2 Conversion of native lignin into lignophenol derivatives and control of their functionality.

Today, lignin or lignin-derived compounds are not used in the neutraceutical/para-pharmaceutical and cosmetic industries. However, it can be imagined that lignin could represent an alternative/substitute for a number of plant polyphenol-containing extracts currently used in such industries, suggesting that lignin could have a more direct protective effect on health, thereby opening up possibilities of using this molecule as a source of bioactive molecules for use in medicine. That is why in this review some examples of the possibilities to use lignins in the biological systems are discussed. However, important biochemical and epidemiological research efforts are necessary before lignin can be used in a medical (as opposed to *para*-pharmaceutical) context. Additionally, it would also be necessary to demonstrate the advantages of using such medicaments/food additives as compared to simply changing to a more fiber-rich diet, containing lignin-rich fiber. Nevertheless, a number of potentially important niches exist for the use of lignin-derived products in the healthcare sector and these aspects will be discussed in the following sections.

21.2 LIGNINS AS ANTIBACTERIALS

The role of lignin in the human and animal metabolism has been mentioned many years ago (Csonka et al. 1929). A loss of lignin in human beings as well as in dogs has been reported. The experiments carried out with dogs and cows point to degradation of lignin. The conclusions are based on the fact that an increase in the hippuric acid eliminated was evidenced after lignin was investigated, as well as a decrease of the methoxyl group content was observed. Thus, experiments performed with a dog showed a loss of 15% of the methoxyl groups initially present in lignin, and in an experiment with a cow, a 37% loss of the methoxyl groups was observed. It was supposed that in these experiments there was direct evidence that lignin is metabolized in the animal body. The results on the benzoic acid eliminated in the urine when lignin was fed support this claim. Experiments were carried out in vitro in which known quantities of lignin mixed with fresh material taken from the four compartments of a cow stomach were incubated. A loss of methoxyl groups was observed, indicating that their partial disappearance of this takes place in vitro in the stomach of the animal and it is not brought about by bacteria but rather by some other agents, possibly an enzyme in the gastric juice of the animal.

As is known, the lignin is taken up as a "dietary fiber" when vegetables and fruits are consumed. It is thought that dietary fiber is useful for the prevention of some diseases (such as cancer), although most is excreted without being digested. To clarify the influence of dietary fiber on an organism, it is important to examine the relationship between the dietary fiber and enterobacteria. In respect to the influence of lignin on enterobacteria, it was reported that woody lignin shows antibacterial activity against *Klebsiella pneumoniae*, *Pseudomonas aeruginosa*, *Escherichia coli*, and *Staphylococcus aureus* (Nelson et al. 1994).

The influence of lignin extracted from rice bran on the growth of enterobacteria and their effect on lignin was also studied. *E. coli*, *Enterococcus faecium*, *Lactobacillus minutus*, *Staphylococcus epidermis*, *Bacteroides vulgatus*, and *Streptococcus intermedius* isolated from human feces were identified and used along with *Bifidobacterium bifidum* (NBRC 13952), *Bifidobacterium adolescentis* (JCM 7044), and *Bifidobacterium longum* (JCM1217). The turbidities (OD_{600}) in the medium including enterobacteria with or without rice bran lignin were determined (Toh et al. 2007). Thus, the OD_{600} values of medium cultivated with *L. acidophilus* and *S. intermedius* in the presence of lignin were significantly greater than those without lignin. The OD_{600} values of *E. coli*, *S. epidermis*, *L. minutus*, and *B. bifidum* in the presence of lignin were significantly lower than those without lignin. Small values of OD_{600} were observed when inhibition of growth of bacteria or bacteriolysis occurred. In the gel permeation chromatography (GPC) of the rice bran lignin in the case of *E. coli*, it was found that molecular mass was reduced from about 5,000 to 10,000 in 24 h.

The change in molecular mass of lignin that was observed in this study means that *E. coli* metabolized lignin and the number of microorganisms was reduced by lignin. Whether the products of the metabolized lignin contributed to the growth of the bacterium is unclear. However, the degradation of rice bran lignin occurred within 24 h after the incubation of *E. coli*, and lignin taken as a meal may be degraded in the intestine by this bacteria. Both herb and wood lignins showed antimicrobial activity for *E. coli*, in spite of the differences in structures and molecular mass between them. As concerning the mechanisms in the case of *E. coli* it is thought that metabolism of lignin by bacteria contributes to its growth. In other cases, there is some possibility that lignin physically functions as the structure of medium, which might contribute to the growth of enterobacteria because lignin is a biopolymer.

The link between uses of growth-promoting antibiotics and antibiotic resistance of pathogenic bacteria in humans has led member countries of the European Union to ban use of these antibiotics in animal agriculture. At the same time, recent findings demonstrate that purified lignin encompasses production and health benefits in animals in the absence of antibiotics. For this reason, purified lignin is one of several natural compounds that have received new interest and scientific consideration as a natural feed additive. Purified lignin may potentially mitigate production and economic losses after the use of antibiotics in livestock production is discontinued. In its purified form, lignin contains several low-molecular mass phenolic monomers that possess biological effects not characteristic of native lignin. The possibility exists that monophenolic fragments of purified lignin may have beneficial effects on productivity of farm animals, safety of animal products, and the environment. Lignin is not digested by monogastric animals, but the rumen bacteria flora facilitates degradation of benzyl-ether linkages of lignin polymers.

On the other hand, the sorption capacity of lignin has been the base in the development of drugs for the therapy of gastroduodenal disorders. Those functional groups allow lignin to adsorb and retain various types of microorganisms and cholesterol, bilirubin, lipid bile acids, and other carcinogenic substances.

21.2.1 In Vitro Studies

The phenolic components of lignin have been reported to inhibit growth of some microorganisms. The comparison of bioactivity toward *E. coli* of nonwood soda-lignin applied in soluble form has revealed that the antimicrobial properties strongly depend on the origin of lignins and, consequently, their structure. The side chain structure and nature of the functional groups of the phenolic compounds are major determinants of the antimicrobial effects of lignin. In general, phenolic fragments with functional groups containing oxygen (–OH,–CO, COOH) in the side chain are less inhibitory, whereas the most inhibitory phenolic fragments are those containing a double bound in α, β positions of the side chain, and a methyl group in the γ position.

The clearest effects were observed for lignins obtained from sisal, abaca, and flax (Telysheva et al. 2005). The same lignins are characterized with the high radical-scavenging activity in tests with DPPH-free radical and superoxide anion radical. This could indicate some relationship between these lignin properties (Figure 21.3).

The antimicrobial effect of lignin products under study was also detected against *Bacillus subtilis* and a range of phytopathogenic bacteria (*Erwinia carotovora, Bacillus polymixa, Pseudomonas syringae, X. vesicatoria*) and fungi (*Fusarium oxysporium, Cladosporium herbarum, Botrytis cinerea, Rhizoctonia solani, Micriodochium nivale*). The results obtained confirmed that the level of microorganism growth inhibition depends on a lignin sample prehistory, its concentration in cultivation medium, and type of microorganisms. Kraft lignin provided a bactericide effect toward *E. carotovora* and *X. vesicatoria* at lower concentration under study (0.25% on nutrition medium). However, in the case of *P. syringe* and *B. polymyxa*, the kraft lignin did not show a bactericide effect even at the 2% concentration. Spruce hydrolysis lignin (HL) showed the bactericide effect toward *P. syringe* and *Xanthomonas vesicatoria* in 2% concentration, so at lower concentration only slight bacteriostatic activity was observed. The modification with quaternary ammonium compound increases bactericidal properties of lignins; e.g., for HL a stable bactericide effect was achieved at the both concentrations used. Fungal species revealed different resistance to lignin products, and in total, the investigated products can be arranged according to their inhibitory effect in the following range: kraft lignin < quaternized HL lignin < quaternized kraft lignin. The inactivation of pathogenic

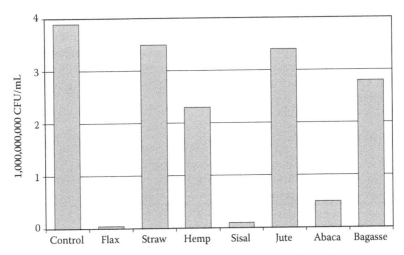

FIGURE 21.3 Influence of different lignin samples on bacteria *Escherichia coli* development.

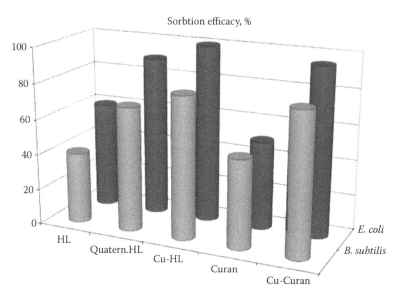

FIGURE 21.4 Influence of different lignin samples on pathogenic bacteria sorption. (Curan is commercial kraft lignin produced by Borregaard Ltd, Norway.)

microorganisms activity may be explained also by their sorption properties onto lignin products: nonmodified HL and kraft lignin are capable of adsorbing 45%–60% of the gram-negative and gram-positive bacteria *E. coli* and *B. subtilis*, respectively, in 1 h (Figure 21.4).

The lignins modified with the quaternary ammonium compounds, well-known antimicrobial agents, revealed the higher sorption capacity (70%–80% in 1 h) toward the same microorganisms. In contrast to the nonmodified lignin samples, quaternized lignin limited the retained microorganisms' vital activity. The Cu-containing modified lignins are characterized with the highest activity toward both the bacteria and besides that totally suppress the vital activity of the bacteria adsorbed.

The polyphenolic compounds of lignin cause cell membrane damage and lysis of bacteria with subsequent release of cell contents. Monophenolic compounds, such as carvacrol and cinnamaldehyde, also possess antibacterial properties. Carvacrol and thymol exhibit antimicrobial effects determining bacterial cell membrane disintegration and releases of cell contents. In contrast, cinnamaldehyde penetrates the bacterial cell membrane to reduce intracellular pH and cause ATP depletion. The antibacterial mechanistic actions seem to vary among phenolic compounds. To date there is no report establishing the mode of action of a product consisting of multiple phenolic compounds.

Some lignin products owing to their microstructure and diversity of sorption centers can realize a possibility of simultaneous sorption of organic pollutants and microorganisms and thus activate the process of pollutant biodegradation on interfaces. This was confirmed by sorption experiments carried out with 2,4-D and its bacteria-degrader *Burkholderia cepacia*, which was used along with HL and lignin modified with Si-containing compounds.

21.2.2 IN VIVO STUDIES

Indulin (40 and 100 g/kg of dry matter [DM]) reduced volatile fatty acids (VFA) concentrations in the ceca and large intestine of chickens, suggesting that purified lignin altered the fermentation pattern of the chicken intestinal tract by inhibiting growth of certain bacteria. In an attempt to provide a better evaluation of the antibacterial effects of lignin, the broiler chickens were orally challenged with pathogenic strains of *E. coli* and fed antibiotic-free diets or one containing Alcell lignin or antibiotics. Alcell lignin reduced cecal concentrations of total *E. coli* after 3 and 9 days. *E. coli* inhibition was more pronounced at higher concentrations (25 versus 12.5 g/kg of DM), suggesting that the antibacterial effects of Alcell lignin mostly occur at higher doses. Further research demonstrated that dietary Alcell lignin reduced concentrations of *E. coli* in poultry litter when compared to antibiotic-free diets or one containing antibiotics. Intestinal *E. coli* contaminate poultry carcasses during slaughter, thereby representing an important cause of food-borne illnesses in humans. At the same time, poultry litter may contribute to the spread of antibiotic-resistant genes into the food chain. Moreover, litter *E. coli* is the major causative pathogen implicated in cellulitis, a major cause of carcass condemnation at processing plants. These findings suggest that purified Alcell lignin may represent a dietary strategy to reduce *E. coli* load in the chicken intestine and litter that could offer an opportunity to improve the safety of poultry products and the control of cellulitis. Effects of purified lignin on these variables remain to be studied.

In a mouse model, Alcell lignin did not alter aerobic bacterial growth in the cecum, but reduced translocation of these bacteria in the lymph nodes and liver after burn injury. In mice, a pine wood lignin was reported to possess antitumor and antibacterial properties against the pathogens *E. coli* and *Pseudomonas*.

Ruminant responses to dietary lignin are variable. Alcell lignin had no effect on rumen fermentation in sheep and did not alter fecal concentrations of anaerobic, aerobic, and coliform microorganisms in calves. This could be the result of degradation of the phenolic compounds of lignin by rumen microbes. Para-coumaric and ferulic acids are extensively converted into reduced phenolics by ruminal microorganisms, as evidenced by their ruminal disappearances. The infusion of phenolic compounds, such as quinic acid, phenolic benzoic, and phenyl acetic acids, into sheep rumen caused large increments in urinary outputs of phenolic acids and phenols. These observations suggest that phenolic monomers are extensively metabolized by the rumen microbial flora to more chemically reduced forms, which are subsequently absorbed and metabolized. It seems, therefore, that the antimicrobial action of lignin may be species dependent and occurs mostly in nonruminants.

21.2.3 PREBIOTIC EFFECTS OF PURIFIED LIGNIN

Prebiotics are indigestible feed ingredients that selectively stimulate growth or metabolic activity of a limited number of intestinal microorganisms in birds and mammals. *Lactobacilli* and *Bifidobacteria* are beneficial bacteria that limit intestinal colonization of pathogens by competing for nutrients and binding sites and by secreting antibacterial substances. However, at higher level (i.e., 25 g/kg of DM), lignin inhibited growth of these bacteria. Prebiotics may also play important roles in improving the intestinal morphology of animal species. Longer villi are generally correlated with better intestinal health and improved efficiency of digestion and absorption. Goblet cells are responsible for the production of mucins, which destroy intestinal pathogens. In broiler chickens, Alcell lignin (12.5 g/kg of DM) increased villi height and goblet cell number. However, there was no benefit at a higher lignin level (i.e., 25 g/kg of DM) suggesting that prebiotic effect of Alcell lignin only occurs within a limited inclusion range (Baurhoo et al. 2008).

21.3   LIGNINS AS ANTIOXIDANTS AND PHOTOPROTECTORS

An antioxidant is defined as any substance that in low concentrations diminishes or inhibits the oxidative damage in a target molecule.

The researches on naturally occurring polyphenols have received more attention in human medicine, versus animal agriculture, as polyphenols exert several health benefits in humans. First, they inhibit oxidation of low-density lipoproteins, thereby decreasing risks of heart diseases. Second, polyphenols possess anti-inflammatory and anticarcinogenic properties and third, they are effective antioxidants for food lipids. Third, the neuronal toxicity is mediated and enhanced by reactive oxygen species (ROS) or reactive nitrogen species (RNS) and thus that the oxidative stress accelerates the cell death of neuronal cells in neurodegenerative disorders. In fact, such neuronal cell death has been reported to be attenuated by antioxidants and free radical scavengers (Catignani and Carter 1982).

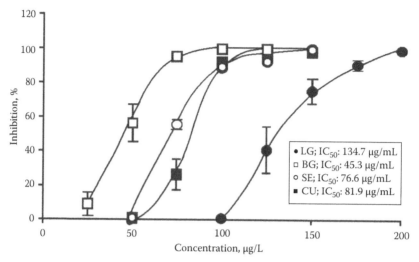

FIGURE 21.5 Inhibitory effects of different lignin solutions on hemolysis induced by AAPH. LG, lignosulfonates; BG, lignin from bagasse; SE, lignin from steam explosion; and CU, Curan, a commercial lignin.

Thus, the phenolic fragments of kraft lignin, obtained as by-products during pulp manufacture process are as effective as vitamin E as an antioxidant in humans.

The antioxidant effect of different types of lignin has been evaluated by their capacity to inhibit human erythrocyte hemolysis induced by AAPH [2,2′-azobis (2-amidopropane) dihydrochloride], a peroxyl radical initiator (Vinardell et al. 2005).

The antioxidant activity is expressed as IC_{50} (concentration determining 50% inhibition on hemolysis induced by AAPH). The values presented in Figure 21.5 are obtained for 45.3–134 µg/mL, bagasse lignin having the highest antioxidant activity, because of its low IC_{50} value.

The photoprotective activity has been determined after incubating the erythrocytes suspension for 30 min in the presence of different lignins and chlorpromazine (CPZ), a photohemolytic compound (Figure 21.6).

A product could be considered as photoprotector if it increases the concentration of chlorpromazine necessary to induce hemolysis. The values of chlorpromazine (CPZ) concentrations that determine hemolysis and photohemolysis range from 60 to 189 mg/mL in the presence of different lignins at a single concentration of 100 µg/mL. The three lignins studied showed no protective activity at this concentration. Nevertheless, the lignins showed a protective effect on the hemolysis induced by CPZ when red blood cells are not irradiated. These results confirm the antioxidant activity of the lignins, since CPZ also induced hemolysis by oxidative stress (Vinardell et al. 2005).

The hemolysis of erythrocytes induced by AAPH was also studied in the presence of Ligmed-A. This product which is composed of about 90% lignin was prepared from sugarcane and is commercially available as antidiarrheal drug (Mitjans et al. 2001). The antioxidant effect of Ligmed-A was expressed as the percentage of hemolysis inhibition at several concentrations. The inhibitory effect was dependent on concentration of lignin, the greatest effect (the IC_{50} was 106.63 µg/mL) being obtained at the highest concentration studied (Garcia et al. 2006).

The antioxidant efficiency of different lignin samples was also studied by Boeriu et al. using two models (Boeriu et al. 2004). The first approach was to measure the ability of lignin samples to react with free radicals in aqueous solution. The $ABTS^{+\cdot}$ [ABTS—2,2′-azino-bis(3-ethylbenzothiazoline-6-sulphonate)] cation radicals were generated by an enzymatic system consisting of peroxidase and hydrogen peroxide. The antioxidant activities of lignins were determined by measuring the rate of disappearance of the ABTS radical in the presence of lignin. All lignin samples tested reacted with ABTS radical (Figure 21.7), but their efficiency was about 20%–30% (on mass) of that measured for the commercial antioxidant BHT (butylated hydroxytoluene). Lignin samples from

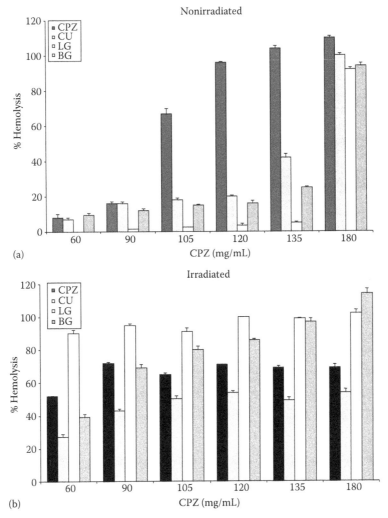

FIGURE 21.6 Hemolysis and photohemolysis of CPZ in the presence and absence of different lignins (for the significance of lignin samples see Figure 21.5). (a) nonirradiated (b) irradiated results are expressed as mean ± SE.

abaca, sisal and jute, and the softwood lignins SW-Kr-1 and SW-Kr-3 show the highest scavenging activity against the ABTS radical.

Further, the effect of lignin on the oxidation of unsaturated compounds by molecular oxygen has been evaluated. The oxidation of linoleic acid in micellar solution initiated with azo-compounds [2,2′-azobis-(2-amidopropane)-dihydrochloride-ABAP] has been studied in the absence and presence of lignins and tocopherol (Vitamin E), which is a commonly used as antioxidant. Antioxidant activities of lignin compounds were determined by measurement of the rate of generation of linoleic acid peroxide during inhibited peroxidation of linoleic acid initiated by ABAP with known rate of initiation. All lignins show some ability to act as chain-breaking antioxidants during lipid peroxidation, but only lignins from abaca, sisal, hardwood, and jute and the sulfur-free softwood lignin SW-SF-2 have significant antioxidant effects (Figure 21.8).

The most effective chain-breaking antioxidant lignin showed about 30% (on mass basis) of the activity of tocopherol in the same system.

The results suggest that the radical scavenging activity of lignins, although induced by hydroxyl phenolic groups, is not linearly related to the concentration of the phenolic groups, but is influenced also by other structure-related factors such as the steric availability of aromatic hydroxyl group and the

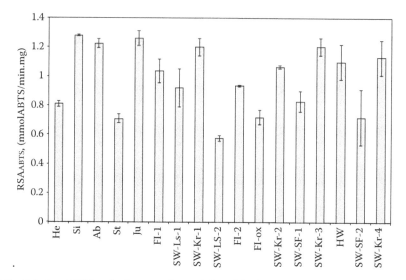

FIGURE 21.7 Relative ABTS-radical scavenging activity of lignin samples. He, hemp; Si, sisal; Ab, abaca; Ju, jute; Fl-1, flax; SW-Ls-1, lignosulfonate from softwood (Boresperse 3A); SW-Kr-1, kraft from softwood (Indulin AT); SW-Ls-2, lignosulfonate from softwood (Wafex P); Fl-2, soda flax (Bioplast); Fl-ox-soda, flax oxidized; SW-Kr-2, kraft from softwood (Curan 100); SW-SF-1, soda from softwood (precipitated at high pH); SW-Kr-3, kraft from softwood; HW, organosolv (Alcell) from mixed hardwoods; SW-SF-2, soda softwood (precipitated at low pH); SW-Kr-4-kraft (Curan 2711P).

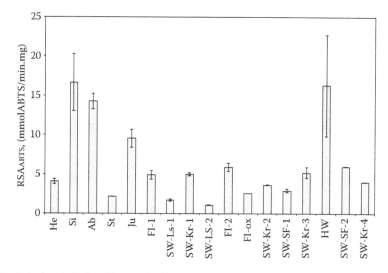

FIGURE 21.8 Relative chain-breaking antioxidant effect of lignin in lipid peroxidation. Significance of the samples is the same as in Figure 21.7.

stabilization of phenolic radicals. Adsorption phenomena at the oil–water interface may be of importance for reactions in micellar systems (e.g., inhibition of oxidation of polyunsaturated fatty acids).

The antioxidant capacity of black liquor lignin was evaluated by the inhibition of formation of free radicals such as superoxide anion, hydroxyl radical, and radicals generated by UV-light (Perez-Perez et al. 2005). In addition, isolated lignin antioxidant activity (AOA) and total antioxidant activity (TAA) values were determined and the effect of isolated lignin on lipid peroxidation process was examined using TBA and FOX methods. In all cases, the antioxidant capacity of lignin was compared to natural antioxidant (melatonin, quercetin, and commercial lignin). In relation to superoxide anion, all the analyzed samples could decrease superoxide anion formation (Table 21.1).

TABLE 21.1

Inhibition Percentages of Superoxide Anion and Hydroxyl Radical Generation

Sample	Superoxide Anion	Hydroxyl Radical
Precipitated lignin	51.44 ± 1.29	33.68 ± 0.91
Commercial lignin	47.15 ± 2.04	27.81 ± 1.30
Melatonin	79.06 ± 0.32	53.89 ± 1.07
Quercetin	71.46 ± 0.85	53.07± 1.13

Precipitated lignin 3 mg/mL; melatonin 1 μM; quercetin 1 μM; commercial lignin 3 mg/mL. Mean ± SE (n = 9). Means within a column sharing the same letter were not significantly different by Newman–Keuls multiple comparison test ($p < 0.05$).

TABLE 21.2

AOA and TAA Values of Precipitated Lignin and Compared Antioxidants

Sample	AOA (mM)	TAA (mM)
Precipitated lignin	0.91 ± 0.03	56.71 ± 0.27
Commercial lignin	0.82 ± 0.01	4.24 ± 0.24
Melatonin	0.83 ± 0.02	6.99 ± 0.17
Quercetin	0.86 ± 0.03	10.55 ± 0.23

Precipitated lignin 3 mg/mL; melatonin 1 μM; quercetin 1 μM; commercial lignin 3 mg/mL. Mean ± SE (n = 9). Means within a column sharing the same letter were not significantly different by Newman–Keuls multiple comparison test ($p < 0.05$).

Precipitated lignin had an inhibition percentage for superoxide anion similar to commercial lignin but lower than melatonin and quercetin. In the case of hydroxyl radical, the same behavior was observed, precipitated lignin being better able to reduce hydroxyl radical formation than commercial lignin, but less effective than melatonin and quercetin. This can be due to the fact that melatonin is a natural antioxidant widely studied and which lost in aging prevents the kidnapping of free radicals and organism protection, and quercetin which has antitumoral, anti-inflammatory, and antioxidant potential.

AOA values of precipitated lignin were determined and the results are summarized in Table 21.2. Precipitated lignin had the highest AOA values, even higher than natural and commercial antioxidants used in these experiments. AOA values of precipitated lignin were high as compared with other samples analyzed by this method (Koracevic et al. 2001), like urine (0.17 mM), saliva (0.84 mM), and brain–spinal fluid (0.095). TAA values of precipitated lignin were determined by ABTS$^{+\cdot}$ TAA values are considered higher according to the results presented by Re et al. (2003), in which a value of TAA superior to 3.0 mM is considered high, and values below 3.0 mM are considered low. Precipitated lignin sample displays TAA values that were comparable and even greater than commercial lignin and lower than melatonin and quercetin. In addition, these values of TAA are high when compared with other samples analyzed by this method, like 1.16 ± 0.18 mM for blood, 0.33 ± 0.02 mM for white onion, and 4.00 ± 0.20 mM for feluric acid. It is important to emphasize that quercetin displayed values of TAA superior to that of the Trolox, which is used as a standard in this method. This behavior has been reported for quercetin, which was approximately 3.3 times

TABLE 21.3

Inhibition Percentages of Formation of MDA and LHPs Determined by FOX and TBA methods

Sample/Methods	Enzymatic TBA	Nonenzymatic TBA	FOX	UV-Light TBA	UV-Light FOX
Precipitated lignin	82.1 ± 0.4	55.4 ± 0.5	40.0 ± 0.8	52.29 ± 0.4	55.3 ± 0.6
Commercial lignin	69.9 ± 0.5	57.4 ± 0.4	61.4 ± 0.5	39.93 ± 0.6	48.5 ± 0.5
Melatonin	82.3 ± 0.4	55.4 ± 0.1	62.4 ± 0.4	51.95 ± 0.8	57.5 ± 0.3
Quercetin	73.6 ± 0.5	58.7 ± 0.6	56.0 ± 0.1	55.69 ± 0.2	58.2 ± 0.5

Precipitated lignin 3 mg/mL; melatonin 1 μM; quercetin 1μM; commercial lignin 3 mg/mL. Mean ± SE (n = 9). Means within a column sharing the same letter were not significantly different by Newman–Keuls multiple comparison test (p < 0.05).

more powerful than Trolox to scavenge radical cation ABTS, whereas the mono- and dichloride derivatives of quercetin were 5.8 and 6.1 times more efficient than the Trolox, respectively.

To verify the antioxidant capacity of precipitated lignin, two methods were used to measure the products of lipid peroxidation. These products were lipid hydroperoxides (LHPs), one of first products of this process, and malondialdehyde (MDA), a final product. The results of these experiments are shown in Table 21.3, in which the inhibition percentages of the formation of each one in different methods are reported. All the studied samples can inhibit formation of MDA and LHPs, in a similar way than melatonin, quercetin, and commercial lignin. The effect of the precipitated lignin was also studied on the radicals generated by UV-light. It was observed clearly that all the studied compounds had the capacity to decrease the formation of MDA and LHPs in a statistically significant way.

For the moment, the mechanism by which lignin acts like antioxidant is not known, but it can be presumed that it can be a scavenger of free radicals thanks to its phenolic structure. Several mechanisms can be proposed: (1) lignin acts like a chelating agent scavenging metals of Fenton's reagent (Fe II and Mn II); (2) lignin acts as a "suicide" antioxidant, receiving the attack of the hydroxyl radical avoiding the action of this radical on target molecules; or (3) lignin inhibits enzymes involved in the metabolic pathways able to generate free radicals. Phenols are efficient in scavenging of peroxyl radicals because of their molecular structures, which include hydroxyl groups that contain mobile hydrogens. In addition, the action of phenolic compounds can be related to their capacity to reduce or to chelate necessary divalent ions for several reactions. On the other hand, phenolic antioxidants can interrupt the oxidation reactions by hydrogen atom transfer or by electron transfer with the formation of the radical cation phenoxy, which is fast and reversible deprotonated forming the phenoxy radical. These two mechanisms always happen in parallel but with different rates. Although these mechanisms have the same net results, the electron transfer is preferable because the radical cation formed by electron transfer can be mutagenic.

To assess the radical scavenging activity of a potential antioxidant, the stable free radical 1,1-diphenyl-2-picryl hydrazyl (DPPH$^{\cdot}$) is used as active radical (Telysheva et al. 2000). The reactivity of DPPH$^{\cdot}$ is far lower than that of oxygen-containing free radicals (HO$^{\cdot}$, RO$^{\cdot}$, ROO$^{\cdot}$, and O$_2^{-}$), and unlike them, the interaction rate is not diffusion controlled, but is ruled by the electronic structure of the antioxidants. To evaluate the radical scavenging activity, potential antioxidants are allowed to react with DPPH, usually in mixtures of water with organic solvent. Due to the presence of an unpaired electron, DPPH$^{\cdot}$ radical gives a strong absorption band having a maximum at 517 nm in the visible region. In the presence of a free radical scavenger, the absorption decreases, and the resulting discoloring is stoichiometric with respect to the number of electrons taken up. Thus, DPPH$^{\cdot}$ test provides information on the capacity of compounds to scavenge free radicals independent from such factors as the presence of initiators or enzymes in a system. The methanol-soluble fractions from commercial softwood kraft and hydrolysis lignin as well as laboratory acid-soluble, alkaline, and ethanol lignin from softwood were studied.

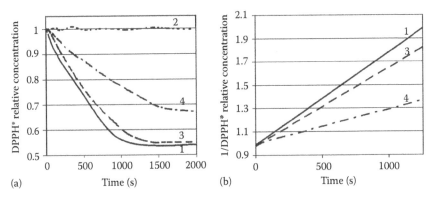

FIGURE 21.9 The kinetics of DPPH decay (a) and the dependence of the reciprocal DPPH˙ concentration on the interaction time (b) in the presence of (1) aspen alkaline lignin; (2) methylated aspen alkaline lignin; (3) birch alkaline lignin; (4) spruce alkaline lignin.

For lignin fractions soluble in methanol, it has been confirmed that the OH_{phen} group content plays a crucial role in radical scavenger activity of lignin in the DPPH˙ assay (Dizhbite et al. 2004). This was proved by the methylation of lignin with diazomethane (Figure 21.9, curve 2), and dimethoxy units (syringyl derivatives) exhibited higher antioxidant efficacy (Figure 21.9 curves 1,3 versus curve 4).

Methanol-soluble fraction of kraft lignin is characterized by a higher antiradical activity than hydrolysis lignin. According to the data obtained, lignin fractions have stoichiometries expressed as a number of reduced DPPH, of 2 and 1.3, respectively, for kraft and hydrolysis lignin, which could be correlated with 240 mg of gallic acid equivalent (GAE) per 1 g for fraction from kraft lignin against 126 mg GAE for that from hydrolysis lignin. These results are in good agreement with the bioactivity of these fractions against phytopathogenic microorganisms, which could be connected, to some extent, with the inhibition of radical processes by mobile lignin fractions (Figure 21.10).

Regarding the antioxidant activity of isolated lignins, the presence of π-conjugation system, for example, stilbenes in kraft lignin, should be taken into account besides the aforementioned factors. These systems are characterized with high ability to form self-donor-acceptor complexes, thus decreasing the energy of polymer molecules excitement from singlet into triplet state and increasing

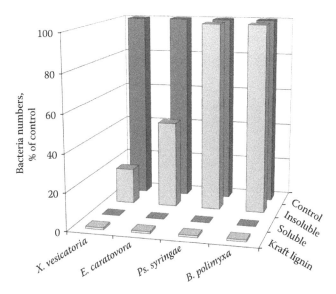

FIGURE 21.10 (See companion CD for color figure.) The influence of lignin on phytopathogenic microorganisms.

the reactivity of polymer. Π-conjugation systems of lignin could act as catalyst or activators of the lignin phenolic structures interaction with DPPH˙, but do not react with DPPH˙ directly. Therefore, rate constant for DPPH˙ disappearance in the case of different lignins depends on both the number of phenolic hydroxyls and degree of π-conjugation systems development. It has been established that the extended π-conjugation systems of polymeric lignins significantly increase rate constants of the second-order one-side process of DPPH˙ decay, whereas the high molecular mass, enhanced heterogeneity and polydispersity are among the factors decreasing the radical scavenging activity.

The ability of various lignophenols derivatives to protect SH-Y5Y cells from hydrogen peroxide (H_2O_2)-induced cell death has been studied (Akao et al. 2004). It was observed that lignocresol derivatives were more potent in the neuroprotective activity compared with other lignophenols. Especially, the carboxymethylated lignocresol (lig-8) from bamboo displayed the most potent apoptosis preventing activity at 20 and 30 μM. This process was explained by blocking the caspase-3 activation, which is responsible by apoptosis (Figure 21.11). Results from in vivo and in vitro experiments suggested that lig-8 possibly blocked the activation of caspase-3.

It is known that mitochondria play a major limiting role in the apoptosis of neuronal cells. H_2O_2-induced apoptosis is considered to be mediated via the mitochondrial pathway, and lignin derivative was shown to ameliorate the apoptosis by preventing depolarization of mitochondrial membrane PT. The results have proved that lignophenol derivatives protected human dopaminergic neuroblastoma SH-SY5Y cells against cell death due to apoptosis induced by oxidative stress.

The properties of lignins, especially the network structures, differ greatly depending on the lignocellulosics, because the precursors building up lignin differ in their species. Generally, softwood lignins consist mainly of guaiacyl aromatic units, and hardwood lignins contain evenly guaiacyl and syringyl units. On the other hand, herb lignins including bamboo have para-hydroxy phenyl units rather than these units. Certainly, the structure containing the para-hydroxy phenyl units in bamboo lignin and additional phenolic groups may contribute to the neuroprotective activity.

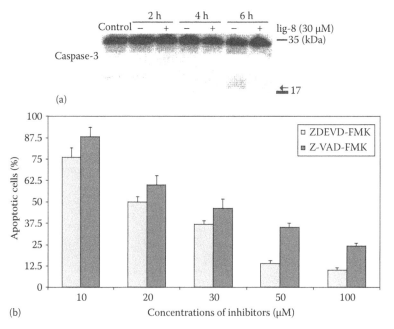

(a)

(b) Concentrations of inhibitors (μM)

FIGURE 21.11 Involvement of caspase-3 activation in the H_2O_2-induced apoptosis and inhibition of caspase-3 activation by lig-8. (a) Western blot analysis of caspase-3. The arrow indicates the active form of caspase-3. **(See companion CD for color figure.)** (b) Apoptosis inhibition assay using caspase-3 inhibitor ZDEVD-FMK or pan-caspase inhibitor Z-VAD-FMK. The apoptotic cell death was evaluated by Hoechst 33342 staining. Results are the mean ± SD of three independent experiments. The percentage of apoptotic cells without the inhibitor is as 100%.

21.4 LIGNINS IN REDUCTION OF CARCINOGENESIS

Damage of deoxyribonucleic acid (DNA) by oxygen radicals and various chemicals is thought to be an important etiologic factor in the development of chronic diseases such as cancer. Carcinogenicity and mutagenicity of chemicals may be modulated by other chemicals mostly prepared by organic synthesis. As to the several drawbacks of synthesis, compounds for human organism examination of preparations of plant origin for this purpose received more attention in the last few years. Lignin component of biomass was derived from kraft pulping and prehydrolysis of wood was investigated with regard to its ability to bind N-nitrosamines, well-known carcinogens and to reduce DNA strand breaks in H_2O_2 and N-methyl-N-nitro-N-nitrosoguanidine (MNNG)-damaged mammalian cells as well as to induce SOS response 4-nitroquinoline-N-oxide (4NQO) in *E. coli* PQ37 (Kosikova et al. 2002).

The most effective lignin sorbents of N-nitrosodiethylamine were modified spruce and beech kraft lignins and these were examined with respect to their influence on the ability of mutagenic compound 4NQO to induce SOS response in the microbial system *E. coli* (Figures 21.12 through 21.14). The *SOS response* is a postreplication DNA repair system that allows DNA replication to bypass lesions or errors in the DNA. The SOS uses the RecA protein. The RecA protein, stimulated

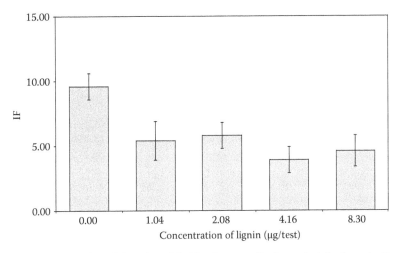

FIGURE 21.12 Antimutagenic activity of modified kraft spruce lignin against 4-nitroquinoline-N-oxide.

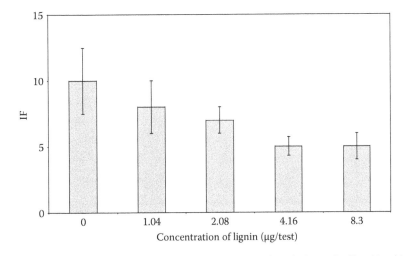

FIGURE 21.13 Antimutagenic activity of prehydrolysis lignin against 4-nitroquinoline-N-oxide.

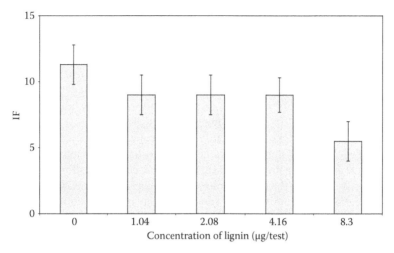

FIGURE 21.14   Antimutagenic activity of modified kraft beech lignin against 4-nitroquinoline-*N*-oxide.

by single-stranded DNA, is involved in the inactivation of the LexA repressor thereby inducing the response. It is an error-prone repair system. *SOS box* is the region in the promoter of various genes to which the LexA repressor binds to repress the transcription in the absence of DNA damage.

The observed antimutagenic activity of lignin samples is probably caused by their low degree of cross-linking density resulting in highly absorptive capacity toward 4NQO. The mutual differences of both modified lignins could be explained in terms of different ratios of guaiacyl and syringyl structures in beech and spruce lignin. The obtained results indicate that antimutagenic potential of lignin is based on its ability to absorb xenobiotics and thus to protect DNA damage.

The antioxidative effect of lignin sample 1 (prehydrolysis lignin) on DNA in human cells VH 10 exposed to oxidative treatment with H_2O_2 was investigated. The protective antioxidative effect of lignin sample 1 results from the distribution of H_2O_2-induced DNA breakage in pretreated (+lignin) and nonpretreated human cells (−lignin) Figure 21.15.

It is clear that in cells preincubated with lignin before oxidative treatment, there are significantly fewer heavily damaged cells (percent tail DNA > 40) than in those treated with H_2O_2 alone. It is known that lignin, which has unique hindered phenolic hydroxyl groups, acts as a stabilizer of reactions induced by oxygen and its radical reduction products. Based on this, it can be suggested that

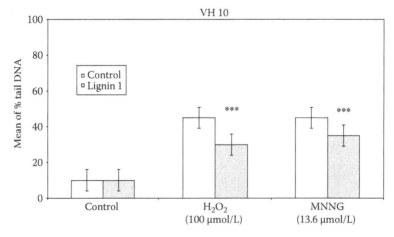

FIGURE 21.15   Influence of lignin 1 on level of DNA strand breaks induced by H_2O_2 and MNNG. Values are means of three independent experiments. Significant difference (***p = 0.001).

phenolic groups of lignin matrix with active dissociable hydrogen are responsible for scavenging of reactive oxygen species and reduction of oxidative DNA damage in lignin-pretreated cells.

In the case of the experiments with human Caco-2 colon carcinoma cells exposed to oxidative damage, sample 1 (prehydrolysis lignin) and sample 2 (condensed kraft lignin) were used as antioxidants. Both lignin samples tested were found to be effective in the reduction of strand breaks.

Further experiments with H_2O_2-treated Chinese hamster cells V79 confirmed the reduction of DNA strand breaks by both lignin samples. Occurrence of oxidative DNA lesions induced by H_2O_2 was reduced more significantly in the case of lignin 1 probably due to its lower molecular mass, which allows a better transfer of lignin macromolecules into cells.

Others experiments were performed using N-methyl-N'-nitro-N-nitrosoguanidine (MNNG). In contrast to H_2O_2, MNNG does not induce any significant level of oxidative damage to DNA but it alkylates several positions on DNA. The level of single strand breaks of DNA induced by MNNG was also reduced by both lignin samples. The protective effect of lignin tested on MNNG-treated cells may be associated with their low cross-linking density resulting in high binding affinity for carcinogenic and mutagenic compounds. The obtained results could be correlated with the suggestion that the reduction of single-strand breaks of DNA by lignin has a different character in MNNG and H_2O_2-treated cells. Reduction of H_2O_2-induced DNA may be caused by scavenging of OH radicals. On the other hand, reduction of alkylating activity of MNNG to cause instability of N-glycosyl bond on DNA lesions may correlate with the adsorption affinity of lignin tested for N-nitrosodiethylamine.

The effect of lignin on cytotoxicity, mutagenicity, and SOS response induced by nitroquinoline-N-oxide (4NQO), 3-(5-nitro-2-furyl) acrylic acid (5NFAA), 2-nitrofluorene (2NF) as well as hydrogen peroxide was investigated in bacterial assay system, i.e., the Ames test with *Salmonella typhimurium* TA 98, TA 100, TA 102, and the SOS chromo test with *E. coli* PQ37 (Mikulasova and Kosikova 2003).

The experiments were performed in order to establish to which extent antimutagenic properties of lignin are influenced by its genetic origin and methods of isolation. As shown in Figure 21.16, all modified lignins were effective in reducing IF by 4NQO as well by H_2O_2 (Figure 21.16).

Effect of wood-derived lignins on 4NQO (13.15 nmol) and H_2O_2 (1.1 mmol) induced SOS response. The data are the mean from at least three independent experiments.

Kraft lignin derivatives 1 and 2 isolated from spruce and beech wood exhibited a similar efficiency (Figure 21.16a and b). Sample 3 derived from beech wood prehydrolysate had an effect comparable with both kraft lignins (Figure 21.16c). The modification of sample 3 yielded lignin

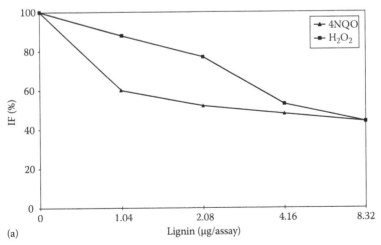

(a)

FIGURE 21.16 Effect of wood-derived lignins on 4NQO (13.15 nmol) and H_2O_2 (1.1 mmol) induced SOS response. The data are the mean from at least three independent experiments: (a) Spruce kraft lignin.

(continued)

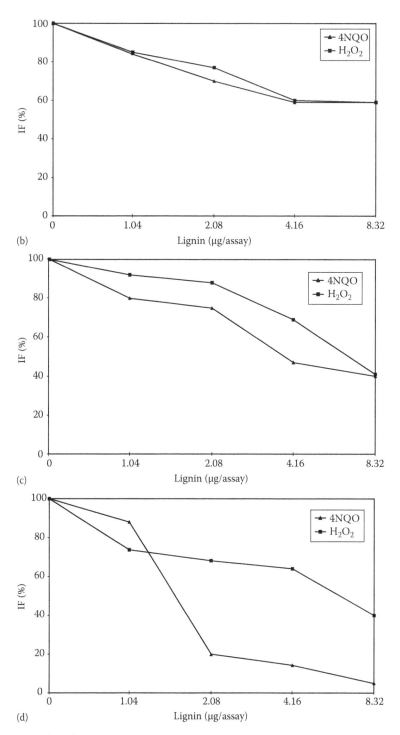

FIGURE 21.16 (continued) Effect of wood-derived lignins on 4NQO (13.15 nmol) and H_2O_2 (1.1 mmol) induced SOS response. The data are the mean from at least three independent experiments: (b) beech kraft lignin, (c) beech prehydrolysis lignin, and (d) beech condensed prehydrolysis lignin.

preparation 4 which exhibited the strongest ability to decrease activity of 4NQO (Figure 21.16d). The determination of molecular parameters of lignins 3 and 4 shows that the modification increased the average molecular mass about 62.5% and decreased cross-links about 10.5%. The increased efficiency of lignin 4 could be explained by degradation of cross-links and linearization of the lignin macromolecule during modification. The samples prepared from steam-treated annual plants showed a relatively lower efficiency than those isolated from wood.

Therefore, the results obtained indicate that tested lignin derivatives decrease mutagenic activity of 4NQO, 2NF, and H_2O_2, and the differences could be correlated with their genetic origin and methods of modification (Mikulasova and Kosikova 2003).

The DNA-protective effects of lignin were examined in rat testicular cells and rat peripheral blood lymphocytes using in vitro and ex vivo experiments. H_2O_2 and visible light-excited methylene blue (MB) were used as DNA-damaging agents. Testicular cells were chosen because the germinal epithelium of tests is one of the most proliferative active tissues potentially susceptible to DNA-damaging effects. As a second target, peripheral blood lymphocytes were chosen because dietary lignin or its metabolites circulate in the animal organism probably through the blood system. For the in vitro experiments, isolated cells were preincubated with lignin for 2 h before treatment with one of the oxidative agents. In ex vivo experiments, the cells were exposed to H_2O_2 or visible light-excited MB after isolation from rats fed either a common diet or a lignin-supplemented diet. The water-soluble, sulfur-free lignin used in experiments was obtained by fractionation of hardwood hydrolysate. The level of direct single-strand DNA breaks in H_2O_2-treated cells was measured by the classical Comet assay, and the level of oxidative DNA lesions in visible light-treated cells was measured by a modified Comet assay. It was found that lignin reduced DNA lesions induced by H_2O_2 or visible light-excited MB both in vitro and ex vivo. Therefore, the lignin polymer obtained by fractionation of hardwood hydrolysate manifested a specific type of antimutagenic effect (Labaj et al. 2004).

Antitumor substances were separated from pinecones of *Pinus parviflora*. Spectral analysis of one isolated fraction revealed that this fraction included a lignin-related structure as its main constituent. This fraction along with various commercially lignins (alkali lignin, dealkali lignin, and lignosulfonates) significantly stimulated iodination of human peripheral blood polymorphonuclear (PMN) cells. The PMN iodination increased with incubation time under the influence of alkali lignin and reached a transient plateau at 150 min after start of incubation with $Na^{125}I$. The PMN iodination again rapidly increased after 180 min. This might be due to some damage of PMN, since PMN aggregation occurred at this time. Optimal concentration of alkali lignin was 50 µg/mL (Sakagami et al. 1991).

But it was observed that the PMN-stimulating activity of alkali lignin was significantly inhibited by the presence of serum. This suggests the importance of structurally modifying lignins to reduce their nonspecific binding to various proteins.

The capacity for stimulation of PMN iodination is substantially related to the lignified or tannin-like polyphenolic structure, although the details of structural requirement need further investigation. It was found that the activity of mentioned products depended more on molecular mass than on the number of hydroxyl groups on each benzene ring in the molecule or on the presence of sugars or hexahydroxyldiphenyl groups. The stimulation of PMN iodination by lignin-related phenolic substances or a tannin-related compound was observed in only myeloperoxidase-containing cells and was significantly reduced by myeloperoxidase inhibitors. These results suggest that the mechanism involved might specifically potentate the active oxygen-mediated host defense capacity against microbial and viral infections and possibly tumor growth and metastasis. The mechanism might have played a role in the antimicrobial, antitumor, antiparasite, and antiviral activities displayed by the pinecone extract and some lignified materials (Sakagami et al 2005).

Lignin derivatives demonstrated differential responses between cells derived from tumor tissues and from normal tissues and strongly inhibited the growth of mouse sarcoma cells (FRUKTO) and rat fetal cells (Ad 12-3Y1-Z-19) transformed with adenovirus type 12 (Sorimachi 1992). The production of tumor necrosis factor (TNF-α), which induced multinucleated giant cell formation, is regulated by polyanions such as lignin derivatives. ELISA for TNF-α showed that polyanion induced TNF-α production by macrophages. The secretion of TNF-α from the cells reached a maximum at 3–6 h, and then showed a slight decline. Northern blotting of TNF-α mRNA evidenced that the amount of TNF-α reached a maximum within 1 h of macrophage culture in the presence of a lignin derivative. On the other hand, TNF-α mRNA was undetectable in the control cells. It was concluded that stimuli such as that provided by lignin derivatives increases the amount of TNF-α mRNA, which is then followed by translation of TNF-α (Sorimachi et al 1995).

It was also observed that the soluble lignin derivatives obtained from an extract of the culture medium of *Lentinus edodes* mycelia induced the secretion of tumor necrosis factor TNF-α. At the same time, lignin derivative markedly induced the secretion interleukin-8, nitric oxide, along with TNF-α in macrophages. The level of these compounds could be negatively and positively modulated by interferon-γ (IFN-γ). It seems that IFN-γ is one of the important cytokines that regulate macrophage functions, particularly in the presence of foreign bodies. In the coculture of macrophages and human urinary bladder carcinoma cells, lignin derivative induced the cytotoxic effect of macrophages on the tumor cells. Therefore, IFN-γ could regulate macrophage–tumor cell interaction (Sorimachi et al. 2003).

21.5   ANTI-HIV PROPERTIES OF LIGNINS

Natural compounds can be used to reduce the signals that tell macrophages they are in the second and third stages, thereby easing entry into the fourth stage. Macrophages play an essential role in the immune response. Natural compounds could be used to stimulate macrophage (and other immune cell) activity and to reduce immune evasion. Natural compounds could promote the antiangiogenic mode of macrophages by reducing the signals that tell macrophages angiogenesis is necessary. Compounds that support the immune system do so by providing proper nutrition to immune cells or by acting as antioxidants or both. (Immune cells such as macrophages require large amounts of antioxidants for protection from the free radicals they generate.) Some natural compounds also help the immune system by providing a regulatory influence, thereby normalizing the extremes in activity found in diseases like cancer. Compounds that stimulate the immune system often do so by increasing the production of cytokines, including IL-2 and interferons, but may act in other ways as well. These natural compounds tend to increase production of a broad range of cytokines.

Macrophages play an important role in the immune system and their function is regulated by substances (lymphokines) produced by sensitized lymphocytes. It is known that human immunodeficiency virus (HIV) induces the acquired immune deficiency syndrome (AIDS), which results in life-threatening opportunistic infection and malignancies. HIV infects not only human T cells but also transformed B cells and macrophages. Indeed HIV particles were detected in macrophages obtained from AIDS patients. In addition, it was shown that receptor protein (CD4) for HIV exists on cell surface in the HIV-infected cells.

It has been reported that several samples of lignin extracted from natural products in land plants and fungi had anti-HIV activities toward cultured cells. Thus, a water-solubilized lignin derivative (EP3), chromatographically purified from water extract of culture medium of *Lentinus edodes* mycelia (LEM), inhibited the cytopathic effect of HIV in AH8 cells derived from human T cells in vitro (Suzuki et al. 1990). The culture medium of *Lentinus edodes* mycelia was extracted with hot water and this extract was called LEM. LEM was further fractionated with ethanol followed by hydrophobic column chromatography. The most active fraction was called EP3. Lignosulfonate (LS) was prepared by ethanol precipitation from waste liquor of the acid sulfite pulping process of wood.

When macrophages were cultured for 24 h in the presence of EP3, the number of macrophages increased. In addition, EP3 accelerated the glucose consumption of macrophages. EP3 increased thymidine uptake and glucose consumption in murine macrophages. Giant cell formation was not observed during the first 24 h of incubation of macrophages with EP 3 at 37°C, but increased markedly during the next 24 h. Multinucleation further proceed between 2 and 3 days of culture. When EP3 was removed from culture medium after macrophages were incubated with the drug for 24 h, giant cell formation still took place over the next 24 h. Thus, macrophages retain the memory of contact with EP3. The giant cells were active in carrying out phagocytosis of India ink.

Lignosulfonate also induced giant cell formation of macrophages. This substance is also known to increase glucose consumption and thymidine uptake in murine macrophages and to inhibit the cytopathic effect of HIV. These suggest that lignin derivatives in general may induce the multinucleation of macrophages in vitro. The plasma membrane of macrophages is presumably altered by EP3 and LS to induce the cell fusion, and this effect seems to be specific for macrophages (Sorimachi et al. 1990, 1997)

However, examining the antiviral activity of EP3 and LS, it was found that these drugs prevented the cytopathic effects of DNA viruses (herpes simplex viruses) and RNA viruses (Western equine encephalitis virus, polio virus, and measles virus) in VERO-317 cells derived from African green monkey kidney, although giant cell formation did not occur. In addition, the oral administration of lignin derivative (LEM) improved hepatic functions in hepatitis-B patients. *These results suggest that lignin derivatives can affect cellular metabolism in cells other than macrophages.*

In other experiments, a purified fraction EPS4 also obtained from LEM by ethanol precipitation followed by hydrophobic chromatography and gel filtration chromatography completely inhibited the HIV-1-induced cytopathic effect in vitro at concentrations of greater than or equal to 10 μg/mL. Chemical and spectral analysis revealed that EPS4 is composed of water-soluble lignins containing minor amounts of protein (3.2%) and sugars (12.2%). Taken together with the previously reported observation that EPS 4 promotes the activation of macrophages and the proliferation of bone marrow cells, the fraction appears to possess both an immunostimulating activity and anti-HIV effect in vitro (Suzuki et al. 1998).

A water-soluble lignin derivative of high molecular mass was found in boiling water extracts of an edible mushroom, *Fuscoporia oblique*. This product inhibited protease of human immunodeficiency virus type 1 (Ichimura et al. 1990).

HIV-1 protease was inhibited by lignin extracted from wheat bran and by synthetic lignin-like compounds, and at the same time, these induced cytopathicity in cultured cells. To examine the molecular mass dependency of the protease inhibition by these materials, synthetic dehydrogenated polymers (DHP) start in from p-coumaric acid and ferulic acid were fractionated into four ranges of molecular mass by passing through ultra-filtration membranes. All the DHP fractions exhibited HIV-1 protease inhibitory activity, although there was a tendency for the inhibitory activity to increase as the molecular mass increased.

The anti-HIV-1 activities of oligomers of p-coumaric and ferulic acids and wheat bran lignin were examined by using MT-4 cells. It was observed, while wheat bran lignin inhibited the cytopathicity induced by HIV-1 replication in MT-4 cells like several other lignins, it showed cytotoxicity at twice the effective dose (Ichimura et al 1999).

When rat bone marrow macrophages were incubated with acetyl lignin in the presence of fetal bovine serum, the macrophages secreted tumor necrosis factor (TNF-α) in a dose-dependent manner. This was followed by macrophage multinucleation. Lignin derivative was found to have a significant effect on TNF-α secretion at a minimum dose of 2 μg/mL and produced no significant further increase at levels above 50 μg/mL, while multinucleation was most active at 10 μg/mL. However, multinucleation did not occur at higher concentration of acetyl lignin (50 and 100 μg/mL). Secretion of TNF-α was significantly reduced in the absence of fetal bovine serum, whereas multinucleation was very active, starting after 6 h of incubation. At a concentration of 100 μg/mL, lignosulfonate only induced low levels of TNF-α secretion from macrophages, but induced active multinucleation. The multinucleation induced by addition of LS was inhibited by further addition of acetyl lignin. Thus, macrophages multinucleation was most active when a low level of TNF-α was secreted from macrophages (Okazaki et al. 1996).

21.6 LIGNIN AS SPERMICIDE

Lignin-derived macromolecules (LDMs) that are highly sulfonated (LSA) are negatively charged compounds that exhibit a number of biological activities. LSA and other LDMs are potent inhibitors of HIV in vitro, possibly through interference with the CD-4 receptor/HIV interaction. In vitro microbicidal activity of LDMs has been demonstrated with several other viruses as well. This antiviral and antipathogen activity is cell surface mediated and is likely related to similar inhibitory effects of other sulfonated and sulfated macromolecules. Selective effects of LDMs on various cell types have also been observed. For example, an LDM has been shown to dramatically inhibit the growth of fibroblast and sarcoma cells, but yet activate murine macrophages and induce proliferation of bone marrow cells. LDMs have also been shown to inhibit fertilization in echinoderms without showing cytotoxic effects through a mechanism that involves inhibition of the sperm acrosome reaction. LSA inhibits the sea urchin sperm acrosome reaction and it competes with the natural sulfated ligand, egg jelly, on the sperm surface. LSA binds to the head of capacitated macaque sperm, a location consistent with its biological activity. LSA significantly inhibited the binding of macaque sperm to macaque zonae pellucidae both when the compound was added to sperm after capacitation and when it was added to sperm before Percoll separation and capacitation (Tollner et al. 2002).

LSA has a complex highly branched structure that may promote "binding" to a large number of sperm surface receptors that recognize anionic domains. LSA may provide a useful tool for elucidating the relationship between capacitation and the sperm functions involved in binding to the zona pellucida. Due to its lack of cytotoxicity and its antifertility effect on noncapacitated sperm, LSA is a strong candidate for development as a vaginal contraceptive. In addition, formulations containing LSA have been shown to prevent infection with sexually transmitted herpes simplex virus in the mouse without causing vaginal irritation.

21.7 MECHANISM

The information concerning the biological properties of lignin could be correlated with the similar well-known properties of lignans. This subject is very interesting having in mind the data existing on the part of lignin and lignans as antibacterial, antioxidants, anti-HIV and in the prevention of breast and prostate cancers, osteoporosis, or cardiovascular diseases. Thus, it was supposed that lignins are major dietary precursors of mammalian lignan. Lignins are structurally related to lignans but differ from them by their polymeric nature and distribution in plants (Begum et al. 2004).

The most common monomers in lignins are guiacyclopropane (G). Therefore, lignins are structurally related to secoisolariciresinol (SECO) and matairesinol (MAT) (Figure 21.17).

The rats were fed with cereal brans that had and had not had lignans removed by solvent extraction to examine the respective contributions of plant lignans and lignins to the formation of mammalian lignans. The rats were also fed with a deuterated synthetic lignin (DHP) and the excretion of deuterated enterolactone (ENL) was measured in urine. Thus, it was demonstrated for the first time that lignins are precursors of lignan phytoestrogens (Nicolle et al. 2002).

Therefore, lignin can be metabolized into ENL in rats. The yield of D_2-ENL was 655 nmol/g DHP (195 μg/g). This yield is ≈10 times higher than that calculated for bran lignins. These differences could be explained either by a limited accessibility to the substrate for bran lignins embedded in the cell wall or to structural differences between native bran lignins and synthetic lignins. Lignins, because of their polymeric nature and of their embedding in the cell wall, are usually considered inert in the digestive tract. However, the results obtained show that they are metabolized by the gut microflora to form part of the ENL excreted in urine. But more data on lignan and lignin contents in foods will be required to determine their respective contributions to the mammalian lignan formation.

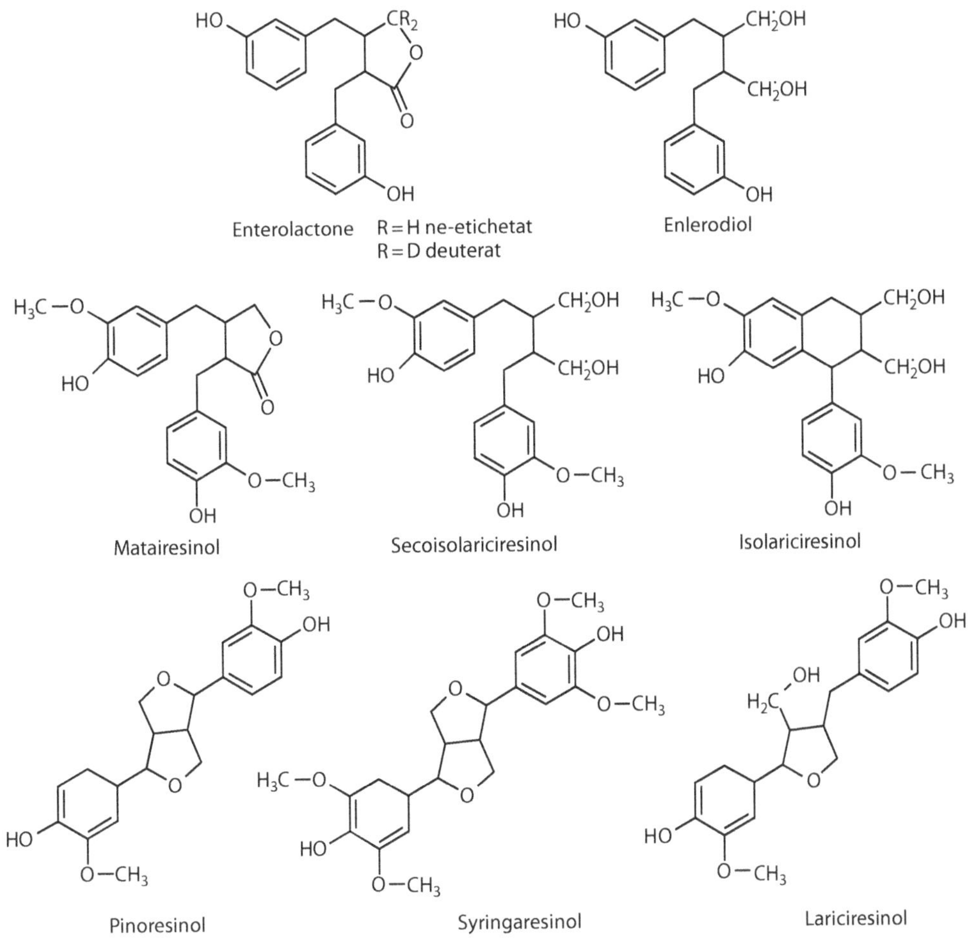

FIGURE 21.17 Chemical structures of mammalian lignans and of the plant lignans. (Adapted from Begun et al. 2004.)

A significantly higher formation of enterolactone was found in pigs fed with the *rye diet* (high in lignans) and higher fecal and urinary excretion and circulating levels of mammalian lignans than pigs fed with *wheat diet* (low in lignans). The conversion of mammalian lignan precursor to entero-lactone was 48% with the wheat diet and 60% with the rye diet. Mammalian lignans are absorbed by passive diffusion from the large intestine and the resulting fraction of the absorbed mammalian lignans undergoes enterohepatic circulation, resulting in low diurnal variation in plasma levels of enterolactone (Bach Knudsen et al. 2003).

One the other hand, the experimental results could suggest a mechanism whereby the free radical-scavenging activity of lignin in dietary fiber may be involved in the fiber–colon cancer interaction. Thus, the ability of dietary fiber to protect against colon cancer may be partly determined by the amount of lignin in dietary fiber as well as the free radical-scavenging ability of lignin. Data showing that lignin is a free radical scavenger were obtained when the NADH-phenazine methosulfate-nitro blue tetrazolium free radical-producing system was used. An alkali lignin solution with a concentration of $46.29\,\mu g/mL$ causes 50% (IC_{50}) inhibition of uric acid production by xanthine oxidase. Alkali lignin with an IC_{50} of $59.08\,\mu g/mL$ inhibits the activity of xanthine oxidase, one of the enzymes related to the production of superoxide anion radicals, and presents a mixed-type noncompetitive inhibition pattern. Using deoxyribose method, it was found that alkali lignin is a hydroxyl radical scavenger with an IC_{50} of $250\,\mu g/mL$, and using the thiobarbituric acid method, it was observed that alkali lignin inhibits nonenzymatic and enzymatic lipid peroxidation with an IC_{50} of 72 and 100 µg/mL, respectively. Alkali-lignin also hinders the activity of glucose-6-phosphate dehydrogenase, another enzyme related to the generation of superoxide anion radicals with IC_{50} of 123.6 µg/mL and obstructs the growth and viability of cancer interaction (HeLa) cells in a dose-dependent manner (Lu et al. 1998).

Various natural and synthetic compounds including alkaloids, terpenoids, and phenolics were tested for inhibition of the cell surface expression of intercellular adhesion molecule-1 (ICAM-1) and vascular cell adhesion molecule-1 (VCAM-1), both of which are crucial in the regulation of immune response and inflammation (Figure 21.18).

Preliminary studies on the structure–activity relationships suggested that *chalcone* and *isoflavone* seem to be superior to *dehydrochalcone*, *flavone*, and *flavanone* in down regulating the cell surface expression of VCAM-1 and ICAM-1:

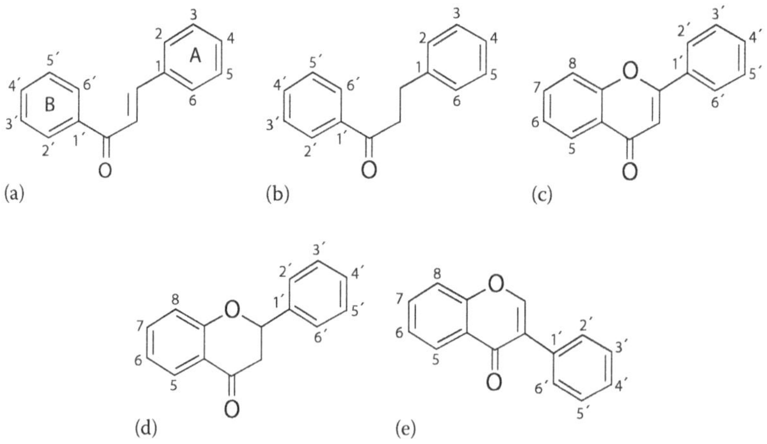

Flavonoids affecting cell expression of VCAM-1 and ICAM-1 are: a-chalcone; b-dehydrochalcone; c-flavone; d-flavanone; e-isoflavone (genistein).

Studies on the relationship between the substitution pattern of the hydroxy groups and the VCAM-1 inhibitory activity in 24 chalcone derivatives have suggested that chalcone bearing one or two hydroxy group at C-2' of the ring B and C-4 of the ring A are potentially active in decreasing the cell surface level of VCAM-1 expression. However, the inhibitory activities of these compounds were restored to the control level by the chemical conversion (acetylation, benzylation, or methylation) of hydroxy groups. These results seem to indicate that the coplanarity formed by the hydrogen bonding between the 2'-hydroxy group of the ring B and the adjacent conjugated ketone may be important in inhibiting the cell surface expression of VCAM-1 (Tanaka et al. 2001).

The relationship between macromolecular and *mutagenic* properties of lignin-containing compounds (LCC) was studied. LCC in water medium and in swollen state (at a moisture ≥ 60%) behave as polyanions and/or neutral molecules according to a value of pH, whose action on biological systems is connected with competition mechanisms. They disturb the structural–functional systems of cells: at different levels (*gene, membrane, enzymes*). The influence of molecular mass value both of natural and technical LCC prepared by different methods (fractionation by solvents, neutron irradiation plasma treatment) on their mutagenic properties was proved by cytogenetical testing. The biological properties of LCC were studied for molecular mass values of 4000–7000. The oligomers with these molecular masses induce the chromosomal aberration in sexual cells of male *Benedictia baicalensis* (Baikal lake mollusc) and root cells of plants at a concentration more than 0.1 and 0.016 g/L, respectively, regardless of the LCC nature and mode of preparation. Relationship between logarithm of molecular mass and mutagenic effect on mollusks is approximately linear and may be used both for estimating and prediction of LCC influence on hydrobionts. LCC with an average molecular mass

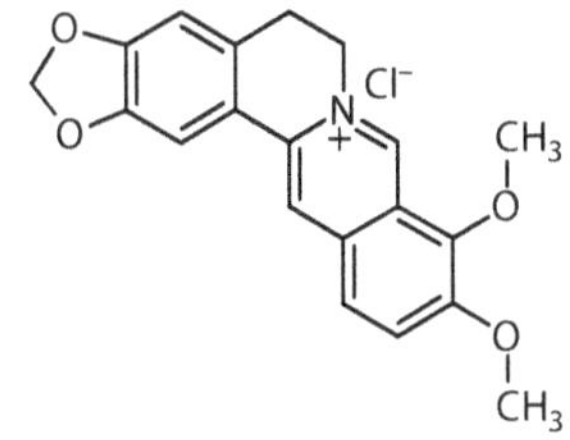

Berberic chloride (1)

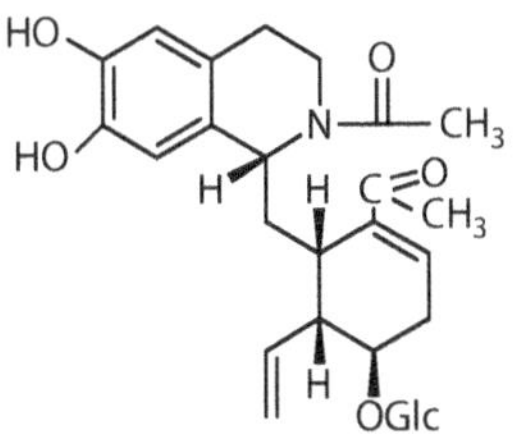

Ipecoside (2)

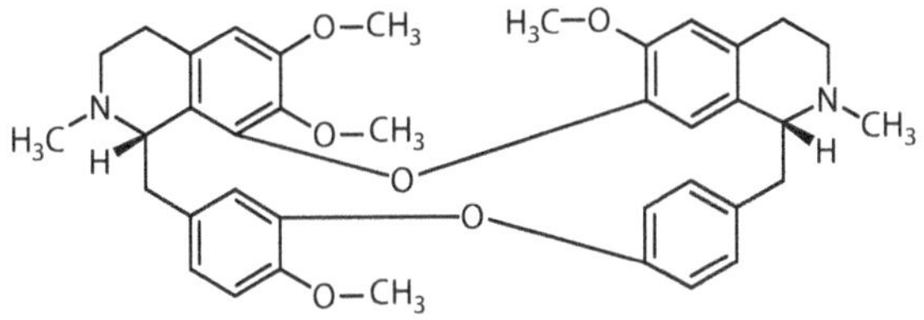

Isotetrandrine (3)

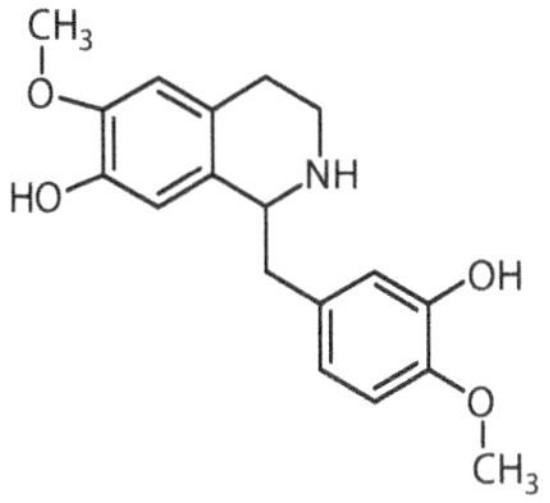

(±)-N-norreticuline (4)

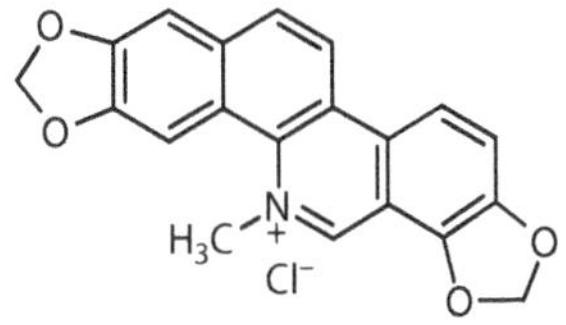

Sanguinarine chloride (5)

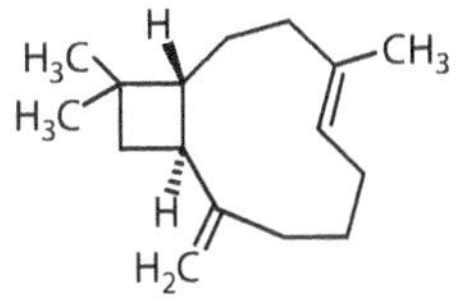

β-Caryophyllene (6)

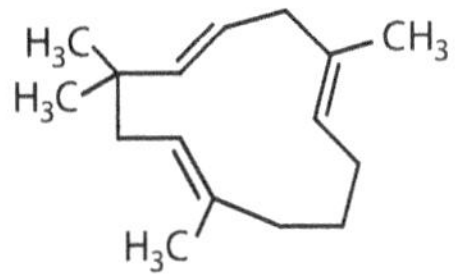

α-Humulene (7)

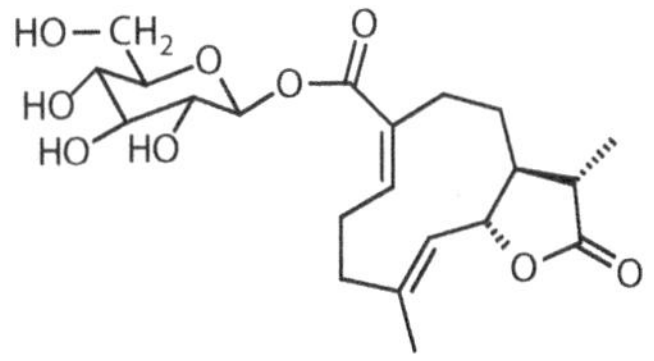

11,13-Dihydro-taraxinic acid
14-O- β-D-glucoside (8)

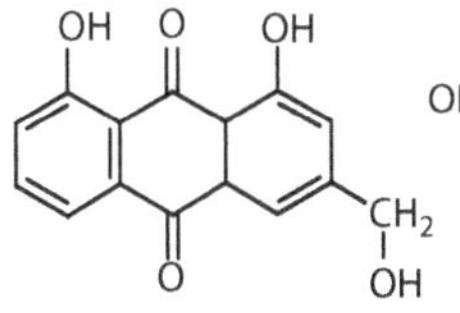

Aloe-emodin (9)

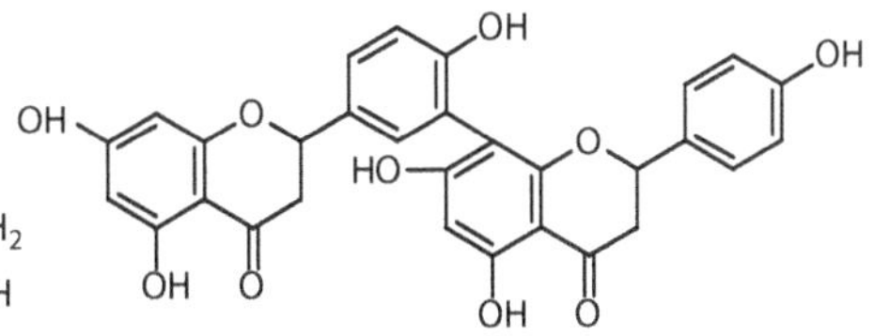

Amentoflavone (10)

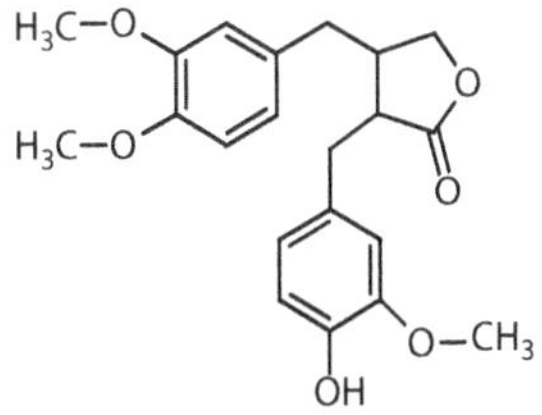

(±)-Arctigenin (11)

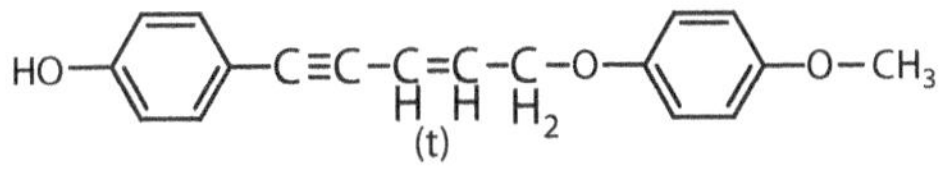

Asparenyol (12)

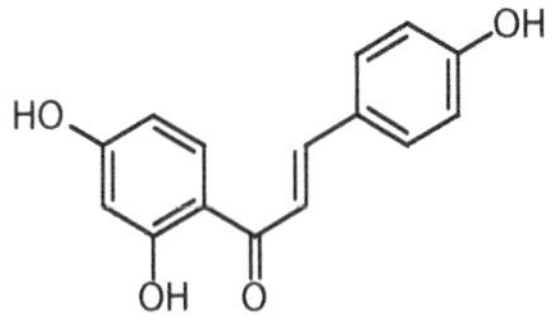

Isoliquiritigenin (13)

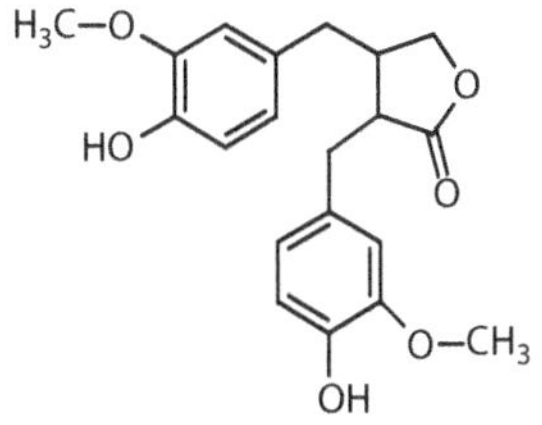

(±)-Matairesinol (14)

FIGURE 21.18 Chemical structure of various compounds affecting the expression of cell adhesion molecules.

below 4000 and above 1000 had no cytogenetical effect on mollusks and plants. The increase of hydrophobicity and toxicity was, however, observed at the relationship $C_{alk}:C_{ar} = 3–5:1$ (content of carbon atoms in aliphatic chain is more than 10). The increasing oxidation value of aromatic rings and increasing content of *quinone* and *methylquinone* structures as result of LCC transformation by the action of biogenic and abiogenic factors causes enhancement of toxic and mutagenic properties. The frequency of chromosomal aberration in sexual cells of mollusks was increased in response to more oxidized high-molecular mass fractions of LCC, but the reduction of LCC (by neutron irradiation) decreased their mutagenic activity. A comparison between microheterogeneity calculated based on potentiometric titration data and biological activity showed its influence on the mutagenic and growth regulating activity. The fractions of LCC with molecular masses of 4000–7000 and narrow distribution factor revealed the high growth-regulating activity depended on content of carboxylic and hydroxylic groups, as well as density of negative charge of macromolecules.

Thus, the biological properties of LCC are determined by the content of labile fractions and active functional groups, density of charges on macromolecules and depend both on value of molecular mass and its distribution and conditions of environment that favored the transformation of macromolecules and inactivation of reactive groups (Novikova et al. 1998).

Plant lignans from sources such as flaxseeds, whole grain cereals, berries, vegetable, and fruits are metabolized in the colon by microflora into enterodiol and enterolactone. Previous research has focused on plant lignans as reducing risk of prostate cancer and improving menopause health, gene health, breast health, heart health, hair loss, acne, and inflammation. The metabolites of lignans may also be beneficial for people with colorectal adenomas growth in the colon and rectum that are considered to be precursors for colorectal cancer. The benefits of lignans appeared to be dose dependent meaning that a higher plasma level of the metabolites enterodiol and enterolactone were associated with a greater benefit. A plasma level of enterolactone of more than 26.3 nmol/L was associated with a reduction in the risk of colorectal adenomas of 37%, compared with people with intakes of less than 4.6 nmol/L.

Enterodiol was similarly associated with colorectal adenomas as enterolactone, although concentrations of enterodiol were 5- to 10-fold lower. This suggests that in the human body, enterodiol might be more active than enterolactone.

The hydroxymatairesinol (HMR lignan) and to a greater extent its human metabolite enterolactone may suppress the growth of prostate cancer cells. The results are reported to show that both enterolactone and hydroxymatairesinol increased the programmed cell death (apoptosis) in a dose-dependent manner. Enterolactone is said to be two times as effective as hydroxymatairesinol but less than half that of cycloheximide (cytotoxic agent) and estradiol.

A high intake of plant lignans could reduce the risk of breast cancer for premenopausal women. In women with high plasma levels of enterolactone (above 12.96 nmol/L), high lignan intake was associated with a 58% reduction of breast cancer risk, while average plasma enterolactone levels of 24.96 nmol/g were associated with a reduction of 62%.

A high intake of enterolignans (enterodiol and enterolactone) on the basis of dietary intake (846 µg/day) combined with a high plasma enterolactone level was associated with a reduction in the risk of breast cancer of 64%. Research has shown enterolactone to stimulate the synthesis and circulating levels of a biochemical called sex hormone-binding globulin (SHBG). Through this activity, enterolactone may reduce the free bioavailable pool of circulating estrogen, thereby reducing estrogen penetration in tissues and the risks of an adverse estrogen balance. There is also evidence that enterolactone may inhibit biosynthesis of estrogen by blocking aromatase, a key enzyme in biosynthesis of estradiol. Collectively through these multiple mechanisms of action, lignans appear to have a positive influence on the estrogen balance in the body.

Flaxseed, the richest known source of plant lignans and omega-3 fatty acid, has been shown to have chemoprotective effects in animal and cell studies. Omega-3 fatty acid suppresses the production of interleukin-1 (IL-1), tumor necrosis factor (TNF) and leukotriene B4 (LTB4), and of oxygen free radicals (OFRs) by polymorphonuclear leukocytes (PMNLs) and monocytes. Lignans possess antiplatelet activating factor (PAF) activity and are antioxidant. PAF, IL-1, TNF, and LTB4 are known to stimulate PMNLs to produce OFRs. Modest dietary flaxseed supplementation is effective in reducing hypocholesterolemic atherosclerosis markedly without lowering serum cholesterol. Its effectiveness against hypercholesterolemic atherosclerosis could be due to suppression of enhanced production of OFRs by PMNL in hypercholesterolemia. Dietary flaxseed supplementation could, therefore, prevent hypercholesterolemia-related heart attack and strokes (Parasad 1997).

Flaxseed and its lignan secoisolariciresinol diglucoside (SDG) inhibit mammary tumor development in rats. At the same time, the increased plasma insulin-like growth factor I (IGF-1) concentrations are associated with increased breast cancer risk. The anticancer effect of flaxseed and SDG may be related, in part, to reductions in plasma IGF-I (Rickard et al. 2000).

SDG is an antioxidant. It scavenges for certain free radicals like hydroxyl (HO·). Our bodies produce free radicals continually as we use (oxidize) fats, proteins, alcohol, and some carbohydrates for energy. Free radicals can damage tissues and have been implicated in the pathology of many diseases like atherosclerosis, cancer, and Alzheimer's disease. The flax lignan secoisolariciresinol (SECO) and the mammalian lignins enterodiol, and enterolactone also act as antioxidants. Indeed the antioxidant action of SECO and enterodiol is greater than that of vitamin E.

The flaxseeds' effects may be mediated through its influence on endogenous hormone production and metabolism. Two competing pathways in estrogen metabolism involve production of the 2-hydroxylated and 16-α-hydroxylated metabolites. Because of the proposed differences in biological activities of these metabolites, the balance of the two pathways has been used as a biomarker for breast cancer risk. The effects of flaxseeds consumption on urinary estrogen metabolite excretion were examined. The results obtained suggest that flaxseed may have chemoprotective effects in postmenopausal women. Lignans bind to estrogen receptors on sex hormone-binding globulin (SHBG), thus blocking the binding of estrogen and testosterone. As SHBG is found in breast cancer cells, the binding of mammalian lignans to SHBG may interfere with cancer processes that are controlled by estrogen (Haggans et al. 1999).

It has been shown that secoisolariciresinol diglycoside (SDG) decreases some early markers of colon cancer risk over the short term. Further studies followed to establish whether over the long term flaxseed still exerts a colon cancer-protective effect, whether its effect may, in part, be due to its high content of SDG and whether any change in β-glucuronidase activity plays a role in the protective effect. Urinary lignan excretion, which is an indicator of mammalian lignan production, was significantly increased in the case of rats that were fed with flaxseed and defated flaxseed. It was concluded that flaxseed has a colon cancer-protective effect, which is due, in part, to SD and that the protective effect of flaxseed is associated with increased β-glucuronidase activity (Jenab and Thompson 1996).

21.8 CONCLUSIONS

Little attention has so far been paid to the utility in biological science of lignins, which have long been regarded as waste from the pulp industry. However, the existing information evidences that lignin and its derivatives have a multifunctional role in biological systems (antibacterials, antioxidants and photoprotectors, anticarcinogen, anti-HIV, and spermicide) influencing cell metabolism. Having in mind the polyphenolic character of lignins, one may suppose that the action of lignins is based on their interactions with cell surface at the level of membrane lipids. Thus, the action of lignins can be correlated with their molecular mass and hydroxyl group content by hydrophilic–hydrophobic interactions. At the same time, lignins can be transformed by different metabolic pathways into products (e.g., lignans) that can be useful for human and animal health.

Since many molecular species of lignin are available, depending on the natural sources and methods of extraction and modification, further investigations are definitely worthwhile for isolating and identifying the principle responsible for their notable biological activity. In this context, we need to have information emanating from studies of the molecular biology of genomics and proteomics in order to enhance our comprehension of intricate roles/functions at the molecular level of the lignin and its derivatives.

REFERENCES

Akao, Y., Seki, N., Nakagawa, Y. et al. 2004. A highly bioactive lignophenol derivative from bamboo lignin exhibits a potent activity to suppress apoptosis induced by oxidative stress in human neuroblastoma SH-SY5Y cells. *Bioorg. Med. Chem.* 12: 4791–4801.

Bach Knudsen, K. E., Serena, A., Kjaer, A. K. et al. 2003. Rye bread in the diet of pigs enhances the formation of enterolactone and increases its levels in plasma, urine and feces. *J. Nutr.* 133: 1368–1375.

Baurhoo, B., Ruiz-Deria, C. A., and Zhao, X. 2008. Purified lignin: Nutritional and health impacts on farm animals—A review. *Anim. Feed Sci. Technol.* 144: 175–184.

Begum, N. A., Nicolle, C., Mila, I. et al. 2004. Dietary lignins are the precursors of mammalian lignans in rats. *J. Nutr.* 134: 120–127.

Boeriu, C. G., Bravo, D., Gosselink, R. J. A. et al. 2004. Characterisation of structure-dependent functional properties of lignin with infrared spectroscopy. *Ind. Crop. Prod.* 20: 205–218.

Catignani, G. L. and Carter, M. E. 1982. Antioxidant properties of lignin. *J. Food Sci.* 47: 1745–1748.

Csonka, F. A., Phillips, M., and Breese Jones, D. 1929. Studies of lignin metabolism. *J. Biol. Chem.* LXXXV: 65–75.

Dizhbite, T., Telysheva, G., Jurkjana, V., and Viesturs, U. 2004. Characterization of the radical scavenging activity of lignins-natural antioxidants. *Bioresour. Technol.* 95: 309–317.

Garcia, L., Abajo, C., del Campo, J. et al. 2006. Antioxidant effect of Ligmed—A on humane erythrocytes *in vitro*. *Pharmacologyonline* 3: 514–519.

Haggans, C. J., Hutchins, A. M., Olson, B. A. et al. 1999. Effect of flaxseed consumption on urinary estrogen metabolites in postmenopausal women. *Nutr. Cancer* 33: 188–195.

Ichimura, T., Otake, T., Mori H. et al. 1999. HIV-1 Protease inhibition and anti-HIV effect of natural and synthetic water soluble lignin-like substances. *Biosci. Biotechnol. Biochem.* 63: 2202–2204.

Ichimura, T., Watanabe, O., and Maruyama, S. 1990. Inhibition of HIV-1 protease by water-soluble lignin-like substance in an edible mushroom, *Fusciporia oblique*. *Agric. Biol. Chem.* 54: 479–487.

Jenab, M. and Thompson, Lu. 1996. The influence of flax seed on colon carcinogenesis and beta-glucuronidase activity. *Carcinogenesis* 17: 1343–1348.

Koracevic, D., Koracevic, G., Djordjevic, V. et al. 2001. Method for measurement of antioxidant activity in human fluids. *J. Clin. Pathol.* 54: 356–361.

Kosikova, B., Slamenova, D., Mikulasova, M. et al. 2002. Reduction of carcinogensis by bio-based lignin derivatives. *Biomass Bioenergy* 23: 153–159.

Labaj, J., Slameoova, D., Lazarova, M., and Kosikova, B. 2004. Lignin-stimulated reduction of oxidative DNA lesions in testicular cells and lymphocytes of sprague-dawley rats *in vitro* and *ex vivo*. *Nutr. Cancer* 50: 198–205.

Le Digabel, F. and Averous, L. 2006. Effects of lignin content on the properties of lignocellulose-based composites. *Carbohydr. Polym.* 66: 537–545.

Lu, F. J., Chu, L. H., and Gau, R. J. 1998. Free radical-scavenging properties of lignin. *Nutr. Cancer* 30: 31–38.

Mikulasova, M. and Kosikova, B. 2003. Modulation of mutagenicity of various mutagens by lignin derivatives. *Mutat. Res.* 535: 171–180.

Mitjans, M., Garcia, L., Marrero, E. et al. 2001. Study of Ligmed-A, an antidiarrheal drug based on lignin, on rat small intestine enzyme activity and morphometry. *J. Vet. Pharmacol. Ther.* 24: 349–351.

Nelson, J. L., Alexander, J. W., Gianotti, L. et al. 1994. Influence of dietary fiber on microbial growth *in vitro* and bacterial translocation after burn injury in mice. *Nutrition* 10: 32–33.

Nicolle, C., Manach, C., Morand, C. et al. 2002. Mammaliann lignan formation in rats fed a wheat bran diet. *J. Agric. Food Chem.* 50: 6222–6226.

Novikova, N. L., Ostrovskaya, M. R., Kozhova, M. O. et al. 1998. *The Biological Properties of Lignin-Contained Compounds*. Abstract of Posters, p. 125, Institute of Organic Chemistry, Novosibirsk, Russia.

Okazaki, M., Akimoto, K., and Sorimaki, K. 1996. Polyanion induced preferential multinucleation in macrophages at low level of TNF-α secretion. *Cell Struct. Funct.* 21: 277–282.

Parasad, K. 1997. Dietary flax seed in prevention of hypercholesterolemic atherosclerosis. *Atherosclerosis* 132: 69–76.

Perez-Perez, E. M., Rodriguez-Malaver, J. A., and Dumitrieva, N. 2005. Antioxidant activity of lignin from black liquor. In *Proceedings of the 7th ILI Forum-Barcelona*, Barcelona, Spain, pp. 191–194.

Popa, V. I. 2007. Lignin and sustainable development. *Cellulose Chem. Technol.* 41: 591–593.

Ranby, B. 2001. Degradation of important polymer materials—An overview of basic reactions. In *Recent Advances in Environmentally Compatible Polymers*, eds. J. F. Kennedy, G. O. Phillips, and P. A. Williams, Guest Editor Hyoe Hatakeyama, pp. 3–14. Cambridge, U.K.: Woodhead Publishing Limited.

Re, R., Proteggente, A. R., Saija, A. et al. 2003. The compositional characterization and antioxidant activity of fresh juices from Sicilian sweet orange (*Citrus sinensis* L. Osbeck) varieties. *Free Radic. Res.* 37: 681–687.

Rickard, S. E., Yuan, Y. V., and Thomson, L. U. 2000. Plasma insulin-like growth factor levels in rats are reduced by dietary supplementation of flaxseed or its lignan seicoisolariciresinol diglycoside. *Cancer Lett.* 161: 47–55.

Sakagami, H., Hashimoto, K., Suzuki, F. et al. 2005. Molecular requirements of lignin-carbohydrate complexes for expression of unique biological activities. *Phytochemistry* 66: 2108–2120.

Sakagami, H., Kawazoe, Y., Toshinari Oh-Hara, T. et al. 1991. Stimulation of human perripheral blood polymorphonuclear cell iodination by lignin—Related substances. *J. Leukoc. Biol.* 49: 277–282.

Sorimachi, K. 1992. Differential response of lignin derivatives between tumor and normal tissue derived cell lines: Effects on cellular adhesion and cell growth. *Cell Biol. Int. Rep.* 16: 249–257.

Sorimachi, K., Akimoto, K., Niwa, A. et al. 1997. Delayed cytocidal effect of lignin derivatives on virally transformed rat fibroblasts. *Cancer Detect. Prev.* 22: 111–117.

Sorimachi, K., Akimoto, K., Tsuru, K. et al. 1995. Secretion of TNF-α from macrophage following induction with induction with a lignin derivatives. *Cell Biol. Int. Rep.* 19: 833–838.

Sorimachi, K., Akimoto, K., Yamazaki, S. et al. 1990. Multinucleation of macrophages with water-solubilized lignin derivatives *in vitro*. *Cell Struct. Funct.* 15: 317–322.

Sorimachi, K., Akimoto, K., and Yamazaki, S. 2003. Modulation of interleukin-8 and nitric oxide synthase mRNA levels by interferon-γ in macrophages stimulated with lignin derivatives and lipopolysaccharides. *Cancer Detect. Prev.* 27: 1–4.

Suzuki, H., Iiyama, K., Yoshida, O. et al. 1990. Structural characterization of the immunoactive and antiviral water solubilized lignin in an extract of the culture medium of *Lentinus edodes* Mycelia (LEM). *Agric. Biol. Chem.* 54: 479–487.

Suzuki, H., Okubo, A., Yamzaki, S. et al. 1998. Inhibition of the infectivity and cytopathic effect of human immunodeficiency virus by water-soluble lignin in an extract of the culture medium of *Lentinus edodes* mycelia (LEM). *Biosci. Biotechnol. Biochem.* 62: 575–577.

Tanaka, S., Sakata, Y., Morimoto, K. et al. 2001. Influence of natural and synthetic compounds on cell surface expression of cell adhesion molecules, ICAm-1 and VCAM-1. *Planta Med.* 67: 108–113.

Telysheva, G., Dizhbite, T., Lebedeva, G. et al. 2005. Lignin products for decontamination of environment objects from pathogenic microorganisms and pollutants. In *Proceedings of the 7th ILI Forum*, Barcelona, Spain, pp. 71–74.

Telysheva, G., Dizhbite, T., Tirzite, D., and Jurjane, V. 2000. Applicability of a free radical (DPPH·) method for estimation of antioxidant activity of lignin and its derivatives. In: *Proceedings of the 5th International Lignin Institute, Forum, Commercial Outlets for New Lignins and Definitions of New Projects*, Bordeaux, France, pp. 153–160.

Toh, K., Yokoyama, H., Takahashi, C. et al. 2007. Effect of herb lignin on the growth of enterobacteria. *J. Gen. Appl. Microbiol.* 53: 201–205.

Tollner, L. T., Overstreet, W. J., Li, W. M. et al. 2002. Lignosulfonic acid blocks *in vitro* fertilization of macaque oocytes when sperm are treated either before or after capacitation. *J. Androl.* 23: 889–898.

Vinardell, M. P., Ugartondo, V., and Mitjans, M. 2005. Antioxidant and photoprotective action of lignins from different sources assessed in human red blood cells. In *Proceedings of the 7th ILI Forum*, Barcelona, Spain, pp. 75–77.

22 Carbohydrate-Derived Self-Crosslinkable In Situ Gelable Hydrogels for Modulation of Wound Healing

Lihui Weng, Christine Falabella, and Weiliam Chen

CONTENTS

22.1 INTRODUCTION

Repair of wounds after trauma or surgery is of significant clinical and research importance. Wound healing is a dynamic interactive process; it typically evolves from an initial inflammatory response, and progresses through various stages to finally achieve complete resolution [1]. The capacity of an implantable to maintain a moist environment is recognized as an important factor in normal wound repair. It is recognized that moisture retention is very important in burn injury as heavy water loss from the wound bed by exudation and evaporation may lead to a fall in body temperature and increase in the metabolic rate. Hydrogels, with their high water retention capacity, appear to be optimal media to assist wound healing [2]; thus, much interest has been focused on developing hydrogel-based wound dressings from biomaterials [3–5].

Hydrogels are three-dimensional, hydrophilic, polymeric networks capable of maintaining large amounts of water or biological fluids [6]. Injectable, flowable, and in situ gelable systems have received much attention in tissue engineering because of several advantages. First, this type of in situ gel-forming matrix, initially a fluid and readily conforms to the surrounding, could be directly introduced through a needle/catheter and thereby obviates surgical implantation [7]. Second, therapeutic agents, such as cells, drugs and growth factors, could be incorporated into the matrix by simple pre-mixing. Third, the aqueous in situ gel-forming matrix is completely devoid of residual solvents as compared with other systems formed by various processing methods [8].

Among all the materials proposed for preparation of hydrogels, natural polysaccharides collectively represent a class of abundant renewable resources; they are advantageous over synthetic polymers in biocompatibility and biodegradability. Moreover, many of these carbohydrates are classified by the U.S. FDA as *GRAS (Generally Regarded As Safe)*. Another widely used *GRAS* polymer for the preparation of hydrogel is gelatin, which is also a renewable resource. In recent years, many of these natural polymers, such as hyaluronan, dextran, and chitosan, have been utilized to formulate hydrogels, and their applications in drug delivery, tissue engineering, and gene therapy have been extensively investigated.

22.1.1 HYALURONIC ACID (HA)

Hyaluronan (HA) is a linear non-sulfated polysaccharide consisting of alternating β-1, 4-linked units of β-1, 3-linked glucuronic acid and *N*-acetyl-D-glucosamine (Figure 22.1a). It presents mainly in human connective tissues such as cartilages and in the vitreous, where its primary role is to lubricate body tissue and block the spread of invading microorganisms [9]. HA is also a main constituent of the extracellular matrix (ECM), which is known to play important roles in both cell differentiation/growth and wound repair [10,11]. Due to the remarkable viscoelastic properties of HA and its lack of immunogenicity, it is widely used in ophthalmologic surgery and in orthopedics for the treatment of joint diseases. In addition, HA inhibits both platelet adhesion and aggregation, and stimulates angiogenesis, making it attractive for vascular applications [12].

Native HA is very soluble and can be degraded rapidly in vivo by free radicals or enzymes such as hyaluronidase [13]. Various crosslinking techniques aimed at enhancing HA's longevity and physicochemical properties without compromising its biocompatibility have been developed to formulate hydrogels, films and microspheres [14–18]. The crosslinkers generally used are diepoxide [14], disulfide [16], glycidyl ether [19], glutaraldehyde [20], and water-soluble carbodiimides [15]. HA hydrogels and their variants have attracted great interest as new biocompatible and biodegradable materials with applications in drug delivery and tissue engineering. Much effort has also been made to develop in situ gelable HA hydrogels; a system was produced by chemical crosslinking of

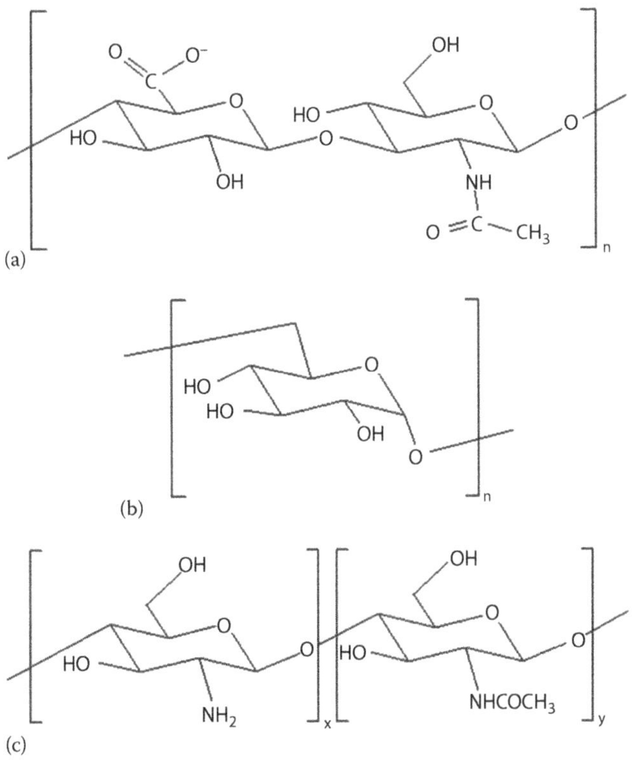

FIGURE 22.1 The structures of (a) hyaluronan, (b) dextran, and (c) chitosan.

adipic dihydrazide-derivatized HA and periodate oxidized HA [21]. Simple mixing of the polymer solutions with pre-dissolved bupivacaine resulted in producing a prolonged local anesthetic effect of the drug [21].

Extracellular matrix HA is prominent in both fetal development and the early phases of wound healing, especially in fetuses where the wound repair is scar-free [22]. In contrast, accumulation of HA is reduced in adult tissues, where wound healing is characterized by scarring, or in keloids characterized by exuberant collagen deposition [23]. The suggestion that extracellular HA might prevent fibrosis have prompted the investigations on the effect of HA derivatives and hydrogels on promoting wound healing [24–26]. Moreover, thiol-modified hyaluronan hydrogels have been developed to slowly release bFGF, and the results showed that the bFGF released from the hydrogel dramatically stimulated blood vessel formation [25].

22.1.2 DEXTRAN

Dextran is a bacterial derived polysaccharide comprising α-1,6 linked D-glucopyranose as well as some α-1,2-, α-1,3- or α-1,4-linked side chains [27,28] (Figure 22.1b). Other than exceptional biocompatibility, dextran is also biodegradable; it has been used as a blood volume expander and for purification of biologics [29]. Dextran has been widely used in the biomedical field to formulate hydrogels for sustained protein (lysozyme, albumin, immunoglobulin G) and drug delivery [30–33]. A methacrylate moiety is the key feature of these dextran derivatives and they are photocrosslinked; the hydrogels formed swell in aqueous media and are partially hydrolyzable by dextranase. Drugs can be incorporated into dextran hydrogels and the versatility of this type of matrices renders it possible to modulate the rate of drug release.

Heparin-like dextran derivatives can interact with heparin-binding growth factors FGF-1, FGF-2 and TGF-b, protecting them from proteolytic and enzymatic degradations; this maneuver also favors the binding to their high-affinity tyrosine kinase receptors [34–36]. They also exhibit both in vitro and in vivo anti-complement properties [37] and anticoagulant activities [38]. Regenerating agent (RGTA) dextran polymers have been shown to promote several tissue-repair processes by stimulating cell growth, differentiation or migration in various in vitro and in vivo models, including muscles [39–41], skin [42], gingival tissue [43,44], and bone [45]. Moreover, part of the biological effects of the RGTA family is attributable to their influence on extracellular matrix remodeling either by modulating collagen biosyntheses [43,46] or by regulating the enzyme activities of matrix metalloproteases [42,43,47], plasmin and cathepsin G [48,49]. For example, film containing glycidyl methacrylate-derivatized dextran/acrylic acid nanoparticles loaded with an antibacterial agent has been prepared and used as an investigational wound dressing material [50]. This film not only maintains a moist environment over a wound bed but also provides a continuous and sustained release of the antibacterial agent.

22.1.3 CHITOSAN

Chitosan is a partially de-acetylated derivative of chitin found in arthropod exoskeletons. It comprises α-1, 4-linked D-glucosamine residues with a variable number of randomly located N-acetyl-glucosamine groups [51] (Figure 22.1c). Chitosan is approved by the FDA as a biocompatible, biodegradable, non-toxic, and bioresorbable biomaterial [52–55], and it has been widely utilized in many fields including pharmaceutics, tissue engineering, food processing, textiles, and so on [56–59].

Chitosan is a semi-crystalline polymer and its crystallinity is dependent upon the degree of deacetylation; the crystallinity is the maximum for both chitin (i.e., 0% deacetylated) and fully deacetylated (i.e., 100%) chitosan, with the minimum crystallinity at intermediate degrees of deacetylation. Due to its stable crystalline structure, chitosan is insoluble in aqueous solutions above pH 7 [51]. Various chemical modifications have been employed to enhance the aqueous solubility of chitosan; some examples are reductive amination with phosphorycholine-glyceraldehyde [60], sulfation [61], carboxymethylation [62], carboxyethylation [63], and N- or O-acylation [64,65]. Amongst these, introducing the highly hydrophilic acrylic acid into the chitosan chains through Michael's reaction is both facile and effective to render it soluble in water [63].

Chitosan has been reported as a wound-healing accelerator [66]. At the inflammatory stage, infiltrating neutrophils cleanse the wound of foreign agents, chitosan accelerates the infiltration of inflammatory cells, consequently accelerating wound cleansing; various chitosan hydrogels have been developed for wound dressing. A photocrosslinked chitosan hydrogel with hemostasis property was formulated and used as a dressing for wound occlusion and tissue adhesive [67]. Moreover, chitosan hydrogel with epidermal growth factor (EGF) incorporated was developed and used in second-degree burn wounds in rat models aiming to accelerate epithelialization [68].

22.1.4 GELATIN

Gelatin is a collagen-derived protein with unique gelation properties; due to a partial recovering of the triple-helix conformation of native collagen [69], gelatin is a liquid above its gelation temperature (t_{gel}) while a gel below it.

Gelatin exhibits several advantages; it is amphipathic and bioresorbable and its physicochemical properties can be easily modified [70–72]. However, gelatin hydrogels melt within the body temperature range, and it is susceptible to in vivo degradation by gelatinases. The rapid degradation of gelatin does not afford sufficient functional longevity to its structural properties in vivo, thereby limiting its utility in biomedical applications. Much effort has been devoted to modulate the physical properties of gelatin hydrogels via chemical crosslinking [73–77]; both glutaraldehyde and formaldehyde have been employed as crosslinking agents to reinforce this protein. This type of treatment increases gelatin's resistance to enzymatic degradation [78–80] by masking the specific enzymatic recognition sites; however, the disadvantage for this approach is a reduction in biocompatibility [81].

Recently, non-cytotoxic crosslinking methods mediated by transglutaminases have been developed. Transglutaminases are a class of natural enzymes capable of catalyzing the acyl-transfer reaction between the amino group of lysine and the carboxyamide group of glutamine on proteins [70,82,83]. In particular, calcium-independent microbial transglutaminase is able to catalyze the formation of covalent crosslinking between glutamine and lysine residues [84,85].

Gelatin is typically used in wound dressing products instead of collagen as the immunogenic potential of the former is generally lower than that of the latter [86–88]. The high absorptive capacity of gelatin hydrogel prevents fluid accumulation; excess wound fluids and cell debris would be absorbed and retained inside the hydrogel applied. Tissue ingrowth could also take place in the biodegradable gelatin matrix with the regenerated tissue and implant fully integrated. Furthermore, EGF has also been incorporated into the gelatin hydrogel wound dressings for efficacy enhancement [89].

In this chapter, we summarize the investigations conducted on several different in situ hydrogel systems composed of the above four biomacromolecules. The physicochemical properties of these hydrogels and their variants are described in detail; the efficacies of these hydrogels in various in vivo experimental models are also revealed.

22.2 ACRYLATED CHITOSAN/OXIDIZED DEXTRAN FOR ACCELERATION OF WOUND HEALING [90]

Native chitosan is insoluble in organic solvents but can be dissolved in aqueous dilute acid; this aspect renders chitosan relatively difficult to process by conventional industrial methods, thereby limiting its versatility in biomedical applications. Introducing carboxyethyl groups into the polymer chain enables chitosan to dissolve in water. Polysaccharides generally degrade relatively rapidly and do not have the endurances of synthetic polymers; crosslinking of polysaccharide chains can enhance their stability and thus durability. Various methods on utilizing small molecule crosslinkers for formulating chitosan hydrogels have been reported [27,91–94]; however, cytotoxicity potential are known to associate with many small molecule crosslinkers. Furthermore, the slow gelation times of some of these hydrogel formulations limits their applicability when rapid in situ gelation is required.

Oxidizing dextran with periodate is a classic method to functionalize dextran, and this maneuver produces a macromolecular crosslinker with multiple functional aldehyde groups capable of reacting with those polymers bearing free amino groups to form hydrogels [95]. Water-soluble carboxyethyl chitosan (CEC) was crosslinked with oxidized dextran (Odex), and the structure, cytotoxicity, biodegradability, and gelation mechanism of the Odex/CEC system were investigated; the gelation time of the hydrogel produced could be modulated and it was both non-cytotoxic and biodegradable. The hydrogel's efficacy in enhancing the healing of murine full-thickness dermal wounds was validated.

22.2.1 Modification of Dextran and Chitosan

The vicinal hydroxyl groups of dextran were modified with periodate to form aldehyde groups. As summarized in Table 22.1, the actual oxidation degrees of periodate-treated dextrans as determined by ^{1}H NMR with the theoretical oxidation degrees of 5%, 15%, and 25% were in actuality 4.6%, 7.4%, and 15.8%, respectively, and the corresponding molecular weights determined by HPLC were 48.1, 38.5, and 36.8 kDa, respectively. The substitution degree of CEC was determined as approximately 45% by ^{1}H NMR. Introducing carboxyl groups into the structure of chitosan partially disrupted the strong hydrogen bonds of native chitosan, rendering it water-soluble. Mixing Odex and CEC readily formed transparent hydrogels that gelled rapidly, and the gelation time could be modulated by varying the formulation parameters, including polymer concentration and temperature. The mechanism was proposed in our previous work [96]; essentially, the aldehyde groups on Odex formed Schiff base with the free amino groups on the CEC, leading to the formation of a 3-D crosslinked network structure (Figure 22.2).

TABLE 22.1

Oxidation Degree of Odex Determined from ¹H NMR and Number Average of Molecular Weight (M_n) Estimated by GPC

Oxidized Dextran	Theoretical Oxidation (%)	Oxidation Degree ¹H NMR (%)[a]	M_n (kDa)	Gelling Time (s)[b]
Odex-I	5	4.6	48.1	359 ± 5
Odex-II	15	7.4	38.5	103 ± 4
Odex-III	25	15.8	36.8	79 ± 4

Source: Modified from *Biomaterials*, 29, Weng, L., Romanov, A., Rooney, J., and Chen, W., Non-cytotoxic, in situ gelable hydrogels composed of N-carboxyethyl chitosan and oxidized dextran, 3905–3913. Copyright 2008, with permission from Elsevier.

[a] Calculated by integrating the anomeric proton peak at δ 4.8 ppm and the tert-butyl proton peak at δ 1.3 ppm.

[b] Gelling time for 2.5% solutions of Odex and 2.5% CEC solutions in 0.01 M PBS at 37°C.

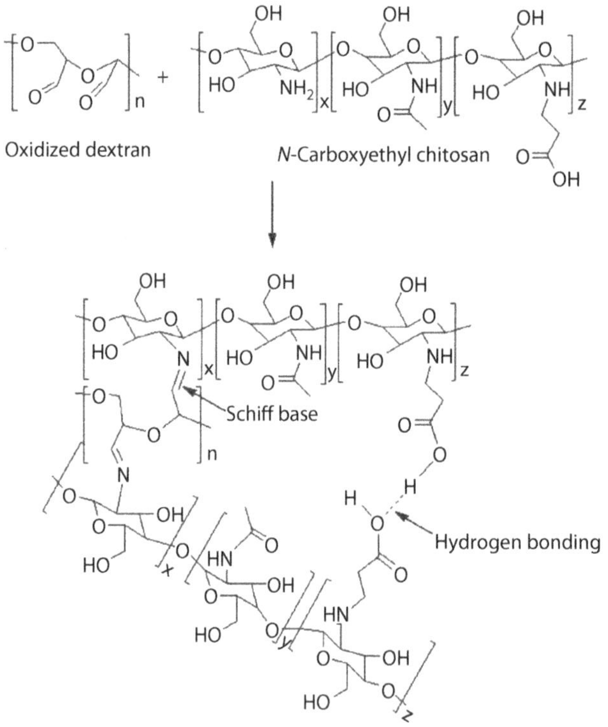

Oxidized dextran crosslinked N-carboxyethyl chitosan

FIGURE 22.2 Schematic presentation of the reaction between Odex and CEC. (Reprinted with permission from Weng, L., Chen, X., and Chen, W., Rheological properties of in situ crosslinkable hydrogels from dextran and chitosan derivatives, *Biomacromolecules*, 8(4), 1109–1115. Copyright 2007 American Chemical Society.)

22.2.2 RHEOLOGICAL ANALYSIS OF HYDROGEL FORMATION

Figure 22.3 showed the variation of the elastic (G′) and viscous moduli (G″) as well as the complex viscosity (η*) versus time for a 2.5% Odex/CEC (ratio 5:5) preparation. With G′ lower than G″, the mix initially exhibited the behavior of a viscous fluid corresponding to a sol. Both moduli elevated rapidly, and the buildup rate of G′ was considerably higher than that of G″ attributable to the rapid

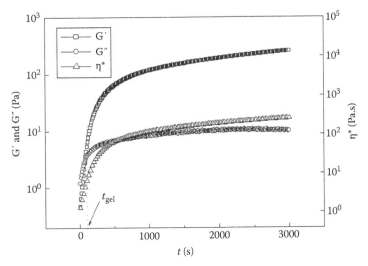

FIGURE 22.3 Time evolution of the storage modulus (G′) and loss modulus (G″) for 2.5% (w/v) Odex-III/CEC (ratio 5:5) at 37°C. The time where G′ and G″ crossover is denoted as t_{gel} (the gelation point). (Reprinted from *Biomaterials*, 29, Weng, L., Romanov, A., Rooney, J., and Chen, W., Non-cytotoxic, in situ gelable hydrogels composed of N-carboxyethyl chitosan and oxidized dextran, 3905–3913. Copyright 2008, with permission from Elsevier.)

occurrence of crosslinking. G′ and G″ eventually intersected ($t = t_{gel}$, where t_{gel} is the gel point), indicative of the transition of the Odex/CEC system from a viscous behavior-dominated liquid-phase to an elastic behavior-dominated solid-phase [97]; eventually, both moduli leveled off, implying the formation of a well-developed three-dimensional network. The η^* also underwent a similar transformation. The t_{gel} for this particular formulation was approximately 78 s and the gelation speed/rate could be modulated [96]. The relationship between the gelation time and the oxidation degree of Odex is shown in Table 22.1. Using Odex solutions and CEC solutions at 2.5% concentration, the gelation time rapidly decreased with increase in the oxidation degree of the Odex (from 5% to 25%), indicating that gelation was strongly correlated with the abundance of aldehyde groups.

22.2.3 MORPHOLOGY OF THE HYDROGELS

The Odex/CEC hydrogel was transparent and yellowish (Figure 22.4a) with its color intensity increasing gradually with time. Figure 22.4b depicts the SEM image of a fractured lyophilized hydrogel prepared

(a)

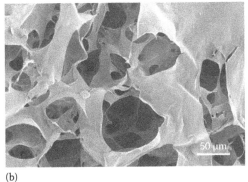

(b)

FIGURE 22.4 **(See companion CD for color figure.)** 2.5% (w/v) Odex-III/CEC (ratio 5:5) hydrogels: (a) the photographic image and (b) the SEM image (b); scale bar: 50 μm. (Reprinted from *Biomaterials*, 29, Weng, L., Romanov, A., Rooney, J., and Chen, W., Non-cytotoxic, in situ gelable hydrogels composed of N-carboxyethyl chitosan and oxidized dextran, 3905–3913. Copyright 2008, with permission from Elsevier.)

from a 2.5% Odex/CEC (ratio 5:5) mixture, revealing a highly porous and interconnected interior structure; the average pore size was approximately 50 μm. It could be inferred that the hydrogel had a high water retention capacity, and both small and macromolecules could diffuse into and out of the structure.

22.2.4 SWELLING PROPERTIES OF THE HYDROGELS

Figure 22.5 showed the relationship between the swelling ratio (q) of the Odex/CEC hydrogels prepared from a 2.5% Odex/CEC (ratio 5:5) mixture. q initially decreased relatively rapidly with an increase in the Odex content from 30% to 50%; this was followed by a slight decline when the Odex content was increased to 60%; and finally, q increased when the Odex content was increased to 70%. This observation was attributed to the variation of the crosslinking density in the hydrogels, with an elevation of the Odex content. As depicted in Table 22.1, when the Odex content was in the range of 50%–60%, the theoretical ratio of the –CHO on the Odex to the –NH$_2$ on the CEC was approximately 1 (i.e., mixing the 2.5% Odex and 2.5% CEC solutions in equal volume); it could be inferred that the crosslinking density increased as the Odex content of the hydrogel increased from 30% to 50% (with –NH$_2$ groups in excess), leading to a decrease in the swelling ratio. When the crosslinking density reached its maximum (50%–60%), q was at the minimum. An excess of –CHO from an increase in the Odex content resulted in the lowering of the degree of crosslinking, assuring a higher swelling ratio.

22.2.5 IN VITRO HYDROLYTIC DEGRADATION OF THE HYDROGELS

Degradation of the Odex/CEC hydrogels was dependent upon the relative amount of Odex and CEC used for the formulation. In vitro degradation of various 2.5% Odex/CEC hydrogels (ratio 3:7, 5:5, and 7:3) was conducted by monitoring their weight losses. The results shown in Figure 22.6 indicates that all hydrogels underwent relatively fast weight loss initially with the rates of degradation moderated thereafter, implying that they were degradable hydrolytically. The 7:3 formulation degraded considerably faster than the 5:5 formulation, which had the lowest degradation rate. The rate of hydrogel degradation was

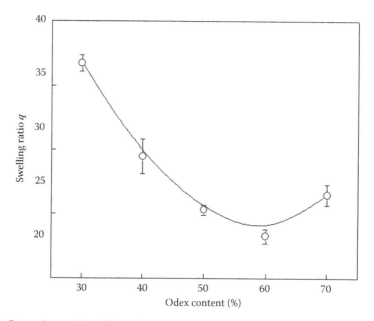

FIGURE 22.5 Dependence of swelling ratio of Odex-III/CEC hydrogels (ratio 5:5) on the Odex content of the hydrogel. (Reprinted from *Biomaterials*, 29, Weng, L., Romanov, A., Rooney, J., and Chen, W., Non-cytotoxic, in situ gelable hydrogels composed of N-carboxyethyl chitosan and oxidized dextran, 3905–3913. Copyright 2008, with permission from Elsevier.)

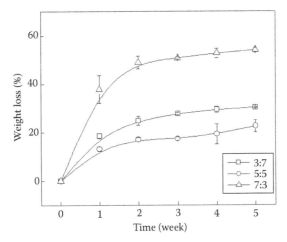

FIGURE 22.6 Hydrolytic degradation of three Odex-III/CEC hydrogels. (Reprinted from *Biomaterials*, 29, Weng, L., Romanov, A., Rooney, J., and Chen, W., Non-cytotoxic, in situ gelable hydrogels composed of N-carboxyethyl chitosan and oxidized dextran, 3905–3913. Copyright 2008, with permission from Elsevier.)

dependent upon the crosslinking density [98]; greater crosslinking typically resulted in forming more stable hydrogels. The 5:5 formulation had the highest crosslinking density and degraded the slowest; the 7:3 formulation degraded slower than the 3:7 formulation despite their comparable theoretical crosslinking densities. The direct implication was hydrogel degradation was dependent upon both the breakdown of the crosslinking as well as the stability of both Odex and CEC.

22.2.6 CYTOTOXICITY OF THE HYDROGELS AND CELL VIABILITY ANALYSIS

The cytotoxicity and cell viability of the Odex/CEC (3:7, 5:5, and 7:3) hydrogels were evaluated by seeding cells in either the presence or absence of the hydrogel. Cell viabilities were determined by MTS assay on days 3, 7, 12, and 30, respectively, and the results were depicted in Figure 22.7. The hydrogel did not appear to affect cell growth as there was no detectable difference in the viabilities between the cells incubated with the hydrogels and the control within 1 week. However, on days 3 and 7, the medium of the 3:7

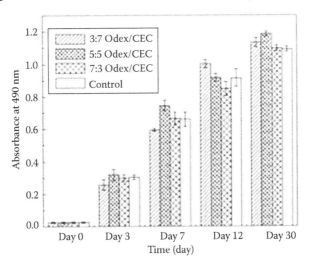

FIGURE 22.7 MTS assay of the cell viabilities in the presence of 3:7, 5:5, and 7:3 Odex-III/CEC hydrogels compared with monolayer cells control by the indirect contact method. (Reprinted from *Biomaterials*, 29, Weng, L., Romanov, A., Rooney, J., and Chen, W., Non-cytotoxic, in situ gelable hydrogels composed of N-carboxyethyl chitosan and oxidized dextran, 3905–3913. Copyright 2008, with permission from Elsevier.)

Odex/CEC hydrogel showed a slightly lower absorbance than those of the other two hydrogels (5:5 and 7:3) as well as the control; this could be attributed to the excess carboxyl functionalities in the presence of excess CEC. After 30 days of culture the obtained results showed that the extent of cell proliferation was comparable to that of the controls, further confirming the hydrogel's non-cytotoxicity; it could also be inferred that the byproducts from hydrogel degradation were nontoxic. Hydrogels were broken down by the enzymes produced by the cells, evidently due to degradation of the polymer network [99].

22.2.7 CELL ENCAPSULATION IN HYDROGELS

Dermal fibroblasts were encapsulated by the Odex/CEC hydrogel. Cells entrapped by the hydrogel initially assumed round shapes (Figure 22.8a), and they were different from the cells residing

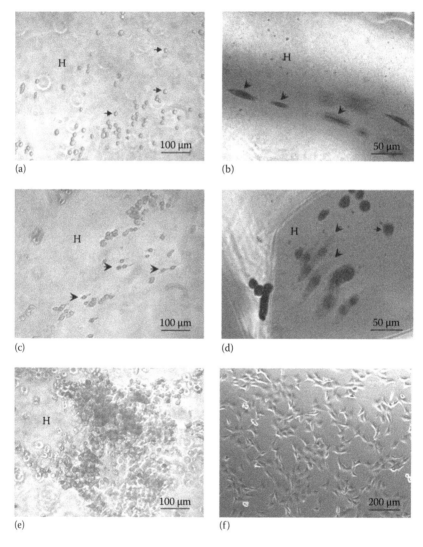

FIGURE 22.8 Morphologies of cells encapsulated in the Odex-III/CEC (ratio 5:5) hydrogel. (a) Day 3, (c) 1 week, (e) 2 week, (b and d) 1 week stained with crystal violet, and (f) 2-D control. ▲: cell assuming round shape, ▼: spread out cells, and **H**: hydrogel. (Reprinted from *Biomaterials*, 29, Weng, L., Romanov, A., Rooney, J., and Chen, W., Non-cytotoxic, in situ gelable hydrogels composed of N-carboxyethyl chitosan and oxidized dextran, 3905–3913. Copyright 2008, with permission from Elsevier.)

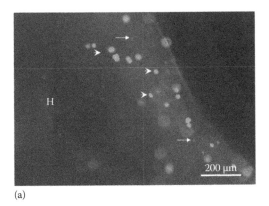
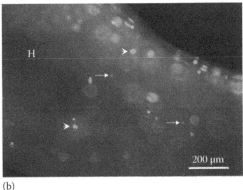

(a)   (b)

FIGURE 22.9 **(See companion CD for color figure.)** Live/Dead™ staining of the cells encapsulated in the Odex-III/CEC hydrogel. (a) Day 3; (b) 1 week. ▼: live cells; ↑: dead cells; *: spread out cells, **H**: hydrogel. (Reprinted from *Biomaterials*, 29, Weng, L., Romanov, A., Rooney, J., and Chen, W., Non-cytotoxic, in situ gelable hydrogels composed of N-carboxyethyl chitosan and oxidized dextran, 3905–3913. Copyright 2008, with permission from Elsevier.)

on the 2D culture with their typical spiny shapes (Figure 22.8f). Cells did not initially adhere to the hydrogel, some of the cells started to adhere after 1 week as indicated by them assuming oval/ or rod shape (Figure 22.8b through d). Figure 22.7c showed the living cells; they were spreading out to form clusters indicating proliferation. Cell-laden hydrogels were fixed with ethanol (70%) and stained by crystal violet (Figure 22.8b and d), the cells and their nuclei could readily be identified ("▼"); they were evidently adapting to the hydrogel surrounding them several days after incubation. Fibroblasts secrete and deposit extracellular matrix (ECM) on polymers [100,101]. The ECM rendered the hydrogel more amenable to cell growth, functioning and cell–cell interactions [102]. The cell density steadily increased with time (Figure 22.8e), and the cells clearly were proliferating in the hydrogel network. Eventually, cells broke down the hydrogel via the degradative enzymes they produced, further confirming the biodegradability of the hydrogel network.

Live/Dead™ staining was performed to differentiate the living cells (green) from the dead cells (red). Figure 22.9 depicted the fluorescent microscopy of stained cells residing in the 2.5% Odex/ CEC (ratio 5:5) hydrogels. Cells maintained their characteristic round shape on day 3 (Figure 22.9a) with nearly 95% cells alive (green color); they started to spread out (marked with "*") by day 7 and this was in good agreement with the results shown in Figure 22.8. The viability of cells was not affected by their entrapment inside the hydrogel.

22.2.8 ACCELERATION OF WOUND HEALING IN MURINE TRANSCUTANEOUS FULL-THICKNESS WOUNDS

In situ forming hydrogels capable of fully conforming to wound beds are advantageous over preformed scaffolds. Chitosan is known to be both biocompatible and bacteriostatic [103] and it has been shown to promote wound healing [104,105]. The precursor of the 2.5% Odex/CEC mix (5:5) was applied to mouse full-thickness transcutaneous wound models (n = 7 per group) to form hydrogel in situ. Figure 22.10 showed the typical wound beds shortly after the procedure and closure with Tegaderm™. The histological sections of two typical wound beds after 7 days were depicted in Figure 22.11. Figure 22.11a showed the histology of a typical wound bed implanted with the Odex/CEC hydrogel, its counterpart receiving PBS was depicted in Figure 22.11b. The implant showed evidence of partial degradation, and the granulation tissues were present in the dermis of both groups; the infiltrated inflammatory cells were more abundant in the group receiving the Odex/CEC hydrogel. The degrees of healing were quantified as% granulation tissue formation and% re-epithelialization. The extent of granulation tissue

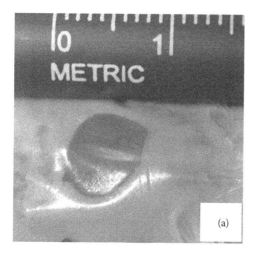 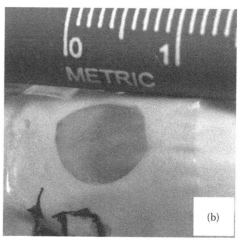

FIGURE 22.10 (**See companion CD for color figure.**) Wound beds of mice covered by Tegaderm™. (a) PBS treated, and (b) Odex/CEC hydrogel treated. (Reprinted from *Biomaterials*, 29, Weng, L., Romanov, A., Rooney, J., and Chen, W., Non-cytotoxic, in situ gelable hydrogels composed of N-carboxyethyl chitosan and oxidized dextran, 3905–3913. Copyright 2008, with permission from Elsevier.)

formation in the wound bed with the hydrogel was 84.9% ± 9.0% (Figure 22.11a) as compared with the 46.3% ± 8.3% for the control (Figure 22.11b) ($p < 0.001$). Granulation tissue fills the wound bed and set the stage for re-epithelialization [2]. Application of the Odex/CEC hydrogel resulted in 100% re-epithelialization, as compared with the counterpart treated with PBS (54.3% ± 8.4%, $p < 0.001$).

22.3 GELATIN/OXIDIZED HYALURONAN (AMENABLE TO CELL MIGRATION)

Both collagen and HA are the major components of the extracellular matrix (ECM); impairment of ECM is a fundamental cause of many soft tissue dysfunctions. Hydrogels composed of HA and gelatin could serve as transient substitutes for impaired soft tissues. HA is inherently rigid due to the presence of β-1,4 backbone linkages and repeated pyranoid rings; therefore, HA hydrogels are brittle. Partial oxidation of HA introduces open rings into its structure which is a strategy to circumvent the chain rigidity issue, thereby improves the elastic properties of the HA hydrogels formed [18]. Gelatin melts at body temperature and this feature limits the potential biomedical applications of pure gelatin hydrogels. Chemical crosslinking has been extensively utilized as a strategy to modulate the physical properties of gelatin hydrogels [69,75,76]. Hydrogels composed of HA and gelatin have been formulated by small molecule crosslinker-mediated reactions [77,106]. Here we describe an approach that does not require any extraneous crosslinker and the hydrogel formulated is very amenable to deep cell infiltration; this is in contrast to other comparable systems where cell attachment occurs mostly on the surface of the hydrogels [77]. These temperature-insensitive self-crosslinkable hydrogels were formulated from partially oxidized HA (oHA) and gelatin in the presence of borax and their physical properties were characterized; dermal fibroblast and mice subdermal models were also used to characterize their in vitro and in vivo interactions, respectively.

22.3.1 PREPARATION OF oHA AND FORMULATION OF oHA/GELATIN HYDROGELS

Figure 22.12a depicted the modifications of HA's hydroxyl groups to aldehydes and the opening of the sugar ring to form a linear chain [18]. The degree of oxidation could be controlled by adjusting the feed ratio and it was estimated by periodate consumption [107]; ¹H NMR on tert-butyl carbazate–treated oHA was performed for validation [18]. The NMR spectra of both HA and oHA were depicted

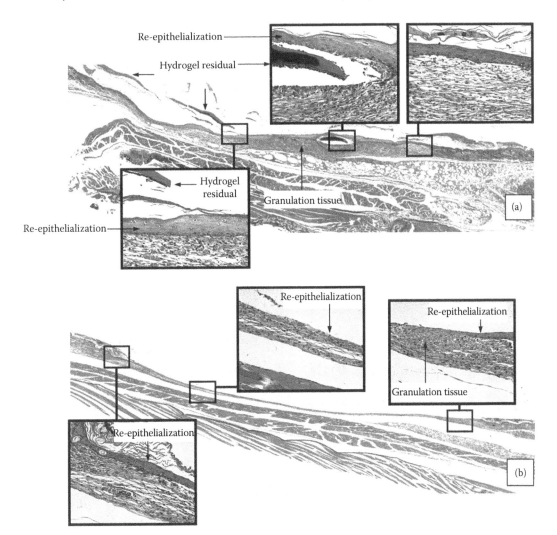

FIGURE 22.11 **(See companion CD for color figure.)** Acceleration of wound healing by Odex/CEC hydrogels in mouse full-thickness dermal wounds. (a) A representative wound bed implanted with hydrogel with both robust epidermal hyperplasia and granulation tissue formation. Part of the implanted hydrogel was fractured and neo-epidermal tissue apparently developed transversely along the side of the hydrogel fragment. (b) A representative wound bed treated with PBS (no hydrogel). Moderate epidermal hyperplasia and granulation tissue formation were observed. The images were captured under 20 × magnification, whereas the insets were captured under 200 × magnification. (Reprinted from *Biomaterials*, 29, Weng, L., Romanov, A., Rooney, J., and Chen, W., Non-cytotoxic, in situ gelable hydrogels composed of N-carboxyethyl chitosan and oxidized dextran, 3905–3913. Copyright 2008, with permission from Elsevier.)

in Figure 22.13, the peaks at 1.87 and 1.32 ppm were from the methyl groups on HA (3H) and tert-butyl groups from oHA (9H, denoted by "*" in Figure 22.13), respectively [18]. The oxidation degree (Table 22.2) increased from 16.7% to 44.4% when the periodate:HA ratio was changed from 20% to 70%; the corresponding average molecular weight of oHA decreased from 183.8 to 35.5 kDa. The FTIR spectra of native HA (Figure 22.14a) and oHA (Figure 22.14b) confirmed oxidation; the peak at 1735 cm^{-1}, emerged as a shoulder on the oHA spectrum, was attributed to the symmetric vibration of aldehyde groups on the oHA [108].

Figure 22.12b depicted the reaction between oHA and gelatin, the presence of borax (a FDA *GRAS* material) favored Schiff base formation [75,109,110]. Various formulations composed of

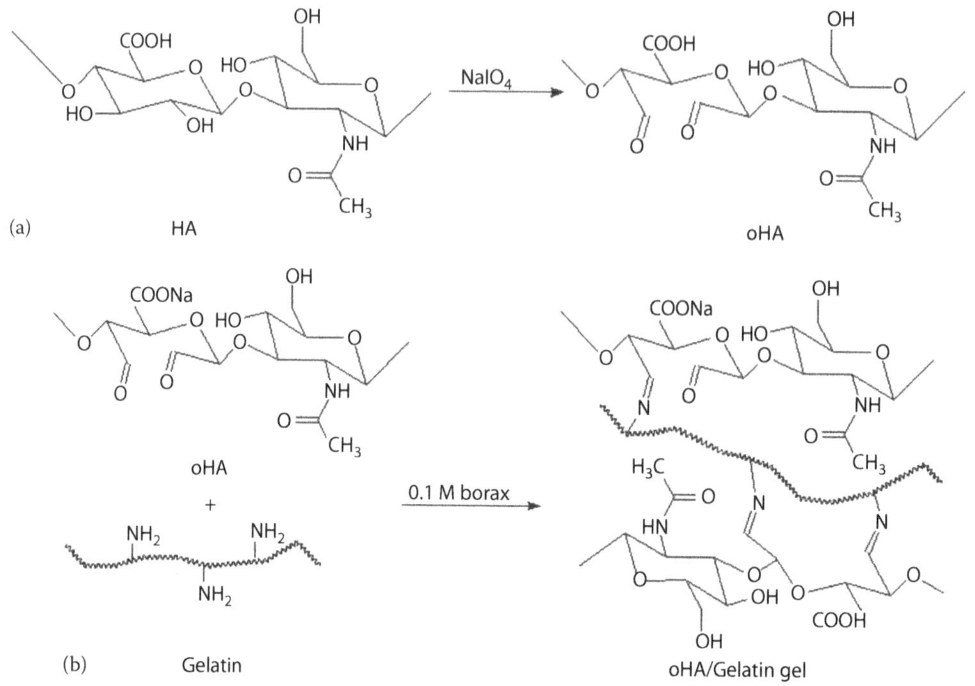

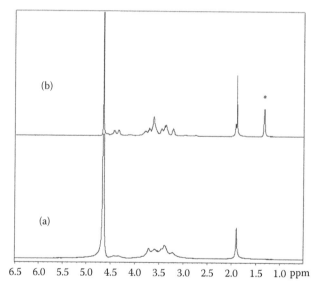

FIGURE 22.12 (a) Oxidation of HA by sodium periodate and (b) crosslinking reaction between oxidized HA and gelatin. (From Weng, L., Pan, H., and Chen, W.: Self-crosslinkable hydrogels composed of partially oxidized hyaluronan and gelatin: In vitro and in vivo responses. *J. Biomed. Mater. Res.* 2008. 85A. 352–365. Copyright Wiley-VCH Verlag GmbH & Co. KGaA. Reprinted with permission.)

FIGURE 22.13 ¹H NMR spectra of (a) HA and (b) oHA treated with tert-butyl carbazate. * denotes the peak of tert-butyl groups (9H). (From Weng, L., Pan, H., and Chen, W.: Self-crosslinkable hydrogels composed of partially oxidized hyaluronan and gelatin: In vitro and in vivo responses. *J. Biomed. Mater. Res.* 2008. 85A. 352–365. Copyright Wiley-VCH Verlag GmbH & Co. KGaA. Reprinted with permission.)

TABLE 22.2
Oxidation Degree of oHA Prepared

Preparation	Oxidation Degree (Theoretical) (%)	Oxidation Degree (Experimental) (%)	M_n (kDa)
1	20	16.7	183.8
2	30	20.3	71.1
3	40	23.4	57.8
4	50	27.8	41.2
5	70	44.4	35.5

Source: Modified from Weng, L., Pan, H., and Chen, W.: Self-crosslinkable hydrogels composed of partially oxidized hyaluronan and gelatin: In vitro and in
vivo responses. *J. Biomed. Mater. Res.* 2008. 85A. 352–365. Copyright
Wiley-VCH Verlag GmbH & Co. KGaA Reproduced with permission.

Note: Calculated by NMR and number average of molecular weight (M_n) measured by HPLC.

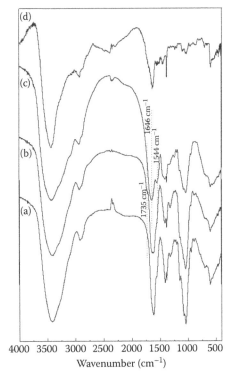

FIGURE 22.14 FT-IR spectra of (a) HA, (b) oHA, (c) oHA/gelatin hydrogel, and (d) gelatin. (From Weng, L.,
Pan, H., and Chen, W.: Self-crosslinkable hydrogels composed of partially oxidized hyaluronan and gelatin:
In vitro and in vivo responses. *J. Biomed. Mater. Res.* 2008. 85A. 352–365. Copyright Wiley-VCH Verlag
GmbH & Co. KGaA. Reprinted with permission.)

oHA of different oxidation degrees and different gelatin contents were prepared (see Table 22.3)
and evaluated. The FTIR spectra of a lyophilized oHA/Gelatin hydrogel and native gelatin were
depicted in Figure 22.14c and d, respectively. The two absorption peaks (3300 and $1650\,cm^{-1}$) in
Figure 22.14d were the N–H and the C=O stretching, respectively [18]. The peaks at 1646 and
$1544\,cm^{-1}$ were stretching of the C=N bonds [111], indicating crosslinking.

TABLE 22.3

Characteristics of the oHA/Gelatin Hydrogels Used

Hydrogel	Oxidation Degree of oHA (%)	oHA:Gelatin[a]
oHG-1	16.7	5:5
oHG-2	20.3	5:5
oHG-3	23.4	5:5
oHG-4	27.8	5:5
oHG-5	44.4	5:5
oHG-6	27.8	4:6
oHG-7	27.8	6:4

Source: Modified from Weng, L., Pan, H., and Chen, W.: Self-crosslinkable hydrogels composed of partially oxidized hyaluronan and gelatin: In vitro and in vivo responses. *J. Biomed. Mater. Res.* 2008. 85A. 352–365. Copyright Wiley-VCH Verlag GmbH & Co. KGaA. Reproduced with permission.

[a] The concentrations of oHA and gelatin solutions used are 20% (w/v) each.

22.3.2   Hydrogel Morphological Analyses

The typical cross-sectional SEM images of three typical lyophilized hydrogel formulations (oHG-4, oHG-7 and oHG-8) (oHA of 27.8% degree of oxidation) were depicted in Figure 22.15. The structures were highly porous (average pore size: ~60 μm), implicating the materials' potential to accommodate cell migration. The hydrogels formulated from oHA:gelatin at the ratios of 4:6 (oHG-7) and

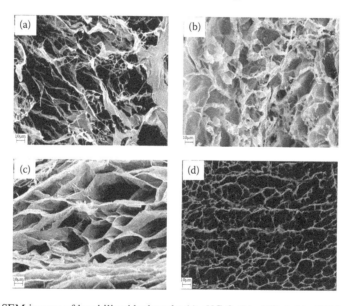

FIGURE 22.15   SEM images of lyophilized hydrogels: (a) oHG-8, (b) oHG-4, (c) oHG-7, and (d) 20% (w/v) pure gelatin hydrogel. Scale bar: 10 μm. (From Weng, L., Pan, H., Chen, W: Self-crosslinkable hydrogels composed of partially oxidized hyaluronan and gelatin: In vitro and in vivo responses. *J. Biomed. Mater. Res.* 2008. 85A. 352–365. Copyright Wiley-VCH Verlag GmbH & Co. KGaA. Reproduced with permission.)

6:4 (oHG-8) resulted in less compact structures (Figure 22.15a and c), as compared with oHG-4 (5:5) (Figure 22.15b), suggested the highest crosslinking density at a ratio 5:5. The structure of pure gelatin hydrogel was shown in Figure 22.15d for comparison with its highly porous network attributable to its triple helical physical crosslinks [69].

22.3.3 SWELLING PROPERTIES OF HYDROGELS

Figure 22.16 depicted the correlation between hydrogel swelling and the oxidation degrees of oHA (Figure 22.16a) and the gelatin contents (Figure 22.16b). The swelling ratio (q) sharply declined with the increase of the oxidation degrees of oHA from 16.7% to 23.4%; this was followed by a rapid moderation when the oxidation degree of oHA was increased to 44.4%. Initial increase in free aldehyde residues resulted in extensive Schiff base formation leading to the rapid drop in water uptake. As a large amount of free amino residues on the gelatin was occupied, subsequent increase in the oxidation degree of oHA resulted in substantially less Schiff base formation and thus the swift decrease of water uptake. Lastly, with the bulk of the amino residues consumed, further increase in the oxidation degree of oHA only resulted in a very modest additional decrease of water uptake. q reached its minimum at a gelatin content of 50%, suggesting the greatest degree of crosslinking.

22.3.4 RHEOLOGICAL ANALYSES OF HYDROGELS

Figure 22.17a depicted the typical frequency sweep profiles of three representative hydrogel formulations (oHG-3, oHG-4, and oHG-5) showing their viscoelasticities under periodic strain. Loss moduli (G″) were weakly dependent on the frequencies in the range of 0.5–100 Hz; storage moduli (G′) appeared to be constant. G″ was considerably larger than G′, indicating the dominance of the elastic properties over the viscous properties, thereby, the presence of well-developed networks [69,112]. G′ elevated from 3,600 to 10,000 Pa with the corresponding increase in the oxidation degree of oHA from 23.4% to 44.4% (oHG-3, oHG-5), this was due to the increase in aldehyde residues and therefore greater Schiff base linkages. The magnitude of elevation in the oxidation degree of oHG-3/oHG-4 was 4.4% as compared with that of oHG-4/oHG-5 at 16.6%; however, the extent of G′ increase for the former was considerably greater than that from the latter, suggesting the

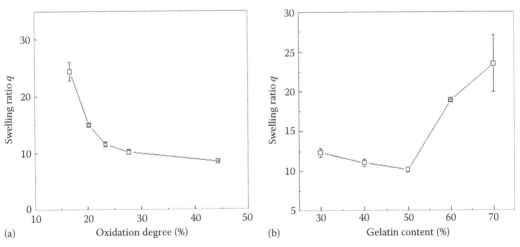

FIGURE 22.16 Correlation of swelling ratio (q) of hydrogels on the oxidation degree of oHA (a) and gelatin content (b). (From Weng, L., Pan, H., and Chen, W.: Self-crosslinkable hydrogels composed of partially oxidized hyaluronan and gelatin: In vitro and in vivo responses. *J. Biomed. Mater. Res.* 2008. 85A. 352–365. Copyright Wiley-VCH Verlag GmbH & Co. KGaA. Reprinted with permission.)

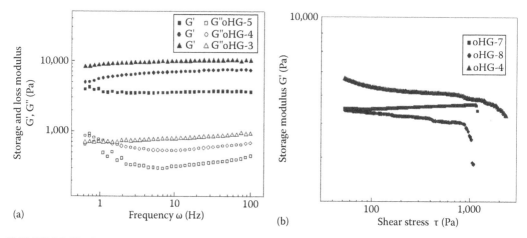

FIGURE 22.17 Rheological spectra of hydrogels: (a) Correlation of the storage modulus G′ and the loss modulus G″ to the oscillating frequency of hydrogels. (b) Correlation of the storage modulus G′ on the oscillatory shear stress of hydrogels. (From Weng, L., Pan, H., and Chen, W.: Self-crosslinkable hydrogels composed of partially oxidized hyaluronan and gelatin: In vitro and in vivo responses. *J. Biomed. Mater. Res.* 2008. 85A. 352–365. Copyright Wiley-VCH Verlag GmbH & Co. KGaA. Reprinted with permission.)

consumption of most of the amino residues on gelatin when the oxidation degree of oHA reached 27.8% (i.e., oHG-4). These results were in strong agreement with those depicted in Figure 22.16.

The oscillation stress sweep profiles of three other hydrogel formulations (oHG-4, oHG-7, and oHG-8) at 1 Hz were depicted in Figure 22.17b. The linear viscoelastic region (LVR) is the range where G′ is independent of the applied stress [113]. Due to a slight slippage at high shear stress, The G′s of oHG-4 and oHG-7 showed a moderate decrease with shearing stress increases; the critical shear stress (i.e., breakdown, at the end of LVR) was different from the slippage-initiated decrease in G′. The G′ was the highest for the oHG-4 hydrogel formulation composed of oHA (at 27.8% oxidation): Gelatin at a 5:5 ratio; the breakdown shear stress was in the order of 5:5(oHG-4) > 4:6(oHG-7) > 6:4 (oHG-8), suggesting that the key contributory role of the oHA:Gelatin ratio to the mechanical strengths of the hydrogels. oHG-4 appeared to have the most optimal ratio for maximizing the interaction of aldehyde and amino groups and thus resulting in a hydrogel with the greatest mechanical strength.

22.3.5 THREE-DIMENSIONAL INFILTRATION AND DISTRIBUTION OF CELLS IN HYDROGELS

Sections of cell-laden hydrogels (~400 μm thick) were stained with crystal violet. Cells attached to all hydrogel formulations 1 day after seeding (Figure 22.18a) with cell numbers increased with time

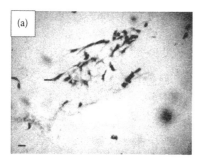
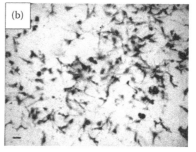

FIGURE 22.18 **(See companion CD for color figure.)** Cell attached to hydrogels (top view). (a) Day 1, and (b) day 3 after cell seeding. Cells were stained with crystal violet. Scale bar: 20 μm. (From Weng, L., Pan, H., and Chen, W.: Self-crosslinkable hydrogels composed of partially oxidized hyaluronan and gelatin: In vitro and in vivo responses. *J. Biomed. Mater. Res.* 2008. 85A. 352–365. Copyright Wiley-VCH Verlag GmbH & Co. KGaA. Reprinted with permission.)

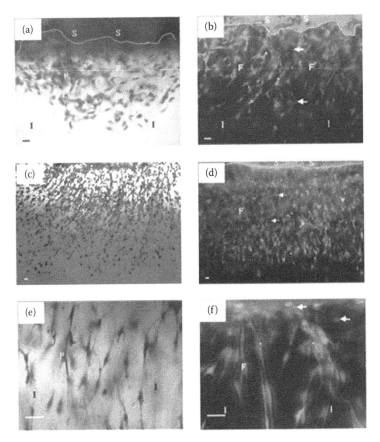

FIGURE 22.19 (See companion CD for color figure.) Cross sections of cell-laden hydrogel (oHG-7): (a, b) 3 days, (c, d) 5 days, and (e, f) 5 days after cell seeding. Samples are stained with crystal violet (a, c, e) and Live/Dead dye (b, d, f), respectively. The demarcation line marked as: **S**: Hydrogel surface, **I**: Hydrogel interior, **F**: Fibroblasts, ⬆: dead cells (red), *: live cells (green). Scale bar: 20 μm. (From Weng, L., Pan, H., and Chen, W.: Self-crosslinkable hydrogels composed of partially oxidized hyaluronan and gelatin: In vitro and in vivo responses. *J. Biomed. Mater. Res.* 2008. 85A. 352–365. Copyright Wiley-VCH Verlag GmbH & Co. KGaA. Reprinted with permission.)

(Figure 22.18b), eventually forming multiple cell layers. The cell attachment rate, the distribution pattern and the numbers of cells attached were comparable in all hydrogel formulations. The hydrogels that were more densely crosslinked had smaller pores and thus, slower cell infiltration. Cell infiltrated the fastest into the hydrogels composed of a 4:6 oHA to Gelatin ratio (oHG-7), deep infiltration was self-evident at 3 and 5 days after seeding (Figure 22.19a, c, and e). The highly elongated cell morphology observed was distinctively different from the typical shapes of cells seeded on culture dishes. The arrays formed were highly organized with the leading cells assuming spherical conformations while the trailing cells were elongated. Both cell infiltration and their corresponding penetration depth increased with time (compare Figure 22.19a and c); there was a corresponding decrease in material cohesiveness signifying cell-mediated degradation and mechanical deterioration.

Both the structure and components of oHA/Gelatin hydrogels bear resemblance to the in vivo environment. Cells attaching to hydrogels were described [114–116], but the robust cell infiltration into the oHA/Gelatin hydrogel had not previously been unambiguously demonstrated. Methods devised to induce cell ingrowth into hydrogels included pre-mixing cells in hydrogel precursor solutions, integrating chemo-attractants into hydrogels, stamping and centrifugation were less than optimal [117–121]. The majority of these studies involved embedding of cell-laden particles into hydrogels and the out-migration of cells was inferred as cell infiltration [122,123].

Despite HA being a major component of ECM, cells do not infiltrate into the structures formed by native HA [124]. Blending oHA with gelatin to form hydrogel formed a structure very amenable to cell infiltration; the morphological/organizational patterns of cells resembled the in vivo counterparts. The implication is that the extent of cell infiltration and the physicomechanical properties of the oHA/Gelatin could be modulated by adjusting the formulation parameters. These hydrogels have the potential to be applied as soft tissue substitutes, as scaffolds for tissue reconstruction, and they could be modified as drug delivery vehicles, particularly for bioactive agents.

22.3.6 LONG-TERM CELL VIABILITY IN THE PRESENCE OF HYDROGELS AND IN VIVO BIOCOMPATIBILITY

The viabilities of the cells infiltrated into the hydrogels were depicted in Figure 22.20. In general, all samples reached full confluence in 1 week and both the hydrogels and their degradation byproducts did not have any apparent effect on cell viability. The results obtained for all oHA/Gelatin hydrogels were virtually identical and they further corroborated with those shown in Figure 22.19b, d, and f.

Fibroblasts produce and maintain ECM [125], and dysfunctional ECM protein distribution/deposition are associated with many pathological conditions [126]. Deposition of ECM on the hydrogel 5 days after cell seeding was demonstrated (Figure 22.21a and b); evidently, the pores inside the hydrogels where cells attached were mostly covered by ECM and the structures where cells had not yet reached were comparable to those of the cell-free controls (Figure 22.21c). The morphological changes of partially degraded hydrogels were also examined (Figure 22.21c and d). The orderly porous structure of the hydrogel surface had turned into fibrous structure suggesting surface erosion (Figure 22.21d), whereas the interior of the hydrogel where cells did not reach remained intact (Figure 22.21c and d). HA and gelatin in their native states are rapidly degraded both in vitro and in vivo [127]. The cell-mediated degradation was a more accurate reflection of in vivo degradation as compared with in vitro degradation studies conducted in either PBS or enzyme solutions [75,119]. Fibroblasts produce many enzymes including

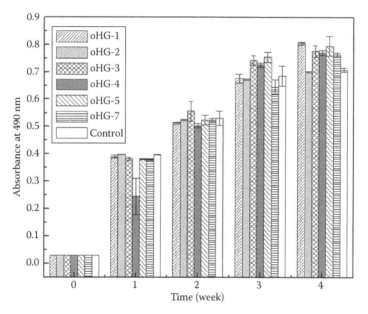

FIGURE 22.20 MTS assay for testing cell viability on various hydrogel formulations compared with monolayer fibroblast control (absence of hydrogel). (From Weng, L., Pan, H., and Chen, W.: Self-crosslinkable hydrogels composed of partially oxidized hyaluronan and gelatin: In vitro and in vivo responses. *J. Biomed. Mater. Res.* 2008. 85A. 352–365. Copyright Wiley-VCH Verlag GmbH & Co. KGaA. Reprinted with permission.)

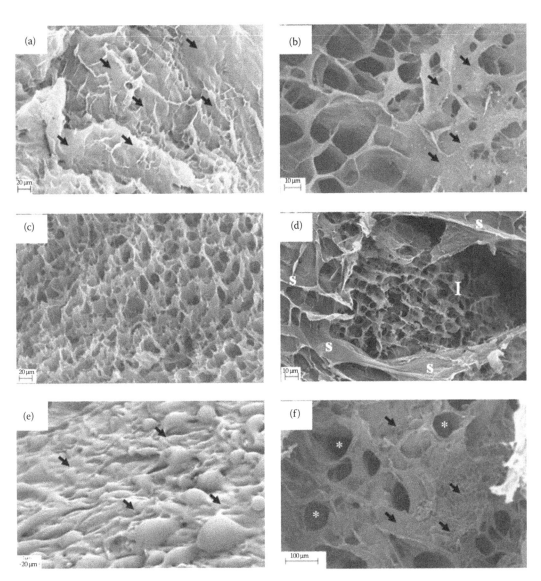

FIGURE 22.21 In vitro (a, b, c) deposition of ECM, and in vitro cell-mediated degradation of oHG-6 hydrogel (d). (a, b) Pores of hydrogel are masked and filled with ECM at day 5 after cell seeding; ♠: deposited ECM. (a) hydrogel surface (scale bar: 20 μm), and (b) cross section of fractured hydrogel (scale bar: 10 μm); right side: pores are masked by ECM; left side: unmasked pores; (c) interior of pristine hydrogel remained intact and are not masked by cell deposited ECM (scale bar: 20 μm), and (d) morphology of hydrogel showing degradation (scale bar: 10 μm). Pore sizes on the surface are larger than those in the interior, indicating cell-mediated degradation. **S**: Hydrogel surface; **I**: Hydrogel interior. In vivo (e, f) deposition of ECM. (e) The hydrogel surface is completely covered by ECM (scale bar: 10 μm), and (f) the explanted hydrogel cross-sectional 1 week post-implantation (scale: 100 μm), *: unmasked pores are the tunnels created by cell infiltration, and ↑: deposited ECM. (From Weng, L., Pan, H., and Chen, W.: Self-crosslinkable hydrogels composed of partially oxidized hyaluronan and gelatin: In vitro and in vivo responses. *J. Biomed. Mater. Res.* 2008. 85A. 352–365. Copyright Wiley-VCH Verlag GmbH & Co. KGaA. Reprinted with permission.)

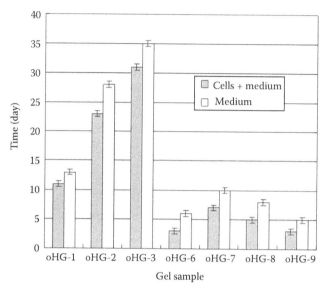

FIGURE 22.22 Cell mediated degradation time of various hydrogel formulations. Hydrogels in cell culture medium are used as controls.

hydrolases such as hyaluronidase and collagenases, capable of degrading oHA/Gelatin. Cell infiltrated into the hydrogel broke down the crosslinks and hastened the degradation leading to disintegration.

The degradation results of the hydrogels after 5 weeks of incubation with fibroblasts were shown in Figure 22.22. Hydrogel degradation was defined as the time span from cell seeding to material disintegration, signified by a loss of cohesiveness. The degradation rates were controlled by the degree of oxidation of oHA and thus the crosslinking densities of the hydrogel. The disintegration time of the hydrogel formulations composed of 5:5 (oHA:gelatin, oHG-1, oHG-2 and oHG-3) increased from 11 to 32 days when the oxidation degree of oHA was elevated from 16.7% (oHG-1) to 23.4% (oHG-3). In contrast, the hydrogels formulated from oHA of 27.8% oxidation degree in different ratios (oHG-6(3:7), oHG-7(4:6), oHG-8(6:4), oHG-9(7:3)) had disintegrated in 7 days. These results suggested that a greater degree of crosslinking was achieved at a ratio of 5:5, which appeared to be the most optimal for material stability and hence resistance to degradation.

Hydrogels were implanted subcutaneously in mice. The gross appearance of the explanted material and the adjoining tissues indicated that the hydrogels and their degradation byproducts were non-cytotoxic; the inflammatory response typically induced by implantation was mild with no sign of tissue necrosis. The implanted hydrogels were evidently degrading (reduced to approximately half the pre-implant size) and appeared to be integrating with the adjoining tissues 3 days post-implant. The hydrogels maintained their cohesiveness after 7 days; their sizes were further reduced to approximately one-fourth of their pre-implant sizes and they were surrounded by very thin fibrous capsules. After 3 weeks, the implants were almost fully resorbed, as indicated by a reduction of their sizes to approximately one-tenth of their initial sizes.

An explanted hydrogel was depicted in Figure 22.23. The hydrogel was surrounded by both neutrophils and macrophages 3 days after implantation (a, b); it was encapsulated in a thin fibrous layer of connective tissue, typically observed in material implanted subdermally [128], 7 days after implantation. There were similarities between the extents of cell infiltration observed in both the in vivo and in vitro studies (depicted in Figure 22.20). One week after implantation (c, d), both the hydrogel and its surrounding capsule were populated by fibroblasts, with inflammatory cells (macrophages, neutrophils and mast cells) scattered throughout. Fibroblasts secrete hydrolases and macrophages secrete both hydrolytic and oxidative enzymes in conjunction with its phagocytic activities were all contributory to material degradation [129]. Three week after implantation (e, f), most of the hydrogel had apparently

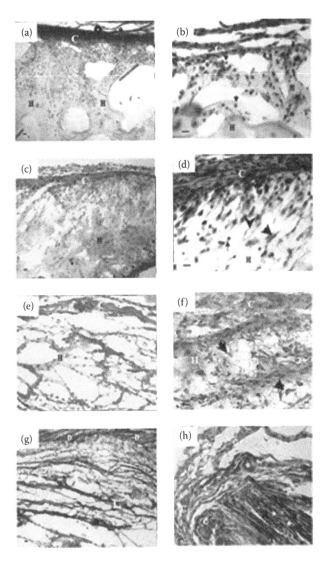

FIGURE 22.23 **(See companion CD for color figure.)** Explanted hydrogels stained with H&E. (a, b) day 3, (c, d) day 7, (e, f) day 21 post-implantation of hydrogels, (g) intact mouse skin, and (h) day 21 post-implantation of Vicryl™ suture. **C**: Collagenous capsule, **H**: Hydrogel, **D**: Dermis, L: Loose connective tissue, ↑: Macrophage, ▼: Fibroblast, and *: Suture. Scale: 20 μm. (From Weng, L., Pan, H., and Chen, W.: Self-crosslinkable hydrogels composed of partially oxidized hyaluronan and gelatin: In vitro and in vivo responses. *J. Biomed. Mater. Res.* 2008. 85A. 352–365. Copyright Wiley-VCH Verlag GmbH & Co. KGaA. Reprinted with permission.)

been replaced by loose connective tissues (g), indicating full resolution of the implant site; the noticeable absence of macrophages implied the inflammation had subsided. A thin collagenous capsule in concert with a lack of foreign body giant cells suggested that the inflammatory response induced by the implanted hydrogel was both mild and transient. For comparison, the tissue adjoining the implanted poly-lactide-co-glycolide Vicryl™ sutures (as the benchmark) showed intense inflammation with conspicuous cell infiltration, they were enclosed by thick fibrous tissues (h).

Consistent with the histology depicted earlier, the SEM images of ECM deposition on the explanted hydrogel (Figure 22.21e and f) suggested a more metabolically active in vivo environment resulted in extensive new ECM formation on hydrogels and the original porous hydrogel structure could not be easily distinguished 1 week after implantation (Figure 22.21e). For comparison, Figure 22.21f depicted the cross section of the hydrogel, some of the "tunnels" ("*") created by cells during their infiltration could be identified. An orderly replacement of the hydrogel by ECM suggested restoration of the damaged implant site. A combination of robust host cell infiltration and ECM deposition resulted in the hydrogels' maintenance of their cohesiveness after implantation.

22.4 CARBOXYETHYL CHITOSAN/OXIDIZED DEXTRAN HYDROGEL FOR PREVENTION OF POST-SURGICAL ABDOMINAL ADHESION

A great majority (>90%) of the patients undergoing abdominal surgery experience some form of adhesions. Adhesions are the results of prolonged inflammation of injured tissues, fibroblast ingrowth and neovascularization leading to the formation of fibrous tissue [130], triggering symptoms including severe pain, bowel obstruction, and infertility [131]. Clinical products targeting reduction of abdominal adhesions generally fall into two categories: non-absorbable and absorbable. Non-biodegradable polytetrafluoroethylene membranes are highly effective in preventing adhesions; however, these membranes have to be surgically removed, thus, biodegradable barriers are appealing alternatives. One of more prominent biodegradable products is a dry uncrosslinked, carboxymethylcellulose/hyaluronan film [132], which forms a gel via absorbing moisture from the surrounding tissues after deployment. The occlusive barrier formed between the site of injury is to reduce adhesion formation without interfering with the repair of tissues under the barrier [133,134]. However, these films are brittle and also relatively difficult to apply as they easily adhere to any moist surface, these drawbacks limit the acceptance of this type of film among surgeons [135].

A resorbable rapid in situ gelable hydrogel barrier that can be applied as a flowable fluid precursor and capable of fully conforming to any complex geometries will be an effective transient barrier, moreover, it is also compatible with endoscopic/laparoscopic procedures. We investigated the feasibility of using the in situ gelable hydrogel formulation composed of partially oxidized dextran (Odex) and N-carboxyethyl chitosan (CEC) [96] as a potential anti-adhesion barrier. There are two main advantages for this system: (1) both chitosan and dextran have long safety records; (2) Odex and CEC readily crosslink to form hydrogels at physiological pH/temperature by electrostatic interaction, hydrogen bonding, and eventually stabilize by Schiff base formation [136]. Odex serves as a macromolecular crosslinker to react with CEC to form a macromolecular network obviating the need of small molecule crosslinking agents, thereby, decreasing the cytotoxicity concerns [136] (see Figure 22.2 for illustration of the hydrogel formation). The efficacy of this hydrogel system was validated in a 21 day, rat cecum abrasion model and commercially available uncrosslinked, carboxymethylcellulose/hyaluronan film was used as a benchmark for comparison.

The results of the rat cecum abrasion study were summarized in Table 22.4. The mean adhesion score for the control animals was 2.1 ± 1.2. Animals treated with the carboxymethylcellulose/hyaluronan

TABLE 22.4
Anti-Adhesion Scores

Score	Control (N = 11)	Odex/CEC Hydrogel (N = 9)	Carboxymethylcellulose/ Hyaluronan (N = 8)
0	2	3	0
1	1	4	7
2	2	1	0
3	6	1	1

Source: Modified from *J. Surg. Res.* Copyright (2009) (in press), with permission from Elsevier.

film and the Odex/CEC hydrogel had adhesion scores of 1.3 ± 0.7 and 1.0 ± 1.0, respectively. The adhesion score of the Odex/CEC hydrogel treated group was significantly lower than that of the control ($p = 0.029$). Animals treated with the carboxymethylcellulose/hyaluronan film also showed a lower adhesion score ($p = 0.056$). There was no statistical difference between the two treatments, although the Odex/CEC hydrogel group showed a slight trend toward a lower score than that of the group treated with the carboxymethylcellulose/hyaluronan film ($p = 0.214$). Twenty one days after intervention, modest amounts of hydrogel remnants (as small fragments), localized along the areas adjoining the cecum and abdominal wall, were identified in 6 out of the 9 animals treated with the Odex/CEC hydrogel (Figure 22.24a and b) [137]. Amber-colored remnants of carboxymethylcellulose/hyaluronan film were also identified in 4 animals (score = 1).

Histological analysis of tissues either received no treatment (control) or Odex/CEC hydrogel were performed. The control specimen was consistent with a severe adhesion score of 3 (Figure 22.25a), a robust and fibrous tissue layer formed between the cecum and abdominal wall. The Odex/CEC treated specimen (adhesion score 1, with visible remnant fragments of hydrogel) showed completely healed abdominal wall (Figure 22.25b) and cecum surface (Figure 22.25c); the ongoing hydrogel degradation was also captured, the presence of multi-nucleated macrophages ingesting the residual hydrogel pieces within the re-epithelialized tissue were self-evident (circled area, Figure 22.25b).

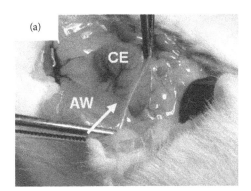
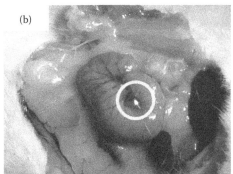
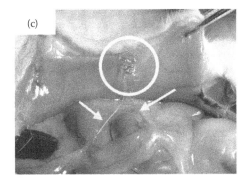

FIGURE 22.24 **(See companion CD for color figure.)** Examination of cecal and abdominal wall tissues from rats 21 days after surgery: (a) An example of a severe adhesion (score = 3) case: a control rat with adhesion of the cecum (CE) to the abdominal wall (AW). The arrow indicates the area where the adhesion is present. Odex/CEC hydrogel remnants found on both (b) the cecum, and (c) the abdominal wall of one animal subject. This animal received an adhesion score of 1 due to some light, filmy adhesions of fat/omentum to the abdominal wall. The circled regions of (b) and (c) highlight the presence of the hydrogel pieces while the arrows in (c) point out the light adhesions that formed inside the rat abdomen after 21 days. (Reprinted from *J. Surg. Res.* 159, Falabella, C.A., Melendez, M.M., Weng, L., and Chen, W., Novel macromolecular crosslinking hydrogel to reduce intra-abdominal adhesions, 772–778. Copyright 2009, with permission from Elsevier.)

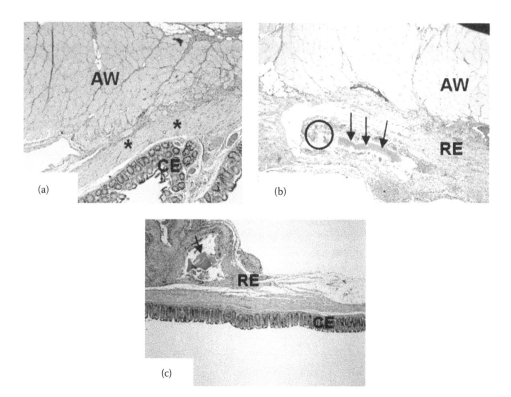

FIGURE 22.25 Photomicrographs of abdominal and cecum tissue excised 21 days after injury. (a) Cross section of rat tissue from untreated, control animals with severe adhesion (*) between the abdominal wall (AW) and the cecum (CE) surface (score = 3). (b) Site of abdominal wall defect in a rat treated with the Odex/CEC hydrogel. Remnants of the hydrogel are highlighted with arrows and located inside the re-epithelialized (RE) abdominal wall defect site. The circled area highlights a fused, giant cell that has ingested a portion of the hydrogel remnant. (c) Site of cecum serosal abrasion from the same rat in (b). Fragments of residual hydrogel are located inside the re-epithelialized tissue on the serosal surface. Magnification 4X for all photomicrographs. (Reprinted from *J. Surg. Res.* Copyright 2009 (in press), with permission from Elsevier.)

The window for efficacy of an adhesion barrier is recognized as between 12 and 36 h after surgery [138,139]. A longer-term endpoint (21 day after surgery) was chosen to observe the adhesion severity in conjunction with validating the biodegradability of the hydrogel. The typical post-operative sacrifice time in comparable studies performed on rat models range from 7 to 14 days [140,141]. Adhesions are normally manifested within the first few days after surgery and more organized, rigid and vascular tissues typically form between 5 and 7 days [132,135]. Any adhesions present after 7 days could persist for several months due to the remodeling occurring inside the adhesion tissue [132].

Analyses of the histological results of the Odex/CEC hydrogel remnants in conjunction with the adjoining tissue at the injured site did not show cellular infiltration into the hydrogel bulk, suggesting a degradation mechanism involving surface erosion [90]. This feature suggested that the hydrogel served as an effective, occlusive barrier to prevent inflammatory cells and fibroblasts from forming aggressive tissue adhesions between the injured cecum and the abdominal wall. Granulation tissue, an indicator of wound healing, was observed at the site of injury with re-epithelialization occurring underneath the hydrogel suggesting tissue repair/restoration. Some residual hydrogel fragments inside the granulation tissue were detected and they were surrounded by macrophages with a noticeable presence of foreign body giant cells, the presence of internalized hydrogel fragments suggesting finalizing of the biodegradation process. This was comparable to the ingestion of other dextran containing biodegradable hydrogel particulates by foreign body giant cells [142], it should be noted that the presence of the multi-nucleated giant cells surrounding any foreign material introduced into the abdominal cavity during surgery at 21 days is not unusual [143–145]. The hydrogel did not appear to interfere with the normal immune response inside the abdomen; therefore, the main mechanism of moderating tissue adhesion was likely through the maintenance of a physical barrier between the injured surfaces.

There was no statistical significance between the severity of adhesion scores between the groups treated by carboxymethylcellulose/hyaluronan film and the Odex/CEC hydrogel; however, the capacity of the hydrogel to conform to complex geometries is very advantageous. The former is typically applied to damaged tissues as a rigid but brittle sheet thus creating a dimensional limitation on open surgeries. It has been reportedly demonstrated that the carboxymethylcellulose/hyaluronan film could be utilized in conjunction with laparoscopic surgeries; nonetheless, the brittle nature of the film could cause difficulties in deployment in addition to the precaution needed to avoid any moisture on the application tools [146]. Conversely, the Odex/CEC hydrogel is a conformal tissue sealant; it could easily be adapted in both conventional surgeries and laparoscopic procedures as either a syringe-borne configuration or through a laparoscopic port. Upon mixing, a steady but rapid increase in both the viscosity and adhesiveness of the precursor mix ensures its localization. The Odex/CEC hydrogel also circumvents the need to dry the bowel surface which could cause further irritation and this is generally regarded as undesirable.

In this study, we have applied an in situ gelable Odex/CEC hydrogel formulation to reduce the incidence of abdominal adhesions in a rat cecum abrasion model. The hydrogel is easy to apply, fully conformal to any geometries, and has achieved comparable, overall efficacy in adhesion prevention as the commercially available carboxymethylcellulose/hyaluronan film.

22.5 SUPPRESSION OF WOUND HEALING BY HYDROGEL-MEDIATED DELIVERY OF AN iNOS INHIBITOR [147]

Macrophage plays significant roles in the post-implant inflammatory response mounted against various biomaterials; it also directs the progress of the wound healing process [148,149]. Activated macrophages impede the growth of microorganisms and malignant cells; they are also involved in ECM production, cell proliferation and angiogenesis [150], moreover, macrophages have been implicated in the pathophysiologies of various inflammatory diseases [151]. Nitric oxide (NO) is one of the major mechanisms for macrophages to mediate inflammatory response and biomaterial degradation [152–154]. NO produced by inducible nitric oxide synthase (iNOS) plays important roles in many physiological and pathophysiological functions [155]; it has also been strongly implicated in dermal wound healing [153–156]. Application of a selective iNOS inhibitor has been shown to reduce inflammatory extravasations [157], a condition known to closely associate with scar or formation of polyposis in wound healing. Therefore, scar formation could theoretically be reduced by timely suppression of iNOS. Guanidinoethyl disulfide (GED, Figure 22.26) is a potent iNOS inhibitor [158] and it has been shown to suppress experimental gingivitis linked to focal increase in iNOS level [159].

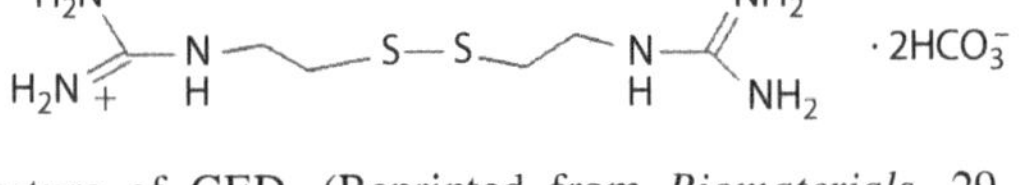

FIGURE 22.26 The structure of GED. (Reprinted from *Biomaterials*, 29, Weng, L., Ivanova, N.D., Zakhaleva, J., and Chen, W., In vitro and in vivo suppression of cellular activity by guanidinoethyl disulfide released from hydrogel microspheres composed of partially oxidized hyaluronan and gelatin, 4149–4156. Copyright 2008, with permission from Elsevier.)

The self-crosslinkable hydrogel prepared by blending oxidized oHA with gelatin we shown previously was used as the basic experimental vehicle for the delivery of GED. It was also previously shown that biodegradable hydrogel microspheres were capable of enhancing the efficacy of various bioactive agents [160–162]; we therefore developed microspheres using the oHA/Gelatin blend as the carrier for GED. The structural and physical properties of the microspheres were characterized and the formulation parameters affecting the characteristics of microspheres and their GED release profiles were investigated. Finally, the in vivo efficacy of the GED microspheres in suppressing macrophages infiltration was validated in mice full-thickness dermal wound models.

22.5.1 MICROSPHERES FOR GED DELIVERY

The GED-carrying microsphere was formulated using a water-in-oil emulsion process. oHA (oxidation degree: 44.7%) and gelatin solutions were blended at approximately 40°C and it was emulsified with soya oil using Span 80 as a surfactant, the emulsion was maintained at 50°C overnight. Microspheres were crosslinked/stabilized by the formation of Schiff base between aldehyde groups and amino groups in the mixture via a similar mechanism revealed previously [99]. GED-loaded oHA/gelatin microspheres were prepared by incorporating GED solutions of various concentrations (0.2%, 1%, and 2%). The encapsulation efficiencies of these microspheres were between 37% and 76%; the amount of GED captured could be adjusted (see Table 22.5).

22.5.2 CHARACTERIZATION OF MICROSPHERES

The SEM images of plain oHA/gelatin microspheres and GED containing microspheres (1% loading) in both dried and swollen state were shown in Figure 22.27. The microspheres were distorted and it was likely the results of the drying process. The surfaces of these microspheres in their dried state were moderately coarse (Figure 22.27a and c) with scattered small pores. Moreover, the diameters of the GED containing microspheres were considerably smaller than their counterparts with no GED. In contrast, the microsphere surfaces in their swollen state exhibited highly porous morphologies.

TABLE 22.5
Particle Size Distribution of Blank and GED Incorporated oHA/Gelatin Microspheres

Microsphere Composition	Mean Diameter (μm) ± SD	GED Encapsulation Yield (%)
oHA/gelatin	91.56 ± 18.76	—
oHA/gelatin/0.2% GED	86.28 ± 19.69	76.64
oHA/gelatin/1% GED	56.39 ± 9.43	70.98
oHA/gelatin/2% GED	31.39 ± 8.67	37.4

Source: Modified from *Biomaterials*, 29, Weng, L., Ivanova, N.D., Zakhaleva, J., and Chen, W., In vitro and in vivo suppression of cellular activity by guanidino-ethyl disulfide released from hydrogel microspheres composed of partially oxidized hyaluronan and gelatin, 4149–4156. Copyright 2008, with permission from Elsevier.

Note: Their GED encapsulation efficiencies were also presented.

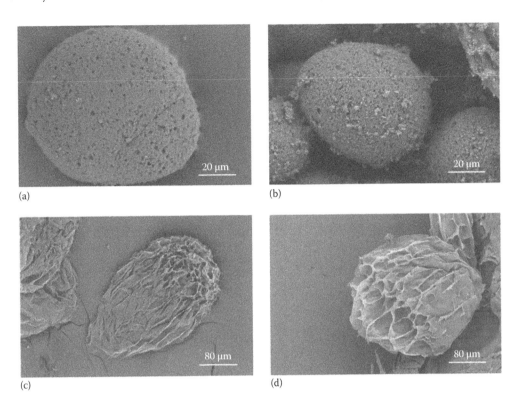

(a)

(b)

(c)

(d)

FIGURE 22.27 SEM photomicrographs of: (a, c) plain oHA/gelatin microspheres, and (b, d) 1% GED-loaded oHA/gelatin microspheres. (a, b): original dry state, and (c, d): lyophilized after fully swollen. (Reprinted from *Biomaterials*, 29, Weng, L., Ivanova, N.D., Zakhaleva, J., and Chen, W., In vitro and in vivo suppression of cellular activity by guanidinoethyl disulfide released from hydrogel microspheres composed of partially oxidized hyaluronan and gelatin, 4149–4156. Copyright 2008, with permission from Elsevier.)

The size distributions of four microspheres formulations were depicted in Figure 22.28; in general, their spreads broadened considerably with decrease in their GED contents. The mean sizes of the plain microspheres, and the GED containing microspheres with 0.2%, 1%, and 2% GED incorporated, were $91.6 \pm 18.8\,\mu m$ (a), $86.3 \pm 19.7\,\mu m$ (b), $56.4 \pm 9.4\,\mu m$ (c), and 31.4 ± 8.7 (d) μm, respectively (as summarized in Table 22.5). The mean size of the former was generally larger than that of the latter; moreover, there was an inverse correlation between the microsphere size and the GED loading. GED-loaded microspheres were generally smaller when compared with its counterpart containing no GED; and it decreased from 91 to 31 μm, with the concentration of GED solution incorporated during the preparation increased from 0% to 2%. Decrease in solution viscosity caused by the increase of GED concentration during emulsification resulted in the formation of smaller emulsion droplets and thus the smaller microspheres formed [162].

Fully swollen oHA/gelatin microspheres (containing 1% GED loading) and the swelling kinetics were depicted in Figure 22.29. All microspheres formulations reached their full swollen state very quickly (< 500 s); the microspheres containing GED had higher equilibrium swelling ratios compared with their counterpart without GED, and this ratio appeared to be linearly related to their GED contents. The presence of GED interfered with the interactions between amino groups and aldehyde groups on the gelatin and oHA, respectively; increase in the microspheres' GED content led to a corresponding reduction of their crosslinking density and thus greater swelling.

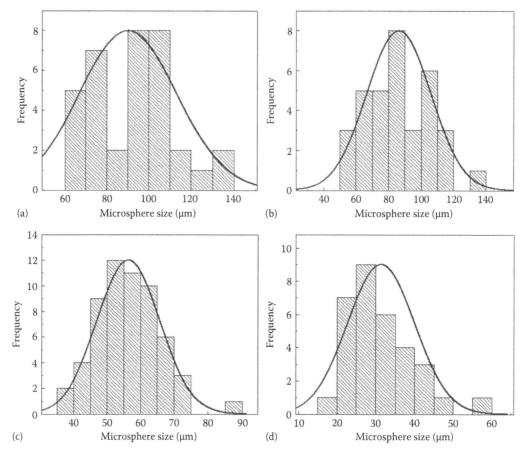

FIGURE 22.28 Frequency and size distribution plots of oHA/gelatin microspheres. (Reprinted from *Biomaterials*, 29, Weng, L., Ivanova, N.D., Zakhaleva, J., and Chen, W., In vitro and in vivo suppression of cellular activity by guanidinoethyl disulfide released from hydrogel microspheres composed of partially oxidized hyaluronan and gelatin, 4149–4156. Copyright 2008, with permission from Elsevier.)

22.5.3 IN VITRO RELEASE OF GED

GED release was performed in 0.01 M PBS at 37°C and the GED HPLC eluent-time profiles were depicted in Figure 22.30. Typical eluent profiles of GED solutions (3 h time-point) were shown in Figure 22.30a and b. The retention time for GED was approximately 13 min and multiple peaks also appeared on the eluent profile of the release samples in the vicinity of 13 min suggesting the release of GED was accompanied by the release of matrix materials from the microspheres.

Figure 22.31 showed the release kinetics of three oHA/gelatin microspheres formulated using 0.2%, 1%, and 2% GED solutions. The initial release kinetics (first 7 h) hi-lighted in the inset indicated that GED release was related to the microspheres' initial GED content; this was consistent with other comparable microspheres prepared by physically entrapping the active principles [163,164]. Typical burst releases signifying the discharge of free residual GED bound to microspheres' surface were observed in all formulations [165,166]. After the initial burst, the release of GED from the microspheres continued at a modest pace.

22.5.4 MACROPHAGE INHIBITION BY GED RELEASED FROM MICROSPHERES

Macrophage play important roles in mediating post-implant tissue/biomaterial inflammatory responses and wound healing [167–169]. In the presence of inflammatory stimuli, the production of NO by

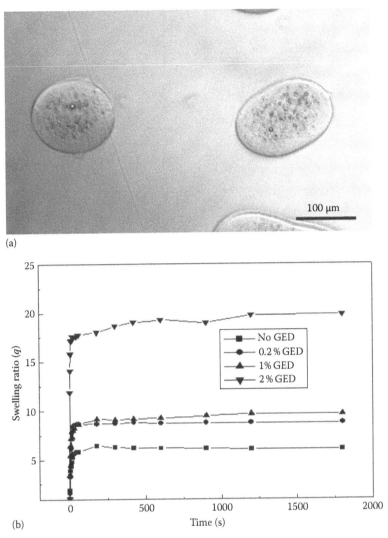

FIGURE 22.29 The effect of GED contents on the swelling dynamics of the oHA/gelatin microspheres. (Reprinted from *Biomaterials*, 29, Weng, L., Ivanova, N.D., Zakhaleva, J., and Chen, W., In vitro and in vivo suppression of cellular activity by guanidinoethyl disulfide released from hydrogel microspheres composed of partially oxidized hyaluronan and gelatin, 4149–4156. Copyright 2008, with permission from Elsevier.)

macrophages rises sharply [170,171]; therefore, suppression of the iNOS activity and thus the NO secretion is a potential strategy to modulate the inflammatory responses evoked by the implant. Macrophage was used as the model cell type for validating the effect of the GED-loaded oHA/Gelatin microspheres.

GED containing microspheres were incorporated into the oHA/Gelatin hydrogels (abbreviated as GOG hydrogel film) and macrophages were cultured directly with them. As controls, macrophages were seeded in either cell culture medium (i.e., background, control) or in the presence of the oHA/Gelatin hydrogel films (abbreviated as OG, as the negative blank control). The effects of GED on macrophages growth and their distributions were observed at 3–6 days after seeding and the results were depicted in Figure 22.32. GED-loaded microspheres evidently suppressed the proliferation of the macrophages; the effect was particularly prominent on day 3, and cells attached to both the culture well (bottom) and the hydrogels. Macrophages residing on the culture wells (Figure 22.32a) and on the OG hydrogel (Figure 22.32c) assumed their characteristic round morphology with a considerable number of them also spread out,

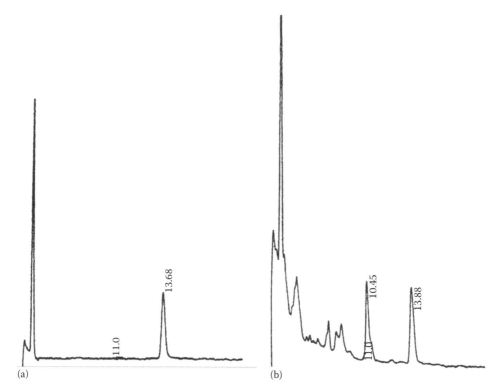

FIGURE 22.30 Typical HPLC elution profiles: (a) A standard GED solution and (b) a sample from the release test of 1% GED-loaded oHA/gelatin microspheres. (Reprinted from *Biomaterials*, 29, Weng, L., Ivanova, N.D., Zakhaleva, J., and Chen, W., In vitro and in vivo suppression of cellular activity by guanidino-ethyl disulfide released from hydrogel microspheres composed of partially oxidized hyaluronan and gelatin, 4149–4156. Copyright 2008, with permission from Elsevier.)

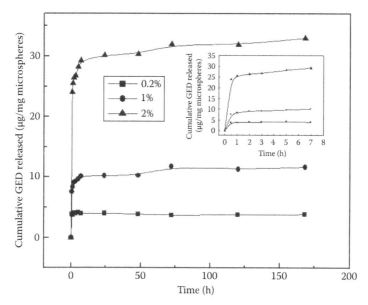

FIGURE 22.31 GED release profiles of oHA/gelatin microspheres formulated with 0.2%, 1%, and 2% GED solutions. (Reprinted from *Biomaterials*, 29, Weng, L., Ivanova, N.D., Zakhaleva, J., and Chen, W., In vitro and in vivo suppression of cellular activity by guanidinoethyl disulfide released from hydrogel microspheres composed of partially oxidized hyaluronan and gelatin, 4149–4156. Copyright 2008, with permission from Elsevier.)

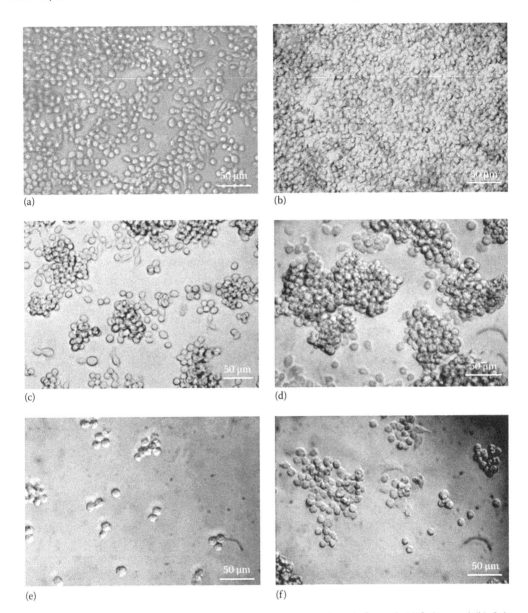

FIGURE 22.32 Morphology of macrophages in the presence hydrogel. Control: (a) 3 days and (b) 6 days. OG hydrogel: (c) 3 days and (d) 6 days. GOG hydrogel: (e) 3 days and (f) 6 days. The GED-loaded microspheres in the GOG hydrogel was formulated with a 1% GED solution. (Reprinted from *Biomaterials*, 29, Weng, L., Ivanova, N.D., Zakhaleva, J., and Chen, W., In vitro and in vivo suppression of cellular activity by guanidinoethyl disulfide released from hydrogel microspheres composed of partially oxidized hyaluronan and gelatin, 4149–4156. Copyright 2008, with permission from Elsevier.)

indicating their activation (i.e., inflammatory response) [168]. Conversely, virtually all the cells co-cultured with the GOG hydrogel maintained their typical non-activated round morphology (Figure 22.32e); GED evidently fully suppressed their activation. Cells also proliferated rapidly in the control group (Figure 22.32b) with a more moderate cell growth in the OG group (Figure 22.32d); macrophage proliferation was suppressed in the presence of the GOG hydrogel (Figure 22.32f). These results were quantitatively summarized in Figure 22.33. The control group had higher cell numbers as compared with both OG and GOG groups, strongly suggesting the release of GED resulted in the suppression of macrophage proliferation. By day

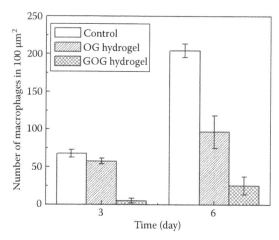

FIGURE 22.33 The densities of macrophages after cultured for 3 days and 6 days, respectively. The GED-loaded microspheres in the GOG hydrogel was formulated with a 1% GED solution. (Reprinted from *Biomaterials*, 29, Weng, L., Ivanova, N.D., Zakhaleva, J., and Chen, W., In vitro and in vivo suppression of cellular activity by guanidinoethyl disulfide released from hydrogel microspheres composed of partially oxidized hyaluronan and gelatin, 4149–4156. Copyright 2008, with permission from Elsevier.)

6, cells in the control group have reached full confluence and they were growing over each other (see Figure 22.32b), whereas the cells in the GOG groups reached 80% confluence (Figure 22.32d); in contrast, the increase in cell numbers in the GOG group was very modest (Figure 22.32f). Continual cell suppression was self-evident despite daily culture medium changes, implicating sustained release of GED from the microspheres entrapped inside the oHA/gelatin hydrogels.

22.5.5 Suppression of Cell Infiltration into Hydrogel in Mice Full-Thickness Transcutaneous Wound Models

The efficacy of the GOG hydrogel in suppressing cell infiltration was validated by mouse full-thickness incisional dermal wound models. The predominant cell types interacting strongly with implants are macrophages and fibroblasts during wound healing; they also interact and modulate the functions of each other [136,172]. As a typical response to injury, the number of macrophages in the wound bed surges almost immediately as the initial inflammatory reaction, their number usually peaks between days 3 and 4, and begins to subside by days 6–7; the number of fibroblasts surges at day 1, peaks by days 5–6, and begins to subside by days 7–8 [173,174]. Days 3 and 7 appeared to be the most optimal time-points for showing GED suppression of cell activities. On day 3, tissue formation, typically manifested as an emerging white ring around the edge of the wound bed, during the early stage of healing was observed in both the unimplanted control and the OG hydrogels. No tissue formation was observed in the wound beds of the mice implanted with the GOG hydrogels, indicating the suppression of wound healing. The histology sections of the hydrogels and the tissues adjacent to them after 3 days of implantation were depicted in Figure 22.34. The capacity of the oHA/gelatin hydrogel to accommodate cell infiltration/proliferation in subdermal implantation was shown earlier [99] (Figure 22.23); the cell infiltration depicted in Figure 22.34a (OG hydrogel) was in good agreement with those shown previously [99] (Figure 22.23). However, the numbers of cells infiltrated into the GOG hydrogel (Figure 22.34b) were substantially lower than their OG counterparts shown in Figure 22.34a; moreover, the maximum penetration depth was approximately half the thickness of the hydrogel. Under high magnifications, the great majority of these infiltrated cells were identified as macrophages ("▼") with some neutrophils and occasionally fibroblasts, which resided predominantly along the fibrous tissues adjacent to the hydrogel. Cell infiltration was suppressed by

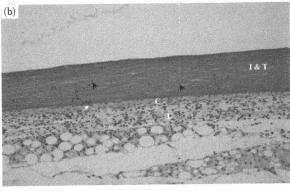

FIGURE 22.34 (See companion CD for color figure.) H&E staining of typical wounds after 3 days. The wound beds were (a) implanted with OG hydrogel, and (b) implanted with GOG hydrogel containing GED-loaded microspheres formulated from 1% GED solution. C: Fibrous tissue formed, I: Implant, L: Loose connective tissue, ▼: Macrophage, ↑: Fibroblast. The images were captured under 20 × magnification. (Reprinted from *Biomaterials*, 29, Weng, L., Ivanova, N.D., Zakhaleva, J., and Chen, W., In vitro and in vivo suppression of cellular activity by guanidinoethyl disulfide released from hydrogel microspheres composed of partially oxidized hyaluronan and gelatin, 4149–4156. Copyright 2008, with permission from Elsevier.)

GED, and by manual cell counting, the ratio of cells in the GOG and OG hydrogels was determined as 4.8:1. There were also clear difference between both the thicknesses and densities (as reflected by the staining intensity) of the two types of hydrogels. Macrophages produce enzymes capable of degrading implanted materials [171], and both hyaluronan and gelatin are very susceptible to these enzymes. Partial digestion of the OG hydrogel caused breakdown of crosslinks, leading to a partial loss of the hydrogel's structural integrity, rendering it less resistant to water uptake and, thus, swelling. Collectively, the net effect is an increased thickness of the OG hydrogel on the histology section and less intense staining. Conversely, GED in the GOG hydrogel suppressed the macrophage activities leading to a greater degree of preservation of the structural integrity of the hydrogel.

22.6 CONCLUSIONS

This chapter summarizes our investigations on natural material derived and crosslinker-free in situ gelable formable/conformable hydrogels; this type of hydrogels is versatile and advantageous over their counterparts requiring extraneous crosslinking reagents or photo-induction. They can be formulated to either prevent or facilitate cell adhesion/migration and their physicomechanical properties could be tailored and optimized for different purposes. The potential biomedical applications (enhance wound healing and prevent post-surgical adhesion) have also been demonstrated. The hydrogel can also be formulated into drug delivery vehicles.

ABBREVIATIONS

CEC	Carboxyethyl chitosan
ECM	Extracellular matrix
EGF	Epidermal growth factor
FGF	Fibroblast growth factor
GED	Guanidinoethyl disulfide
GRAS	Generally regarded as safe
iNOS	Inducible nitric oxide synthase
HA	Hyaluronan
Odex	Oxidized dextran
oHA	Oxidized hyaluronan
TGF	Transforming growth factor
LVR	Linear viscoelastic region

REFERENCES

1. Wilgus, T.A. 2008. Immune cells in the healing skin wound: influential players at each stage of repair. *Pharm Res* 58:112–116.
2. Balakrishnan, B., Mohanty, M., Umashankar, P.R., Jayakrishnan, A. 2005. Evaluation of an in situ forming hydrogel wound dressing based on oxidized alginate and gelatin. *Biomaterials* 26:6335–6342.
3. Yusof, N., Hafiza, A.H.A., Zohdi, R.M., Bakar, Z.A.M. 2007. Development of honey hydrogel dressing for enhanced wound healing. *Rad Phys Chem* 76:1767–1770.
4. Garcia, Y., Wilkins, B., Collighan, R.J., Griffin, M., Pandit, A. 2008. Towards development of a dermal rudiment for enhanced wound healing response. *Biomaterials* 29:857–868.
5. Rayment, E.A., Dargaville, T.R., Shooter, G.K., George, G.A., Upton, Z. 2008. Attenuation of protease activity in chronic wound fluid with bisphosphonate-functionalised hydrogels. *Biomaterials* 29:1785–1795.
6. Peppas, N.A., Bures, P., Leobandung, W., Ichikawa, H. 2000. Hydrogels in pharmaceutical formulations. *Eur J Pharm Biopharm* 50:27–46.
7. Shu, X.Z., Liu, Y., Palumbo, F.S., Luo, Y., Prestwich, G.D. 2004. In situ crosslinkable hyaluronan hydrogels for tissue engineering. *Biomaterials* 25:1339–1348.
8. Klouda, L., Hacker, M.C., Kretlow, J.D., Mikos, A.G. 2009. Cytocompatibility evaluation of amphiphilic, thermally responsive and chemically crosslinkable macromers for in situ forming hydrogels. *Biomaterials* 30:4558–4566.
9. Coviello, T., Matricardi, P., Marianecci, C., Alhaique, F. 2007. Polysaccharide hydrogels for modified release formulations. *J Control Release* 119:5–24.
10. Tammi, M.I., Day, A.J., Turley, E.A. 2002. Hyaluronan and homeostasis: a balancing act. *J Biol Chem* 277:4581–4584.
11. Morra, M. 2005. Engineering of biomaterials surfaces by hyaluronan. *Biomacromolecules* 6: 1205–1223.
12. Slevin, M., Krupinski, J., Gaffney, J., Matou, S., West, D., Delisser, H., Savani, R.C., Kumar, S. 2007. Hyaluronan-mediated angiogenesis in vascular disease: uncovering RHAMM and CD44 receptor signaling pathways. *Matrix Biol* 26:58–68.
13. Hornebeck W. 2003. Down-regulation of tissue inhibitor of matrix metalloprotease-1 (TIMP-1) in aged human skin contributes to matrix degradation and impaired cell growth and survival. *Pathol Biol* 51:569–573.
14. Segura, T., Anderson, B.C., Chung, P.H., Webber, R.E., Shull, K.R., Shea, L.D. 2005. Crosslinked hyaluronic acid hydrogels: a strategy to functionalize and pattern. *Biomaterials* 26:359–371.
15. Baumann, M. D., Kang, C.E, Stanwick, J.C., Wang, Y., Kim, H., Lapitsky, Y., Shoichet, M.S. 2009. An injectable drug delivery platform for sustained combination therapy. *J Control Release* 138:205–213.
16. Shu, X.Z., Liu, Y., Palumbo, F., Prestwich, G.D. 2003. Disulfide-crosslinked hyaluronan-gelatin hydrogel films: a covalent mimic of the extracellular matrix for in vitro cell growth. *Biomaterials* 24:3825–3834.
17. Esposito, E., Menegatti, E., Cortesi, R. 2005. Hyaluronan-based microspheres as tools for drug delivery: a comparative study. *Inter J Pharm* 288:35–49.
18. Jia, X.Q., Burdick, J.A., Kobler, J., Clifton, R.J., Rosowski, J.J., Zeitels, S.M. et al. 2004. Synthesis and characterization of in situ cross–linkable hyaluronic acid-based hydrogels with potential application for vocal fold regeneration. *Macromolecules* 37:3239–3248.

19. Kogan, G., Šoltés, L., Stern, R., Schiller, J., Mendichi, R. 2008. Hyaluronic acid: its function and degradation in in vivo systems. *Stud Nat Prod Chem* 34:789–882.
20. Tomihata, K., Ikada, Y. 1997. Crosslinking of hyaluronic acid with glutaraldehyde, *J Polym Sci Part A Polym Chem* 35:3553–3559.
21. Jia, X., Colombo, G., Padera, R., Langer, R., Kohane, D.S. 2004. Prolongation of sciatic nerve blockade by in situ cross-linked hyaluronic acid. *Biomaterials* 25:4797–4804.
22. Croce, M.A., Dyne, K., Boraldi, F., Quaglino, D.Jr., Cetta, G., Tiozzo, R., Pasquali, R.I. 2001. Hyaluronan affects protein and collagen synthesis by in vitro human skin fibroblasts. *Tissue Cell* 33:326–331.
23. Meyer, L.J., Russell, S.B., Russell, J.D., Trupin, J.S., Egbert, B.M., Shuster, S., Stern, R. 2000. Reduced hyaluronan in keloid tissue and cultured keloid fibroblasts. *J Invest Dermatol* 114:953–959.
24. Xu, H., Ma, L., Shi, H., Gao, C., Han, C. 2007. Chitosan–hyaluronic acid hybrid film as a novel wound dressing: in vitro and in vivo studies. *Polym Adv Technol* 18:869–875.
25. Cai, S., Liu, Y., Shu, X.Z., Prestwich, G.D. 2005. Injectable glycosaminoglycan hydrogels for controlled release of human basic fibroblast growth factor. *Biomaterials* 26:6054–6067.
26. Witte, S.H., Olaifa, A.K., Lewis, A.J., Eggleston, R.B., Halper, J., Kietzmann, M., Baeumer, W., Mueller, E. 2009. Application of exogenous esterified hyaluronan to equine distal limb wounds. *J Equine Veter Sci* 29:197–205.
27. Cana, H.K., Denizlia, B.K., Gunera, A., Rzaev, Z.M.O. 2005. Effect of functional crosslinking agents on preparation and swelling properties of dextran hydrogels. *Carbohydr Polym* 59:51–56.
28. Nowakowska, M., Zapotoczny, S., Sterzel, M., Kot, E. 2004. Novel water-soluble photosensitizers from dextrans. *Biomacromolecules* 5:1009–1014.
29. Suda, E.J., Thomas, K.E., Pabst, T.M., Mensah, P., Ramasubramanyan, N., Gustafson, M.E., Hunter, A.K. 2009.Comparison of agarose and dextran-grafted agarose strong ion exchangers for the separation of protein aggregates. *J Chromatogr A* 1216:5256–5264.
30. Hiemstra, C., Zhong, Z., van Steenbergen, M.J., Hennink, W.E., Feijen, J. 2007. Release of model proteins and basic fibroblast growth factor from in situ forming degradable dextran hydrogels. *J Control Release* 122:71–78.
31. Pitarresi, G., Palumbo, F.S., Giammona, G., Casadei, M.A., Moracci, F.M. 2003. Biodegradable hydrogels obtained by photocrosslinking of dextran and polyaspartamide derivatives. *Biomaterials* 24:4301–4313.
32. Cadee, J.A., de Groot, C.J., Jiskoot, W., den Otter, W., Hennink, W.E. 2002. Release of recombinant human interleukin-2 from dextran-based hydrogels. *J Control Release* 78:1–13.
33. Hornig, S., Bunjes, H., Heinze, T. 2009. Preparation and characterization of nanoparticles based on dextran–drug conjugates. *J Colloid Interface Sci* 338:56–62.
34. Vercoutter-Edouart, A.S., Dubreucq, G., Vanhoecke, B., Rigaut, C., Renaux, F., Correia, L.D., Lemoine, J., Bracke, M., Michalski, J.C., Correia, J. 2008. Enhancement of PDGF-BB mitogenic activity on human dermal fibroblasts by biospecific dextran derivatives. *Biomaterials* 29:2280–2292.
35. Rouet, V., Meddahi-Pelle, A., Miao, H.Q., Vlodavsky, I., Caruelle, J.P., Barritault, D. 2006. Heparin-like synthetic polymers, named RGTAs, mimic biological effects of heparin in vitro. *J Biomed Mater Res A* 78:792–797.
36. Logeart-Avramoglou, D., Huynh, R., Chaubet, F., Sedel, L., Meunier, A. 2002. Interaction of specifically chemically modified dextrans with transforming growth factor beta-1: potentiation of its biological activity. *Biochem Pharmacol* 63:129–137.
37. Banz, Y., Gajanayake, T., Matozan, K., Yang, Z., Rieben, R. 2009. Dextran sulfate modulates MAP kinase signaling and reduces endothelial injury in a rat aortic clamping model. *J Vasc Surg* 50:161–170.
38. Huynh, R., Chaubet, F., Jozefonvicz, J. 2001. Anticoagulant properties of dextran methylcarboxylate benzylamide sulfate (DMCBSu): a new generation of bioactive functionalized dextran. *Carbohydr Res* 332:75–83.
39. Meddahi, A., Bree, F., Papy-Garcia, D., Gautron, J., Barritault, D., Caruelle, J.P. 2002. Pharmacological studies of RGTA, a heparan sulfate mimetic polymer, efficient on muscle regeneration. *J Biomed Mater Res* 62:525–531.
40. Papy-Garcia, D., Barbosa, I., Duchesnay, A., Saadi, S., Caruelle, J.P., Barritault, D. et al. 2002. Glycosaminoglycan mimetics (RGTA) modulate adult skeletal muscle satellite cell proliferation in vitro. *J Biomed Mater Res* 62:46–55.
41. Barbosa, I., Morin, C., Garcia, S., Duchesnay, A., Oudghir, M., Jenniskens, G. et al. 2005. A synthetic glycosaminoglycan mimetic (RGTA) modifies natural glycosaminoglycan species during myogenesis. *J Cell Sci* 118:253–264.
42. Garcia-Filipe, S., Barbier-Chassefiere, V., Alexakis, C., Huet, E., Ledoux, D., Kerros, M.E. et al. 2007. RGTA OTR4120, a heparan sulfate mimetic, is a possible long-term active agent to heal burned skin. *J Biomed Mater Res A* 80:75–84.

43. Escartin, Q., Lallam-Laroye, C., Baroukh, B., Morvan, F.O., Caruelle, J.P., Godeau, G. et al. 2003. A new approach to treat tissue destruction in periodontitis with chemically modified dextran polymers. *FASEB J* 17:644–651.

44. Lallam-Laroye, C., Escartin, Q., Zlowodzki, A.S., Barritault, D., Caruelle, J.P., Baroukh, B. et al. 2006. Periodontitis destructions are restored by synthetic glycosaminoglycan mimetic. *J Biomed Mater Res A* 79:675–683.

45. Blanquaert, F., Carpentier, G., Morvan, F., Caruelle, J.P., Barritault, D., Tardieu, M. 2003. RGTA modulates the healing pattern of a defect in a monolayer of osteoblastic cells by acting on both proliferation and migration. *J Biomed Mater Res A* 64:525–532.

46. Alexakis, C., Mestries, P., Garcia, S., Petit, E., Barbier, V., Papy-Garcia, D. et al. 2004. Structurally different RGTAs modulate collagen-type expression by cultured aortic smooth muscle cells via different pathways involving fibroblast growth factor-2 or transforming growth factor-beta1. *FASEB J* 18:1147–1149.

47. Mestries, P., Alexakis, C., Papy-Garcia, D., Duchesnay, A., Barritault, D., Caruelle, J.P. et al. 2001. Specific RGTA increases collagen V expression by cultured aortic smooth muscle cells via activation and protection of transforming growth factor-beta1. *Matrix Biol* 20:171–181.

48. Ledoux, D., Papy-Garcia, D., Escartin, Q., Sagot, M.A., Cao, Y., Barritault, D. et al. 2000. Human plasmin enzymatic activity is inhibited by chemically modified dextrans. *J Biol Chem* 275:29383–29390.

49. Ledoux, D., Merciris, D., Barritault, D., Caruelle, J.P. 2003. Heparin-like dextran derivatives as well as glycosaminoglycans inhibit the enzymatic activity of human cathepsin G. *FEBS Lett* 537:23–29.

50. Zhang, H., Gu, C., Wu, H., Fan, L., Li, F., Yang, F., Yang, Q. 2007. Immobilization of derivatized dextran nanoparticles on konjac glucomannan/chitosan film as a novel wound dressing. *BioFactors* 30:227–240.

51. Bhatnagar, A., Sillanpää, M. 2009. Applications of chitin- and chitosan-derivatives for the detoxification of water and wastewater — A short review. *Adv Colloid Interface Sci* 152:26–38.

52. Wang, Q., Zhang, J., Wang, A. 2009. Preparation and characterization of a novel pH-sensitive chitosan-g-poly(acrylic acid)/attapulgite/sodium alginate composite hydrogel bead for controlled release of diclofenac sodium. *Carbohydr Polym* 78:731–737.

53. Majeti, N.V., Kumar, R. 2000. A review of chitin and chitosan applications. *React Funct Polym* 46:1–27.

54. Boucard, N., Viton, C., Agay, D., Mari, E., Roger, T., Chancerelle, Y., Domard, A. 2007. The use of physical hydrogels of chitosan for skin regeneration following third-degree burns. *Biomaterials* 28:3478–3488.

55. Chang, Y.Y., Chen, S.J., Liang, H.C., Sung, H.W., Lin, C.C., Huang, R.N. 2004. The effect of galectin 1 on 3T3 cell proliferation on chitosan membranes. *Biomaterials* 25:3603–3611.

56. Dutkiewicz, J.K. 2002. Superabsorbent materials from shellfish waste—A review. *J Biomed Mater Res* 63:373–381.

57. Kandile, N.G., Nasr, A.S. 2009. Environment friendly modified chitosan hydrogels as a matrix for adsorption of metal ions, synthesis and characterization. *Carbohydr Polym* 78:753–759.

58. Felix, L., Hernandez, J., Arguelles-Monal, W.M., Goycoolea, F.M. 2005. Kinetics of gelation and thermal sensitivity of N-isobutyryl chitosan hydrogels. *Biomacromolecules* 6:2408–2415.

59. Huang, R., Chen, G., Yang, B., Gao, C. 2008. Positively charged composite nanofiltration membrane from quaternized chitosan by toluene diisocyanate cross-linking. *Sep Purif Technol* 61:424–429.

60. Tiera, M.J., Qiu, X.P., Bechaouch, S., Shi, Q., Fernandes, J.C., Winnik, F.M. 2006. Synthesis and characterization of phosphorylcholine-substituted chitosans soluble in physiological pH conditions. *Biomacromolecules* 7:3151–3156.

61. Zhang, C., Qu, G., Sun, Y., Wu, X., Yao, Z., Guo, Q., Ding, Q., Yuan, S., Shen, Z., Ping, Q., Zhou, H. 2008. Pharmacokinetics, biodistribution, efficacy and safety of N-octyl-O-sulfate chitosan micelles loaded with paclitaxel. *Biomaterials* 29:1233–1241.

62. Lu, G., Kong, L., Sheng, B., Wang, G., Gong, Y., Zhang, X. 2007. Degradation of covalently cross-linked carboxymethyl chitosan and its potential application for peripheral nerve regeneration. *Eur Polym J* 43:3807–3818.

63. Jiang, H., Wang, Y., Huang, Q., Li, Y., Xu, C., Zhu, K., Chen, W. 2005. Biodegradable hyaluronic acid/N-carboxyethyl chitosan/protein ternary complexes as implantable carriers for controlled protein release. *Macromol Biosci* 5:1226–1233.

64. Badawy, M.E.I., Rabea, E.I., Rogge, T.M., Stevens, C.V., Smagghe, G., Steurbaut, W., Hofte, M. 2004. Synthesis and fungicidal activity of new N, O-acyl chitosan derivatives. *Biomacromolecules* 5:589–595.

65. Zhu, A., Shan, B., Yuan, Y., Shen, J. 2003. Preparation and blood compatibility of phosphorylcholine-bonded O-butyrylchitosan. *Polym Int* 52:81–85.

66. Ueno, H., Yamada, H., Tanaka, I., Kaba, N., Matsuura, M., Okumura, M., Kadosawa, T., Fujinaga, T. 1999. Accelerating e!ects of chitosan for healing at early phase of experimental open wound in dogs. *Biomaterials* 20:1407–1414.

67. Ishiharaa, M., Nakanishi, K., Ono, K., Sato, M., Kikuchi, M., Saito,Y., Yura, H., Matsuia, T., Hattoria, H., Uenoyama, M., Kurita, A. 2002. Photocrosslinkable chitosan as a dressing for wound occlusion and accelerator in healing process. *Biomaterials* 23:833–840.

68. Alemdaroglu, C., Degim, Z., Celebi, N.C., Zor, F., Ozturk, S., Erdogan, D. 2006. An investigation on burn wound healing in rats with chitosan gel formulation containing epidermal growth factor. *Burns* 32:319–327.

69. Boudet, C., Lliopoulos, L., Poncelet, O., Cloitre, M. 2005. Control of the chemical cross-linking of gelatin by a thermosensitive polymer: example of switchable reactivity. *Biomacromolecules* 6:3073–3078.

70. Broderick, E.P., O'Halloran, D.M., Rochev, Y.A., Griffin, M., Collighan, R.J., Pandit, A.S. 2005. Enzymatic stabilization of gelatin-based scaffolds. *J Biomed Mater Res Part B* 72B:37–42.

71. Hurley, P.A., Clarke, M., Crook, J.M., Wise, A.K., Shepherd, R.K. 2003. Cochlear immunochemistry: a new technique based on gelatin embedding. *J Neurosci* 129:81–86.

72. Gaspard, S., Oujja, M., Abrusci, C., Catalina, F., Lazare, S., Desvergne, J.P., Castillejo, M. 2008. Laser induced foaming and chemical modifications of gelatine films. *J Photochem Photobiol A* 193:187–192.

73. Liang, H.C., Chang, W.H., Liang, H.F., Lee, M.H., Sung, H.W. 2004.Crosslinking structures of gelatin hydrogels crosslinked with genipin or a water-soluble carbodiimide. *J Appl Polym Sci* 91: 4017–4026.

74. Senthilkumar, K.S., Saravanan, K.S., Chandra, G., Sindhu, K.M., Jayakrishnan, A., Mohanakumar, K.P. 2007. Unilateral implantation of dopamine-loaded biodegradable hydrogel in the striatum attenuates motor abnormalities in the 6-hydroxydopamine model of hemi-parkinsonism. *Behav Brain Res* 184:11–18.

75. Balakrishnan, B., Jayakrishnan, A. 2005. Self-crosslinkable biopolymers as injectable in situ forming biodegradable scaffolds. *Biomaterials* 26: 3941–3951.

76. Balakrishnan, B., Jayakrishnan, A. 2005. Oxidized chondroitin sulfate-crosslinked gelatin matrixes: a new class of hydrogels. *Biomaterials* 6: 2040–2048.

77. Hong, S.R., Chong, M.S., Lee, S.B., Lee, Y.M., Song, K.W., Park, M.H., Hong, S.H. 2004. Biocompatibility and biodegradation of crosslinked gelatin/hyaluronic acid sponge in rat subcutaneous tissue. *J Biomater Sci: Polym Ed* 15: 201–214.

78. Patel, Z.S., Yamamoto, M., Ueda, H., Tabata, Y., Mikos, A.G. 2008. Biodegradable gelatin microparticles as delivery systems for the controlled release of bone morphogenetic protein-2. *Acta Biomater* 4:1126–1138.

79. Huang, Y., Onyeri, S., Siewe, M., Moshfeghian, A., Madihally, S.V. 2005. In vitro characterization of chitosan–gelatin scaffolds for tissue engineering. *Biomaterials* 26:7616–7627.

80. Yamamoto, M., Takahashi, Y., Tabata, Y. 2003. Controlled release by biodegradable hydrogels enhances the ectopic bone formation of bone morphogenetic protein. *Biomaterials* 24:4375–4383.

81. van Luyn, M.J.A., van Wachem, P.B., Dijkstra, P.J., Olde Damink, L.H.H., Feijen, J. 1995. Calcification of subcutaneously implanted collagen in relation to cytotoxicity, cellular interactions and crosslinking. *J Mater Sci Mater Med* 11:169–175.

82. Griffin, M., Casadio, R., Bergamini, C.M. 2002. Transglutaminases: Nature's biological glues. *Biochem J* 368:377–396.

83. Crescenzi, V., Francescangeli, A., Taglienti, A. 2005. New gelatin-based hydrogels via enzymatic networking. *J Biomed Mater Res Part B*: 72B:37–42.

84. Gu, Y.S., Matsumura, Y., Yamaguchi, S., Mori, T. 2001. Action of proteinglutaminase on alpha-lactalbumin in the native and molten globule states. *J Agric Food Chem* 49:5999–6005.

85. Kang, Y.N., Kim, H., Shin, W.S., Woo, G., Moon, T.W. 2003. Effect of disulfide bond reduction on bovine serum albumin-stabilized emulsion gel formed by microbial transglutaminase. *J Food Sci* 68:2215–2220.

86. Deng, C.M., He, L.Z., Zhao, M., Yang, D., Liu, Y. 2007. Biological properties of the chitosan-gelatin sponge wound dressing. *Carbohydr Polym* 69: 583–589.

87. Wang, T.W., Sun, J.S., Wu, H.C., Tsuang, Y.H., Wang, W., Lin, F.H. 2006. The effect of gelatin–chondroitin sulfate–hyaluronic acid skin substitute on wound healing in SCID mice. *Biomaterials* 27:5689–5697.

88. Balakrishnan, B., Mohanty, M., Fernandez, A.C., Mohananc, P.V., Jayakrishnan, A. 2006. Evaluation of the effect of incorporation of dibutyryl cyclic adenosine monophosphate in an *in situ*-forming hydrogel wound dressing based on oxidized alginate and gelatin. *Biomaterials* 27:1355–1361.

89. Ulubayram, K.A., Cakar, N., Korkusuz, P., Ertan, C., Hasirci, N. 2001. EGF containing gelatin-based wound dressings. *Biomaterials* 22:1345–1356.

90. Weng, L., Romanov, A., Rooney, J., Chen, W. 2008. Non-cytotoxic, in situ gelable hydrogels composed of N-carboxyethyl chitosan and oxidized dextran. *Biomaterials* 29, 3905–3913.

91. Yin, L., Fei, L., Cui, F., Tang, C., Yin, C. 2007. Superporous hydrogels containing poly(acrylic acid-co-acrylamide)/*O*-carboxymethyl chitosan interpenetrating polymer networks. *Biomaterials* 28:1258–1266.

92. Roughley, P., Hoemann, C., DesRosiers, E., Mwale, F., Antoniou, J., Alini, M. 2006. The potential of chitosan-based gels containing intervertebral disc cells for nucleus pulposus supplementation. *Biomaterials* 27:388–396.

93. Chen, L., Tian, Z., Du, Y. 2004. Synthesis and pH sensitivity of carboxymethyl chitosan-based polyampholyte hydrogels for protein carrier matrices. *Biomaterials* 25:3725–3732.

94. Crompton, K.E., Prankerd, R.J., Paganin, D.M., Scott, T.F., Horne, M.K., Finkelstein, D.I., Gross, K.A., Forsythe, J.S. 2005. Morphology and gelation of thermosensitive chitosan hydrogels. *Biophys Chem* 117:47–53.

95. Draye, J., Delaey, B., Van de Voorde, A., Van Den Bulcke, A., Bogdanov, B., Schacht, E. 1998. In vitro release characteristic of bioactive molecules from dextran dialdehyde crosslinked gelatin hydrogels films. *Biomaterials*19:99–107.

96. Weng, L., Chen, X., Chen, W. 2007. Rheological properties of in situ crosslinkable hydrogels from dextran and chitosan derivatives. *Biomacromolecules* 8:1109–1115.

97. Winter, H.H., Chambon, F. 1986. Analysis of linear viscoelasticity of a crosslinking polymer at the gel point. *J Rheol* 30:367–382.

98. Lee, K.Y., Bouhadir, K.H., Mooney, D.J. 2000. Degradation behavior of covalent crosslinked poly(aldehyde guluronate) hydrogels. *Macromolecules* 33:97–101.

99. Weng, L., Pan, H., Chen, W. 2008. Self-crosslinkable hydrogels composed of partially oxidized hyaluronan and gelatin: in vitro and in vivo responses. *J Biomed Mater Res* 85A: 352–365

100. Tamada, Y., Ikada, Y. 2004. Fibroblast growth on polymer surface and biosynthesis of collagen. *J Biomed Mater Res* 28:783–789.

101. Ohno, T., Yoo, M.J., Swanson, E.R., Hirano, S., Ossoff, R.H., Rousseau, B. 2009. Regenerative effects of basic fibroblast growth factor on extracellular matrix production in aged rat vocal folds. *Ann Otol Rhinol Laryngol* 118:559–564.

102. Mo, X.M., Xu, C.Y., Kotaki, M., Ramakrishna, S. 2004. Electrospun P(LLA-CL) nanofiber: a biomimetic extracellular matrix for smooth muscle cell and endothelial cell proliferation. *Biomaterials* 25:1883–1890.

103. Liu, X.F., Guan, Y.L., Yang, D.Z., Li, Z., Yao, K.D. 2001. Antibacterial action of chitosan and carboxymethylated chitosan. *J Appl Polym Sci* 79:1324–1335.

104. Ueno, H., Mori, T., Fujinaga, T. 2001. Topical formulations and wound healing applications of chitosan. *Adv Drug Deliv Rev* 52:105–115.

105. Berger, J., Reist, M., Mayer, J.M., Felt, O., Gurny, R. 2004. Structure and interactions in chitosan hydrogels formed by complexation or aggregation for biomedical applications. *Eur J Pharm Biopharm* 57:35–52.

106. Shu, X.Z., Ghosh, K., Liu, Y., Palumbo, F.S., Luo, Y., Clark, R.A., Prestwich, G.D. 2004. Attachment and spreading of fibroblasts on an RGD peptide–modified injectable hyaluronan hydrogel. *J Biomed Mater Res A* 68:365–375.

107. Bouhadir, K.H., Hausman, D.S., Mooney, D.J. 1999. Synthesis of cross-linked poly(aldehyde guluronate) hydrogels. *Polymer* 40:3575–3584.

108. Guan, Y.L., Shao, L., Yao, K.D. 1996. A study on correlation between water state and swelling kinetics of chitosan-based hydrogels. *J Appl Polym Sci* 61:2325–2335.

109. Strauss, G., Kral, H. 1982. Borate complexes of amphotericin-B: polymeric species and aggregates in aqueous solutions. *Biopolymers* 21:459–470.

110. Suzuki, Y., Nishimura, Y., Tanihara, M., Suzuki, K., Nakamura, T., Shimizu, Y., Yamawaki, Y., Kakimaru, Y. 1998. Evaluation of a novel alginate gel dressing: cytotoxicity to fibroblasts in vitro and foreign-body reaction in pig skin in vivo. *J Biomed Mater Res* 39:317–322.

111. Aivd, B., Bogdanov, B., Rooze, N.D., Schacht, E.H., Cornelissen, M., Berghmans, H. 2000. Structural and rheological properties of methacrylamide modified gelatin hydrogels. *Biomacromolecules* 1:31–38.

112. Rudraraju, V.S., Wyandt, C.M. 2005. Rheology of microcrystalline cellulose and sodium carboxymethyl cellulose hydrogels using a controlled stress rheometer: part II. *Inter J Pharm* 292:63–73.

113. Santoveña, A., Álvarez-Lorenzo, C., Concheiro, A., Llabrés, M., Fariña, J.B. 2004. Rheological properties of PLGA film-based implants: correlation with polymer degradation and SPf66 antimalaric synthetic peptide. *Biomaterials* 25:925–931.

114. Santiago, L.Y., Nowak, R.W., Rubin, J.P., Marra, K.G. 2006. Peptide-surface modification of poly(caprolactone) with laminin-derived sequences for adipose-derived stem cell applications. *Biomaterials* 27:2962–2969.

115. Stabenfeldt, S.E., Garcia, A.J., Laplaca, M.C. 2006. Thermoreversible laminin-functionalized hydrogel for neural tissue engineering. *J Biomed Mater Res A* 77: 718–725.

116. Seo, S.J., Choi, Y.J., Akaike, T., Higuchi, A., Cho, C.S. 2006. Alginate/Galactosylated chitosan/heparin scaffold as a new synthetic extracellular matrix for hepatocytes. *Tissue Eng* 12: 33–44.

117. Hwang, N.S., Kim, M.S., Sampattavanich, S., Baek, J.H., Zhang, Z., Elisseeff, J. 2006. Effects of three-dimensional culture and growth factors on the chondrogenic differentiation of murine embryonic stem cells. *Stem Cells* 24: 284–291.

118. Sontjens, S.H., Nettles, D.L., Carnahan, M.A., Setton, L.A., Grinstaff, M.W. 2006. Biodendrimer-based hydrogel scaffolds for cartilage tissue repair. *Biomacromolecules* 7: 310–316.

119. DeLong, S.A., Gobin, A.S., West, J.L. 2005. Covalent immobilization of RGDS on hydrogel surfaces to direct cell alignment and migration. *J Control Release* 109:139–148.

120. Stevens, M.M., Mayer, M., Anderson, D.G., Weibel, D.B., Whitesides, G.M., Langer, R. 2005. Direct patterning of mammalian cells onto porous tissue engineering substrates using agarose stamps. *Biomaterials* 26: 7636–7641.

121. Mironov, V., Kasyanov, V., Shu, X.Z., Eisenberg, C., Eisenberg, L., Gonda, S., Trusk, T., Markwald, R.R., Prestwich, G.D. 2005. Fabrication of tubular tissue constructs by centrifugal casting of cells suspended in an in situ crosslinkable hyaluronan-gelatin hydrogel. *Biomaterials* 26: 7628–7635.

122. Mahoney, M.J., Anseth, K.S. 2006. Three-dimensional growth and function of neural tissue in degradable polyethylene glycol hydrogels. *Biomaterials* 27:2265–2274.

123. Alberts, B., Johnson, A., Lewis, J., Raff, M., Roberts, K., Walter, P. 2002. Cell junctions, cell adhesion, and the extracellular matrix. In: *Molecular Biology of the Cell* (4th edn.), Chapter 19, Garland Publishing, New York.

124. Tian, W.M., Hou, S.P., Ma, J., Zhang, C.L., Xu, Q.Y., Lee, I.S., Li, H.D., Spector, M., Cui, F.Z. 2005. Hyaluronic acid-poly-D-lysine-based three-dimensional hydrogel for traumatic brain injury. *Tissue Eng* 11:513–525.

125. Clark, R.A.F. 1996. Wound repair: overview and general considerations. In: R.A.F. Clark, Ed., *The Molecular and Cellular Biology of Wound Repair* (2nd edn.), Plenum, New York, pp. 3–35.

126. Silva, E.A., Mooney, D.J. 2004. Synthetic extracellular matrices for tissue engineering and regeneration. *Curr Top Dev Biol* 64:181–205.

127. Young, J.J., Cheng, K.M., Tsou, T.L., Liu, H.W., Wang, H.J. 2004. Preparation of cross-linked hyaluronic acid film using 2-chloro-1-methylpyridinium iodide or water-soluble 1- ethyl - (3, 3- dimethylaminopropyl) carbodiimide. *J Biomater Sci Polym Ed* 15:767–780.

128. Danielsson, C., Ruault, S., Basset-Dardare, A., Frey, P. 2006. Modified collagen fleece, a scaffold for transplantation of human bladder smooth muscle cells. *Biomaterials* 27:1054–1060.

129. Ratner, B.D., Hoffman, A.S., Schoen, F.J., Lemons, J.E. 2004. *Biomaterials Science: An Introduction to Materials in Medicine* (2nd edn.), Elsevier, Amsterdam, the Netherlands.

130. Vural, B., Cantürk, N.Z., Esen, N., Solakoglu, S., Cantürk, Z., Kirkali, G., Sökmensüer, C. 1999. *Hum Reprod* 14:49–54.

131. Boland, G.M., Weigel, R.J. 2006. Formation and prevention of postoperative abdominal adhesions. *J Surg Res* 132:3–12.

132. Becker, J.M., Dayton, M.T., Fazio, V.W. et al. 1996. Prevention of postoperative abdominal adhesions by a sodium hyaluronate-based bioresorbable membrane: a prospective, randomized, double-blind multicenter study. *J Am Coll Surg* 183:297–306.

133. Gago, L.A., Saed, G.M., Chauhan, S. et al. 2003. Seprafilm (modified hyaluronic acid and carboxymethylcellulose) acts as a physical barrier. *Fertil Steril* 80:612–616.

134. Liakakos, T., Thomakos, N., Fine, P.M. et al. 2001. Peritoneal adhesions: etiology, pathophysiology, and clinical significance. Recent advances in prevention and management. *Dig Surg* 18:260–273.

135. Liu, Y., Shu, X.Z., Prestwich, G.D. 2007. Reduced postoperative intraabdominal adhesions using Carbylan-SX, a semisynthetic glycosaminoglycan hydrogel. *Fertil Steril* 87:940–948.

136. Lundorff, P., van Geldorp, H., Tronstad, S.E. et al. 2001. Reduction of postsurgical adhesions with ferric hyaluronate gel: a European study. *Hum Reprod* 16:1982–1988.

137. Falabella, C.A., Melendez, M.M., Weng, L., Chen, W. 2009. Novel macromolecular crosslinking hydrogel to reduce intra-abdominal adhesions, *J. Surg. Res.*, 159:772–778.

138. DiZerega, G.C.J.D. 2001. Peritoneal repair and post-surgical adhesion formation. *Hum Reprod Update* 7:547–555.

139. Harris, E.S., Morgan, R.F., Rodeheaver, G.T. 1995. Analysis of the kinetics of peritoneal adhesion formation in the rat and evaluation of potential antiadhesive agents. *Surgery* 117:663–669.

140. Arnold, P.B., Green, C.W., Foresman, P.A. et al. 2000. Evaluation of resorbable barriers for preventing surgical adhesions. *Fertil Steril* 73:157–161.

141. Dunn, R., Lyman, M.D., Edelman, P.G. et al. 2001. Evaluation of the SprayGel adhesion barrier in the rat cecum abrasion and rabbit uterine horn adhesion models. *Fertil Steril* 75:411–416.

142. Cadee, J.A., van Luyn, M.J., Brouwer, L.A. et al. 2000. In vivo biocompatibility of dextran-based hydrogels. *J Biomed Mater Res* 50:397–404.

143. Numanoglu, V., Cihan, A., Salman, B. et al. 2007. Comparison between powdered gloves, powder-free gloves and hyaluronate/carboxymethylcellulose membrane on adhesion formation in a rat caecal serosal abrasion model. *Asian J Surg* 30:96–101.

144. Risberg, B. 1997. Adhesions: preventive strategies. *Eur J Surg Suppl* 577:32–39.

145. Anderson, J.M. 2004. Inflammation, wound healing, and the foreign body response. In: Ratner, B.D., Hoffman, A.S., Schoen, F.J., Lemons, J.E., Eds., *Biomaterials Science: An Introduction to Materials in Medicine*, Elsevier Academic Press, San Diego, CA, p. 296.

146. Chuang, Y.C., Fan, C.N., Cho, F.N. et al. 2008. A novel technique to apply a Seprafilm (hyaluronate-carboxymethylcellulose) barrier following laparoscopic surgeries. *Fertil Steril* 90:1959–1963.

147. Weng, L., Ivanova, N.D., Zakhaleva, J., Chen, W. 2008. In vitro and in vivo suppression of cellular activity by guanidinoethyl disulfide released from hydrogel microspheres composed of partially oxidized hyaluronan and gelatin. *Biomaterials* 29:4149–4156.

148. Suzuki, E., Umezawa, K. 2006. Inhibition of macrophage activation and phagocytosis by a novel NF-κB inhibitor, dehydroxymethylepoxyquinomicin. *Biomed Pharmacother* 60:578–586.

149. Amy, C., Gao, Q., John, K.W. 2007. Macrophage matrix metalloproteinase-2/-9 gene and protein expression following adhesion to ECM-derived multifunctional matrices via integrin complexation. *Biomaterials* 28:285–298.

150. Maresz, K., Ponomarev, E.D., Barteneva, N., Tan, Y., Manna, M.K., Dittel, B.N. 2008. IL-13 induces the expression of the alternative activation marker Ym1 in a subset of testicular macrophages. *J Reprod Immunol* 78:140–148.

151. Coleman, J.W. 2001. Nitric oxide in immunity and inflammation. *Inter Immunopharmacol* 1:1397–1406.

152. Cantoni, O., Palomba, L., Guidarelli, A., Tommasini, I., Cerioni, L., Sestili, P. 2002. Cell signaling and cytotoxicity by peroxynitrite. *Environ Health Perspect* 110:823–825.

153. Lee, R.H., Efron, D., Tantry, U., Barbul, A. 2001. Nitric oxide in the healing wound: a time-course study. *J Surg Res* 101:104–108.

154. Schwentker, A., Vodovotz, Y., Weller, R., Billiar, T.R. 2002. Nitric oxide and wound repair: role of cytokines. *Nitric Oxide* 7:1–10.

155. Kibbe, M., Billiar, T., Tzeng, E. 1999. Inducible nitric oxide synthase and vascular injury. *Cardiovasc Res* 43:650–657.

156. Isenberg, C., Kolmann, H.L. 2005. Neurocutaneous angiomatosis: manifestation with cystic tumors. *Nervenarzt* 76:202–204.

157. Mabley, J.G., Pacher, P., Bai, P., Wallace, R., Goonesekera, S., Virag, L. et al. 2004. Suppression of intestinal polyposis in Apcmin/+ mice by targeting the nitric oxide or poly(ADP-ribose) pathways. *Mut Res* 548:107–116.

158. Suarez-Pinzon, W.L., Mabley, J.G., Strynadka, K., Power, R.F., Szabo, C., Rabinovitch, A. 2001. An inhibitor of inducible nitric oxide synthase and scavenger of peroxynitrite prevents diabetes development in NOD mice. *J Autoimmunity* 16:449–455.

159. Paquette, D.W., Rosenberg, A., Lohinai, Z., Southan, G.J., Williams, R.C., Offenbacher, S. et al. 2006. Inhibition of experimental gingivitis in beagle dogs with topical mercatoalkylguanidines. *J Periodontol* 77:385–391.

160. Wong, H.L., Wang, M.X., Cheung, P.T., Yao, K.M., Chan, B.P. 2007. A 3D collagen microsphere culture system for GDNF-secreting HEK293 cells with enhanced protein productivity. *Biomaterials* 28:5369–5380.

161. Yun, Y.H., Goetz, D.J., Yellen, P., Chen, W. 2004. Hyaluronan microspheres for sustained gene delivery and site-specific targeting. *Biomaterials* 25:147–157.

162. Agnihotri, S.A., Aminabhavi, T.M. 2006. Novel interpenetrating network chitosan-poly(ethylene oxide-g-acrylamide) hydrogel microspheres for the controlled release of capecitabine. *Int J Pharm* 324:103–115.

163. Cortesi, R., Esposito, E., Osti, M., Squarzoni, G., Menegatti, E., Davis, S.S. et al. 1999. Dextran cross-linked gelatin microspheres as a drug delivery system. *Eur J Pharm Biopharm* 47:153–160.

164. Dinarvand, R., Rahmania, E., Farboda, E. 2003. Gelatin microspheres for the controlled release of all-trans-retinoic acid topical formulation and drug delivery evaluation. *Iranian J Pharm Res* 2:47–50.

165. Patil, S.D., Papadmitrakopoulos, F., Burgess, D.J. 2007. Concurrent delivery of dexamethasone and VEGF for localized inflammation control and angiogenesis. *J Control Release* 117:68–79.

166. O'Donnell, P.B., McGinity, J.W. 1997. Preparation of microspheres by the solvent evaporation technique. *Adv Drug Deliv Rev* 28:25–42.
167. Bernatchez, S.F., Parks, P.J., Gibbons, D.F. 1996. Interaction of macrophages with fibrous materials *in vitro*. *Biomaterials* 17:2077–2086.
168. Esposito, A., Sannino, A., Cozzolino, A., Quintiliano, S.N., Lamberti, M., Ambrosio, L. et al. 2005. Response of intestinal cells and macrophages to an orally administered cellulose-PEG based polymer as a potential treatment for intractable edemas. *Biomaterials* 26:4101–4110.
169. Gamal-Eldeen, A.M., Amer, H., Helmy, W.A., Talaat, R.M., Ragab, H. 2007. Chemically-modified polysaccharide extract derived from *Leucaena leucocephala* alters Raw 264.7 murine macrophage functions. *Int Immunopharm* 7:871–888.
170. Bredt, D.S., Snyder, S.H. 1994. Nitric oxide: a physiologic messenger molecule. *Annu Rev Biochem* 63:175–195.
171. Vijayasekaran, S., Fitton, J.H., Hicks, C.R., Chirila, T.V., Crawford, G.J., Constable, I.J. 1998. Cell viability and inflammatory response in hydrogel sponges implanted in the rabbit cornea. *Biomaterials* 19:2255–2267.
172. Xa, Z., Triffitt, J.T. 2006. A review on macrophage responses to biomaterials. *Biomed Mater* 1:1–9.
173. Witte, M.B., Barbul, A. 1997. General principles of wound healing. *Surg Clin North Am* 77:509–528.
174. Enoch, S., Leaper, D.J. 2005. Basic science of wound healing. *Surgery* 23:37–42.

23 Dental and Maxillofacial Surgery Applications of Polymers

E.C. Combe

CONTENTS

23.1 INTRODUCTION

23.1.1 POLYMERS IN THE ORAL ENVIRONMENT

There is a long history of the application of polymers in dentistry. In the 1800s, before the advent of synthetic polymers, India-rubber and gutta-percha were employed; the latter material is still in use in endodontic therapy.

Given the current vast range of available synthetic polymers, and the ability to modify polymer properties—for example, by co-polymerization, cross-linking, incorporation of plasticizers and the inclusion of reinforcing constituents—it is not surprising that polymers are playing an essential major role in oral and maxillofacial treatments. To demonstrate the broad scope of the subject, Table 23.1 lists a wide range of types of polymer-containing materials, giving their principal dental and maxillofacial surgery application(s).

The fundamental question to be asked is this—how will a synthetic material behave in the oral and maxillofacial environment? In contrast to other parts of the body, materials in the mouth are exposed to saliva, bacteria, and components of drinks and foodstuffs. Extremes of temperature (0°C–90°C) and pH (gastric secretions have pH values from 1.0 to 3.5) can occur [1].

In addition to, or as well as biological effects, materials may suffer chemical or mechanical breakdown under the aggressive conditions of the oral environment. This includes corrosion, fatigue, fracture, and surface wear. Thus, there must be consideration for the safety, durability and efficacy of biomedical polymers. Figure 23.1 is a general illustration of the principal concerns about the placement of a synthetic material in a living environment.

Thus, there are stringent requirements that apply to polymers designed to be placed in the oral cavity. In certain cases there may be poor biocompatibility. Biocompatibility may be defined as "the ability of a material to perform with an appropriate host response in a specific situation" [2,3]. Problems of biocompatibility of polymers may occur from a number of sources, including the possibility of elution of unreacted monomers or co-monomers, polymerization initiators, stabilizers, contaminants, and products of chemical decomposition [4]. In addition to chemical considerations, surface features—for example, roughness, surface charge, porosity—may influence host response. Potentially, a material may be irritant, toxic, allergenic, teratogenic, mutagenic, or carcinogenic. Biocompatibility is a recurring theme of this review.

Before detailed consideration of each polymer application, an overview of polymers and their principal applications is given. This is desirable because classification of the subject is made difficult

TABLE 23.1

Classification and Principal Dental Applications of Polymers

Type of Polymer		Polymer Name	Application	Hardening Mechanism
Synthetic polymers	1. Condensation polymers	Polyamides—e.g., nylon Polycarbonates	Prosthodontics	Polymerization
	2. Free-radical addition, linear polymers	Poly(methyl methacrylate) — acrylic resins — PMMA	Prosthodontics	Polymerization
	3. Free-radical addition, cross-linked	Dimethacrylates	Resin composites, sealants and cements	Polymerization
	4. Flexible polymers	Plasticized acrylics	Denture soft linings	Polymerization
		Elastomers:	Impressions	Polymerization
		Polysulfides	Soft linings	
		Siloxanes	Mouth protectors	
		Polyether	Maxillofacial polymers	
	5. Water soluble polymers	Aqueous poly(acrylic acid) and related co-polymers	Adhesive dental cements	Acid-base reaction
Naturally occurring polymers	6. Polyisoprenes	Latex	Orthodontic elastics	
		Gutta percha	Endodontics	
	7. Polysaccharides	Alginates	Impressions	Gel formation
	8. Polysaccharide	Hyaluronic acid	Soft tissue augmentation	
	9. Protein	Collagen	Soft tissue augmentation	

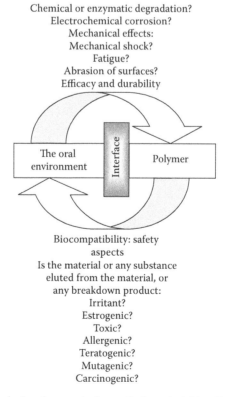

Chemical or enzymatic degradation?
Electrochemical corrosion?
Mechanical effects:
Mechanical shock?
Fatigue?
Abrasion of surfaces?
Efficacy and durability

The oral environment — Interface — Polymer

Biocompatibility: safety
aspects
Is the material or any substance
eluted from the material, or
any breakdown product:
Irritant?
Estrogenic?
Toxic?
Allergenic?
Teratogenic?
Mutagenic?
Carcinogenic?

FIGURE 23.1   Considerations in the placement of a synthetic material in a living environment.

by the fact that a given polymer type may have multiple biomedical applications, and a given application may have the choice of a number of materials. In particular, there are some dental procedures (e.g., placement of sealants, orthodontic bonding, restoring teeth materials and adhesives), where there is "competition" between two fundamentally different types of polymer-containing material—that is, those based on the free radical polymerization of dimethacrylates and those containing aqueous solutions of acidic synthetic polyelectrolytes that react with basic powders.

In addition to being structural materials (e.g., components of tooth restoration and prosthetic materials), polymers also make a significant contribution to cements and adhesives. A review of three adhesive mechanisms is given, all of which utilize polymers, and are widely applicable in contemporary dental practice.

23.1.2 OVERVIEW OF APPLICATIONS

The extensive range of applications of dental and maxillofacial applications of polymers is listed in a logical sequence, in general working from the least invasive procedures to the most invasive. Thus, the prevention of disease and trauma is considered, followed by applications such as orthodontics and esthetic treatments, then with the repair of diseased teeth, followed by the replacement of teeth and then maxillofacial surgery.

Dental and maxillofacial applications of polymers and polymer-containing materials include

- The prevention of decay—pit and fissure sealants for teeth.
- The prevention of the effects of trauma—resilient polymers can be used for mouth guards for sports players.
- Orthodontic treatment—polymer-containing cements for bonding brackets and bands to teeth can be employed. In some cases, orthodontic appliances will include polymeric elastics.
- Esthetic treatment—the cementation of translucent veneers on to tooth enamel.
- The direct restoration of teeth.
- Endodontic therapy.
- The recording of impressions, for the preparation of inlays, crowns and fixed partial dentures and removable complete dentures.
- Indirect restorative materials, that is, those that are fabricated outside of the mouth and cemented into place.
- Provisional (temporary) crown and bridge materials.
- Complete dentures.
- Denture teeth.
- Resilient polymers as linings for dentures.
- Periodontology—guided tissue regeneration.
- Soft tissue augmentation.
- Maxillofacial prosthetic materials.
- Obturators.
- Wound closure materials (sutures and tissue adhesives).

One critically important factor to consider is that in many cases bonded interfaces are required, thus adhesion to tooth enamel and dentin, and to ceramics and alloys is essential for success. There exists a wide range of bonding agents, many of which are polymer-based.

23.2 OVERVIEW OF MECHANISMS AND MATERIALS

In classifying the synthetic polymers, it is important to distinguish between two categories of polymer-containing material, which are tooth-colored and which harden directly in the mouth. Some materials harden by a polymerization reaction (e.g., number 3 in Table 23.1), and in others an acidic polyelectrolyte in aqueous solution reacts with a basic or amphoteric powder to form a set material (number 5 in Table 23.1). As stated earlier, in a number of applications there is a choice between examples of these general material types; where relevant the different types of material will be compared and contrasted.

23.2.1 Free Radical Addition Polymerization*

Many dental biomaterials set by a mechanism of free radical addition polymerization (though some contemporary materials employ a cationic mechanism). Three different mechanisms of activation/initiation are employed—namely, thermal, photochemical and redox methods.

23.2.1.1 Activation/Initiation Systems

23.2.1.1.1 Thermal Activation

Some materials, such as denture materials based on poly(methyl methacrylate)—see Section 23.10—contain benzoyl peroxide. On heating, this compound forms free radicals by the homolytic decomposition of the weak –O–O– bond.

23.2.1.1.2 Photochemical Activation

Photoactivated initiators can be employed. Frequently the diketone camphorquinone and an amine are used [5], and free radicals are generated in the presence of light of wavelength approximately 470 nm (Figure 23.2). A more recent alternative diketone that is sometimes employed is 1-phenyl-1,2-propanedione [6].

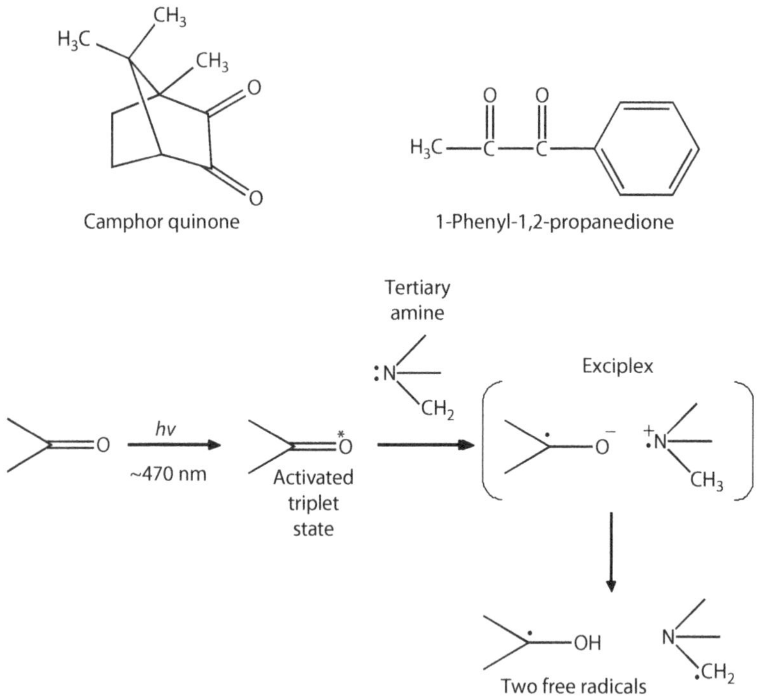

Camphor quinone

1-Phenyl-1,2-propanedione

FIGURE 23.2 Photochemical activation.

* Condensation polymerization reactions are also used (Section 23.8.3), but not for tooth-colored materials that harden in the mouth.

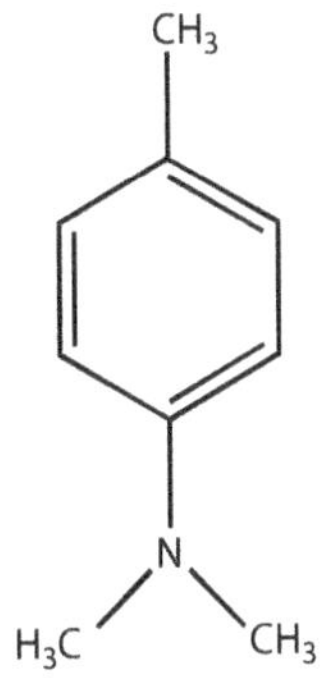

FIGURE 23.3 *N,N*-dimethyl-*p*-toluidine.

23.2.1.1.3 Redox Activation

Alternatively, a redox system be employed—for example, when free radicals to initiate the polymerization are formed by the reaction of benzoyl peroxide with a tertiary amine such as *N,N*-dimethyl-*p*-toluidine (Figure 23.3). These systems are sometimes referred to as "self-curing," "autopolymerizing," or "chemical curing."

23.2.1.1.4 Dual Cure

In some instances, dual cure materials are available—that is, they have both a photochemical reaction and redox activation.

23.2.1.2 Methacrylates and Dimethacrylates

An extensive list of dentally used methacrylates and dimethacrylates monomers has been given by Van Landuyt et al. [7]. These monomers are classified as follows:

- Monofunctional acrylates and methacrylates—used in adhesive and prosthodontic applications (Section 23.10).
- Cross-linking monomers based on dimethacrylates (Figure 23.4)—widely used as the basis of sealants (Section 23.4.1), orthodontic adhesives (Section 23.5.1), restorative materials (Section 23.7.1) and temporary inlay, crown and bridge materials (Section 23.9). Of these, the most widely used are Bis-GMA, UDMA, and TEGDMA.
- Methacrylate monomers with functional group(s) to promote adhesion. The functional groups have been classified as (i) sulfonic acid (e.g., in Figure 23.5), (ii) phosphate (Figure 23.6), phosphonate (Figure 23.7), carboxyl (Figure 23.8), and alcohol (Figure 23.9) [7]. These are discussed in more detail in Section 23.3.2.

23.2.1.3 Factors Associated with Polymerization

23.2.1.3.1 Reaction Exotherm

The fact that free radical addition polymerization reactions are exothermic has a number of dental ramifications. First, a polymer-based restorative material should not generate enough hear to damage the sensitive pulp of the tooth [8] (Section 23.7.1). Second, the substantial temperature during the polymerization of methyl methacrylate has to be taken into account into the processing of acrylic dentures (Section 23.10.1) to avoid volatilization of unpolymerized monomer [9].

23.2.1.3.2 Polymerization Shrinkage

Polymerization reactions are usually accompanied by shrinkage. For example, the volumetric shrinkage during the polymerization of methyl methacrylate is 21%. This is an important factor in the fabrication of complete acrylic dentures (Section 23.10). Contraction is also critical in polymer-based

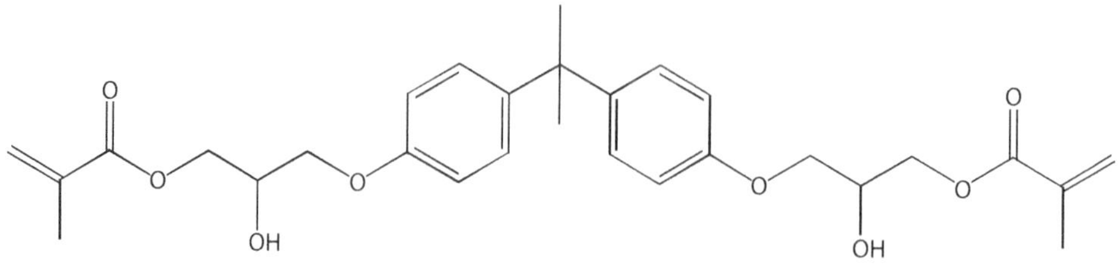

Bis-GMA (Bowen's resin): Bisphenol A diglycidyl methacrylate

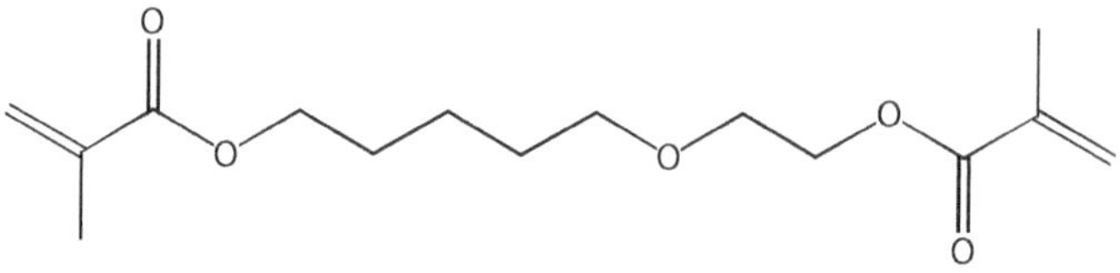

TEGDMA: triethylene glycol dimethacrylate

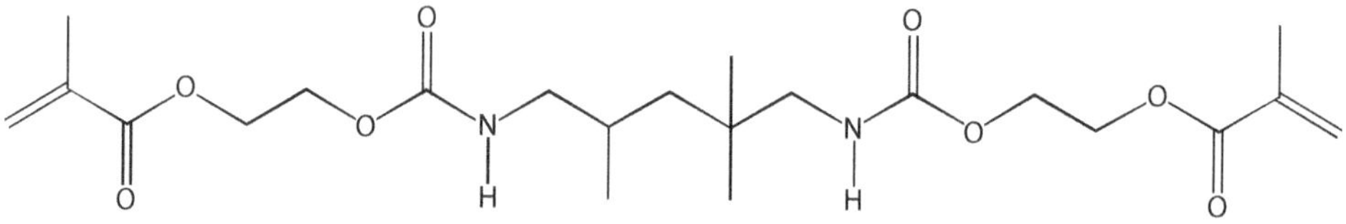

UDM: urethane dimethacrylate

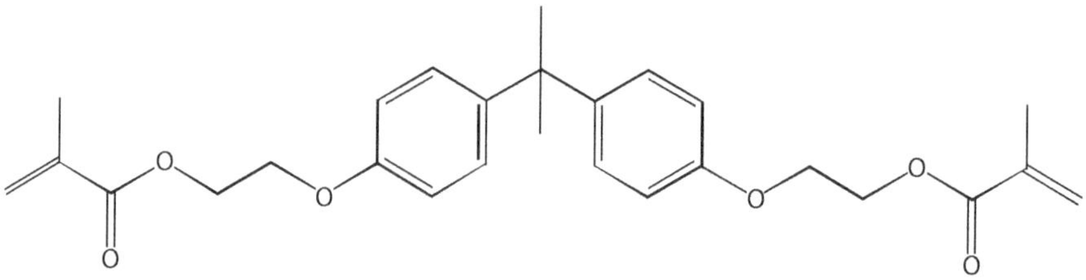

Bis EMA: ethoxylated bisphenol A glycol dimethacrylate

FIGURE 23.4   Dimethacrylate monomers.

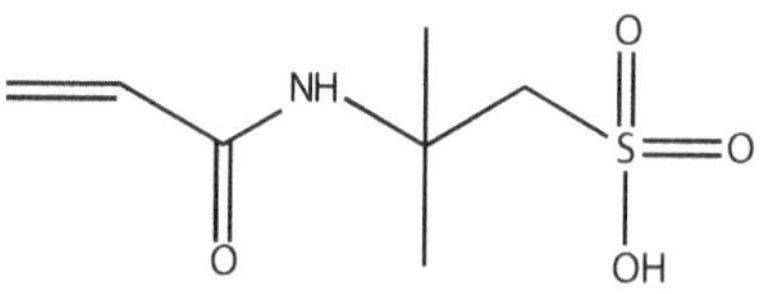

AMPS: 2-acrylamido-2-methyl-1
propanesulfonic acid

FIGURE 23.5   Methacrylate monomer with sulfonic acid functional group.

tooth restorative materials—the stress resulting from shrinkage during polymerization may compromise bonding of the material to tooth substance [10] (Section 23.7.1.1.4). As will be discussed later, there is intensive research to overcome this concern.

23.2.1.3.3   Degree of Conversion of Monomer to Polymer

It needs to be recognized that not all double bonds react in free radical addition polymerization. In the case of the polymerization of methyl methacrylate, there is always a certain amount of residual

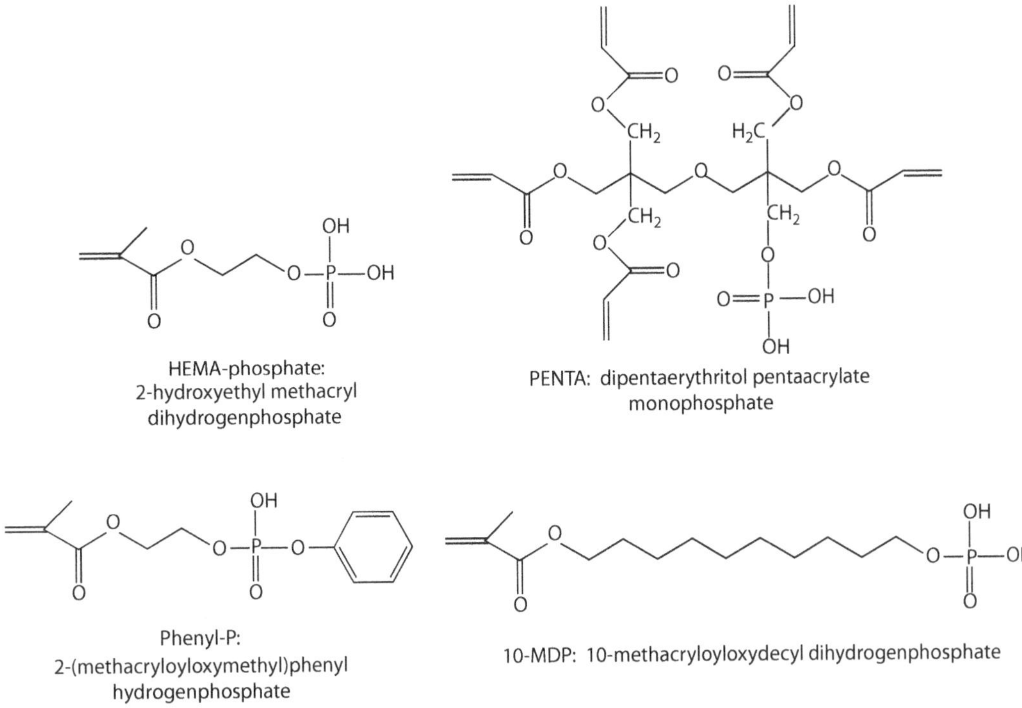

FIGURE 23.6 Monomers with phosphate groups.

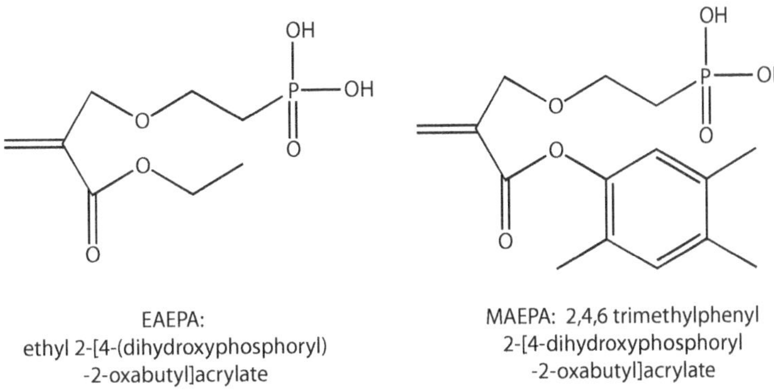

FIGURE 23.7 Monomers with phosphonate groups.

monomer. This may have deleterious biological consequences. In the case of dimethacrylates, the degree of conversion of –C=C– bonds may influence biocompatibility, mechanical properties and durability of the material [11].

23.2.1.4 Inhibition of Polymerization

23.2.1.4.1 Oxygen

Free radical addition polymerization is inhibited in the presence of oxygen [12]. As shown in Figure 23.10, oxygen can form an adduct with a free radical; this will slow the rate of polymerization. This may have implications for the clinical placement of fissure sealants (Section 23.4.1) and polymer-containing tooth restorative materials (Section 23.7.1).

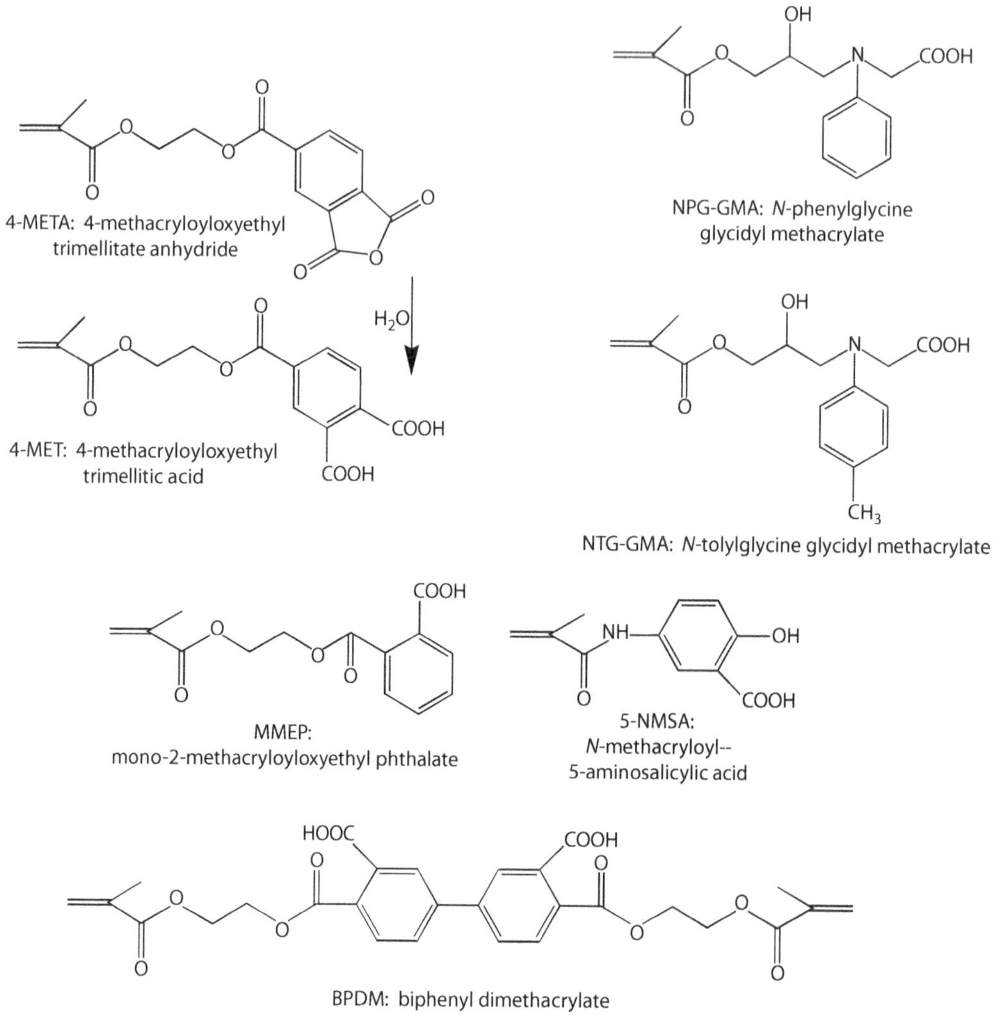

4-META: 4-methacryloyloxyethyl
trimellitate anhydride

H_2O

4-MET: 4-methacryloyloxyethyl
trimellitic acid

NPG-GMA: *N*-phenylglycine
glycidyl methacrylate

NTG-GMA: *N*-tolylglycine glycidyl methacrylate

MMEP:
mono-2-methacryloyloxyethyl phthalate

5-NMSA:
N-methacryloyl--
5-aminosalicylic acid

BPDM: biphenyl dimethacrylate

FIGURE 23.8 Monomers with carboxyl groups.

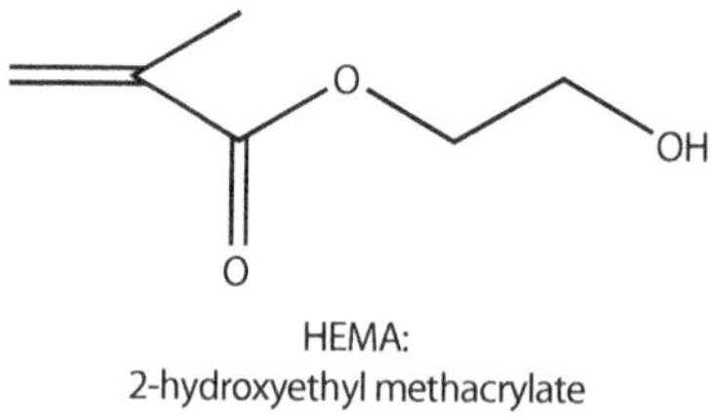

HEMA:
2-hydroxyethyl methacrylate

FIGURE 23.9 Monomer with an –OH group.

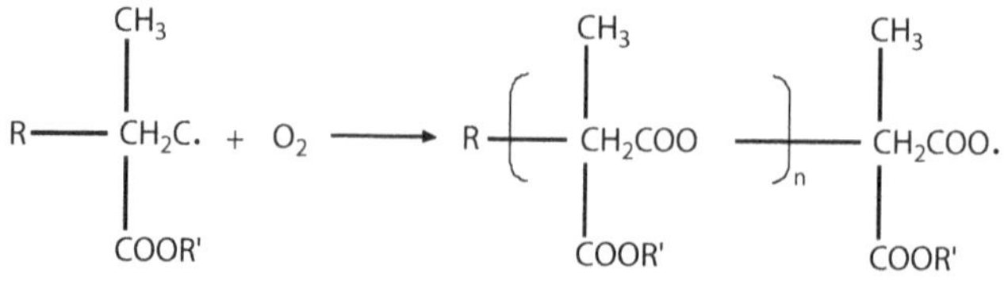

FIGURE 23.10 Oxygen inhibition of polymerization.

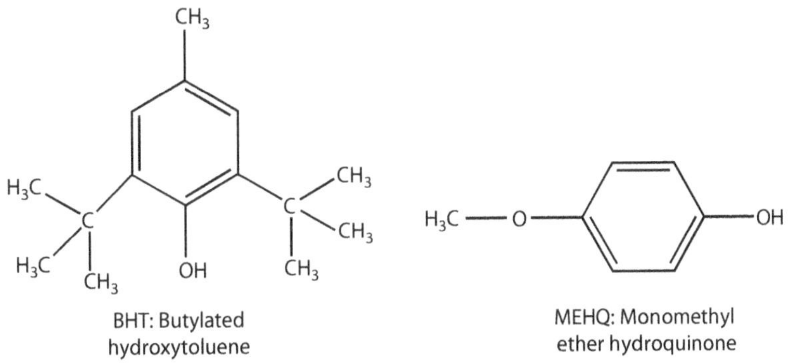

FIGURE 23.11   Inhibitors of polymerization.

23.2.1.4.2 *Eugenol*
3-(4-hydroxy-3-methoxyphenyl)-1-propene (eugenol) is used as a constituent of some dental cements. Since this compound is a scavenger for free radicals, it is imperative that it should not come into contact with a material that is polymerizing by a free radical mechanism [13].

23.2.1.4.3 *Butylated Hydroxyl Toluene (BHT) and Monomethyl Ether Hydroquinone (MEHQ)*
Because of the tendency of monomers to polymerize under unfavorable storage conditions, minute quantities of free radical scavengers such as BHT or MEHQ (Figure 23.11) are added to dental materials to confer stability on storage [4]. When the polymerization is activated, there is an induction period during which the scavenger reacts with free radicals. When the scavenger is consumed, the polymerization will commence [14]. Given that these compounds can be eluted from resins, the biocompatibility implications need to be considered (Section 23.7.1) [4,15,16].

23.2.2 REACTION OF A BASIC POWDER WITH AN ACIDIC POLYELECTROLYTE IN AQUEOUS SOLUTION

Some dental biomaterials contain an aqueous solution of a synthetic polyelectrolyte acid, which is designed to undergo an acid–base reaction with an inorganic solid. For example, zinc oxide reacts with aqueous poly(acrylic acid)—PAA—to give zinc polycarboxylate. This was the first chemically adhesive dental adhesive cement [17].

Subsequently, glass-ionomer cements were developed using similar polyelectrolyte solutions, but with a powdered fluoro-aluminosilicate (FAS) glass [18]. Early materials used an acrylic acid homopolymer, but today, copolymers with itaconic or maleic acid are often employed (Figure 23.12). The general term for these polymers and copolymers is polyalkenoic acids, and glass-ionomers materials are sometimes referred to as polyalkenoates. Material properties are dependent on molecular weight (MW) of the polymer, higher MWs giving stronger material for a given concentration of the acid.

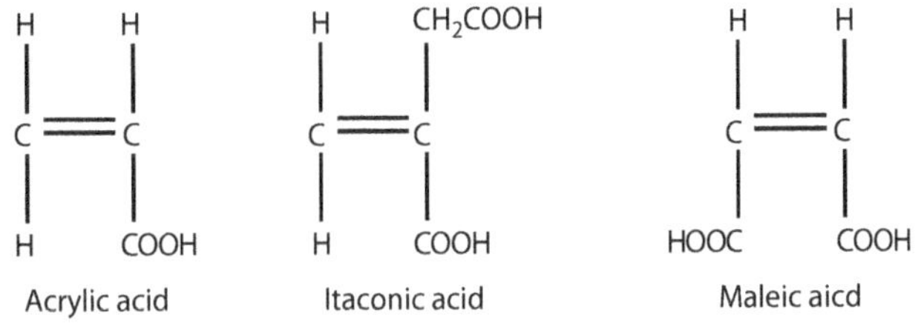

FIGURE 23.12   Component monomers of water-soluble polyelectrolytes used in glass-ionomer cements.

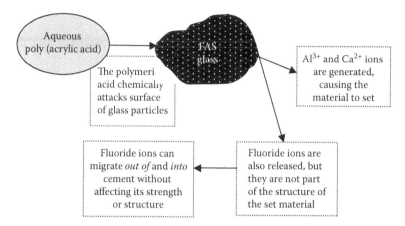

FIGURE 23.13 Setting and fluoride release of glass-ionomer cements.

However, there are limitations—viscosity of the solution increases with molecular weight for a given concentration. In practice, typical figures are a 47.5% solution of MW 23,000, given that if the viscosity of the cement liquid was too great, it would be impossible to mix with the FAS glass powder [19].

After mixing the polyalkenoate acid with FAS glass, a number of events occur, as depicted in Figure 23.13. The acid attacks the surfaces of the glass particles, resulting in extraction of ions from the glass. Three of these are of great significance. Ca^{2+} and Al^{3+} can crosslink the polyalkenoate anions, contributing to the development of strength and rigidity of the material. F^- ions are also released, but remain in the aqueous phase. They are not part of the structure of the material, and they can diffuse through the aqueous medium, without alteration in mechanical properties of the material. They can migrate both out of and into the cement; the clinical significance of this will be discussed later. It should be noted that the structure of the set material is composite (although dental terminology usually restricts the term "composite" to refer to resin composite materials—see the following text), since only the outer portion of the FAS glass particles react. The set material is composed of unreacted glass surrounded by a siliceous hydrogel, and held together by the cross-linked polyalkenoate chains (Figure 23.14) [20].

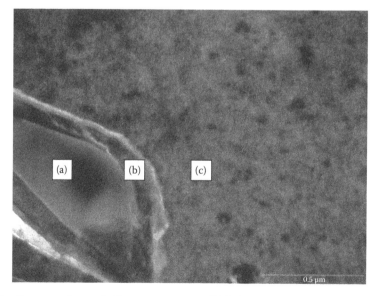

FIGURE 23.14 Structure of set polyalkenoate cement. a, FAS glass particle; b, siliceous hydrogel; c, matrix of reaction products. (Courtesy of J. Perdigão.)

23.2.3 MATERIAL TYPES

23.2.3.1 Cements and Luting Agents

A luting cement is defined as a material used to attach indirect restorations to prepared teeth [21]. There is a wide range of such materials, both for temporary and permanent luting (Table 23.2) [22,23], many of which are polymer-containing.

23.2.3.2 "Continuum"

Based on the earlier discussion, we can now give a preliminary discussion of four types of dental biomaterial, which are all capable of hardening in the mouth, and all of which are polymer-containing and tooth-colored. They all harden by free radical polymerization and/or an acid-base reaction. They are used for a host of dental applications, including orthodontics, pedodontics, restorative dentistry and prosthodontics, often with several materials being advocated for any one application.

The four types of material have been described as a "continuum" [24]. Because there is often a choice of material for a given application, it is essential that both the similarities and differences between these various materials be understood. Figure 23.15 shows the main types of material, and describes their principal characteristics. The general principles of these materials are discussed here, and in later sections of the review (e.g., tooth restorations) more details will be given, including detailed chemical composition, current concerns, and potential research developments.

The two original materials of the "continuum" are the resin composites and the glass-ionomers. These will be considered first. The other two materials were derived from these, as shown by the arrows on Figure 23.15.

TABLE 23.2
Dental Luting Cements

Material	Main Constituents as Supplied	Setting Mechanism(s)	Comments
Zinc oxide-eugenol	Zinc oxide + eugenol	Acid-base	Not polymer-containing; for temporary cementation
Zinc phosphate	Zinc oxide + buffered aq. H_3PO_4	Acid-base	Not polymer-containing
Zinc polycarboxylate	Zinc oxide + aq. poly(acrylic acid)	Acid-base	Chemically adhesive to tooth substance – see Section 23.2.2
Glass-ionomer	FAS glass + aq. polyalkenoic acid	Acid-base	Chemically adhesive – see Section 23.3.2
Resin-modified glass-ionomer	FAS glass + aq. polyalkenoic acid with, e.g., pendant methacrylate groups (Figure 23.17)	Acid-base + polymerization	Improved mechanical properties compared to glass-ionomer
Polyacid-modified resin composite ("compomer")	Resin composite + unreacted glass-ionomer type constituents	Free radical addition polymerization	See Section 23.2.3.6
Polymer-based adhesives	Monomers (dimethacrylate) and inorganic fillers	Free radical addition polymerization	Similar to tooth restorative materials, but with lower filler content (Section 23.7.1)
Adhesive polymers	Contains bifunctional adhesive monomer such as 4-META (Figure 23.8) or 10-MDP (Figure 23.6)	Free radical addition polymerization	Adhesion to metals as well as tooth substance (Section 23.3.2.5)

Source: Hill, E.E., *Dent. Clin. N. Am.*, 51, 643, 2007.

	Resin composites	Polyacid-modified resin composites (compomers)	Resin-modified glass-ionomers	Glass-ionomer cements
Setting Mechanism(s)	Polymerization	Polymerization	Polymerization and Acid-Base	Acid-Base Reaction
Inorganic constituent	Chemically inert during setting	Chemically inert during setting	Involved in acid-base reaction	Involved in acid-base reaction
Structure of set material	Composite	Composite	Composite	Composite
Aqueous phase?	No	No aqueous phase as placed	Yes	Yes
Fluoride release	Low potential for release of fluoride[a]	Very limited scope for fluoride release	Fluoride release	Fluoride release
Mechanical properties	Quite tough materials	Not as tough as resin composites	Not as brittle as glass-ionomers	Quite brittle materials
Applications	Tooth restorations, fissure sealants, adhesive resins	Some tooth restorations	Luting, some tooth restorations	Luting, some tooth restorations

[a] Note that some resin composites have been developed with special additives (e.g., YbF_3), claimed to release fluoride in vivo.

FIGURE 23.15 The "continuum" of tooth-colored restorative materials.

23.2.3.3 Resin Composites

In many applications, such as sealants, restorative materials, orthodontic bonding and adhesive cements, a dimethacrylate polymer is employed, following the pioneering work of Bowen [25]. The main difunctional monomers are BisGMA (or derivatives thereof) or UDMA (Figure 23.4). These monomers are very viscous, so are diluted with TEGDMA (Figure 23.4) to facilitate the incorporation of particulate inorganic filler, which may comprise up to 86 mass% of the material. The types of filler particles and their functions are classified in Section 23.7.1. To aid bonding between the two phases of the composite, the inorganic fillers are treated with a silane coupling agent (typically 3-methacryloxypropyl trimethoxysilane—Figure 23.16). Other constituents include polymerization initiators and stabilizers, as discussed earlier. Materials for restoring teeth usually contain pigments to enhance the esthetic appeal of the material.

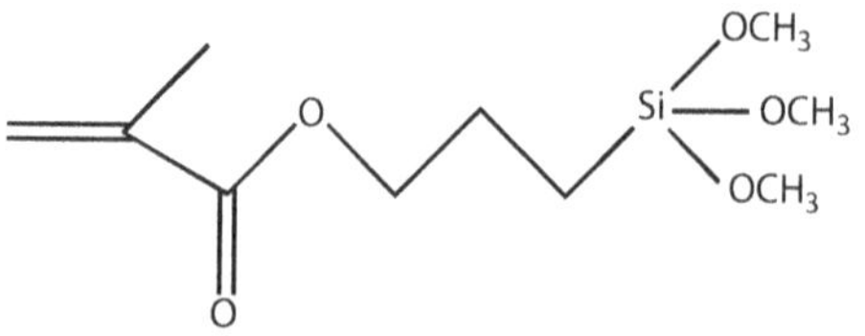

3-Methacryloxypropyl
trimethoxy silane

FIGURE 23.16 Silane coupling agent.

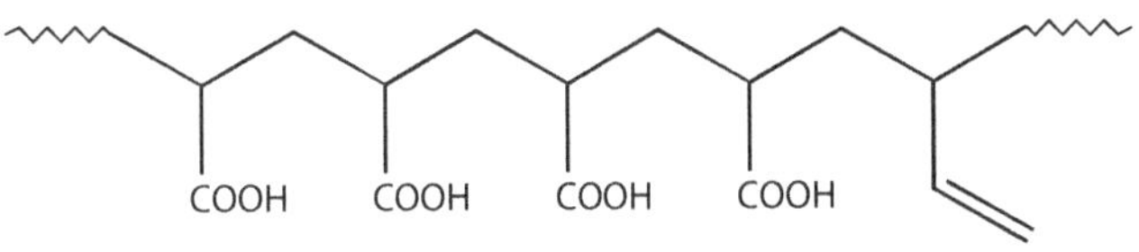

FIGURE 23.17 Modified poly(alkenoic acid) with pendant methacrylate group.

23.2.3.4 Glass-Ionomer (Polyalkenoate) Cements

These materials have two potentially useful properties for dentistry. First, the ability to leach fluoride to the surrounding tooth tissue can have great potential benefit for the prevention or reduction of tooth decay (caries), although the relevance of in vitro studies to long-term clinical benefits is not well established [26]. Second, they are capable of chemical adhesion to many materials, including tooth material, as discussed later (Section 23.3.2).

23.2.3.5 Resin-Modified Glass-Ionomer Cements

Subsequently, materials have been developed in which *both* an acid-base reaction and a polymerization reaction occur—these are termed resin-modified glass-ionomers. They can contain 2-hydroxyethyl methacrylate (HEMA, Figure 23.9), which can be polymerized because of the incorporation of a photo-sensitive initiator. Additionally, some materials contain poly(acrylic acid) on to which are grafted branches with vinyl groups (Figure 23.17). In such cases, co-polymerizaton with HEMA can occur, in addition to the acid-base reaction (Figure 23.18). This modification has the effect of producing tough materials, which retain the merits of fluoride leaching and chemical adhesion [27].

23.2.3.6 Polyacid-Modified Resin Composites ("Compomers")

Compomers are essentially a subset of resin composites, which have been modified by the incorporation of additional monomers with acidic functional groups [28]—for example, the di-ester of 2-hydroxyethyl methacrylate with 1,2,3,4-butanetetracarboxylic acid or a cycloaliphatic dicarboxylic dimethacrylate [29]. These additional monomers comprise less than 10 wt% of the total material. In addition, some reactive glass powder, as used in glass-ionomers, is included. When compomers were introduced they were sometimes wrongly confused with "glass-ionomers" [30].

The materials harden by free radical addition polymerization only, usually with visible light activation (as for composites, but contrast with resin-modified glass-ionomers). One product uses a redox system [31]. Once the material has polymerized, it can absorb water, and in the long term, some glass-ionomer type reaction can occur, giving some potential for the release of fluoride.

23.2.3.7 General Comparison of Material Properties

Given that the aforementioned four classes of material are often recommended for multiple applications, it is helpful to compare and contrast their properties, before considering their dental applications. Some very general comparisons can be given:

- In terms of adhesion to tooth substance, the materials containing polyelectrolytes (glass-ionomer and resin-modified glass-ionomer) are chemically adhesive (Section 23.3.2.1).
- Materials with an aqueous phase (glass-ionomer and resin-modified glass-ionomer) have the potential for fluoride release.
- Mechanical properties such as fracture toughness are usually superior for the resin composites compared with the polyelectrolyte materials [32].

23.2.3.8 Critique of the Concept of a "Continuum"

Although the concept of a continuum is helpful in understanding the origin of each type of material, and in comparing and contrasting properties, it needs to be noted that there is a discontinuity in the diagram, represented by a vertical line in Figure 23.15. To the right of this line the materials contain an aqueous phase, and to the left they do not; this distinction is significant in terms of potential for fluoride release.

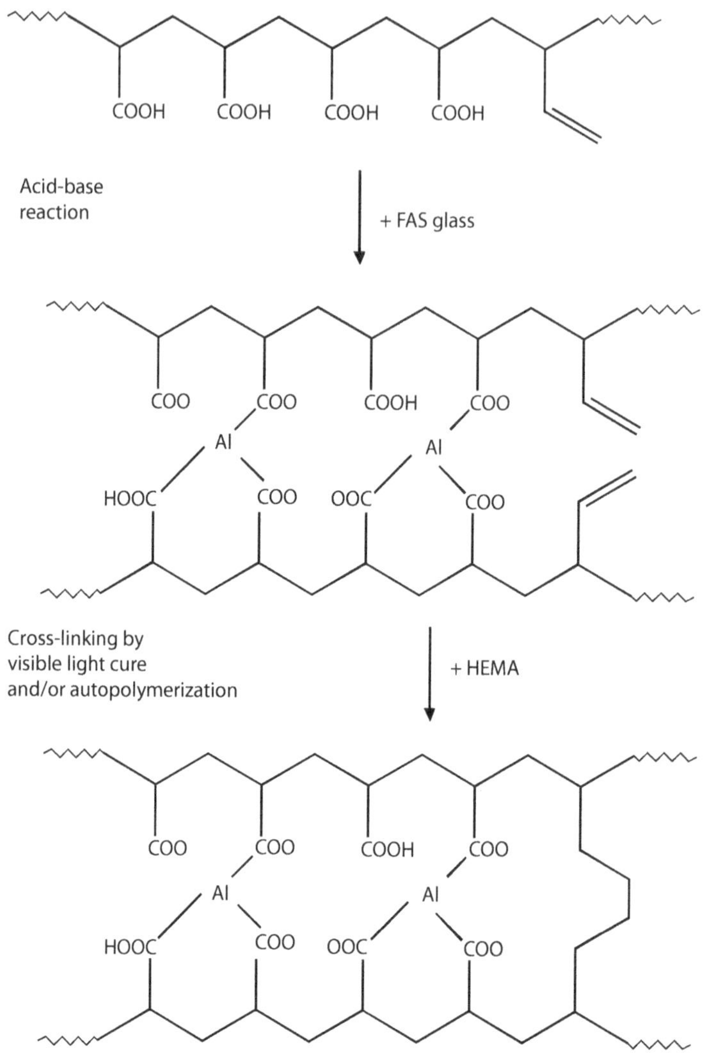

FIGURE 23.18 Mechanisms of setting of a resin-modified glass-ionomer.

23.3 INTERFACES AND ADHESION

23.3.1 APPLICATIONS AND DENTAL SIGNIFICANCE

In the last 50 years, there has been considerable emphasis on the development of adhesive materials for dentistry [33]. In relation to bonding to tooth tissue it was realized, for example, that non-adhesive dental restorations could not seal the tooth cavity, resulting in "microleakage" of saliva, bacteria and bacterial toxins to the pulp of the tooth. Thus postoperative sensitivity, recurrent decay and staining at the margins of the restorations can occur.

In addition to bonding to teeth, the following applications of adhesion technology involving polymerizing materials are important:

- Bonding to surfaces of ceramic and alloy restorations and prostheses that are fabricated in a laboratory before being cemented in place.
- Adhesion between the phases of a composite material—for example, by the use of a silane compound as already discussed.

TABLE 23.3
Principal Dental Applications of Adhesion

1. Chemical adhesion
   a. Glass-ionomer cements to tooth substance
   b. Adhesive resins to alloys
   c. Silane agents to ceramics
   d. Soft tissue adhesives
2. Micromechanical adhesion
   a. Adhesion to etched enamel
   b. Contributes to dentin bonding
   c. Abraded alloy surfaces
   d. Etched ceramics (HF)
3. Hybridization bonding
   a. Bonding to dentin

There are essentially three mechanisms of adhesion, and each is applied to the bonding to tooth tissue:

- Chemical adhesion, where is chemical attraction between adhesive and adherend.
- Micromechanical adhesion, where a material is applied to a roughened or porous surface; on setting a micromechanical bond is formed.
- Hybridization bonding, in which the adhesive is absorbed by the adherend; the adhesive sets to form a hybrid of adhesive and adherend.

It should be noted that (1) surface preparation is critical to good bonding and (2) in some instances more than one of the mechanisms may be responsible for the observed bonding between materials. Table 23.3 lists the dental applications of adhesion in terms of mechanism of action; in very many instances a polymeric material has been developed for this purpose.

23.3.2 MECHANISMS AND MATERIALS

23.3.2.1 Chemical Adhesion to Tooth Tissue

The mechanism of the chemical bonding of the polyelectrolyte cements has been explored by Wilson et al. [34]. They studied the interaction between polyacrylate ions and hydroxyapatite (the main constituent of tooth enamel and a significant component of dentin), and postulated that polyelectrolyte chains are embedded in the enamel or dentin surface with the release of calcium and phosphate ions (Figure 23.19). The pretreatment of dentin with a dilute poly(acrylic acid) conditioning solution did not improve clinical bonding [35].

23.3.2.2 Micromechanical Adhesion to Tooth Enamel

This depends on roughening of the surface of enamel, followed by the application of a monomeric material that can flow into pores on the rough surface, and polymerize in situ, resulting in a micromechanical lock between the materials. This is widely used in sealants, orthodontics and restorative dentistry. It is based on the discovery by Buonocore in 1955 [36] that enamel can be roughened by the application of phosphoric acid (Figure 23.20). Resins applied to such a surface form "prism-like" tags both in vitro [37] and in vivo [38]. Though this technique was not developed or applied for quite a number of years, it is now very widely applied (see Sections 23.4.1, 23.5.1, and 23.6).

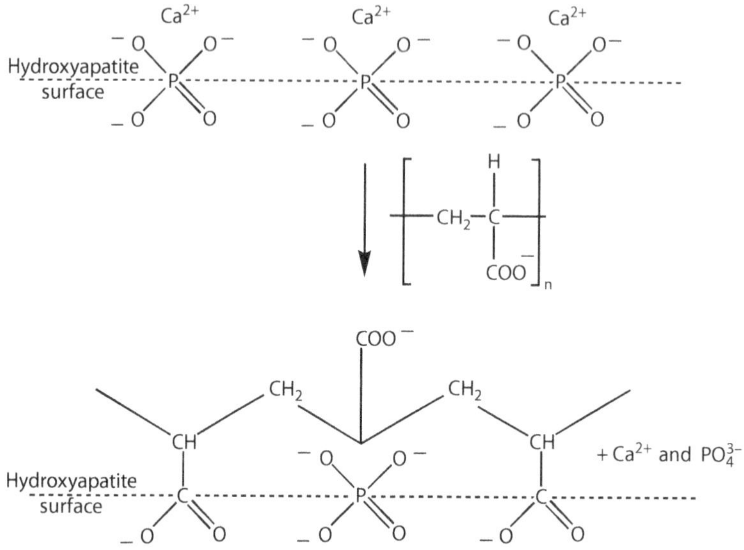

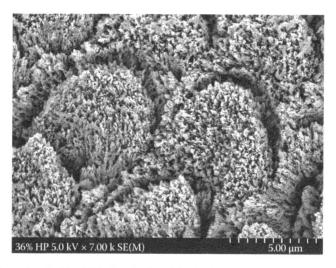

FIGURE 23.19 Suggested mechanism for the adhesion of polyelectrolyte cements. (After Wilson, A.D. et al., *J. Dent. Res.*, 62, 590, 1983.)

FIGURE 23.20 Enamel etched by 36% phosphoric acid gel. (Courtesy of J. Perdigão.)

23.3.2.3 Hybridization Bonding to Tooth Dentin

23.3.2.3.1 Nature of Dentin

A number of approaches have been used to bond to dentin, which has been the subject of intensive research for many years [33]. The first factor to be considered is how to treat the dentin surface for bonding. Dentin consists of a composite of hydroxyapatite particles in a matrix of collagen. It has fluid-filled tubules that converge toward the pulp. It is imperative that any adhesive system should ensure that the tubules are sealed to prevent fluid movement in either direction. Following dental instrumentation of dentin, a smear layer is formed, composed of denatured collagen and mineral. This penetrates into the dentinal tubules to form smear plugs [39]—see Figure 23.21. An important fundamental question is whether the smear layer should be removed or retained before bonding.

Another factor to be considered for a bonding system is that it should be hydrophilic, given that the dentin contains water, but also hydrophobic to be compatible with a polymer-containing

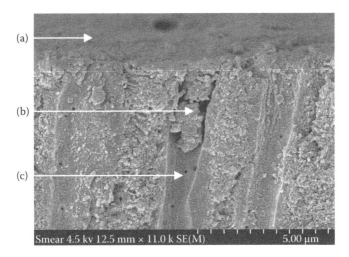

(a)

(b)

(c)

Smear 4.5 kv 12.5 mm × 11.0 k SE(M)   5.00 μm

FIGURE 23.21   Smear layer and smear plugs. a, smear layer; b, smear plug; c, dentinal tubule. (Courtesy of J. Perdigão.)

restorative material. Thus, many monomers are used in the different products, which contain a polymerizable group (e.g., acrylate or methacrylate) and also hydrophilic functional groups—which are shown and classified later in the chapter in Figures 23.5 through 23.9. However, if a material is too hydrophilic, there may be increased degradation of the resin–dentin bond [40].

23.3.2.3.2 Historical Developments

Early work in dentin bonding proposed glycerophosphoric acid dimethacrylate as a dental adhesive [41,42] studied the interaction of this compound with dentin. The first commercial adhesive materials contained phosphorylated methacrylate monomers [43]. However, the resultant shear bond strengths were low (typically 5–7 MPa, about one-third of the strength of the bond to etched enamel) [43] because the hydrophobic monomer did not penetrate the water-containing smear layer [44].

Fusayama [45] introduced the idea of preparing dentin by etching it with phosphoric acid (the enamel of the cavity can be simultaneously etched), and this led to adhesive systems that are classified as "etch and rinse"; that is etching (or conditioning) is followed by copious rinsing to remove any reaction products. This procedure has a significant effect on the smear layer; it removes the smear layer and demineralizes the most superficial hydroxyapatite crystals [46]. As a result, the conditioned dentin can be readily infiltrated by monomers. Nakabayashi et al. etched dentin with a 10% citric acid and 3% ferric chloride mixture and applied 4-META (see Figure 23.8), a monomer with both hydrophobic and hydrophilic groups [47]. These authors demonstrated the infiltration of monomers into the treated dentin; these monomers polymerized in situ to form a hybrid layer composed of resin, collagen, residual hydroxyapatite and a small quantity of water. The resultant tensile bond strength was 18 MPa [47]. This hybrid layer (Figure 23.22) was reported to be about 10 μm in thickness, following etching with 20% H_3PO_4 for 30 s [48]. In contrast, Hashimoto et al. reported a hybrid layer of <5 μm after 30 s etching with 35% H_3PO_4 [49]. Subsequently "self-etching primers" were developed—for example, it was shown that an aqueous solution of 20% phenyl-P (Figure 23.6) and 30% 2-hydroxyehtyl methacrylate (Figure 23.9) creates diffusion channels in the dentin, which can be infiltrated by monomers [50]. However, durability of bonding was a concern—if the dentinal collagen is not completely protected by resin it can be hydrolyzed, causing defects in bonding.

23.3.2.3.3 Current Materials: Classification and Constituents

Today, there is a wide range of available dentin bonding systems and the details of their composition have been reported [7,46]. They may be classified into four types, which are illustrated in Figure 23.23 together with the practical stages involved in the application of each. The first two

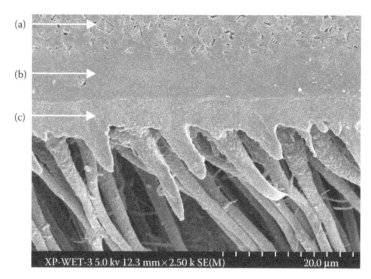

FIGURE 23.22   Bonding to dentin. a, resin composite restorative material; b, adhesive layer; c, hybrid layer. (Courtesy of J. Perdigão.)

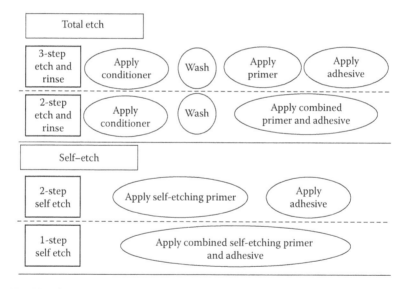

FIGURE 23.23   Classification of dentin bonding systems.

types involve removal of the smear layer ("etch and rinse" systems), and the second two do not ("self etch"). One of the objectives of these developments has been to reduce the number of treatment steps, with the potential to reduce treatment errors (but see the following text). Using the classification of Van Landuyt et al. [7] the materials are

- Three-step etch and rinse adhesives; a conditioner (typically a 35% phosphoric acid gel), similar to the etchant for enamel. After washing off the debris, a primer is applied, followed by an adhesive.
- Two-step etch and rinse adhesives; the primer and adhesive are combined, thus eliminating one step of the practical procedure.

- Two-step self-etch adhesives, in which the conditioner and primer are combined ("self-etching primer").
- One-step self-etch adhesives—a single material is applied to the dentin.

It should be noted that self-etch systems are also being recommended for application to enamel. Dentin bonding systems contain the following types of constituents, though not all products utilize all these components types [7]:

- Dimethacrylate monomers(s) such as Bis-GMA, UDMA, and TEGDMA (Figure 23.4)—these are the same monomer types as used in resin composites (see the following text). These monomers enable the bonding material to covalently bond to the composite restorative material.
- A hydrophilic monomer that is polymerizable, particularly HEMA (Figure 23.9).
- A few products contain polyalkenoic acids, capable of chemical bonding to the calcium of hydroxyapatite.
- Methacrylate monomers with functional groups—including those detailed in Figures 23.5 through 23.8. Note that these monomers contain both hydrophilic and hydrophobic groups.
- Inorganic fillers (also used in resin composite restorative materials).
- Solvating agent (alcohol/water, acetone).
- Polymerization initiators, for example, for visible light curing.

23.3.2.3.4 Durability
Reporting in 1988, Van Meerbeek et al. claimed that there had been important improvements in materials, particular in terms of retention [51]. However, no material could prevent microleakage over a long time period. More recently, it was claimed that the materials with fewer practical steps were not as effective [52]. To quote, "Simplification so far appears to induce loss of effectiveness" [53]. De Munck et al. [54] regard three-step etch and rinse systems to be the "gold standard."

23.3.2.3.5 Current Concerns and Developments
A major problem is that when polymerizing, a resin composite undergoes shrinkage, and this places stress on the bond to dentin. In the later discussion on composites (Section 23.7.1.1.4), the attempts to minimize this problem are discussed.

The long-term stability of bonding is also an issue, particularly the hydrolytic stability of chemical bonding to hydroxyapatite. Inoue et al. [55] and Van Landuyt et al. [56] showed that durability of bonding depends on the chemical bonding of the functional monomer. Yoshida et al. [57] determined that 10-MDP formed very stable bonds with hydroxyapatite, superior to the performance of 4-MET and phenyl-P.

There are some issues of concern in relation to one-step self etch adhesives. Those that do not contain HEMA are prone to display droplets within the material, attributed to phase separation between monomer and solvent; this can compromise bond durability [58]. In adhesives that were HEMA-rich, droplets were also observed due to water absorption from dentin, caused by osmosis [59]. The term nanoleakage has been introduced to describe nanometer-sized spaces filled with water [60]—distinguish from microleakage described earlier.

Although the presence of HEMA can prevent phase separation, its allergenic potential is a concern. Due to low molecular weight, it can penetrate through gloves to cause contact dermatitis [61].

Currently, the problem of degradation of the hybrid layer is being investigated. When the dentin bonding system used acetone as solvent, extensive degradation is observed; this can be reduced by the application of chlorhexidine gluconate [62]. It is believed that this compound increases the stability of the collagen fibrils in the hybrid layer against endogeneous metaloproteinases [63].

In the etch-and-rinse adhesives, primer or combined primer-adhesive is applied to moist dentin. Replacing water with ethanol has been reported to increase the durability of bond strength [64].

23.3.2.4 Bonding to Ceramics

Ceramics based on feldspathic porcelain can be etched with hydrofluoric acid, then treated with a silane agent (Figure 23.16), which bonds both to the ceramic surface and to the resin-based composite cement adhesive [65].

However, many modern tougher dental ceramics are based on alumina or zirconia, which are not capable of being etched by HF. For glass-infiltrated alumina restorations, currently there is a choice of bonding technique—either coat the surface with silica, apply a silane compound, and use an adhesive based on bis-GMA resin or roughen the surface with airborne Al_2O_3 particle abrasion, followed by application of a phosphate-containing polymer cement. For zirconia-based materials, this latter procedure appears to be the best available at this time [66].

23.3.2.5 Bonding to Metallic Materials

Silica coating, as described earlier, can also be applied to cast dental alloys, followed by silane treatment. Alternatively, tin plating has been applied. An adhesive resin cement, containing 4-META or 10-MDP may be employed [67].

23.4 PREVENTION OF DISEASE AND THE EFFECTS OF TRAUMA

23.4.1 Pit and Fissure Sealants

23.4.1.1 Purpose

Dental decay (caries) has been described as the most common disease of school children [68]. Rough pit and fissure surfaces of teeth are difficult to clean, and are therefore more susceptible to caries. In addition, fluoride protection is less beneficial in protecting pits and fissures, compared with smooth enamel surfaces. Epidemiological research has emphasized the need for some means of protection of pits and fissures [69].

This problem has been tackled by the application of sealants. A fissure sealant has been defined as "a material that is placed in pits and fissures of teeth in order to prevent or arrest the development of dental caries" [70]. These polymeric materials are applied to the pits and fissures that are prone to decay. They provide a physical barrier that protects pits and fissures from microorganisms and particles of food [71]. However, although sealants have been available for over 30 years, they were described in 2005 as being "perhaps the most tested yet most underused technique in clinical preventive dentistry" [72].

Early attempts at pit and fissure sealing included the use of materials containing methyl-2-cyanoacrylate. However, clinical tests showed "almost total failure of the fissure sealant" [73].

Representatives of all the groups of material within the "continuum" of tooth-colored materials (Section 23.2.3) have been advocated as sealants. However, materials based on dimethacrylates are the most widely used and thus will be considered first, followed by a critique of alternative materials.

23.4.1.2 Classification of Dimethacrylate-Based Sealants

There is a plethora of different materials, often confusingly designated as "first generation," "second generation" etc. Differences have occurred in activation system. Some early materials were activated by UV radiation. However, there were two grounds for their criticism. In the first place, the safety of using 365-nm radiation was called into question because of potential biological damage such as skin cancer, eye damage and mutagenic affects [74]. Furthermore, the efficiency of the curing lights was sometimes poor because of insufficient intensity of radiation [75].

In today's materials, there are many different choices—for example,

- There is a choice of "self-cure" vs. visible light cure (VLC) materials.
- Filled and unfilled polymers are available; the filled materials resemble resin composite restorative materials (Section 23.7.1), but usually with lower filler concentration. In fact, one type of composite restorative material (flowable composites) is being suggested for sealant use.

- Sealants may be clear or opaque or tinted. Some change color—for example, from clear to a green color, or from pink to opaque white, after polymerization, in order to ensure adequate placement of sealant.
- Some sealants contain fluoride.

23.4.1.3 Biocompatibility

Free-radical addition polymerized sealants often have an oxygen-inhibited layer, estimated to be from 7–84 μm in depth [76]. Is this harmful? A major concern has been the claim that sealants may leach compounds that are estrogenic, such as bisphenol-A (2,2-bis[4-hydroxyphenyl]propane (BPA), a component in the synthesis of Bis-GMA. Olea et al. [77] claimed that "the use of bis-GMA resins in dentistry, and in particular the use of sealants in children, appears to contribute to human exposure to xenoestrogens." Joskow [78] reported that one brand of sealant leached BPA. However, more recently, a systematic review has shown that there is no risk of exposure to BPA from sealants, though removal of the soft layer of un- or under-polymerized material is recommended [79]. It has also been shown that the formation of an oxygen-inhibited layer can be prevented by placing a thin layer of oil to form an air barrier, and then shining the polymerizing light through this barrier [80].

23.4.1.4 Bonding to Enamel and Retention Rates

Having looked at the safety of sealants, it is now necessary to consider efficacy. We consider how well they are retained on enamel and whether there has been success in caries prevention. A Cochrane evidence-based review showed that caries reduction from 87% at 1 year to 60% at 4–4½ years can be achieved. Retention of the sealant on the enamel was from 79% to 92% after 1 year, 71% to 85% at 2 years, and 61% to 80% at 3 years [81].

23.4.1.5 Effect on Bacteria Levels in Caries Lesions

The efficacy of sealants has assumed their application to sound enamel. What if inadvertent sealing over dental caries is carried out? Consistent with the finding that bacteria levels are reduced if covered by sealants [82], a review by Griffin et al. supported the technique of placing sealants over caries lesions that were not cavitated [83].

23.4.1.6 Application of Self-Etch Adhesives

Since self-etch adhesives have been introduced for bonding to dentin (Section 23.3.2), it was suggested that they could be employed for bonding sealants to enamel [84–87].

23.4.1.7 Application of Compomers

In his review on compomers, Nicholson [28] claimed that these materials performed well in a number of applications, including fissure sealing, despite a decrease of strength in vivo associated with water uptake. Two short-term clinical trials appear to confirm this [88,89], but the outcomes of longer term clinical data are awaited.

23.4.1.8 Application of Polyelectrolyte Materials

Glass-ionomer and resin-modified glass-ionomer cements have two features that render them potentially attractive as sealants—namely, their adhesion to enamel and their ability to release fluoride. Thus, they have been tested as fissure sealants. A systematic review showed that neither resin-based or glass-ionomer was superior to the other in the prevention of dentin lesion development [90]. However, the retention of resin-based sealants (by micromechanical adhesion) has been shown to be greater than that of chemically adhered glass-ionomers [91].

23.4.1.9   Summary of Current Clinical Recommendations
Evidence-based clinical recommendations have been published by the American Dental Association Council on Scientific Affairs [92].

- Sealants are recommended for pits and fissures of both primary and permanent teeth if there is risk of caries development. They can also be applied to early carious lesions, to reduce the probability of progression of the lesion [83,92].
- Although glass-ionomer cement can be used as interim preventive agent, particularly if moisture control during placement is difficult, resin-based materials should be the first choice of material [93].
- In relation to bonding to enamel, if self-etching bonding agents are used without the separate step of enamel etching, there may be less retention that the standard acid-etch procedure. If such bonding agents are used on already etched enamel, sealant retention would be enhanced [92].
- For maximum effectiveness, it is important that sealant be monitored regularly, and reapplied as necessary [92].

23.4.1.10   Research Needs
Beauchamp et al. [92] have also listed numerous research needs, including studies of preventive effectiveness and cost-effectiveness, timing of sealant application, and exploration and evaluation of new materials.

23.4.2   Mouth Protectors

23.4.2.1   Purpose and Requirements
Mouth protectors (sometimes referred to as mouth guards or gum shields) are recommended for participants in many types of sport to prevent orofacial injuries and concussion. A mouth protector have been defined by the American Society for Testing and Materials as a "resilient device or appliance placed inside the mouth (or inside and outside), to reduce mouth injuries, particularly to teeth and surrounding structures" [94]. The American Dental Association has listed some 29 sporting activities that pose hazards resulting from a blow to the head or mouth [95].

The ideal requirements for a mouth protector include its function, its fit and its durability. In terms of function, the material must have the ability to absorb impact. In terms of fit, there should be accurate adaptation to the individual's mouth structures and should have smooth well-contoured edges. The protector should be comfortable to wear and not interfere with speech or breathing. The material should resist tearing and should be easily cleanable to avoid unpleasant odors or tastes.

23.4.2.2   Materials
Over the years, five types of polymer have been advocated for mouth protector construction [96,97]. These include

- Ethylene-vinyl acetate copolymer (EVA)—this is currently the most widely used material, owing to its ease of fabrication.
- Poly(vinyl chloride)—noting that there are concerns relating to the biocompatibility of phthalate esters used as plasticizers—these materials have been banned in the European Union [98].
- Natural rubber.
- Soft acrylic resin.
- Polyurethane.

23.4.2.3 Classification

Mouth protectors can be classified into three types [94]:

1. Type I—Thermoplastic type. These may be either mouth-formed or vacuum-formed.
2. Type II—Thermosetting type.
3. Type III—Stock type.

It should be noted that alternative classifications have been given by the American National Standards Institute and Standards Australia International [99].

Of these three types, Type III is generally considered to be the least satisfactory. Such materials do not provide good fit and comfort, may not stay in place comfortably, as they have no retention and they can interfere with mouth breathing and speech [97]. Currently, the choice is between mouth-formed custom-made protectors.

23.4.2.3.1 Mouth-Formed Protectors

There are two varieties of mouth-formed protectors. Some are thermoplastic "boil and bite" materials, usually an EVA copolymer. The material is softened in hot water, briefly cooled and then shaped in the mouth [100]. A problem with this approach to fabrication is that the material may become over-thinned in some regions, thus giving insufficient protection to the wearer [97].

Alternative mouth-formed protectors are the shell-liner materials, in which a PVC shell contains an inner lining of a resilient material such as a plasticized acrylic or a silicone rubber [100].

23.4.2.3.2 Custom-Made Protectors

These are frequently prepared on a gypsum model or cast, using low heat and vacuum, using EVA. Sometimes, there may be insufficient thickness of material in the formed protector. This problem can be minimized by using a pressure lamination technique [100].

23.4.2.4 Properties

EVA is a random co-polymer produced by high-pressure free radical polymerization. Properties are dependent on the vinyl acetate content—as this is increased it disrupts the polymer crystallinity, which has important consequences for the properties of the co-polymer. Increase vinyl acetate content (up to 28%) linearly decreases the hardness (Figure 23.24), flexural modulus (Figure 23.25), and softening temperature (Figure 23.26), and increases the impact strength (Figure 23.27) [101].

23.4.2.5 Conclusion

A recent extensive meta-analysis by Knapik et al. [102] has concluded that the "overall risk of an orofacial injury is 1.6–1.9 times higher when a mouthguard is not worn, relative to wearing a mouthguard." It is as yet unknown if mouthguards afford protection from concussion.

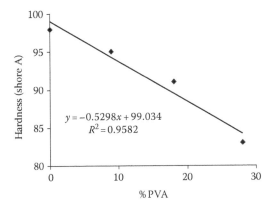

FIGURE 23.24 Effect of quantity of PVA in EVA copolymer on hardness.

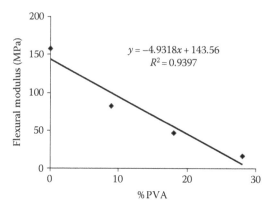

FIGURE 23.25 Effect of quantity of PVA in EVA copolymer on flexural modulus.

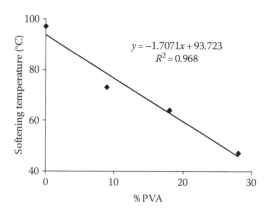

FIGURE 23.26 Effect of quantity of PVA in EVA copolymer on softening temperature.

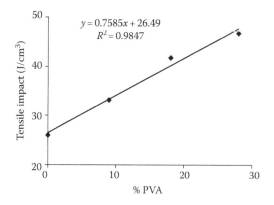

FIGURE 23.27 Effect of quantity of PVA in EVA copolymer on impact energy.

23.5 ORTHODONTICS

23.5.1 ORTHODONTIC ADHESIVES

The role of polymers in orthodontic treatment is now considered. Orthodontics deals with the treatment of malocclusion and other irregularities of the teeth and jaws. Polymer-containing materials are used in orthodontics as agents to bond brackets to tooth enamel (Figure 23.28), and also as cements to retain bands around teeth (Figure 23.29). The requirements for orthodontic adhesives are quite stringent, and include good biocompatibility and durable adhesion to enamel.

As with pit and fissure sealants (see the earlier text), cyanoacrylates were tested and found to lack bonding durability [103]. Today, all four classes of material in the "continuum" (see the earlier text) are employed.

For cementing orthodontic bands, a recent systematic review has found insufficient evidence to establish which adhesive is the most effective adhesive for bands, though polyelectrolyte

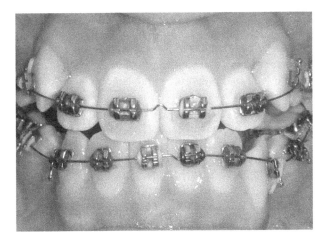

FIGURE 23.28   Orthodontic brackets cemented to tooth enamel. (Courtesy of J. Beyer.)

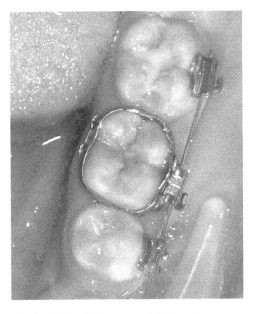

FIGURE 23.29   Cemented orthodontic band. (Courtesy of J. Beyer.)

FIGURE 23.30   Orthodontic elastics. (Courtesy of J. Beyer.)

materials can adhere to both enamel and stainless steel, can release and take up fluoride, and have antibacterial properties [104].

By way of contrast, in relation to the bonding of brackets, there is widespread use of polymerizing materials, using a range of bonding techniques (acid etch, self etch, etc.). However it has been predicted that in the future use of compomer and glass-ionomer based materials will increase [105].

The potential iatrogenic effects of direct orthodontic bonding need to be considered.

- Is there danger from BPA, as was once alleged for fissure sealants? Eliades et al. [106] have measured BPA release from orthodontic adhesives and demonstrated the absence of any estrogenicity [107].
- One problem is that decalcification of enamel (white spot lesions) can occur during orthodontic treatment [108]. Reportedly this can be reduced by the application of a fluoride-containing varnish [109] or a fissure sealant [110].
- Is there permanent damage to enamel after debonding? Studies have shown that enamel cracking [111] and facture [112] can occur.

23.5.2   ORTHODONTIC ELASTICS

Orthodontic elastics are used in some treatments to aid tooth movement (Figure 23.30). Traditionally these have been made of cis-polyisoprene (latex). However there is concern about hypersensitivity [113,114]. As a result synthetic non-latex elastics are available, reportedly containing styrene-butadiene with mineral oil and small amounts of stearic acid, magnesium carbonate, silicic acid, and an antioxidative stabilizer [115]. As a result of dynamic mechanical testing it was shown that the synthetic materials need more frequent replacement than the latex products because of force loss [115,116].

23.6   ESTHETIC DENTISTRY: VENEERS

A veneer is a thin layer of an esthetic material, which is placed over a tooth surface, to mask tooth defects or discoloration (e.g., tetracycline staining), to correct the appearance of uneven or crooked teeth, to cover chips in teeth and to close diastemas. They play a very significant role in elective dental esthetics [117]. A merit of this treatment approach is that there is little or no removal of tooth structure—up to about 0.5 or 0.75 mm [117,118].

Veneers may be made from translucent ceramics such as feldspathic porcelain, cemented into place with a visible light cured polymeric material, in conjunction with enamel etching and silane treatment of the ceramic. In terms of clinical durability, long-term clinical success has been claimed [119]. Survival rates of veneers up to 16 years have been determined [120].

Alternatively, tooth bleaching followed by application of a resin composite (see Section 23.7.1.) may be used in certain circumstances, as a more conservative approach to esthetic dentistry [121].

23.7 TOOTH RESTORATION

23.7.1 TOOTH FILLING MATERIALS

For many years, the main options for the restoration of teeth were silver amalgam for posterior teeth and silicate cements for anterior teeth. Polymer-containing materials have now completely replaced silicates for anterior restorations, and have also partially displaced amalgam. The early history of usage of polymers for tooth restoration has been recorded by Puckett et al. [122].

The requirements for materials for restoring teeth are stringent. Such materials are in an aggressive environment, encountering dietary acids, plaque-forming bacteria, and enzymes; in addition, masticatory forces have the potential to cause degradation. In addition to biocompatibility, and the ability to adhere durably to tooth tissue, chemical stability, mechanical toughness, surface wear resistance, esthetic properties, and ease of handling are also important.

In this section, each of the four classes in the "continuum" (Section 23.2.3.2) is considered and critiqued, with the greatest emphasis on resin composites. Emphasis is placed on the potential limitations of these materials, and the research efforts to bring about improvements. Also some new types of material that have become recently available are then reviewed.

23.7.1.1 Resin Composites

The basic principles of these materials have already been presented (Section 23.2.3.3). As supplied they consist of dimethacrylate monomers mixed with silane-coated inorganic particles, initiators, stabilizers and pigments. Early materials were supplied as two pastes with a peroxide/tertiary amine redox system. Most current products are visible light cured. It should also be noted that indirect composite resins are available for laboratory fabrication [123].

23.7.1.1.1 Classification by Inorganic Filler Content

Fillers have included silica, barium glass, strontium glass, and zirconia/silica. Composites are often classified according to their filler content [124]. Table 23.4 classifies current materials. Different particle size and size distribution have been used, including macrofilled materials (Figure 23.31), microfillers (Figure 23.32), a combination of different filler sizes (Figure 23.33) and nanofillers (Figure 23.34). The functions of these fillers include: improvement in mechanical properties such as compressive strength and hardness, reduction in shrinkage associated with polymerization and conferment of radiopacity on the material if a heavy metal (Ba, Sr, Zr) is

TABLE 23.4

Classification of Resin Composite Tooth Restorative Materials according to Their Filler Content

Class of Composite	Particle Size of Filler
Macrofilled	1–50 μm (e.g., Ba glass, SiO_2)
Large particle hybrid	1–20 μm glass + 0.04 μm SiO_2
Midifilled hybrid	0.1–10 μm glass + 0.04 μm SiO_2
Minifilled hybrid	0.1–2 μm glass + 0.04 μm SiO_2
Continuum filled	0.01–3.5 μm zirconia/silica (mean 0.6 μm)
Microfilled	0.04 μm SiO_2
Nanofilled (translucent)	75 nm SiO_2 + clusters of agglomerated 75 nm particles; cluster size range = 0.6–1.4 μm
Nanofilled (radiopaque)	Non-agglomerated 20 nm SiO_2 + 0.6–1.4 μm clusters of zirconia/silica particles of size 5–20 nm

Concise 5.0 kV 14.1 mm × 5.00 k SE(M)   10.0 μm

FIGURE 23.31   Macrofilled composite—see Table 23.4. (Courtesy of J. Perdigão.)

A110 5.0 kV 14.0 mm × 20.0 k SE(M)   2.00 μm

FIGURE 23.32   Microfilled composite—see Table 23.4. (Courtesy of J. Perdigão.)

present. Good esthetics depends on translucency of the material, but a mismatch of refractive index between polymer and filler can be detrimental [125]. Developments in filler technology have included the incorporation of fluoride, for example, ytterbium trifluoride.

23.7.1.1.2   Classification by Rheology

Composites have been produced with a range of consistencies. Flowable materials, with 52–68 wt% filler are capable of being injected through a syringe. They may contain colloidal silica which confers thixotrpy on the material [126] so that it is easy to syringe but will stay in place after syringing, until it is adapted to the tooth cavity. There is considerable variation in available products in terms of composition and properties [127]. Some of the lower viscosity products are being used as fissure sealants. Others are being recommended for Class III and Class V restorations, as liners under other restorations and for repairing small defects in indirect restorations.

In contrast, materials with higher viscosity have been produced; they are termed "packable composites." They were designed to mimic the handling properties of dental amalgam. Placement and

FIGURE 23.33 Hybrid filled composite—see Table 23.4. (Courtesy of J. Perdigão.)

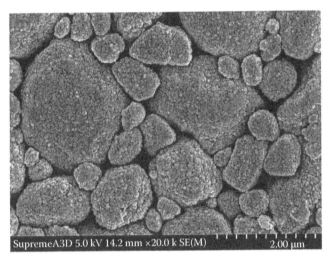

FIGURE 23.34 Nanofilled composite, showing clusters of nanoparticles—see Table 23.4. (Courtesy of J. Perdigão.)

handling properties may be better than that of other composites for posterior restorations [128] but they are not necessarily superior to them in mechanical properties [129].

23.7.1.1.3 Issues Relating to Biostability, Biocompatibility, and Chemical and Mechanical Durability

It needs to be realized that the conversion of monomer to polymer is not a chemically efficient reaction since 25%–60% of methacrylate groups remain unreacted [130]. Thus, there is concern about the elution of monomers; lower molecular weight monomers are more likely to be extracted than higher molecular weight ones [131]. TEGDMA has been shown to be extracted from composites [132,133], with potential toxic effects on gingival fibroblasts [134]. TEGDMA and HEMA have been shown to express cyclooxygenase-2 (COX-2) which is involved in inflammatory reactions [135].

Another issue is degradation of polymers in the oral environment—that is—biostability. Santerre et al. [136] have listed the potential types of polymer degradation processes. Of particular interest is the possibility for breakdown of the silane coupling agent and also enzymatic degradation of polymer components.

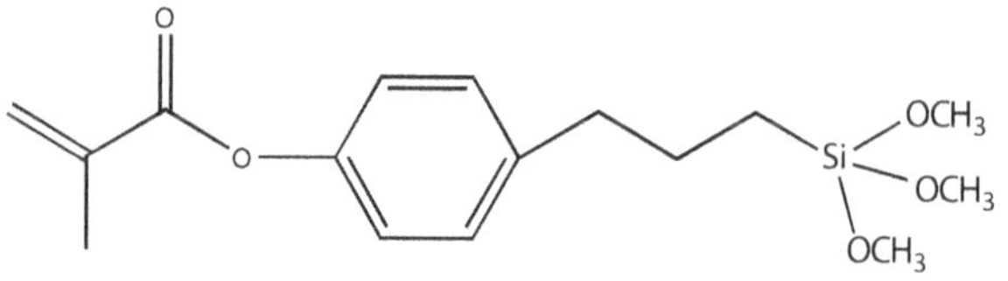

3-(3-methoxy-4-methacryloylphenyl)propyl-
trimethoxysilane

FIGURE 23.35 Alternative silane coupling agent (cf. Figure 23.16).

To counteract the possibility of hydrolytic breakdown of 3-methacryloxypropyl trimethoxysilane (Figure 23.16), Nihei et al. [137] have synthesized a silane with a hydrophobic phenyl group—3-(3-methoxy-4-methacryloyloxyphenyl)propyltrimethoxysilane (Figure 23.35). They demonstrated that composites with this silane had better wear resistance, which was attributed to the formation of a hydrophobic layer on the filler surface.

Enzymatic degradation of the resin components of composites has been extensively studied, given that saliva contains esterases [138] which can catalyze the hydrolysis of resin composites [139]. The biochemical stability of a composite is dependent on the material's composition [140]. Following incubation in cholesterol esterase, degradation products such as methacrylic acid, TEGMA and bishydroxypropoxyphenylpropane can be produced [140].

Chemical and/or biochemical breakdown can cause physical breakdown of the material. This has been extensively reviewed by Drummond [141], including the effects of aging in different chemical media, and cyclic loading.

23.7.1.1.4 Polymerization Shrinkage

It is well documented that composites contract during polymerization. Smaller monomer molecules show the greatest shrinkage on polymerization. For example, the volumetric shrinkage of pure bis-GMA and TEGDMA is 6.1% and 14.3%, respectively [142], or 5.2% and 12.5%, respectively [143]. For dental resin composites, the volumetric shrinkage is typically around 2% [144]. On polymerization, a shrinkage stress results. Of particular importance is the stress developed after the material has become solid—the so-called "post-gel" shrinkage [145] The stress depends not only on the magnitude of the shrinkage but also on the viscoelastic properties of the material and the configuration of the tooth cavity [143]. Deleterious outcomes can result from this phenomenon. These include tooth deformation, debonding of the restoration, propagation of cracks in enamel and post-operative sensitivity [146].

Substantial research has been, and is being, carried out to solve this problem. Approaches include:

- The development of polyfluorinated prepolymer multifunctional urethane methacrylate—reported to give low shrinkage and a high degree of conversion of monomer [147].
- Cyclopolymerizable monomers (oxy bis-methacrylates) have been synthesized (Figure 23.36) and found to be capable of producing cross-linked polymers from bulk polymerization. These have been tested in combination with Bis-GMA; less contraction and good mechanical properties were claimed [148].
- The ring-opening polymerization of spiro orthocarbonates has been suggested as a means of providing expansion to counteract the shrinkage associated with polymerizing dimethacrylates [149–151]. By using bis-GMA derivatives (in place of bis-GMA) with spiro orthocarbonates, Moon et al. [152] claimed to achieve reduction in shrinkage without compromise of mechanical properties.

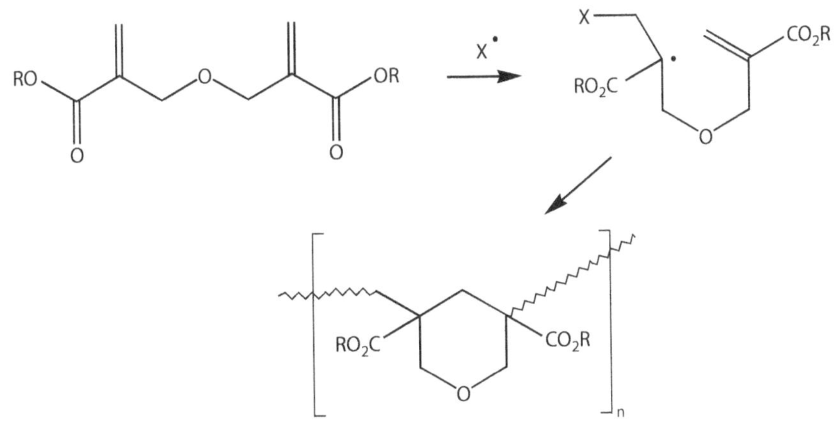

FIGURE 23.36 Cyclopolymerization of an oxy bis-methacrylate monomer. (After Stansbury, J.W., *J. Dent. Res.*, 69, 844, 1990.)

- New high molecular weight dimethacrylate monomers, based on hydrogenated dimer acid have been developed. Although these materials have low modulus of elasticity and mechanical strength, they are claimed to low water absorption and polymerization shrinkage [153].
- Urethane derivatives of bis-GMA have been shown to have favorable properties compared with bis-GMA—namely, higher degree of polymerization and lower polymerization shrinkage. However, lower flexural strength was reported [154].

Fluorinated monomers have been developed in attempts to improve the hydrophobicity of the composite, hence potentially reducing its liability to degrade under oral conditions. Early work led to hydrophobic polymers with poor mechanical properties [155]. Later research [156] claims the development of fluorinated polymers with a high degree of polymerization conversion.

The chemistry of the monomers used in both conventional and experimental materials has been extensively reviewed by Peutzfeldt [130] and Moszner et al. [157].

Two chemical developments in composites have led to the production of clinically used dental materials—namely, the so-called siloranes and ormocers. These are considered separately later.

23.7.1.2 Siloranes

The word *silorane* is derived from a combination of *silox*ane and oxi*rane*. The silorane monomer is shown in Figure 23.37. The siloxane component was chosen for its well known hydrophobicity, and oxiranes are known to give stable polymers with low shrinkage on polymerization. Siloranes can undergo cationic addition polymerization, in contrast to (di)methacrylate polymerization which is usually a free radical reaction. The cationic initiator is a combination of camphorquinone (as used in visible light cure free radical systems), an iodonium salt and an electron donor. In the presence of visible light, cations are generated. The reaction proceeds as shown in Figure 23.38 [158]. Volumetric shrinkage on polymerization is 1%—about half of that for typical composites. This is attributed to the effect of the ring opening reaction. Fillers include quartz and radiopaque yttrium fluoride [159]. Mechanical properties are similar to those of conventional composites [160].

An adhesive system has been developed for use with the silorane composite. This includes a self-etch primer containing phosphorylated methacrylates and a hydrophobic adhesive [161] aimed at bonding the hydrophobic restorative material to the more hydrophilic tooth material.

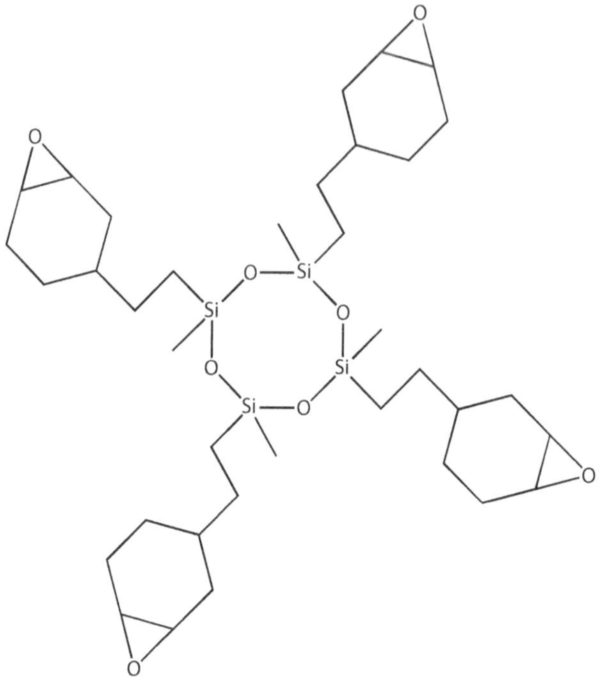

FIGURE 23.37 Silorane monomer.

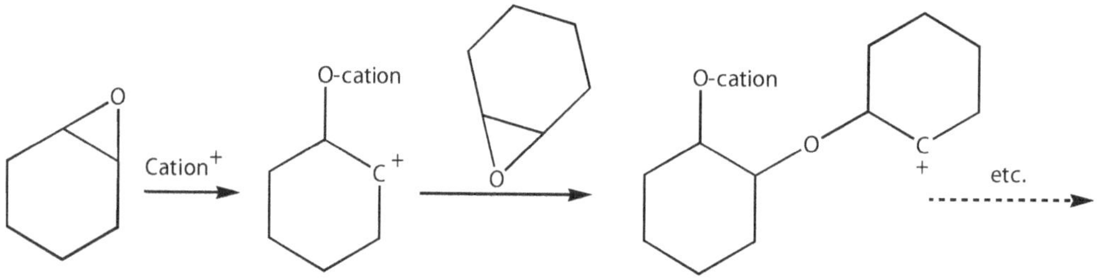

FIGURE 23.38 Cationic polymerization of siloranes.

23.7.1.3 Ormocers*

An *ormocer* is an *org*anically *mo*dified *cer*amic. These materials were developed by Wolter, Storch and Ott [162], who report the synthesis of thioether- and urethane-(meth)acrylate silanes. The general features of the chemistry of the latter compounds are outlined in Figure 23.39. In addition to an ormocer, the commercially available dental composite includes a quantity of traditional diethacrylates (e.g., bis-GMA) and inorganic filers [159]. The material demonstrated good wear resistance in in vitro testing [163].

23.7.1.4 Compomers

The chemistry of these has been discussed earlier (Section 23.2.3.6) [28]. Given that they are not as tough as resin composites, their application is mainly in areas of the mouth where they will be subjected to little or no stress—for example, in deciduous teeth and cervical and proximal restorations [164].

---

* Ormocer® is a trademark of the Fraunhofer Gesellschaft, Germany.

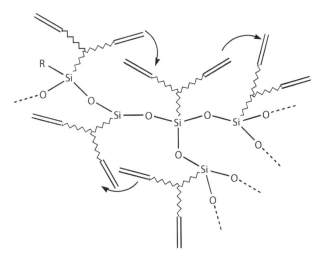

FIGURE 23.39 Ormocer structure.

23.7.1.5 Glass-Ionomers and Resin-Modified Glass-Ionomers

As recorded earlier (Section 23.2.3.7), these polyelectrolyte-based materials are more brittle than resin composites and thus cannot withstand high forces [32]. However, different formulations of the material type are very versatile, being used for provisional restorations, tooth cavity lining and base materials, luting restorations, fissure sealants (Section 23.4.1) and bonding agents [165].

23.7.1.6 Summary of Data on Clinical Longevity of Tooth-Colored Restorations

The data for clinical longevity of polymer-containing tooth restorative materials present a challenge to both clinician and research worker. It has been shown in clinical studies that resin composite is not as durable as dental amalgam, and that "routine clinical practice may be producing suboptimal results" [166].

23.7.2 ENDODONTIC MATERIALS

When trauma or oral disease affects the pulp of the tooth, endodontic treatment needs to be carried out. A root filling is placed with the aim of obturating the root canal system. Sealing of the root canal is important to prevent micro-organisms and/or toxins from tracking from the coronal aspect of the root to the root apex.

Traditionally, gutta percha [*trans*-poly(isoprene)] has been employed. Often this core material is used in conjunction with a "sealer" composed of zinc oxide-eugenol cement. However, there is no adhesion, either between the sealer and the dentin or between the sealer and the gutta percha. Thus, there has been extensive work in attempts to develop adhesive systems to eliminate undesirable leakage.

Alternative sealers include an epoxy-resin-based material and glass-ionomer cement [167]. The former material shows some toxicity in the unset state [168,169]. The latter sealer is hard and adhesive, but makes retreatment difficult if this should be necessary [170].

More recent developments include the development of endodontic restorations where the components act as a single unit—this is termed a "monoblock." One example is a composite core material consisting of thermoplastic polycaprolactone with bioactive glass and bismuth and barium salts to confer radiopacity. This is used in conjunction with a sealer which is a composite resin, based on filled dimethacrylates—not dissimilar from the restorative materials discussed earlier (Section 23.7.1). This system comes with a self-etch primer for dentin, composed of a sulfonic acid terminated monomer, HEMA, water and initiator for polymerization. It is claimed that adhesion, both to the dentin and between sealer and core, is obtained. However, in vitro fluid filtration tests on this system have shown that it does not eliminate leakage [171]. Moreover, it has been shown this system, in common with others poses "significant cytotoxic risks and that cytotoxicity generally increased with time" [172].

An alternative self-etch sealing material, based on 4-META technology, has been designed for use with either gutta percha or polycaprolactone cores. However, it has been shown to have limited etching potential, which may reduce the ability to seal [173].

A recent development is gutta percha coated with methacrylate resin, designed to be used with a urethane dimethacrylate-based sealer [174].

Polyelectrolyte-containing materials have also been proposed. A new commercial system uses gutta percha coated with glass-ionomer particles to be used in conjunction with a glass-ionomer sealer [174]. An experimental system incorporated high-molecular-weight poly(acrylic acid) particles into both gutta percha and a sealer and was shown to have low leakage [175].

23.8 IMPRESSION MATERIALS

23.8.1 INTRODUCTION AND CLASSIFICATION

Discussion now turns to dental situations where a restoration, prosthesis or appliance is fabricated outside the mouth. A clinical impression of the tissues is required. There are stringent demands for such materials, including appropriate rheological properties, a high degree of dimensional accuracy and stability, elastic recovery on displacement from the tissues, and hydrophilicity. Currently used impression materials, which set to form an elastic solid, are polymer-containing. They may be classified as follows:

1. Hydrocolloid materials:
   a. Agar (reversible)
   b. Alginate (irreversible)
2. Synthetic (non-aqueous) elastomers:
   a. Polysulfides (condensation polymerization)
   b. Condensation silicones (condensation polymerization)
   c. Addition silicones (addition polymerization)
   d. Polyethers (addition polymerization)

23.8.2 HYDROCOLLOIDS

23.8.2.1 Agar: Reversible Hydrocolloids

Agar is a reversible hydrocolloid. The material is heated to form a sol, and it gels in the mouth (with an appropriate cooling technique) to form the impression. The principal constituent is agar, which is a sulfuric ester of a linear polymer of galactose. The composition of this material is given in Table 23.5. This material is seldom used, as it involves use of an elaborate procedure and equipment for controlling the heating and cooling.

TABLE 23.5
Constituents of Agar Hydrocolloid Impression Materials

Constituent	wt. %	Function
Agar	14	Colloid
Borax	0.2	Strengthens the gel but retards the setting of dental stone model and die materials
Potassium sulfate	2	Accelerates the setting of dental stone
Water	83.8	Dispersion medium

TABLE 23.6
Constituents of Alginate Impression Materials

Constituent	wt. %	Function
Soluble salt of alginic acid (sodium, potassium or triethanol amine salt)	12	Reacts with calcium ions to give calcium alginate gel
Slowly soluble calcium salt (for example, calcium sulfate)	12	Releases calcium ions to reacts with salt of alginic acid
Trisodium phosphate	2	Reacts with calcium ions go give calcium phosphate and delay gel formation
Filler (for example, diatomaceous earth)	70	Increases cohesion of mix and strengthens gel
Silicofluorides or fluorides	Trace	Improves surface of stone model
Flavoring agents	Trace	Improves patient acceptability
Chemical indicator (present in some materials)	Trace	Changes color with pH change during setting to indicate correct timings for the impression procedure

23.8.2.2 Alginates: Irreversible Hydrocolloids

Alginates are irreversible hydrocolloids, that is, they form a gel irreversibly. They have been in use for over 60 years, and their chemistry is well known [176,177]. They are supplied as powders to be mixed with water. A typical composition is given in Table 23.6. Three of the components take part in two chemical reactions that occur sequentially. The gel is formed by the reaction of the alginic salt with calcium sulfate:

$$Na_nAlg + n/2\ CaSO_4 \longrightarrow n/2\ Na_2SO_4 + Ca_{n/2}Alg$$

However, this reaction would lead to very rapid gel formation, and not permit sufficient time for recording the impression before gelation. Thus, trisodium phoshate is included as a retarder. It reacts preferentially with calcium sulfate until it has all reacted:

$$2Na_3PO_4 + 3CaSO_4 \longrightarrow Ca_3(PO_4)_2 + 3Na_2SO_4$$

There are some important limitations to this material. The gel is not tough, and is liable to tear, although a constituent such as diatomaceous earth is included to strengthen the material. A more major limitation is the fact that the gel is dimensionally unstable and is susceptible to changes in fluid content. Three things can occur depending on the conditions: (1) syneresis, which is the contraction of the gel which causes a separation of liquid, (2) loss of moisture—a diffusion controlled process [178]—which causes shrinkage, and (3) imbibition or uptake of moisture, which can cause swelling.

Alginates are most commonly used for impressions for complete dentures, and in orthodontics for the preparation of study casts. However, they are not usually considered to have adequate dimensional accuracy and stability for impressions for cast alloy restorations or prostheses.

23.8.3 NON-AQUEOUS ELASTOMERS

23.8.3.1 Introduction

As noted earlier, two of the elastomers set by a condensation reaction and two by an addition reaction. Loss of volatile reaction products from a set impression can result in shrinkage, hence loss of dimensional stability. As a result, addition polymers are now the more widely used materials.

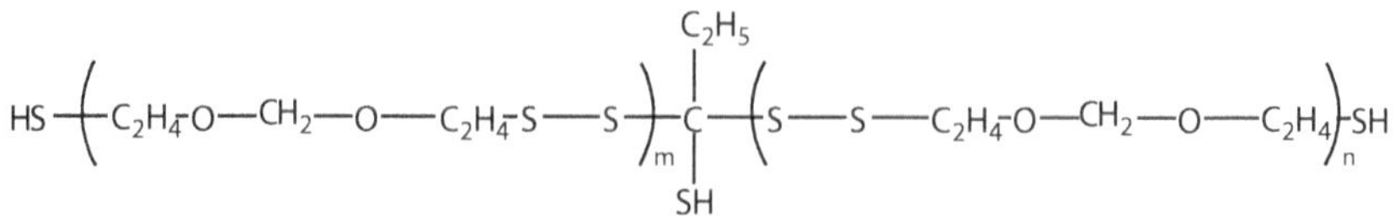

FIGURE 23.40   Polysulfide molecule with one pendant and two terminal –SH groups.

23.8.3.2 Polysulfides

These materials are supplied as two pastes. The "base" paste contains a polysulfide, the formula of which is given in Figure 23.40. This molecule contains three –SH groups; two of these groups are terminal, one is pendant. The base paste also contains a filler, between 11% and 54% for different materials, for example, TiO_2. This paste is usually colored white, because of the color of the filler. The reactor paste (sometimes called "accelerator" or "catalyst" paste) contains lead dioxide (PbO_2) that causes polymerization and cross-linking, by oxidation of –SH groups (Figure 23.40). Sulfur is also present. An oil (an ester or chlorinated paraffin) is included to form a paste of the correct consistency. This paste is brown colored due to the lead dioxide. On mixing the pastes, the –SH groups can be oxidized by PbO_2, giving S-S linkages, resulting in both chain lengthening and cross-linking.

23.8.3.3 Condensation Silicones

These are also two paste materials. The base paste contains a silicone polymer with terminal hydroxyl groups and a filler. This is mixed with a reactor paste containing a cross-linking agent such as an alkoxy ortho-silicate or a polymer thereof. In the reactor paste there is also an activator, usually an organo-tin compound, such as dibutyl-tin dilaurate. On mixing, cross-linking occurs with the elimination of alcohol (Figure 23.41).

23.8.3.4 Addition Silicones (Vinyl Polysiloxane)

Silicones that set by an addition reaction are now widely used, and are often termed vinyl polysiloxanes. These materials contain an organo-hydrogen siloxane, which can undergo an addition reaction with a siloxane compound containing a terminal vinyl group in the presence of a precious metal catalyst such as H_2PtCl_6. This type of reaction can lead to the formation of a cross-linked silicone rubber (Figure 23.42) [179].

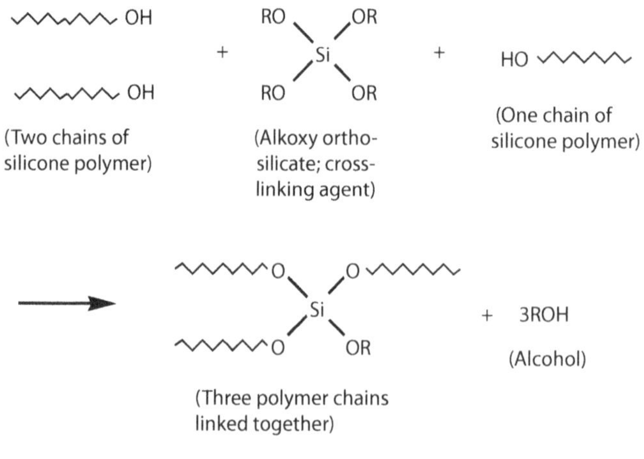

FIGURE 23.41   Condensation polymerization of silicones.

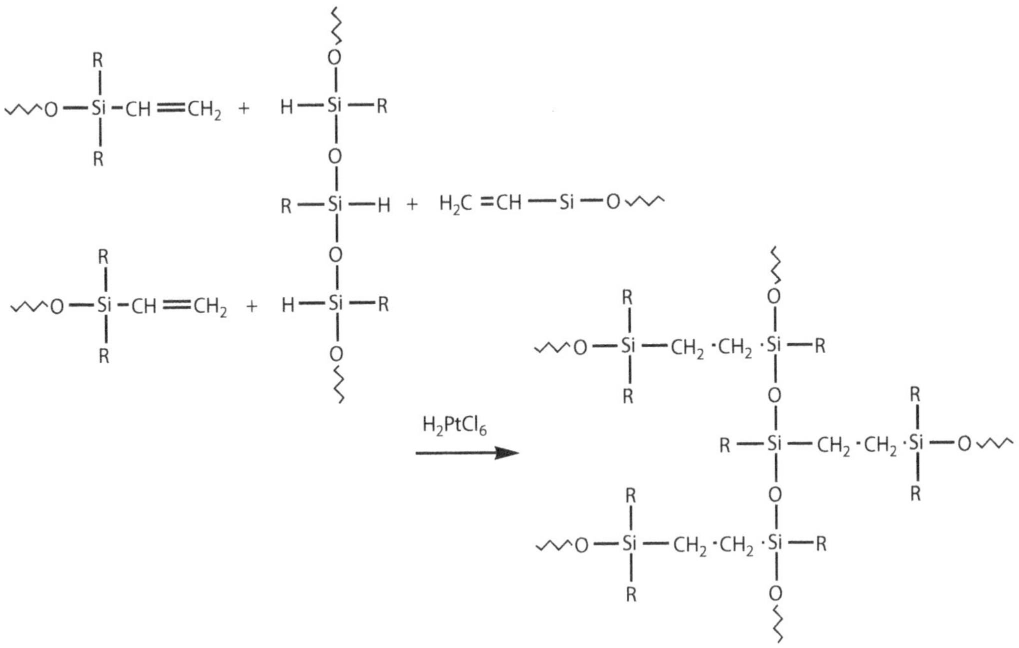

FIGURE 23.42 Addition polymerization of silicones.

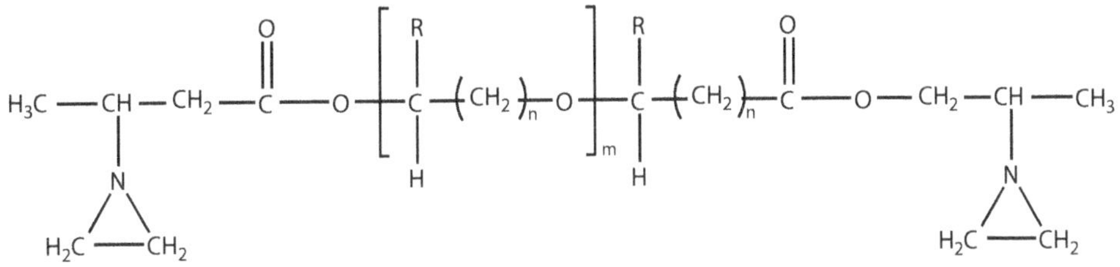

FIGURE 23.43 A polyether with imine end groups.

23.8.3.5 Polyether Materials

The base paste contains a polyether with imine end groups (Figure 23.43), a plasticizer and a filler. The reactor paste contains an aromatic sulfonate, a plasticizer and a filler. Setting is by cross-linking reaction of the imine groups by cationic addition polymerization [180].

23.8.3.6 General Comparison of Elastomers

In comparing and contrasting vinyl polysiloxane and polyether materials, some general points can be made:

- Both yield impressions of high accuracy and dimensional stability
- Vinyl polysiloxanes are not hydrophilic, which can be problematic if the material does not adequately wet moist tissues during impression taking. Some products have added surfactants in attempts to minimize this problem [181].
- Polyether materials are intrinsically hydrophilic, in contrast to the silicone-based materials.

23.8.3.7 Disinfection of Impression Materials

It is now becoming mandatory for impressions to be disinfected before transmission to the dental laboratory. Numerous experiments have been carried out on chemical disinfection by spraying or immersing impression materials. Usually disinfection does not affect the dimensional stability of an impression, though there are occasional exceptions to this [182]. In general non-aqueous elastomeric impressions are less likely to be adversely affected by immersion than the hydrocolloid-based materials.

23.9 PROVISIONAL (INTERIM) RESTORATIONS

23.9.1 PURPOSE AND REQUIREMENTS

Interim restorations are essential in fixed prosthodontic treatment. During the interval between tooth preparation and delivery of the definitive restoration, protection from thermal changes, margin breakage, caries and pulpal irritation is essential. An interim restoration also provides tooth stability and aids the patient in terms of oral function.

The requirements for such materials are essentially similar to those for definitive materials, although strength and esthetics may not be so important. One factor to consider is reaction exotherm—if this is too high, pulpal damage could ensue.

23.9.2 CRITIQUE OF MATERIALS

The available polymer-based materials include

- Poly(methyl methacrylate), supplied as a polymer powder to be mixed with monomer, and initiated by a redox reaction
- Materials based on higher methacrylates such as poly(ethyl methacrylate)
- Filled dimethacrylate polymers, similar to resin composites, but with less inorganic filler

Of these materials, the latter are probably the most widely used, given that the acrylic materials have the highest exotherm and the lowest flexural strength. The materials have been reviewed by Gratton et al. [183].

23.10 PROSTHODONTIC MATERIALS

23.10.1 COMPLETE DENTURES

23.10.1.1 Poly(Methyl Methacrylate): Techniques of Polymerization

Complete dentures are usually fabricated from acrylic polymers-poly(methyl methacrylate) or PMMA. Traditionally the procedure involves preparing a dough from monomer and beads of polymer, with benzoyl peroxide. This dough is packed under pressure in a two part mold and heated to activate the polymerization of the monomer. Temperature control is important. Too little heating will result in a denture with excessive residual monomer (see the following text). However very rapid initial heating can cause the dough temperature to exceed the boiling point of the monomer (100.9°C), because of the exothermic heat. This can result in a polymer with gaseous porosity.

Alternative polymerization methods have been explored, including autopolymerizing resins used in a "fluid resin" technique, for example, for emergency treatment [184], visible light curing [185], including a methyl methacrylate-free urethane dimethacrylate [186] and microwave heating [187].

23.10.1.2 Limitations of Acrylic Denture Materials

There are several areas of concern about PMMA as a denture material.

23.10.1.2.1 Biocompatibility and Residual Monomer Content

The residual monomer (RM) content of dental PMMA is related to the method of polymerization. For example,

RM of fluid resin > RM of heat cured resin ≥ RM of microwave cures resin [188].

The RM of a heat-cured material has been reported to be 0.045%–0.106%, compared with a figure of 0.185% for an autopolymerizing material [189].

The reasons for concern about RM are twofold. Firstly it may act as a plasticizer and compromise the mechanical properties of the material. Secondly, RM can be a problem in relation to biocompatibility. RM can cause irritation, inflammation and an allergic response [190]. Autopolymerizing acrylics have been found to be the most cytotoxic [190].

23.10.1.2.2 Mechanical Properties

PMMA dentures can fracture from flexural fatigue and impact forces [191]. As a result alternative polymers have been tested, including epoxy resins, polystyrene, and polycarbonate [192].

Modifications to PMMA have been advocated, including rubber-acrylic graft co-polymers, and the incorporation of up to 30% by weight low molecular weight butadiene styrene rubber [193].

Fiber reinforcement studies have included the use of carbon fibers [194], high modulus poly(p-phenylene terephthalamide) fibers (Kevlar®) [195], glass fibers [196–198], ultra-high molecular weight polyethylene fibers [199–201], and PMMA fibers [202].

Metallic reinforcement of PMMA has met with problems, including stress concentration around the metallic material which can cause weakening [191]. Also, bonding between phases has been difficult to ensure. In attempts to overcome this, the following have been tested:

- Surface roughening treatment of the metal [203]
- Silanization techniques [204]
- An adhesive resin containing 4-META [205]

23.10.1.2.3 Radiopacity

A problem with PMMA dentures is that if fracture occurs, there may be ingestion or aspiration of broken fragments. Radiological detection of such fragments is impossible, and deaths have resulted [206]. Attempts have been made to solve this problem by including constituents with elements such as iodine, barium or bismuth, but as yet there is no radiopaque denture material that can match PMMA in terms of esthetics and mechanical properties [206].

23.10.1.3 Hypoallergenic Denture Polymers

For patients who are allergic to methacrylates there are two possible alternative chemistries available. An injection molded nylon denture base material is available [207]. Also, a number of urethane-based materials are being produced [208].

23.10.2 Denture Teeth

Although porcelain is sometimes used as a synthetic tooth material, most dentures are fabricated with polymeric teeth. Options include conventional PMMA, cross-linked PMMA with conventional PMMA coating, or an interpenetrating polymer network (IPN) of cross-linked and uncrosslinked PMMA [209]. Composite teeth (nano-filled and micro-filled) are also produced [210]. It appears that the conventional PMMA teeth have the lowest fracture toughness [209] and least resistance to abrasive wear [211].

Good bonding between PMMA denture and teeth is essential. It appears that heat-cured polymers bond better to teeth than visible light cured resins [212]. Further, when heat curing is used, higher polymerization temperature results in greater diffusion of monomer from the denture base into the tooth, resulting in better bond strength [213].

23.10.3 RESILIENT POLYMERS AS DENTURE LININGS

23.10.3.1 Temporary Lining Materials

Temporary resilient lining materials for dentures may be used as tissue conditioners in order to condition inflamed and distorted soft tissues underlying poorly fitting dentures. They may also be used for recording functional impressions [214].

These materials contain a polymer powder which is mixed with a liquid plasticizer. Traditional materials have been based on poly(ethyl methacrylate) with an ethanol/ester mixture. On mixing, the ethanol increases the rate of gel formation. However, it is lost from the material within 24 h. Thus a product has been developed without ethanol; it is based on a *n*-butyl methacrylate/*i*-butyl methacrylate copolymer [215]. It should be noted that some materials whose composition has been reported by Murata et al. [215] contain phthalate esters—compared to mouth protectors (Section 23.4.2), where these compounds have been banned [98].

23.10.3.2 Permanent Resilient Lining Materials

These can be used when the tissues of the denture beading area are atrophied and pressure from the denture during mastication would cause pain. They can also be used for utilizing undercut areas to achieve maximum retention of the denture.

Requirements for resilient materials are stringent, and at this time there is no ideal material. Obviously the materials should not be toxic or irritant, and should ideally be free from odor and taste. Secure permanent bonding to the PMMA denture base is essential. There should be no adverse effect from the oral environment; thus loss of plasticizers (if present) is undesirable, as also is absorption of water causing swelling. The material should be permanently resilient. The composition and properties of these materials have been extensively reviewed [216].

23.10.3.2.1 Methacrylate-Based Materials

Heat-activated higher methacrylate polymers are used, in which poly(ethyl methacrylate) or co-polymers are mixed with ethyl, butyl or 2-ethoxyethyl methacrylate. Butyl phthalyl butyl glycolate or a phthalate ester plasticizer is incorporated. In general, these materials adhere well to PMMA dentures, but may leach plasticizer components, causing hardening and dimensional change.

23.10.3.2.2 Siloxane-Based Materials

These are usually chemically similar to the condensation silicone impression materials discussed earlier (Section 23.8.3). One addition polymerizable material, requiring heat at 100°C has been produced. In general, it is difficult to secure durable bonding of a silicone to PMMA. In attempts to overcome this, a silicone product with pendant methacrylate groups is available. A major concern with silicone soft linings is that they support the growth of *Candida albicans* [217].

23.10.3.2.3 Suggested Modified and Alternative Materials

In the development of modified or new materials, the following experimental approaches, among others, have been tested:

- The use of polymerizable plasticizers [218,219]
- The application of a fluoroalkyl methacrylate polymer [220]
- The use of a poly(fluoroalkoxy) phosphazine elastomeric system [216]

23.10.4 DENTURE ADHESIVES

These products are usually a blend of polymer salts such as sodium carboxymethyl cellulose and zinc and calcium salts of poly(methyl vinyl ether-co-maleic anhydride), as well as oils, flavors, binding materials and preservatives [221,222].

23.11 POLYMER APPLICATIONS IN PERIODONTOLOGY

23.11.1 Membranes for Guided Tissue Regeneration

Guided tissue regeneration (GTR) is a surgical procedure that utilizes barrier membranes to direct the growth of new bone and soft tissue. It is necessary to exclude unwanted cell lines from healing sites to allow growth of desired tissues. A barrier membrane provides primary closure and stability of the wound that promotes undisturbed and uninterrupted healing, creates and maintains space to facilitate bone in-growth and migration of undifferentiated mesenchymal cells [223–225]. Nyman et al. [226] demonstrated the potential of this technique in dentistry.

Two membrane types are currently used in GTR. These are

1. Synthetic non-resorbable PTFE membranes such as Gore-tex®
2. Synthetic resorbable membranes formed from gylcolide and lacide copolymers, collagen, etc.

23.11.2 Other Constituents

A membrane may be used alone or with a bone graft (an autograft or allograft) or bone substitute.

Growth factors like bone morphogenic proteins (BMP) and human recombinant platelet derived growth factors (rhPDGF) can also be used in conjunction with a barrier membrane. BMPs can be produced using recombinant DNA technology. BMP-2 and BMP-7 have FDA approval for human use. The biological basis of bone morphogenesis was shown by Urist [227]. A recent review has been presented by Bashutski et al. [228].

23.12 DENTAL IMPLANTS

Titanium-based implants are finding increasing use in dentistry, for the functional replacement of missing teeth. They have achieved a considerable measure of success, related to the process of osseointegration, in which newly formed bone develops and integrates with the implant surface [229]. Extensive research is being carried out on surface modification to enhance the interaction with bone. Such surface treatments include (but are not limited to) the use of polymers, which may be naturally occurring or synthetic materials. Research has included collagen and biomimetic polymers similar to the extracellular matrix of bone [230].

23.13 SOFT TISSUE AUGMENTATION

23.13.1 Objectives and Requirements

Numerous materials have been employed for facial rejuvenation, in attempts to correct the effects of aging. Such procedures are increasing in popularity. Material requirements include: non-allergenic, minimal side effects, longevity, yet with the ability to reverse the treatment and ease of use [231].

23.13.2 Materials

Materials may be classified according to (a) chemical composition—naturally occurring vs. synthetic polymers [232], (b) temporary vs. permanent materials [233], or (c) reversible vs. nonreversible materials [234]. Cross-linked hyaluronic acid (HA) is the most popular of the polymeric materials [234]; this may be avian-derived or bacterial-cultured [232]. HA materials are reversible because treatment can be reversed by the injection of hyaluronidase [234]. Collagen-containing materials (from bovine, cadaveric or cell-cultured sources) are available [232]. Other polymers include poly-L-lactic acid microparticles, PMMA microspheres in 3.5% bovine collagen, silicone liquids, poly(alkyl imide), and polyacrylamide hydrogel. Not all of these materials have received FDA approval [232].

23.14　MAXILLOFACIAL PROSTHESES

23.14.1　REQUIREMENTS

Facial defects resulting from surgery, trauma or malformation require prosthetic rehabilitation when reconstructive surgery is not feasible or advisable [235]. The polymers employed for such treatment include silicone and polyurethane elastomers and methacrylates [236,237] though no type of material shows all the ideal properties. Applications include the use of rigid materials (e.g., for orbital prostheses) and materials that can mimic the flexibility of living tissue. Some prostheses are fabricated from a silicone bonded to a rigid acrylic base.

The requirements for such materials are stringent, including the obvious ones of biocompatibility and durability. From the patient's viewpoint, color, texture and hygiene considerations are paramount. The material must also have suitable mechanical properties including flexibility and resistance to tearing.

23.14.2　MATERIALS

23.14.2.1　Silicone Polymers

Medical grade silicones—both heat vulcanizing and autopolymerizing—are used, but the latter are preferred. Their chemistry is somewhat similar to the impression materials discussed earlier (Section 23.8.3), consisting of filled siloxanes. They may be polymerized by a condensation reaction or an addition reaction, as for the impression materials. Fillers may include silica or kieselguhr; though these increase strength, they may reduce the polymer's translucency [237].

Aziz et al. [238] have compared the mechanical properties, such as hardness, tensile strength, tear strength and water absorption of a range of silicone products. In a separate study, Hulterström et al. [239] claimed that addition-type silicones showed little or no water absorption, in contrast to two condensation materials. However, this may be related to the silica filler; –OH groups on the surface of the silica may facilitate water absorption, whereas surface-treated hydrophobic fillers may inhibit this [238].

In attempts to achieve good esthetics, many pigments and coloring techniques have been suggested [240], including intrinsic and extrinsic coloration [237]. However, color stability is difficult to achieve, given that factors such as ultraviolet light exposure, application of cosmetics and cleaning agents may operate. Inorganic pigments appear to be more stable over time compared with organic colorants [241].

A methacryloxypropyl-terminated silicone has been developed to facilitate adhesion to non-silicone-based adhesives [242]. Improvements in physical properties of maxillofacial silicones have been achieved by the incorporation of nanosized oxides (TiO_2, ZnO, CeO_2) at concentrations of 2%–2.5% [243].

23.14.2.2　Methacrylates

PMMA can be used for restoration of defects with tissue beds that are relatively non-mobile [237], for ocular prostheses [244] and for hollow obturator prostheses. For this latter application, visible light curing denture materials are being advocated [245].

23.14.2.3　Other Polymers

Other polymeric maxillofacial materials include polyurethanes [246] and chlorinated polyethylene [247]. Glass fiber-reinforced composite have been proposed for a reinforcing substructure to reinforce a silicone elastomer [248].

23.15　WOUND CLOSURE MATERIALS

The requirements for suturing in dentistry are exacting, given the presence of saliva, and functions such as speech, mastication and swallowing [249]. Non-resorbable sutures such as silk resist the effects of oral fluids, but require a clinical appointment for removal. Resorbable sutures include gut, chromic gut, polyglactin, and poly(glycolic acid) [250]. Gassner [251] has extensively reviewed wound closure materials.

23.16 CONCLUSION

It has been shown that there is extensive ongoing research in biomedical polymers for dental and maxillofacial applications. The limitations of current materials present great challenges to the researcher. Providers of treatment are challenged to make wise decisions on the selection and utilization of materials. The future is exciting as the possibilities of new technologies are demanding radically new approaches to the subject [252].

ACKNOWLEDGMENTS

The author is indebted to colleagues for helpful discussion, particularly, Dr. C. Aparicio, Dr. J. Beyer, Dr. Yasser F. Gomaa, Dr. V. Thumbigere Math and Dr. J. Perdigão. Also Drs. Perdigão and Beyer kindly supplied some of the illustrations.

REFERENCES

1. K. R. St. John. Biocompatibility of dental materials. *Dent. Clin. N. Am.* 51:747–760 (2007).
2. J. C. Wataha. Principles of bicompatibility for dental practitioners. *J. Prosthet. Dent.* 86:203–209 (2001).
3. D. F. Williams. On the mechanisms of biocompatibility. *Biomater.* 29:2941–2953 (2008).
4. V. B. Michelsen, H. Lygre, R. Shålevik, A. B. Tveit, E. Solheim. Identification of organic eluates from four polymer-based dental filling materials. *Eur. J. Oral Sci.* 111:263–271 (2003).
5. M. Taira, H. Urabe, T. Hirose, K. Wakasa, M. Yamaki. Analysis of photo-initiators in visible-light-cured dental composite resins. *J. Dent. Res.* 67:24–28 (1988).
6. G. J. Sun, K. H. Chae. Properties of 2,3-butanedione and 1-phenyl-1,2-propanedione as new photosensitizers for visible light cured dental resin composites. *Polymer* 41:6205–6212 (2000).
7. K. L. Van Landuyt, J. Snauwaert, J. De Munck, M. Peumans, Y. Yoshida, A. Poitevin, E. Coutinho, K. Suzuki, P. Lambrechts, B. Van Meerbeek. Systematic review of the chemical composition of contemporary dental adhesives. *Biomater.* 28:3757–3785 (2007).
8. D. N. Raith, J. E. Palamara, H. H. Messer. Temperature change, dentinal fluid flow and cuspal displacement during resin composite restoration. *J. Oral Rehabil.* 34:693–701 (2007).
9. S. A. Faraj, B. Ellis. The effect of processing temperatures on the exotherm, porosity and properties of acrylic denture base. *Br. Dent. J.* 147:209–212 (1979).
10. L. G. Lopes, E. B. Franco, J. C. Pereira, R. F. L. Mondelli. Effect of light-curing units and activation mode on polymerization shrinkage and shrinkage stress of composite resins. *J. Appl. Oral Sci.* 16:35–42 (2008).
11. J. L. Ferracane, J. C. Mitchem, J. R. Condon, R. Todd. Wear and marginal breakdown of composites with various degrees of cure. *J. Dent. Res.* 76:1508–1516 (1997).
12. L. Feng, B. I. Suh, Acrylic resins resisting oxygen inhibition during free-radical photocuring. I. Formulation attributes. *J. Appl. Pol. Sci.* 112:1565–1571 (2009).
13. S. Fujisawa, Y. Kadoma. Action of eugenol as a retarder against polymerization of methyl methacrylate by benzoyl peroxide. *Biomaterials* 18:701–703 (1997).
14. W. D. Cook, P. M. Standish. Polymerization kinetics of resin-based restorative materials. *J. Biomed. Mater. Res.* 17:275–282 (1983).
15. C. A. Lapp, G. S. Schuster. Effects of DMAEMA and 4-methoxyphenol on gingival fibroblast growth, metabolism, response to interleukin-1. *J. Biomed. Mater. Res.* 60:30–35 (2002).
16. M. Reed, H. Fujiwara, D. C. Thompson. Comparative metabolism, covalent binding and toxicity of BHT congeners in rat liver slices. *Chem. Biol. Interact.* 138:155–170 (2001).
17. D. C. Smith. A new dental cement. *Br. Dent. J.* 124:381–384 (1968).
18. A. D. Wilson, B. E. Kent. A new translucent cement for dentistry. The glass ionomer cement. *Br. Dent. J.* 132:133–135 (1972).
19. A. D. Wilson, S. Crisp, G. Abel. Characterization of glass-ionomer cements. 4. Effect of molecular weight on physical properties. *J. Dent.* 5:117–120 (1977).
20. P. V. Hatton, I. M. Brook. Characterisation of the ultrastructure of glass-ionomer (poly-alkenoate) cement. *Br. Dent. J.* 173:275–277 (1992).
21. The Academy of Prosthodontics. The glossary of prosthodontic terms. *J. Prosthet. Dent.* 94:21–38 (2005).

22. E. E. Hill. Dental cements for definitive luting: a review and practical clinical considerations. *Dent. Clin. N. Am.* 51:643–658 (2007).

23. D. Edelhoff, M. Özcan. To what extent does the longevity of fixed dental prostheses depend on the function of the cement? *Clin. Oral Impl. Res.* 18(supplement 3):193–204 (2007).

24. J. O. Burgess, B. K. Norling, H. R. Rawls, J. L. Ong. Directly placed esthetic restorative materials—the continuum. *Comp. Contin. Ed. Dent.* 17:731–732,734 (1996).

25. R. L. Bowen. U. S. Patent 3,066,112 (1962).

26. F. M. Burke, N. J. Ray, R J. McConnell. Fluoride-containing restorative materials. *Int. Dent. J.* 56:33–43 (2006).

27. A. D. Wilson. Resin-modified glass-ionomer cements. *Int. J. Prosthodont.* 3:425–429 (1990).

28. J. W. Nicholson. Polyacid-modified composite resins ("compomers") and their use in clinical dentistry. *Dent. Mater.* 23:615–622 (2007).

29. G. Eliades, A. Kakaboura, G. Palaghias. Acid-base reaction and fluoride release profiles in visible light-cured polyacid-modified composite restoratives (compomers). *Dent. Mater.* 14:57–63 (1998).

30. J. W. McLean, J. W. Nicholson, A. D. Wilson. Proposed nomenclature for glass-ionomer dental cements and related materials. *Quint. Int.* 25:587–589 (1994).

31. J. W. Nicholson, M. A. McKenzie. The properties of polymerizable luting cements. *J. Oral Rehabil.* 26:767–774 (1999).

32. C. H. Lloyd, L. Mitchell. The fracture toughness of tooth coloured restorative materials. *J. Oral Rehabil.* 11:257–272 (1984).

33. R. W. Phillips. Advancements in adhesive restorative dental materials. *J. Dent. Res.* 45:1662–1667 (1966).

34. A. D. Wilson, H. J. Prosser, D. M. Powis. Mechanism of adhesion of polyelectrolyte cements to hydroxy-apatite. *J. Dent. Res.* 62:590–592 (1983).

35. M. J. Tyas. The effect of dentine conditioning with polyacrylic acid on the clinical performance of glass ionomer cement. *Aust. Dent. J.* 38:46–48 (1993).

36. M. G. Buonocore. A simple method of increasing the adhesion of acrylic filling materials to enamel surfaces. *J. Dent. Res.* 34:849–853 (1955).

37. M. G. Buonocore, A. Matsui, A. J. Gwinnett. Penetration of resin dental materials into enamel surfaces with reference to bonding. *Archs. Oral Biol.* 13:61–70 (1968).

38. A. J. Gwinnett, L. W. Ripa. Penetration of pit and fissure sealants into conditioned human enamel *in vivo*. *Archs. Oral Biol.* 18:435–439 (1973).

39. G. W. Marshall Jr, S. J. Marshall, J. H. Kinney, M. Balooch. The dentin substrate: structure and properties related to bonding. *J. Dent.* 25:441–458 (1997).

40. F. R. Tay, D. H. Pashley. Have dentin adhesives become too hydrophilic? *J. Can. Dent. Assoc.* 69:726–731 (2003).

41. Gebr. De Trey Aktiengesellschaft. Swiss Patent 278,946 (1951).

42. I. R. H. Kramer, J. W. McLean. Alterations in the staining reactions of dentine resulting from a constituent of a new self-polymerizing resin. *Brit. Dent. J.* 93:150–153 (1952).

43. W. H. Douglas. Clinical status of dentine bonding agents. *J. Dent.* 17:209–215 (1989).

44. I. Watanabe, N. Nakabayashi, D. H. Pashley. Bonding to ground dentin by a phenyl-P self-etching primer. *J. Dent. Res.* 73:1212–1220 (1994).

45. T. Fusayama. *New Concepts in Operative Dentistry*. p. 118. Quintessence Publishing, Chicago, IL (1980).

46. J. Perdigão. New developments in dental adhesion. *Dent. Clin. N. Am.* 51:333–357 (2007).

47. N. Nakabayashi, K. Kojima, E. Masuhara. The promotion of adhesion by the infiltration of monomers into tooth substrates. *J. Biomed. Mater. Res.* 16:265–273 (1982).

48. S. Uno, W. J. Finger. Effects of acidic conditioners on dentine demineralization and dimension of hybrid layers. *J. Dent.* 24:211–216 (1996).

49. M. Hashimoto, H. Ohno, K. Endo, M. Kaga, H. Sana, H. Oguchi. The effect of hybrid layer thickness on bond strength: demineralized dentin zone of the hybrid layer. *Dent. Mater.* 16:406–411 (2000).

50. N. Nakabayashi, T. Saimi. Bonding to intact dentin. *J. Dent. Res.* 75:1706–1715 (1996).

51. B. Van Meerrbeck, J. Perdigão, P. Lambrechts, G. Vanherle. The clinical performance of adhesives. *J. Dent.* 26:1–20 (1998).

52. L. Breschi, A. Mazzoni, A. Ruggeri, M. Cadenaro. Dental adhesion review: aging and stability of the bonded interface. *Dent. Mater.* 24:90–101 (2008).

53. M. Peumans, P. Kanumilli, J. De Munck, K. Van Landuyt, P. Lambrechts, B. Van Meerbeek. Clinical effectiveness of contemporary adhesives: a systematic review of current clinical trials. *Dent. Mater.* 21:864–881 (2005).

54. J. De Munck, K. Van Landuyt, M. Peumans, A. Poitevin, P. Lambrechts, M. Braem, B. Van Meerbeek. A critical review of the durability of adhesion to tooth tissue: methods and results. *J. Dent. Res.* 84:118–132 (2005).

55. S. Inoue, K. Koshiro, Y. Yoshida, J. De Munck, K. Nagakane, K. Suzuki, H. Sano, B. Van Meerbeek. Hydrolytic stability of self-etch adhesives bonded to dentin. *J. Dent. Res.* 84:1160–1164 (2005).

56. K. L. Van Landuyt, Y. Yoshida, I. Hirata, J. Snauwaert, J. De Munck, M. Okazaki, K. Suzuki, P. Lambrechts, B. Van Meerbeck. Influence of the chemical structure of functional monomers on their adhesive performance. *J. Dent. Res.* 87:757–761 (2008).

57. Y. Yoshida, K. Nagakane, R. Fukuda, Y. Nakayama, M. Okazaki, H. Shintani, S. Inoue et al. Comparative study on adhesive performance of functional monomers. *J. Dent. Res.* 83:454–458 (2004).

58. K. L. Van Landuyt, J. De Munck, J. Snauwaert, E. Continho, A. Poitevin, Y. Yoshida, S. Inoue et al. Monomer-solvent phase separation in one-step self-etch adhesives. *J. Dent. Res.* 184:183–188 (2005).

59. K. L. Van Landuyt, J. Snauwaert, J. De Munck, E. Continho, A. Poitevin, Y. Yoshida, K. Suzuki, P. Lambrechts, B. Van Meerbeek. Origin of interfacial droplets with one-step adhesives. *J. Dent. Res.* 86:739–744 (2007).

60. A. F. Reis, M. Giannini, P. N. R. Pereira. Long-term TEM analysis of the nanoleakage patterns in resin-dentin interfaces produced by different bonding strategies. *Dent. Mater.* 23:1164–1172 (2007).

61. K. L. Van Landuyt, J. Snauwaert, M. Peumans, J. De Munck. The role of HEMA in one-step self-etch adhesives. *Dent. Mater.* 24:1412–1419 (2008).

62. M. G. Brackett, F. R. Tay, W. W. Brackett, A. Dib, F. A. Dipp, S. Mai, D. H. Pashley. In vivo chlorhexidine stabilization of hybrid layers of an acetone-based dentin adhesive. *Oper. Dent.* 34:379–383 (2009).

63. R. Stanislawczuk, R. C. Amaral, C. Zander-Hrande, D. Gagler, A. Reis, A. D. Loguercio. Chlorhexidine-containing acid conditioner preserves the longevity of resin-dentin bonds. *Oper. Dent.* 34:481–490 (2009).

64. H. Hosaka, Y. Nishitani, J. Tagami, M. Yoshiyama, W. W. Brackett, K. A. Agee, F. R. Tay, D. H. Pashley. Durability of resin-dentin bonds to water- vs. ethanol-saturated dentin. *J. Dent. Res.* 88:146–151 (2009).

65. J. P. Matinlinna, L. V. J. Lassila. M. Özcan, A. Yli-Urpo, P. K. Vallittu. An introduction to silanes and their clinical applications in dentistry. *Int. J. Prosthodont.* 17:155–164 (2004).

66. M. B. Blatz, A. Sadan, M. Kern. Resin-ceramic bonding: a review of the literature. *J. Prosthet. Dent.* 89:268–274 (2003).

67. R. L. Bertolotti. Adhesion to porcelain and metal. *Dent. Clin. N. Am.* 51:433–451 (2007).

68. U.S. Department of health and Human Services. *Oral Health in America: A Report of the Surgeon General*. U. S. Department of Health and Human Services, National Institute of Dental and Craniofacial Research, National Institutes of Health: Washington, DC; Rockford, IL; Bethesda, MD (2000).

69. S. M. Adair. The role of sealants in caries prevention programs. *J. Calif. Dent. Assoc.* 31:221–227 (2003).

70. R. Welbury, M. Raadal, N. A. Lygidakis. EAPD guidelines for the use of pit and fissure sealants. *Eur. J. Paediatric Dent.* 5:179–184 (2004).

71. L. W. Ripa. Sealants revisited: an update of the effectiveness of pit-and-fissure sealants. *Caries Res.* 27 (supplement 1):77–82 (1993).

72. R. J. Simonsen. Preventive resin restorations and sealants in light of current evidence. *Dent. Clin. N. Am.* 49:815–823 (2005).

73. R. C. Parkhouse, G. B. Winter. A fissure sealant containing methyl-2-cyanoacrylate as a caries preventive agent. A clinical evaluation. *Br. Dent. J.* 130:16–19 (1971).

74. D. C. Birdsell, P. J. Bannon, R. B. Webb. Harmful effects of near-ultraviolet radiation used for polymerization of a sealant and a composite resin. *J. Am. Dent. Assoc.* 94:311–314 (1977).

75. K. C. Young, M. Hussey, F. C. Gillespie, K. W. Stephen. The performance of ultraviolet lights used to polymerize fissure sealants. *J. Oral Rehabil.* 4:181–191 (1977).

76. I. E. Ruyter. Unpolymerized surface layers on sealants. *Acta Odont. Scan.* 39:27–32 (1981).

77. N. Olea, R. Pulgar, P. Pérez, F. Olea-Serrano, A. Rivas, A. Novillo-Fertrell, V. Pedraza, A. M. Soto, C. Sonnenschein. Estrogenicity of resin-based composites and sealants used in dentistry. *Environ. Health Perspect.* 104:298–305 (1996).

78. R. Joskow, D. B. Barr, J. R. Barr, A. M. Calafat, L. L. Needham, C. Rubin. Exposure to bisphenol A from bis-glycidyl dimethacrylate-based dental sealants. *J. Am. Dent. Assoc.* 137:353–362 (2006).

79. A. Azarpazhooh, P. A. Main. Is there a risk of harm or toxicity in the placement of pit and fissue sealant materials? A systematic review. *J. Can. Dent. Assoc.* 74:179–183 (2008).

80. J. H. Warford II, J. H. Warford III, E.C. Combe. U. S. Patent 6,620,859 (2003).

81. A. Ahovuo-Saloranta, A. Hiiri, A. Norblad, M. Mäkelä, H. V. Worthington. Pit and fissure sealants for preventing dental decay in the permanent teeth of children and adolescents. *Cochrane Database of Systematic Reviews* 2008, Issue 4. Art No.: CD001830. DOI: 10.1002/14651858.CD001830.pub3.

82. E. M. Oong, S. O. Griffin, W. G. Kohn, B. F. Gooch, P. W. Caufield. The effect of dental sealants on bacteria levels in caries lesions. A review of the evidence. *J. Am. Dent. Assoc.* 139:271–278 (2008).

83. S. O. Griffin, E. Oong, W. Kohn, B. Vidakovic, B. F. Gooch, J. Bader, J. Clarkson, M. R. Fontana, D. M. Meyer, R. G. Roozier, J. A. Weintraub, D. T. Zero. The effectiveness of sealants in managing caries lesions. *J. Dent. Res.* 87:169–174 (2008).

84. R. J. Feigal, I. Quelhas. Clinical trial of a self-etching adhesive for sealant application: success at 24 months with Prompt L-Pop. *Am J. Dent.* 16:249–251 (2003).

85. A. Peutzfeldt, L. A. Nielsen. Bond strength of a sealant to primary and permanent enamel: phosphoric acid versus self-etching adhesive. *Pediatr. Dent.* 26:240–244 (2004).

86. J. Perdigão, J. W. Fundingsland, S. Duarte Jr, M. Lopes. Microtensile adhesion of sealants to intact enamel. *Int. J. Paediatric Dent.* 15:342–348 (2005).

87. R. J. Feigal. Self-etch adhesives for sealants? *J. Esthetic Restorative Dent.* 19:69–70 (2007).

88. R. M. Puppin-Rontani, M. E. Baglioni-Gouvea, M. F. deGoes, F. Garcia-Godoy. Compomer as a pit and fissure sealant: effectiveness and retention after 24 months. *J. Dent. Child.* 73:31–36 (2006).

89. N. Yakut, H. Sönmez. Resin composite sealant vs. polyacid-modified resin composite applied to post eruptive mature and immature molars: two year clinical study. *J. Clin. Pediatr. Dent.* 30:215–218 (2006).

90. N. Beiruti, J. E. Frencken, M. A. van't Hof, W. H. Van Palenstein Helderman. Caries-preventive effect of resin-based and glass ionomer sealants over time: a systematic review. *Commun. Dent. Oral Epidemiol.* 34:403–409 (2006).

91. R. J. Simonsen. Pit and fissure sealant: review of the literature. *Pediatr. Dent.* 24:393–414 (2002).

92. J. Beauchamp, P. W. Caufield, J. J. Crall, K. Donly, R. Feigal, B. Gooch, A. Ismail, W. Kohn, M. Siegal, R. Simonsen. Evidence-based clinical recommendations for the use of pit-and-fissure sealants. *J. Am. Dent. Assoc.* 139:257–267 (2008).

93. A. Azarpazhooh, P. A. Main. Pit and fissure sealants in the prevention of dental caries in children and adolescents: a systematic review. *J. Can. Dent. Assoc.* 74:171–177 (2008).

94. American Society for Testing and Materials. *Standard Practice for Care and Use of Athletic Mouth Protectors.* American Society for Testing and Materials, Philadelphia, PA, F 697–00 (2006).

95. American Dental Association. The importance of using mouthguards. *J. Am. Dent. Assoc.* 135:1061 (2004).

96. R. E. Going, R. E. Loehman, M. S. Chan. Mouthguard materials: their physical and mechanical properties. *J. Am. Dent. Assoc.* 89:132–138 (1974).

97. J. B. Park, K. L. Shaull, B. Overton, K. J. Donly. *J. Prosthet. Dent.* 72:373–380 (1994).

98. J. B. Fontelles, C. Clarke. Directive 2005/84/EC of the European Parliament. *Off. J. Eur. Union*, 48:40–43 (2005).

99. T. E. Gould, S. G. Piland, J. Shin, C. E. Hoyle, S. Nazarenko. Characterization of mouthguard materials: physical and mechanical properties of commercialized products. *Dent. Mater.* 25:771–780 (2009).

100. ADA Council on Access, Prevention and Interprofessional relations and ADA Council on Scientific Affairs. Using mouthguards to reduce the incidence and severity of sports-related oral injuries. *J. Am. Dent. Assoc.* 137:1712–1720 (2006).

101. A. Picozzi. Mouth protectors. *Dent. Clin. N. Am.* 19:385–387 (1975).

102. J. J. Knapik, S. W. Marshall, R. B. Lee, S. S. Darakjy, S. B. Jones, T. A. Mitchener, G. C. delaCruz, B. H. Jones. Mouthguards in sport activities. *Sports Med.* 37:117–144 (2007).

103. D. J. Howells, P. Jones. In vitro evaluation of a cyanoacrylate bonding agent. *Br J. Ortho.* 16:75–78 (1989).

104. D. Millett, N. Mandall, J. Hickman, R. Mattick, A.-M. Glenny. Adhesives for fixed orthodontic bands. A systematic review. *Angle Orthod.* 79:193–199 (2009).

105. T. Eliades. Orthodontic materials research and applications: part 1. Current status and projected future developments in bonding and adhesives. *Am J. Orthod. Dentofacial Orthop.* 130:445–451 (2006).

106. T. Eliades, A. Hiskia, G. Eliades, A. E. Athanasiou. Assessment of bisphenol-A release from orthodontic adhesives. *Am J. Orthod. Dentofacial Orthop.* 131:72–75 (2007).

107. T. Eliades, V. Gioni, D. Kletsas, A. E. Athanasiou, G. Eliades. Oestrogenicity of orthodontic adhesive resins. *Eur. J. Orthod.* 29:404–407 (2007).

108. L. Gorelick, A. M. Geiger, A. J. Gwinnett. Incidence of white spot formation after bonding and banding. *Am. J. Orthod.* 81:93–98 (1982).

109. S. Imran. Fluoride varnish reduces white spot lesions during orthodontic treatment. *Evid. Based Dent.* 9:81 (2008).

110. A. W. Benham, P. M. Campbell, P. H. Buschang. Effectiveness of pit and fissure sealants in reducing white spot lesions during orthodontic treatment. A pilot study. *Angle Orthod.* 79:338–345 (2009).

111. F. Heravi, R. Rashed, L. Raziee. The effects of bracket removal on enamel. *Aust. Orthod. J.* 24:110–115 (2008).

112. C. S. Chen, M. L. Hsu, K. D. Chang, S. H. Kuang, P. T. Chen, Y. W. Gung. Failure analysis: enamel fracture after debonding orthodontic brackets. *Angle Orthod.* 78:1071–1077 (2008).

113. C. Nattrass, A. J. Ireland, C. R. Lovell. Latex allergy in an orthognathic patient and implications for clinical management. *Br. J. Oral Maxillofac. Surg.* 37:11–13 (1999).

114. C. Bertoncini, E. Cioni, B. Grampi, P. Gandini. In vitro properties' change of latex and non-latex orthodontic elastics. *Prog. Orthod.* 7:76–84 (2006).

115. P. Gandini, R. Gennai, C. Bertoncini, S. Massironi. Experimental evaluation of latex-free orthodontic elastics' behaviour in dynamics. *Prog. Orthod.* 8:88–99 (2007).

116. M. L. Kersey, K. E. Glover, G. Heo, D. Raboud, P. W. Major. A comparison of dynamic and static testing of latex and nonlates orthodontic elastics. *Angle Orthod.* 73:181–186 (2003).

117. M. Pcumans, B. Van Meerbeek, P. Lambrechts, G. Vanherle. Porcelain veneers: a review of the literature. *J. Dent.* 28:163–177 (2000).

118. G. J. Christensen. Have porcelain veneers arrived? *J. Am. Dent. Assoc.* 122:81 (1991).

119. J. R. Calamia, C. S. Calamia. Porcelain laminate veneers: reasons for 25 years of success. *Dent. Clin. N. Am.* 51:399–417 (2007).

120. D. Layton, T. Walton. An up to 16-year prospective study of 304 porcelain veneers. *Int. J. Prosthod.* 20:389–396 (2007).

121. F. J. T. Burke, M. G. D. Kelleher. Perspectives. The "daughter test" in elective esthetic dentistry. *J. Esthet. Restor. Dent.* 21:143–146 (2009).

122. A. D. Puckett, J. G. Fitchie, P. C. Kirk, J. Gamblin. Direct composite restorative materials. *Dent. Clin. N. Am.* 51:659–675 (2007).

123. M. N. Mandikos, G. P. McGivney, E. Davis, P. J. Bush, J. M. Carter. A comparison of the wear resistance and hardness of indirect composite resins. *J. Prosthet. Dent.* 85:386–395 (2001).

124. F. Lutz, R. W. Phillips. A classification and evaluation of composite resin systems. *J. Prosthet. Dent.* 50:480–488 (1983).

125. A. C. Shortall, W. M. Palin, P. Burtscher. Refractive index mismatch and monomer reactivity influence composite curing depth. *J. Dent. Res.* 87:84–88 (2008).

126. R. W. Hasel. U. S Patent 5,944,527 (1999).

127. A. Nuray, L. E. Tam, D. McComb. Flow, strength, stiffness and radiopacity of flowable resin composites. *J. Can. Dent. Assoc.* 69:516–521 (2003).

128. R. D. Jackson, M. Morgan. The new posterior resins and a simplified placement technique. *J. Am. Dent. Assoc.* 131:375–383 (2000).

129. K. F. Leinfelder, S. C. Bayne. E. J. Swift Jr. Packable composites: overview and technical considerations. *J. Esthet. Dent.* 11:234–249 (1999).

130. A. Peutzfeldt. Resin composites in dentistry: the monomer systems. *Eur. J. Oral Sci.* 105:97–116 (1997).

131. K. Tanaka, M. Taira, H. Shintani, K. Wakasa, M. Yamaki. Residual monomers (TEGDMA and Bis-GMA) of a set visible-light-cured dental composite resin when immersed in water. *J. Oral Rehabil.* 18:353–362 (1991).

132. W. Spahl, H. Budzikiewicz, W. Guersten. Determination of leachable components from four commercial dental composites by gas and liquid chromatography/mass spectrometry. *J. Dent.* 26:137–145 (1998).

133. U. Ortengren, H. Wellendorf, S. Karlsson, I. E. Ruyter. Water sorption and solubility of dental composites and identification of monomers released in an aqueous environment. *J. Oral Rehabil.* 28:1106–1115 (2001).

134. Y. Issa, D. C. Watts, P. A. Brunton, C. M. Waters, A. J. Duxbury. Resin composite monomers alter MTT and LDH activity of human gingival fibroblasts *in vitro*. *Dent. Mater.* 20:12–20 (2004).

135. D. H. Lee, N. R. Kim, B.-S. Lim, Y.-K. Lee, H.-C. Yang. Effects of TEGDMA and HEMA on the expression of COX-2 and iNOS in cultures murine macrophage cells. *Dent. Mater.* 25:240–246 (2009).

136. J. P. Santerre, L. Shajii, B. W. Leung. Relation of dental composite formulations to their degradation and the release of hydrolyzed polymeric-resin-derived products. *Crit. Rev. Oral Biol. Med.* 12:136–151 (2001).

137. T. Nihei, A. Dabanoglu, T. Teranaka, S. Kurata, K. Ohashi, Y. Kondo, N. Yoshino, R. Hickel, K.-H. Kunzelmann. Three-body-wear resistance of the experimental composites containing filler treated with hydrophobic silane coupling agents. *Dent. Mater.* 24:760–764 (2008).

138. L. Lindqvist, C. E. Nord, P. O. Söder. Origin of esterase in human whole saliva. *Enzyme* 22:165–175 (1977).

139. Y. Finer, J. P. Santerre. Salivary esterase activity and its association with the biodegradation of dental composites. *J. Dent. Res.* 83:22–26 (2004).

140. Y. Finer, J. P. Santerre. Influence of silanated filler content on the biodegradation of bisGMA/TEGDMA dental composite resins. *J. Biomed. Mater. Res.* 81A:75–84 (2007).

141. J. L. Drummond. Degradation, fatigue, and failure of resin dental composite materials. *J. Dent. Res.* 87:710–719 (2008).

142. N. Moszner, U. Salz. New developments of polymeric dental composites. *Prog. Polym. Sci.* 26:535–576 (2001).

143. R. R. Braga, R. Y. Ballester, J. L. Ferracane. Factors involved in the development of polymerization shrinkage stress in resin-composites: a systematic review, *Dent. Mater.* 21:962–970 (2005).

144. M. J. M. Coelho Santos, G. C. Santos Jr, H. Nagem Filho, R. F. L. Mondelli, O. El-Mowafy. Effect of light curing method on volumetric polymerization shrinkage of resin composites. *Oper. Dent.* 29:157–161 (2004).

145. R. L. Sakaguchi, A. Versluis, W. H. Douglas. Analysis of strain gage method for measurement of post-gel shrinkage in resin composites. *Dent. Mater.* 13:233–239 (1997).

146. D. Tantbirojn, A. Versluis, M. R. Pintado, R. DeLong, W. H. Douglas. Tooth deformation patterns in molars after composite restoration. *Dent. Mater.* 20:535–542 (2004).

147. J. M. Antonucci, J. W. Stansbury, S. Venz. Synthesis and properties of polyfluorinated prepolymer multifunctional urethane methacrylate. *Poly. Mater. Sci. Eng.* 59:388–396 (1988).

148. J. W. Stansbury. Cyclopolymerizable monomers for use in dental resin composites. *J. Dent. Res.* 69:844–848 (1990).

149. V. P. Thompson, E. F. Williams, W. J. Bailey. Dental resins with reduced shrinkage during hardening. *J. Dent. Res.* 58:1522–1532 (1979).

150. J. W. Stansbury, W. J. Bailey. Evaluation of spiro orthocarbonate monomers capable of polymerizing with expansion as ingredients in dental composite materials. In: C. G. Gebelein, R. L. Dunn (eds.). *Progress in Biomedical Polymers,* Plenum Press, New York, pp. 133–139 (1990).

151. J. W. Stansbury. Synthesis and evaluation of new oxaspiro monomers for double ring-opening polymerization. *J. Dent. Res.* 71:1408–1412 (1992).

152. E. J. Moon, J. Y. Lee, C. K. Kim, B. H. Cho. Dental restorative composites containing 2,2-Bis-[4-(2-hydroxy-3-methacryloyloxy propoxy) phenyl] propane derivatives and spiro orthocarbonates. *J. Biomed. Mater. Res. Part B: Appl. Biomater.* 73B:338–346 (2005).

153. M. Trujillo-Lemon, J. Ge, H. Lu, J. Tanaka, J. W. Stansbury. Dimethacrylate derivatives of dimer acid. *J. Polym. Sci. Part A Polym. Chem.* 44:3921–3929 (2006).

154. C. A. Khatri, J. W. Stansbury, C. R. Schultheisz, J. M. Antonucci. Synthesis, characterization and evaluation of urethane derivatives of bis-GMA. *Dent. Mater.* 19:584–588 (2003).

155. J. W. Stansbury, J. M. Antonucci. Dimethacrylate monomers with various fluorine contents and distributions. *Dent. Mater.* 15:166–173 (1999).

156. S. Kurata, N. Yamazaki. New matrix polymers for photo-activated resin composites using di-α-fluoroacrylic acid derivatives. *Dent. Mater. J.* 27:434–540 (2008).

157. N. Moszner, U. Salz. Chapter 2: Composites for dental restoratives. In: S. S. Shalaby, U. Salz (eds.). *Polymers for Dental and Orthopedic Applications,* CRC Press, Boca Raton, FL, pp. 13–67 (2007).

158. W. Weinmann, C. Thalacker, R. Guggenberger. Siloranes in dental composites. *Dent. Mater.* 21:68–74 (2005).

159. V. Miletic, V. Ivanovic, B. Dzeletovic, M. Lezaja, Temperature changes in silorane-, ormocer-, and dimethacrylate-based composites and pulp chamber roof during light-curing. *J. Esther. Restor. Dent.* 21:122–132 (2009).

160. N. Ilie, R. Hickel. Macro-, micro- and nano-mechanical investigations on silorane and methacrylate-based composites. *Dent. Mater.* 25:810–819 (2009).

161. S. Duarte Jr, J.-H. Park, F. M. Varjão, A. Sadan. Nanoleakage, ultramorphological characteristics, and microtensile bond strengths of a new low-shrinkage composite to dentin after artificial aging. *Dent. Mater.* 25:589–600 (2009).

162. H. Wolter, W. Storch, H. Ott. New inorganic/organic copolymers (Ormocer®s) for dental applications. In: A. K. Cheetham, C. J. Brinker, M. L. Mecartney, C. Sanchez (eds.). *Material Research Society Symposium Proceedings,* San Francisco, CA, April 4–8, 1994, Vol. 346, pp. 143–149. Materials Research Society, Pittsburgh, PA (1994).

163. D. A. Tagtekin, F. C. Yanikoglu, F. O. Bozkurt, B. Kogoglu, H. Sur. Selected characteristics of an Ormocer and a conventional hybrid resin composite. *Dent. Mater.* 20:487–497 (2004).

164. J. M. Meyer, M. A. Cattani-Lorente, V. Dupuis. Compomers: between glass-ionomer cements and composites. *Biomaterials* 19:529–539 (1998).

165. J. W. Nicholson. Glass-ionomers in medicine and dentistry. *Proc. Instn. Mech. Engrs. Part H*, 212:121–126 (1998).

166. B. L. Chadwick, P. M. H. Dummer, F. D. Dunstan, A. S. M. Gilmour, R. J. Jones, C. J. Phillips, J. Rees, S. Richmond, J. Stevens, E. T. Treasure. What type of filling? Best practice in dental restorations. *Quality in Health Care* 8:202–207 (1999).

167. R. S. Gatewood. Endodontic materials. *Dent. Clin. N. Am.* 51:695–712 (2007).

168. L. Spångberg. Biological effects of root end filling materials. II. Effect in vitro of water-soluble components of root canal filling materials in HeLa cells. *Odontol. Revy.* 20:133–145 (1969).

169. L. Spångberg. Biological effects of root end filling materials. IV. Effect in vitro of solubilized root canal filling materials in HeLa cells. *Odontol. Revy.* 20:289–299 (1969).

170. J. Moshonov, M. Trope, S. Friedman. Retreatment efficacy 3 months after obturation using glass ionomer cement, zinc oxide-eugenol, and epoxy resin sealers. *J. Endod.* 20:90–92 (1994).

171. R. Raina, R. J. Loushine, R. N. Weller, F. R. Tay, D. H. Pashley. Evaluation of the quality of the apical seal in Resilon/Epiphany and Gutta-Percha/AH plus-filled root canals by using a fluid filtration approach. *J. Endod.* 33:944–948 (2007).

172. S. Bouillaguet, J. C. Wataha, F. R. Tay, M. G. Brackett, P. E. Lockwood. Initial in vitro biological response to contemporary endodontic sealers. *J. Endod.* 32:989–992 (2006).

173. S. Mai, Y. K. Kim, N. Hirashi, J. Ling, D. H. Pashley, F. R. Tay. Evaluation of the true self-etching potential of a fourth generation self-adhesive methacrylate resin-based sealer. *J. Endod.* 35:870–874 (2009).

174. F. R. Tay, D. H. Pashley. Monoblocks in root canals: a hypothetical or a tangible goal. *J. Endod.* 22:391–398 (2007).

175. E. S. Reeh, E. C. Combe. New core and sealer materials for root canal obturation and retrofilling. *J. Endod.* 28:520–523 (2002).

176. S. Buchan, R.W. Peggie. Role of ingredients in alginate impression compounds. *J. Dent. Res.* 45:1120–1129 (1966).

177. W. D. Cook. Alginate dental impression materials: chemistry, structure and properties. *J. Biomed. Mater. Res.* 20:1–24 (1986).

178. N. Nallamuthu, M. Braden, M. P. Patel. Dimensional changes of alginate dental impression materials. *J. Mater. Sci.: Mater. Med.* 17:1205–1210 (2006).

179. J. G. Stannard, R. G. Craig. Modifying the setting rate of an addition-type silicone impression material. *J. Dent. Res.* 58:1377–1382 (1979).

180. C. Shen, Chapter 9 of *Phillips' Science of Dental Materials.* 11th edn., In: K. J. Anusavice (eds.). Saunders, Philadelphia, PA, pp. 216–217 (2003).

181. A. Sadan. Hydrophilic vinyl polysiloxane impression materials. *Pract. Proc. Aesthet. Dent.* 17:310 (2005).

182. E. Kotsiomiti, A. Tzialla, K. Hatjivasiliou. Accuracy and stability of impression materials subjected to chemical disinfection – a literature review. *J. Oral Rehabil.* 35:291–299 (2008).

183. D. G. Gratton, S. A. Aquilino. Interim restorations. *Dent. Clin. N. Am.* 48:487–497 (2004).

184. C. M. Becker, C. C. Swoope, C. A. Schwalm. Emergency dentures. *J. Prosthet. Dent.* 32:514–519 (1974).

185. R. E. Ogle, S. E. Sorensen, E. A. Lewis. A new visible light-cured resin system applied to removable prosthodontics. *J. Prosthet. Dent.* 56:497–504 (1986).

186. A. M. Diaz-Arnold, M. A. Vargas, K. L. Shuall, J. E. Laffoon, F. Qian. Flexural and fatigue strengths of denture base resin. *J. Prosthet. Dent.* 100:47–51 (2008).

187. R. Diwan, E. C. Combe, A. A. Grant. The use of microwave energy to cure acrylic resins. *Clin. Mater.* 12:117–120 (1993).

188. S. Sadamori, T. Ganefiyanti, T. Hamada, T. Arima. Influence of thickness and location on the residual monomer content of denture base cured by three processing methods. *J. Prosthet. Dent.* 72:19–22 (1994).

189. J. F. McCabe, R. M. Basker. Tissue sensitivity to acrylic resin. *Br. Dent. J.* 140:347–350 (1976).

190. J. H. Jorge, E. T. Giampaolo, A. L. Machado, C. E. Vergani. Cytotoxicity of denture base acrylic resins: a literature review. *J. Prosthet. Dent.* 90:190–193 (2003).

191. D. C. Jagger, A. Harrison, K. D. Jandt. The reinforcement of dentures. *J. Oral Rehabil.* 26:185–194 (1999).

192. G. D. Stafford, J. F. Bates, R. Huggett, R. W. Handley. A review of the properties of some denture base polymers. *J. Dent.* 8:292–306 (1980).

193. R. A. Rodford. Further development and evaluation of high-impact-strength denture base materials. *J. Dent.* 18:151–157 (1990).

194. C. K. Schreiber. The clinical application of carbon fibre/polymer denture bases. *Br. Dent. J.* 137:21–22 (1974).

195. J. M. Berrong, R. M. Weed, J. M. Young. Fracture resistance of Kevlar-reinforced poly(methyl methacrylate) resin: a preliminary study. *Int. J. Prosthodont.* 3:391–395 (1990).

196. P. K. Vallittu, V. P. Lassila, R. Lappalainen. Acrylic resin fibre composite. Part I: The effect of fibre concentration on fracture resistance. *J. Prosthet. Dent.* 71:607–612 (1994).

197. P. K. Vallittu, V. P. Lassila, R. Lappalainen. Transverse strength and fatigue of denture acrylic glass fiber composite. *Dent. Mater.* 10:116–121 (1994).

198. P. K. Vallittu, K. Narva. Impact strength of a modified glass fiber poly(methyl methacrylate). *Int J. Prosthodont.* 10:142–148 (1997).

199. M. Braden, K. W. M. Davy, S. Parker, N. H. Ladizesky, I. M. Ward. Denture base poly (methyl methacrylate) reinforced with ultra-high molulus polyethylene fibres. *Brit. Dent. J.* 164:109–113 (1988).

200. D. L. Gutteridge. Reinforcement of poly (methyl methacrylate) with ultra-high-modulus polyethylene fibre. *J. Dent.* 20:50–54 (1992).

201. N. H. Ladizesky, T. W. Chow. Reinforcement of complete denture bases with continuous high performance polyethylene fibres. *J. Prosthet. Dent.* 68:934–939 (1992).

202. J. L. Gilbert, D. S. Ney, E. P. Lautenschlager. Self reinforced composite poly (methyl methacrylate): static and fatigue properties. *Biomaterials* 16:1043–1055 (1995).

203. P. K. Vallittuil, V. P. Lassila. Effect of metal strengthener's surface roughness on fracture resistance of acrylic denture base material. *J. Oral Rehabil.* 19:385–391 (1992).

204. P. K. Vallittu. Effect of some properties of metal strengtheners on the fracture resistance of acrylic denture base material. *J. Oral Rehabil.* 20:241–248 (1993).

205. G. L. Polyzois. Reinforcement of denture acrylic resin. The effect of metal inserts and denture resin type on fracture resistance. *Eur. J. Prosthodont. Rest. Dent.* 3:275–278 (1995).

206. K. E. Bloodworth, P. J. Render. Dental acrylic resin radiopacity: literature review and survey of practitioners' attitudes. *J. Prosthet. Dent.* 67:121–123 (1992).

207. N. Yunus, A. A. Rashid, L.L. Azmi, M. I. Abu-Hassan. Some flexural properties of a nylon denture base polymer. *J. Oral Rehabil.* 32:65–71 (2005).

208. P. Pfeiffer, N. An, P. Schmage. Repair strengths of hypoallergenic denture base materials. *J. Prosthet. Dent.* 100:292–301 (2008).

209. A. C. Murphy, R. G. Hill. Fracture toughness of tooth acrylics. *J. Mater. Sci. Mater. Med.* 14:1011–1015 (2003).

210. S. Suzuki. In vitro wear of nano-composite denture teeth. *J. Prosthodont.* 13:238–243 (2004).

211. K. R. Reis, G. Bonfante, L. F. Pegoraro, P. C. Conti, P. C. Oliveira, O. B. Kaizer. In vitro wear resistance of three types of poly(methyl methacrylate) denture teeth. *J. Appl. Oral Sci.* 16:176–180 (2008).

212. J. M. Clancy, D. B. Boyer. Comparative bond strengths of light-cured, heat-cured, and autopolymerizing denture resins to denture teeth. *J. Prosthet. Dent.* 61:457–462 (1989).

213. P. K. Vallittu, I. E. Ruyter, R. Nat. The swelling phenomenon of acrylic resins polymer teeth at the interface with denture base polymers. *J. Prosthet. Dent.* 78:194–199 (1997).

214. W. W. Chase. Tissue conditioning utilizing dynamic adaptive stress. *J. Prosthet. Dent.* 11:804–815 (1961).

215. H. Murata, Y. Narasaki, T. Hamada, F. F. McCabe. An alcohol-free tissue conditioner—A laboratory study. *J. Dent.* 34:307–315 (2006).

216. M. Braden, P. S. Wright, S. Parker. Soft lining materials—A review. *Eur. J. Prosthodont. Rest. Dent.* 3:163–174 (1995).

217. D. R. Radford, S. J. Challacombe, J. D. Walter. Denture plaque and adherence of *Candida albicans* to denture-base materials in vivo and in vitro. *Crit. Rev. Oral Biol. Med.* 10:99–116 (1999).

218. S. Parker, D. Martin, M. Braden. Soft acrylic resin materials containing a polymerisable plasticiser I: Mechanical properties. *Biomater.* 19:1695–1701 (1988).

219. S. Parker, D. Martin, M. Braden. Soft acrylic resin materials containing a polymerisable plasticiser II: Water absorption characteristics. *Biomaterials* 20:55–60 (1999).

220. I. Hayakawa, N. Akiba, E. Keh, Y. Kasuga. Physical properties of a new denture lining material containing a fluoroalkyl methacrylate polymer. *J. Prosthet. Dent.* 96:53–58 (2006).

221. J. E. Grasso. Denture adhesives: changing attitudes. *J. Am. Dent. Assoc.* 127:90–96 (1996).

222. J. E. Grasso. Denture adhesives. *Dent. Clin. N. Am.* 48:721–733 (2004).

223. A. H. Melcher. On the repair potential of periodontal tissues. *J. Periodont.* 256–260 (1976).

224. A. H. Melcher. Healing of wounds in the periodontium. In: A. H. Melcher, W. H. Bowen (eds.). *Biology of the Periodontium*, Academic Press, London, U.K., pp. 497–529 (1969).

225. F. Isidor, T. Karring, S. Nyman. J. Lindhe. The significance of coronal growth of periodontal ligament tissue for new attachment formation. *J. Clin. Periodont.* 13:145–150 (1986).

226. S. Nyman, J. Lindhe, T. Karring, H. Rylander. New attachment following surgical treatment of human periodontal disease. *J. Clin. Periodont.* 9:290–296 (1982).

227. M. R. Urist. Bone: formation by autoinduction. *Science* 150(698):893–899 (1965).

228. J. D. Bashutski, H.-M. Wang. Periodontal and endodontic regeneration. *J. Endod.* 35:321–328 (2009).

229. P. I. Branemark, B. O. Hannson, R. Adell, U. Breine, J. Lindstrom, O. Hallen, A. Ohman. Osseointegrated implants in the treatment of the edentulous jaw. Experience from a 10-year period. *Scand. J. Plast. Reconstr. Surg. Suppl.* 16:1–132 (1977).

230. H. Schliephake, D. Scharnweber. Chemical and biological functionalization of titanium for dental implants. *J. Mater. Chem.* 18:2404–2414 (2008).

231. A.Tezel, G. H. Fredrickson. The science of hyaluronic acid dermal fillers. *J. Cosmetic and Laser Ther.* 10:35–42 (2008).

232. D. W. Buck, A. Alam, J. Y. S. Kim. Injectable fillers for facial rejuvenation: a review. *J. Plast. Reconstr. Aesth. Surg.* 62:11–18 (2008).

233. J. L. Cohen. Understanding, avoiding and managing dermal filler complications. *Derm. Surg.* 34:S92–S99 (2008).

234. K. C. Smith. Reversible vs. nonreversible fillers in facial aesthetics: concerns and considerations. *Derm. Online J.* 14:3 (2008).

235. A. J. Valauri. Maxillofacial prosthetics. *Aesth. Plast. Surg.* 6:159–164 (1982).

236. J. C. Lemon, S. Kiat-amnuay, L. Gettleman, J. W. Martin, M. S. Chambers. Fixed prosthetic rehabilitation: preprosthetic surgical techniques and biomaterials. *Curr. Opin. Otolaryngol. Head Neck Surg.* 13:255–262 (2005).

237. H. Huber, S. P. Studer. Materials and techniques in maxillofacial prosthodontic rehabilitation. *Oral Maxillofac. Surg. Clin. N. Am.* 14:73–93 (2002).

238. T. Aziz, M. Waters, R. Jagger. Analysis of the properties of silicone rubber maxillofacial prosthetic materials. *J. Dent.* 31:67–74 (2003).

239. A. K. Hulterström, A. Berglund, I. E. Ruyter. Wettability, water sorption and water solubility of seven silicone elastomers used for maxillofacial prostheses. *J. Mater. Sci Mater. Med.* 19:225–231 (2008).

240. J. J. Gary, C. T. Smith. Pigments and their application in maxillofacial elastomers: a literature review. *J. Prosthet. Dent.* 80:204–208 (1998).

241. S. P. Haug, B. K. Moore, C. J. Andres. Color stability and colorant effect on maxillofacial elastomers. Part II: Weathering effect on physical properties. *J. Prosthet. Dent.* 81:423–430 (1999).

242. J. H. Lai, L. L. Wang, C. C. Ko, R. L. DeLong, J. S. Hodges. New organosilicon maxillofacial prosthetic materials. *Dent. Mater.* 18:281–286 (2002).

243. Y. Han, S. Kiat-amnuay, J. M. Powers, Y. Zhao. Effect of nano-oxide concentration on the mechanical properties of a maxillofacial silicone elastomer. *J. Prosthet. Dent.* 100:465–473 (2008).

244. S. O. Bartlett, D. J. Moore. Ocular prosthesis: a physiologic system. *J. Prosthet. Dent.* 29:450–459 (1973).

245. Y. Grossman, I. Savion. The use of a light-polymerized resin-based obturator for the treatment of the maxillofacial patient. *J. Prosthet. Dent.* 94:289–292 (2005).

246. G. T. Grant, R. M. Taft, S. T. Wheeler. Practical application of polyurethane and Velcro in maxillofacial prosthetics. *J. Prosthet. Dent.* 85:281–283 (2001).

247. S. Kiat-amnuay, P. J. Waters, D. Roberts, L. Gettleman. Adhesive retention of silicone and chlorinated polyethylene for maxillofacial prostheses. *J. Prosthet. Dent.* 99:483–488 (2008).

248. H. Kurunmäki, R. Kantola, M. M. Hatamleh, D. C. Watts, P. K. Vallittu. A fiber-reinforced composite prosthesis restoring a lateral midfacial defect: a clinical report. *J. Prosthet. Dent.* 100:348–352 (2008).

249. A. Vasanthan, K. Satheesh, W. Hoopes, P. Lucaci, K. Williams, J. Rapley. Comparing suture strengths for clinical applications: a novel in vitro study. *J. Periodontol.* 80:618–624 (2009).

250. R. J. Shaw, T. W. Negus, T. K. Mellor. A prospective evaluation of the longevity of resorbable sutures in oral mucosa. *Brit. J. Oral Maxillofac. Surg.* 34:252–254 (1996).

251. R. Gassner. Wound closure materials. *Oral Maxillofac. Surg. Clin.* 14:95–104 (2002).

252. D. F. Williams. On the nature of biomaterials. *Biomaterials* 30:5897–5909 (2009).

24 Biomaterials as Platforms for Topical Administration of Therapeutic Agents in Cutaneous Wound Healing

Rhiannon Braund and Natalie J. Medlicott

CONTENTS

24.1 INTRODUCTION

Biomaterials have found application in cutaneous wound healing either as the basis of wound dressings or as platforms for the delivery of therapeutic agents into wounds. Increased wound epithelialization was first reported in pigs following application of occlusive dressings made from polyethylene films by Winter in 1962 and has led to the current practice of maintaining a moist wound environment to optimize wound healing. Since this time, much has been discovered on the mechanisms of wound healing to support this practice. This together with advances in biomaterial science and increased availability of a range of biocompatible materials has led to the development of modern products used for treatment of wounds today (Yudanova and Reshetov 2006, Singer and Dagum 2008). Still remaining, however, is the fundamental challenge of improving the healing of chronic wounds such as foot ulcers in diabetic patients (Greenhalgh 2003, Mekkes et al. 2003).

The aim of this chapter is to review the current use of biomaterials in wound healing and, from this, to gain an appreciation of the potential of biomaterials to enhance wound healing outcomes through their promotion of an optimal environment for healing and use as drug delivery platforms.

24.2 CUTANEOUS WOUNDS

The skin is a vital organ in all animal species, it covers the body and provides protection to the underlying tissues from the environment in which we live. However, periodic damage to the skin can and does occur leaving the affected individual vulnerable. Many reviews are available on the

Polymeric Biomaterials: Structure and Function, Volume 1

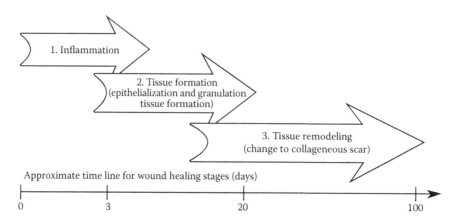

FIGURE 24.1 Stages of normal wound healing. Approximate timeline is from Clark. (From Clark, R.A.F., Wound repair: Overview and general considerations, in R.A.F. Clark (ed.), *The Molecular and Cellular Biology of Wound Repair*, 2nd edn., Plenum Press, New York, 1995.)

physiological, cellular, and biochemical response to cutaneous wounds and readers are directed to these reviews for a greater depth of discussion of the wound healing process while a brief overview follows (Kiritsy et al. 1993, Scaffer and Nanney 1996, Singer and Clark 1999, Yamaguchi and Yoshikawa 2001, Monaco and Lawrence 2003, Bao et al. 2009).

The normal physiological response to wounds is termed the "wound healing response." It aims to reduce loss of blood, by initiation of the clotting cascade, then to trigger an inflammation response with the formation of granulation tissue and ultimately results in re-epithelialization with tissue maturation and remodeling (Singer and Clark 1999, Monaco and Lawrence 2003). These stages of wound healing are generally considered as sequential, but overlap in time as shown in Figure 24.1. The cellular and biochemical responses associated with each stage are becoming increasingly understood (Clark 1995) and this information is being used to propose therapeutic targets to enhance wound healing outcomes.

Wounds in general do not represent a single well-defined entity, but rather cover a range of presentations classified by both cause and severity from minor cuts and abrasions to extensive loss of large areas of skin in, for example, burns (Singer and Dagum 2008). The body's ability to recover unaided appears highly dependent on the type and severity of the wound and on the existence of underlying diseases and/or infection. The importance of these factors is reflected in the often high mortality associated with burn injury, where patient age >60 years, burn area >40% of the body surface, and coexisting inhalation injury have been identified as risk factors for mortality (Ryan et al. 1998). At the time of any cutaneous wounding, there is a dramatic increase in skin permeability with potential for entry of toxic substances and infective organisms. From transdermal drug delivery literature, the importance of the stratum corneum (outmost layer of the skin) in preventing entry of compounds applied to the skin is well understood (Walker and Smith 1996, Menon 2002, Magnusson et al. 2004). Microorganisms also cannot easily penetrate intact skin but see no such barrier once wounding has occurred. Colonization of wounds by microorganisms can both delay healing and increase the risk of systemic infection (Thomsen 1994, Heggers 2003). A committee of the Wound Healing Society in 1994 developed definitions of wounds, wound healing, and wound attributes to facilitate comparison between wound healing studies. In this they defined an acute wound as "one that proceeds through an orderly and timely reparative process to establish sustained anatomic and functional integrity" and a chronic wound as "one that has failed to proceed through an orderly and timely reparative process to produce anatomic and functional integrity or has proceeded through the repair process without establishing a sustained anatomic and functional result" (Lazarus et al. 1994).

Modern biomaterials and tissue-engineered skin substitutes form an integral part of today's wound treatments. They are used firstly to assist in the reduction of blood loss through hemostatic mechanisms and action as a replacement physical barrier to prevent entry of microorganisms. Then if the mechanisms of wound repair are well understood, these same materials may be developed into treatments that actively promote wound healing. As part of this advance, biomaterial dressings are becoming increasingly recognized for their potential to be used as platforms for the local delivery of therapeutic agents (Lawrence and Diegelmann 1994, Sakiyama-Elbert and Hubbell 2000, Wong et al. 2003, Braund et al. 2007a, 2009, Bader and Kao 2009, Mulder et al. 2009).

24.2.1 Acute Wounds and the Normal Wound Healing Response

In response to clean, acute, cutaneous wounds, there is a reasonably orderly interplay between the vascular and cellular responses and the chemical mediators such as growth factors and cytokines to control the progression of wound healing (Kiritsy et al. 1993, Scaffer and Nanney 1996, Monaco and Lawrence 2003, Frokjaer and Otzen 2005). Risk factors and underlying diseases that impede successful wound healing are recognized, with diabetes mellitus and peripheral vascular disease being two of the most well recognized (Greenhalgh 2003, Brem and Tomic-Canic 2007, Milic et al. 2009). The goals of acute wound management include ensuring rapid and complete wound closure, optimal return of skin functions with a low level of tissue scarring (Singer and Dagum 2008). In order to achieve this, risk factors recognized to delay healing are identified and reduced, thereby creating a wound environment that has an increased chance of optimal healing. A number of references give guidelines for optimizing wound healing, and some common approaches that are summarized in Figure 24.2 include removal of dead tissue, prevention of wound infection, pressure off-loading of the wounded area, and correction of any underlying vascular deficits (Franz et al. 2008). Efforts are also directed at reducing the extent of scarring through maintenance of a moist wound environment during healing. To achieve this, occlusive and moisture controlling dressings are commonly used (Field and Kerstein 1994, Singer and Dagum 2008). Burns constitute a significant group of acute wounds, with the body area affected being highly variable. Outcomes are dependent on surface area, depth of injury, and whether or not the lungs are involved in the burn injury (Ryan et al. 1998). For burn wounds of moderate or lower severity, treatment may focus on the local area of the burn and includes skin grafting and use of biomaterial dressings.

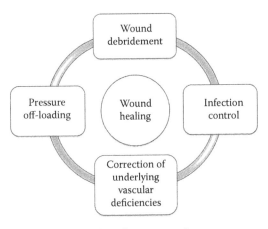

FIGURE 24.2 Factors contributing to resolution of acute wounds.

TABLE 24.1
Some Systemic and Local Factors That Contribute to Delayed Wound Healing

Systemic Factors	Local Factors
Aging	Arterial occlusion/insufficiency
Alcoholism	Foreign bodies
Diabetes mellitus	Ischemia
Drugs, e.g., steroids	Pressure (e.g., bed-ridden patients)
Immunodeficiency	Tissue necrosis
Liver disease	Ulcerating tumors
Nutritional deficiencies	Ulcerating skin diseases
Peripheral vascular disease	Wound infection
	Vasculitis
	Venous insufficiency

Source: Burns, J.L. et al., *Clin. Plast. Surg.*, 30, 47, 2003.

24.2.2 CHRONIC WOUNDS: IMPAIRED WOUND HEALING

The mechanisms by which acute wounds turn into chronic wounds are not completely understood, but as research on the biochemistry and cellular biology of chronic wounds continues, new understandings are gained on the differences between acute and chronic wound environments so that new therapeutic targets may be identified in the future. In the meantime, the high financial and patient morbidity costs associated with treatments for chronic wounds make this an important health-care area especially as the population ages and diseases such as diabetes mellitus increase in prevalance (Sen et al. 2009). Treatments described in many published guidelines aim to minimize the risk factors identified as impeding wound healing (Robson and Barbul 2006, Franz et al. 2007, Hinchliffe et al. 2008). The conditions and coexisting diseases known to make a significant contribution to delayed wound healing are listed in Table 24.1 (Burns et al. 2003) and although treatment of the underlying cause of impaired wound healing would seem the logical course of action to improve wound healing outcomes in patients, complete resolution is often difficult if these underlying diseases cannot be well controlled. Newer treatments for chronic wounds have included the use of biomaterial dressings to maintain the moist wound environment together with active treatments with antiseptics and antibiotic agents to control wound infection. Topical delivery of growth factors has also been investigated with mixed success. These areas are discussed further in the next sections.

24.3 BIOMATERIALS USED IN WOUND HEALING APPLICATIONS

Modern wound dressings are usually presented in the form of gels, thin films, or foam sheets (Boateng et al. 2008). Properties of an ideal wound dressing have been suggested by many including Zhang et al. (2005). These include provision of hemostatic action, sufficient bacterial barrier, adsorbant to exudates, ability to maintain a moist wound environment, flexibility to cover wound surface, adherent to health tissue while nonadherent to wound tissue, easy and painless to remove, and low cost. Along with these properties, materials used should exhibit tissue biocompatibility so they do not promote negative inflammatory changes in the wound tissue (Anderson et al. 2008) including development of foreign body reactions (Rosenberg 2006). There is also an optimal size and shape requirement for individual wounds so that biomaterials whose size and shape can be easily manipulated to fill and/or cover non-uniform shaped wounds have potential advantage.

The most appropriate dressing to use is often determined by the type and extent of the wound and some general guidelines are available in the literature (Singer and Dagum 2008). Numerous polymeric materials have been reported as the basis of modern dressings in wound healing literature and many of these have successfully reached the market (Yudanova and Reshetov 2006, Boateng et al. 2008). Yudanova and Reshetov (2006) suggested a need for systematic classification of wound dressings according to their characteristic features. Boateng et al. (2008) gave three classifications: firstly, according to the dressing function; secondly, according to the type of material used; and thirdly, according to the physical form of product. Jeffcoate also gave a similar classification of wound dressings into films, foams, hydrogels, hydrocolloids, alginates, hydrofibers, and medicated dressings (Jeffcoate et al. 2004).

Synthetic polymeric materials have a variety of physical properties and those used in wound dressings are generally processed into gels, thin sheets or films, or foams. The resulting dressings have different fluid absorbencies, shapes, and water and vapor permeabilities to allow selection of appropriate dressings to maintain moist, but not too wet, wound environments (Table 24.2).

The biological dressings are prepared from materials that have been reported to play an active part in the wound healing process and include alginates, collagen, hyaluronic acid, and fibrin. Alginate polymers have formed the basis of wound dressing products for many years. Both sodium and calcium alginates are used and the degree of alginate cross-linking and hence physical properties of the material are dependent on the calcium concentration within the alginate (Ichioka et al. 1998). Following application to wounds, sodium and calcium ions can exchange between the wound fluid and the applied alginate so that physical properties can be influenced by the biological environment. Comparative studies have shown that alginates typically remain on the wound for longer than hydrocolloid materials (Ichioka et al. 1998). Calcium alginate also appears to increase fibroblast proliferation in the wound (Doyle et al. 1996) and calcium ions may help with clotting through a hemostatic effect (Blair et al. 1988).

Collagen is a major constituent of connective tissue and forms a large portion of the wound granulation tissue. Its use in wound dressings aims to provide a ready supply of collagen for incorporation into granulation tissue. Sources of collagen with low immunogenicity and good biodegradability are available and have been used in wound dressings and for dermal reconstruction (Macleod et al. 2005, 2008, Benito-Ruiz et al. 2006, Shevchenko et al. 2008). Arul et al. (2007) showed that collagen alone gave an increased intra-wound collagen (compared with saline-treated controls) 21 days after experimental wounds in diabetic rats. Then, if collagen was used as a platform for delivery of a therapeutic agent, biotinylated tripeptide Gly-His-Lys, wounds showed complete re-epithelialization and densely packed collagen tissue at day 21. Shevchenko et al. (2008) reported that a paste form of collagen allowed better cellular infiltration over sheet collagen in full-thickness wounds. Hence, the biomaterial physical form as well as chemical composition can influence the wound healing outcome.

Chitosan is a versatile polysaccharide material; it is made up of linear chains of D-glucosamine and N-acetyl-D-glucosamine units. Hydrogels can be prepared using a variety of cross-linking agents to produce materials with varying physical properties and geometries, for example, gels, powders, beads, films, tablets, capsules, microspheres, microparticles, sponges, and nano and microfibers (Bhattarai et al. 2010). Dressings composed of chitosan (Rossi et al. 2007, Liu et al. 2008, Meng et al. 2010) or chitosan with alginates (Khor and Lim 2003, Liu et al. 2008, Ho et al. 2009, Meng et al. 2010) or hyaluronic acid (Choi et al. 2001) have been reported widely for their beneficial effects on wound healing.

Tissue-engineered artificial skin can be acellular where the biomaterial provides a scaffold architecture into which surrounding cells migrate and proliferate. Alternatively, cells (usually fibroblasts or keratinocytes) can be seeded into the biomaterial scaffold to initiate tissue regeneration. Nondegrading and biodegradable polymers have been used as the scaffold forming materials and they are especially useful for replacing lost tissue in full-thickness wounds (Miller and Patrick 2003, Kamolz et al. 2008, Priya et al. 2008, Dieckmann et al. 2010). Bota et al. (2010) showed that for expanded polytetrafluoroethylene (ePTFE) implants, a porous structure with an

TABLE 24.2

Biomaterial Dressings Used in Cutaneous Wound Healing

	Polymers	Properties
Hydrocolloids	Carboxymethyl cellulose	Colloidal materials
		Prepared as films or thin sheets
	Gelatin	Adhere to dry and moist areas
	Pectin	Initially impermeable to water vapor
		Absorb fluid to form a gel, which becomes increasingly permeable to air and water as water content increases
		Useful for moderately exudating wounds
Alginates	Sodium or calcium alginate (polysaccharide monomers: mannuronate and guluronic acid)	Prepared as porous sheets or flexible fibers. Fibers can be packed into wounds
		Form gel structure on contact with water. High mannuronic content gives softer gels. Higher guluronic content gives firmer gels
		Can be cross-linked with calcium ions to give a material with a stronger barrier function
		Useful for moderate to heavily exudating wounds due to high absorbency
		May dehydrate drier wounds
		Often require frequent changing
Hydrogels	Poly(methacrylates)	Insoluble, swellable hydrophilic materials
	polyvinylpyrrolidine	Prepared as amorphous gels or elastic solid sheets or films
		Cross-linking produces matrices that can entrap water
		Application of amorphous gels usually requires covering with a secondary dressing
		Sheets may be polymer backed with adhesive edges to minimize water loss through the hydrogel
		Due to preparation with high water content, they do not adsorb much exudate
		Fluid accumulation under dressing can cause tissue maceration
		Useful for deep abrasions and wounds with crusted surface exudate
Semipermeable films	Nylon derivatives	Permeable to water vapor but not liquid exudate
	Polyurethane	Thin, often transparent films
	Acrylic derivatives	Highly flexible and elastic so that they can conform easily to the skin contours
		Useful for relatively shallow wounds
Foam dressings	Polyurethane	Prepared with an open porous polyurethane structure
		High absorbency
		Provide a protective padding
		May be cut into shaped to fit within wounds
		Useful for partial or full-thickness wound as well as heavily exudating wounds
Biological dressings	Collagen	Made from biomaterials that play an active part in the wound healing process (also called bioactive dressings)
	Hyaluronic acid	
	Chitosan	See text for further properties
	Alginates	
	Elastin	
	Fibrin	

TABLE 24.2 (continued)
Biomaterial Dressings Used in Cutaneous Wound Healing

	Polymers	Properties
Tissue-engineered artificial skin	Synthetic or biological polymers, e.g., polyesters, collagen, silicone, hyaluronic acid derivatives, fibrin	May be acellular or cell-containing matrices. The cellular component is usually fibroblasts or keratinocytes Polymeric material forms a scaffold that replicates dermal architecture through which the regenerating tissue can form The scaffolds should have optimal mechanical properties initially to support regenerating tissue, and then may degrade once tissue is established

Sources: Boateng, J.S., *J. Pharm. Sci.*, 97, 2892, 2008; Singer, A.J. and Dagum, A.B., *N. Engl. J. Med.*, 359, 1037, 2008.

internodal distance of 4.4 μm allowed cellular invasion and resulted in the formation of a thinner fibrous capsule surrounding the implant. Smaller pore sizes (internodal distances of 1.2 and 3.0 μm) and nonporous PTFE in contrast were associated with little cellular infiltration and thicker fibrous capsules surrounding the implanted material. In vitro studies were also reported and showed an increased secretion of early proinflammatory cytokines for the larger pore sized ePTFE incubated with monocyte/macrophage cells (Bota et al. 2010). Results were suggestive of an increased macrophage activation in response to ePTFE topography and an ability to modify the tissue foreign body response through scaffold topography (Bota et al. 2010). Chang et al. (2008) showed that surface hydrophobicity affected interactions with lymphocytes and macrophages. They measured cytokines, chemokines, and matrix proteins produced on culture of monocytes and lymphocytes with polyethylene terephthalate–based surfaces. Hydrophilic anionic or neutral surfaces produced greater pro-inflammatory responses with less breakdown of extracellular matrix than hydrophilic cationic materials, where a more anti-proinflammatory response was generated with increased ratio of metalloproteinases (MMP-9) to tissue inhibitors of MMP (TIMP) observed. Since MMPs normally show a 1:1 stoichiometry with their inhibitors (TIMP), an increase in the ratio of MMP/TIMP could be expected to result in increased extracellular matrix breakdown in the wound.

24.4 TOPICAL DELIVERY OF DRUGS TO WOUNDS

A further consideration in the development of an improved wound dressing relies on integration of biomaterial and drug delivery sciences. The goal is to use the biomaterial as a platform for drug release, and then future dressings may enhance wound healing both through moist wound environment and pharmacological mechanisms. Two classes of therapeutic agents that have received considerable attention for topical administration are the anti-infective agents and growth factors. While the topical application of anti-infective agents has been part of wound management for many years, growth factors have more recently been investigated (Kiritsy et al. 1993, Lawrence and Diegelmann 1994, Werner and Grose 2003, Wong et al. 2003, Robson et al. 2004, Amery 2005, Braund et al. 2007a,b, Davidson 2007). Becaplermin, a recombinant platelet-derived growth factor (PDGF-BB), is available for topical applications under the name Regranex® in many countries (Papanas and Maltezos 2010). Clinical trials have shown becaplermin to be beneficial in the treatment of diabetic foot ulcers with adequate blood supply (Embil et al. 2000, Papanas and Maltezos 2010). Akopian et al. (2006) reported benefit in the treatment of larger, higher grade wounds, following topical application of growth factors (platelet-derived wound healing factor or recombinant platelet-derived growth factor-BB) at the early stage of treatment. Routine use of becaplermin in wounds has, however, been somewhat limited. Proposed reasons for this have been the product's relatively high cost and its perceived reduced efficacy outside the well-controlled clinical trial environment. More recently, an increased cancer risk has been reported for patients treated with three or more tubes (i.e., 45 g total) of Regranex® (Papanas and Maltezos 2010).

24.4.1 Anti-Infective Agents Delivery from Biomaterials

Topical application of anti-infective agents aims to reduce wound colonization by microorganisms and subsequent infection, thereby exerting a positive effect on wound healing.

Silver preparations have been used as semi-solid creams and ointments in patients with burns (Boateng et al. 2008, Singer and Dagum 2008). Incorporation into wound dressing materials has also occurred with products such as Aquacoat®, which produces a slow release of silver into the wound. The silver in this product is deposited onto the surface of the material as a 1 μm thick layer using a physical vapor deposition method. The potential advantage suggested for this mode of delivery of silver to wounds was that the total silver dose could be reduced thereby lowering systemic exposure (Tredget et al. 1998). A further advantage of application of silver in the wound dressing is that it is easier to remove than a cream formulation at the dressing change. Leaper reviewed the use of silver-containing dressings in wounds and recognized the emergence of an increasing number of silver-containing dressings (Leaper and Durani 2008). The incorporation of nanocrystallized silver seems to be associated with good sustained-release properties and lower risk of systemic toxicity.

Other antiseptic agents incorporated into wound dressings include povidone iodine (Reimer et al. 2000) and chlorhexidine (Rossi et al. 2007, Sulea et al. 2011). Both these agents have be applied to wounds as solutions in conventional treatment protocols, but incorporation into biomaterials offers the advantage of sustained release and reduced frequency of application. Antibiotic agents, for example, gentamicin (Zilberman et al. 2009) and minocycline (Aoyagi et al. 2007), have also been incorporated into biomaterials for direct application to wounds to reduce bacterial infection. While the topical application of antiseptic agent has not usually been associated with the occurrence of resistant microorganism (Leaper and Durani 2008, Milstone et al. 2008), release of topical antibiotics needs to be carefully controlled so that treatments prevent wound infection but do not increase the emergence of antibiotic-resistant strains of microorganisms.

24.4.2 Growth Factors Delivery from Biomaterials

Topically delivered growth factors to promote wound healing have long been proposed in the scientific literature (Kiritsy et al. 1993, Werner and Grose 2003, Robson et al. 2004, Fu et al. 2005, Braund et al. 2007a, Davidson 2007, Papanas and Maltezos 2007). Whether topically applied growth factors can be used for the treatment of poorly resolving chronic wounds remains to be seen. Because these compounds are protein molecules, processing into biomaterials will need consideration of the effects of processing variables on protein physical stability, that is, secondary and tertiary structures (Arakawa et al. 2001, Frokjaer and Otzen 2005). Protein instability with generation of aggregated protein can result in both decreased biological activity and increased risk of adverse immunological reactions (Rosenberg 2006).

It is known that chemokines, cytokines, and growth factors, when produced in an orderly sequence in response to wounds, mediate and control the biochemical and cellular events of the healing response (Kiritsy et al. 1993, Frokjaer and Otzen 2005). How exogenously applied growth factors modify this process is unknown so that further research is needed to improve our understanding in this area. Benefits have been shown in wound closure rates in animal models following application of growth factors (Chan et al. 2006). Chan et al. showed in diabetic mice that although application of recombinant platelet-derived growth factor (Regranex®) did not increase the rate of wound closure of 1.5 cm² wounds, it did increase the amount of granulation tissue in the wound. Positive reports have also been described in human clinical trials (Smiell 1998, Rees et al. 1999, Smiell et al. 1999, Embil et al. 2000, Papanas and Maltezos 2010). A recent review has examined the potential of growth factor delivery to wounds (Braund et al. 2007a). From this, it appears that mixed clinical outcomes have been observed. Still little research has been done on the optimal time-course of exogenous growth factor application. If this is important, then research into biomaterials that can precisely control release of incorporated growth factors may provide a key to future development of pharmacologically active wound dressings.

24.5 PHARMACOKINETIC MODELS OF DRUG DELIVERY TO WOUNDS

A simple model is shown in Figure 24.3 where the concentration of free drug in a wound environment is determined by its release from the applied biomaterial and loss through systemic absorption, binding to local tissues and degradation within the wound. As the wound heals, the barrier function of the outermost layer will change. Cross and Roberts (1999) reported the change in permeability of full-thickness excisional wounds to solutes of different molecular sizes. In this work, they showed that large molecules (FGF-2 and epidermal growth factor) could only penetrate to superficial layers of the developing granulation tissue. Braund et al. have shown that fluorescent-labeled FGF-2 applied as single dose (0.3 μg) in a hypromellose gel or film could be detected in wound tissue sections by confocal microscopy at least 8 h after application. They also measured intra-wound FGF-2 tissue concentration as a function of depth into the wound (Braund et al. 2009).

Fibroblast growth factors belong to the heparin-binding family of growth factors (Burgess and Maciag 1989). The interaction between FGF and heparin is reported to improve physical stability of the growth factor (Gospodarowicz and Cheng 1986, Zakrewska et al. 2009), and heparin has been used to stabilize this protein in the development of some FGF-releasing biomaterials (Tanihara et al. 2001). There is some question as to whether heparin is needed in the interaction between FGF and its receptor (FGFR) in vivo. Zakrewska et al. (2009) suggested that heparin was not necessary if the physical stability of FGF was maintained in some other way. Hence, it may be possible that interaction with polymeric biomaterials may fulfill this function.

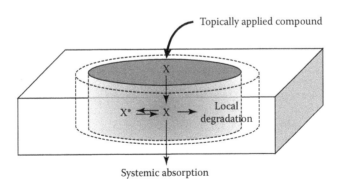

FIGURE 24.3 Schematic illustrating the potential fate of topically administered drugs in a model wound. X = amount of drug in the wound, X* = tissue-bound drug. (Modified from Braund, R. et al., *Curr. Drug Deliv.*, 4, 195, 2007a.)

24.6 CONCLUDING COMMENTS

Both synthetic and biological polymeric materials have shown application in wound healing treatments. The greatest advances have been in the development of materials with ranges of physical forms, fluid absorbencies, and water vapor transmission rates. These modern wound dressings address the clinical needs of maintaining a moist, but not wet, wound environment and acting as an artificial replacement for lost dermal tissue. Release of biologically active agents from wound dressing is receiving increased attention and may offer future enhancements in the area of difficult to treat chronic wounds. Extended release of antimicrobial agents from wound dressings can help to control wound infection, whereas the topical application of growth factors may have the potential to directly influence the biochemical course of the wound healing response.

REFERENCES

Akopian, G., Nunnery, S. P., Piangenti, J. et al. 2006. Outcomes of conventional wound treatment in a comprehensive wound center. *Am. Surg.*, 72, 314–317.

Amery, C. M. 2005. Growth factors and the management of the diabetic foot. *Diabet. Med.*, 22, 12–14.

Anderson, J. M., Rodriguez, A. and Chang, D. T. 2008. Foreign body reaction to biomaterials. *Semin. Immunol.*, 20, 86–100.

Aoyagi, S., Onishi, H. and Machida, Y. 2007. Novel chitosan wound dressing loaded with minocycline for the treatment of severe burn wounds. *Int. J. Pharm.*, 330, 138–145.

Arakawa, T., Prestrelski, S. J., Kenney, W. C. and Carpenter, J. F. 2001. Factors affecting short-term and long-term stabilities of proteins. *Adv. Drug Deliv. Rev.*, 46, 307–326.

Arul, V., Kartha, R. and Jayakumar, R. 2007. A therapeutic approach for diabetic wound healing using biotinylated GHK incorporated collagen matrices. *Life Sci.*, 80, 275–284.

Bader, R. A. and Kao, W. J. 2009. Modulation of the keratinocyte-fibroblast paracrine relationship with gelatin-based semi-interpenetrating networks containing bioactive factors for wound repair. *J. Biomater. Sci. Polym. Ed.*, 20, 1005–1030.

Bao, P., Kodra, A., Tomic-Canic, M. et al. 2009. The role of vascular endothelial growth factor in wound healing. *J. Surg. Res.*, 153, 347–358.

Benito-Ruiz, J., Guisantes, E. and Serra-Renom, J. M. 2006. Porcine dermal collagen: A new option for soft-tissue reconstruction of the lip. *Plast. Reconstr. Surg.*, 117, 2517–2519.

Bhattarai, N., Gunn, J. and Zhang, M. 2010. Chitosan-based hydrogels for controlled, localized drug delivery. *Adv. Drug Deliv. Rev.*, 62, 83–99.

Blair, S. D., Backhouse, C. M., Harper, R., Matthews, J. and McCollum, C. N. 1988. Comparison of absorbable materials for surgical haemostasis. *Br. J. Surg.*, 75, 69–71.

Boateng, J. S., Matthews, K. H., Stevens, H. N. E. and Eccleston, G. M. 2008. Wound healing dressings and drug delivery systems: A review. *J. Pharm. Sci.*, 97, 2892–2923.

Bota, P. C. S., Collie, A. M. B., Puolakkainen, P. et al. 2010. Biomaterial topography alters healing in vivo and monocyte/macrophage activation in vitro. *J. Biomed. Res. A*, 95A, 649–657.

Braund, R., Hook, S. M., Greenhill, N. and Medlicott, N. J. 2009. Distribution of fibroblast growth factor-2 (FGF-2) within model excisional wounds following topical application. *J. Pharm. Pharmacol.*, 61, 193–200.

Braund, R., Hook, S. and Medlicott, N. J. 2007a. The role of topical growth factors in chronic wounds. *Curr. Drug Deliv.*, 4, 195–204.

Braund, R., Tucker, I. G. and Medlicott, N. J. 2007b. Hypromellose films for the delivery of growth factors for wound healing. *J. Pharm. Pharmacol.*, 59, 367–372.

Brem, H. and Tomic-Canic, M. 2007. Cellular and molecular basis of wound healing in diabetes. *J. Clin. Invest.*, 117, 1219–1222.

Burgess, W. H. and Maciag, T. 1989. The heparin-binding (fibroblast) growth factors family of proteins. *Annu. Rev. Biochem.*, 58, 575–606.

Burns, J. L., Mancoll, J. S. and Phillips, L. G. 2003. Impairments to wound healing. *Clin. Plast. Surg.*, 30, 47–56.

Chan, R. K., Liu, P. H., Pietramaggiori, G. et al. 2006. Effect of recombinant platelet-derived growth factor (Regranex®) on wound closure in genetically diabetic mice. *J. Burn Care Res.*, 27, 202–205.

Chang, D. T., Jones, J. A., Meyerson, H. et al. 2008. Lymphocyte/macrophage interactions: Biomaterial surface-dependent cytokine, chemokine and matrix protein production. *J. Biomed. Mater. Res.*, 87, 676–687.

Choi, Y. S., Lee, S. B., Hong, S. R. et al. 2001. Studies on gelatin-based sponges. Part III: A comparative study of cross-linked gelatin/alginate, gelatin/hyaluronate and chitosan/hyaluronate sponges and their application as a wound dressing in full-thickness skin defect of rat. *J. Mater. Sci. Mater. Med.*, 12, 67–73.

Clark, R. A. F. 1995. Wound repair: Overview and general considerations. In: R.A.F. Clark (ed.) *The Molecular and Cellular Biology of Wound Repair*, 2nd edn. New York: Plenum Press.

Cross, S. and Roberts, M. S. 1999. Defining a model to predict the distribution of topically applied growth factors and other solutes in excisional full-thickness wounds. *J. Invest. Dermatol.*, 112, 36–41.

Davidson, J. M. 2007. Growth factors: The promise and the problems. *Int. J. Lower Extrem. Wounds*, 6, 8–10.

Dieckmann, C., Renner, R., Milkova, L. and Simon, J. C. 2010. Regenerative medicine in dermatology: Biomaterials, tissue engineering, stem cells, gene transfer and beyond. *Exp. Derm.*, 19, 697–706.

Doyle, J. W., Roth, T., Smith, R. M., Li, Y. Q. and Dunn, R. M. 1996. Effects of calcium alginate on cellular wound healing processes modelled in vitro. *J. Biomed. Mater. Res.*, 32, 561–568.

Embil, J. M., Papp, K., Sibbald, G. et al. 2000. Recombinant human platelet-derived growth factor-BB (becaplermin) for healing chronic lower extremity diabetic ulcers: An open-label clinical evaluation of efficacy. *Wound Repair Regen.*, 8, 162–168.

Field, C. K. and Kerstein, M. D. 1994. Overview of wound healing in a moist environment. *Am. J. Surg.*, 167 (Suppl 1A), 2S–6S.

Franz, M. G., Robson, M. C., Steed, D. L. et al. 2008. Guidelines to aid healing of acute wounds by decreasing impediments of healing. *Wound Repair Regen.*, 16, 723–748.

Franz, M. G., Steed, D. L. and Robson, M. C. 2007. Optimizing healing of the acute wound by minimizing complications. *Curr. Prob. Surg.*, 44, 691–763.

Frokjaer, S. and Otzen, D. E. 2005. Protein drug stability: A formulation challenge. *Nat. Rev. Drug Disc.*, 4, 298–306.

Fu, X., Li, X., Cheng, B., Chen, W. and Sheng, Z. 2005. Engineered growth factors and cutaneous wound healing: Success and possible questions in the past 10 years. *Wound Repair Regen.*, 13, 122–130.

Gospodarowicz, D. and Cheng, J. 1986. Heparin protects basic and acidic FGF from inactivation. *J. Cell. Physiol.*, 128, 475–484.

Greenhalgh, D. G. 2003. Wound healing and diabetes mellitus. *Clin. Plast. Surg.*, 30, 37–45.

Heggers, J. P. 2003. Assessing and controlling wound infection. *Clin. Plast. Surg.*, 30, 25–35.

Hinchliffe, R. J., Valk, G. D., Apelqvist, J. et al. 2008. Specific guidelines on wound and wound-bed management. *Diabet. Met. Res. Rev.*, 24 (Suppl. 1), S188–S189.

Ho, Y. C., Mi, F. L., Sung, H. W. and Kuo, P. L. 2009. Heparin-functionalized chitosan-alginate scaffolds for controlled release of growth factor. *Int. J. Pharm.*, 376, 69–75.

Ichioka, S., Harii, K., Nakahara, M. and Sato, Y. 1998. An experimental comparison of hydrocolloid and alginate dressings, and the effect of calcium ions on the behaviour of alginate gel. *Scand. J. Plast. Reconstr. Surg. Hand Surg.*, 32, 311–316.

Jeffcoate, W. J., Price, P. and Harding, K. G. 2004. Wound healing and treatments for people with diabetic foot ulcer. *Diabetes. Metab. Res. Rev.*, 20, S78–S89.

Kamolz, L. P., Lumenta, D. B., Kitzinger, H. B. and Frey, M. 2008. Tissue engineering for cutaneous wounds: An overview of current standards and possibilities. *Eur. Surg.*, 40, 19–26.

Khor, E. and Lim, L. Y. 2003. Implantable applications of chitin and chitosan. *Biomaterials*, 24, 2339–2349.

Kiritsy, C. P., Lynch, A. B. and Lynch, S. E. 1993. Role of growth factors in cutaneous wound healing: A review. *Crit. Rev. Oral Biol. Med.*, 4, 729–760.

Lawrence, W. T. and Diegelmann, R. F. 1994. Growth factors in wound healing. *Clin. Dermatol.*, 12, 157–169.

Lazarus, G. S., Cooper, D. M., Knighton, D. R. et al. 1994. Definitions and guidelines for assessment of wounds and evaluation of healing. *Arch. Dermatol.*, 130, 489–493.

Leaper, D. J. and Durani, P. 2008. Topical antimicrobial therapy of chronic wounds healing by secondary intention using iodine products. *Int. Wound J.*, 5, 361–368.

Liu, B. S., Yao, C. H. and Fang, S. S. 2008. Evaluation of a non-woven fabric coated with a chitosan bi-layer composite for wound dressing. *Macromol. Biosci.*, 8, 432–440.

MacLeod, T. M., Cambrey, A., Williams, G., Sanders, R. and Green, C. J. 2008. Evaluation of Permacol™ as a cultured skin equivalent. *Burns*, 34, 1169–1175.

Macleod, T. M., Williams, G., Sanders, R. and Green, C. J. 2005. Histological evaluation of Permacol™ as a subcutaneous implant over a 20-week period in the rat model. *Br. J. Plast. Surg.*, 58, 518–532.

Magnusson, B. M., Anissimov, Y. G., Cross, S. E. and Roberts, M. S. 2004. Molecular size as the main determinant of solute maximum flux across the skin. *J. Invest. Dermatol.*, 122, 993–999.

Mekkes, J. R., Loots, M. A. M., Van der Wal, A. C. and Bos, J. D. 2003. Causes, investigation and treatment of leg ulceration. *Br. J. Dermatol.*, 148, 388–401.

Meng, X., Tian, F., Yang, J. et al. 2010. Chitosan and alginate polyelectrolyte complex membranes and their properties for wound dressing application. *J. Mater. Sci. Mater. Med.*, 21, 1751–1759.

Menon, G. K. 2002. New insight into skin structure: Scratching the surface. *Adv. Drug Deliv. Rev.*, 54 (Suppl. 1), S3–S17.

Milic, D. J., Zivic, S. S., Bogdanovic, D. C., Karanovic, N. D. and Golubovic, D. C. 2009. Risk factors related to the failure of venous ulcers to heal with compression treatment. *J. Vasc. Surg.*, 49, 1242–1247.

Miller, M. J. and Patrick, C. W. 2003. Tissue engineering. *Clin. Plast. Surg.*, 30, 91–103.

Milstone, A. M., Passaretti, C. L. and Perl, T. M. 2008. Chlorhexidine: Expanding the armamentarium for infection control and prevention. *Clin. Infect. Dis.*, 46, 274–281.

Monaco, J. L. and Lawrence, W. T. 2003. Acute wound healing: An overview. *Clin. Plast. Surg.*, 30, 1–12.

Mulder, G., Tallis, A. J., Marshall, V. T. et al. 2009. Treatment of nonhealing diabetic foot ulcers with a platelet derived growth factor gene-activated matrix (GAM501): Results of a phase 1/2 trial. *Wound Rep. Regen.*, 17, 772–779.

Papanas, N. and Maltezos, E. 2007. Growth factors in the treatment of diabetic foot ulcers: New technologies, any promises? *Int. J. Low. Extrem. Wounds*, 6, 37–53.

Papanas, N. and Maltezos, E. 2010. Benefit-risk assessment of becaplermin in the treatment of diabetic foot ulcers. *Drug Saf.*, 33, 455–461.

Priya, S. G., Jungvid, H. and Kumar, A. 2008. Skin tissue engineering for tissue repair and regeneration. *Tissue Eng. Part B Rev.*, 14, 105–118.

Rees, R. S., Robson, M. C., Smiell, J. M., Perry, B. H. and The Pressure Ulcer Study, G. 1999. Becaplermin gel in the treatment of pressure ulcers: A phase II randomized, double-blind, placebo-controlled study. *Wound Repair Regen.*, 7, 141–147.

Reimer, K., Vogt, P. M., Broegmann, B. et al. 2000. An innovative topical drug formulation for wound healing and infection treatment: In vitro and in vivo investigations of a povidone-iodine liposome hydrogel. *Dermatology*, 201, 235–241.

Robson, M. C. and Barbul, A. 2006. Guidelines for the best care of chronic wounds. *Wound Repair Regen.*, 14, 647–648.

Robson, M. C., Dubay, D. A., Wang, X. and Franz, M. G. 2004. Effect of cytokine growth factors on the prevention of acute wound failure. *Wound Repair Regen.*, 12, 38–43.

Rosenberg, A. S. 2006. Effects of proteins aggregates: An immunological perspective. *AAPS J.*, 8, E501–E507.

Rossi, S., Marciello, M., Sandri, G. et al. 2007. Wound dressings based on chitosans and hyaluronic acid for the release of chlorhexidine diacetate in skin ulcer therapy. *Pharm. Dev. Technol.*, 12, 415–422.

Ryan, C. M., Schoenfeld, D. A., Thorpe, W. P. et al. 1998. Objective estimates of the probability of death from burn injuries. *N. Engl. J. Med.*, 338, 362–366.

Sakiyama-Elbert, S. E. and Hubbell, J. A. 2000. Development of fibrin derivatives for controlled release of heparin-binding growth factors. *J. Control. Release*, 65, 389–402.

Scaffer, C. J. and Nanney, L. B. 1996. Cell biology of wound healing. *Int. Rev. Cytol.*, 169, 151–158.

Sen, C. K., Gordillo, G. M., Roy, S. et al. 2009. Human skin wounds: A major and snowballing threat to public health and the economy. *Wound Repair Regen.*, 17, 763–771.

Shevchenko, R. V., Sibbons, P. D., Sharpe, J. R. and James, S. E. 2008. Use of a novel porcine collagen paste as a dermal substitute in full-thickness wounds. *Wound Repair Regen.*, 16, 198–207.

Singer, A. J. and Clark, R. A. F. 1999. Cutaneous wound healing. *N. Engl. J. Med.*, 341, 734–746.

Singer, A. J. and Dagum, A. B. 2008. Current management of acute cutaneous wounds. *N. Engl. J. Med.*, 359, 1037–1046.

Smiell, J. M. 1998. Clinical safety of becaplermin (rhPDGF-BB) gel. *Am. J. Surg.*, 176 (Suppl 2A), 68S–73S.

Smiell, J. M., Wieman, J., Steed, D. L. et al. 1999. Efficacy and safety of becaplermin (recombinant human platelet-derived growth factor-BB) in patients with nonhealing, lower extremity diabetic ulcers: A combined analysis of four randomized studies. *Wound Repair Regen.*, 7, 335–346.

Sulea, D., Albu, M. G., Ghica, M. V. et al. 2011. Characterization and in vitro release of chlorhexidine digluconate contained in type I collagen porous matrices. *Rev. Roum. Chim.*, 56, 65–71.

Tanihara, M., Suzuki, Y., Yamamoto, E., Noguchi, A. and Mizushima, Y. 2001. Sustained release of basic fibroblast growth factor and angiogenesis in a novel covalently crosslinked gel of heparin and alginate. *J. Biomed. Mater. Res.*, 56, 216–221.

Thomsen, P. D. 1994. What is Infection? *Am. J. Surg.*, 167 (Suppl 1), 7S–11S.

Tredget, E. E., Shankowsky, H. A., Groeneveld, A. and Burrell, R. 1998. A matched-pair, randomised study evaluating the efficacy and safety of Acticoat silver-coated dressing for the treatment of burn wounds. *J. Burn Care Rehabil.*, 19, 531–537.

Walker, R. B. and Smith, E. W. 1996. The role of percutaneous penetration enhancers. *Adv. Drug Deliv. Rev.*, 18, 295–301.

Werner, S. and Grose, R. 2003. Regulation of wound healing by growth factors and cytokines. *Physiol. Rev.*, 83, 835–870.

Winter, G. D. 1962. Formation of the scab and the rate of epithelialization of superficial wounds in the skin of the young domestic pig. *Nat. Biotechnol.*, 193, 293–294.

Wong, C., Inman, E., Spaethe, R. and Helgerson, S. 2003. Fibrin-based biomaterials to deliver human growth factors. *Thromb. Haemost.*, 89, 573–582.

Yamaguchi, Y. and Yoshikawa, K. 2001. Cutaneous wound healing: An update. *J. Dermatol.*, 28, 521–534.

Yudanova, T. N. and Reshetov, I. V. 2006. Modern wound dressings: Manufacturing and properties. *Pharm. Chem. J.*, 40, 24–31.

Zakrewska, M., Wiedlocha, A., Szlachcic, A., Krowarsch, D. and Otlewski, J. 2009. Increased protein stability of FGF-1 can compensate for its reduced affinity for heparin. *J. Biol. Chem.*, 284, 25388.

Zhang, Y., Lim, C. T., Ramakrishna, R. and Huang, Z. M. 2005. Recent developments of polymer nanofibres for biomedical and biotechnological applications. *J. Mater. Sci. Mater. Med.*, 16, 933–946.

Zilberman, M., Golerkansky, E., Elsner, J. J. and Berdicevsky, I. 2009. Gentamicin-eluting bioresorbable composite fibres for wound healing applications. *J. Biomed. Mater. Res.*, 89A, 654–666.

25 Polymers for Artificial Joints

Masayuki Kyomoto, Toru Moro, and Kazuhiko Ishihara

CONTENTS

25.1 ARTIFICIAL JOINT REPLACEMENT

25.1.1 HIP JOINT REPLACEMENT AND ITS CLINICAL PERFORMANCE

A normal joint in our body is made up of bones that are lined by surface cartilage. The joint is surrounded by a capsule with a thin lining of synovial cells that produce a thin layer of lubrication film. The lubrication film (synovial fluid) together with the surface cartilage (articular cartilage) acts as a shock absorber and allows the joint to move smoothly; this protective action endures for many years (such as 50–60 years). If the surface cartilage is badly damaged or if the joint surfaces are misaligned (e.g., hip dysplasia), then the cartilage will wear out much quicker than in normal wear and tear, and as a result, the bone under the cartilage layer becomes exposed. The exposed bone starts to rub against the other and the process of osteoarthritis (wear and tear) is established. Osteoarthritis is, therefore, the result of mechanical wear and tear on a joint. Its main feature is a loss of surface cartilage with bone rubbing on bone and this may include joint pain, tenderness, stiffness, creaking, and locking of joints. This process produces pain and local inflammation. In osteoarthritis, a variety of potential forces—hereditary, developmental, metabolic, and the mechanical—may initiate processes leading to loss of cartilage. As the body struggles to contain ongoing damage, immune and regrowth processes can accelerate damage. Sometimes the body tries to relieve this pain by increasing the amount of fluid in the joint. This is why joints are swollen. The formation of bone spurs and cysts around the joint is another hallmark of osteoarthritis.

The most common type of arthritis leading to total hip replacement is degenerative arthritis (e.g., osteoarthritis) of the hip joint. This type of arthritis is generally seen with aging, trauma, or congenital abnormality (dysplasia) of the hip joint. Other conditions leading to total hip replacement include bone fractures, rheumatoid arthritis, and bone death (aseptic necrosis) of the femoral head. Bone necrosis can be caused by fracture of the hip, alcohol and drugs (such as prednisone and prednisolone), diseases (such as systemic lupus erythematosus), and conditions (such as kidney transplantation).

Total hip arthroplasty (THA) or hemi-arthroplasty is a surgical procedure whereby the diseased cartilage and the bone of the hip joint are surgically replaced with an artificial joint to restore joint movement (Figure 25.1). In general, total hip joint replacement consists of three (cement fixation) or four (cementless fixation) parts as follows:

- A plastic cup and metal shell that replaces the hip socket (acetabulum)
- A metal or ceramic ball that replaces the fractured femoral head
- A femoral metal stem that is attached to the shaft of the bone to add stability to the prosthesis

If a hemi-arthroplasty is performed, either the femoral head or the hip socket (acetabulum) will be replaced with a prosthetic device.

Upon inserting the prosthesis into the central core of the femur, it is fixed with a bone cement of poly(methyl methacrylate). Alternatively, a "cementless" prosthesis is used, which has microscopic

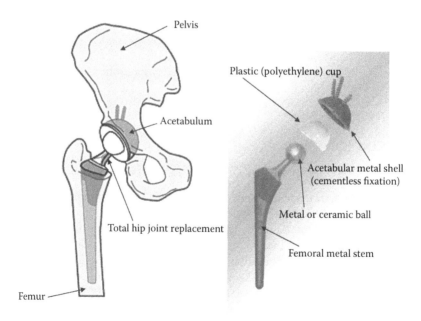

FIGURE 25.1 Schematic model of total hip joint replacement and typical product.

pores that allow bone ingrowth from the normal femur into the prosthesis stem. This "cementless" hip is considered to have a longer duration and will be chosen especially for younger patients.

THA is one of the most successful joint surgeries performed today. The operation relieves pain and stiffness symptoms, and most patients (over 80%) need no help in walking. In well-selected patients who are appropriate candidates for total hip replacements, the effects of the procedure last for at least 10 years in nearly 95% of patients [1,2]. However, with time, many problems have been observed due to the limited long-term fixation of the replacement. Hence, improvements with new devices and techniques are necessary. The future will provide newer devices that will further improve patient outcomes and lessen the potential for complications. Moreover, with improved devices and techniques, the operation could be recommended for younger individuals.

25.1.2 KNEE JOINT REPLACEMENT AND ITS CLINICAL PERFORMANCE

The most common type of arthritis leading to knee replacement is also degenerative arthritis (i.e., osteoarthritis) of the knee joint. For patients with mild arthritis, which is confined to one of the condyles of the knee, the surgeon might decide to perform unicondylar knee arthroplasty (UKA). If the arthritis is more serious and both the condyles of the knee are diseased, the surgeon may perform a bicondylar total knee arthroplasty (TKA) (Figure 25.2). Finally, in the case of extreme circumstances, such as a revision operation or in the event of tumor resection, semi-constrained hinged knee design with metal or ceramic materials might be employed.

TKA includes a prosthesis consisting of three or four parts as follows:

1. A metal or ceramic femoral component
2. A plastic insert or all-plastic tibial component
3. A metal or ceramic tibial tray
4. A plastic patellar component

The plastic ultra high molecular weight polyethylene (UHMWPE) insert or all-plastic tibial component plays a primary role: articulating either against a metallic or ceramic femoral component; in several cases in patellar resurfacing, the UHMWPE may articulate against cartilage.

TKA continues to be a remarkably successful operation for pain relief. Cementless fixation of the artificial knee joint is much less common than cement fixation. Today, TKA yields beneficial and predictable results with a survival rate of over 90% at 10 years after surgery [3].

On the other hand, the number of primary TKA has continuously increased. Although the frequency of revision surgery has not changed, the failure mechanism that necessitates revision TKA appears to be increasingly related to UHMWPE wear and tear along with osteolysis. Rand et al.

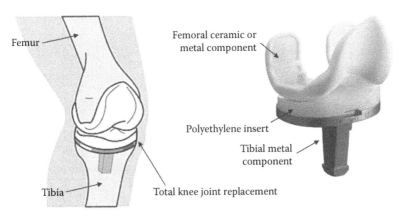

FIGURE 25.2 Schematic model of total knee joint replacement and typical product.

reported that implant loosening was the major mode of failure [4]. Aseptic loosening was the leading reason for revision arthroplasty, followed by osteolysis and polyethylene wear. However, the true wear and tear rate of polyethylene in TKA is not exactly known for a number of reasons as follows: no radiographic methods for measuring it have been established, the complex geometry of knee implants makes this task far more difficult than in THA, and the activity levels of the patient are largely unreported. In clinical outcome studies, the wear and tear of the tibial UHMWPE insert should be correlated with not only the period of clinical use but also the condition and type of use. The orthopedic community has also recognized that oxidative degradation can adversely affect the wear properties of UHMWPE. Oxidative degradation occurs due to prolonged shelf life and clinical use. An oxidatively degraded UHMWPE insert generally shows bad results with failure by wear and tear or osteolysis.

Wear and tear in TKA is far more dependent on alignment and ligament balancing techniques than wear and tear in THA is. The understanding of alterations in knee kinematics in TKA has markedly improved, but the surgical technique for TKA remains largely unchanged. Improvements in surgical technique with experience and teaching should reduce the frequency of component positioning that tends to accelerate wear and tear.

25.1.3 INCREASE IN HIP JOINT ARTHROPLASTY PROCEDURES

THA is widely recognized as a successful and effective treatment for degenerative hip joint disease. Worldwide, the number and rate of artificial hip joints used for primary and revised THA are substantially increasing every year. For example, the rate of primary THA per 100,000 persons increased by 46%, and the rates of revision THA increased by 60%, respectively, during 1990–2002 in the United States [1]. The number of primary THA increased from 119,000 in 1990 to 193,000 in 2002. Taking into account the population changes according to the United States Census Bureau, the overall rate of primary THA is 15 procedures per 100,000 persons per decade. On the other hand, in Japan, the number of primary THA and hemi-arthroplasties increased from 15,040 and 27,916 in 1994 to 35,793 and 49,315 in 2006, by 138% and 77%, respectively (Figure 25.3) [5]. Hip joint arthroplasties have important implications for health costs in Japan. For example, if the 85,108 hip joint arthroplasties performed in 2006 were to increase by 1%, the potential increase in cost could be 581 million (≈$5.81 million) based on recent procedural cost estimates of 0.7 million (≈$0.007 million) for each hip joint arthroplasty.

A general trend pointing to an increase in both the number and the rate of revision arthroplasties has also been observed. The number of revision THA increased from approximately 24,000 in 1990 to 43,000 in 2002 in the United States [1]. Further, the mean revision burden for THA was noted to be 17.5% (15.2%–20.5%). Taking into account the population changes according to the U.S. Census Bureau, the overall rate of revision THA was 3.7 procedures per 100,000 persons per decade.

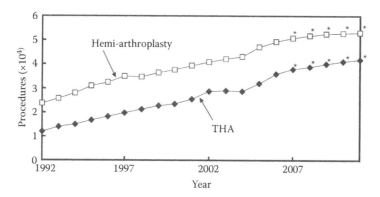

FIGURE 25.3   Procedures of primary THA and hemi-arthroplasty in Japan. *Forecast value.

TABLE 25.1
Revision Burdens for Hip Arthroplasty in Various Countries

Country	Period	Revision Burden (%)	Comments
Australia	1999–2002	18.2	—
Canada	2002–2003	13.1	—
Finland	1980–2001	15.7	—
Finland	1990–2001	18.3	—
Norway	1987–1998	15.0	—
Norway	1994–1998	16.4	—
Sweden	1979–2000	7.7	—
Sweden	1992–2000	11.0	—
Sweden	1992–2000	6.4	$\geq$65 years old
United States	1990–2002	17.5	—
United States	1990–2002	16.9	$\geq$65 years old

With the exception of THA performed in Sweden, the revision burden in the United States compared favorably with that in several countries with established total joint registries (Table 25.1) [1,2]. Overall, the THA revision burden of 17.5% in the United States from 1990 through 2002 fell within the range of revision burdens of 15.0%–18.3% observed in Norway, Finland, and Australia. In Canada, the revision burden for THA was lower (13.1% for 2002–2003). The overall revision burden for THA in the United States was substantially greater than the revision burden reported for Sweden (7%–11%).

25.1.4 PROBLEMS OF JOINT REPLACEMENT: OSTEOLYSIS

Table 25.2 illustrates the reasons for revision in the 14,081 first revisions for THA performed in the previous study [2]. The majority (75.3%) of the revision surgeries were performed because of aseptic loosening with or without focal osteolysis, 7.6% were performed to treat primary or secondary infection, and 8.8% were performed for technical reasons and dislocation that could have been mainly related to misalignment of the implants. Periprosthetic fractures (5.1%), implant fractures (1.5%), and a number of less prevalent reasons constituted the balance of the reasons.

TABLE 25.2
Reasons for Revision THA

Reason	Number	Share (%)
Aseptic loosening	10610	75.3
Primary deep infection	948	6.7
Dislocation	810	5.8
Fracture only	716	5.1
Technical error	425	3.0
Implant fracture	215	1.5
Secondary infection	128	0.9
Polyethylene wear	126	0.9
Pain	46	0.3
Miscellaneous	56	0.4
Missing	1	<0.1
Total	14081	100

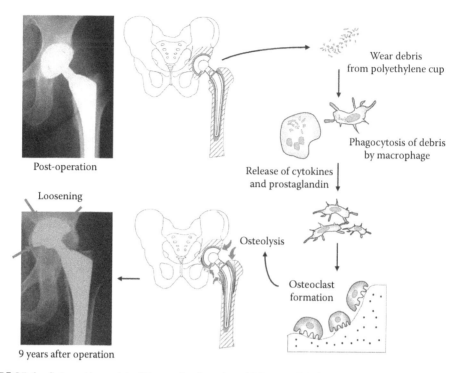

Post-operation

Loosening

9 years after operation

Wear debris
from polyethylene cup

Phagocytosis of debris
by macrophage

Release of cytokines
and prostaglandin

Osteolysis

Osteoclast
formation

FIGURE 25.4 Schematic model of the mechanisms by which wear debris leads to osteolysis.

As shown in Table 25.2, a consensus statement on total hip joint replacement concluded that the major remaining issues of concern included long-term fixation of the acetabular component, osteolysis due to wear debris, the biological response to debris, and problems related to revision surgery. Although acetabular fixation is no longer a problem, wear and related complications continue to be the major issue affecting the longevity of total hip joint replacements. The bone loss associated with osteolysis can result in pelvic dissociation and instability and major segmental cortical defects in the femur. Young active patients are most at risk for wear and osteolysis.

The precise mechanisms by which wear debris leads to osteolysis will ultimately be determined by defining how specific types of particles combine with environmental factors to permit interactions with specific types of cells that then communicate with each other through the release of soluble mediators (Figure 25.4). Generation of wear debris occurs immediately after implant insertion and ultimately results in a profile of particles that includes all total hip joint replacement materials [6,7]. The extent of bone resorption at the implant–bone interface varies with the severity of the granulomatous tissue response to wear debris and determines the time lapsed before implant loosening occurs. Wear particle-induced macrophage activation plays a role in periprosthetic osteolysis. Essentially, this occurs by two biological mechanisms. First, wear particle-associated macrophages release proinflammatory factors (e.g., cytokines, growth factors, prostaglandins) that enhance the activity of osteoclasts, the cells that carry out bone resorption. Second, osteoclasts are formed from mononuclear precursors that are present in the wear particle-induced macrophage infiltrate. These processes are not mutually exclusive; other stromal and inflammatory cell elements found at the bone–implant interface likely influence both the extent of osteoclast formation and bone resorption.

The contribution of the cells present within the macrophage-rich inflammatory tissue to the induction of bone resorption and implant loosening involves multiple cellular mechanisms. The macrophages are activated by the particles and subsequently release proinflammatory cytokines and other agents that induce bone resorption. Macrophage products capable of inducing bone resorption include interleukin (IL)-1α, IL-1β, IL-6, tumor necrosis factor (TNF)-α, arachidonic acid metabolites, and degradative enzymes. The existence of multiple factors at one site is likely to accelerate bone destruction. IL-1α, IL-1β, and TNF-α may also induce secondary effects on other cell types (such as osteoblasts) in the interfacial membrane, resulting in the release of matrix-degrading enzymes, including collagenase, stromelysin, gelatinases, and plasminogen activators. Granulocyte/macrophage colony-stimulating factor (GM-CSF) has also been implicated in cellular proliferation in the interfacial membrane around implants. Other cytokines that may exhibit immunomodulatory roles include IL-12, which is increased in the pseudosynovial fluid in patients with aseptic loosening of hip joint replacement. A primary response of macrophages to particulate debris is the increased release of TNF-α. TNF-α release results in part from the exposure of macrophages to particles, which activates the transcription factor NF-κB; this reaction is related to membrane receptor events. Alteration of the bone surface by these proteases may stimulate osteoclast bone-resorbing activity and may influence the recruitment and adhesion of mononuclear phagocyte osteoclast precursors at the bone–implant interface.

A second important mechanism relevant to the role of macrophages in implant loosening is revealed by data demonstrating that wear particle-associated macrophages are capable of differentiating into multinucleated cells that exhibit all the phenotypic features of osteoclasts. Osteoclasts are highly specialized multinucleated cells that are uniquely capable of carrying out lacunar resorption. Osteoclasts are formed by fusion of bone marrow-derived mononuclear precursors that circulate in the monocyte fraction. A number of cellular and humoral factors are known to influence RANKL and osteoprotegerin (OPG) expression. Osteoclast formation in periprosthetic tissues can effectively be viewed as a balance between the productions of these two factors. Various cytokines and growth factors (apart from macrophage CSF) abundant in periprosthetic tissues in aseptic loosening, such as IL-1 and TNF-α, increase the OPG mRNA expression by osteoblasts, suggesting that these factors that stimulate osteoclastic bone-resorbing activity appear to act conversely to downregulate osteoclast formation. Prostaglandins such as PGE2 have also been shown to increase RANKL production and to decrease OPG release, thus stimulating osteoclast formation and bone resorption. Inflammatory cells, such as T-cells, are present in the arthroplasty membrane and may influence osteoclast differentiation and periprosthetic osteolysis by modulating RANKL expression and OPG production. Recent studies have also highlighted the role of certain cytokines (e.g., TNF-α, IL-1β, and IL-1) in inducing osteoclast formation both in the presence and absence of RANKL.

25.2 BEARING MATERIALS FOR JOINT REPLACEMENTS

25.2.1 POLYETHYLENE BEARING MATERIAL

Polyethylene is a polymer formed from ethylene (C_2H_4), which is a gas with a molecular weight of 28. The generic chemical formula for polyethylene is $-(C_2H_4)_n-$, where n is the degree of polymerization. For UHMWPE, the molecular chain can consist of as many as 0.2×10^6 ethylene repeat units, i.e., the molecular chain of UHMWPE contains up to 0.4×10^6 carbon atoms.

There are several kinds of polyethylene, which are synthesized with different molecular weights and chain architectures. Low-density polyethylene (LDPE) and linear low-density polyethylene (LLDPE) generally have branched and linear chain architectures, respectively, each with a molecular weight of typically less than 5×10^4 g/mol. High-density polyethylene (HDPE) is a linear polymer with a molecular weight of up to 0.2×10^6 g/mol. In comparison, UHMWPE has a molecular weight of up to 6×10^6 g/mol. In fact, the molecular weight is so ultra-high that it cannot be measured directly by conventional methods and must instead be inferred by its intrinsic viscosity. Table 25.3 summarizes the physical and chemical properties of LDPE, LLDPE, HDPE, and UHMWPE.

TABLE 25.3

Typical Physical and Chemical Properties of LDPE, LLDPE, HDPE, and UHMWPE

Property	LDPE	LLDPE	HDPE	UHMWPE
Molecular weight (10^6 g/mol)	—	—	0.05–0.25	2–6
Melting temperature (°C)	110–115	110–125	130–137	125–138
Poisson's ratio	—	—	0.40	0.46
Specific gravity	0.910–0.930	0.910–0.925	0.952–0.965	0.932–0.945
Tensile modulus of elasticity (GPa)	0.1–0.4	0.1–1.6	0.4–4.0	0.8–1.6
Tensile yield strength (MPa)	7–14	7–42	26–33	21–28
Tensile ultimate strength (MPa)	3–57	8–46	22–31	39–48
Elongation (%)	145–1000	460–1100	10–1200	350–525
Crystallinity (%)	<50	—	60–80	39–75

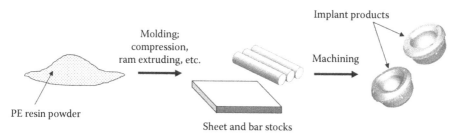

FIGURE 25.5 Typical processing steps in the manufacture of UHMWPE implants.

Three industrial steps are needed to manufacture orthopedic implants. First, UHMWPE must be polymerized from ethylene gas. Second, the polymerized UHMWPE, in the form of resin powder, needs to be consolidated into a sheet (i.e., compression molding), rod (i.e., ram extrusion), or near-net shaped implant (i.e., direct compression molding). Finally, in most instances, the UHMWPE implant needs to be machined into its final shape (Figure 25.5).

Since the 1950s, UHMWPE powders have been produced by Ruhrchemie (currently known as Ticona GmbH, Oberhausen, Germany) using the Ziegler process. The main ingredients for processing UHMWPE are reactive ethylene gas, hydrogen, and titanium tetrachloride catalyst. The polymerization takes place in a solvent used for mass and heat transfer. The requirements for medical-grade UHMWPE powder are specified in the American Society for Testing and Materials (ASTM) standard F648 and the International Organization for Standardization (ISO) standard 5834-1 [8,9].

Historically, the UHMWPE powder has been converted by compression molding since the 1950s, because the industries in the area around Ruhrchemie already were experienced in this processing technique. Today, compression-molded sheets of the UHMWPE are produced commercially by two companies (Orthoplastics, Ltd., Lancashire, United Kingdom, and Meditech Poly Hi Solidur, Ltd., Fort Wayne, IN). One UHMWPE sheet is pressed between the upper and middle plates, and the second is produced between the middle and lower plates. The plates are oil heated and hydraulically actuated from below. The heating and loading systems are all computer controlled. Finally, the entire press is contained in a clean room, to reduce the contamination of extraneous matter into the sheet. In contrast, ram extrusion of UHMWPE was developed by converters in the United States during the 1970s. Today, only few converters supply medical-grade ram extrusion UHMWPE to the orthopedic industry. Medical-grade extrusion facilities are owned by Orthoplastics, Ltd., Meditech Poly Hi Solidur, Ltd., and Westlake Plastics Co. Ltd. (Lenni, PA). The process is as follows. UHMWPE powder is fed continuously into an extruder. The extruder itself consists essentially of a hopper that allows powder to enter a heated receiving chamber, a horizontal reciprocating ram, a heated die, and an outlet. Within the extruder, the UHMWPE is maintained under pressure by the ram as well as by the backpressure of the molten UHMWPE. The backpressure is caused by frictional forces of the molten resin against the heated die wall surface as it is forced horizontally though the outlet. Beyond the outlet, the UHMWPE rod is slowly cooled in a series of electric heating mantles.

25.2.1.1 History of Polyethylene in the Orthopedic Field

The load-bearing articulating surface materials used in total joint arthroplasty comprise metallic alloys, ceramics, and polymers. The articulating couples of primary concern—those that generate considerable amounts of wear leading to osteolysis—include UHMWPE cups and inserts. Accordingly, in the past several decades, most research and development have been focused on improving the wear resistance of UHMWPE (Table 25.4).

Introduced clinically in November 1962 by Charnley, UHMWPE articulating against a metallic femoral ball remains the gold standard bearing surface combination in total hip joint replacement [11]. Considering how rapidly technology can change in the field of orthopedics, the long-term role that UHMWPE has played in hip joint replacement since the 1960s is fairly remarkable.

In the 1970s, the properties of UHMWPE were modified by including carbon fibers within the matrix of polyethylene, thereby creating a carbon fiber-reinforced UHMWPE, known as Poly II (Zimmer, Inc., Warsaw, IN) [13]. However, this UHMWPE composite was not found to exhibit consistent and improved clinical results relative to the conventional UHMWPE introduced by Charnley. The material was designed with orthopedic bearing applications in mind, under the assumption that increasing the modulus and ultimate tensile strength of the bearing as well as decreasing its creep properties would increase its longevity. This assumption was reasonable, since bearing surfaces are subject to high contact stresses, conditions under which conventional UHMWPE had often been observed to be pitted or delaminated. The inclusion of short chopped carbon fibers in a UHMWPE matrix resulted in a composite material with improved mechanical properties in vitro. Thus, the expectation was that Poly II would be more resistant to the pitting and delamination often seen in

TABLE 25.4
History of UHMWPE Development for Joint Replacement

Year	Comments
1958	Clinical use of polytetrafluoroethylene as bearing material of implants in hip arthroplasty by Charnley et al. [10]
1962	Charnley et al. adopts UHMWPE for use in hip arthroplasty [11]
1969	UHMWPE was gamma-ray sterilized in air with a minimum dose of 25 kGy [12]
1970	Commercial release of the Poly II-carbon fiber-reinforced UHMWPE for hip arthroplasty by Zimmer, Inc. [13]
1971	Clinical introduction of the 100 Mrad PE—extremely highly CLPE by more than 1000 kGy of gamma-ray irradiation in air by Oonishi et al. [16]
1982	Commercial release of alumina ceramic balls articulating against UHMWPE by Kyocera, Corp. [19]
1986	Clinical introduction of silane cross-linked HDPE by Wroblewski et al. [20]
1991	Commercial release of the Hylamer—highly crystalline UHMWPE for hip arthroplasty by DePuy Orthopedics, Inc. [21]
1997	Commercial release of highly CLPE with an energy-ray irradiation of 50–105 kGy by several orthopedic product manufacturers [13,24]
2006	Clinical use of the vitamin E-blended UHMWPE in knee arthroplasty produced by Nakashima Medical, Co. Ltd., as a trial [44]
2007	Clinical use of the PMPC-grafted CLPE in hip arthroplasty produced by Japan Medical Materials Corp. as a trial [42]

joint replacements. Further, wear testing of Poly II conducted by the manufacturer revealed it to have lower wear than conventional UHMWPE, suggesting that the strength benefits would result in longer lasting hip arthroplasties. However, unfortunately, the promise shown by Poly II in vitro was not borne out in the clinical setting, and within a short time after implantation, many patients presented with osteolysis and complete mechanical failure of their bearing surfaces [14]. One possible explanation for the mechanical failure was that the poor crack propagation resistance of Poly II was due to the carbon fibers not bonding with the UHMWPE matrix, instead serving as stress concentrators and crack nucleation sites [15].

In Japan, during the 1970s, an important technological advancement occurred: the clinical introduction of an extremely highly cross-linked polyethylene (CLPE) with more than 1000 kGy of gamma-ray irradiation in air by Oonishi et al., the so-called 100 Mrad PE [16,17]. A similar advancement in extremely highly CLPE also occurred in South Africa during the 1970s, where researchers in Pretoria clinically introduced a UHMWPE that was gamma-ray irradiated with up to 700 kGy in the presence of acetylene [18]. During the 1980s, two other noteworthy developments occurred relative to polyethylene in joint replacements. Chas F. Thackray-DePuy International Ltd. (Leeds, United Kingdom) began the development of an injection-molded HDPE that could be cross-linked by silane coupling. Only 22 of these implants were produced and implanted by Wroblewski et al. starting in 1986 [20]. After an initial wear period (initial bedding-in period), these cross-linked HDPE components have been found to exhibit very low clinical wear rates.

In 1991, a highly crystalline UHMWPE known as Hylamer was patented by Li et al. from E. I. Du Pont de Nemours and Company (Wilmington, DE) and marketed by the DePuy-DuPont Orthopedics joint venture (Newark, DE) [21]. Hylamer is a hot isostatically pressed UHMWPE, leading to the formation of an extended-chain crystallite morphology with thick (200–500 nm) lamellae and higher crystallinity (65%–71%) [13]. In contrast, conventional low-pressure sintered UHMWPE displays a folded-chain crystalline morphology with much thinner lamellae (10–50 nm in thickness) and a crystallinity of 50%–55%. By varying the postconversion heating, pressure, and cooling sequence, a family of materials was developed with varying crystalline morphologies and sizes. Hylamer has a higher density and crystallinity than conventional UHMWPE. Although the yield and ultimate strength of Hylamer are slightly higher, the most noticeable change occurs in the elastic modulus, which is nearly double for Hylamer as compared to conventional UHMWPE. The clinical results for the highly crystalline UHMWPE, which were clarified in the 1990s, have been mixed and are therefore controversial. Although several studies reported worse clinical performance using the Hylamer compared with conventional UHMWPE, other studies reported several satisfactory or even improved performances [13].

25.2.1.2 Cross-Linked Polyethylene

High-energy ray irradiation cross-linking and thermal treatment of UHMWPE has aroused intense scientific and commercial interest within the orthopedic field since the late 1990s (Figure 25.6). For decades, the cross-linking of polyethylene has been known to improve the abrasion resistance of the polymer for industrial applications. However, only a few applications of this technology have been reported in orthopedics literature [13,22,23]. All high-energy ray irradiation, including the standard 25- to 40-kGy dose of gamma-ray irradiation used for sterilization, leads to the formation of free radicals in polymeric materials through homolytic chain cleavage. In UHMWPE, some of these free radicals recombine with each other to form cross-links or trans-vinylene bonds, while others remain as highly reactive species in the structure for extended periods of time. Although the gel content of UHMWPE may be increased to 80% by an average gamma-ray radiation dose of 25 kGy, the polymer becomes highly cross-linked (corresponding to a gel content of 90%–100%) after an absorbed dose of 50 kGy [13,24]. Despite the plateau in gel content, the cross-linking density in UHMWPE did not reach an asymptotic value until a dose of 100–150 kGy had been absorbed. Therefore, several CLPEs, irradiated with 50–105 kGy, have been launched since 1998 and used extensively.

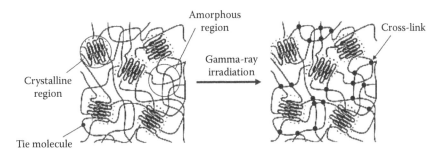

FIGURE 25.6   Schematic illustration of cross-linking induced by gamma-ray irradiation.

TABLE 25.5
Wear Reductions in Early- and Mid-Term Clinical Studies of CLPE Cups Compared with Conventional UHMWPE Cups

Manufacturing Process for CLPE	Mean Follow-Up Period (Years)	Wear Reduction (%)	Reference
Cold-irradiated and annealed	2.0	85	[27]
	2.3	42	[28]
	2.3	94	[23]
	4.0	58	[29]
	4.9	60	[30]
Cold-irradiated and remelted	2.8	72	[31]
	3.2	45	[32]
	5.3	73	[24]
	5.5	95	[33]
Warm-irradiated and remelted	2.0	54	[34]
	2.6	94	[35]
	2.9	44	[36]
	3.0	23	[37]
	3.8	83	[38]
	5.0	55	[39]

In several independent reviews of the literature, it was found that osteolysis is rare in patients in whom the UHMWPE cup is wearing at a rate of less than about 0.1 mm/year, but osteolysis becomes much more frequent and extensive as the wear rate increases substantially above this "threshold" value [25,26]. In several studies with a mean duration of follow-up of ~5 years or longer, the mean rates of wear of CLPE cups (Table 25.5) were all well below 0.1 mm/year [23,24,27–39].

On the other hand, the osteolysis threshold of 0.1 mm/year was established for hip joints with conventional UHMWPE cups—i.e., those that either were not cross-linked or were moderately cross-linked during gamma-ray sterilization. Some investigators have reported that the mean particle size is smaller with CLPE and that, in equivalent volumes, smaller particles tend to be more likely to cause osteolysis [40]. If that is correct, these factors could lead to the osteolysis threshold being somewhat lower for CLPE. We are aware of only one published case report of clinically relevant osteolysis in a hip with a CLPE cup [41]. However, the hip in question also had a forged-steel surface-grit-blasted femoral component that, at revision, was found to be loose at the stem–cement interface. Because the osteolysis in this hip joint occurred endosteally around the loosened stem, with no acetabular osteolysis, it is highly possible that the lesions were primarily due to debris produced at the stem–cement interface rather than from the CLPE cups. Continued close monitoring of patients with CLPE cups is essential to determine if the improved wear resistance that has been observed in the mid-term, as summarized here, will translate into a substantial reduction in the prevalence and severity of osteolysis at long-term follow-up.

25.2.1.3 Antioxidants for Polyethylene

Recently, there has been an explosion of interest in the research and development of vitamin E as an antioxidant for UHMWPE in the orthopedic field. The primary role of vitamin E (α-tocopherol) is to stabilize the active free radicals resulting from oxidation. The antioxidant activity of vitamin E is due to hydrogen abstraction from the –OH group on the chroman ring by a peroxy free radical, which can combine with another free radical (Figure 25.7). In a gamma-ray irradiated UHMWPE with vitamin E, peroxy free radicals abstract a hydrogen from vitamin E, forming hydroperoxides. The oxidative degradation cascade in the gamma-ray irradiated UHMWPE is hindered in the presence of vitamin E.

The idea of vitamin E-blended UHMWPE is popular in the industrial field: the first widespread applications of the vitamin E-blending technology actually appeared in food packaging since the 1980s. In the orthopedic field, Tomita et al. demonstrated the use of vitamin E-blended UHMWPE in order to prevent delamination by reducing crack formation at the grain boundaries of UHMWPE in 1998 [43]. Then, they demonstrated that the vitamin E-blended UHMWPE with gamma-ray sterilization exhibited a higher resistance to oxidation and fatigue wear compared with conventional UHMWPE. In light of its acceptance as an effective antioxidant, the vitamin E-blended UHMWPE

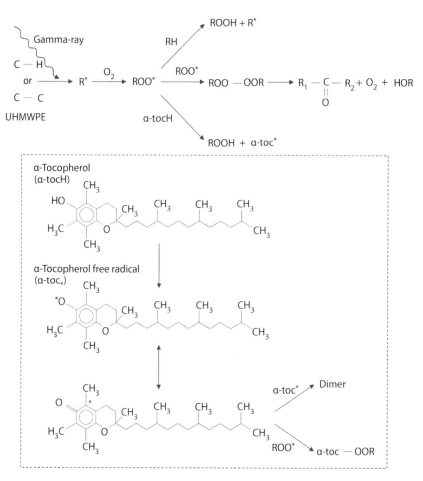

FIGURE 25.7 Schematic illustration of reaction of vitamin E (α-tocopherol).

insert in TKR was produced by Nakashima Medical Co. Ltd. (Okayama, Japan) and is being used in a clinical trial in Japan since 2006 [44]. Although this trial has taken place, the clinical results have not been published yet.

Subsequently, many orthopedic manufacturers have developed CLPE with vitamin E for joint replacements. However, several new problems have arisen, in particular for the procedures of introduction of vitamin E into the polyethylene as follows: (1) blending during compression molding or extrusion before the cross-linking and (2) diffusion after the cross-linking and machining [45]. The disadvantages of the former are that the cross-link density is suppressed to a low value during the cross-linking procedure with (e.g., gamma-ray) irradiation (Figure 25.8). On the other hand, those of the latter are that it is difficult to control the concentration and distribution of diffused vitamin E.

In both the cases, the hypothesized advantage of the vitamin E-blended/diffused CLPE is that the vitamin E protects the CLPE against oxidative degradation (Figure 25.9).

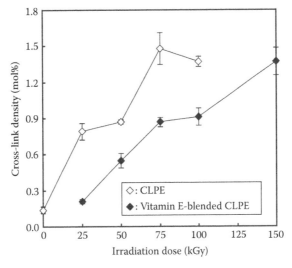

FIGURE 25.8 Cross-link density of vitamin E-blended CLPE as a function of the gamma-ray irradiation. Bar: Standard deviations.

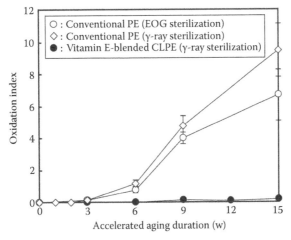

FIGURE 25.9 Oxidative degradation (oxidation index) of vitamin E-blended CLPE as a function of the accelerated aging duration in air at 80°C. Bar: Standard deviations.

25.2.2 Phospholipid Polymers for Mimicking Articular Cartilage

25.2.2.1 Hydration Lubrication

Water attracted by hydrophilic macromolecules in the surface layer plays an important role in lubrication. As macromolecules are flexible, they cannot support a load by themselves. The water in the surface layer would support most of the load because the water is attracted by the macromolecules. Frictional forces arise due to the adhesion of macromolecules to the counter surface. The time-dependent properties of friction forces can be interpreted as follows (Figure 25.10) [46]. Under a load, water exudes slowly from the surface layer with or without sliding. As the result of water loss, the thickness of the surface layer reduces and the water content of the surface layer decreases. Consequently, the degree of adhesion to the opposite bearing surface increases and the frictional force also increases. Therefore, it may be concluded that friction depends essentially on the water content of the surface layer. This hydration would lead to low friction and wear, by acting as "hydration lubrication."

25.2.2.2 Articular Cartilage and Material Design

Although the lubrication mechanism of human joints has been studied since the 1930s, it has not yet been understood clearly. However, it is well known that the composition elements of the articular cartilage surface consist of the collagen network, hyaluronic acid, and proteoglycan subunits. The proteoglycan subunits form a gel-like surface layer due to hydration along with the joint synovia. Although the binding between the proteoglycan subunits and hyaluronic acid can be visible [47], the binding between hyaluronic acid and the collagen network has not yet been confirmed. It was reported by Obara et al. [48] that the friction coefficient of joints increases when the gelled material on the cartilage surface is removed by gauze. After this, the joint surface is lubricated only by joint synovia or hyaluronic acid, i.e., following the loss of the gel comprising proteoglycan subunits, the friction coefficient of the joint cannot be lowered again. This fact indicates that the proteoglycan aggregates are not combined with the collagen network by physical adsorption and that the hydrophilic macromolecules on the joint surface play an important role in keeping the friction at low levels. A previous study reported that the hydrophilic macromolecules of the cartilage surface are assumed to have a brush-like structure: a part of the proteoglycan aggregate brush is bonded with the collagen network on the cartilage surface (Figure 25.11) [49]. The rest of the proteoglycan aggregate floats freely in joint synovia.

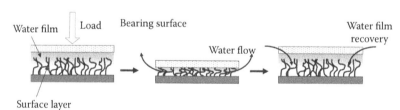

FIGURE 25.10   Schematic model of hydration lubrication.

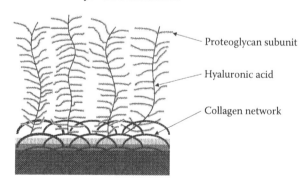

FIGURE 25.11   Schematic model of the brush-like structure of the cartilage surface.

Longfield et al. [50] and Ikeuchi et al. [51] reported that the lubrication mechanism of joints mainly comprises hydration lubrication. Hydrophilic macromolecules induce low friction by promoting the formation of a fluid film that is retained by the water molecules' attraction forces, indicating that the water molecules' attraction forces are important for realizing low friction. The lubrication of the human joint appears to occur by hydration lubrication because the surface layer of the joint resembles the structure of the gelled material in the human joint. Sasada et al. proposed a new idea for joint lubrication, named "Surface gel hydration lubrication" [49] for a lubrication mechanism peculiar to such hydrophilic macromolecules. A bearing surface with a brush-like structure comprising hydrophilic macromolecules in artificial hip joints was therefore assumed to be similar to that of articular cartilage. The hydration lubrication interface can also be regarded to mimic the natural joint cartilage in vivo. The novel material design with this hydration lubrication should be necessary.

25.2.2.3 Photo-Induced Surface "Grafting from" Polymerization

The grafting of polymers onto various surfaces has been studied for over 50 years and has played an important role in many areas of biomaterial science and technology, e.g., colloidal stabilization, adhesion, lubrication, tribology, and rheology. Recent work has focused on the synthesis of so-called polymer brushes whereby the polymer chains stretch out away from the surface or substrate [52,53]. There are three primary methods for modifying a planar substrate with an organic polymer: (a) physical coating, (b) chemical coating and/or "grafting to," and (c) "grafting from" (Figure 25.12). This includes physical coating such as spin or dip coating; however, the polymer is merely adsorbed onto the substrate and may diffuse away when the substrate is immersed into a solvent in which it is soluble. Chemical coating utilizes the functional group of the polymer to chemically attach onto the substrate via several coating techniques. Robust layers may be created by utilizing a self-assembled monolayer (SAM) in order to immobilize a reactive functionality. Thus, the polymer can be attached to the surface, provided the preformed polymer possesses a functional group that is capable of bonding with the surface (e.g., a polymer containing a primary amine could form an amide bond with a carboxylic acid-terminated SAM). This approach

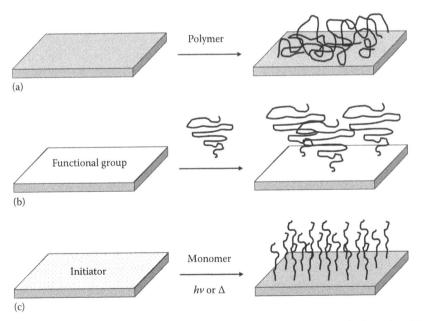

FIGURE 25.12 Approaches for modifying a substrate with a polymer: (a) physical coating (adsorption), (b) chemical coating (immobilization) and/or "grafting to," and (c) "grafting from."

is known as the "grafting to" technique and has been considerably successful at synthesizing robust layers of 1–50 nm in thickness. However, the "grafting to" technique is limited by diffusion barriers that prevent the preformed polymer from intercalating through the tethered polymer to the reactive substrate. Therefore, the "grafting to" method yields a low-density brush. In contrast, the "grafting from" approach has been utilized to synthesize high-density polymer brushes [54]. The conformation of these polymer brushes in a solvent can dramatically change with the graft density. At low-graft density, they will assume a "mushroom" conformation with a coil dimension similar to that of free chains. With increasing graft density, the graft chains will be obliged to stretch away from the substrate, forming a "polymer brush." These high-density brushes can be much thicker and range in size from a nanometer-scale to greater than a micrometer-scale. The great increase in thickness for "grafting from" layers is due to much higher grafting densities compared to those of "grafting to" layers.

Photochemical initiation has several advantages over thermal initiation. First, certain functional groups are not thermally stable; therefore, it is desirable to activate polymerization at room temperature. This also simplifies manufacturing processes. Furthermore, most alkylthiolate SAMs are not stable above 70°C and may begin to degrade at the temperatures required for most thermal initiations. Second, photoinitiation is generally faster than thermal initiation. Third, the initiation process may be activated at almost every temperature; this yields great flexibility when controlling the reactivity and processability of a layer. Surface-initiated polymerization has been carried out with a variety of initiators, and Figure 25.13 describes some of the most common photoinitiators [55]. For surface-initiated polymerization applications, these initiators are typically modified and covalently bonded to the substrate to yield a "grafting from" polymerization. Alternatively, photo-sensitizers can be added to bulk solutions in order to abstract hydrogen atoms from the substrate. For example, benzophenone (BP, Figure 25.13d) is converted to a reactive triplet state after ultraviolet (UV) irradiation; this triplet is capable of abstracting hydrogen atoms from various moieties. Tertiary amines or thiol-ene systems have been activated with photo-sensitizers, but until recently have not been used for surface-initiated polymerization. Other photoinitiators include peroxides (Figure 25.13a) and benzoin derivatives (Figure 25.13c); of the two, only peroxides have been used for surface-initiated polymerization. The most common free radical photoinitiators are derivatives of 2,2′-azobisisobutyronitrile (AIBN, Figure 25.13b), and these have been used by several research groups for "grafting from" polymerizations from various substrates. Recently, controlled free radical polymerizations have gained much recognition owing to their low polydispersities and "living"-like properties [55]. Indeed, "living" polymerizations have a tremendous advantage for surface-initiated polymerization since it is possible to grow block copolymers and to terminate the polymerization with specific end-groups. However, most living free radical polymerizations utilize thermal initiation; for example, atom transfer radical polymerization may be

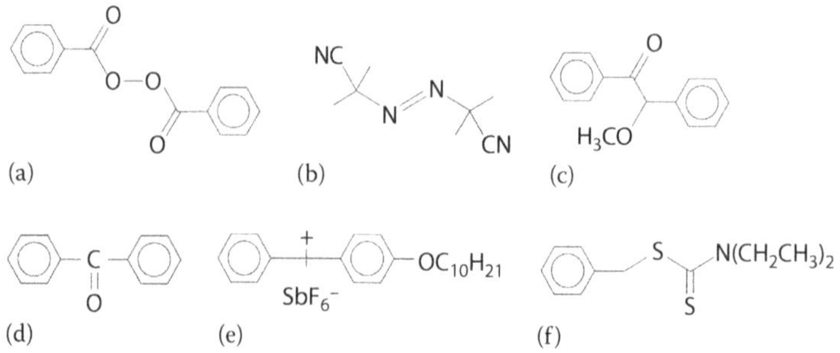

FIGURE 25.13 Various types of photoinitiators: (a) benzoylperoxides, (b) 2,2′-azobisisobutyronitrile compounds based on AIBN, (c) benzoin methylethers, (d) triplet photo-sensitizers, benzophenone (e) onium salts for cationic polymerization, and (f) controlled free radical polymerization with photoiniferters.

used for surface-initiated polymerization, but the rate of initiation and propagation is relatively slow as compared to that of photoinitiation and typically leads to layers that are less than 50 nm thick. Thus, there is a rich variety of photopolymerization strategies that may be utilized in the future, although very few examples of photosurface-initiated polymerization have been reported to date.

25.2.2.4 Poly(MPC)-Grafted Polyethylene

Surface modification is important for improvements in bearing materials. Moro et al. have demonstrated the creation of an artificial hip joint based on the novel concept of "hydration lubrication" by using poly(MPC) (PMPC)-grafted onto the surface of CLPE (PMPC-grafted CLPE); this device is designed to reduce wear and suppress bone resorption [56–58]. A previous study has reported that the hydrogel cartilage surface is assumed to have a brush-like structure: a part of the proteoglycan aggregate brush is bonded with the collagen network on the cartilage surface [49]. Therefore, the bearing surface with PMPC in artificial hip joints is assumed to have a brush-like structure similar to that of articular cartilage (Figure 25.14). The hydration lubrication interface can therefore be regarded to mimic the natural joint cartilage in vivo.

MPC, a methacrylate monomer with a phospholipid polar group in the side chain, is a novel biomaterial designed and developed by Ishihara et al. that mimics the neutral phospholipids of cell membranes [59]. MPC polymers are one of the most common biocompatible and hydrophilic polymers studied thus far, which have potential applications in a variety of fields, such as biology, biomedical science, and surface chemistry, because they possess the unique properties of good biocompatibility, high lubricity and low friction, anti-protein adsorption, and cell membrane-like surface [59–62]. Hence, MPC is hydrophilic and can form a thin film of free water under physiological conditions. Several medical devices have already been developed by utilizing the MPC polymers. These devices have been subjected to clinical use with the approvals of the Ministry of Health, Labour and Welfare (MHLW) of Japan and the Food and Drug Administration (FDA) of the United States; therefore, the efficacy and safety of the MPC polymer as a biomaterial are well established (Table 25.6) [63–82].

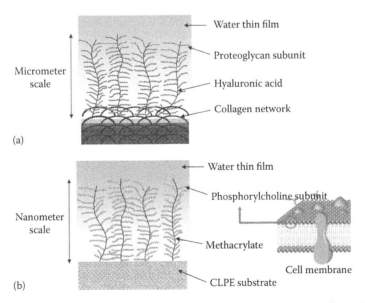

FIGURE 25.14 Schematic model of PMPC-grafted CLPE surface mimicking cartilage. (a) Cartilage. (b) PMPC-grafted CLPE.

TABLE 25.6
Medical Devices with MPC Polymer

Medical Device	Product Name	Manufacturer	Clinical Introduction	Reference
Artificial heart	Evaheart	Sun Medical	Current trial	[63]
Artificial joint	Aquala	Japan Medical Materials	Current trial	[42]
Artificial lung	Mimesys	Sorin Biomedica	2002	[64]
	Synthesis	Sorin Biomedica	2003	[65]
	Physio	Sorin Italia	2005	—
Catheter	Eliminate	Clinical Supply	—	MHLW approval
Contact lens	Proclear	Cooper Vision	1998	FDA approval
Guide wire	Aqua diver	Clinical Supply	—	MHLW approval
	Inter through	Clinical Supply	—	—
	Hunter	Biocompatible	1997	FDA approval
Micro catheter	Londis	Clinical Supply	2005	MHLW approval
Stent	Endeavor	Medtronic	Current trial	[66]
	Endeavor I	Medtronic	Current trial	[67,68]
	Endeavor II	Medtronic	Current trial	[67–69]
	Endeavor II CA	Medtronic	Current trial	[67,68]
	Endeavor III	Medtronic	Current trial	[67,68,70]
	TriMaxx	Abbott Laboratories	2005	[71]
	ZoMaxx	Abbott Laboratories	—	[72]
	Biodiv Ysio	Biocompatible	2000	[73–82]
Tympanostomy tube	—	Gyrus, Grace Medical	2000	FDA approval

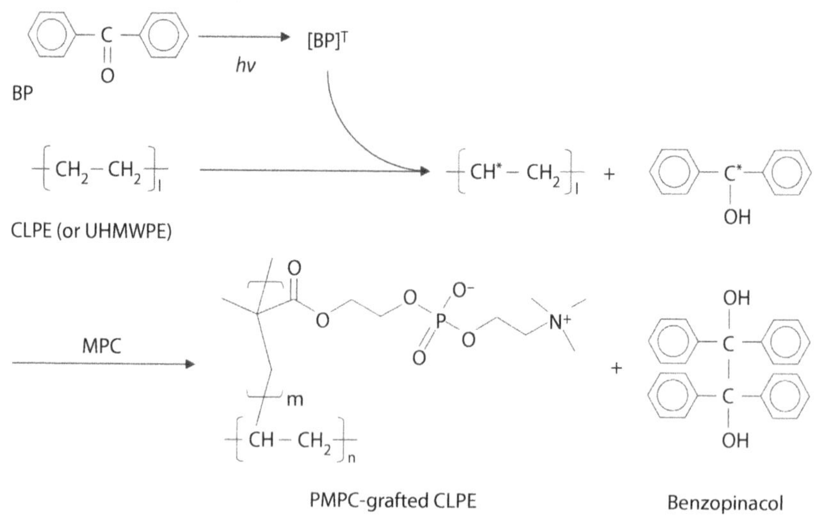

FIGURE 25.15 Schematic illustration of MPC graft polymerization by using the BP system.

The grafting of biocompatible and hydrophilic PMPC with CLPE has been accomplished by using a photo-initiated "grafting-from" polymerization. The photo-initiated "grafting-from" polymerization reaction by using a typical BP photoinitiator is shown in Figure 25.15. First, the physically coated BP on CLPE is excited by UV irradiation. The BP excited to the triplet state extracts a hydrogen atom from the –CH$_2$– group and then generates a radical that is capable of initiating the graft polymerization of MPC.

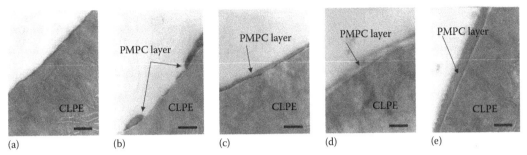

FIGURE 25.16 Cross-sectional TEM images of PMPC-grafted CLPE obtained with a 0.5 mol/L MPC concentration and various photo-irradiation times. Bar: 200 nm. (a) 11 min. (b) 23 min. (c) 45 min. (d) 90 min. (e) 180 min.

This technique has several important benefits as follows: direct grafting of PMPC to CLPE—thereby forming C–C covalent bonding between the PMPC and CLPE substrate, high mobility of the chains of the PMPC, a high density, and controlling the length of the introduced PMPC.

Figure 25.16 shows cross-sectional transmission electron microscope (TEM) images of PMPC-grafted CLPE produced with various photo-irradiation times during polymerization [83]. With photo-irradiation times longer than 45 min, a 100- to 200-nm-thick PMPC-grafted layer was clearly observed on the surface of the CLPE substrate. The MPC-covered region was coexistent with uncovered regions after a photo-irradiation time of 23 min, although the thickness of the covered region on the PMPC layer remained the same (100–200 nm). With photo-irradiation for 11 min, no PMPC layer was observed on the surface of the CLPE. These results indicate that the density of the grafted PMPC can be controlled by the polymerization time. This is attributable to the fact that the number of polymer chains produced in a radical polymerization reaction is generally correlated with the photo-irradiation time.

Figure 25.17 shows the static water-contact angle of PMPC-grafted CLPE as a function of the photo-irradiation time used for polymerization (0.50 mol/L MPC concentration) [83]. The static water-contact angle of untreated CLPE was 90° and decreased markedly with a decrease in the photo-irradiation time. The static water-contact angle decreased as the irradiation time was increased.

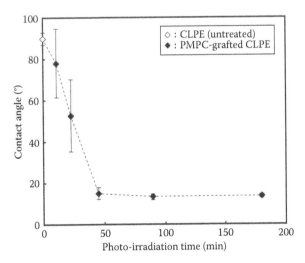

FIGURE 25.17 Static water-contact angle of PMPC-grafted CLPE as a function of the photo-irradiation time with a 0.5 mol/L MPC concentration. Bar: Standard deviations.

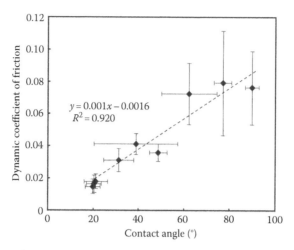

FIGURE 25.18 Relationship between dynamic coefficient of friction and contact angle in the PMPC-grafted CLPE surface. Bar: Standard deviations.

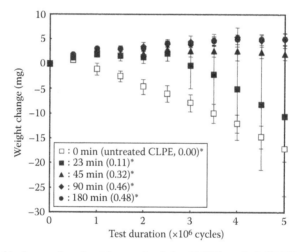

FIGURE 25.19 Weight change (gravimetric wear) of the PMPC-grafted CLPE cups obtained with a 0.5 mol/L MPC concentration and various photo-irradiation times in the hip joint simulator wear test. Bar: Standard deviations. *P–O group indexes are in parentheses.

Figure 25.18 shows the relationship between the dynamic coefficient of friction and the contact angle [84]. The dynamic coefficient of friction tended to increase with the contact angle. This increase was linear to a degree of accuracy, and the correlation coefficient was 0.920.

Figure 25.19 shows the gravimetric wear of PMPC-grafted CLPE with various photo-irradiation times during the hip joint simulation test. The PMPC-grafted CLPE cups were found to wear significantly less than the untreated CLPE cups. The wear of the PMPC-grafted CLPE cups subjected to 23-min photo-irradiation time started to increase after 2.5×10^6 cycles. The PMPC-grafted CLPE cups exhibited a slight increase in weight. This was partially attributable to enhanced fluid absorption in the tested cups than in the load-soak controls. When using the gravimetric method, the weight loss in the tested cups is corrected by subtracting the weight gain in the load-soak controls; however, this correction cannot be perfectly achieved because only the tested cups are continuously subjected to motion and load. Fluid absorption in the tested cups is generally slightly higher than that in the load-soak controls. Consequently, the

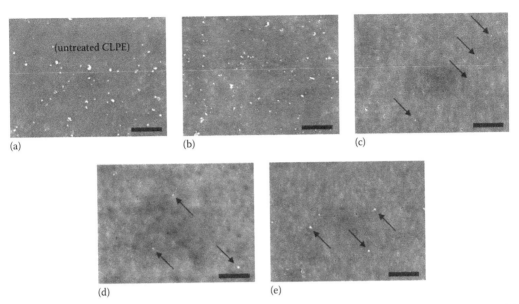

FIGURE 25.20 FE-SEM images of wear particles of the untreated CLPE and PMPC-grafted CLPE with various photo-irradiation times during the 4.5×10^6–5.0×10^6 cycles of the hip joint simulation test. Arrows: wear particles. Bar: 5 μm. (a) 0 min. (b) 23 min. (c) 45 min. (d) 90 min. (e) 180 min.

correction for fluid absorption by using the load-soak data as the correction factor leads to a slight underestimation of the actual weight loss. The initial wear rate is defined as that from the start to 0.5×10^6 cycles, and the steady wear rate is considered as that from 4.0×10^6 to 5.0×10^6 cycles. All the untreated CLPE and PMPC-grafted CLPE cups showed low initial wear rates of -1.42 to -3.74 mg/10^6 cycles. The steady wear rate of the untreated CLPE cups and the PMPC-grafted CLPE cups with a low P–O group index of 0.11 (23-min photo-irradiation time) increased to 5.11 and 5.48 mg/10^6 cycles, respectively. In contrast, the wear rates of the PMPC-grafted CLPE cups with high P–O group indexes, i.e., 0.46 (90-min photo-irradiation time) and 0.48 (180-min photo-irradiation time), were markedly lower at 0.32 and -0.02 mg/10^6 cycles, respectively.

Figure 25.20 shows field emission scanning electron microscope (FE-SEM) images of wear particles of the untreated CLPE and PMPC-grafted CLPE with various photo-irradiation times during the 4.5×10^6–5.0×10^6 cycles of the hip joint simulation test. The wear particles of the untreated CLPE and PMPC-grafted CLPE cups, as characterized by FE-SEM, were predominantly submicrometer-sized granules. The wear particles of the PMPC-grafted CLPE cups with 45-, 90-, and 180-min photo-irradiation times were found to be significantly lesser than those for the untreated CLPE cups and the PMPC-grafted CLPE cups with 23-min photo-irradiation time.

In summary, an artificial hip joint based on the novel concept of "hydration lubrication" was created by using PMPC grafted onto the surface of CLPE for reducing the wear debris of UHMWPE. The approach using "hydration lubrication" is surely novel in the field of orthopedic biomaterials science, and joint replacement with hydration lubrication can pioneer the "next generation" artificial joint. Furthermore, these joint replacements have the potential to be applied in the orthopedic field in the near future [85]. The clinical trial for such joint replacements with hydration lubrication (i.e., PMPC-grafted CLPE acetabular cup) has been started at the University of Tokyo and other hospitals in Japan since 2007. For this novel PMPC-grafted CLPE material, close monitoring of clinical performance and accurate quantification of wear rates would be essential for the early recognition of unforeseen problems.

25.2.3   POLY(ETHER-ETHER KETONE) BEARING MATERIALS

25.2.3.1   Structure and Properties

Poly(aryl-ether-ketone) (PAEK), including poly(ether-ether-ketone) (PEEK), is a new family of high-performance thermoplastic polymers, consisting of an aromatic backbone molecular chain interconnected by ketone and ether functional groups, i.e., a BP unit is included in its molecular structure. Polyaromatic ketones exhibit enhanced mechanical properties, and their chemical structure is stable, resistant to chemical and radiation damage, and compatible with several reinforcing agents (such as glass and carbon fibers; carbon fiber-reinforced PEEK [CFR-PEEK], Figure 25.21). Therefore, they are considered to be promising materials for not only industrial applications but also biomedical applications.

In the 1980s, the in vivo stability of various PAEK materials and the tissue response to the same were investigated [86]. Recently, PEEK has emerged as the leading high-performance super-engineering plastic candidate for replacing metal implant components, especially in the field of orthopedics and spinal surgery (Table 25.7) [87]. In recent studies, the tribological and bioactive properties of PEEK, which is used as a bearing material and flexible implant in orthopedic and spinal surgeries, has been investigated [88–90]. However, conventional single-component PEEK cannot satisfy these requirements (e.g., antibiofouling, wear resistance, and fixation to a bone) for use as an artificial joint or intervertebral body fusion cage [87]. For further improving the capabilities of PEEK as an implant biomaterial, various studies have focused upon the lubricity and antibiofouling of the polymer, either via reinforcing agents or surface modifications [91,92]. Therefore, multicomponent polymer systems have been designed in order to synthesize new multifunctional

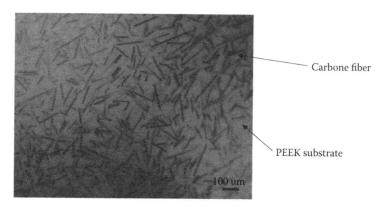

Carbone fiber

PEEK substrate

100 um

FIGURE 25.21   Fluorescence microscopic images of CFR-PEEK. Bar: 100 μm.

TABLE 25.7
Typical Physical and Chemical Properties of PEEK and CFR-PEEK

Property	PEEK	30% CFR-PEEK	60% CFR-PEEK
Molecular weight (10^6 g/mol)	0.08–0.12	0.08–0.12	0.08–0.12
Melting temperature (°C)	343	343	—
Poisson's ratio	0.36–0.40	0.40–0.44	0.38–0.44
Specific gravity	1.3	1.4	1.6
Flexural modulus (GPa)	4	20	135
Tensile ultimate strength (MPa)	93–97	170–228	>2000
Elongation (%)	30–40	1–2	1
Crystallinity (%)	30–35	30–35	30–35

biomaterials. In order to use PEEK and related composites in the implant applications, they can be engineered to have a wide range of physical, mechanical, and surface properties.

25.2.3.2 Tribological Properties

In the 1990s, CFR-PEEK was evaluated as a bearing material for hip and knee joint replacement [88]. Wang et al. carried out a more comprehensive tribological investigation of PEEK composites for both hip and knee joint replacements. CFR-PEEK formulations were blended with 20%–30% mass discontinuous polyacrylonitrile (PAN) or pitch fibers. Under higher stress, cylinder-on-flat loading conditions, the PEEK composites exhibited higher wear rates than conventional UHMWPE. In contrast, under the lower stress hip simulator test conditions, all the PEEK composites had substantially lower wear rates than conventional UHMWPE, with the lowest wear observed between 30% pitch CFR-PEEK against ceramics. In contrast, unreinforced PEEK wore at six times the rate of UHMWPE. The results of this study underscored the importance of fiber reinforcement in lowering stress and conforming contact applications and provided a further basis for exploring PEEK composites for hip joint replacements, especially in combination with ceramics as opposed to Co-Cr heads. Therefore, alumina became the femoral head material of choice for THA applications with CFR-PEEK. Co-Cr heads, when used in conjunction with CFR-PEEK liners, exhibited substantially higher wear, with observations of scratching of the metallic surface by the carbon fibers. On the other hand, this study also suggested that PEEK composites were unsuitable for knee applications, regardless of the fiber content of the composite or the type of the counter surface. The authors recommended that the composite materials should not be used as a tibial insert for knee joint replacement.

To validate the in vivo wear behavior and compatibility of CFR-PEEK wear debris, a clinical study was initiated in Italy starting in 2001 using the ABG II total hip system (Stryker SA, Montreux, Switzerland). The CFR-PEEK liners were fabricated from injection-molded PEEK blended with 30% pitch fibers, and the bearing surfaces were machined to achieve the desired final tolerance. After a mean follow-up period of 3 years, none of the liners needed to be revised due to aseptic loosening. This clinical trial is still ongoing, and the detailed results have not yet been published. Overall, the available preliminary clinical data support the short-term effectiveness of CFR-PEEK as a bearing material for hip joint replacement. However, in a conventional hip joint replacement design, the current data do not yet demonstrate a long-term clinical advantage of CFR-PEEK over other well-established bearing alternatives, such as CLPE.

25.2.3.3 Surface Modification

On the other hand, surface modification is one of the most important technologies for the preparation of new multifunctional biomaterials for satisfying several requirements. Surface modifications used today include coating, blending, and grafting.

It is well known that when BP is exposed to photo-irradiation such as ultraviolet-ray (UV)-irradiation, a pinacolization reaction is induced; this results in the formation of semi-benzopinacol radicals (i.e., ketyl radicals) that act as photo-initiators. Therefore, in this study, we have focused upon a BP unit in PEEK and formulated a novel self-initiated surface-graft polymerization method that utilizes the BP unit in "graft from" polymerization (Figure 25.22) [93,94]. This polymerization reaction involving free radicals is photoinduced by UV-irradiation. Under UV-irradiation, a BP unit in PEEK can undergo the following reactions in monomeric aqueous solutions [95–101] as follows: the pinacolization reaction (photoreduction by H-abstraction of a BP unit in PEEK) results in the formation of a semi-benzopinacol radical, which can initiate the graft-from polymerization of the feed monomer as the main reaction, and the graft-to polymerization (the radical chain end of the active-polymer couples with the semi-benzopinacol radical of the PEEK surface) as a subreaction. In addition, a photoscission reaction occurs as a subreaction, which may not need a hydrogen (H) donor. The cleavage reaction induces recombination and graft-from polymerization. When water polymerization is performed in the presence of an H-donor, a phenol unit may be subsequentially formed due to H-abstraction. This technique enables the direct grafting of the functional polymer

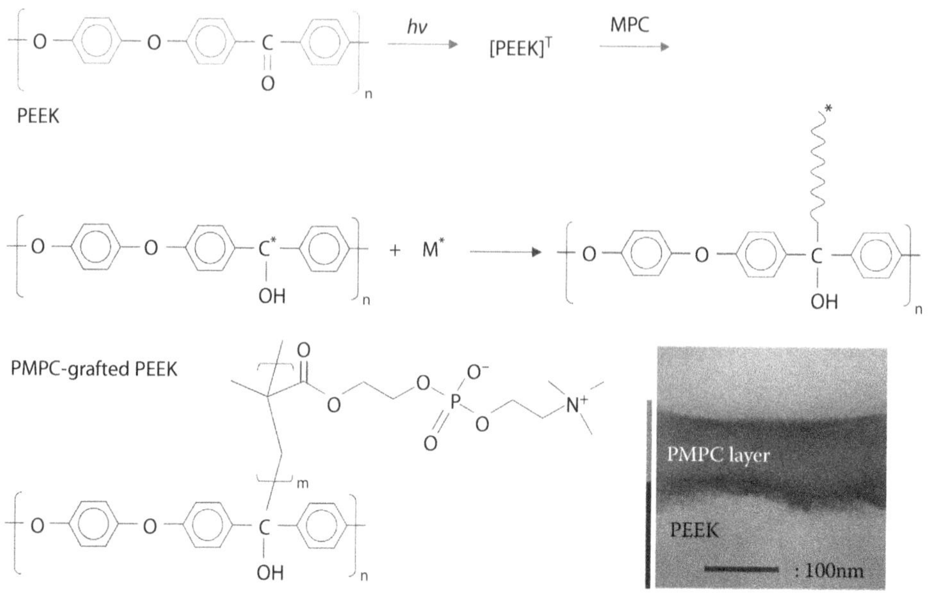

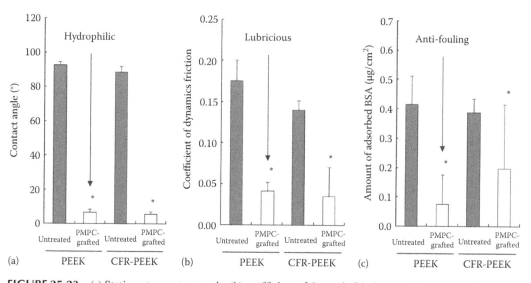

FIGURE 25.22 Schematic illustration for the preparation and cross-sectional TEM image of PMPC-grafted PEEK.

onto the PEEK surface in the absence of a photoinitiator, thereby resulting in the formation of a C–C covalent bond between the functional polymer and PEEK substrate.

Kyomoto et al. demonstrated the fabrication of a biocompatible and highly hydrophilic nanometer-scale-modified surface by PMPC-grafting onto the self-initiated PEEK surface using a photo-induced pinacolization reaction (Figure 25.22) [93,94].

This novel and simple self-initiated surface-graft polymerization on the PEEK surface induces unique properties such as lubricity and anti-protein adsorption by PMPC grafting, which are novel phenomena in the field of orthopedic and spinal surgery (Figure 25.23). Moreover, the fabrications of the PMPC-grafted PEEK and CFR-PEEK can result in next-generation orthopedic and spinal applications.

FIGURE 25.23 (a) Static water-contact angle, (b) coefficient of dynamic friction, and (c) amount of adsorbed BSA of PMPC-grafted PEEK and CFR-PEEK. *t-test, significant difference ($p < 0.05$) as compared to the untreated PEEK and CFR-PEEK, respectively.

25.3 FIXATION MATERIALS FOR JOINT REPLACEMENTS

25.3.1 POLY(METHYL METHACRYLATE) BONE CEMENT

Bone cements based on poly(methyl methacrylate) (PMMA) are important polymer materials in joint replacement surgery: PMMA bone cements are primarily used for the fixation of joint replacement. In the fixation of joint replacement, the self-curing bone cement fills the free space between the joint replacement and bone and constitutes a very important interface.

The PMMA bone cements are two-component systems, comprising a polymer powder and a liquid methyl methacrylate (MMA) monomer. The polymer powder component is composed of PMMA and/or methacrylate copolymer. The polymer powder contains benzoyl peroxide (BPO), which acts as an initiator for radical polymerization. The polymer powder also contains an X-ray contrast agent. In the liquid monomer, MMA is the main constituent; however, at times, other methacrylates—such as butyl methacrylate—are used. In order for the MMA to be used in bone cements, it must be polymerizable, i.e., it must contain a carbon double bond that can be broken. Furthermore, the liquid monomer contains an aromatic amine, such as N,N-dimethyl-p-toluidine (DMT), as an activator of radical formation. Additionally, it contains an inhibitor (e.g., hydroquinone) to avoid premature polymerization in storage aging.

PMMA bone cements are polymeric materials produced by the radical polymerization of MMA (Figure 25.24). The polymerization process starts when the polymer powder and liquid monomer are mixed, resulting in a reaction between the BPO initiator and DMT activator, forming radicals. Consequently, the DMT causes a breakdown of the BPO in the reaction process by electron transfer, resulting in the formation of benzoyl radicals. The C=C of the MMA monomer has a pair of electrons that is attacked by the free radical to form a new chemical bond between the initiator fragment and one of the C=C bond of the monomer molecule. The other electron of the C=C bond stays on the C atom that is not bonded to the initiator fragment, creating a new free radical. This unpaired electron is capable of attacking the C=C bond of a new monomer unit. This process, with the breakdown of the initiator molecule to form radicals, is called the "initiation reaction" of the radicals' polymerization. The new radical reacts with another MMA molecule in the same way as the initiator fragment reaction. Another radical is always formed when this reaction takes place, over and over again. This process of the growing polymer chain is called a "propagation reaction." As the polymerization continues, the rate of termination (termination reaction) is decreased,

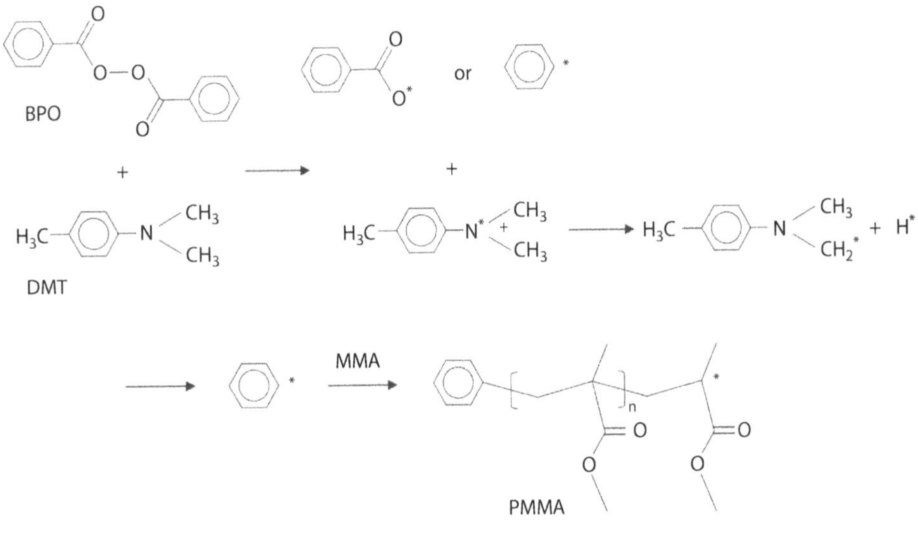

FIGURE 25.24 Schematic illustration of formation of radical polymerization chains in PMMA bone cement.

because the diffusion of chain growth and the combination of chain ends is reduced. The system becomes depleted of free radicals by recombination of the two radical chains and the polymerization ceases.

25.3.2 BONE CEMENT HISTORY

In 1936, Heraeus Kulzer GmbH & Co. KG (Wehrheim, Germany) found that dough material could be produced by mixing PMMA powder and a liquid MMA monomer, which was cured when BPO was added and the blend was heated to 100°C [102]. In 1958, Charnley et al. successfully fixed a femoral stem prosthesis by using PMMA as bone cement [103]. Since the introduction of PMMA bone cement by Charnley et al., many people have had their lives dramatically and remarkably improved by this innovation. At the beginning of the 1970s, in an effort to alleviate periprosthetic infection—the most feared complication after joint replacement—Buchholz et al. advocated the addition of antibiotics to bone cement [104]. Their idea was to add antibiotics to the bone cement in order to reduce the incidence of infection. The PMMA bone cement with a gentamicin powder was demonstrated to be stable and offered suitable antibiotic activity.

25.3.3 PROBLEMS WITH PMMA BONE CEMENT

Since the introduction of PMMA bone cement in the orthopedic field, bone cement is considered to have become one of the effective polymer materials for fixation in joint replacement. However, aseptic loosening remains a serious issue on a multifactorial basis as follows. Aseptic loosening is suggested to be a result of monomer-mediated bone damage; during end-polymerization, there is volumetric shrinkage of the cement, potentially compromising the bone–cement interface. Using a fluid displacement model, Charnley observed that the volume of cement increases to a maximum during polymerization, before shrinking slightly, although not to its initial volume. This may be a largely theoretical concern, but there is a conflict between the stiffness of cement and the adjacent bone. The Young's modulus is 0.5–1.0 GPa for cancellous bone; 15–20 GPa, for cortical bone; 2 GPa, for bone cement; 1 GPa, for titanium alloy; and 220 GPa, for cobalt–chromium–molybdenum alloy. The bone cement may provide a shock-absorbing layer between elastic bone and a stiff implant. The conflict between degrees of stiffness is therefore much greater for cementless implants, and in some instances, the cement mantle and its interfaces may be the weak link in the construct. The bone–cement interface is the key to the survival of a THR. The combination of matte-surfaced collared femoral stems and poor cementing technique are intrinsic to the failure of some implants. Polished collarless tapered stems generally give better fixation, while cement particles were once considered a biological cause of aseptic loosening. However, wear particles of polyethylene are now seen as primary initiators of the biological reactions in aseptic osteolysis.

25.3.4 SOLUTIONS FOR THE PROBLEMS OF PMMA BONE CEMENT

25.3.4.1 Antibiotic-Loaded Acrylic Cement

In 1969, Buchholz et al. incorporated gentamicin in PMMA bone cement for the treatment of infection in prosthetic joints [104]. Initially, the antibiotic was added during the operation, and subsequently during manufacture, making antibiotic-loaded PMMA bone cement widely available as part of antimicrobial prophylaxis in primary arthroplasty. There is valid evidence to support the prophylactic use of antibiotic-loaded PMMA bone cement, which remains a standard practice in Europe, and is in transition in the United States. In 2003, the Food and Drug Administration accepted the use of three commercial antibiotic-loaded PMMA bone cements in the second stage of revision surgery for prosthetic joint infection. Their use in primary THA has not been authorized. The use of antibiotic-loaded PMMA bone cement in joint replacement provides short- to medium-term protection against prosthetic infection. It aims to overlap with, and then replace, the prophylaxis provided by perioperative intravenous antibiotics. To achieve this, the antibiotic must be released from cement in adequately high concentrations that exceed the minimum inhibitory concentration of potential colonizing bacteria. Gentamicin is the most common additive because it has, among other features, a good spectrum of concentration-dependent bactericidal activity, thermal stability, and high water solubility. In 1980, Wahlig et al. gave robust evidence of gentamicin release from PMMA bone cement for up to 5.5 years in patients who had undergone hip joint replacement [105]. Others have also confirmed the reliable release of gentamicin from PMMA bone cement. However, concerns regarding antibiotics in cement still persist, including concerns regarding the induction of antibiotic resistance. In 1989, Hope et al. found that 90% of Staphylococcal strains isolated from infected hip joint replacements were resistant to gentamicin, but if plain cement had been used at the initial operation, the resistance rate was only 16% [106]. Other studies have confirmed that antibiotic-loaded PMMA bone cement reduces infection in total joint replacement at the price of increasing bacterial resistance. These problems have not been demonstrated clinically, although they have been postulated. Despite the aim of achieving early and total release, all in vitro studies show that only 5%–8% of the added antibiotic is ever freed. Clinical studies have shown a low concentration of the release of gentamicin in failing hip joint replacements up to 25 years after the primary operation, a potent stimulus for antibiotic resistance. In summary, we need to design guidelines regarding the use of antibiotic-loaded PMMA bone cement and advise its use only in cases where the patient has significant risk factors for infection.

25.3.4.2 Bone Bioactive Organic–Inorganic Hybrid Bone Cement

Recently, Ishihara et al. reported that PMMA bone cement containing hydroxyapatite as a bone bioactive-filler was developed using 4-methacryloyloxyethyl trimellitate anhydride (4-META) to promote adhesion to both bone and hydroxyapatite [107]. The mechanical properties of this PMMA cement with 4-META and hydroxyapatite did not decrease significantly with increasing hydroxyapatite filler content in the cement. In contrast, the mechanical properties decreased with increasing hydroxyapatite filler content in the absence of 4-META. The fracture surface of the cement clearly showed that there are no gaps between hydroxyapatite filler and PMMA matrix resin; this means that the hydroxyapatite filler adhered to the PMMA matrix resin by addition of 4-META [108]. The hydroxyapatite filler along the surface increased with increased hydroxyapatite filler content in the cement. The PMMA cement with 4-META and hydroxyapatite adhered also to bone with a tensile bond strength of higher than 10 MPa [109–111].

The sol–gel process is popular for the preparation of nanometer-scaled composites of organic–inorganic components. It has been reported that organically modified silicates can be synthesized through hydrolysis and polycondensation of tetraethoxysilane and poly(dimethylsiloxane). The nanometer-scaled organic–inorganic composite has the potential to show the properties of both organic and inorganic components. Based on the finding that a $CaO–SiO_2$ glass is effective in producing bone bioactivity, Ohtsuki et al. applied organic modification with Si–OH and Ca^{2+} to PMMA bone cement in order to induce the bone bioactivity in the bone cement (Figure 25.25) [112,113]. This bone bioactive PMMA cement has the potential to demonstrate bone-bonding properties in the bone–cement interface (Figure 25.25d).

25.4 FUTURE PERSPECTIVES

Every year, the number and prevalence of primary and revision hip, knee, and other joint replacements are increasing substantially worldwide. As a result, the quality of all artificial joints is becoming increasingly important. Therefore, we can see that functional, durable, and natural joint-like artificial joint replacements will be necessary for all artificial joints. We consider that the important research goal for the future is the creation of the ultimate artificial joint interface that mimics the

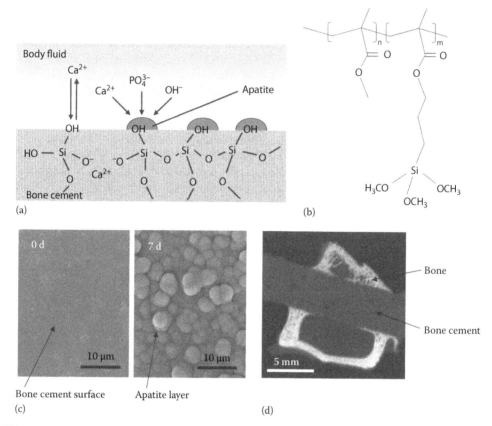

FIGURE 25.25 Schematic illustration of bone bioactive PMMA bone cement: (a) concept of bone bioactive cement, (b) chemical structure, (c) apatite formation in simulated body soaking fluid by SEM observation, and (d) pQCT image after 9 weeks implantation, Animal experiment.

natural joint cartilage. Additionally, a functional, durable, and natural surface not only would be of great service to applications as medical devices such as artificial joints but also would be important to biomaterial and bioengineering sciences. For example, it is well known that the composition elements of the articular cartilage surface consist of the collagen network, hyaluronic acid, and proteoglycan subunits. However, the functions of the articular cartilage surface have not been well explained. We consider that a bioengineering surface with new polymeric biomaterial would help in clarifying these phenomena; further, researches in biotribological science can elucidate the functions of articular cartilage surface. We believe that the designs of polymeric biomaterials and bioengineering surfaces will act as key technologies for further evolving biomaterial and bioengineering sciences, and we hope that this issue will be addressed by future scientists in polymeric biomaterial and bioengineering sciences.

REFERENCES

1. Kurtz S, Mowat F, Ong K, Chan N, Lau E, Halpern M. 2005. Prevalence of primary and revision total hip and knee arthroplasty in the United States from 1990 through 2002. *J Bone Joint Surg Am* 87(7):1487–1497.
2. Malchau H, Herberts P, Eisler T, Garellick G, Söderman P. 2002. The swedish total hip replacement register. *J Bone Joint Surg Am* 84 (Suppl 2):2–20.
3. Sierra RJ, Cooney WP 4th, Pagnano MW, Trousdale RT, and Rand, JA. 2004. Reoperations after 3200 revision TKAs: Rates, etiology, and lessons learned. *Clin Orthop Relat Res* 425:200–206.

4. Rand JA, Trousdale RT, Ilstrup DM, Harmsen WS. 2003. Factors affecting the durability of primary total knee prostheses. *J Bone Joint Surg Am* 85(2):259–265.

5. Orthopedic medical device market in Japan 2007. In: *Medical Bionics (Artificial Organ) Market 2007*. Tokyo, Japan: Yano Research Institute, Ltd., pp. 275–344.

6. Harris WH. 1995. The problem is osteolysis. *Clin Orthop Relat Res* 311:46–53.

7. Sochart DH. 1999. Relationship of acetabular wear to osteolysis and loosening in total hip arthroplasty. *Clin Orthop Relat Res* 363:135–150.

8. ASTM F648-07, 2007. Standard specification for ultra-high-molecular-weight polyethylene powder and fabricated form for surgical implants. West Conshohocken, PA: ASTM International.

9. ISO 5834-1, 2007. Implants for surgery—Ultra-high-molecular-weight polyethylene—Part 1: Powder form., Geneva, Switzerland: International Organization for Standardization.

10. Charnley J. 1961. Arthroplasty of the hip. A new operation. *Lancet* 27;1(7187):1129–1132.

11. Dupont JA, Charnley J. 1972. Low-friction arthroplasty of the hip for the failures of previous operations. *J Bone Joint Surg Br* 54(1):77–87.

12. Isaac GH, Dowson D, Wroblewski BM. 1996. An investigation into the origins of time-dependent variation in penetration rates with Charnley acetabular cups—Wear, creep or degradation? *Proc Inst Mech Eng H* 210(3):209–216.

13. Kurtz SM, Muratoglu OK, Evans M, Edidin AA. 1999. Advances in the processing, sterilization, and crosslinking of ultra-high molecular weight polyethylene for total joint arthroplasty. *Biomaterials* 20(18):1659–1688.

14. Wright TM, Bartel DL. 1986. The problem of surface damage in polyethylene total knee components. *Clin Orthop Relat Res* 205:67–74.

15. Connelly GM, Rimnac CM, Wright TM, Hertzberg RW, Manson JA. 1984. Fatigue crack propagation behavior of ultrahigh molecular weight polyethylene. *J Orthop Res* 2(2):119–125.

16. Oonishi H, Takayama Y, Tsuji E. 1992. Improvement of polyethylene by irradiation in artificial joints. *Radiat Phys Chem* 39(6):495–504.

17. Kyomoto M, Ueno M, Kim SC, Oonishi H, Oonishi H. 2007. Wear of '100 Mrad' cross-linked polyethylene: Effects of packaging after 30 years real-time shelf-aging. *J Biomater Sci Polym Ed* 18(1):59–70.

18. Grobbelaar CJ, Du Plessis TA, Marais F. 1978. The radiation improvement of polyethylene prostheses: A preliminary study. *J Bone Joint Surg Br* 60:370–374.

19. Oonishi H, Wakitani S, Murata N, Saito M, Imoto K, Kim S, Matsuura M. 2000. Clinical experience with ceramics in total hip replacement. *Clin Orthop Relat Res* 379:77–84.

20. Wroblewski BM, Siney PD, Dowson D, Collins SN. 1996. Prospective clinical and joint simulator studies of a new total hip arthroplasty using alumina ceramic heads and cross-linked polyethylene cups. *J Bone Joint Surg Br* 78(2):280–285.

21. Li S, Burstein AH. 1994. Ultra-high molecular weight polyethylene. The material and its use in total joint implants. *J Bone Joint Surg Am* 76(7):1080–1090.

22. Muratoglu OK, Bragdon CR, O'Connor DO, Jasty M, Harris WH. 2001. A novel method of cross-linking ultra-high-molecular-weight polyethylene to improve wear, reduce oxidation, and retain mechanical properties. *J Arthroplasty* 16(2):149–160.

23. Oonishi H, Kim SC, Takao Y, Kyomoto M, Iwamoto M, Ueno M. 2006. Wear of highly cross-linked polyethylene acetabular cup in Japan. *J Arthroplasty* 21(7):944–949.

24. McKellop H, Shen FW, Lu B, Campbell P, Salovey R. 2000. Effect of sterilization method and other modifications on the wear resistance of acetabular cups made of ultra-high molecular weight polyethylene. A hip-simulator study. *J Bone Joint Surg Am* 82(12):1708–1725.

25. Dowd JE, Sychterz CJ, Young AM, Engh CA. 2000. Characterization of long-term femoral-head-penetration rates. Association with and prediction of osteolysis. *J Bone Joint Surg Am* 82(8):1102–1107.

26. Oparaugo PC, Clarke IC, Malchau H, Herberts P. 2001. Correlation of wear debris-induced osteolysis and revision with volumetric wear-rates of polyethylene: A survey of 8 reports in the literature. *Acta Orthop Scand* 72(1):22–28.

27. Röhrl S, Nivbrant B, Mingguo L, Hewitt B. 2005. In vivo wear and migration of highly cross-linked polyethylene cups a radiostereometry analysis study. *J Arthroplasty* 20(4):409–413.

28. Martell JM, Verner JJ, Incavo SJ. 2003. Clinical performance of a highly cross-linked polyethylene at two years in total hip arthroplasty: A randomized prospective trial. *J Arthroplasty* 18(7 Suppl 1):55–59.

29. Krushell RJ, Fingeroth RJ, Cushing MC. 2005. Early femoral head penetration of a highly cross-linked polyethylene liner vs a conventional polyethylene liner. A case controlled study. *J Arthroplasty* 20(7 Suppl 3):73–76.

30. D'Antonio JA, Manley MT, Capello WN, Bierbaum BE, Ramakrishnan R, Naughton M, Sutton K. 2005. Five-year experience with Crossfire highly cross-linked polyethylene. *Clin Orthop Relat Res* 441:143–150.

31. Heisel C, Silva M, dela Rosa MA, Schmalzried TP. 2004. Short-term in vivo wear of cross-linked polyethylene. *J Bone Joint Surg Am* 86:748–751.

32. Sychterz CJ, Engh CA Jr, Engh CA. 2004. A prospective, randomized clinical study comparing Marathon and Enduron polyethylene acetabular liners: 3 year results. *J Arthroplasty* 19:258.

33. Engh CA Jr, Stepniewski AS, Ginn SD, Beykirch SE, Sychterz-Terefenko CJ, Hopper RH Jr, Engh CA. 2006. A randomized prospective evaluation of outcomes after total hip arthroplasty using cross-linked Marathon and non-cross-linked Enduron polyethylene liners. *J Arthroplasty* 21(6 Suppl 2):17–25.

34. Digas G, Karrholm J, Thanner J, Malchau H, Herberts P. 2003. Highly cross-linked polyethylene in cemented THA: Randomized study of 61 hips. *Clin Orthop Relat Res* 417:126–138.

35. Manning DW, Chiang PP, Martell JM, Galante JO, Harris WH. 2005. In vivo comparative wear study of traditional and highly cross-linked polyethylene in total hip arthroplasty. *J Arthroplasty* 20:880–886.

36. Hopper RH Jr, Young AM, Orishimo KF, McAuley JP. 2003. Correlation between early and late wear rates in total hip arthroplasty with application to the performance of Marathon cross-linked polyethylene liners. *J Arthroplasty* 18(7 Suppl 1):60–67.

37. Digas G, Karrholm J, Thanner J, Malchau H, Herberts P. 2004. Highly cross-linked polyethylene in total hip arthroplasty: Randomized evaluation of penetration rate in cemented and uncemented sockets using radiostereometric analysis. *Clin Orthop Relat Res* 429:6–16.

38. Bragdon CR, Barrett S, Martell JM, Greene ME, Malchau H, Harris WH. 2006. Steady-state penetration rates of electron beam-irradiated, highly cross-linked polyethylene at an average 45-month follow-up. *J Arthroplasty* 21:935–943.

39. Dorr LD, Wan Z, Shahrdar C, Sirianni L, Boutary M, Yun A. 2005. Clinical performance of a Durasul highly cross-linked polyethylene acetabular liner for total hip arthroplasty at five years. *J Bone Joint Surg Am* 87:1816–1821.

40. Ries MD, Scott ML, Jani S. 2001. Relationship between gravimetric wear and particle generation in hip simulators: Conventional compared with cross-linked polyethylene. *J Bone Joint Surg Am* 83 (Suppl 2):116–122.

41. Bradford L, Kurland R, Sankaran M, Kim H, Pruitt LA, Ries MD. 2004. Early failure due to osteolysis associated with contemporary highly cross-linked ultra-high molecular weight polyethylene. A case report. *J Bone Joint Surg Am* 86(5):1051–1056.

42. Moro T, Takatori Y, Ishihara K, Kyomoto M, Nakamura K, Kawaguchi H. 2009. Progress of research in osteoarthritis. Invention of longer lasting artificial joints. *Clin Calcium* 19(11):1629–1637.

43. Shibata N, Tomita N. 2005. The anti-oxidative properties of α-tocopherol in γ-irradiated UHMWPE with respect to fatigue and oxidation resistance. *Biomaterials* 26(29):5755–5762.

44. Suzuki M, Lee T, Miyagi J, Kobayashi T, Sasho T, Nakagawa K, Fujiwara K, Nishimura N, Kuramoto K, Moriya H, Takahashi K. 2010. Evaluation of vitamin E added ultra high molecular weight polyethylene in total knee arthroplasty. Joint fluid concentrations of tocopherol and matrix metalloproteinase 9. *J Bone Joint Surg Br* 92-B (Suppl I): 131.

45. Brach Del Prever EM, Bistolfi A, Bracco P, Costa L. 2009. UHMWPE for arthroplasty: Past or future? *J Orthop Traumatol* 10(1):1–8.

46. Yoshida H, Morita Y, Ikeuchi K. 2003. Biological lubrication of hydrated surface layer in small intestine. In: Dowson D, editor. *Tribological Research and Design for Engineering Systems*. Amsterdam, the Netherlands: Elsevier, pp. 425–428.

47. Buckwalter JA, Rosenberg L. 1983. Structural changes during development in bovine fetal epiphyseal cartilage. *Coll Relat Res* 3(6):489–504.

48. Obara T, Mabuchi K, Iso T, Yamaguchi T. 1997. Increased friction of animal joints by experimental degeneration and recovery by addition of hyaluronic acid. *Clin Biomech (Bristol, Avon)* 12(4):246–252.

49. Ishikawa Y, Hiratsuka K, Sasada T. 2006. Role of water in the lubrication of hydrogel. *Wear* 261:500–504.

50. Longfield MD, Dowson D, Walker PS, Wright V. 1969. "Boosted lubrication" of human joints by fluid enrichment and entrapment. *Biomed Eng* 4(11):517–522.

51. Ikeuchi K, Kusaka J, Yamane D, Fujita S. 1999. Time-dependent wear process between lubricated soft materials. *Wear* 229:656–659.

52. Milner ST. 1991. Polymer brushes. *Science* 251:905–914.

53. Nagasaki Y, Kataoka K. 1996. An intelligent polymer brush. *Trends Polym Sci* 4(2):59–64.

54. Edmondson S, Osborne VL, Huck WTS. 2004. Polymer brushes via surface-initiated polymerizations. *Chem Soc Rev* 33:14–22.

55. Dyer DJ. 2006. Photoinitiated synthesis of grafted polymers. In: Abe A et al., editors. *Advances in Polymer Science*. Berlin, Germany: Springer-Verlag, pp. 47–65.

56. Moro T, Takatori Y, Ishihara K, Konno T, Takigawa Y, Matsushita T, Chung UI, Nakamura K, Kawaguchi H. 2004. Surface grafting of artificial joints with a biocompatible polymer for preventing periprosthetic osteolysis. *Nat Mater* 3:829–837.

57. Moro T, Takatori Y, Ishihara K, Nakamura K, Kawaguchi H. 2006. 2006 Frank Stinchfield Award: Grafting of biocompatible polymer for longevity of artificial hip joints. *Clin Orthop Relat Res* 453:58–63.

58. Moro T, Kawaguchi H, Ishihara K, Kyomoto M, Karita T, Ito H, Nakamura K, Takatori Y. 2009. Wear resistance of artificial hip joints with poly(2-methacryloyloxyethyl phosphorylcholine) grafted polyethylene: Comparisons with the effect of polyethylene cross-linking and ceramic femoral heads. *Biomaterials* 30(16):2995–3001.

59. Ishihara K, Ueda T, Nakabayashi N. 1990. Preparation of phospholipid polymers and their properties as polymer hydrogel membranes. *Polym J* 22(5):355–360.

60. Brash JL. 2000. Exploiting the current paradigm of blood-material interactions for the rational design of blood-compatible materials. *J Biomater Sci Polym Ed* 11(11):1135–1146.

61. Zwaal RF, Comfurius P, van Deenen LL. 1977. Membrane asymmetry and blood coagulation. *Nature* 268(5618):358–360.

62. Hayward JA, Chapman D. 1984. Biomembrane surfaces as models for polymer design: The potential for haemocompatibility. *Biomaterials* 5(3):135–142.

63. Yamazaki K, Saito S, Tomioka H, Miyagishima M, Kobayashi K, Miyake T, Ishii H, Kawai A, Aomi S, Tagusari O, Niwaya K, Nakatani T, Kobayashi J, Kitamura S, Kihara S, Kurosawa H. 2006. Clinical trial of EVAHEART: Next generation left ventricular assist device. *Circulation J* 70 (Suppl 1):59.

64. Myers GJ, Gardiner K, Ditmore SN, Swyer WJ, Squires C, Johnstone DR, Power CV, Mitchell LB, Ditmore JE, Cook B. 2005. Clinical evaluation of the Sorin Synthesis oxygenator with integrated arterial filter. *J Extra Corpor Technol* 37(2):201–206.

65. Myers GJ, Johnstone DR, Swyer WJ, McTeer S, Maxwell SL, Squires C, Ditmore SN et al. 2003. Evaluation of Mimesys phosphorylcholine (PC)-coated oxygenators during cardiopulmonary bypass in adults. *J Extra Corpor Technol* 35(1):6–12.

66. Fajadet J, Wijns W, Laarman GJ, Kuck KH, Ormiston J, Münzel T, Popma JJ, Fitzgerald PJ, Bonan R, Kuntz RE; ENDEAVOR II Investigators. 2006. Randomized, double-blind, multicenter study of the Endeavor zotarolimus-eluting phosphorylcholine-encapsulated stent for treatment of native coronary artery lesions: Clinical and angiographic results of the ENDEAVOR II trial. *Circulation* 114(8):798–806.

67. Gershlick A, Kandzari DE, Leon MB, Wijns W, Meredith IT, Fajadet J, Popma JJ, Fitzgerald PJ, Kuntz RE; ENDEAVOR Investigators. 2007. Zotarolimus-eluting stents in patients with native coronary artery disease: Clinical and angiographic outcomes in 1,317 patients. *Am J Cardiol* 100(8B):45M–55M.

68. Kandzari DE, Leon MB. 2006. Overview of pharmacology and clinical trials program with the zotarolimus-eluting endeavor stent. *J Interv Cardiol* 19(5):405–413.

69. Sakurai R, Hongo Y, Yamasaki M, Honda Y, Bonneau HN, Yock PG, Cutlip D, Popma JJ, Zimetbaum P, Fajadet J, Kuntz RE, Wijns W, Fitzgerald PJ; ENDEAVOR II Trial Investigators. 2007. Detailed intravascular ultrasound analysis of Zotarolimus-eluting phosphorylcholine-coated cobalt-chromium alloy stent in de novo coronary lesions (results from the ENDEAVOR II trial). *Am J Cardiol* 100(5):818–823.

70. Kandzari DE, Leon MB, Popma JJ, Fitzgerald PJ, O'Shaughnessy C, Ball MW, Turco M et al. 2006. Comparison of zotarolimus-eluting and sirolimus-eluting stents in patients with native coronary artery disease: A randomized controlled trial. *J Am Coll Cardiol* 48(12):2440–2447.

71. Abizaid A, Popma JJ, Tanajura LF, Hattori K, Solberg B, Larracas C, Feres F, Costa Jde R Jr, Schwartz LB. 2007. Clinical and angiographic results of percutaneous coronary revascularization using a trilayer stainless steel-tantalum-stainless steel phosphorylcholine-coated stent: The TriMaxx trial. *Catheter Cardiovasc Interv* 70(7):914–919.

72. Abizaid A, Lansky AJ, Fitzgerald PJ, Tanajura LF, Feres F, Staico R, Mattos L et al. 2007. Percutaneous coronary revascularization using a trilayer metal phosphorylcholine-coated zotarolimus-eluting stent. *Am J Cardiol* 99(10):1403–1408.

73. Han SH, Ahn TH, Kang WC, Oh KJ, Chung WJ, Shin MS, Koh KK, Choi IS, Shin EK. 2006. The favorable clinical and angiographic outcomes of a high-dose dexamethasone-eluting stent: Randomized controlled prospective study. *Am Heart J* 152(5):887, e1–e7.

74. Kwok OH, Chow WH, Law TC, Chiu A, Ng W, Lam WF, Hong MK, Popma JJ. 2005. First human experience with angiopeptin-eluting stent: A quantitative coronary angiography and three-dimensional intravascular ultrasound study. *Catheter Cardiovasc Interv* 66(4):541–546.

75. Airoldi F, Di Mario C, Ribichini F, Presbitero P, Sganzerla P, Ferrero V, Vassanelli C et al. 2005. 17-Beta-estradiol eluting stent versus phosphorylcholine-coated stent for the treatment of native coronary artery disease. *Am J Cardiol* 96(5):664–667.

76. Rodriguez A, Rodríguez Alemparte M, Fernández Pereira C, Sampaolesi A, da Rocha Loures Bueno R, Vigo F, Obregón A, Palacios IF; LASMAL investigators. 2005. Latin American randomized trial of balloon angioplasty vs coronary stenting for small vessels (LASMAL): Immediate and long-term results. *Am J Med* 118(7):743–751.

77. Bakhai A, Booth J, Delahunty N, Nugara F, Clayton T, McNeill J, Davies SW, Cumberland DC, Stables RH; SV Stent Investigators. 2005. The SV stent study: A prospective, multicentre, angiographic evaluation of the BiodivYsio phosphorylcholine coated small vessel stent in small coronary vessels. *Int J Cardiol* 102(1):95–102.

78. Shinozaki N, Yokoi H, Iwabuchi M, Nosaka H, Kadota K, Mitsudo K, Nobuyoshi M. 2005. Initial and follow-up results of the BiodivYsio phosphorylcholine coated stent for treatment of coronary artery disease. *Circ J* 69(3):295–300.

79. Hausleiter J, Kastrati A, Mehilli J, Schühlen H, Pache J, Dotzer F, Glatthor C, Siebert S, Dirschinger J, Schömig A; ISAR-SMART-2 Investigators. 2004. A randomized trial comparing phosphorylcholine-coated stenting with balloon angioplasty as well as abciximab with placebo for restenosis reduction in small coronary arteries. *J Intern Med* 256(5):388–397.

80. Boland JL, Corbeij HA, Van Der Giessen W, Seabra-Gomes R, Suryapranata H, Wijns W, Hanet C et al. 2000. Multicenter evaluation of the phosphorylcholine-coated biodivYsio stent in short de novo coronary lesions: The SOPHOS study. *Int J Cardiovasc Intervent* 3(4):215–225.

81. Kuiper KK, Nordrehaug JE. 2000. Early mobilization after protamine reversal of heparin following implantation of phosphorylcholine-coated stents in totally occluded coronary arteries. *Am J Cardiol* 85(6):698–702.

82. Grenadier E, Roguin A, Hertz I, Peled B, Boulos M, Nikolsky E, Amikam S, Kerner A, Cohen S, Beyar R. 2002. Stenting very small coronary narrowings (<2 mm) using the biocompatible phosphorylcholine-coated coronary stent. *Catheter Cardiovasc Interv* 55(3):303–308.

83. Kyomoto M, Moro T, Konno T, Takadama H, Yamawaki N, Kawaguchi H, Takatori Y, Nakamura K, Ishihara K. 2007. Enhanced wear resistance of modified cross-linked polyethylene by grafting with poly(2-methacryloyloxyethyl phosphorylcholine). *J Biomed Mater Res A* 82(1):10–17.

84. Kyomoto M, Moro T, Miyaji F, Hashimoto M, Kawaguchi H, Takatori Y, Nakamura K, Ishihara K. 2008. Effect of 2-methacryloyloxyethyl phosphorylcholine concentration on photo-induced graft polymerization of polyethylene in reducing the wear of orthopaedic bearing surface. *J Biomed Mater Res A* 86(2):439–447.

85. Kyomoto M, Moro T, Miyaji F, Hashimoto M, Kawaguchi H, Takatori Y, Nakamura K, Ishihara K. 2009. Effects of mobility/immobility of surface modification by 2-methacryloyloxyethyl phosphorylcholine polymer on the durability of polyethylene for artificial joints. *J Biomed Mater Res A* 90(2):362–371.

86. Brown SA, Hastings RS, Mason JJ, Moet A. 1990. Characterization of short-fibre reinforced thermoplastics for fracture fixation devices. *Biomaterials* 11(8):541–547.

87. Kurtz SM, Devine JN. 2007. PEEK biomaterials in trauma, orthopedic, and spinal implants. *Biomaterials* 28(32):4845–4869.

88. Wang A, Lin R, Stark C, Dumbleton JH. 1999. Suitability and limitations of carbon fiber reinforced PEEK composites as bearing surfaces for total joint replacements. *Wear* 225–229:724–727.

89. Joyce TJ, Rieker C, Unsworth A. 2006. Comparative in vitro wear testing of PEEK and UHMWPE capped metacarpophalangeal prostheses. *Biomed Mater Eng* 16(1):1–10.

90. Latif AM, Mehats A, Elcocks M, Rushton N, Field RE, Jones E. 2008. Pre-clinical studies to validate the MITCH PCR Cup: A flexible and anatomically shaped acetabular component with novel bearing characteristics. *J Mater Sci: Mater Med* 19(4):1729–1736.

91. Yu S, Hariram KP, Kumar R, Cheang P, Aik KK. 2005. In vitro apatite formation and its growth kinetics on hydroxyapatite/polyetheretherketone biocomposites. *Biomaterials* 26(15):2343–2352.

92. Fan JP, Tsui CP, Tang CY, Chow CL. 2004. Influence of interphase layer on the overall elasto-plastic behaviors of HA/PEEK biocomposite. *Biomaterials* 25(23):5363–5373.

93. Kyomoto M, Ishihara K. 2009. Self-initiated surface graft polymerization of 2-methacryloyloxyethyl phosphorylcholine on poly(ether-ether-ketone) by photo-irradiation. *ACS Appl Mater Interfaces* 1(3):537–542.

94. Kyomoto M, Moro T, Takatori Y, Kawaguchi H, Nakamura K, Ishihara K. 2010. Self-initiated surface grafting with poly(2-methacryloyloxyethyl phosphorylcholine) on poly(ether-ether-ketone). *Biomaterials* 31(6):1017–1024.

95. Giancaterina S, Rossi A, Rivaton A, Gardette JL. 2000. Photochemical evolution of poly(ether ether ketone). *Polym Degrad Stab* 68(1):133–144.

96. Wang H, Brown HR, Li Z. 2007. Aliphatic ketones/water/alcohol as a new photoinitiating system for the photografting of methacrylic acid onto high-density polyethylene. *Polymer* 48(4):939–948.

97. Yang W, Rånby B. 1999. Photoinitiation performance of some ketones in the LDPE–acrylic acid surface photografting system. *Eur Polym J* 35(8):1557–1568.

98. Qiu C, Nguyen QT, Ping Z. 2007. Surface modification of cardo polyetherketone ultrafiltration membrane by photo-grafted copolymers to obtain nanofiltration membranes. *J Membr Sci* 295(1–2):88–94.

99. Nguyen HX, Ishida H. 1986. Molecular analysis of the melting behaviour of poly(aryl-ether-ether-ketone). *Polymer* 27(9):1400–1405.

100. Cole KC, Casella IG. 1992. Fourier transform infrared spectroscopic study of thermal degradation in films of poly(etheretherketone). *Thermochim Acta* 211:209–228.

101. Qiu KY, Si K. 1996. Grafting reaction of macromolecules with pendant amino groups via photoinitiation with benzophenone. *Macromol Chem Phys* 197:2403–2413.

102. Kuehn KD, Ege W, Gopp U. 2005. Acrylic bone cements: Composition and properties. *Orthop Clin North Am* 36(1):17–28.

103. Charnley J. 1964. The bonding of prostheses to bone by cement. *J Bone Joint Surg Br* 46:518–529.

104. Buchholz HW, Elson RA, Engelbrecht E, Lodenkämper H, Röttger J, Siegel A. 1981. Management of deep infection of total hip replacement. *J Bone Joint Surg Br* 63(3):342–353.

105. Wahlig H, Dingeldein E. 1980. Antibiotics and bone cements. Experimental and clinical long-term observations. *Acta Orthop Scand* 51(1):49–56.

106. Hope PG, Kristinsson KG, Norman P, Elson RA. 1989. Deep infection of cemented total hip arthroplasties caused by coagulase-negative staphylococci. *J Bone Joint Surg Br* 71(5):851–855.

107. Ishihara K, Arai J, Nakabayashi N, Morita S, Furuya K. 1992. Adhesive bone cement containing hydroxyapatite particle as bone compatible filler. *J Biomed Mater Res* 26(7):937–945.

108. Lee RR, Ogiso M, Watanabe A, Ishihara K. 1997. Examination of hydroxyapatite filled 4-META/MMA-TBB adhesive bone cement in vitro and in vivo environment. *J Biomed Mater Res* 38(1):11–16.

109. Morita S, Kawachi S, Yamamoto H, Shinomiya K, Nakabayashi N, Ishihara K. 1999. Total hip arthroplasty using bone cement containing tri-n-butylborane as the initiator. *J Biomed Mater Res* 48(5):759–763.

110. Sakai T, Morita S, Shinomiya K, Watanabe A, Nakabayashi N, Ishihara K. 2000. Prevention of fibrous layer formation between bone and adhesive bone cement: In vivo evaluation of bone impregnation with 4-META/MMA-TBB cement. *J Biomed Mater Res* 52(1):24–29.

111. Sakai T, Morita S, Shinomiya K, Watanabe A, Nakabayashi N, Ishihara K. 2000. In vivo evaluation of the bond strength of adhesive 4-META/MMA-TBB bone cement under weight-bearing conditions. *J Biomed Mater Res* 52(1):128–134.

112. Ohtsuki C, Miyazaki T, Kyomoto M, Tanihara M, Osaka A. 2001. Development of bioactive PMMA-based cement by modification with alkoxysilane and calcium salt. *J Mater Sci: Mater Med* 12(10–12):895–899.

113. Miyazaki T, Ohtsuki C, Kyomoto M, Tanihara M, Mori A, Kuramoto K. 2003. Bioactive PMMA bone cement prepared by modification with methacryloxypropyltrimethoxysilane and calcium chloride. *J Biomed Mater Res A* 67(4):1417–1423.

Index